S0-AKJ-302

Special Volume: Computational Chemistry
Guest Editor: C. Le Bris

Handbook of Numerical Analysis

General Editor:

P.G. Ciarlet

Laboratoire Jacques-Louis Lions
Université Pierre et Marie Curie
4 Place Jussieu
75005 PARIS, France

and

Department of Mathematics
City University of Hong Kong
Tat Chee Avenue
KOWLOON, Hong Kong

ELSEVIER
Amsterdam • Boston • London • New York • Oxford • Paris
San Diego • San Francisco • Singapore • Sydney • Tokyo

Volume X

Special Volume: Computational Chemistry

Guest Editor:

C. Le Bris

CERMICS
Ecole Nationale des Ponts et Chaussées
77455 Marne La Vallée, France

2003
ELSEVIER
Amsterdam • Boston • London • New York • Oxford • Paris
San Diego • San Francisco • Singapore • Sydney • Tokyo

Chemistry Library

ELSEVIER SCIENCE B.V.
Sara Burgerhartstraat 25
P.O. Box 211, 1000 AE Amsterdam, The Netherlands

© 2003 Elsevier Science B.V. All rights reserved.

This work is protected under copyright by Elsevier Science, and the following terms and conditions apply to its use:

Photocopying:
Single photocopies of single chapters may be made for personal use as allowed by national copyright laws. Permission of the Publisher and payment of a fee is required for all other photocopying, including multiple or systematic copying, copying for advertising or promotional purposes, resale, and all forms of document delivery. Special rates are available for educational institutions that wish to make photocopies for non-profit educational classroom use.

Permissions may be sought directly from Elsevier Science Global Rights Department, PO Box 800, Oxford, UK; phone: (+44) 1865 843830, fax: (+44) 1865 853333, e-mail: permissions@elsevier.co.uk. You may also complete your request online via the Elsevier Science homepage (http://www.elsevier.com), by selecting 'Customer Support' and then 'Obtaining Permissions'.

In the USA, users may clear permissions and make payments through the Copyright Clearance Center, Inc., 222 Rosewood Drive, Danvers, MA 01923, USA; phone: (+1) 978 7508400, fax: (+1) 978 7504744, and in the UK through the Copyright Licensing Agency Rapid Clearance Service (CLARCS), 90 Tottenham Court Road, London W1P 0LP, UK; phone: (+44) 207 631 5555; fax: (+44) 207 631 5500. Other countries may have a local reprographic rights agency for payments.

Derivative Works:
Tables of contents may be reproduced for internal circulation, but permission of Elsevier Science is required for external resale or distribution of such material.
Permission of the Publisher is required for all other derivative works, including compilations and translations.

Electronic Storage or Usage:
Permission of the Publisher is required to store or use electronically any material contained in this work, including any chapter or part of a chapter.

Except as outlined above, no part of this work may be reproduced, stored in a retrieval system or transmitted in any form or by any means, electronic, mechanical, photocopying, recording or otherwise, without prior written permission of the Publisher.
Address permissions requests to: Elsevier Science Global Rights Department, at the phone, mail, fax and e-mail addresses noted above.

Notice:
No responsibility is assumed by the Publisher for any injury and/or damage to persons or property as a matter of products liability, negligence or otherwise, or from any use or operation of any methods, products, instructions or ideas contained in the material herein. Because of rapid advances in the medical sciences, in particular, independent verification of diagnoses and drug dosages should be made.

First edition 2003

Library of Congress Cataloging in Publication Data
A catalog record from Library of Congress has been applied for.

British Library Cataloguing in Publication Data
A catalog record from British Library has been applied for.

For information on published and forthcoming volumes URL = http://www.elsevier.com/locate/series/hna

ISBN: 0-444-51248-9
ISSN (Series): 1570-8659

⊗ The paper used in this publication meets the requirements of ANSI/NISO Z39.48-1992 (Permanence of Paper).

Printed in Great Britain by MPG Books Ltd, Bodmin, Cornwall

QA 297 H287 1990 V.10 CHEM

General Preface

In the early eighties, when Jacques-Louis Lions and I considered the idea of a *Handbook of Numerical Analysis*, we carefully laid out specific objectives, outlined in the following excerpts from the "General Preface" which has appeared at the beginning of each of the volumes published so far:

> During the past decades, giant needs for ever more sophisticated mathematical models and increasingly complex and extensive computer simulations have arisen. In this fashion, two indissociable activities, *mathematical modeling* and *computer simulation*, have gained a major status in all aspects of science, technology and industry.
>
> In order that these two sciences be established on the safest possible grounds, mathematical rigor is indispensable. For this reason, two companion sciences, *Numerical Analysis* and *Scientific Software*, have emerged as essential steps for validating the mathematical models and the computer simulations that are based on them.
>
> *Numerical Analysis* is here understood as the part of *Mathematics* that describes and analyzes all the numerical schemes that are used on computers; its objective consists in obtaining a clear, precise, and faithful, representation of all the "information" contained in a mathematical model; as such, it is the natural extension of more classical tools, such as analytic solutions, special transforms, functional analysis, as well as stability and asymptotic analysis.
>
> The various volumes comprising the *Handbook of Numerical Analysis* will thoroughly cover all the major aspects of Numerical Analysis, by presenting accessible and in-depth surveys, which include the most recent trends.
>
> More precisely, the Handbook will cover the *basic methods of Numerical Analysis*, gathered under the following general headings:
>
> – Solution of Equations in $\mathbb{R}^n$,
> – Finite Difference Methods,
> – Finite Element Methods,
> – Techniques of Scientific Computing.

It will also cover the *numerical solution of actual problems of contemporary interest in Applied Mathematics*, gathered under the following general headings:

- Numerical Methods for Fluids,
- Numerical Methods for Solids.

In retrospect, it can be safely asserted that Volumes I to IX, which were edited by both of us, fulfilled most of these objectives, thanks to the eminence of the authors and the quality of their contributions.

After Jacques-Louis Lions' tragic loss in 2001, it became clear that Volume IX would be the last one of the type published so far, i.e., edited by both of us and devoted to some of the general headings defined above. It was then decided, in consultation with the publisher, that each future volume will instead be devoted to a single "*specific application*" and called for this reason a "*Special Volume*". "*Specific applications*" will include Mathematical Finance, Meteorology, Celestial Mechanics, Computational Chemistry, Living Systems, Electromagnetism, Computational Mathematics etc. It is worth noting that the inclusion of such "specific applications" in the *Handbook of Numerical Analysis* was part of our initial project.

To ensure the continuity of this enterprise, I will continue to act as Editor of each Special Volume, whose conception will be jointly coordinated and supervised by a Guest Editor.

P.G. CIARLET
July 2002

Foreword

It is my pleasure and an honor to act as a Guest Editor of this special volume of the *Handbook of Numerical Analysis* devoted to *Computational Chemistry*.

Computational Chemistry is a very active field of Scientific Computing, at the crossroads between Physics, Chemistry, Applied Mathematics, Computer Science. It is indeed unbelievably difficult to simulate large systems of interacting particles at the microscopic scale. A deep understanding of the phenomena at play is necessary, clever algorithms are to be used, advanced techniques need to be resorted to, huge computational task forces are needed in order to successfully treat the complex and sophisticated equations that rule the interactions and model the behaviour of matter at this scale. Despite huge efforts and real progress made by the communities of applied scientists involved in the field, it is true that, to date, the field is unsatisfactorily explored by numerical analysts and experts in scientific computing. Numerical strategies and algorithmic techniques have been mostly designed by theoretical chemists and physicists. In this respect, the situation is radically different from the situation in computational mechanics, e.g., where the two communities of experts in mechanics and experts in mathematics, that is to say the community that designs the models and that which designs the numerical codes have made progress, during the second half of the past century, progressively enriching one another by constant interaction. The development of the numerical techniques in Computational Chemistry has followed another road, more application driven, in a way.

It is now time (or at least I modestly think so) for the two communities to meet each other, exchange ideas and compare viewpoints; in one word, to interact. Over the last ten years we have witnessed some efforts in this direction, and in particular efforts of the applied mathematicians to get involved into this scientific adventure. Some of the leading experts of this effort are contributors to this volume. On the other hand, some chemists and physicists have welcomed interactions with applied mathematicians, making fruitful collaborations possible by patiently explaining the foundations of their models, the context and the goals of their scientific activity, to mathematicians. Clearly, the two communities have benefited from such experiences.

Of course, the present volume cannot give an account of all the existing collaboration between the two camps. I have collected some contributions issued from some groups over the world which have played the game of collaborating. The reason why these contributions appear in this book is primarily that all of them are instances of works of outstanding quality and impact, that cannot be ignored. But, of course, as the work of editing such a book is also a reflection of a personal trajectory, many works of equal outstanding quality might unfortunately have been forgotten, just because the author of these lines has not come across them. Anyway, I hope that this book presents a fair, if not complete, account of the scientific field. Sincere apologies are expressed for not making it better.

The book is intended to have a true pedagogical content. It is designed to be a useful guide, for experts of one (or both) of the two fields, mathematics and chemistry, as well as for researchers who wish to discover both fields. In order to try and reach this goal, the following friendly rule of the game was established: each contribution of the volume is authored either by a team of researchers, or by a single author working at the interface, so that in either case both communities (mathematics and chemistry/physics) involved should be represented. In doing so, I hope the accessibility of the book has been improved.

The first four contributions aim at introducing the field. They present the basics that will be used through the volume. The modelling aspects, together with the mathematical aspects, are rather well detailed. This has been done at the price of not going very far into the description and analysis of the most sophisticated methods. On the other hand, the subsequent contributions detail more advanced aspects, both from the viewpoint of methods and of applications. They aim at showing how lively and ever creative this scientific domain is. From the point of view of the applied mathematician and that of the numerical analyst, these contributions may in particular be seen as a rich mine of open problems, most of them of outstanding difficulty. But, as always is the case with risky investments, a high pay-off can be hoped for.

The book opens with four expository surveys in a (hopefully) pedagogical style. The first one, *Computational quantum chemistry: a primer*, by Eric Cancès, Mireille Defranceschi, Werner Kutzelnigg, Claude Le Bris, and Yvon Maday is the bottom line for the book, and introduces the most commonly used models and techniques for the molecular case, hardly approaching the more complicated cases. Consequently, the next two contributions deal with the condensed phases, respectively, liquid and solid: *The modelling and simulation of the liquid phase*, by Jacopo Tomasi, Benedetta Mennucci and Patrick Laug; and *An introduction to the first-principles simulations of extended systems*, by Fabio Finocchi, Jacek Goniakowski, Xavier Gonze and Cesare Pisani. The fourth contribution of this introductory part addresses the very specific aspects of the relativistic modelling: *Computational approaches of relativistic methods in quantum chemistry*, by Jean-Paul Desclaux, Jean Dolbeault, Maria Jesus Esteban, Paul Indelicato, Eric Séré.

As mentioned, the second part of the volume consists of contributions focusing on special techniques and/or applications. This part is organized as follows: a series of five contributions dealing with advanced aspects of numerical techniques for various systems and various applications, a number of contributions making the link with the very important field of biological applications, and a final sequence of three contributions addressing topics related to the laser control of chemical reactions.

Stochastic-like methods are of growing importance for computational chemistry, and this is why Alan Asparu-Guzik and William A. Lester Jr. were asked to write the contribution *Quantum Monte Carlo methods for the solution of the Schrödinger equation for molecular systems*. Advanced numerical techniques showing the current trends of the community are next detailed in the contributions *Linear scaling methods for the solution of Schrödinger's equation*, by Stefan Goedecker, *Finite-difference methods for ab initio electronic structure and quantum transport calculations of nanostructures*, by Jean-Luc Fattebert and Marco Buongiorno Nardelli, *Using real space pseudopotentials for the electronic structure problem*, by James R. Chelikowski, Leeor Kronik, Igor Vasiliev, Manish Jain, and Yousef Saad, *Scalable multiresolution algorithms for classical and quantum molecular dynamics simulations of nanosystems* by Aiichiro Nakano, Timothy J. Campbell, Rajiv K. Kalia, Sanjay Kodiyalam, Shuji Ogata, Fuyuki Shimojo, Xiaotao Su and Priya Vashistha. All these leading experts provide the reader with an extensive view of their respective fields of competence. The variety of these contributions shows that, apart from the obvious interest *per se* for the modelling and simulation of molecular systems, the techniques of Computational Chemistry become relevant, and are likely to become a *must* within the next few years, in other fields of the engineering sciences. The reason is (at least) twofold. First, as the size of many technological devices shrinks, sometimes down to the nanoscale, a microscopic description of their physical behaviour becomes necessary. Second, as the microscopic structure of the materials used for macroscopic purposes becomes richer and richer, their modelling at the pure macroscopic level is difficult, and their macroscopic description needs to be fed by models at lower scales.

In the same spirit, with a view to showing the connections between computational chemistry and other fields of science, the next two contributions deal with applications related to biology, which is a companion field of science for chemistry: *Simulating chemical reactions in complex systems*, by Martin A. Field, and *Biomolecular conformations can be identified as metastable sets of molecular dynamics*, by Christof Schütte and Wilhelm Huisinga, the focus on the latter being more on mathematical aspects.

The volume ends with contributions, by outstanding researchers of the field, on the control of chemical reactions, in the sense "laser control of the evolution of quantum systems", one of the major and the most promising field of applications related to theoretical chemistry: *Theory of intense laser-induced molecular dissociation: from simulation to control*, by Osman Atabek, Roland Lefebvre and Thanh-Tunh Nguyen-Dang, *Numerical methods for molecular time-dependent Schrödinger equations – Bridging the perturbative to non-perturbative regime*, by André Bandrauk and Hui-Zhong Lu, *Control of quantum dynamics: concepts, procedures and future prospects*, by Herschel Rabitz, Gabriel Turinici and Eric Brown.

Ending this preface, and thereby my work as a guest editor, I would like to give my sincere thanks to some colleagues from whom I benefited a lot: first, from the chemistry world, my long-term colleague Mireille Defranceschi, and second my companion colleagues who participate into the endeavor of improving the mathematical techniques for computational chemistry: Eric Cancès, Isabelle Catto, Maria Esteban, Pierre-Louis Lions, Yvon Maday, Eric Séré and Gabriel Turinici, to cite only the close relatives.

Let me finally express my deep gratitude to Philippe Ciarlet, and have a respectful and commemorative thought for the greatly missed Jacques-Louis Lions, who together invited me to edit this volume, a day of January 2000.

CLAUDE LE BRIS
Paris, June 2002

Contents of Volume X

Contents of the Handbook

VOLUME VII

SOLUTION OF EQUATIONS IN $\mathbb{R}^n$ (PART 3)

TECHNIQUES OF SCIENTIFIC COMPUTING (PART 3)

VOLUME VIII

SOLUTION OF EQUATIONS IN $\mathbb{R}^n$ (PART 4)

TECHNIQUES OF SCIENTIFIC COMPUTING (PART 4)

NUMERICAL METHODS FOR FLUIDS (PART 2)

VOLUME IX

NUMERICAL METHODS FOR FLUIDS (PART 3)

VOLUME X

SPECIAL VOLUME: COMPUTATIONAL CHEMISTRY

Special Volume:
Computational Chemistry

Computational Quantum Chemistry: A Primer

Eric Cancès

C.E.R.M.I.C.S., Ecole Nationale des Ponts et Chaussées, 6 & 8 Avenue Blaise Pascal, Cité Descartes, Champs sur Marne, 77455 Marne La Vallée Cedex 2, France
e-mail: cances@cermics.enpc.fr
url: cermics.enpc.fr/~cances

Mireille Defranceschi

Commissariat à l'Energie Atomique, CE-Saclay, DEN/DSOE/RB, 91191 Gif-sur-Yvette Cedex, France
e-mail: mireille.defranceschi@cea.fr

Werner Kutzelnigg

Lehrstuhl für Theoretische Chemie Ruhr-Universität Bochum, D-44780 Bochum, Germany
e-mail: Werner.Kutzelnigg@ruhr-uni-bochum.de

Claude Le Bris

C.E.R.M.I.C.S., Ecole Nationale des Ponts et Chaussées, 6 & 8 Avenue Blaise Pascal, Cité Descartes, Champs sur Marne, 77455 Marne La Vallée Cedex 2, France
e-mail: lebris@cermics.enpc.fr
url: cermics.enpc.fr/~lebris

Yvon Maday

Laboratoire Jacques-Louis Lions, Université Paris VI, 175 rue du Chevaleret, Boite Courrier 187, 75252 Paris Cedex 05, France
e-mail: maday@ann.jussieu.fr

Computational Chemistry
Special Volume (C. Le Bris, Guest Editor) of
HANDBOOK OF NUMERICAL ANALYSIS, VOL. X
P.G. Ciarlet (Editor)
© 2003 Elsevier Science B.V. All rights reserved

Contents

We present a broad overview of the models in use in Computational quantum chemistry (typically N-body Schrödinger, Hartree–Fock type, density functional type models), the hierarchy of the different approximations (ab initio, semi-empirical, ...), of their physical foundations. We shall emphasize the mathematical difficulties these models rise, and the way they are treated (or not ...). We explain the most standard discretization schemes (the most advanced aspects being treated in other contributions below) such as the LCAO approximation etc., the algorithms in use for their resolution (SCF, ...), and the related numerical analysis. At the end of the chapter, it is expected that the reader has a good knowledge of the basic techniques, and of the mathematical and numerical analysis needed to make rigorously founded computations.

CHAPTER I

The Basic Modelling

1. Introduction

Quantum chemistry aims at understanding the properties of matter through the modelling of its behaviour at a subatomic scale, where matter is described as an assembly of nuclei and electrons.

At this scale, the equation that rules the interactions between these constitutive elements is the Schrödinger equation. It can be considered (except in few special cases notably those involving relativistic phenomena or nuclear reactions) as a universal model for at least three reasons. First it contains all the physical information of the system under consideration so that any of the properties of this system can be deduced in theory from the Schrödinger equation associated to it. Second, the Schrödinger equation does not involve any empirical parameter, except some fundamental constants of Physics (the Planck constant, the mass and charge of the electron, ...); it can thus be written for any kind of molecular system provided its chemical composition, in terms of natures of nuclei and number of electrons, is known. Third, this model enjoys remarkable predictive capabilities, as confirmed by comparisons with a large amount of experimental data of various types.

On the other hand, using this high quality model requires working with space and time scales which are both very tiny: The typical size of the electronic cloud of an isolated atom is the Angström (10^{-10} m), and the size of the nucleus embedded in it is 10^{-15} m; the typical vibration period of a molecular bond is the femtosecond (10^{-15} s), and the characteristic relaxation time for an electron is 10^{-18} s. Consequently, quantum chemistry calculations concern very short time (say 10^{-12} s) behaviors of very small size (say 10^{-27} m^3) systems. The underlying question is therefore whether information on phenomena at these scales is or not of some help to understand, or better predict, macroscopic properties of matter.

It is certainly not true that *all* macroscopic properties can be simply upscaled from the consideration of the short time behavior of a tiny sample of matter. Many of them proceed (also) from ensemble or bulk effects, that are far from being easy to understand and to model. Striking examples are found in the solid state or biological systems. Cleavage, the ability minerals have to naturally split along crystal surfaces (e.g., mica yields to thin flakes) is an ensemble effect. Protein folding is also an ensemble effect which originates in the presence of the surrounding medium; it is responsible for peculiar properties (e.g., unexpected acidity of some reactive site enhanced by special interactions) on which rely vital processes.

However, it is undoubtedly true that on the other hand *many* macroscopic phenomena originate from elementary processes which take place at the atomic scale. Let us mention for instance the fact that the elastic constants of a perfect crystal or the color of a chemical compound (which is related to the wavelengths absorbed or emitted during optic transitions between electronic levels) can be evaluated by atomic scale calculations. In the same fashion, the lubrifying properties of graphite are essentially due to a phenomenon which can be entirely modelled at the atomic scale. Other examples will be given later on.

It is therefore founded to simulate the behaviour of matter at the atomic scale in order to understand what is going on at the macroscopic one. The journey is however a long one. Starting from the basic principles of quantum mechanics to model the matter at the subatomic scale, one finally uses statistical mechanics to reach the macroscopic scale. It is often necessary to rely on intermediate steps to deal with phenomena which take place on various *mesoscales*. Possibly, one couples one approach to the others within the so-called *multiscale* models. In this contribution, we shall indicate how this journey can be done, focusing rather on the first scale (the subatomic one), than on the latter ones.

It has already been mentioned that at the subatomic scale, the behavior of nuclei and electrons is governed by the Schrödinger equation, either in its time dependent form

$$i\frac{\partial}{\partial t}\Psi(t) = H\Psi(t), \tag{1.1}$$

or in its time independent form

$$H\Psi = E\Psi. \tag{1.2}$$

The physical meaning of these two equations will be clarified later on. Let us only mention at this point that

- in both equations H denotes the quantum Hamiltonian of the molecular system under consideration; from a mathematical viewpoint, it is a self-adjoint operator on some Hilbert space $\mathcal{H}$; *both* the Hilbert space $\mathcal{H}$ and the operator H depend on the nature of the system;
- Ψ is the wavefunction of the system; it completely describes its state; it is a vector of the unit sphere of $\mathcal{H}$.

Eq. (1.1) is a first order linear evolution equation, whereas Eq. (1.2) is a linear eigenvalue equation.

For the reader more familiar with numerical analysis than with quantum mechanics, the linear nature of the problems stated above may look auspicious. What makes in fact extremely difficult the numerical simulation of Eqs. (1.1) and (1.2) is essentially the huge size of the Hilbert space $\mathcal{H}$: Indeed, $\mathcal{H}$ is roughly some symmetry constrained subspace of $L^2(\mathbb{R}^d)$, with $d = 3(M+N)$, M and N, respectively, denoting the number of nuclei and the number of electrons the system is made of. The parameter d is already 39 for a single water molecule and reaches rapidly 10^6 for polymers or biological molecules. In addition, a consequence of the universality of the model is that one has to deal at the same time with several energy scales. In molecular systems indeed, the basic elementary interaction between nuclei and electrons (the two-body Coulomb interaction) shows itself in various complex physical and chemical phenomena whose characteristic energies cover several orders of magnitude: The binding energy of core electrons in heavy atoms is 10^4 times as large as a typical covalent bond energy, which is itself around 20 times as large as the energy of a hydrogen bond. High precision or at least controlled error cancellations are thus required to reach chemical accuracy when starting from the Schrödinger equation.

Clever approximations of problems (1.1) and (1.2) are therefore needed. The main two approximation strategies, namely the Born–Oppenheimer–Hartree–Fock and the Born–Oppenheimer–Kohn–Sham strategies, end up with large systems of coupled *nonlinear* partial differential equations, each of these equations being posed on $L^2(\mathbb{R}^3)$. The size of the underlying functional space is thus reduced at the cost of a dramatic increase of the mathematical complexity of the problem: Nonlinearity. The mathematical and numerical analysis of the resulting models is one of the major concern of the present contribution.

Before we get to the heart of the matter, we would like to provide the reader with a list of bibliographical references that can advantageously be utilized as points of entry in the huge bibliography devoted to computational chemistry. We have deliberately chosen *treatises*, and in addition only *a few* of them. Of course our choice is biased, and the subsequent short list should not be considered else than a personal view on the topic. General treatises dealing with the models of quantum chemistry are, e.g., LEVINE [1991], SZABO and OSTLUND [1982], MCWEENY [1992], HEHRE, RADOM, SCHLEYER and POPLE [1986], DREIZLER and GROSS [1990], MARCH [1992], PARR and YANG [1989]. A practical field of interest is that of the models for the condensed phase, which are pretty well exposed in ALLEN and TILDESLEY [1987] (liquid phase) or ASHCROFT and MERMIN [1976] (solid phase). Let us also mention that some references in the same spirit will be given in Chapter V to cover the slightly different field of *molecular dynamics*. In addition to these treatises, we would like to mention the very interesting and useful series entitled *Reviews in computational chemistry* (14 volumes to date) featuring high level state of the art review articles.

Let us end this introduction by saying we gratefully acknowledge the help and advice of X. Blanc, M. Lewin and G. Turinici, who carefully read the manuscript and suggested many improvements.

2. Some historical milestones

Theoretical chemistry originates in the first theoretical considerations of the XVIII century. In the course of the XIX century theoretical arguments have grown into a wide body of concepts. Finally the fundaments of quantum chemistry were settled down in the 1920s. The first step in this direction was taken by Bohr (1913) when he made two postulates. The first one was that an atomic system can only exist in particular stationary states with energies varying in a discrete set and the second that transitions between two states by a radiation correspond to an energy equal to the product of the frequency of the radiation and the Planck constant. A more general quantization rule was discovered independently by Wilson (1915) and Sommerfeld (1916), thus making possible the application of Bohr's postulates to a wider variety of atomic systems. In parallel, attempts to understand qualitatively the physical properties of metals were made by Drude (1900) who proposed the free electron model and in 1910 Madelung gave a first theory for ionic solids. After Schrödinger wrote down his famous equation in SCHRODINGER [1926], quantitative aspects were investigated. But it rapidly appeared that an exact solution for the Schrödinger equation cannot be found when the system contains three particles or more. Consequently approximations had to be introduced. The first one historically speaking and on which rely almost all quantum calculations is the so-called Born–Oppenheimer approximation (BORN and OPPENHEIMER [1927], BORN [1951]). The fact that the mass of a nucleus is at least 1836 times larger than the electron mass is used to consider the electrons in a stationary state in the field of the nuclei in their positions at a given time. This leads to an effective electronic energy which is a parametric function of the nuclear motions. This approximation is at the basis of the concept of potential energy surface from which are extracted geometries of chemical structures. Because the Schrödinger equation does not have separate solutions due to the interelectronic interactions (physically this corresponds to the fact that electrons do not move independently of one another and it is said that they are *correlated*) a first crude approximation was proposed by HARTREE [1928] who looked for separable functions which represented the best approximations to the true N-body wavefunction. In parallel HEITLER and LONDON [1927] published a paper on the hydrogen molecule which is considered as the conventional date for the beginning of the new discipline quantum chemistry. The Hartree approximation was rapidly improved in FOCK [1930] and SLATER [1930a], SLATER [1930b] as follows; first, due to the Pauli exclusion principle no more than two electrons can be in the same orbital state; secondly, electrons are physically indistinguishable from one another and the wavefunction should be antisymmetric to the interchange of any two sets of coordinates. Numerical solutions of the atomic Hartree–Fock equations were obtained in a systematic way and are collected in the book by HARTREE [1957]. The late twenties and thirties saw pioneering calculations on the Helium ground state in HYLLERAAS [1928] and on the hydrogen ground state in JAMES and COOLIDGE [1933]. By 1935, quantum chemistry was already delineated as a distinct sub-discipline due to the contributions of pioneers such as HUND [1928], DIRAC [1926], HUCKEL [1931a], HUCKEL [1931b], HUCKEL [1932], MULLIKEN [1928], PAULING [1931] or SLATER [1936a], SLATER [1936b]. These people grasped the new possibilities offered by quantum mechanics for chemical problems. They are credited with showing that the applications of quantum chemistry were, indeed, possible, especially so after the introduction of a number of new concepts and the adoption of certain approximation methods. For instance whereas the first idea might be to use, as basis functions, the exact eigenfunctions of the hydrogen-like Hamiltonian, which turned out to be practically intractable, it was proposed in 1930 to use Slater-type-orbitals (STO). They are still used nowadays even if following the suggestion of BOYS [1950], most of the present calculations are done using Gaussian-type-orbitals (GTO). The corresponding functions for solids are the Bloch functions or their alternative form the Wannier functions (see PARMENTER and WANNIER [1937]) constructed from a limited number of local functions multiplied by a periodic exponential term to obey the Bloch theorem which reflects the translational periodicity of the system. Instead of the Bloch functions it is also convenient for periodic solids to use plane waves which have been improved later as augmented plane waves or linear augmented plane waves (see ANDERSEN [1975]) which are still largely used. In COULSON [1935] was also tackled the problem of the hydrogen molecule whilst in D.R. HARTREE, W. HARTREE and SWIRLES [1939] was developed the multiconfiguration self-consistent field approach. Relativistic effects were introduced as early as 1935 in SWIRLES [1935], SWIRLES [1936] via the Dirac–Hartree–Fock self-consistent field method.

From the pure conceptual point of view, two different roads have been pursued towards a qualitative and quantitative understanding of chemical bonding. The first one is the molecular orbital theory based on molecules constructed as an assembly of atoms and independent electrons; it was first developed by Hund, Mulliken and elaborated by Slater, Hückel and others. Coulson extended the molecular orbital theory and the self-consistent field to polyatomic molecules in COULSON [1937], COULSON [1938]. On the other hand, PAULING [1931] introduced and developed the concept of resonance using essentially Heitler and London picture.

Contemporaneously, solids were also studied in SOMMERFELD and BETHE [1933] considering the free electron model of metals as described by Drude but corrected with the right statistics for electrons (the Fermi–Dirac statistics instead of the bosonic one). It is only much later that a theory of strong electron interactions in solid appeared with the work by LANDAU [1957a]. For this purpose the homogeneous gas is the natural reference point of density-functional theories which appeared in 1964 with the famous papers by HOHENBERG and KOHN [1964], KOHN and SHAM [1965]. Some authors (HEDIN and LUNDQVIST [1971], LUNDQVIST and MARCH [1983]) modifying the exchange functionals have also deeply influenced solid state physics. Prior to the density functional theory period, solid state physicists mainly used tight-binding methods (i.e., semi-empirical methods) which can describe the band structure of a large variety of crystalline systems, see SLATER and KOSTER [1954]. At the beginning of 1932, TAMM [1932] showed the existence of specific electronic states localized near the crystal boundary and this was the starting point of the theory of solid surfaces. Tamm's idea immediately drew the attention of other theoretical physicists and, as a consequence, many papers appeared in the 1930s; FOWLER [1933], RIJANOV [1934] and others rather rapidly developed the theory of surface states. All of the above papers were more methodological in nature than philosophical, and none of them explained how surface states originate from the atomic levels as the crystal is formed. The calculations performed in SHOCKLEY [1939] gave conditions of existence of surface levels. Finally the advanced theory really developed in the 1950s; a step towards a sophisticated general theory of localized states was taken in KOUTECKÝ [1957] based on perturbation theory of the Hamiltonian of the electron in an infinite ideal crystal. Appreciable progress has been made during the last decades of the twentieth century in theoretical studies of surface states of metals or semiconductors. However, the mathematical difficulties arising from the boundary conditions at the crystal surface, and the consequent drastic approximations make this field an active area of development today.

In spite of the efforts of all these brilliant scientists, it is only when computers were available that computations effectively began. In the early 1950s, HALL [1951] and ROOTHAAN [1951] presented the matrix Hartree–Fock equations for molecular systems in a form suitable for implementation on a computer. Finally the development of softwares began in the sixties with the paper by MULLIKEN and ROOTHAAN [1959] describing an automatic program for the determination of self-consistent field studies on diatomic molecules. Then appeared the first considerations on correlation with the works by SINANOĞLU [1964], SINANOĞLU [1966], SILVER, MEHLER and RUEDENBERG [1968], MILLER and RUEDENBERG [1968], BENDER and DAVIDSON [1969], ČIŽEK [1966], ČIŽEK [1969]. Works on the calculations of derivatives also began. In the seventies was developed the many-body perturbation theory, within the algebraic approximation and using the Møller–Plesset partition of the Hamiltonian (see MØLLER and PLESSET [1934], POPLE, KRISHNAN, SCHLEGEL and BINKLEY [1978b]), for the treatment of correlation. Finally during the past twenty years, there was an explosion of calculations relying on the use of more and more powerful computers. Already in the sixties the theoretical chemists have succeeded in developing ab initio programs, such as POLYATOM (see BARNETT [1963]) and IBMOL (see CLEMENTI and DAVIS [1966]), and the first release of the universally used program GAUSSIAN 70 (still in evolution) appeared as early as 1970, developed by Pople and coworkers.

A much less standardization is present in solid state physics because of historical reasons. The homogeneous electron gas has for long played a central role as the prototype of condensed many-electron systems modeling metals. Due to its limited efficiency it has prevented calculations from sophistication even if the basic models of solid state physics have been refined so as to describe more and more accurately a wide spectrum of observable quantities of real physical systems. It is only with the development of density functional theory that it has become a computational science. Although in many respects the determinations of electronic structures are very similar for molecules and bulk solids or surfaces (the basic problem is the study of a many-electron system in the Coulomb field of fixed nuclei), the theoretical developments in the two fields have diverged to such an extent that today they often do not share a common language. One main obstacle has been the fact that the methods applied in quantum chemistry cannot be carried over to solids, in particular when electrons are well delocalized (with Hartree–Fock methods the density of states of the electron gas is zero at the Fermi level whatever the degree of refinement of the method). On the other side, quantum chemists have been suspicious about density functional theory because it manipulates one electron wavefunctions that are not simple to interpret physically. The development of quantum chemistry has proven that ab initio calculations based on controlled approximations capable of systematic improvements have made simpler computational schemes based on uncontrolled simplifications obsolete. Whether or not the same will eventually hold true for solid-state theory remains to be seen. However given their overlapping concerns, quantum chemistry and solid-state theory should be considered together. Common features of methods as well as links between the two fields will be stressed.

3. Current status of computational chemistry

Starting from pure academic knowledge quantum chemistry has gained the rank of full partner in most chemical and physical researches carried out today, from organic chemistry to materials science and chemical engineering.

At the beginning of the XXI century it is more commonly named *atomistic modelling* than theoretical chemistry because there has been a remarkable shift; in addition to the classical 0 K or temperature independent quantum chemistry and solid state physics, contributions from side sciences allow now the evaluation of more significantly temperature-dependent properties and time-dependent evolutions. The word "modelling" has a broad sense, including a wide range of computer-based approaches with various strategies which can even overlap with computer-aided design of molecules or materials. Thanks to the improvement of methodologies and the efficiency of computational algorithms (joined to the explosion of computers) calculations are now accurate enough so as to be used as predictive data.

We are now at a time when theory and experiments are about to compete for demonstrating the existence of compounds not yet synthesized. There are several examples in the recent literature of convincing numerical determination of the structure of unknown compounds. Such examples remain marginal but modelling appears more and more as an additional tool for chemists among the wide spectrum of available analytical methods during the process of elaboration.

The role of modelling depends greatly on the industry involved. In drug design, the research part of the design of the molecule is essential and modelling appears as a guide for pre-selecting some molecules as potential good candidates for a given application. Modelling allows to spare experiments and money. In chemical industries, the part of design is less important and modelling often refers to other tools such as thermodynamic data. In materials industry, modelling includes determination of specific properties, strain–layer phenomena, electronic processes. Materials science is more technological than academic condensed matter science. For instance in industries prevails the adequacy of the product for some purpose (materials with the right polarizabilities, with the expected color and so on), the long-term behavior (evolution of materials under creep conditions, breakdown of ceramics under electric stress or irradiation) but also the economic and environmental aspects (in food industries).

Atomistic modelling is now acting even out of the restricted academic research area thanks to widely distributed or commercial codes preventing a waste of time in the definition of processes and in the analysis of new synthesized compounds. Among the numerous existing codes let us quote just a few: GAUSSIAN, ADF or MOLPRO for molecular purposes and CASTEP, VASP or ABINIT for solid calculations. They are used to such an extent that electronic structures can be determined using these "black boxes"; the applied fields where calculations can be ran go from astrochemistry to materials science (ceramics, polymers, dispersed mesosystems) to biology. With the computerization of sciences theoretical chemists are now divided into two groups, the first and far the smallest one is composed of theoreticians still dealing with concepts who try to elaborate concepts and to theoretisize chemistry and the second one who benefiting from the explosion of the power of computers, want to get specific information about structures, energetics, observables as obtained from the calculations. In the next future it can be guessed that atomic modelling will evolve in order to be able to treat complex systems: multicomponent chemical situations, multiscale effects, ... Fields which are now in progress.

4. A hierarchy of models

Only systems of extremely small size (which can still be of interest for very peculiar applications, mostly related to fundamental physics) can be directly tackled by the simulation of the Schrödinger equation (1.1) or (1.2). Most systems of interest for quantum chemistry are thus out of reach without any approximation, the most commonly used being the Born–Oppenheimer approximation. This approximation consists in taking into account the separation of scales between the masses of the nuclei and the masses of the electrons (as mentioned above, the lightest nucleus is 1836 times heavier than the electron) to treat nuclei and electrons on different levels. The Born–Oppenheimer approximation gives rise to various models, depending on the way the asymptotic expansion in the mass ratio is carried out and cut off.

In most cases encountered in molecular simulation, it is relevant to use the simplest model, namely the Born–Oppenheimer approximation,

(H1) nuclei are *classical point-like particles*,

(H2) the state of the electrons (represented by some electronic wavefunction ψ_e) only depends on the positions in space $(\bar{x}_1, \dots, \bar{x}_M)$ of the M nuclei,

(H3) this state is in fact the *electronic ground state*, i.e., the one that minimizes the energy of the electrons for the nuclear configuration $(\bar{x}_1, \dots, \bar{x}_M)$ under consideration.

This model is often referred to as *the* Born–Oppenheimer approximation of the Schrödinger equation in the chemical literature; for convenience, we will also adopt this abuse of the language in the sequel. Although the latter model is usually valid in practice, it is however necessary to weaken the assumptions (H1)–(H3) in some cases; we shall come back to this point in Section 6, and Chapters IV and V.

Quantum chemistry calculations performed in the Born–Oppenheimer setting described above mainly consist

- either in solving the *geometry optimization* problem, that is to compute the equilibrium molecular configuration that minimizes the energy of the system; mathematically, this problem reads

$$\inf\{W(\bar{x}_1, \dots, \bar{x}_M),\ (\bar{x}_1, \dots, \bar{x}_M) \in \mathbb{R}^{3M}\}; \tag{4.1}$$

  finding the most stable molecular configuration is an interesting result *per se* but is also a first stage in the computation of numerous properties like for instance infrared spectrum or elastic constants;
- or in performing an *ab initio molecular dynamics* simulation, that is to simulate the time evolution of the molecular structure according to the Newton law of classical

mechanics

$$m_k \frac{d^2\bar{x}_k}{dt^2}(t) = -\nabla_{\bar{x}_k} W\big(\bar{x}_1(t), \ldots, \bar{x}_M(t)\big), \tag{4.2}$$

where m_k denotes the mass of the kth nucleus. Molecular dynamics simulations allow to compute various transport properties (thermal conductivity, viscosity, ...) as well as some nonequilibrium properties. This will be briefly addressed in Chapter V of the present contribution.

In both cases, $W(\bar{x}_1, \ldots, \bar{x}_M)$ denotes the N-body potential the nuclei are subjected to. It can be split into two terms

$$W(\bar{x}_1, \ldots, \bar{x}_M) = U(\bar{x}_1, \ldots, \bar{x}_M) + \sum_{1 \leqslant k < l \leqslant M} \frac{z_k\, z_l}{|\bar{x}_k - \bar{x}_l|},$$

where z_k is the charge of the kth nucleus. The latter term has a clear physical meaning: it is the internuclear electrostatic repulsion energy. The former term stands for the contribution of the electrons to the interaction between the nuclei; it is a function of $(\bar{x}_1, \ldots, \bar{x}_M)$ since the electronic state is assumed to depend only on the positions of the nuclei.

For a given nuclear configuration, it follows from the Born–Oppenheimer approximation that $U(\bar{x}_1, \ldots, \bar{x}_M)$ can be in turn obtained by solving a time-independent Schrödinger equation of type (1.2) but of lower complexity than the original one since it concerns electrons only. Such a procedure permits to obtain an analytical calculation of the potential U for the H_2^+ ion, which contains a unique electron (see LANDAU and LIFCHITZ [1977] for instance); however, it is again not amenable (neither analytically nor computationally) for systems with more than a few electrons (say 3). One therefore resorts to further approximations of this problem. These approximations can typically be classified into two categories.

The first category is that of the so-called ab initio methods. These methods remain based only upon the first principles of quantum mechanics without any further approximations. In particular, they do not involve any empirical parameter (except some fundamental constants of Physics). The typical example for such methods is the Hartree–Fock approximation and its improvements, we shall come back to later on. Another wide class of instances of this category of methods consists of the ones issued from the density functional theory. They will also be dealt with in the sequel.

Depending on the specific choice of the method, on the accuracy required, and on the computer facility available, the ab initio methods allow today for the simulations of systems up to one hundred or one thousand atoms. In time dependent simulations, they are only convenient for small-time simulations, say not more than a picosecond (10^{-12} s). We shall concentrate on this type of methods in the rest of this chapter, mainly because they are well founded, physically, mathematically, and numerically, and in addition because they are also of course very much used. However, their limitation to systems of "small" size has created the need for less accurate methods that could be used for the simulation of larger systems, at the evident price of a loss of accuracy and control on the quality of the result. We now briefly turn to them.

The second category of approximation methods for the Schrödinger equation is that of the *parameterized methods*. This category is in turn divided into two subcategories: The *semi empirical methods* and the *empirical methods*. The *semi empirical methods* are based upon the observation along which it is possible to speed up the ab initio methods by inserting inside some results known from experiments or previous computations. The algorithmic complexity of the ab initio method is consequently reduced and larger systems may therefore be treated. *Empirical methods* go one step further, by considering explicitly only the nuclei. The whole system is consequently ruled by the classical laws of Newtonian mechanics. One therefore ends up with a so-called *molecular mechanics* problem (4.1) in the stationary setting, or a *classical molecular dynamics* problem (4.2). Most of the times, the form of the potential W is postulated, and its parameters are fitted on the basis of experimental data and/or ab initio numerical computations. Of course, the amount of numerical computation needed to simulate (4.1) or (4.2) with empirical potentials is reduced of many orders of magnitude with respect to the same computation when the potential W is simultaneously calculated at the ab initio level. On the other hand, as the electronic part of the system is to some extent frozen, it is not possible to simulate with such parameterized methods situations when the system under study experiences deep changes in the electronic structure. The main instance is that of a chemical reaction. Another drawback of these methods is the so-called *transferability* difficulty: A given empirical potential is well parameterized for one or several given system (or at least one hopes so), and *consequently* has no reason to be good at predicting properties for other ones. On the other hand, there are plenty of situations when simulating the dynamics of the system with empirical potential is enough to understand the physical phenomena at the leading order: The protein folding is one of them, of challenging importance. Indeed, empirical potentials seem to work quite well for standard biological compounds, at least in the absence of chemical reactions. There are also the situations when the process under investigation involves several millions of atoms (crack propagation, e.g.); such phenomena can only be simulated through empirical methods at this day. For us, who focus on the ab initio level, another useful application of empirical methods is to give at a low price a reasonable initial first evaluation (initial guess) of a quantity we aim at evaluating more precisely on a "small system".

As always, there is therefore room for a large variety of methods, depending on the application targeted (see Fig. 4.1).

In particular, it is to be remarked that, although we have clearly mentioned that ab initio methods are restricted to small systems (both in terms of space and in terms of time), they have a twofold interest:

- They can be indeed used to simulate very precisely small systems, which has a great interest per se.
- They can also be used in a coupled simulation of a far larger system. By saying "coupled simulation", we have in mind at least two situations. Firstly, one may use the *sequential* coupling where the ab initio method is used in a preliminary step to compute precisely, e.g., N-body interaction potentials between atoms. These potentials, which are parameterized in a next step (for instance, using a least square fitting), are finally used in an empirical simulation such as a molecular dynamics

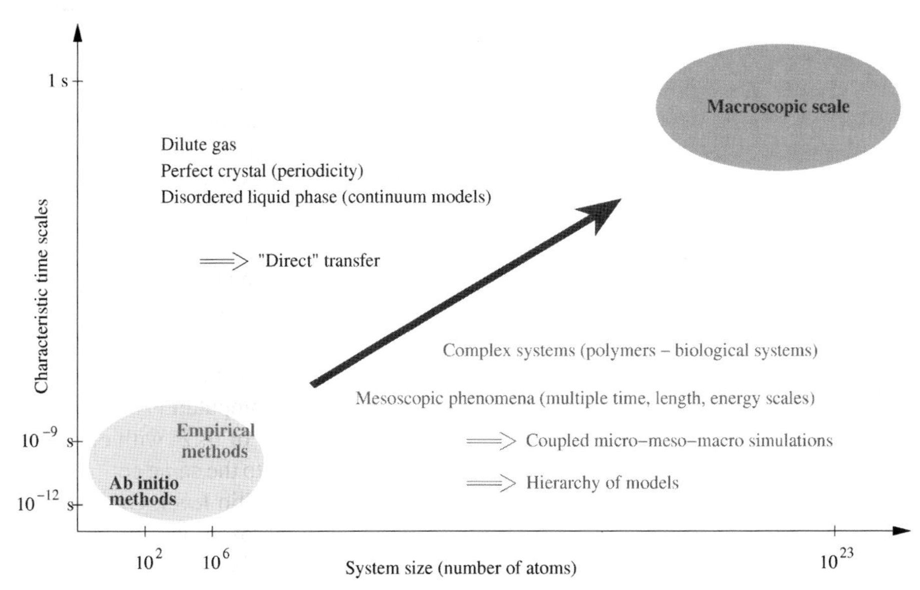

FIG. 4.1. From the micro-scale to the macro-scale.

simulation. Secondly, one may perform a *parallel* coupling, i.e., a simulation that couples two levels of approximation, namely the ab initio one and the empirical one. This is for instance the case when one studies the dynamics of a large molecule for a particular reaction. It is clear that not all atoms the molecular system is composed of are chemically reactive at the same time. Some of them are, most of them are not. The first ones are then treated with an ab initio method explicitly accounting for the change in the electronic configurations, whilst the latter ones are treated as inactive objects that are simply evolved through an empirical molecular dynamics simulation.

One must therefore bear in mind that, in that respect, the field of application of ab initio methods is *not* restricted to small systems.

Computational chemistry models follow from theoretical considerations; they need then to be tested by systematic comparisons of their outputs with known experimental results. Regarding ab initio methods, the success of a model is associated with its sophistication (see Section 8) and the quality of the basis sets used (see Section 24). Given the present state of the art of the theory, where there is a mosaic of options, and with the scope of reaching the status of predictive tool, one has to know the range of applicability of each model (to be able to use it even for computing systems for which experimental data are unavailable). Up to now, the accuracy reached with theoretical results is lower than that obtained with experimental ones, except for very small molecules. Even if improvements are appearing very fast, a complete virtual laboratory is not about to exist. However, if quantitative values are not straightfully obtainable with computations, trends can be obtained with enough confidence so as to allow to select potentially good molecules or materials for a given application out of various candidates. Successes or failures of a given model are known and published, but they are not always well enough documented and a know-how is generally necessary to obtain results in which one may be confident. Thus, even if black box codes are commercially available, accurate results can hardly be obtained by a nonspecialist. Nevertheless general trends are known. For instance, the Hartree–Fock model cannot describe metals and overestimates energy gaps in insulators; besides, Hartree–Fock infrared absorption frequencies are 10% lower than the correct ones. As far as density functional methods are concerned, they are known, among other trends, to poorly describe weak interactions (hydrogen bonds, Van der Walls interactions, ...). For more precise values the interested reader can refer to HEHRE, RADOM, SCHLEYER and POPLE [1986] or BARONI, GIANNOZZI and TESTA [1987].

As regards parameterized models (molecular dynamics), their range of applicability is more restricted and they usually can be used as interpolation tools rather than extrapolation ones. Finally let us quote that one of the successes of these methods is that they allow a vision of the atomic structure which is not attainable by any experimental method. It is now possible to virtually travel in a 3D chemical system at the atomistic level (NAKANO [2002]).

5. The basic model: Quantum description of a molecular system

We now get to the heart of the matter, and present the basic models of quantum chemistry. It is to be emphasized that we shall only deal here with *nonrelativistic* models. In the case when the molecular system under study involves one or many heavy atoms (atoms belonging to the bottom-half of the periodic table of chemical elements, such as Uranium), the relativistic effects do have some major importance. Such a situation requires specific models, issued from relativistic quantum chemistry, that will not be dealt with here. An entire contribution of the present volume will be devoted to the physical and mathematical foundations of these models, together with the current state of the art of the computations based upon these models. We refer the interested reader to this contribution, or also to the few lines at the end of the present chapter. Let us only mention here that, neglecting the relativistic effects for heavy atoms and modelling them in the nonrelativistic framework we shall detail in the sequel, may lead to noncorrect physical conclusions. For instance, modelling a heavy atom such as Gold in a nonrelativistic setting leads to the conclusion that gold is *not* yellow or that Mercury is not a liquid metal! In order to remedy that, one can use fully relativistic models. One can also resort to an hybrid approach that combines relativistic and nonrelativistic techniques, that we wish to mention here for it can be used efficiently. One partitions the set of electrons into two subsets: The subset of core electrons, that in a classical model evolve close to the nucleus at relativistic speeds and for which relativistic effects are to be accounted for, and the subset of outer shell electrons. To treat the latter ones, a nonrelativistic model (such as one of those which will be detailed below) is then built. It treats the effects of the core electrons through a so-called *pseudopotential* that is added to the electrostatic potential created by the nucleus. This pseudopotential is in turn known from a preliminary relativistic computation.

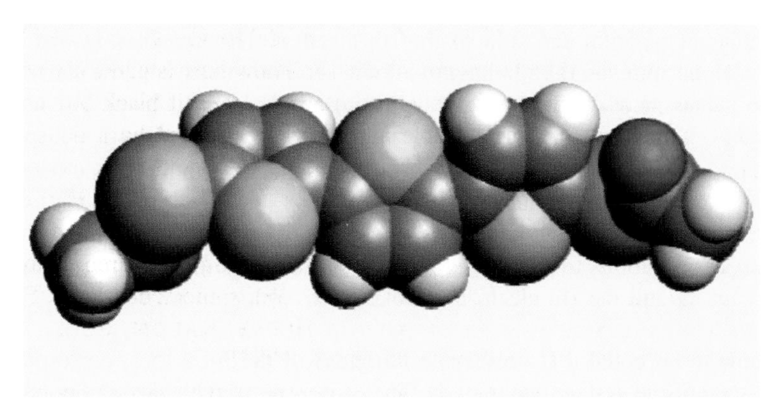

FIG. 5.1. Ground state of a molecular system: Bis-acetyl-pentathiophen (carbon is in grey, oxygen in red, sulfur in yellow and hydrogen in white).

We begin with an exposition of the quantum description of a molecular system. This gives us the opportunity to rapidly recall, for the convenience of the reader, the basics of quantum mechanics required to understand and manipulate the models of computational chemistry. The reader who is familiar with the framework of quantum mechanics can easily skip this part and proceed directly to the next section. We next present the central problem of computational quantum chemistry, namely the determination of the electronic ground state of a molecular system. We explain why it is a crucial problem which in most cases can only be attacked after some approximations. The search for this ground state through various approximations will be the purpose of Section 6. But before we get to that section, we briefly describe a direct attack of the problem, without any further approximation.

Let us consider an isolated molecular system composed of M nuclei and N electrons. In the sequel, we shall never detail further the structure of a nucleus, which is the domain of nuclear physics, and is not of our concern here (because energies involved in nuclear physics are far larger than those involved in the chemical bonds we are mainly interested in). Each nucleus will be treated here as a whole. We shall only keep trace of the substructure of the nucleus by remembering of how many protons and nucleons it is composed of. This will influence its total charge, the values its spin can take, and the properties of symmetry of the wavefunction. We shall even get rid of the latter two facts in most circumstances, as will be seen below.

For pedagogical purposes, this introductory presentation is limited to the so-called *pure states* (we shall introduce a more general setting later on). In this setting, the state of the molecular system under consideration is entirely described by a wavefunction of the form

$$\Psi(t; \bar{x}_1, \bar{\sigma}_1; \dots; \bar{x}_M, \bar{\sigma}_M; x_1, \sigma_1; \dots; x_N, \sigma_N) \tag{5.1}$$

valued in $\mathbb{C}$. Here, t denotes the time variable, $\bar{x}_k$ and $\bar{\sigma}_k$, respectively, denote the variables of position and spin of the kth nucleus, while x_i and σ_i, respectively, denote the variables of position and spin of the ith electron. The variables $\bar{x}_k$ and x_i vary in $\mathbb{R}^3$, while the spin variables belong to a finite set. For an electron, the spin can only have two values, here denoted by $|+\rangle$ (*spin up*) and $|-\rangle$ (*spin down*). For a nucleus composed of K nucleons, the spin variable can take $\frac{1}{4}(K+2)^2$ values if K is even, and $\frac{1}{4}(K+1)(K+3)$ values if K is odd. Let us denote henceforth by A_k the finite set of values for the spin of the kth nucleus.

From a physical standpoint, $|\Psi(t;\bar{x}_1,\bar{\sigma}_1;\ldots;\bar{x}_M,\bar{\sigma}_M;x_1,\sigma_1;\ldots;x_N,\sigma_N)|^2$ denotes the density of probability to simultaneously measure at time t the kth nucleus at position $\bar{x}_k$ with spin $\bar{\sigma}_k$ and the ith electron at position x_i with spin σ_i, for all $1\leqslant k\leqslant M$, $1\leqslant i\leqslant N$.

It is now to be noted that only *some* functions of the form (5.1) correspond to a physical state of the system under study. They are those which enjoy at any time t the following two properties

- *Normalization*: the function $\psi(t,\cdot)$ needs to have L^2 norm 1, i.e.,

$$\begin{aligned}\Psi(t,\cdot)&\in L^2\big(\big(\mathbb{R}^3\times A_1\big)\times\cdots\times\big(\mathbb{R}^3\times A_M\big)\times\big(\mathbb{R}^3\times\big\{|+\rangle,|-\rangle\big\}\big)^N,\mathbb{C}\big)\\&\sim L^2\big(\big(\mathbb{R}^3\times A_1\big),\mathbb{C}\big)\otimes\cdots\otimes L^2\big(\big(\mathbb{R}^3\times A_M\big),\mathbb{C}\big)\\&\quad\otimes L^2\big(\mathbb{R}^3\times\big\{|+\rangle,|-\rangle\big\}\big),\mathbb{C}\big)^{\otimes N}\end{aligned}$$

  with

$$\begin{aligned}\big\|\Psi(t,\cdot)\big\|^2=\int_{\mathbb{R}^{3M}}d\bar{x}_1\cdots d\bar{x}_M\sum_{\bar{\sigma}_1\cdots\bar{\sigma}_M}\int_{\mathbb{R}^{3N}}dx_1\cdots dx_N\\\times\sum_{\sigma_1\cdots\sigma_N}\big|\Psi(t;\bar{x}_1,\bar{\sigma}_1;\ldots;\bar{x}_M,\bar{\sigma}_M;x_1,\sigma_1;\ldots;x_N,\sigma_N)\big|^2=1.\end{aligned}$$

  This property is natural from a physical standpoint, for $|\Psi(t,\cdot)|^2$ represents a probability density.
- *Indistinguishability of identical particles*: This property stems from the fact that the measures operated on the system cannot depend on the specific way we have chosen to label the particles. In our setting, one must not be able to discriminate between two electrons, or to discriminate between two nuclei composed of the same number of protons and neutrons. It is simple to see that, mathematically, this translates into the property along which, for any time t, the function $\Psi(t,\cdot)$ is enforced to be either symmetric or antisymmetric with respect to the exchange of the coordinates of both space and spin (treated together) of two identical particles. More precisely, by definition of bosons and fermions, the function $\Psi(t,\cdot)$ has to be
  - *symmetric* with respect to the exchange of the coordinates of both space and spin of two identical particles which are bosons. In our framework, the bosons are exactly the nuclei composed of an even number of nucleons;
  - *antisymmetric* with respect to the exchange of the coordinates of both space and spin of two identical particles which are fermions. In our framework, the fermions are the electrons, and the nuclei composed of an odd number of nucleons.

In particular, the antisymmetry with respect to permutations of electrons reads

$$\begin{aligned} &\Psi\big(t; \{\bar{x}_k, \bar{\sigma}_k\}; x_{p(1)}, \sigma_{p(1)}; x_{p(2)}, \sigma_{p(2)}; \ldots; x_{p(N)}, \sigma_{p(N)}\big) \\ &\quad = (-1)^{\varepsilon(p)} \Psi\big(t; \{\bar{x}_k, \bar{\sigma}_k\}; x_1, \sigma_1; x_2, \sigma_2; \ldots; x_N, \sigma_N\big), \end{aligned} \tag{5.2}$$

where p denotes a permutation of the indices for electrons $\{1, 2, \ldots, N\}$, and $\varepsilon(p)$ denotes the signature of the permutation p. From the above property (5.2), one can deduce the so-called *Pauli exclusion principle*, stating that two electrons cannot be in the same state of spin and position, namely

$$\Psi\big(t; \{\bar{x}_k, \bar{\sigma}_k\}; x_1, \sigma_1; \ldots; x_N, \sigma_N\big) = 0$$

as soon as there are two indices $i \neq j$ such that $x_i = x_j$ and $\sigma_i = \sigma_j$.

A crucial consequence of the above properties is that the space of wavefunctions $\Psi(t, \cdot)$ will not be the entire L^2 space of functions of $M + N$ space variables, plus the spin variables. At this stage, the space of wavefunctions to be considered is

$$\mathcal{H} = \mathcal{H}_n \otimes \mathcal{H}_e, \tag{5.3}$$

where

$$\mathcal{H}_n = L^2_{\text{as}}\big((\mathbb{R}^3 \times A_1) \times \cdots \times (\mathbb{R}^3 \times A_M), \mathbb{C}\big), \tag{5.4}$$

$$\mathcal{H}_e = \bigwedge_{i=1}^{N} L^2\big(\mathbb{R}^3 \times \{|+\rangle, |-\rangle\}, \mathbb{C}\big). \tag{5.5}$$

The subscript "as" underlines the fact that the function has to enjoy specific antisymmetric or symmetric properties depending on the structure of the nuclei. For brevity, we do not make these properties more explicit in this general presentation, but it is a simple exercise to state them while considering particular cases of molecular systems.

In (5.5), let us indicate that the symbol $\bigwedge$ means the usual tensorial product $\otimes$ *with the additional assumption* that one only keeps the antisymmetrized products. This fact obviously refers to the antisymmetry requirement that we have described above.

In order to make more precise the functional spaces $\mathcal{H}_n$ and $\mathcal{H}_e$ we shall work with, let us now introduce the Hamiltonian of the system. For now on, we assume, unless otherwise mentioned, that the molecular system under study is *isolated*. In particular, the system is not subjected to any external electromagnetic field.

The Hamiltonian of the molecular system reads

$$\begin{aligned} H = &- \sum_{k=1}^{M} \frac{1}{2m_k} \Delta_{\bar{x}_k} - \sum_{i=1}^{N} \frac{1}{2} \Delta_{x_i} - \sum_{i=1}^{N} \sum_{k=1}^{M} \frac{z_k}{|x_i - \bar{x}_k|} \\ &+ \sum_{1 \leqslant i < j \leqslant N} \frac{1}{|x_i - x_j|} + \sum_{1 \leqslant k < l \leqslant M} \frac{z_k z_l}{|\bar{x}_k - \bar{x}_l|} \end{aligned} \tag{5.6}$$

where we have denoted by m_k and z_k the mass and the charge of the kth nucleus. In order to write H, we have chosen a special system of units, that we shall adopt henceforth.

This special system of units, called the *atomic unit system*, is commonly used in quantum chemistry. In this system, one has

$$m_e = 1, \qquad e = 1, \qquad \hbar = 1, \qquad \frac{1}{4\pi\varepsilon_0} = 1,$$

where m_e, e, $\hbar$, ε_0, respectively, denote the electron mass, the elementary charge, the reduced Planck constant, and the dielectric permittivity of vacuum. Accordingly, the unit of mass is 9.11×10^{-31} kg, the unit of length (denoted by a_0 and called the Bohr radius) 5.29×10^{-11} m, the unit of time 2.42×10^{-17} s, and the unit of energy, called the Hartree, 4.36×10^{-18} J, that is 27.2 eV, or also 627 kcal/mol. In this particular system, the mean value of the electron–nucleus distance in the Hydrogen atom is of the order of one, while the ground state energy of this atom is -0.5. As well, in this system, the speed of light is $1/\alpha$ (where α is the fine structure constant), i.e., roughly 137.

The above Hamiltonian (5.6) can be deduced from the one in the classical mechanics framework

$$H_{\text{cl}} = \sum_{k=1}^{M} \frac{p_{\bar{x}_k}^2}{2m_k} + \sum_{i=1}^{N} \frac{p_{x_i}^2}{2} - \sum_{i=1}^{N}\sum_{k=1}^{M} \frac{z_k}{|x_i - \bar{x}_k|} + \sum_{1\leqslant i<j\leqslant N} \frac{1}{|x_i - x_j|} + \sum_{1\leqslant k<l\leqslant M} \frac{z_k z_l}{|\bar{x}_k - \bar{x}_l|}$$

by replacing the x component by the operator consisting in a multiplication by x, and the momentum p_x along the x component by the operator $-i\nabla_x$. The meaning of each term in (5.6) is then clear: The first two terms correspond to the kinetic energy, respectively, of the nuclei and of the electrons, the latter three ones model the electrostatic energy between the particles, between nuclei and electrons first, between electrons secondly, and between nuclei finally.

It is to be remarked that in the case we are dealing with, namely that of an isolated molecular system, the Hamiltonian does not act on the spin variable $\bar{\sigma}_k$ of the nuclei (it does not act on the spin variable of the electrons σ_i either). In addition, it is physically relevant to consider the nuclei as very localized particles in space (we shall most often treat them as point particles), that can be distinguish from one another, in most of the cases at least. The combination of the two facts allows one to make the approximation consisting of forgetting the spin variable for the nuclei, which amounts to replacing the definition (5.4) of $\mathcal{H}_n$ by

$$\mathcal{H}_n = L^2(\mathbb{R}^{3M}, \mathbb{C}).$$

6. The basic model: Determination of the ground state

As announced above, let us now concentrate on the determination of the ground state of the molecular system. In the natural environment indeed, chemical and physical systems are usually found in their most stable state; for instance the iron element is abundantly found as iron oxides (which are chemical combinations of iron and oxygen atoms) and native metallic iron does not exist. To obtain metallic iron (i.e., only iron atoms) one has to provide energy to the oxides to transform them (it is called a reduction in chemistry). More generally speaking any chemical system A reacts, spontaneously or with a compound X, to give products $B, C, \ldots$ according to a chemical reaction:

$$A(+X) \rightarrow B + C + \cdots$$

if the variation of energy of the previous reaction corresponds to a stabilization of the whole system. For instance metallic sodium is stable in the absence of oxygen, but spontaneously burns in its presence. The above thermodynamic consideration does not suffice however to explain all the observations; for instance, it can be shown that the most stable form of the carbon element is graphite (at least in room temperature and pressure) whereas exists another less stable form: Diamond. However diamonds are usually considered as eternal! And they are indeed because of kinetic arguments. From the strict thermodynamic point of view, graphite is more stable but the rate of the spontaneous transformation of diamond into graphite is so low, in usual conditions, that the transformation does not occur in practice.

At this stage, the minimization problem to be solved in order to determine the ground state is

$$E_0 = \inf\{\langle\psi, H\psi\rangle, \ \psi \in \mathcal{H}, \ \|\psi\|_{L^2} = 1\}. \tag{6.1}$$

Let us first notice that any ground state ψ_0, that is any minimizer of (6.1) satisfies the time independent Schrödinger equation

$$H\psi_0 = E_0\psi_0.$$

As the Hamiltonian H defined by (5.6) is a *real* operator, both the real part and the imaginary part of ψ_0 are themselves ground states (provided they are not zero and up to normalization). Consequently, *it suffices to consider real valued wavefunctions.*

It is also clear from the form of the Hamiltonian H that, if we want to only consider states of finite energy, we need to impose the L^2 integrability of the first derivative of the wavefunctions with respect to the space variables. This is the necessary and sufficient condition to give a proper sense to the kinetic energy terms. In view of standard results in functional analysis, this also allows one to properly define the electrostatic Coulombian interaction term.

The above two arguments make it possible to further restrict the space $\mathcal{H}$ of wavefunctions we shall consider henceforth. We define $\mathcal{H}$ (keeping the same notations as there is no risk of ambiguity) as the tensor product of

$$\mathcal{H}_n = H^1(\mathbb{R}^{3M}),$$

and

$$\mathcal{H}_e = \bigwedge_{i=1}^{N} H^1\big(\mathbb{R}^3 \times \{|+\rangle, |-\rangle\}\big). \tag{6.2}$$

In the notation H^1, it is understood that only the differentiability with respect to the space variable $\bar{x}_k$, respectively x_i, and of course not with respect to the discrete spin variables, is required.

Problem (6.1) cannot be tackled as such, for it suffers for many drawbacks. From the theoretical standpoint, it is to be remarked that, although the energy, i.e., the value of (6.1) is well defined, the minimization problem itself is *not* attained. Indeed there is no minimizer to (6.1): The Hamiltonian can be written in terms of the coordinates of the barycenter of the system, plus the coordinates that are invariant by translation as follows

$$H = -\frac{1}{2m_G}\Delta_{x_G} + H_I,$$

where $m_G = \sum_{k=1}^{M} m_k + N$ is the total mass of the system; one next remarks that the operator $-\frac{1}{2m_G}\Delta_{x_G}$ has a purely continuous spectrum, which implies, in view of the form of H_I, that H has also a purely continuous spectrum, and therefore it has no ground state. From the practical viewpoint, even if we leave the difficulty related to the existence of a ground state aside, the size of the space $\mathcal{H}$ makes it impossible to find a value for the energy. We shall therefore need to cure both disadvantages in the sequel.

Let us now briefly present a simple way to formally recover what is called today the Born–Oppenheimer approximation of problem (6.1). Let us first assume that an arbitrary wavefunction for the molecular system is not only a *sum* of products of wavefunctions for the nuclei and wavefunctions for the electrons, as stated in (5.3), but is a *single* product of *one* wavefunction ψ_n for the nuclei and *one* wavefunction ψ_e for the electrons. The minimization problem (6.1) is then approximated by

$$\inf\big\{\langle\psi, H\psi\rangle,\ \psi = \psi_n\psi_e,\ \psi_n \in \mathcal{H}_n,\ \|\psi_n\|_{L^2} = 1,\ \psi_e \in \mathcal{H}_e,\ \|\psi_e\|_{L^2} = 1\big\},$$

which also reads

$$\inf\left\{\sum_{k=1}^{M} \frac{1}{2m_k}\int_{\mathbb{R}^{3M}} |\nabla_{\bar{x}_k}\psi_n|^2 + \int_{\mathbb{R}^{3M}} W|\psi_n|^2,\ \psi_n \in \mathcal{H}_n,\ \|\psi_n\|_{L^2} = 1\right\} \tag{6.3}$$

with

$$W(\bar{x}_1, \dots, \bar{x}_M) = U(\bar{x}_1, \dots, \bar{x}_M) + \sum_{1 \leqslant k < l \leqslant M} \frac{z_k z_l}{|\bar{x}_k - \bar{x}_l|}, \tag{6.4}$$

$$U(\bar{x}_1, \dots, \bar{x}_M) = \inf\big\{\big\langle\psi_e, H_e^{(\bar{x}_1,\dots,\bar{x}_M)}\psi_e\big\rangle,\ \psi_e \in \mathcal{H}_e,\ \|\psi_e\|_{L^2} = 1\big\}, \tag{6.5}$$

$$H_e^{(\bar{x}_1,\dots,\bar{x}_M)} = -\sum_{i=1}^{N} \frac{1}{2}\Delta_{x_i} - \sum_{i=1}^{N}\sum_{k=1}^{M} \frac{z_k}{|x_i - \bar{x}_k|} + \sum_{1 \leqslant i < j \leqslant N} \frac{1}{|x_i - x_j|}. \tag{6.6}$$

The Hamiltonian $H_e^{(\bar{x}_1,\dots,\bar{x}_M)}$ is called the electronic Hamiltonian, for it only acts on the electron coordinates, the nuclei being seen as parameters. The second term of it, namely the attraction of the nuclei on the electrons will be henceforth denoted by

$$\mathcal{V}(x_1,\dots,x_N)=\sum_{i=1}^{N}V(x_i)=-\sum_{i=1}^{N}\sum_{k=1}^{M}\frac{z_k}{|x_i-\bar{x}_k|}.$$

We next let the mass m_k of each nucleus go to infinity. It is easy to see, letting simultaneously ψ_n concentrate on the points $(\bar{x}_1,\dots,\bar{x}_M)$ where W attains its minimum, that the infimum of (6.3) goes to

$$\inf\{W(\bar{x}_1,\dots,\bar{x}_M),\ (\bar{x}_1,\dots,\bar{x}_M)\in\mathbb{R}^{3M}\}. \tag{6.7}$$

Solving (6.1) therefore amounts to minimizing W on $\mathbb{R}^{3M}$. In computational quantum chemistry, one refers to (6.7) as the Born–Oppenheimer approximation to (6.1). Notice that each value of W is a minimization problem, namely (6.4)–(6.6). Solving (6.5) is *solving the electronic problem* for a given configuration of nuclei $(\bar{x}_1,\dots,\bar{x}_M)\in\mathbb{R}^3$. Solving (6.7) is *solving the geometry optimization problem*. We shall deal with the latter in Section 25.

Let us mention that it is possible to improve the basic Born–Oppenheimer approximation by a post-treatment of the solution of the geometry optimization problem: Indeed, it can be considered that the points of $\mathbb{R}^{3M}$ where the modulus of the minimizer ψ_n of (6.3) is not negligible is a small neighborhood of the equilibrium geometry $(\bar{x}_1^{\text{opt}},\dots,\bar{x}_M^{\text{opt}})$; in these points, the potential W can be approximated by the second order Taylor expansion of W around $(\bar{x}_1^{\text{opt}},\dots,\bar{x}_M^{\text{opt}})$; putting these two arguments together, one can replace the potential W in (6.3) by its second order Taylor expansion around $(\bar{x}_1^{\text{opt}},\dots,\bar{x}_M^{\text{opt}})$. The resulting minimization problem, which reduces to calculating the ground state of a collection of independent harmonic oscillators, can be solved analytically: Its minimum is the result of (6.7) corrected by a (usually small) positive contribution, the so-called *zero point energy* (see, e.g., SZABO and OSTLUND [1982]), which accounts for the quantum nature of the nuclei.

Let us now concentrate on the solution to (6.5). Searching for the value of the infimum in (6.5) and the states where it is attained (in most cases, there is at least one such state, contrary to the situation encountered in (6.1)) is the central task of computational quantum chemistry. As announced above, the knowledge of the ground state energy and that of the ground state wavefunction is interesting per se, and is in addition a preliminary task for many, if not all, computations. In the static setting, it is for instance needed for the computation of the response of the system to any kind of solicitations. Such a computation sometimes also requires the knowledge of the excited states, but in any case requires that of the ground state before all (see Section 4). In the time dependent setting, the knowledge of (6.5) allows, e.g., to use it in an adiabatic simulation. We shall come back to it in Chapter V into more details.

From the theoretical viewpoint, (6.5) is a well posed problem enjoying very convenient properties. We will not detail these; let us however mention the following properties of the minimizer Ψ to (6.5), because they will have some impact on the discretization issues we shall examine in Chapter II. It indeed turns out that properly accounting for the qualitative behavior of the exact ground state wavefunction is a crucial fact for designing a numerical approximation of it. T. Kato has, in his fundamental study on the solution of the n-electron Schrödinger equation in the field of *clamped point nuclei* with charge Z_μ at $\vec{R}_\mu$ and with a *Coulomb interaction* between the electrons (KATO [1980]), derived some important theorems, among others that the wavefunction is bounded everywhere and that the Schrödinger equation is satisfied in any neighborhood of the Coulomb singularities. A consequence is that the wavefunction has a discontinuous first derivative at the Coulomb singularities and has to satisfy so-called cusp relations. That for the *nuclear cusp* is

$$\lim_{r_{k\mu}\to 0}\left(\frac{\partial\Psi}{\partial r_{k\mu}}\right)_{\mathrm{av}} = -Z_\mu\Psi(r_{k\mu}=0) \tag{6.8}$$

with $r_{k\mu}=|\vec{r}_k-\vec{R}_\mu|$, $\vec{r}_k$ the position of the kth electron, and av meaning spherical averaging.

The *correlation cusp* is characterized by

$$\lim_{r_{ij}\to 0}\left(\frac{\partial\Psi}{\partial r_{ij}}\right)_{\mathrm{av}} = \frac{1}{2}\Psi(r_{ij}=0) \tag{6.9}$$

with $r_{ij}=|\vec{r}_i-\vec{r}_j|$.

Also the coalescence of *three particles*, e.g., a nucleus and two electrons has been studied. The Fock expansion given in FOCK [1958], ABBOTT and MASLEN [1987], which contains logarithmic terms in $(r_1^2+r_2^2)$, is not yet understood in all details (MORGAN III [1986]). Inclusion of terms that appear in a Fock expansion, has for practical variational calculations of the He atom lead to a significantly improved convergence (FRANKOWSKI and PEKERIS [1966], FREUND, HUXTABLE and MORGAN III [1984]), but has not been applied beyond the He ground state. A similarly good convergence could later also be achieved in a much simpler way without any logarithmic terms, see DRAKE and YAN [1994].

In the same vein, one also knows the behavior of the exact square root of the electron density ϱ for large distances from any nucleus (AHLRICHS, M. HOFFMANN-OSTENHOFF, TH. HOFFMANN-OSTENHOFF and MORGAN III [1981], THIRRING [1983]). Actually,

$$\sqrt{\varrho(x)}\leqslant C(1+r)^{(Z-n+1)/\sqrt{2\varepsilon}-1}\exp\bigl(-\sqrt{2\varepsilon}\,r\bigr),$$

where $r=|x|$, ε is the first ionization potential, n the number of electrons, and Z the nuclear charge (or the sum of the nuclear charges for a molecule). There is an exponential decay, determined by the first ionization potential. The discretization strategy will have to account for it. We do not go into further details here.

Let us now turn to the practical standpoint: Problem (6.5) cannot be attacked as such, unless for very special cases we deal with in Section 7. From the practical standpoint, Problem (6.5) indeed suffers from two main drawbacks

- the space $\mathcal{H}_e$ of wavefunctions to be explored is a subspace of $L^2(\mathbb{R}^{3N})$, which is of too large a size to be discretized as soon as $N \geqslant 3$,
- the calculation of the energy $\langle \psi_e, H_e^{(\bar{x}_1,\ldots,\bar{x}_M)} \psi_e \rangle$ for an arbitrary ψ_e involves integrals of the form

$$\int_{\mathbb{R}^{3N}} \frac{1}{|x_i - x_j|} |\psi_e|^2$$

due to the presence of the electrostatic repulsion between electrons in (6.6). The calculation of these integrals (called in chemistry the *bielectronic* integrals, for they involve the positions of *two* electrons simultaneously) is an overwhelming practical task (even when $N = 2$ one encounters serious difficulties, for integrating on a 6-dimensional domain is not an easy task).

7. The basic model: Direct attacks

For systems involving a few (say today six or seven) electrons, a direct variational approximation of problem (6.5) can be envisaged. This consists in considering a finite-dimensional subspace $V_h = \operatorname{span}(\chi_1, \ldots, \chi_{N_b})$ of $H^1(\mathbb{R}^3 \times \{|+\rangle, |-\rangle\})$, and in approximating (6.5) by the finite dimensional problem

$$\inf\left\{\left\langle \psi_e, H_e^{(\bar{x}_1,\ldots,\bar{x}_M)} \psi_e \right\rangle,\ \psi_e \in \bigwedge_{i=1}^{N} V_h,\ \|\psi_e\|_{L^2} = 1\right\}. \tag{7.1}$$

The dimension of the vector space $\bigwedge_{i=1}^{N} V_h$ is $\binom{N_b}{N}$; it is therefore clear that for an exact resolution of problem (7.1) being possible, it is necessary that

(1) the dimension N_b of V_h is small enough;
(2) for a given test wavefunction $\psi_e \in \bigwedge_{i=1}^{N} V_h$, the computation of the energy (especially that of the bielectronic integral) is not too costly.

The first condition notably eliminates finite element like methods. On the other hand, both conditions can be satisfied for well chosen atomic orbitals basis sets (see Chapter II). From a numerical viewpoint, problem (7.1) is seen as a special occurrence of the configuration interaction (CI) methods that will be described in Chapter II. For this reason, this variational method is referred to as the *Full CI* method in the literature.

This method gives excellent results when applicable (i.e., for few electron systems), but is unfortunately out of reach for larger systems due to the intrinsic combinatoric complexity of the Full CI methodology.

A completely different approach consists in using stochastic (Monte Carlo) methods to directly attack problem (6.5). A whole contribution of this volume being devoted to these methods, we do not develop further here.

8. The approximations

The most commonly used approximations to the minimization problem (6.5) can schematically be classified into two main classes (Fig. 8.1)

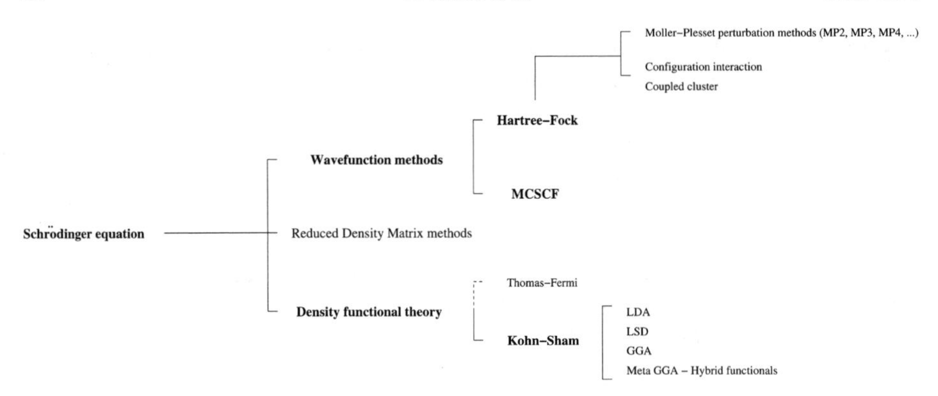

FIG. 8.1. Classification of electronic ground state calculation methods.

- *wavefunction methods* aim at finding an approximation of the ground state electronic wavefunction, i.e., of the minimizer of (6.5). Basically, a variational approximation is used: The variational space $\mathcal{H}_e$ is reduced (for it is too large in most cases to be numerically tractable) but the "exact" form of the energy $\langle \psi_e, H_e^{(\bar{x}_1,\ldots,\bar{x}_M)} \psi_e \rangle$ is kept. The famous Hartree–Fock approximation is one of them, as well as configuration interaction (CI) and multi-configuration self-consistent field (MCSCF) methods. Nonvariational perturbation methods have also been developed in order to improve the Hartree–Fock results, notably the Møller–Plesset perturbation methods (MP2, MP3, MP4, ...) and the coupled cluster (CC) expansions;
- *density functional methods* are issued from the density functional theory. They are based on a reformulation of problem (6.5) in such a way that the main variable is the electronic density

$$\rho(x) = N \int_{\mathbb{R}^{3(N-1)}} \sum_{\sigma_1,\ldots,\sigma_N} \left|\psi_e(x, \sigma_1, x_2, \sigma_2, \ldots, x_N, \sigma_N)\right|^2 dx_2 \cdots dx_N$$

  (i.e., a scalar field on $\mathbb{R}^3$) rather than the wavefunction (i.e., a scalar field on $\mathbb{R}^{3N}$) as in the original problem (6.5). The density functional theory indeed states that the ground state energy can be obtained by minimizing some functional of the density. Unfortunately this theory does not provide a tractable way to compute this energy functional, which therefore must be approximated.

From the standpoint of scientific computing, this classification roughly amounts to a discrimination between on the one hand variational approaches, such as the finite element methods used to attack variational problems by approaching the set of test functions by finite-dimensional approximations while keeping the energy functional (or the equation) unchanged, and on the other hand finite difference like methods that approximate the operator itself (or the energy functional).

A third class consists of the methods dealing with *density matrices* instead of wavefunctions or densities. These methods are clearly less commonly used than the first two ones at this day, but on the other hand they seem very promising.

We shall devote the next three sections to the presentation of all these methods.

Loosely speaking, one may say that a schematic answer to the natural question: "*Which approximation for which community and which applications*?" is the following. Wavefunction methods are the methods of choice for those of the chemists who are interested in the precise simulations of systems of small size. The reason is that there is not that many parameters that are free to fix in the wavefunction setting (the size of the basis set, its nature, and a few items like these). Wavefunction type calculations are thus likely to be the most intrinsic ones. In addition, they can provide results as accurate as wanted ... provided computational time is unlimited. On the other hand, when computational time is a bottleneck (i.e., for larger size systems) density functional theory methods are more convenient: In general, they offer a much better accuracy than the basic wavefunction method (i.e., the Hartree–Fock approximation) for a given computational cost. This is why these methods are widely used by those of the chemists who are interested in large molecular systems (e.g., biological systems) as well as most solid-state physicists. The fact that various "parameters" need to be arbitrarily chosen for these methods, for instance the form of the energy functional itself, does make the method particularly efficient for some situations (the above parameters are tuned for a specific calculation for the system they are dedicated to) but is sometimes seen as a lack of rigour by chemists. The poor universality of DFT computation is recognized as a clear drawback. From our formal viewpoint, the two approaches deserve an equal attention.

As we focus in the following section on the electronic problem, we shall henceforth fix the nuclei positions $(\bar{x}_1, \dots, \bar{x}_M)$ and therefore forget the superscript over the electronic Hamiltonian H_e defined in (6.6). The problem under consideration is to calculate an approximation to the right hand side of (6.5), namely

$$\inf\{\langle \psi_e, H_e\psi_e\rangle,\ \psi_e \in \mathcal{H}_e,\ \|\psi_e\|_{L^2} = 1\}.$$

9. Wavefunction methods: The Hartree–Fock approximation

We detail in this section and in the next three sections one of the most commonly used approximations of the stationary Schrödinger equation, namely the *Hartree–Fock approximation*.

For the sake of simplicity, we shall first expose the models *without* taking into account the spin variables. We proceed in this fashion for clarity and in order to focus on the mathematical features. The equations we shall derivate are therefore those mathematicians are used to deal with for abstract studies.

Next, we shall present the same approximation, but with another language, that of the density matrices, which, apart from its interest for formal studies, turns out to be the most efficient language to build and analyze the numerical algorithms in use. We refer to Chapter II for such issues.

In a third stage, we detail the so-called *post Hartree–Fock methods* that are improvements of the basic setting. In practice, it is indeed rare that one stops at the Hartree–Fock level. Improvements are most often necessary.

Finally, for the sake of completeness, we shall detail the different versions of the Hartree–Fock approximation explicitly accounting for the spin variable. They are of course the approximations used in the numerical practice.

Let us already mention that a huge bibliography has been devoted to the Hartree–Fock model and the models it gives rise to. Among other excellent references, we wish to point out SZABO and OSTLUND [1982], HEHRE, RADOM, SCHLEYER and POPLE [1986], LEVINE [1991], MCWEENY [1992].

As announced above, we forget for a while the spin variables. The electronic wavefunction is therefore a function of the $3N$ space coordinates of the N electrons.

The Hartree–Fock approximation, abbreviated in HF, consists in restricting in the variational problem (6.5) (here written without the spin variable),

$$U(\bar{x}_1, \ldots, \bar{x}_M) = \inf\left\{ \langle \psi_e, H_e \psi_e \rangle,\ \psi_e \in \bigwedge_{i=1}^{N} H^1(\mathbb{R}^3),\ \|\psi_e\|_{L^2} = 1 \right\},$$

the variational space $\mathcal{H}_e$ to that of functions of variables $(x_1, \ldots, x_N) \in \mathbb{R}^{3N}$ that can be written as a *single* determinant (i.e., an antisymmetrized product) of N functions defined on $\mathbb{R}^3$. In the whole generality, an arbitrary element of $\mathcal{H}_e$ only is a converging infinite sum of such determinants. The Hartree–Fock approximation is therefore defined as

$$U^{\mathrm{HF}}(\bar{x}_1, \ldots, \bar{x}_M) = \inf\left\{ \langle \psi_e, H_e \psi_e \rangle,\ \psi_e \in \mathcal{S}_N \right\}, \tag{9.1}$$

with

$$\begin{aligned} \mathcal{S}_N &= \left\{ \psi_e \in \mathcal{H}_e \mid \exists \Phi = \{\phi_i\}_{1 \leqslant i \leqslant N} \in \mathcal{Y}_N,\ \psi_e = \frac{1}{\sqrt{N!}} \det\bigl(\phi_i(x_j)\bigr) \right\}, \\ \mathcal{Y}_N &= \left\{ \Phi = \{\phi_i\}_{1 \leqslant i \leqslant N},\ \phi_i \in H^1(\mathbb{R}^3),\ (\phi_i, \phi_j) = \int_{\mathbb{R}^3} \phi_i \phi_j = \delta_{ij},\ 1 \leqslant i, j \leqslant N \right\}. \end{aligned} \tag{9.2}$$

In order to make the form of the wavefunction more explicit for the reader, we expand it as follows:

$$\psi_e(x_1, \ldots, x_N) = \frac{1}{\sqrt{N!}} \det\bigl(\phi_i(x_j)\bigr) = \frac{1}{\sqrt{N!}} \begin{vmatrix} \phi_1(x_1) & \ldots & \phi_1(x_N) \\ \cdot & & \cdot \\ \cdot & & \cdot \\ \cdot & & \cdot \\ \phi_N(x_1) & \ldots & \phi_N(x_N) \end{vmatrix}. \tag{9.3}$$

In the language of quantum chemistry, a function of the form (9.3) is called a *Slater determinant* (thus the notation $\mathcal{S}_N$). The ϕ_i are called *molecular orbitals*; other names such as *monoelectronic* (or *single-particle*) *wavefunctions* are also found in the literature.

Of course, restricting the minimization to *some* specific forms of functions in (9.1) provides only with an upper bound of the energy (6.5).

The main advantage, with respect to other types of approximations such as those of the density functional theory that we shall see later on, is that the Hartree–Fock ground state energy precisely is *always* an upper bound for the exact exergy (6.1) (one speaks of a *variational* approximation). In addition, when comparing the energy of different configurations of the same chemical system (geometry optimization, rotation barriers) or that of different molecular systems (in order to compute reaction and activation energies of chemical reactions), cancellations of errors are observed in practice. They are however not explained mathematically.

The main drawback of the approximation is that the simplification basically consists in considering that the electrons are independent from one another, which is obviously not true. Indeed, forgetting for a while about the antisymmetry, which is due to the Pauli principle, the Hartree–Fock approximation heuristically consists in writing that the probability density

$$|\psi|^2(x_1, \dots, x_N)$$

of finding the N electrons at positions $(x_1, \dots, x_N)$ can be written as the product

$$|\phi_1|^2(x_1) \cdots |\phi_N|^2(x_N).$$

In terms of the theory of probabilities, this amounts to considering the positions of the electrons as independent stochastic variables.

This simplification causes a loss of *correlation* between the positions of the electrons. The word "correlation" can be understood again in terms of the theory of probabilities, but it is also meaningful in terms of quantum chemistry: Under this approximation, one electron evolves independently from the way the other ones do. It is however to be remarked that the formal argument we have given above has to be somewhat weakened in the following fashion. Due to the antisymmetry requirement, the Hartree–Fock state (9.3) is not a direct product of one electron wavefunctions, but a determinant of them. Consequently, *some* correlation (in the sense of probability theory) is in fact embodied in the Hartree–Fock setting. Indeed, the probability density $|\psi_e(x_1, \dots, x_N)|^2$ that two electrons are exactly at the same position ($x_i = x_j$ for some $i \neq j$) vanishes since two columns of the determinant (9.3) are then identical. By continuity, we may therefore admit that the probability to find two electrons almost in the same position is small. The positions of two different electrons are therefore correlated because of the fermionic nature of the electrons taken into account in the antisymmetry properties of the wavefunction. In fact, this formal argument must be itself slightly weakened by the fact that in the simplified picture we give here, we do not take the spin into account. Anticipating on the sequel (see (12.1)), the complete precise argument implies that, in the Hartree–Fock model, only two electrons *sharing the same spin* are in some sense correlated to one another, while two electrons of different spin are not (one speaks of a *Fermi hole* around a given electron). Obviously, this is not sufficient to account for reality where positions of electrons are also correlated, whatever their spins, because of the electrostatic repulsion which prevents two electrons to get too close (one speaks of a *Coulomb hole*). A direct consequence is that uncorrelated wavefunctions are too much concentrated around the nuclei; introducing correlation allows a more diffuse character to the functions.

The price to pay for the loss of correlation is that the Hartree–Fock energy, i.e., the scalar value defined by (9.1) is *strictly* larger than the true energy (6.5). In the language of quantum chemistry, the *correlation energy* is precisely defined as the difference between the two. Some corrections of the Hartree–Fock model can then be implemented in order to cure the disease. They will be dealt with later on.

Let us now write the Hartree–Fock approximation in a more explicit fashion. For $\Phi = \{\phi_i\}_{1\leqslant i\leqslant N} \in \mathcal{Y}_N$ and $\psi_e \in \mathcal{S}_N$ the Slater determinant built with Φ, we denote by

$$\tau_\Phi(x, x') = \sum_{i=1}^{N} \phi_i(x)\phi_i(x'), \tag{9.4}$$

and

$$\rho_\Phi(x) = \rho_{\psi_e}(x) = \sum_{i=1}^{N} \left|\phi_i(x)\right|^2.$$

The function τ_Φ and ρ_Φ are, respectively, called the *density matrix* and the *density* associated to the state ψ_e. The function τ_Φ is in fact a special occurrence for a Slater determinant ψ_e of the general formula

$$\gamma_1(x, x') = \int_{\mathbb{R}^{3(N-1)}} \psi_e(x, x_2, x_3, \ldots, x_N)\psi_e(x', x_2, x_3, \ldots, x_N)\, dx_2 \cdots dx_N \tag{9.5}$$

which defines in the whole generality the *first order reduced density matrix* (1-RDM in short) calculated from the state ψ_e (objects of this type will be manipulated in Section 16 below). The density ρ_Φ is then deduced from τ_Φ.

This allows us to compute

$$\begin{aligned} E^{\mathrm{HF}}(\Phi) &= \langle \psi_e, H_e \psi_e \rangle \\ &= \sum_{i=1}^{N} \frac{1}{2}\int_{\mathbb{R}^3} |\nabla\phi_i|^2 + \int_{\mathbb{R}^3} \rho_\Phi\, V + \frac{1}{2}\int_{\mathbb{R}^3}\int_{\mathbb{R}^3} \frac{\rho_\Phi(x)\rho_\Phi(x')}{|x - x'|}\, dx\, dx' \\ &\quad - \frac{1}{2}\int_{\mathbb{R}^3}\int_{\mathbb{R}^3} \frac{|\tau_\Phi(x, x')|^2}{|x - x'|}\, dx\, dx'. \end{aligned} \tag{9.6}$$

We therefore write

$$U^{\mathrm{HF}}(\bar{x}_1,\ldots,\bar{x}_M)$$
$$= \inf\Bigg\{\sum_{i=1}^{N}\frac{1}{2}\int_{\mathbb{R}^3}|\nabla\phi_i|^2 + \int_{\mathbb{R}^3}\rho_\Phi V + \frac{1}{2}\int_{\mathbb{R}^3}\int_{\mathbb{R}^3}\frac{\rho_\Phi(x)\rho_\Phi(x')}{|x-x'|}\,dx\,dx'$$
$$-\frac{1}{2}\int_{\mathbb{R}^3}\int_{\mathbb{R}^3}\frac{|\tau_\Phi(x,x')|^2}{|x-x'|}\,dx\,dx',$$
$$\phi_i\in H^1(\mathbb{R}^3),\ \int_{\mathbb{R}^3}\phi_i\phi_j=\delta_{ij},\ 1\leqslant i,j\leqslant N\Bigg\}. \tag{9.7}$$

The above *spinless Hartree–Fock model* (9.7) has been extensively studied by mathematicians and experts at mathematical physics. For reasons that will be made clear in Section 17, it is not a simple task to prove that such a minimization problem has a full mathematical meaning. Issues of mathematical interest are to know

(i) whether the infimum U^{HF} defined by the right hand side of (9.7) is indeed a finite scalar and not $-\infty$,

(ii) whether the infimum is attained, i.e., whether there exists at least one N-tuple of functions ϕ_i such that the energy of this N-tuple is exactly the infimum, in which case the Slater determinant Φ built with this N-tuple is called a Hartree–Fock ground state,

(iii) whether the Hartree–Fock ground state is indeed unique, up to trivial invariance properties of the Hartree–Fock energy functional we shall come back to.

The essential mathematical background needed to understand the difficulty in answering the above questions will be detailed in Section 17. Let us only mention here the crucial mathematical results known at this day, and refer the reader to the bibliography for the details on the results and proofs.

The proof of an existence of at least one minimizer to (9.7), namely of an Hartree–Fock ground state, is due to LIEB and SIMON [1977a]. It holds under the condition that the total nuclear charge $Z=\sum_{k=1}^{M}z_k$ of the molecular system satisfies $Z>N-1$, where we recall that N is the number of electrons. This mathematical condition translates the physical property that the nuclei should be sufficiently positively charged in order to be able to bind the N electrons at their vicinity. The proof makes use of the following two properties.

First, the infimum (9.7) is not changed if one replaces the orthonormality constraint $\int_{\mathbb{R}^3}\phi_i\phi_j=\delta_{ij}$, $1\leqslant i,j\leqslant N$, into the so-called *relaxed constraint* that the matrix $[\int_{\mathbb{R}^3}\phi_i\phi_j]$ is less than or equal to the identity matrix in the sense of Hermitian matrices, namely that

$$\sum_{i,j=1}^{N}\left(\int_{\mathbb{R}^3}\phi_i\phi_j\right)\xi_i\xi_j\leqslant\sum_{i=1}^{N}\xi_i^2,\quad \forall\{\xi_i\}_{1\leqslant i\leqslant N}\in\mathbb{R}^N.$$

Therefore

$$U^{\mathrm{HF}}(\bar{x}_1, \dots, \bar{x}_M) = \inf\left\{ E^{\mathrm{HF}}(\Phi),\ \phi_i \in H^1(\mathbb{R}^3),\ \left[\int_{\mathbb{R}^3} \phi_i \phi_j\right] \leqslant [\delta_{ij}] \right\}. \tag{9.8}$$

One then proves, when possible, the existence of a minimizer to (9.8). Secondly, any minimizer to the above problem with relaxed constraint (9.8) is indeed "on the boundary", i.e., satisfies $\int_{\mathbb{R}^3} \phi_i \phi_j = \delta_{ij}$, $1 \leqslant i, j \leqslant N$. This yields a Hartree–Fock ground state.

These properties will have fundamental consequences on the numerical practice, as will be seen in Chapter II.

It is simple to see that such a minimizer satisfies the Euler–Lagrange equations associated to (9.7), namely the following system of N coupled partial differential equations

$$\begin{cases} -\dfrac{1}{2}\Delta\phi_i + V\phi_i + \left(\displaystyle\sum_{j=1}^{N} |\phi_j|^2 \star \frac{1}{|x|}\right)\phi_i - \displaystyle\sum_{j=1}^{N}\left(\phi_i\phi_j \star \frac{1}{|x|}\right)\phi_j = \displaystyle\sum_{j=1}^{N} \lambda_{ij}\phi_j, \\ \displaystyle\int_{\mathbb{R}^3} \phi_i\phi_j = \delta_{ij}, \end{cases} \tag{9.9}$$

where the matrix λ_{ij} is the Hermitian matrix of Lagrange multipliers associated to the orthonormality constraints.

Next, it is easy to see that both the Hartree–Fock energy (9.6) and the so-called *Fock operator*

$$\begin{aligned} F_\Phi &= -\frac{1}{2}\Delta + V + \left(\sum_{j=1}^{N} |\phi_j|^2 \star \frac{1}{|x|}\right) - \sum_{j=1}^{N}\left(\cdot\phi_j \star \frac{1}{|x|}\right)\phi_j \\ &= -\frac{1}{2}\Delta + V + \left(\rho_\Phi \star \frac{1}{|x|}\right) - \int_{\mathbb{R}^3} \frac{\tau_\Phi(x, x')}{|x - x'|} \cdot (x')\, dx' \end{aligned} \tag{9.10}$$

appearing in the left hand side of (9.9), are invariant with respect to unitary transforms of N-tuple $\Phi = (\phi_1, \dots, \phi_N)$: More precisely, for any $U \in U(N)$

$$\tau_{\Phi U} = \tau_\phi, \quad \rho_{\Phi U} = \rho_\Phi, \quad E^{\mathrm{HF}}(\Phi U) = E^{\mathrm{HF}}(\Phi) \quad \text{and} \quad F_{\Phi U} = F_\phi. \tag{9.11}$$

Let us mention that this unitary invariance originates from the fact that a determinant remains unchanged when its columns are transformed through a unitary transform.

We therefore may use this invariance to diagonalize the matrix of Lagrange multipliers and obtain some new N-tuple, still denoted by $(\phi_1, \dots, \phi_N)$ with a slight abuse of notation, such that

$$\begin{cases} -\frac{1}{2}\Delta\phi_i + V\phi_i + \left(\sum_{j=1}^{N} |\phi_j|^2 \star \frac{1}{|x|}\right)\phi_i - \sum_{j=1}^{N}\left(\phi_i\phi_j \star \frac{1}{|x|}\right)\phi_j = \lambda_i\phi_i, \\ \int_{\mathbb{R}^3} \phi_i\phi_j = \delta_{ij} \end{cases} \tag{9.12}$$

for N scalars λ_i. As the unitary transform we have operated has also left the Hartree–Fock energy unchanged, the Slater determinant built with the new N-tuple is also a minimizer. The λ_i, which mathematically arise as the diagonal form of the matrix of Lagrange multipliers associated to the orthonormality constraint, are called the *one-electron energies* of the *occupied orbitals* ϕ_i, $1 \leqslant i \leqslant N$. They can be interpreted in terms of ionization energies in theoretical chemistry (the result is known under the name of the Koopmans' theorem, see, e.g., SZABO and OSTLUND [1982]).

Eqs. (9.12), which can be written in the more compact form

$$\begin{cases} F_\Phi\phi_i = \lambda_i\phi_i, \\ \int_{\mathbb{R}^3} \phi_i\phi_j = \delta_{ij}, \end{cases} \tag{9.13}$$

are known as the *Hartree–Fock equations*. They take the form of a *nonlinear eigenvalue problem*. Indeed, the λ_i appear as eigenvalues, associated to the eigenfunctions ϕ_i. This nonlinear eigenvalue problem is known in quantum chemistry as a *self consistent field* problem, abbreviated in SCF, in order to emphasize that the electrostatic field the electrons are experiencing (namely the last two terms of the Fock operator (9.10)) indeed depends on the state of the electrons, i.e., on the ϕ_is. Any minimizer to (9.7) is, up to an orthogonal transform, a solution to (9.12). Moreover, it has been proved in LIONS [1987] that if Ψ is a Hartree–Fock ground state, and therefore is a solution to (9.12) then the λ_i are not any, but necessarily are the lowest N eigenvalues of the Fock operator F_Φ. Another interesting property, established in BACH, LIEB, LOSS and SOLOVEJ [1994], is that, for any minimizer again, the Nth energy level λ_N is not degenerated.

It should now be emphasized that it is not known whether such a minimizer to (9.7) is unique or not (up to an orthogonal transformation). In particular, the article by PAYNE [1979] which is sometimes cited in theoretical chemistry as a reference for the proof of uniqueness of such a ground state is not satisfactory from the mathematical standpoint. The question remains unsolved and, in our opinion, is likely to be very difficult to settle. Likewise, it is also an open problem to know whether the electronic density ρ_Φ built from any Hartree–Fock ground state Φ is itself unique. It is also important to note that *it is not known* whether any solution to (9.12) conversely is a minimizer to (9.7). In particular, it has been proven in LIONS [1987] that there are infinitely many solutions to (9.12), that can be in some sense interpreted as excited states. It is not known either which additional property we must impose on a solution to (9.12) to make it a minimizer to (9.7). The property along which the λ_i are the lowest N eigenvalues of the Fock operator F_Φ has not been proven, nor disproven, to be sufficient.

Apart from the above mathematical questions (i)–(iii) which obviously are crucial, there are other mathematical properties useful to judge for the validity and the physical relevance of a mathematical minimization problem of the type (9.7). Let us mention a few of them.

When a solution to (9.12) exists, and in particular for the Hartree–Fock ground state, it is possible to show that the ϕ_i solutions to (9.12) are necessarily smooth functions (except at the point nuclei), that in addition decrease exponentially fast at infinity. This latter property implies the exponential fall-off of the electronic density, which agrees with experimental observations and with the theoretical properties of the exact ground-state. Some upper bounds on the decrease may also be estimated. Such regularity results may be useful in the numerical practice when deciding the way these functions are discretized.

We have claimed that a sufficient condition for a ground state to (9.7) to exist is that $Z > N - 1$. It is also known from LIEB [1984] that when $2Z + M \leqslant N$ (we recall that M is the number of nuclei the system is composed of), then no ground state exists. This is coherent with the physical observation along which a given number of positive charges cannot bind an arbitrary large number of electrons to form a stable molecular system.

A natural question consists of asking whether one can or cannot recover the exact Schrödinger problem (6.5) from the Hartree–Fock approximation, even in some weak, formal, sense. We have indeed mentioned that the Hartree–Fock energy is a bound from above for the true energy (due to the variational nature of the Hartree–Fock approximation). Can the difference between the two be made arbitrarily small in some well chosen regime? The answer is affirmative. It indeed turns out that the relative difference between the Hartree–Fock ground state energy and the exact ground state energy of an atom goes to zero as the nuclear charge Z goes to infinity. For this result and studies in the same vein, we refer to BACH [1992] and the references therein.

All the above mathematical results contribute to give confidence in the ability of the Hartree–Fock approximation (and further of its improvements) to fit to experimental reality. They justify a posteriori and from the theoretical standpoint the success of this approximation in the numerical practice.

At this point, it is important to wonder about the impact of the state of the mathematical knowledge described so far on the practice of quantum chemistry.

From the standpoint of theoretical chemistry, it is to be underlined that the knowledge of the existence of the infimum, and the knowledge on the existence and possible uniqueness of the ground state in some given approximation, such as the present Hartree–Fock one is not only of an academical concern. Indeed, we have mentioned previously that the existence of the ground state of a compound is fundamental to apprehend its chemical reactivity. It is worth stressing at this point that the search of the chemical stability is one of the main purpose of computer-aided designs of new materials. The same is true for intermediate species such as highly negative anions. However, one has to be aware that for practical instances the proof of the existence and possible uniqueness of a ground state in some given approximation is not sufficient. The compound under study has to be stable for any possible approximation. Moreover, one has to *compute* the ground state, but it is another story which we shall consider in Chapter II.

Likewise, the impact on the numerical practice is very important.

It is of course useful to know that a minimizer exists when one tries to approximate it numerically, although in most situations the numerics (fortunately) does not wait for such an existence result; indeed finite dimensional problems issued from numerical approximations are simpler in their behavior. We have already mentioned that the regularity and the decreasing property of the one-electron wavefunctions may be useful in a convenient choice for its discretization, e.g., truncation of the computational domain. The uniqueness of the minimizer, or of any quantity built from it (such as the electronic density), would be an extremely useful property for numerical purposes. In particular, it is necessary to expect the unconditional convergence of the minimization algorithms. Being in such an incomplete knowledge of the mathematical properties of the minimizers and of the solutions to (9.12) is therefore embarrassing for the numerics. This is all the more true as the direct attack of the minimization problem (9.7) with minimization algorithms is not the most usual way to proceed (although it is indeed possible, contrarily to what is sometimes erroneously claimed, we shall come back to that important point in Chapter II). The most commonly used approach to attack (9.7) is indeed to iteratively solve Eqs. (9.12). Therefore proving the convergence of such algorithms at first, and next proving the convergence to a (at least) local minimizer to (9.7) are likely to be difficult. And we shall see it is indeed the case. This is in fact the best result one can hope, for proving the convergence of such algorithms to a global minimizer is out of reach without knowing about its uniqueness.

The slightest information one can get from the mathematical study can be very useful in the numerical practice. For instance, the property along which the Nth energy level is, or is not, degenerated at the ground state has important consequences on the algorithms. Likewise, the fact that the λ_i are the lowest N eigenvalues is used explicitly in the practice. Finally, the fact that one can relax the constraint by using the expression (9.8) of the Hartree–Fock energy, still knowing that the minimizer is necessarily orthonormal, has been extensively used to construct new algorithms that have proven to be extremely efficient. All this will be detailed in Chapter II.

10. Wavefunction methods: Density matrix formulation of the Hartree–Fock model

Let us now introduce the *first order density operator*

$$\mathcal{D}_\Phi = \sum_{i=1}^{N} |\phi_i\rangle\langle\phi_i| \tag{10.1}$$

associated with some $\Phi = \{\phi_i\} \in \mathcal{Y}_N$. In formula (10.1) the classical *bra-ket* notations have been used; with more standard mathematical notations, (10.1) reads

$$\mathcal{D}_\Phi = \sum_{i=1}^{N} (\phi_i, \cdot)_{L^2}\phi_i.$$

The density operator associated with the state Φ therefore is the orthogonal projector from $L^2(\mathbb{R}^3)$ onto the N-dimensional vector space generated by the ϕ_i's; it can be checked that unitary transforms of the ϕ_i also leave the density operator unchanged:

$$\forall U \in U(N),\ \mathcal{D}_{\Phi U} = \mathcal{D}_\Phi.$$

Let us also remark that the density matrix τ_Φ defined by (9.4) is the Green kernel of the density operator $\mathcal{D}_\Phi$.

It is possible, and indeed very convenient for numerical purposes as will be seen below, to rewrite the Hartree–Fock problem by considering the density operator as the unknown variable.

On the one hand, the Hartree–Fock energy functional is a function of the density operator only:

$$\langle \psi_e, H_e \psi_e \rangle = E^{\mathrm{HF}}(\Phi) = \mathcal{E}^{\mathrm{HF}}(\mathcal{D}_\Phi),$$

where

$$\mathcal{E}^{\mathrm{HF}}(\mathcal{D}) = \mathrm{Trace}(h\mathcal{D}) + \frac{1}{2}\mathrm{Trace}\big(\mathcal{G}(\mathcal{D}) \cdot \mathcal{D}\big).$$

The Hamiltonian

$$h = -\frac{1}{2}\Delta + V$$

is called the *core Hamiltonian* and for all $\phi \in H^1(\mathbb{R}^3)$ and $x \in \mathbb{R}^3$ we have denoted by

$$\big(\mathcal{G}(\mathcal{D}) \cdot \phi\big)(x) = \left(\rho_\mathcal{D} \star \frac{1}{|y|}\right)(x)\phi(x) - \int_{\mathbb{R}^3} \frac{\tau_\mathcal{D}(x, x')}{|x - y|}\phi(x')\,dx',$$

where $\tau_\mathcal{D}(x, y)$ is the Green kernel of $\mathcal{D}$ and $\rho_\mathcal{D}(x) = \tau_\mathcal{D}(x, x)$.

On the other hand, the variational space one minimizes upon can also be expressed in terms of the density matrix. It is indeed easy to show that the density operators $\mathcal{D}$ that originate from a Slater determinant of finite energy, i.e., with columns belonging to $H^1(\mathbb{R}^3)$, are exactly the orthogonal projectors of rank N defined from $L^2(\mathbb{R}^3)$ onto $H^1(\mathbb{R}^3)$. It follows that (9.7) also reads

$$\inf\big\{\mathcal{E}^{\mathrm{HF}}(\mathcal{D}),\ \mathcal{D}^2 = \mathcal{D} = \mathcal{D}^*,\ \mathrm{Trace}(\mathcal{D}) = N\big\}. \tag{10.2}$$

Because of the property stating that one may also relax the orthonormality constraint ((9.7) is equivalent to (9.8)), it is immediate to see that (10.2) also reads

$$\inf\big\{\mathcal{E}^{\mathrm{HF}}(\mathcal{D}),\ \mathcal{D}^2 \leqslant \mathcal{D} = \mathcal{D}^*,\ \mathrm{Trace}(\mathcal{D}) = N\big\}.$$

The Hartree–Fock equation can then be written

$$\begin{cases} \mathcal{F}(\mathcal{D}) \cdot \phi_i = \lambda_i \phi_i, \\ \displaystyle\int_{\mathbb{R}^3} \phi_i \phi_j = \delta_{ij}, \\ \mathcal{D} = \displaystyle\sum_{i=1}^{N} |\phi_i\rangle\langle\phi_i|, \end{cases}$$

with

$$\mathcal{F}(\mathcal{D}) = h + \mathcal{G}(\mathcal{D}), \tag{10.3}$$

where the Fock operator reads as in (9.10)

$$\begin{aligned} \big(\mathcal{F}(\mathcal{D}) \cdot \phi\big)(x) = {} & -\frac{1}{2}\Delta\phi(x) + V(x)\phi(x) + \left(\rho_{\mathcal{D}} \star \frac{1}{|y|}\right)(x)\phi(x) \\ & - \int_{\mathbb{R}^3} \frac{\tau_{\mathcal{D}}(x, x')}{|x - y|} \phi(x')\, dx'. \end{aligned}$$

11. Wavefunction methods: Post Hartree–Fock methods

Five classes of post Hartree–Fock methods can be identified

(1) *Møller–Plesset perturbation* (MP2, MP3, MP4, ...) methods;
(2) *configuration interaction* (CI) methods;
(3) *coupled cluster* (CC) methods;
(4) *explicitly correlated wavefunction* methods;
(5) *multi-configuration self consistent field* (MCSCF) methods.

The methods of the first four classes consist in improving on the energy or wavefunction generated by the Hartree–Fock approximation; they are post-processing of Hartree–Fock calculations. The description of the Møller–Plesset, configuration interaction, and coupled cluster methods, whose formulations in terms of *sum over states* are based upon the discretized form of the Hartree–Fock equations, is postponed until Chapter II, once the discretization of the Hartree–Fock problem has been detailed. Explicitly correlated wavefunction methods are analyzed in Section 32.

We thus focus in the present section on MCSCF methods.

As mentioned several times above, the Hartree–Fock approximation is a variational approximation of (6.5), i.e., an approximation constructed by restricting the variational space to a smaller one, and that therefore provides an upper bound to the exact result. As a consequence, improving the approximation amounts to enlarging the variational space.

The *multiconfiguration self consistent field method* (abbreviated in MCSCF) aims at recovering more generality on the wavefunction ψ_e than the Hartree–Fock method. In order to recover the lost correlation, one remembers that any antisymmetric wavefunction of N variables is *not* of the form of a *single* determinant, but it is indeed of the form of an *infinite sum* of such determinants. This is nothing else but writing explicitly any arbitrary element of the space $\mathcal{H}_e = \bigwedge_{i=1}^{N} H^1(\mathbb{R}^3 \times \{|+\rangle, |-\rangle\})$. By enlarging the variational space to *some* or *any* finite sum of determinants

$$\psi_e(x_1,\ldots,x_N)=\sum_{q=1}^{Q} c_q \frac{1}{\sqrt{N!}}\det\bigl(\phi_i^q(x_j)\bigr) \tag{11.1}$$

for some integer Q, instead of considering only the functions of the form (9.3), one gets a better approximation to (6.5). We shall make precise below the integer Q, the coefficients c_q and the wavefunctions ϕ_i.

In the MCSCF method, the energy is minimized upon some set of functions (11.1), for functions ϕ_i^q satisfying convenient orthonormality constraints, and for coefficients c_q subject to a convenient constraint so that the functions ψ_e defined by (11.1) are normalized. Of course, the set of values $c_1=1$, $c_k=0$ for $k\geqslant 2$, for the coefficients of the linear combination (11.1) corresponds to a Hartree–Fock state. Formula (11.1) however covers more general functions.

More precisely, the MCSCF problem can be stated as follows

$$E_N^K=\inf\Biggl\{\langle\psi_e,H_e\psi_e\rangle,\ \psi_e=\sum_{I=\{i_1,\ldots,i_N\}\subset\{1,\ldots,K\}} c_I\frac{1}{\sqrt{N!}}\det(\phi_{i_1},\ldots,\phi_{i_N}),$$
$$\phi_i\in H^1(\mathbb{R}^3),\ \int_{\mathbb{R}^3}\phi_i\phi_j=\delta_{ij},\ \sum_I c_I^2=1\Biggr\}, \tag{11.2}$$

where $K\geqslant N$ is some given fixed integer. The constraints the ϕ_i and the c_I are subject to amount to enforcing $\|\psi_e\|_{L^2}=1$ and $\mathrm{Rank}(\mathcal{D}_{\psi_e})\leqslant K$, where $\mathcal{D}_{\psi_e}$ is the density operator built from ψ_e, that is to say the operator on $L^2(\mathbb{R}^3)$ whose kernel is the first order reduced density matrix (1-RDM) $\gamma_1(x,x')$ defined by (9.5). Of course, $K=N$ corresponds to the Hartree–Fock case.

The Euler–Lagrange equations of this minimization problem can be derived in the same fashion as they are in the Hartree–Fock setting, and one then obtains the so-called *MCSCF equations*, which basically are of the same form as the Hartree–Fock equations (9.12), apart from tedious technical details.

The MCSCF model (11.2) gives rise to the same theoretical interesting questions as the HF model does. The first mathematical result on the MCSCF model is due to one of us who has studied in LE BRIS [1994] the existence of a minimizer for two particular forms of the general problem (11.2) (when $K=N+2$), and also has shown

$$U(\bar{x}_1,\ldots,\bar{x}_M)\leqslant E_N^{N+2}(\bar{x}_1,\ldots,\bar{x}_M)<E_N^N=U^{\mathrm{HF}}(\bar{x}_1,\ldots,\bar{x}_M). \tag{11.3}$$

The strict inequality theoretically justifies the need to resort to the MCSCF method. It indeed shows that the HF ground state cannot be the correct one. It is corroborated by the numerical observation, which shows a great improvement of the HF approximation with the MCSCF model (improvement of energies up to 10%).

This preliminary result has been considerably complemented in FRIESECKE [2002] where the existence of a minimizer is proven (for any $K\geqslant N$, under the natural condition $Z > N - 1$) for a model of the form (11.2), that is however too general to be tackled in the numerical practice. Generalizing (11.3), the same author has also proven the strict inequality $E_N^{K+2} < E_N^K$. The proofs have recently been simplified and the results further complemented and improved in LEWIN [2002]. This latter work puts the mathematical knowledge on the MCSCF model on the same level as that on the HF model. Such a parallel between the two models reflects the heuristic property that they both come from successive degrees of variational approximation of the one and only original problem (6.5). It is therefore natural that they basically share the same mathematical feature.

For large systems, solving the whole generality of the MCSCF model (11.2) would represent an overwhelming practical task. In the numerical practice, only a limited number of configurations (i.e., of determinants) is therefore usually considered, cleverly selected to account for the main part of the correlation; this is the basis of the *Complete Active Space* (CASSCF) methods. CASSCF methods require some know-how and a deep a priori understanding of the correlation phenomena occurring in the system under study. With such physical ingredients, CASSCF methods reveal as very efficient methods, that greatly improve the approximation of the energy (6.5). From the practical viewpoint, CASSCF methods are heavier to implement than the Hartree–Fock method. It is easily understandable as the optimization here runs on two sets of variables, namely both the functions ϕ_i and the coefficients c_q. The contemporary mathematical understanding of the MCSCF model allows one to be optimistic and foresee improvements of the efficiency of the numerical algorithms in use for the MCSCF model directly coming from a better mathematical understanding of this complicated model. This is likely to make the MCSCF method all the more popular.

12. Wavefunction methods: Spin dependent models

Now that we have a good understanding of the spinless Hartree–Fock model on the formal standpoint, let us resume our presentation while explicitly accounting for the spin. This is of course the physically relevant case. We only detail the wavefunction formalism. Adaptations to the density matrix formalism are easy.

The *General Hartree–Fock* (GHF) approximation consists in restricting in (6.5) the variational space $\mathcal{H}_e$ to antisymmetrized products of functions of $H^1(\mathbb{R}^3 \times \{|+\rangle, |-\rangle\})$. We thus obtain

$$U^{\mathrm{GHF}}(\bar{x}_1, \ldots, \bar{x}_M) = \inf\{\langle \psi_e, H_e \psi_e \rangle,\ \psi_e \in \mathcal{S}_N\},$$

with

$$\mathcal{S}_N = \left\{ \psi_e \in \mathcal{H}_e \mid \exists \Phi = \{\phi_i\}_{1 \leqslant i \leqslant N} \in \mathcal{Y}_N^S,\ \psi_e = \frac{1}{\sqrt{N!}} \det\bigl(\phi_i(x_j, \sigma_j)\bigr) \right\}, \tag{12.1}$$

$$\mathcal{Y}_N^S = \Biggl\{ \Phi = \{\phi_i\}_{1 \leqslant i \leqslant N},\ \phi_i \in H^1\bigl(\mathbb{R}^3 \times \{|+\rangle, |-\rangle\}\bigr), \\ (\phi_i, \phi_j) = \sum_{\sigma \in \{|+\rangle, |-\rangle\}} \int_{\mathbb{R}^3} \phi_i(x, \sigma) \phi_j(x, \sigma)\, dx = \delta_{ij},\ 1 \leqslant i, j \leqslant N \Biggr\}.$$

Denoting by

$$\tau_\Phi(x,\sigma;x',\sigma')=\sum_{i=1}^N \phi_i(x,\sigma)\phi_i(x',\sigma'),$$

and

$$\rho_\Phi(x)=\sum_\sigma \tau_\Phi(x,\sigma;x,\sigma)=\sum_{i=1}^N\sum_\sigma\left|\phi_i(x,\sigma)\right|^2,$$

this allows us to compute

$$\begin{aligned}E^{\mathrm{GHF}}(\Phi)&=\langle\psi_e,H_e\psi_e\rangle\\&=\sum_{i=1}^N\frac{1}{2}\int_{\mathbb{R}^3}\sum_\sigma|\nabla\phi_i|^2+\int_{\mathbb{R}^3}\rho_\Phi V+\frac{1}{2}\int_{\mathbb{R}^3}\int_{\mathbb{R}^3}\frac{\rho_\Phi(x)\rho_\Phi(x')}{|x-x'|}\,dx\,dx'\\&\quad-\frac{1}{2}\int_{\mathbb{R}^3}\int_{\mathbb{R}^3}\sum_{\sigma,\sigma'}\frac{|\tau_\Phi(x,\sigma;x',\sigma')|^2}{|x-x'|}\,dx\,dx'.\end{aligned}$$

We therefore write the general Hartree–Fock approximation as follows

$$\begin{aligned}&U^{\mathrm{GHF}}(\bar{x}_1,\ldots,\bar{x}_M)\\&\quad=\inf\Bigg\{\sum_{i=1}^N\frac{1}{2}\int_{\mathbb{R}^3}\sum_\sigma|\nabla\phi_i|^2+\int_{\mathbb{R}^3}\rho_\Phi V+\frac{1}{2}\int_{\mathbb{R}^3}\int_{\mathbb{R}^3}\frac{\rho_\Phi(x)\rho_\Phi(x')}{|x-x'|}\,dx\,dx'\\&\qquad-\frac{1}{2}\int_{\mathbb{R}^3}\int_{\mathbb{R}^3}\sum_{\sigma,\sigma'}\frac{|\tau_\Phi(x,\sigma;x',\sigma')|^2}{|x-x'|}\,dx\,dx'\;\Big|\;\phi_i\in H^1\big(\mathbb{R}^3\times\{|+\rangle,|-\rangle\}\big)\\&\qquad\sum_{\sigma\in\{|+\rangle,|-\rangle\}}\int_{\mathbb{R}^3}\phi_i(x,\sigma)\phi_j(x,\sigma)\,dx=\delta_{ij},\;1\leqslant i,j\leqslant N\Bigg\}.\end{aligned}\tag{12.2}$$

The one-electron wavefunctions $\phi_i(x,\sigma)$ are usually called *molecular spin-orbitals*.

This general model is not very much used in the numerical practice (except in some computations of spectroscopic spectra or magnetic properties). It is usually replaced by one of the three particular forms of it, which all consist in further restricting the space of wavefunctions by prescribing some specific dependence with respect to the spin variables.

The first of these three models is the *Restricted Hartree–Fock* model (abbreviated in RHF). It is a Hartree–Fock type model that stems from the fundamental concept of Lewis electron pairs. This model can only be used on molecular systems composed of an *even* number N of electrons, the so-called *closed shells* systems. To each of the $p=N/2$ pairs of electrons, a *spacial* wavefunction $\phi_i\in H^1(\mathbb{R}^3)$ is associated. The two electrons of the pair share this same spacial wavefunction. One is of *spin up*, the other one of *spin down*. The wavefunctions of the RHF model are Slater determinants of N-tuples of the form

$$\{\phi_1\alpha, \phi_1\beta, \dots, \phi_p\alpha, \phi_p\beta\},$$

where $\phi_i \in H^1(\mathbb{R}^3)$ and $\int_{\mathbb{R}^3} \phi_i\phi_j = \delta_{ij}$, $1 \leqslant i, j \leqslant p$. The functions α and β are functions of the spin variable defined as follows

$$\alpha\big(|+\rangle\big) = 1, \qquad \alpha\big(|-\rangle\big) = 0, \qquad \beta\big(|+\rangle\big) = 0, \qquad \beta\big(|-\rangle\big) = 1. \tag{12.3}$$

It is simple to see that

$$\tau_\Phi\big(x, |+\rangle; x', |-\rangle\big) = \tau_\Phi\big(x, |-\rangle; x', |+\rangle\big) = 0,$$

$$\tau_\Phi\big(x, |+\rangle; x', |+\rangle\big) = \tau_\Phi\big(x, |-\rangle; x', |-\rangle\big) = \sum_{i=1}^{p} \phi_i(x)\phi_i(x').$$

Thus the RHF model reads

$$U^{\mathrm{RHF}}(\bar{x}_1, \dots, \bar{x}_M) = \inf\big\{E^{\mathrm{RHF}}(\Phi),\ \Phi \in \mathcal{Y}_p\big\},$$

$$E^{\mathrm{RHF}}(\Phi) = \sum_{i=1}^{p} \int_{\mathbb{R}^3} |\nabla\phi_i|^2 + \int_{\mathbb{R}^3} \rho_\Phi V + \frac{1}{2}\int_{\mathbb{R}^3}\int_{\mathbb{R}^3} \frac{\rho_\Phi(x)\rho_\Phi(x')}{|x - x'|}\,dx\,dx' - \frac{1}{4}\int_{\mathbb{R}^3}\int_{\mathbb{R}^3} \frac{|\tau_\Phi(x, x')|^2}{|x - x'|}\,dx\,dx',$$

with

$$\tau_\Phi(x, x') = \sum_\sigma \tau_\Phi(x, \sigma; x', \sigma) = 2\sum_{i=1}^{p} \phi_i(x)\phi_i(x'),$$

$$\rho_\Phi(x) = \tau_\Phi(x, x) = 2\sum_{i=1}^{p} \big|\phi_i(x)\big|^2,$$

$$\mathcal{Y}_p = \bigg\{\Phi = \{\phi_i\}_{1\leqslant i\leqslant p},\ \phi_i \in H^1\big(\mathbb{R}^3\big),\ (\phi_i, \phi_j) = \int_{\mathbb{R}^3} \phi_i\phi_j = \delta_{ij},\ 1 \leqslant i, j \leqslant p\bigg\}. \tag{12.4}$$

Let us also mention that the Restricted Hartree–Fock model as defined above *must not* be confused with a different one, unfortunately known in the mathematical literature under the same name and sometimes used by mathematicians for theoretical purposes.

The latter model is a spinless model which is constructed from the spinless Hartree–Fock model (9.6) by erasing the exchange term

$$-\frac{1}{2}\int_{\mathbb{R}^3}\int_{\mathbb{R}^3}\frac{|\tau_\Phi(x,x')|^2}{|x-x'|}\,dx\,dx'.$$

The second model constructed from (12.2) is the *Unrestricted Hartree–Fock model*, abbreviated in UHF model. It is appropriate in order to deal with open shell molecular systems, which are systems with a odd number of electrons such as radicals, and systems with an even number of electrons but whose ground state is not a spin singlet state (see, e.g., SZABO and OSTLUND [1982]). In this model, the wavefunction is the form of a determinant built with functions that are *either* with spin up *or* with spin down. In other words, the spin dependence of each function cannot be else than a simple dependence consisting of the multiplication by either the function α or the function β defined above in (12.3). More explicitly, the minimization is performed upon the set of wavefunctions that are determinants of the form

$$\{\phi_1^\alpha\alpha,\dots,\phi_{N_\alpha}^\alpha\alpha;\phi_1^\beta\beta,\dots,\phi_{N_\beta}^\beta\beta\},$$

where the two integers N_α and N_β are such that $N_\alpha+N_\beta=N$. Denoting by

$$\tau_{\Phi^\alpha}(x,x')=\sum_{i=1}^{N_\alpha}\phi_i^\alpha(x)\phi_i^\alpha(x'),\qquad \tau_{\Phi^\beta}(x,x')=\sum_{i=1}^{N_\beta}\phi_i^\beta(x)\phi_i^\beta(x'),$$

$$\rho_{\Phi^\alpha}=\sum_{i=1}^{N_\alpha}\left|\phi_i^\alpha(x)\right|^2,\qquad \rho_{\Phi^\beta}=\sum_{i=1}^{N_\beta}\left|\phi_i^\beta(x)\right|^2,\qquad \rho=\rho_{\Phi^\alpha}+\rho_{\Phi^\beta},$$

the UHF problem reads

$$\begin{aligned}&U^{\mathrm{UHF}}(\bar{x}_1,\dots,\bar{x}_M)\\&\quad=\inf_{N_\alpha+N_\beta=N}\inf\left\{E^{\mathrm{UHF}}\left(\Phi^\alpha,\Phi^\beta\right),\ \left(\Phi^\alpha,\Phi^\beta\right)\in\mathcal{Y}_{N_\alpha}\times\mathcal{Y}_{N_\beta}\right\},\end{aligned}\tag{12.5}$$

where

$$\begin{aligned}&E^{\mathrm{UHF}}\left(\Phi^\alpha,\Phi^\beta\right)\\&\quad=\sum_{i=1}^{N_\alpha}\frac{1}{2}\int_{\mathbb{R}^3}|\nabla\phi_i^\alpha|^2+\sum_{i=1}^{N_\beta}\frac{1}{2}\int_{\mathbb{R}^3}|\nabla\phi_i^\beta|^2+\int_{\mathbb{R}^3}\rho V\\&\qquad+\frac{1}{2}\int_{\mathbb{R}^3}\int_{\mathbb{R}^3}\frac{\rho(x)\rho(x')}{|x-x'|}\,dx\,dx'-\frac{1}{2}\int_{\mathbb{R}^3}\int_{\mathbb{R}^3}\frac{|\tau_{\Phi^\alpha}(x,x')|^2}{|x-x'|}\,dx\,dx'\\&\qquad-\frac{1}{2}\int_{\mathbb{R}^3}\int_{\mathbb{R}^3}\frac{|\tau_{\Phi^\beta}(x,x')|^2}{|x-x'|}\,dx\,dx'.\end{aligned}$$

In the numerical practice, the integers N_α and N_β that yield the minimum value in (12.5) are determined upon the basis of chemical arguments and one only solves for this particular pair of integers

$$\inf\{E^{\mathrm{UHF}}(\Phi^\alpha, \Phi^\beta),\ (\Phi^\alpha, \Phi^\beta) \in \mathcal{Y}_{N_\alpha} \times \mathcal{Y}_{N_\beta}\}.$$

$$\begin{aligned} \phi_3 &= \phi_3^\alpha \uparrow + \phi_3^\beta \downarrow \\ \phi_2 &= \phi_2^\alpha \uparrow + \phi_2^\beta \downarrow \\ \phi_1 &= \phi_1^\alpha \uparrow + \phi_1^\beta \downarrow \end{aligned}$$

General Hartree–Fock
$(N = 3)$

$$\begin{aligned} &- \phi_4 \\ &- \phi_3 \\ &- \phi_2 \\ &- \phi_1 \end{aligned}$$

Spinless Hartree–Fock
$(N = 4)$

$$\begin{aligned} &\uparrow \phi_3^\alpha \\ &\uparrow \phi_2^\alpha \quad \downarrow \phi_2^\beta \\ &\uparrow \phi_1^\alpha \quad \downarrow \phi_1^\beta \end{aligned}$$

Unrestricted Hartree–Fock
$(N_\alpha = 3,\ N_\beta = 2)$

$$\begin{aligned} &\uparrow\downarrow \phi_3 \\ &\uparrow\downarrow \phi_2 \\ &\uparrow\downarrow \phi_1 \end{aligned}$$

Restricted Hartree–Fock
$(N = 6,\ N_p = 3)$

The main disadvantage of the UHF model is that the resulting wavefunction is not necessarily a pure spin state. Let us comment on this point. As the Hamiltonian commutes with the spin observable S^2 (see, e.g., MCWEENY [1992]), the ground state is an eigenvector of S^2 (or may be chosen so in case of degeneracy). The problem is that the UHF ground state often is *not* an eigenvector of S^2.

This disadvantage is remedied in a third model, the so-called *Restricted Open-shell Hartree–Fock* (ROHF) model; basically, the ROHF model is a mixing of the RHF and UHF models (the lowest energy molecular orbitals are populated by two electrons as in the RHF setting whereas upper energy molecular orbitals are populated by only one electron like in the UHF setting) for which all permutations of the spin variables corresponding to a given eigenstate of S^2 and S_z are considered; consequently, the ROHF model is a multideterminantal method (test wavefunctions are sum of several Slater determinants) such that for any ROHF wavefunction, the spatial parts of the ith molecular spin-orbitals is the same for all the determinants forming it; only the spin parts varies to reproduce the desired spin state.

Of course, in the same manner as it has been done above on the spinless Hartree–Fock model, Hartree–Fock equations can be derived for the spin dependent models. The language of density matrices also allows to conveniently formulate the problems and the optimality equations. Likewise, post Hartree–Fock spin dependent models are developed and used, in the same vein as the spinless extensions we have described. We skip such easy extensions that we leave to the reader.

What about the mathematical properties of these spin-dependent Hartree–Fock models? The existence proof developed on the spin-less Hartree–Fock model (9.7) we have defined above for pedagogical purposes can easily be transposed to the four spin constrained Hartree–Fock models defined in the present section. The fundamental reason is that the spin which takes value in a finite set has no impact on the existence results for which the main difficulty comes from the fact that the space variables vary in an unbounded domain $\mathbb{R}^3$. This will be made clear in Section 17.

The same existence results hold, the same questions related to uniqueness are open.

Lastly, the important (also for numerical purposes) mathematical properties of the Hartree–Fock minimizers stating that

- the occupied molecular orbitals necessarily are the lowest N eigenstates of the Fock operator;
- the Nth eigenstate of the Fock operator is not degenerate

still hold for the general spin-dependent Hartree–Fock model (12.2) and can be adapted to the Unrestricted Hartree–Fock model (12.5). On the other hand, there is no mathematical argument to decide whether the above two results are also valid for the Restricted Hartree–Fock model; they are believed to be true for no numerical counterexample seems to have been met yet. The situation is opposite for the Restricted open-shell Hartree–Fock model: In some ROHF numerical simulations, the first of the above two properties seems to be violated. It is not proved however that these counterexamples "pass to the limit" when the discretization is refined to make the numerical solution converge to a solution of the continuous model, nor *a fortiori* that the latter solution is actually a ground state of the continuous model.

13. The density functional theory: The idea of Hohenberg and Kohn

Let us go back to the original electronic minimization problems (6.5) and (6.6) that we reproduce here for convenience

$$U(\bar{x}_1, \dots, \bar{x}_M) = \inf\{\langle \psi_e, H_e \psi_e \rangle,\ \psi_e \in \mathcal{H}_e,\ \|\psi_e\|_{L^2} = 1\}, \tag{13.1}$$

$$H_e = -\sum_{i=1}^{N} \frac{1}{2}\Delta_{x_i} - \sum_{i=1}^{N}\sum_{k=1}^{M} \frac{z_k}{|x_i - \bar{x}_k|} + \sum_{1 \leqslant i < j \leqslant N} \frac{1}{|x_i - x_j|}, \tag{13.2}$$

where the variational space is given by (6.2)

$$\mathcal{H}_e = \bigwedge_{i=1}^{N} H^1(\mathbb{R}^3 \times \{|+\rangle, |-\rangle\}). \tag{13.3}$$

We now attack this problem by a completely different approach than the variational Hartree–Fock one. In order to lighten the notations, the spin variables will now be omitted throughout this section, except when spin dependence plays a specific role.

As above, the positions of the nuclei are fixed, and we denote by

$$V(x) = -\sum_{k=1}^{M} \frac{z_k}{|x - \bar{x}_k|}$$

the electrostatic potential generated by the nuclei.

We shall denote in this section the electronic Hamiltonian H_e by H_V to emphasize the special role that the external potential V the electrons are subjected to will play in the sequel. Problems (13.1)–(13.3) is thus embodied into the general class of problems

$$E(V) = \inf\{\langle\psi_e, H_V\psi_e\rangle,\ \psi_e \in \mathcal{H}_e,\ \|\psi_e\|_{L^2} = 1\}, \tag{13.4}$$

$$H_V = H_1 + \sum_{i=1}^{N} V(x_i), \tag{13.5}$$

$$H_1 = -\sum_{i=1}^{N} \frac{1}{2}\Delta_{x_i} + \sum_{1\leqslant i<j\leqslant N} \frac{1}{|x_i - x_j|}. \tag{13.6}$$

The subscript 1 in the notation H_1 originates from the adiabatic connection theory where the notation H_0 refers a model Hamiltonian in which the Coulomb interaction has been turn off

$$H_0 = -\sum_{i=1}^{N} \frac{1}{2}\Delta_{x_i}$$

and where $(H_\lambda)_{\lambda\in[0,1]}$ denotes a family of Hamiltonians linking smoothly H_0 and H_1 (for instance $H_\lambda = -\sum_{i=1}^{N} \frac{1}{2}\Delta_{x_i} + \sum_{1\leqslant i<j\leqslant N} \frac{\lambda}{|x_i-x_j|}$); the adiabatic connection is a useful tool in density functional theory (see, e.g., DREIZLER and GROSS [1990]).

As the functions in $\mathcal{H}_e$ are H^1 with respect to the space variables, it is easy to see that one can give a rigorous meaning to all the terms of $\langle\psi_e, H_V\psi_e\rangle$ as soon as (for instance) $V \in L^{3/2} + L^\infty$. We assume henceforth that any V we deal with belongs to this class of potentials.

Denoting by

$$\mathcal{X}_N = \{\psi_e \in \mathcal{H}_e,\ \|\psi_e\|_{L^2} = 1\}$$

the set of admissible wavefunctions, we define, for an arbitrary $\psi_e \in \mathcal{X}_N$, the corresponding electronic density

$$\rho(x) = N \int_{\mathbb{R}^{3(N-1)}} \left|\psi_e(x, x_2, \ldots, x_N)\right|^2 dx_2 \cdots dx_N. \tag{13.7}$$

It is easy to see that the conditions for ψ_e belonging to $\mathcal{X}_N$ imply that ρ at least varies in the following convex set

$$\mathcal{I}_N = \left\{\rho \geqslant 0,\ \sqrt{\rho} \in H^1(\mathbb{R}^3),\ \int_{\mathbb{R}^3} \rho = N\right\}. \tag{13.8}$$

Conversely, it can be shown that for any ρ in this set we may construct a wavefunction ψ_e with density ρ given by (13.7) that belongs to $\mathcal{X}_N$. One says that the densities $\mathcal{I}_N$ are *N-representable* with wavefunctions of finite energy.

The purpose of the density functional theory, abbreviated in DFT, is to replace the minimization problem (6.5) defined in terms of the unknown wavefunction ψ_e by a minimization problem set on the unknown density ρ. This latter problem, together with the class of admissible ρ to be considered, will be determined with the help of the class of problems (13.4)–(13.6). We shall now briefly see how. For more details on this fascinating theory, we refer the reader to the reviews by LIEB [1983], JONES and GUNNARSSON [1989], KOHN [1999], and also to the four monographs by ESCHRIG [1996], DREIZLER and GROSS [1990], PARR and YANG [1989] and MARCH [1992].

The first observation is that, depending on V, the problem (13.4) may or may not admit a minimizer. In addition, this minimizer may either be unique, or not.

We therefore introduce the class

$$\mathcal{V}_N = \{V \text{ such that (13.4) admits a minimizer}\}.$$

The seminal idea that founded the density functional theory was the following observation due to HOHENBERG and KOHN [1964]: *Suppose that ψ_1 is a ground state for the potential V_1 (i.e., a minimizer to problem* (13.4) *with $V = V_1$); suppose as well that ψ_2 is a ground state for V_2; suppose finally that $V_1 - V_2$ is not a constant; then the electronic density ρ_i built from* (13.7) *with ψ_i, $i = 1, 2$, are different from one another.*

The proof of this result, known as the *Hohenberg and Kohn theorem*, is surprisingly simple and we reproduce it here for consistency.

By definition of ψ_i, we have $H_1\psi_i + V_i\psi_i = E(V_i)\psi_i$ for $i = 1, 2$. For V_i convenient (which is indeed the case if we constrained V_i to belong to the class $L^{3/2} + L^\infty(\mathbb{R}^3)$), this implies that ψ_i cannot vanish on a set of positive measures (otherwise, thanks to a mathematical result known as the unique continuation theorem, ψ_i would vanish everywhere, and could not be of unit norm). Consequently, ψ_1 cannot satisfy $H_1\psi_1 + V_2\psi_1 = \lambda\psi_1$ whatever λ. Should it be the case, we would have $(V_1 - V_2 - E(V_1) + \lambda)\psi_1 \equiv 0$, which would imply that $V_1 - V_2$ is a constant and would contradict the hypothesis. Therefore, ψ_1 cannot be a minimizer to $E(V_2)$ and thus $E(V_2) < \langle\psi_1, H_{V_2}\psi_1\rangle$. Likewise, $E(V_1) < \langle\psi_2, H_{V_1}\psi_2\rangle$. These two strict inequalities can be respectively written $E(V_2) < E(V_1) - \int \rho_1 V_1 + \int \rho_1 V_2$ and $E(V_1) < E(V_2) - \int \rho_2 V_2 + \int \rho_2 V_1$. Adding the two of them, we get $0 < \int(\rho_1 - \rho_2)(V_1 - V_2)$, which shows that $\rho_1 \neq \rho_2$. The theorem is proven.

This result means that, up to the irrelevant addition of a constant, the knowledge of the density ρ completely determines the potential V the system is subjected to, and therefore the ground state wavefunction(s) of it. From the heuristic viewpoint, such a result is reasonable, for the density, by the location and the specific form of its cusps (formula (6.8)), determines the location of the nuclei together with their charge. This in turn determines the potential V, and thus the ground state, which is a *functional* of the density. Let us examine a mathematical form of the consequences of that discussion.

Consider now ρ in the set

$$\mathcal{A}_N = \{\rho \mid \exists\, V \in \mathcal{V}_N \text{ such that } E(V) \text{ is attained at some } \psi_e \text{ that has density } \rho\} \tag{13.9}$$

For such a ρ (which is said to be *pure-state* V *representable*), the above theorem now allows to define the *Hohenberg–Kohn density functional*

$$F_{\mathrm{HK}}(\rho) = E(V) - \int_{\mathbb{R}^3} \rho V, \tag{13.10}$$

where V is *one* of the potentials of (13.9). Let us notice that

$$F_{\mathrm{HK}}(\rho) = \langle \psi_e, H_1 \psi_e \rangle,$$

where ψ_e is any wavefunction defined by (13.9). Indeed, it is straightforward to see that the Hohenberg and Kohn theorem shows that the right hand side does not depend on the choice of this potential V, as all the convenient V differ necessarily only from a constant, which indeed cancels out.

Consider conversely $V \in \mathcal{V}_N$. It is easy to see that $E(V)$ defined by (13.4) is also

$$E(V) = \inf\left\{ F_{\mathrm{HK}}(\rho) + \int_{\mathbb{R}^3} \rho V,\ \rho \in \mathcal{A}_N \right\}. \tag{13.11}$$

This property *theoretically* fulfills the goal we have fixed at the beginning of the section, as it replaces the search for a minimizing ψ_e by a search for a minimizing ρ.

However, it is not fully satisfactory for

- the set of potentials $\mathcal{V}_N$ for which it holds is not known,
- the set $\mathcal{A}_N$ one minimizes upon is not known either,
- the energy functional F_{HK} is not known explicitly, otherwise than through the somewhat obscure formula (13.10).

Therefore, some alternative approaches to determine some minimization problem of the form (13.11) that could be more explicitly computable have been developed. We now present two of them.

A very natural idea, due to Levy and Lieb, is to set

$$F_{\mathrm{LL}}(\rho) = \inf\{\langle \psi_e, H_1 \psi_e \rangle,\ \psi_e \in \mathcal{X}_N,\ \psi_e \text{ has density } \rho\}, \tag{13.12}$$

and to simply remark that

$$\begin{aligned} E(V) &= \inf\{\langle \psi_e, H_V \psi_e \rangle,\ \psi_e \in \mathcal{X}_N\} \\ &= \inf\left\{ \inf\{\langle \psi_e, H_1 \psi_e \rangle,\ \psi_e \in \mathcal{X}_N,\ \psi_e \text{ has density } \rho\} + \int_{\mathbb{R}^3} \rho V,\ \rho \in \mathcal{I}_N \right\} \\ &= \inf\left\{ F_{\mathrm{LL}}(\rho) + \int_{\mathbb{R}^3} \rho V,\ \rho \in \mathcal{I}_N \right\}. \end{aligned} \tag{13.13}$$

The functional F_{LL} is called the *Levy–Lieb density functional*. Then one can show that $F_{\mathrm{LL}}(\rho) = F_{\mathrm{HK}}(\rho)$ for all $\rho \in \mathcal{A}_N$.

It is to be mentioned at this point that the explicit construction of $F_{\mathrm{LL}}(\rho)$ for a given ρ is of course an open problem, as it is for $F_{\mathrm{HK}}(\rho)$. In order to fulfill this goal, what seems to us an interesting approach, that improves previous existing ones, is a mathematical approach based upon a convenient description of the set of ψ_e which have density ρ. The approach has been thoroughly studied mathematically in BOKANOWSKI and GREBERT [1996a], BOKANOWSKI and GREBERT [1996b], BOKANOWSKI and GREBERT [1998] and other works by the same authors. Basically, it consists in decomposing the information contained in a state ψ_e into one information contained in the density ρ itself and one contained in a state with constant density. The density being fixed, one therefore only optimizes upon the latter to evaluate $F_{\mathrm{LL}}(\rho)$. This gives rise to a promising numerical strategy to compute $F_{\mathrm{LL}}(\rho)$. More definite conclusions about the practical efficiency of this approach are yet to be obtained.

It is also possible to define another density functional which enjoys better mathematical properties (LIEB [1983]). For this purpose, we need to introduce the concept of *mixed states*. In quantum mechanics, states that can actually be described by a single wavefunction ψ_e are called *pure state*. The N-particle density operator associated to ψ_e is defined by

$$\Gamma_N = |\psi_e\rangle\langle\psi_e|.$$

It is a rank-1 operator on $\mathcal{H}_e$ (in fact the orthogonal projector on $\mathbb{R}\psi_e$ for the L^2 norm). Its kernel is the so called N-particle density matrix

$$\gamma_N(x_1, x_2, \ldots, x_N; x_1', x_2', \ldots, x_N') = \psi_e(x_1, x_2, \ldots, x_N)\psi_e(x_1', x_2', \ldots, x_N').$$

The density $\rho(x)$ and the first order reduced density matrix (1-RDM) $\gamma_1(x, x')$ associated to a pure state N-particle density operator Γ are, respectively, the density and the 1-RDM associated with the wavefunction ψ_e, namely

$$\rho(x) = N \int_{\mathbb{R}^{3(N-1)}} |\psi_e(x, x_2, \ldots, x_N)|^2 \, dx_2 \cdots dx_N,$$

$$\gamma_1(x, x') = \int_{\mathbb{R}^{3(N-1)}} \psi_e(x, x_2, x_3, \ldots, x_N)\psi_e(x', x_2, x_3, \ldots, x_N) \, dx_2 \cdots dx_N.$$

In general, quantum systems are not in pure states but in *mixed states*: they must be described by an *ensemble N-particle density operator*, that is to say by a convex combination of pure states N-particle density operators

$$\Gamma = \sum_{i=1}^{+\infty} p_i |\psi_e^i\rangle\langle\psi_e^i|, \quad 0 \leqslant p_i \leqslant 1, \ \sum_{i=1}^{+\infty} p_i = 1, \ \psi_e^i \in \mathcal{X}_N.$$

The coefficient p_i can be interpreted in Statistical Physics as the probability for the system being in the pure state ψ_e^i. Let us denote by $\mathcal{D}_N$ the set of *ensemble N-particle density operators*; clearly, $\mathcal{D}_N$ is a convex set (it is in fact the convex hull of the set of pure state N-particle density operators).

The density $\rho(x)$ and 1-RDM $\gamma_1(x, x')$ associated to an ensemble N-particle density operator Γ are, respectively,

$$\rho(x) = \sum_{i=1}^{+\infty} p_i \, \rho^i(x)$$

and

$$\gamma_1(x, x') = \sum_{i=1}^{+\infty} p_i \gamma_1^i(x, x'), \tag{13.14}$$

where $\rho_i(x)$ and $\gamma_1^i(x, x')$ are the density and 1-RDM associated with the wavefunction ψ_e^i. By linearity,

$$\mathrm{Trace}(\Gamma) = 1, \qquad \mathrm{Trace}(H_1 \Gamma) = \sum_{i=1}^{+\infty} p_i \langle \psi_e^i, H_1 \psi_e^i \rangle,$$

$$\mathrm{Trace}(H_V \Gamma) = \sum_{i=1}^{+\infty} p_i \langle \psi_e^i, H_1 \psi_e^i \rangle + \int_{\mathbb{R}^3} \rho V.$$

It is then easy to see that on the one hand

$$E(V) = \inf\{\mathrm{Trace}(H_V \Gamma), \ \Gamma \in \mathcal{D}_N\},$$

and on the other hand

$$\{\rho \mid \exists \Gamma \in \mathcal{D}_N \text{ such that } \Gamma \text{ has density } \rho\} = \mathcal{I}_N.$$

The latter property follows from the convexity of $\mathcal{I}_N$. Consequently,

$$E(V) = \inf\left\{ F_{\mathrm{L}}(\rho) + \int_{\mathbb{R}^3} \rho V, \ \rho \in \mathcal{I}_N \right\}, \tag{13.15}$$

where the *Lieb density functional* $F_{\mathrm{L}}(\rho)$ is defined by

$$F_{\mathrm{L}}(\rho) = \inf\{\mathrm{Trace}(H_1 \Gamma), \ \Gamma \text{ is an ensemble } N\text{-particle density operator with density } \rho\}. \tag{13.16}$$

The functional $F_{\mathrm{L}}(\rho)$ is in fact equal to the Legendre transform of $E(V)$, namely

$$F_{\mathrm{L}}(\rho) = \sup\left\{ E(V) - \int_{\mathbb{R}^3} \rho V, \ V \in L^{3/2} + L^{\infty} \right\},$$

the latter expression making sense for any $\rho \in L^1 \cap L^3$. One therefore also has

$$E(V) = \inf\left\{ F_{\mathrm{L}}(\rho) + \int_{\mathbb{R}^3} \rho V,\ \rho \in L^1 \cap L^3 \right\}.$$

It can be proved that for any $\rho \in \mathcal{A}_N$, $F_{\mathrm{L}}(\rho) = F_{\mathrm{LL}}(\rho) = F_{\mathrm{HK}}(\rho)$ and that F_{L} is the convex hull of the functionals F_{HK} and F_{LL}.

At this stage no approximation has been made. But, as neither the Hohenberg–Kohn density functional F_{HK} nor its extensions F_{LL} and F_{L} can be easily evaluated for the real system of interest (N interacting electrons), approximations are needed to make of the density functional theory a practical tool for computing electronic ground states. Approximations rely on exact or very accurate evaluations of the density functional for reference systems "close" to the real system:

- In Thomas–Fermi and related models, the reference system is the *uniform noninteracting electron gas*;
- In Kohn–Sham models (by far the most commonly used), it is a system of N *noninteracting* electrons.

14. The density functional theory: Thomas–Fermi and related models

Far earlier than the groundbreaking work by Hohenberg and Kohn, the idea of replacing the problem of determining the full ground state wavefunction by that of determining only the ground state electronic density was attacked, independently by Thomas and Fermi in the 1920s. We now briefly review their method and the developments of it. For more details, we refer to the review article by LIEB [1981], which collects the theoretical results known in the early 1980s and in particular the work by LIEB and SIMON [1977b]. Another interesting review article is SPRUCH [1991]. In addition, we refer to the fourth volume of THIRRING [1983], and also to LIEB and LOSS [2001]. An account for more recent results may be found in CANCÈS [1998].

In the perspective of the modern density functional theory, the Thomas–Fermi theory can be seen as one way, among others, to derive a functional of ρ to model the kinetic energy term, exactly as Kohn and Sham, and Janak did years later (see below).

In Thomas–Fermi and related models, the homogeneous noninteracting electron gas is the reference system. The main property of this system is that the first order reduced density matrix of the ground state is known analytically:

$$\gamma_1(x, x') = 3\rho \left(\frac{\sin(k_{\mathrm{F}}|x-x'|) - (k_{\mathrm{F}}|x-x'|)\cos(k_{\mathrm{F}}|x-x'|)}{(k_{\mathrm{F}}|x-x'|)^3} \right), \tag{14.1}$$

where ρ is the (uniform) density of the noninteracting electron gas and k_{F} the Fermi wavevector modulus defined as $k_{\mathrm{F}} = (3\pi^2\rho)^{1/3}$. The proof of this result can be read in any basic textbook of Statistical Physics.

This allows one to compute the kinetic energy density of the system:

$$t = \left(-\frac{1}{2} \Delta \gamma_1(x, x') \right)\Big|_{x=x'} = C_{\mathrm{TF}} \rho^{5/3},$$

where $C_{\mathrm{TF}} = \frac{10}{3}(3\pi^2)^{2/3}$ denotes the Thomas–Fermi constant.

The idea of Thomas and Fermi was to approximate

- the kinetic energy term for a molecular system by

$$T_{\mathrm{TF}}(\rho) = C_{\mathrm{TF}} \int_{\mathbb{R}^3} \rho(x)^{5/3}\,dx,$$

- the interelectronic repulsion energy by the electrostatic self-interaction energy of a *classical* charge distribution of density ρ

$$J(\rho) = \frac{1}{2} \int_{\mathbb{R}^3} \int_{\mathbb{R}^3} \frac{\rho(x)\rho(y)}{|x-y|}\,dx\,dy.$$

The Thomas–Fermi model therefore consists in considering the minimization of the functional

$$F_{\mathrm{TF}}(\rho) = C_{\mathrm{TF}} \int_{\mathbb{R}^3} \rho^{5/3} + \int_{\mathbb{R}^3} \rho V + \frac{1}{2} \int_{\mathbb{R}^3} \int_{\mathbb{R}^3} \frac{\rho(x)\rho(y)}{|x-y|}\,dx\,dy$$

over all admissible densities ρ, which here describe the set

$$\left\{\rho \geqslant 0,\ \rho \in L^1 \cap L^{5/3}(\mathbb{R}^3),\ \int_{\mathbb{R}^3} \rho = N\right\}.$$

The Thomas–Fermi model is much less accurate than the Hartree–Fock model for two reasons: On the one hand the kinetic energy modelling is more crude, and on the other hand there is no exchange term accounting for the fermionic nature of the electrons. As the exchange term is a function of the first order reduced density matrix it is possible to use again the uniform noninteracting electron gas as a reference system to compute it; indeed, the exchange energy density of this system reads

$$-\frac{1}{4} \int_{\mathbb{R}^3} \frac{|\gamma_1(x,x')|^2}{|x-x'|}\,dx' = -C_{\mathrm{D}}\rho^{4/3},$$

where $C_{\mathrm{D}} = \frac{3}{4}(\frac{3}{\pi})^{1/3}$ denotes the Dirac constant (the closed-shell assumption has been used, giving rise to the factor $1/4$ in the above expression). We thus end up with the so-called Thomas–Fermi–Dirac functional

$$F_{\mathrm{TFD}}(\rho) = C_{\mathrm{TF}} \int_{\mathbb{R}^3} \rho^{5/3} + \int_{\mathbb{R}^3} \rho V + \frac{1}{2} \int_{\mathbb{R}^3} \int_{\mathbb{R}^3} \frac{\rho(x)\rho(y)}{|x-y|}\,dx\,dy - C_{\mathrm{D}} \int_{\mathbb{R}^3} \rho^{4/3}.$$

The Thomas–Fermi and Thomas–Fermi–Dirac models therefore read

$$I_N^{\mathrm{TF,TFD}} = \inf\left\{F_{\mathrm{TF,TFD}}(\rho),\ \rho \geqslant 0,\ \rho \in L^1 \cap L^{5/3}(\mathbb{R}^3),\ \int_{\mathbb{R}^3} \rho = N\right\}. \tag{14.2}$$

In addition, a correction term, due to von Weizsäcker and expressed in terms of $\int_{\mathbb{R}^3} |\nabla\sqrt{\rho}|^2$, can be added to the kinetic energy; this term is obtained by studying perturbations of formula (14.1) generated by small heterogeneities of the density. This term leads to the Thomas–Fermi–von Weizsäcker model

$$F_{\mathrm{TFW}}(\rho) = C_{\mathrm{W}} \int_{\mathbb{R}^3} \left|\nabla\sqrt{\rho}\right|^2 + C_{\mathrm{TF}} \int_{\mathbb{R}^3} \rho^{5/3} + \int_{\mathbb{R}^3} \rho V + \frac{1}{2} \int_{\mathbb{R}^3} \int_{\mathbb{R}^3} \frac{\rho(x)\rho(y)}{|x-y|}\, dx\, dy$$

and the Thomas–Fermi–Dirac–von Weizsäcker model,

$$F_{\mathrm{TFDW}}(\rho) = C_{\mathrm{W}} \int_{\mathbb{R}^3} \left|\nabla\sqrt{\rho}\right|^2 + C_{\mathrm{TF}} \int_{\mathbb{R}^3} \rho^{5/3} + \int_{\mathbb{R}^3} \rho V + \frac{1}{2} \int_{\mathbb{R}^3} \int_{\mathbb{R}^3} \frac{\rho(x)\rho(y)}{|x-y|}\, dx\, dy - C_{\mathrm{D}} \int_{\mathbb{R}^3} \rho^{4/3},$$

respectively, thus the minimization problem for

$$I_N^{\mathrm{TFW,TFDW}} = \inf\left\{ F_{\mathrm{TFW,TFDW}}(\rho),\ \rho \geqslant 0,\ \sqrt{\rho} \in H^1(\mathbb{R}^3),\ \int_{\mathbb{R}^3} \rho = N \right\}, \tag{14.3}$$

where $C_{\mathrm{W}} = \frac{1}{72}$. We refer the reader to the bibliography for more details on the modelling. Let us only mention that, contrary to what is sometimes erroneously stated, the two coefficients C_{W} and C_{TF} of, respectively, the von Weizsäcker and Thomas–Fermi terms, both modelling the kinetic energy, are of the same order (namely $\hbar^2$ in universal units).

From the standpoint of physics, Thomas–Fermi like models allow to recover *qualitatively* many fundamental physical properties of the microscopic matter. This has justified a constant interest over the last years of the 20th century. We send the reader to the bibliography quoted above for more information.

From the standpoint of mathematics, the Thomas–Fermi type models really deserve a huge interest because they constitute an excellent toy model, or test problem, to elaborate mathematical arguments in a simple setting. These arguments can then be extended, in a second stage, to more sophisticated and more commonly used models such as the Hartree–Fock type models or the Kohn–Sham like models. This is particularly true for the Thomas–Fermi models that include the von Weizsäcker term. They are formally alike the Hartree–Fock (or Kohn–Sham) model for *one* monoelectronic wavefunction, the role of which is here played by $\sqrt{\rho}$. The recent history of the mathematical progress on the models issued from quantum chemistry has shown that the analogy is not fortuitous and can be exploited very efficiently.

As far as the state of the mathematical knowledge on Thomas–Fermi type models is concerned, let us only say the following. The minimization problems (14.2) and (14.3) are all well posed, in the sense they admit (at least) a minimizer under the sufficient condition $N \leqslant Z$. Some of the models also allow for the existence of some negative ions, i.e., admit a minimizer for $N \leqslant N_c$, with some $N_c > Z$. This minimizer is unique in the TF, TFD, and TFW models, because of convexity properties. The most difficult model to deal with is the TFDW model. Many open questions remain on that latter model, which in a sense is really close to the Kohn–Sham model. For brevity, we only consider further the TFW model, which shares some common features with the Hartree–Fock model. For any minimizer ρ, $u = \sqrt{\rho}$ is a nonnegative solution to an equation of the form

$$-\Delta u + Vu + \rho^{2/3}u + \left(\rho \star \frac{1}{|x|}\right)u = \lambda u$$

for some λ that can be shown to be negative for $Z \leqslant N$ (and in fact for more general N than that). This equation is a nonlinear eigenvalue problem. From this equation follow a lot of qualitative properties of $\sqrt{\rho}$ such as the regularity, and the asymptotic behaviour at the point nuclei and at infinity. Some symmetry can also be inferred from it depending on the geometry of the set of nuclei. As instances of mathematical works dealing with Thomas–Fermi type theories, we wish to quote BENGURIA, BREZIS and LIEB [1981], BENGURIA and LIEB [1985], BENGURIA and YARUR [1990], BENILAN, GOLDSTEIN and RIEDER [1991], BENILAN, GOLDSTEIN and RIEDER [1992], GOLDSTEIN and RIEDER [1991], J.A. GOLDSTEIN, G.R. GOLDSTEIN and WENYIAO JIA [1995], SOLOVEJ [1990], LE BRIS [1993b], LE BRIS [1995].

What we have claimed for the mathematical features is also true for the numerical aspects. Some algorithms can be advantageously tested on these TF like models before being adapted to the case of interest.

Thomas–Fermi like models thus have the twofold advantage of being good test theories both from the viewpoint of the theoretical physics and from the viewpoint of mathematics.

In the numerical practice however, they are not used that much, for they are most often considered out-of-date since the appearance of Kohn–Sham models in the 1960s; the latter are indeed by far more accurate. It is nevertheless a strange phenomenon that Thomas–Fermi type models periodically come back to the general landscape of computational physics and chemistry, with some publications that are devoted to them and appear in the leading journals of the communities. For instance, some groups are now developing the so-called *orbital free models* which are refinements of Thomas–Fermi like models (WANG and CARTER [2000]). In these models indeed, the electronic state is represented by *one* scalar field on $\mathbb{R}^3$ (the density, or often in practice its square root), whereas in Kohn–Sham models it is represented by N scalar fields on $\mathbb{R}^3$ (the Kohn–Sham orbitals, see next section). Orbital free models thus permit to simulate much larger systems than Kohn–Sham models . . . at the cost of a loss of accuracy.

15. The density functional theory: Kohn–Sham models

For the sake of simplicity, we mainly deal here with spinless Kohn–Sham models just as we proceeded in the presentation of the Hartree–Fock approximation in Section 9. Spin-dependent Kohn–Sham models used in practice are build along with the method detailed in Section 12 in the Hartree–Fock setting.

The groundbreaking contribution to practical implementations of the density functional theory is due to Kohn and Sham, who stated up the idea of taking as the reference system a system of N noninteracting electrons.

For such a system, the density functional is obtained by replacing the interacting Hamiltonian H_1 of the real system (formula (13.6)) by the Hamiltonian of the reference system

$$H_0 = -\sum_{i=1}^{N} \frac{1}{2} \Delta_{x_i}. \tag{15.1}$$

The analogue of the Levy–Lieb density functional (13.12) then is the Kohn–Sham kinetic energy functional

$$T_{\mathrm{KS}}(\rho) = \inf\bigl\{\langle \psi_e, H_0 \psi_e\rangle,\ \psi_e \in \mathcal{X}_N,\ \psi_e \text{ has density } \rho\bigr\}. \tag{15.2}$$

When (15.2) admits a minimizer of the form of a Slater determinant, (15.2) may clearly be simplified into

$$T_{\mathrm{KS}}(\rho) = \inf\left\{\frac{1}{2}\sum_{i=1}^{N}\int_{\mathbb{R}^3} |\nabla\phi_i|^2,\ \phi_i \in H^1(\mathbb{R}^3),\ \int_{\mathbb{R}^3} \phi_i\phi_j = \delta_{ij},\ \sum_{i=1}^{N} |\phi_i|^2 = \rho\right\}. \tag{15.3}$$

The difficulty with this formulation is that it is not true that for any arbitrary $\rho \in \mathcal{A}_N$, (15.2) admits some Slater determinant for minimizer. Some counterexamples are known. However, the above formula (15.3) can be *formally* extended to any $\rho \in \mathcal{A}_N$, or even to any $\rho \in \mathcal{I}_N$ (the set $\mathcal{I}_N$ is defined by (13.8)).

This representability problem is avoided when using the Lieb density functional associated with the noninteracting Hamiltonian H_0. Let us indeed define the *Janak kinetic energy functional* as

$$T_{\mathrm{J}}(\rho) = \inf\bigl\{\mathrm{Trace}(H_0\Gamma),\ \Gamma \in \mathcal{D}_N,\ \Gamma \text{ has density } \rho\bigr\}.$$

It turns out indeed that

- for any ensemble N-particle density operator Γ, one has

$$\mathrm{Trace}(H_0\Gamma) = -\int_{\mathbb{R}^3} \left(\frac{1}{2}\Delta_x \gamma_1(x,x')\right)\bigg|_{x=x'} dx,$$

  where γ_1 is the 1-RDM associated to Γ along formula (13.14);
- the set of all 1-RDM that originate from *ensemble* N-particle density operators is known; it reads

$$\left\{\gamma_1(x,x') = \sum_{i=1}^{+\infty} n_i \phi_i(x)\phi_i(x'),\ \phi_i \in H^1(\mathbb{R}^3),\right.$$
$$\left.\int_{\mathbb{R}^3} \phi_i\phi_j = \delta_{ij},\ 0 \leqslant n_i \leqslant 1,\ \sum_{i=1}^{+\infty} n_i = 1\right\}.$$

This important result is known as the *ensemble first order reduced density matrices representability theorem*. No similar representability result holds for *pure state N-particle density operators*.

An alternative way to define a Janak kinetic energy functional, which is valid for any $\rho \in \mathcal{I}_N$, therefore reads

$$T_{\mathrm{J}}(\rho) = \inf\left\{ \frac{1}{2}\sum_{i=1}^{+\infty} n_i \int_{\mathbb{R}^3} |\nabla\phi_i|^2,\ \phi_i \in H^1(\mathbb{R}^3),\ \int_{\mathbb{R}^3} \phi_i\phi_j = \delta_{ij},\right.$$
$$\left. 0 \leqslant n_i \leqslant 1,\ \sum_{i=1}^{+\infty} n_i = 1,\ \sum_{i=1}^{+\infty} n_i|\phi_i|^2 = \rho \right\}.$$

We shall see below that this functional yields some improvements not only in the theory, but also in the numerical practice with respect to the Kohn–Sham functional. For the time being, let us only notice that, for an arbitrary $\rho \in \mathcal{I}_N$, one only has

$$T_{\mathrm{J}}(\rho) \leqslant T_{\mathrm{KS}}(\rho)$$
$$\leqslant \inf\left\{ \frac{1}{2}\sum_{i=1}^{N} \int_{\mathbb{R}^3} |\nabla\phi_i|^2,\ \phi_i \in H^1(\mathbb{R}^3),\ \int_{\mathbb{R}^3} \phi_i\phi_j = \delta_{ij},\ \sum_{i=1}^{N} |\phi_i|^2 = \rho \right\}.$$

The density functionals T_{KS} and T_{J} associated with the noninteracting Hamiltonian H_0 are likely to provide rather good approximations of the kinetic energy of the real (interacting) system.

On the other hand, the Coulomb energy

$$J(\rho) = \frac{1}{2}\int_{\mathbb{R}^3}\int_{\mathbb{R}^3} \frac{\rho(x)\,\rho(y)}{|x-y|}\,dx\,dy$$

representing the electrostatic energy of a *classical* charge distribution of density ρ is a reasonable guess for the interelectronic interaction energy in a system of N electrons of density ρ.

The errors on both the kinetic energy and the electrostatic interaction are put together in the *exchange-correlation energy* defined as the difference

$$E_{\mathrm{xc}}(\rho) = F_{\mathrm{LL}}(\rho) - T_{\mathrm{KS}}(\rho) - J(\rho),$$

or

$$E_{\mathrm{xc}}(\rho) = F_{\mathrm{L}}(\rho) - T_{\mathrm{J}}(\rho) - J(\rho),$$

depending on the choices for the interacting and noninteracting density functionals. The construction of approximated exchange-correlation functionals rendering numerical simulations possible will be dealt with below.

When the Kohn–Sham kinetic energy functional is chosen, the abstract problem (13.13) can then be replaced by the celebrated *Kohn–Sham model*

$$E^{\mathrm{KS}}(V) = \inf\left\{ \frac{1}{2}\sum_{i=1}^{N}\int_{\mathbb{R}^3}|\nabla\phi_i| + \int \rho V + \frac{1}{2}\int_{\mathbb{R}^3}\int_{\mathbb{R}^3}\frac{\rho(x)\rho(y)}{|x-y|}\,dx\,dy + E_{\mathrm{xc}}(\rho),\right.$$
$$\left. \phi_i \in H^1(\mathbb{R}^3),\ \int_{\mathbb{R}^3}\phi_i\phi_j = \delta_{ij} \right\}, \tag{15.4}$$

where ρ is a notation for $\sum_{i=1}^{N}|\phi_i|^2$, *if one assumes that* (15.3) *holds for some minimizer* ρ *of* (13.13). The trouble is that this assumption is not always valid. The Kohn–Sham approximation consists in *assuming* that we can forget this restriction.

On the other hand, no such assumption is required if the Janak kinetic energy functional is used to reformulate problem (13.15) as an *extended Kohn–Sham model*

$$E^{\mathrm{EKS}}(V)$$
$$= \inf\left\{ \frac{1}{2}\sum_{i=1}^{+\infty} n_i\int_{\mathbb{R}^3}|\nabla\phi_i|^2 + \int \rho V + \frac{1}{2}\iint_{\mathbb{R}^3\times\mathbb{R}^3}\frac{\rho(x)\rho(y)}{|x-y|}\,dx\,dy + E_{\mathrm{xc}}(\rho)\right.$$
$$\left. \phi_i \in H^1(\mathbb{R}^3),\ \int_{\mathbb{R}^3}\phi_i\phi_j = \delta_{ij},\ 0 \leqslant n_i \leqslant 1,\ \sum_{i=1}^{+\infty} n_i = N \right\}, \tag{15.5}$$

where ρ is a notation for $\sum_{i=1}^{+\infty} n_i|\phi_i|^2$.

One can then write the Euler–Lagrange equations associated to the above two minimization problems. This requires some differentiability properties of the functional $E_{\mathrm{xc}}(\rho)$. Assuming for a while this differentiability, we may write that the ϕ_i which minimize (15.4) are solutions to the *Kohn–Sham equations*

$$\begin{cases} -\dfrac{1}{2}\Delta\phi_i + V\phi_i + \left(\rho \star \dfrac{1}{|x|}\right)\phi_i + v_{\mathrm{xc}}(\rho)\phi_i = \lambda_i\phi_i, \\ \displaystyle\int_{\mathbb{R}^3}\phi_i\phi_j = \delta_{ij}, \end{cases} \tag{15.6}$$

where λ_i denote the Lagrange multiplier (the matrix of Lagrange multiplier has been again diagonalized, in the same manner as it was in the Hartree–Fock setting), and where $v_{\mathrm{xc}}(\rho)$ denotes the derivative of $E_{\mathrm{xc}}(\rho)$ with respect to ρ. In view of what has been said above, E_{xc} can be considered as a functional on $L^1 \cap L^3$, so that its derivative is a vector of the dual space, namely $v_{\mathrm{xc}}(\rho) \in L^{3/2} + L^{\infty}$.

Along the same differentiability assumption for $E_{\mathrm{xc}}(\rho)$, the Euler–Lagrange equations associated to (15.5) read

$$\begin{cases} -\frac{1}{2}\Delta\phi_i + V\phi_i + \left(\rho \star \frac{1}{|x|}\right)\phi_i + v_{\mathrm{xc}}(\rho)\phi_i = \lambda_i\phi_i, \\ \int_{\mathbb{R}^3} \phi_i\phi_j = \delta_{ij}, \\ n_i = 1 \quad \text{if } \lambda_i < \varepsilon_{\mathrm{F}}, \\ 0 \leqslant n_i \leqslant 1 \quad \text{if } \lambda_i = \varepsilon_{\mathrm{F}}, \\ n_i = 0 \quad \text{if } \lambda_i > \varepsilon_{\mathrm{F}}, \\ \sum_{i=1}^{+\infty} n_i = N. \end{cases} \tag{15.7}$$

The parameter ε_F is the *Fermi level* of the system; it is the Lagrange multiplier of the constraint $\sum_{i=1}^{+\infty} n_i = N$.

From the theoretical viewpoint, it is to be remarked that the issue of differentiability of $E_{\mathrm{xc}}(\rho)$ at the ρ we look for (namely the ρ which is the density of a minimizer to (6.5) or (13.4)) is indeed closely related to the difficult questions of V representability we have mentioned above. We shall not comment further on this property and refer the reader to the bibliography for more details.

Let us come back to the Kohn–Sham equations (15.6). The first comment to make is that, whereas the theoretical argument we have made to derive them is completely different, we end up with a system of equations that is *formally* almost the same as the Hartree–Fock system (9.12). The object we manipulate still is a N-tuple of functions defined on $\mathbb{R}^3$ (times the spin variable). The form of the system still is that of a nonlinear eigenvalue problem. However, some differences are to be pointed out.

From the standpoint of theoretical chemistry, the ϕ_i cannot be interpreted as the monoelectronic wavefunctions of the ground state at hand. Indeed, they are some corresponding electronic wavefunctions in the noninteracting system with the same density as the interacting system under study (we recall such a noninteracting system is *assumed* to exist here).

Again from the standpoint of theoretical chemistry, the problem to be solved is not aimed at being a simple approximation of the original problem, like the Hartree–Fock is: On the contrary, it is theoretically the *exact* translation of it. It is only *in the practice*, when the potential v_{xc} is given an explicit (and thus approximate) form, which in turn requires to "forget" about the differentiability difficulty, that this system of equations becomes an approximation.

From the mathematical viewpoint, the arguments needed to prove that a minimizer to (15.2) exists, and thus that a normalized solution to (15.6) exists, are somewhat different from those developed in the Hartree–Fock framework. The reason is that this model does not originate from a variational approximation to (6.5). The proof of the existence of a minimizer, for commonly used approximations of $v_{\mathrm{xc}}(\rho)$ is due to one of us, see LE BRIS [1993a].

Let us conclude with three theoretical differences that are of paramount importance for numerical simulations:

(1) the Hartree–Fock energy (9.6) is a fourth order polynomial in the ϕ_is whereas the dependency of the Kohn–Sham energy (see Eq. (15.4)) on the ϕ_is is more complex, even within the simple LDA framework (see formula (15.8));

(2) under the differentiability assumption stated above, the *Kohn–Sham operator* appearing in (15.2) reads

$$-\frac{1}{2}\Delta + V_{\text{eff}}(\rho),$$

where the effective potential $V_{\text{eff}}(\rho) = V + \rho \star \frac{1}{|x|} + V_{\text{xc}}(\rho)$ is local (it is a multiplicative operator); the Fock operator has not such a simple form because of the nonlocal exchange term (the fourth term of (9.12)). In the numerical practice, i.e., when a convenient form of the exchange-correlation functional is postulated, the differentiability *with respect to* ρ is satisfied for the LDA, LSD and GGA functionals (see formulae (15.8)–(15.13)), but not for the hybrid functionals which are not explicit in ρ for they also include some part of the Hartree–Fock exchange; on the other hand, the latter functionals are differentiable with respect to the ϕ_is so that Kohn–Sham equations, but with a nonlocal effective potential, can nevertheless be written;

(3) it has already been underlined that at the Hartree–Fock ground state, the Lagrange multipliers λ_i appearing in the Hartree–Fock equations (9.13) are the lowest N eigenvalues of the Fock operator; this property does not hold (or at least is not proven to be true) for the Kohn–Sham model even if it is most often supposed to hold for numerical purposes; on the other hand it is clear, in view of the Euler–Lagrange equations (15.7), that any critical point of the generalized Kohn–Sham model is such that all the orbitals of the Kohn–Sham operator which are below the Fermi level are fully occupied whereas all the orbitals which are above the Fermi level are virtual (i.e., unoccupied); lastly, at the Fermi level, occupation is not a priori prescribed so that partial occupancy may occur.

We shall come back to these three points in Chapter II.

The question is now to find approximations for E_{xc}.

The simplest approximation of $E_{\text{xc}}(\rho)$, already suggested in the Kohn and Sham reference article, is the *local density approximation* (LDA); it consists in taking

$$E_{\text{xc}}^{\text{LDA}}(\rho) = \int_{\mathbb{R}^3} \rho(x)\varepsilon_{\text{xc}}^{\text{LDA}}\big(\rho(x)\big)\,dx, \tag{15.8}$$

where $\varepsilon_{\text{xc}}^{\text{LDA}}(\bar{\rho})$ is the exchange-correlation energy per particle in a uniform electron gas of density $\bar{\rho}$. The function $E_{\text{xc}}(\rho)$ is usually split into

$$E_{\text{xc}}^{\text{LDA}}(\rho) = E_{\text{x}}^{\text{LDA}}(\rho) + E_{\text{c}}^{\text{LDA}}(\rho), \tag{15.9}$$

where $E_{\text{x}}^{\text{LDA}}(\rho)$ is the *LDA exchange functional* corresponding to the Dirac term computed in the previous section

$$E_{\text{x}}^{\text{LDA}}(\rho) = \int_{\mathbb{R}^3} \rho(x)\varepsilon_{\text{x}}^{\text{LDA}}\big(\rho(x)\big)\,dx \quad \text{with } \varepsilon_{\text{x}}^{\text{LDA}}(\bar{\rho}) = -C_{\text{D}}\bar{\rho}^{1/3}, \tag{15.10}$$

and $E_c^{\mathrm{LDA}}(\rho)$ is the *LDA correlation functional*

$$E_c^{\mathrm{LDA}}(\rho) = \int_{\mathbb{R}^3} \rho(x)\varepsilon_c^{\mathrm{LDA}}\big(\rho(x)\big)\,dx. \tag{15.11}$$

No analytical formulation for $\varepsilon_c^{\mathrm{LDA}}$ is known; on the other hand, the function $\bar{\rho} \mapsto \varepsilon_c^{\mathrm{LDA}}(\bar{\rho})$ (from $\mathbb{R}^+$ to $\mathbb{R}$) can be sampled once and for all by very accurate Quantum Monte Carlo simulations; interpolation formulae can then be drawn from this sampling, as for instance the widely used (PERDEW and ZUNGER [1981]) interpolation formula

$$\varepsilon_c^{\mathrm{PZ81}}(\bar{\rho}) = \begin{cases} 0.0311\,\mathrm{Log}(r_s) - 0.0480 + 0.0020 r_s\,\mathrm{Log}(r_s) - 0.0116 r_s & \text{if } r_s < 1, \\ -0.1423/\big(1 + 1.0529\sqrt{r_s} + 0.3334 r_s\big) & \text{if } r_s \geqslant 1, \end{cases}$$

where $r_s = (3/4\pi\bar{\rho})^{1/3}$.

Let us underline that both in Thomas–Fermi like and in Kohn–Sham LDA models, the uniform electron gas is used as a reference system; what makes the difference between the two approaches is that this very crude reference system is used to evaluate the *whole* density functional in the former models whereas it is only used for a *small contribution* to it (around 10% of the total energy in practice) in the latter models. This explains the better accuracy of Kohn–Sham LDA models with respect to Thomas–Fermi like models and the wide range of use in practical calculations.

Let us now briefly describes how to go beyond the LDA approximation.

The first step consists in introducing spin dependent densities:

$$\rho_+(x) = N \int_{\mathbb{R}^{3(N-1)}} \sum_{\sigma_2,\ldots,\sigma_N} \big|\psi_e\big(x,|+\rangle,x_2,\sigma_2,\ldots,x_N,\sigma_N\big)\big|^2\,dx_2\cdots dx_N,$$

$$\rho_-(x) = N \int_{\mathbb{R}^{3(N-1)}} \sum_{\sigma_2,\ldots,\sigma_N} \big|\psi_e\big(x,|-\rangle,x_2,\sigma_2,\ldots,x_N,\sigma_N\big)\big|^2\,dx_2\cdots dx_N$$

in order to discriminate between spin-up and spin-down electrons and in building *local spin density* (LSD) exchange-correlation functionals of the form

$$E_{\mathrm{xc}}^{\mathrm{LSD}}(\rho_+,\rho_-) = \int_{\mathbb{R}^3} \rho(x)\varepsilon_{\mathrm{xc}}^{\mathrm{LSD}}\big(\rho_+(x),\rho_-(x)\big)\,dx. \tag{15.12}$$

From a theoretical viewpoint, there is no need to introduce explicit spin dependence in the picture. The only reason why LSD (usually) give better results than LDA is that spin-dependent *approximate local* exchange correlation functionals are better than spin-independent ones.

The second step is to take into account density inhomogeneities in the evaluation of E_{xc} *via* a local expansion of the exchange-correlation energy density in terms of $\rho_\pm$ and $\nabla\rho_\pm$

$$E_{\mathrm{xc}}^{\mathrm{GGA}}(\rho_+,\rho_-) = \int_{\mathbb{R}^3} f\big(\rho_+(x),\rho_-(x),\nabla\rho_+(x),\nabla\rho_-(x)\big)\,dx \tag{15.13}$$

the acronym GGA standing for *Generalized Gradient Approximation*. The most commonly used two representatives of this category are the exchange-correlation energy of PERDEW and WANG [1986], and PERDEW, BURKE and ERNZERHOF [1996]; both of them are non-empirical in the sense that the determination of the function f does not use any experimental data. They are based on a fine analysis of the so-called exchange-correlation hole, see PERDEW, BURKE and ERNZERHOF [1996].

Additional improvements involve dependence of the exchange-correlation functional on the Kohn–Sham density matrix which reads

$$\tau(x,\sigma;x',\sigma') = \sum_{i=1}^{N} \phi_i(x,\sigma)\phi_i(x',\sigma')$$

in the standard Kohn–Sham model and

$$\tau(x,\sigma;x',\sigma') = \sum_{i=1}^{+\infty} n_i\phi_i(x,\sigma)\phi_i(x',\sigma')$$

in the generalized Kohn–Sham model. This dependence usually occurs through the kinetic energy density

$$t(x,\sigma) = \left(-\frac{1}{2}\Delta_x \tau(x,\sigma;x',\sigma)\right)\bigg|_{x=x'}$$

in which case one speaks of *meta-GGA* functionals, or through the Hartree–Fock exchange term

$$E_{\mathrm{X}}^{\mathrm{HF}}(\tau) = -\frac{1}{2}\int_{\mathbb{R}^3}\int_{\mathbb{R}^3}\sum_{\sigma,\sigma'} \frac{\tau(x,\sigma;x',\sigma')}{|x-x'|}\,dx\,dx',$$

in which case one speaks of *hybrid* functionals; a widely used hybrid functional is the B3LYP functional (BECKE [1993]).

The story is far from being over. The construction of more sophisticated exchange-correlation functionals able to describe fine phenomena such as hydrogen bonding or van der Waals forces, is still an active field of research.

16. Direct calculation of reduced density matrices

In addition to the two main approaches addressed so far, namely the wavefunction based methods (whose most popular representative is the Hartree–Fock approximation) and those related to density functional theory, a third possibility consists in formulating the N-electron Schrödinger equation in terms of *reduced density matrices*.

Let us come back to (6.5), (6.6)

$$U(\bar{x}_1,\ldots,\bar{x}_M) = \inf\bigl\{\langle\psi_e, H_e\psi_e\rangle,\ \psi_e \in \mathcal{H}_e,\ \|\psi_e\|_{L^2} = 1\bigr\},$$

$$H_e = -\sum_{i=1}^{N} \frac{1}{2}\Delta_{x_i} - \sum_{i=1}^{N}\sum_{k=1}^{M} \frac{z_k}{|x_i - \bar{x}_k|} + \sum_{1 \leqslant i < j \leqslant N} \frac{1}{|x_i - x_j|}$$

and notice that, due to the fact that the Hamiltonian only contains one- and two-electrons operators, the energy $\langle \psi_e, H_e \psi_e \rangle$ can be expressed in terms of the so-called *two-particle density matrix*, often also called *second order reduced density matrix* (2-RDM)

$$\gamma_2(x_1, x_2, x_1', x_2') = \int_{\mathbb{R}^{3(N-2)}} dx_3 \cdots dx_N \sum_{\sigma_1 \cdots \sigma_N} \psi_e(x_1, \sigma_1, x_2, \sigma_2, x_3, \sigma_3, \ldots, x_N, \sigma_N)$$

$$\times\ \psi_e(x_1', \sigma_1, x_2', \sigma_2, x_3, \sigma_3, \ldots, x_N, \sigma_N). \tag{16.1}$$

We have normalized γ_2 here as $\operatorname{Trace}\gamma_2 = 1$; other authors prefer a normalization according to $\operatorname{Trace}\gamma_2 = N(N-1)$. Let us rewrite H_e as

$$H_e = \sum_{i=1}^{N} h_1(i) + \sum_{1 \leqslant i < j \leqslant N} h_2(i, j),$$

where $h_1(i)$ denotes the one-body Hamiltonian $-\frac{1}{2}\Delta_{x_i} - \sum_{k=1}^{M} \frac{z_k}{|x_i - \bar{x}_k|}$ and $h_2(i, j)$ the two-body Hamiltonian $\frac{1}{|x_i - x_j|}$. The energy can then be written as

$$\langle \psi_e, H_e \psi_e \rangle = \frac{N}{2} \operatorname{Trace}(K_e \Gamma_2),$$

where Γ_2 denotes the two-particle density operator, namely the operator $L^2(\mathbb{R}^3 \times \mathbb{R}^3)$ whose kernel is γ_2, and where K_e is the so-called *reduced Hamiltonian operator*. It is a self-adjoint operator on $L^2(\mathbb{R}^3 \times \mathbb{R}^3)$ formally defined by

$$K_e = h_1(1) + h_1(2) + (N-1)h_2(1, 2).$$

What makes this expression attractive is that the energy in now an integral on $\mathbb{R}^6$ (and not on $\mathbb{R}^{3N}$ as before).

One could now try to find the minimum of the energy as functional of γ_2. However, this must be done under the restriction that γ_2 is derivable from an (antisymmetric) N-electron wavefunctions ψ_e, or at least from an ensemble state. The formulation of this restriction is known as the *N-representability problem.* More precisely, one speaks of *pure state N-representability* or *ensemble N-representability*, depending on whether one wants that γ_2 originates from a wavefunction according to Eq. (16.1), or from a N-particle density operator in the sense of

$$\gamma_2 = \sum_{i=1}^{+\infty} n_i \gamma_2^i, \quad 0 \leqslant n_i \leqslant 1, \ \sum_{i=1}^{+\infty} n_i = 1, \ \gamma_2^i \text{ of the form (16.1).}$$

Unfortunately, to ensure N-representability of a given γ_2, one only knows

- necessary conditions, that are known to be not sufficient,
- and sufficient conditions that are too restrictive in practice.

Apart from trivial necessary conditions, such as hermiticity and antisymmetry, as well as the correct trace, the most important necessary conditions are related to the nonnegativity of γ_2 itself, and of operators related to γ_2, such as the *two-hole matrix* and the *particle–hole matrix* (COLEMAN and YUKALOV [2000], KUMMER [1977], KUTZELNIGG and MUKHERJEE [1999], MUKHERJEE and KUTZELNIGG [2001]). The first attempt to solve the N electron problem in terms of γ_2 is probably due to BOPP [1959], who did not care about N-representability, and even assumed an incorrect upper bound for the eigenvalues of γ_2. The search for a solution of the N-representability problem is much associated with the name of COLEMAN [1963], and a good account of this problem and its history is found in his recent monograph (together with Yukolov) COLEMAN and YUKALOV [2000]. This has the subtitle: *Coulson's challenge*, to honor the first chemist who has suggested to follow this track. The most recent results of Coleman on the N-representability problem can be found in COLEMAN [2002].

It is interesting to compare a theory in terms of γ_2 with density functional theory (DFT). While the energy as a functional of the density is *unknown*, the energy as functional of γ_2 is *exactly known*. On the other hand, while one does not know the restrictions for a variation of γ_2, there appear to be no restrictions on the variations of the density, i.e., no N-representability problem in DFT. Strictly speaking, in DFT the (unknown) functional has to take care of N-representability, i.e., of the fermionic nature of the electrons, so the N-representability problem is somewhat hidden, rather than really absent. In Kohn–Sham type approaches, the construction of the Kohn–Sham determinant cares much for approximate N-representability.

By hindsight we can say that the community of the 2-particle density matrix theory was much more reluctant to rely on only approximate N-representability conditions, than was the DFT community to rely on approximate density functionals, and that this different attitude of the two communities has strongly influenced the history of the quantum theory of many-electron systems (KUTZELNIGG and HERIGONTE [1999]). Actually long ago GARROD and MILHAILOVIĆ [1975] has done successful calculations with approximate N-representability conditions, and recent applications by Nakatsuji (NAKATA, EHARA and NAKATSUJI [2002]) look encouraging.

Somewhat on the way between DFT and 2-RDM theory is the formulation of the energy as a *functional of the one-particle density matrix* 1-RDM. Unlike for the 2-RDM theory the energy as a functional of the 1-RDM is not known, but a larger part of it is known than for the corresponding functional of the density. Actually both the kinetic energy and the exchange energy are known functionals of the 1-RDM, so only the part referring to the correlation energy is unknown. Some preliminary formulations of a *one-particle density matrix functional theory* (GOEDECKER and UNMRIGAR [1998], CIOSLOWSKI and PERNAL [1999], YASUDA [2002]) suggest that methods of this kind may become competitive with DFT.

There is an alternative to attempts to formulate the theory entirely in terms of the 2-RDM and care for stationarity of its variations, and where one rather considers stationarity conditions with respect to variations of the wavefunction, but formulates these in terms of density matrices.

The oldest approach on these lines is the hierarchy of *contracted Schrödinger equations* (NAKATSUJI [1976], COHEN and FRESHBERG [1976]). One can formally start with the Schrödinger equation

$$H_e \psi_e = E \psi_e, \tag{16.2}$$

multiply this from the left by ψ_e and integrate over a subset of the variables $(x_1, \sigma_1, \ldots, x_N, \sigma_N)$. So one is lead to a *hierarchy* of equations involving the reduced matrices of different particle rank, i.e., the matrices built in the same fashion as (16.1) with integration over a subset of the (x_i, σ_i). Integrating over $(\sigma_1, \sigma_2, \ldots, \sigma_p, x_{p+1}, \sigma_{p+1}, \ldots, x_N, \sigma_N)$ defines in this way γ_p, the p-particle reduced matrix.

The hierarchy obtained from (16.2) by successive integrations is of the form

$$\text{Left-hand side in } \gamma_p, \gamma_{p+1}, \gamma_{p+2} = E\gamma_p$$

for $1 \leqslant p \leqslant N-2$ (while the equations $p = N-1$ and $p = N$ are slightly different). More precisely, one obtains

$$\begin{aligned} &\left(\sum_{i=1}^{p} h_1(i) + \sum_{1 \leqslant i < j \leqslant p} h_{12}(i,j)\right)\gamma_p \\ &\quad + (N-p)\int_{\mathbb{R}^3}\left(h_1(p+1) + \sum_{1 \leqslant i \leqslant p} h_2(i, p+1)\right)\gamma_{p+1}\, dx_{p+1} \\ &\quad + \frac{(N-p)(N-p+1)}{2}\int_{\mathbb{R}^3}\int_{\mathbb{R}^3} h_2(p+1, p+2)\gamma_{p+2}\, dx_{p+1}\, dx_{p+2} \\ &= E\gamma_p. \end{aligned} \tag{16.3}$$

Any member of this hierarchy relates γ_p to γ_{p+1} and γ_{p+2}. Since the sequence of the γ_p does not terminate before γ_N, a truncation of the hierarchy at some $p < N$ is not possible, which does not appear to make these equations very useful.

The same is true for the p-particle *Brillouin conditions* (KUTZELNIGG [1979]). These can formally be written as

$$\langle \psi_e | [H, X] | \psi_e \rangle,$$

where X is an *excitation operator*, such that one has a one-particle Brillouin condition for X a one-particle excitation operator etc. In the same compact notation the contracted Schrödinger equations are

$$\langle \psi_e | X(H-E) | \psi_e \rangle.$$

The Brillouin conditions have the advantage that γ_p is only related to γ_{p+1} (and not γ_{p+2}), but they leave the *diagonal elements* of the γ_p undetermined.

A breakthrough came with the work of VALDEMORO [1985], VALDEMORO [1992], VALDEMORO, TEL and PEREZ-ROMERO [2000] who succeeded, in a rather tricky way, to construct approximations of the γ_p in terms of the γ_k with $k < p$, and could so define a hierarchy of *contracted Schrödinger equations*, that can be truncated at any particle rank p. This *reconstruction* was refined by NAKATSUJI and YASUDA [1996], while MAZZIOTTI [1998a] gave a rationalization of the reconstruction in terms of the *cumulants* of the density matrices, noting that the suggested 'reconstruction' of the γ_k corresponds essentially to a truncation of the sequence of the k-particle cumulants. MUKHERJEE and KUTZELNIGG [2001] gave a formulation of the theory entirely in terms of the cumulants, making use of a generalization of *normal ordering of operators* and Wick's theorem with respect to an arbitrary reference functions. The *irreducible k-particle contracted Schrödinger equations* (ICSE_k) and *k-particle irreducible Brillouin conditions* (IBC_k) only involve separable (additively separable) quantities – like the cumulants – which can be represented by *connected* diagrams, which is an important paradigm of many-body quantum mechanics.

The cumulants of the density matrices can be defined via a generating function (KUTZELNIGG and MUKHERJEE [1999], MAZZIOTTI [1998a]). A cumulant λ_k, corresponding to a density matrix γ_k, represents, so to say, that part of γ_k, which goes beyond what one expects for independent particles (KUTZELNIGG and MUKHERJEE [1997], KUTZELNIGG and MUKHERJEE [1999]). The two-particle cumulant $\lambda_2(1,2;1',2')$ is, e.g., defined as

$$\lambda_2(1,2;1',2') = \gamma_2(1,2;1',2') - \gamma_1(1;1')\gamma_1(2;2') + \gamma_1(1;2')\gamma_1(2;1').$$

For a single Slater determinant wavefunction all cumulants (except $\lambda_1 = \gamma_1$) vanish. The cumulants have a lot of very nice properties (KUTZELNIGG and MUKHERJEE [1999]). In particular the trace of λ_k is of $O(N)$, when in the same normalization that of γ_k is of $O(N^k)$. An essential property of the cumulants is that they are *additively separable*, which means that for a supersystem AB, consisting of two noninteracting subsystems A and B, the λ_k are decomposable as

$$\lambda_k(AB) = \lambda_k(A) + \lambda_k(B).$$

Neither the wavefunction, nor any of the γ_k, except γ_1 have this separability property. While truncation of the γ_k at some k is not allowed, a similar truncation of the λ_k can be justified. Truncation at $k = 1$ automatically leads to Hartree–Fock theory, which furnishes a first approximation of γ_k, while truncation at $k = 2$ defines a *two-particle approximation*, from which one gets a first approximation of λ_2.

Clearly, a lot of work remains to be done to give proper theoretical, mathematical and numerical foundations to the approaches just discussed. Nevertheless, there is currently a great activity in this field, and optimism appears to be justified.

17. The mathematical background

This section is devoted to a brief presentation of the mathematical background needed to put the models of quantum chemistry on a sound ground. On purpose, we focus on the question of existence of minimizers, i.e., ground states, for the various minimization problems we have encountered so far.

Clearly, this section is written in a simplified mathematical language that is particularly aimed at a reader that has no specific competence in advanced mathematics. It is a revised version of some parts of two survey articles written in the same spirit (DEFRANCESCHI and LE BRIS [1997], DEFRANCESCHI and LE BRIS [1999]). We refer the reader from the chemistry community to these two articles for more details. On the other hand, experts at mathematics can easily skip this section, and proceed to the following one. For more details on the arguments and on the techniques, we refer to REED and SIMON [1975], THIRRING [1983], SCHECHTER [1981], CYCON, FROESE, KIRSCH and SIMON [1987], BLANCHARD and BRUNING [1982], together with LE BRIS [1993a], LIONS [1987] and many chapters of DEFRANCESCHI and LE BRIS [2000].

The reason why we focus on the question of existence of a ground-state, while there exists a lot of other questions that are both interesting from the viewpoint of quantum chemistry and challenging for the mathematician is directly related to the main purpose of this Handbook, namely numerics. There exists material that extensively comment upon the proof of asymptotic exactness of the different approximations described above with respect to the original exact Schrödinger problem. One could also examine the various attempts to determine the maximal negative charge of a stable anion in a given model. All the works devoted to such questions feature very fine mathematical arguments and results that have important theoretical consequences. The results also have some impact on the numerics, but this impact is less evident.

We prefer to concentrate here on the question of existence of a minimizer, which is indeed a central issue for the understanding of the convergence of numerical algorithms, and therefore the construction of efficient discretizations of the problem. It turns out that the proof of existence of a minimizer in infinite dimension (since the problem we address are indeed posed, at the theoretical level, on an infinite-dimensional space such as $H^1(\mathbb{R}^3)$) is really alike the first step of the proof of the convergence of any algorithm used to compute the finite-dimensional approximation of this minimizer. This will become clear in Chapter II. We shall make below an attempt to let the reader feel why it is so. In the same fashion, knowing about the uniqueness of the minimizer would be interesting, for the same purpose. However, we only rarely know about this uniqueness, and the situations when we indeed know the uniqueness holds are too academic to deserve any real attention in this text aimed at practical purposes.

For pedagogical purposes, it is convenient to reformulate the search for a ground state for all the models we have introduced above under the form of the following abstract problem

Find a function ϕ_0 in a function space X that satisfies some given constraint $J(\phi_0) = N$ (translating the fact that the number of electrons is fixed) and that minimizes the

energy functional $E(\phi)$ on the convenient set of states

$$I_N = E(\phi_0) = \inf\{E(\phi),\ \phi \in X,\ J(\phi) = N\}. \tag{17.1}$$

Let us at once mention that this formulation is not *strictly speaking* equivalent to *all* the models described above, but it gives at least a unified way of seeing all of them from a mathematical standpoint.

Our goal is to mathematically understand the difficulties encountered in trying to establish the existence of a convenient ϕ_0.

Let us consider a minimizing sequence of (17.1), namely a sequence satisfying

$$\begin{cases} \phi^n \in X, \\ J(\phi^n) = N, \\ \lim_{n\to+\infty} E(\phi^n) = I_N. \end{cases}$$

The question under study is to know whether this sequence ϕ^n can be said to converge, possibly up to the extraction of a subsequence, in a convenient topology, toward a function ϕ_0, and whether this ϕ_0 is a minimizer or not.

As we want the sequence to converge, the less we can ask is that it is bounded. The first step is therefore to check that the sequence ϕ^n is bounded in the strongest possible topology. As we deal here with wavefunctions that have finite kinetic energy, it is easily understandable that the functional space X will have a close relationship with H^1, and therefore that the sequence ϕ^n is bounded for the H^1 norm (this has to be proven, but it is most often the case).

We then come across a usual difficulty in analysis: We have at hand a sequence of functions that is bounded and we want to make it converge. The game we play is to weaken the topology, in order to give a better chance to this sequence to converge. In the present case, it will be customary to prove that the sequence ϕ^n converges for the weak topology in H^1. This creates a limit, ϕ_0, which is a limit only in a weak sense, but has the enormous advantage to exist. Of course, it is not at all clear that the limit will also be a limit in the usual (strong) sense. To fix the ideas of the reader, let us mention that ϕ^n is said to weakly converge in L^2 to ϕ_0 when $\int \phi^n \psi$ goes to $\int \phi_0 \psi$ for any L^2 function ψ, while as usual it is said to converge strongly to ϕ_0 when $\int |\phi^n - \phi_0|^2$ goes to zero. It is clear that the latter implies the former, but the converse implication is false (consider $\phi^n(x) = \sin(nx)$).

The price to pay for the weakening of topology is that we now miss information on the limit ϕ_0. Indeed, most of the energy functionals we manipulate are continuous for the usual (strong) topology, but not necessarily for the weak topology we have now. A corollary is that we do not know whether

$$\text{weak-}\lim_{n\to+\infty} \phi^n = \phi_0 \quad \text{implies or not that}$$

$$E(\phi_0) = \lim_{n\to+\infty} E(\phi^n) = I_N \text{ and } I(\phi_0) = N.$$

This will have to be proven in each case.

Therefore, one has to examine the so-called possible lack of compactness (another word for strong convergence) for a given weak converging sequence. Typically, these lack of compactness are of three possible natures (or of course any combination of these)

- oscillation: The functions oscillate more and more rapidly as n goes to infinity, such as a sine function with frequency n,
- concentration: The functions concentrate around a given point, such as a sequence of regular functions approximating a Dirac mass in the limit,
- escape at infinity: The functions have support that go to infinity, either while breaking into pieces (one speaks of a dichotomy) or while spreading over the space (one speaks of a vanishing).

In the type of problems we deal with, the first two behaviours are not allowed, because of the kinetic energy bound: The fact that the sequence is bounded in H^1 impedes such behaviours. On the contrary, because the domain the functions are defined upon is the whole space $\mathbb{R}^3$, we may observe the third behaviour. It is indeed possible, that with a view to minimize further the energy, the functions ϕ^n begin to escape at infinity. Chemically, this has to be thought of as some loss of charge at infinity like in a ionization. Starting with N electrons, one ends up, e.g., with $N-1$ because the molecular system with N electrons is not stable, and has energy strictly larger than the system with $N-1$ electrons, thus the ionization. This is to be related with the fact that, e.g., the Hartree–Fock problem with relaxed constraint (9.8) is not necessarily equal to the original Hartree–Fock problem (9.7). This explains the difficulty the mathematician has in trying to prove that the constraint $J(\phi_0) = N$ is satisfied in the limit. This also explains why a condition on the positive charge contained in the system will play a role (see the condition $Z > N - 1$ for the Hartree–Fock approximation, e.g.). Considering the fact that the only possible lack of compactness is at infinity, the problems we deal with here are called *locally compact problems*. This has the following sense: If we abruptly restrict the domain of functions to a fixed bounded domain of $\mathbb{R}^3$ instead of the whole of it, then the existence of a minimizer becomes a simple fact to prove. Of course, the so obtained minimizer depends on the chosen domain, and in particular on its size. If we now let the bounded domain progressively increase its size, the sequence of minimizers may or may not converge to a minimizer on the whole space. The question of asking whether the sequence of minimizers may or may not converge to a minimizer on the whole space is exactly of the same difficulty as the original one.

This simple observation is not naive. It has a close relationship with the numerics. Suppose indeed that we discretize our functions ϕ on a finite-dimensional space. The problem of finding a minimizer in this finite-dimensional space is a trivial one (in the above sophisticated mathematical language, it suffices to say that in finite dimension, the weak and the strong topology coincide). The question of asking whether this minimizer converges to an intrinsic limit when the basis approaches a basis of, say, $L^2(\mathbb{R}^3)$ is the real issue. Answering it is again of the same difficulty as proving the existence of the minimizer above. By enlarging the basis set, one will indeed allow the ϕ^n to put some mass at larger and larger distances, and we end up with the same question.

The reader familiar with computational quantum chemistry will advantageously relate the present discussion with questions such as: Basis sets with diffuse functions, or basis sets for molecular systems under electric fields. In such situations, some charge is explicitly allowed to go far from the nuclei (by putting some basis functions far away from them). The existence of a minimizer in such settings can be a pure numerical artefact: Pushing the basis function further may result in a change of the minimizer, proving that the result is not intrinsic in terms of the basis, and therefore that it corresponds to a minimizer that does not exist.

This being said, there remains to give some indications on how one indeed proves that the minimizing sequence does not escape at infinity, and therefore that there exists a minimizer.

In order to give a rough idea of the argument, let us come back to a simple quadratic case

$$I=\inf\left\{\left\langle(-\Delta+W)\phi,\phi\right\rangle,\ \phi\in H^1(\mathbb{R}^3),\ \int|\phi|^2=1\right\},\tag{17.2}$$

which is a particular case of (17.1), not far from the original Schrödinger case. In (17.2), we assume being given a potential W, say in $L^{3/2}+L^\infty$, that goes to zero at infinity. Note that in this quadratic case, the scalar I is both the value of the infimum (17.2) and the first eigenvalue of the linear operator involved $-\Delta+W$. It is a simple exercise to see that, up to an extraction, a minimizing sequence ϕ^n for (17.2) converges to some ϕ_0 that satisfies $\int_{\mathbb{R}^3}|\phi_0|^2\leqslant 1$ and $\langle(-\Delta+W)\phi_0,\phi_0\rangle\leqslant I$. Assume now $I<0$ and $\int_{\mathbb{R}^3}|\phi_0|^2<1$. Then we have

$$\frac{\langle(-\Delta+W)\phi_0,\phi_0\rangle}{\int_{\mathbb{R}^3}|\phi_0|^2}<\left\langle(-\Delta+W)\phi_0,\phi_0\right\rangle\leqslant I,$$

which yields a contradiction, for $(\int_{\mathbb{R}^3}|\phi_0|^2)^{-1/2}\phi_0$ is a test function for (17.2). Therefore the conclusion is: If we choose W so that $I<0$, then a minimizer exists. Conversely, using spectral theory arguments it can be shown that $I<0$ is indeed a necessary condition for the existence of a minimizer, for the first eigenvalue of $-\Delta+W$ (for such a W) cannot be nonnegative (see LIEB [1981]).

As a minimizer to (17.2) is a ground state for the operator $-\Delta+W$, it is easily understandable that spectral theory (here however simple) has played a role for the proof of existence of a minimizer to (17.2). For the same reason, spectral theory will also play a role in the proof of existence of a minimizer to (17.1). Loosely speaking, the strategy there consists in proving that the weak limit ϕ_0 of the minimizing sequence ϕ^n is an eigenfunction of an operator of the type $-\Delta+W$. For instance, to fix the idea, in the simple Thomas–Fermi–von Weizsäcker setting, ϕ_0 is shown to be a solution to

$$-\Delta\phi_0+V\phi_0+\frac{5}{3}C_{\mathrm{TF}}\phi_0^{4/3}\phi_0+\left(\phi_0^2\star\frac{1}{|x|}\right)\phi_0=\lambda_0\phi_0$$

which can be written as $(-\Delta+W)\phi_0=\lambda_0\phi_0$ where

$$W=V+\frac{5}{3}C_{\mathrm{TF}}\phi_0^{4/3}+\phi_0^2\star\frac{1}{|x|}.$$

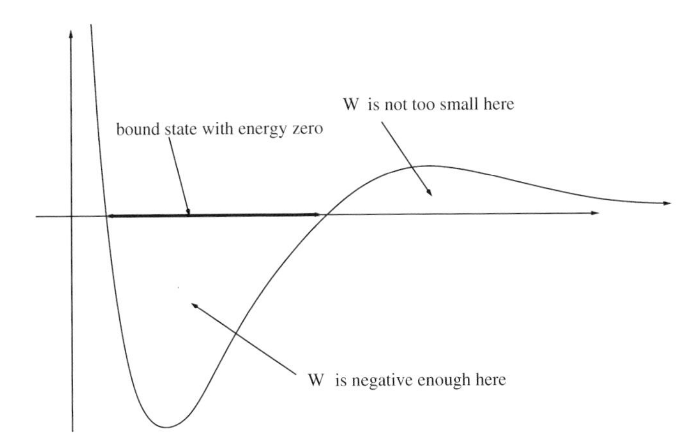

FIG. 17.1. "Heuristic" conditions for existence of a bound state of zero energy.

Then the question is: Can the associated eigenvalue (which is indeed the Lagrange multiplier associated to the constraint $J(\phi) = N$) vanish or not? exactly as it was questioned whether the ground state eigenvalue of (17.2) can vanish or not. This will in turn be studied either by exploiting ready-to-use results of spectral theory, or by proving on the case of interest some scalar condition analogous to $I < 0$ for (17.2). In both cases, the essential feature is to know whether the potential W is sufficiently negative (i.e., attractive) at finite distance, and sufficiently positive at infinity (i.e., prevents diffusion), see Fig. 17.1. If it is the case, then it will allow $-\Delta + W$ to have an eigenvalue below its essential spectrum, and this eigenvalue will be associated to a normalized eigenvector, indeed ϕ_0. In terms of physics, it is exactly proving that the Hamiltonian has at least a bound state below the states of diffusion. Let us mention that the condition *analogous to* $I < 0$ may take for the different occurrences of (17.1) some more complicated form. One speaks of the strict-subadditivity conditions (see LIONS [1987], or the monograph by STRUWE [1990] where many related techniques are exposed). We shall come back to them in a more intuitive framework in the next section.

The main message of this section that the mathematical arguments needed to prove the existence of a ground state for a given model are *not* straightforward. They involve very fine mathematical results and approaches, that need to be combined in a specific way to address each case as a particular case. No general existence theorem exists (for we put ourselves here in a nonlinear framework). The main body of the mathematical work has only been done in the late twentieth century, when nonlinear analysis arrived to a mature stage. Despite all the efforts, a lot remains to be done.

To be kept in mind is also the fact that the existence proofs will have enormous counterparts in the proofs of convergence of algorithms.

Another point we want to briefly address in this section concerns the hypothesis we have made that we only consider nonrelativistic models. It is to be mentioned for completeness that the relativistic models are far more difficult to study and manipulate than the nonrelativistic ones. Let us mention that the central difficulty comes from the fact that the kinetic operator, which is the Laplacian in the nonrelativistic setting, is now replaced by the Dirac operator. This amounts to replacing a nonnegative operator with purely continuous spectrum $[0, +\infty[$ by an operator with purely continuous spectrum $]-\infty, -1] \cup [1, +\infty[$, that is an operator which is not bounded from below, and moreover has an infinite (continuous) number of negative eigenvalues. In other words, one leaves the standard minimization framework. All the arguments we have made collapse at once, and completely different techniques must be resorted to. For instance, one cannot make the Dirac operator a bounded from below operator simply by projecting it onto the orthogonal to a finite dimensional subspace, for one cannot eliminate the part of the spectrum $]-\infty, -1]$. The techniques to be used are much less naive.

Let us only quote the following noteworthy result (due to M. Esteban and E. Séré) that we formulate somewhat vaguely. Of concern is the relativistic analogue to the Hartree–Fock model, namely the Dirac–Fock model, built by simply replacing in (9.1) the Laplacian operator in H_e by the Dirac operator and by considering that the test functions ϕ_i in (9.2) are spinnors, i.e., test functions valued in $\mathbb{C}^4$. The result states that eigenstates may be defined properly, with eigenvalues in the spectral gap $[-1, +1]$ of the Dirac operator. The eigenstates corresponding to eigenvalues in $[-1, 0]$ model electrons. One eigenstate among them is particular. In some sense it corresponds to the smallest possible eigenvalue. In the nonrelativistic limit, this eigenstate converges to a Hartree–Fock ground state. Therefore, it can be seen as the ground state of the Dirac–Fock model. Of particular importance is that the mathematical theory provides with a variational characterization of this ground state, which is completely new and opens new tracks for its numerical computation. For a complete view on the modelling, the mathematical properties and the various numerical approximations in the relativistic framework, we refer the reader to another Contribution of this Handbook.

18. Geometry optimization

Let us briefly approach in this section the geometry optimization problem that have been left aside so far, for we wanted to concentrate on the search for the electronic ground state (6.5) and its approximations, the positions of the nuclei being considered as known and fixed.

Let us reproduce it here

$$\inf\left\{W(\bar{x}_1, \dots, \bar{x}_M),\ (\bar{x}_1, \dots, \bar{x}_M) \in \mathbb{R}^{3M}\right\},$$

where W is given by (6.4), i.e.,

$$W(\bar{x}_1, \dots, \bar{x}_M) = U(\bar{x}_1, \dots, \bar{x}_M) + \sum_{1 \leqslant k < l \leqslant M} \frac{z_k\, z_l}{|\bar{x}_k - \bar{x}_l|}$$

and U denotes the exact electronic ground state energy. It is clear that once U has been approximated by any of the approximations presented above, we may define the analogous approximation to W. For instance, to fix the ideas, one can consider the geometry optimization for the Hartree–Fock approximation

$$\inf\left\{U^{\mathrm{HF}}(\bar{x}_1,\ldots,\bar{x}_M)+\sum_{1\leqslant k<l\leqslant M}\frac{z_k\,z_l}{|\bar{x}_k-\bar{x}_l|},\ (\bar{x}_1,\ldots,\bar{x}_M)\in\mathbb{R}^{3M}\right\} \tag{18.1}$$

with U^{HF} defined by (9.1).

From the theoretical viewpoint, it is not at all mathematically clear whether this problem (18.1) admits a minimizer, i.e., whether there exists a collection $(\bar{x}_1,\ldots,\bar{x}_M)$ of position of nuclei, together with the Hartree–Fock ground state associated to this particular set of positions, that minimizes (18.1). The state of the art of the mathematical knowledge is indeed that, for the Hartree–Fock approximation, it is an open question. On the other hand, for the analogue minimization problem to (18.1) set with simpler models than the Hartree–Fock model for the electronic state, the answer is known. For the Thomas–Fermi model, it is known that the answer is negative. Moreover, it is proven (and known as *Teller no binding theorem*) that in this theory, no molecular system is stable, which amounts to claiming that no minimizer to the geometry optimization problem exists (as soon as there are at least two nuclei, of course). The reason is that, for any partition of the set of $M\geqslant 2$ nuclei, the energy (in the sense of (18.1)) of the two parts together is larger than the energy of the two subsystems taken at infinity from one another. On the contrary, for improvements of the crude Thomas–Fermi model (TFW, e.g.), and for the bosonic analogue to the Hartree–Fock model, namely the Hartree model (which we have not described above, for it is of almost no practical use in computational chemistry), then it is known that there exists a minimizer for the geometry optimization problem. The proof of this existence is very difficult, and due to CATTO and LIONS [1992]. The idea is precisely to show that, contrary to Teller no binding theorem for the TF case, a so-called strict subadditivity condition of the type

$$I_{M=M_1+M_2,N=N_1+N_2} < I_{M_1,N_1} + I_{M_2,N_2}$$

holds. It means that the energy of the system with M nuclei and N electrons is strictly less than the sum of the energies of the subsystems, no matter the way the original system is split into subsystems. The same methodology could have been applied to the pure electronic problem (at a fixed configuration of nuclei) to study the possible escape of an electron at infinity. We alluded to that in the previous section. On the basis of these mathematical results (and also on the basis of the billions of numerical experiments done so far in computational chemistry), it is *believed* that the geometry optimization problem for the Hartree–Fock case (18.1) admits a minimizer, but at this day no mathematical proof has confirmed this belief. For sophisticated models of density functional theory, such as the Kohn–Sham model, the geometry optimization problem is also an open issue from the mathematical viewpoint.

Maybe this discussion is the good opportunity to emphasize one point. By showing that a geometry optimization problem admits a minimizer in the setting of a given model, the mathematician of course does not aim at proving that the system under consideration exists. For instance, he perfectly knows that the molecule of water exists! He only aims at testing the ability of the model under consideration to reproduce this reality. This is a step towards more confidence in the model, and thus more confidence in the numerical prediction of it. What has just been said on the geometry optimization problem could be stated analogously on many mathematical issues in the context of chemistry.

Despite the lack of exact theoretical results, the geometry optimization problem is of course explicitly solved in computational chemistry. We shall see the numerical techniques for this purpose in the next section. Let us only say here that they all make use of a very peculiar mathematical property of the problem, that we detail now. This property is related to the so-called *analytical derivatives* calculations.

Let us for this purpose recast the geometry optimization problems (18.1) in the following general mathematical form

$$\inf\{W(\bar{x}),\ \bar{x} \in \Omega\}, \qquad W(\bar{x}) = \inf\{E(\bar{x},\phi),\ \phi \in \mathcal{H},\ g(\bar{x},\phi) = 0\}, \tag{18.2}$$

where W is defined in some domain $\Omega \subset \mathbb{R}^n$, where $\mathcal{H}$ is a Hilbert space (of finite or infinite dimension) and where

$$E : \Omega \times \mathcal{H} \to \mathbb{R}, \qquad g : \Omega \times \mathcal{H} \to \mathcal{K}$$

are regular enough ($\mathcal{K}$ is a finite-dimensional Hilbert space). It is indeed easy to recognize in this formal setting $\bar{x}$ as the collection of coordinates of the nuclei, ϕ as the electronic wavefunction, $E(\bar{x},\phi)$ as the energy functional depending both on the nuclear coordinates and the electronic wavefunction, and $g(\bar{x},\phi)$ as the orthonormality conditions on the molecular orbitals. We make it depend explicitly on $\bar{x}$, although it does not in theory, because we have practical purposes and in the numerics it will often depend on $\bar{x}$ (through the basis set, see Chapter II). The function W represents the electronic ground state energy.

Suppose we now want to search for a minimum to (18.2) and use for this purpose a deterministic gradient-like minimization method (see Chapter II). Then $\frac{\partial W}{\partial \bar{x}_k}$ is needed. For $\phi(\bar{x})$ denoting a ground state for $\bar{x}$ (uniqueness of $\phi(\bar{x})$ and regularity for the map $\bar{x} \mapsto \phi(\bar{x})$ are assumed), we may write formally

$$\begin{aligned}\frac{\partial W}{\partial \bar{x}_k}(\bar{x}) &= \frac{\partial}{\partial \bar{x}_k}\big(E(\bar{x},\phi(\bar{x}))\big)\\ &= \frac{\partial E}{\partial \bar{x}_k}(\bar{x},\phi(\bar{x})) + \left\langle \nabla_\phi E(\bar{x},\phi(\bar{x})), \frac{\partial \phi}{\partial \bar{x}_k}(\bar{x})\right\rangle_{\mathcal{H}},\end{aligned}$$

by the chain rule. Next, as $\phi(\bar{x})$ is a minimizer, it satisfies

$$\nabla_\phi E(\bar{x},\phi(\bar{x})) = d_\phi g(\bar{x},\phi(\bar{x}))^T \cdot \lambda(\bar{x})$$

for some Lagrange multiplier $\lambda(\bar{x}) \in \mathcal{K}$. And finally, by differentiation of the constraint, we have

$$\frac{\partial g}{\partial \bar{x}_k}\big(\bar{x}, \phi(\bar{x})\big) + d_\phi g\big(\bar{x}, \phi(\bar{x})\big) \cdot \frac{\partial \phi}{\partial \bar{x}_k}(\bar{x}) = 0.$$

Inserting the latter two informations into the first formula yields

$$\frac{\partial W}{\partial \bar{x}_k}(\bar{x}) = \frac{\partial E}{\partial \bar{x}_k}(\bar{x}) - \left\langle \lambda(\bar{x}), \frac{\partial g}{\partial \bar{x}_k}\big(\bar{x}, \phi(\bar{x})\big) \right\rangle_{\mathcal{K}}.$$

The crucial point is that $\frac{\partial \phi}{\partial \bar{x}_k}(\bar{x})$ has been eliminated. Therefore, the gradient of W can be directly computed from $(\bar{x}, \phi(\bar{x}), \lambda(\bar{x}))$ (that is to say the set of positions of nuclei considered, the electronic ground state, and the monoelectronic energies) without any further calculations. This property makes the gradient a very easily computable quantity in this setting, which is highly unusual in other contexts of scientific computing for the engineering sciences. Let us remark that the manipulations above are only formal, and do not have a sound mathematical ground in most settings (giving them a rigorous meaning indeed in particular requires the uniqueness of $\phi(\bar{x})$ for each $\bar{x}$, which is not known).

As will be seen in the sequel, this property will be used both for the solution to the geometry optimization problem (Section 37), and in the context of ab initio molecular dynamics (Chapter V).

CHAPTER II

Ground State Calculations

19. Introduction

In the spirit of the previous chapter, we mainly concentrate here again on the electronic problem (6.5), i.e., on the determination of the electronic ground state wavefunction and energy of the molecular system for a given nuclear configuration. It has already been mentioned that further calculations, such as those involving electronic excited states, will be dealt with later on. Geometry optimization will be briefly dealt with in Section 37.

What we have at hand at this stage are various models for the approximation of the electronic ground state. The question under study here is the following: *How do we build numerical approximations of these models*?

Basically, the two methods of preference are the Hartree–Fock approximation (together with the improvements of it, namely the so-called post Hartree–Fock methods), and the density functional type methods.

Both types of methods lead to a minimization problem, either (9.7) or (15.4), which formally read

$$\inf\left\{\mathcal{E}(\phi_1,\dots,\phi_N),\ \phi_i\in H^1(\mathbb{R}^3),\ \int_{\mathbb{R}^3}\phi_i\phi_j=\delta_{ij},\ 1\leqslant i,j\leqslant N\right\} \tag{19.1}$$

for a given, rather intricate, energy functional $\mathcal{E}$.

This problem involves one electron wavefunctions, typically denoted by ϕ_i, defined on the space $\mathbb{R}^3$ (forgetting again the spin variable). The ground state and the ground state energy are then constructed from the knowledge of these ϕ_i. It has also been mentioned that the ϕ_i, which originally are defined through a minimization problem, can also be determined on the basis of the following observation: They are a solution to the Euler–Lagrange equations associated to the original problem. In order to fix the ideas, let us figure out we work either on the Hartree–Fock problem (9.7), or on the Kohn–Sham problem (15.4). Then instead of directly attacking the minimization of the energy, we know that we may try to solve the nonlinear eigenvalue problem (9.13), or respectively (15.6), both equations being of the form

$$\left\{\begin{aligned} &H_\Phi\phi_i=\lambda_i\phi_i,\\ &\int_{\mathbb{R}^3}\phi_i\phi_j=\delta_{ij}.\end{aligned}\right. \tag{19.2}$$

It has indeed been underlined that although the two theories originate from radically different viewpoints, and although the ϕ_i do not have the same meaning in the two theories, the mathematical forms of the equations are basically the same. This being said, two issues are to be addressed. First, which strategy do we choose to solve the problem? Second, how do we construct a finite dimensional approximation of the objects we manipulate?

The first issue is: *Is it better to directly attack the minimization problem* (19.1), *or to try to solve Eqs.* (19.2)?

Theoretically, there is one and only one possible answer. In a nonconvex setting, such as the one we have here, many solutions of Eqs. (19.2) which do not correspond to the ground state we search for are likely to exist: Local but nonglobal minima, local maxima, and critical points. Even if we exhibit and impose second order conditions, we have no way to be sure that the solution of the equation we shall compute is indeed the ground state since any local minimum satisfies the latter conditions. Therefore, considering the minimization problem is the only acceptable solution.

In the numerical practice, the landscape is radically different. Precisely because a nonconvex minimization problem such as (19.1) is likely to have many local minimizers (and it indeed has in practice) performing a direct numerical minimization with classical deterministic algorithms has long been considered as risky, or even hopeless. Consequently, the codes aim at solving the nonlinear eigenvalue problem (19.2), even if it is not ensured that the solution, which surely is a critical point of the energy, is a ground state, in view of the previous theoretical warning. To our opinion, another factor has biased the choice between a direct minimization and a solution to the critical point equation. In the early days of computational chemistry, the only quantities that were both computationally available and experimentally measurable were the ionization energies, namely the λ_i in (19.2) (we were then in the framework of the Hartree–Fock approximation). As a consequence, all implementations done in computational chemistry firstly aim at computing these quantities, which naturally arise in the solution of the nonlinear eigenvalue problem. This was a natural behaviour. And the contemporary state of the art inheritates from this.

Such a state of the art however deserves a comment.

Today, in the engineering sciences, many, if not all, of the minimization problems of interest are indeed nonconvex problems. Some of them are nevertheless treated by optimization techniques. Think of the numerous instances provided by the optimal control theory, although it is true that in the latter setting a "good" solution, and not necessarily the "best" solution, is usually searched for. To say in what sense the minimization problems arising in computational chemistry are or are not tractable by direct minimization techniques is not an easy task. But our strong belief is that the tracks consisting in a direct minimization of the problem, for instance by stochastic techniques, must not be forgotten. For the time being, apart from this somewhat speculating statement, there is a clear and immediate consequence.

Most of the efficient techniques we are aware of to solve the problem are based upon the following observation: One does not search for *a* solution of the nonlinear eigenvalue equation (19.2) but for a solution *that is a ground state*. Therefore, within the nonlinear cycle, the information along which the equation indeed comes from a minimization problem is inserted, more or less intensively (see Section 28). This translates in particular into the *aufbau* principle. Alternatively, one may cast the most efficient algorithms under the form of minimization algorithms that only use the solution of the eigenvalue problem as an intermediate step to find a better iterate. This alternative, rather recent, viewpoint will be detailed in Section 30.

The conclusion of the above discussion is therefore: Rather than directly attacking the minimization problem, or directly solving the critical point equation, one should take benefit of the two viewpoints (minimization *and* search for a critical point) in order to build efficient algorithms.

But, as Eq. (19.2) is a nonlinear eigenvalue problem, a new, twofold, question arises. One first needs to understand *how to iterate on the nonlinearity*, and second, the nonlinearity being fixed at a given value, one needs to understand *how to solve the linear eigenvalue problem*. Both issues are of challenging difficulty. For the first one, we know from the previous discussion that this solution procedure has to keep track of the minimization nature of the problem. This will be a useful guide for the construction of efficient algorithms, as will be seen in Section 30. As this requires a good knowledge of up-to-date techniques of numerical optimization, we recall in Sections 26 and 27 the basics on such techniques. For the second one, namely the diagonalization issue, one most often uses well established techniques (such as the Jacobi–Davidson algorithm, see, e.g., SAAD [1992]). It is to be emphasized that the diagonalization of the Hamiltonian matrix is the core of a computational chemistry code, in the same respect as the solution procedure of the Laplace equation is the core of most codes of incompressible fluid mechanics, or the solution to the Riemann problem is the core of a code for compressible fluid dynamics. It is therefore particularly important to use efficient techniques for the routine devoted to that, and also to cleverly insert this routine in the general chain of computations. A particularly challenging subject is the by-pass of such diagonalization techniques by the so-called *order N methods*; the standard diagonalization methods are indeed known to be especially time consuming as they require of the order of N^3 elementary operations for a matrix of size $N \times N$. This timely topic is exposed in Section 36.

Let us now turn to the second main issue: *How do we discretize the objects*? This will actually be the first issue dealt with in this chapter (Sections 20–25). For the Hartree–Fock approximation as well as for the Kohn–Sham equations, or any density functional type equations, the main objects one needs to manipulate are the monoelectronic wavefunctions ϕ_i. Alternatively, one may also wish to manipulate the density matrices. Whatever the objects, they basically are a solution to an equation that involves partial derivatives. As usual then, the discretization may follow two roads: Either one discretizes the differential operators, which essentially gives rise to the finite differences type methods, or one approximates the space of test functions, which gives rise to Galerkin-like techniques. The former methods, although clearly less commonly used in computational chemistry at the present time, deserve attention; they are the main topic of an entire chapter in this volume, see the contribution by Fattebert and Nardelli, and will not be addressed further in the present chapter. The latter techniques, those of Galerkin type, are extensively examined in Section 20. The different possible basis sets will be presented and commented on. For the sake of clarity, we shall henceforth focus on the Hartree–Fock approximation. Unless otherwise mentioned, all what we detail on the numerical discretization techniques can be straightforwardly adapted to treat the other cases of interest. When this adaptation involves new developments or deserves specific comments, we shall make remarks (see in particular Sections 30 and 35).

20. Galerkin approximation

As announced, we begin by dealing with the discretization of the objects manipulated, and for clarity, we first concentrate ourselves on the discretization of the spinless Hartree–Fock (HF in short) approximation (9.7); for convenience, let us recall that this model reads

$$\inf\left\{E^{\mathrm{HF}}(\Phi),\ \Phi=\{\phi_i\}\in\left(H^1(\mathbb{R}^3)\right)^N,\ \int_{\mathbb{R}^3}\phi_i\phi_j=\delta_{ij},\ 1\leqslant i,j\leqslant N\right\},$$

with

$$\begin{aligned}E^{\mathrm{HF}}(\{\phi_i\}) &= \sum_{i=1}^N \frac{1}{2}\int_{\mathbb{R}^3}|\nabla\phi_i|^2+\int_{\mathbb{R}^3}\rho_\Phi\, V+\frac{1}{2}\int_{\mathbb{R}^3}\int_{\mathbb{R}^3}\frac{\rho_\Phi(x)\rho_\Phi(x')}{|x-x'|}\,dx\,dx'\\ &\quad -\frac{1}{2}\int_{\mathbb{R}^3}\int_{\mathbb{R}^3}\frac{|\tau_\Phi(x,x')|^2}{|x-x'|}\,dx\,dx',\\ \tau_\Phi(x,x') &= \sum_{i=1}^N\phi_i(x)\phi_i(x'),\\ \rho_\Phi(x) &= \sum_{i=1}^N\left|\phi_i(x)\right|^2.\end{aligned}$$

The Galerkin approximation procedure consists in approaching the *infinite-dimensional* HF problem by a *finite-dimensional* HF problem as follows: The HF energy is minimized over the set of molecular orbitals ϕ_i that can be expanded on a given finite basis set $\{\chi_\mu\}_{1\leqslant\mu\leqslant N_b}$:

$$\phi_i=\sum_{\mu=1}^{N_b}C_{\mu i}\chi_\mu.$$

The ith column of the rectangular matrix $C\in\mathcal{M}(N_b,N)$ contains the N_b coefficients of the molecular orbital ϕ_i in the basis $\{\chi_\mu\}_{1\leqslant\mu\leqslant N_b}$.

Denoting by $S\in\mathcal{M}_S(N_b)$ ($\mathcal{M}_S(N_b)$ is the vector space of the symmetric square matrices of size $N_b\times N_b$) with the so-called *overlap* matrix

$$S_{\mu\nu}=\int_{\mathbb{R}_3}\chi_\mu\chi_\nu, \tag{20.1}$$

the constraints $\int_{\mathbb{R}^3} \phi_i \phi_j = \delta_{ij}$ read

$$\delta_{ij} = \int_{\mathbb{R}^3} \phi_i \phi_j = \int_{\mathbb{R}^3} \left(\sum_{\nu=1}^{N_b} C_{\nu i} \chi_\nu \right) \left(\sum_{\mu=1}^{N_b} C_{\mu j} \chi_\mu \right) = \sum_{\mu=1}^{N_b} \sum_{\nu=1}^{N_b} C_{\mu j} S_{\mu\nu} C_{\nu i},$$

or in matrix form

$$C^* S C = I_N,$$

where I_N denotes the identity matrix of rank N. In addition,

$$\begin{aligned}
\sum_{i=1}^{N} \frac{1}{2} \int_{\mathbb{R}^3} |\nabla \phi_i|^2 + \int_{\mathbb{R}^3} \rho_\Phi V &= \sum_{i=1}^{N} \left(\frac{1}{2} \int_{\mathbb{R}^3} |\nabla \phi_i|^2 + \int_{\mathbb{R}^3} V |\phi_i|^2 \right) \\
&= \sum_{i=1}^{N} \left(\frac{1}{2} \int_{\mathbb{R}^3} \left| \nabla \sum_{\mu=1}^{N_b} C_{\mu i} \chi_\mu \right|^2 + \int_{\mathbb{R}^3} V \left| \sum_{\mu=1}^{N_b} C_{\mu i} \chi_\mu \right|^2 \right) \\
&= \sum_{i=1}^{N} \sum_{\mu=1}^{N_b} \sum_{\nu=1}^{N_b} h_{\mu\nu} C_{\nu i} C_{\mu i} \\
&= \operatorname{Trace}(hCC^*),
\end{aligned}$$

where $h \in \mathcal{M}_S(N_b)$ denotes the matrix of the core Hamiltonian $-\frac{1}{2}\Delta + V$ in the basis $\{\chi_\mu\}$:

$$h_{\mu\nu} = \frac{1}{2} \int_{\mathbb{R}_3} \nabla \chi_\mu \cdot \nabla \chi_\nu + \int_{\mathbb{R}^3} V \chi_\mu \chi_\nu. \tag{20.2}$$

Denoting by

$$(\mu\nu|\kappa\lambda) = \int_{\mathbb{R}^3} \int_{\mathbb{R}^3} \frac{\chi_\mu(x) \chi_\nu(x) \chi_\kappa(x') \chi_\lambda(x')}{|x - x'|} \, dx \, dx', \tag{20.3}$$

and by

$$J(X)_{\mu\nu} = \sum_{\kappa,\lambda=1}^{N_b} (\mu\nu|\kappa\lambda) X_{\kappa\lambda}, \qquad K(X)_{\mu\nu} = \sum_{\kappa,\lambda=1}^{N_b} (\mu\lambda|\nu\kappa) X_{\kappa\lambda},$$

where X is any $N_b \times N_b$ matrix, the Coulomb and exchange terms respectively read

$$\begin{aligned}
\int_{\mathbb{R}^3} \int_{\mathbb{R}^3} \frac{\rho_\Phi(x) \rho_\Phi(x')}{|x - x'|} \, dx \, dx' &= \sum_{\mu,\nu,\kappa,\lambda=1}^{N_b} \sum_{i,j=1}^{N} (\mu\nu|\kappa\lambda) C_{\mu i} C_{\nu i} C_{\kappa j} C_{\lambda j} \\
&= \operatorname{Trace}\big(J(CC^*) CC^*\big)
\end{aligned}$$

and

$$\int_{\mathbb{R}^3}\int_{\mathbb{R}^3} \frac{|\tau_\Phi(x,x')|^2}{|x-x'|}\,dx\,dx' = \sum_{\mu,\nu,\kappa,\lambda=1}^{N_b} \sum_{i,j=1}^{N} (\mu\lambda|\kappa\nu) C_{\mu i} C_{\nu i} C_{\kappa j} C_{\lambda j}$$
$$= \operatorname{Trace}\big(K(CC^*)CC^*\big).$$

We thus end up with the following minimization problem

$$\inf\big\{E^{\mathrm{HF}}(CC^*),\ C \in \mathcal{W}_N\big\}, \tag{20.4}$$

where

$$\mathcal{W}_N = \big\{C \in \mathcal{M}(N_b, N),\ C^*SC = I_N\big\},$$

and where for any $D \in \mathcal{M}_S(N_b)$

$$E^{\mathrm{HF}}(D) = \operatorname{Trace}(hD) + \frac{1}{2}\operatorname{Trace}\big(J(D)D\big) - \frac{1}{2}\operatorname{Trace}\big(K(D)D\big).$$

Denoting by

$$G(D) = J(D) - K(D),$$

we may also write

$$E^{\mathrm{HF}}(D) = \operatorname{Trace}(hD) + \frac{1}{2}\operatorname{Trace}\big(G(D)D\big).$$

The HF energy can be written in terms of the so-called *density matrix*

$$D = CC^*$$

for the matrix $D = CC^*$ is in fact the representation of the (infinite-dimensional) density matrix $\tau_\Phi(x,x')$ associated to $\Phi = \{\phi_i\}$, $\phi_i(x) = \sum_{\mu=1}^{N_b} C_{\mu i}\chi_\mu(x)$, in the basis $\{\chi_\mu\}$:

$$\tau_\Phi(x,x') = \sum_{i=1}^{N} \phi_i(x)\phi_i(x') = \sum_{i=1}^{N}\left(\sum_{\mu=1}^{N_b} C_{\mu i}\chi_\mu(x)\right)\left(\sum_{\nu=1}^{N_b} C_{\nu i}\chi_\nu(x')\right)$$
$$= \sum_{\mu,\nu=1}^{N_b}\left(\sum_{i=1}^{N} C_{\mu i}C_{\nu i}\right)\chi_\mu(x)\chi_\nu(x') = \sum_{\mu,\nu=1}^{N_b} D_{\mu\nu}\chi_\mu(x)\chi_\nu(x').$$

In addition, it is easy to see that the matrices $D = CC^*$ with $C \in \mathcal{M}(N_b, N)$ and $C^*SC = I_N$ are the symmetric matrices which satisfy $\operatorname{Trace}(SD) = N$ and $DSD = D$. An alternative formulation of the finite dimension HF problem (20.4) then reads

$$\inf\big\{E^{\mathrm{HF}}(D),\ D \in \mathcal{P}_N\big\} \tag{20.5}$$

with

$$\mathcal{P}_N = \left\{ D \in \mathcal{M}_S(N_b),\ DSD = D,\ \text{Trace}(SD) = N \right\}.$$

Let us remark that when the overlap matrix S is the identity matrix, that is to say when the basis $\{\chi_\mu\}$ is orthonormal, $\mathcal{P}_N$ turns out to be the set of rank-N orthogonal projectors on $\mathbb{R}^{N_b}$.

The other HF type models and the Kohn–Sham models are discretized in the same way. We now turn to them for the sake of completeness. The reader less interested in these details can easily skip the sequel and directly proceed to Section 21.

The spin-dependent models used in practice, namely the Restricted Hartree–Fock (RHF) and the Unrestricted Hartree–Fock (UHF) (see Section 9) are discretized as follows

$$I^{\text{RHF}} = \inf\left\{ E^{\text{RHF}}\left(CC^*\right),\ C \in \mathcal{W}_{N_p} \right\} \tag{20.6}$$

with

$$E^{\text{RHF}}(D) = 2\,\text{Trace}(hD) + 2\,\text{Trace}\big(J(D)D\big) - \text{Trace}\big(K(D)D\big),$$

and

$$I^{\text{UHF}} = \inf\left\{ E^{\text{UHF}}\left(C^\alpha C^{\alpha *}, C^\beta C^{\beta *}\right),\ C^\alpha \in \mathcal{W}_{N_\alpha},\ C^\beta \in \mathcal{W}_{N_\beta} \right\} \tag{20.7}$$

with

$$\begin{aligned} E^{\text{UHF}}\left(D^\alpha, D^\beta\right) &= \text{Trace}\left(hD^\alpha\right) + \text{Trace}\left(hD^\beta\right) \\ &\quad + \frac{1}{2}\,\text{Trace}\big(J\left(D^\alpha + D^\beta\right)\left(D^\alpha + D^\beta\right)\big) \\ &\quad - \frac{1}{2}\,\text{Trace}\big(K\left(D^\alpha\right)D^\alpha\big) - \frac{1}{2}\,\text{Trace}\big(K\left(D^\beta\right)D^\beta\big), \end{aligned}$$

or, in the density matrix formalism

$$I^{\text{RHF}} = \inf\left\{ E^{\text{RHF}}(D),\ D \in \mathcal{P}_{N_p} \right\}$$

and

$$I^{\text{UHF}} = \inf\left\{ E^{\text{UHF}}\left(D^\alpha, D^\beta\right),\ D^\alpha \in \mathcal{P}_{N_\alpha},\ D^\beta \in \mathcal{P}_{N_\beta} \right\}.$$

Let us now turn to the Kohn–Sham and extended Kohn–Sham models which have been defined in Section 15.

The Kohn–Sham models are also discretized accordingly; closed-shell and open-shell KS models look like the RHF and the UHF model, respectively. It is necessary to resort to open-shell models when the number of electrons is odd and also when spin dependent approximated exchange-correlation functionals (LSD functionals for instance) are used.

For brevity, we only deal here with closed-shell KS models. After discretization, they read

$$I^{\mathrm{KS}} = \inf\{E^{\mathrm{KS}}(CC^*),\ C \in \mathcal{W}_{N_p}\} \tag{20.8}$$

with

$$E^{\mathrm{KS}}(D) = 2\,\mathrm{Trace}(hD) + 2\,\mathrm{Trace}\big(J(D)D\big) + E_{\mathrm{xc}}(D),$$

$E_{\mathrm{xc}}(D)$ denoting the exchange-correlation energy. If, for instance, a LDA functional is used (see Eqs. (15.9)–(15.11)),

$$E_{\mathrm{xc}}(D) = \int_{\mathbb{R}^3} \rho(x)\varepsilon_{\mathrm{xc}}^{\mathrm{LDA}}\big(\rho(x)\big)\,dx \quad \text{with } \rho(x) = 2\sum_{\mu,\nu=1}^{N_p} D_{\mu\nu}\chi_\mu(x)\chi_\nu(x).$$

In the extended Kohn–Sham setting, the discretized closed-shell models read

$$\begin{aligned} I^{\mathrm{EKS}} = \inf\{&E^{\mathrm{KS}}(C\Delta C^*),\ C \in \mathcal{M}(N_b, N_b),\ C^*SC = I_{N_b},\\ &\Delta = \mathrm{Diag}(n_1, \ldots, n_{N_b}),\ 0 \leqslant n_i \leqslant 1,\ \mathrm{Trace}(\Delta) = N_p\} \end{aligned} \tag{20.9}$$

or, more simply, in the density matrix formalism

$$I^{\mathrm{EKS}} = \inf\{E^{\mathrm{KS}}(D),\ D \in \widetilde{\mathcal{P}}_{N_p}\},$$

where

$$\widetilde{\mathcal{P}}_{N_p} = \{D \in \mathcal{M}_S(N_b),\ DSD \leqslant D,\ \mathrm{Trace}(SD) = N_p\},$$

and where $E^{\mathrm{xc}}(D)$ has the same expression as in the standard KS setting. For any N, the set $\widetilde{\mathcal{P}}_N$ is the convex hull of the set $\mathcal{P}_N$; from a physical viewpoint it is the set of trace-N density matrices for which occupation numbers all are in the range [0, 1].

21. First order optimality conditions

As said above, in the numerical practice, the Hartree–Fock approximation of the ground state is rarely computed through a direct attack of the above minimization problem, in order to prevent the iterates from being trapped in one of the possibly numerous local minima. On the contrary, the problem is solved through the resolution of the Euler–Lagrange equation associated with the minimization problem. The fact that one does not seek for any arbitrary solution to this equation, that is to say any arbitrary critical point of the Hartree–Fock energy functional, but precisely for a global minimizer, will however be extensively used in the building of numerical algorithms. This is where the variational nature of the Hartree–Fock approximation will come back into the picture. We shall deal with this below. For the time being, let us make precise the equation we are going to solve.

The spinless HF, the RHF and the UHF models, as well as the standard KS model, all are of the generic form

$$\inf\{E(x),\ x \in \mathcal{M}\}, \tag{21.1}$$

where E is a real-valued function on a n-dimensional Euclidean space $\mathcal{H}$ and $\mathcal{M}$ a m-dimensional submanifold of $\mathcal{H}$. Denoting by $T_x\mathcal{M}$ the tangent space at $x \in \mathcal{M}$ to the submanifold $\mathcal{M}$, the first order optimality conditions that are satisfied by a minimizer x_* of (21.1) read

$$\begin{cases} \nabla E(x_*) \in (T_{x_*}\mathcal{M})^\perp, \\ x_* \in \mathcal{M}. \end{cases} \tag{21.2}$$

Let $\mathcal{K}$ be an Euclidean space and $g : \mathcal{H} \to \mathcal{K}$ a mapping such that

$$\mathcal{M} = \{x \in \mathcal{H},\ g(x) = 0\}.$$

It is assumed that E and g are regular enough (say C^1) and that g is a submersion at each point $x \in \mathcal{M}$ (which means that for all $x \in \mathcal{M}$, $g'(x)$ is surjective); a consequence of the latter assumption is that $\dim(\mathcal{K}) = n - m$ and $T_x\mathcal{M} = \operatorname{Ker}(g'(x))$. The optimality conditions (21.2) thus also read

$$\begin{cases} \nabla E(x_*) \in \operatorname{Ker}(g'(x_*))^\perp, \\ g(x_*) = 0. \end{cases}$$

As $\operatorname{Ker}(g'(x_*))^\perp = \operatorname{Ran}(g'(x_*)^T)$, the above conditions are equivalent to the following ones: There exists $\lambda_* \in \mathcal{K}$ such that

$$\begin{cases} \nabla E(x_*) = g'(x_*)^T \cdot \lambda_*, \\ g(x_*) = 0. \end{cases} \tag{21.3}$$

Eqs. (21.3) are the *Euler–Lagrange equations* associated with the constrained optimization problem

$$\inf\{E(x),\ x \in \mathcal{H},\ g(x) = 0\}, \tag{21.4}$$

which is obviously equivalent to problem (21.1); λ_* is the *Lagrange multiplier* of the constraints $g(x) = 0$.

Problem (20.4) is of the form (21.4) if one equips the vector space $\mathcal{H} = \mathcal{M}(N_b, N)$ with the scalar product defined by $(C, C')_{\mathcal{H}} = \operatorname{Trace}(C^*C')$, the vector space $\mathcal{K} =$

$\mathcal{M}_S(N)$ with the scalar product $(\Sigma, \Sigma')_{\mathcal{K}} = \mathrm{Trace}(\Sigma\Sigma')$, and if one sets $E(C) = E^{\mathrm{HF}}(CC^*)$ and $g(C) = C^*SC - I_N$. One thus obtains

$$\nabla E(C) = 2F(C)C, \qquad g'(C)\cdot W = C^*SW + W^*SC,$$
$$g'(C)^T \cdot \Sigma = 2SC\Sigma \tag{21.5}$$

with $D = CC^*$ and $F(C) = h + G(D)$. The matrix $F(C)$ corresponds to the discretization of the Fock operator denoted by F_Φ and defined by (9.10): For $\Phi = \{\phi_i\}$, $\phi_i(x) = \sum_{\mu=1}^{N_b} C_{\mu i}\chi_\mu(x)$, the representation of F_Φ in the basis $\{\chi_\mu\}$, namely the symmetric $N_b \times N_b$ matrix defined $\langle \chi_\mu, F_\Phi \chi_\nu\rangle$, is the matrix $F(C)$. It is by definition the *Fock matrix* associated with the molecular orbital matrix $C \in \mathcal{M}(N_b, N)$. With a slight abuse of notations, we also denote by

$$F(D) = h + G(D)$$

the Fock matrix associated with some $D \in \mathcal{M}_S(N_b)$. Let us notice that the finite-dimensional counterpart of the unitary invariance (9.11) reads: for any $U \in U(N)$,

$$D_{CU} = D_C, \qquad E(CU) = E(C), \qquad F(CU) = F(C), \tag{21.6}$$

where $D_C = CC^*$ and $D_{CU} = (CU)(CU)^*$ are the density matrices associated with C and CU, respectively.

The linear map $g'(C)$ is clearly surjective if $\Sigma = C^*SC$ is invertible (then $g'(C)\cdot(\frac{1}{2}\Sigma^{-1}\Lambda) = \Lambda$, for any $\Lambda \in \mathcal{M}_S(N)$); this is the case in particular for any C such that $g(C) = 0$. The Euler–Lagrange equations (21.3) therefore read

$$\begin{cases} F(D)C = SC\Lambda, \\ C^*SC = I_N, \\ D = CC^*. \end{cases} \tag{21.7}$$

These are the spinless HF equations.

The Euler–Lagrange equations for the RHF model (20.6), the UHF model (20.7), the KS model (20.8), and the extended KS model (20.9) can be derived in the same manner. We now mention them for the sake of completeness. As above, this part can be skipped in a first reading.

For the RHF model, one has

$$\begin{cases} F^{\mathrm{RHF}}(D)C = SC\Lambda, \\ C^*SC = I_{N_p}, \\ D = CC^*, \end{cases}$$

with $F^{\mathrm{RHF}}(D) = h + 2J(D) - K(D)$ and, for the UHF model,

$$\begin{cases} F^\alpha\big(D^\alpha, D^\beta\big)C^\alpha = SC^\alpha \Lambda^\alpha, \\ C^{\alpha *}SC^\alpha = I_{N_\alpha}, \\ D^\alpha = C^\alpha C^{\alpha *}, \\ F^\beta\big(D^\alpha, D^\beta\big)C^\beta = SC^\beta \Lambda^\beta, \\ C^{\beta *}SC^\beta = I_{N_\beta}, \\ D^\beta = C^\beta C^{\beta *}, \end{cases} \tag{21.8}$$

with $F^\alpha(D^\alpha, D^\beta) = h + J(D^\alpha + D^\beta) - K(D^\alpha)$ and $F^\beta(D^\alpha, D^\beta) = h + J(D^\alpha + D^\beta) - K(D^\beta)$. Let us remark that if $N_\alpha = N_\beta = N_p$ and if $D^\alpha = D^\beta = D$, $E^{\mathrm{UHF}}(D^\alpha, D^\beta) = E^{\mathrm{RHF}}(D)$ and $F^\alpha(D^\alpha, D^\beta) = F^\beta(D^\alpha, D^\beta) = F^{\mathrm{RHF}}(D)$. For a system with $N = 2N_p$ electrons, the RHF ground state therefore is at least a critical point of the UHF energy for $N_\alpha = N_\beta = N_p$.

Likewise, the KS equations read

$$\begin{cases} F^{\mathrm{KS}}(D)C = SC\Lambda, \\ C^*SC = I_{N_p}, \\ D = CC^*, \end{cases}$$

with $F^{\mathrm{KS}}(D) = h + J(D) + F_{\mathrm{xc}}(D)$ and $F_{\mathrm{xc}}(D) = \frac{1}{2}\nabla E_{\mathrm{xc}}(D)$. We have assumed here that the function $C \mapsto E_{\mathrm{xc}}(CC^*)$ is differentiable; this is always the case *in practice* (i.e., for commonly used *approximated* exchange-correlation functionals). Theoretical questions related to the differentiability of the *exact* exchange-correlation functional with respect to the density are left aside (see H. ENGLISCH and R. ENGLISCH [1983]).

Finally, the first order optimality conditions associated with the extended KS model (20.9) have a more complicated structure:

$$\begin{cases} F^{\mathrm{KS}}(D)C\Delta = SC\Lambda, \\ C^*SC = I_{N_b}, \\ D = C\Delta C^*, \quad \Delta = \mathrm{Diag}(n_1, \cdots, n_{N_b}), \\ n_i = 0 \quad \text{if } \Phi_i^T F^{\mathrm{KS}}(D)\Phi_i > \varepsilon_{\mathrm{F}}, \\ n_i = 1 \quad \text{if } \Phi_i^T F^{\mathrm{KS}}(D)\Phi_i < \varepsilon_{\mathrm{F}}, \\ 0 \leqslant n_i \leqslant 1 \quad \text{if } \Phi_i^T F^{\mathrm{KS}}(D)\Phi_i = \varepsilon_{\mathrm{F}}, \\ \displaystyle\sum_{i=1}^{N_b} n_i = N_p, \end{cases} \tag{21.9}$$

where Φ_i denotes the ith column of the matrix C and ε_{F} the Fermi level. The first three optimality conditions are associated with a variation of C under the equality constraints

$C^* SC = I_{N_b}$; they can be established exactly as above. The latter four conditions are related to a variation of the n_i with respect to the inequality constraints $0 \leqslant n_i \leqslant 1$ and $\mathrm{Trace}(\Delta) = N_p$. They result from the fact that transferring δn electron pairs from the orbital Φ_i to the orbital Φ_j involves a first order variation of the energy equal to

$$\Delta E^{\mathrm{EKS}} = 2\delta n\big(\Phi_j^T F^{\mathrm{KS}}(D)\Phi_j - \Phi_i^T F^{\mathrm{KS}}(D)\Phi_i\big).$$

22. The *aufbau* principle and its mathematical foundations

Using the unitary invariance (21.6) and noticing that for any $U \in U(N)$ and any $C \in \mathcal{M}(N_b, N)$, $CU \in \mathcal{M}(N_b, N)$ and

$$C^* SC = I_N \ \Rightarrow \ (CU)^* S(CU) = I_N,$$
$$(CU)(CU)^* = CC^*,$$

it is clear that the spinless HF model (20.4) is invariant with respect to the unitary transform $C \mapsto CU$. It is a standard result that this invariance can be used to diagonalize the matrix Λ in (21.7). One obtains

$$\begin{cases} F(D)C = SCE, \quad E = \mathrm{Diag}(\varepsilon_1, \varepsilon_2, \ldots, \varepsilon_N), \\ C^* SC = I_N, \\ D = CC^*. \end{cases} \tag{22.1}$$

In particular, still denoting by Φ_i the ith column of C,

$$F(D)\Phi_i = \varepsilon_i S\Phi_i. \tag{22.2}$$

It is to be noticed that solving (22.1) provides *some* solutions to (20.4) (and also local minima and saddle points we are not interested in) but not *all* solutions; on the other hand, all solutions to (20.4) can be recovered from solutions to (22.1) by applying the unitary transform $C \leftarrow CU$ for all $U \in U(N)$. Another equivalent viewpoint is that among the density matrices solutions to (22.1), there are *all* the solutions to (20.5). The same analysis can exactly be conducted for the KS model.

It has been underlined in Section 9 that for a minimizer of the infinite-dimensional spinless HF problem (9.13), the eigenvalues of the Fock operator are the lowest N eigenvalues of the Fock operator F_Φ. This property is preserved in the finite-dimensional setting: If C is a minimizer of (20.4), $(\varepsilon_1, \varepsilon_2, \ldots, \varepsilon_{N_p})$ are the lowest N_p eigenvalues of the generalized eigenvalue problem (22.2). It is said that C and $D = CC^*$ satisfy the *aufbau* principle. This property, which is strongly exploited in the building of self-consistent field (SCF) algorithms, is also fulfilled for the UHF and GHF models. Let us prove it for instance in the UHF setting. After diagonalization of the matrices Λ^α and Λ^β in (21.8) with the help of the unitary invariance already mentioned, the UHF equations become

$$
\begin{cases}
F^\alpha\big(D^\alpha, D^\beta\big)C^\alpha = SC^\alpha E^\alpha, & E^\alpha = \mathrm{Diag}(\varepsilon_1^\alpha, \ldots, \varepsilon_{N_\alpha}^\alpha),\\
C^{\alpha *} S C^\alpha = I_{N_\alpha}, & \\
D^\alpha = C^\alpha C^{\alpha *}, & \\
F^\beta\big(D^\alpha, D^\beta\big)C^\beta = SC^\beta E^\beta, & E^\beta = \mathrm{Diag}\big(\varepsilon_1^\beta, \ldots, \varepsilon_{N_\beta}^\beta\big),\\
C^{\beta *} S C^\beta = I_{N_\beta}, & \\
D^\beta = C^\beta C^{\beta *}. &
\end{cases}
\tag{22.3}
$$

In this setting, it is true that at the UHF minimizer, $(\varepsilon_1^\alpha, \ldots, \varepsilon_{N_\alpha}^\alpha)$ are the lowest N_α eigenvalues of $F^\alpha(D^\alpha, D^\beta)$ and that $(\varepsilon_1^\beta, \ldots, \varepsilon_{N_\beta}^\beta)$ are the lowest N_β eigenvalues of $F^\beta(D^\alpha, D^\beta)$. Let us prove this assertion. For this purpose let us denote by (C^α, C^β) with $C^\alpha = (\Phi_1^\alpha|\Phi_2^\alpha|\cdots|\Phi_{N_\alpha-1}^\alpha|\Phi_{N_\alpha}^\alpha)$ a minimizer of (20.7) satisfying (22.3) and consider $C'^\alpha = (\Phi_1^\alpha|\Phi_2^\alpha|\cdots|\Phi_{N_\alpha-1}^\alpha|\Phi_a^\alpha)$ where $a \in |[N_\alpha + 1, N_b]|$. A simple algebraic manipulation leads to

$$
\begin{aligned}
&E^{\mathrm{UHF}}\big(C'^\alpha C'^{\alpha *}, C^\beta C^{\beta *}\big) - E^{\mathrm{UHF}}\big(C^\alpha C^{\alpha *}, C^\beta C^{\beta *}\big)\\
&\quad = \varepsilon_a^\alpha - \varepsilon_{N_\alpha}^\alpha - \frac{1}{2}\int_{\mathbb{R}^3}\int_{\mathbb{R}^3} \frac{|\phi_{N_\alpha}^\alpha(x)\phi_a^\alpha(y) - \phi_{N_\alpha}^\alpha(y)\phi_a^\alpha(x)|^2}{|x-y|}\,dx\,dy,
\end{aligned}
$$

where

$$
\phi_{N_\alpha}^\alpha(x) = \sum_{k=1}^{N_b}\big[\Phi_{N_\alpha}^\alpha\big]_k \chi_k(x), \qquad \phi_a^\alpha(x) = \sum_{k=1}^{N_b}\big[\Phi_a^\alpha\big]_k \chi_k(x).
$$

If (C^α, C^β) is a minimizer of the UHF energy,

$$
E^{\mathrm{UHF}}\big(C'^\alpha C'^{\alpha *}, C^\beta C^{\beta *}\big) - E^{\mathrm{UHF}}\big(C^\alpha C^{\alpha *}, C^\beta C^{\beta *}\big) \geqslant 0.
$$

Therefore

$$
\varepsilon_a^\alpha - \varepsilon_{N_\alpha}^\alpha \geqslant \frac{1}{2}\int_{\mathbb{R}^3}\int_{\mathbb{R}^3} \frac{|\phi_N(x)\phi_a(y) - \phi_N(y)\phi_a(x)|^2}{|x-y|}\,dx\,dy > 0.
$$

We can thus conclude that

- $(\varepsilon_1^\alpha, \ldots, \varepsilon_{N_\alpha}^\alpha)$ are the *lowest N_α eigenvalues* of F^α (see also LIONS [1987]);
- there is always a *positive gap* between the α-component lowest unoccupied molecular orbital (LUMO) and the α-component highest occupied molecular orbital (HOMO); see also BACH, LIEB, LOSS and SOLOVEJ [1994].

Of course the same result holds for the β-components.

On the other hand, it is not proven (to our knowledge at least) that the RHF and the KS minimizers satisfy the *aufbau* principle; it is nevertheless always assumed that they do so in the numerical practice.

Let us now turn to the extended KS model; it is easy to see that if the unitary invariance is used to diagonalize the matrix Λ in (21.9), we obtain

$$\begin{cases} F^{\mathrm{KS}}(D)C = SCE, & E = \mathrm{Diag}(\varepsilon_1, \ldots, \varepsilon_{N_b}), \\ C^*SC = I_{N_b}, & \\ D = C\Delta C^*, & \Delta = \mathrm{Diag}(n_1, \cdots, n_{N_b}), \\ n_i = 0 & \text{if } \varepsilon_i > \varepsilon_{\mathrm{F}}, \\ n_i = 1 & \text{if } \varepsilon_i < \varepsilon_{\mathrm{F}}, \\ 0 \leqslant n_i \leqslant 1 & \text{if } \varepsilon_i = \varepsilon_{\mathrm{F}}, \\ \sum_{i=1}^{N_b} n_i = N_p. & \end{cases} \tag{22.3$'$}$$

It follows that *for any critical point* of (20.9), the orbitals which correspond to eigenvalues that are below the Fermi level are fully occupied, whereas those which correspond to eigenvalues that are above the Fermi level are empty; if the Fermi level is degenerated, the occupancies of the corresponding orbitals are not a priori prescribed: Transferring some electrons from one orbital to another one with the same ε leaves the energy unchanged *at the first order*; on the other hand, the energy changes at higher orders so that the occupancies at the Fermi level also have to be optimized.

Let us conclude this section by giving useful characterizations of the solutions to

$$\begin{cases} FC = SCE, & E = \mathrm{Diag}(\varepsilon_1, \varepsilon_2, \ldots, \varepsilon_N), \\ C^*SC = I_N, & \\ D = CC^* & \end{cases} \tag{22.4}$$

(where F is a given matrix) which only involve density matrices. As will be proven shortly, the solutions to (22.4) are the same as the solutions in $\mathcal{P}_N$ to the equation

$$[F, D] = 0, \tag{22.5}$$

where $[\cdot,\cdot]$ denotes the "commutator" defined by $[A, B] = ABS - SBA$. In addition, the solutions to (22.4) which satisfy the *aufbau* principle are the same as the solutions to the problem

$$D = \arg\inf\{\mathrm{Trace}(FD'),\ D' \in \mathcal{P}_N\}.$$

If in addition there is a positive gap between the Nth and the $(N+1)$th eigenvalue of F (i.e., if $\varepsilon_1 \leqslant \cdots \leqslant \varepsilon_N < \varepsilon_{N+1} \leqslant \cdots \leqslant \varepsilon_{N_b}$), then the *aufbau* solutions D to (22.4) are also the solutions to

$$D = \arg\inf\{\mathrm{Trace}(FD'),\ D' \in \widetilde{\mathcal{P}}_N\}. \tag{22.6}$$

Let us prove these statements in the case when $S = I_N$ (the general case is obtained by making the changes of $F \leftarrow S^{-1/2} F S^{-1/2}$ and $D \leftarrow S^{1/2} D S^{1/2}$). It is straightforward to check that any solution D to (22.4) is solution to (22.5). Conversely, if $D \in \mathcal{P}_N$ satisfies (22.5), then F and D commute (recall that we have assumed that $S = I_N$) and therefore any eigenspaces of D is F-invariant. In particular, F can be diagonalized on $\mathrm{Ran}(D)$; system (22.4) follows. Let us now prove that *aufbau* solutions to (22.4) are in fact the solutions to (22.6) in the case when $\gamma = \varepsilon_{N+1} - \varepsilon_N > 0$. Let us denote by D a solution to (22.4), by D' a current matrix such that $D' = D'^*$, $\mathrm{Trace}(D') = N$, $D'^2 \leqslant D'$ and by D'_{ij} its coefficients in an orthonormal basis in which $F = \mathrm{Diag}(\varepsilon_1, \varepsilon_2, \ldots, \varepsilon_{N_b})$ with $\varepsilon_1 \leqslant \varepsilon_2 \leqslant \cdots \leqslant \varepsilon_{N_b}$. In such a basis $D = \mathrm{Diag}(1, \ldots, 1, 0, \ldots, 0)$. As in addition $\mathrm{Trace}(D') = \sum_{i=1}^{N_b} D'_{ii} = N$, we get first

$$
\begin{aligned}
\|D - D'\|^2 &= \mathrm{Trace}\big((D - D') \cdot (D - D')\big) \\
&= \mathrm{Trace}\big(D^2\big) + \mathrm{Trace}\big(D'^2\big) - 2\,\mathrm{Trace}\big(DD'\big) \\
&\leqslant \mathrm{Trace}(D) + \mathrm{Trace}\big(D'\big) - 2\,\mathrm{Trace}\big(DD'\big) \\
&= 2N - 2\sum_{i=1}^{N} D'_{ii} = 2\sum_{i=N+1}^{N_b} D'_{ii}.
\end{aligned}
$$

In addition, $\mathrm{Trace}(FD) = \sum_{i=1}^{N} \varepsilon_i$, $\mathrm{Trace}(FD') = \sum_{i=1}^{N_b} \varepsilon_i D'_{ii}$ and

$$
0 \leqslant D'_{ii} \leqslant 1 \quad \text{for any } 1 \leqslant i \leqslant N_b,
$$

for $D'^2 \leqslant D' = D'^*$ implies $|D'_{ii}|^2 + \sum_{j \neq i} |D'_{ij}|^2 \leqslant D'_{ii}$. Collecting the above results, we obtain

$$
\begin{aligned}
\mathrm{Trace}\big(FD'\big) &= \sum_{i=1}^{N_b} \varepsilon_i D'_{ii} \geqslant \sum_{i=1}^{N} \varepsilon_i D'_{ii} + \sum_{i=N+1}^{N_b} (\varepsilon_N + \gamma) D'_{ii} \\
&= \sum_{i=1}^{N} \varepsilon_i D'_{ii} + \varepsilon_N \sum_{i=N+1}^{N_b} D'_{ii} + \gamma \sum_{i=N+1}^{N_b} D'_{ii} \\
&= \sum_{i=1}^{N} \varepsilon_i D'_{ii} + \varepsilon_N \left(N - \sum_{i=1}^{N} D'_{ii} \right) + \gamma \sum_{i=N+1}^{N_b} D'_{ii} \\
&= \sum_{i=1}^{N} \varepsilon_i + \sum_{i=1}^{N} (\varepsilon_N - \varepsilon_i)\big(1 - D'_{ii}\big) + \gamma \sum_{i=N+1}^{N_b} D'_{ii}.
\end{aligned}
$$

As for any $1 \leqslant i \leqslant N$, $0 \leqslant D'_{ii} \leqslant 1$ and $\varepsilon_N \geqslant \varepsilon_i$, we finally obtain

$$
\mathrm{Trace}\big(FD'\big) \geqslant \mathrm{Trace}(FD) + \frac{\gamma}{2} \|D - D'\|^2, \tag{22.7}
$$

which proves the equivalence between (22.4) and (22.6).

23. The hydrogen-like atom

As a preliminary step before the building of efficient basis sets to solve the electronic problem, it is useful to study in detail the simple case of the *hydrogen-like* ion, that is a system consisting of a single nucleus with charge Z together with a single electron. This system will indeed serve as a paradigm for the computations of more complicated molecular systems. In addition, and this is our primary goal here, it will provide efficient basis sets for them.

In the Born–Oppenheimer approximation, the Hamiltonian for the hydrogen-like atom reads

$$H_Z = -\frac{1}{2}\Delta - \frac{Z}{|x|}$$

and acts on $L^2(\mathbb{R}^3)$. Due to the fact that there is only one electron, no further approximation is necessary in order to tackle the system, both from a theoretical and a numerical viewpoint.

Standard general results of spectral theory indeed allow one to have a very detailed description of the spectrum of this operator. The potential $-\frac{Z}{|x|}$ being a compact perturbation of the Laplacian, the essential spectrum of H_Z is that of $-\frac{1}{2}\Delta$, namely $\sigma_{\text{ess}}(H_Z) = [0, +\infty[$. In addition, H_Z has no nonnegative discrete eigenvalue, and has a countable infinite number of negative eigenvalues, forming an increasing sequence of eigenvalues, all of which with finite multiplicity, that converges to zero. It is moreover possible to obtain the precise value of these eigenvalues and a complete description of the corresponding eigenspaces, as will now be seen.

Concerning the first eigenvalue, namely the ground state energy in terms of quantum mechanics, standard arguments of spectral theory again allow one to claim the following. In view of some specific property of the Coulomb potential $-\frac{Z}{|x|}$ (a property that is shared by a large class of potentials but not by any arbitrary compact perturbation of the Laplacian), it can be proven that the minimizer to

$$\inf\left\{ \langle \psi, H_Z \psi \rangle = \frac{1}{2}\int_{\mathbb{R}^3} |\nabla \psi|^2 - Z\int_{\mathbb{R}^3} \frac{|\psi|^2}{|x|},\ \psi \in H^1(\mathbb{R}^3),\ \|\psi\|_{L^2} = 1 \right\}$$

is unique up to a change of sign. More precisely, there exist exactly two minimizers to the above problem. One is some $\psi_1^Z > 0$ and the other one is $-\psi_1^Z$. In addition, it can be claimed that any eigenfunction of H_Z which is positive is necessarily equal to ψ_1^Z. In other words, up to a change of sign, the normalized ground state of H_Z is unique, and is the only eigenfunction enjoying the property of being of constant sign over $\mathbb{R}^3$.

The uniqueness property is at once used to claim that ψ_1^Z necessarily is a radially symmetric function, for any $\psi_1^Z(R \cdot x)$ is also a positive ground state when R is a rotation in $\mathbb{R}^3$. Consequently, the knowledge of ψ_1^Z is reduced to that of the function $f(r) > 0$ solution to

$$-\frac{1}{2}f''(r) - \frac{f'(r)}{r} - \frac{Zf(r)}{r} = E_1 f(r).$$

It is straightforward to see that $f(r) = \exp(-Zr)$ is a solution with $E_1^Z = -Z^2/2$. As it is positive, we can conclude it yields the unique positive ground state wavefunction ψ_1^Z. Therefore

$$\psi_1^Z(x) = \frac{Z^{3/2}}{\sqrt{\pi}} e^{-Z|x|}$$

and the first eigenvalue of H_Z (the ground state energy) is $E_1^Z = -Z^2/2$.

Let us now continue with the other eigenvalues. We begin by writing H_Z in terms of the spherical coordinates (r, θ, ϕ)

$$H_Z = -\frac{1}{2r}\frac{\partial^2}{\partial r^2} r - \frac{1}{2r^2}\Delta_S - \frac{Z}{r},$$

where Δ_S denotes the Laplace–Beltrami operator:

$$\Delta_S = \frac{1}{\sin\theta}\frac{\partial}{\partial\theta}\left(\sin\theta\frac{\partial}{\partial\theta}\right) + \frac{1}{\sin^2\theta}\frac{\partial^2}{\partial\phi^2}.$$

It is to be noted at this stage that the operators H_Z, $L_z = (-ix \times \nabla) \cdot e_z = -i\frac{\partial}{\partial z}$, and $L^2 = (-ix \times \nabla)^2$ (where $L = -ix \times \nabla$ denotes the kinetic moment operator) commute, and therefore that one may construct a basis set of eigenfunctions of H_Z that are also eigenfunctions of L_z and L^2. Now, the eigenfunctions of L_z and L^2 are explicitly known as the functions

$$\psi(r, \theta, \phi) = R(r)\, Y_l^m(\theta, \phi),$$

where R is an arbitrary square integrable function on $L^2(\mathbb{R}^+)$ equipped with the measure $d\mu = 4\pi r^2\, dr$, and Y_l^m denotes a spherical harmonic. These latter functions are defined as the solutions to

$$\Delta_S \cdot Y_l^m = -l(l+1)Y_l^m \quad \text{and} \quad -i\frac{\partial}{\partial\phi}Y_l^m(\theta, \phi) = mY_l^m(\theta, \phi). \tag{23.1}$$

Their explicit expression in terms of the variables (θ, ϕ) is

$$Y_l^m(\theta, \phi) = \frac{1}{\sqrt{2\pi}} P_l^m(\cos\theta) e^{im\phi},$$

where the $P_l^m(\cos\theta)$ denote the *Legendre functions*, which are in turn obtained from the mth derivative $P_l^{(m)}$ of the *Legendre polynomials* $P_l(x) = \frac{(-1)^l}{2^l l!}\frac{d^l}{dx^l}(1-x^2)^l$ as follows

$$\begin{cases} P_l^m(\cos\theta) = (\sin\theta)^m P_l^{(m)}(\cos\theta) & \text{for } 0 \leqslant m \leqslant l, \\ P_l^{-m}(\cos\theta) = (-1)^m \dfrac{(l-m)!}{(l+m)!} P_l^m(\cos\theta) & \text{for } 0 \leqslant m \leqslant l. \end{cases}$$

To fix the ideas, let us give the first spherical harmonics:

$$Y_0^0(\theta,\phi) = \frac{1}{\sqrt{4\pi}},$$

$$Y_1^0(\theta,\phi) = \sqrt{\frac{3}{4\pi}}\cos\theta, \qquad Y_1^{\pm 1}(\theta,\phi) = \pm\sqrt{\frac{3}{8\pi}}\sin\theta e^{\pm i\phi},$$

$$Y_2^0(\theta,\phi) = \sqrt{\frac{5}{16\pi}}(3\cos^2\theta - 1), \qquad Y_2^{\pm 1}(\theta,\phi) = \mp\sqrt{\frac{15}{8\pi}}\sin\theta\cos\theta\, e^{\pm i\phi},$$

$$Y_2^{\pm 2}(\theta,\phi) = \sqrt{\frac{15}{32\pi}}\sin^2\theta e^{\pm 2i\phi}.$$

The spherical harmonics enjoy the following properties:

- $(Y_l^m)_{l\geqslant 0, -l\leqslant m\leqslant l}$ is a basis of the Hilbert space $L^2(S^2)$;
- for any (l,m), $l \geqslant 0$, $-l \leqslant m \leqslant l$, and any (l',m'), $l' \geqslant 0$, $-l' \leqslant m' \leqslant l'$,

$$\begin{aligned}(Y_l^m, Y_{l'}^{m'})_{L^2(S^2)} &= \int_{S^2} Y_l^{m*} Y_{l'}^{m'} \\ &= \int_0^{\pi}\int_0^{2\pi} Y_l^m(\theta,\phi)^* Y_{l'}^{m'}(\theta,\phi)\sin\theta\, d\theta\, d\phi = \delta_{ll'}\delta_{mm'};\end{aligned}$$

- any $f \in L^2(S^2)$ can be uniquely decomposed along this basis as:

$$f = \sum_{l=0}^{+\infty}\sum_{m=-l}^{l} (Y_l^m, f)_{L^2(S^2)} Y_l^m.$$

If we now come back to the question of determining the eigenfunctions of H_Z, we see that, without loss of generality, we may search for $\psi(r,\theta,\phi)$ in the form

$$\psi(r,\theta,\phi) = R(r)Y_l^m(\theta,\phi),$$

which amounts to solving

$$-\frac{1}{2r}\frac{d^2}{dr^2}\big(rR(r)\big) + \frac{l(l+1)}{2r^2}R(r) - \frac{Z}{r}R(r) = ER(r). \tag{23.2}$$

For each fixed integer l, it turns out that Eq. (23.2) has a countable infinite number of normalized solutions (i.e., such that $\int_0^{+\infty} r^2|R(r)|^2\,dr = 1$), that are in fact well known. The description of these solutions indeed allows one to obtain the following precise description of the eigenstates of H_Z:

- the eigenvalues of H_Z form an increasing sequence $(E_n^Z)_{n\geqslant 1}$ defined by

$$E_n^Z = -\frac{Z^2}{2n^2};$$

  each of which is of multiplicity n^2;

- the n^2 eigenfunctions associated to E_n^Z, that are also eigenstates of L^2 and L_z, are

$$\psi_{nlm}^Z(r,\theta,\phi) = Q_{nl}(Zr)e^{-Zr/n}Y_l^m(\theta,\phi),$$
$$0 \leqslant l \leqslant n-1, \ -l \leqslant m \leqslant l, \tag{23.3}$$

  where $Q_{nl}(x) = x^l(C_0 + C_1x + \cdots + C_{n-l-1}x^{n-l-1})$ denotes the polynomial with coefficients defined by the induction formula $C_i = -\frac{2}{n}\frac{n-l-i}{i(2l+i+1)}C_{i-1}$, the value of C_0 being fixed (up to a phase) by the normalization condition $\|\psi_{nlm}^Z\|_{L^2} = 1$;
- one has

$$H_Z\psi_{nlm}^Z = E_n^Z\psi_{nlm}^Z, \qquad L^2\psi_{nlm}^Z = l(l+1)\psi_{nlm}^Z, \qquad L_z\psi_{nlm}^Z = m\psi_{nlm}^Z.$$

The integers n, l, and m are, respectively, called the *principal*, *azimuthal* and *magnetic quantum numbers*. It is customary in the language of quantum chemistry to denote ψ_{nlm}^Z by an integer (n), followed by a letter corresponding to the value of l through the glossary

$$l = 0 \rightarrow s, \qquad l = 1 \rightarrow p, \qquad l = 2 \rightarrow d, \qquad l = 3 \rightarrow f.$$

For instance, ψ_{100}^Z, formerly denoted by ψ_1^Z, is denoted by 1s, while the three functions ψ_{21m}^Z for $m = -1, 0, 1$ are the 2p states.

24. Basis sets: LCAO approximation

Recall that our goal is to build an efficient finite-dimensional approximation of the space of wavefunctions to be considered for the determination of the electronic structure of the molecular system under study. In quantum chemistry packages, several basis sets are at the user's disposal; it is the user's task to select the basis set which is best adapted to the molecular system and to the properties he intends to compute.

As the molecular system is formally made of an assembly of atoms, it is a natural idea to choose as finite-dimensional space the vectorial space generated by a finite number of atomic orbitals (AO). One speaks of a *LCAO approximation*, the acronym LCAO standing for *linear combination of atomic orbitals*.

Atomic orbital basis sets are built as follows

(1) to each chemical element A of the periodic table, one associates a collection $\{\xi_\mu^A\}_{1 \leqslant \mu \leqslant n_A}$ of linearly independent functions of $H^1(\mathbb{R}^3)$ (these functions will be made precise in the sequel),
(2) one is then able to construct a basis set for any molecular system by collecting all the ξ_μ^A for the different atoms the system is composed of. As a matter of example, let us consider a water molecule H_2O: The so-obtained basis set reads

$$\{\chi_\mu\} = \big\{\xi_1^H(x - \bar{x}_{H_1}), \ldots, \xi_{n_H}^H(x - \bar{x}_{H_1}); \xi_1^H(x - \bar{x}_{H_2}), \ldots, \xi_{n_H}^H(x - \bar{x}_{H_2});$$
$$\xi_1^O(x - \bar{x}_O), \ldots, \xi_{n_O}^O(x - \bar{x}_O)\big\},$$

where $\bar{x}_{H_1}$, $\bar{x}_{H_2}$ et $\bar{x}_O$, respectively, denote the locations in $\mathbb{R}^3$ of the two hydrogen atoms and the oxygen atom. This forms a basis containing $N_b = 2n_H + n_O$ functions.

The definition of a basis of atomic orbitals is that of a set collecting the $\{\xi_\mu^A\}_{1 \leqslant \mu \leqslant n_A}$ for various chemical elements A, possibly for all elements of the periodic table. For each A involved, an optimization is performed so as to choose the most efficient choice of orbitals that yield the best approximation for some properties of the isolated atom and possibly of various small molecules involving the element A. The basis sets that are obtained in this way are surprisingly efficient. The consideration of only a small number of AO allows for a very fine approximation of the Hartree–Fock ground state. For instance in the Gaussian atomic orbital basis 6-31++G(3df,3pd), which is yet considered as a large size basis and is only used to obtain high accuracy results, there are only 18 AO for Hydrogen (1 electron), 39 for Carbon (6 electrons) and 47 for Magnesium (12 electrons). Apart from the heuristic remark along which it is "natural" that a basis built with atoms yields a good basis for a molecule, there is no deep understanding, and *a fortiori* no mathematical argument, for the noticeable efficiency of such basis sets.

Atomic orbital basis sets are very efficient and are therefore very much used in quantum chemistry calculations. Two shortcomings can nevertheless be identified:

(1) The construction and use of AO basis sets is more an art than a science; contrary to the situation encountered with finite element or plane wave basis sets, for which the quality of the basis is directly related to a characteristic length (or energy) scale, there is no simple way to evaluate a priori the performances of a given AO basis. The choice of an AO basis for solving a given problem mostly relies upon the physical intuition and on some practical know-how; this problem is all the more acute as the determination of some molecular properties is very sensitive to the choice of the basis set;

(2) in the study of interactions between two or more chemical compounds, the results are biased by the so-called *Basis Set Superposition Errors* (BSSE). Let us illustrate BSSE on a simple example consisting in determining the binding energy of some dimer A–A. The binding energy can be computed by first calculating the energy E_A of the mononer A alone on the one hand, the energy $E_{A\text{–}A}$ of the dimer A–A on the other hand, and second making the difference $\Delta E = E_{A\text{–}A} - 2E_A$. The difficulty is that the calculations of E_A and $E_{A\text{–}A}$ are not performed in the same AO basis since in the second calculation the basis is twice as large as in the first one, thus an error in the estimate of ΔE. An alternative consists in using in both cases the largest basis but it is not clear that this is the right way to proceed. For details on this issue which mainly remains open, the reader is referred to KESTNER and COMBARIZA [1999].

The following lines mainly focus on *Hydrogen-like* and *Slater-type atomic orbitals* on the one hand, and on *Gaussian-type atomic orbitals* on the other hand. The latter type of atomic orbitals is by far the most commonly used in practice. For completeness, we also briefly mention the *numerical atomic orbitals* that are used in some quantum chemistry DFT codes.

(A) Hydrogen-like and Slater type orbitals

On the ground of the developments made in Section 23, we have a good knowledge of the orbitals for the hydrogen-like atoms, and it is the only case we have at hand at this stage. Therefore, in the first row of candidates for forming a collection of atomic orbitals $\{\xi_\mu^{\mathrm{A}}\}_{1\leqslant\mu\leqslant n_{\mathrm{A}}}$ stand the hydrogen-like orbitals $\psi_{nlm}^{Z}(r,\theta,\phi)$ defined by (23.3). The parameter Z can be tuned to fit experimental results; for example, the orbitals 1s, 2s and 2p of the carbon atom are well described by the hydrogen-like orbitals $\psi_{100}^{Z_{1\mathrm{s}}}$, $\psi_{200}^{Z_{2\mathrm{s}}}$ and $\psi_{21m}^{Z_{2\mathrm{p}}}$, with $Z_{1\mathrm{s}}=6.0$, $Z_{2\mathrm{s}}=4.0$ and $Z_{2\mathrm{p}}=3.1$, respectively. Hydrogen-like orbitals provide high accuracy results but computations are very tedious.

Slater type orbitals (STO) are simplified forms of hydrogen-like orbitals: The exponential factor $e^{-Zr/n}$ is left unchanged but the polynomial Q_{nl} is replaced by the monomial r^l. Up to a normalization constant, the first STO then read in Cartesian coordinates

$$\begin{aligned}
&\xi_{1s}(x,y,z)=e^{-\alpha_{1s}r},\\
&\xi_{2s}(x,y,z)=re^{-\alpha_{2s}r},\\
&\xi_{2p_x}(x,y,z)=xe^{-\alpha_{2p}r},\\
&\xi_{2p_y}(x,y,z)=ye^{-\alpha_{2p}r},\\
&\xi_{2p_z}(x,y,z)=ze^{-\alpha_{2p}r},
\end{aligned}$$

where $r=(x^2+y^2+z^2)^{1/2}$. STO have been introduced in SLATER [1930b] and widely used in the early days of quantum chemistry. They still serve for theoretical studies, or as intermediate steps for computations. However, they are almost no longer used in the numerical practice, because one encounters a terrible bottleneck when using them: The computational cost of multicenter bielectronic integral calculations.

At this stage, it is indeed useful to recall the following fact. The basis set functions χ_μ being chosen, they are included in the actual computations of the energy (or of the Fock operator) through the calculations of the matrix coefficients $h_{\mu\nu}$ of (20.2), and of the tensor $(\mu\nu|\kappa\lambda)$ in (20.3). The latter calculation is the most demanding one, as it indeed requires of the order of N_b^4 calculations (we recall N_b denotes the number of basis set elements) of 6-dimensional integrals:

$$(\mu\nu|\kappa\lambda)=\int_{\mathbb{R}^3}\int_{\mathbb{R}^3}\frac{\chi_\mu(x)\chi_\nu(x)\chi_\kappa(x')\chi_\lambda(x')}{|x-x'|}\,dx\,dx'.$$

This calculation is a bottleneck of the whole computation. Without any further simplification, it is illusion to hope to calculate all these integrals for a reasonable number N_b of elements in the basis set. The computation time required would be prohibitive. This is why the Slater type orbitals need to be replaced by another type of atomic orbitals that allows for a more efficient calculation of the above bielectronic integrals.

(B) Gaussian type orbitals

The groundbreaking idea by BOYS [1950] in the early fifties, that all of a sudden has changed the whole landscape of quantum chemistry, was to use Gaussian-type orbitals (GTO), that is to say Gaussian functions or successive derivatives of Gaussian functions:

$$\xi(x, y, z) = C x^{n_x} y^{n_y} z^{n_z} e^{-\alpha r^2}. \tag{24.1}$$

In the above formula, C is a normalization constant, n_x, n_y, n_z are nonnegative integers, α a positive real number and $r = (x^2 + y^2 + z^2)^{1/2}$.

The crucial advantage in considering such functions is that the calculation of the overlap matrix (20.1), of the core Hamiltonian matrix (20.2), and above all of the bielectronic integrals (20.3) can then be greatly simplified. In particular, the computations of the six-dimensional integrals (20.3) are brought down to the numerical computations of one-dimensional integrals of the form $F(w) = \int_0^1 e^{-ws^2}\, ds$. Let us outline the main steps of these simplifications.

One begins by noticing that it is sufficient to know how to efficiently calculate the integrals when the χ_μ involved are of the form $\chi_\mu = \xi_\mu(\cdot - \bar{x}_\mu)$ with $\xi_\mu = C_\mu e^{-\alpha_\mu r^2}$ a Gaussian function. Indeed, this being done, it is a simple exercise that uses successive integrations by parts to establish recursion formulae that relate integrals for functions of the form (24.1) with integrals for functions of the same form but with a polynomial of inferior degree.

One next remarks that the product of two Gaussian functions still is a Gaussian function:

$$\forall x \in \mathbb{R}^3,\ C_\mu e^{-\alpha_\mu |x - \bar{x}_\mu|^2} C_\nu e^{-\alpha_\nu |x - \bar{x}_\nu|^2} = C_{\mu\nu} e^{-\gamma_{\mu\nu} |x - \bar{y}_{\mu\nu}|^2},$$

where

$$C_{\mu\nu} = C_\mu C_\nu e^{-\frac{\alpha_\mu \alpha_\nu}{\alpha_\mu + \alpha_\nu} |\bar{x}_\mu - \bar{x}_\nu|^2}, \quad \gamma_{\mu\nu} = \alpha_\mu + \alpha_\nu, \quad \alpha_{\mu\nu} = \frac{\alpha_\mu \alpha_\nu}{\alpha_\mu + \alpha_\nu} \quad \text{and}$$

$$\bar{y}_{\mu\nu} = \frac{\alpha_\mu \bar{x}_\mu + \alpha_\nu \bar{x}_\nu}{\alpha_\mu + \alpha_\nu}.$$

This allows for a straightforward calculation of the overlap integrals:

$$S_{\mu\nu} = \int_{\mathbb{R}^3} \chi_\mu \chi_\nu = C_{\mu\nu} \left(\frac{\pi}{\gamma_{\mu\nu}} \right)^{3/2}$$

as well as of the kinetic energy integrals

$$\int_{\mathbb{R}^3} \nabla \chi_\mu \cdot \nabla \chi_\nu = 2\alpha_{\mu\nu} (3 - 2\alpha_{\mu\nu}) S_{\mu\nu}.$$

The bielectronic integrals (20.3) for Gaussian functions are now computed as follows. One has

$$(\mu\nu|\kappa\lambda) = C_{\mu\nu}C_{\kappa\lambda}\int_{\mathbb{R}^3}\int_{\mathbb{R}^3} e^{-\gamma_{\mu\nu}|x-\bar{y}_{\mu\nu}|^2}\frac{1}{|x-x'|}e^{-\gamma_{\kappa\lambda}|x'-\bar{y}_{\kappa\lambda}|^2}\,dx\,dx'.$$

Using the equalities

$$e^{-\gamma|x|^2} = \frac{1}{(2\pi)^3}\left(\frac{\pi}{\gamma}\right)^{3/2}\int_{\mathbb{R}^3} e^{-k^2/4\gamma}e^{ik\cdot x}\,dk \quad \text{and}$$

$$\frac{1}{|x|} = \frac{1}{(2\pi)^3}\int_{\mathbb{R}^3}\frac{4\pi}{|k|^2}e^{ik\cdot x}\,dk,$$

it can then be shown that

$$(\mu\nu|\kappa\lambda) = \frac{2\pi^2 C_{\mu\nu}C_{\kappa\lambda}}{(\gamma_{\mu\nu}\gamma_{\kappa\lambda})^{3/2}|\bar{y}_{\mu\nu}-\bar{y}_{\kappa\lambda}|}\int_0^{+\infty}\frac{\sin u}{u}e^{-u^2/4w_{\mu\nu\kappa\lambda}}\,du,$$

with

$$w_{\mu\nu\kappa\lambda} = \frac{\gamma_{\mu\nu}\gamma_{\kappa\lambda}}{\gamma_{\mu\nu}+\gamma_{\kappa\lambda}}|\bar{y}_{\mu\nu}-\bar{y}_{\kappa\lambda}|^2.$$

Finally, as

$$\int_0^{+\infty}\frac{\sin u}{u}e^{-u^2/4w}\,du = \sqrt{\pi w}\int_0^1 e^{-ws^2}\,ds,$$

one solely expresses the bielectronic integral in terms of the function $F(w) = \int_0^1 e^{-ws^2}\,ds$:

$$(\mu\nu|\kappa\lambda) = \frac{2\pi^{5/2}C_{\mu\nu}C_{\kappa\lambda}}{(\gamma_{\mu\nu}\gamma_{\kappa\lambda})^{3/2}|\bar{y}_{\mu\nu}-\bar{y}_{\kappa\lambda}|}\sqrt{w_{\mu\nu\kappa\lambda}}\,F(w_{\mu\nu\kappa\lambda}). \tag{24.2}$$

A sampling of the function F is stored in memory so that the values of F at the various $w_{\mu\nu\kappa\lambda}$ involved can be computed by interpolation. Lastly, let us remark that the integrals

$$\int_{\mathbb{R}^3} V\chi_\mu\chi_\nu = -\sum_{k=1}^{M} z_k\int_{\mathbb{R}^3}\frac{\chi_\mu(x)\chi_\mu(x)}{|x-\bar{x}_k|}$$

needed for the assembling of the core Hamiltonian matrix $h_{\mu\nu}$ can easily be deduced from formula (24.2) by taking $\bar{y}_{\kappa\lambda} = \bar{x}_k$, $C_{\kappa\lambda} = (\gamma_{\kappa\lambda}/\pi)^{3/2}$ and letting $\gamma_{\kappa\lambda}$ go to infinity. We thus obtain

$$\int_{\mathbb{R}^3} V\chi_\mu\chi_\nu = -\frac{2\pi C_{\mu\nu}}{\gamma_{\mu\nu}}\sum_{k=1}^{M} z_k F\big(\gamma_{\mu\nu}|\bar{y}_{\mu\nu}-\bar{x}_k|^2\big).$$

The main advantage of GTO with respect to STO lays in formula (24.2). On the other hand, the drawback of the former is that they do not describe correctly the shape of the molecular orbitals both near the nuclei and at infinity. Indeed, the HF electronic density has a cusp at each nucleus and fall-off like $e^{-\varepsilon|x|}$ at infinity; a few STO, but many GTO, are necessary to represent correctly such a density shape. If *primitive* Gaussian functions such as (24.1) are used, the basis set has to be much larger than a typical STO basis set for the same accuracy. This problem can be remedied by the use of basis sets made of *contracted Gaussian functions*, which are linear combinations of primitive Gaussian functions:

$$\xi(x, y, z) = \sum_{k=1}^{K} C_k x^{n_x^k} y^{n_y^k} z^{n_z^k} e^{-\alpha_k r^2}, \tag{24.3}$$

in which the C_k are optimized once and for all in order to accurately represent the cusps and the fall-off. In general the K primitive Gaussian functions all belong to a prescribed symmetry class (i.e., all the n_x^k are equal to one another, all the n_y^k are equal, all the n_z^k are equal).

Many Gaussian basis sets are available in the literature. Computational quantum chemistry packages propose a large choice of already implemented basis sets. Some users however prefer to define their own basis. Gaussian basis sets are characterized by the following terminology:

- In *single zeta* basis sets, only one contracted Gaussian function per occupied orbital is used; for instance five contracted Gaussian functions (1s, 2s, $2p_x$, $2p_y$ and $2p_z$) are associated to each atom from Boron to Neon; the STO-3G basis set, in which each occupied orbital is described by a contraction of 3 primitive Gaussians, belongs to this class;
- in *double zeta* (resp. in *triple zeta*) basis sets, two (resp. three) contracted Gaussian functions per occupied orbital are used;
- *split-valence* basis sets consist in taking one basis function for each core orbital (like in single zeta basis sets) and two basis functions for each valence orbital (like in double zeta basis sets). In the 6-31G basis set for instance,
  - one basis function which is itself the contraction of 6 primitive Gaussians is associated to each core orbital;
  - two basis functions are associated to each valence orbital, one of them being the contraction of 3 primitive Gaussians, the other one being a single primitive Gaussian.

  Thus, each atom from Boron to Neon carries 9 basis functions (1 for the 1s core orbital, 2 for each of the 2s, $2p_x$, $2p_y$ and $2p_z$ valence orbitals) collecting 22 primitive Gaussians.

In some cases, it is necessary to incorporate to the basis set *polarization* and/or *diffuse* basis functions:

- *Polarization functions* are functions whose angular momentum quantum number l is above the highest l of the occupied orbitals. For instance, polarization functions are of p type at least for Hydrogen, of d type at least for Carbon or Oxygen, ...

Polarization functions are necessary to describe hybridation and polarization phenomena (for instance to compute polarizabilities);

- *diffuse functions* are primitive or contracted Gaussian functions for which the parameters α_k in expression (24.3) are smaller than those of the valence orbitals; they are therefore more *diffuse*. By analogy with the hydrogen-like atoms whose orbitals are given by (23.3), diffuse functions are in some way atomic orbitals whose principal quantum number n is above the highest n of the occupied orbitals. Diffuse functions are needed to compute excited states, as well as negative ions.

In the Hartree–Fock setting, the computational effort necessary to build the mean-field Hamiltonian matrix (i.e., the Fock matrix) in a basis containing N_b element a priori scales as N_b^4 because of the calculation of the bielectronic integrals. For large systems however, the scaling is much lower in practice due to the fact that the overlap of two Gaussian atomic orbitals attached to two nuclei far away from one another is negligible. Various algorithms based on a priori estimates of the integrals have been developed in the late 70s and the 80s (see GILL [1994] and the references therein); it is estimated that the scaling of these algorithms is around $N_b^{2.7}$ in practice. The prefactor mainly depends on the degree of contraction of the Gaussian atomic orbitals, i.e., of the parameter K in (24.3). A much better scaling ($O(N_b)$) can be obtained with linear scaling algorithms based on Greengard and Rokhlin Fast Multipole Method (see GREENGARD and ROKHLIN [1997]). The latter method, initially developed to compute the self-energy of a large set of interacting point particles in a gravitational (or Coulomb) field, has been adapted to Coulomb and exchange matrix computations by SCHWEGLER and CHALLACOMBE [1997], SCHWEGLER and CHALLACOMBE [2000], SCHWEGLER and CHALLACOMBE [1999], SCHWEGLER, CHALLACOMBE and HEAD-GORDON [1997]. When combined with alternative to diagonalization methods (see Section 36), FMM based Fock matrix computations allow HF calculations of about one thousand atoms on today's available workstations.

The above discussion was focused on the Hartree–Fock case. In Kohn–Sham calculations the exchange-correlation matrix also has to be built and there is no hope to find out analytical formulae for computing its coefficients. For instance, the component of the closed shell mean-field Hamiltonian matrix corresponding to the simple LDA exchange term (15.10) reads

$$\left[F_x^{\mathrm{LDA}}\right]_{\mu\nu} = -\frac{4}{3}C_{\mathrm{D}}\int_{\mathbb{R}^3}\rho(x)^{1/3}\chi_\mu(x)\chi_\nu(x)\,dx$$

$$\text{with } \rho(x) = 2\sum_{\mu,\nu=1}^{N_b} D_{\mu\nu}\chi_\mu(x)\chi_\nu(x).$$

Due to the noninteger power $1/3$, the integral cannot be performed analytically. Most often, the integral is simply evaluated by numerical integration on a grid. Linear scaling integration methods have been introduced recently (see SCUSERIA [1999], CHALLACOMBE [2000] and references therein).

(C) Numerical atomic orbitals

Numerical atomic orbitals are defined by their numerical values on a grid. They are in general compactly supported in spheres centered at the nuclei and whose radii do not exceed a few atomic units. Both the Kohn–Sham mean-field Hamiltonian and the overlap matrix S are thus sparse and the locations of the nonzero elements is a priori known (the coefficients associated to two atomic orbitals which do not overlap vanishes). This is of greatest interest in the perspective of linear scaling algorithms (see Section 36 and the contribution of S. Goedecker to this volume).

The main problem with numerical atomic orbitals is the computation of the Coulomb contribution

$$\int_{\mathbb{R}^3} V\chi_\mu\chi_\nu + \int_{\mathbb{R}^3}\left(\rho\star\frac{1}{|x|}\right)\chi_\mu\chi_\nu$$

to the mean-field Hamiltonian. In general, the total Coulomb potential W solution to

$$-\Delta W = 4\pi(\rho_N - \rho) \tag{24.4}$$

(where ρ_N is the distribution of the nuclear charge) is first computed on the regular grid by solving (24.4) with a fast Fourier transform technique; the integrals

$$\int_{\mathbb{R}^3} W\chi_\mu\chi_\nu$$

are then evaluated by numerical integration (the integrand is in fact compactly supported). This assembling technique scales as $\mathrm{O}(N_b)$. However, it is not completely satisfactory for two reasons:

(1) Eq. (24.4) is set on the whole space $\mathbb{R}^3$. In grid methods based on FFT, boundary conditions have to be specified. Homogeneous Dirichlet conditions are not satisfactory because of the slow decay of the Coulomb potential (especially for ionic systems for which W decays as $|x|^{-1}$). Periodic boundary conditions are not satisfactory either. Lastly, coupling grid methods with integral equation methods leads to cumbersome numerical techniques;

(2) In the models presented so far, nuclear charges are supported by Dirac distributions:

$$\rho_N = \sum_{k=1}^{N} z_k\delta_{x_k}.$$

FFT based grid methods are not designed to deal with such singularities. The nuclear charge therefore has to be smeared out; for that pseudopotentials are needed (even for the Hydrogen atom!).

25. Basis sets: Plane waves and other basis sets

Atomic orbital basis sets are used in most of the gas or liquid phase calculations. Some calculations are also done with plane waves basis sets or more generally with basis sets originally dedicated to crystal phase calculations. Computing an isolated molecule in such a basis set amounts to computing a periodic system containing one of the molecule under study in each cell. One has then to strike a balance between two incompatible requirements:

- On the one hand, the smaller the cell the stronger the interactions between the molecule and its images; this may induce major changes of the molecular properties;
- On the other hand, the larger the cell the larger the number of plane waves required for reaching a given accuracy.

Let us also add that when plane waves are used, the size of the basis set is in any case much larger than the number of basis functions necessary to perform the same calculation with the same accuracy with atomic orbitals basis sets. The size of the plane wave basis set may even become completely prohibitive; as for numerical basis sets, pseudopotentials are needed to smear out the nuclear Coulomb singularity. For all these reasons, plane wave basis set are generally not recommended for calculations on isolated molecules. On the other hand, they are one of the methods of choice for the solid-state.

Incidentally, let us also point out that wavelet basis sets have been successfully tested on simple cases (DEFRANCESCHI and FISCHER [1998], FISCHER [1994]) and that new developments in this field are in progress (GOEDECKER and IVANOV [1998], MARKVOORT, PINO and HILBERS [2001]).

26. Basics on optimization techniques: The unconstrained case

We present in this section and in the following one some basics of numerical optimization. By doing so, of course we do not claim at any originality. Needless to say, the present section is not a substitute to a detailed treatise about numerical optimization. Our only purpose is to provide the reader with a rapid overview of the methods, so that he could approach the next Sections 28–30, specifically devoted to the optimization in the peculiar context of computational chemistry, with a knowledge of the general techniques. Our viewpoint is rather schematic, and we shall only give a rapid and vague overview, focusing on the concepts rather than on the techniques.

For a complete description, we rather refer to BONNANS, GILBERT, LEMARECHAL and SAGASTIZABAL [1997] and also to KELLEY [1999], NOCEDAL and WRIGHT [1999]. An interesting reference, mainly dedicated for a readership coming from the chemistry community, is the excellent review by SCHLICK [1992], which may serve as a useful introduction before reading more advanced treatises.

We mainly concentrate ourselves on the deterministic methods, leaving aside the stochastic like methods.

The simplest problem of optimization is to find the minimum of a function

$$E : \mathbb{R}^N \to \mathbb{R}$$

over the whole space $\mathbb{R}^N$. For simplicity, we assume that the function E is sufficiently regular so that we can deal at least with its first and when needed with its second derivatives. Adaptations of the algorithms described below to the nondifferentiable cases can be read in the bibliography. They can be really intricate.

All optimization algorithms consist in building a sequence of iterates x_k, that hopefully converges to a global minimizer of E. In practice, it is at best *one* of the global minimizer, and in fact most of the times, it is only a *local* minimizer and not a *global* one.

An optimization method is said to be *locally convergent* if it is able to find a given local minimum if the initial guess is "close enough" to this local minimum and *globally convergent* if the method converges toward some local minimum, whatever the initial guess. The reader has to keep in mind that a method which globally converges is not necessarily a method which converges toward a global minimum.

A practical way to classify the optimization algorithms is to classify them with respect to the number of derivatives of E they take into account for building the new iterate x_{k+1} from the current one x_k. Only the zero order information provided by $E(x_k)$ is used for algorithms such as the downhill simplex method. Such algorithms are essentially very slow. In order to accelerate the convergence, it is natural to insert at each step more information on the function to be minimized. This can be done by accounting for the first derivative $\nabla E(x_k)$ of E at the current iterate, which gives rise to the gradient-like methods. One step further is to also take into account the second derivative $\nabla^2 E(x_k)$, which is the Hessian of E. Accordingly to the number of derivatives taken into account, the search is accelerated in the zones where E is regular (think of a parabolic basin of E). The price to pay is however the risk to remain trapped around in a local minimizer within a large basin of E, and not being able to escape from that basin. This can only be remedied for by inserting some stochasticity in the algorithm, which is for instance the purpose of stochastic gradient algorithms.

The fundamental guideline for the construction of an optimization algorithm is the notion of *a local model*.

At the vicinity of the current iterate x_k, the form of the function E is not known. However, it is reasonable to approximate it by a given model. Typically, this model is based upon a Taylor expansion of E in the neighborhood of x_k, the degree of which is related to the number of derivatives we have decided above to take into account. This model being fixed, it is next natural to hope that the minimizer of a local model provides with a reasonable guess for a minimizer to E, at least around x_k. For instance, if one chooses to approximate E by a second order Taylor expansion

$$E(x) \simeq \widetilde{E}(x) = E(x_k) + \nabla E(x_k)(x - x_k) + \frac{1}{2}\nabla^2 E(x_k)(x - x_k)^2$$

and if the next iterate x_{k+1} is accordingly defined as the (supposedly existing) minimizer of the right-hand side, then one ends up with the famous Newton algorithm

$$x_{k+1} = x_k - \left(\nabla^2 E(x_k)\right)^{-1} \nabla E(x_k). \tag{26.1}$$

The clear limitation of the notion of local model is precisely its locality. As soon as the Taylor expansion is legal, i.e., in the immediate vicinity of x_k, it is clear that the model reproduces the function E. However, it is not at all sure, and there is of course no reason for it, that this is a good approximation of E further than *infinitely close to* x_k. Therefore, the minimizer of the local model, which might be far away from x_k, may be a poor approximation of the minimizer of E, and even lead to a wrong track. Incidentally, let us mention that this observation is at the roots of a class of methods known as trust region methods which make use of the comparison between the model and the exact function.

A way to improve the approach is to divide the construction of the next iterate x_{k+1} in two steps, that are usually done sequentially (but may also be done simultaneously in very peculiar algorithms that will not be dealt with here). The construction of x_{k+1} is made of two steps

- (Step 1) the choice of a (hopefully) *descent direction* d_k;
- (Step 2) the choice of an optimal length t_k along this direction.

For the choice of the direction, possibilities are given, as before, by the local model of E, namely the gradient, the Newton direction, ... For the second step, called the *linesearch*, the optimal length t_k is chosen with the help of a local model but the true function E may be itself explicitly used. Let us briefly define this step.

Once the direction d_k is chosen, the problem to be answered is that of the choice of the length. Basically, one is back to the same original problem of minimizing E, but this time only on the real line (aligned with the descent direction). Of course, a naive approach is to exactly search for the minimizer of the local model of E along this direction. In the "genuine" Newton algorithm, this is the strategy chosen, and it leads to the unit length, appearing in the right-hand side of (26.1). But as the local model is only a model, and as the direction may not be the good descent direction a posteriori, there is no strict need to minimize exactly the model along the direction. This observation is the basis for the linesearch algorithm, which aims at only finding a point on the line where the function E itself has "sufficiently" decreased. The word "sufficiently" is given a precise meaning below.

A linesearch algorithm consists itself of a two step iteration

- (Step (i)) a range being given, one chooses a good point in this range;
- (Step (ii)) a range and a point therein being given, one decides to restrict the range to the left part or the right part of it.

In this fashion, one iteratively restricts the size of the range where the approximate minimizer is searched for, until convergence. It is to be mentioned that, again, Step (i) may be done accordingly to a local model for E. Once the convergence is reached, this defines t_k thus the new iterate $x_{k+1} = x_k + t_k d_k$ and one returns to the main iteration loop.

This main iteration loop is stopped on the basis of a criterion, that mostly expresses the fact that the derivative ∇E at the current iterate is zero within a given threshold fixed in advance. Hopefully, this means that the current iterate is a critical point. It can then be tested whether it is, in some vague sense at least, a local minimum. On the other hand, there is no way to test by a (necessarily) local criterion, whether it is a global minimum.

So far we have only described the general principles. In order to fix the ideas, let us now overview a few algorithms that originate from these principles and that implement them. We first present the various algorithms classified along the way they define the descent direction (Step 1 above). It is to be understood that all of them are followed by a linesearch, which is more explicitly described next.

As rapidly said above, the natural idea for a descent direction is that of the first derivative of E, $d_k = -\nabla E(x_k)$, which amounts to make use of the first order local model

$$E(x) \simeq \widetilde{E}(x) = E(x_k) + \nabla E(x_k) \cdot (x - x_k).$$

With an optimized length t_k along this direction, this method is the well known *steepest descent* method. Actually, it must not be advised, because the crude first order approximation leads to a very slow convergence. However, the steepest descent direction is sometimes used in quantum chemistry calculations for it is easy to compute.

It can however be greatly improved on the basis of the following observation. The steepest descent method is slow in a twofold sense. First, the gradient direction may be good at infinitesimal distance but very bad at noninfinitesimal ones. Clearly this cannot be improved simply. Second, the method has no memory of the previous iterations (mathematically speaking, it is *Markovian*). Therefore, zones of the minimization space that have already been explored may be explored again, with as a counterpart a great loss in efficiency. The *nonlinear conjugate gradient method* aims at curing this disease. It is one of the simplest representative of a class of methods that are not based upon the exploration of a single direction (as in Step 1 above), but rather on *a set of directions*, thus a subspace of finite dimension. Here the subspace chosen is

$$K_k = x_k + \mathrm{Vect}\big(\nabla E(x_0), \ldots, \nabla E(x_k)\big).$$

Therefore the new iterate x_{k+1} is defined as a solution to

$$\inf_{x \in K_k} E(x), \tag{26.2}$$

or at least, more practically speaking, as a local minimizer to this problem, in the neighborhood of x_k.

In addition to the obvious increasing memory requirement, as the size of the space one minimizes upon increases through the iterations, the solution to this problem may be tedious to find when k is large. Therefore, a simplification is needed. The foundation of such a simplification comes from the examination of the case when E is quadratic (which again amounts to considering a local model for E). When E reads $E(x) = \frac{1}{2} x^T A x - b^T x$, where A is a symmetric definite positive matrix and b is a given vector, it turns out that the problem (26.2) has a simple solution: The minimizer x_{k+1} to (26.2) is given by the induction formulae

$$x_{k+1} = x_k + t_k d_k,$$

$$d_k = -\nabla E(x_k) + c_{k-1} d_{k-1}, \tag{26.3}$$

$$c_{k-1} = \frac{|\nabla E(x_k)|^2}{|\nabla E(x_{k-1})|^2}, \tag{26.4}$$

$$t_k = -\frac{(\nabla E(x_k) \cdot d_k)}{(d_k, A \cdot d_k)}.$$

Therefore the minimization step (26.2), which is to be done at each iteration loop of the main minimization algorithm, may be done explicitly with a simple calculus. In fact, this property that an apparently complicated sequence of minimization problems over a particular subspace reduces to an induction formula (possibly with more than one level as above), is a special case of a more general property for a class of methods dedicated to the iterative solution procedure for linear systems $Ax = b$, namely that of the Krylov spaces methods. A complete numerical analysis of these methods, with the proof of convergence of algorithms and an estimate for their speed of convergence, may be read in the literature.

On the basis of this quadratic case, the idea is to replace, even for a nonquadratic function E, the search for a minimizer to (26.2) by an induction formula of the type (26.3). More precisely, the descent direction is defined as

$$d_k = -\nabla E(x_k) + c_{k-1} d_{k-1},$$

where the length c_{k-1} is fixed somewhat arbitrarily (and in fact in view of simple toy situations). In the practice, the most commonly used choice is that of the *Polak–Ribière* formula

$$c_{k-1} = \frac{(\nabla E(x_k) - \nabla E(x_{k-1})) \cdot \nabla E(x_k)}{|\nabla E(x_{k-1})|^2}, \tag{26.5}$$

which turns out to be equivalent to formula (26.4) when E is quadratic. Next, the length t_k that defines the new iterate $x_{k+1} = x_k + t_k d_k$ is found as in Step 2 by a linesearch. The nonlinear conjugate gradient algorithm that has been built this way is an efficient method to solve unconstrained minimization problems. It is in fact the most powerful first order method. There are some cases of divergencies, but in most cases the algorithm works pretty well, in spite of the lack of a theoretical numerical analysis.

In the difficult cases that are not satisfactorily solved by the nonlinear conjugate gradient or in order to accelerate convergence, there is the opportunity to resort to a second order algorithm, namely a Newton type algorithm. As said above, the descent direction is evaluated from a second order local model of the function, and thus reads as a solution to

$$\nabla^2 E(x_k) \cdot d_k + \nabla E(x_k) = 0. \tag{26.6}$$

In the case when E is quadratic, the optimal length (Step 2) is of course unity, but for arbitrary E, the determination of d_k is next followed by a linesearch to find the optimal length along this direction.

The clear bottleneck of the method is the need to solve (26.6), which is indeed a linear system that may be of too large a size. Therefore, the Newton method as described so far is only used when (26.6) is "easy" to solve (which in particular requires that the Hessian $\nabla^2 E(x_k)$ is easy to compute itself, this being far from obvious in the cases of interest). A natural simplification is provided by the *quasi-Newton* methods, that replace the solution to (26.6) by

$$d_k = -B_k \nabla E(x_k),$$

where B_k aims at approximating $(\nabla^2 E(x_k))^{-1}$ and is given by an induction formula, such as the famous BFGS formula (that acronym standing for Broyden, Fletcher, Goldfarb, Shanno)

$$B_{k+1} = B_k - \frac{s_k y_k^T B_k + B_k y_k s_k^T}{(y_k, s_k)} + \left[1 + \frac{(y_k, B_k \cdot y_k)}{(y_k, s_k)}\right] \frac{s_k s_k^T}{(y_k, s_k)}$$

(with $s_k = x_{k+1} - x_k$, $y_k = \nabla E(x_{k+1}) - \nabla E(x_k)$, and $B_0 = I$, except if some better initial guess can be obtained by a specific analysis of the problem under consideration). The foundation of this formula is that it is both imposed that B_k is positive definite and that $B_{k+1} \cdot (\nabla E(x_{k+1}) - \nabla E(x_k)) = x_{k+1} - x_k$, which in view of Taylor formula traduces that B_k approximates $(\nabla^2 E(x_k))^{-1}$ in the average along the segment line linking x_k and x_{k+1}.

Let us now examine the linesearch problem. A descent direction being given, the goal is now to find some $x_{k+1} = x_k + t_k d_k$ on the line $x_k + t d_k$ such that the function $q(t) = E(x_k + t d_k)$ *almost attains* a minimum at t_k. Let us underline that "almost" is enough: It is not worth losing time in solving very accurately the linesearch problem at each step since the minimum of E that we search for is not (in general) on the line $x_k + t d_k$.

For clarity, let us assume that $q'(0) < 0$ (this corresponds to the case when d_k is actually a descent direction). The desired t_k is determined through the following twofold constraint. First, $E(x_{k+1})$ is imposed to be sufficiently inferior to $E(x_k)$, which is expressed by stating that the function $q(t)$ needs to satisfy at $t = t_k$ the condition

$$q(t) \leqslant q(0) + m_1 t q'(0). \tag{26.7}$$

Simultaneously, it is useful to impose that, among the points t that satisfy the above condition, the one we search for must satisfy

$$q'(t) \geqslant m_2 q'(0), \tag{26.8}$$

which expresses the fact that in some weak sense, the derivative of E at t is close to zero. In the above formulae, it is convenient to set $0 < m_1 < m_2 < 1$. Incidentally, it is to be remarked that condition (26.7) actually implies that t is not too large, and that condition (26.8) actually implies that t is not too small. Of course, the specific choice of conditions (26.7), (26.8) made above is arbitrary, and known as the *Wolfe selection rule*. Other choices may be made, that express mathematically the same ideas as (26.7), (26.8).

As soon as these two conditions are satisfied for some t, the linesearch algorithm stops at t, sets $t = t_k$ and returns to the main iteration loop. To reach this situation, the algorithm obeys to the following two step iterations, starting from a given range, say $[t_- = 0, t_+ = 1]$,

- Step (i): Select some t in $[t_-, t_+]$: this is done either by an arbitrary "universal" law, such $t = \frac{t_- + t_+}{2}$, or by a "more specific" law that accounts for a local model of E, say use $q(0)$, $q(1)$, $q'(0)$, etc., to approximate E by interpolation with a polynomial and choose t as the minimizer of this polynomial on the line,

- Step (ii): Test if t satisfies the conditions; if not, decide whether to continue the linesearch on $[t_-, t]$ or $[t, t_+]$, depending on which of the two conditions (26.7) or (26.8) is satisfied.

For details on the possible choice for the implementation of the two steps, and the improvements of the basic ideas we have given here, we refer to the literature.

An efficient linesearch is a crucial requirement for an efficient algorithm, as it is easily understandable in view of the large number of occurrences of the linesearch in the main optimization loop.

27. Basics on optimization techniques: The constrained case

Let us now turn to the constrained optimization, that is the situation when the function E considered so far is to be minimized *only on a subset* of $\mathbb{R}^N$.

Let us consider again the generic equality constrained problem

$$\inf\{E(x),\ x \in \mathcal{H},\ g(x) = 0\} \tag{27.1}$$

with $E : \mathcal{H} \to \mathbb{R}$ and $g : \mathcal{H} \to \mathcal{K}$ are C^2 ($\mathcal{H}$ and $\mathcal{K}$ are Euclidean spaces) whose Euler–Lagrange equations have been derived in Section 21: They read

$$\begin{cases} \nabla E(x_*) - g'(x_*)^T \cdot \lambda_* = 0, \\ g(x_*) = 0. \end{cases}$$

Let us now introduce the *Lagrangian* of the optimization problem (27.1) defined by

$$\mathcal{L}(x, \lambda) = E(x) - \big(\lambda, g(x)\big)_{\mathcal{K}}, \tag{27.2}$$

where $(\cdot, \cdot)_{\mathcal{K}}$ denotes the Hermitian scalar product on $\mathcal{K}$. The Lagrangian formulation is very convenient to state the first and the second order optimality conditions, that is the purpose of the following theorem, the proof of which is standard:

THEOREM 27.1.

(1) First order optimality condition: *A necessary condition for x_* being a local minimum of* (27.1) *is that there exists $\lambda_* \in \mathcal{K}$ such that*

$$\begin{cases} \nabla_x \mathcal{L}(x_*, \lambda_*) = 0, \\ \nabla_\lambda \mathcal{L}(x_*, \lambda_*) = 0. \end{cases} \tag{27.3}$$

(2) Necessary second order optimality condition: *A necessary condition for x_* satisfying* (27.3) *being a local minimum of* (27.1) *is that*

$$\forall h \in V_*,\ h^T \nabla^2_{xx} \mathcal{L}(x_*, \lambda_*) h \geqslant 0,$$

where $V_* := T_{x_*}\mathcal{M} = \mathrm{Ker}(g'(x_*))$.

(3) Sufficient second order optimality condition: *A sufficient condition for* x_* *satisfying* (27.3) *being a strict local minimum of* (27.1) *is that*

$$\forall h \in V_*,\ h^T \nabla^2_{xx}\mathcal{L}(x_*,\lambda_*)h > 0. \tag{27.4}$$

A natural idea in order to solve the minimization problem (27.1) is to directly attack it by a standard algorithm for unconstrained minimization, and to slightly modify the algorithm to account for the constraint. An instance of such an approach is the projected gradient algorithm which consists in the iterations

$$x_{k+1} = P_{\mathcal{M}}\big(x_k - \nabla E(x_k)\big),$$

where $P_{\mathcal{M}}$ denotes the projection operator on the manifold $\mathcal{M} = \{x \in \mathcal{H};\ g(x) = 0\}$. Such an approach is most often too naive to work well. In particular, it experiences the worst difficulties when $\mathcal{M}$ is not convex, which is the case for computational chemistry ($\mathcal{M}$ is typically the set of idempotent matrices). Therefore, such approaches need to be complemented by approaches truly dedicated to the constrained cases.

Most of the efficient numerical methods for solving the optimization problem (27.1) are in fact dedicated to finding a primal–dual solution (x_*, λ_*) to the Euler–Lagrange equations (27.3). The simplest method of this kind is the celebrated Newton method, which consists in using the basic Newton algorithm (see, e.g., BURDEN and FAIRES [1993]) for solving the nonlinear equations (27.3). More precisely, the Newton algorithm generates a sequence (x_k, λ_k) in the following way: The current iterate (x_k, λ_k) is updated by first solving the linearized equations

$$\begin{cases} \text{Search } (d_k,\mu_k) \in \mathcal{H}\times\mathcal{K} \text{ such that} \\ E''(x_k)\cdot d_k - \big(g''(x_k)\cdot d_k\big)^T\cdot\lambda_k - g'(x_k)^T\cdot\mu_k + \nabla E(x_k) - g'(x_k)^T\cdot\lambda_k = 0, \\ g'(x_k)\cdot d_k + g(x_k) = 0, \end{cases}$$

then setting $x_{k+1} = x_k + d_k$, $\lambda_{k+1} = \lambda_k + \mu_k$. Let us now denote by $L_k = \nabla^2_{xx}\mathcal{L}(x_k,\lambda_k)$ the second derivative of the Lagrangian with respect to x at (x_k, λ_k), and by $g_k = g(x_k)$, $s_k = -\nabla E(x_k)$ and $B_k = g'(x_k)$. It is easy to see that the pair (d_k, λ_{k+1}) can be obtained by solving

$$\begin{cases} \text{Search } (d_k,\lambda_{k+1}) \in \mathcal{H}\times\mathcal{K} \text{ such that} \\ L_k\cdot d_k - B_k^T\cdot\lambda_{k+1} - s_k = 0, \\ B_k\cdot d_k + g_k = 0 \end{cases} \tag{27.5}$$

or, in matrix form,

$$\begin{pmatrix} L_k & B_k^T \\ B_k & 0 \end{pmatrix}\cdot\begin{pmatrix} d_k \\ -\lambda_{k+1} \end{pmatrix} = \begin{pmatrix} s_k \\ -g_k \end{pmatrix}. \tag{27.6}$$

The existence and uniqueness of the solution of the equivalent problems (27.5), (27.6) is guaranteed if the two assumptions below are fulfilled

(1) $g_k \in \text{Ran}(B_k)$;

(2) L_k is positive definite on $V_k := \text{Ker}(B_k)$.

If x_* is a local minimum of (27.1) which satisfies condition (27.4) and if the current iterate (x_k, λ_k) is close enough to the solution (x_*, λ_*) of (27.3), then the two conditions are fulfilled by continuity. The Newton step is therefore perfectly defined in the neighborhood of a strict local minimum of (27.1). It can be proved that the Newton method is locally superlinearly convergent in the pair (x, λ) (and quadratically provided E'' and g'' are Lipschitz). On the other hand, the Newton algorithm may diverge if the initial guess is not in the neighborhood of a solution. Here also global convergence can be obtained with a linesearch along the Newton direction aiming at minimizing some *penalty function* taking into account both the criterion E and the deviation to the constraints g. For more details, we refer to NOCEDAL and WRIGHT [1999].

In addition, under the above two assumptions, the primal solution d_k is also solution to the quadratic programming problem

$$\inf\left\{\frac{1}{2}(d, L_k d) - (s_k, d),\ d \in \mathcal{H},\ B_k d + g_k = 0\right\}. \tag{27.7}$$

The Newton method is thus a special case of Sequential Quadratic Programming (SQP): The fully nonlinear constrained problem (27.1) is replaced by a sequence of quadratic programs, for the resolution of which efficient numerical methods are available. More sophisticated SQP algorithms involve approximations of the second derivative (quasi-Newton methods), and extensions to inequality constraints. Some of them tackle the additional difficulty when g is not a submersion (in which case the constraints $B_k \cdot d + g_k = 0$ may be inconsistent, that is to say that this equation may have no solution). Lastly, let us mention that the condition number of the problem can be improved by augmented Lagrangian techniques as in FORTIN and GLOWINSKI [1983].

Let us now briefly present some classical methods for computing the primal–dual solution (d_k, λ_{k+1}) to system (27.5).

The first strategy consists in the *range–space methods* that compute *first* the primal solution d_k, *then* the dual solution λ_{k+1}. Let us assume that are known

- some η_k satisfying $B_k \cdot \eta_k + g_k = 0$;
- a linear surjective mapping P_k from $\mathcal{H}$ onto V_k;
- a left-hand side inverse of B_k^T denoted by Q_k^l.

The solution d_k to (27.7) is of the form $d_k = \eta_k + \delta_k$ where $\delta_k \in V_k$. As P_k is surjective, $\delta_k = P_k u_k$ where u_k is a minimizer of the problem

$$\inf\left\{\frac{1}{2}(u, A_k u) - (b_k, u),\ u \in \mathcal{H}\right\} \tag{27.8}$$

with $A_k = P_k^* L_k P_k$ and $b_k = P_k^* \cdot (s_k + L_k \eta_k)$. The conjugated gradient algorithm is perfectly adapted to solving problem (27.8) provided the matrix A_k is positive definite on $\text{Ran}(P_k^*) = (\text{Ker}\, P_k)^\perp$, that is to say if L_k is positive definite on $V_k = \text{Ran}(P_k)$. As any onto mapping P_k can be used, various choices are possible. They are algebraically, but not numerically equivalent: P_k can indeed be seen as a preconditioner of the linear system. Once d_k is known, the updated Lagrange multiplier λ_{k+1} can be obtained as

$$\lambda_{k+1} = Q_k^l \cdot (L_k \cdot d_k - s_k).$$

The second strategy, followed by the *null-space methods*, consists in proceeding the other way round, that is to say in computing *first* the dual solution λ_{k+1} as a solution to the symmetric linear system

$$\left(B_k L_k^{-1} B_k^T\right) \cdot \lambda_{k+1} = -B_k L_k^{-1} \cdot s_k - g_k,$$

then the primal solution as a solution to the symmetric linear system

$$L_k d_k = s_k + B_k^T \cdot \lambda_{k+1}.$$

Lastly, the third method, known as the *Lagrangian method*, is to compute *simultaneously* d_k and λ_{k+1} by solving problem (27.6). Problems such as (27.6) have been widely studied in numerical analysis because they play a central role in many fields of scientific computing, for instance, in computational fluid mechanics; efficient algorithms have thus been designed to solve them (QUARTERONI and VALLI [1994]).

28. The SCF cycles in the Hartree–Fock setting: Generalities

We now concentrate on the strategy to solve the discretized spinless Hartree–Fock problem (20.4). We again focus on the spinless setting for the sake of simplicity. The algorithms detailed below can be easily transposed to the RHF, UHF and GHF settings. All the following theoretical convergence results are valid for the UHF and GHF cases. On the other hand, we shall see that only some of them are proven for the RHF setting at the present time.

We shall both describe the algorithms in use and provide the reader with a numerical analysis of them, that is to say with a proof of the main results of convergence available at this day. We shall also point out the lack of rigorous results, should the occasion arise. It is to be noted that the analysis we reproduce here is due to two of us. To the best of our knowledge, there is no other works dealing with a rigorous analysis of the algorithms in use for the SCF cycle in electronic structure computations. The only mathematical work we are aware of is the very interesting one by AUCHMUTY and WENYAO JIA [1994], who have studied the convergence of a toy algorithm which is *not* used in practice. In addition, some studies, such as, e.g., SCHLEGEL and MCDOUALL [1991], SEEGER and POPLE [1976], STANTON [1981a], STARIKOV [1993], ZERNER and HEHENBERGER [1979], KOUTECKY and BONACIC [1971], DOUADY, ELLINGER, SUBRA and LEVY [1980], NATIELLO and SCUSERIA [1984], FISCHER and ALMLÖF [1992], CHABAN, SCHMIDT and GORDON [1997], have appeared in the chemical literature and are sometimes referred to. Although they provide with enlightening (numerical) experimental results, comprehensive numerical tests of ideas, and clever recipes to remedy for possible lacks of convergence in the numerical codes, they cannot be considered as being rigorous from the standpoint of numerical analysis. The results described below *complement* these studies from the theoretical standpoint.

Before getting to the heart of the matter, let us also make precise that we shall give below the proofs of convergence in the setting of the *finite-dimensional* approximation. Actually, the same results hold, with more complicated proofs and sometimes with additional hypotheses, in the *infinite-dimensional* setting of the equations. From the implementation viewpoint, the finite-dimensional is enough to ensure convergence the basis set being considered as fixed. However, when one aims at evaluating the impact of the growing size of the basis set (up to infinity), it is useful to have at hand the result for the infinite dimension. We claim that these results hold, refer to CANCÈS and LE BRIS [2000a] for the details, and only mention their finite-dimensional versions for brevity.

Various strategies for reaching SCF convergence are studied in the sequel. They can be classified in three groups:

(1) Fixed point iterations on the Euler–Lagrange equations;
(2) Direct minimization methods;
(3) Relaxed constrained algorithms.

The methods of the first group consist in using a fixed point algorithm to find out a solution to the HF equations (22.1). The *aufbau* principle is taken into account in order to try and get a minimizer of the HF energy. As stated in Section 22, any spinless HF, UHF, or GHF minimizer satisfies the *aufbau* principle; on the other hand, the same property is only conjectured for RHF minimizers. Fixed point iterations have long been the methods of reference. They are unfortunately built on unsafe foundations since they all proceed from the Roothaan algorithm, which is known to lead to convergence failures in many cases. The Roothaan algorithm is the simplest fixed point algorithm one can think of. It is a natural idea that dates back to ROOTHAAN [1951]. It is now obsolete, and not used any longer in computations (and we shall indeed explain below why), but as it serves for a basis for more sophisticated algorithms, it nevertheless deserves our attention; its study is the purpose of Paragraph (A) of Section 29. We shall see that this algorithm either converges or oscillates between two states which are not solutions to (22.1). Basic attempts to stabilize the Roothaan algorithm include the *level-shifting* algorithm (SAUNDERS and HILLIER [1973]) (see Section 29) and simple mixing techniques. Although they cure some theoretical drawbacks of the Roothaan algorithm, these algorithms are not yet extremely efficient, for they are very slow and too often converge toward critical points which are only local minima or even, as far as the level-shifting algorithm is concerned, saddle points. Therefore, with a view to both enforcing and accelerating the convergence of the iterations, the *Direct Inversion in the Iterated Subspace* (DIIS) algorithm has been introduced by PULAY [1982]. It is still universally used in nowadays calculations. The basic idea of the algorithm is to take benefit of the characterization (22.5) of the solutions to systems such as (22.4) in order to insert damping into the iterations. It will be overviewed in Paragraph (C) of Section 29. It turns out that the DIIS algorithm works extremely well: In many cases, it typically converges in a dozen of iterations, although there is *no theoretical reason* why it should even converge. However, the DIIS algorithm suffers from a qualitative drawback: it is not ensured that the Hartree–Fock energy decreases throughout the iterations. In practice, there exist cases where this algorithm does not converge. Such cases are not numerous for molecular systems containing first and second-row atoms only, but are much more numerous for systems containing atoms with d or f electrons (like transition metals for instance). We believe both observations justify to consider alternative strategies.

Direct minimization methods, when properly implemented, converge whatever the initial guess. Unfortunately the risk to converge toward "the closest" local (nonglobal) minimum is very high so that they are not appropriate for performing the first steps of the optimization procedure. On the other hand, they are the methods of choice to get quadratic convergence once in the neighborhood of (what is hoped to be) a global minimum. The most popular direct minimization methods are Newton-like methods. They actually provide quadratic convergence, which dramatically reduces the number of SCF steps; unfortunately, as we shall see below, each Newton-like SCF step may be very costly. Several quasi-Newton methods have also been suggested and tested. The description of such algorithms is the matter of Section 30.

Relaxed constrained algorithms (RCA) have been introduced in CANCÈS and LE BRIS [2000b], CANCÈS [2001a], CANCÈS [2001b]. They are proven to converge in the UHF setting and also in the less commonly used GHF setting; no such rigorous result has been proved for the RHF model but all the calculations performed to date show convergence. Paragraph (B) of Section 30 is devoted to the description and the tests of this new family of algorithms. In particular, these algorithms ensure the decreasing of the Hartree–Fock energy throughout the iterations, contrary to DIIS. In addition, although one can only prove that the obtained solution is a local minimizer of the HF energy, numerical tests show that

- the solution obtained by RCA is always the same whatever the initial guess chosen in the list of commonly used initial guesses (core Hamiltonian diagonalization, optimum of a semi-empirical method, ...); this very important robustness property is not fulfilled by the SCF algorithms of the first two groups;
- the solution obtained by RCA is always at least as good as the solution obtained by any other method (the energy at convergence is the lowest one observed).

RCA available today do not converge as fast as the DIIS algorithm when the latter does converge, but on the other hand they always converge. It is in fact likely that the best way to obtain very efficient SCF algorithms is to combine several of the elementary strategies described above. A step forward in this direction is detailed in KUDIN, SCUSERIA and CANCÈS [2002]. The numerical analysis we now give for the sake of consistency is reproduced from CANCÈS and LE BRIS [2000a], CANCÈS and LE BRIS [2000b], CANCÈS [2001a], and CANCÈS [2001b].

Before we begin our analysis of the algorithms, we need to make a few remarks and give some definitions. We shall see that all the SCF algorithms presented below, except the direct minimization techniques of Section 30, can be recast in the following form

$$\begin{cases} \widetilde{F}_k C_{k+1} = S C_{k+1} E_{k+1}, \quad E_{k+1} = \mathrm{Diag}\big(\varepsilon_1^{k+1}, \dots, \varepsilon_N^{k+1}\big), \\ C_{k+1}^* S C_{k+1} = I_N, \\ D_{k+1} = C_{k+1} C_{k+1}^*, \end{cases} \tag{28.1}$$

where $\varepsilon_1^{k+1} \leqslant \varepsilon_2^{k+1} \leqslant \cdots \leqslant \varepsilon_N^{k+1}$ are the smallest N eigenvalues of the linear generalized eigenvalue problem

$$\widetilde{F}_k \phi = \varepsilon\, S\phi,$$

and where C_{k+1} collects N orthonormal eigenvectors associated with $\varepsilon_1^{k+1}, \varepsilon_2^{k+1}, \ldots, \varepsilon_N^{k+1}$. The expression of the current Fock matrix $\widetilde{F}_k$ characterizes the algorithm. We have for instance

- $\widetilde{F}_k = F(D_k)$ for the Roothaan algorithm (Section 29);
- $\widetilde{F}_k = F(D_k) - bD_k$ where b is a positive constant for the level-shifting algorithm (Section 29);
- $\widetilde{F}_k$ is a linear combination of $F(D_0), F(D_1), \ldots, F(D_k)$ for the DIIS algorithm (Section 30);
- $\widetilde{F}_k = F(\widetilde{D}_k)$ for the optimal damping algorithm (Section 30), where $\widetilde{D}_k$ is a pseudo-density matrix which satisfies the *relaxed* constraints $\widetilde{D}_k^2 \leqslant \widetilde{D}_k$ and is defined so that the HF energy $E^{\mathrm{HF}}(\widetilde{D}_k)$ decreases at each iteration.

The hope is that C_k, D_k and $\widetilde{F}_k$ converge, respectively to C, D and $F(D)$, so that we get from (28.1) a solution to (22.1) in the limit $k \to +\infty$.

By analogy with the terminology defined in Section 22, the procedure consisting in assembling the matrix $D_{k+1} \in \mathcal{P}_N$ by populating the N molecular orbitals of lowest energy of the current Fock matrix $\widetilde{F}_k$ is referred to as the *aufbau principle*. It is justified by the results stated in Section 22. A necessary and sufficient condition for the matrix D_{k+1} being defined in a unique way is that $\varepsilon_N^{k+1} < \varepsilon_{N+1}^{k+1}$. Degeneracies in the spectrum are in general related to the symmetries of the system: In the cases when the system does not exhibit any symmetry, numerical experiments show that the eigenvalues of $\widetilde{F}_k$ are generically nondegenerate for any k, whereas it may not be the case when the system does exhibit symmetries (consider for instance the spherical symmetry of the Hamiltonian in the atomic case). Degeneracies create technical difficulties which complicate the theoretical studies on SCF convergence. For the sake of simplicity, we therefore assume that the *uniform well-posedness* (UWP) property introduced in CANCÈS and LE BRIS [2000a] is satisfied:

UWP PROPERTY. *A SCF algorithm of the form* (28.1) *with initial guess* D_0 *will be said to be uniformly well-posed if there exists some positive constant* γ *such that*

$$\forall k \in \mathbb{N},\ \varepsilon_{N+1}^{k+1} \geqslant \varepsilon_N^{k+1} + \gamma.$$

Let us point out that some convergence results can be obtained without resorting to the UWP assumption. In particular, it turns out that the level-shifting algorithm is automatically UWP as soon as the shift parameter is large enough.

Before turning to the study of SCF algorithms, the notion of convergence has to be made precise. All the algorithms under examination below consist in building from an initial guess D_0 a sequence of density matrices (D_k) which hopefully converges toward a solution D of the HF problem. To date, we are in fact not able to prove mathematical convergence results of the form "the sequence (D_k) generated by the algorithm converges toward a minimizer D of the HF problem (20.5)" for at least two reasons. The first one is a consequence of the strategy itself: We are solving the Euler–Lagrange equations associated with the HF minimization problem, namely the HF equations (22.1); even in case of convergence we have no argument to conclude that the so-obtained critical point is actually a minimum (even local) of the HF energy. The second one is technical and therefore is likely to be overcome in the future: We have no precise description of the topology of the set of the critical points of (20.5); this lack of information prevents us from proving the convergence of the whole sequence (D_k) toward a solution D to the HF equations. We can at best obtain that $D_{k+1} - D_k$ goes to zero, that for "large" k, that D_k is "close to" a solution to the HF equations (20.5) satisfying the *aufbau* principle, and that a subsequence converges to a such a solution of the HF equations. For instance, it may happen that the HF problem admits a connected manifold of minima; this phenomenon is observed in particular for open-shell atoms because the spherical symmetry of the problem is broken by the HF approximation (this can be related to the mathematical result of BACH, LIEB, LOSS and SOLOVEJ [1994] stating that "there are no unfilled shell" in the HF ground states). We cannot then discriminate between the case when the sequence (D_k) converges toward a point of the manifold and the case when the sequence (D_k) is attracted by the manifold together with a slow drift parallel to the manifold.

We shall consequently adopt here the following two convergence criteria, which are sufficient in practice. We shall say that a SCF algorithm of the form (28.1) *numerically converges toward a solution to the HF equations* if the sequence (D_k) satisfies

(1) $D_{k+1} - D_k \to 0$;

(2) $[F(D_k), D_k] \to 0$;

with as above $[A, B] = ABS - SBA$, and that it *numerically converges toward an aufbau solution to the HF equations* if the sequence (D_k) satisfies

(1′) $D_{k+1} - D_k \to 0$;

(2′) $\mathrm{Trace}(F(D_k)D_k) - \inf\{\mathrm{Trace}(F(D_k)D),\ D \in \mathcal{P}_N\} \to 0$.

These convergence criteria are directly related to the characterizations (22.5) and (22.6) of solutions and *aufbau* solutions to problems such as (22.4), respectively. As all norms are equivalent in finite dimension, we do not need to specify the matrix norm. Let us remark that (2′) implies (2).

29. The SCF cycles in the Hartree–Fock setting: A variety of classical strategies

(A) Roothaan fixed-point algorithm

The Roothaan algorithm (also called *simple SCF* or *pure SCF* or *conventional SCF* in the literature), that has been introduced in ROOTHAAN [1951], consists in generating a sequence (D_k^{Rth}) in $\mathcal{P}_N$ satisfying

$$\begin{cases} F(D_k^{\mathrm{Rth}})C_{k+1} = SC_{k+1}E_{k+1}, \quad E_{k+1} = \mathrm{Diag}(\varepsilon_1^{k+1}, \ldots, \varepsilon_N^{k+1}), \\ C_{k+1}^* SC_{k+1} = I_N, \\ D_{k+1}^{\mathrm{Rth}} = C_{k+1}C_{k+1}^*, \end{cases}$$

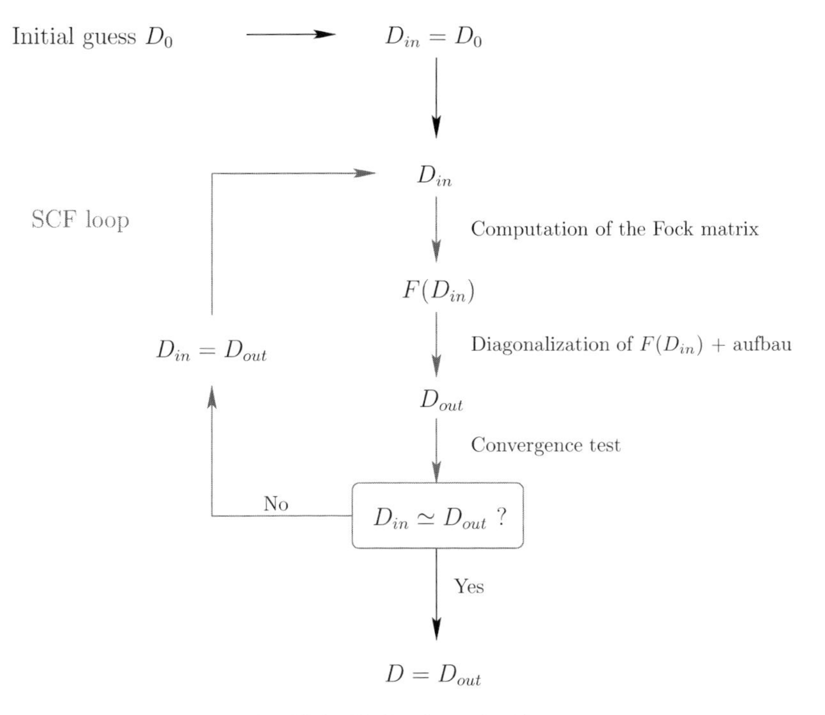

FIG. 29.1. The Roothaan algorithm.

where $\varepsilon_1^{k+1} \leqslant \varepsilon_2^{k+1} \leqslant \cdots \leqslant \varepsilon_N^{k+1}$ are the smallest N eigenvalues of the linear generalized eigenvalue problem

$$F(D_k^{\mathrm{Rth}})\phi = \varepsilon S\phi$$

and where the $N_b \times N$ matrix C_{k+1} collects N orthonormal eigenvectors of $F(D_k^{\mathrm{Rth}})$ associated with $\varepsilon_1^{k+1}, \varepsilon_2^{k+1}, \ldots, \varepsilon_N^{k+1}$. The iteration procedure of the Roothaan algorithm can therefore be summarized by the diagram of Fig. 29.1.

The convergence properties of the Roothaan algorithm are not satisfactory: although the Roothaan algorithm sometimes numerically converges toward a solution to the HF equations, it frequently oscillates between two states, none of them being solution to the HF equations:

$$D_{k+2}^{\mathrm{Rth}} - D_k^{\mathrm{Rth}} \to 0, \quad \text{but } D_{k+1}^{\mathrm{Rth}} - D_k^{\mathrm{Rth}} \not\to 0.$$

The behavior of the Roothaan algorithm can be explained by introducing the auxiliary function

$$E(D, D') = \mathrm{Trace}(hD) + \mathrm{Trace}(hD') + \mathrm{Trace}(G(D)D'),$$

(the notations are defined in Section 20), which is symmetric since $\mathrm{Trace}(G(D)D') = \mathrm{Trace}(G(D')D)$, and which satisfies $E(D, D) = 2E^{\mathrm{HF}}(D)$. Let us indeed minimize E

alternatively with respect to each of the two arguments D and D':

$$\begin{aligned}
&D_1 = \arg\inf\{E(D_0, D),\ D \in \mathcal{P}_N\},\\
&D_2 = \arg\inf\{E(D, D_1),\ D \in \mathcal{P}_N\},\\
&D_3 = \arg\inf\{E(D_2, D),\ D \in \mathcal{P}_N\},\\
&\ldots
\end{aligned}$$

This minimization procedure is usually called *relaxation* or *alternative direction minimization* in the mathematical literature, and should not be confused with relaxed constrained algorithms, that we shall see below. For the first two steps, we obtain

$$\begin{aligned}
D_1 &= \arg\inf\{E(D_0, D),\ D \in \mathcal{P}_N\}\\
&= \arg\inf\{\mathrm{Trace}(hD_0) + \mathrm{Trace}(hD) + \mathrm{Trace}\big(G(D_0)D\big),\ D \in \mathcal{P}_N\}\\
&= \arg\inf\{\mathrm{Trace}\big(F(D_0)D\big),\ D \in \mathcal{P}_N\}\\
&= D_1^{\mathrm{Rth}},
\end{aligned}$$

and, since E is symmetric on $\mathcal{P}_N \times \mathcal{P}_N$,

$$\begin{aligned}
D_2 &= \arg\inf\{E\big(D, D_1^{\mathrm{Rth}}\big),\ D \in \mathcal{P}_N\}\\
&= \arg\inf\{E\big(D_1^{\mathrm{Rth}}, D\big),\ D \in \mathcal{P}_N\}\\
&= \arg\inf\{\mathrm{Trace}(hD) + \mathrm{Trace}\big(hD_1^{\mathrm{Rth}}\big) + \mathrm{Trace}\big(G\big(D_1^{\mathrm{Rth}}\big)D\big),\ D \in \mathcal{P}_N\}\\
&= \arg\inf\{\mathrm{Trace}\big(F\big(D_1^{\mathrm{Rth}}\big)D\big),\ D \in \mathcal{P}_N\}\\
&= D_2^{\mathrm{Rth}}.
\end{aligned}$$

It follows by induction that the sequences generated by the relaxation algorithm on the one hand, and by the Roothaan algorithm on the other hand, are the same. The functional E, which decreases at each iteration of the relaxation procedure can therefore be interpreted as a Lyapunov functional of the Roothaan algorithm. This basic remark is the foundation of the proof of the following result.

THEOREM 29.1. *Let $D_0 \in \mathcal{P}_N$ such that the Roothaan algorithm with initial guess D_0 is UWP. Then the sequence (D_k^{Rth}) generated by the Roothaan algorithm satisfies one of the following two properties*

- *either (D_k^{Rth}) numerically converges toward an aufbau solution to the HF equations*
- *or (D_k^{Rth}) numerically oscillates between two states, none of them being an aufbau solution to the HF equations.*

PROOF. For the sake of simplicity, we assume here that $S = I_{N_b}$. For any $k \in \mathbb{N}$, we deduce from inequality (22.7) that

$$\mathrm{Trace}\big(F\big(D_{k+1}^{\mathrm{Rth}}\big)D_{k+2}^{\mathrm{Rth}}\big) + \frac{\gamma}{2}\big\|D_{k+2}^{\mathrm{Rth}} - D_k^{\mathrm{Rth}}\big\|^2 \leqslant \mathrm{Trace}\big(F\big(D_{k+1}^{\mathrm{Rth}}\big)D_k^{\mathrm{Rth}}\big).$$

Adding $\operatorname{Trace}(hD_{k+1}^{\mathrm{Rth}})$ to both terms of the above inequality, we obtain

$$E\big(D_{k+1}^{\mathrm{Rth}}, D_{k+2}^{\mathrm{Rth}}\big) + \frac{\gamma}{2}\big\|D_{k+2}^{\mathrm{Rth}} - D_k^{\mathrm{Rth}}\big\|^2 \leqslant E\big(D_k^{\mathrm{Rth}}, D_{k+1}^{\mathrm{Rth}}\big).$$

We then sum up the above inequalities for $k \in \mathbb{N}$ and we get $\sum_{k\in\mathbb{N}} \|D_{k+2}^{\mathrm{Rth}} - D_k^{\mathrm{Rth}}\|^2 < +\infty$, which involves in particular that

$$D_{k+2}^{\mathrm{Rth}} - D_k^{\mathrm{Rth}} \to 0.$$

Now, either $D_{k+1}^{\mathrm{Rth}} - D_k^{\mathrm{Rth}}$ converges to zero or it does not. In the former case, we deduce from the characterization of D_{k+1}^{Rth} by

$$\operatorname{Trace}\big(F\big(D_k^{\mathrm{Rth}}\big)D_{k+1}^{\mathrm{Rth}}\big) = \inf\big\{\operatorname{Tr}\big(F\big(D_k^{\mathrm{Rth}}\big)D\big),\ D \in \mathcal{P}_N\big\}$$

that

$$\operatorname{Trace}\big(F\big(D_k^{\mathrm{Rth}}\big)D_k^{\mathrm{Rth}}\big) - \inf\big\{\operatorname{Tr}\big(F\big(D_k^{\mathrm{Rth}}\big)D\big),\ D \in \mathcal{P}_N\big\} \to 0.$$

Convergence toward an *aufbau* solution to the HF equations is thus established. In the latter case

$$\begin{aligned}
&\operatorname{Trace}\big(F\big(D_{2k}^{\mathrm{Rth}}\big)D_{2k}^{\mathrm{Rth}}\big) - \inf\big\{\operatorname{Tr}\big(F\big(D_{2k}^{\mathrm{Rth}}\big)D\big),\ D \in \mathcal{P}_N\big\}\\
&\quad = \operatorname{Tr}\big(F\big(D_{2k}^{\mathrm{Rth}}\big)D_{2k}^{\mathrm{Rth}}\big) - \operatorname{Trace}\big(F\big(D_{2k}^{\mathrm{Rth}}\big)D_{2k+1}^{\mathrm{Rth}}\big)\\
&\quad \geqslant \frac{\gamma}{2}\big\|D_{2k}^{\mathrm{Rth}} - D_{2k+1}^{\mathrm{Rth}}\big\|^2 \not\to 0.
\end{aligned}$$

Convergence of (D_{2k}) toward an *aufbau* solution to the HF equations is therefore excluded; the same argument holds for (D_{2k+1}). □

Mimicking the proof of Theorem 29.1, it is easy to establish in addition that (D_{2k}, D_{2k+1}) converges up to an extraction to a critical point $(D, D') \in \mathcal{P}_N \times \mathcal{P}_N$ of the functional E, which satisfies

$$\begin{cases}
F(D')C = SCE,\\
C^*C = I_N,\\
D = CC^*,\\
F(D)C' = SC'E',\\
C'^*C' = I_N,\\
D' = C'C'^*,
\end{cases}$$

where E and E' are diagonal matrices collecting the smallest N eigenvalues of the generalized eigenvalue problems $F(D')\phi = \varepsilon S\phi$ and $F(D)\phi = \varepsilon S\phi$, respectively. Besides, as $E(D_{2k}, D_{2k+1})$ is decreasing, *the whole* sequence (D_{2k}, D_{2k+1}) converges to (D, D') if this critical point is a strict (local) minimum. In this case, the alternatives are

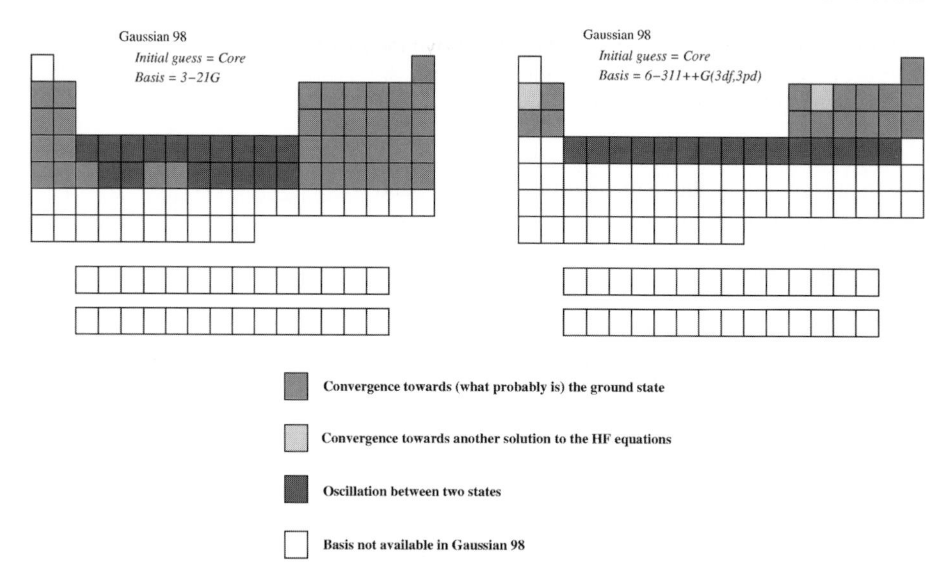

FIG. 29.2. Searching the ground state of some atoms with the Roothaan algorithm. Results are shown on the periodic table of the elements. The original figure was published in E. Cancès, SCF algorithms for HF electronic calculations, Fig. 2.2, p. 31, in: Mathematical Models and Methods for Ab Initio Quantum Chemistry, Lecture Notes in Chemistry, Vol. 74, M. Defranceschi and C. Le Bris Eds., Springer-Verlag (Berlin, Heidelberg) 2000.

- either (D, D') is on the diagonal of $\mathcal{P}_N \times \mathcal{P}_N$ (i.e., $D = D'$) and (D_k^{Rth}) converges toward an *aufbau* solution to the HF equations;
- or (D, D') is not on the diagonal of $\mathcal{P}_N \times \mathcal{P}_N$ (i.e., $D \neq D'$) and (D_k^{Rth}) oscillates between two states which are not *aufbau* solutions to the HF equations.

Oscillations can be observed even for simple chemical systems. As a matter of example, we have tested the Roothaan algorithm in the UHF setting for the atoms of the periodic table and for two sets of atomic orbitals, namely the Gaussian basis sets 3-21G and 6-311++G(3df,3pd) (see GAUSSIAN [1998] user's guide). The initial guess is obtained by diagonalization of the core Hamiltonian. Calculations have been performed with Gaussian 98 (GAUSSIAN [1998]). The results are reported on Fig. 29.2; they indicate that

(1) Both alternatives (convergence *and* oscillation) are met in practice.
(2) Convergence toward a critical point of the HF problem which is not a global minimum can sometimes be observed.
(3) For the same system, we can get convergence for one basis set and oscillation for another basis set.

(B) Level shifting algorithm

In view of the analysis of the Roothaan algorithm, one can guess that convergence can be obtained by adding to the functional E a penalty term E_p which prevents critical points of $E + E_p$ from being located out of the diagonal $\{(D, D),\ D \in \mathcal{P}_N\}$.

It turns out that the very simple penalty functional $E_p = b\|D - D'\|^2$, where b is a positive constant and where $\|\cdot\|$ denotes as above the Hilbert–Schmidt norm, perfectly suits. Let us indeed define

$$E^b(D, D') = \mathrm{Trace}(hD) + \mathrm{Trace}(hD') + \mathrm{Trace}\big(G(D)D'\big) + b\big\|D - D'\big\|^2.$$

The relaxation algorithm associated with the minimization problem

$$\inf\big\{E^b(D, D'),\ (D, D') \in \mathcal{P}_N \times \mathcal{P}_N\big\}$$

generates the sequence (D_k^b) defined by (28.1) with

$$\widetilde{F}_k = F\big(D_k^b\big) - bD_k^b.$$

The sequence (D_k^b) is in fact the sequence generated by the level-shifting algorithm with shift parameter b. This algorithm, introduced in SAUNDERS and HILLIER [1973], on the basis of completely different arguments, has satisfactory theoretical convergence properties:

THEOREM 29.2. *There exists a positive constant b_0 such that for any $D_0 \in \mathcal{P}_N$ and for any level-shift parameter $b \geqslant b_0$,*

(1) *the level-shifting algorithm with shift parameter b is UWP;*
(2) *the sequence of the energies $E^{\mathrm{HF}}(D_k^b)$ decreases toward some stationary value $\mathcal{E}$ of E^{HF};*
(3) *the sequence (D_k^b) numerically converges toward a solution to the HF equations.*

The proof of this theorem can be read in CANCÈS [2001b].

Unfortunately, the level-shift parameters $b \geqslant b_0$ which guarantee convergence are large so that convergence is very slow; in addition, many numerical observations show that the level-shifting algorithm often converges toward solutions to the HF equation which do not satisfy the *aufbau* principle: At convergence, the density matrix D is by construction the projector on the lowest N orbitals of some *shifted* operator $F(D) - bD$; these orbitals are also orbitals of the Fock matrix $F(D)$, but not necessarily the lowest N ones. This means that the energy of the so-obtained critical point of the HF energy functional is necessarily above that of the HF ground state.

(C) DIIS algorithm

The most commonly used fixed point method is the *Direct Inversion in the Iterative Subspace* (DIIS) algorithm (PULAY [1982]). This algorithm turns out to be very efficient in most cases, even if no theoretical argument can provide a rigorous explanation for that. It consists in keeping in memory all (or some of) the density matrices computed at the previous steps and in choosing the updated density matrix in the vector subspace generated by all the density matrices stored in memory, in such a way that some criterion is minimized. In the DIIS algorithm, the criterion

$$\left\| \sum_{i=0}^{k+1} c_i \big[F(D_i), D_i\big] \right\|^2,$$

is minimized under the constraint $\sum_{i=0}^{k+1} c_i = 1$. As above, the bracket $[\cdot,\cdot]$ denotes the "commutator" $[A, B] = ABS - SBA$; the norm $\|\cdot\|$ is the Frobenius (or Hilbert–Schmidt) norm ($\|A\| = \mathrm{Trace}(AA^*)^{1/2}$). It is natural to try to zero the commutator since the condition $[F(D), D] = 0$ is equivalent to the HF equations (see Eq. (22.5)). However, as already pointed out, minimizing the latter criterion does not always force convergence. The DIIS algorithm finally reads

$$\begin{cases} F(\widetilde{D}_k)C_{k+1} = SC_{k+1}E_{k+1}, \quad E_{k+1} = \mathrm{Diag}(\varepsilon_1^{k+1}, \ldots, \varepsilon_N^{k+1}), \\ C_{k+1}^* SC_{k+1} = I_{N_p}, \\ D_{k+1} = C_{k+1}C_{k+1}^*, \\ \widetilde{D}_{k+1} = \displaystyle\sum_{i=0}^{k+1} c_i^{\mathrm{opt}} D_i, \quad \{c_i^{\mathrm{opt}}\} = \arg\inf\left\{ \left\| \sum_{i=0}^{k+1} c_i\big[F(D_i), D_i\big] \right\|^2, \sum_{i=0}^{k+1} c_i = 1 \right\}, \end{cases}$$

where $\varepsilon_1^{k+1} \leqslant \varepsilon_2^{k+1} \leqslant \cdots \leqslant \varepsilon_N^{k+1}$ are the smallest N eigenvalues of the linear generalized eigenvalue problem

$$F(\widetilde{D}_k)\phi = \varepsilon S\phi.$$

Despite its nice performances, there are some cases when the DIIS algorithm does not converge; difficulties raise up in particular when the molecular system contains atoms of the third raw and beyond with d or f electrons. In that case, other SCF algorithms have to be resorted to.

30. The SCF cycles in the Hartree–Fock setting: Some new strategies

(A) Direct minimization techniques

Direct minimization techniques consist in adapting some general optimization schemes briefly reviewed in Sections 26 and 27 to the HF setting.

The first Newton-like algorithm for computing HF ground states appeared in the early 60s with Bacskay's quadratic convergent (QC) method (BACSKAY [1961]). The basic idea was to make a change of variable to remove the constraints in order to be able to use a standard Newton algorithm for *unconstrained* optimization (see Section 26). The convenient parametrization of the manifold $\mathcal{P}_N$ used by Bacskay is the following: For any $C \in \mathcal{M}(N_b, N_b)$ such that $C^* S C = I_{N_b}$,

$$\mathcal{P}_N = \left\{ C \exp(A) D_0 \exp(-A) C^*, \ D_0 = \begin{bmatrix} I_N & 0 \\ 0 & 0 \end{bmatrix}, \right.$$
$$\left. A = \begin{bmatrix} 0 & -A_{\mathrm{vo}}^* \\ A_{\mathrm{vo}} & 0 \end{bmatrix}, \ A_{\mathrm{vo}} \in \mathcal{M}(N_b - N, N) \right\};$$

the suffix *vo* denotes the "*virtual-occupied*" off-diagonal block of the matrix A. Let us now denote by

$$E^C(A_{\mathrm{vo}}) = E^{\mathrm{HF}}\big(C \exp(A) D_0 \exp(-A) C^*\big).$$

The problem becomes now to minimize $E^C(A_{\mathrm{vo}})$. Starting from some reference matrix C, Bacskay QC algorithm proposes to apply to this unconstrained minimization problem, *one* Newton step starting from $A_{\mathrm{vo}} = 0$, then to update the reference matrix C by setting $C = C \exp A$ with

$$A = \begin{bmatrix} 0 & -A_{\mathrm{vo}}^* \\ A_{\mathrm{vo}} & 0 \end{bmatrix},$$

where A_{vo} denotes the result of the Newton step.

Denoting by $\nabla^2 E^C(0)$ the Hessian of E^C at $A_{\mathrm{vo}} = 0$, Bacskay QC algorithm therefore reads

$$\begin{cases} \text{Compute the solution } A_{\mathrm{vo}}^k \text{ of the Newton equation} \\ \nabla^2 E^{C_k}(0) \cdot A_{\mathrm{vo}} + \nabla E^{C_k}(0) = 0, \\ \text{Set } C_{k+1} = C_k \exp(A_k) \quad \text{with } A_k = \begin{bmatrix} 0 & -A_{\mathrm{vo}}^{k*} \\ A_{\mathrm{vo}}^k & 0 \end{bmatrix}. \end{cases}$$

Let us remark that $\nabla E^{C_k}(0) \in \mathcal{M}(N_b - N, N)$ ($N_b - N$ denotes the number of virtual orbitals) is the off-diagonal left-bottom block of the "commutator" $[F(D_k), D_k]$, with $D_k = C_k C_k^*$, in the basis of the molecular orbitals.

A natural alternative to Bacskay QC is to use Newton-like algorithm for *constrained* optimization (see Section 27) in order to directly attack problem (20.4). The Lagrangian of the constrained minimization problem (20.4) is defined by

$$\begin{aligned} \mathcal{L}(C, \Lambda) &= E^{\mathrm{HF}}\big(CC^*\big) - \big(C^* S C - I_N, \Lambda\big)_{\mathcal{K}} \\ &= \mathrm{Trace}\big(hCC^*\big) + \frac{1}{2}\mathrm{Trace}\big(G\big(CC^*\big)CC^*\big) - \mathrm{Trace}\big(\big(C^* S C - I_N\big)\Lambda\big). \end{aligned}$$

For a given iterate (C_k, Λ_k), the operators $L_k = \nabla^2_{CC}\mathcal{L}(C_k, \Lambda_k)$, $B_k = g'(C_k)$ and $B_k^T = g'(C_k)^T$ introduced in Section 27 are then defined by

$$\begin{aligned} &\forall W \in \mathcal{M}(N_b, N), \quad L_k W = F_k W + G\big(D_W^{C_k}\big) C_k - S W \Lambda_k, \\ &\forall W \in \mathcal{M}(N_b, N), \quad B_k W = C_k^* S W + W S C_k^*, \\ &\forall M \in \mathcal{M}_S(N), \quad B_k^T M = 2 S C_k M, \end{aligned}$$

where (C_k, Λ_k) denotes the current iterate, $F_k = F(D_k)$ with $D_k = C_k C_k^*$ and $D_{W_k}^{C_k} = W_k C_k^* + C_k W_k^*$.

The Newton step to solve the Euler–Lagrange equations (22.1), reads

$$\begin{cases} \text{Search } (W_k, M_k) \in \mathcal{M}(N_b, N) \times \mathcal{M}_S(N) \text{ such that} \\ F_k W_k + G\big(D_{W_k}^{C_k}\big) C_k - S W_k \Lambda_k - S C_k M_k + F_k C_k - S C_k \Lambda_k = 0, \\ C_k^* S W_k + W_k^* S C_k + C_k^* S C_k - I_N = 0, \end{cases} \tag{30.1}$$

where (C_k, Λ_k) denotes the current iterate, $F_k = F(D_k)$ with $D_k = C_k C_k^*$ and $D_{W_k}^{C_k} = W_k C_k^* + C_k W_k^*$. Unfortunately, due to the unitary invariance (21.6), the operator L_k (using the notations of Section 26) is not positive definite on V_k. More precisely,

- we already know that at convergence, the restriction of L_k to V_k is only positive semi-definite and its kernel is the set $\{W = C_k A,\ A \in \mathcal{M}_A(N)\}$;
- in the neighborhood of a solution, the restriction of L_k to V_k is even not positive in general. Its spectrum is made of $N(N_b - N)$ positive eigenvalues and $N(N-1)/2$ eigenvalues very close to zero which can either be positive or negative.

It is however possible to remedy for the difficulty encountered in the Newton method by modifying the algorithm in order to drop out the unitary invariance (21.6), see, e.g., SHEPARD [1993a]. To begin with, let us notice that the matrix

$$\eta_k = \frac{1}{2} C_k \big(\Sigma_k^{-1} - I_N\big), \tag{30.2}$$

with $\Sigma_k = C_k^* S C_k$, satisfies $C_k^* S \eta_k + \eta_k^* S C_k + C_k^* S C_k - I_N = 0$. Let us then define

$$Z_k = F_k C_k - S C_k \Lambda_k + F_k \eta_k + G\big(D_{\eta_k}^{C_k}\big) C_k - S \eta_k \Lambda_k, \tag{30.3}$$

and let us consider the following system

$$\begin{cases} \text{Search } \big(W_k', M_k'\big) \in \mathcal{M}(N_b, N) \times \mathcal{M}(N) \text{ such that} \\ F_k W_k' + G\big(D_{W_k'}^{C_k}\big) C_k - S W_k' \Lambda_k - S C_k M_k' + Z_k = 0, \\ C_k^* S W_k' = 0, \end{cases} \tag{30.4}$$

where Z_k is given by formula (30.3). It is not difficult to prove that the above problem has a unique solution (W_k', M_k) and it is straightforward to see that (W_k, M_k) with $W_k = W_k' + \eta_k$ is solution to (30.1). Finally, a possible second-order algorithm reads:

- *Step* 0 (initialization). Choose an initial guess (C_0, Λ_0) and set $k = 0$.
- *Step* 1. Compute $D_k = C_k C_k^*$ and $F_k = F(D_k)$. If the residues $F_k C_k - S C_k \Lambda_k$ and $C_k^* S C_k - I_N$ are "small enough" then stop else continue.
- *Step* 2. Compute $\Sigma_k = C_k^* S C_k$, Σ_k^{-1}, η_k and Z_k given by (30.2) and (30.3), respectively.
- *Step* 3. Compute the solution (W_k', M_k') of problem (30.4) by a range–space method in which the linear system is solved by conjugated gradient; for that use the projector P_k, its adjoint P_k^* and the left-hand side inverse of $B_k'^T$ defined for any

$X \in \mathcal{M}(N_b, N)$ by

$$P_k \cdot X = \left(I - C_k \Sigma_k^{-1} C_k^* S\right) X, \qquad P_k^* \cdot X = \left(I - S C_k \Sigma_k^{-1} C_k^*\right) X,$$
$$Q_k^l \cdot X = \Sigma_k^{-1} C_k^* X.$$

- *Step* 4. Set $C_{k+1} = C_k + \eta_k + W_k'$ and $\Lambda_{k+1} = \Lambda_k + M_k'$.
- *Step* 5. Set $k = k + 1$ and go to Step 1.

The leak point of the above algorithm is that the matrix Σ_k^{-1} is needed. Fortunately, the matrix Σ_k is close to identity (C_k is assumed to be close to a local minimum) so that the matrix Σ_k^{-1} can be well approximated

- either by $\Sigma_k^{-1} \simeq 2 - \Sigma_k$,
- or by the result of a few Hotelling's iterations $\Sigma_k^{-1} \simeq X_{n_0}$ where the matrices X_n are defined by $X_0 = I$ and $X_{n+1} = 2X_n - X_n \Sigma_k X_n$.

Another possibility to bypass the computation of Σ_k^{-1} is to modify the algorithm in such a way that $\Sigma_k = C_k^* S C_k = I_N$ at each step. This can be done as follows

- *Step* 0 (initialization). Choose an initial guess $C_0 \in \mathcal{W}$ and set $k = 0$.
- *Step* 1. Compute $D_k = C_k C_k^*$, $F_k = F(D_k)$ and $\Lambda_k = C_k^* F_k C_k$. If the residue $F_k C_k - S C_k \Lambda_k$ is "small enough" then stop else
- *Step* 2. Compute $P_k = I - D_k S$ and search by deflated conjugated gradient the solution $W_k \in \mathcal{M}(N_b, N)$ to the linear system

$$\begin{cases} P_k^*\left(F_k P_k W_k + G\left(D_{W_k}^{C_k}\right) C_k - S W_k \Lambda_k\right) = -P_k^*(F_k C_k - S C_k \Lambda_k), \\ W_k \in \operatorname{Ran} P_k. \end{cases}$$

- *Step* 3. Compute $M_k = (I - W_k^* S W_k)^{1/2}$ and $C_{k+1} = C_k M_k + W_k$.
- *Step* 4. Set $k = k + 1$ and go to Step 2.

Here again, the matrix $M_k = (I - W_k^* S W_k)^{1/2}$ is close to identity and can therefore be approximated by $M_k \simeq I - \frac{1}{2} W_k^* S W_k$.

The computational costs of the various Newton type algorithms described above are very high for applying the HF Hessian to some vector W consists in assembling the bi-electronic part $G(D_W^{C_k})$ of the Fock matrix $F(D_W^{C_k})$ for the matrix $D_W^{C_k} = C_k W^* + W C_k^*$. One Fock matrix construction is thus needed for each step of the conjugated gradient algorithm which is itself the inner loop of the Newton algorithm. Various attempts have been made to build quasi-Newton versions of Bacskay QC algorithm (see for instance FISCHER and ALMLÖF [1992], CHABAN, SCHMIDT and GORDON [1997]). We are not aware of any work on quasi-Newton methods for solving the constrained optimization problem (20.4), but we are convinced that this is a promising track.

(B) Relaxed constraint algorithms

RCA are based on the following remarkable property of the HF energy functional: All the local minima of $E^{\mathrm{HF}}(D)$ in

$$\widetilde{\mathcal{P}}_N = \left\{D \in \mathcal{M}_S(N_b),\ DSD \leqslant D,\ \operatorname{Trace}(SD) = N\right\}$$

are in fact on

$$\mathcal{P}_N = \{D \in \mathcal{M}_S(N_b),\ DSD = D,\ \mathrm{Trace}(SD) = N\}.$$

The proof of this statement relies upon simple arguments that we outline here. Any critical point of

$$\inf\{E^{\mathrm{HF}}(D),\ D \in \widetilde{\mathcal{P}}_N\} \tag{30.5}$$

satisfies Euler–Lagrange equations similar to (22.3′). Assume now that the critical point does not belong to $\mathcal{P}_N$ (i.e., does not satisfy the constraints $DSD = D$); in that case, at least two orbitals $\Phi \in \mathbb{R}^{N_b}$ and $\Phi' \in \mathbb{R}^{N_b}$ associated to the same eigenvalue ε_{F} of the Fock matrix $F(D)$, are only partially occupied (their occupation numbers n and n' are in the range $]0, 1[$). It is easy to see that transferring δn electrons from Φ to Φ' yields the density matrix

$$\widetilde{D} = D + \delta n\big(|\Phi'\rangle\langle\Phi'| - |\Phi\rangle\langle\Phi|\big)$$

which belongs to the minimization set $\widetilde{\mathcal{P}}_N$ if δn is small enough, and which generates a variation of the energy equal to

$$\Delta E = E(\widetilde{D}) - E(D) = -\frac{\delta n^2}{2}\int_{\mathbb{R}^3}\int_{\mathbb{R}^3}\frac{|\phi(x)\phi'(y) - \phi(y)\phi'(x)|^2}{|x-y|}\,dx\,dy < 0,$$

where $\phi(x) = \sum_{\mu=1}^{N_b} \Phi_\mu \chi_\mu(x)$ and $\phi'(x) = \sum_{\mu=1}^{N_b} \Phi'_\mu \chi_\mu(x)$. Therefore any local minimizer of problem (30.5) is in fact a local minimizer of the HF model (20.5).

RCA aim at searching a minimizer to problem (30.5) by direct minimization techniques; due to the property stated above, this minimizer is a solution to (20.5). Problem (30.5) is easier to attack with direct minimization algorithm than problem (20.5) for the set $\widetilde{\mathcal{P}}_N$ is convex. With convex variational sets indeed, mixing or iterative subspace methods are based on a sound ground since any convex combination of previously computed iterates still is in the variational set.

Let us now present the simplest RCA, called the *optimal damping algorithm* (ODA). For the sake of simplicity, we detail here the spinless HF version of the algorithm. The ODA is defined by the following iterative procedure

$$\begin{cases} F(\widetilde{D}_k)C_{k+1} = SC_{k+1}E_{k+1}, \quad E_{k+1} = \mathrm{Diag}(\varepsilon_1^{k+1}, \ldots, \varepsilon_N^{k+1}), \\ C_{k+1}^* SC_{k+1} = I_{N_p}, \\ D_{k+1} = C_{k+1}C_{k+1}^*, \\ \widetilde{D}_{k+1} = \arg\inf\{E^{\mathrm{HF}}(\widetilde{D}),\ \widetilde{D} \in \mathrm{Seg}[\widetilde{D}_k, D_{k+1}]\}, \end{cases}$$

where $\mathrm{Seg}[\widetilde{D}_k, D_{k+1}] = \{(1-\lambda)\widetilde{D}_k + \lambda D_{k+1},\ \lambda \in [0,1]\}$ denotes the line segment linking $\widetilde{D}_k$ and D_{k+1}. As E^{HF} is a second degree polynomial in the density matrix, this step consists in minimizing a quadratic function of λ in $[0, 1]$, what can be done analytically. The procedure is initialized with $\widetilde{D}_0 = D_0$, $D_0 \in \mathcal{P}_N$ being the initial guess.

The ODA thus generates two sequences of matrices:

- The main sequence of density matrices $(D_k)_{k\in\mathbb{N}}$ which will be proven to numerically converge toward an *aufbau* solution to the HF equations;
- A secondary sequence $(\widetilde{D}_k)_{k\geqslant 1}$ of pseudo-density matrices which belong to $\widetilde{\mathcal{P}}_N$.

The latter statement is a direct consequence of the convexity of $\widetilde{\mathcal{P}}_N$. The properties of the ODA are collected in the following theorem, which is valid for the GHF and the UHF models (in the latter case D_k denotes the pair (D_k^α, D_k^β)). It is not proven for the RHF model but as far as we know, no numerical counterexample has been found out yet.

THEOREM 30.1. *Let us consider an initial guess $D_0 \in \mathcal{P}_N$ such that the optimal damping algorithm is UWP. Then,*

(1) *the sequence $E^{\mathrm{HF}}(\widetilde{D}_k)$ decreases toward a stationary value of the HF energy;*
(2) *the sequence (D_k) numerically converges toward an* aufbau *solution to the HF equations.*

The reason why the ODA works can be understood as follows: let us first consider $\widetilde{D}_k \in \widetilde{\mathcal{P}}_N$ and $D' \in \mathcal{P}_N$, and let us compute the variation of the HF energy on the line segment

$$\mathrm{Seg}\big[\widetilde{D}_k, D'\big] = \big\{(1-\lambda)\widetilde{D}_k + \lambda D',\ \lambda \in [0,1]\big\}.$$

We obtain for any $\lambda \in [0,1]$,

$$E^{\mathrm{HF}}\big((1-\lambda)\widetilde{D}_k + \lambda D'\big) = E^{\mathrm{HF}}\big(\widetilde{D}_k\big) + \lambda\,\mathrm{Tr}\big(F\big(\widetilde{D}_k\big)\cdot\big(D' - \widetilde{D}_k\big)\big) + \frac{\lambda^2}{2}\mathrm{Tr}\big(G\big(D' - \widetilde{D}_k\big)\cdot\big(D' - \widetilde{D}_k\big)\big).$$

The "steepest descent" direction, i.e., the density matrix D for which the slope $s_{\widetilde{D}_k\to D} = \mathrm{Tr}(F(\widetilde{D}_k)\cdot(D - \widetilde{D}_k))$ is minimum, is given by the solution to the minimization problem

$$D = \arg\inf\big\{\mathrm{Tr}\big(F\big(\widetilde{D}_k\big)\big(D' - \widetilde{D}_k\big)\big),\ D' \in \mathcal{P}_N\big\},$$

which also reads

$$D = \arg\inf\big\{\mathrm{Tr}\big(F\big(\widetilde{D}_k\big)D'\big),\ D' \in \mathcal{P}_N\big\}.$$

This is precisely the direction D_{k+1} selected by the *aufbau* principle. As $E^{\mathrm{HF}}(\widetilde{D}_k)$ decreases and as $\widetilde{\mathcal{P}}_N$ is compact, the sequence $(\widetilde{D}_k)$ is forced to converge toward a critical point of E^{HF}, necessarily located on $\mathcal{P}_N$. This is the essence of the proof of the above theorem.

The practical implementation of the ODA is detailed in CANCÈS and LE BRIS [2000b], where numerical tests are also reported. The cost of one ODA iteration is approximatively the same as that of one iteration of the Roothaan algorithm. Numerical tests performed so far demonstrate that the ODA is particularly efficient in the early steps of the iteration procedure.

An attempt of improvement of the ODA for the latest steps of the procedure consists, in the spirit of iterative subspace methods, in keeping in memory all (or some of) the density matrices computed at the previous steps and in minimizing the HF energy in the convex set generated by all the density matrices stored in memory; the following procedure by CANCÈS and LE BRIS [2000b], KUDIN, SCUSERIA and CANCÈS [2002] is referred to as the EDIIS algorithm, for *Energy* Direct Inversion in the Iterative Subspace:

$$\begin{cases} F(\widetilde{D}_k)C_{k+1} = SC_{k+1}E_{k+1}, \\ C_{k+1}^* SC_{k+1} = I_{N_p}, \\ D_{k+1} = C_{k+1}C_{k+1}^*, \\ \widetilde{D}_{k+1} = \arg\inf\left\{ E^{\mathrm{HF}}(\widetilde{D}),\ \widetilde{D} = \sum_{i=0}^{k+1} c_i D_i,\ 0 \leqslant c_i \leqslant 1,\ \sum_{i=0}^{k+1} c_i = 1 \right\}. \end{cases}$$

This algorithm is similar to the DIIS algorithm, except that in the DIIS algorithm, the coefficients $\{c_i^{\mathrm{opt}}\}$ minimize the residual

$$\left\| \sum_{i=0}^{k+1} c_i \big[F(D_i), D_i \big] \right\|^2$$

under the constraint $\sum_{i=0}^{k+1} c_i = 1$. Minimizing the latter residual does not force convergence, whereas minimizing the energy (as in EDIIS) does.

PROOF OF THEOREM 30.1. Let us denote by $\widetilde{F}_k = F(\widetilde{D}_k)$ and by $s_{k+1} = \mathrm{Trace}(\widetilde{F}_k \times (D_{k+1} - \widetilde{D}_k))$. In view of inequality (22.7),

$$s_{k+1} = \mathrm{Trace}\big(\widetilde{F}_k D_{k+1}\big) - \mathrm{Trace}\big(\widetilde{F}_k \widetilde{D}_k\big) \leqslant -\frac{\gamma}{2} \| D_{k+1} - \widetilde{D}_k \|^2.$$

Let us denote by b_0 a positive constant such that

$$\forall (\widetilde{D}, \widetilde{D}') \in \mathcal{M}_S(N_b) \times \mathcal{M}_S(N_b), \\ \mathrm{Trace}\big(G(\widetilde{D}' - \widetilde{D}) \cdot (\widetilde{D}' - \widetilde{D})\big) \leqslant b_0 \| \widetilde{D} - \widetilde{D}' \|^2.$$

For any $\lambda \in [0, 1]$,

$$E^{\mathrm{HF}}\big((1-\lambda)\widetilde{D}_k + \lambda D_{k+1}\big) \\ \leqslant E^{\mathrm{HF}}(\widetilde{D}_k) - \frac{\gamma}{2} \| D_{k+1} - \widetilde{D}_k \|^2 \lambda + \frac{b_0}{2} \| D_{k+1} - \widetilde{D}_k \|^2 \lambda^2.$$

Therefore

$$
\begin{aligned}
&E^{\mathrm{HF}}(\widetilde{D}_{k+1})\\
&\quad = \inf\{E^{\mathrm{HF}}((1-\lambda)\widetilde{D}_k + \lambda D_{k+1}),\ \lambda \in [0,1]\}\\
&\quad \leqslant \inf\left\{E^{\mathrm{HF}}(\widetilde{D}_k) - \frac{\gamma}{2}\|D_{k+1} - \widetilde{D}_k\|^2\lambda + \frac{b_0}{2}\|D_{k+1} - \widetilde{D}_k\|^2\lambda^2,\ \lambda \in [0,1]\right\}\\
&\quad = E^{\mathrm{HF}}(\widetilde{D}_k) - \alpha\|D_{k+1} - \widetilde{D}_k\|^2
\end{aligned}
$$

with $\alpha = \gamma^2/8b_0$ if $\gamma \leqslant 2b_0$, $\alpha = (\gamma - b_0)/2$ otherwise. We then add up the above inequalities for $k \in \mathbb{N}$, and we get $\sum \|D_{k+1} - \widetilde{D}_k\|^2 < +\infty$, which implies that

$$
D_{k+1} - \widetilde{D}_k \to 0. \tag{30.6}
$$

As $\widetilde{D}_{k+1} \in [\widetilde{D}_k, D_{k+1}]$, it follows that

$$
\widetilde{D}_{k+1} - \widetilde{D}_k \to 0,
$$

and then that

$$
D_{k+1} - D_k \to 0.
$$

Besides

$$
\mathrm{Trace}(F(\widetilde{D}_k)D_{k+1}) = \inf\{\mathrm{Tr}(F(\widetilde{D}_k)D),\ D \in \mathcal{P}_N\}. \tag{30.7}
$$

Putting together (30.6) and (30.7), we finally obtain

$$
\mathrm{Trace}(F(D_{k+1})D_{k+1}) - \inf\{\mathrm{Trace}(F(D_{k+1})D),\ D \in \mathcal{P}_N\} \to 0. \qquad \square
$$

Let us make a remark on the standard mixing algorithm defined by

$$
\begin{cases}
F(\widetilde{D}_k)C_{k+1} = SC_{k+1}E_{k+1}, \quad E_{k+1} = \mathrm{Diag}(\varepsilon_1^{k+1}, \ldots, \varepsilon_N^{k+1}),\\
C_{k+1}^* SC_{k+1} = I_{N_p},\\
D_{k+1} = C_{k+1}C_{k+1}^*,\\
\widetilde{D}_{k+1} = (1-\alpha)\widetilde{D}_k + \alpha D_{k+1},
\end{cases}
$$

where α is a fixed damping parameter. In view of the above analysis, the standard mixing algorithm can be viewed as a steepest descent procedure with a fixed step. This explains both its robustness (it converges if the parameter α is small enough), and its relatively poor efficiency. In practice, the parameter α is often chosen equal to 0.25, which is not enough to force convergence.

Let us end this section by briefly mentioning the state of the art for the SCF cycles in the Kohn–Sham setting. Basically the discussion related to fixed point iterations and direct minimization can be applied to the KS setting of CANCÈS [2001a], KUDIN, SCUSERIA and CANCÈS [2002]; the only difference is that we do not have yet a mathematical proof of the results contained in Theorem 29.1 in this framework (for any standard exchange-correlation functional), although this theorem actually seems to perfectly describe what is observed in the numerical practice.

On the other hand, the situation is radically different as far as RCA are concerned. The reason is that relaxing the constraints $DSD = D$ in the KS model leads to the *extended* Kohn–Sham model. RCA thus provide solutions to the extended Kohn–Sham equations. In many cases, the KS orbitals obtained at convergence are in fact either fully occupied or empty, so that one recovers a solution to the KS equations. On the other hand, some numerical examples demonstrate that it is not always the case.

31. Rate of convergence of expansion in a Gaussian basis set

Having examined the full solution procedure for a given basis, we now go into the analysis of the impact of the choice of the basis. The general framework we have is the following: We search for the solution Φ_0 of a given minimization problem set on a space X. When a space X_δ of finite dimension is fixed, with a view to approximate the space X, we may formally split the error between the exact solution of the problem set in infinite dimension and the solution Φ_δ found numerically into two components.

One component of this error comes from the fact that a space of finite dimension cannot in whole generality contain any arbitrary function of the infinite-dimensional space. The best one can hope is, instead of the exact function Φ_0, obtain the function in X_δ that is as close as possible to Φ_0. This latter function, denoted $\pi_\delta \Phi_0$ here, is called the best approximation of Φ_0.

The second component of the error is due to the fact that what we obtain when running the algorithm is not $\pi_\delta \Phi_0$ itself, but only an approximation of it, namely the result Φ_δ of the algorithm. Therefore one can (formally) estimate the global error as follows:

$$\|\Phi_0 - \Phi_\delta\| \leqslant \|\Phi_0 - \pi_\delta \Phi_0\| + \|\pi_\delta \Phi_0 - \Phi_\delta\|.$$

Clearly the first component of the error only depends on the quality of the approximation space X_δ, that is to say of the basis set of X_δ, while the second component depends on the quality of the algorithm itself (an issue which has been examined in Sections 28–30). We shall concentrate on the first component in the present section, at least in the simple case of one-electron systems, postponing the many electron case to Section 32. Both Sections 31 and 32 are written in a style easily accessible for a reader familiar with the literature of chemistry. Sections 33 and 34 then focus on more detailed mathematical aspects.

For evaluating the quality of the basis set, we come back to the problem of finding the solution of the exact Schrödinger equation (in fact for the hydrogen-like case).

Let us therefore consider a *family of basis sets*, with each set characterized by its dimension d, and let us assume that there is a prescription to construct all members of this family. Let us further assume that for $d \to \infty$ we arrive at a *complete* basis, with the meaning of *complete* yet to be specified. We will, of course, like that in the limit $d \to \infty$ the considered matrix eigenvalues converge to the corresponding eigenvalues of the Schrödinger equation.

This has, for a long time been taken for granted. Only KLAHN and BINGEL [1977] investigated the conditions for the convergence of the sequence of matrix eigenvalues to the exact eigenvalues. They found that completeness *in the L^2 Lebesgue space* is not sufficient, but that rather completeness in the *first Sobolev space H^1* is required. From the mathematical standpoint, such a result could indeed be expected, as the H^1 norm is the right norm for the continuity, and further the coercitivity, of the energy functional.

These authors were also able to show, that the basis sets used conventionally, in particular *Gaussian* basis sets fulfill this completeness requirement, provided that some, rather weak precautions are taken. So convergence of conventional basis expansions is guaranteed, although a numerical extrapolation to $d \to \infty$ may turn out to be rather difficult, in particular if the convergence is *slow*. This is a hint that the *rate of convergence* is of more practical importance than just convergence. Actually, even more important that the rate of convergence is in practice the ratio between the convergence and the complexity of the computations.

If a one-electron basis $\{\chi_p\}_d$ of *spin-orbitals* is given, then the set of all n-electron Slater determinants constructable from the χ_p is a *basis* for an n-electron system. The expansion of the n-electron Hamiltonian in this basis leads to what is usually called *full CI*, where we recall that CI stands for *configuration interaction*. Extrapolation to $d \to \infty$, i.e., from *full CI* to *complete CI*, which is hardly feasible in practice, leads to the exact result under the same condition as for one-electron systems, i.e., for completeness of the basis in H^1.

To the attention of the reader who is not familiar with the mathematical issues, let us now briefly give a flavor of the questions of approximation examined below by considering very naive situations.

Let us begin with the approximation of real numbers. Nobody would seriously evaluate $\ln(2)$ from the series $\ln(1 + x) = 1 - \frac{1}{2}x + \frac{1}{3}x^2 + \cdots$. The error due to truncating the series after n terms is of $O(\frac{1}{n})$, and $n \approx 10^6$ is needed to get $\ln(2)$ with 6-figure accuracy. An *inverse-power law* for the truncation error means usually *poor convergence*. Much better is the (geometric) series $2 = 1 + \frac{1}{2} + \frac{1}{4} + \frac{1}{8} + \cdots$. Here the truncation error $(\frac{1}{2})^{n+1}$ decreases *exponentially* with the number n of terms included. Generally exponential convergence is desirable, but it cannot always be achieved. There is a third type of convergence behavior, that appears to gain importance, namely with a truncation error that goes as $f - f_n \sim A\exp(-b\sqrt{n})$ and that will play an important role below. After the real numbers, let us now turn to approximations of functions. One of the most popular basis expansions is realized in the Fourier series. Let $f(x) = f(x + 2\pi)$ be a periodic function. Then the expansion $f_n(x) = f_0 + \sum_{k=1}^{n} a_k \cos(kx) + \sum_{k=1}^{n} a_k \sin(kx)$ converges in the L^2_{per}-norm, provided that f is L^2_{per}. However the convergence is *exponential*, i.e., the truncation error goes as $\|f - f_n\| \sim A\exp(-cn)$ *only if* $f(x)$ and all of its derivatives are continuous functions. If some derivative is discontinuous, there is only an inverse-power law convergence $\|f - f_n\| \sim Bn^{-k}$, with k determined by the order of the first discontinuous derivative (MORGAN III [1984]). The rate of convergence of a Fourier-type expansion depends sensitively on the *singularities* of the function to be expanded. One can usually speed up the convergence, if one complements the basis by one or more functions (called *comparison functions* in HILL [1985]), that contain the "correct" singularity (thus making the basis formally "*overcomplete*"). An alternative is, as we shall see, to care for an expansion that is *less* sensitive to the singularities.

Let us now come back to the context of quantum chemistry. We begin by considering explicitly atomic one-electron systems, leaving the more difficult many-electron case for Section 32 below. In addition, we consider essentially the *ground state of the hydrogen atom*, but the results are valid for molecular systems as well, and even for approximate solutions of n-electron systems in terms of effective one-electron Hamiltonians, such as in Hartree–Fock theory. In all these situations the critical aspect, as far as convergence of a basis expansion is concerned, is the *representation of the nuclear cusp*, in terms of a basis of functions that are analytic at the position of a nucleus. From a strict mathematical standpoint, we may describe the situation as follows. Suppose the singularity of the exact wavefunction Φ_0 is not properly reproduced by its approximation Φ_δ. Then $\|\Phi_0 - \Phi_\delta\|_{L^\infty}$ is large, thus so is $\|\Phi_0 - \Phi_\delta\|_{H^{3/2+\varepsilon}}$ (in space dimension 3). Thus, using an inverse inequality, $\|\Phi_0 - \Phi_\delta\|_{H^1}$ is not small.

It is convenient to have a look at three types of Gaussian basis sets (BOYS [1950])

(a) $\tilde{\psi}_{nlm} = N r^{n-1} e^{-\eta_l r^2} Y_l^m(\vartheta, \varphi)$; $n - l - 1 = 0, 2, 4, 6, \ldots$;

(b) as (a) but $n - l - 1 = 0, 1, 2, 3, 4, \ldots$;

(c) $\psi_{klm} = N r^l e^{-\eta(l,k) r^2} Y_l^m(\vartheta, \varphi)$.

In order to test our methodology, we want to expand the (normalized) wavefunction ψ of the ground state of the hydrogen atom, or rather of hydrogen-like ions with nuclear charge Z

$$\psi = Z^{-\frac{3}{2}} \pi^{-\frac{1}{2}} \exp(-Zr)$$

in any of the three basis sets, noting that all these Gaussian basis functions are unable to describe both the correct ψ near the nucleus (no cusp) and far from a nucleus (too fast decay). All three basis sets are *complete* (in the first Sobolev space) in the limit of an infinite basis (for case (c) if the $\eta(l, k)$ increase sufficiently fast with k), but the rate of convergence is quite different.

KLAHN and MORGAN III [1984] have studied the convergence of expansions of the ground state of the H atom in a basis of type (a) i.e., a basis of 1s, 3s, 5s etc. Gaussians with the same exponential factor. This is equivalent to a basis of Hermite functions. The result was frustrating. The error of the energy goes as $\sim d^{-3/2}$ if d is the dimension of the basis. The rate of convergence can somewhat be improved, if one optimizes the factor η in the exponent as function of the dimension d (KLOPPER and KUTZELNIGG [1986], HILL [1995]). The conjecture in KLOPPER and KUTZELNIGG [1986] that the error goes as d^{-2}, could not be confirmed, and the asymptotic behavior is rather complicated (HILL [1995]).

If one adds a 2s Gaussian function ($n = 2$), the error is reduced to $\sim d^{-3}$, and it is further diminished if one includes 4s, 6s functions etc. For basis set (b) the convergence becomes exponential and this is the best that one can get. This is rather noteworthy, since basis (a) is complete and (b) *overcomplete*.

Why is the basis with just even powers of r in front of $\exp(-\alpha r^2) r^l Y_l^m(\vartheta, \varphi)$ so much poorer than the basis with even and odd powers (although both basis sets are complete)? The reason is – as first pointed out in SCHWARTZ [1962] in a different context and worked out in detail in HILL [1985] based on the work by KLAHN and MORGAN III [1984] – that exponential (and hence fast) convergence is only possible (at least heuristically) if the basis set describes the *singularities* of the function to be expanded.

Obviously the correct ground state wavefunction of the H-atom has a cusp at $r = 0$ (KATO [1980]), while Gaussians of type (a) and (c) have a vanishing first derivative at $r = 0$. Applying arguments very similar to those current in the theory of Fourier series (MORGAN III [1984]), KLAHN and MORGAN III [1984] were able to demonstrate the origin of the slow convergence in terms of basis (a). Basis (b) describes the nuclear cusp correctly. However, the molecular integrals for the functions of basis (b) are *considerably more complicated* than for the sets (a) or (c), that basis (b) is of no practical interest.

Obviously basis set (c) behaves quite differently. The convergence turned out – surprisingly – to be effectively much better than for basis sets (a), although not as fast as exponential. There were some numerical studies on the convergence, e.g., rather extensive ones in SCHMIDT and RUEDENBERG [1979], FELLER and RUEDENBERG [1979], but these did not lead to very clear-cut conclusions. The great merit of the works by SCHMIDT and RUEDENBERG [1979], FELLER and RUEDENBERG [1979] has been the introduction of even-tempered basis sets. For these the orbital exponents $\eta(l,k)$ are constructed as

$$\eta(l,k) = \alpha(d,l)\beta(d,l)^{k-1}, \tag{31.1}$$

i.e., for each *angular quantum number* l and each dimension d, the basis is characterized by only two parameters α and β. Such basis sets perform a little less well than basis sets with individually optimized orbital coefficients $\eta(l,k)$ (HUZINAGA [1965]), but are surprisingly good in view of their compactness. KLOPPER and KUTZELNIGG [1986] studied the error ε of the energy of the H atom ground state with Huzinaga basis sets of dimension d and found an excellent numerical fit of the form $\varepsilon = A\exp(-b\sqrt{d})$.

(A) Preliminary on the wavefunction

In order to evaluate the quality of the approximation in a given basis set of the exact ground state function of the H-atom let us start from the following identity

$$e^{-\alpha r} = \frac{\alpha}{2\sqrt{\pi}} \int_0^\infty s^{-3/2} \exp\left\{-\frac{\alpha^2}{4s} - sr^2\right\} ds \tag{31.2}$$

which expresses an exponential function as a *Gaussian integral transform* (SHAVITT and KARPLUS [1962]).

To arrive at a basis expansion of the exponential function, the first step is to restrict the integration domain in (31.2) from s_1 to s_2 (KUTZELNIGG [1994], KUTZELNIGG [1989b]). This gives rise to two cut off errors

$$\varepsilon_{c1}(r) = \int_0^{s_1} f(s,r)\,ds, \tag{31.3}$$

$$\varepsilon_{c2}(r) = \int_{s_2}^{\infty} f(s,r)\,ds. \tag{31.4}$$

In the second step one transforms the integration variable s to another variable q, and in a final step the integral over q is replaced by a sum over an equidistant grid. The variable transformation is chosen so that the grid points sample the integrand in an effective way. We shall, in fact, choose such a transformation that the grid points correspond to an *even-tempered basis*. From now on n denotes the number of grid points.

$$\int_{s_1}^{s_2} f(s,r)\,ds = \int_{q(s_1)}^{q(s_2)} f\big(s[q],r\big)\frac{ds}{dq}\,dq = \int_{q_1}^{q_2} g(q,r)\,dq \approx h\sum_{k=1}^{n} g(q_k,r),$$

$$q_k = q(s_1) + \left(k - \frac{1}{2}\right)h; \quad h = \frac{q(s_2) - q(s_1)}{n}; \quad s(q) = \exp\left(\frac{q}{2}\right).$$

The total error in the wavefunction is then the sum of the discretization error $\varepsilon_d(r)$ and of the two cut-off errors $\varepsilon_{c1}(r)$ and $\varepsilon_{c2}(r)$. We define the error as the difference between the exact and the approximate results:

$$\varepsilon_d(r) = \int_{s_1}^{s_2} f(s,r)\,ds - h\sum_{k=1}^{n} g(q_k,r).$$

In order to make $\varepsilon_{c1}(r)$ and $\varepsilon_{c2}(r)$ as small as possible, one should make s_1 as small and s_2 as large as possible. For the discretization error ε_d, one expects that it becomes smaller as the interval length h becomes smaller. For a given number n of points a big h reduces the cut-off errors, and a small h reduces the discretization error. The optimum h will then be some compromise.

We want to measure the error by a single number rather than by a function of r. There are various possibilities, e.g.:

(a) The mean square error, i.e., the distance of the expansion to the exact function in a Hilbert space norm.
(b) The maximum error in a Chebyshev norm.
(c) The error of the energy expectation value.
(d) The variance of the energy expectation value.
(e) The error of the density at the position of the nucleus.

In view of the relation to variational calculations the option (c) looks particularly promising, as expected in such a context and even if we have seen so that for (e) is also a useful guide.

Before we go into details we note that for any of these norms one finds that the sum of the two cut-off errors goes asymptotically as $\varepsilon_c \sim A(hn)^k \exp(-hn)$ where h is the interval length, n the number of intervals, and A and k are constants. The discretization error, i.e., the error of using a trapezoid formula has the asymptotic behavior $\varepsilon_d \sim Bh^{-l}\exp(-\frac{b}{h})$ with B, l, and b constants. As expected, a small h is good for ε_d, while a large h makes ε_c small. The best compromise is achieved for $h \sim \sqrt{b/n}$; $\varepsilon \sim Dn^m \exp(-\sqrt{bn})$. While the estimation of the truncation error is trivial, that of the trapezoid approximation looks somewhat unorthodox. It does not at all remind the traditional asymptotic expansion of the error (STOER [1972], KUTZELNIGG [1994]), based on the Taylor expansion in each interval (KUTZELNIGG [1994], KUTZELNIGG [1989b]).

$$\varepsilon_d = -\frac{h^2}{24}\left[g'(q_2) - g'(q_1)\right] + \frac{7h^4}{5760}\left[g^{(3)}(q_2) - g^{(3)}(q_1)\right] + \mathrm{O}\left(h^6\right). \tag{31.5}$$

Of course this error estimate is only valid if the integrand allows a power series expansion in h, but this is obviously not the case here. The discretization error has even an *essential singularity* at $h = 0$. On the other hand an error quadratic in h would be unacceptably poor anyway. The special kind of a bell-shaped integrand decaying very fast both at $-\infty$ and $+\infty$ allows a different and more powerful error estimate.

One arrives at a useful asymptotic expansion of the discretization error, if one formulates the discretization as a convolution with a periodic δ-function and then uses the Fourier expansion of the periodic δ-function. The result is (KUTZELNIGG [1994])

$$\varepsilon_d = -2\sum_{l=1}(-1)^l \int_{q_1}^{q_2} g(q)\cos\frac{2l\pi(q-q_1)}{h}\,dq. \tag{31.6}$$

For the cases that we are interested in, the term with $l = 1$ dominates (KUTZELNIGG [1994]) and determines the leading term in the asymptotic expansion

$$\varepsilon_d \sim 2\int_{q_1}^{q_2} g(q)\cos\frac{2\pi(q-q_1)}{h}\,dq. \tag{31.7}$$

One can actually extend the integration domain to the full range (usually $-\infty < q < \infty$), i.e., consider

$$\varepsilon_{d\infty} = 2\int_{-\infty}^{\infty} g(q)\cos\frac{2\pi(q-q_1)}{h}\,dq. \tag{31.8}$$

On one hand this often simplifies the evaluation of the integral, and moreover in this way we have completely separated cut-off and discretization errors. The 'coupling' between the two errors, i.e., the difference between (31.6) and (31.8), in other words the cut-off error of the discretization error, is usually of higher order and negligible.

The discretization error $\varepsilon_{d\infty}$ corresponds formally to replacing the integral over an infinite domain by a (countable) infinite sum of terms. There is one problem with $\varepsilon_{d\infty}$.

For finite values of q_1 and q_2 the division of the integration domain into intervals of length h is unique. This is no longer the case if both integration limits are infinite. Then the 'phase' $\eta = 2\pi q_1$ becomes indefinite and there is no unique limit for $q_1 \to -\infty$. An indefinite phase can be avoided, if one places one grid point at the maximum of the integrand (provided it has a single maximum). We shall see that the expression (31.8) is useful, nevertheless. The indefinite phase has, on one hand, some numerical reality. On the other hand, it is possible to average over the indefinite phase in some simple and meaningful way.

(B) Error on the energy expectation value

We want to minimize the error ε of the energy expectation value for the hydrogen atom

$$\varepsilon = \int f(r)\{H - E_0\} f(r) r^2\, dr, \tag{31.9}$$

where $H = -\frac{1}{2}\Delta - \frac{Z}{r}$, $E_0 = -\frac{1}{2}Z^2$, and $f(r)$ is an approximation to

$$\psi(r) = 2Z^{3/2} e^{-Zr} = \int_0^\infty \phi(t,r)\, dr, \quad \phi(t,r) = \frac{2}{\sqrt{Z\pi}} t^{-2} \exp\left(-\frac{Z^2}{4t^2} - r^2 t^2\right).$$

The cut-off error of the expectation value is

$$\begin{aligned} \varepsilon_c &= \int_{t_1}^{t_2} dt \int_{t_1}^{t_2} ds \int_0^\infty dr\, r^2 \phi(t,r)\big(\widehat{H} - E_0\big)\phi(s,r) \\ &= \int_{t_1}^{t_2} dt \int_{t_1}^{t_2} ds\, F(t,s). \end{aligned} \tag{31.10}$$

The integration over r can be performed in closed form

$$\begin{aligned} F(t,s) &= \int_0^\infty dr\, r^2 \Phi(t,r)\big(\widehat{H} - E_0\big)\Phi(s,r) \\ &= \frac{\exp\left(-\frac{Z^2(s^2+t^2)}{4s^2t^2}\right)}{s^2t^2} \left\{ \frac{Z^7}{2\sqrt{\pi}} \frac{1}{(s^2+t^2)^{3/2}} - \frac{2Z^6}{\pi(s^2+t^2)} + \frac{3Z^5 s^2 t^2}{\sqrt{\pi}(s^2+t^2)^{5/2}} \right\}. \end{aligned} \tag{31.11}$$

The final integration is best performed in spherical polar coordinates

$$\begin{aligned} \tilde{t} &= \frac{1}{t} = p\cos\varphi = \frac{2u}{Z}\cos\varphi, \\ \tilde{q} &= \frac{1}{q} = p\sin\varphi = \frac{2u}{Z}\sin\varphi, \\ dt\, dq &= \tilde{t}^{-2}\tilde{q}^{-2}\, d\tilde{t}\, d\tilde{q}; \qquad d\tilde{t}\, d\tilde{q} = p\, dp\, d\varphi = \frac{4}{Z^2} u\, du\, d\varphi. \end{aligned} \tag{31.12}$$

$F(t, q)$ is then replaced by

$$G(u, \varphi) = -Z^2 e^{-u^2} \left\{ \frac{3u^2}{\sqrt{\pi}} \sin^3 2\varphi - \frac{8u^3}{\pi} \sin^2 2\varphi + \frac{2u^4}{\sqrt{\pi}} \sin^3 2\varphi \right\}. \tag{31.13}$$

Since

$$\int_0^\infty dt \int_0^\infty dq\, F(t, q) = \int_0^\infty dt \int_{t_1}^{t_2} dq\, F(t, q) = \int_{t_1}^{t_2} dt \int_0^\infty dq\, F(t, q) = 0,$$

the cut-off error (31.10) can be rewritten as

$$\begin{aligned} \varepsilon_c = &- \int_{t_1}^\infty dt \int_{t_1}^\infty dq\, F(t, q) - \int_{t_2}^\infty dt \int_{t_2}^\infty dq\, F(x, q) \\ &- \int_0^{t_1} dt \int_{t_2}^\infty dq\, F(t, q) - \int_{t_1}^\infty dt \int_{t_2}^\infty dq\, F(t, q) \end{aligned} \tag{31.14}$$

the first term in (31.14) can be regarded as the *lower cut-off error*

$$\varepsilon_{c1} = - \int_{t_1}^\infty dt \int_{t_1}^\infty dq\, F(t, q) \tag{31.15}$$

and the second term in (31.14) as the *upper cut-off error*

$$\varepsilon_{c2} = - \int_{t_2}^\infty dt \int_{t_2}^\infty dq\, F(t, q) \tag{31.16}$$

while the last two terms in (31.14) represent a coupling between upper and lower cut-off. These are small of higher order and do not contribute to the leading term in the asymptotic expansion of the cut-off error. One gets

$$\begin{aligned} \varepsilon_{c2} = &\frac{Z^5}{\sqrt{\pi}} \left(\frac{4}{3} - \frac{5\sqrt{2}}{6} \right) t_2^{-3} + Z^6 \left(\frac{1}{4} - \frac{1}{\pi} \right) t_2^{-4} \\ &+ \frac{Z^7}{\sqrt{\pi}} \frac{4 - 3\sqrt{2}}{20} t_2^{-5} + \mathrm{O}\left(t_2^{-6}\right), \end{aligned} \tag{31.17}$$

$$\varepsilon_{c1} = \frac{Z^3}{\sqrt{2\pi}} t_1^{-1} e^{-Z^2/2t_1^2} \left[1 + \mathrm{O}(t_1) \right]. \tag{31.18}$$

Note that the upper-cut-off error ε_{c2} is proportional to t_2^{-3}, while the lower cut-off error behaves as $\exp[-a/t_1^2]$. Hence, there is a much more sensitive dependence on the upper cut-off parameter t_2 than on the lower cut-off parameter t_1. For an even tempered basis t_1 and t_2 are related as $t_2^2 = t_1^2 e^{nh}$. This relation and the minimization $\varepsilon_c = \varepsilon_{c1} + \varepsilon_{c2}$ with respect to t_1 for fixed n and h lead to

$$t_1 = \frac{\gamma Z}{\sqrt{6hn}}\left\{1 + \mathrm{O}\left(\frac{\ln(hn)}{hn}\right)\right\}, \qquad \varepsilon_c = A(hn)^{\frac{3}{2}} e^{\frac{-3hn}{2}}\{1 + \mathrm{O}(hn)\}$$

with A a numerical constant.

The leading term of the discretization error is

$$\varepsilon_d = 4\int_0^\infty dt \int_0^\infty dq\, F(t,s)\cos\frac{2\pi g(t)}{h}\cos\frac{2\pi g(s)}{h}. \tag{31.19}$$

For an even-tempered basis with the mapping function

$$g(t) = 2\ln t \tag{31.20}$$

the discretization error (31.19) becomes

$$\begin{aligned}\varepsilon_d &= 2\int_0^\infty dt \int_0^\infty dq\, F(t,q)\cos\frac{4\pi \ln t + \eta}{h}\cos\frac{4\pi \ln q + \eta}{h}\\ &= \mathrm{Re}\int_0^\infty du \int_0^{\pi/2} d\varphi\, G(u,\varphi)\left\{\left(\frac{2u}{Z}\right)^{8\pi i/h}\left(\frac{\sin 2\varphi}{2}\right)^{4\pi i/h} e^{2i\eta/h}\right.\\ &\qquad\left. + (\tan\varphi)^{4\pi i/h}\right\}\end{aligned} \tag{31.21}$$

with $G(u,\varphi)$ defined by (31.13) and with η the indefinite phase mentioned previously.

There are two contributions to ε_d corresponding to the two terms in braces in the last expression in (31.21), both of which can be expressed in terms of the Gamma function for a complex variable. While the first of these contributions is a rapidly oscillating function of h, the second contribution depends monotonically on h. Fortunately the second one dominates, going as $h^{-3}\exp(-\frac{2\pi^2}{h})$ for small h, while the absolute value (rather than the real part) of the first contribution goes as $h^{-3/2}\exp(-\frac{2\pi^2}{h})$. So for the leading contribution up to $\mathrm{O}(h^{-2}\exp[-\frac{2\pi^2}{h}])$ we only need to consider the second contribution. For this one we get

$$\begin{aligned}\varepsilon_d \approx \varepsilon_{d2} &= 8Z^2 \operatorname{cosech}\left(\frac{2\pi^2}{h}\right)\left\{\frac{\pi^2}{2h} + \frac{2\pi^4}{h^3}\right\} - 4Z^2 \operatorname{sech}\left(\frac{2\pi^2}{h}\right)\left\{1 - \frac{4\pi^2}{h^2}\right\}\\ &= 8Z^2 e^{-2\pi^2/h}\left\{\frac{4\pi^4}{h^3} - \frac{8\pi^2}{h^2} + \frac{\pi^2}{h} - \frac{1}{2}\right\} + \mathrm{O}\left(e^{-6\pi^2/h}\right)\\ &= \frac{8\pi^4 Z^2}{h^3} e^{-2\pi^2/h}\left[1 + \mathrm{O}(h)\right].\end{aligned}$$

We can now minimize the total error $\varepsilon = \varepsilon_c + \varepsilon_d$ with respect to h and obtain the following optimal values

$$h = \frac{2\pi}{\sqrt{3n}}\left[1 + \mathrm{O}\left(\frac{\ln n}{\sqrt{n}}\right)\right], \tag{31.22}$$

$$\varepsilon = -\pi(3n)^{3/2}\exp\left\{-\pi\sqrt{3n}\left[1 + \mathrm{O}\left(n^{-1/2}\right)\right]\right\}, \tag{31.23}$$

$$t_1 = \frac{Z}{\sqrt{3nh}}\left[1 + \mathrm{O}(n)\right]. \tag{31.24}$$

The asymptotically (i.e., for large n) optimized parameters α and β of the even-tempered basis in the sense of (31.1) are

$$\beta = \exp h = \exp\left(2\pi/\sqrt{3n}\right), \tag{31.25}$$

$$\alpha = t_1^2\beta^{-1/2} = \frac{Z^2}{2\pi\sqrt{3n}}\exp\left(-\pi\sqrt{3n}\right). \tag{31.26}$$

In Table 31.1 the quantity $\frac{\ln|\varepsilon|}{\sqrt{n}}$ is displayed for various values of the basis dimension n, and ε obtained in different ways. Specifically, ε is chosen

(a) from the asymptotic formula (31.23),
(b) from an explicit application of the trapezoid formula to the integral (31.19) using t_1 and h as estimated by (31.24) and (31.22), respectively,
(c) from a brute force variational calculation using an even tempered basis with the asymptotically optimized parameters (31.25), (31.26),
(d) a variational calculation with the even-tempered basis sets optimized by SCHMIDT and RUEDENBERG [1979] individually for various small dimensions n and extrapolated numerically for large n,

TABLE 31.1

$\frac{\ln|\varepsilon|}{\sqrt{n}}$ for the H atom ground state; $\varepsilon = E_0 - \langle H\rangle$

n	(a)	(b)	(c)	(d)	(e)
10	−3.46607	−3.437	−3.53	−3.87	−4.28
20	−3.81214	−3.845	−3.96	−4.07	−4.65
30	−4.00008	−4.055	−4.17	−3.83	−4.84
40	−4.12495	−4.190	−4.27	−3.64	−5.02
50	−4.21659	−4.289	−4.38	−3.49	
∞	−5.4414				

(a) $\varepsilon = (3n)^{3/2}\pi\exp(-\pi\sqrt{3n})$;
(b) explicit discretization of the integral;
(c) variational calculation with the asymptotically exact even tempered basis;
(d) variational calculation with the Schmidt–Ruedenberg basis;
(e) variational calculation with the Morgan–Haywood basis. All variational calculations where done in 128-bit arithmetic.

(e) a variational calculation with the Morgan–Haywood basis, which is not of even-tempered type (see below).

One first observes the good agreement between the prediction of the error by means of (31.23) and the error that one gets from explicit use of the trapezoid approximation. Noting that (31.23) may also be written as

$$\varepsilon = \exp\left\{-\pi\sqrt{3n} + \ln\left[\pi 3^{3/2}\right] + \frac{3}{2}\ln n + \mathrm{O}\left(n^{-1/2}\right)\right\}$$

and hence

$$\frac{\ln|\varepsilon|}{\sqrt{n}} = \pi\sqrt{3} - \frac{\ln[\pi 3^{3/2}]}{\sqrt{n}} - \frac{3}{2}\frac{\ln n}{\sqrt{n}} + \mathrm{O}\left(n^{-1}\right),$$

one sees that $\frac{\ln n}{\sqrt{n}}$ is not negligible, for the n-value considered here. Even for $n = 50$ one is still far from the asymptotic value $-\pi\sqrt{3}$ for $\ln|\varepsilon|/\sqrt{n}$. Actually for n between $n = 10$ and $n = 50$, $\ln n/\sqrt{n}$ varies only between 0.73 and 0.55.

The errors of the results for the variational calculation with the asymptotically optimized basis sets are still rather close to those of an explicit discretization and to the estimate (31.23). In view of the $n - 1$ linear variational parameters, it is not astonishing that the errors of the variational calculation are somewhat smaller, but the effect of this additional flexibility is not very large. The comparison with the Schmidt–Ruedenberg basis reveals that the basis optimized individually for each n gains over the asymptotically optimized basis as long as n is small. However, for large n (30 to 50) the extrapolation from the Schmidt–Ruedenberg fit is rather poor.

We shall comment later on the errors obtained with the Morgan–Haywood basis.

(C) Error on the wavefunction, and other errors

We have optimized the parameters α and β of an even-tempered basis such that the error of the energy expectation value is minimized asymptotically for large n. There are other possible criteria. One can, e.g., try to minimize the distance in Hilbert space between the exact and the approximate (expanded) wavefunction. The integral to be approximated instead of (31.9) is then

$$\varepsilon = \int \left[f(r) - \psi(r)\right]^2 r^2\, dr.$$

For this quantity the cut-off-errors ε_{c1} and ε_{c2} behave as

$$\varepsilon_{c1} \sim t_1^{-1} e^{-Z^2/(2t_1^2)}, \tag{31.27}$$

$$\varepsilon_{c2} \sim t_2^{-4}, \tag{31.28}$$

instead of (31.17), (31.18). While the lower cut-off error has again a very weak dependence on t_1 as for the energy expectation value, the upper cut-off-error is less sensitive to t_2 as in the example of the energy expectation value (namely $\sim t_2^{-4}$ rather than $\sim t_2^{-3}$). As a consequence, the optimum t_1 is now $t_1 \sim Z/\sqrt{8hn}$, i.e., the factor 6 in (31.24) is now replaced by 8: The lower cut-off is somewhat reduced. For the interval h one now gets $h \sim \pi/\sqrt{n}$ to be compared with (31.22), i.e., h and hence β is reduced. Therefore also the highest exponent η_n is smaller than for minimization of the energy expectation value. Basis functions with large exponents are required less to minimize the distance in Hilbert space than to minimize the error of the energy.

Another possible criterion is to minimize the variance of the energy expectation value. Now the integral that we want to approximate is

$$\varepsilon = \int f(r)(H - E_0)^2 f(r) r^2 \, dr.$$

Instead of (31.16) we now get $\varepsilon_{c2} \sim t_2^{-1}$, i.e., the variance is extremely sensitive to the upper cut-off, i.e., to the inclusion of large orbital exponents. This means the optimum α and β are much larger than those needed to minimize the error of the energy.

One may even want to choose the basis such that certain expectation values, say of r^k are best approximated. This is a different situation from the three ones considered so far.

The error of the energy, the distance in Hilbert space, and the variance are all bounded. A consequence of this is that the terms which are linear in the error of the wavefunction vanish, and the errors of the respective expressions are determined by terms quadratic in the error of the wavefunction. This implies that the indefinite phase has no effect on the dominating error terms.

All this is no longer the case for a simple expectation value, say that of the nuclear attraction potential $-Z/r$ or the kinetic energy T. The errors of these expectation values have qualitatively a similar behavior as the error of the energy (as suggested by the virial theorem). However, the discretization error depends on the indefinite phase, and it can be either positive or negative, such that the overall error does not decrease monotonically with n, but has a somewhat irregular dependence on the basis dimension n, especially for large n. This has also be found in computer experiments, and is obviously not a consequence of numerical instabilities.

An important expectation value is that of the delta function $\delta(r)$, i.e., the density at the nucleus. We cannot go into details, but we mention the following observation.

For the expectation value of $\delta(r)$ the error depends on the upper cut-off as $\sim t_2^{-1}$ (somewhat like the variance of the energy). For the expectation value of $\delta'(r)$ one finds that the cut-off error ε_{c2} goes as $\sim t_2^0$, i.e., *becomes independent* of t_2, and for $\delta''(r)$ it even goes as $\sim t_2^1$, i.e., *increases with* t_2. This means that the error of $\langle \delta(r) \rangle$, i.e., the *value* of the wavefunction at $r = 0$ converges to 0 with extension of the basis, while the error of $\langle \delta'(r) \rangle$, i.e., of the first derivative of the wavefunction at $r = 0$, does not go to zero for $t_2 \to \infty$, and the error of $\langle \delta''(r) \rangle$ even diverges.

Although even-tempered basis sets are quite good, they are obviously not the best possible choice.

The even tempered basis follows automatically from the mapping (31.20). It appears therefore recommended to study a more general class of mappings and to find an optimum one. In fact one should try to find a mapping such that for an integration domain specified by t_1 and t_2 a smaller number of intervals leads to the same (or a smaller) discretization error.

At first glance it is just a matter of imagination how to choose a more general class of mapping functions. In practice the problem arises that for most mappings that one can think of the integrals which arise cannot be obtained in closed form. However, we do need the explicit dependence on the interval length and the basis dimension in order to derive asymptotic estimates for the optimum basis parameters and the overall error.

The most interesting proposal for a basis determined by a small number of parameters, that is definitely superior to an even-tempered basis is that of Morgan and Haywood. For this basis the large orbital exponents rise faster than exponentially. This is also what one finds if one optimizes the orbital exponents individually. The analytic expression given by Morgan and Haywood looks, nevertheless, somewhat puzzling. In fact the orbital exponents are chosen according to the prescription $\eta_k^{(n)} = ck^{-d}\exp[a(n^b - k^b)]$. The basis contains 4 parameters (a, b, c, d), which all should asymptotically be independent of n.

There is some evidence, but no proof that the error of the energy expectation value obtained with this basis goes as $\varepsilon \sim 3n^{1/2}\exp\{-\pi\sqrt{3n}\}$.

A convergence behavior as found here, is probably less exotic than believed not too long ago. In a recent paper by STAHL [1992] the error of a rational Chebyshev approximation of order (n, n) – i.e., for the quotient of two polynomials of degree n – to the function $|x|$ in the interval $[-1, 1]$ has been shown to go asymptotically as $8\exp(-\pi\sqrt{n})$.

Using similar techniques BRAESS [1995] studied a problem related to that occurring in the basis expansion reported here. He was able to get rigorous bounds for the error. Unfortunately the problems were not identical, so a direct comparison is not possible. Actually Braess studied the expansion of r^{-1} and $\exp(-\alpha r)$ in a weighted L_1 norm, with an exponential weight factor and found a $\exp(-a\sqrt{n})$ behavior of the error as in KUTZELNIGG [1994].

The same kind of behavior is quite common if one tries to approximate an integral over the entire real axis by a finite sum making use of the theory of Whittaker's cardinal functions or sinc-functions (WHITTAKER [1915], WHITTAKER [1935], STENGER [1981], STENGER [1993]). The works by STENGER [1981], STENGER [1993] point out that the expansion based on the theory of Whittaker's cardinal functions is inferior to an expansion in orthogonal polynomials if the function to be expanded is *nonsingular*, but by far superior in the presence of singularities at the boundaries, to which it is little sensitive.

32. On the convergence of the correlated wavefunction

In the vein of what has been done in Section 31 in the simple case of the hydrogen-like atom, we come back to the question of evaluating the best approximation of the exact ground state wavefunction, this time with a special emphasis on many-electron systems. We indeed consider the effect of the *correlation cusp* on the convergence of basis expansions. This first arises for two-electron systems, to which we limit our attention here, but it is virulent for all n-electron systems.

Let us emphasize that as we are dealing here with highly technical aspects, this section may easily be skipped at a first reading.

(A) The configuration interaction (CI) and the partial wave expansion (PWE) for two-electron systems

Like any n-electron wavefunction, that for $n = 2$ can be expanded in the n-electron Slater determinants constructable from the given one-electron basis. However, for $n = 2$, space and spin can be completely separated, and the wavefunction is either a symmetric function in space multiplied by an antisymmetric spin function (for a *singlet state*) or vice versa (for a *triplet state*).

The spatial wavefunction for a singlet can then be expanded in products of spinfree orbitals

$$\Psi(1,2) = \sum_{k,l} c_{k,l}\big\{\varphi_k(\vec{r}_1)\varphi_l(\vec{r}_2) + \varphi_k(\vec{r}_2)\varphi_l(\vec{r}_1)\big\}.$$

A spinfree two-electron wavefunction for an atom in an ^1S*-state* can be written as $\psi(r_1, r_2, r_{12})$ and can be expanded in a partial wave expansion, i.e., in Legendre polynomials of the cosine of the angle $\vartheta \equiv \vartheta_{12}$ between the electronic position vectors $\vec{r}_1$ and $\vec{r}_2$

$$\psi(\vec{r}_1, \vec{r}_2) = \sum_{l=0}^{\infty} \psi_l(r_1, r_2)\ P_l(\cos\vartheta). \tag{32.1}$$

The partial wave expansion is closely related to an expansion in one-electron states. One sees this as one further expands

$$\psi_l(r_1, r_2) P_l(\cos\vartheta) = \frac{4\pi}{2l+1} \sum_{p,q} c_{pq}\varphi_{lp}(r_1)\varphi_{lq}(r_2) \times \sum_{m=-l}^{l} Y_l^m(\vartheta_1, \varphi_1) Y_l^{m*}(\vartheta_2, \varphi_2) \tag{32.2}$$

In fact the lth term in the partial wave expansion (PWE) (32.1) is equal to the contribution of one-electron functions with angular quantum number l in a CI-type expansion of ψ.

(B) The formulas of Schwartz and Hill

SCHWARTZ [1962] was the first to study the rate of convergence of the expansion of correlated wavefunctions in a one-electron basis. He considered the perturbative treatment of the He ground state, with the bare nuclear Hamiltonian as H_0 and the electron interaction as the perturbation.

The first order wavefunction ψ was expanded in a partial wave expansion (32.1),

$$H_0 = \frac{1}{2}\nabla_1^2 - \frac{1}{2}\nabla_2^2 - \frac{Z}{r_1} - \frac{Z}{r_2}; \quad V = r_{12}^{-1}, \tag{32.3}$$

$$(H_0 - E_0)\phi = 0; \quad (H_0 - E_0)\psi = -(V - E_1)\phi, \tag{32.4}$$

$$E_2 = \sum_{l=0}^{\infty} E_l^{(2)}; \quad E_l^{(2)} = \langle\phi|V|\psi_l P_l\rangle = \langle\phi|V_l P_l^2|\psi_l\rangle, \tag{32.5}$$

$$V_l = \frac{r_\langle^{l-1}}{r_\rangle^l}.$$

Schwartz was able to show that the partial wave increments $E_l^{(2)}$ given by (32.5) obey the asymptotic formula (SCHWARTZ [1962])

$$E_l^{(2)} = -\frac{45}{256}\left(l + \frac{1}{2}\right)^{-4} + \frac{225}{1024}\left(l + \frac{1}{2}\right)^{-6} + \mathrm{O}\left(\left[l + \frac{1}{2}\right]^{-8}\right). \tag{32.6}$$

The perturbative treatment just outlined gives the perturbation expansion of the energies of all ions isoelectronic with He as

$$E = Z^2\left(E_0 + \frac{1}{Z}E_1 + \frac{1}{Z^2}E_2 + \frac{1}{Z^3}E_3 + \cdots\right)$$

known as the $1/Z$-expansion (LAYZER [1959]) which one easily derives by appropriate scaling ($\vec{r} \to \vec{r}/Z$) in (32.3), (32.4).

Schwartz's derivation is based on the fact that the exact wavefunction displays a correlation cusp (6.9) (KATO [1980]) which implies that the 1st order wavefunction behaves for small r_{12} as

$$\psi = \left(1 + \frac{1}{2}r_{12}\right)\phi + \mathrm{O}(r_{12}^2),$$

$$\psi_k \sim \frac{1}{2}(r_{12})_k\phi; \quad (r_{12})_k = \frac{1}{2l+3}\frac{r_\langle^{l+2}}{r_\rangle^{l+1}} - \frac{1}{2l-1}\frac{r_\langle^l}{r_\rangle^{l-1}},$$

where ϕ is the unperturbed (zeroth-order) wavefunction. The correlation cusp itself is a direct consequence of the singularity of the Coulomb interaction r_{12}^{-1} at $r_{12} = 0$.

More recently HILL [1985], taking up prior work by KLAHN and MORGAN III [1984], studied the rate of convergence of variational calculations from a rigorous mathematical point of view (for a review of this problem see MORGAN III [1984]). One of his examples was the partial wave expansion of the Helium ground state, for which he showed that the error ΔE_L of a variational calculation in a basis truncated at some angular quantum number L goes as $\Delta E_L = C_1(L+1)^{-3} + C_2 L^{-4} + O(L^{-5})$. Hill was able to derive closed form expressions of C_1 and C_2 in terms of the wavefunction $\Psi(r_1, r_2, r_{12})$ of the ground state of He

$$C_1 = 2\pi^2 \int_0^\infty |\Psi(r, r, 0)|^2 r^5 \, dr,$$

$$C_2 = \frac{12\pi}{5} \int_0^\infty |\Psi(r, r, 0)|^2 r^6 \, dr.$$

The result of HILL [1985] implies that the partial wave increments due to basis functions with a given l in a CI calculation are asymptotically proportional to $(l + \frac{1}{2})^{-4}$, like the partial wave increments to $E^{(2)}$ studied by Schwartz, just with a different coefficient. This behavior of a CI expansion had been conjectured in LAKIN [1965] and had been demonstrated in a numerical calculation of the ground state of He in CARROLL, SILVERSTONE and METZGER [1979], which led to very pessimistic conclusions as to the applicability of CI to accurate calculations.

A $(l + \frac{1}{2})^{-4}$ behavior of the partial wave increments also holds in perturbation theory based on the Hartree–Fock–Hamiltonian as H_0, again with a different coefficient (KUTZELNIGG and MORGAN III [1992]).

The results of SCHWARTZ [1962] and HILL [1985] are limited to the ground states of He-like ions. Generalizations to arbitrary states of two-electron atoms were derived in KUTZELNIGG and MORGAN III [1992]. These will be given below. To prepare this, we must first have a careful look at the cusp condition for general two-electron systems.

(C) The generalized correlation cusp for two-electron systems

We now consider the solution of the Schrödinger equation near $r_{12} = 0$ explicitly for a *two-electron system*, and we separate off the spin. The eigenfunctions can then be classified

(a) as *singlet* or *triplet* depending on symmetry or antisymmetry with respect to exchange of $\vec{r}_1$ and $\vec{r}_2$;

(b) as *gerade* (g) or *ungerade* (u) depending on symmetry or antisymmetry with respect to $\vec{r}_k \to -\vec{r}_k$;

(c) by the total angular momentum $\mathcal{L}$;

(d) as of *natural* or *unnatural* parity (KUTZELNIGG and MORGAN III [1992]). Since for one-electron states *l even* always corresponds to *gerade* parity and *l odd* to *ungerade* parity, while for n-electron states *any* L can be associated with overall *gerade or ungerade* parity, it does make sense to define that a state has *natural* parity, if its parity is equal to $(-1)^L$, and *unnatural* parity, if the parity of the state is equal to $(-1)^{(L+1)}$, So S_g, P_u, D_g etc., have natural parity, S_u, P_g, D_u have unnatural parity.

We express the spinfree wavefunction Ψ of a two-electron system in terms of center-of-mass and relative coordinates $\vec{R}$ and $\vec{r}$: $\Psi(\vec{R},\vec{r})$, $\vec{R}=\frac{1}{2}(\vec{r}_1+\vec{r}_2)$, $\vec{r}=\vec{r}_1-\vec{r}_2$.

$$\Psi(-\vec{R},-\vec{r})=\pm\Psi(\vec{R},\vec{r}); \quad + \text{for gerade}, \quad - \text{for ungerade states},$$

$$\Psi(\vec{R},-\vec{r})=\pm\Psi(\vec{R},\vec{r}); \quad + \text{for singlet}, \quad - \text{for triplet states}.$$

We expand Ψ in spherical harmonics of the center of mass and relative coordinates $\vec{R}=(R,\Theta,\phi)$, $\vec{r}=(r,\vartheta,\varphi)$,

$$\Psi=\sum_{L,M}\sum_{k,m}\Psi_{Lk}^{Mm}(R,r)Y_L^M(\Theta,\phi)Y_k^m(\vartheta,\varphi).$$

Obviously

L even, k even for singlet gerade,
L odd, k even for singlet ungerade,
L odd, k odd for triplet gerade,
L even, k odd for triplet ungerade,

L and k couple to the total angular momentum $\mathcal{L}$. We have

$\mathcal{L}-L-k$ even for natural parity states,
$\mathcal{L}-L-k$ odd for unnatural parity states.

This leads to the classification

natural parity singlet: k even, $\mathcal{L}-L$ even,
natural parity triplet: k odd, $\mathcal{L}-L$odd,
unnatural parity singlet: k even, $\mathcal{L}-L$ odd,
unnatural parity triplet: k odd, $\mathcal{L}-L$ even.

The minimum allowed even/odd k is $0/1$, respectively; however, $k=0$ is impossible for $\mathcal{L}-L$ odd. Hence the leading partial waves in the relative coordinate are

$k=0$ for natural parity singlet states,

$k=1$ for arbitrary triplet states,

$k=2$ for unnatural parity singlet states.

We now consider the Schrödinger equation at $r_{12}\equiv r\to 0$

$$\left[-\frac{1}{2r^2}\frac{\partial}{\partial r}r^2\frac{\partial}{\partial r}+\frac{k(k+1)}{2r^2}+\frac{\lambda}{r}+\mathrm{O}(r^0)\right]\Psi_k^m(\vec{R},\vec{r})=0. \tag{32.7}$$

In order to satisfy (32.7) Ψ_k must behave as

$$\Psi_k^m(\vec{R},\vec{r}) \sim \left\{ r^k \left(1 + \frac{\lambda}{2(k+1)} r \right) + \mathrm{O}\left(r^{k+2}\right) \right\} Y_k^m(\vartheta,\varphi). \tag{32.8}$$

For unnatural-parity singlet states where the leading partial wave of the relative motion is that with $k=0$, (32.8) is essentially Kato's correlation cusp condition (6.9). For triplet states with $k=1$ (32.8) implies the generalized cusp condition (PACK and BYERS BROWN PACK [1966]), while the result with $k=2$ for natural parity singlet states has only rather recently been found, see KUTZELNIGG and MORGAN III [1992], MORGAN III and KUTZELNIGG [1993], KUTZELNIGG and MORGAN III [1996].

If we set $\lambda = 1$ in (32.7, 32.8) we get the correlation cusp relation for the exact wavefunction. For $\lambda = 1/Z$ the relation between the unperturbed wavefunction ϕ and the 1st order function in the $1/Z$ expansion is obtained: $\psi = \frac{1}{2(k+1)} r_{12}\phi + \mathrm{O}(r_{12}^2)$. For $\lambda = 0$ we get the behavior of the unperturbed wavefunction ϕ at $r_{12} \to 0$.

(D) Convergence of the partial wave expansion

In KUTZELNIGG and MORGAN III [1992] were derived the leading terms of the PWE contributions to the second (and also third order) in Z^{-1} for general states of atoms isoelectronic with He. The second-order results are summarized in Table 32.1, with the corresponding numerical values in Table 32.2.

To illustrate Table 32.1 we consider the $(1s)^1S$ ground state of He-like ions, that has already been studied long ago in SCHWARTZ [1962]. Here one has

$$N = \frac{1}{2}, \quad F_0 = 1; \qquad R_1(r) = R_2(r) = 2Z^{3/2}e^{-Zr}; \qquad \varrho_+(r) = 32Z^6e^{-4Zr};$$

$$R_+^{(5)} = 32\int_0^\infty Z^6 e^{-4Zr} r^5\, dr = \frac{15}{16}$$

and the value for the coefficient of $(l+\frac{1}{2})^{-4}$ is $-\frac{45}{256}$, in agreement with (32.6).

The most noteworthy message of KUTZELNIGG and MORGAN III [1992] is that the leading term is of $\mathrm{O}([l+\frac{1}{2}]^{-4})$ only for *natural-parity singlet* states, it is of $\mathrm{O}([l+\frac{1}{2}]^{-6})$ for *triplet* states of either natural or unnatural parity (but with a different origin in the two cases), and it is of $\mathrm{O}([l+\frac{1}{2}]^{-8})$ for *unnatural-parity singlet* states.

To second order in perturbation theory in Z^{-1} only even orders in $([l+\frac{1}{2}]^{-1})$ arise. Odd orders in $([l+\frac{1}{2}]^{-1}$ appear in third and higher orders in Z^{-1}, and are therefore present in HILL [1985] for the nonexpanded wavefunction.

The generalization to many-electron systems involves the two-particle density matrices (KUTZELNIGG and MORGAN III [1992]).

(E) Wavefunctions with explicit r_{ij} dependence

The slow convergence with expansions in a basis of Slater determinants is avoided, if one uses variational trial wavefunctions that depend explicitly on the interelectronic

TABLE 32.1

Expressions for the leading terms in the partial wave expansion of the 2nd order energy $E^{(2)}$ of He-like ions, for an unperturbed state characterized by the quantum numbers n_1, l_1, n_2, l_2, L with the radial one-electron functions $R_1(r), R_2(r)$

Natural parity singlet	$-\frac{3}{8} N R_+^{(5)} F_0 \{(l+\frac{1}{2})^{-4} - \frac{5}{4}(l+\frac{1}{2})^{-6}\}$
Natural parity triplet	$-N\{\frac{5}{64} F_0 R_-^{(7)} + \frac{15}{256}(F_1 - G_1) R_+^{(5)}\}(l+\frac{1}{2})^{-6}$
Unnatural parity triplet	$-\frac{15}{128} N F_1 R_+^{(5)} (l+\frac{1}{2})^{-6}$
Unnatural parity singlet	$-N\{\frac{35}{384} F_1 R_-^{(7)} + \frac{77}{3072}(F_2 - G_2) R_+^{(5)}\}(l+\frac{1}{2})^{-8}$

Auxiliary expressions

$N = (1 + \delta_{n_1 n_2} \delta_{l_1 l_2})^{-1} \qquad M = (-1)^L (2l_1 + 1)(2l_2 + 1)$

$R_+^{(5)} = \int_0^\infty \varrho_+(r, r) r^5 dr, \ \varrho_\pm(r, r') = \frac{1}{2} |R_1(r) R_2(r') \pm R_2(r) R_1(r')|^2$

$R_-^{(7)} = \int_0^\infty \varrho'_-(r, r) r^7 dr, \ \varrho'_-(r, r') = \partial^2 [\varrho_-(r, r')] / \partial r'^2$

$F_p = M \sum_{n=0}^{2\min(l_1, l_2)} (2n+1)[n(n+1)]^p \begin{Bmatrix} l_1 & l_1 & n \\ l_2 & l_2 & L \end{Bmatrix} \begin{pmatrix} l_1 & l_1 & n \\ 0 & 0 & 0 \end{pmatrix} \begin{pmatrix} l_2 & l_2 & n \\ 0 & 0 & 0 \end{pmatrix}$

$G_p = M \sum_{n=|l_1 - l_2|}^{l_1 + l_2} (2n+1)[n(n+1)]^p \begin{Bmatrix} l_1 & l_2 & n \\ l_2 & l_1 & L \end{Bmatrix} \begin{pmatrix} l_1 & l_2 & n \\ 0 & 0 & 0 \end{pmatrix}^2$

$F_0 = G_0 = (2l_1 + 1)(2l_2 + 1) \begin{pmatrix} l_1 & l_2 & L \\ 0 & 0 & 0 \end{pmatrix}^2$

Expressions like $\begin{pmatrix} l_1 & l_2 & L \\ 0 & 0 & 0 \end{pmatrix}$ and $\begin{Bmatrix} l_1 & l_2 & n \\ l_2 & l_1 & L \end{Bmatrix}$ are Wigner $3j$ and $6j$ symbols, respectively.

distances r_{ij}, in particular that contain terms linear in the r_{ij} and allow to describe the correlation cusp (6.9) correctly. This has been realized long ago in HYLLERAAS [1929], who obtained an excellent variational energy for the ground state of He-like ions with a compact ansatz $\Psi(r_1, r_2, r_{12})$ with only a small number of parameters. This was improved in KINOSHITA [1957], while the elaborate calculation done in PEKERIS [1958] reminded a landmark for quite some time.

A similar ansatz was used successfully for the H_2-molecule in JAMES and COOLIDGE [1933] and mainly in KOLOS and ROOTHAAN [1960], KOLOS and WOLNIEWICZ [1964a], WOLNIEWICZ [1993], KOLOS and RYCHLEWSKI [1993].

The generalization of this concept, known as *Hylleraas–CI* to n-electron systems has for a long time been prohibited by the need to evaluate many-electron integrals of the type

$$\int f_1(\vec{r}_1) f_2(\vec{r}_2) \cdots r_{12} r_{13} r_{23}^{-1} \, d\vec{r}_1 \, d\vec{r}_2 \, d\vec{r}_3$$

and even more complicated ones. The evaluation of each single integral is quite time consuming, but even more problematic is the large number of these integrals. Only rather recently some progress has been achieved for 3- and 4-electron atoms

TABLE 32.2

Numerical values of the coefficients of $(l+\frac{1}{2})^{-n}$ of the 2nd order energy (in the $1/Z$ expansion) for various states of the He isoelectronic series

State	$n=4$	$n=6$	$n=8$
$^1S_g(1s^2)$	−0.17578125	0.21972656	
$^1S_g(1s2s)$	−0.04115226	0.05144033	
$^1S_g(1s3s)$	−0.01158714	0.01448392	
$^1S_g(1s4s)$	−0.00480707	0.00600883	
$^3S_g(1s2s)$	0.0	−0.06001372	
$^3S_g(1s3s)$	0.0	−0.01457445	
$^3S_g(1s4s)$	0.0	−0.00585253	
$^1P_u(1s2p)$	−0.09602195	0.12002744	
$^1P_u(1s3p)$	−0.02433300	0.03041625	
$^1P_u(1s4p)$	−0.00978518	0.01223148	
$^3P_u(1s2p)$	0.0	−0.02667276	
$^3P_u(1s3p)$	0.0	−0.00137296	
$^3P_u(1s4p)$	0.0	−0.00024087	
$^1D_g(1s3d)$	−0.00486660	0.00608325	
$^1D_g(1s4d)$	−0.00253919	0.00317399	
$^3D_g(1s3d)$	0.0	−0.00601988	
$^3D_g(1s4d)$	0.0	−0.00251752	
$^1F_u(1s4f)$	−0.00009732	0.00012165	
$^3F_u(1s4f)$	0.0	−0.00027007	
$^1D_g(2p^2)$	−0.27685547	0.34606934	
$^3P_g(2p^2)$	0.0	−0.43258667	
$^3P_g(2p3p)$	0.0	−0.18811699	
$^1P_g(2p3p)$	0.0	0.0	−1.25536739
$^1D_u(2p3d)$	0.0	0.0	−0.09012894
$^1D_u(3p3d)$	0.0	0.0	0.76076507
$^3D_u(2p3d)$	0.0	−0.41385738	
$^1F_u(2p3d)$	−0.28378792	0.35473490	
$^3F_u(2p3d)$	0.0	−0.08750128	
$^1F_g(2p4f)$	0.0	0.0	−0.19227989
$^3F_g(2p4f)$	0.0	−0.06054294	

(LÜCHOW and KLEINDIENST [1994], KLEINDIENST, LÜCHOW and MERCKENS [1994], BUESSE and KLEINDIENST [1995], BUESSE, KLEINDIENST and LÜCHOW [1998], THAKKAR and KOGA [1994], NEWMAN [1997]). It is unlikely that Hylleraas–CI has a serious chance to be practicable for systems with more than, say, 4 electrons.

In atomic calculations one can – alternatively to r_{12}, with some success, use $r_<$ and $r_>$, i.e., the smaller and the greater of r_1 and r_2 (with a generalization to n-electron systems) as independent variables (GOLDMAN [1998]).

One is obviously faced with the problem that traditional CI-type calculations in a one-electron basis do not allow a high accuracy, because of the slow convergence with the basis size, while calculations like Hylleraas–CI that allow a fast convergence in principle, are impracticable, except for very small systems.

As commented in Section 31 the slow convergence of an expansion in a one-electron basis is due the poor representation of the correlation cusp in such a basis. This correlation cusp is an exact relation that the wavefunction has to satisfy at the coalescence of two electrons, while the nuclear cusp is related to the coalescence of the position of an electron and a nucleus.

There are also some general relations available for the coalescence of *three particles*. The keyword to this is *Fock expansion* (FOCK [1958]) and it requires that the wavefunction of the He ground state contains terms in $(r_1^2 + r_1^2)^{p/2}\{\ln(r_1^2 + r_1^2)\}^q$. Including logarithmic terms FRANKOWSKI and PEKERIS [1966] needed only a 246 term wavefunction to be more accurate than PEKERIS [1958] in his original work with more than 1000 terms. The performance was even superseded in FREUND, HUXTABLE and MORGAN III [1984] who took especially care to describe the three-particle coalescence correctly, and obtained the nonrelativistic He ground state energy with 13-figure accuracy. More recently has been obtained in DRAKE and YAN [1994] an even higher accuracy with a rather simple-minded ansatz using a linear combination of two Hylleraas type wavefunctions with different orbital exponents,

$$\Psi = \sum_{i,j,k} r_1^i r_2^j r_{12}^k \big[c_{ijk}^{(1)} \exp(-\alpha_1 r_1 - \beta_1 r_2) + c_{ijk}^{(2)} \exp(-\alpha_2 r_1 - \beta_2 r_2)\big]. \tag{32.9}$$

So the question whether the presence of logarithmic terms in the wavefunction is really helpful, is open again. It is hard to believe that they will ever play a possible role beyond the He isoelectronic series.

Meanwhile an 18-figure accuracy has been achieved by KLEINDIENST and EMRICH [1990], THAKKAR and KOGA [1994], NEWMAN [1997] and this does not appear to be the end (see GOLDMAN [1998], where the ansatz which involves $r_\langle$ and $r_\rangle$, looks particularly promising).

Still another approach is the promising *R*12 *method* by KLOPPER and KUTZELNIGG [1987], NOGA, KLOPPER and KUTZELNIGG [1998], KLOPPER [1998]. Its basic idea is to combine a conventional CI-like basis expansion approach like MP2 or CC with the inclusion of terms that take care of the linear r_{ij} dependence of the wavefunction near the coalescence of two electrons. If one starts from a single Slater determinant reference function Φ with the orbitals φ_I occupied, one constructs *additional n-electron basis functions*, obtained from Φ on replacing an orbital pair $\varphi_I(1)\varphi_J(2)$ by a pair function

$$r_{12}\varphi_I(1)\varphi_J(2) \tag{32.10}$$

and includes these additional n-electron basis functions in a variational calculation. Actually one uses different pair functions for singlet and triplet pairs, and one even includes pair functions, in which $\varphi_K(1)\varphi_L(2)$ is replaced by (32.10) (with K, L different from I, J) (KLOPPER [1991]). In practice one first orthogonalizes these pair functions to the *excited* Slater determinants that one includes in the CI part, such that the remaining pair functions correct for the *incompleteness of the one-electron basis* to describe the correlation cusp. An essential feature of the R12 method is that only two-electron integrals are needed. However, in addition to the classical electron repulsion integrals $\langle\varphi_p(1)\varphi_q(2)|r_{12}^{-1}|\varphi_r(1)\varphi_s(2)\rangle$ also $\langle\varphi_p(1)\varphi_q(2)|r_{12}|\varphi_r(1)\varphi_s(2)\rangle$ and $\langle\varphi_p(1)\varphi_q(2)|[T, r_{12}]|\varphi_r(1)\varphi_s(2)\rangle$ need to be evaluated.

An essential ingredient of the R12 method is the systematic introduction of completeness insertions in such a way that (i) 3- and 4-electron integrals do not arise explicitly, (ii) the results become exact in the limit of a complete one-electron basis, (iii) the basis truncation error decreases much faster with the size of the basis than in conventional calculations, such that much higher accuracy is achieved with less computational effort. The only disadvantage is that no strict upper-bond property holds, but this is not even the case for conventional coupled-cluster calculations. There is no stronger limitation of the size of the system to be computed than for conventional CC calculations.

A certain drawback is that relatively large basis sets have to be used, they should be sufficiently close to complete for the *low angular momenta*, i.e., essentially for s, p, d, and to some extent f. This can be avoided if one introduces an *auxiliary basis* to satisfy completeness relations, while for the actual calculations a smaller basis can be used (KLOPPER and SAMSON [2002]). If one includes explicit linear r_{12}-dependent terms in the wavefunctions and expands only the remainder (KLOPPER and KUTZELNIGG [1987]), the l-increments of the remainder go essentially as $(l+\frac{1}{2})^{-8}$, at least at the level of 2nd order perturbation theory, which is a substantial improvement.

It is somewhat surprising that in the R12 method there appears no need for a *damping* of the r_{12} factor for large r_{12}. It appears that on one hand the decay of the one-electron functions, and on the other hand the linear variational coefficient, takes care of this to a sufficient extent.

An alternative to the use of linear r_{ij}-terms as in the R12 method, is the method of *Gaussian geminals*. Here one introduces correlation factors of the form $\exp(-\gamma r_{12}^2)$. With such functions it is *not* possible to satisfy the correlation cusp exactly, but all integrals that arise, including three and four-electron integrals, can be evaluated in closed form. The convergence of this kind of expansion is much faster than that of a CI (or a partial-wave expansion), but by far not as fast as Hylleraas–CI. It is possibly similar to that of the expansion of a 1s hydrogen wavefunction in a Gaussian basis.

Recently RYCHLEWSKI, CENCEK and KOMASA [1994] have been able to achieve for the ground state of the He atom a similar accuracy as in PEKERIS [1958] even with a wavefunction of Gaussian geminal type

$$\Psi = \sum_i c_i \exp\bigl(-\alpha_{1i}|\vec{r}_1 - \vec{A}_i|^2 - \alpha_{2i}|\vec{r}_2 - \vec{B}_i|^2 - \beta_i|(\vec{r}_1 - \vec{r}_2)|^2\bigr).$$

For He 12-figure accuracy was reported in RYCHLEWSKI, CENCEK and KOMASA [1994], CENCEK and KUTZELNIGG [1996]. This is surprising since this ansatz neither fulfills the nuclear cusp nor the correlation cusp conditions. Although it is not yet fully understood why this works, some preliminary comments can be made.

The Gaussian geminal method has been implemented for many-electron atoms or molecules in combination with, e.g., MP2 and CCSD (SZALEWICZ, JEZIORSKI, MONKHORST and ZABOLITZKY [1983]). A rather difficult practical problem is that of the choice of the optimum nonlinear parameters γ. This has so far inhibited the application of the Gaussian geminal method beyond HF (WENZEL, ZABOLITZKY, SZALEWICZ, JEZIORSKI and MONKHORST [1986]) or H_2O (BUKOWSKI, JEZIORSKI, RYBAK and SZALEWICZ [1995]). The technique of RYCHLEWSKI, CENCEK and KOMASA [1994] to optimize nonlinear parameters has to the authors' knowledge not yet been applied in this context.

(F) Extrapolation methods

The first method is the exact extrapolation to the basis limit. If one is able to compute the individual partial wave increments E_l to the energy accurately, and knows at the same time the correct asymptotic expansion of these increments, one can combine these two sources of information in an effective way. Take as an example the 2nd order energy $E^{(2)}$ of the ground state of He-like ions in the $1/Z$ expansion. We know that in view of (32.6) the truncation error of $E^{(2)}$ – for limiting the series at $l = L$ is given as

$$\Delta E_L^{(2)} = \frac{15}{256}(L+1)^{-3} - \frac{45}{1024}(L+1)^{-5} + \mathrm{O}\big([L+1]^{-7}\big). \tag{32.11}$$

Using this formula we can estimate that $L = 38$ is required to get an accuracy of 1 microhartree from a brute force partial wave expansion. Let us now assume that we calculate some truncated $E_L^{(2)}$, but correct it for the first term of the truncation error (32.11). Then the error is dominated by the second term of (32.11). To make this remaining error smaller than 1 microhartree we need only $L = 8$. If we correct also for the second term the maximum required L is reduced to $L = 6$. This means that it is sufficient to obtain the partial wave increments for $l = 0, 1, 2, \ldots, 6$ in order to obtain $E^{(2)}$ accurate to 1 microhartree, provided that one knows the coefficients in the asymptotic error formula (32.11). This is confirmed by taking the partial-wave increments up to $l = 6$ from SCHMIDT and HIRSCHHAUSEN [1983], which sum to -157499.67 μE_h; the first term in (1.11) corrects this to -157670.49 μE_h, and the second to -157667.88 μE_h, compared with the exact value (BYRON and JOACHAIN [1967]) of 157666.43 μE_h. Of course, this procedure requires that the low partial wave increments are sufficiently accurate, since errors in them cannot be corrected by the asymptotic error formula.

A second method consists of numerical extrapolations. If one is able to evaluate the partial wave increments E_l for low l, but does *not* know the exact asymptotic behavior of the E_l, one can try to extrapolate numerically. This is dangerous, because usually one evaluates E_l increasingly less accurately for increasing l, which can result in an overestimation of the rate of convergence. Some extrapolations were reported, in which a wrong general analytic form of the asymptotic error was assumed. It is, in this context, at least helpful to know that the truncation error goes as $(L+1)^{-3}$. *Educated* numerical extrapolations are starting to play an important role in practical calculations striving at high accuracy: PETERSON, WOON and DUNNING Jr. [1994], DUNNING Jr. [1989], PETERSON, WILSON, WOON and DUNNING Jr. [1997], WILSON and DUNNING Jr. [1997], MARTIN and TAYLOR [1997], PETERSSON and BRAUNSTEIN [1985].

Thirdly, there is the so-called principle quantum number expansion. Although the asymptotic behavior of the partial wave expansion is well understood, an extrapolation based on the E_l is difficult, because for increasing l an increasing size of the basis is needed (unless one uses expansions in $r_\rangle$ and $r_\langle$ (GOLDMAN [1998]), which is only feasible for atoms), so good E_l values for moderately large l are quite expensive. An interesting alternative is the *principal quantum number expansion* of HELGAKER, KLOPPER, KOCH and NOGA [1997]. There is striking evidence (although an analytic proof is still missing) that the error of truncating, say, a CI expansion of the He ground state, after a certain *principle quantum number* n, i.e., after 2s, 2p, or 3s, 3p, 3d etc., goes (like the error of the l-expansion) as $(n_{\max}+1)^{-3}$. The evaluation of the *principle quantum number increments* E_n is much easier than that of the partial waver increments E_l.

33. Convergence analysis

Let us first make another incursion in mathematical considerations. Indeed we have explained to which extent the models are stated on firm mathematical and phenomenological ground, we have also presented some basis sets for the approximation and argued on the way they are chosen; we denote hereafter by X_δ a generic discrete space. It was clarified also in the previous sections, that there exists a little bit of tune-up in the definition of the basis sets, in order to achieve better accuracies. Hence at least two questions have to be raised now

(1) are the hopes of the developers sound, in the sense that the solution that gets out from the computations actually reveals the accuracy that is available in the associated discrete spaces X_δ;

(2) to which extent the final results, i.e., the outputs or observables, that are derived from a given computation, are close to the exact value.

The first question is related to the *optimality* of the approximation, another way to state it is: What can be said on the ratio – denoted as ρ – of the error between the exact solution and the computed solution by the error between the exact solution and the closest element in the discrete space. The approximation is said optimal if this ratio is not parameter dependent.

The second question is related to the *certification* of the results. Indeed, most often, decisions are made after one computation (and generally a coarse one) and not after a series of computations with finer and finer approximations. It is thus not known whether the asymptotic rates, taking into account the previous ratio ρ, allows to tell if the current results are reliable or not.

These two issues will be developed in what follows. Before going into more details in this analysis, more has to be said about the solutions to the problem. Let us focus our attention on the Hartree–Fock problem and more precisely to the ground state of these equations. In order to be in the best theoretical framework, let us also consider that the computations are done in the case of a positive ion, hence $N+1 \leqslant \sum_{k=1}^{M} z_k$ so that it is known that the eigenvalue problem (9.12) admits an infinite sequence of negative eigenvalues converging to zero that can be ranked in increasing order. The ground state is then associated with the evaluation of the smallest N eigenvalues and related eigenfunctions or related atomic orbitals. We denote by $\Phi_0 = (\phi_{i,0})_{i=1}^N$ the solution to the minimization problem that is

$$\min_{\Psi \in \mathcal{K} \cap \mathcal{H}} \mathcal{E}^{\mathrm{HF}}(\Psi),$$

where $\mathcal{K}$ is defined by

$$\mathcal{K} = \{(\Phi_1, \ldots, \Phi_N) \in [L^2(\mathbb{R}^3)]^N;\ \langle \Phi_i, \Phi_j \rangle = \delta_{ij}\}.$$

Actually it is interesting at this stage to introduce the notation

$$\forall \Phi = (\Phi_i)_{i=1}^N,\ F_{ij}(\Phi) = \langle \Phi_i, \Phi_j \rangle - \delta_{ij} \tag{33.1}$$

so that for instance $\mathcal{K}$ appear as the intersection of the kernels of all F_{ij}.

This minimization solution is associated with Euler–Lagrange equations that take the form of the nonlinear eigenvalue problem (9.12). It has already been stated that it is not known whether this problem has a unique solution. What is also sure is that if Φ_0 is a solution then $U\Phi_0$ is also a solution for any unitary $N \times N$ matrix U where it appears useful to consider the elements of $\mathcal{H}$ as columns $(N \times 1)$ vectors.

Due to the previous invariance and in order to get a better idea on the error that can exist between the solution Φ_0 and an approximation, denoted as Φ_δ hereafter, it is more appropriate to consider the minimum between Φ_0 and any $U\Phi_\delta$ where U is a unitary matrix. This leads to the introduction, for any element $\Psi_1, \Psi_2 \in \mathcal{H} = [H^1(\mathbb{R}^3)]^N$ of the notation

$$U_{\Psi_1,\Psi_2} = \arg\min\{\|U\Psi_1, \Psi_2\|_{[L^2(\mathbb{R}^3)]^N};\ U \in \mathcal{U}(N)\}, \tag{33.2}$$

where $\mathcal{U}(N)$ represents the set of all unitary $N \times N$ matrices.

In the same spirit, the decomposition

$$\mathcal{H} = \mathcal{M}_\Phi \oplus \Phi^{\perp\perp}$$

has been introduced in MADAY and TURINICI [2002a], with

$$\mathcal{M}_\Phi = \{M\Phi;\ M \in \mathbb{R}^{N\times N}\},$$
$$\Phi^{\perp\perp} = \{\Psi = (\psi_i)_{i=1}^N \in \mathcal{H};\ \langle \psi_i, \phi_j \rangle = 0;\ i, j = 1, \ldots, N\}.$$

Indeed, it is quite obvious that by setting $M_{ij} = \langle \psi_i, \phi_j \rangle$, we have $\Psi - M\Phi \in \Phi^{\perp\perp}$. More precisely, we also have

$$\mathcal{H} = \mathcal{A}_\Phi \oplus \mathcal{S}_\Phi \oplus \Phi^{\perp\perp},$$

where

$$\mathcal{A}_\Phi = \{C\Phi;\ M \in \mathbb{R}^{N\times N},\ C^t = C\},$$
$$\mathcal{S}_\Phi = \{S\Phi;\ M \in \mathbb{R}^{N\times N},\ S^t = -S\}$$

by decomposing the matrix M in its antisymmetric part, denoted as C, and its symmetric part, denotes as S. With these notations applied to $\Psi - \Phi$, with $\Phi, \Psi \in \mathcal{H} \cap \mathcal{K}$, we thus have the decomposition

$$\Psi = \Phi + C\Phi + S\Phi + W, \quad W \in \Phi^{\perp\perp}. \tag{33.3}$$

In what follows, we will denote, for any $\Psi_1, \Psi_2 \in [L^2(\mathbb{R}^3)]^N$: $\Psi_1 \perp\perp \Psi_2$ if, for any $i, j = 1, \dots, N$, $i \neq j$, we have $\langle \psi_{1i}, \psi_{2j} \rangle = 0$. Let us recall now the following interesting result (see MADAY and TURINICI [2002a] Lemma 1). For any $\Phi, \Psi \in \mathcal{H} \cap \mathcal{K}$, the matrix $U_{\Psi,\Phi}$ solution of (33.2) has the property

$$U_{\Psi,\Phi}\Psi - \Phi \in \mathcal{S}_\Phi \oplus \Phi^{\perp\perp} \tag{33.4}$$

which means that through some appropriate rotation, the antisymmetric component C in (33.3) can be set to zero. If not zero, the symmetric part, is nevertheless not the main part of the decomposition since (see MADAY and TURINICI [2002a] Lemma 2) there exists constants C_1, C_2 depending only on N such that

$$\|S\Phi\|_{[L^2(\mathbb{R}^3)]^N} \leqslant C_1 \|\Psi - \Phi\|^2_{(L^2(\mathbb{R}^3))^N}, \tag{33.5}$$

$$\|S\Phi\|_{\mathcal{H}} \leqslant C_2 \|\Psi - \Phi\|^2_{\mathcal{H}} \|\Phi\|_{\mathcal{H}}. \tag{33.6}$$

Let us consider now $\pi_\delta(\Phi_0)$ as being the best fit of Φ_0 by elements of the discrete space X_δ say in the $L^2(\mathbb{R}^3)$-norm. It has been explained, by looking at the hydrogen-like atom that for any i, $1 \leqslant i \leqslant N$, $\lim \|\pi_\delta \phi_{i,0} - \phi_{i,0}\| = 0$, the norm being either the $L^2(\mathbb{R}^3)$- or the $H^1(\mathbb{R}^3)$-norm, actually the proof goes through other atomic orbitals (see MADAY and TURINICI [2002a] for more details). It has also been stated that the convergence rate may be quite rapid, depending on the basis set that is chosen. The first point is to prove that a solution for the discrete problem exists in a neighborhood of the solution we want to approximate and immediately will follow the proof that the distance between the exact solution and the approximated one is of the same asymptotic order as the best fit. The first point goes through the remark that, for any $\Phi \in \mathcal{K}$ and any $W \in \Phi^{\perp\perp}$, there exists a diagonal $N \times N$ matrix T such that

$$\Phi + T\Phi + W \in \mathcal{K} \tag{33.7}$$

this is done in a standard manner by the same trick as for the Gramm–Schmidt orthonormalization process and from (33.5) it follows easily that

$$\|T\Phi\|_{[L^2(\mathbb{R}^3)]^N} \leqslant C\|W\|^2_{(L^2(\mathbb{R}^3))^N}. \tag{33.8}$$

In addition, T is a regular function of the argument W.

Next, for any $\Phi \in \mathcal{H} \cap \mathcal{K}$, we denote by

$$\mathcal{E}^\Phi(\cdot) = \mathcal{E}^{\mathrm{HF}}(\cdot) + \sum_{i,j=1}^{N} \Lambda_{ij} F_{ij}(\cdot), \tag{33.9}$$

where $\Lambda_{ij} = \langle \mathcal{F}_\Phi \Phi_i, \Phi_j \rangle$, $i, j = 1, \dots, N$, and $\mathcal{F}_\Phi$ denotes the Fock operator. It is then straightforward to note that, Φ_0 is a solution to the Hartree–Fock equation leads to

$$D\mathcal{E}^{\Phi_0} = 0.$$

In addition, being a minimum it follows that $a_{\Phi_0} = D^2\mathcal{E}^{\Phi_0}$ is positive. It is proven in MADAY and TURINICI [2002a] a more precise result: Let X_{Φ_0} be the orthogonal space (in $\mathcal{H}$) of the elements of $\mathcal{A}_{\Phi_0} \oplus \Phi_0^{\perp\perp}$ over which a_{Φ_0} vanishes, then

$$\forall \Psi \in X_{\Phi_0},\ \Psi \neq 0\colon\ a_{\Phi_0}(\Psi, \Psi) > 0$$

and there exists a constant $\alpha > 0$ depending only on X_{Φ_0} such that a_{Φ_0} is α-elliptic over X_{Φ_0}, that, for simplicity, we assume to coincide with $\Phi_0^{\perp\perp}$.

We are now in a position to establish the existence in a neighborhood of Φ_0 of a minimizer of the discrete Hartree–Fock problem. This is done by writing, for any Ψ_δ in $\mathcal{K} \cap [X_\delta]^N$

$$\begin{aligned}
&\mathcal{E}^{\mathrm{HF}}(\Psi_\delta) - \mathcal{E}^{\mathrm{HF}}\big(\pi_\delta(\Phi_0)\big) = \mathcal{E}^{\Phi_0}(\Psi) - \mathcal{E}^{\Phi_0}\big(\pi_\delta(\Phi_0)\big)\\
&\quad = D\mathcal{E}^{\Phi_0}(\Psi - \Phi_0) - D\mathcal{E}^{\Phi_0}\big(\Phi_0 - \pi_\delta(\Phi_0)\big) + \frac{1}{2}D^2\mathcal{E}^{\Phi_0}(\Psi - \Phi_0, \Psi - \Phi_0)\\
&\qquad - \frac{1}{2}D^2\mathcal{E}^{\Phi_0}\big(\Phi_0 - \pi_\delta(\Phi_0), \Phi_0 - \pi_\delta(\Phi_0)\big)\\
&\qquad + \mathcal{O}\big(\|\Psi - \Phi_0\|^3 + \big\|\Phi_0 - \pi_\delta(\Phi_0)\big\|^3\big).
\end{aligned}$$

Reminding first that $D\mathcal{E}^{\Phi_0}$ is zero, using the decomposition $\Psi - \pi_\delta(\Phi_0) = S\pi_\delta(\Phi_0) + W$, where $S \in \mathcal{S}$ and $W \in \pi_\delta(\Phi_0)^{\perp\perp} \cap [X_\delta]^N = (\Phi_0)^{\perp\perp} \cap [X_\delta]^N$ and taking into account the fact that $\|W\| = \mathcal{O}(\|\Psi - \pi_\delta(\Phi_0)\|)$ together with (33.5), this simply leads to

$$\begin{aligned}
&\mathcal{E}^{\mathrm{HF}}(\Psi) - \mathcal{E}^{\mathrm{HF}}\big(\pi_\delta(\Phi_0)\big)\\
&\quad = \frac{1}{2}D^2\mathcal{E}^{\Phi_0}\big(W - \Phi_0 - \pi_\delta(\Phi_0), W - \Phi_0 - \pi_\delta(\Phi_0)\big)\\
&\qquad - \frac{1}{2}D^2\mathcal{E}^{\Phi_0}\big(\Phi_0 - \pi_\delta(\Phi_0), \Phi_0 - \pi_\delta(\Phi_0)\big)\\
&\qquad + \mathcal{O}\big(\big\|\Psi - \pi_\delta(\Phi_0)\big\|^3 + \big\|\Phi_0 - \pi_\delta(\Phi_0)\big\|^3\big)\\
&\quad = \frac{1}{2}D^2\mathcal{E}^{\Phi_0}\big(W - \Phi_0 - \pi_\delta(\Phi_0), W - \Phi_0 - \pi_\delta(\Phi_0)\big)\\
&\qquad - \frac{1}{2}D^2\mathcal{E}^{\Phi_0}\big(\Phi_0 - \pi_\delta(\Phi_0), \Phi_0 - \pi_\delta(\Phi_0)\big) + \mathcal{O}\big(\|W\|^3 + \big\|\Phi_0 - \pi_\delta(\Phi_0)\big\|^3\big),
\end{aligned}$$

where $\mathcal{O}(\|W\|^3)$ is actually a regular function of W. From the ellipticity property stated above, there exists a unique element W in $X_\delta \cap X_{\Phi_0}$ in a small enough neighborhood of 0, that minimizes $\mathcal{E}^{\mathrm{HF}}(\Psi) - \mathcal{E}^{\mathrm{HF}}(\pi_\delta(\Phi_0))$ – hence $\mathcal{E}^{\mathrm{HF}}(\Psi)$ – and

$\Psi \in \mathcal{K} \cap X_\delta$ is recovered from W thanks to (33.7). From the same ellipticity property it is straightforward to get that

$$\|W\| \leqslant c\|\Phi_0 - \pi_\delta \Phi_0\|$$

and the optimality of the approximation is obtained.

We have just proven that, in a small enough neighborhood of Φ_0 (or $\pi_\delta \Phi_0$), there exists a discrete solution Φ_δ (that is unique to some extent, the W) and such that the error, between Φ_0 and Φ_δ is of the same order as $\Phi_0 - \pi_\delta \Phi_0$.

This is quite reassuring in the sense that if the dimension of discrete space increases, the accuracy of the approximation will improve but it is not enough to certify the result. This is the object of the next section.

34. Analysis of the certification of the results

We have now to get place in a different philosophy. Indeed, we do not want to prove that the discrete solution exists and can be computed, we do not want either to state that if the discrete space is large enough we obtain an accurate result (even optimal) which is *a priori* considerations. What we want now consists in a final stage where we have computed one approximate solution so that we are able to validate the result and understand its pertinency a posteriori. We cannot use any more the reference to Φ_0 explicitly since this is the unknown object from which we have an approximation named Φ_δ. The only thing that we accept, and this is justified from the previous analysis, is that, we are close to Φ_0 and the question is: "How close are we?" This question enters in what in known in numerical analysis to be the *a posteriori* analysis and more precisely, for what we shall present here, in the general paradigm of the definition of explicit lower and upper bounds for outputs introduced in MADAY, PATERA and PERAIRE [1999], PARASCHIVOIU and PATERA [1998] and first analyzed in MADAY and PATERA [2000]. The output of interest can be either linear or nonlinear in terms of the solution. Here, as a particular output, we consider the Hartree–Fock energy as in MADAY and TURINICI [2002a]. In the previous section, the analysis showed that the bilinear form a_{Φ_0} is elliptic. Such an information is important as it will allow to prove that, provided Φ_δ is close enough to Φ_0, then a_{Φ_δ} is also elliptic with an ellipticity constant independent of δ. Let us report here the analysis of MADAY and TURINICI [2002a] where, for the sake of simplicity, we assume again that $X_{\Phi_0} = \Phi_0^{\perp\perp}$. It then follows that, for any $\Xi \in \Phi_\delta^{\perp\perp}$, the decomposition (33.3) based on Φ_0 leads to $\Xi = M\Phi_0 + \widetilde{\Xi}$ where $M_{ij} = \langle \xi_i, \Phi_0 \rangle$. It also follows that

$$|M_{ij}| = \left|\langle \xi_i, \Phi_0 - \Phi_\delta \rangle\right| \leqslant \|\Xi\|_{L^2(\mathbb{R}^3)} \|\Phi_0 - \Phi_\delta\|_{L^2(\mathbb{R}^3)}$$

so that, for $\|\Phi_0 - \Phi_\delta\|_{\mathcal{H}} \leqslant \varepsilon$, small enough,

$$\begin{aligned} a_{\Phi_\delta}(\Xi, \Xi) &= a_{\Phi_\delta}\big(M\Phi_0 + \widetilde{\Xi}, M\Phi_0 + \widetilde{\Xi}\big) \\ &\geqslant a_{\Phi_\delta}\big(\widetilde{\Xi}, \widetilde{\Xi}\big) - C\|\Xi\|_{\mathcal{H}} \big\|\widetilde{\Xi}\big\|_{\mathcal{H}} \|\Phi_0 - \Phi_\delta\|_{\mathcal{H}} \\ &\geqslant a_{\Phi_\delta}\big(\widetilde{\Xi}, \widetilde{\Xi}\big) - C\|\Xi\|_{\mathcal{H}}^2 \|\Phi_0 - \Phi_\delta\|_{\mathcal{H}}. \end{aligned}$$

Since $|\Lambda_{ij}^{\delta} - \Lambda_{ij}^{0}| \leqslant C\|\Phi_0 - \Phi_\delta\|_{\mathcal{H}}$, it follows that

$$a_{\Phi_\delta}(\widetilde{\Xi}, \widetilde{\Xi}) - a_{\Phi_0}(\widetilde{\Xi}, \widetilde{\Xi}) \leqslant c\|\widetilde{\Xi}\|_{\mathcal{H}}^2 \|\Phi_0 - \Phi_\delta\|_{\mathcal{H}}$$

so that, there exists a constant $\gamma > 0$, depending only on Φ_0 such that

$$a_{\Phi}(\Xi, \Xi) \geqslant \frac{2\alpha}{3}\|\widetilde{\Xi}\|_{\mathcal{H}}^2 \geqslant \frac{\alpha}{2}\|\Xi\|_{\mathcal{H}}^2 \tag{34.1}$$

for $\|\Phi_0 - \Phi_\delta\|_{\mathcal{H}} \leqslant \gamma$ small enough.

We can now follow the same lines as in the previous section using the ellipticity of $D^2\mathcal{E}^{\Phi_\delta}$ and the decomposition $\Phi_0 - \Phi_\delta = S\Phi_\delta + W$ where W belongs to $\Phi_\delta^{\perp\perp}$. We get similarly

$$\begin{aligned}
&\mathcal{E}^{\mathrm{HF}}(\Phi_0) - \mathcal{E}^{\mathrm{HF}}(\Phi_\delta) = \mathcal{E}^{\Phi_\delta}(\Phi_0) - \mathcal{E}^{\Phi_\delta}(\Phi_\delta) \\
&\quad = \mathcal{E}^{\Phi_\delta}(\Phi_\delta + S\Phi_\delta + W) - \mathcal{E}^{\Phi_\delta}(\Phi_\delta) = D\mathcal{E}^{\Phi_\delta}(S\Phi_\delta + W) \\
&\qquad + \frac{1}{2}D^2\mathcal{E}^{\Phi_\delta}(S\Phi_\delta + W, S\Phi_\delta + W) + \mathcal{O}(\varepsilon^3).
\end{aligned} \tag{34.2}$$

From the definition of the discrete minimizer Φ_δ, we derive that $D\mathcal{E}^{\Phi_\delta}(S\Phi_\delta) = 0$, in addition, the following bounds are simple or derived from (33.6): $\|W\|_{\mathcal{H}} \leqslant C\varepsilon$ and $\|S\Phi_\delta\|_{\mathcal{H}} \leqslant C\varepsilon^2$, so that

$$\mathcal{E}^{\mathrm{HF}}(\Phi_0) - \mathcal{E}^{\mathrm{HF}}(\Phi_\delta) = D\mathcal{E}^{\Phi_\delta}(W) + \frac{1}{2}D^2\mathcal{E}^{\Phi_\delta}(W, W) + \mathcal{O}(\varepsilon^3). \tag{34.3}$$

Let us now introduce the problem of finding $\widehat{W} \in \Phi_\delta^{\perp\perp}$ such that

$$D^2\mathcal{E}^{\Phi_\delta}(\widehat{W}, \Psi) + D\mathcal{E}^{\Phi_\delta}(\Psi) = 0, \quad \forall \Psi \in \Phi_\delta^{\perp\perp} \tag{34.4}$$

that has, due to assumed ellipticity over $\Phi_\delta^{\perp\perp}$, a unique solution. The element $\widehat{W}$ is named *the reconstructed error* for reasons that will be sketched afterwards. This allows to derive from (34.2) that

$$\begin{aligned}
\mathcal{E}^{\mathrm{HF}}(\Phi_0) &= \mathcal{E}^{\mathrm{HF}}(\Phi_\delta) - D^2\mathcal{E}^{\Phi_\delta}(\widehat{W}, W) + \frac{1}{2}D^2\mathcal{E}^{\Phi_\delta}(W, W) + \mathcal{O}(\varepsilon^3) \\
&= \mathcal{E}^{\mathrm{HF}}(\Phi_\delta) - \frac{1}{2}D^2\mathcal{E}^{\Phi_\delta}(\widehat{W}, \widehat{W}) + \frac{1}{2}D^2\mathcal{E}^{\Phi_\delta}(W - \widehat{W}, W - \widehat{W}) + \mathcal{O}(\varepsilon^3).
\end{aligned}$$

We thus get the explicit lower bound

$$\mathcal{E}^{\mathrm{HF}}(\Phi_0) \geqslant \mathcal{E}^{\mathrm{HF}}(\Phi_\delta) - \frac{1}{2}D^2\mathcal{E}^{\Phi_\delta}(\widehat{W}, \widehat{W}) + \mathcal{O}(\varepsilon^3)$$

so that certainly

$$\mathcal{E}^{\mathrm{HF}}(\Phi_0) \geqslant \mathcal{E}^{\mathrm{HF}}(\Phi_\delta) - D^2\mathcal{E}^{\Phi_\delta}\big(\widehat{W}, \widehat{W}\big). \tag{34.5}$$

We can remind again that, the variational principle always provides a discrete minimum that is larger than the global one so that

$$\mathcal{E}^{\mathrm{HF}}(\Phi_\delta) \geqslant \mathcal{E}^{\mathrm{HF}}(\Phi_0) \geqslant \mathcal{E}^{\mathrm{HF}}(\Phi_\delta) - D^2\mathcal{E}^{\Phi_\delta}\big(\widehat{W}, \widehat{W}\big). \tag{34.6}$$

The previous equation thus provides an explicit upper and lower bound on the Hartree–Fock energy.

Four remarks are in order. The first one is related to the information that is in this two-sided bound. It is obvious that it would be of no interest if $D^2\mathcal{E}^{\Phi_\delta}(\widehat{W}, \widehat{W})$ would be large. On the contrary, as proven in MADAY and TURINICI [2002a] there exits a constant such that

$$\|\widehat{W}\|_{\mathcal{H}} \leqslant c\|W\|_{\mathcal{H}}, \tag{34.7}$$

so that the width of the bound is small and of the same order as $\|\mathcal{E}^{\mathrm{HF}}(\Phi_\delta) - \mathcal{E}^{\mathrm{HF}}(\Phi_0)\|_{\mathcal{H}}$.

The second remark is that the computation of $\widehat{W}$ involves a direct problem and not an eigenvalue problem. In addition, all operators in this problem are linear. The resolution of the (34.4) is thus much more simple than the original minimization (or eigenvalue) problem. This is the reason why it is feasible, instead of solving (34.4), to solve a discrete version of it (actually the only possible one) on a discrete space $X_{\tilde{\delta}}$ much finer than X_δ. The two-sided bound proposed in (34.6) is then estimating $\mathcal{E}^{\mathrm{HF}}(\Phi_{\tilde{\delta}})$ instead of $\mathcal{E}^{\mathrm{HF}}(\Phi_0)$. Nevertheless, the dimension of $X_{\tilde{\delta}}$ makes $\mathcal{E}^{\mathrm{HF}}(\Phi_{\tilde{\delta}})$ and $\mathcal{E}^{\mathrm{HF}}(\Phi_0)$ very close, even though the computation of $\Phi_{\tilde{\delta}}$ itself is never required. This strategy has been numerically illustrated in MADAY and TURINICI [2002a] on various STOnG basis with increasing n's.

The third remark concerns another application of the validation method. This method has currently been presented in order to certify the output after the calculation of the discrete solution that, we remind, is solution of a nonlinear problem. We remind also that the resolution of this nonlinear problem is obtained through an iterative process that enters under the SCF strategies. The other application that can be proposed for this a posteriori process is the validation of the iterative process per se. Let now consider that the solution of the discrete problem Φ_δ is obtained as the limit of a sequence Φ_δ^k as k goes to infinity. Of course k never goes to infinity but up to some $k_{\max}$ and the certification can be used to validate the chosen step $k_{\max}$. Indeed, the reconstructed error can be computed by finding $\widehat{W}^k \in \Phi_\delta^{k,\perp\perp} \cap X_\delta$ such that

$$D^2\mathcal{E}^{\Phi_\delta^k}\big(\widehat{W}^k, \Psi\big) + D\mathcal{E}^{\Phi_\delta^k}(\Psi) = 0, \quad \forall \Psi \in \Phi_\delta^{k,\perp\perp} \cap X_\delta, \tag{34.8}$$

instead of (34.4). This strategy has been implemented in the code Asterix (ERNENWEIN, ROHMER and BÉNARD [1990], ROHMER, DEMUYNCK, BÉNARD, WIEST, BACHMANN, HENRIET and ERNENWEIN [1990], WIEST, DEMUYNCK, BÉNARD, ROHMER and ERNENWEIN [1991]) on the carbyne molecule $Cr(CO)_4C\ell CH$ and has proved of be efficient and has well illustrated the validation property of this estimator (cf. MADAY and TURINICI [2002a]).

The last remark goes further in this particular context of the Hartree–Fock problem since the reconstructed error $\widehat{W}$ can actually be shown to be very close to the actual (main part of) the error W. Indeed it is proven in (cf. MADAY and TURINICI [2002a]) that $\widehat{W} = W + \mathcal{O}(\varepsilon^2)$ so that an improvement on the solution Φ_δ can be proposed by setting $\tilde{\Phi}_\delta = \Phi_\delta + \widehat{W}$. Note that this allows to name $\widehat{W}$ as *reconstructed error*. Actually further slight improvement can be added in this direction to propose, from this ε approximation to Φ_0, a ε^2 approximation to Φ_0 and a ε^4 approximation of the energy.

35. Post-Hartree–Fock methods

As already mentioned in Section 11, post-Hartree–Fock methods include the Møller–Plesset perturbation (MP2, MP3, ...), the configuration interaction (CI), the coupled cluster (CC), the multi-configuration self-consistent field (MCSCF) methods, and finally methods using explicitly correlated wavefunctions (see Section 32).

(A) Møller–Plesset perturbation method

The Møller–Plesset perturbation methods originate from standard perturbation theory, which is a commonly used tool in quantum mechanics. For convenience, let us briefly describe it. Let us consider an isolated quantum system modeled by the Hamiltonian H_0 and a corresponding stationary state ψ_0 of energy E_0. Let us apply to this system a time-independent external perturbation modeled by the interaction Hamiltonian $\mathcal{V}$, and search for a stationary state ψ of the global Hamiltonian $H = H_0 + \mathcal{V}$ "close" to ψ_0. For this purpose, one searches for $\lambda \in \mathbb{C}$ a solution $(\psi(\lambda), E(\lambda))$ to

$$\begin{cases} (H_0 + \lambda\mathcal{V})\psi(\lambda) = E(\lambda)\psi(\lambda), \\ \|\psi(\lambda)\| = 1 \end{cases} \tag{35.1}$$

of the form $\psi(\lambda) = \sum_{n=0}^{+\infty} \lambda^n \psi_n$, $E(\lambda) = \sum_{n=0}^{+\infty} \lambda^n E_n$. Reporting the latter expressions in (35.1), one obtains for any $n \in \mathbb{N}^*$

$$(RS_n) \begin{cases} (H_0 - E_0) \cdot \psi_n = E_n \psi_0 + f_n, \\ (\psi_0, \psi_n) = \alpha_n, \end{cases}$$

where f_n and α_n depends only on $\{\psi_k\}_{0 \leqslant k \leqslant n-1}$ and $\{E_k\}_{0 \leqslant k \leqslant n-1}$. When the triangular system (RS) consisting in all the systems (RS_n) together has a unique solution, one ends up with two formal series in λ

$$\sum_{n=0}^{+\infty} \lambda^n \psi_n \quad \text{and} \quad \sum_{n=0}^{+\infty} \lambda^n E_n$$

usually called *Rayleigh–Schrödinger expansions*. If the convergence radii of both of these expansions are greater than one, it is clear that $\psi = \psi(1) = \sum_{n=0}^{+\infty} \psi_n$ and

$E = E(1) = \sum_{n=0}^{+\infty} E_n$ satisfy

$$H\psi = E\psi,$$

which provides a solution to the problem. In practice, one hardly estimates the radius of convergence of the perturbation expansion rigorously (although this has sometimes be done, e.g., in the so-called Z^{-1}-expansion of atomic correlation energies). Even if the radius of convergence of the perturbation series is zero, which is often the case (e.g., for an atom in an external electric field), its leading orders may still be physically useful.

The mathematical foundations of the perturbation theory for linear operators can be read in the reference textbook by KATO [1980].

The Møller–Plesset perturbation method consists in applying the general technique described above to the search for a ground state of the electronic Hamiltonian H_e taking the HF ground state $\mathcal{D} = \sum_{i=1}^{N} |\phi_i\rangle\langle\phi_i|$ as a reference point. More precisely, one first defines on the Hilbert space $\mathcal{H}_e$ the unperturbed Hamiltonian (corresponding to a HF ground state)

$$H_0 = \sum_{i=1}^{N} \mathcal{F}(\mathcal{D})_{x_i},$$

where $\mathcal{F}(\mathcal{D})$ is the Fock operator associated with $\mathcal{D}$ (see Eq. (10.3)) and $\mathcal{F}(\mathcal{D})_{x_i}$ denotes the action of this one-particle on particle i. Next, one defines the perturbed Hamiltonian as the true electronic Hamiltonian ($H = H_e$). The perturbation is therefore given by

$$\mathcal{V} = H - H_0 = \sum_{1 \leqslant i < j \leqslant N} \frac{1}{|x_i - x_j|} - \sum_{i=1}^{N} \mathcal{G}(\mathcal{D})_{x_i}.$$

Let us then remark that the Slater determinant $\psi_0 = \frac{1}{\sqrt{N!}} \det(\phi_i(x_i))$ corresponding to the HF ground state is an eigenstate (in fact the ground state) of H_0 on $\mathcal{H}_e$; indeed,

$$H_0\psi_0 = E_0\,\psi_0$$

with $E_0 = \sum_{i=1}^{N} \varepsilon_i$, the ε_i denoting the smallest N eigenvalues (with multiplicities) of the Fock operator $\mathcal{F}(\mathcal{D})$ (cf. Section 9). The perturbation method can now be used.

The nth order Møller–Plesset perturbation allows to compute the energy

$$E^{\mathrm{MP}n} = \sum_{k=0}^{n} E_k.$$

At the first order, one recovers the Hartree–Fock energy; indeed

$$E^{\mathrm{MP1}} = E_0 + E_1 = E_0 + \langle\psi_0, \mathcal{V}\psi_0\rangle = \sum_{i=1}^{N} \varepsilon_i - \frac{1}{2}\mathrm{Trace}\big(\mathcal{G}(\mathcal{D})\mathcal{D}\big) = E^{\mathrm{HF}}.$$

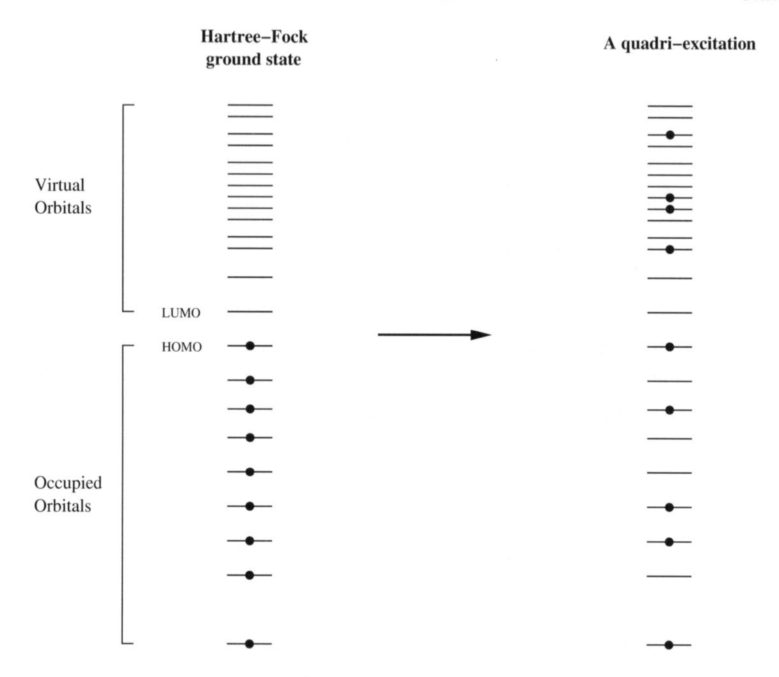

FIG. 35.1. Schematic representation of a quadri-excitation of a spinless HF ground state.

At the second order, one obtains

$$E^{\mathrm{MP2}} = E_0 + E_1 + E_2 = E^{\mathrm{HF}} - \langle \psi_0, \mathcal{V}(H_0 - E_0)^{-1} Q\mathcal{V}\psi_0 \rangle,$$

where Q denotes the projector $I - |\psi_0\rangle\langle\psi_0|$.

Within a Galerkin approximation, the diagonalization of the Fock matrix $F(D) \in \mathcal{M}_S(N_b)$ yields a set of N *occupied* orbitals and a set of $N_v = N_b - N$ *virtual* orbitals.

According to the aufbau principle the occupied orbitals are those of lowest energy. The *Highest Occupied Molecular Orbital* and the *Lowest Unoccupied Molecular Orbital* are referred to by the acronyms HOMO and LUMO, respectively. A Slater determinant built with $N - k$ occupied and k virtual orbitals is called an *excited configuration*, or a k-excitation of the HF ground state.

The term E_2 can be easily computed by *sum over states*, for among all the Slater determinants built with eigenvectors of $F(D)$, only the di-excitations have a nonzero contribution. After calculation (see, e.g., MCWEENY [1992]),

$$E_2 = \frac{1}{4} \sum_{i,j=1}^{N} \sum_{a,b=1}^{n-N} \frac{|\langle \psi_{ij}^{ab}, \mathcal{V}\psi_0 \rangle|^2}{\varepsilon_i + \varepsilon_j - \varepsilon_a - \varepsilon_b} = \frac{1}{4} \sum_{i,j=1}^{N} \sum_{a,b=1}^{n-N} \frac{|(ij||ab)|^2}{\varepsilon_i + \varepsilon_j - \varepsilon_a - \varepsilon_b},$$

where ψ_{ij}^{ab} is the di-excitation obtained by replacing in the HF ground state the occupied orbitals ϕ_i and ϕ_j by the virtual orbitals ϕ_a and ϕ_b.

For the spinless HF model,

$$
\begin{aligned}
(ij\|ab) &= (ij|ab) - (ia|jb) \\
&= \int_{\mathbb{R}^3}\int_{\mathbb{R}^3} \frac{\phi_i(x)\phi_j(x)\phi_a(x')\phi_b(x')}{|x-x'|}\,dx\,dx' \\
&\quad - \int_{\mathbb{R}^3}\int_{\mathbb{R}^3} \frac{\phi_i(x)\phi_a(x)\phi_j(x')\phi_b(x')}{|x-x'|}\,dx\,dx'.
\end{aligned}
$$

The computation of the terms $E_3, E_4, \ldots$ can be performed in the same way but calculations become heavier and heavier as the order of the perturbation increases. Only MP2, MP3 and MP4 are commonly used in practice.

(B) Configuration interaction

Let us assume that the Hartree–Fock problem has already been solved in a given basis set $\{\chi_\mu\}_{1\leqslant\mu\leqslant N_b}$, and that the Fock matrix corresponding to the HF ground state has been diagonalized; one thus is provided with N *occupied* spin orbitals $(\phi_i)_{1\leqslant i\leqslant N}$ (N orthonormal eigenvectors associated with the lowest N eigenvalues of the Fock matrix) and with $N_b - N$ *virtual* orbitals $(\phi_i)_{N+1\leqslant i\leqslant N_b}$ ($N_b - N$ orthonormal eigenvectors associated with the highest $N_b - N$ eigenvalues of the Fock matrix).

For any subset $I = (i_1, \ldots, i_N)$ of $|[1, N_b]|$ containing N elements, we denote by ψ_I the Slater determinant

$$
\psi_I(x_1, \ldots, x_N) = \frac{1}{\sqrt{N!}} \det\big(\phi_{i_k}(x_j)\big). \tag{35.2}
$$

It is possible to build in this way $\binom{N_b}{N}$ Slater determinants which form an orthonormal basis of the vector space $\bigwedge_{i=1}^N V_h$ where $V_h = \operatorname{span}(\chi_1, \ldots, \chi_{N_b})$. The fact that ψ_I and ψ_J are orthogonal if $I \neq J$ is a direct consequence of the orthonormality of the $(\phi_i)_{1\leqslant i\leqslant N_b}$.

The configuration interaction (CI) method consists in searching the lowest energy wavefunction in the vector space spanned by a subset of the ψ_I. Let us denote by $\mathcal{I}$ a subset of the $\binom{N_b}{N}$ partitions of $|[1, N_b]|$ containing N elements. For a given wavefunction

$$
\psi_e = \sum_{I\in\mathcal{I}} c_I \psi_I, \tag{35.3}
$$

one has

$$
\langle\psi_e, H_e\psi_e\rangle = \sum_{I,J\in\mathcal{I}} c_I c_J \langle\psi_I, H_e\psi_J\rangle \quad \text{and} \quad \|\psi_e\|_{L^2} = \Big(\sum_{I\in\mathcal{I}} c_I^2\Big)^2.
$$

A generic CI problem therefore reads

$$\inf\left\{ \sum_{I,J\in\mathcal{I}} c_I c_J \langle \psi_I, H_e \psi_J \rangle,\ (c_I) \in \mathbb{R}^{\mathrm{Card}\,\mathcal{I}},\ \sum_{I\in\mathcal{I}} c_I^2 = 1 \right\}. \tag{35.4}$$

Two extreme cases can be identified:

- When $\mathcal{I}$ contains the single element $\{(1, 2, \ldots, N)\}$, one is left with the HF ground state;
- When $\mathcal{I}$ contains all the $\binom{N_b}{N}$ partitions of $|[1, N_b]|$ containing N elements, one is led to what is usually called the *Full CI* method.

The number of terms in a full-CI expansion increases essentially exponentially with the size of the one-electron basis. This limits the practical use of full CI to very small basis sets, which also means very small atoms or molecules. One is therefore obliged to resort to a *selection of the excited configurations to be included* in the CI expansion. The most obvious selection consists in considering only those excited configurations, which have a nonvanishing matrix element of the Hamiltonian with the reference configuration ψ_0. These configurations define the *first-order interacting space*. For ψ_0 a single Slater determinant, the first-order interacting space consists of all *double excitations*

$$\psi_{ij}^{ab} = a_{ij}^{ab} \psi_0,$$

where a_{ij}^{ab} is the *excitation operator* which replaces the occupied spin orbitals ϕ_i, ϕ_j by the formerly unoccupied ones ϕ_a, ϕ_b. A CI with all doubly excited configurations is called CID. Single excitations

$$\psi_i^a = a_i^a \psi_0,$$

do not belong to the first-order interacting space, since their matrix elements with ψ_0 vanish due to the *Brillouin Theorem*, provided that ψ_0 is solution of the Hartree–Fock equations. It is, nevertheless, recommended to include the ψ_i^a, i.e., to use CISD rather than CID. This gives an improvement at a rather low extra cost.

The next step is to include triply excited configurations

$$\psi_{ijk}^{abc} = a_{ijk}^{abc} \psi_0,$$

which is much more expensive, since the number of the ψ_{ijk}^{abc} is $\frac{1}{12}N(N-1)(N-2) \times (N_b - N)(N_b - N - 1)(N_b - N - 2)$ compared to $N(N_b - N)$ single excitations and $\frac{1}{4}N(N-1)(N_b - N)(N_b - N - 1)$ double excitations.

Problem (35.4) has a very simple mathematical structure: Its solution is the lowest eigenvalue of the symmetric matrix $[\langle \psi_I, H_e \psi_J \rangle]_{I,J\in\mathcal{I}}$. The only difficulty comes from the size $\mathrm{Card}\,\mathcal{I} \times \mathrm{Card}\,\mathcal{I}$ of the matrix, which scales as N^4 for CID and CISD, N^6 for CISDT, and so on. Even more serious is the scaling with N_b for large basis sets. Despite the sparsity of the matrix $[\langle \psi_I, H_e \psi_J \rangle]$ (if two determinants ψ_I and ψ_J have less than $N-2$ orbitals in common $\langle \psi_I, H_e \psi_J \rangle = 0$), the CI method therefore requires a huge computational effort for large systems.

All selected CI schemes (unlike full CI) have a serious drawback, that makes them practically useless. The energy obtained by CISD or CISDT etc., is not *extensive*, or as it is often stated not *size-consistent* (POPLE, KRISHNAN, SCHLEGEL and BINKLEY [1978a]). This means that for a supersystem AB, consisting of two *noninteracting* subsystems A and B, the energy $E(AB)$ is *not* equal to $E(A)+E(B)$. One even finds that for a supersystem consisting of n noninteracting 2-electron systems the energy scales with $\sqrt{n}$ rather than with n (AHLRICHS [1974]). This unphysical behavior is *not* shared by Møller–Plesset perturbation theory, which is extensive.

Attempts to remedy this defect of CI lead us immediately to coupled-cluster theory (COESTER and KÜMMEL [1960]).

(C) Coupled clusters

Operators which describe physical observables are usually *additively separable*, which means that the operator $\Omega(AB)$ for a supersystem AB, consisting of two noninteracting subsystems A and B, is related to the corresponding operators $\Omega(A)$, $\Omega(B)$ for the subsystems as

$$\Omega(AB) = \Omega(A) + \Omega(B).$$

Consider now the *wave operator* W, which formally relates the correlated (e.g., full CI) wavefunction ψ to the reference wavefunction ψ_0.

$$\psi = W\psi_0.$$

One easily sees that W is multiplicatively separable (PRIMAS [1961], PRIMAS [1963], KUTZELNIGG [1989a])

$$W(AB) = W(A)W(B).$$

If one only admits double excitations in W, one cannot satisfy this multiplicative separability. If one allows independent double excitations in either subsystem, one must allow quadruple excitations in the supersystem. A way out of this dilemma consists in writing W in exponential form (COESTER and KÜMMEL [1960], BARTLETT [1989])

$$W = e^S,$$

and to expand the *cluster amplitude* S, rather than the wave operator W, in terms of excitation operators

$$\begin{aligned} S &= S_1 + S_2 + S_3 + \cdots, \\ S_1 &= S^i_a a^a_i, \\ S_2 &= S^{ij}_{ab} a^{ab}_{ij}, \quad \text{etc.,} \end{aligned} \tag{35.5}$$

where S_a^i etc., are expansion coefficients, and where the Einstein summation convention over repeated indices is implied (KUTZELNIGG [1982], KUTZELNIGG and KOCH [1983]). Since S is an additively separable operator, the expansion (35.5) is justified. Like in CI, one can define a hierarchy of approximations CCSD, CCSDT etc., with increasing excitation rank, but now depending correctly on the number of particles.

Unfortunately the energy expectation value for a CC wavefunction

$$E = \frac{\langle\psi_0|e^{S^\dagger}He^S|\psi_0\rangle}{\langle\psi_0|e^{S^\dagger}e^S|\psi_0\rangle}$$

is hardly manageable. One must expand the exponential, and neither the expansion of the numerator, nor that of the denominator terminate. One therefore renounces usually on a variational formulation and rather uses an approach related to the method of moments.

One starts with the Schrödinger equation

$$He^S\psi_0 = Ee^S\psi_0$$

and multiplies this from the left by e^{-S}

$$e^{-S}He^S\psi_0 = E\psi_0.$$

One then takes the scalar products from the left with ψ_0^* and $\psi_0^* X^\dagger$, where $X^\dagger$ is some deexcitation operator. The energy is then

$$E = \langle\psi_0|e^{-S}He^S|\psi_0\rangle = \langle\psi_0|H + [H,S] + \frac{1}{2}\big[[H,S],S\big] + \cdots + |\psi_0\rangle$$

while for any $X^\dagger$ we get

$$0 = \langle\psi_0|X^\dagger e^{-S}He^S|\psi_0\rangle = \langle\psi_0|X^\dagger H + X[H,S] + \frac{1}{2}X^\dagger\big[[H,S],S\big] + \cdots + |\psi_0\rangle.$$

Fortunately the Hausdorff expansion terminates, if ψ_0 is a single Slater determinant. For a CCSD ansatz we get

$$\begin{aligned}
E &= \langle\psi_0|e^{-S_2}He^{S_2}|\psi_0\rangle = \langle\psi_0|H + [H,S_2]|\psi_0\rangle,\\
0 &= \langle\psi_0|X_2^\dagger e^{-S_2}He^{S_2}|\psi_0\rangle\\
&= \langle\psi_0|X_2 H + X^\dagger[H,S_2] + \frac{1}{2}X_2^\dagger\big[[H,S_2],S_2\big]|\psi_0\rangle,
\end{aligned}$$

where $X_2^\dagger$ is a two-particle deexcitation operator of the type a_{ab}^{ij}. Choosing as $X_2^\dagger$ exactly the Hermitian conjugates of the operators into which S_2 is expanded, one gets a nonlinear system of as many equations as there are unknown expansion coefficients S_{ab}^{ij}.

Unlike CI, CC does not provide a strict upper bound to the ground state energy, but one can, at least, show (KUTZELNIGG [1991], KUTZELNIGG [1998]) that the error of the energy is quadratic in the error of the wavefunction. The lack of an upper bound is outweighed by the size-consistency (extensivity) of the results.

At present CC methods represent the state-of-the-art for high-precision calculations for medium-sized atoms or molecules. There is some evidence that the CC hierarchy converges rather rapidly with the particle rank, i.e., via CCSD, CCSDT etc., to full CI. In this respect CC is definitely superior to both MP and selected CI. A drawback is the rather unfortunate scaling with the particle number. However, considerable progress has been achieved in recent years on the way to linear scaling, mainly by taking advantage of the localizability of wavefunctions.

CCSD can be regarded as a kind of standard method. CCSDT is already so demanding, that it is hardly used in complete form. One rather uses approximations to CCSDT, in which the triple excitations are only treated at a perturbative level. Particularly popular approximations are referred to as CCSD(T) (RAGHAVACHARI, POPLE and HEAD-GORDON [1989]) and CCSD[T] (NOGA, BARTLETT and URBAN [1987]). CCSDTQ (including quadruple excitation) has only been used occasionally (BARTLETT [1989]). Rather promising is the combination of CCSD or CCSD[T] with the R12-method, to take care of the correlation cusp (NOGA, TUNEGA, KLOPPER and KUTZELNIGG [1995], NOGA, KLOPPER and KUTZELNIGG [1998]).

(D) Multi-configuration self-consistent field

The MCSCF method has been described in Section 11. MCSCF are CI methods in which *both* the orbitals and the weights of the configurations (i.e., of the determinants) are optimized.

After discretization in a finite basis set $\{\chi_\mu\}_{1\leqslant\mu\leqslant N_b}$, problem (11.2) reads

$$\inf\left\{\sum_{i,j=1}^{K}\sum_{\mu,\nu=1}^{N_b} h_{\mu\nu}\gamma_{ij}C_{\mu j}C_{\nu i}+\frac{1}{2}\sum_{i,j,k,l=1}^{K}\sum_{\mu,\nu,\kappa,\lambda=1}^{N_b}(\mu\nu|\kappa\lambda)\Gamma_{ijkl}C_{\kappa k}C_{\lambda l}C_{\mu i}C_{\nu j},\right.$$
$$C\in\mathcal{M}(N_b,K),\ C^*SC=I_K,\ (c_I)\in\mathbb{R}^{\mathrm{Card}\,\mathcal{I}},\ \sum_{I\in\mathcal{I}}c_I^2=1,$$
$$\left.\gamma_{ij}=\sum_{I,J\in\mathcal{I}}c_Ic_J\gamma_{ij}^{IJ},\ \Gamma_{ijkl}=\sum_{I,J\in\mathcal{I}}c_Ic_J\Gamma_{ijkl}^{IJ}\right\}, \tag{35.6}$$

where $\mathcal{I}$ denotes the set of the $\binom{K}{N}$ partitions of $|[1,K]|$ containing N elements, where $h_{\mu\nu}$ and $(\mu\nu|\kappa\lambda)$ are, respectively, defined by (20.2) and (20.3) and where the coefficients γ_{ij}^{IJ} and Γ_{ijkl}^{IJ} which take their values in the set $\{-1,0,1\}$, are such that

$$\langle\psi_I,H_e\psi_J\rangle=\sum_{i,j=1}^{K}\sum_{\mu,\nu=1}^{N_b}h_{\mu\nu}\gamma_{ij}^{IJ}C_{\mu j}C_{\nu i}$$
$$+\frac{1}{2}\sum_{i,j,k,l=1}^{K}\sum_{\mu,\nu,\kappa,\lambda=1}^{N_b}(\mu\nu|\kappa\lambda)\Gamma_{ijkl}^{IJ}C_{\kappa k}C_{\lambda l}C_{\mu i}C_{\nu j},$$

where ψ_I is defined by (35.2).

Complete Active Space (CASSCF) methods, which have been already mentioned in Section 11, consist in restricting the set $\mathcal{I}$. For this purpose, a number N_a of *active orbitals* is selected; the set $\mathcal{I}$ is then chosen such that for any $I \in \mathcal{I}$, $|[1, K - N_a]| \subset I$. Admissible configurations are therefore built by

- populating all of the $K - N_a$ *inactive orbitals* $(\Phi_i)_{1 \leqslant i \leqslant K-N_a}$;
- selecting $N - K + N_a$ orbitals among the N_a *active orbitals* $(\Phi_i)_{K-N_a+1 \leqslant i \leqslant K}$.

CAS-SCF corresponds to performing full CI within the active space.

In MCSCF it is customary to apply a second-order quadratically convergent (Newton–Raphson type) iteration scheme, which converges fast, but which requires the construction of the Hessian of the Hamiltonian in each iteration (SHEPARD [1993b]). Quasi-Newton methods may also be used (CHABAN, SCHMIDT and GORDON [1997]).

MCSCF and coupled cluster (CC) are, to some extent, complementary. While CC takes care mainly of the so-called *dynamic* correlation effects (which are characterized by a mixture of excited configurations with small coefficients, but large contributions to the energy), MCSCF mainly deals with situations of near-degeneracy, where some excited configurations enter with large coefficients. One refers then to *nondynamic* correlation. In those cases where it is necessary to use MCSCF rather than Hartree–Fock as a starting point, it is usually still necessary to built a CC type calculation upon MCSCF, to take care of the dynamic correlation effects as well.

36. The linear subproblem

In Sections 28–30 devoted to the description and analysis of the SCF algorithm, we have considered the linear problem as a subroutine. It is now time to focus on the inner loop of the SCF cycle. At each cycle, the current mean-field Hamiltonian, here denoted by F, is used to build a new density matrix on the basis of the *aufbau* principle.

According to characterization (22.6) of *aufbau* solutions to systems such as (22.4), this amounts to solving the minimization problem

$$\inf\left\{\mathrm{Trace}(FD),\ D \in \mathcal{M}_S(N_b),\ DSD = D,\ \mathrm{Trace}(SD) = N\right\} \tag{36.1}$$

or in the language of molecular orbitals

$$\inf\left\{\mathrm{Trace}\left(FCC^*\right),\ C \in \mathcal{M}(N_b, N),\ C^*SC = I_N\right\}, \tag{36.2}$$

the two forms being equivalent through the transform $D = CC^*$. Here, F and S, respectively, denote the mean-field Hamiltonian matrix and the overlap matrix, which are symmetric and sparse (at least for large systems, in LCAO approximation). The matrix S is in addition positive definite, for it is the Gram matrix $(\chi_\mu, \chi_\nu)_{L^2}$ for the basis set $\{\chi_\mu\}_{1 \leqslant \mu \leqslant N_b}$. The integer N, defining one constraint in (36.1), or fixing the size of the matrices in (36.2), is the number of electrons, that should not be confused with the number N_b of basis set functions. Typically, for atomic orbitals basis sets, N_b is of the order of mN, where, say, $2 \leqslant m \leqslant 10$, which means that the issue we want to address is basically to *find the lowest N eigenvalues and associated eigenvectors of a matrix of size* $(mN) \times (mN)$.

Let us remark that in the special case when $N = 1$ and $S = I_{N_b}$, (36.2) simply reads

$$\inf\{(\Phi, F\Phi),\ \Phi \in \mathbb{R}^{N_b},\ \|\Phi\|^2 = 1\}. \tag{36.3}$$

Our aim here is to report on various techniques allowing to solve problem (36.1), or alternatively (36.2) in $O(N)$ operations (recall that within LCAO approximation, N and N_b are of the same order of magnitude).

The direct approach to solve these problems is to diagonalize F (or more precisely to solve the generalized eigenvalue problem $F\phi = \varepsilon S\phi$) and choose the lowest N eigenvectors among the N_b ones. Essentially, the algorithmic complexity of diagonalization methods is of the order of N^3 (see, e.g., DEMMEL [1997]). The real drawback of the standard diagonalization techniques is however not that large scaling. Rather, it is the following observation: *Even* if the matrix F to be diagonalized is sparse, the time needed for the computations of the eigenvectors cannot scale linearly, for it cannot be $O(1)$ for each vector. Therefore, linear scaling cannot be achieved with such diagonalization techniques.

In fact, linear scaling methods are founded on the following simple remark: Actually, SCF algorithms only require the projector onto the subspace generated by the smallest N eigenstates of the matrix, and not the precise knowledge of the eigenstates themselves. Diagonalization can be avoided, which significantly reduces the algorithmic complexity, basically from N^3 to N (in some cases at least). These methods, known as the *linear scaling methods*, therefore allow to compute systems of greater size. Let us now examine the basic principles of these.

We emphasize in the sequel the mathematical foundations of the methods, presenting what we hope to be a pedagogic and synthetic overview. For a complete and more detailed description, with emphasis on the physicochemical aspects, as well as for comparative studies of the different methods, we refer to the survey articles by GOEDECKER [1999], GALLI [2000], DANIELS and SCUSERIA [1999], as well as to another contribution in the present volume.

Let us however give a necessarily schematic state of the art for these methods aimed at reaching the linear scaling. Essentially, one may say that these methods are remarkably efficient for the modelling of insulators, namely of systems for which there is a noninfinitesimal gap between the HOMO and the LUMO of F ($\varepsilon_{N+1} - \varepsilon_N = \gamma > 0$). On the other hand they experience the worst difficulties when dealing with metallic systems ($\gamma \simeq 0$). For the latter, they clearly are in a nonsatisfactory state and need to be further developed and adapted. None of the above methods efficiently work on this kind of systems. From the numerical standpoint, the difficulty of metallic systems is twofold. First, the problem of finding the eigenvalues (which is more or less the problem that is addressed here) is ill-conditioned in the sense that the eigenvalues for the Hamiltonian of such systems are very close to one another. Second, the density matrix D for the ground state is basically a full matrix, and not a sparse one as it is the case for insulators and as it is basically required to reach the linear scaling. The two difficulties together are an overwhelming task for the current algorithms. Of course, huge efforts are directed toward the improvement of methods so that they can also treat such systems.

The purpose is now to see major instances of approaches that are alternatives to diagonalization. The situation we regard is the following. The mean-field Hamiltonian F is assumed to be known and fixed (and from the implementation viewpoint already assembled). We assume in addition that F is sparse, which is the case for LCAO calculations on large systems. We shall see below that some additional assumptions on the situation we tackle are necessary to hope for an order N method.

In order to reach the linear scaling, there are three steps

(1) as pointed out above, we first remark that *strictly speaking*, the lowest N eigenvectors are not needed; only the subspace they generate is to be found out;
(2) the problems (36.1) and (36.2) are reformulated in such a way that the constraint $C^*SC = I_N$, or equivalently $DSD = D$, disappears;
(3) an algorithm is constructed, which scales cubically in the whole generality, but which scales linearly when F is sparse and when we make the additional assumption that *the density matrix D we seek is also sparse*. This is a crucial additional assumption, which clearly restricts the systems that can be attacked, in a nontrivial and nonexplicit way; it can be proven, at least for simplified models of electronic structures, that the matrix D is actually sparse for insulators, whereas it is not for metallic systems. Under the assumption that D is sparse, the algorithm we build therefore involves only products of sparse matrices, which cost $O(N)$ operations. As the object we search for is also sparse, the number of such products is small and therefore we hopefully never have to do products of full matrices.

In practice, the sparsity of the various matrices involved in the calculation (essentially various polynomials in F) results either from some locality assumption (a matrix element is set to zero if it corresponds to two atomic orbitals carried by atoms far away from other) or from some cut-off rule (a matrix element is set to zero if its modulus is below some threshold). It is to be underlined that the numerical analysis of the different linear scaling methods detailed below without cut-off or locality assumption is easy (the main results are reported below), but that a numerical analysis of these methods taking into account some cut-off rule or locality assumption, is not yet available. The efficiencies of the different methods have therefore only been investigated on a few benchmark calculations which are far from reproducing all the situations met in practice.

Let us mention that, as far as problem (36.1) is concerned, we have focused so far on the orthonormality (or idempotency) constraint, and not on the constraint fixing the number of electrons. In fact, the latter is far easier, since it can be associated with a scalar Lagrange multiplier μ, called the *Fermi level* or the *chemical potential*. It indeed defines the *occupied levels* (for $\lambda_i < \mu$) and the *unoccupied* ones (for $\lambda_i > \mu$), respectively corresponding to the lowest N eigenvectors and the $N_b - N$ following ones. For simplicity, we shall mostly assume in the sequel that this multiplier μ is known. In practice, it can be determined simultaneously by an external optimization loop. In variational methods, it is also possible to take into account the *affine* constraint $\mathrm{Trace}(SD) = N$ by projecting the descent direction on the vector space $\{M \in \mathcal{M}_S(N_b), \mathrm{Trace}(SM) = 0\}$.

The linear scaling methods we present are collected into two families, namely the penalization techniques and the approximation techniques.

On purpose, a family of linear scaling methods is left aside, that arising from the *divide and conquer* paradigm. Loosely speaking, the idea is to partition the molecular system into subsystems, next compute the "partial" density matrices, and finally make some merging to build the global density matrix. The idea is inheritated from the very efficient domain decomposition methods that are commonly used in the numerical solution for variational problems in engineering sciences. However, the crucial difference, to our opinion at least, is that in the engineering sciences, these techniques have been thoroughly studied and optimized through decades of numerical analysis. They are therefore in a mature age. On the other hand, in computational chemistry, these techniques seem to be still in their conception age. It is our hope that they will grow up fast and efficiently. At this day, however, we prefer to concentrate on other techniques and refer to the bibliography for a complete account for the state of the art of the divide and conquer approach.

(A) Penalization approaches

The penalization method consists in eliminating the constraint of orthonormality (respectively of idempotency) by constructing an *exact penalized functional*. Mathematically, this means that the constrained optimization problem

$$\inf\{f(x),\ x\in\mathbb{R}^n,\ c(x)=0\} \tag{36.4}$$

is replaced by the unconstrained optimization problem

$$\inf\{h(x),\ x\in\mathbb{R}^n\} \tag{36.5}$$

in such a way that any local minimizer to (36.4) is a minimizer to (36.5) (the converse being rarely possible). Then a standard algorithm of unconstrained numerical minimization, such as the nonlinear conjugate gradient algorithm for instance, is performed on the latter problem.

The prototype of such a method is the *orbital minimization method*, abbreviated in OM method, and due to ORDEJON, DRABOLD and MARTIN [1995]. It arises not as a penalization method but rather as follows. Rather than directly attacking (36.2), one remarks that a genuine way to obtain a matrix $C'\in\mathcal{M}(N_b,N)$ satisfying $C'^{*}SC'=I_N$ from a generic rank-N matrix $C\in\mathcal{M}(N_b,N)$ is to set

$$C'=C\Sigma^{-1/2},\quad \text{where } \Sigma=C^{*}SC.$$

This transform is known as the Löwdin orthonormalization procedure in the chemical literature (LÖWDIN [1950]). Problem (36.2) therefore also reads

$$\inf\{\mathrm{Trace}\big(FC(C^{*}SC)^{-1}C^{*}\big),\ C\in\mathcal{M}(N_b,N),\ C^{*}SC \text{ nonsingular}\}.$$

In the simplified setting when $N=1$ and $S=I_{N_b}$, this amounts to saying that instead of minimizing $(\Phi,F\Phi)$ on the euclidean unit sphere of $\mathbb{R}^{N_b}$ (as in (36.3)), one sets

$\Phi = \Psi(\Psi, \Psi)^{-1/2}$, and then minimizes the Rayleigh quotient

$$\inf\left\{\frac{(\Psi, F\Psi)}{(\Psi, \Psi)},\ \Psi \in \mathbb{R}^{N_b},\ \Psi \neq 0\right\}.$$

In order to improve efficiency, it is now needed to *approximate* $\Sigma^{-1} = (C^* SC)^{-1}$ and not explicitly calculate it. A standard way is to replace the inverse by the series

$$\Sigma^{-1} = \big(I - (I - \Sigma)\big)^{-1} = \sum_{k=0}^{+\infty} (I - \Sigma)^k$$

and truncate the expansion at a finite (small) order. The simplest way to proceed is to set

$$\Sigma^{-1} = I + (I - \Sigma) = 2 - \Sigma = 2 - C^* SC.$$

We finally obtain the OM optimization problem

$$\inf\big\{\mathrm{Trace}\big(FC\big(2 - C^* SC\big)C^*\big),\ C \in \mathcal{M}(N_b, N)\big\}. \tag{36.6}$$

In this form, the quantity $2I - \Sigma = I + (I - \Sigma)$ can be interpreted as a penalty method: One penalizes the fact that Σ is not the identity matrix. In the simplest form, it replaces (36.3) by

$$\inf\big\{(\Psi, F\Psi)\big(2I - (\Psi, \Psi)\big),\ \Psi \in \mathbb{R}^{N_b}\big\}.$$

Let us now check that one obtains in this way an *exact* penalization method. The gradient of the functional

$$\Omega(C) = \mathrm{Trace}\big(FC\big(2I - C^* SC\big)C^*\big)$$

with respect to the Hilbert–Schmidt scalar product (defined on $\mathcal{M}(N_b, N)$ by $(A, B) = \mathrm{Trace}(AB^*)$), reads

$$\nabla\Omega(C) = 4FC - 2FCC^* SC - 2SCC^* FC.$$

It is thus clear that any solution to the Euler–Lagrange equations $FC = SC\Lambda$, $C^* SC = I_N$, is a critical point of Ω. In addition, if C is solution to (36.2), and if all the occupied orbitals of F are negative, then the sufficient condition

$$\begin{aligned}&\forall M \in \mathcal{M}(N_b, N),\ M \neq 0,\\ &\mathrm{Trace}\big(M^* FM\big(2I - C^* SC\big) - C^* FCM^* SM - 2C^* FM\big(M^* SC + C^* SM\big)\big) > 0\end{aligned}$$

for C being a local minimum of Ω is fulfilled.

From a chemical viewpoint, the occupied orbitals of the mean-field Hamiltonian must be negative. It may happen however that it is not the case in practice (for instance, in some steps of the SCF procedure); in this case, it is always possible to shift the Hamiltonian, and replace F by $F - \varepsilon_{\mathrm{F}} I$ where ε_{F} is the Fermi level.

The OM method is known to have satisfactory convergence properties when the initial guess of the nonlinear conjugated gradient is close to the solution; for initial guesses which are not that good, thousands of conjugated gradient iterations may be required so that the OM method cannot compete with the other methods described below. It has to be noticed that initial guesses of good quality are available in the latest steps of the SCF procedure: Suffices to choose the result of the previous iteration. In the same way, extrapolation techniques are used to build good initial guesses, on and after the second step in geometry optimization or molecular dynamics iterations.

Let us mention that another viewpoint to obtain the minimization problem (36.6) is that of Lagrange multipliers. Minimizing (36.2) yields the Euler–Lagrange equation

$$FC = SC\Lambda,$$

so that at any minimizer C of (36.2), the corresponding Lagrange multiplier Λ satisfies

$$\Lambda = C^* FC. \tag{36.7}$$

For obtaining this expression, we have simply multiplied the Euler–Lagrange equation by C^* in the left-hand side and used the orthonormality condition $C^*SC = I_N$. Let us now consider the Lagrangian

$$\mathcal{L}(C, \Lambda) = \operatorname{Trace}\bigl(FCC^*\bigr) - \bigl(\Lambda, C^*SC - I_N\bigr)$$

associated with (36.2) and the related saddle point problem

$$\inf_C \sup_\Lambda \mathcal{L}(C, \Lambda).$$

If one replaces in this problem the *independent variable* Λ by the expression (36.7) of Λ as a function of C which is valid at the saddle points, one recovers problem (36.6).

Let us now examine the penalization method when applied to the problem written in terms of density matrices rather than in terms of molecular orbitals.

Let us assume for a while that $S = I_{N_b}$. The foundation of the penalization methods for dealing with problem (36.1) is to penalize the constraint $D^2 = D$. In doing that, one needs to be careful to only build penalized functionals that are indeed exact, in the sense defined above, in order not to create spurious minimizers or on the other hand not to lose some. A strategy to remedy to this difficulty is to use functionals of the type

$$\operatorname{Trace}(FD) + \operatorname{Trace}\bigl(Fg(D)\bigr) \tag{36.8}$$

rather than functionals of the more naive style

$$\operatorname{Trace}(FD) + \lambda \bigl\| D^2 - D \bigr\|.$$

Here, $g(D)$ is a function that is bound to enjoy some particular properties:

- It has to satisfy $g(D)=0$ for any D such that $D^2=D$, which means that the penalty term vanishes when the constraint is fulfilled;
- it has to satisfy $g'(D)=-I$ when $D^2=D$ so that the derivative of (36.8) $h\to \mathrm{Trace}(Fh+Fg'(D)h)$ vanishes at the minimizer.

In addition, the calculation of $g(D)$ for a given D must be easy and a simple way to fulfill this condition is to take g as a polynomial; in this case the two above conditions imply that the polynomial g must satisfy

- $g(0)=g(1)=0$;
- $g'(0)=g'(1)=-1$.

Depending on the choice of function g, various methods can be designed. We now examine the simplest of them, corresponding to the choice $g(D)=3D^2-2D^3-D$. This method is referred to as the *Density Matrix Minimization* method, abbreviated in DMM, or in LNV from the authors of the original article (LI, NUNES and VANDERBILT [1993]). The DMM functional therefore reads

$$\Omega_\mu(D)=\mathrm{Trace}\big(F_\mu\big(3D^2-2D^3\big)\big),$$

where $F_\mu=F-\mu I$, μ denoting the Fermi level. The reason why one has to shift the Hamiltonian spectrum will be clarified below. Let us mention that in the simple case when $N=1$, this method simply replaces (36.3) by

$$\inf\big\{(\Phi,F_\mu\Phi)\big(3\|\Phi\|^2-2\|\Phi\|^4\big),\ \Phi\in\mathbb{R}^{N_b}\big\}.$$

The gradient of Ω_μ with respect to the Hilbert–Schmidt scalar product on $\mathcal{M}_S(N_b)$ reads

$$\nabla\Omega_\mu(D)=3(F_\mu D+DF_\mu)-2\big(F_\mu D^2+DF_\mu D+D^2F_\mu\big).$$

The solution D to (36.1) with $S=I_{N_b}$ commutes with F_μ and satisfies $D^2=D$. It is therefore clear that it is a critical point of F_μ.

Let us now examine the second-order optimality condition: In an orthonormal basis where both F_μ and the solution D to (36.1) are diagonal, one has for any $d\in\mathcal{M}_S(N_b)$,

$$\begin{aligned}\Omega_\mu(D+d)=\Omega(D)&-\sum_{i=1}^{N}\varepsilon_i\left(\sum_{j=1}^{N_b}d_{ij}^2+2\sum_{j=1}^{N}d_{ij}^2\right)\\&+\sum_{i=N+1}^{N_b}\varepsilon_i\left(\sum_{j=1}^{N}d_{ij}^2+2\sum_{j=N+1}^{N_b}d_{ij}^2\right)+\mathrm{o}\big(d^2\big),\end{aligned}$$

where $\varepsilon_1\leqslant\varepsilon_2\leqslant\cdots\leqslant\varepsilon_N<0<\varepsilon_{N+1}\leqslant\cdots\leqslant\varepsilon_n$ denote the eigenvalues of F_μ. This proves that D is a strict local minimum of Ω_μ. We clearly see on the above formula the role of the Fermi level: In order to ensure that D is a local minimum of Ω_μ, the parameter μ must be such that all the occupied eigenvalues are negative and all the virtual eigenvalues are positive.

It is lastly possible to show that the solution to (36.1) is in fact the unique local minimizer to Ω_μ. Let us indeed assume that Ω_μ has two local minima D and D'. The restriction of Ω_μ to the line containing D and D' is a third degree polynomial, which admits two local minima; we reach a contradiction.

The DMM optimization problem

$$\inf\{\mathrm{Trace}(F_\mu(3D^2 - 2D^3)),\ D \in \mathcal{M}_S(N_b)\} \tag{36.9}$$

therefore has a unique local minimizer, which is the solution to (36.1). Let us however remark that the infimum of the above problem is $-\infty$ (the function Ω_μ is a third order polynomial along each line of $\mathcal{M}_S(N_b)$). For the solution of (36.1) being obtained by solving (36.9) with a nonlinear conjugated gradient algorithm, it is necessary that the initial guess of the conjugated gradient is in the "attraction basin" of the local minimum. As for the OM method, the density matrices obtained in the previous steps of the calculation (inside a SCF, geometry optimization, or molecular dynamics procedure) can be used to define a "good" initial guess. Let us also remark that in the nonlinear conjugated gradient procedure, linesearches (which consist in minimizing a third order polynomial) can be performed analytically.

In the general case when the overlap matrix S is not the identity, the formulae must be modified along

$$\Omega_\mu(D) = \mathrm{Trace}(F_\mu(3DSD - 2DSDSD)).$$

Another possibility to penalize the constraint $D^2 = D$ is to penalize not with the trace norm but with the Hilbert–Schmidt norm of operators. It gives rise to the *Kohn method* (see KOHN [1996]). This method consists in minimizing

$$\widetilde{\Omega}_\mu(D) = \mathrm{Trace}(F_\mu D^2) + \alpha \| D^2 - D \|_{\mathrm{HS}}.$$

Note that D has been replaced by D^2 in the first term, and in addition $D^2 = D$ is penalized through the second term.

A simple picture for this method can be given in terms of the (unknown) eigenvalues ε_i of F. Suppose that the matrix D along an orthonormal basis made of the eigenvectors χ_i of F reads $D = \mathrm{Diag}(\lambda_i)$. Then we may write the above minimization under the following form

$$\inf \sum_{i=1}^{N_b} \lambda_i^2(\varepsilon_i - \mu) + \alpha \sqrt{\sum_{i=1}^{N_b} \lambda_i^2(1 - \lambda_i)^2}. \tag{36.10}$$

In order to ensure consistency, it is required that the minimizer is exactly given by $\lambda_i = 1$ when $\varepsilon_i \leqslant \mu$, and $\lambda_i = 0$ when $\varepsilon_i > \mu$. This in turn can be ensured by choosing the penalization weight α large enough. More precisely the condition on α is

$$\alpha \geqslant 2 \sqrt{\sum_{\varepsilon_i \leqslant \mu} (\varepsilon_i - \mu)^2},$$

which amounts to roughly estimating the ε_i before setting the algorithm.

The case $N = 1$ allows to understand why the method works. To fix the ideas, suppose we search for the ground state of a Hamiltonian F which has exactly one negative eigenvalue ε below μ. The minimization in (36.10) leads to $\lambda_i = 0$ for $i \geqslant 2$, so that we end up with

$$\inf \lambda^2(\varepsilon - \mu) + \alpha\lambda|\lambda - 1|$$

for some $\varepsilon < \mu$. It is easily seen that a minimizer exists if and only if α and large enough, and then the value is exactly $\lambda = 1$. Basically, the case $N = 1$ therefore amounts to replacing (36.3) by

$$\inf \langle \Phi, F_\mu \Phi \rangle \|\Phi\|^2 + \alpha\big(\|\Phi\|^4\big(\|\Phi\|^2 - 1\big)^2\big).$$

(B) Nonvariational approximations

An alternative to the above techniques is provided by nonvariational approximations. They consist in approaching the solution D to the problem (36.1), which reads

$$D = \mathcal{H}(\mu - F)$$

in the special case when $S = I_N$ and

$$D = S^{-1/2}\mathcal{H}\big(\mu - S^{-1/2} F S^{-1/2}\big) S^{-1/2} = \mathcal{H}\big(\mu - S^{-1} F\big) S^{-1} \tag{36.11}$$

in the general case, through a "standard" approximation of the function $\mathcal{H}$. In the above formulae, $\mathcal{H}$ denotes the Heaviside function ($\mathcal{H}(x) = 0$ if $x < 0$ and $\mathcal{H}(x) = 1$ if $x > 0$) and μ is the Fermi level.

A first possibility is to resort to the *Fermi Operator Expansion* (FOE) which consists in approaching the Heaviside function $\mathcal{H}$ by a Chebyshev polynomial approximation.

Let us assume in a first stage that $S = I_{N_b}$ and suppose that we know both an estimate of the spectral window of F (i.e., a lower bound $\varepsilon_{\min}$ of the lowest eigenvalue and an upper bound $\varepsilon_{\max}$ of the highest eigenvalue) and the exact value of the Fermi level μ. Let us consider the rescaled Hamiltonian

$$F' = \alpha F + \beta \quad \text{with } \alpha = \frac{1}{\max(\varepsilon_{\max} - \mu, \mu - \varepsilon_{\min})} \text{ and } \beta = -\alpha\mu.$$

The coefficients α and β have been chosen such that the eigenvectors of F corresponding to the eigenvalues of F belonging to the range $[-\infty, \mu[$ (resp. to the range $]\mu, +\infty[$) are the eigenvectors to F' associated to the eigenvalues of F' belonging to the range $[-1, 0[$ (resp. to the range $]0, 1]$). The eigenvalues of F' all are in the range $[-1, 1]$ and it is then easy to check that the minimizer D of (36.1) (with $S = I_{N_b}$) satisfies

$$D = \mathcal{H}(-F').$$

Let us denote by T_j the jth Chebyshev polynomial and by $(c_j)_{0 \leqslant j \leqslant +\infty}$ the Chebyshev coefficients of the function $\mathcal{H}(-x)$ on the range $[-1, 1]$. One has

$$\mathcal{H}(-x) = \sum_{j=0}^{+\infty} c_j T_j(x)$$

and therefore

$$D = \sum_{j=0}^{+\infty} c_j T_j(F').$$

The FOE method consists in truncating the above expansion to a given order k which depends both on the spectral gap and on the required accuracy (see GOEDECKER [1999]). Because of the discontinuity of the Heaviside function at zero, it is well known that the Chebyshev expansion (truncated at finite order) yields Gibbs oscillations, which can be cured by replacing the Chebyshev expansion by the Chebyshev–Jackson expansion. This simply involves a slight modification of the coefficients c_j (we refer to JAY, KIM, SAAD and CHELIKOWSKI [1999] for more details).

The computation of the truncated expansion

$$D_k = \sum_{j=0}^{k} c_j T_j(F')$$

is done by taking advantage of the recursion formula

$$T_{j+1}(F') = 2F'T_j(F') - T_{j-1}(F'), \quad T_0(F') = I_{N_b}, \ T_1(F') = F',$$

which allows to compute *independently* each column of the matrix; this method therefore is easily parallelizable.

As in practice the Fermi level μ is not known a priori, it has to be computed by an outer loop. Iterations are driven by the fact that the trace of D_k is lower than N if μ is too small and larger than N in the opposite case; the search for μ can be accelerated by using the relation

$$\frac{\partial \operatorname{Trace}(D_k)}{\partial \mu} = -\operatorname{Trace}\left(\sum_{j=0}^{k} c_j \frac{dT_j}{dx}(F')\right).$$

The FOE method can be adapted to deal with nonorthonormal basis sets, i.e., to the general case when $S \neq I_{N_b}$. One way to proceed is to replace in the above developments the mean-field Hamiltonian F by the (non-Hermitian) matrix $\widetilde{F} = S^{-1}F$ (see formula (36.11)). It has indeed been remarked (see GIBSON, HAYDOCK and LAFEMINA [1993]) that the matrix $\widetilde{F}$ is also sparse (although S^{-1} is not). The matrix $\widetilde{F}$ can be assembled column by column by solving independently the N_b linear systems generated by the matrix equation $S\widetilde{F} = F$.

A second instance of nonvariational approximation technique is provided by the method of purification of the density matrix. This method was introduced in PALSER and MANOLOPOULOS [1998] following the earlier MCWEENY [1992] work. The idea is the following one. Consider the function defined on the real line $f(x) = 3x^2 - 2x^3$. Then, if $x_0 \in]-1/2, 3/2[$, the algorithm defined by the induction formula $x_{k+1} = f(x_k)$ converges (quadratically) to $\mathcal{H}(x_0 - 1/2)$, where $\mathcal{H}$ is the Heaviside function.

Let us again assume in a first stage that $S = I_{N_b}$ and that we know both an estimate of the spectral window of F and the exact value of the Fermi level μ. Let us this time consider the rescaled Hamiltonian

$$F' = \alpha F + \beta \quad \text{with } \alpha = \frac{-1}{2\max(\varepsilon_{\max} - \mu, \mu - \varepsilon_{\min})} \text{ and } \beta = \frac{1}{2} - \alpha\mu,$$

where the coefficients α and β have been chosen such that the eigenvectors of F corresponding to the eigenvalues of F belonging to the range $[-\infty, \mu[$ (resp. to the range $]\mu, +\infty[$) are the eigenvectors to F' associated to the eigenvalues of F' belonging to the range $]1/2, 1]$ (resp. to the range $[0, 1/2[$). The eigenvalues of F' all belong to the range $[0, 1]$ and the minimizer D of (36.1) (with $S = I_{N_b}$) satisfies

$$D = \mathcal{H}(F' - 1/2);$$

indeed in an eigenvectors basis, the matrix $F' - 1/2$ is a diagonal matrix where the minimal modes of F (in the range $]-\infty, \mu]$) correspond to positive entries and the maximal modes of F to negative entries. In such a basis, D is precisely the diagonal projector with entries 0 and 1 on the corresponding modes. This equality is obviously independent of the basis.

The sequence defined by

$$D_0 = F', \qquad D_{k+1} = f(D_k) = 3D_k^2 - 2D_k^3$$

therefore converges toward D. This algorithm can be interpreted as the approximation of the Heaviside function by a polynomial, namely that defined by the nth iteration of the function f (which is called the *McWeeny purification function*). It can also be interpreted as an iteration to obtain the orthogonal projection of $D_0 = F'$ onto the set $\{D \in \mathcal{M}_S(N_b), D^2 = D, \operatorname{Trace}(D) = N\}$.

The above algorithm can be adapted to the case when $S \neq I_{N_b}$ in the following way:

$$D_0 = \alpha S^{-1} F S^{-1} + \beta S^{-1}, \qquad D_{k+1} = 3D_k S D_k - 2D_k S D_k S D_k.$$

In practice, it is easy to compute good estimates for $\varepsilon_{\min}$ and $\varepsilon_{\max}$ (for instance by the shifted Lanczos method). On the other hand, it is difficult to get the exact value of the Fermi level; for this reason, the following variant of the above algorithm is used

$$D_{k+1} = \begin{cases} \dfrac{1}{1-c_k}\big((1-2c_k)D_k + (1+c_k)D_kSD_k - D_kSD_kSD_k\big) & \text{if } c_k \leqslant 1/2, \\ \dfrac{1}{c_k}\big((1+c_k)D_kSD_k - D_kSD_kSD_k\big) & \text{if } c_k > 1/2, \end{cases}$$

with

$$c_k = \frac{\operatorname{Trace}(D_kSD_k - D_kSD_kSD_k)}{\operatorname{Trace}(D_k - D_kSD_k)}.$$

Let us notice that, contrary to what occurs in the previously presented three methods, the whole available information on the matrix to be diagonalized is inserted in the initial D_0, and does not appear in the recursion formula. Consequently, although this method is much better than the others in some cases (for instance if the trivial case when F' is diagonal), the cut-off rule or the localization assumption used to guarantee linear scaling unfortunately often makes the method miss the target.

37. Geometry optimization techniques

Let us now give some hints on the numerical techniques to deal with the geometry optimization problem (6.7) that we reproduce here for convenience:

$$\inf\big\{W(\bar{x}_1,\ldots,\bar{x}_M),\ (\bar{x}_1,\ldots,\bar{x}_M) \in \mathbb{R}^{3M}\big\},$$

where $W(\bar{x}_1,\ldots,\bar{x}_M)$ is the sum of the internuclear Coulomb repulsion energy $\sum_{1\leqslant k<l\leqslant M} \frac{z_k z_l}{|\bar{x}_k - \bar{x}_l|}$ and of the electronic ground state energy, defined by (6.5), the latter being computed by one of the approximation methods previously presented (Hartree–Fock, post-Hartree–Fock, Kohn–Sham, ...). Theoretical questions related to the geometry optimization problem have already been briefly addressed in Section 18.

From a numerical viewpoint, the problem is difficult for, on the one hand, the dimension of the minimization space may be very large (M may reach several hundreds for ab initio models, several millions for empirical models), and, on the other hand each computation of a single value of the potential W may be a computationally demanding task (at least for ab initio models).

In most cases, it is not efficient to solve the geometry optimization problem for molecules in Cartesian coordinates. The conditioning of the problem is indeed much better when the optimization problem is written in terms of *internal coordinates* (bond lengths and angles). When the molecular system is very flexible, or when it exhibits cycles (like the benzene molecule), it has been observed that *redundant internal coordinates* (all bonds, all valence angles between bonded atoms, and all dihedral angles between bonded atoms are considered as optimization parameters) are more appropriate than nonredundant ones. When needed, redundancies can be removed by using a generalized inverse of the quasi-Newton matrix; constraints can be added by using appropriate projectors (PENG, AYALA, SCHLEGEL and FRISCH [1995]).

Usually, deterministic methods are resorted to when a good guess of the geometry can be found out by theoretical or experimental considerations. The goal then is only to determine more precisely geometrical parameters (bond lengths and angles). The most commonly used deterministic methods are of quasi-Newton types (see Section 26). Analytical derivatives formulae (see Section 18) are indeed available for most of the commonly used ab initio methods (Hartree–Fock, Kohn–Sham, Møller–Plesset, configuration interaction, CASSCF, ...), so that the first order derivatives of W at some $(\bar{x}_1, \ldots, \bar{x}_M)$ are easily obtained once the corresponding electronic ground state (necessary to get the value W at this point) has been computed. For example, let us detail how analytical derivative formulae are obtained for the spinless Hartree–Fock model: Under the HF-LCAO approximation, the derivative of the potential

$$W = \operatorname{Trace}(hD) + \frac{1}{2}\operatorname{Trace}\bigl(G(D)D\bigr) + V_{\text{nuc}}$$

(where $V_{\text{nuc}} = \sum_{1 \leqslant k < l \leqslant M} \frac{z_k z_l}{|\bar{x}_k - \bar{x}_l|}$ denotes the internuclear potential) with respect to some parameter λ reads

$$\frac{\partial W}{\partial \lambda} = \operatorname{Trace}\left(\frac{\partial h}{\partial \lambda} D\right) + \frac{1}{2}\operatorname{Trace}\left(\frac{\partial G}{\partial \lambda}(D)D\right) + \operatorname{Trace}\left(F(D)\frac{\partial D}{\partial \lambda}\right) + \frac{\partial V_{\text{nuc}}}{\partial \lambda}.$$

Besides, using the Hartree–Fock equation $F(D)C = SCE$

$$\begin{aligned}
\operatorname{Trace}\left(F(D)\frac{\partial D}{\partial \lambda}\right) &= \operatorname{Trace}\left(F(D)\frac{\partial}{\partial \lambda}(CC^*)\right) \\
&= \operatorname{Trace}\left(F(D)\frac{\partial C}{\partial \lambda}C^*\right) + \operatorname{Trace}\left(F(D)C\frac{\partial C^*}{\partial \lambda}\right) \\
&= \operatorname{Trace}\left(C^*F(D)\frac{\partial C}{\partial \lambda}\right) + \operatorname{Trace}\left(F(D)C\frac{\partial C^*}{\partial \lambda}\right) \\
&= \operatorname{Trace}\left(EC^*S\frac{\partial C}{\partial \lambda}\right) + \operatorname{Trace}\left(SCE\frac{\partial C^*}{\partial \lambda}\right) \\
&= \operatorname{Trace}\left(E\left(C^*S\frac{\partial C}{\partial \lambda} + \frac{\partial C^*}{\partial \lambda}SC\right)\right).
\end{aligned}$$

Inserting in the latter expression the relation

$$C^*S\frac{\partial C}{\partial \lambda} + \frac{\partial C^*}{\partial \lambda}SC = -C^*\frac{\partial S}{\partial \lambda}C$$

obtained by differentiating with respect to λ the orthonormality condition $C^*SC = I_N$, one gets

$$\frac{\partial W}{\partial \lambda} = \text{Trace}\left(\frac{\partial h}{\partial \lambda} D\right) + \frac{1}{2}\text{Trace}\left(\frac{\partial G}{\partial \lambda}(D)D\right) - \text{Trace}\left(D_E \frac{\partial S}{\partial \lambda}\right) + \frac{\partial V_{\text{nuc}}}{\partial \lambda},$$

where $D_{\text{E}} = CEC^* = S^{-1}F(D)D$ denotes the *energy weighted density matrix*. The derivative $\frac{\partial C}{\partial \lambda}$ has thus been eliminated.

As atomic orbitals are centered on the nuclei, the AO basis functions move during the geometry optimization procedure, so that the derivatives $\frac{\partial h}{\partial \lambda}$, $\frac{\partial G}{\partial \lambda}$ and $\frac{\partial S}{\partial \lambda}$ do not vanish. On the other hand, they can be easily computed for Gaussian atomic orbitals since the derivative of a Gaussian-type function still is a Gaussian-type function.

Stochastic methods have to be resorted to for large systems because of the potentially huge number of local minima; all the standard stochastic optimization techniques, such as evolutionary strategies (MICHALEWICZ [1999]), simulated annealing, smoothing techniques, basin hoping, have been tested in the field of molecular simulation (NEUMAIER [1997]). Let us lastly mention that both Monte Carlo and finite temperature molecular dynamics algorithms (and in particular the Car–Parinello method, see Chapter V and PAYNE, TETER, ALLAN, ARIAS and JOANNOPOULOS [1992]) can be used to generated pseudo-random trajectories.

More information on geometry optimization as well as numerous references are provided in the review article by NEUMAIER [1997].

CHAPTER III

The Condensed Phases

38. Introduction

So far we have been concerned with *isolated* atoms, ions or molecules for which various models have been mentioned. The reasons for solving such equations should be obvious: Equilibrium geometries, potential energy surfaces, heats of reaction, excitation energies, optical, electrical or magnetic properties, activation barriers, to name just a few, can all be calculated from first principles. Doing so the underlying assumption is that a molecular system can be regarded as the set of atoms it is composed with, making abstraction of its environment.

This is the point of view currently used in quantum chemistry for dealing with low density gases (although from the strict physical point of view, it cannot be taken for granted (WOOLEY [1978])). For molecular systems in condensed phases on the other hand, it is well known that dealing with an isolated molecule does not always make sense. For instance, iodine, I_2, in the standard thermodynamic conditions (temperature 298 K and pressure 1 atmosphere) is a dark purple solid. When a small amount of iodine is dissolved into water the solution takes a very pale yellowish colour, whereas in tetrachloride carbon, CCl_4, the solution is pink. In the three cases the central entity is the I_2 molecule, but its environment varies: in the solid the environment is made of other I_2 molecules regularly ordered on a cubic lattice; in an aqueous solution the I_2 molecule is surrounded of H_2O molecules and in CCl_4 solutions, the I_2 molecule is surrounded of CCl_4 molecules. Obviously a single calculation on an isolated I_2 molecule cannot explain the three colours for the above mentioned three situations. Another striking example is the study of the two common solid phases of carbon, one being graphite and the other one diamond; they are two different periodic arrangements of carbon atoms, that can by no means be described by a calculation on a single Carbon atom. These examples convincingly demonstrate the need for more sophisticated models taking into account the environment of the molecular system under study in order to correctly describe the physico-chemical properties on the system *in situ*.

Our task in the present chapter is to study electronic ground states in solids and in the liquid phase. It is worth stressing at this point that we limit ourselves to some specific (and somewhat simplistic) cases. As far as the solid phase is concerned, we shall only be concerned with the perfect crystals. In fact, defects in crystalline solids are known to control atomic transport processes and reactivity as well as many aspects of the crystal thermodynamic, spectroscopic and mechanical properties; the contrasting states of order and disorder provide one of the central and most fascinating themes in solid state science. Even more dramatic are the formidable problems posed by the atomic and electronic structures and by the transport properties of amorphous materials. Also quasi-crystals are ignored in the sequel. Unfortunately these basic challenges in the study of solid state materials represent a program which has been only slightly worked out so far, and hence lies largely out of the scope of the present chapter. Nevertheless even the study of perfectly ordered crystals is a field vigorously pursued and there are still enormously many facts and properties which has been uncovered by the theoreticians in their work during the years which have elapsed since the fundamental works of the beginning of the XX century. The traditional main interest of solid state physics is toward the response properties and dynamical behaviour of solids, whereas chemists are mostly interested in their chemical properties (in particular the total energy of the ground state and the charge distributions), surface properties and reactivity. Initially two main categories of solids were considered (metals and insulators) but now problems under consideration cover a wide spectrum ranging in complexity from simple solids to high temperature superconductors, catalysts, microporous materials, molecular solids and very recently fullerenes and other nanostructures.

As concerns liquid phases, we will limit in the present chapter to systems which do not interact with solvents from the *chemical* viewpoint (e.g., a biological molecule in an aqueous solution). Situations when solvation competes with complexation (i.e., solvent molecules undergo chemical bonds with the central species) are therefore ignored in the sequel despite they are of great interest in chemistry. The more sophisticated models needed to cope with them are indeed out of the scope of the models presented here. We shall only focus on the current thinking on how best to model long length-scale electrostatic phenomena in fluids. Even doing so we are far from being generic because of the enormous variety of solvents. Long length-scale electrostatic phenomena in simple solvents (e.g., water, benzene or alcohols) basically depend on their macroscopic dielectric constant only: Such solvents enter plainly in the models here described. But most of the chemical reactions occur in complex fluids, such as ionic solutions (salted water) or even mixtures of solvents. Now industrial processes deal with molten salts (e.g., aluminium extraction), cryogenic solvents (e.g., lyophilization), micro emulsions (depolluting processes) or liquid crystals. An understanding of the basic science governing these solvations would permit one to thoroughly describe phase diagrams, interfacial properties, etc., but unfortunately, this represents a program which lies largely in the future.

39. The solid phase

The rest of this chapter (but its last section) is concerned with ab initio numerical solid state chemistry, that is, the computation of electronic structures in a solid. We will only consider the case of perfect crystals, i.e., an infinite periodic arrangement of classical nuclei, in which electrons are supposed to satisfy the Schrödinger equation.

This presentation, though by no means complete, will give an idea of the state of the art and the trends in the area. A more detailed contribution in the present volume will give more details on the subject.

The implementation of computer programs for the study of the electronic structure of molecules and crystals has followed almost independent patterns, and the state of the art in the two fields of research presents markedly distinct features, although the basic problem is the same: the study of a many-electron system in the Coulomb field of fixed point-like nuclei.

A great variety of techniques are currently employed for the description of the electronic properties of crystals, resulting in much less standardization. This is for both intrinsic and historical reasons. The problem of crystal stability is not of primary importance in solid state physics, at least when chemically simple crystalline compounds are considered. Well-established empirical or semi-empirical schemes exist for correlating crystal structure and stability with the chemical nature of the atoms involved. For these reasons the problem of describing the ground-state of a real crystal and calculating its energy has historically not been the object of much concern for solid state physicists. What they are normally interested in is what happens to the system when some experimental probe disturbs it, for instance, transitions between states. An extraordinary variety of approaches have therefore flourished in solid state physics: a wide spectrum of sophisticated mathematical tools are employed, specific Hamiltonians are used for the different problems, leading to a mosaic of options for the treatment of crystalline solids.

40. Lattice structures

First of all let us introduce the standard language of lattice structure. Considering a perfect crystal, that is, a system of infinite size consisting of a periodically repeated set of nuclei, together with an associated set of electrons. Assuming that the nuclei are fixed, we may describe their positions by a motif m repeated on the nodes of a lattice $\mathcal{R}$:

$$\mathcal{R} = \{ua + vb + wc,\ (u, v, w) \in \mathbb{Z}^3\}.$$

The vectors a, b, c are assumed to be linearly independent vectors of $\mathbb{R}^3$, and are called a basis of $\mathcal{R}$. Using the approximation of classical nuclei, the motif m represents N point particles of positive charges, that is, mathematically speaking, a set of positive Dirac masses:

$$m = \sum_{k=1}^{N} z_k \delta_{\bar{x}_k},$$

where the vectors $\bar{x}_k$ may be chosen to be in the set:

$$\mathcal{Q} = \left\{xa + yb + zc,\ (x, y, z) \in \left[-\frac{1}{2}, \frac{1}{2}\right[^3\right\}.$$

We call a *unit cell* of the lattice $\mathcal{R}$ any semi-open polyhedron such that its translations along a subset of the vectors of the lattice fill in the whole space $\mathbb{R}^3$ without overlapping.

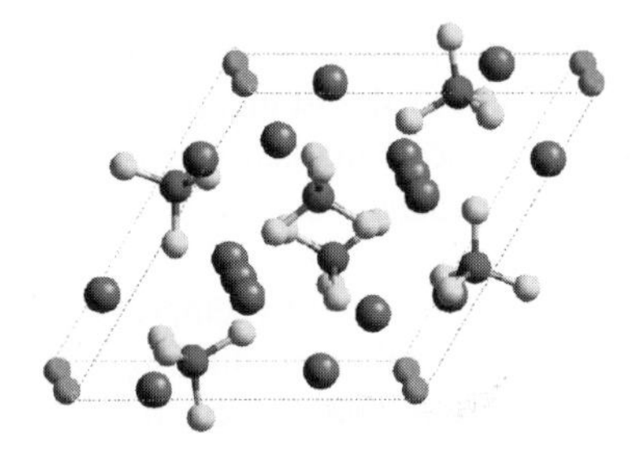

FIG. 40.1. Unit cell of fluoroapatite $Ca_{10}(PO_4)_6F_2$.

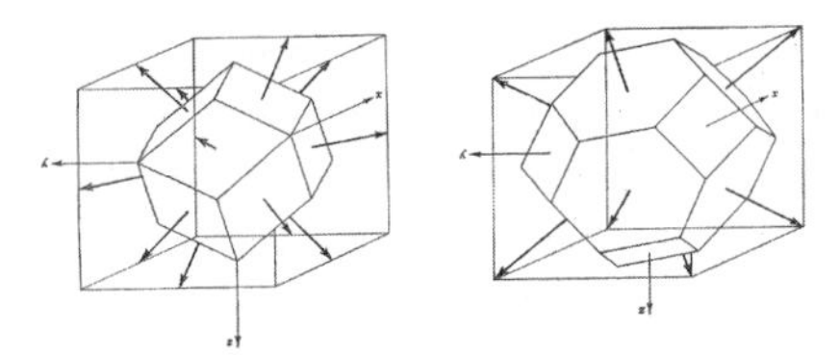

FIG. 40.2. Wigner–Seitz cells for the face-centered and body-centered structures.

An example of a unit cell is given in Fig. 40.1. The set $\mathcal{Q}$ defined previously is thus a unit cell of $\mathcal{R}$. Many choices of unit cell are possible. If its volume is minimal, which is the case of $\mathcal{Q}$, we call the cell a *primitive unit cell*. Such a cell satisfies the property that when translated by all the vectors of the lattice, it constitutes a partition of $\mathbb{R}^3$. The cell, however, does not show the symmetry of the point group, and it would be more satisfactory if we could find a unit cell, which does show this symmetry. The way to do this was pointed out in the early years of crystallography (WIGNER and SEITZ [1933]). A particular primitive unit cell is the Wigner–Seitz cell, or Dirichlet zone, which is constructed as follows (see Fig. 40.2):

(i) The point $0 \in \mathcal{R}$ is connected to all of its neighbours by segments;
(ii) The mediator planes of these segments are traced;
(iii) Each of these planes define a half-space containing 0, and the intersection of these half-spaces is a convex polyhedron, which is the Wigner–Seitz cell.

This cell presents the advantage of being unique.

The aim of the ab initio calculations is to describe the electronic structure of a crystal, the lattice on which the nuclei are set being given. These calculations handle many $\mathcal{R}$-periodic functions, that can be treated by Fourier analysis. This is why we now introduce the notion of reciprocal lattice. Given a basis (a, b, c) of $\mathcal{R}$, we set:

$$\mathcal{R}^\star = \{ua' + vb' + wc',\ (u, v, w) \in \mathbb{Z}^3\},$$

where the basis (a', b', c') is defined as follows:

$$a.a' = b.b' = c.c' = 2\pi,$$
$$b.a' = c.a' = a.b' = c.b' = a.c' = b.c' = 0.$$

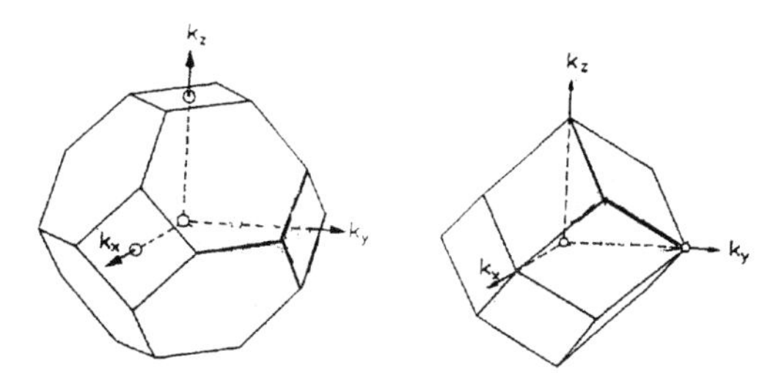

FIG. 40.3. Brillouin zones for the face-centered and body-centered structures.

In other words, the matrices of columns, respectively, (a, b, c) and $\frac{1}{2\pi}(a', b', c')$ are inverse of each other. The point is that this definition of $\mathcal{R}^\star$ does not depend on the chosen basis. The Wigner–Seitz cell of the reciprocal lattice is called the first Brillouin zone (BZ) of the crystal. Two examples of BZ are given in Fig. 40.3.

In many cases, the relevant parameter of the lattice is its invariance group, rather than the lattice itself. Crystals may have more symmetries than the translation symmetry. The operations, which result from these additional symmetries, are rotations or reflections, and they form groups. They have the characteristics that they are operations which leave one point, say, the origin unchanged. If we denote such an operation by R_i, it performs a linear operation on the rectangular coördinates (x_1, x_2, x_3) of any atom:

$$R_i \Psi(x_1, x_2, x_3) = \Psi(x'_1, x'_2, x'_3),$$

where the quantities x'_p are linear functions of the x_p's, given by

$$x'_p = \sum_q \alpha^i_{pq} x_q, \tag{40.1}$$

and Ψ any function of the rectangular coordinates.

The matrix α^i of quantities α^i_{pq} describes the operation R_i of the point group. These are real quantities, satisfying orthogonality conditions. If the operation of the point group corresponds to a rotation, the determinant of α^i equals to 1. If it corresponds to an improper rotation, namely, a rotation plus an inversion, or a reflection, the determinant of α^i equals to -1. We may denote Eq. (40.1) in a matrix form as follows:

$$r' = \alpha^i r,$$

where r and r' are vectors and α^i is a tensor. Finally the crystalline space group is obtained by combining the operations of translation present in the crystal (those defining its periodicity) and the point group. The operators of such a space group are noted $\{R_i, \tau_T\}$ and defined by:

$$\{R_i, \tau_T\}\Psi(r) = \Psi(\alpha^i r + T),$$

where $T \in \mathcal{R}$ and where τ_T denotes the associated translation operator. These operators must fulfill some requirements to form a group; there must be a multiplicative table (the successive application of two operations of a group must give an operation of the group). Or in other words, the point group must transform the vectors τ_T into the same set of vectors.

In crystallography it has been shown (SEITZ [1936]) that there are just 230 possible space groups. These are made up by combining 14 different sets of vectors t_1, t_2, t_3, leading to lattices of points called the 14 Bravais lattices, with 32 possible point groups. Each of these 32-point groups is said to form a crystal class. Finally, out of the 230 space groups above mentioned, a relatively small number is found in the structures of a great majority of the elements and compounds (INTERNATIONAL TABLES FOR X-RAY CRYSTALLOGRAPHY [1952]). All these symmetries allow to limit the Brillouin zone to an irreducible zone where eigenvalues of the operators are unique. For instance, for cubic structures, the irreducible zone (for a spinless calculation) is only 1/48 of the total Brillouin zone.

41. Band theory

The first step in studying the electronic structure of a solid is the examination of the properties of a Schrödinger operator with periodic potential: $H = -\frac{1}{2}\Delta + V$, V being $\mathcal{R}$-periodic. Such an operator may either come from a crude *linear* semi-empirical model, or consists of the current iterate of the inner loop of a self-consistent field iteration algorithm. Then the question is: *Can we find a spectral decomposition of* H?, which leads to the following determination of stationary states of H:

$$-\frac{1}{2}\Delta\psi + V\psi = E\psi. \tag{41.1}$$

For any $T \in \mathcal{R}$, we denote by τ_T the translation operator

$$\psi \mapsto \psi(\cdot + T). \tag{41.2}$$

The point is that H is invariant under such translations, which means, in terms of operators, that

$$\tau_T H = H \tau_T \tag{41.3}$$

for all $T \in \mathcal{R}$. As in addition all the $(\tau_T)_{T\in\mathcal{R}}$ commute with each others, one can impose to the generalized eigenvectors of H to be also generalized eigenvectors of all the τ_T:

$$\begin{cases} H\psi = E\psi, \\ \forall T \in \mathcal{R},\ \tau_T\psi = \gamma(T)\psi, \text{ for some } \gamma(T) \in \mathbb{C}. \end{cases} \tag{41.4}$$

As $\tau_{T+T'} = \tau_T\tau_{T'} = \tau_{T'}\tau_T$, we infer from (41.4) that $\gamma(T+T') = \gamma(T)\gamma(T')$. On the other hand, since τ_T is norm-preserving, we necessarily have $|\gamma(T)| = 1$. It follows that there exists a vector k such that:

$$\gamma(T) = e^{ik\cdot T}. \tag{41.5}$$

From the fact that $T \in \mathcal{R}$, the addition of a vector of the reciprocal lattice $\mathcal{R}^*$ to k does not change $\gamma(T)$, so that we may impose that k is in some primitive unit cell of $\mathcal{R}^*$, for example the first Brillouin zone (BZ). Eq. (41.4) thus becomes: There exists $k \in BZ$, such that

$$\begin{cases} H\psi = E\psi, \\ \psi(x+T) = e^{ik\cdot T}\psi(x), \quad \forall T \in \mathcal{R}, \end{cases} \tag{41.6}$$

or equivalently

$$\begin{cases} H\psi = E\psi, \\ \psi(x)e^{-ik\cdot x}, \quad \mathcal{R}\text{-periodic}. \end{cases} \tag{41.7}$$

For $k \in BZ$, let us denote by $H_k = -\frac{1}{2}\Delta - ik\cdot\nabla + \frac{1}{2}|k|^2 + V$. It is easy to check that

$$\begin{cases} H\psi = E\psi, \\ \psi(x)e^{-ik\cdot x} \quad \mathcal{R}\text{-periodic} \end{cases} \Leftrightarrow \begin{cases} H_k u = Eu, \\ u \quad \mathcal{R}\text{-periodic}, \end{cases} \quad \text{where } u(x) = \psi(x)e^{-ik\cdot x}.$$

The operator H_k acting on $L^2(\mathcal{Q})$ with periodic boundary conditions is self-adjoint and its inverse is compact: Therefore there exists an increasing sequence $(\varepsilon_j^k)_{j\in\mathbb{N}}$ and $(u_j^k)_{j\in\mathbb{N}}$ such that

$$\begin{cases} H_k u_j^k = \varepsilon_j^k u_j^k, \\ u_j^k \quad \mathcal{R}\text{-periodic}, \\ (u_j^k)_{j\in\mathbb{N}} \quad \text{is an orthonormal basis of } L^2_{\text{per}}(\mathcal{Q}). \end{cases}$$

Denoting by $\psi_j^k(x) = u_j^k(x)\, e^{ik\cdot x}$, one obtains

$$\begin{cases} H\psi_j^k = \varepsilon_j^k \psi_j^k, \\ \psi_j^k e^{-ik\cdot x} \quad \mathcal{R}\text{-periodic}, \\ \int_{\mathcal{Q}} \psi_j^k \psi_j^{k*} = \delta_{jj'}. \end{cases}$$

The continuity of the function $k \mapsto \varepsilon_j^k$ follows the continuity of the mapping $k \mapsto H_k$.

This leads us to the celebrated Bloch theorem (ASHCROFT and MERMIN [1976]), which precisely describes the spectrum $\sigma(H)$ of this type of Schrödinger operators:

THEOREM 41.1. *Let $H = -\frac{1}{2}\Delta + V$ be a Schrödinger operator on $L^2(\mathbb{R}^n)$, with $V \in L^2_{\text{loc}}(\mathbb{R}^n)$ $\mathcal{R}$-periodic. Then for all $k \in BZ$, there exists an increasing sequence of reals $\{\varepsilon_j^k\}_{j\in\mathbb{N}}$ and a sequence of L^2_{loc} functions $\{\psi_j^k\}_{j\in\mathbb{N}}$ such that:*

(1) $H\psi_j^k = \varepsilon_j^k \psi_j^k$;
(2) $\psi_j^k(x)e^{-ik\cdot x}$ *is* $\mathcal{R}$*-periodic*;
(3) $\int_Q \psi_j^k \psi_{j'}^{k'} = \delta_{jj'}\delta_0(k-k')$;
(4) $I = \frac{1}{|BZ|}\int_{BZ}\sum_{j=0}^{+\infty}|\psi_j^k\rangle\langle\psi_j^k|\,dk$ *and* $H = \frac{1}{|BZ|}\int_{BZ}\sum_{j=0}^{+\infty}\varepsilon_j^k|\psi_j^k\rangle\langle\psi_j^k|\,dk$;
(5) $k \mapsto \varepsilon_j^k$ *is continuous on* BZ*, for all* $j \in \mathbb{N}$;
(6) $\sigma(H) = \bigcup_{j\in\mathbb{N}}[\inf_{k\in BZ}\varepsilon_j^k, \sup_{k\in BZ}\varepsilon_j^k]$.

The above argument is by no means a rigorous proof of this result. The interested reader may find more detailed proofs in REED and SIMON [1975].

Essentially, we see that the spectrum of such a Schrödinger operator consists of intervals, which are called *bands* of the operator. This gives us an intuition on the behavior of electrons in a crystal: Assuming that they are represented by the above eigenstates, preferably the lowest ones, we see that if there are exactly enough electrons to fill in an entire number of bands, we have an energetical gap between the highest occupied states and the lowest unoccupied ones. On the other hand, if a band is only partially filled, there is no such a gap. In the first case, one needs a great amount of energy to take an electron from its stationary state to an excited state, and the crystal behaves like an insulator. In the second case, a small amount of energy is sufficient for an electron to reach an excited state: This is a metallic behavior.

If one simulates a crystal at zero-temperature, the above Bloch states are filled in using the *aufbau* principle, that is, starting from the lowest energy state, then the next lowest, and so on. Of course, this is a much clearer concept in the molecular case, where there is only a countable infinite number of eigenstates, than in the crystal case, where the energy levels are indexed by the continuum. We however know that we have a fixed number of electrons per unit cell, although an electron associated to a precise unit cell need not be localized in this cell. Thus, one eigenstate does not represent an electron by itself, but the set of all similar electrons repeated through the translations by the lattice vectors (physically, this corresponds to delocalized electrons).

More precisely, the first-order reduced density operator τ obtained by populating the eigenstates of H according to the *aufbau* principle is formally defined by

$$\tau = \frac{1}{|BZ|}\int_{BZ}\sum_{j=0}^{+\infty}\mathcal{H}(\varepsilon_F - \varepsilon_j^k)|\psi_j^k\rangle\langle\psi_j^k|\,dk,$$

where $\mathcal{H}$ denotes the Heaviside function and where ε_F is the Fermi level. Denoting by $\tau(x,y)$ the Green kernel of τ, namely the first-order density matrix (1-RDM), the number of electrons per cell reads

$$\begin{aligned}\int_Q \tau(x,x)\,dx &= \int_Q \frac{1}{|BZ|}\int_{BZ}\sum_{j=0}^{+\infty}\mathcal{H}(\varepsilon_F - \varepsilon_j^k)|\psi_j^k(x)|^2\,dx\,dk \\ &= \frac{1}{|BZ|}\int_{BZ}\sum_{j=0}^{+\infty}\mathcal{H}(\varepsilon_F - \varepsilon_j^k)\,dk.\end{aligned}$$

The function

$$f(\varepsilon) = \frac{1}{|BZ|} \int_{BZ} \sum_{j=0}^{+\infty} \mathcal{H}\big(\varepsilon - \varepsilon_j^k\big)\, dk \tag{41.8}$$

is continuous, nondecreasing, goes to zero at $-\infty$ and to $+\infty$ at $+\infty$; the Fermi level (or Fermi energy) ε_F is therefore defined by

$$f(\varepsilon_F) = N, \tag{41.9}$$

where N is the number of electrons per cell. The Fermi level is univoquely defined for metals; for insulators, $f(\varepsilon)$ is constant equal to N in the gap between the highest filled band (the valence band) and the lowest empty band (the conducting band) and it is of common use to define the Fermi level as the center of the gap. Let us notice that filling one band entirely amounts to putting one electron (or two electrons in closed-shell models) per unit cell.

Lastly, let us mention that the derivative of $f(\varepsilon)$, denoted by $n(\varepsilon)$, is called the density of states and plays an important role in solid state physics (ASHCROFT and MERMIN [1976]).

42. Early electronic structure models for the crystal phase

Simple models of periodic solids have been developed in the early years of quantum mechanics, in particular, the so-called *muffin-tin*, *tight-binding* and *augmented plane waves* methods. All these methods in their most simple form have somewhat disappeared, but the central physical ideas are still used and the corresponding vocabulary is still valid even if it is with more sophisticated methods. This is why it is worth accounting for them.

(A) The muffin-tin method

Placing an atom in a solid we may expect that the potential remains spherically symmetric in the core region which is unlikely to be affected. On the contrary the wavefunctions outside the core region are expected to be considerably distorted due to the repulsion of the electrons of the neighbouring atoms. As far as the potential of an electron is concerned however, we could argue that, when it is roughly equidistant from all neighbouring atoms, it would not be particularly attracted to any of them, which means that it would be moving in an approximately zero field or constant potential. The muffin-tin potential which simply involves taking the potential V spherically symmetrical about the nuclei at their centers and equal to a constant V_0 everywhere outside of the spherical regions is the simplest case. This has the great advantage that the one-electron equation (41.1) can be solved exactly, because the equation has separate solutions in polar coordinates, the so-called spherical harmonics, and the equation has plane waves as solutions for the radial part of the wavefunctions (FLETCHER [1971]).

(B) The tight-binding method

The previous method is based on the assumption that electrons are not closely attached to any particular nucleus in the solid, as is the case of metals. When it is not so, e.g., when the outer electrons of neighbouring atoms adopt bonding wavefunctions concentrated along the lines joining the nuclei (covalent solids) or when the outer electrons of some atoms are essentially transferred into orbitals concentrated about other atoms (ionic solids), the above approximation does not hold. In such cases it is convenient to consider that the electrons spend most of their time close to the same nucleus in orbit about the nucleus and that their orbitals can be described as a linear combination of the wavefunctions not very different from those in which they would be in the free atom or ion. Taking into account the periodicity of the system, the electronic wavefunctions in the solid writes as Bloch functions:

$$\chi_i(k,r) = N^{-1/2} \sum_n e^{ik\cdot R_n} \phi_i(r - R_n), \tag{42.1}$$

where R_n are the nucleus positions and the ϕ_i are the atomic orbitals. The approach is often referred to as tight-binding (TB) method since it works particularly well for electrons which are tightly bound to their nuclei. As such, this choice of basis functions does not introduce any simplification in solving the set of one-electron equations as defined in (42.1). However in the early years, the separation of atoms was considered as large compared with the spread of the atomic orbitals, i.e., V is varying only slowly between the atoms (as in the nearly-free-electron approximation) and this leads to simplifications analogous to those previously described (FLETCHER [1952]).

(C) The augmented plane wave method

Clearly taking advantages of both methods would give a better representation of the situation found in a crystal. This is why, in 1937 an alternative method, which has become known as the augmented-plane-wave method was proposed (SLATER [1937]). This starts by assuming that we are dealing with a potential energy function which is spherically symmetrical within spheres centered on the nuclei and which is constant outside these spheres. The idea of the method is that in the spheres we use the same sort of expansion as in the muffin-tin method (namely, the solution of the Schrödinger equation for the spherical potential) and outside the spheres, we use an expansion in plane waves. This is not a perfect representation of the potential actually found in a crystal, but it is not a bad one since it has the advantage that we can set up a really rigorous solution of the Schrödinger equation. Having found these solutions, the small departures of the potential from the assumed form can be treated as perturbations.

43. Ab initio computations for the crystal phase: Hartree–Fock and Kohn–Sham type models

So far, we have dealt with the Schrödinger equation satisfied by *one* electron in a periodic potential. Let us now consider the general Schrödinger equation as written for a finite system and assume that it holds for an infinite system. Doing so we implicitly use the currently made approximations, the so-called Born–Oppenheimer and the neglect of the relativistic effects. However, new difficulties arise due to the infinite number of particles (electrons and nuclei). We shall address them below. As in the case for a finite system, it is necessary for calculations of electronic structures in the solid state to resort to approximations of the Schrödinger equation.

For the sake of clarity and brevity, we mainly restrict ourselves here to the Hartree–Fock approximation. The treatment of the density functional type models basically follows the same lines. As the latter models are clearly the most commonly used ones in the solid state science community, they will deserve a complete expository survey, given in a following contribution of the present volume. For the vague ideas we give here, the Hartree–Fock setting is a convenient framework.

Recall that the Hartree–Fock equations for a finite molecular system read

$$-\frac{1}{2}\Delta\phi_j + V\phi_j + \left(\rho \star \frac{1}{|x|}\right)\phi_j - \int_{\mathbb{R}^3} \frac{\tau(x,x')}{|x-x'|}\phi_j(x')\,dx' = \varepsilon_j\phi_j, \tag{43.1}$$

with

$$\tau(x,x') = \sum_{j=1}^{N} \phi_j(x)\phi_j(x')^*, \qquad \rho(x) = \tau(x,x),$$

and $V = -\sum_{k=1}^{M} \frac{z_k}{|x-\bar{x}_k|}$ represents the electrostatic potential created by the nuclei. The number M is the number of nuclei, and $\bar{x}_k$ and z_k, respectively, their positions and charges; $\star$ denotes the convolution product over $\mathbb{R}^3$. The Hartree–Fock equations are the Euler–Lagrange equation of the energy

$$\begin{aligned} E^{\mathrm{HF}}(\{\phi_j\}) &= \frac{1}{2}\sum_{j=1}^{N}\int_{\mathbb{R}^3} |\nabla\phi_j|^2 + \int_{\mathbb{R}^3} \rho V + \frac{1}{2}\int_{\mathbb{R}^3}\int_{\mathbb{R}^3} \frac{\rho(x)\rho(x')}{|x-x'|}\,dx\,dx' \\ &\quad - \frac{1}{2}\int_{\mathbb{R}^3}\int_{\mathbb{R}^3} \frac{|\tau(x,x')|^2}{|x-x'|}\,dx\,dx' + \frac{1}{2}\sum_{k\neq l} \frac{z_k z_l}{|\bar{x}_k - \bar{x}_l|}, \end{aligned} \tag{43.2}$$

under some orthonormality constraints on the molecular orbitals ϕ_j. We see here that although the above energy clearly makes sense in the case where N is finite, the solid state case is *not that obvious*. It can at once be understood by the examination of the nuclei-nuclei repulsion term

$$\frac{1}{2}\sum_{k\neq l} \frac{z_k z_l}{|\bar{x}_k - \bar{x}_l|}$$

which is infinite when the sites $\bar{x}_k$ vary in an infinite lattice. In order to give a proper meaning to the energy, one needs to group the electrostatic terms in a convenient way, so as to make appear electrostatic potentials created by *neutral* infinite systems. This difficulty translates into the need of a proper definition for the interaction law of an infinite system of particles when these particles are periodically arranged. Obviously, the good candidate is the Green function G_{per} of the Laplacian in a periodic setting, in the same vein as the Coulomb interaction law is the Green function of the Laplacian in the whole space.

A rigorous mathematical definition of the Hartree–Fock energy will therefore involve the potential G_{per} satisfying

$$\begin{cases} -\Delta G_{\mathrm{per}} = 4\pi \left(\delta_0 - \dfrac{1}{|\mathcal{Q}|} \right), \\ G_{\mathrm{per}} \quad \mathcal{R}\text{-periodic}, \end{cases}$$

where $|\mathcal{Q}|$ is the volume of the unit cell $\mathcal{Q}$.

A large body of mathematical work has been devoted to the enterprise of giving a sound mathematical ground to the Hartree–Fock model for infinite solids, which enterprise is essentially fulfilled to date. As this is not the main purpose of this chapter to focus on such theoretical issues, we prefer not to elaborate further on these issues, and refer the reader to the bibliography, e.g., the treatise by CATTO, LE BRIS and LIONS [1998], the articles by CATTO, LE BRIS and LIONS [2002], CATTO, LE BRIS and LIONS [2001] and the review article by Catto et al. in DEFRANCESCHI and LE BRIS [2000]. Let us only mention that the procedure that recasts the Hartree–Fock model for infinite periodic crystals (as usually manipulated in solid state physics) as a limit of the same problem for finite molecular systems, when the size of the latter goes to infinity is referred to as the *thermodynamic limit problem* (even if the procedure holds at zero temperature). Mathematically, one shows that, for simple models, as the nuclei placed at points $\bar{x}_k$ all are of the same nature and progressively fulfill the whole lattice $\mathbb{Z}^3$, the minimizer of the molecular problem converges to the minimizer of a minimization problem of the same form, but set on the unit cubic cell of the lattice. Simultaneously, the energy of the molecular system *divided by its size* converges also to the periodic energy thus obtained. For the Hartree–Fock problem, which is more intricate, the mathematical proof is not complete, but on the basis of partial arguments, it is strongly believed that the minimization problem obtained through such a limit is precisely that manipulated in solid state physics, namely a problem of the following form: Minimize an energy of the type (43.2) where all integrals on $\mathbb{R}^3$ are replaced by 6-dimensional integrals on the unit cell of the lattice times the Brillouin zone, and the Coulombian interaction potential $\frac{1}{|x|}$ is replaced by the potential G_{per}, or an analogue to it. The functions one minimizes upon are now Bloch functions of the form $\phi_j^k(x)$, where x varies in the unit cell, k in the Brillouin zone and $j \in \mathbb{N}^*$. The whole mathematical study of such a problem can be conducted. We refer to the bibliography.

Let us consider this issue as being settled, and now concentrate ourselves not on the Hartree–Fock energy functional, but on the Hartree–Fock equations themselves, that is to say on the sense (43.1) should be given in the case of a solid. We only make formal manipulations in the sequel, that can in fact be given a sound ground through the complete mathematical analysis we eluded to above.

Formally, the Fock Hamiltonian may be written in the following form:

$$F = -\frac{1}{2}\Delta + V + \rho \star \frac{1}{|x|} - \int_{\mathbb{R}^3} \frac{\tau(x,x')}{|x-x'|} \bullet (x')\,dx'. \tag{43.3}$$

Bearing in mind our comments in G_{per}, we introduce $V_{\text{tot}} = V + \rho \star \frac{1}{|x|}$ which is to be understood as a solution of the Poisson equation:

$$\begin{cases} -\Delta V_{\text{tot}} = -4\pi\Big(\sum_{T\in\mathcal{R}} m(\cdot + T) - \rho\Big), \\ V_{\text{tot}} \quad \mathcal{R}\text{-periodic} \end{cases} \tag{43.4}$$

with the measure m defining the nuclei in the primitive unit cell. Let us assume that ρ is periodic (which will indeed be the case *a posteriori*). This equation has a periodic solution as soon as $m - \rho$ is neutral over the unit cell of $\mathcal{R}$. A similar treatment is possible for the exchange term (last term in (43.1)). In the case of a solid, the formal Hartree–Fock equations are derived through Bloch's theorem: All sums over j involved in the definition of ρ and τ are replaced by sums over j and integrals over BZ, and the wavefunctions and energies are labelled by $(j,k) \in \mathbb{N}^* \times BZ$. More precisely, setting

$$\tau(x,x') = \sum_{j\in\mathbb{N}} \int_{BZ} \phi_j^k(x)\phi_j^{k^*}(x')\big(\varepsilon_{\text{F}} - \varepsilon_j^k\big)_+\,dk, \tag{43.5}$$

where $t_+ = \max(t,0)$, (the term $(\varepsilon_{\text{F}} - \varepsilon_j^k)_+$ selects only the states having lower energy than the Fermi energy ε_{F} defined in (41.9)), then we define

$$F\phi = -\frac{1}{2}\Delta\phi + V_{\text{tot}}\phi - \int_{\mathbb{R}^3} \frac{\tau(x,x')}{|x-x'|}\phi(x')\,dx', \tag{43.6}$$

where V_{tot} is solution to (43.4) with

$$\rho(x) = \tau(x,x) = \sum_{j\in\mathbb{N}} \int_{BZ} \big|\phi_j^k(x)\big|^2 \big(\varepsilon_F - \varepsilon_j^k\big)_+\,dk.$$

The Hartree–Fock wavefunctions ϕ_j^k are defined as the solutions to

$$\begin{cases} F\phi_j^k = \varepsilon_j^k \phi_j^k, \\ \forall j \in \mathbb{N},\ \forall k \in BZ,\ e^{-ikx}\phi_j^k(x)\ \mathcal{R}\text{-periodic}, \\ \int_{\mathcal{Q}} \phi_j^k(x)\,\phi_{j'}^{k'}(x)^*\,dx = \delta(k-k')\delta_{jj'}. \end{cases} \tag{43.7}$$

It is one of the purposes of the mathematical analysis of the problem to recast these equations, only *postulated* above and introduced as the infinite analogue to the finite version (43.1), as the rigorous Euler–Lagrange equations of a well-posed periodic minimization problem, set on the unit cell of the lattice, and involving an energy functional of Hartree–Fock type, formally reminiscent of (43.2).

In principle, one therefore needs to solve an infinite number of Hartree–Fock equations, for system (43.7) is indeed an infinite collection (indexed by the points k of the Brillouin zone) of molecular-like Hartree–Fock type systems

$$F\phi_j = \varepsilon_j \phi_j.$$

At the discrete level, *a set of* Hartree–Fock type systems needs to be solved, the number of which is the number of points k taken into account in the discretization of the Brillouin zone. In practice, it turns out that, fortunately, a limited number of points k is generally enough to obtain accurate results.

For DFT Kohn–Sham models, the equations for a finite number of electrons are of the form

$$\begin{cases} -\frac{1}{2}\Delta\phi_j + V_{\mathrm{eff}}\phi_j = \varepsilon_j\phi_j, \quad \forall j, \\ V_{\mathrm{eff}} = V_{\mathrm{tot}}(\rho) + v_{\mathrm{xc}}(\rho), \\ \rho = \sum_j |\phi_j|^2, \end{cases} \tag{43.8}$$

where the ε_j are the lowest N eigenvalues of the operator $-\frac{1}{2}\Delta + V_{\mathrm{eff}}$, where $V_{\mathrm{tot}}(\rho) = V + \rho \star \frac{1}{|x|}$, and where $v_{\mathrm{xc}} = \frac{1}{2}\frac{\partial E_{\mathrm{xc}}}{\partial \rho}$ denotes the exchange-correlation potential (see Section 13). In the case of a solid, one needs to group the electrostatic terms together in V_{tot} so as to be able to compute it through Poisson's equation (43.4). Moreover, the same dependency on the wavevector k as in HF theory appears here. Therefore, (43.8) becomes:

$$\begin{cases} -\frac{1}{2}\Delta\phi_j^k + V_{\mathrm{eff}}\phi_j^k = \varepsilon_j^k\phi_j^k, \quad \forall j, k, \\ V_{\mathrm{eff}} = V_{\mathrm{tot}}(\rho) + v_{\mathrm{xc}}(\rho), \\ \rho = \sum_j \int_{BZ} |\phi_j^k|^2 (\varepsilon_{\mathrm{F}} - \varepsilon_j^k)_+ \, dk. \end{cases} \tag{43.9}$$

The equations are then treated analogously to the Hartree–Fock case.

The success of density functional calculations is closely related to the choice of the approximation for the exchange-correlation energy. Many possibilities are presently available in the literature. Starting with the simplest one: The Local Density Approximation (LDA) which assumes that the exchange-correlation energy depends only on the local electron density around each volume element or in other words it is assumed that the evaluation of the exchange-correlation effects, the real electron density surrounding each volume element can be replaced by a constant electron density of the same value as at the reference point (see Section 15). This is an excellent approximation for metallic systems, but it is found to be also a surprisingly good approximation for a wide variety of systems. For systems with strongly varying electron density it represents a severe simplification revealed by systematic errors (bond lengths within 0.05 Å, too large binding energies), but gradient-corrected density functionals (Generalized Gradient Approximation (GGA)) seem to offer a remedy. The basic idea in these schemes is the inclusion of terms in the exchange-correlation expressions that depend on the gradient of the electron density and not only on its value at each point in space (see again Section 15).

44. Basis sets

In both DFT and HF formalisms, one chooses a basis to expand the one-electron wavefunctions ϕ_i^k and solves Eqs. (43.7) or (43.9). The first crucial choice is that of the basis functions. Plane waves (PW) are by far the preferred bases. The reason is that they are Bloch functions, they form an orthogonal set, they make most algebraic manipulations very simple (the kinetic operator is diagonal), Fast Fourier Transform (FFT) algorithms can be systematically adopted. The disadvantage of using PWs is due to the oscillating behaviour of crystalline orbitals near the nuclei (core region); the price to pay is the huge number of PWs to describe accurately the details of the wavefunctions (their Fourier coefficients are not decaying very fast).

The use of Gaussian-Type Orbitals (GTO), e.g., basis functions made of Gaussian functions multiplied by a spherical harmonics centred on the nuclei, permits to describe accurately electronic distributions both in the valence and in the core region with a limited number of basis functions. The advantage of GTOs is that the Fourier transform of a Gaussian is a Gaussian and they make the computation of bielectronic integrals easier. Fast Multipole Methods (FMM) have recently been introduced in crystal phase calculations with GTO in order to reach linear scaling of the Fock matrix building (KUDIN and SCUSERIA [1998]). Among the disadvantages associated with the choice of GTOs there is their nonorthogonality, the need to take into account the so-called basis set superposition error and the risk of pseudo-linear dependence catastrophes when very diffuse functions are used (overcompleteness of the basis set).

Considering the advantages and disadvantages of PWs and GTOs, the completion of GTO basis with PWs is computationally convenient (BARONI, PASTORI-PARRARRICINI and PEZZICA [1985]). GTOs are more particularly used to represent core electrons while valence electrons are represented by a mixed combination of PWs and GTOs. Indeed the additional problems raised by the computation of overlap and bi-electronic integrals are solved by the fact that (in the case of a PW and a GTO) the first ones are simply the Fourier transforms of Gaussians, and the second ones the Fourier transforms of Gaussians multiplied by the Fourier transform of $1/|x|$.

In the frame of DFT there is also the possibility to calculate numerically all needed integrals (TE VELDE and BAERENDO [1991]) . With this choice, much greater freedom is possible in the selection of the representative functions: They may include PWs, GTOs, Slater-type orbitals, numerical bases or any combination of these

Similarly to the idea originally proposed by Slater (see Section 42), the APW method is the characteristic feature of a wide variety of techniques. It is originating in the fact that for an atom placed in a solid, one can assume to a good approximation that the potential is still spherically symmetrical in the core. The wavefunctions outside the core of its electrons are expected to be considerably distorted, particularly due to the repulsion of the outer electrons of neighbouring atoms. It would not be particularly attracted to any of them, i.e., it would be moving in an approximately constant potential.

Relying on the general ideas just discussed the Muffin-Tin orbitals (MTO) are composed of two regions: The first one is a union of nonintersecting spheres around the nuclei, and a second one is the remaining one called the interstitial region. The muffin-tin approximation consists essentially in assuming that the potential and electron density are radially symmetric in the atomic spheres (spherical harmonics multiplied by radial functions are used as a representative state), and in the interstitial region a PW expansion is used. A whole family of techniques rely on this representation: APW (augmented PWs) (SLATER [1937]), LAPW (linearized APWs) (ANDERSEN [1975]), FLAPW (full-potential LAPWs) (WIMMER, KRAKAUER, WEINERT and FREEMAN [1981]). With the latter improvement, very accurate descriptions of wavefunctions can be obtained for all-electron DFT. Alternatively in the KKR method (KORRINGA [1947], KOHN and ROSTOKER [1954]) the potential in the interstitial region is supposed to be flat and the wavefunctions are expanded in phase shifted spherical waves.

45. Pseudopotentials

The concept of pseudopotential is closely related to the choice of the basis set for its role consists mainly in reducing the size of the basis set used in the calculations. It has been observed long ago that the nearly free electron approximation provides a good description of the behavior of electrons in a metal. On the other hand the metal potential is approximately a superposition of atomic potentials which oscillates strongly and more in each atomic core region. This paradox was first used in 1959 (PHILLIPS and KLEINMAN [1959]) and other pseudopotentials were invented next which are largely employed in codes because apart for being less time consuming than all electron methods, they also provide an efficient way of considering relativistic effects. The physical and chemical properties of an assembly of interacting atoms depend mainly upon the motion of the loosely bound valence electrons and only slightly upon the ionic core electrons motion. When the system is deformed, the valence electron density is strongly distorted, but each ionic core density follows almost rigidly the displaced atomic nucleus. On the other hand, the effect of the core states upon the valence electron motion is important and must be taken into account. The pseudopotential method is a technique which eliminates the explicit consideration of the core states, while treating almost exactly their effect upon the valence electrons: Inside spheres centered on the various nuclei the highly oscillating potential is replaced by a smooth pseudopotential while out of the spheres the correct potential is considered. The pseudopotential has to satisfy some obvious numerical constraints (e.g., continuity of the functions and of their first two derivatives onto the spheres) but also some physical constraints (the pseudoatom and the real atom must have the same behavior regarding the chemical bonds and for some cases their chemical hardness – i.e., capability of their electronic clouds to be polarized – have to be identical

(HEINE [1970])). Physically speaking using a pseudopotential is equivalent to have a pseudoatom built from a positive ion with part of its electrons (the core electrons) frozen and the other electrons (the valence electrons) free. An important property of pseudopotentials is their transferability. A pseudopotential is said to be transferable if the same ionic potential can be used for different atomic arrangements. Accordingly the pseudoatoms are described by pseudowavefunctions (for the various levels s, p, d, f, etc.) and the system is described by a pseudoenergy which has no physical meaning whereas differences between pseudoenergies can be compared with physical quantities. The first task for generating pseudopotentials is to select for each type of atom the radii of the spheres for the s, p, d, f, etc., levels which have to include all the oscillations of the potential but leaving out of the sphere the last extremum. Then after selecting a parametrized expression for the pseudopotential, the various constraints are applied yielding to different categories of pseudopotentials. A few words enter the vocabulary of pseudopotentials which have to be known by the user of pseudopotentials who has to select from physical considerations, which type of pseudopotential his problem needs. Norm Conserving (NC) pseudopotentials are such that the norm of the wavefunction is preserved for a given energy and approximate for the nearby energies, soft and ultra-soft mean that a relatively small basis set can be used. Over the years, numerous pseudopotentials of increasingly better quality have appeared and are now widely spread (among many others, see TROULLIER and MARTINS [1990], VANDERBILT [1990]).

46. Specific numerical features of solid state calculations

The selection of a given theoretical model opens a spectrum of technical possibilities the choice of which is crucial for the success of the calculations.

Choice of a plane wave basis sets

For periodic systems, one-electron wavefunctions are written as Bloch functions

$$\chi_\beta(r)e^{ikr},$$

where $\chi_\beta(r)$ is a periodic function and where e^{ikr} involves a Bloch vector k belonging to the Brillouin zone. The periodic term $\chi_\beta(r)$ can be expressed in terms of plane waves

$$\chi_\beta(r) = \sum_G c^k_{\beta,G} e^{iGr},$$

where the sum is expanded on the whole reciprocal space and is infinite. Consequently wavefunctions appear as being developed on an infinite number of plane waves and on an infinite number of k-points; practically infinite calculations are not runnable, but two physical observations allow to circumvent the difficulty:

(i) For close values of k electronic wavefunctions do not differ significantly; the infinite number of wavefunctions corresponding to a given small zone of the Brillouin Zone can thus be approximatively described by the wavefunction associated to the mean value of k in this small zone;

(ii) coefficients $c_{\beta,k+G}$ of the plane waves with a low kinetic energy (given by $\frac{1}{2}|k+G|^2 < E_c$) are significantly higher than those corresponding to plane waves with a high kinetic energy. The plane wave basis can be truncated keeping only those with a kinetic energy smaller than a selected cut-off energy, E_c. The electronic wavefunctions are therefore determined for a finite number of k-points of the Brillouin zone.

Choice of the k-point sampling

The calculations described above involve integrating periodic functions over values of k. The k-partition of the periodic problem made possible by translational periodicity requires the choice of a suitable sampling set of k-points. Selected values of k lie in the first Brillouin zone, but to minimize the calculations the set should be as small as possible, yet representative enough to allow accurate interpolation throughout the reciprocal space. Methods have been developed for the selection of special k-points which rely partly on symmetry considerations (see, for instance, ZICOVICH-WILSON and DOVESI [1998], MONKHORST and PACK [1976]). From the practical point of view, the only thing to do for the user is to determine how many points are necessary to obtain accurate results. The general trend is that for metals many points are needed while for insulators a few k-points yield realistic values. As a matter of fact calculations on insulating materials can even be run with one k-point.

Ewald sums

Due to the summations appearing on the Coulomb potential (and the exchange term in HF theory), a special attention has to be brought to give a sense to the expressions. This problem already present in the molecular case is much more delicate to deal with in the periodic case, because the truncation of the series cannot be based only on a criterion of range of distance between the centers of the interacting functions. Since these series are only conditionally convergent, efficient methods have been implemented to handle these sums. In many textbooks this problem is addressed, but a recent overview can be found in BLANC [2000].

Many other points need specific attentions but they cannot all be listed in the present section. More details will be given in the contribution dedicated to the study of solid phase and we also refer the interested reader to the dedicated literature (see, for instance, CATLOW [1998], DOVESI, ORLANDO, ROETTI, PISANI and SAUNDERS [2000], GALE, CATLOW and MACKRODT [1992], GILLAN [1991]).

47. The liquid phase

Let us now turn to the liquid phase, and to the related modelling issues in view of the numerical simulations of phenomena occurring in this particular state of matter. The purpose of this section is only to briefly introduce the reader to the subject, as an entire contribution of the present volume will give a comprehensive and detailed survey of the different strategies of modelling and the numerical techniques.

For clarity, we (again) concentrate ourselves on the spinless Hartree–Fock model. In order to pedagogically introduce the models for the liquid phase, let us for a moment rewrite the problem of determining the nuclei configuration together with the electronic structure of the Hartree–Fock ground state in the slightly unusual following form:

$$\inf\left\{E^{\mathrm{el}}\big(\{\bar{x}_k\}\big)+\sum_{1\leqslant k<l\leqslant M} z_k z_l G(\bar{x}_k-\bar{x}_l),\ \{\bar{x}_k\}\in\mathbb{R}^{3M}\right\}, \tag{47.1}$$

where

$$E^{\mathrm{el}}\big(\{\bar{x}_k\}\big)=\inf\left\{E^{\mathrm{HF}}_{\{\bar{x}_k\}}\big(\{\phi_i\}\big),\ \phi_i\in H^1\big(\mathbb{R}^3\big),\ \int_{\mathbb{R}^3}\phi_i\phi_j=\delta_{ij},\ 1\leqslant i,j\leqslant N\right\} \tag{47.2}$$

and

$$\begin{aligned} E^{\mathrm{HF}}_{\{\bar{x}_k\}}\big(\{\phi_i\}\big) &= \frac{1}{2}\int_{\mathbb{R}^3}\sum_{i=1}^{N}|\nabla\phi_i|^2-\int_{\mathbb{R}^3}\sum_{k=1}^{M} z_k G(x-\bar{x}_k)\rho(x)\,dx \\ &\quad+\frac{1}{2}\int_{\mathbb{R}^3}\int_{\mathbb{R}^3} G(x-y)\rho(x)\rho(y)\,dx\,dy \\ &\quad-\frac{1}{2}\int_{\mathbb{R}^3}\int_{\mathbb{R}^3} G(x-y)\big|\tau(x,y)\big|^2\,dx\,dy \end{aligned} \tag{47.3}$$

with as usual $\tau(x,y)=\sum_{i=1}^{N}\phi_i(x)\phi_i(y)$ and $\rho(x)=\tau(x,x)$. In the above formulae, $\bar{x}_k$ denotes the position in $\mathbb{R}^3$ of the kth nucleus, $z_k>0$ its charge and $G(x-y)=\frac{1}{|x-y|}$ the Green kernel in $\mathbb{R}^3$ of the operator $-\frac{1}{4\pi}\Delta$.

The associated Euler–Lagrange equations (the Hartree–Fock equations) for the electronic part read

$$-\Delta\phi_i+V_{\mathrm{tot}}\phi_i-\int_{\mathbb{R}^3}\tau(\cdot,y)G(\cdot-y)\phi_i(y)\,dy=\varepsilon_i\phi_i,\quad 1\leqslant i\leqslant N, \tag{47.4}$$

$$-\Delta V_{\mathrm{tot}}=-4\pi\left(\sum_{\nu=1}^{M} z_\nu\delta_{\bar{x}_\nu}-\rho\right). \tag{47.5}$$

The reason why we have isolated an electrostatic (classical) subproblem (47.5) apart from the quantum equations (47.4) in the above system will clearly appear below. Let us already mention that if the Hartree–Fock setting we use here is replaced by any other frameworks, like those of density-functional theory or Molecular Mechanics, only Eqs. (47.4) are changed, while (47.5) remains valid.

Most of the physical and chemical phenomena of interest in Chemistry and Biology take place in the liquid phase and it is well known from experimental evidences that solvent effects play a crucial role in these processes (POLTIZER and MURRAY [1994]). It is thus important to extend the model presented above to solvated molecules. The first

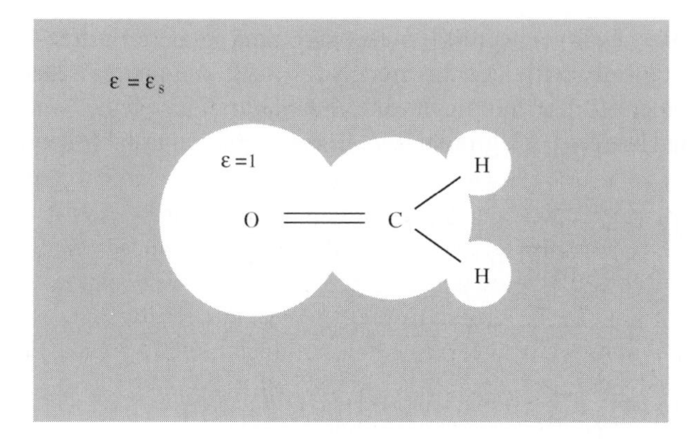

FIG. 47.1. A H_2CO molecule surrounded by a continuum medium.

idea is to enlarge the system under study: The technique used to model a single isolated molecule can be applied to a "supermolecule" consisting of the solvated molecule under study plus several neighboring solvent molecules. Despite its conceptual simplicity, this method suffers from two major drawbacks: It does not take into account long-range effects of the solute–solvent interaction and, above all, it significantly increases the size of the system, thus the time needed for computations. The latter argument turns out to be prohibitive when the solvated molecule already contains several dozens of atoms.

Coupling the Hartree–Fock model with a solvation continuum model (TOMASI and PERSICO [1994]) constitutes an economical alternative. This consists in locating the solute molecule under study inside a cavity Ω_i, standing for a solvent excluding volume, surrounded by a continuous medium modelling the solvent (Fig. 47.1).

In the standard model that goes back to Born, Kirkwood and Onsager (see TOMASI and PERSICO [1994] for a historical review), the continuous medium behaves as a homogeneous isotropic dielectric of relative permittivity ε_s ($\varepsilon_s > 1$). The electrostatic interactions between the charge distributions which compose the solute (points nuclei and electronic cloud) are affected by the presence of the solvent: The standard Coulomb potential $\frac{1}{|x-y|}$, which is the Green kernel $G(x-y)$ of $-\frac{1}{4\pi}\Delta$ in $\mathbb{R}^3$ must be replaced in (47.3) by that of the operator $-\frac{1}{4\pi}\mathrm{div}\,(\varepsilon\nabla\cdot)$, with $\varepsilon(x) = 1$ inside the cavity Ω_i and $\varepsilon(x) = \varepsilon_s$ outside. This should not be surprising for the reader who remembers of the above treatment of the interaction law for solids. The electrostatic subproblem (47.5) is thus modified and becomes:

$$-\mathrm{div}\big(\varepsilon(x)\nabla V_{\mathrm{tot}}(x)\big) = -4\pi\rho_t(x), \tag{47.6}$$

denoting $\rho_t = \sum_{\nu=1}^{M} z_\nu \delta_{\bar{x}_\nu} - \rho$ the total charge of the molecule. It is however to be noticed that in practice (with a view to keep the efficiency of the computations of bielectronic integrals in vacuum) the exchange term in (47.3) is left unchanged.

In the earliest calculations, the cavity was a sphere and Eq. (47.6) was solved by expansion on spherical harmonics. In order to better represent the solvent excluding volume, successive refinements of the cavity definition have been used: ellipsoidal cavities, for which (47.6) may still be solved by expansion on a basis, then molecular shape cavities, for which (47.6) is solved in general by integral equation methods. The notion of molecular shape cavity is rather vague, and actually several definitions are in use at the present time (TOMASI and PERSICO [1994]). Anyway, most of the cavities have more or less the same shape and volume: They are variations around the basic molecular shape cavity made of a union of M spheres, each of them centered on a nucleus and of radius αR_{VdW}, R_{VdW} denoting the Van der Walls radius of the corresponding atom and α being a fixed coefficient ($\alpha = 1.2$ for neutral solutes).

Over the last few years, some extensions of the standard solvation continuum model have been proposed to cover the cases when the solvent is either an ionic solution or a liquid crystal. In the former case (HONIG and NICHOLLS [1995]), Eq. (47.6) is replaced by the so-called Poisson–Boltzmann equation

$$-\operatorname{div}\big(\varepsilon(x)\nabla V(x)\big) + \varepsilon(x)k_{\mathrm{B}}T\kappa^2(x)\operatorname{sh}\big(V(x)/k_{\mathrm{B}}T\big) = -4\pi\rho_t(x), \tag{47.7}$$

where T is the temperature, k_{B} the Boltzmann constant and where the field $\kappa(x)$ is constant, equal to κ_s ($1/\kappa_s$ is the Debye length and accounts for the ion screening), outside the cavity and vanishes inside the cavity. Eq. (47.7) can be approximated in many cases by the *linearized* Poisson–Boltzmann equation

$$-\operatorname{div}\big(\varepsilon(x)\nabla V(x)\big) + \varepsilon(x)\kappa^2(x)V(x) = -4\pi\rho_t(x), \tag{47.8}$$

with $\varepsilon(x) = 1$ and $\kappa(x) = 0$ inside the cavity Ω_i and $\varepsilon(x) = \varepsilon_s > 1$ and $\kappa(x) = \kappa_s > 0$ outside. In the latter case (MENNUCCI, COSSI and TOMASI [1995]), Eq. (47.6) keeps the same formal expression,

$$-\operatorname{div}\big(\underline{\underline{\varepsilon}}(x)\cdot\nabla V(x)\big) = -4\pi\rho_t(x) \tag{47.9}$$

but the dielectric constant $\underline{\underline{\varepsilon}}(x)$ is no longer a scalar: It is a 3×3 anisotropic symmetric tensor so that $\underline{\underline{\varepsilon}}(x) = \underline{\underline{I}}_3$ inside the cavity Ω_i and $\underline{\underline{\varepsilon}}(x) = \underline{\underline{\varepsilon}}_s$ outside ($\underline{\underline{I}}_3$ denotes the 3×3 unit tensor).

In practice, Eq. (47.6) is most often solved by an integral equation method (see the contribution of B. Mennucci et al. in the present volume). Recently (see, e.g., CANCÈS, LE BRIS, MENNUCCI and TOMASI [1999]), integral equation methods has been suggested for also solving (47.7) and (47.9) (three-dimensional finite element or finite difference methods were previously used).

In practical calculations, the potential ϕ and the field $\nabla\phi$ generated by ρ_t *in the vacuum* are needed to compute the left-hand side of the integral equations; they are easy to obtain, as soon as the basis functions χ_p are of Gaussian type. This is the reason why the computational time needed to perform a Hartree–Fock calculation on a solvated molecule is not much longer than that needed to perform the same calculation in the vacuum.

Let us now turn to the geometry optimization problem (47.1). As computing the function $E^{\mathrm{el}}(\{\bar{x}_k\})$ for a given nuclear conformation $\{\bar{x}_k\}$ is time consuming, it is crucial to perform the geometry optimization with as few steps as possible. This amounts to saying that the information provided by the derivatives of the energy $E^{\mathrm{el}}(\{\bar{x}_k\})$ with respect to the nuclear coordinates need to be incorporated in the iterations. In the vacuum, analytical derivatives of $E^{\mathrm{el}}(\{\bar{x}_k\})$ are known for Hartree–Fock and related models, as above mentioned. The so-obtained estimations of the gradient are thus of great accuracy for a low computational cost. An extension of these analytical derivative formulae for solvated molecules, including cases when the solvent is an anisotropic dielectric or an ionic solution, is now also known for the continuum models, see CANCÈS and MENNUCCI [1998]. The main technical difficulty lies on the fact that the cavity is moving with the nuclei. It is shown that two new terms appear, the first one stemming for the motion of the charges, the second one for the motion of the cavity itself, both being easy to deal with. Experimental measures as well as numerical results have clearly shown that the molecular geometry is affected by solvent effects. These two terms cannot therefore be ignored in the geometry optimization phase.

In fact, other attempts to extend the computation of analytical derivatives to solvated molecules have been carried out for molecular shape cavities and standard isotropic dielectrics. The basic idea underlying these previous attempts consisted in differentiating the integral equation *after* a boundary element approximation. Proceeding in this way requires to derive the mesh drawn on the interface Γ with respect to the motion of the nuclei, which is time consuming and may lead to uncontrolled numerical errors. The alternative approach suggested in CANCÈS and MENNUCCI [1998] avoids computing the derivative of the apparent surface charge, and is furthermore also adapted for anisotropic dielectrics or ionic solutions: Globally speaking, it consists in deriving *first* the integral equation so as to write $\frac{\partial V_{\mathrm{tot}}}{\partial \bar{x}_\nu}$ as a solution to the derived equation, and next inserting it in the derivative of (47.1). Numerical tests have confirmed the efficiency of the formulae obtained with this latter approach: The computational time is reduced for the standard case and good results are provided in nonstandard cases out of reach so far.

To be as comprehensive as possible, let us point out that other terms, with physical counterparts (cavity formation, entropy, dispersion and repulsion corrections), must be added to the energy computed via the geometry optimization procedure to obtain a numerical value which can be compared with experiment. We refer the reader to CANCÈS and MENNUCCI [1998], TOMASI and PERSICO [1994] for further details.

Let us conclude this section by mentioning that solvation continuum models seem very crude and seem to contrast with the concern of high accuracy usually adopted in quantum chemistry. However, despite their apparent simplicity, they have brought up lots of interesting results, and remain at the present time the most efficient way to take into account solvent effects, mainly because they are a natural place to make use of the powerful machinery of integral equation methods. Several numerical and mathematical studies associated with refinements of the solvation continuum models remain to be done.

CHAPTER IV

Quantum Chemistry and its Relation to Experiment

So far we have been concerned solely with Born–Oppenheimer ground state electronic structure calculations (which are used, e.g., to determine equilibrium molecular structures or formation energy of chemical compounds). It is the scope of the present chapter to examine the relations of other kinds of quantum chemistry calculations to various experimentally observed electronic properties.

First of all, let us underline that any experimental observation is performed neither on a single isolated molecule nor on a perfect solid but on a macroscopic aggregate of molecules or on partially doped solids. In addition, any observation involves some kind of interaction with the system. None of these facts have been taken into account in setting up the Schrödinger equation or the Hamiltonian, but both must be allowed for when we try to interpret physical and chemical observations using results reached by solving any approximation of the Schrödinger equation. For instance, we may need to consider the electronic excitations which play a significant role for a huge number of technological applications or even biological processes, e.g., they are at the basis of photosynthesis or chemical reactions, they govern the behavior of the electronic devices.... We may also need to consider the interaction of a molecule with an external electromagnetic field and also the internal electromagnetic interactions arising from spin and orbital motion. The present chapter will address two main topics; the first one will be the description of excited states (Section 48) and the second one will consider two types of perturbation: external electric F or magnetic B field (Section 49). The two fields may either be time independent, leading to static properties, or time dependent, leading to dynamic properties. In the latter case, we will mainly treat electromagnetic radiations characterized by a frequency. Many other physical quantities may be derived from quantum chemical calculations. They are not included in the present volume because they rise almost no mathematical issues (the only point to be considered is the way their accuracy is monitored, a striking example is found for infra-red calculations (POPLE, SCOTT, WONG and RADOM [1993])). We conclude this chapter by examining briefly the study of reaction mechanisms (Section 50).

Since this chapter mainly concentrates on the physical aspects, it is primely dedicated to a reader familiar with the concepts and applications of theoretical chemistry. The reader more interested by mathematical aspects may easily skip it at a first reading and directly proceed to the next one.

48. Excited states

Let us begin with the determination of excited wavefunctions. Ab initio methods allow for the determination of ground state structure or properties of a molecule or a solid but the calculation of excited states is still a challenge for theoreticians, especially as regards materials. There are various processes responsible for excitations in a quantum system, the two basic ones are the *single-particle* and *two-particle excitations*. A first example of the first category is *photoemission*: A beam of light irradiates a sample, one electron is ejected from the materials and its kinetic energy is analyzed, it is then possible to determine what are the energies of the occupied levels (the levels involved might be either core or valence levels). Immediately after excitation, the system undergoes relaxation. One class of relaxations is of radiative nature (such as fluorescence or phosphorescence, i.e., hopping from higher to lower excited states or to the ground state with emission of radiations) or nonradiative nature (the energy is transformed into vibrations, hence an increase of the temperature of the sample). Instead of being ejected, the electron may be promoted from an occupied to an empty level (this is called a two-electron process); calculations provide information on the occupied levels (where in the final state there is a hole) and the empty levels (which are singly occupied at the end). It is called *photo-ionization*. The section will review the many ways in which the excited states of molecules or solids and the transition energies can be calculated. The very fact that there are several ways attests to the observation that none are particularly standard for all occasions.

This is still an area of quantum theory of research interest. There is no real "black-box" solution that will hold much hope for great accuracy whatever the situation. Rather individual cases will need individual attention, and this is especially true for large systems, systems containing heavy elements, solids, or systems in which the ground state is of open-shell character. Owing to the importance of correlations for excited states and in order to obtain high-quality results, it often proves desirable to have a quantitative treatment of electron correlations. The point is that the correlation-energy contribution to the ground-state energy may be small compared with the dominating contributions of the self-consistent field; however, when energy differences with respect to the ground state are calculated, the changes in the correlation energy may become equal to or even larger than those resulting from changes in the self-consistent field. For example, in diamond, the energy gap for exciting an electron from the valence into the conduction band is reduced by a factor of $1/2$ due to correlations.

Let us quote that depending on the purpose of the calculation, the methodology to select is different. For instance in chemistry, where people are most often interested in the nature of the excited state, one has to obtain both the eigenvalues and the eigenfunctions. Eigenfunctions are somewhat more difficult to obtain than eigenvalues. It is useful to distinguish between two cases, depending on whether the excited state has the same or a different symmetry than the lower state.

Let us first deal with wavefunction methods. The different symmetry case is easy to handle, as the lowest energy state of a given symmetry may be handled completely analogously to the ground state. A HF wavefunction may be obtained by a proper specification of the occupied orbitals, and the resulting wavefunction improved by adding electron correlation by for example Configuration Interaction, Møller–Plesset or Coupled Cluster methods. The only possible caveat may be that the state is an open-shell, which often requires a small MCSCF wavefunction for an adequate zero-order description. Excited states having energy solutions of the same symmetry are more difficult to treat. It will in general be difficult to generate a HF type wavefunction for such states, as the variational optimization will collapse to the lowest energy solution of the given symmetry. The lack of a proper HF solution means that perturbation and coupled cluster methods are not well suited to calculating excited states, although excited state properties may be calculated directly using propagator methods. It is however, relatively easy to generate higher energy states by Configuration Interaction methods, this simply corresponds to using the $(n+1)$th eigenvalue from the diagonalization of the Configuration Interaction matrix as a description of the nth excited state (the first, i.e., the lowest, eigenvalue is the ground state, the second one is the first excited state, etc.). Such a Configuration Interaction procedure will normally employ a set of HF orbitals from a calculation on the lowest energy state, and the Configuration Interaction procedure is therefore biased against the excited states.

We can expect no better in density-functional theory: An important and contemporary research area is still to find out a workable density-functional for the excited states. One might ask whether the Euler–Lagrange equation of the ground-state for the density-functional theory can be used to define excited states. What is usually done is to freeze the mean-field Hamiltonian at the ground state, and consider solutions other than the ground-state as corresponding to true excited states. In a formal mathematical language, the ground state equation reading

$$H_{\Psi_0}\Psi_0 = E_0\Psi_0,$$

the excited states Ψ_n are thus defined by

$$H_{\Psi_0}\Psi_n = E_n\Psi_n.$$

We could be optimistic about this, first because this is legal in the linear case and second because the ground-state density determines the Hamiltonian and hence all of the eigenstates. It was shown by PERDEW and LEVY [1985] that, indeed, non-ground-state solutions of the ground-state Euler–Lagrange equation, if there are any, do correspond to some excited state, but one certainly would not expect general solutions for excited states to be obtained in this way.

For the lowest excited states of a given symmetry different from that of the ground state, one just restricts trial functions in the variational procedure to functions of a given symmetry. This subspace theory (GUNNARSON and LUNDQVIST [1976], VON BARTH [1979], H. ENGLISCH and R. ENGLISCH [1983], KUTZLER and PAINTER [1987]) in spite of the unhappy fact that the Hohenberg–Kohn functional will in general differ from symmetry to symmetry is also used as a procedure in DFT. A next question is, given a particular excited state of interest, whether there might exist a variational principle whereby its density and energy could be found, and in which its density is the basic variable. Such a principle does in fact exist, based on a principle proved in wavefunction theory (MACDONALD [1934]). In addition, it has also be proved in LIEB [1985] that there exists no universal variational density functional procedure yielding an individual excited state. Following some authors (THEOPHILOU [1979], KOHN [1986]) a whole computational scheme can be obtained in the Kohn–Sham scheme.

Let us now turn to the determination of excited energies. In studying excited states we distinguish between well-separated excitation energies with no or small degeneracies of the eigenstates on the one hand, and eigenvalue clusters or continuous spectra on the other hand. Excitations of valence electrons in solids usually belong to the latter category. When the excitation energies are well separated, we can sometimes directly calculate the eigenstates and their energies by starting from a single-reference SCF eigenstates and treating the effects of the residual interactions by suitable approximations. In almost all cases however, the use of a multi-reference SCF calculation as a starting point becomes vital. Depending on the orders of magnitude of the degree of degeneracy of the energy values, we can characterize the appropriate methods as:

- Koopmanns' theorem,
- Δ(SCF) methods, methods in which two separate self-consistent field (SCF) calculations are performed, and the transition energies are obtained as the difference between the two results,
- Configuration Interaction (CI) methods,
- Direct methods, in which the transition energies are obtained directly, as in Green's function or propagator calculations.

There are detailed in the sequel.

Before entering the details, let us mention that solvation can affect electronic molecular spectroscopy and major shifts may accompany solvation. There are several ways to treat them. One way is to perform a classical simulation, and on top of it, a spectroscopic calculation every fixed number of molecular dynamics or Monte Carlo moves, using the structures obtained from the simulation. If the interactions are weak, the solvent molecules might be well represented by point charges, dipoles or higher momenta in the method used to calculate the spectroscopic quantities, or alternatively by a dielectric continuum. If some of the interactions are strong, or if there is charge transfer then the quantum calculation must include the tightly bound molecules along with the central species.

(A) Koopmans' theorem

As far as 1930, the Hartree–Fock study of the electronic structure of atoms was settled down and one of the main theorem of quantum chemistry was "demonstrated" (KOOPMANS [1933]). The *ionization potential* $\mathrm{IP}(i)$ of an atom or a molecule is defined as the difference between the ground state energy E_N of the system with N electrons and the energy E^i_{N-1} of the ionized state with $N-1$ electrons obtained by ionization of the ith level. This means that the molecular orbital of the ionized molecule is represented by a Slater determinant constructed from the same orbitals ϕ_i (doubly occupied) as the initial species but a single electron is removed from orbital i. Since in spectroscopy, observable transitions are said to be vertical, which means that the ionized species is well described by the geometry of the initial molecule, it is usual as a first approximation to neglect the relaxation of the molecular orbitals in the ionization process. Assuming the molecular orbitals of the neutral species are unchanged by the ionization process, the HF energy E^i_{N-1} of the ith ionized state is related to the energy E_N of the ground state by

$$E^i_{N-1} = E_N - \varepsilon_i,$$

where ε_i is the HF energy of the ith level ϕ_i ($F_\Phi \phi_i = \varepsilon_i \phi_i$, where F_Φ is the Fock operator associated with the ground state). Therefore the ionization potentials IP(i) read

$$\mathrm{IP}(i) = E^i_{N-1} - E_N = -\varepsilon_i.$$

This result is known under the name of Koopmans' theorem.

This theorem therefore allows evaluations of ionization potentials from the values of the orbital energies in the framework of the HF theory. In fact the success of the Koopmans' theorem (which in many cases reproduces experimental data) results from error compensation. However, this "theorem" fails in some cases: Main failures originate in the neglect of the geometrical relaxations and in the different correlation energies in the neutral and the ionized species. The failures of Koopmans' theorem have been analyzed in BERTHIER [1989]. Let us also stress that Koopmans' theorem has no equivalent in the Kohn–Sham framework of density-functional theory and should not be used.

(B) The Δ(SCF) method

A better evaluation of the ionization potentials can be obtained by the difference between the correct energies of both the initial and the final species, this is the Δ(SCF) method. In a Δ(SCF) calculation, starting from a closed-shell ground state for ease of example, a Restricted Open-shell Hartree–Fock (ROHF) or an Unrestricted Hartree–Fock (UHF) calculation has to be done for the excited state. When proceeding in this way, the Slater determinants representing the excited states are composed of orbitals that are not the same as those composing the ground state. The advantage of a ROHF calculation is that it yields to a wavefunction which is by construction an eigenfunction of the operator S^2 whereas in a UHF calculation contamination of spin is observed which is eliminated either by a projection onto the exact eigenfunction of spin or by annihilation of the spin contaminant (ZERNER [1997]). The advantage of such a method is that it allows to access excited states, but it is appropriate for certain excited states only, namely those that are of different spatial or spin symmetry to all lying below them. As concerns the transition energy, it is obtained from the subtraction of two large numbers obtained from two separate calculations for which it is difficult to insure a similar quality. Therefore it is necessary to perform balanced calculations on the initial and the final states of interest in order to obtain reliable predictions. This method is not appropriate for solids. Additionally, if several ionization energies are required, then it is necessary to carry out calculations on the initial state and each of the ionized states of interest.

(C) The CI methods

One possibility to circumvent some of the difficulties encountered in the Δ(SCF) methods is to use a CI method. Typically, we recall from Section 35 that in ground state CI calculations, one defines a set of basis functions, performs an SCF calculation to find occupied and virtual MO's (spin-orbitals), uses these molecular spin-orbitals to form configuration functions (i.e., Slater determinants), writes the total molecular wavefunction as a linear combination of the configuration functions (only the configuration functions that have the symmetry properties of the state being calculated are included) and uses the variational principle to find the optimal coefficients of the expansion. The latter step corresponds to finding the ground state eigenvalue of the Configuration Interaction matrix; this gives the clue to defining the nth eigenstate as the $(n+1)$th eigenmode of the configuration interaction matrix. The number of configurations increases rapidly and for practical purposes one has to reduce it. The lowest level of theory for a qualitative description of excited states is a CI including only the singly excited determinants, denoted CIS (Configuration Interaction including Single excitations). CIS gives wavefunctions of roughly HF quality for excited states, since no orbital optimization is involved. For valence excited states, for example those arising from excitations between orbitals in an unsaturated molecular system, this may be a reasonable description. For low-lying states, inclusion of the double excitations is necessary yielding CID (Configuration Interaction including Double excitations); in some other cases CISD (Configuration Interaction including Single and Double excitations) is required. In general CISD is not a successful level of theory for spectroscopy, and triple excitations are needed (CISDT). Obviously the CI method presents the advantage of being done on one calculation and give accurate results. However the choice of configurations is an "art", and results are not size-consistent. This means that, for noninteracting subsystems, the amount of correlation energy introduced depends on the number of noninteracting subsystems (except for full CI calculations). Obviously this method is not suitable for solids.

(D) The direct methods

An alternative is to use direct methods, which are able to compute directly in a single calculation the transition energies and the transition probabilities. When an electron is ejected from some level the electronic cloud reacting to the created hole relaxes; what we need to know is the response of the electronic cloud to this modification. Analytically this can be done in terms of response functions. The names usually associated with such techniques are the *propagator methods* also named *Green's function methods*. They arise from time-dependent perturbation theory. They were first introduced in theoretical physics (FEYNMAN [1948]) but they were really employed (ÖHRN and LINDERBERGH [1965]) in quantum chemistry some twenty years later. For two different operator P and Q a representation of a propagator may be (FRIEDMAN [1956]):

$$\langle\langle P; Q\rangle\rangle_\omega = \lim_{\eta\to 0+} \sum_{i\neq 0}\left(\frac{\langle\psi_0|P|\psi_i\rangle\langle\psi_i|Q|\psi_0\rangle}{\omega - E_i + E_0 + i\eta} \pm \frac{\langle\psi_0|Q|\psi_i\rangle\langle\psi_i|P|\psi_0\rangle}{\omega - E_i - E_0 - i\eta}\right), \tag{48.1}$$

where the $\pm$ sign depends on whether P and Q are number conserving operators or not (i.e., for Bose- or Fermi-type operators). In the above formula, Ψ_i denotes the ith eigenstate of the system, and E_i the associated energy. By suitable choices for P and Q operators, a whole variety of properties may be calculated (ODDERSHEDE and SABIN [1991]) such as excitation energies, polarizabilities or magnetic properties. If both operators P and Q are any function of the Hamiltonian of the system, they can be expressed in terms of its spectral resolution as:

$$H = \sum_{n,N} E_n^N |\psi_n^N\rangle\langle\psi_n^N| \tag{48.2}$$

in which the sum includes all numbers of particles N and all possible states n. In this formula, we have denoted by Ψ_n^N the nth excited state of the system consisting of N electrons. The associated energy is denoted by E_n^N. Thus (48.1) can now be written:

$$\langle\langle P; Q\rangle\rangle_\omega = \lim_{\eta\to 0+} \sum_{n,N'} \left(\frac{\langle\psi_0^N|P|\psi_n^{N'}\rangle\langle\psi_n^{N'}|Q|\psi_0^N\rangle}{\omega + E_0^N - E_n^{N'} + i\eta} - \frac{\langle\psi_0^N|Q|\psi_n^{N'}\rangle\langle\psi_n^{N'}|P|\psi_0^N\rangle}{\omega - E_0^N + E_n^{N'} - i\eta} \right). \tag{48.3}$$

Information is contained in the poles of the propagator, for instance when:

$$\omega = \pm\left(E_0^N - E_n^{N'}\right). \tag{48.4}$$

This quantity corresponds to the one we are looking for. Note that the above expression is valid in either a HF framework or in DFT. Two of the more commonly used propagators are the polarization propagator and the electron propagator. In the polarization propagator, the number of particles in the system remains constant, and the operators correspond to the excitation process, the Green's function has poles at the eigenvalues of the Hamiltonian if it is a "one-body" Green's function. In the "electron propagator", the operators correspond to the removal or addition of an electron, that is ionization or electron attachment processes. When expanded in terms of a Hartree–Fock reference function, the ionization energies and electron affinities can be considered in terms of orbital relaxation and electron correlation corrections to Koopmans' theorem. As already stated if one wants to calculate the ionization potential correctly he has to calculate the exact energy difference between the N and $N - 1$ particle systems. This can be done by using the many-body Green's function which has poles at differences between eigenvalues. Due to this difficulty of treating many-body Green's functions, it was proposed by Dyson to circumvent the difficulty by introducing an effective potential $\Sigma(E)$ which is energy dependent, it is called the *self-energy* where the exact Green's function can be obtained iteratively from a "one-body" Green's function of the system. Many possible iterative schemes are present in the literature (let us just stress that the starting point can either be obtained from HF and post HF calculations but also from DFT calculations). One of them the so-called *GW method* is more particularly used for solids (HEDIN [1965]) which basic implementation is described in several papers by HYBERTSEN and LOUIE [1985], HYBERTSEN and LOUIE [1986], GODBY, SCHLÜTER and SHAM [1986], GODBY, SCHLÜTER and SHAM [1988] and actual evolutions are given in a recent publication by REINING, ONIDA and GODBY [1997]. However, spectroscopic properties involving two-particle excitations are often only poorly described at this particle level and higher level theory is needed but they lie out of the scope of the present section.

49. Chemical systems in the presence of external fields

Modifications of electronic structures under the influence of purposely applied electric fields are the subject of many research areas: Obtaining of piezoelectric materials by relaxation of the molecular structure of polymers films, orientation of biological macromolecules, removing of centrosymmetry in active side chain polymer materials for nonlinear optics, enhancing reactivity and selectivity by confining molecules in zeolite channels, catalytic and orientation effects in the electric double layer in electrochemical reactions, etc. The complex chemical and physical modifications are now a main field of interest for quantum calculations.

(A) Particles in electromagnetic fields

We develop here the procedures for calculating the time-dependent equations describing the action of an external electromagnetic field. Starting from the time-dependent Schrödinger equation

$$i\frac{\partial \Psi}{\partial t} = H\Psi \tag{49.1}$$

a complete treatment would need to employ field theory (in particular to be able to deal with spontaneous emission). However, in most cases it is possible to consider the incorporation in a purely classical way of an external field. From classical electromagnetic theory, it is known that the electric field strength F, and the magnetic flux density B, may be derived from a scalar potential ϕ and vector potential A according to:

$$\begin{cases} F = -\nabla\phi - \dfrac{\partial A}{\partial t}, \\ B = \nabla \times A. \end{cases} \tag{49.2}$$

Replacement of the potentials by new potentials according to

$$\begin{cases} A \to A' = A + \nabla f, \\ \phi \to \phi' = \phi - \dfrac{\partial f}{\partial t}, \end{cases} \tag{49.3}$$

where f is an arbitrary function, leads to exactly the same fields. These equations define a change of *gauge*. It is clear that whenever the potentials occur in a Hamiltonian they must do so in such a way that all physical quantities, such as the energy and the electron density, are gauge-invariant. To set up the Hamiltonian, we must return to the axiomatic foundations of quantum mechanics and find a set of canonically conjugate coordinates in the classical problem, in order to set up a quantum-mechanical Hamiltonian operator. The classical Lagrangian for a particle with charge q and mass m moving in a field specified by potentials ϕ and A, with position x and velocity v, is:

$$L(x, v) = \frac{1}{2}mv^2 - q\phi(x) + q\,v \cdot A(x), \tag{49.4}$$

so that

$$p = \nabla_v L = mv + qA(x),$$
$$H(x, p) = \frac{1}{2m}\big(p - qA(x)\big)^2 + q\phi(x). \tag{49.5}$$

Converting this to operator form, summing over all the particles indexed by j, and including the potential energy of interaction, we get for the classical Hamiltonian operator

$$H_{\mathrm{cl}}\big(\{x_j\}, \{p_j\}\big) = \sum_j \frac{1}{2m_j}\big(p_j - q_j A(x_j)\big)^2 + \sum_j q_j\phi(x_j), \tag{49.6}$$

from which we deduce the quantum Hamiltonian

$$H = \sum_j \frac{1}{2m_j}\big(-i\nabla_{x_j} - q_j A(x_j)\big)^2 + \sum_j q_j\phi(x_j). \tag{49.7}$$

We must indeed verify that the energies and densities obtained from H are gauge-invariant. To do this, we first verify that the momentum operator

$$\pi_j = p_j - q_j A(x_j) \tag{49.8}$$

has an invariant expectation value if a gauge transformation of the electromagnetic field accordingly to (49.3) is accompanied with the following gauge transformation of the local phase of the wavefunction

$$\Psi\big(\{x_j\}\big) \to \Psi'\big(\{x_j\}\big) = \Psi\big(\{x_j\}\big)e^{i\lambda(\{x_j\})},$$

where

$$\lambda\big(\{x_j\}\big) = \sum_j q_j f(x_j). \tag{49.9}$$

It is indeed straightforward to check that for any j

$$\langle \Psi' | \pi_j' | \Psi' \rangle = \langle \Psi | \pi_j | \Psi \rangle \tag{49.10}$$

and that it follows immediately that both $\pi = \sum_j \pi_j$ and the kinetic energy operator $T = \sum_j \frac{1}{m_j} \pi_j^2$ are gauge-invariant. Doing so we have shown that the time-independent Schrödinger equation is gauge-invariant. In the above-mentioned equations, time has been omitted for simplicity; time can be reintroduced and the invariance of the time-dependent case can be shown as well, even if it requires further discussion (COHEN-TANNOUDJI, DUPONT-ROC and GRYNBERG [1988]).

To proceed to the calculation of field-dependent effects, it is usual to expand the term π_j^2 in the Hamiltonian (49.6), leading to

$$\pi_j^2 = \frac{1}{2m_j}\big[-\Delta_{x_j} + iq_j \operatorname{div}(A)(x_j) + 2iq_j A(x_j) \cdot \nabla_{x_j} + q_j^2 A^2(x_j)\big]. \tag{49.11}$$

It is usually convenient to take advantage of the physical property that any part of A that is expressible as the gradient of some scalar field f (see (49.3)) makes zero contribution to B, to choose A divergence free (i.e., such that $\operatorname{div} A = 0$). This choice defines the co-called *Coulomb gauge*. Assuming Coulomb gauge throughout, the Hamiltonian including applied fields is thus

$$H = H_0 + \sum_j q_j \phi(x_j) + 2i \sum_j \frac{q_j}{m_j} A(x_j) \cdot \nabla_{x_j} + \sum_j \frac{q_j^2}{2m_j} A^2(x_j), \tag{49.12}$$

where H_0 is the field-free Hamiltonian; the second term corresponds to the action of an electric field and the last two terms to the action of a magnetic field. Due to the intensities of the light sources usually employed the last term of (49.12) is negligible compared with the others and will not be considered in the sequel. Let us add that including spin effects, an additional term describing the interaction of the spin magnetic moments with the magnetic field appears in the previous equation, leading to

$$H = H_0 + \sum_j q_j \phi(x_j) + 2i \sum_j \frac{q_j}{m_j} A(x_j) \cdot \nabla_{x_j} - \sum_j \mathcal{M}_j \cdot B(x_j), \tag{49.13}$$

where $\mathcal{M}_j$ is the magnetic dipole moment operator associated with particle j (see below).

In nearly every case, solutions for a molecule subjected to static fields or to radiation fields are obtained from the solutions of the stationary state problem involving the field-free Hamiltonian H_0. Radiation effects, provided that they are weak ($\leqslant 10^{-1}$ W/m^2) and of fairly long wavelength ($\geqslant 1$ nm to be compared with chemical bonds of the order of magnitude 0.1 nm) are expected to unaffect the energy levels of the molecule, but only induce transitions between them. Therefore the properties of a system in an external electromagnetic field are considered as static ones and the interaction with the field is treated as a small term added to the initial Hamiltonian. In many instances, it is expedient to approximate solutions of the time-dependent equation using time-dependent variational methods or perturbation theory assuming the Hamiltonian differs only slightly from an operator H_0, whose eigenvalues and eigenvectors are supposed to be known.

(B) Nonlinear electric properties

As an illustration, we consider the effect of a static uniform external electric field (more commonly denoted F) on a molecular system. The strategy we detail is largely used for nonlinear optical property calculations. The assumption of uniformness corresponds physically to the fact that almost all fields can be considered as uniform at the atomic scale. The Hamiltonian H, derived from the expression given in (49.13) reduces to the first two terms

$$H_0 + H', \tag{49.14}$$

and H' is regarded as "small" compared to the unperturbed Hamiltonian H_0. Therefore both the eigenvalues and the eigenfunctions can be expanded in a power series. The expansion of the energy E (for H) in a power series of the electric field with components F_i, close to the unperturbed value $F_i = 0$, writes (BUCKINGHAM [1967], BUCKINGHAM and ORR [1967], BOGAARD and ORR [1975], FRANKEN and WARD [1963], MILLER, ORR and WARD [1977], BISHOP [1987], BISHOP [1988], BISHOP [1989], BISHOP [1991])

$$E(F_i) = E_0 - \mu_i F_i - \frac{1}{2}\alpha_{ij} F_i F_j - \frac{1}{6}\beta_{ijk} F_i F_j F_k - \frac{1}{24}\gamma_{ijkl} F_i F_j F_k F_l \tag{49.15}$$

(the Einstein convention of summation over repeated indices has been used). In the above formula E_0 is the ground state energy of the unperturbed Hamiltonian H_0, the vector

$$\mu_i = -\left.\frac{\partial E}{\partial F_i}\right|_{F=0}$$

is the *permanent electric dipole moment*, the second-order tensor

$$\alpha_{ij} = -\left.\frac{\partial^2 E}{\partial F_i \partial F_j}\right|_{F=0}$$

is the static *electric polarizability tensor*, and the third-order and fourth-order tensors

$$\beta_{ijk} = -\left.\frac{\partial^3 E}{\partial F_i \partial F_j \partial F_k}\right|_{F=0}$$

and

$$\gamma_{ijkl} = -\left.\frac{\partial^4 E}{\partial F_i \partial F_j \partial F_k \partial F_l}\right|_{F=0}$$

are known as the static *electric hyperpolarizabilities and second order hyperpolarizabilities tensors*, respectively. The permanent electric dipole moment as well as the polarizability and hyperpolarizability tensors can be experimentally measured in the gas phase.

The microscopic relation (49.15) can be parallel to a macroscopic one describing the response of a macroscopic material submitted to an electric field. The electrons and the nuclei are displaced, the first ones in the opposite direction of the field the others in the direction of the field, the material is polarized. For anisotropic materials, the polarization can be expressed as

$$P_i = P_i^0 + \chi_{ij}^{(1)} F_j + \chi_{ijk}^{(2)} F_j F_k + \chi_{ijkl}^{(3)} F_j F_k F_l, \tag{49.16}$$

where P^0 is the macroscopic polarization of the material in the absence of external electric field and where $\chi^{(n)}$ are tensors of order n related to the electric susceptibility of the material: $\chi^{(1)}$ is the *linear electric susceptibility*, $\chi^{(2)}$ and $\chi^{(3)}$ are, respectively, the *second order and third order nonlinear electric susceptibilities* of the material.

Deriving the energy (49.15) with respect to the field, the parallel between the macroscopic and microscopic levels leads to an equivalence between the macroscopic polarization of the material and the molecular dipole moment μ, the linear susceptibility of the material and the molecular polarizability α. The polarizability is associated with the capacity of a system to reorganize its charges under the influence of an external electric field; macroscopically it is directly related to the refraction index of a substance which is a complex quantity for conductive materials. Depending on the frequency range, the index is responsible of the so-called "skin effect" where high frequency waves propagate only on the external part of a conductor. For even higher frequencies the index is responsible of the metallic reflexion of visible light.

More recently, it has been largely investigated for the hyperpolarizability β describes the first nonlinear effects, for example, Second Harmonic Generation – SHG – (i.e., due to second-order terms, an intense radiation of wavelength ω going through a crystal has a wavelength $\omega/2$ at the exit) (FRANKEN, HILL, PETERS and WEINREICH [1961]), dc-Pockels effect (responsible for amplitude modulation (TARASSOV [1979]), optical switches (WILSON and HAWKES [1983]) and Optical Rectification – OR – (i.e., similar phenomenon as SHG but for two different wavelengths ω_1 and ω_2 yielding to a radiation of wavelength $\omega_3 = \omega_1 \pm \omega_2$) and the second hyperpolarizability γ is associated with phenomena such as the Third Harmonic Generation – THG –, dc-Kerr effect (the refractive index in presence of an electric field depends on the square of the intensity of the field due to third order terms), Electric Field-Induced Second Harmonic Generation – EFISH, etc. The polarizability tensor is also important in the description of molecular interactions since it appears in the expressions of the Debye and London dispersion energies.

Although the contributions originating from the field-induced nuclear relaxation and the distortion of the vibrational motions can be nonnegligible (KIRTMAN and HASAN [1992], BISHOP and PIPIN [1995]), the study of the electronic distribution is generally restricted to the reorganization of the electronic distribution. The reason is partially due to the slow motion of the nuclei with respect to the electrons which adapt quasi-instantaneously to the positions of the nuclei.

Within this context calculating $E(F_i)$ with formula (49.15) is equivalent to find out the stabilization of the energy of the system upon switching on the electric field. There are three main methods for calculating the effect of a perturbation: Derivative techniques, perturbation theory and propagator methods, they are detailed in the sequel. Since such calculations are much time consuming efforts have been done to reduce the computational costs. Besides the elementary properties of index permutational symmetry and intrinsic point group symmetry of a given tensor accounted for, much more powerful group-theoretical tools have been developed to speed up coupled Hartree–Fock calculations (LAZZERETTI and ZANASI [1979], LAZZERETTI, MALAGOLI and ZANASI [1996]).

In recent years effects of the surroundings have been introduced either for the liquid phase (CAMMI and TOMASI [1994a], CAMMI and TOMASI [1994b], DI BELLA, MARKS and RATNER [1994]) or for the solid phase (LEVINE and BETHEA [1975], LEVINE and BETHEA [1978]).

The first type of method called *Finite-field (FF) method* consists in minimizing the total energy of the system undergoing the electric perturbation, by calculating the derivatives of the energy or the dipole moment with respect to the field amplitude using the finite difference approximation as first proposed for closed-shell atoms in COHEN and ROOTHAN [1965] and further developed in POPLE, MCIVER and OSTLUND [1968]. Derivative techniques consider the energy in presence of the perturbation W

$$\inf\{\mathcal{E}(\Psi) + \langle\Psi, W\Psi\rangle,\ \Psi \text{ convenient}\}, \tag{49.17}$$

where $\mathcal{E}$ is a given energy functional, and perform an analytical (in the simplest case) or numerical differentiation of the energy n times to derive a formula for the nth order property.

For example, the elements of the polarizability tensor can be calculated either as the second derivative of the energy with respect to the field. The choice of the amplitude of the field needs some skill to do the numerical differentiation correctly, and the precision of the method becomes worse and worse as the order of the derivative increases. Nevertheless it has been widely used because it does not require any additional methodology than the classical quantum chemical codes. The presence of the electric field does not modify the form of the Hartree–Fock or Kohn–Sham equations: It just adds a term representing the electron-field interaction to the core Hamiltonian; this terms writes as

$$\int_{\mathbb{R}^3} (-F \cdot x)\chi_\mu(x)\chi_\nu(x)\,dx, \tag{49.18}$$

where $\{\chi_\mu\}$ are the atomic basis functions. Through the iterative process the dependence of the density matrix with the electric field appears leading to the reorganization of the electronic cloud.

From the numerical point of view (BISHOP and SOLUNAC [1985]), convergence problems may be encountered originating on the difficulty of running numerical derivatives but foremost from the nonbounded character of the operator $-q.F$. Indeed, it turns out that, due to the presence of the electric field, the infimum defined mathematically by (49.17) is $-\infty$, which of course implies a fortiori that there is no ground state in this setting. Furthermore, it can also be shown that there is even no solution to the Euler–Lagrange equation associated to (49.17) (see CANCÈS and LE BRIS [1998]), namely the SCF equations in this framework. In other words, the mathematical foundation of the computations done in practice cannot be found in a variational setting. In the linear case, the resonance theory provides with a way to circumvent this difficulty, but in the present nonlinear case, it is an open mathematical question to give a meaning to the computations. Of course, due the use of compactly supported functions, or in other terms, due to the use of a finite basis, computations do converge, to some well defined *local* minimizer ("local" is to be understood in the spatial sense). Practical calculations are therefore very routinely done. In order to circumvent the above difficulty, some authors (POUCHAN and BISHOP [1984], BISHOP, CHAILLET, LARRIEU and POUCHAN [1985], MAROULIS and BISHOP [1985]) have tried to put around the molecule a set of charges supposed to annihilate the electric field in the surroundings of the molecule. But most of the literature ignores this method and just select "correctly" the basis set. There is a huge literature regarding the influence of the basis set on the calculated values (see, for example, WERNER and MEYER [1976], ZEISS, SCOTT, SUZUKI, CHONG and LANGHOFF [1979], VAN DUIJNEVELDT-VAN DEN RIJDT and VAN DUIJNEVELDT [1982], CHRISTIANSEN and MCCULLOUGH Jr. [1977], AMOS and WILLIAMS [1979], DIERCKSEN, KELLÖ and SADLEJ [1983], ALMLÖF and FAEGRI JR. [1983], LIU and DYKSTRA [1985], SEKINO and BARTLETT [1986])) from which it appears that a basis of good quality for an energy calculation is not necessarily good for its derivative calculations due to the fact that long distances in the wavefunction are involved by the action of the field.

In certain situations, however, it is expedient to use some form of perturbation theory in which the Hamiltonian can be regarded as the sum of an operator H_0, whose eigenvalues and eigenfunctions are known, and a small perturbation H' to the unperturbed Hamiltonian H_0. The assumption that H' is small suggests that we can expand the perturbed eigenfunctions and eigenvalues as power series in H' (this strategy, which also founds the Møller–Plesset perturbation method, is described in Chapter II). The perturbed wavefunctions and energy levels read

$$\begin{aligned} \Psi_n &= \Psi_n^{(0)} + \Psi_n^{(1)} + \Psi_n^{(2)} + \cdots, \\ E_n &= E_n^{(0)} + E_n^{(1)} + E_n^{(2)} + \cdots, \end{aligned} \tag{49.19}$$

where $\Psi_n^{(0)}$, $E_n^{(0)}$ and Ψ_n, E_n, respectively, are the nth eigenstates of the Hamiltonians H_0 and H.

It is implicit in the present treatment that the unperturbed states form a discrete set (which is always the case once the model has been discretized); we assume in addition that the unperturbed state is nondegenerated (the degenerate case can be found in classic textbooks). The expansions up to second order (wavefunctions) and third order (energy), respectively, write

$$E_i^{(1)} = \langle \Psi_i^{(0)} | H' | \Psi_i^{(0)} \rangle, \tag{49.20}$$

$$\Psi_i^{(1)} = \sum_{j \neq i} \frac{\langle \Psi_j^{(0)} | \Psi_i^{(0)} \rangle}{E_i^{(0)} - E_j^{(0)}} \Psi_j^{(0)}, \tag{49.21}$$

$$E_i^{(2)} = \sum_{j \neq i} \frac{|\langle \Psi_j^{(0)} | H' | \Psi_i^{(0)} \rangle|^2}{E_i^{(0)} - E_j^{(0)}} \Psi_j^{(0)}, \tag{49.22}$$

$$\begin{aligned} \Psi_i^{(2)} &= \sum_{j \neq i} \frac{\langle \Psi_j^{(0)} | H' | \Psi_i^{(1)} \rangle - E_i^{(1)} \langle \Psi_j^{(0)} | \Psi_i^{(1)} \rangle}{E_i^{(0)} - E_j^{(0)}} \Psi_j^{(0)} \\ &= \sum_{j \neq i, k \neq i} \frac{\langle \Psi_j^{(0)} | H' | \Psi_k^{(0)} \rangle \langle \Psi_k^{(0)} | H' | \Psi_i^{(0)} \rangle}{(E_i^{(0)} - E_j^{(0)})(E_i^{(0)} - E_k^{(0)})} \Psi_j^{(0)} \\ &\quad - E_i^{(1)} \sum_{j \neq i} \frac{\langle \Psi_j^{(0)} | H' | \Psi_i^{(0)} \rangle}{(E_i^{(0)} - E_j^{(0)})^2} \Psi_j^{(0)}, \end{aligned} \tag{49.23}$$

$$\begin{aligned} E_i^{(3)} &= \sum_{j \neq i, k \neq i} \frac{\langle \Psi_i^{(0)} | H' | \Psi_j^{(0)} \rangle \langle \Psi_j^{(0)} | H' | \Psi_k^{(0)} \rangle \langle \Psi_k^{(0)} | H' | \Psi_i^{(0)} \rangle}{(E_i^{(0)} - E_j^{(0)})(E_i^{(0)} - E_k^{(0)})} \\ &\quad - E_i^{(1)} \sum_{j \neq i} \frac{\langle \Psi_i^{(0)} | H' | \Psi_j^{(0)} \rangle \langle \Psi_j^{(0)} | H' | \Psi_i^{(0)} \rangle}{(E_i^{(0)} - E_j^{(0)})^2}. \end{aligned} \tag{49.24}$$

Due to the summation over all the excited states this method is called the *Sum Over States (SOS) method.* Implementation of such formula requires truncations. In the framework of the molecular orbital theory, the various excited wavefunctions are constructed by replacing successively one or more occupied orbital φ_n by virtual orbitals φ_m in the Slater determinant; only mono, di, tri or quadri excitations are taken most of the time. The SOS method has been widely used by several authors (ANDRE, BARBIER, BODART and DELHALLE [1987]) as described above but also in an improved way by using a CI procedure in order to describe more correctly the excited states (PIERCE [1989]).

Let us point out that this formalism can be extended to the case of crystalline compounds (GONZE and VIGNERON [1989]).

The so-called *propagator methods* can also be used for the calculation of polarizabilities. Starting from (48.3) giving the spectral representation of the propagator another commonly used propagator is obtained for the case where the operator P and Q introduced above are both equal to the position operator, which corresponds to a number of particles conserved in the system ($N = N'$). The spectral representation of the polarization propagator is:

$$\langle\langle r; r\rangle\rangle_\omega = \lim_{\eta\to 0+} \sum_n \left(\frac{\langle\psi_0|r|\psi_n\rangle\langle\psi_n|r|\psi_0\rangle}{\omega + E_0 - E_n + i\eta} - \frac{\langle\psi_0|r|\psi_n\rangle\langle\psi_n|r|\psi_0\rangle}{\omega - E_0 + E_n - i\eta} \right) \tag{49.25}$$

the propagator describes the response of the dipole moment to a linear electric field; it corresponds to the exact frequency-dependent polarizability tensor. In the limit where $\omega \to 0$ (i.e., where the perturbation is time-independent), the propagator is identical to the second-order perturbation formula for a constant field. Choosing a nonzero value for ω corresponds to a time-dependent field with a frequency ω. The propagator defines the frequency-dependent polarizability of the system. The time-dependent density response can be obtained through a self-consistent equation, its representation is not computationally tractable because it presumes complete knowledge of many-electron wavefunctions and energies. However, the response function for a noninteracting ground state, has a much simpler spectral representation (ZANGWILL and SOVEN [1980]). This technique works either for Hartree–Fock and post Hartree–Fock wavefunctions or DFT calculations. The density-functional theory of dynamic linear response has been applied to small molecules (LEVINE and SOVEN [1984]) and to solids (BARONI, GIANNOZZI and TESTA [1987], GONZE [1995]) and has been further extended to nonlinear response (SENATORE and SUBBASWAMY [1987]).

(C) The magnetic field perturbation

Considering now the terms of Eq. (49.13) corresponding to the presence of an external uniform magnetic field, B, and assuming there is no electric field, one obtains:

$$H = H_0 - \sum_j \frac{q_j}{m_j} A(x_j) p_j - \sum_j \mathcal{M}_j \cdot B,$$

where index j runs for all particles of the system, q_j, m_j $\mathcal{M}_j$ are, respectively, the magnitude of the charge, the mass and the magnetic dipole moment operator of the jth particle. As the magnetic field is uniform,

$$A(x) = \frac{1}{2} B \times x$$

is a vector potential satisfying the Coulomb gauge. Using this expression to rewrite the second term of the Hamiltonian one obtains

$$H = H_0 - \sum_j \frac{q_j}{2m_j} B \cdot L_j - \sum_j \mathcal{M}_j \cdot B,$$

where

$$L_j = x_j \times p_j$$

denotes the angular momentum operator of particle j.

Let us now deal with the third term of the Hamiltonian. Quantum Field Theory allows to relate the magnetic dipole moment operator $\mathcal{M}_j$ to the spin operator S_j. For spin-1/2 particles, one has for instance

$$\mathcal{M}_j = \gamma S_j,$$

where S_j is the spin operator of the particle and where γ is some scalar depending on the physical nature of the particle: $\gamma = -e/m_e$ for an electron (up to some tiny relativistic corrections), $\gamma \simeq 2.79e/M_p$ for a proton, $\gamma \simeq -1.91e/M_n$ for a neutron where e is the unit charge ($e = 1$ in atomic units) and m_e, M_p and M_n the masses of the electron, of the proton and of the neutron, respectively ($m_e = 1$ in atomic units).

The electronic Hamiltonian therefore reads

$$H_e = H_e^0 + \frac{1}{2}\sum_{j=1}^{N} B \cdot (L + 2S)$$

where H_e^0 is the electronic Hamiltonian in absence of magnetic field and where L and S are the total angular momentum and spin operators:

$$L = \sum_{j=1}^{N} L_j, \qquad S = \sum_{j=1}^{N} S_j.$$

We see that the magnetic field interacts with both the angular momentum and spin operators.

Effects of magnetic fields are used in resonance spectroscopies; in most practical cases the magnetic field effects are small. From the practical point of view, experiments consist in applying static magnetic fields of a certain strength (typically 10 Gauss) together with a weaker time-dependent harmonic magnetic field, whose frequency is tuned to scan some frequency range (typically 10^5 to 10^7 Hz) in order to track the resonances. For instance let us consider first the case of a single electron in a state with no orbital angular momentum, in presence of the field: each energetical level is split into two components (because the spin has two eigenvalues $S_z = \pm 1/2$) as shown in Fig. 49.1.

This small splitting of the energy levels revealed by the field is the so-called Zeeman effect.

Let us stress also that the above equations are based on perturbation theory, this means that it is assumed that the magnetic field is weak, i.e., the Hamiltonian considers only the electronic degrees of freedom and accordingly neglects the nuclei. When the magnetic field is strong, the situation may be markedly different, and strong effects may show up because nuclei may possess spins with which the magnetic field may interact. The

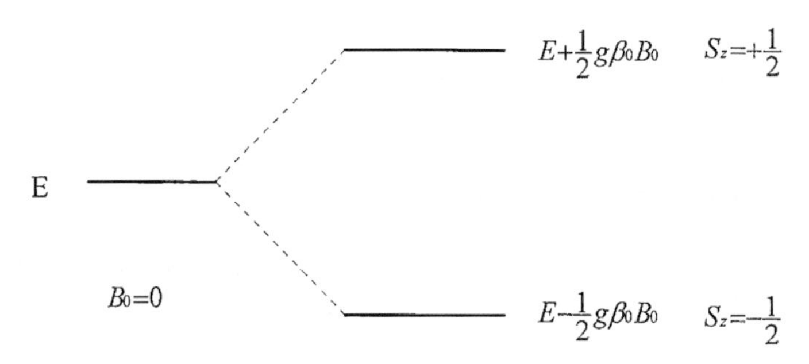

FIG. 49.1. Splitting of the two components of the spin (B_0 is the oscillating field).

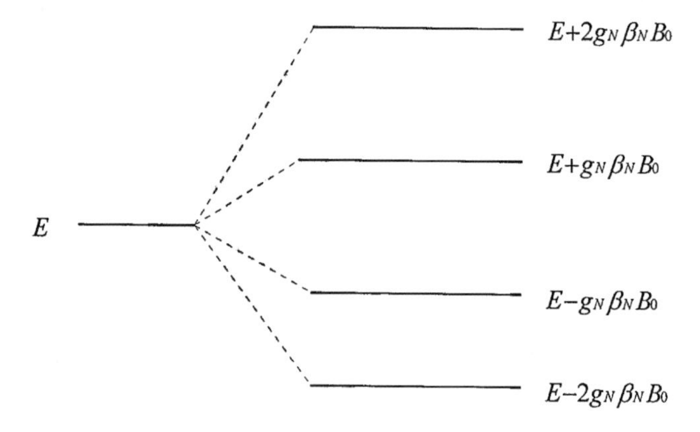

FIG. 49.2. Nuclear Zeeman effect for $I = 3/2$.

system we have just studied represents a simplification. In particular the classically inferred Hamiltonian, Eq. (49.13), is incomplete, for it omits internal electromagnetic interactions associated with the intrinsic spin of the particles. Also this is studied in some experiments.

Similarly, magnetic fields induce a splitting (nuclear Zeeman effect) of the $(2I + 1)$ spin states of a nucleus of spin I (Fig. 49.2)

Also this is used in some experiments of magnetic resonance (NMR: Nuclear magnetic resonance). Not all the nuclei can be seen in NMR experiments because some of them have zero nuclear moments, interesting atoms for chemistry are: Hydrogen, carbon, nitrogen, oxygen, fluorine, phosphorus, sulfur, etc., which exhibit g factors of various values and signs. The interest of the method originates in that a given nucleus occurring in different chemical environments, shows slightly different splittings because the actual field experienced by the nucleus is dependent of its surroundings due to the presence of induced currents in the electronic distribution. Each nucleus n is found to behave as if the external field B it experiences were replaced by $B_n = (1 - \sigma_n)B$ where σ_n is a shielding constant. This phenomenon is called the "chemical shift" in NMR signals and although usually very small it may be measured accurately and important chemical information about chemical compositions of a given compound can be obtained. The energies of these resonances are first of all determined by that part of the system that is studied. Finally it is important to notice that NMR experiments can be run also for solids.

50. Elucidation of reaction mechanisms

The elucidation of reaction mechanisms is a major challenge for chemical theory. Although the geometries of reactants and products may generally be obtained experimentally using a wide range of spectroscopic methods, these same techniques provide little if any detailed information about connecting pathways. The scarcity of experimental data greatly enhances the overall value of the calculations. The major objective is therefore a survey of the potential energy surface that is the location of stationary points and the determination of transition structures and of the related activation energies. In Figs. 50.1 and 50.2 are given examples of reaction profiles for the addition of a molecule A to a molecule B, leading to the aggregate species AB, or of transfer of an atom or a fragment B between fragments A and C. Computational methods may also be employed to characterize short-lived reactive intermediates, which correspond to shallow

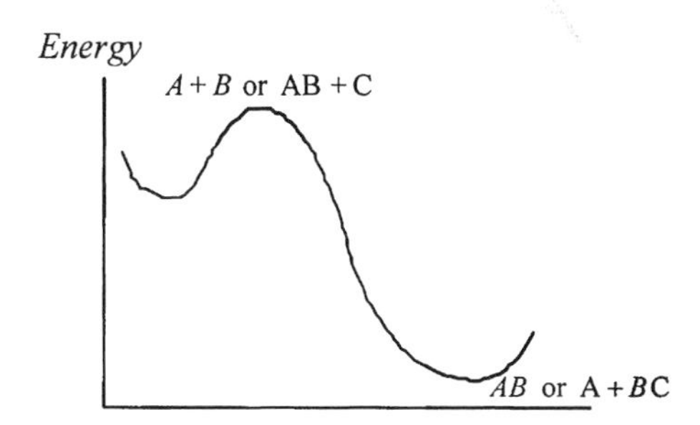

FIG. 50.1. Reaction profile for addition of A to B or transfer of fragment B with activation energy.

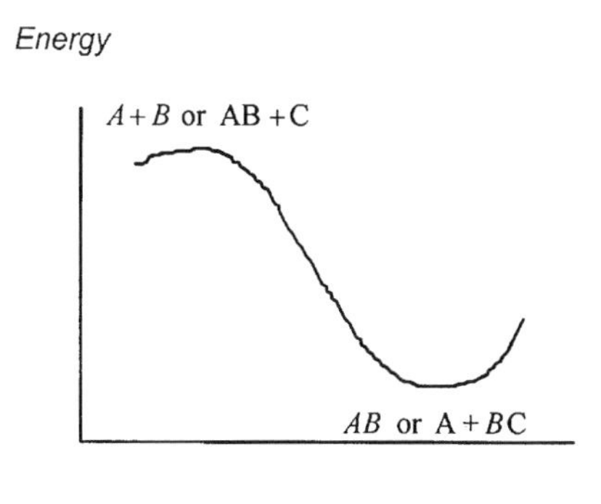

FIG. 50.2. Reaction profile for addition of A to B or transfer of fragment B without activation energy.

local minima on potential surfaces. While such species may sometimes be detected experimentally, detailed information regarding their structures is usually difficult to obtain.

Quantitative molecular orbital theory can be used to investigate the properties of short-lived reactive intermediates just as those of more stable molecules. However the success of simple theoretical models in determining the properties of stable molecules may not carry over into the description of reaction pathways, because transition species are normally characterized by weak partial bonds, which are readily to be broken or formed.

The application of theory to an intimate description of the way chemical reaction proceed is a difficult task. Having found a possible pathway, it is difficult to rule out completely any other possibility through other transition states. Furthermore potential surfaces may be very sensitive to the level of theory employed; structures obtained using low levels of theory with small basis sets, may differ significantly from those obtained with large basis sets and higher levels of theory. The way to obtain a good mapping of a potential surface is not straightforward and a lot of efforts have to be done, and a good knowledge of chemistry is often a necessary condition for the success of the enterprise.

Regarding the computational procedures, things differ significantly depending on the nature of the chemical intermediates one is looking for. Numerically speaking, stationary points have an energy gradient equal to zero for the set of $(3N-6)$ internal coordinates of an N-atom molecule. A stationary point may correspond either to a local energy minimum or a local energy maximum or even to an energy saddle point. These may be distinguished from one another by the knowledge of the second derivatives of the energy with respect to the internal coordinates, the number of which is $(3N-6)(3N-7)/2$. The matrix of these second derivatives has to be diagonalized. If the eigenvalues of the matrix are all positive, the stationary point is a minimum and corresponds to a stable structure. If all the eigenvalues are negative, the point is a maximum and corresponds to a physically uninteresting state. If the point presents one and only one negative eigenvalue, it is a saddle point which corresponds to a transition state along the reaction path related to a peculiar reaction coordinate.

There are several numerical methods for locating saddle points and reaction paths. We will not detail these metods here since they are addressed in two contributions of the present volume, FIELD [2003] and SCHÜTTE and HUISINGA [2003], respectively. We also refer the reader to the bibliography.

In the formalism so far developed the notion of temperature has been ignored, an energy calculation in quantum chemistry is essentially a zero-degree (0 K) calculation. As far as one is concerned with electronic structures, ignoring the temperature is correct since a pattern of nuclei and electrons is not a macroscopic system and as such the notion of temperature does not exist. However in the modeling of real systems where one is interested in energetical quantities, temperature might play a role and has to be introduced. For instance let us consider a chemical reaction where various reactants react (in stoichiometric proportions) to yield to products (in stoichiometric proportions), an estimate of the energy (at 0 K) of the reaction is obtained by

$$\Delta U_0 = \sum_{\text{products}} \alpha_p E_p^{\text{tot}} - \sum_{\text{reactants}} \alpha_r E_r^{\text{tot}}$$

from the total energies of the products and the reactants. The energy so-obtained is the free enthalpy of the reaction at 0 K. Depending on the sign of the free enthalpy a chemist knows if a reaction is possible or not. Most of the time temperature influences the reaction by changing the proportions of the products, and it needs to be introduced in the calculation through the determination of the entropy (S) of the reaction. A correct thermodynamical quantity for the knowledge of reactions is the variation of the free enthalpy (G) defined as

$$\Delta G_T = \Delta H_T - T \Delta S_T$$

for a given temperature T.

Coupling statistical thermodynamics and quantum calculations, it is possible from the total energy of the system at 0 K (ΔU_0) to evaluate the variation of the free enthalpy of the reaction for finite temperatures

$$\Delta H_T = \Delta U_0 + \Delta ZPE + \Delta[H_0 - H_T],$$

where ZPE is the residual energy due to vibrations at 0 K, and the last term $H_0 - H_T$ contains three terms: H_{trans}, H_{vib} and H_{rot} which originate in the translational, vibrational and rotational motions of the molecules. All these quantities can be obtained either from atomic calculations (ZPE is obtained from the vibrational modes of the system) or from thermodynamical calculations (which are derived in many textbooks of statistical thermodynamics). When these quantities are known, it is then possible to access physical characteristics of compounds such the specific heat at constant pressure or at constant volume, etc. The calculations of all these thermodynamical quantities are implemented for molecules in standard codes of quantum chemistry. As regards solids things are somewhat more difficult to implement even if similar physical fundaments are involved, for instance because of the difficulty to correctly describe the vibrations of crystals an approximation is most often used which consists in considering only the harmonic vibrations of the nuclei. In the case of solids containing intrinsic defects, the methodology just given which relies on the assumption that solids are made of perfect crystals cannot explain experimental data. Physically, the introduction of defects changes both the total energy of the system and the entropy of the crystal. The method used to treat such cases lies out of the scope of the present volume and readers interested in such aspects can refer to specialized books (see, for instance, CATLOW [1994]).

CHAPTER V

The Time Dependent Models and their Discretization

51. Introduction

Let us now deal with time dependent problems, and therefore recall here for convenience the time dependent Schrödinger equation

$$i\frac{\partial}{\partial t}\Psi = H\Psi, \tag{51.1}$$

that we have briefly quoted in Chapter I and then mostly forgotten, rather focusing on its stationary analogue. The Hamiltonian H is supposed to be that of a molecular system, the wavefunction Ψ defines its state at current time. Both the Hamiltonian and the space of wavefunctions are the same as those defined in Chapter I. We shall recall them below. In the same fashion, we choose units where $\hbar = 1$.

As in the stationary case, the above time dependent equation can only (and indeed need only) be simulated as such in rare situations, that will be briefly addressed in Section 52. In almost all cases, one deals with *approximations* of this equation, all based upon the decoupling of the degrees of freedom for the electrons and for the nuclei. One of the enormous difficulty of simulating (51.1) is indeed its *multiscale* nature. We have already mentioned that the nuclei and the electrons are very different objects in size and mass. This has influenced the modelling in the static setting. Here, in the time-dependent one, the difficulty is enhanced by the account for time: The timescales also are very different from one another. This separation of lengthscales and timescales is a drawback when trying to simulate (51.1) directly (when needed), but is also an advantageous property to take benefit from when building approximations of the equation. The observation that founds the decoupling, thus the simplification, is indeed the following: The nuclei can often be treated as classical point particles that correspond to the slow timescales in the problem, while the electrons are quantum objects that evolve at rapid timescales. This decoupling roughly amounts to doing the Born–Oppenheimer approximation in the static setting. Once decoupled, the system still needs to be approximated. To define further the type of approximation requires more information on the situation at hand and the application targeted. The main two families of approximations, namely the adiabatic and the nonadiabatic one, will be overviewed below. The difference lies in the treatment of the electrons. As for the nuclei, a Newtonian dynamics has to be simulated. This is the domain of the very famous *molecular dynamics*, which is the topic of a brief overview in Section 55. It is indeed out of the scope (and out of the pretension) of such an introductory article to describe in detail the techniques of molecular dynamics, and a fortiori its applications. We shall therefore provide the reader with a list of bibliographical references to approach the vast literature devoted to this rich subject. In Section 56, we rapidly evoke one of the representatives of nonadiabatic models, namely the *time dependent Hartree–Fock approximation*. Finally, in Section 57, we give one possible application of these time dependent models: The simulation of laser control at the molecular scale. This very active field of today's Theoretical chemistry and Physics is indeed a formidable field of application for the techniques of scientific computing. In order to prepare the reader to the reading of the contributions of the present volume specifically devoted to this issue, we briefly introduce the basic concepts.

52. Simulations of the original equation

For tiny systems, Eq. (51.1) can be simulated in its original form. We refer, e.g., to BANDRAUK and CHELKOWSKI [2000] for such simulations. These are the only cases when the wavefunction Ψ in (51.1) really depends on all the electrons and nuclei of the molecular system (5.1):

$$\Psi(t; \bar{x}_1, \bar{\sigma}_1; \ldots; \bar{x}_M, \bar{\sigma}_M; x_1, \sigma_1; \ldots; x_N, \sigma_N) \tag{52.1}$$

and the Hamiltonian, as introduced in (5.6) in Chapter I, fully depends on all these variables:

$$H = -\sum_{k=1}^{M} \frac{1}{2m_k} \Delta_{\bar{x}_k} - \sum_{i=1}^{N} \frac{1}{2} \Delta_{x_i} - \sum_{i=1}^{N} \sum_{k=1}^{M} \frac{z_k}{|x_i - \bar{x}_k|} + \sum_{1 \leqslant i < j \leqslant N} \frac{1}{|x_i - x_j|} + \sum_{1 \leqslant k < l \leqslant M} \frac{z_k z_l}{|\bar{x}_k - \bar{x}_l|}. \tag{52.2}$$

On the other hand, it may happen that the evolution under study is essentially that of the nuclei of the system, that this evolution cannot be properly treated without considering the nuclei as quantum objects, and that the effects of the electrons can be modelled through an effective potential $U(\bar{x}_1, \ldots, \bar{x}_M)$ acting on the nuclei. In some sense, the viewpoint is the opposite to the one we have had in the previous sections by focusing on the quantum nature of the electrons and treating the nuclei classically. The cases where the genuine quantum nature of the nuclei play a role are mostly related to situations encountered in fundamental physics, an example of which is the laser control of molecular evolutions, or also the quantum tunneling of protons. For the simulation, the wavefunction (52.1) is simplified into a function of the nuclear variables only

$$\Psi(t; \bar{x}_1, \bar{\sigma}_1; \ldots; \bar{x}_M, \bar{\sigma}_M)$$

and the Hamiltonian is modified accordingly

$$H = -\sum_{k=1}^{M} \frac{1}{2m_k} \Delta_{\bar{x}_k} + U(\bar{x}_1, \ldots, \bar{x}_M) + \sum_{1 \leqslant k < l \leqslant M} \frac{z_k z_l}{|\bar{x}_k - \bar{x}_l|}.$$

Both these situations will be addressed in the present volume, in a contribution by A. Bandrauk and collaborators, and in a contribution by O. Atabek and collaborators. Further details on the modelling and the numerical treatment of these equations will be given there.

Leaving aside the above two (very) specific situations, the main trend of time dependent simulations in computational chemistry consists of models that consider the nuclei as classical objects and the electrons as quantum ones. The nuclei, each of which is described by a charge z_k at a point position $\bar{x}_k(t) \in \mathbb{R}^3$, evolve through the classical law of Newtonian dynamics

$$m_k \frac{d^2 \bar{x}_k}{dt^2}(t) = -\nabla_{\bar{x}_k} W\big(\bar{x}_1(t), \ldots, \bar{x}_M(t), t\big). \tag{52.3}$$

The potential W is given by

$$W(\bar{x}_1, \ldots, \bar{x}_M, t) = U(\bar{x}_1, \ldots, \bar{x}_M, t) + \sum_{1 \leqslant k < l \leqslant M} \frac{z_k z_l}{|\bar{x}_k - \bar{x}_l|}, \tag{52.4}$$

where the electronic contribution U will be more precisely described below. On the other hand, as in the static Born–Oppenheimer approximation, the wavefunction is consequently reduced to a function of the electronic variables

$$\psi_e(t; x_1, \sigma_1; \ldots; x_N, \sigma_N), \tag{52.5}$$

subject to the time-dependent Hamiltonian

$$\begin{aligned} &H_e^{(\bar{x}_1(t), \ldots, \bar{x}_M(t))} \\ &\quad = -\sum_{i=1}^{N} \frac{1}{2} \Delta_{x_i} - \sum_{i=1}^{N} \sum_{k=1}^{M} \frac{z_k}{|x_i - \bar{x}_k(t)|} + \sum_{1 \leqslant i < j \leqslant N} \frac{1}{|x_i - x_j|}, \end{aligned} \tag{52.6}$$

and bound to evolve as a normalized element in the space

$$\mathcal{H}_e = \bigwedge_{i=1}^{N} H^1\big(\mathbb{R}^3 \times \{|+\rangle, |-\rangle\}, \mathbb{C}\big). \tag{52.7}$$

The central issue is then to determine at which level of approximation the *evolution* of electrons needs to be treated.

53. Nonadiabatic simulations

In full generality, and in the vein of what has been described so far, the electronic wavefunction satisfies the time-dependent Schrödinger equation

$$i\frac{\partial\psi_e}{\partial t}=H_e^{(\bar{x}_1(t),\ldots,\bar{x}_M(t))}\psi_e. \tag{53.1}$$

Accordingly, the potential U in (52.4) is set to the value

$$U(\bar{x}_1,\ldots,\bar{x}_M,t)=-\sum_{k=1}^{M}\int_{\mathbb{R}^3}\frac{z_k\rho(t,x)}{|x-\bar{x}_k|}\,dx, \tag{53.2}$$

where the electronic density $\rho(t,x)$ is evaluated from the current electronic state defined by

$$\rho(t,x)=N\sum_{\sigma_1,\sigma_2,\ldots,\sigma_N}\int_{\mathbb{R}^{3(N-1)}}|\psi_e|^2(t;x,\sigma_1;x_2,\sigma_2;\ldots;x_N,\sigma_N)\,dx_2\cdots dx_N. \tag{53.3}$$

The above equations mean that each nucleus moves according to the Newton dynamics (52.3) in the electrostatic potential (52.4) created by the other nuclei and by the mean electronic density ρ. The potential U contributing to W is called the *Hellmann–Feynman potential*; its expression is connected with the *Ehrenfest theorem* (see LANDAU and LIFCHITZ [1977] and also BORNEMANN, NETTERSHEIM and SCHÜTTE [1996] for a mathematical argument).

The state of the system at time t is thus defined by the knowledge of

$$\left(\left\{\bar{x}_k(t),\frac{d\bar{x}_k}{dt}(t)\right\}_{1\leqslant k\leqslant M},\psi_e(t)\right)\in\mathbb{R}^{6M}\times\mathcal{H}_e,$$

with $(\bar{x}_k(t),\psi_e(t))$ obeying to the full system (52.3)–(52.4)–(53.1)–(53.2)–(53.3), namely

$$\begin{cases} m_k\dfrac{d^2\bar{x}_k}{dt^2}(t)=-\nabla_{\bar{x}_k}W\big(\bar{x}_1(t),\ldots,\bar{x}_M(t),t\big),\\ W(\bar{x}_1,\ldots,\bar{x}_M,t)=U(\bar{x}_1,\ldots,\bar{x}_M,t)+\displaystyle\sum_{1\leqslant k<l\leqslant M}\frac{z_kz_l}{|\bar{x}_k-\bar{x}_l|},\\ U(\bar{x}_1,\ldots,\bar{x}_M,t)=-\displaystyle\sum_{k=1}^{M}\int_{\mathbb{R}^3}\frac{z_k\rho(t,x)}{|x-\bar{x}_k|}\,dx,\\ \rho(t,x)=N\displaystyle\sum_{\sigma_1,\sigma_2,\ldots,\sigma_N}\int_{\mathbb{R}^{3(N-1)}}|\psi_e|^2(t;x,\sigma_1;x_2,\sigma_2;\ldots;x_N,\sigma_N)\,dx_2\cdots dx_N,\\ i\dfrac{\partial\psi_e}{\partial t}=H_e^{(\bar{x}_1(t),\ldots,\bar{x}_M(t))}\psi_e,\\ H_e^{(\bar{x}_1,\ldots,\bar{x}_M)}=-\displaystyle\sum_{i=1}^{N}\frac{1}{2}\Delta_{x_i}-\sum_{i=1}^{N}\sum_{k=1}^{M}\frac{z_k}{|x_i-\bar{x}_k|}+\sum_{1\leqslant i<j\leqslant N}\frac{1}{|x_i-x_j|}. \end{cases} \tag{53.4}$$

One speaks in this case of the *nonadiabatic approximation*, as opposed to the adiabatic approximation, which is the matter of next section. Let us just mention at this stage that the adiabatic approximation consists in assuming that for any time t the electrons are in the ground-state (or in some well-defined excited state) of the electronic Hamiltonian $H_e^{(\bar{x}_1(t),\ldots,\bar{x}_M(t))}$. The adiabatic approximation is valid in many cases (see next section), but electronic nonadiabatic effects are however important in some fundamental molecular processes; examples are notably encountered in collisions involving electronic excited states whose energy surfaces are intersecting (level crossings).

From the practical numerical viewpoint, it turns out that Eq. (53.1) can only be handled for small systems. Therefore, most often, a *time-dependent* approximation of the electronic *time-dependent* Schrödinger equation (53.1) is chosen, in the spirit of those which have been developed above in the stationary case. Section 56 below will be mostly devoted to one of them, the *time-dependent Hartree–Fock approximation*. We shall emphasize the mathematical aspects and skip the numerical ones.

54. Adiabatic simulations

As mentioned in the previous section, it can be considered in many applications that the electrons stay in a well-defined Born–Oppenheimer energy surface (most often the ground-state energy surface). This approximation rely on physical arguments (the characteristic relaxation time of the electrons being so small with respect to that of the nuclei, it can be considered that the electronic wavefunction reacts *adiabatically* to a change in the position of the nuclei), and can be justified by mathematical arguments BORNEMANN, NETTERSHEIM and SCHÜTTE [1996] under the assumptions that on the one hand the velocity of the nuclei is "small enough" and that on the other hand the Born–Oppenheimer energy surface under consideration is "isolated" from the rest of the spectrum (let us notice that this assumption is never fulfilled for metals).

In practice, the adiabatic approximation is valid for the simulation of *physical* properties (phase diagrams, surface reconstruction, diffusion in alloys, ...) of ordered or disordered systems in their liquid of amorphous states, as well as for the simulation of most chemical reactions.

In adiabatic simulations, the time-dependent Schrödinger equation can thus be replaced by a stationary one; the main advantage of this replacement is that the time step for numerical integration of the dynamics can now be chosen of the same order of magnitude as the characteristic evolution time of the *nuclei* rather than as the one of the *electrons* (the latter is roughly 100 times as small as the former). If the state of reference considered for the electronic part is the electronic ground state (it could as well be an excited state, and the modifications of the lines below are straightforward), then (53.1) is replaced by

$$\psi_e(t) = \arg\inf\big\{\langle\psi_e, H_e^{(\bar{x}_1(t),\ldots,\bar{x}_M(t))}\psi_e\rangle,\ \psi_e \in \mathcal{H}_e,\ \|\psi_e\|_{L^2} = 1\big\}, \tag{54.1}$$

and it can be proven that in these conditions the potential U has to be set to

$$U(\bar{x}_1, \ldots, \bar{x}_M) = \inf\big\{\langle\psi_e, H_e^{(\bar{x}_1,\ldots,\bar{x}_M)}\psi_e\rangle,\ \psi_e \in \mathcal{H}_e,\ \|\psi_e\|_{L^2} = 1\big\}. \tag{54.2}$$

The coupled problem to simulate is therefore

$$\left\{\begin{array}{l} m_k \dfrac{d^2\bar{x}_k}{dt^2}(t) = -\nabla_{\bar{x}_k} W\big(\bar{x}_1(t), \ldots, \bar{x}_M(t)\big), \\ W(\bar{x}_1, \ldots, \bar{x}_M) = U(\bar{x}_1, \ldots, \bar{x}_M) + \displaystyle\sum_{1 \leqslant k < l \leqslant M} \frac{z_k z_l}{|\bar{x}_k - \bar{x}_l|}, \\ U(\bar{x}_1, \ldots, \bar{x}_M) = \inf\big\{\big\langle \psi_e, H_e^{(\bar{x}_1, \ldots, \bar{x}_M)} \psi_e \big\rangle,\ \psi_e \in \mathcal{H}_e,\ \|\psi_e\|_{L^2} = 1\big\}, \\ H_e^{(\bar{x}_1, \ldots, \bar{x}_M)} = -\displaystyle\sum_{i=1}^N \frac{1}{2}\Delta_{x_i} - \sum_{i=1}^N \sum_{k=1}^M \frac{z_k}{|x_i - \bar{x}_k|} + \sum_{1 \leqslant i < j \leqslant N} \frac{1}{|x_i - x_j|}. \end{array}\right. \tag{54.3}$$

In this case, one speaks of the *adiabatic approximation* (here on the Born–Oppenheimer ground state energy surface). Note that the explicit dependence upon time has disappeared in the expression of W and U.

Let us remark that the adiabatic approximation is in fact the generalization of the Born–Oppenheimer approximation to a time-dependent setting. The electronic state is bound to evolve on a given energy surface. It is not easy at all to study the mathematical properties of this approximation together with the relationship it has with the nonadiabatic setting. Particularly intricate features are involved when the so-called Born–Oppenheimer energy surfaces happen to cross each others for some given configuration of nuclei. For a mathematical study of this approximation, we refer to HAGEDORN [1996], HAGEDORN [1980].

In practice, the minimization problem (54.2) has to be approximated, as in the pure time-independent case, by one of the standard (Hartree–Fock or density functional) methods. Doing so, the system obtained by through the adiabatic approximation reads

$$\left\{\begin{array}{l} m_k \dfrac{d^2\bar{x}_k}{dt^2}(t) = -\nabla_{\bar{x}_k} W(\bar{x}_1(t), \ldots, \bar{x}_M(t)), \\ W(\bar{x}_1, \ldots, \bar{x}_M) = U(\bar{x}_1, \ldots, \bar{x}_M) + \displaystyle\sum_{1 \leqslant k < l \leqslant M} \frac{z_k\, z_l}{|\bar{x}_k - \bar{x}_l|}, \\ U(\bar{x}_1, \cdots, \bar{x}_M) \quad \text{evaluated in a given static model.} \end{array}\right. \tag{54.4}$$

If the model is the Hartree–Fock approximation, one needs to find at each time step the Hartree–Fock ground state, which in practice (following the remarks we have made in Chapter II) amounts to solving the SCF equations

$$\left\{\begin{array}{l} F_\Phi \phi_i = \lambda_i \phi_i, \\ \displaystyle\int_{\mathbb{R}^3} \phi_i \phi_j = \delta_{ij}, \end{array}\right. \tag{54.5}$$

where

$$F_\Phi = -\frac{1}{2}\Delta - \sum_{k=1}^M \frac{z_k}{|\cdot - \bar{x}_k(t)|} + \left(\sum_{j=1}^N |\phi_j|^2 \star \frac{1}{|x|}\right) - \sum_{j=1}^N \left(\cdot \phi_j \star \frac{1}{|x|}\right)\phi_j, \tag{54.6}$$

and then computing the gradient of the energy

$$U(\bar{x}_1, \ldots, \bar{x}_M) = U^{\mathrm{HF}}(\bar{x}_1, \ldots, \bar{x}_M) = E^{\mathrm{HF}}(\phi_1, \ldots, \phi_N)$$

using analytical derivatives formulae as explained in Section 37.

Likewise, if the static model is, e.g., a Kohn–Sham model, then, from (15.6) of Chapter I, the equations

$$\begin{cases} K(\rho_\Phi)\phi_i = \lambda_i \phi_i, \\ \displaystyle\int_{\mathbb{R}^3} \phi_i \phi_j = \delta_{ij}, \end{cases} \tag{54.7}$$

where

$$K(\rho_\Phi) = -\frac{1}{2}\Delta - \sum_{k=1}^{M} \frac{z_k}{|\cdot - \bar{x}_k(t)|} + \rho_\Phi \star \frac{1}{|x|} + v_{\mathrm{xc}}(\rho_\Phi) \tag{54.8}$$

and $\rho_\Phi = \sum_{i=1}^{N} |\phi_i|^2$ need to be solved at each time step.

Even within this approximation, the coupled problem (54.4) remains very time-consuming since a minimization problem has to be solved for each time step in order to compute ∇U. It must indeed be remembered that the work we have done in Chapter II is now to be done at each time step of the discretization in time of the Newtonian dynamics of the nuclei. In other words, the solution procedure of the static electronic problem is now the *inner loop* of the dynamics. Therefore, from a numerical viewpoint, the task for simulating an adiabatic-type model is a sequence of 3-step iterations on the time variable:

(i) Solve the electronic minimization problem (which more practically means solving the nonlinear eigenvalue problem (SCF problem)),
(ii) compute the gradient of the interaction potential W,
(iii) integrate in time the Newtonian dynamics.

We have largely commented upon the first two steps: (i) is the central issue discussed in Chapter II, while (ii) is done by the analytical derivative method introduced mathematically in Chapter I and numerically in Chapter II. Therefore, it remains to now focus on step (iii). This will be the topic of Section 55. But before we get to that, let us briefly mention an alternative method, that is in some sense intermediate between the adiabatic and the nonadiabatic approximation.

With a view to circumvent the difficulty of solving a minimization problem (i.e., in practice a nonlinear eigenvalue problem) at each time step, which turns out to be more difficult that advancing forward in time a time-dependent electronic Schrödinger equation, the idea has arisen to replace the minimization by a *virtual* time evolution. This is the so-called *Car–Parrinello dynamics*, introduced in CAR and PARRINELLO [1985]. It consists in replacing the minimization problem by a fictitious (nonphysical) electronic dynamics which makes the electronic wavefunction evolve in the neighbourhood of the adiabatic state. Heuristically, the method is close to the notion of relaxing an holonomous constraint in classical dynamics. From a mathematical point of view, the Car–Parrinello method is investigated in BORNEMANN and SCHÜTTE [1998]. Explicitly, system (54.4) together with the Kohn–Sham LDA minimization at each time step (which amounts to solving (54.7)) is replaced by

$$\begin{cases} m_k \dfrac{d^2 \bar{x}_k}{dt^2}(t) = -\nabla_{\bar{x}_k} W\big(\bar{x}_1(t), \ldots, \bar{x}_M(t), t\big), \\ W(\bar{x}_1, \ldots, \bar{x}_M, t) = E^{\mathrm{KS}}_{\bar{x}_1, \ldots, \bar{x}_M}\big(\phi_1(t), \ldots, \phi_N(t)\big) + \displaystyle\sum_{1 \leqslant k < l \leqslant M} \frac{z_k z_l}{|\bar{x}_k - \bar{x}_l|}, \\ \mu \dfrac{\partial^2 \phi_i}{\partial t^2}(t) = -K(\rho_{\Phi(t)})\phi_i(t) + \displaystyle\sum_{j=1}^{N} \Lambda_{ij}(t)\phi_j(t), \\ \Lambda_{ij}(t) = \big\langle \phi_j(t), K(\rho_{\Phi(t)})\phi_j(t) \big\rangle - \mu \displaystyle\int_{\mathbb{R}^3} \frac{\partial \phi_j}{\partial t}(t) \frac{\partial \phi_i(t)}{\partial t}(t), \end{cases} \tag{54.9}$$

where μ is a fictitious mass, that needs to be properly fitted to reach efficiency in the computations. In particular, it is chosen much larger than the electron mass ($m_e = 1$). This allows one to numerically integrate (54.9) with a larger timestep than that for the nonadiabatic simulations, thus giving a more efficient method. On the other hand, although the timestep used for (54.9) needs to be smaller than that used for the adiabatic simulation, the Car–Parrinello method is also more efficient than the adiabatic ones because no minimization is required. Formally, the limit $\mu \to 0$ allows one to recover the adiabatic approximation, as it may be understood looking at the last two lines of (54.9) which then reduce to (54.7).

55. Molecular dynamics

As announced, we concentrate ourselves in this section on the numerical simulation of the Newtonian dynamics, which is a part of any of the above coupled simulations, the nonadiabatic one (53.4), the adiabatic one (54.3), or even the Car–Parrinello simulation (54.9).

The focus is therefore on the simulation of the Newtonian equation (52.3) that we reproduce here

$$m_k \frac{d^2 \bar{x}_k}{dt^2}(t) = -\nabla_{\bar{x}_k} W\big(\bar{x}_1(t), \ldots, \bar{x}_M(t), t\big), \tag{55.1}$$

complemented by initial data on the positions and velocities at time 0. The system may be casted into the form of a *Hamiltonian system*

$$\begin{cases} \dfrac{dq_k}{dt} = \dfrac{\partial H}{\partial p_k}(q_1, p_1, \ldots, q_M, p_M, t), & k = 1, \ldots, M, \\ \dfrac{dp_k}{dt} = -\dfrac{\partial H}{\partial q_k}(q_1, p_1, \ldots, q_M, p_M, t), & k = 1, \ldots, M, \end{cases} \tag{55.2}$$

where we have used the standard notations of the field of mathematics devoted to the study of Hamiltonian systems, namely p_k and q_k for the momentum and position of particle k, respectively. We have introduced the *Hamiltonian*

$$H(q_1, p_1, \dots, q_M, p_M, t) = \frac{1}{2} \sum_{k=1}^{M} p_k^2 + W(q_1, \dots, q_M, t). \tag{55.3}$$

which yields

$$\begin{cases} \dfrac{\partial H}{\partial p_k}(q_1, p_1, \dots, q_M, p_M, t) = p_k, \\ \dfrac{\partial H}{\partial q_k}(q_1, p_1, \dots, q_M, p_M, t) = -\nabla_{q_k} W(q_1, \dots, q_M, t). \end{cases} \tag{55.4}$$

In the language of physics, H is of course the energy of the system (kinetic plus potential energy). In the case when H does not depend explicitly upon the time variable t (which in our context happens for an adiabatic simulation), the Hamiltonian system is said to be *autonomous*.

Here, we do not make precise the origin of the potential W, as it may come

- either from a *formula* (for instance the very famous *Lennard-Jones interatomic potential* ruling the interaction between two particles at distance r by a law in $\frac{1}{r^{12}} - \frac{1}{r^6}$, the coefficients of which are fitted on each specific case, or any more sophisticated form of this); in the same vein, a lot of efforts in the domain of molecular dynamics are devoted toward the derivation of efficient and realistic *force fields* that can be used in the right hand side of the Newton equations,
- or from a calculation done with analytical derivatives (in the adiabatic case), or an evaluation of an integral depending on the current electronic density (in the nonadiabatic case),
- or from an expression of interatomic potential obtained by any other means (numerical data, ...).

The methods for the time integration will essentially not depend on the nature of this potential W. *The point* is however that, whatever the nature of W, the time needed for the calculation of the gradient in the right-hand side of (55.1) is *the main time consuming part* of the simulation. The reason is twofold: (a) The calculation of W itself together with its derivatives, which task is already huge when simulating small systems at the quantum level, and (b) the number of $\bar{x}_k$ involved, which increases further the combinatorial complexity when simulating, even with empirical force fields, very large systems. In the case of analytical formula for W, it is of course less time consuming, but it must however not be looked upon.

For simplicity, we shall only consider in the sequel the case when the system is autonomous, i.e., when W does not depend (explicitly) on time, which covers the case of empirical force fields and of the adiabatic approximation.

A complete detailed description of the algorithms in use for the simulation of general dynamics such as (55.1), of the specific adaptations needed to address the molecular dynamics problem, and of the overwhelming numbers of applications of the molecular dynamics techniques would require (at least) an entire volume. Molecular dynamics is indeed the most popular and commonly used field of molecular simulation. Calculations on millions of atoms are routinely done either on ensemble of molecules or on solids. Clearly, this is out of our scope here. We prefer to give here a brief overview in the following spirit: *We emphasize the peculiarities of the problem of simulating* (55.1) (*in the autonomous case*) *in the context of computational chemistry, with respect to simulating a dynamics in general*. It is therefore understood that the reader has a basic knowledge of the finite difference schemes used to simulate ordinary differential equations.

For a more complete description, we refer to

- HAIRER, NORSETT and WANNER [1993], HAIRER and WANNER [1996], HAIRER, LUBICH and WANNER [2002] for treatises of reference on the numerical simulation of ordinary differential equations,
- SANZ-SERNA and CALVO [1994] for the specific numerical analysis needed in the case when the equations are those of an Hamiltonian system,
- FRENKEL and SMIT [1996] for a complete treatise on molecular dynamics, as well as HAILE [1992] and RAPAPORT [1995] for more details on the numerical implementation,
- TUCKERMAN and MARTYNA [2000] for a recent state of the art survey in the field (see also NEUMAIER [1997] for the general context).
- DEUFLHARD ET AL. [1999] for a book showing various advanced aspects of the field of molecular dynamics.

The first thing to be said on numerical algorithms in use for the simulation of (55.1) is that these algorithms are in fact not dedicated to simulate one trajectory obeying to (55.1) for a given *precise* initial condition, but for an initial condition that may be only vaguely determined. This holds either because the initial condition is not that well-known, or because one is not interested in it, or both. Consequently, the properties that have drawn the attention of practitioners are not necessarily the ones that are usually addressed by numerical analysis in other fields of scientific computing, where the initial condition is pretty well known and controlled. The reason for this is that the primary use of a molecular dynamics simulation is to sample the phase space in order either to compute statistical averages (with a view to evaluate macroscopic quantities at thermodynamic equilibrium) or to search for minima and saddle points of the energy surface (as for protein folding). Some other applications of molecular dynamics require a knowledge of the time evolution itself (crack propagation in crystalline or amorphous materials, nonequilibrium thermodynamics) but for these kinds of calculations to make sense, results should not be really influenced by small changes in the choice of the initial condition.

As said above, the usual purpose of a molecular dynamics simulation is to evaluate the value of an observable, A, on a given system, by simulating a trajectory of this system which samples the phase space and computing the average value of A along this trajectory. The foundation of the method is ergodicity. More explicitly, assume we want to compute the average value $\langle A\rangle$ of A. This average value reads as the following integral over the phase space of the position/impulsion of the M particles

$$\langle A\rangle = \int_{\mathbb{R}^{6M}} A(q_1, p_1, \ldots, q_M, p_M) f(q_1, p_1, \ldots, q_M, p_M)\, dq_1\, dp_1 \cdots dq_M\, dp_M,$$

where f is the distribution function in the phase space (we do not make precise here the whole context coming from the foundations of statistical mechanics, and rather refer to the bibliography). This multidimensional integral can be evaluated as such through a Monte–Carlo sampling method, but an alternative method also exists. Indeed, if one knows how to exhibit a particular ergodic dynamics for the M particles

$$\begin{cases} \dfrac{dq_i}{dt} = p_i, & i = 1, \ldots, M, \\ \dfrac{dp_i}{dt} = F(q_j, p_j), & i = 1, \ldots, M, \end{cases} \tag{55.5}$$

where F is some force-field, so that the measure

$$f(q_1, p_1, \ldots, q_M, p_M)\, dq_1\, dp_1 \cdots dq_M\, dp_M$$

is precisely the invariant measure of the dynamics, then the average value of A may *also* be evaluated from the integral

$$\langle A \rangle = \lim_{T \to +\infty} \frac{1}{T} \int_0^T A\big(q_1(t), p_1(t), \ldots, q_M(t), p_M(t)\big)\, dt \tag{55.6}$$

along a trajectory of the system. This is where molecular dynamics comes into play.

Consequently, the integration of a system of the form (55.2) can be either done to simulate the exact evolution of a given molecular system obeying to (55.1), or (and it is the most general situation) to simulate a possibly fictitious trajectory (55.5), only aiming at evaluating an average value of a given observable on the system.

For simplicity, let us suppose (55.1) (or equivalently (55.5)) are Hamiltonian systems of the form

$$\begin{cases} \dfrac{dq}{dt} = \dfrac{\partial H}{\partial p}, \\ \dfrac{dp}{dt} = -\dfrac{\partial H}{\partial q}. \end{cases} \tag{55.7}$$

What are we aiming at? Typically, we want to know the mean state of the system for long time evolutions (think of formula (55.6)).

What is our main practical constraint? We know that the force field $\frac{\partial H}{\partial q}$ may be very time consuming to evaluate, and therefore that we must evaluate it as rarely as possible. It draws the attention to algorithms where large time steps δt are possible.

From the numerical viewpoint, the purpose is, as usual, to build algorithms that reproduce the theoretical mathematical properties of the system (55.7) to be simulated. These properties are the following.

First, there is *reversibility* in time: One may change the sign of time at a given instant and follow back the trajectory exactly. As long as possible, it is therefore important to choose an algorithm that reproduces this time reversibility at the discrete level.

Second, there is *energy conservation*. It is a simple fact to check that the Hamiltonian $H(q(t), p(t))$ is conserved by the dynamics. This energy conservation is not that important at the scale of a few time steps, but is a crucial issue for large times. Therefore, for a given algorithm, the emphasis will be laid upon the absence of too much a large energy drift after a huge number of time steps, rather than conservation of energy at a high order in δt. The reason is also that, accordingly to what has been just said on large time steps, a numerical analysis of energy conservation at the order $(\delta t)^p$ in the limit $\delta t \to 0$ is not very convincing in our context where large δt are to be used.

Third, there is *symplecticity*. This is a general property of flows of Hamiltonian systems (the *flow* is the application $\Phi(t,\cdot)$ that maps the initial condition of the system onto the position in the phase space at time t); we do not want to detail symplecticity here, and will only mention that it always imply that the flow keeps the volume constant in the phase space (and in fact it is equivalent to this conservation property in the case of a one-dimensional evolution), which conservation is indeed related to the conservation of energy. Indeed, it is understandable that an algorithm that does not preserve the volume in the phase space will have difficulties in preserving the volume of the domain {energy less than a given threshold}, or even (asymptotically) the curves of constant energy. We argue here somewhat vaguely, but this discussion can be given a sound mathematical meaning. Indeed, it can be shown, by a numerical analysis called *backward analysis*, that algorithms enjoying symplecticity at the discrete level have the following property: Their *numerical flow*, say Φ_n (i.e., the application mapping, through the discrete evolution, the initial data to the position/impulsion at discrete time t_n) is close to the exact flow $\widetilde{\Phi}$ of an Hamiltonian system (in fact close at the order $e^{-1/\delta t}$ over time intervals of length $e^{1/\delta t}$ if δt denotes the discretization time-step). This latter system is not the original system, but its energy $\widetilde{H}$ is close to the energy H of the original system (in fact close at the order $(\delta t)^p$ if the numerical scheme is of order p). Consequently, $\widetilde{\Phi}$, being the exact flow of an Hamiltonian system, preserves the energy $\widetilde{H}$ of it. This energy $\widetilde{H}$ is close to that of the original system H; thus $\widetilde{\Phi}$ almost preserves H. Finally, Φ_n, being close to $\widetilde{\Phi}$, behaves accordingly. In conclusion, the symplecticity is the property of choice for the simulation of Hamiltonian systems on large times.

Another property, that we wish to isolate, is the fact that the dynamics depends continuously upon the initial data. However, it is also well known that this dependence, although continuous, is very sharp. In other words, variating the initial condition from a small amount leads to a total modification of the position at "final" time T. This property of *chaotic behaviour* is a plague from the numerical viewpoint, but will be less of concern here as we are not too much demanding on the precision of the initial data.

From all the previous considerations, it follows that we need to use an algorithm that is *symplectic*, *reversible*, that *allows for large time steps* (which in other words means quite robust). The typical example of such an algorithm for Hamiltonians of the form $H(q,p) = \frac{p^2}{2} + V(q)$ in use for molecular simulation is the *Verlet algorithm* (VERLET [1967]).

In coordinates (q,p), denoting q_n, p_n, respectively, the approximations of $q(t_n)$, $p(t_n)$, at time $t_n = n\delta t$, this algorithm reads

$$\begin{cases} q_{n+1} + q_{n-1} = 2q_n - (\delta t)^2 \nabla V(q_n), \\ p_n = \dfrac{q_{n+1} - q_{n-1}}{2\delta t}. \end{cases} \tag{55.8}$$

It is obviously an explicit algorithm, which may be written in the alternative forms

$$\begin{cases} q_{n+1} = q_n + \delta t p_{n+1/2}, \\ p_{n+1/2} = p_{n-1/2} + (\delta t)\nabla V(q_n), \end{cases} \tag{55.9}$$

which is that of the *leap-frog algorithm*, or

$$\begin{cases} q_{n+1} = q_n + \delta t p_n + \dfrac{(\delta t)^2}{2}\nabla V(q_n), \\ p_{n+1} = p_n + \dfrac{\delta t}{2}\big(\nabla V(q_{n+1}) + \nabla V(q_n)\big), \end{cases} \tag{55.10}$$

which is called *velocity Verlet*.

This algorithm works remarkably well in the context of molecular dynamics, and is the reference choice, unless other specific requirements are to be accounted for. Of course, there are plenty of other, more sophisticated, algorithms, for which we refer to the bibliography.

So far, we have only considered simulation of systems of Hamiltonian nature, formally written under the form (55.7). If the system under study comes from a dynamics (55.1), the system is genuinely Hamiltonian and therefore treating it as (55.7) is legal. But we also have mentioned that most of the evolutions encountered in molecular dynamics are systems of the type (55.5), built on the specific purpose to compute average values such as (55.6). It is not at all clear that (55.5) is Hamiltonian. It is actually the case when the statistical ensemble on which the average $\langle A\rangle$ is defined is the microcanonical ensemble, which means the evolution lies on the set where the number of particles N, the volume of the system V, and its energy E are kept constant. On the other hand, it is false when the ensemble is another statistical average, as it is often the case in applications. For instance, the N, V, T ensemble, where the temperature T, instead of the energy E, is kept constant, is often to be explored. A typical dynamics, in the sense of (55.5), to generate this ensemble, is that of the Nosé–Hoover thermostat (NOSE [1984], HOOVER [1985], NOSE [1986]). In its simplest case, for one particle in 1D, this dynamics reads

$$\begin{cases} \dfrac{dq}{dt} = \dfrac{p}{m}, \\ \dfrac{dp}{dt} = F - \dfrac{p_\xi}{Q}p, \\ \dfrac{d\xi}{dt} = \dfrac{p_\xi}{Q}, \\ \dfrac{dp_\xi}{dt} = \dfrac{1}{m}p^2 - k_B T, \end{cases} \tag{55.11}$$

where Q is a coupling constant and T denotes the temperature that needs to be fixed (k_B is the Boltzmann constant). The evolution of the additional pair of variables

(ξ, p_ξ) aims at measuring to what extent the constraint of fixed temperature is obeyed.

This form is in fact only a trivial case of a general form of the so-called *Nosé–Hoover chain*, where more than one additional pair of variables is used.

In order to efficiently discretize *non-Hamiltonian* dynamics such as (55.11), the strategy follows the same lines as the one we have described above for a *Newtonian* dynamics. One tries to establish the list of quantities that are conserved via the dynamics (like the energy H is in the Hamiltonian dynamics), and then the same conservation property is imposed on the discretization scheme. Likewise, when possible, the measure preserved by the flow of the system is determined (like the Lebesgue measure of the phase space which is preserved by the Newtonian dynamics). Again, the algorithm is bound to preserve the same measure. The numerical analysis then follows the same spirit as above. We refer to the bibliography for the details.

56. Mathematical analysis of some nonadiabatic models

We study here system (53.4). This system is a perfect example of a multiscale problem. As mentioned above, it is indeed to be remembered that the characteristic times of the nuclear dynamics and that of the electronic dynamics are different from three orders of magnitude at least. Special techniques will therefore be required. It is easily understandable that the same time step is not to be used for the two equations. These special techniques for multiscale evolution problems, involving operator splitting methods, multiscale schemes, ... are not in the scope of this introductory survey. We only focus again on the basics. Considering that we have seen how to integrate the Newtonian dynamics (we have actually detailed only the autonomous case, which does not apply *stricto sensu* here, but the reader will admit the techniques basically follow the same lines), we concentrate ourselves on the electronic part. For the sake of brevity, we only concentrate on the mathematical properties, and refer the reader to the bibliography for more details on the numerics.

Let us begin with a brief overview of some mathematical results known on the time-dependent Schrödinger equation in its general form (53.1). If the electronic system is isolated (erase the term modelling the attraction of the nuclei in (52.6)), the Hamiltonian H_e is independent on the variable t. In this case, the well-posedness of the Cauchy problem

$$\begin{cases} i\dfrac{\partial \psi_e}{\partial t} = H_e \psi_e, \\ \psi_e(0) = \psi_e^0, \end{cases}$$

with $\psi_e^0 \in \mathcal{H}_e$ is guaranteed by the Stone theorem. More precisely, the evolution of the system is governed by a group of unitary operators on $\mathcal{H}_e$, the so-called propagator $(U(t,s))_{(t,s)\in\mathbb{R}^2}$, which satisfies

$$\psi_e(t) = U(t,s)\psi_e(s) \quad \text{for all } (t,s) \in \mathbb{R}^2,$$

and which enjoys the following properties:

(1) $U(t,s)U(s,r) = U(t,r)$ for all $(t,s,r) \in \mathbb{R}^3$;

(2) $U(t,s)$ is unitary on $\mathcal{H}_e$ for all $(t,s) \in \mathbb{R}^2$ and $(t,s) \mapsto U(t,s)$ is strongly continuous from $\mathbb{R}^2$ to $\mathcal{L}(\mathcal{H}_e)$, the space of bounded linear maps from $\mathcal{H}_e$ onto itself;

(3) denoting by D the domain of the operator H_e, $U(t,s) \in \mathcal{L}(D)$ for all $(t,s) \in \mathbb{R}^2$ and $(t,s) \mapsto U(t,s)$ is strongly continuous from $\mathbb{R}^2$ to $\mathcal{L}(D)$;

(4) the equalities $i\frac{dU(t,s)}{dt} = H_e U(t,s)$ and $i\frac{dU(t,s)}{ds} = -U(t,s)H_e$ hold strongly as equalities between operators from D to $\mathcal{H}_e$.

In view of the time-invariance, we have in addition here $U(t,s) = U(t-s) = e^{-i(t-s)H_e}$.

On the other hand, when the Hamiltonian explicitly depends on t, which happens in particular when the presence of the moving nuclei is accounted for in the Hamiltonian H_e, or when an external time-dependent electric field is turned on, the existence of a propagator may be difficult to establish. A few general results exist, in particular the Kato's theorem (KATO [1980]), but they do not apply in all cases. For instance, for $\mathcal{H} = L^2(\mathbb{R}^3)$ and $H(t) = -\Delta + \mathcal{V} + \mathcal{E}(t) \cdot x$ with $\mathcal{V}(x) = -\frac{Z}{|x|}$ (a particle in a fixed Coulomb potential subjected to a time-dependent electric field $\mathcal{E}(t)$) the Kato's theorem permits to conclude provided the time-dependent part of the potential is regular enough, which in the above setting means $t \mapsto \mathcal{E}(t)$ is continuous (IORIO JR. and MARCHESIN [1984]). On the contrary, it is not sufficient to conclude for $\mathcal{H} = L^2(\mathbb{R}^3)$ and $H(t) = -\Delta + \mathcal{V} + \mathcal{W}(t)$ with $\mathcal{V}(x) = -\frac{Z}{|x|}$ and $W(t,x) = \frac{1}{|x-\bar{x}(t)|}$ (which for instance models a particle placed in a fixed attractive Coulomb potential $\mathcal{V}$ and interacting with a moving charged particle through the interaction potential $\mathcal{W}$, which is precisely our case here) because of the singularity of the time-dependent potential. In this latter situation, which also occurs for example in the study of collision processes, the existence of the propagator may be established either as in WULLER [1986] by locally deforming the set of coordinates so that the moving particle (thus the singularity) remains fixed in this frame (let us note that such an approach allows one to conclude only when the time-dependent part of the potential is of the general form $W(t,x) = W(x - \bar{x}(t))$), or as in YAJIMA [1987] by resorting to $L^p - L^q$ estimates provided the time-dependent part of the potential is not too singular (the Coulomb singularity being convenient in $\mathbb{R}^3$).

As said above, as in the time-independent setting, the electronic Schrödinger equation cannot be solved directly and additional approximations are necessary. We have chosen here to focus on the Hartree–Fock method which we briefly present here in a spinless context to simplify, so that here $\mathcal{H}_e = \bigwedge_{i=1}^N L^2(\mathbb{R}^3, \mathbb{C})$. Taking the spin into account make the notations more cumbersome but does not raise any additional mathematical nor numerical difficulties.

Of course, there exist other methods of approximation for the time-dependent Schrödinger equation. For instance, a theory of time-dependent density functional has been developed. However, to the best of our knowledge, such a theory is not mature enough to be discussed in such a volume. We therefore prefer to concentrate on an Hartree–Fock type approximation. It is to be noted that this time-dependent Hartree–Fock approximation is not flawless either. It has the enormous advantage, from our formal viewpoint, to be mathematically well posed and defined. In addition, it provides rather satisfactory results on the chemical side. It is however clear that it is less well understood (and therefore less commonly used) than its stationary counterpart.

The *time-dependent Hartree–Fock approximation* is of variational nature: It consists in forcing the wavefunction ψ_e to evolve on the manifold

$$\mathcal{A} = \left\{ \psi_e(x_1, \ldots, x_n) = \frac{1}{\sqrt{N!}} \det\bigl(\phi_i(x_j)\bigr),\ \phi_i \in H^1(\mathbb{R}^3, \mathbb{C}),\ \int_{\mathbb{R}^3} \phi_i \phi_j^* = \delta_{ij} \right\}$$

of $\mathcal{H}_e$ and in replacing Eq. (53.1) by the stationarity condition for the action

$$\int_0^T \bigl\langle \psi_e(t), \bigl(i\partial_t \psi_e(t) - H_e(t)\psi_e(t)\bigr)\bigr\rangle \, dt.$$

The associated Euler–Lagrange equations (McWEENY [1992]) read

$$i \frac{\partial \phi_i}{\partial t} = F_\Phi \phi_i + \sum_{j=1}^N \lambda_{ij} \phi_j,$$

where F_Φ is the Fock Hamiltonian

$$F_\Phi = -\frac{1}{2}\Delta + \sum_{k=1}^M \frac{z_k}{|\cdot - \bar{x}_k|} + \left(\sum_{i=1}^N |\phi_i|^2 \star \frac{1}{|x|} \right) - \sum_{j=1}^N \left(\phi_j^* \cdot \star \frac{1}{|x|} \right) \phi_j$$

and where the matrix $(\lambda_{ij}(t))$ is Hermitian for any t. We emphasize that, as in the stationary case, the Hartree–Fock approximation has created nonlinearity: Indeed the Hartree–Fock Hamiltonian depends on the electronic wavefunction. Contrary to what is sometimes claimed in the chemical literature, the $\lambda_{ij}(t)$ should not be interpreted as Lagrange multipliers associated with the constraints $\int_{\mathbb{R}^3} \phi_i \phi_j^* = \delta_{ij}$ (these constraints are automatically propagated by the dynamics because of the self-adjointness of the Hartree–Fock Hamiltonian), but rather as degrees of freedom associated with the gauge invariance $\phi_i(t) \to U_{ij}(t)\phi_j(t)$ for any regular unitary $N \times N$ matrix valued function $t \mapsto U(t)$. In particular, this gauge invariance can be used to set the $\lambda_{ij}(t)$ to zero for all t so that the above system can be transformed into the simpler one

$$i \frac{\partial \phi_i}{\partial t} = F_\Phi \phi_i. \tag{56.1}$$

Let us notice that the usual time-independent Hartree–Fock equations can be easily deduced from (56.1), like the time-independent Schrödinger equation is deduced from the time-dependent Schrödinger equation: Indeed, let us search for solutions of the form $\phi_i(t, x) = \phi_i(x) e^{-i\lambda_i t}$; we thus obtain

$$F_\Phi \cdot \phi_i = \lambda_i \phi_i.$$

The time-dependent Hartree–Fock model has been mathematically studied by CHADAM and GLASSEY [1975] who proved the well-posedness of the Cauchy problem for fixed nuclei $\bar{x}_k$. Clearly, this assumption is too restrictive for the case we have at hand where the nuclei move accordingly to the Newton equations (52.3), and this requires extending the Chadam and Glassey's result by including *also* the nuclear dynamics into the evolution system.

Here, the system under study couples the electronic Hartree–Fock evolution equation with the Hellmann–Feynmann nuclear dynamics and reads:

$$\begin{cases} m_k \dfrac{d^2\bar{x}_k}{dt^2}(t) = -\nabla_{\bar{x}_k} W\big(t; \bar{x}_1(t), \dots \bar{x}_M(t)\big), \\ \bar{x}_k(0) = \bar{x}_k^0, \\ \dfrac{d\bar{x}_k}{dt}(0) = \bar{v}_k^0, \\ W(t; \bar{x}_1, \dots, \bar{x}_M) = -\displaystyle\sum_{k=1}^{M}\sum_{i=1}^{N} z_k \int \frac{|\phi_i(t,x)|^2}{|x-\bar{x}_k|}\,dx + \sum_{1\leqslant k<l\leqslant M} \frac{z_k z_l}{|\bar{x}_k - \bar{x}_l|}, \\ i\dfrac{\partial \phi_i}{\partial t} = -\dfrac{1}{2}\Delta\phi_i - \displaystyle\sum_{k=1}^{M} \frac{z_k}{|\cdot - \bar{x}_k(t)|}\phi_i + \left(\sum_{j=1}^{N} |\phi_j|^2 \star \frac{1}{|x|}\right)\phi_i \\ \qquad -\displaystyle\sum_{j=1}^{N}\left(\phi_j^*\phi_i \star \frac{1}{|x|}\right)\phi_j, \\ \phi_i(0) = \phi_i^0. \end{cases} \tag{56.2}$$

The following result, due to two of us in CANCÈS and LE BRIS [1999] shows that this Cauchy problem for (56.2) is well-posed, i.e., that system (56.2) has a unique global solution in a functional space made precise below, provided the ϕ_i^0 are chosen regular enough. This assumption is not restrictive because in practice the initial data are often eigenstates, thus regular functions.

THEOREM 56.1. *Suppose that $\phi_i^0 \in H^2(\mathbb{R}^3, \mathbb{C})$ for all $1 \leqslant i \leqslant N$ and that all the $\bar{x}_k^0$ are different from one another. Then, the system (56.2) has a unique global solution $(\{\phi_i\}_{1\leqslant i\leqslant N}, \{\bar{x}_k\}_{1\leqslant k\leqslant M})$ in*

$$\big(C^1\big([0,+\infty[, L^2(\mathbb{R}^3,\mathbb{C})\big) \cap C^0\big([0,+\infty[, H^2(\mathbb{R}^3,\mathbb{C})\big)\big)^N \times \big(C^2\big([0,+\infty[, \mathbb{R}^3\big)\big)^M.$$

Let us mention the possible extension of the above results for the case when the molecular system is subjected to an external uniform time-dependent electric field $\mathcal{E}(t)$. The molecular Hamiltonian H given by (52.2) is modified by the addition of the external electric potential $\mathcal{V}(t,x) = -\sum_{k=1}^{M} z_k \mathcal{E}(t)\cdot\bar{x}_k + \sum_{i=1}^{N} \mathcal{E}(t)\cdot x_i$ created by the field. It can be shown that the system

$$\begin{cases} m_k \dfrac{d^2\bar{x}_k}{dt^2}(t) = -\nabla_{\bar{x}_k} W\big(\bar{x}_1(t), \dots \bar{x}_M(t), t\big) + z_k \mathcal{E}(t), \\ \bar{x}_k(0) = \bar{x}_k^0, \\ \dfrac{d\bar{x}_k}{dt}(0) = \bar{v}_k^0, \\ W\big(\bar{x}_1, \dots \bar{x}_M(t), t\big) = -\displaystyle\sum_{k=1}^{M}\sum_{i=1}^{N} z_k \int \frac{|\phi_i(t,x)|^2}{|x-\bar{x}_k(t)|}\,dx \\ \qquad\qquad + \displaystyle\sum_{1\leqslant k<l\leqslant M} \frac{z_k z_l}{|\bar{x}_k(t)-\bar{x}_l(t)|}, \\ i\dfrac{\partial\phi_i}{\partial t} = -\dfrac{1}{2}\Delta\phi_i - \displaystyle\sum_{k=1}^{M} \frac{z_k}{|\cdot - \bar{x}_k(t)|}\phi_i + \left(\sum_{j=1}^{N} |\phi_j|^2 \star \frac{1}{|x|}\right)\phi_i \\ \qquad - \displaystyle\sum_{j=1}^{N}\left(\phi_j^*\phi_i \star \frac{1}{|x|}\right)\phi_j + \mathcal{E}(t)\cdot x\phi_i, \\ \phi_i(0) = \phi_i^0, \end{cases} \tag{56.3}$$

is also well-posed.

To state the result, we must indicate that the domain of the self-adjoint operator

$$-\frac{1}{2}\Delta + \mathcal{E}\cdot x$$

contains $H^2_{\text{ef}} = \{\phi \in H^2(\mathbb{R}^3)/\sqrt{1+|x|^2}\phi \in L^2(\mathbb{R}^3)\}$ if $\mathcal{E} \neq 0$ and equals H^2 in the special case when $\mathcal{E} = 0$. The space H^2_{ef} is a Hilbert space when equipped with the norm

$$\|\phi\|_{H^2_{\text{ef}}} = \left(\left\|\sqrt{1+|x|^2}\,\phi\right\|^2_{L^2} + \|\Delta\phi\|^2_{L^2}\right)^{1/2}.$$

Let us now state a result proven in CANCÈS and LE BRIS [1998].

THEOREM 56.2. *Let* $\mathcal{E} \in C^0([0,+\infty[,\mathbb{R}^3)$. *If* $\phi_i^0 \in H^2_{\text{ef}}$ *for all* i, *and all the* $\bar{x}_k^0$ *are different from one another, then the system* (56.3) *has a unique global solution* $(\phi_1,\dots,\phi_N,\bar{x}_1,\dots,\bar{x}_M)$ *in*

$$\left(C^1([0,+\infty[,L^2)\cap C^0([0,+\infty[,H^2_{\text{ef}})\right)^N \times \left(C^2([0,+\infty[,\mathbb{R}^3)\right)^M.$$

As in the stationary setting, it is to be mentioned that improvements of the time-dependent Hartree–Fock equations can be used. For instance, the time-dependent

MCSCF method exists. We again refer to the bibliography for such methods, which are implemented in practice but clearly lack of a mathematical analysis.

Alternatively to the standard form (56.1), the time dependent Hartree–Fock approximation may be written as an evolution equation of the kernel $\gamma_1(x,x')$ of the reduced density matrix introduced in Chapter I:

$$\begin{aligned} i\frac{\partial\gamma_1}{\partial t}(t,x,x') &= \left(-\frac{1}{2}\Delta_x + V(x) + \frac{1}{2}\Delta_{x'} - V(x')\right)\gamma_1(t,x,x') \\ &\quad + \int_{\mathbb{R}^3}\left(\frac{1}{|x-z|} - \frac{1}{|x'-z|}\right)\big(\gamma_1(t,x,x')\gamma_1(t,z,z) \\ &\qquad - \gamma_1(t,x,z)\gamma_1(t,z,x')\big)\,dz. \quad (56.4) \end{aligned}$$

This equation is the *Liouville equation*. Actually, this equation may be derived as follows.

Starting from the time-dependent electronic Schrödinger equation (53.1) with Hamiltonian (52.6), one integrates partially with respect to *some* of the variables x_i, namely the last $N-n$ ones, in the spirit of what has been done in the static case in Chapter I. Denoting $X_n = (x_1,\ldots,x_n)$, $\Delta_{X_n} = \sum_{x_i\in X_n}\Delta_{x_i}$, $V(X_n) = \sum_{x_i\in X_n} V(x_i)$ and

$$\begin{aligned} &\gamma_n\big(X_n, X_n'\big) = \frac{N!}{n!(N-n)!}\int_{\mathbb{R}^{3(N-n)}} dx_{n+1}\cdots dx_N \\ &\psi_e(x_1,\ldots,x_n,x_{n+1},\ldots,x_N)\psi_e\big(x_1',\ldots,x_n',x_{n+1},\ldots,x_N\big)^* \quad (56.5) \end{aligned}$$

the reduced matrix of order n (we have omitted the spin variables for simplicity), we obtain from (53.1) the *hierarchy* of equations

$$\begin{aligned} i\frac{\partial\gamma_n}{\partial t} &= \left(-\frac{1}{2}\Delta_{X_n} + V(X_n) + \frac{1}{2}\Delta_{X_n'} - V\big(X_n'\big)\right)\gamma_n \\ &\quad + \sum_{1\leqslant i<j\leqslant n}\left(\frac{1}{|x_i-x_j|} - \frac{1}{|x_i'-x_j'|}\right)\gamma_n \\ &\quad + (n+1)\sum_{1\leqslant i\leqslant n}\int_{\mathbb{R}^3}\left(\frac{1}{|x_i-z|} - \frac{1}{|x_i'-z|}\right)\gamma_{n+1}\big(X_n,z,X_n',z\big)\,dz \quad (56.6) \end{aligned}$$

if $n \leqslant N-1$, and

$$\begin{aligned} i\frac{\partial\gamma_N}{\partial t} &= \left(-\frac{1}{2}\Delta_{X_N} + V(X_N) + \frac{1}{2}\Delta_{X_N'} - V\big(X_N'\big)\right)\gamma_N \\ &\quad + \sum_{1\leqslant i<j\leqslant N}\left(\frac{1}{|x_i-x_j|} - \frac{1}{|x_i'-x_j'|}\right)\gamma_N. \quad (56.7) \end{aligned}$$

System (56.6) is called a hierarchy because the nth line is an evolution equation for the reduced density matrix of order n (n-RDM) that involves (in the right-hand side) the matrix of order $n+1$.

It turns out that making the Hartree–Fock approximation exactly amounts to assuming that the reduced density of order 2 can be expressed in terms of the first one γ_1 as a determinant of a 2×2 matrix

$$\gamma_2(X_2, X_2') = \frac{1}{2} \det(\gamma_1(x_i, x_j')). \tag{56.8}$$

Consequently, *all* reduced density matrices of order $1 \leqslant n \leqslant N$ can simply be evaluated by the determinant of a $n \times n$ matrix

$$\gamma_n(X_n, X_n') = \frac{1}{n!} \det(\gamma_1(x_i, x_j')). \tag{56.9}$$

The first line of hierarchy (56.6) is then equivalent to (56.4), and the following lines only reduce to the "pointwise" algebraic evaluation (56.9). In this language, improving the time-dependent Hartree–Fock approximation is exactly avoiding closure formulae such as (56.8), (56.9), or adopting closure formulae at higher orders. In this perspective, the time-dependent Hartree–Fock approximation is nothing else but the simplest possible way to close the hierarchy at the lowest possible order.

57. Control of molecular systems

We wish to devote the last section to the modeling of the control of chemical reactions by laser fields. The laser control of chemical reactions is a very active field of laser physics, at the crossroads between computational chemistry, quantum mechanics, and theoretical and experimental femtophysics. Manipulation of molecular systems using laser fields is today an experimental reality (ASSION ET AL. [1998]), provided one restricts his aims to reasonable goals. This leads to a mostly unexplored field for mathematical analysis and numerical simulation. Numerical simulations can indeed efficiently complement the experimental strategy, both by explaining the deep nature of the phenomena involved and by optimizing the parameters to be used experimentally. Let us at once make it clear that apart from the ultimate goal of controlling chemical reactions, the laser control is also a way to isolate and identify the underlying crucial phenomena in some evolution, and therefore to deeply understand the physics at work. In addition, it should be emphasized that from the strict viewpoint of the control itself, it is not claimed that, in the end, the control of the given reaction will necessarily be done industrially or even experimentally with a laser. Once the physical driving phenomena are understood and shown to be controllable, the control can then be made with more classical devices (use of temperature, pressure, and so on).

This topic will be addressed more thoroughly elsewhere in this volume, but we wish to give here a short introduction to the field, with some emphasis on the mathematical aspects, as usual. For a general introduction to the problematic, we refer, e.g., to

KOBAYASHI [1998] and to the more general REPORT OF THE PANEL ON FUTURE DIRECTIONS IN CONTROL, DYNAMICS, AND SYSTEMS [2002] (and the references therein), which mentions the control of quantum evolutions as one of the most promising fields of application for mathematical control.

The use of laser technology in chemistry goes back to the 1960s, with the *mode selective chemistry*. Among the numerous problems that were attacked with this new born technology was the challenge to break a given bond in a given molecular system, so as to make it split into two desired products. Typically, the scheme is to discriminate in

$$ABC \to \begin{cases} A + BC, \\ AB + C \end{cases}$$

by making the first dissociation more likely to happen that the second one (see BRUMER and SHAPIRO [1995] for an introduction to this problematic). The standard example for dissociation is that of a linear molecule, that one wishes to break at a given bond. The natural idea is to choose as the frequency of the laser that of the targeted bond, with the hope to get a resonance effect, and therefore to break the bond. This idea leads to a failure. Indeed, what has been erroneously neglected is the connection of bonds: In such a situation, the energy provided by the laser field only spreads over the whole molecule, without giving rise to any local concentration of energy on the targeted bond. The conclusion follows: The electric field to be used is not to be fitted on a given frequency, but is rather more complicated than that. An intelligent systematic way to design the control should come into the picture: Mathematics.

Today, the usual way to divide practitioners of laser control is the following. There is the strategy of *time-resolved* (or pump dump) schemes TANNOR, KOSLOFF and RICE [1986], which is based upon the idea of using two (or more) lasers, and next of playing on the dephasing between the two lasers to create interferences that will drive the system to the targeted state. Second, there is the so-called *frequency-resolved* (or coherent control) schemes. The strategy, followed, e.g., in SHAPIRO and BRUMER [1993], is then to fit the frequencies of the lasers in order to reach the goal. The two above strategies require a very deep understanding of the phenomena at work, and can only be applied to very small toy-systems, to identify fundamental behaviours. An alternative to these strategies is the somewhat more mathematical *optimal control* strategy, which makes use of both the frequencies and the dephasings to reach the goal. Of course, it is a strategy less based upon the physical intuition, although we shall see that a physical understanding is of course also useful and even necessary to reach efficiency. Examples of this strategy are the works by ZHU and RABITZ [1998], JUDSON, LEHMANN, RABITZ and WARREN [1990], HOKI and FUJIMURA [2001], KANAI and SAKAI [2001]. This is the strategy we shall concentrate on in the sequel.

In such a scientific domain, it is absolutely necessary to bear in mind the physical orders of magnitude and the technological data. Typically, the space scale is that of a molecule, namely a few angströms (10^{-10} m), and the time scale is that of the vibration of a molecular bond, namely ten femtoseconds (10^{-14} s). The total time of a simulation for a Schrödinger like equation is thus typically the picosecond (10^{-12} s). In other words, the whole process of control is over after a time of the order of one picosecond.

One of us has presented in LE BRIS [2000] a rapid account of the practical parameters for a laser field. We also refer to the comprehensive RAPPORT DE L'ACADEMIE [2000] for more details. This latter reference presents a broad overview of the domain, indicating current approaches, both theoretical and experimental, and gives trends for the future. One other useful reference in the same spirit is TOURNOIS ET AL. [1999]. Let us only say that a trade-off has to be made between the power of the laser, its time resolution, its repetition frequency, and also its price and its size. The laser fields that are commonly available have intensities in the range $[10^{12}, 10^{13}]\ \mathrm{W/cm^2}$, are able to have a risetime of the order of 10^{-14} s, and the light they create has frequency around 10^{14} Hz.

Let us now give the typical models in use in this framework.

The evolution of a molecular system subjected to a laser field $\vec{E}$ is most often modeled by the time-dependent Schrödinger equation

$$i\frac{\partial\psi}{\partial t} = H_0\psi + \vec{\mathcal{E}}(t)\cdot\vec{\mathcal{D}}\big(\vec{\mathcal{E}}(t)\big)\psi, \tag{57.1}$$

complemented with the initial condition $\psi(t=0)=\psi_0$. In this equation, the wavefunction ψ is assumed to depend only on the coordinates of the various nuclei the molecular system is composed of. The presence of the electrons is accounted for through an effective potential acting on the nuclei, and contained in the Hamiltonian H_0 of the free system (when the laser is turned off). We denote by $\vec{\mathcal{D}}(\vec{\mathcal{E}})$ the dipole moment of the molecule in presence of an external electric field $\vec{\mathcal{E}}$; at the first order perturbation theory, one can use the form $\vec{\mathcal{D}}(\vec{\mathcal{E}})=\vec{\mu}_0+\overline{\overline{\alpha}}\vec{\mathcal{E}}$. More sophisticated models would feature higher order expansions of $\vec{\mathcal{D}}(\vec{\mathcal{E}})$, still in the perturbation setting, or even a true dependence of the wavefunction ψ and the Hamiltonian H with respect to the coordinates of *all nuclei and electrons* of the molecular system. To the present day, the latter model is out of reach of a numerical treatment.

In order to state a control problem, we need, in addition to the direct equation (57.1) modeling the evolution of the system, to define a *target* (in the sense of *exact control*) or a *cost function* (in the sense of *optimal control*). Reaching the target, or minimizing the cost function will give a formal sense to the physical goal.

The goal usually is to drive the system from one initial state to some other specific state through a controlled time-dependent evolution. Therefore, the problem can be set in the standard mathematical fashion: Reach a target state Ψ_T at final time T. As such, a mathematical study, and a lot of numerics can be done. However, this direct form of the control problem rises a lot of (open) questions, as will be briefly mentioned below. Loosely speaking, let us say for the time being that such an exact control problem may be too difficult to solve in this framework. An alternative to this is to set a control problem with a much more modest goal, or at least a more vague goal. It is then chosen to formulate the mathematical goal not as a distance to a target state, but as the mean value of an observable (a measure of the orientation of the molecular system with the field, for instance). The ultimate goal of the manipulations we want to model is, as we said, the control of chemical reactions. Succeeding in making a chemical reaction possible does not necessarily mean driving the initial state to the final one, but sometimes (and in fact most of the times) only succeeding in *preparing* the initial system in a good way so that afterwards the desired reaction spontaneously happens. This gives some room to define a sufficiently modest goal, translated into a cost function $J(E)$, so that it can be efficiently reached. An instance that will be briefly addressed below is provided by the notion of orientating a molecule in space.

The simple setting we have just indicated above suffices to now underline the peculiarities of the optimal control problem we have to tackle, in comparison with other optimal control problems that the reader may have in mind and that come from more usual domains of the engineering world (aeronautics, ...). Let us now emphasize these peculiarities.

From the standpoint of the mathematical theory, this problem is *bilinear* (the control $\vec{\zeta}$ multiplies the state ψ) which at once puts the problem on a very high level of mathematical difficulty. Indeed, the mathematical theoretical results on bilinear control are very rare. In infinite dimension, i.e., for the partial differential equation (57.1), since the celebrated work by BALL, MARSDEN and SLEMROD [1982], no real progress has been made, to the best of our knowledge. For the finite-dimensional approximation of (57.1), there exist some results that can also be extended to the infinite-dimensional case but that are not very easy to exploit (so far). We refer to the work of TURINICI and RABITZ [2001] for some recent progress on the theory of exact controllability for systems such as those we deal with here. Despite all these efforts, it must however be emphasized that a good notion of exact controllability in the framework of quantum mechanics remains to be defined, to the best of our knowledge. On the other hand, for the optimal control problem, *some* minor things can be done. We refer in particular to CANCÈS, LE BRIS and PILOT [2000] where some of us have proven the existence of an optimal field in a very academic and simplified setting.

A noticeable peculiarity is the fact that, in most cases, the control $\vec{\zeta}$ is *distributed in time*, and not in space. This means that, at the scale of the molecular system under study, it can be considered that the electric field does not vary. Clearly, this is valid under the restriction that the system is not too large (it is an atom, or a small molecule). For a larger system, electric fields of the more general form $\vec{\zeta}(t, x)$ should be accounted for. It is not a crucial fact for the sequel (cases when $\vec{\zeta}$ depends both on time and space could be treated in the same fashion, however with slightly more tedious computations) but it is rather convenient and constitutes a very reasonable approximation. Such a distributed in time control is not that usual for a partial differential equation such as (57.1).

Two other peculiarities come from the very unusual orders of magnitude for the physical quantities that have to be manipulated here. First, the control needs to be an *open-loop* control, since it is clear that one cannot update the field in real time with electronic devices. In other words, the only system that can react as fast as the molecular system is precisely the system itself. The second consequence is that we must think of this problem in a completely different way from the way we think of usual control problems: We are here in a framework where we can do thousands of experiments within a minute (while we cannot launch a rocket thousands times). This ability to make many experiments has in turn two consequences. First, one can imagine, and it is indeed done, to couple the numerical search for the optimal field not with the numerical simulation of Eq. (57.1), but with the experiment itself (ASSION ET AL. [1998], JUDSON and RABITZ [1992]). The experimental solution of (57.1) is indeed much faster than its resolution on

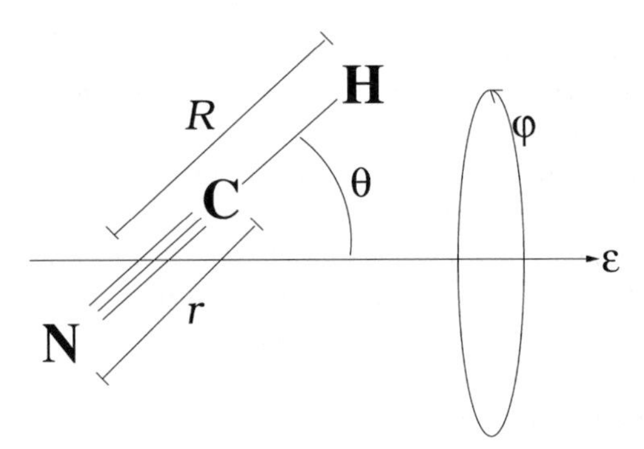

FIG. 57.1. Jacobi coordinates.

a computer, as can easily be understood from the orders of magnitude mentioned above. But in this coupled (experimental *plus* numerics) approach arises the very important question: *How to measure the success*?, for this is an information that must be inserted, with the maximum of accuracy, inside the optimization loop. This is not at all an easy question to answer. On the other hand, one of the major problems of this field is the tremendous amount of data that are at our disposal. A challenge is to find a way to exploit them in the optimization cycle.

Another very peculiar feature appears considering the data that we have mentioned on the laser technology above. One can ask the question whether it is better to optimize upon *only* the fields that are today experimentally feasible or to consider all fields without taking into account any contemporary technological constraint. Both approaches may be useful. In particular, and this is where the peculiarity is, the second one may help in designing the lasers physicists do need for the next generation.

As an instance of a numerical study for an optimal control problem in the framework of laser control, we wish to end this section by rapidly mentioning the optimal control of orientation of a linear molecular system. The work is detailed in AUGER, DION, BEN HAJ YEDDER, KELLER, CANCÈS, LE BRIS and ATABEK [2002], AUGER, DION, BEN HAJ YEDDER, CANCÈS, LE BRIS, KELLER and ATABEK [2003], DION, BEN HAJ YEDDER, CANCÈS, LE BRIS, KELLER and ATABEK [2002], BEN HAJ YEDDER, CANCÈS and LE BRIS [2001], and follows earlier works by DION, KELLER and ATABEK [1999a], DION, KELLER and ATABEK [1999b], DION, KELLER and ATABEK [2001]. It will be addressed in more details in the contribution by O. Atabek and collaborators in the present volume. Of course, as other contributions of the volume will show it, it is by no means the only example of optimal control applied to laser control.

As mentioned above, trying to align and orientate a molecule in space is both a modest and reachable goal. Indeed, once the molecule is conveniently "geometrically" prepared, the aim of controlling the chemical reaction is almost reached. Nature will do the rest of the job. In addition, there are today experimental evidences showing that aligning a molecule (orientation is one step forward alignment) with a laser *is* feasible, and constitutes a significant step that can be used to efficiently control reactions (see the works by SAKAI ET AL. [1999], LARSEN ET AL. [1999]). The problem of orientation is therefore a good problem to look at.

The molecular system under study is the linear HCN molecule (hydrogen cyanide). In the so-called Jacobi coordinates ($\mathbf{R} = (R, r), \theta, \varphi$) used to parameterize the state of the molecule, the free Hamiltonian H_0 can be written as $H_0 = H_{\text{vib}}(\mathbf{R}) + H_{\text{rot}}(\mathbf{R}, \theta, \varphi) + V(\mathbf{R})$ and the dipole moment is written as $D(\zeta(t)) = -\mu_0(R, r)\cos\theta - \frac{\zeta(t)}{2}[\alpha_{\|}(R, r)\cos^2\theta + \alpha_{\perp}(R, r)\sin^2\theta]$. We have denoted by μ_0 the permanent dipole moment. The coefficients $\alpha_{\|}$ and $\alpha_{\perp}$ are, respectively, the parallel and the perpendicular components of the diagonal polarizability tensor $\overline{\overline{\alpha}}$ given by $\alpha_{\|} = \alpha_{zz}$ and $\alpha_{\perp} = \alpha_{xx} = \alpha_{yy}$ when (Oz) is the molecular axis. For simplicity, the case of a rigid rotor can be considered: The variables R and r are fixed at their equilibrium value, and the problem thus only depends on the angular variables θ, ϕ. Furthermore, symmetry conservation around the laser polarization axis allows us to separate the motion in ϕ from the motion in θ, and consider only the latter in the calculations. The Hamiltonian of the system therefore reads

$$H = H(\theta, t) = H_{\text{rot}}(\theta) + H_{\text{laser}}(\theta, t) \tag{57.2}$$

with

$$H_{\text{rot}}(\theta) = -\frac{1}{2(\mu_{HCN}R^2 + \mu_{CN}r^2)}\frac{1}{\sin\theta}\frac{\partial}{\partial\theta}\left(\sin\theta\frac{\partial}{\partial\theta}\right)$$

and

$$H_{\text{laser}}(\theta, t) = -\mu_0(R, r)\zeta(t)\cos\theta - \frac{\zeta^2(t)}{2}\big[\alpha_{\|}(R, r)\cos^2\theta + \alpha_{\perp}(R, r)\sin^2\theta\big].$$

In the former formulae, $\mu_{CN} = \frac{m_C m_N}{m_C + m_N}$ and $\mu_{HCN} = \frac{m_H(m_C + m_N)}{m_H + m_C + m_N}$ represent the reduced masses.

The objective function $J(\zeta)$ typically is either $J = \frac{1}{T}\int_0^T j(t)\,dt$, when it is desired that the orientation is kept as long as possible, or $J = \min_{t\in[0,T]} j(t)$, when the maximum orientation is needed at one time. Both forms make use of an instantaneous criterion $j(t)$ which is the measure of the orientation at time t, namely in this simple case $j(t) = \int_0^\pi \cos\theta\,\|\psi\|_{\mathbb{C}}^2 \sin\theta\,d\theta$. The instantaneous criterion $j(t)$ takes its values in the range $[-1, 1]$, the values -1 and 1 corresponding, respectively, to a molecule pointing in the direction of the laser field polarization axis and in the opposite direction.

The Schrödinger equation

$$\begin{cases} i\dfrac{\partial\psi}{\partial t} = H\psi, \\ \psi(t=0) = \psi_0 \end{cases} \tag{57.3}$$

depending only on the variable θ is numerically solved with an operator splitting method coupled with a FFT for the kinetic part.

As far as the set of laser fields one optimizes upon are concerned, it is to be noticed that, as said above, both strategies of restricting oneself to the experimental state of the art or of considering the most general laser fields are of some interest. In the latter case, it can be considered that the electric field $\zeta(t)$ is the sum of N ($\leqslant 10$) individual linearly-polarized pulses:

$$\zeta(t) = \sum_{n=1}^{N} E_n(t) \sin(\omega_n t + \phi_n).$$

The envelope functions $E_n(t)$ are of given sine-square form,

$$E_n(t) = \begin{cases} 0 & \text{if } t \leqslant t_{0n}, \\ E_{0n} \sin^2\left[\frac{\pi}{2}\left(\frac{t - t_{0n}}{t_{1n} - t_{0n}}\right)\right] & \text{if } t_{0n} \leqslant t \leqslant t_{1n}, \\ E_{0n} & \text{if } t_{1n} \leqslant t \leqslant t_{2n}, \\ E_{0n} \sin^2\left[\frac{\pi}{2}\left(\frac{t_{3n} - t}{t_{3n} - t_{2n}}\right)\right] & \text{if } t_{2n} \leqslant t \leqslant t_{3n}, \\ 0 & \text{if } t \geqslant t_{3n}, \end{cases} \tag{57.4}$$

each pulse being characterized by a set of 7 adjustable parameters, namely its frequency ω_n, relative phase ϕ_n, maximum field amplitude E_{0n}, together with 4 times determining its shape (origin t_{0n}, rise time $t_{1n} - t_{0n}$, plateau $t_{2n} - t_{1n}$, and extinction time $t_{3n} - t_{2n}$).

All beams are polarized along the same axis. This makes a total of $7 \times 10 = 70$ parameters. It should be emphasized that by considering such a superposition, the goal is not to model a situation that is experimentally feasible, but only to generate a "generic" form of signal $\zeta(t)$.

Using such a generic field has one main disadvantage (in addition to the obvious huge difficulty to minimize over $\mathbb{R}^{70}$!): The optimized laser field that is obtained through minimization is likely to be too difficult to analyze! Indeed, the aim is to provide the experimenter with a well identified field to generate. Obviously, a typical field obtained by such a minimization is likely to be difficult to analyze. Therefore, the other strategy consisting in restrict oneself to a superposition of two, or at most three, different lasers is also convenient. Apart from sticking to experimental reality (for instance a system of two lasers with the same pulsation but with two different phases is nothing else than the same laser with different optical paths), it greatly simplifies the post-treatment of results.

In view of the complexity of the cost function to optimize and of the number of parameters, the domain is a field of choice for stochastic like algorithms. In addition, in view of the fact that disposing of many solutions for the problem is a clear advantage (for some of them may be more difficult to realize in practice than others), the stochastic like algorithms to be used are preferably genetic algorithms or evolutionary strategies (see MICHALEWICZ [1999], BÄCK, FOGEL and MICHALEWICZ [1997]) that both handle a *set* of solutions, than a single solution, at each iteration. In order to accelerate these algorithms, that are known to be very time consuming, hybrid methods based upon a combination of stochastic like algorithms with algorithms of deterministic nature can be useful. For instance, genetic algorithms can be coupled with gradient like algorithms. We refer to the articles mentioned above for details upon the algorithms, the methodology and the results that can be obtained by applying an efficient numerical toolbox to this type of optimal control problems.

References

ABBOTT, P.C. and E.N. MASLEN (1987), Coordinate system and analytic expansions for three-body atomic wavefunctions, I. Partial summation for the Fock expansion in hyperspherical coordinates, *J. Phys. A* **20**, 2043–2075.

AHLRICHS, R. (1974), The influence of electron correlation on reaction energies. The dimerization energies of BH_3 and LiH, *Theoret. Chim. Acta* **35**, 59–68.

AHLRICHS, R., M. HOFFMANN-OSTENHOFF, TH. HOFFMANN-OSTENHOFF and J.D. MORGAN III (1981), Bounds on the decay of electron densities with screening, *Phys. Rev. A* **23**, 2106–2117.

ALEXANDER, S.A., H.J. MONKHORST and K. SZALEWICZ (1986), Random tempering of Gaussian-type geminals, I. Atomic systems, *J. Chem. Phys.* **85**, 5821–5825.

ALEXANDER, S.A., H.J. MONKHORST and K. SZALEWICZ (1987), Random tempering of Gaussian-type geminals, II. Molecular systems, *J. Chem. Phys.* **87**, 3976–3980.

ALLEN, M.P. and D.J. TILDESLEY (1987), *Computer Simulation of Liquids* (Oxford Science Publications).

ALMLÖF, J. and K. FAEGRI JR. (1983), Basis set effects in Hartree–Fock studies on aromatic molecules: Hartree–Fock calculations of properties in benzene and hexafluorobenzene, *J. Chem. Phys.* **79**, 2284–2294.

AMOS, R.D. and J.H. WILLIAMS (1979), Accurate SCF calculations of the multipole moments and polarizabilities of acetylene, ethylene and ethane, *Chem. Phys. Lett.* **66**, 471–474.

ANDERSEN, O.K. (1975), Linear methods in band theory, *Phys. Rev. B* **12**, 3060–3083.

ANDRÉ, J.M., C. BARBIER, V.P. BODART and J. DELHALLE (1987), in: D.S. Chemla and J. Zyss, eds., *Nonlinear Optical Properties of Organic Molecules and Crystals* (Academic Press, New York).

ASHCROFT, N.W. and N.D. MERMIN (1976), *Solid-State Physics* (Saunders, Philadelphia).

ASSION, A. ET AL. (1998), Control of chemical reactions by feedback-optimized phase-shaped femtosecond laser pulses, *Science* **282**, 919–922 .

AUCHMUTY, G. and WENYAO JIA (1994), Convergent iterative methods for the Hartree eigenproblem, *Math. Model. and Numer. Anal.* **28**, 575–610.

AUGER, A., C. DION, A. BEN HAJ YEDDER, A. KELLER, E. CANCÈS, C. LE BRIS and O. ATABEK (2002), Optimal laser control of chemical reactions: methodology and results, *Math. Models Methods Appl. Sci.* **12**, 1281–1315.

AUGER, A., C. DION, A. BEN HAJ YEDDER, A. KELLER, E. CANCÈS, C. LE BRIS and O. ATABEK (2003), *Optimal Laser Control of Orientation: The Train of Kicks*, in preparation.

AYALA, P.Y. and P.B. SCHLEGEL (1997), A combined method for determining reaction paths, minima and transition state geometries, *J. Chem. Phys.* **107**, 375–384.

BACH, V. (1992), Error bound for the Hartree–Fock energy of atoms and molecules, *Comm. Math. Phys.* **147**, 527–548.

BACH, V., E.H. LIEB, M. LOSS and J.P. SOLOVEJ (1994), There are no unfilled shells in unrestricted Hartree–Fock theory, *Phys. Rev. Lett.* **72**, 2981–2983.

BÄCK, TH., D.B. FOGEL and Z. MICHALEWICZ, eds. (1997), *Handbook of Evolutionary Computation*, (Oxford Univ. Press, New York).

BACSKAY, G.B. (1961), A quadratically convergent Hartree–Fock (QC-SCF) method. Application to closed shell systems, *Chem. Phys.* **61** 385–404.

BALL, J.M., J.E. MARSDEN and M. SLEMROD (1982), Controllability for distributed bilinear systems, *SIAM J. Control Optim.* **20**, 575–597.

BANDRAUK, A.D. and S. CHELKOWSKI (2000), Asymmetric electron-nuclear dynamics in two-color laser fields: laser phase directional control of photofragments in H_2^+, *Phys. Rev. Lett.* **84**, 3562–3565.

BARNETT, M.P. (1963), Mechanized molecular calculations. The POLYATOM system, *Rev. Mod. Phys.* **35**, 571–579.

BARONI, S., P. GIANNOZZI and A. TESTA (1987), Green's function approach to linear response in solids, *Phys. Rev. Lett.* **58**, 1861–1864.

BARONI, S., G. PASTORI-PARRARRICINI and G. PEZZICA (1985), Quasiparticle band structure of lithium hybride, *Phys. Rev. B* **32**, 4077–4087.

BARTLETT, R.J. (1989), Coupled-cluster approach to molecular structure and spectra: A step forward predictive Quantum Chemistry, *J. Phys. Chem.* **93**, 1697–1708.

BARTLETT, R.J. and G.D. PURVIS (1979), Molecular hyperpolarizabilities, I. Theoretical calculations including correlation, *Phys. Rev. A* **20**, 1313–1322.

BECKE, A.D. (1988), Density-functional exchange-energy approximation with correct asymptotic behavior, *Phys. Rev. A* **38**, 3098–3100.

BECKE, A.D. (1993), Density-functional thermochemistry, III. The role of exact exchange, *J. Chem. Phys.* **98**, 5648–5652.

BENDER, C.F. and E.R. DAVIDSON (1969), Studies in configuration interaction: the first-row diatomic hydrides, *Phys. Rev.* **183**, 23–30.

BENGURIA, R., H. BREZIS and E.H. LIEB (1981), The Thomas–Fermi–von Weizsäcker theory of atoms and molecules, *Comm. Math. Phys.* **79**, 167–180.

BENGURIA, R. and E.H. LIEB (1985), The most negative ion in the Thomas–Fermi–von Weizsäcker theory of atoms and molecules, *J. Phys. B* **18**, 1045–1059.

BENGURIA, R., and C. YARUR (1990), Sharp condition on the decay of the potential for the absence of a zero-energy ground state of the Schrödinger equation, *J. Phys. A* **23**, 1513–1518.

BEN HAJ YEDDER, A., E. CANCÈS and C. LE BRIS (2001), Optimal laser control of chemical reactions using automatic differentiation, in: G. Corliss, Ch. Faure, A. Griewank, L. Hascoët and U. Naumann, eds., *Proceedings of Automatic Differentiation 2000: From Simulation to Optimization* (Springer-Verlag, New York) 203–213.

BENILAN, PH., J.A. GOLDSTEIN and G.R. RIEDER (1991), The Fermi–Amaldi correction in spin polarized Thomas–Fermi theory, in: C. Bennewitz, ed., *Differential Equations and Mathematical Physics* (Academic Press) 25–37.

BENILAN, PH., J.A. GOLDSTEIN and G.R. RIEDER (1992), A nonlinear elliptic system arising in electron density theory, *Comm. Partial Differential Equations* **17**, 2079–2092.

BERTHIER, G. (1989), The three theorems of the Hartree–Fock theory and their extensions, in: R. Carbó, ed., *Quantum Chemistry – Basic Aspects, Actual Trends. Studies in Physical and Theoretical Chemistry*, Vol. 62 (Elsevier, Amsterdam) 91–102.

BISHOP, D.M. (1987), Nuclear motion and electric hyperpolarizabilities, *J. Chem. Phys.* **86**, 5613–5616.

BISHOP, D.M. (1988), Dispersion formulas for certain nonlinear optical processes, *Phys. Rev. Lett.* **61**, 322–324.

BISHOP, D.M. (1989), General dispersion formulas for molecular third-order nonlinear properties, *J. Chem. Phys.* **90**, 3192–3195.

BISHOP, D.M. (1991), Dispersion formula for the average first hyperpolarizabilities $\overline{\beta}$, *J. Chem. Phys.* **95**, 5489–5489.

BISHOP, D.M., M. CHAILLET, C. LARRIEU and C. POUCHAN (1985), Charge perturbation approach to the calculation of molecular polarizabilities. Application to Li2, *Phys. Rev. A* **31**, 2785–2793.

BISHOP, D.M. and J. PIPIN (1995), Theoretical investigation of the Kerr effect for CH_4, *J. Chem. Phys.* **103**, 4980–4984.

BISHOP, D.M. and S.A. SOLUNAC (1985), Aspects on finite field calculations of polarizabilities for H_2^+, *Chem. Phys. Lett.* **122**, 567–571.

BLANC X. (2000), A mathematical insight into ab initio simulations of solid phase, in: M. Defranceschi, C. Le Bris, eds., *Mathematical Models and Methods for Ab Initio Quantum Chemistry*, Lecture Notes in Chemistry, Vol. 74 (Springer) 133–158.

BLANCHARD, PH. and E. BRUNING (1982), *Variational Methods in Mathematical Physics* (Springer).

BOGAARD, M.P. and B.J. ORR (1975), Electric dipole polarizabilities of atoms and molecules, in: A.D. Buckingham, ed., *Molecular Structures and Properties*, Int. Rev. Sci.: Phys. Chem., Series 2, **2** (Butterworths, London) 149–194.

BOKANOWSKI, O.M. and B. GREBERT (1996a), A decomposition theorem for wavefunctions in molecular quantum chemistry, *Math. Models Methods Appl. Sci.* **6**, 437–466.

BOKANOWSKI, O.M. and B. GREBERT (1996b), Deformations of density functions in molecular quantum chemistry, *J. Math. Phys.* **37**, 1553–1557.

BOKANOWSKI, O.M. and B. GREBERT (1998), Utilization of deformations in molecular quantum chemistry and application to density functional theory, *Int. J. Quant. Chem.* **68**, 221–231.

BONNANS, J.-F., J.-C. GILBERT, C. LEMARÉCHAL and C. SAGASTIZABAL (1997), *Optimisation Numérique: aspects théoriques et pratiques* (Springer, Berlin).

BOPP, F. (1959), Ableitung der Bindungsenergie von N-Teilchen-systemen aus 2-Teilchen Dichtematrizen, *Z. Phys.* **156**, 348–359.

BORN, M. (1951), Kopplung der Elektronen und Kernbewegung in Molekeln und Kristallen, *Gött. Nachr. Akad. Wiss. Math. Phys. Kl.* **6**, 1–3.

BORN, M. and J.R. OPPENHEIMER (1927), Zur Quantentheorie der Molekeln, *Ann. Physik* **84**, 457–484.

BORNEMANN, F.A., P. NETTERSHEIM and Ch. SCHÜTTE (1996), Quantum-classical molecular dynamics as an approximation to full quantum dynamics, *J. Chem. Phys.* **105**, 1074–1083.

BORNEMANN, F.A. and Ch. SCHÜTTE (1998), A mathematical investigation of the Car–Parrinello method, *Numer. Math.* **78**, 359–376.

BOWLER, D. ET AL. (1997), A comparison of linear scaling tight-binding methods, *Model. Simul. Mater. Sci. Engrg.* **5**, 199–202.

BOWLER, D. and M. GILLAN (1999), Density matrices in O(N) electronic structure calculations, *Comp. Phys. Comm.* **120**, 95–108.

BOYS, S.F. (1950), Electronic wavefunctions, I. A general method of calculation for the stationary states of any molecular system, *Proc. Roy. Soc. A* **200**, 542–554.

BRAESS, D. (1995), Asymptotics for the approximation of wave functions by sums of exponential sums, *J. Approx. Theory* **83**, 93–103.

BRUMER, P. and M. SHAPIRO (1995), Laser control of chemical reactions, *Sci. Amer.*, March 1995, 34–39.

BUCKINGHAM, A.D. (1967), Permanent and induced molecular moments and long-range intermolecular forces, *Adv. Chem. Phys.* **12**, 107–142.

BUCKINGHAM, A.D. and B.J. ORR (1967), Molecular hyperpolarizabilities, *Quart. Rev.* **21**, 195–212.

BUESSE, G. and H. KLEINDIENST (1995), Double-linked Hylleraas configuration-interaction calculation for the nonrelativistic ground-state energy of the Be atom, *Phys. Rev. A* **51**, 5019–5020.

BUESSE, G., H. KLEINDIENST and A. LÜCHOW (1998), Non-relativistic energies for the Be atom: double-linked Hylleraas-CI calculation, *Int. J. Quant. Chem.* **66**, 241–247.

BUKOWSKI, R., B. JEZIORSKI, S. RYBAK and K. SZALEWICZ (1995), Second-order correlation energy for H_2O using explicitly correlated Gaussian geminals, *J. Chem. Phys.* **102**, 888–897.

BURDEN, R.L. and J.D. FAIRES (1993), *Numerical Analysis*, 5th edn. (PWS Publishing Company).

BYRON, F.W. and C.J. JOACHAIN (1967), Correlation effects in atoms, II. Angular correlations between electrons, *Phys. Rev.* **157**, 1–6.

CAMMI, R. and J. TOMASI (1994a), Analytical derivatives for molecular solutes, I. Hartree–Fock energy first derivatives with respect to external parameters in the polarizable continuum model, *J. Chem. Phys.* **100**, 7495–7502.

CAMMI, R. and J. TOMASI (1994b), Analytical derivatives for molecular solutes, II. Hartree–Fock energy first derivatives and second derivative with respect to nuclear coordinates, *J. Chem. Phys.* **101**, 3888–3897.

CANCÈS, E. (1998), *Simulation moléculaire et effets d'environnement, une perspective mathématique et numérique*, Ph.D. thesis, Ecole Nationale des Ponts et Chaussées, http://cermics.enpc.fr/theses/98/cances.ps.

CANCÈS, E. (2001a), SCF algorithms for Kohn–Sham models with fractional occupation numbers, *J. Chem. Phys.* **114**, 10616–10623.

CANCÈS, E. (2001b), SCF algorithms for Hartree–Fock electronic calculations, in: M. Defranceschi and C. Le Bris, eds., *Mathematical Models and Methods for Ab Initio Quantum Chemistry*, Lecture Notes in Chem., Vol. 74 (Springer), 17–43.

CANCÈS, E. and C. LE BRIS (1998), On the perturbation method for some nonlinear Quantum Chemistry models, *Math. Models Methods Appl. Sci.* **8**, 55–94.

CANCÈS, E. and C. LE BRIS (1999), On the time-dependent electronic Hartree–Fock equations coupled with a classical nuclear dynamics, *Math. Models Methods Appl. Sci.* **9**, 963–990.

CANCÈS, E. and C. LE BRIS (2000a), On the convergence of SCF algorithms for the Hartree–Fock equations, *Math. Model. Num. Anal.* **34**, 749–774.

CANCÈS, E. and C. LE BRIS (2000b), Can we outperform the DIIS approach for electronic structure calculations, *Int. J. Quant. Chem.* **79**, 82–90.

CANCÈS, E., C. LE BRIS, B. MENNUCCI and J. TOMASI (1999), Integral equation methods for molecular scale calculations in the liquid phase, *Math. Models Methods Appl. Sci.* **9**, 35–44.

CANCÈS, E., C. LE BRIS and M. PILOT (2000), Contrôle optimal bilinéaire sur une équation de Schrödinger, *Comptes Rendus de l'Académie des Sciences*, Série I **330**, 567–571.

CANCÈS, E. and B. MENNUCCI (1998), Analytical derivatives for geometry optimization in solvation continuum models, *J. Chem. Phys.* **109**, 249–259.

CAR, R. and M. PARRINELLO (1985), Unified approach for molecular dynamics and density functional theory, *Phys. Rev. Lett.* **55**, 2471–2474.

CARROLL, D.P., H.J. SILVERSTONE and R.M. METZGER (1979), Piecewise polynomial configuration interaction natural orbital study of $1s^2$ helium, *J. Chem. Phys.* **71**, 4142–4163.

CATLOW, C.R.A., ed. (1994), *Defects and Disorder in Crystalline and Amorphous Solids*, NATO ASI Series C, Vol. 418 (Kluwer, Dordrecht).

CATLOW, C.R. (1998), Computer modelling, in: P.v.R. Schleyer, N.L. Allinger, T. Clark, J. Gasteiger, P.A. Kollman, H.F. Schaeffer III and P.R. Schreiner, eds., *The Encyclopaedia of Computational Chemistry* (Wiley, Chichester).

CATTO, I., C. LE BRIS and P.-L. LIONS (1998), *Mathematical Theory of Thermodynamic Limits: Thomas–Fermi Type Models* (Oxford Univ. Press).

CATTO, I., C. LE BRIS and P.-L. LIONS (2001), On the thermodynamic limit for Hartree–Fock type models, *Ann. Inst. H. Poincaré Anal. Non Linéaire* **18**, 687–760.

CATTO, I., C. LE BRIS and P.-L. LIONS (2002), On some periodic Hartree-type models for crystals, *Ann. Inst. H. Poincaré, Anal. Non Linéaire* **19**, 143–190.

CATTO, I. and P.-L. LIONS (1992), Binding of atoms and stability of molecules in Hartree and Thomas–Fermi type theories. Part I: A necessary and sufficient condition for the stability of general molecular systems, *Comm. Partial Differential Equations* **17**, 1051–1110.

CATTO, I. and P.-L. LIONS (1993a), Binding of atoms and stability of molecules in Hartree and Thomas–Fermi type theories. Part 2: Stability is equivalent to the binding of neutral subsystems, *Comm. Partial Differential Equations* **17**, 305–354.

CATTO, I. and P.-L. LIONS (1993b), Binding of atoms and stability of molecules in Hartree and Thomas–Fermi type theories. Part 3: Binding of neutral subsystems, *Comm. Partial Differential Equations* **18**, 381–429.

CATTO, I. and P.-L. LIONS (1993c), Binding of atoms and stability of molecules in Hartree and Thomas–Fermi type theories. Part 4: Binding of neutral systems for the Hartree model, *Comm. Partial Differential Equations* **18**, 1149–1159.

CENCEK, W. and W. KUTZELNIGG (1996), Accurate relativistic energies of one and two electron systems using Gaussian wave functions, *J. Chem. Phys.* **105**, 5878–5885.

CENCEK, W. and W. KUTZELNIGG (1997), Accurate adiabatic connection for the hydrogen molecule using the Gorn–Handy formula, *Chem. Phys. Lett.* **266**, 383–387.

CENCEK, W. and J. RYCHLEWSKI (1993), Many-electron explicitly correlated Gaussian functions, I. General Theory and test results, *J. Chem. Phys.* **98**, 1252–1261.

CEPERLEY, D.M. and B.J. ALDER (1980), Ground state of the electron gas by a stochastic method, *Phys. Rev. Lett.* **45**, 566–569.

CHABAN, G., M.W. SCHMIDT and M.S. GORDON (1997), Approximate second order method for orbital optimization of SCF and MCSCF wavefunctions, *Theor. Chem. Acc.* **97**, 88–95.

CHADAM, J.M. and R.T. GLASSEY (1975), Global existence of solutions to the Cauchy problem for time-dependent Hartree equations, *J. Math. Phys.* **16**, 1122–1230.

CHALLACOMBE, M. (2000), Linear scaling computation of the Fock matrix, V. Hierarchical cubature for numerical integration of the exchange-correlation matrix, *J. Chem. Phys.* **113**, 10037–10043.

CHRISTIANSEN, P.A. and E.A. MCCULLOUGH Jr. (1977), Numerical coupled Hartree–Fock parallel polarizabilities for FH and CO, *Chem. Phys. Lett.* **51**, 468–472.

CIOSLOWSKI, J. and K. PERNAL (1999), Topology of electron–electron interaction sin atoms and molecules, III. Morphology of electron intracule density in two $^1\Sigma_g^+$ states of the hydrogen molecule, *J. Chem. Phys.* **111**, 3401–3409.

ČIŽEK, J. (1966), On the correlation problem in atomic and molecular systems. Calculation of wavefunction components in Ursell-type expansion using quantum-field theoretical methods, *J. Chem. Phys.* **45**, 4256–4266.

ČIŽEK, J. (1969), On the use of the cluster expansion and the technique of diagrams in calculations of correlation effects in atoms and molecules, *Adv. Chem. Phys.* **14**, 35–89.

CLEMENTI, E. and D.R. DAVIS (1966), Electronic structure of large molecular systems, *J. Comp. Phys.* **1**, 223–244.

COESTER, F. and H. KÜMMEL (1960), Short-range correlations in nuclear wave functions, *Nucl. Phys.* **17**, 477–485.

COHEN, L. and C. FRESHBERG (1976), Hierarchy equations for reduced density matrices, *Phys. Rev.* **A13**, 927–930.

COHEN, H.D. and C.C.J. ROOTHAAN (1965), Electric dipole polarizability of atoms by the Hartree–Fock method, I. Theory for closed-shell systems, *J. Chem. Phys.* **43**, S34-S39.

COHEN-TANNOUDJI, C., J. DUPONT-ROC and G. GRYNBERG (1988), *Processus d'interaction entre photons et atomes* (InterEditions/Editions du CNRS, Paris).

COLEMAN, A.J. (1963), Structure of Fermion density matrices, *Rev. Mod. Phys.* **35**, 668–689.

COLEMAN, A.J. (2002), Kummer variety, geometry of N-representability and phase transitions, *Phys. Rev. A* **66**, 22503.

COLEMAN, A.J. and V.I. YUKALOV (2000), *Reduced Density Matrices*, Lecture Notes in Chemistry, Vol. 72 (Springer, Berlin).

COULSON, C.A. (1935), The electronic structure of H_3^+, *Proc. Cambridge Phil. Soc.* **31**, 244.

COULSON, C.A. (1937), The evolution of certain integrals occurring in studies of electronic structures, *Proc. Cambridge. Phil. Soc.* **33**, 104.

COULSON, C.A. (1938), Self-consistent field for molecular hydrogen, *Proc. Camb. Phil. Soc.* **34**, 204–212.

CRAWFORD, T.D. and H.F. SCHAEFER III (2000), An introduction to coupled cluster theory for computational chemists, in: K.B. Lipkowitz and D.B. Boyd, eds., *Rev. Comput. Chemistry*, Vol. 14 (VCH, Weinheim) 33–136.

CYCON, H.L., R.G. FROESE, W. KIRSCH and B. SIMON (1987), *Schrödinger Operators with Applications to Quantum Mechanics and Global Geometry* (Springer, New York).

DANIELS, A. and G. SCUSERIA (1999), What is the best alternative to diagonalization of the Hamiltonian in large scale semiempirical calculations?, *J. Chem. Phys.* **110**, 1321–1328.

DEFRANCESCHI, M. and P. FISCHER (1998), Numerical solution of the Schrödinger equation in a wavelet basis for hydrogen-like atoms, *SIAM J. Numer. Anal.* **35**, 1–12.

DEFRANCESCHI, M. and C. LE BRIS (1997), Computing a molecule: a mathematical viewpoint, *J. Math. Chem.* **21**, 1–30.

DEFRANCESCHI, M. and C. LE BRIS (1999), Computing a molecule in its environment: a mathematical viewpoint, *Int. J. Quant. Chem.* **71**, 257–250.

DEFRANCESCHI, M. and C. LE BRIS, eds. (2000), *Mathematical Models and Methods for Ab Initio Quantum Chemistry*, Lecture Notes in Chemistry, Vol. 74 (Springer).

DEMMEL, J.W. (1997), *Applied Numerical Linear Algebra* (SIAM Press, Philadelphia, PA).

DEUFLHARD, P. ET AL., eds. (1999), *Computational Molecular Dynamics: Challenges, Methods, Ideas*, Lecture Notes in Comput. Sci. Engrg. (Springer).

DI BELLA, S., T.J. MARKS and M.A. RATNER (1994), Environmental effects on nonlinear optical chromophore, Performance. Calculation of molecular quadratic hyperpolarizabilities in solvating media, *J. Amer. Chem. Soc.* **116**, 4440–4445.

DIERCKSEN, G.H.F., V. KELLÖ and A.J. SADLEJ (1983), Perturbation theory of the electron correlation effect for atomic and molecular properties, VII. Complete fourth-order MBPT study of the dipole moment and dipole polarizability of H_2O, *J. Chem. Phys.* **79**, 2918–2923.

DIERCKSEN, G.H.F. and A.J. SADLEJ (1981), Perturbation theory of the electron correlation effect for atomic and molecular properties. Second and third-order correlation corrections to molecular dipole moments and polarizabilities, *J. Chem. Phys.* **75**, 1253–1266.

DION, C., A. BEN HAJ YEDDER, E. CANCÈS, C. LE BRIS, A. KELLER and O. ATABEK (2002), Optimal laser control of orientation: the kicked molecule, *Physical Rev. A* **65**, 063408.

DION, C., A. KELLER and O. ATABEK (1999a), Laser-induced alignment dynamics of HCN: Roles of the permanent dipole moment and the polarizability, *Phys. Rev. A* **59**, 1382–1391.

DION, C., A. KELLER and O. ATABEK (1999b), Two-frequency IR laser orientation of polar molecules. numerical simulations for HCN, *Chem. Phys. Lett.* **302**, 215–223.

DION, C., A. KELLER and O. ATABEK (2001), Orienting molecules using half-cycle pulses, *Eur. Phys. J. D* **14**, 249–255.

DIRAC, P.A.M. (1926), Quantum mechanics of many electron systems, *Proc. Roy. Soc. London A* **113**, 621–641.

DOUADY, J., Y. ELLINGER, R. SUBRA and B. LEVY (1980), Exponential transformation of molecular orbitals: a quadratically convergent SCF procedure, I. General formulation and application to closed-shell ground states, *J. Chem. Phys.* **72**, 1452–1462.

DOVESI, R., R. ORLANDO, C. ROETTI, C. PISANI and V.R. SAUNDERS (2000), The periodic Hartree–Fock method and its implementation in the Crystal code, *Phys. Stat. Sol. (b)* **217**, 63–88.

DRAKE, G.W.F. (1987), New variational techniques for the states of Helium, *Phys. Rev. Lett.* **59**, 1549–1552.

DRAKE, G.W.F. and Z.-C. YAN (1994), Variational eigenvalues for the S states of Helium, *Chem. Phys. Lett.* **229**, 486–490.

DREIZLER, R.M. and E.K.U. GROSS (1990), *Density Functional Theory* (Springer).

DUNNING JR., T.H. (1989), Gaussian basis sets for use in correlated molecular calculations, I. The atoms boron through neon and hydrogen, *J. Chem. Phys.* **90**, 1007–1023.

DYKSTRA, C.E. (1993), Electrostatic interaction potentials in molecular force fields, *Chem. Rev.* **93**, 2339–2353.

DYKSTRA, C.E. and P.G. JASIEN (1984), Derivative Hartree–Fock theory to all orders, *Chem. Phys. Lett.* **109**, 388–393.

ENGLISCH, H. and R. ENGLISH (1983), Hohenberg–Kohn theorem and non-V-representable densities, *Physica A* **121**, 253–268.

ERNENWEIN, R., M.-M. ROHMER and M. BÉNARD (1990), A program system for ab initio MO calculations on vector and parallel processing machines, I. Evaluation of integrals, *Comput. Phys. Comm.* **58**, 305–328.

ESCHRIG, H. (1996), *The Fundamentals of Density Functional Theory* (B.G. Teubner Verlagsgesellschaft).

FELLER, D.F. and K. RUEDENBERG (1979), Systematic approach to extended even-tempered orbital bases for atomic and molecular calculations, *Theoret. Chim. Acta* **52**, 231–251.

FEYNMAN, R.P. (1948), Space-time approach to non-relativistic quantum mechanics, *Rev. Mod. Phys.* **20**, 367–387.

FIELD, M.J. (2003), Simulating chemical reactions in complex systems, in: P.G. Ciarlet and C. Le Bris, eds., *Computational Chemistry*, Handbook of Numerical Analysis, Vol. X (Elsevier, Amsterdam), 667–697.

FISCHER, T.H. and J. ALMLÖF (1992), General methods for geometry and wave function optimization, *J. Phys. Chem.* **96**, 9768–9774.

FISCHER, P. and M. DEFRANCESCHI (1994), The wavelet transform: a new mathematical tool for Quantum Chemistry, in: E.S. Kryachko and J.L. Calais, eds., *Conceptual Trends in Quantum Chemistry* (Kluwer), 227–247.

FLETCHER, G.C. (1952), Density of states curves for the 3d electrons in Nickel, *Proc. Phys. Soc. London A* **65**, 192–202.

FLETCHER, G.C. (1971), *Band Theory of Solids* (North-Holland, Amsterdam).

FOCK, V. (1930), Näherungsmethode zur Lösung des quantenmechanischen Mehrkörperproblems, *Z. Physik* **61**, 126–148.

FOCK, V. (1958), On Schrodinger equation for the Helium atom, *K. Norske Vidensk. Selsk. Forhandl.* **31**, 138–152.

FORESMAN, J.B. and A. FRISCH (1996), *Exploring Chemistry with Electronic Structure Methods*, 2nd edn. (Gaussian, Pittsburgh).

FORTIN, M. and R. GLOWINSKI (1983), *Augmented Lagrangian Methods: Applications to the Numerical Solution of Boundary-Value Problems* (North-Holland, Amsterdam).

FOWLER, R.H. (1933), Note on some electronic properties of conductors and insulators, *Proc. Roy. Soc. London A* **141**, 56–71.

FRANKEN, P.A., A.E. HILL, C.W. PETERS and C.W. WEINREICH (1961), Generation of optical harmonics, *Phys. Rev. Lett.* **7**, 118–119.

FRANKEN, P.A. and J.F. WARD (1963), Optical harmonics and nonlinear phenomena, *Rev. Mod. Phys.* **35**, 23–39.

FRANKOWSKI, K. and C.L. PEKERIS (1966), Logarithmic terms in the wave functions of the ground state of two-electron atoms, *Phys. Rev.* **146**, 46–49.

FRENKEL, D. and B. SMIT (1996), *Understanding Molecular Simulation* (Academic Press, San Diego).

FREUND, D.E., B.D. HUXTABLE and J.D. MORGAN III (1984), Variational calculation of the helium isoelectronic sequence, *Phys. Rev. A* **29**, 980–982.

FRIEDMAN, B. (1956), *Principles and Techniques of Applied Mathematics* (Wiley, New York).

FRIESECKE, G. (2002), The multiconfiguration equations for atoms and molecules: charge quantization and existence of solutions, *Arch. Rat. Mech. Anal.*, to appear.

GALE, J., C.R.A. CATLOW and W.C. MACKRODT (1992), Determination of interatomic potential for alumina, *Modell. Simul. Mater. Sci. Engrg.* **1**, 73–81.

GALLI, G. (2000), Large scale electronic structure calculations using linear scaling methods, *Phys. Stat. Sol (b)* **217**, 231–249.

GARROD, C. and M.V. MILHAILOVIĆ (1975), The variational approach to the two-body density matrix, *J. Math. Phys.* **16**, 868–874.

GASPAR, R. (1954) Über die Approximation des Hartree-Fockschen Potentials durch eine universelle Potential Funktion, *Acta Phys. Hungar.* **3**, 263–286.

GIBSON, A., R. HAYDOCK and J.P. LAFEMINA (1993), *Ab initio* electronic-structure computations with recursion methods, *Phys. Rev. B* **47**, 9229–9237.

GILL, P.M.W. (1994), Molecular integrals over Gaussian basis functions, in: P.-O. Löwdin, J.R. Sabin and M.C. Zerner, eds., *Advances in Quantum Chemistry*, Vol. 25 (Academic Press, San Diego) 141–205.

GILLAN, M.J. (1991), Calculating the properties of materials from scratch, in: M. Meyer and V. Pontikis, eds., *Computer Simulation*, NATO ASI Series E, Vol. 205 (Computer Simulations in Materials Science), 257–281.

GODBY, R.W., M. SCHLÜTER and L.J. SHAM (1986), Accurate exchange-correlation potential for silicon and its discontinuity on addition of an electron, *Phys. Rev. Lett.* **56**, 2415–2418.

GODBY, R.W., M. SCHLÜTER and L.J. SHAM (1988), Self-energy operators and exchange-correlation potentials in semi-conductors, *Phys. Rev. B* **37**, 10159–10175.

GOEDECKER, S. (1998), *Wavelets and their Application for the Solution of Partial Differential Equations* (Presses Polytechniques Universitaires et Romandes, Lausanne).

GOEDECKER, S. (1999), Linear scaling electronic structure methods, *Rev. Modern Phys.* **71**, 1085–1123.

GOEDECKER, S. and O. IVANOV (1998), Linear Scaling solution of the classical Coulomb problem using wavelets, *Sol. State Comm.* **105**, 665–670.

GOEDECKER, S. and C.J. UMRIGAR (1998), Natural orbital functional for the many-electron problem, *Phys. Rev. Lett.* **81**, 866–869.

GOLDMAN, S.P. (1994), Modified configuration interaction method for accurate calculations with small basis sets, *Phys. Rev. Lett.* **73**, 2547–2550.

GOLDMAN, S.P. (1998), Uncoupling correlated calculations in atomic physics: very high accuracy and ease, *Phys. Rev. A* **57**, R677–R680.

GOLDSTEIN, J.A., G.R. GOLDSTEIN and WENYIAO JIA (1995), Thomas–Fermi theory with magnetic fields and the Fermi–Amaldi correction, *Differential Integral Equations* **8**, 1305–1316.

GOLDSTEIN, J.A. and G.R. RIEDER (1991), Thomas–Fermi theory with an external magnetic field, *J. Math. Phys.* **32**, 2907–2917.

GONZE, X. (1995), Adiabatic density-functional perturbation theory, *Phys. Rev. A* **52**, 1096–1114.

GONZE, X. and J.-P. VIGNERON (1989), Density-functional approach to nonlinear response coefficients of solids, *Phys. Rev.* **39**, 13120–13128.

GREENGARD, R. and V. ROKHLIN (1997), A new version of the fast multipole method for the Laplace equation in three dimensions, *Acta Numerica* **6**, 229–269.

GUNNARSON, O. and B.I. LUNDQVIST (1976), Exchange and correlation in atoms, molecules, and solids by the spin-density-functional formalism, *Phys. Rev. B* **13**, 4274–4298.

HAGEDORN, G.A. (1980). A time-dependent Born–Oppenheimer approximation, *Commun. Math. Phys.* **77**, 77–93.

HAGEDORN, G.A. (1996), Crossing the interface between Chemistry and Mathematics, *Notices of the AMS*, March 1996.

HAILE, J.M. (1992), *Molecular Dynamics Simulations* (Wiley, New York).

HAIRER, E., CH. LUBICH and G. WANNER (2002), *Geometric Numerical Integration. Structure-Preserving Algorithms for Ordinary Differential Equations* (Springer, Berlin).

HAIRER, E., S.P. NORSETT and G. WANNER (1993), *Solving Ordinary Differential Equations I*, 2nd edn. (Springer, Berlin).

HAIRER, E. and G. WANNER (1996), *Solving Ordinary Differential Equations II*, 2nd edn. (Springer, Berlin).

HALL, G.G. (1951), The molecular orbital theory of chemical valency, VIII. A method of calculating ionisation potential, *Proc. Roy. Soc. A* **205**, 541–552.

HARTREE, D.R. (1928), The wave mechanics of an atom with the non-Coulomb central field, I. Theory and methods, *Proc. Cambridge Phil. Soc.* **24**, 89.

HARTREE, D.R. (1957), *The Calculation of Atomic Structures* (Wiley, New York).

HARTREE, D.R., W. HARTREE and B. SWIRLES (1939), Self-consistent field, including exchange and superposition of configurations, with some results for oxygen, *Phil. Trans. Roy. Soc. A* **238**, 229–247.

HEDIN, L. (1965), New method for calculating the one-particle Green's function with application to the electron-gas problem, *Phys. Rev.* **139**, A796–A823.

HEDIN, L. and S. LUNDQVIST (1971), Explicit local exchange-correlation potential, *J. Phys. C* **4**, 2064–2083.

HEHRE, W.J., J.A. POPLE ET AL. (1973), GAUSSIAN 70: *Ab initio* SCF calculations on organic molecules, QCPE **236**.

HEHRE, W.J., L. RADOM, P.v.R. SCHLEYER and J.A. POPLE (1986), *Ab Initio Molecular Orbital Theory* (Wiley).

HEINE, V. (1970), The pseudopotential concept, *Sol. State Phys.* **24**, 1–36.

HEITLER, W. and F. LONDON (1927), Wechselwirkung neutraler Atome und homöopolare Bindung nach der Quantenmechanik, *Z. Physik* **44**, 455–472.

HELGAKER, T., W. KLOPPER, H. KOCH and J. NOGA (1997), Basis-set convergence of correlated calculations on water, *J. Chem. Phys.* **106**, 9639–9646.

HILL, S.C. (1985), Rates of convergence and error estimation formulas for the Rayleigh–Ritz variational method, *J. Chem. Phys.* **83**, 1173–1196.

HILL, R.N. (1995), Dependence of the rate of convergence of the Rayleigh–Ritz method on a nonlinear parameter, *Phys. Rev. A* **51**, 4433–4471.

HOHENBERG, P. and W. KOHN (1964), Inhomogeneous electron gas, *Phys. Rev. B* **136**, 864–871.

HOKI, K. and Y. FUJIMURA (2001), Quantum control of alignment and orientation of molecules by optimized laser pulses, *Chem. Phys.* **267**, 187–193.

HONIG, B. and A. NICHOLLS (1995), Classical electrostatics in Biology and Chemistry, *Science* **268**, 1144.

HOOVER, W.H. (1986), Canonical dynamics: equilibrium phase-space distributions, *Phys. Rev. A* **31**, 1695–1697.

HÜCKEL, E. (1931a), Quantentheoretische Beiträge zum Benzolproblem, I. Die Elecktronenkonfiguration des Benzols und vervandter Verbindungen, *Z. Phys.* **70**, 204–286.

HÜCKEL, E. (1931b), Quantentheoretische Beiträge zum Benzolproblem, II. Quantentheorie der induzierten Polaritäten, *Z. Phys.* **72**, 310–337.

HÜCKEL, E. (1932), Quantentheoretische Beiträge zum Benzolproblem, III. Um aromatischen and ungesättigten Verbindungen, *Z. Phys.* **76**, 628–648.

HUND, F. (1928), Über Zuordnungfragen insbesondere über die Zuordnung von Multiplettermen zu Seriengrenzen, *Z. Phys.* **52**, 601–609.

HUZINAGA, S. (1965), Gaussian-type functions for polyatomic systems I, *J. Chem. Phys.* **42**, 1293–1302.

HYBERTSEN, M.S. and S.G. LOUIE (1985), First-principles theory of quasiparticules: calculation of band gaps in semiconductors and insulators, *Phys. Rev. Lett.* **55**, 1418–1420.

HYBERTSEN, M.S. and S.G. LOUIE (1986), Electron correlation in semiconductors and insulators: band gaps and quasiparticle energies, *Phys. Rev. B* **34**, 5390–5413.

HYLLERAAS, E.A. (1928), Über den Grundzustand des Helium-Atoms, *Z. Phys.* **48**, 469–494.

HYLLERAAS, E.A. (1929), Neue Berechnung der Energie des Heliums im Grundzustande, sowie des tiefsten Terms von Ortho-Helium, *Z. Phys.* **54**, 347–366.

HYLLERAAS, E.A. (1930), Über den Grundterm der Zweielektronenprobleme von H^-, He, Li^+, be^{++} usw., *Z. Phys.* **65**, 209–225.

IORIO JR., R.J. and R.D. MARCHESIN (1984), On the Schrödinger equation with time-dependent electric fields, *Proc. Roy. Soc. Edinburgh* **96**, 117–134.

JAMES, H.M. and A.S. COOLIDGE (1933), The ground state of the hydrogen molecule, *J. Chem. Phys.* **1**, 825–835.

JAY, L.O., H. KIM, Y. SAAD and J.R. CHELIKOWSKI (1999), Electronic structure calculations for plane wave codes without diagonalization, *Comp. Phys. Comm.* **118**, 21–30.

JONES, R.O. and O. GUNNARSSON (1989), The density functional formalism, its applications and prospects, *Rev. Mod. Phys.* **61**, 689–746.

JUDSON, R.S., K.K. LEHMANN, H. RABITZ and W.S. WARREN (1990), Optimal design of external fields for controlling molecular motion: Application to rotation, *J. Molecular structure* **223**, 425–456.

JUDSON, R.S. and H. RABITZ (1992), Teaching lasers to control molecules, *Phys. Rev. Lett.* **68**, 1500–1503.

KANAI, T. and H. SAKAI (2001), Numerical simulation of molecular orientation using strong, nonresonant, two-color laser fields, *J. Chem. Phys.* **115**, 5492–5497.

KATO, T. (1957), On the eigenfunctions of many-particle systems in quantum mechanics, *Comm. Pure Appl. Math.* **10**, 151–177.

KATO, T. (1980), *Perturbation Theory for Linear Operators* (Springer, Berlin).

KELLEY, C.T. (1999), *Iterative Methods for Optimization* (SIAM, Philadelphia).

KESTNER, N.R. and J.E. COMBIRAZA (1999), Basis set superposition errors: theory and practice, in: K.B. Lipkowitz and D.B. Boyd, eds., *Reviews in Computational Chemistry*, Vol. 13 (VCH, New York) 99–132.

KINOSHITA, T. (1957), Ground state of the helium atom, *Phys. Rev.* **105**, 1490–1502.

KIRTMAN, B. and M.J. HASAN (1992), Linear and nonlinear polarizabilities of a trans-polysilane from ab initio oligomer calculations, *Chem. Phys.* **96**, 470–479.

KLAHN, B. and W.A. BINGEL (1977), The convergence of the Rayleigh–Ritz method in Quantum Chemistry, *Theor. Chim. Acta* **44**, 26–43.

KLAHN, B. and J.D. MORGAN III (1984), Rates of convergence of variational calculations of expectation values, *J. Chem. Phys.* **81**, 410–433.

KLEINDIENST, H. and R. EMRICH (1990), The atomic three-body problem. An accurate lower bound calculation using wave functions with logarithmic terms, *Int. J. Quant. Chem.* **37**, 257–270.

KLEINDIENST, H., A. LÜCHOW and H.P. MERCKENS (1994), Accurate upper and lower bounds for some excited S-states of the He atom, *Chem. Phys. Lett.* **218**, 441–444.

KLOPPER, W. (1991), Orbital-invariant formulation of the MP2-R12 method, *Chem. Phys. Lett.* **186**, 583–585.

KLOPPER, W. (1998), r_{12}-Dependent wavefunctions, in: P.v.R. Schleyer et al., eds., *Encyclopedia of Computational Chemistry* (Wiley, Chichester) 2351–2379.

KLOPPER, W. and W. KUTZELNIGG (1986), Gaussian basis sets and the nuclear cusp problem, *J. Mol. Struct. Theochem.* **135**, 339–356.

KLOPPER, W. and W. KUTZELNIGG (1987), Møller–Plesset calculations taking care of the correlation cusp, *Chem. Phys. Lett.* **134**, 17–22.

KLOPPER, W. and W. KUTZELNIGG (1991), Wavefunctions with terms linear in the interelectronic coordinates to take care of the correlation cusp, II. Second-order Møller–Plesset (MP2-R12) calculations on molecules of first row atoms, *J. Chem. Phys.* **94**, 2020–2030.

KLOPPER, W. and C.C.M. SAMSON (2002), Explicit correlated second-order Møller–Plesset methods for auxiliary basis sets, *J. Chem. Phys.* **116**, 6397–6410.

KOBAYASHI, M. (1998), Mathematics make molecules dance, *SIAM News* **31**, 1–5.

KOHN, W. (1986), Density-functional theory for excited states in a quasi-local-density approximation, *Phys. Rev. A* **34**, 737–741.

KOHN, W. (1996), Density functional and density matrix method scaling linearly with the number of atoms, *Phys. Rev. Lett.* **76**, 3168–3171.

KOHN, W. (1999), Nobel Lecture: Electronic structure of matter-wave functions and density functionals, *Rev. Mod. Phys.* **71**, 1253–1266.

KOHN, W. and N. ROSTOKER (1954), Solution of the Schrödinger equation in periodic lattices with an application to metallic Lithium, *Phys. Rev.* **94**, 1111–1120.

KOHN, W. and L.J. SHAM (1965), Self-consistent equations including exchange and correlation effects, *Phys. Rev.* **140**, A1133–A1138.

KOLOS, W. and C.C.J. ROOTHAAN (1960), Accurate electronic wave functions for the H_2 molecule, *Rev. Mod. Phys.* **32**, 219–232.

KOLOS, W. and J. RYCHLEWSKI (1993), Improved theoretical dissociation energy and ionization potential for the ground-state of the hydrogen molecule, *J. Chem. Phys.* **98**, 3960–3967.

KOLOS, W. and L. WOLNIEWICZ (1964a), Accurate adiabatic treatment of the ground-state of the hydrogen molecule, *J. Chem. Phys.* **41**, 3663–3673.

KOLOS, W. and L. WOLNIEWICZ (1964b), Accurate computation of vibronic energies and some expectation values for H_2, D_2 and T_2, *J. Chem. Phys.* **41**, 3674–3678.

KOLOS, W. and L. WOLNIEWICZ (1968), Improved theoretical ground-state energy of the hydrogen molecule, *J. Chem. Phys.* **49**, 404–410.

KOOPMANS, T. (1933), Über die Zuordnung von Wellenfunktionen und Eigenwerten zu den einzelnen Elektronen eines Atoms, *Physika* **1**, 104–113.

KORRINGA, J. (1947), On the calculation of the energy of a Bloch wave in a metal, *Physica* **13**, 392–400.

KOUTECKÝ, J. (1957), Contribution to the theory of surface electronic states in the one-electron approximation, *Phys. Rev.* **108**, 13–18.

KOUTECKÝ, J. and V. BONACIC (1971), On convergence difficulties in the iterative Hartree–Fock procedure, *J. Chem. Phys.* **55**, 2408–2413.

KUDIN, K. and G.E. SCUSERIA (1998), A fast multipole algorithm for the efficient treatment of the Coulomb problem in electronic structure calculations of periodic systems with Gaussian orbitals, *Chem. Phys. Lett.* **289**, 611–616.

KUDIN, K., G.E. SCUSERIA and E. CANCÈS (2002), A black-box self-consistent field convergence algorithm: one step closer, *J. Chem. Phys.* **116**, 8255–8261.

KUMMER, H. (1977), About the relationship between some necessary conditions for N-representability, *Int. J. Quant. Chem.* **12**, 1033–1038.

KUTZELNIGG, W. (1979), Generalized k-particle Brillouin conditions and their use for the construction of correlated electronic wavefunctions, *Chem. Phys. Lett.* **64**, 383–387.

KUTZELNIGG, W. (1980), New derivation and a k-particle generalization of SCF-type theories, *Int. J. Quant. Chem.* **18**, 3–10.

KUTZELNIGG, W. (1982), Quantum chemistry in Fock space, I. The universal wave and energy operators, *J. Chem. Phys.* **77**, 3081–3097.

KUTZELNIGG, W. (1989a), Quantum chemistry in Fock space, in: D. Mukherjee, ed., *Aspects of Many-Body Effects in Molecules and Extended Systems*, Lecture Notes in Chemistry, Vol. 50 (Springer, Berlin).

KUTZELNIGG, W. (1989b), Convergence expansions in Gaussian basis, in: *Strategies and Applications in Quantum Chemistry*, A Tribute to G. Berthier, M. Defranceshi and Y. Ellinger, eds. (Kluwer, Dordrecht), 79–102.

KUTZELNIGG, W. (1991), Error analysis and improvements of coupled-cluster theory, *Theoret. Chim. Acta* **80**, 349–386.

KUTZELNIGG, W. (1994), Theory of the expansion of wave functions in a Gaussian basis, *Int. J. Quant. Chem.* **51**, 447–463.

KUTZELNIGG, W. (1998), Almost variational coupled-cluster theory, *Mol. Phys.* **94**, 65–72.

KUTZELNIGG, W. and P.v. HERIGONTE (1999), Electron correlation at the dawn of the 21st century, *Adv. Quant. Chem.* **36**, 185–229.

KUTZELNIGG, W. and W. KLOPPER (1991), Wave functions with terms linear in the interelectronic coordinates to take care of the correlation cusp, I. General theory, *J. Chem. Phys.* **94**, 1985–2001.

KUTZELNIGG, W. and S. KOCH (1983), Quantum Chemistry in Fock space, I & II, *J. Chem. Phys.* **79**, 4315–4335.

KUTZELNIGG, W. and J.D. MORGAN III (1992), Rate of convergence of the partial-wave expansions of atomic correlation energies, *J. Chem. Phys.* **96**, 4484–4508.

KUTZELNIGG, W. and J.D. MORGAN III (1996), Hund's rules, *Z. Phys. D* **36**, 197–214.

KUTZELNIGG, W. and D. MUKHERJEE (1997), Normal order and extended Wick theorem for a multiconfiguration reference wave function, *J. Chem. Phys.* **107**, 432–449.

KUTZELNIGG, W. and D. MUKHERJEE (1999), Cumulant expansion of the reduced density matrices, *J. Chem. Phys.* **110**, 2800–2809.

KUTZLER, F.W. and G.S. PAINTER (1987), Energies of atoms with nonspherical charge densities calculated with non-local density-functional theory, *Phys. Rev. Lett.* **59**, 1285–1288.

LAKIN, W. (1965), On singularities in eigenfunctions, *J. Chem. Phys.* **43**, 2954–2956.

LANDAU, L.D. (1957a), Theory of a Fermi liquid, *Sov. Phys. JETP* **3**, 920–925.

LANDAU, L.D. (1957b), Oscillations of a Fermi liquid, *Sov. Phys. JETP* **5**, 101–108.

LANDAU, L.D. and E. LIFCHITZ (1977), *Quantum Mechanics* (Pergamon Press, Oxford).

LARSEN, J.J. ET AL. (1999), Controlling the branching ratio of photodissociation using aligned molecules, *Phys. Rev. Lett.* **83**, 1123–1126.

LAYZER, D. (1959), On the screening theory of atomic spectra, *Ann. Phys. (NY)* **8**, 271–296.

LAZZERETTI, P., M. MALAGOLI and R. ZANASI (1996), Coupled Hartree–Foch approach to electric hyperpolarizability tensors in benzene, in: Y. Ellinger and M. Defranceschi, eds., *Strategies and Applications in Quantum Chemistry* (Kluwer, Dordrecht).

LAZZERETTI, P. and R. ZANASI (1979), On the use of symmetry in first-order perturbed Hartree–Fock theory. II, *Int. J. Quant. Chem.* **15**, 645–653.

LE BRIS, C. (1993a), *Quelques problèmes mathématiques en chimie quantique moléculaire*, PhD thesis, Ècole Polytechnique.

LE BRIS, C. (1993b), Some results on the Thomas–Fermi–Dirac–von Weizsäcker model, *Differential Integral Equations* **6**, 337–353.

LE BRIS, C. (1994), A general approach for multiconfiguration methods in quantum molecular chemistry, *Ann. Inst. H. Poincaré Anal. Non Linéaire* **11**, 441–484.

LE BRIS, C. (1995), On the spin polarized Thomas–Fermi model with the Fermi–Amaldi correction, *Nonlinear Anal.* **25**, 669–679.

LE BRIS, C. (2000), Control theory applied to quantum chemistry: Some tracks, in: *8th International Conference on Control of Systems Governed by PDEs*, ESAIM PROC 2000, 77–94.

LEVINE, I.N (1991), *Quantum Chemistry*, 4th edn. (Prentice Hall).

LEVINE, B.F. and C.G. BETHEA (1975), Second and third order hyperpolarizabilities of organic molecules, *J. Chem. Phys.* **63**, 2666–2682.

LEVINE, B.F. and C.G. BETHEA (1978), Ultraviolet dispersion of the donor-acceptor charge transfer contribution to the second-order hyperpolarizability, *J. Chem. Phys.* **69**, 5240–5245.

LEVINE, B.F. and P. SOVEN (1984), Time-dependent local-density theory of dielectric effects in small molecules, *Phys. Rev. A* **29**, 625–635.

LEWIN, M. (2002), The multiconfiguration methods in Quantum Chemistry: Palais–Smale condition and existence of minimizers, *C. R. Acad. Sci. Paris Ser. I* **334**, 299–304.

LI, X.P., R.W. NUMES and D. VANDERBILT (1993), Density matrix electronic structure method with linear system-size scaling, *Phys. Rev B.* **47**, 10891–10894.

LIEB, E.H. (1981), Thomas–Fermi and related theories of atoms and molecules, *Rev. Mod. Phys.* **53**, 603–641.

LIEB, E.H. (1983), Density functionals for Coulomb systems, *Int. J. Quant. Chem.* **24**, 243–277.

LIEB, E.H. (1984), Bound of the maximum negative ionization of atoms and molecules, *Phys. Rev. A* **29**, 3018–3028.

LIEB, E.H. (1985), Density functionals for Coulomb systems, in: R.M. Drieizler and J. da Providencia, eds., *Density Functional Methods in Physics* (Plenum, New York) 31–80.

LIEB, E.H. and M. LOSS (2001), *Analysis*, 2nd edn., Grad. Stud. Math., Vol. 14 (American Mathematical Society).

LIEB, E.H. and B. SIMON (1977a), The Hartree–Fock theory for Coulomb systems, *Comm. Math. Phys.* **53**, 185–194.

LIEB, E.H. and B. SIMON (1977b), The Thomas–Fermi theory of atoms, molecules and solids, *Adv. Math.* **23**, 22–116.

LIONS, P.-L. (1985), Hartree–Fock and related equations, in: *Nonlinear Partial Differential Equations and Their Applications*, Lect. Coll. de France Semin., Vol. IX, Pitman Res. Notes Math. Ser., Vol. 181, 304–333.

LIONS, P.-L. (1987), Solutions of Hartree–Fock equations for Coulomb systems, *Comm. Math. Phys.* **109**, 33–97.

LIU, S. and C.E. DYKSTRA (1985), Polarizabilities and hyperpolarizabilities of methane. The importance of valence charge polarization in polyatomic molecules, *Chem. Phys. Lett.* **119**, 407–411.

LÖWDIN, P.-O. (1950), On the non-orthogonality problem connected with the atomic wavefunction in the theory of molecules and crystals, *J. Chem. Phys.* **18**, 365–375.

LÜCHOW, A. and H. KLEINDIENST (1994), Accurate upper and lower bounds to the ^{2}S states of the Lithium atom, *Int. J. Quant. Chem.* **51**, 211–224.

LUNDQVIST, S. and N.H. MARCH, eds. (1983), *Theory of the Inhomogeneous Electron Gas* (Plenum, New York).

MADAY, Y. and A.T. PATERA (2000), Numerical analysis of a posteriori finite element bounds for linear functional outputs, *Math. Models Methods Appl. Sci.* **10** 5, 785–799.

MADAY, Y., A.T. PATERA and J. PERAIRE (1999), A general formulation for a posteriori bounds for output functionals of partial differential equations; application to the eigenvalue problem, *C. R. Math. Acad. Sci. Paris, Série I.* **328**, 9, 823–828.

MADAY, Y. and G. TURINICI (2002a), A posteriori numerical analysis for the Hartree–Fock equations and quadratically convergent methods, *Numer. Math.*, in press.

MADAY, Y. and G. TURINICI (2002b), Convergence analysis of the discretization for the Hartree–Fock problem by Gaussian type basis sets, *C. R. Math. Acad. Sci. Paris*, in preparation.

MARCH, N.H. (1992), *Electron Density Theory of Atoms and Molecules* (Academic Press).

MARKVOORT, A.J., R. PINO and P.A.J. HILBERS (2001), in: V.N. Alexandrov et al., eds., *Interpolation Wavelets in Kohn–Sham Electronic Structure Calculations*, ICCS 2001 (Springer) 541–550.

MAROULIS, G. and D.M. BISHOP (1985), On the dipole and higher polarizabilities of Ne(^{1}S), *Chem. Phys. Lett.* **114**, 182–186.

MARTIN, J.M.L. and P.R. TAYLOR (1997), Benchmark quality total atomization energies of small polyatomic molecules, *J. Chem. Phys.* **106**, 8620–8623.

MAZZIOTTI, D. (1998a), Approximate solution for electron correlation through the use of Schwinger probes, *Chem. Phys. Lett.* **289**, 419–427.

MAZZIOTTI, D. (1998b), 3,5-contracted Schrödinger equation: determining quantum energies and reduced density matrices without wave functions, *Int. J. Quant. Chem.* **70**, 557–570.

MAZZIOTTI, D. (1998c), Contracted Schrödinger equation: determining quantum energies and two-particle density matrices without wave functions, *Phys. Rev. A* **57**, 4219–4234.

MAZZIOTTI, D. (1999), Comparison of contracted Schrödinger and coupled-cluster theories, *Phys. Rev. A* **60**, 4396–4408.

MCDONALD, J.K.L (1934), On modified Ritz method, *Phys. Rev.* **46**, 828–828.

MCWEENY, R. (1992), *Methods of Molecular Quantum Mechanics*, 2nd edn. (Academic Press, San Diego).

MENNUCCI, B., M. COSSI and J. TOMASI (1995), A theoretical model of solvation in continuum anisotropic dielectrics, *J. Chem. Phys.* **102**, 6837–6845.

MICHALEWICZ, M. (1999), *Genetic Algorithms + Data Structure = Evolution Programs* (Springer).

MIKKELSEN, K.V., P. JØRGENSEN and H.J. JENSEN (1994), A multiconfiguration self-consistent reaction field response method, *J. Chem. Phys.* **100**, 6597–6607.

MILLER, C.K., B.J. ORR and J.F. WARD (1977), An interactive segment model of molecular electronic tensor properties: theory and application to electric dipole moments of the halogenated methanes, *J. Chem. Phys.* **67**, 2109–2218.

MILLER, K.J. and K. RUEDENBERG (1968), Electron correlation and separated-pair approximation. An application to Beryllium-like atoms, *J. Chem. Phys.* **48**, 3415–3430.

MØLLER, C. and M.S. PLESSET (1934), Note on an approximation treatment for many-electron systems, *Phys. Rev.* **46**, 618–622.

MONKHORST, H.J. and J.D. PACK (1976), Special points for Brillouin zone integration, *Phys. Rev. B* **13**, 5188–5192.

MORGAN III, J.D. (1984), in: M. Defranceschi and J. Delhalle, eds., *Numerical Determination of the Electronic Structure of Atoms, Diatomic and Polyatomic Molecules* (Kluwer, Dordrecht) 49–84.

MORGAN III, J.D. (1986), Convergence properties of Fock's expansions for s-state eigenfunctions of the helium atom, *Theoret. Chim. Acta* **69**, 181–224.

MORGAN III, J.D. and W. KUTZELNIGG (1993), Hund's rules, the alternating rule, and symmetry holes, *J. Phys. Chem.* **97**, 2425–2434.

MUKHERJEE, D. and W. KUTZELNIGG (2001), Irreductible Brillouin conditions and contracted Schrödinger equations for n-electron systems, I. The equations satisfied by the density cumulants, *J. Chem. Phys.* **114**, 2047–2061; Erratum *ibid.* **114**, 8226.

MÜLLER, H. and W. KUTZELNIGG (1997), A CCSD(T)-R12 study of the ten-electron systems Ne, F^-, FH, H_2O, NH_3, NH^{2-} and CH_4, *Mol. Phys.* **92**, 535–546.

MULLIKEN, R.S. (1928), The assignment of quantum numbers for electrons in molecules, *Phys. Rev.* **32**, 186–222.

MULLIKEN, R.S. and C.C.J. ROOTHAAN (1959), Broken bottlenecks and the future of molecular quantum mechanics, *Proc. U.S. Natl. Acad. Sci.* **45**, 394–398.

NAKANO, A. (2002), http://www.cclms.lsu.edu/cclms/group/.

NAKATA, M., M. EHARA and H. NAKATSUJI (2002), Density-matrix variational theory: Application to the potential energy surfaces and strongly correlated systems, *J. Chem. Phys.* **116**, 5432–5439.

NAKATSUJI, H. (1976), Equation for the direct determination of the density matrix, *Phys. Rev.* **14**, 41–50.

NAKATSUJI, H. and K. YASUDA (1996), Direct determination of the quantum-mechanical density matrix using the density equation, *Phys. Rev. Lett.* **76**, 1039–1042.

NATIELLO, M.A. and G.E. SCUSERIA (1984), Convergence properties of Hartree–Fock SCF molecular calculations, *Int. J. Quant. Chem.* **24**, 1039–1049.

NEUMAIER, A. (1997), Molecular modeling of proteins and mathematical prediction of protein structure, *SIAM Rev.* **39**, 407–460.

NEWMAN, F.T. (1997), Power series with rational coefficients for two-electron atoms energies, *Int. J. Quant. Chem.* **63**, 1065–1078.

NOCEDAL, J. and S.J. WRIGHT (1999), *Numerical Optimization* (Springer).

NOGA, J., R.J. BARTLETT and M. URBAN (1987), Towards a full CCSDT model for electron correlation. CCSDT-n models, *Chem. Phys. Lett.* **134**, 126–132.

NOGA, J., W. KLOPPER and W. KUTZELNIGG (1998), in: R.J. Bartlett, ed., *Recent Advances in Coupled-Cluster Methods* (World Scientific, Singapore) 1.

NOGA, J., D. TUNEGA, W. KLOPPER and W. KUTZELNIGG (1995), The performance of the explicit correlated coupled-cluster method, I. The four-electron systems Be, Li^- and LiH, *J. Chem. Phys.* **103**, 309–320.

NOSÉ, S. (1984), A molecular dynamics method for simulation in the canonical ensemble, *Mol. Phys.* **52**, 255–268.

NOSÉ, S. (1986), An extension of the canonical ensemble molecular dynamics method, *Mol. Phys.* **57**, 187–191.

ODDERSHEDE, J. and J.R. SABIN (1991), Polarization propagator calculation of a spectroscopic properties of molecules, *Int. J. Quant. Chem.* **39**, 371–386.

ÖHRN, Y. and J. LINDERBERG (1965), Propagators for alternant hydrocarbon molecules, *Phys. Rev.* **139**, A1063–A1068.

ORDEJON, P., D.A. DRABOLD and R.M. MARTIN (1995), Linear system-size scaling methods for electronic structure calculations, *Phys. Rev. B* **51**, 1456–1476.

ORR, B.J. and J.F. WARD (1971), Perturbation theory of nonlinear optical polarization of an isolated system, *Mol. Phys.* **20**, 513–526.

PACK, R.T. and W. BYERS BROWN PACK (1966), Cusp conditions for molecular wavefunctions, *J. Chem. Phys.* **45**, 556–559.

PALSER, A. and D. MANOPOULOS (1998), Canonical purification of the density matrix in electronic structure theory, *Phys. Rev. B* **58**, 12704–12711.

PARASCHIVOIU, M. and A.T. PATERA (1998), A hierarchical duality approach to bounds for the outputs of partial differential equations, *Comput. Meth. Appl. Mech. Engrg.* **1158**, 389–407.

PARMENTER, R.H. and G.H. WANNIER (1937), Electronic energy bands in crystals, *Phys. Rev.* **86**, 552–559.

PARR, R.G. and W. YANG (1989), *Density Functional Theory of Atoms and Molecules* (Oxford Univ. Press).

PAULING, L. (1931), Quantum mechanics and the chemical bonds, *Phys. Rev.* **37**, 1185–1186.

PAYNE, P.W. (1979), Density functionals in unrestricted Hartree–Fock theory, *J. Chem. Phys.* **71**, 490–496.

PAYNE, M.C., M.P. TETER, D.C. ALLAN, T.A. ARIAS and J.D. JOANNOPOULOS (1992), Iterative minimization techniques for ab initio total-energy calculations: molecular dynamics and conjugate gradients, *Rev. Mod. Phys.* **64**, 1045–1097.

PEKERIS, C.L. (1958), Ground state of two-electron atoms, *Phys. Rev.* **112**, 1649–1658.

PEKERIS, C.L. (1962), $1\,^1S$, $2\,^1S$, and $2\,^3S$ states of H^- and of He, *Phys. Rev.* **126**, 1470–1476.

PENG, C., P.Y. AYALA, H.B. SCHLEGEL and M.J. FRISCH (1995), Using redundant internal coordinates to optimize equilibrium geometries and transition states, *J. Comp. Chem.* **16**, 49–56.

PERDEW, J.P. (1986), Density-functional approximation for the correlation energy of the inhomogeneous electron gas, *Phys. Rev. B* **33**, 8822–8824.

PERDEW, J.P., K. BURKE and M. ERNZERHOF (1996), Generalized gradient approximation made simple, *Phys. Rev. Lett.* **77**, 3865–3868.

PERDEW, J.P. and M. LEVY (1985), Extrema for the density-functional theory for the energy: Excited states from the ground-state theory, *Phys. Rev. B* **31**, 6264–6272.

PERDEW, J.P. and Y. WANG (1986), Accurate and simple density functional for the electronic exchange energy: generalized gradient approximation, *Phys. Rev. B* **33**, 8800–8802.

PERDEW, J.P. and A. ZUNGER (1981), Self-interaction correction to density-functional approximations for many-electron systems, *Phys. Rev. B* **23**, 5048–5079.

PETERSON, K.A., A.K. WILSON, D.E. WOON and T.H. DUNNING JR. (1997), Benchmark calculations with correlated molecular wave functions, 12. Core correlation effects on the homonuclear diatomic molecules B_2, F_2, *Theoret. Chim. Acta* **97**, 251–259.

PETERSON, K.A., D.E. WOON and T.H. DUNNING JR. (1994), Benchmark calculations with correlated molecular wave functions, IV. The classical barrier height of the $H + H_2 \to H_2 + H$ reaction, *J. Chem. Phys.* **100**, 7410–7415.

PETERSSON, G.A. and M. BRAUNSTEIN (1985), Complete basis set correlation energies, IV. The total correlation energy of the water molecule, *J. Chem. Phys.* **83**, 5129–5134.

PHILLIPS, J.C. and L. KLEINMAN (1959), New method for calculating wave functions in crystals and molecules, *Phys. Rev.* **116**, 287–293.

PIERCE, B.M. (1989), A theoretical analysis of third-order nonlinear optical properties of linear polyenes and benzene, *J. Chem. Phys.* **91**, 791–811.

POLITZER, P. and J.S. MURRAY, eds. (1994), *Quantitative Treatments of Solute/Solvents Interactions* (Elsevier).

POPLE, J.A., R. KRISHNAN, H.B. SCHLEGEL and J.S. BINKLEY (1978a), Electron correlation theories and their application to the study of simple reaction potential surfaces, *Int. J. Quant. Chem.* **14**, 545–560.

POPLE, J., R. KRISHNAN, H.B. SCHLEGEL and J.S. BINKLEY (1978b), Electron correlation theories and their application to the study of simple reaction potential surfaces, *Int. J. Quant. Chem.* **14**, 545–560.

POPLE, J.A., L.W. MCIVER and N.S. OSTLUND (1968), Self-consistent perturbation theory, I. Finite perturbation methods, *J. Chem. Phys.* **49**, 2960–2864.

POPLE, J., A.P. SCOTT, M.H. WONG and L. RADOM (1993), Scaling factors for obtaining fundamental vibrational frequencies and zero-point energies from HF/6-31G* and MP2/6-31G* harmonic frequencies, *J. Chem. Phys.* **33**, 345–350.

POUCHAN, C. and D.M. BISHOP (1984), Static dipole polarizability of the lithium atom, cation and anion, *Phys. Rev. A* **29**, 1–5.

PRIMAS, H. (1961), Eine verallgemeinerte Störungstheorie für quantenmechanische Mehrtielchenprobleme, *Helvet Phys. Acta* **34**, 331–351.

PRIMAS, H. (1963), Generalized perturbation theory in operator form, *Rev. Mod. Phys.* **35**, 710–712.

PULAY, P. (1969), Ab initio calculation of force constants and equilibrium geometries in polyatomic molecules, I. Theory, *Molecular Phys.* **17**, 197–204.

PULAY, P. (1982), Improved SCF convergence acceleration, *J. Comp. Chem.* **3**, 556–560.

QUARTERONI, A. and A. VALLI (1994), *Numerical Approximation of Partial Differential Equations* (Springer).

RAGHAVACHARI, K., J.A. POPLE and M. HEAD-GORDON (1989), A fifth order perturbation comparison of electron correlation theories, *Chem. Phys. Lett.* **157**, 479–483.

RAPAPORT, D.C. (1995), *The Art of Molecular Dynamics Simulation* (Cambridge Univ. Press).

REED, M. and B. SIMON (1975–1980), *Methods of Modern Mathematical Physics*, in 4 vols. (Academic Press).

REINING, L., G. ONIDA and R.W. GODBY (1997), Elimination of unoccupied-state summation in ab initio self-energy calculations for large supercells, *Phys. Rev. B* **56**, R4301–R4304.

RIJANOV, S. (1934), Zur Frage nach der "Elektronenbewegung" im beschräkten Kristallgitter, *Z. Physik* **89**, 806–819.

ROHMER, M.-M., J. DEMUYNCK, M. BÉNARD, R. WIEST, C. BACHMANN, C. HENRIET and M. ERNENWEIN (1990), A program system for ab initio MO calculations on vector and parallel processing machines, II SCF closed shell and open shell iterations, *Comput. Phys. Comm.* **60** , 127–144.

ROOS, B.O., P.R. TAYLOR and E.M. SIEGHBAN (1980), A complete active space SCF method (CASSCF) using a density matrix formulated super CI approach, *Chem. Phys.* **48**, 157–173.

ROOTHAAN, C.C.J. (1951), New developments in molecular orbital theory, *Rev. Mod. Phys.* **23**, 69–89.

RYCHLEWSKI, J., W. CENCEK and J. KOMASA (1994), The equivalence of explicitly correlated Slater and Gaussian functions in variational Quantum Chemistry computations. The ground state of H_2, *Chem. Phys. Lett.* **229**, 657–660.

SAAD, Y. (1992), *Numerical Methods for Large Eigenvalue Problems* (Wiley, New York).

SAKAI, H. ET AL. (1999), Controlling the alignment of neutral molecules by a strong laser field, *J. Chem. Phys.* **110**, 10235–10238.

SANZ-SERNA, J.M. and M.P. CALVO (1994), *Numerical Hamiltonian Problems* (Chapman and Hall).

SAUNDERS, V.R. and I.H. HILLIER (1973), A "level-shifting" method for converging closed shell Hartree–Fock wavefunctions, *Int. J. Quant. Chem.* **7**, 699–705.

SCHECHTER, M. (1981), *Operator Methods in Quantum Mechanics* (North-Holland, Amsterdam).

SCHLEGEL, H.B. and J.J.W. MCDOUALL (1991), Do you have SCF stability and convergence problems?, in: *Computational Advances in Organic Chemistry* (Kluwer Academic) 167–185.

SCHLICK, T. (1992), Optimization methods in computational chemistry, in: K.B. Lipkowitz and D.B. Boyd, eds., *Reviews in Computational Chemistry*, Vol. 3 (VCH Publishers) 1–71.

SCHMIDT, H.M. and H.v. HIRSCHHAUSEN (1983), Perturbation theory in $1/Z$ for atoms: First order pair functions in an l-separated Hylleraas basis-set, *Phys. Rev. A* **28**, 3179–3183.

SCHMIDT, M.W. and K. RUEDENBERG (1979), Effective convergence to complete orbital bases and to the atomic Hartree–Fock limit through systematic sequences of Gaussian primitives, *J. Chem. Phys.* **71**, 3951–3962.

SCHRÖDINGER, E. (1926), Quantiesierung als Eigenwertproblem, *Ann. Physik* **79**, 489–527.

SCHULMAN, J. and D.N. KAUFMAN (1970), Application of many-body perturbation theory to the hydrogen molecule, *J. Chem. Phys.* **53**, 477–484.

SCHÜTTE, CH. and W. HUISINGA (2003), Biomolecular conformations can be identified as metastable sets of molecular dynamics, in: P.G. Ciarlet and C. Le Bris, eds., *Computational Chemistry*, Handbook of Numerical Analysis, Vol. X (Elsevier, Amsterdam), 699–744.

SCHWARTZ, C. (1962), Importance of angular correlation between atomic electrons, *Phys. Rev.* **126**, 1015–1019.

SCHWEGLER, E. and M. CHALLACOMBE (1997), Linear scaling computation of the Fock matrix, *J. Chem. Phys.* **106**, 5526–5536.

SCHWEGLER, E. and M. CHALLACOMBE (1999), Linear scaling computation of the Fock matrix, IV. Multipole accelerated formation of the exchange matrix, *J. Chem. Phys.* **106**, 6223–6229.

SCHWEGLER, E. and M. CHALLACOMBE (2000), Linear scaling computation of the Fock matrix, *Theor. Chem. Acc.* **104**, 344–349.

SCHWEGLER, E., M. CHALLACOMBE and M. HEAD-GORDON (1997), Linear scaling computation of the Fock matrix, II. Rigorous bounds on exchange integrals and incremental Fock bluid, *J. Chem. Phys.* **106**, 9708–9717.

SCUSERIA, G.E. (1999), Linear scaling density functional calculations with Gaussian orbitals, *J. Phys. Chem. A* **103**, 4782–4790.

SEEGER, R. and J.A. POPLE (1976), Self-consistent molecular orbital methods, XVI. Numerically stable direct energy minimization procedures for solution of Hartree–Fock equations, *J. Chem. Phys.* **65**, 265–271.

SEITZ, F. (1936), Matrix-algebraic development of Crystallographic groups, *Zeits. f. Krist.* **94**, 100–130.

SEKINO, H. and R.J. BARTLETT (1986), Hyperpolarizabilities of the hydrogen fluoride molecule: a discrepancy between theory and experiment?, *J. Chem. Phys.* **84**, 2726–2733.

SEKINO, H. and R.J. BARTLETT (1993), Molecular hyperpolarizabilities, *J. Chem. Phys.* **98**, 3022–3037.

SENATORE, G. and K.R. SUBBASWAMY (1987), Nonlinear response of closed-shell atoms in the density-functional formalism, *Phys. Rev. A* **35**, 2440–2447.

SHAPIRO, M. and P. BRUMER (1986), Laser control of product quantum state populations in unimolecular reactions, *J. Chem. Phys.* **84**, 4103–4104.

SHAVITT, I. and M. KARPLUS (1962), Multicenter integrals in molecular Quantum Mechanics, *J. Chem. Phys.* **36** 550–551.

SHAVITT, I. and M. KARPLUS (1965), Gaussian-transform method for molecular integrals, I. Formulation for energy integrals, *J. Chem. Phys.* **43**, 398–414.

SHEPARD, R. (1993a), Elimination of the diagonalization bottleneck in parallel direct-SCF methods, *Theor. Chim. Acta* **84**, 343–351.

SHEPARD, R. (1993b) The MCSCF method, *Adv. Chem. Phys.* **69**, 63–200.

SHOCKLEY, W. (1939), On the surface states associated with a periodic potential, *Phys. Rev.* **56**, 317–323.

SINANOĞLU, O. (1964), Electron correlation in atoms and molecules, *Adv. Chem. Phys.* **6**, 315–412.

SINANOĞLU, O. (1966), Many electron theory of atoms, molecules, and theirs interactions, *Adv. Chem. Phys.* **14**, 237–282.

SILVER, D.M., E.L. MEHLER and K. RUEDENBERG (1968), Atomic orbital overlap integrals; Coulomb integrals between Slater-type atomic orbitals, *J. Chem. Phys.* **49**, 4301–4311.

SLATER, J.C. (1930a), Note on Hartree's method, *Phys. Rev.* **35**, 210–211.

SLATER, J.C. (1930b), Atomic shielding constants, *Phys. Rev.* **36**, 57–64.

SLATER, J.C. (1936a), The ferromagnetism of Nickel, *Phys. Rev.* **49**, 537–545.

SLATER, J.C. (1936b), The ferromagnetism of Nickel, II. Temperature effects, *Phys. Rev.* **49**, 931–937.

SLATER, J.C. (1937), Wavefunctions in a periodic potential, *Phys. Rev.* **51**, 846–851.

SLATER, J.C. and G.F. KOSTER (1954), Simplified LCAO method for the periodic potential problem, *Phys. Rev.* **94**, 1498–1524.

SOLOVEJ, J.P. (1990), Universality in the Thomas–Fermi–von Weizsäcker model of atoms and molecules, *Comm. Math. Phys.* **129**, 561–598.

SOMMERFELD, A. and H. BETHE (1933), Elektronentheorie der Metalle, in: *Handbuch der Physik*, Vol. 24, 2nd edn. (Springer).

SPRUCH, L. (1991), Pedagogic notes on Thomas–Fermi theory (and on some improvements): atoms, stars and the stability of bulk matter, *Rev. Mod. Phys.* **63**, 151–209.

STAHL, H. (1992), Best uniform rational approximation of x on $[-1, 1]$, *Mat. Sbornik* **183**, 85–112.

STANTON, R.E. (1981a), The existence and cure of intrinsic divergence in closed shell SCF calculations, *J. Chem. Phys.* **75**, 3426–3432.

STANTON, R.E. (1981b), Intrinsic convergence in closed-shell SCF calculations. A general criterion, *J. Chem. Phys.* **75**, 5416–5422.

STARIKOV, E.B. (1993), On the convergence of the Hartree–Fock selfconsistency procedure, *Mol. Phys.* **78**, 285–305.

STENGER, F. (1981), Numerical methods based on Whittaker cardinal or sine functions, *SIAM Rev.* **23**, 165–224.

STENGER, F. (1993), *Numerical Methods Based on Sinc and Analytical Functions* (Springer, New York).

STOER, J. (1972), *Einführung in die Numerische Mathematik*, Vol. 1 (Springer, Berlin).

STRUWE, M. (1990), *Variational Methods. Applications to Nonlinear Partial Differential Equations and Hamiltonian Systems* (Springer).

SVENSON, M., S. HUMBEL, R.D.J. FROESE, T. MATSUKARA, S. SIEBER and K. MOROKUMA (1995), ONIOM: a multilayered integrated MO + MM method for geometry optimization and single point energy predictions; a test for Diels–Alder reactions and $Pt(P(t\text{–}Bu)_3)_2 + H_2$ oxidative addition, *J. Chem. Phys.* **100**, 19357–19363.

SWIRLESS, B. (1935), The relativistic self-consistent field, *Proc. Roy. Soc. A* **152**, 625–649.

SWIRLESS, B. (1936), The relativistic interaction of two electrons in the self-consistent field method, *Proc. Roy. Soc. A* **157**, 680–696.

SZABO, A. and N.S. OSTLUND (1982), *Modern Quantum Chemistry: An Introduction to Advanced Electronic Structure Theory* (MacMillan).

SZALEWICZ, K., B. JEZIORSKI, H.J. MONKHORST and J.G. ZABOLITZKY (1983), Atomic and molecular correlation energies with explicitly correlated Gaussian geminals, I and II, *J. Chem. Phys.* **78**, 1420–1430 and **79**, 5543–5542.

TAMM, I.E. (1932), Possible types of electron-binding on crystal surfaces, *Zeits. f. Physik* **76**, 849–850.

TANNOR, D.J., R. KOSLOFF and S.A. RICE (1986), Coherent pulse sequence induced control of selectivity of reactions: exact quantum mechanics calculations, *J. Chem. Phys.* **85**, 5805–5820.

TARASSOV, I. (1979), *Bases physiques de l'électronique quantique* (Editions Mir, Moscow).

TERSOFF, J. (1988), New empirical approach for the structure and energy of covalent systems, *Phys. Rev. B* **37**, 6981–7000.

THAKKAR, A.J. and T. KOGA (1994), Ground state energies for the helium isoelectronic series, *Phys. Rev. A* **50**, 854–856.

THEOPHILOU, A. (1979), The energy density functional formalism for excited states, *J. Phys. C* **12**, 5419–5430.

THIRRING, W. (1983), *A Course in Mathematical Physics*, in 4 vols. (Springer).

TOMASI, J. and M. PERSICO (1994), Molecular interactions in solution: An overview of methods based on continuous distribution of solvent, *Chem. Rev.* **94**, 2027–2094.

TOURNOIS, T. ET AL. (1998), Les impulsions lasers ultra-brèves, in: *Compte rendu des* 3e *Entretiens de la physique* (EDP Sciences).

TROULLIER, N. and J.L. MARTINS (1990), A straightforward method for generating soft transferable pseudopotentials, *Solid State Comm.* **74**, 613–616.

TUCKERMAN, M.E. and G.J. MARTYNA (2000), Understanding modern molecular dynamics: techniques and applications, *J. Phys. Chem.* **104**, 159–178.

TURINICI, G. and H. RABITZ (2001), Quantum wavefunction controllability, *Chem. Phys.* **267**, 1–9.

VALDEMORO, C. (1985), Spin-adapted reduced Hamiltonian, I. Elementary excitations, *Phys. Rev. A* **31**, 2114–2122.

VALDEMORO, C. (1992), Approximating the second-order reduced density matrix in terms of the first-order one, *Phys. Rev. A* **45**, 4462–4467.

VALDEMORO, C., L.M. TEL and E. PEREZ-ROMERO (2000), N-representability problem within the framework of the contracted Schrödinger equation, *Phys. Rev. A* **61**, 032507–032700.

VANDERBILT, D. (1990), Soft self-consistent pseudopotentials in a generalized eigenvalue formalism, *Phys. Rev. B* **41**, 7892–7895.

VAN DUIJNEVELDT-VAN DEN RIJDT, J.G.C.M. and F.B. VAN DUIJNEVELDT (1982), Gaussian basis sets which yield accurate Hartree–Fock electric moments and polarizabilities, *J. Mol. Struct. (Theochem)* **89**, 185–201.

TE VELDE, G. and E.J. BAERENDO (1991), Precise density functional method for periodic structures, *Phys. Rev. B* **44**, 7888–7903.

VERLET, L. (1967), Computer "experiments" on classical fluids, I. Thermodynamical properties of Lennard-Jones molecules, *Phys. Rev.* **159**, 98–103.

VON BARTH, U. (1979), Local-density theory of multiplet structure, *Phys. Rev. A* **20**, 1693–1703.

VON BARTH, J. and L. HEDIN (1972), A local exchange-correlation potential for the spin polarized case, *J. Phys. C* **5**, 1629–1642.

VOSKO, S.H., L. WILK and M. NUSAIR (1980), Accurate spin-dependent electron liquid correlation energy for local spin density calculations: a critical analysis, *Canad. J. Phys.* **58**, 1200–1211.

WANG, Y.A. and E.A. CARTER (2000), Orbital-free kinetic-energy density functional theory, in: S.D. Schwartz, ed., *Theoretical Methods in Condensed Phase Chemistry* (Kluwer).

WENZEL, K.B., J.G. ZABOLITZKY, V. SZALEWICZ, B. JEZIORSKI and H.J. MONKHORST (1986), Atomic and molecular correlation energies with explicitly correlated Gaussian geminals, V. Cartesian Gaussian geminals and the neon atom, *J. Chem. Phys.* **85**, 3964–3974.

WERNER, H.J. and W. MEYER (1976), PNO-CI and PNO-CEPA methods of electron correlation effects, V. Static dipole polarizabilities of small molecules, *Mol. Phys.* **31**, 855–872.

WHITTAKER, E.T. (1915), *Proc. Roy. Soc. Edinburgh* **35**, 183.

WHITTAKER, J.M. (1935), *Interpolatory Function Theory* (Cambridge, London).

WIEST, R., J. DEMUYNCK, M. BÉNARD, M.-M. ROHMER and M. ERNENWEIN (1991), A program system for ab initio MO calculations on vector and parallel processing machines, III. Integral reordering and fourth index transformation, *Comput. Phys. Comm.* **62**, 107–124.

WIGNER, E. and F. SEITZ (1933), On the constitutions of metallic sodium, *Phys. Rev.* **43**, 804–810.

WILSON, W. (1915).

WILSON, A.K. and T.H. DUNNING JR. (1997), Benchmark calculations with correlated molecular wave functions. X. Comparison with "exact" MP2 calculations on Ne, HF, H_2O and N_2, *J. Chem. Phys.* **106**, 8718–8726.

WILSON, J. and J.F.B. HAWKES (1983), *Optoelectronics: An Introduction*, Prentice-Hall International Series in Optoelectronics (Prentice-Hall).

WIMMER, E., H. KRAKAUER, M. WEINERT and A.J. FREEMAN (1981), Full-potential self consistent linearized augmented plane wave method for calculating the electronic structure of molecules and surfaces: O_2 molecule, *Phys. Rev. B* **24**, 864–875.

WOLNIEWICZ, L. (1993), Relativistic energies of the ground-state of the hydrogen molecule, *J. Chem. Phys.* **99**, 1851–1868.

WOOLEY, R.G. (1978), Must a molecule have a shape?, *J. Amer. Chem. Soc.* **100**, 1073–1078.

WULLER, W. (1986), Existence of time evolution for Schrödinger operators with time dependent singular potentials, *Ann. Inst. H. Poincaré Sect. A.* **44**, 155–171.

YAJIMA, K. (1987), Existence of solutions for Schrödinger evolution equations, *Commun. Math. Phys.* **110**, 415–426.

YASUDA, K. (2002), Local approximation of the correlation energy functional in the density matrix functional theory, *Phys. Rev. Lett.* **88**, 053001.

ZANGWILL, A. and P. SOVEN (1980), Density-functional approach to local-field effects in finite systems: Photoabsorption in the rare gases, *Phys. Rev. A* **21**, 1561–1572.

ZEISS, G.D., W.R. SCOTT, N. SUZIKI, D.P. CHONG and S.R. LANGHOFF (1979), Finite-field calculations of molecular polarizabilities using field induced polarization functions: second and third order perturbation correlation corrections to the coupled HF polarizability of water, *Mol. Phys.* **37**, 1543–1572.

ZERNER, M.C. (1997), On calculating the electronic spectroscopy of very large molecules, in: S. Wilson and G.H.F. Diercksen, eds., *Problem Solving in Computational Molecular Science: Molecules in Different Environments*, NATO ASI Series, Vol. 500 (Kluwer Academic, Dordrecht) 249–289.

ZERNER, M.C. and M. HEHENBERGEN (1979), A dynamical damping scheme for converging molecular SCF calculations, *Chem. Phys. Lett.* **62**, 550–554.

ZHU, W. and H. RABITZ (1998), A rapid monotonically convergent iteration algorithm for quantum optimal control over the expectation value of a positive definite operator, *J. Chem. Phys.* **109**, 385–391.

ZICOVICH-WILSON, J. and R. DOVESI (1998), On the use of symmetry-adapted crystalline orbitals in SCF-LCAO periodic calculations, I. The construction of the symmetrized orbitals, *Int. J. Quant. Chem.* **67**, 299–309.

FRISCH, M.J., G.W. TRUCKS, H.B. SCHLEGEL, G.E. SCUSERIA, M.A. ROBB, J.R. CHEESEMAN, V.G. ZAKRZEWSKI, J.A. MONTGOMERY, R.E. STRATMANN, J.C. BURANT, S. DAPPRICH, J.M. MILLAM, A.D. DANIELS, K.N. KUDIN, M.C. STRAIN, O. FARKAS, J. TOMASI, V. BARONE, M. COSSI, R. CAMMI, B. MENNUCCI, C. POMELLI, C. ADAMO, S. CLIFFORD, J. OCHTERSKI, G.A. PETERSSON, P.Y. AYALA, Q. CUI, K. MOROKUMA, D.K. MALICK, A.D. RABUCK, K. RAGHAVACHARI, J.B. FORESMAN, J. CIOSLOWSKI, J.V. ORTIZ, B.B. STEFANOV, G. LIU, A. LIASHENKO, P. PISKORZ, I. KPMAROMI, G. GOMPERTS, R.L. MARTIN, D.J. FOX, T. KEITH, M.A. AL-LAHAM, C.Y. PENG, A. NANAYAKKARA, C. GONZALEZ, M. CHALLACOMBE, P.M.W. GILL, B.G. JOHNSON, W. CHEN, M.W. WONG, J.L. ANDRES, M. HEAD-GORDON, E.S. REPLOGLE and J.A. POPLE (1998), *Gaussian 98* (Revision A.7) (Gaussian Inc., Pittsburgh, PA).

INTERNATIONAL TABLES FOR X-RAY CRYSTALLOGRAPHY (Kynoch Press, Birmingham), 1952–1962.

RAPPORT DE L'ACADÉMIE, ACADÉMIE DES SCIENCES, SCIENCES AUX TEMPS ULTRACOURTS ; DE L'ATOSECONDE AUX PETAWATTS, Rapport sur la science et la technologie, 2000, September.

REPORT OF THE PANEL ON FUTURE DIRECTIONS IN CONTROL, DYNAMICS, AND SYSTEMS (2002), Richard M. MURRAY (chair), http://www.cds.caltech.edu/~murray/cdspanel.

The Modeling and Simulation of the Liquid Phase

J. Tomasi, B. Mennucci

Dipartimento di Chimica e Chim. Indus., Via Risorgimento 35, University of Pisa, Pisa, Italy

Patrick Laug

INRIA Rocquencourt, Gamma project, BP 105, 78153 Le Chesnay cedex, France

Computational Chemistry
Special Volume (C. Le Bris, Guest Editor) of
HANDBOOK OF NUMERICAL ANALYSIS, VOL. X
P.G. Ciarlet (Editor)
© 2003 Elsevier Science B.V. All rights reserved

Contents

An overview on the theoretical and computational methodologies developed so far to study liquids and solutions is presented. The main characteristics of the different methods are outlined and the advantages and shortcomings of the computational approaches are discussed. Particular attention is focused on a specific class of methods, known as continuum solvation models, for which a more detailed description of theoretical and computational aspects is presented. For this class of methods, the concept of molecular cavity is introduced together with a description of the numerical techniques developed to mesh the corresponding surface. For selected methods representing those of larger use, an overview on their applications to the evaluation of energies and properties of liquid systems and molecules in solution is also presented.

CHAPTER I

Introduction to Liquids

For very long tradition chemistry has to deal with liquids. The chain of chemical manipulations performed in laboratories and in factories are mostly performed in a liquid medium and one of the first questions asked when a new chemical is introduced in the use is what is its solubility in solvents. There is a large number of solvents in use in the chemical practice, more than two thousand, and this number is not sufficient: often it is convenient to use solvent mixtures. Chemical reactions, the core of chemistry, are very sensitive to the solvent used to put in intimate contact the reagent species: often a delicate reaction fails if the solvent was not properly chosen. Important areas of chemical research are inherently tied to a solvent. An important example is that of biological systems. All the complex machinery of biological systems works only if the system is in water: without this solvent event the intimate structures of biological molecules collapse and loose their activity.

Chemists are so accustomed to consider many properties of liquids, from the basic physical properties to others, more specific and playing a role in specific cases in the large variety of problems that chemistry has to face. We cannot give here an overview of the properties of interest for liquids, many among which could be the object of computer modeling and we limit ourselves to indicate a few points more directly related to what will be exposed in the following of this chapter.

Liquids often occur in the chemical practice in large quantities. The effects due to the spatial limitations of the liquid sample have not great influence on the phenomena occurring at the interior, which in the current practice is called the bulk solvent. On large scale the bulk pure liquids and solutions exhibit isotropy. Liquids can be however dispersed. A typical example are the water droplets dispersed in the atmosphere, as for example in the fog. Other examples occur in liquid mixtures that present a miscibility lacuna. In all cases the ratio "surface region/bulk" can be relatively large and phenomena occurring at the interface may present properties different with respect to the same occurring in the bulk.

The region of separation between a bulk liquid and a solid body has in some cases a decisive importance. One example is the separation between a salt solution and a metal electrode. All the electrochemical phenomena are influenced by the behavior of this thin portion of the liquid. Another example are the phenomena leading to dissolution of solid bodies in the liquid phase, and to the deposition of material components of the liquid into solid particles. The surface of separation of a liquid and a solid surface presents features of different type when the liquid is in a limited amount. The liquid phase can be present under the form of separate drops, or of a thin liquid film on the solid surface: both occurrences have a great importance in some chemical problems.

Surfaces of bulk liquids can be covered by a thin layer of an immiscible liquid. This is a phenomenon frequently observed in everyday life. Such thin layers can be organized into a structure that maintains some aspects of normal liquids but presents a high local order. With an appropriate selection of the chemical composition of this second phase, it is easy to form ordered layers of well-defined molecular thickness and membranes. Membranes, on their part, preserves fluid-like properties at their interior which can be exploited, for example, in the machinery of biological systems. A liquid can contain small solid particles within the bulk. Many phenomena and many practical applications are based on this particular type of liquid systems. A well-known phenomenon that has played an important role in the development of science is the Brownian motion. The solvent eases the dispersion in the bulk of such small particles, that are however subjected to gravity forces leading to sedimentation. Gravity, one of the basic forces in the universe, plays a little role in the microscopic approach to the study of material systems, but this example shows that there are reasons to forecast that in the future also gravity will be included into some computational models. A liquid can also contain at its interior portion of a second phase organized with specific shapes. This is the domain of micelles, vesicles and other similar structures, all subject of intense research both of basic type and addressing practical applications.

The ordering of specific chemical systems into layers and vesicles has a counterpart in bulk liquids. A special category of liquids are called liquid crystals. They combine macroscopic properties shared by all liquids, in particular the properties of assuming the shape of the vessel in which they are put and the capability of dissolving other chemicals, with a long range order making them similar to crystals. This long range order may assume different forms, to which corresponds specific names for the liquid crystal phase, smectic, nematic, cholesteric. In general the liquid crystal are specific phases of a liquid that can also have an isotropic phase without long range order. For many substances the changes in physical parameters temperature and pressure (T, P) rule the transformation from a solid, to a liquid and then to a gaseous phase (in the order on increasing T and decreasing P), while in the substances giving origin to liquid crystals there are some additional phases between solid and isotropic liquid. The liquid crystals are a subject of intensive study, because of their quite peculiar optical and electric properties.

1. Physical approaches to the study of liquids

The distinction we introduce here between physical and chemical approaches to the study of liquid systems is mainly justified by exposition reasons. The basic principles are the same in both approaches and there is a mutual interchange of methods and procedures: both approaches grow up in harmony. Actually there is a basic difference in the motivations of the two approaches: chemistry is directly interested to the details due to differences in the chemical composition of the fluid (the study of pure liquids could be viewed in this context as an extreme case of solutions), physics arrives to consider effects due to chemical composition (and of specific solutions in particular) at the end of a longer route. The physical approaches we shall consider here do not play much attention to the chemical composition of the system (but some aspects have still to be considered).

Macroscopic approaches

Historically, and also logically, the first contribution of the physical understanding of liquids was obtained with macroscopic approaches. We shall not consider here the mechanical studies that come first in the physical enquiry about fluids, but we directly shift to the thermodynamic approach elaborated in the nineteenth century. The science of heat, thermodynamics, regards all the matter in general, and considers liquids as a specific state of aggregation of the matter, to be treated on the same footing as the others. This emphasis on the uniformity of thermodynamic laws actually is at the basis of our understanding of the phenomena of phase transformation we have quoted above, and of the transfer of models (that we shall quote in the following) elaborated for the gas phase, being simpler to study, to liquids.

Thermodynamics is a rigorous discipline, and the definition of thermodynamic functions (at the equilibrium and out of the equilibrium) must be always reminded in performing studies on liquids with theoretical tools. Even when the model is reduced to a molecular model with attention paid to the details of quantum mechanical calculations on a reduced portion of the liquid, it is wise (often necessary) to keep in mind what is the thermodynamic status of the system, and to what selection of fixed macroscopic variables (pressure, volume, temperature, energy, chemical potential) is made in assessing the model.

A second macroscopic approach of interest for us regards the electric properties of the liquid. Here again we have to go back to the nineteenth century to find the elaboration of the macroscopic theory. A large portion of liquids are poor conductors of electricity (while there are notable exceptions, the whole electrochemistry is based on the conducting properties of ionic solutions), and for this reason the dielectric behavior plays the prominent role. It is convenient to give a short summary because we shall need it later. An external electric field induces a polarization of continuum dielectric media. In the standard version of the Maxwell elaboration, the attention is focused on the dipole polarization with respect to a homogeneous electric field. A vector field, $\mathbf{P}$, is defined giving the value of the dipole density; this vector is related to the two other vector fields defined by Maxwell to satisfy the two basic constitutive relations for electrostatic fields in vacuo: the electric field $\mathbf{E}$ and the displacement vector $\mathbf{D}$ (by definition $\mathbf{E} - \mathbf{D} = 4\pi\mathbf{P}$).

In the simplest case (homogeneous linear dielectrics, constant electric field) the relationship $\mathbf{P} = \chi\mathbf{E}$ with $\chi = (1 - \varepsilon)/4\pi$ holds. The macroscopic dielectric description of the liquids is not limited to the basic homogeneous and isotropic description we have here recalled. For anisotropic fluids (as liquid crystals) a tensorial definition of χ and of the dielectric constant ε must be introduced. There are cases in which the linear regime is not sufficient and the polarization must be described with the aid of higher order terms in the electric field. Models for specific cases call for specific modifications of the basic dielectric equation (a notable example of wide interest in chemistry is given by the Debye–Hückel model for ionic solutions). Other modifications are used for liquids spread on a metallic body.

All the examples we have given here refer to specific applications that shall be considered more in detail later, but of wider occurrence are the dynamic aspects introduced by the time dependence of the external field $\mathbf{E}(t)$, which gives rise to a large number of important phenomena. Relations paralleling those defined for all the static cases we have mentioned can be derived when changes in the electric field are not excessively fast (for the definition of such limits one has to make reference to a microscopic picture of matter). At a basic level, the dynamic case can be studied with the help of sinusoidally varying electric fields: $E(t) = E^0 \cos \omega t$ giving so origin to a complex dielectric response function that can be split into a real part, $\varepsilon'(\omega)$, and an imaginary part, $\varepsilon''(\omega)$. The first is a generalization of the static dielectric constant, $\varepsilon'(0) = \varepsilon$, and the second is called the *loss factor* (BÖTTCHER [1973], BÖTTCHER and BORDEWIJK [1978]). Both functions play an important role in the following of this chapter.

Microscopic approaches

The use of microscopic models based on molecules to describe liquids was introduced at the beginning of the past century. The impact has been enormous pervading all the research fields on liquids. The theory of liquid is at present a molecular theory. Two are the reasons of a so large impact: first, microscopic approaches have opened the way to the use of statistical mechanics, second, they have introduced explicit consideration of molecular interaction potentials, a subject on which chemists have a lot to say. For more than fifty years the research has been led by physicists, following the strategy to which physicists are more inclined, namely discarding the details of the molecular interaction potentials, spending more effort to establish models based on statistical mechanics and taking the model for ideal gases as starting point. Only in the late sixties of the past century aspects of more direct interest for chemistry were taken into consideration. In this chapter we are more interested to the chemical way of describing liquids and so we shall deserve only few sentences to the physical approach, especially to introduce terms that will be used in the following chapters (among the vast literature on the argument, we quote here few titles: BALESCU [1991], REICHL [1980], MCQUARRIE [1976]).

The basic ingredient in this approach is the probability density P in the $6N$-dimensional phase space of a liquid system which satisfies the invariance of a selected set of three macroscopic quantities (the number of particles N, the temperature T, the energy E, or the chemical potential μ). In the formulation of the theory of statistical thermodynamics use is made of another concept, that of ensemble of systems, an ideal collection of numerous systems, all defined by the same invariants, in close contact and exchanging among them what permitted by the invariants. These ensemble are usually indicated as *microcanonical*, *canonical* and *grand canonical*. We report in the table names and invariants of the systems to which are associated. In the following we shall indicate single systems with the name of the ensemble, to avoid proliferation of names. The systems that have been selected correspond, respectively, to closed and isolated, to closed but not isolated and to open systems. Several other systems may be defined, according to the experimental setting used for the study of the properties shown in Table 1.1.

TABLE 1.1

Name of the system	Physical invariants
Microcanonical	N, V, E
Canonical	N, V, T
Grand canonical	V, T, μ

The time evolution of the probability density $P(t)$ can be described in terms of a Liouville equation. The description given by the full P function is too detailed as in practice we only need the probability density to find expectation values or correlation functions for various observables. The observables generally corresponds to one- and two-body operators and so it is sufficient to use reduced densities limited at these two orders: P_1 and P_2, formally obtainable by integration of P on the other variables. The equations of motion for reduced densities must be however expressed in terms of a hierarchy of equations, called BBGKY, where P_1 is given in terms of P_2, P_2 in terms of P_3, and so on, until the so-called thermodynamic limit (quite large). To use the BBGKY hierarchy there is so the need of defining a closure relation, and the various physical methods differ in the choice of it. The BBGKY hierarchy was defined within the classical picture of the fluid; in quantum systems, where the Hamiltonian replaces the Liouvillian operator, use is made of the Wigner functions which lead to a similar hierarchy.

Another approach is available, based on the correlation functions. The two approaches are connected. For brevity we shall limit ourselves to stationary states of the Liouville equation, namely to systems in equilibrium. P can be expressed as a product of uncorrelated one-molecule distribution functions $P_1(i)$ modified by a correlation function g_N:

$$P_N(1, 2, \ldots, N) = P_1(1)P_1(2)\cdots P_1(N)g_N(1, 2, \ldots, N).$$

We have indicated with i the set of coordinates of the ith molecule (remark that implicitly, we have here discarded the grand canonical system, that can be recovered later, and that we have not paid attention to the possible differences among the N molecules of the system). For isotropic and homogeneous fluids each $P_1(i)$ is independent on the coordinates of i, so we may express it in terms of the numeral density ρ

$$P_N(1, 2, \ldots, N) = \rho g_N(1, 2, \ldots, N).$$

The correlation function is then subjected to a cluster expansion:

$$g_N(1, 2, \ldots, N) = \sum_{i<j} g_2(i, j) + \sum_{i<j<k} g_3(i, j, k) + \cdots.$$

Let us consider the two-body correlation function $g_2(i, j)$. It can be divided into two parts, the first called *direct correlation function* $c(i, j)$, describing a direct effect only depending on the interaction potential $V(i, j)$, and the second describing many-body effects, in which i influences j through other molecules. The Ornstein–Zernicke equation, set in 1914 for spherical rigid molecules (nonadditive repulsive potential), and generalized in 1972 to rigid nonspherical molecules, is used as an approximation to connect $g_2(i, j)$ to the direct correlation function. After some manipulations (use is made of an intermediate function indicated as $h(i, j)$) one obtains an expansion into powers of the numeral density, in which the coefficients are given by the integration of appropriate products of direct correlation functions:

$$g_2(i, j) = 1 + c(i, j) + \rho \int c(i, k)c(j, k)\, dr_k + \rho^2 \iint c(i, k)c(j, k)c(k, l)\, dr_k\, dr_l + \cdots.$$

Clearly this approach presents the same problems of the *equation of motion* method. The same set of possible closure relations are in fact used in the two approaches. It is worth to remark that the use of these approaches are limited, with a very few exceptions regarding very simple model fluids (low pressure gases), to a calculation of $g_2(i, j)$ discarding higher terms in the cluster expansion. The same limitation is used in the computer simulations on liquids we shall present in the following of this chapter.

To close this chapter on the physical approach we consider some formal aspects of the calculation of properties. The number of properties that can be treated with physical approaches is large, but does not include many properties of chemical interest (mostly related to the behavior of specific molecules in the liquid). In parallel several properties computed with the physical approach are of little interest in chemistry. It may be worth recalling that properties object of scrutiny in the physical approach can be divided into three categories. First, the properties having a microscopic counterpart and that be defined as averages on the microscopic dynamical functions (they are often called mechanical quantities), examples are the internal energy E and the pressure. Second, the properties that have no microscopical meaning and must be defined in terms of the whole probability density (called thermal quantities); examples are the temperature T and the entropy S. Third, quantities that are fixed in value by the experiment without reference to the internal state of the system; they are called external parameters and typical examples are the number of particles N, the volume V, and the strength of external fields.

The method we are sketching here below can be applied to properties of the first and second type. We limit ourselves to consider time-independent properties in equilibrium systems. The value of a property is given as the average of the values the property has for each distribution, each multiplied by the appropriate statistical weight. The normalization factor of this sum is denoted as the *partition function* Z. Each kind of systems has its own partition function. There is no complete uniformity in the literature about the symbols used to indicate the corresponding partition function, here we shall use Ω, Q, and Ξ for the microcanonical, canonical and grand canonical systems, respectively. The system of reference is the canonical one. The steps leading from these systems to the other will be here omitted. In classical thermodynamics the complex (and complete) set of differential relations connecting the various thermodynamical functions is expressed taking one among them as the leading term (characteristic function). In the (N, V, T) systems the characteristic function is the Helmholtz free energy A. Its expression in statistical thermodynamics is quite simple:

$$A(N, V, T) = -kT \ln Q.$$

The most important thermodynamic functions for canonical systems are reported here below:

Pressure	$P = kT(\partial \ln Q/\partial V)_T$;
Entropy	$S = kT(\partial \ln Q/\partial T)_V + k \ln Q$;
Internal energy	$E = kT^2(\partial \ln Q/\partial T)_V$;
Enthalpy	$H = -(\partial \ln Q/\partial \beta)_V + kTV(\partial \ln Q/\partial V)_T$;
Free energy	A (or G) $= -kT \ln Q$.

Among the five energy functions reported in the table the most important in chemistry is G, also called Gibbs free energy, or free energy tout court. G is the characteristic function for another type of systems: (NPT) called isobaric–isothermal, and it also plays the leading role in the definition of grand canonical systems, through a related thermodynamic function, the chemical potential (in the chemical literature the term chemical potential is often used instead of Gibbs free energy). The function A plays a limited role in chemistry, it has been introduced because the analysis in terms of the canonical ensemble is easier than using other invariants. Remark, however, that when the attention is limited to liquid systems near the standard conditions there is no difference in practice between A and G.

2. Chemical approaches to the study of liquids

The simple model of liquids used in the chemical approach considers liquids as composed by an assembly of molecules at close contact, undergoing incessant collisions. Macroscopic conditions (T and P) rule the kinetic aspects of these collisions, that are however strongly modulated by molecular interactions depending on the chemical nature of the molecules. These interactions are responsible of rare events (with respect to the number of collisions) which give origin to reactions and other phenomena of chemical interest. The same interactions are also responsible of the creation of a partial ordering of local nature, which fades away at larger distances. Chemists tend to mentally average the thermal collisions and to pay more attention to the effects of chemical interactions, both for reactions and for the partial local ordering, which affects the properties of the composing molecules, the solute in particular.

Molecular interactions

The mutual molecular interactions are of different type, some are of very short range, others exert their effects at a longer distance. In general they are nonadditive, but with some exceptions. Chemical practice has permitted to learn a lot about the effects of these

TABLE 2.1

Component	Acronym	Physical meaning
Coulombian	COUL	Electrostatic interactions between rigid charge distributions
Induction	IND	Electrostatic deformation of the charge distributions
Exchange	EXC	Quantum repulsion effect due to the Pauli exclusion principle
Dispersion	DISP	Interactions between fluctuations in the charge distributions
Charge transfer	CHTR	Transfer of electronic charge among partners

forces in the various solvents and obtained some hint about their nature. A more precise classification and definition came from the study with theoretical tools of the behavior of small molecular clusters, composed by two, three or little more molecules (MARGENAU and KESTNER [1971], HOBZA and ZAHRADNIK [1988]). The classification in use for the decomposition of these interactions is reported in Table 2.1.

The whole set of interactions can be described at the quantum mechanical level using the usual electrostatic Hamiltonian, other terms not present in it, like spin-orbit effects or relativistic corrections play a little role in the study of liquid systems.

The complete interaction energy within a given cluster can be easily computed with standard Quantum Mechanical (QM) tools. It is sufficient to compute the energy of the whole system, considered as a *supermolecule*, and then subtract the energies of the separate partners. This interaction energy can be decomposed with the help of some additional QM calculations performed with the same procedure but with deletion of some terms and/or of some steps in the solution of the resulting computational problem. The details are not reported here, suffice it to say that the analysis of such decompositions performed on small clusters at different geometries has permitted to gain a good confidence on the relative importance of these terms, and on their spatial anisotropy.

There is another approach that do not introduce the calculation of the energy of the supermolecule, it was the only approach in use before the advances in electronic computers made possible calculations on supermolecules. Use is made in this approach of the perturbation theory (ARRIGHINI [1981]). The unperturbed Hamiltonian is defined as the sum of the Hamiltonians of the isolated monomers, and the perturbation is just given by the interaction. The perturbative corrections to the energy are expressed order by order and within each order a separation in additive terms corresponding to interactions of different physical origin is performed. There are problems, however, due to the fact that the full Hamiltonian of the cluster has a higher permutational symmetry than the unperturbed one due to the exchange of electrons among partners: the wavefunction of the cluster has to be fully antisymmetrized with respect to all the electrons. This lack of antisymmetry in the exchange of electrons among partners greatly complicates the formulation of the method, and the quality of the results. Here again we do not give details about the procedures elaborated in the years to partly overcome these problems. Suffice it to say that the results obtained for lower orders of the perturbation expansion on small clusters are in fairly good agreement with those obtained with the decomposition of the supermolecule description. The basic point is that neither approaches can be directly used to get a description of the liquid around a given solute.

There are two reasons, strictly connected. The first is that experience has learned that models composed by small size clusters give a very poor description of local solvent effects (and almost nothing on the properties of the whole solution). The second is that the larger aggregates that should be used in modeling require an accurate description of the thermal motions, properly averaged to reach a consistent thermodynamic status. QM supermolecule treatments are ruled out mostly for the difficulty of getting the thermal average. QM calculations at single geometries of a large cluster are feasible with some efforts, but the spanning of the portion of interest of the conformational space, and the following determination and use of the system partition function are out of question. The device often used in the study of isolated chemical systems of looking only at minima and other topological critical points on the potential energy surface (PES) cannot be applied here. The PES presents an exceedingly number of local minima, separated by very low barriers and with flat portions of the surface, hard to treat with statistical mechanics. In practice, the concept of PES is of little utility in treating large molecular aggregates. The direct use of the explicit expressions of the various components of molecular interactions obtained either via decomposition of the supermolecule energy or perturbation theory analysis are out of question for similar reasons. The computational cost is nowadays similar to that of getting the whole supermolecule energy (in addition, the extension of such formulas to the case of many body clusters presents serious difficulties).

There is so the need of reconsidering the problem, paying attention to the physical effects produced by the various types of interactions. We shall give here a short summary of this analysis (TOMASI, MENNUCCI and CAPPELLI [2001]).

COUL. It is the interaction having the larger long range effect. It is strictly nonadditive (i.e., limited to two-body contributions) and strongly anisotropic. In most cases it is the more important contribution (for example in water solutions). It determines the most favorable orientation among partners having asymmetric charge distributions (e.g., dipoles). The formal expression for a two body interaction

$$E_{\mathrm{COUL}} = \iint \rho_A(r_1)\frac{1}{r_{12}}\rho_B(r_2)\,dr_1\,dr_2 = \int \rho_A(r_1)V_B(r_1)\,dr_1$$

suggests that simplifications of this expression can be searched with an opportune modeling of the total charge distribution $\rho(r)$ or of the electrostatic molecular potential $V(r)$. This last presents, as a rule for neutral molecules, positive and negative regions, and the same holds for the integrand giving origin to COUL. Acceptable simplifications must preserve the anisotropy of the property, which, as already said, is essential. One-center multipole expansions work badly if the molecule has a complex shape: better it is to pass to many center expansions. One formulation of large use in chemical computations (especially for solutions of large biological systems) consist in keeping only the first term (a point charge) of multipole expansions centered on all the atoms of the molecule. This approximation is rather grossly, and should be avoided, if possible.

IND. This term is decidedly less anisotropic than the preceding one, but it exhibits a strong nonadditivity. The integrand giving origin to IND is everywhere negative. The modeling of the interaction is generally based on multipole expansions of the electric molecular polarizability. The effect of the other molecules on the charge distribution is in general far from that of a uniform electric field assumed in the definition of the first order polarizability tensor. In spite of it, a formulation often used in computations consists in placing a single isotropic value for the polarizability placed at the center of the molecule. It is a poor approximation, partially justified for a very common solvent, water. The nonadditivity of the contribution is active also at relatively long distances: the electric field of distant molecules must be considered in the contribution at each polarizable center. In turn polarization effects, even when reduced to a single contribution, affects the total polarization. IND must be computed in an iterative way, when these models are used.

EXCH. To satisfy antisymmetry, closed electronic shells have to repel each other. This contribution is everywhere positive. It is nonadditive, but only the effect of nearby molecules play a role, being the contribution considerably short-ranged. Exchange forces can be relatively well described by a set of repulsive stiff potentials, centered on the various nuclei of the molecule, each with a spherical symmetry (typically with a R^{-12} decay law). The exchange forces are the interaction terms left in the oversimplified physical models, in which the molecule is replaced by a sphere, hard or with a soft repulsive potential.

DISP. This term is nonadditive, with quite moderate anisotropy. It may be formally treated as IND, making use here of dynamic multipole polarizabilities. The simple approximation used in chemical computations uses the first term in the development of the first polarizability in terms of the inverse powers of the distance (it is a R^{-6} term) applied again to each atom of the molecule. Dispersion forces are relatively weak and have a moderate long range effect, by far smaller that of COUL and IND. It happens however that the long range effects of the two "classical" electrostatic terms, coupled to their anisotropy, tend to orient molecules in the liquid in the more appropriate orientation, producing so a screening of the global effect. The screening is not active for dispersion terms, and so happens that for several of the examples of liquids we have done in the introduction, the long range interactions are almost solely ruled by dispersion forces. This effect is more evident in the presence of massive bodies.

CHTR. The effects due to the transfer of electronic charge during the molecular encounter are harder to model than the preceding ones. Of course these effects belong to the category of "rare events" but a large part of chemistry is based just on these "rare events". In the approximated formulation of interaction effects we have outlined, electron transfer effects have not been inserted. This is a limitation in the computational procedures using this modelistic elaboration. To consider the effect of charge transfer other methods must be devised.

We have outlined a reasonable way of computing molecular interactions without passing through QM calculations. This procedure can be inserted as a basic element into other procedures performing the systematic scan of the conformational space on the desired thermodynamic ensemble, to get the properties of the liquid. This scan is given via molecular simulations, Monte Carlo (according to a Gibbs picture of thermodynamic averages) or to Molecular Dynamics (according to a Boltzmann picture). These topics will be considered later in this chapter. Such simulations, largely used in the last 20 years, have given a considerable wealth of information about the properties of liquids, by far more detailed and more precise than those obtainable with the physical approach. All the types of liquids we mentioned in the introduction have been at least partially examined with these approaches, and it may be said that what we know today about liquids derives from, or has been confirmed by, molecular simulations. It has been a big effort, quite rewarding. The magnitude of this effort, that has been at a good extent of methodological type, can be appreciated by looking at the final computational costs. To perform a simulation on a liquid system there is to repeat many times the computation of the geometry and of the interactions of a relatively large cluster. For this reason people is compelled to use descriptions of the molecular interactions as simple as possible. In addition, this approach does not give the essential information for which chemists undertook studies on liquids: the effect of a solvent on a chemical reaction and on the molecular properties of the solute. To have them a quantum mechanical description at high level is in fact necessary.

An approach alternative to the simulation on simplified semiclassical interaction potentials exists, and it is able to reach the requested high quality in the description of the molecular units of interest (TOMASI and PERSICO [1994]). This approach is the main subject of this chapter. It is still based on the analysis of the various interaction terms, but in a different way. The basic consideration is that the thermal average performed as the final step in the above described methods can be avoided by replacing the molecular discreteness of the molecular distribution with a continuous distribution function. It is possible to formally define continuous solvent response functions corresponding to the various interaction terms, and satisfying the macroscopic thermodynamic conditions: density, temperature, pressure. It is also possible to formulate equilibrium and nonequilibrium expressions, and to introduce boundary conditions to treat systems different from the bulk liquid. These response functions are then applied to a small molecular system, including the molecule (or molecules) of direct interest, using a QM approach that can be, when desired, of high accuracy. We shall call it the continuum approach, in contrast to the discrete approach of which we have given in the preceding pages a short summary of its basic definition.

CHAPTER II

Continuum Models

As said above, continuum models tend to simplify the problem by introducing solvent response functions describing its interaction with the focused model (we shall call it for brevity the "solute"). The main advantage of this approach consists in a very large reduction of the internal degrees of freedom of the system one has to consider. In passing from the whole solution to the solute + continuum model, a detailed monitoring of all the degrees of freedom of the solvent molecules is no more necessary. The solvent response functions are specific for the various types of interactions occurring in liquids. Among them, that carrying more information is the electrostatic response. For several years also modern continuum methods were limited to the electrostatic interactions only. We shall consider now the origin and evolution of this electrostatic model starting from the seminal paper by ONSAGER [1936] in which he presented several concepts that are the basis of modern continuum methods.

The model used in Onsager's paper was quite simple. A solute, reduced to a point dipole, μ, but provided of polarizability α, is placed into a spherical cavity of appropriate radius; the solvent, placed out of the cavity is described as a continuous isotropic dielectric. The solute charge distribution induces a polarization of the dielectric, and in turns, this polarized medium polarizes the solute charge distribution, via an electric field, called by Onsager the *solvent reaction field* $\mathbf{R}$. As a consequence the solute dipole changes from μ to μ^*, this last depending on μ, its polarizability α, the dielectric constant ε, and the radius a of the cavity. The elaboration of this formula, quite simple, was done on the basis of classical electrostatics.

We have anticipate that this simple model introduces some basic concepts, among them we first quote the cavity. Onsager's definition of cavity is a physical entity: it corresponds to a portion of the physical space in which the solvent is not allowed, because already occupied by the solute molecule. Kirkwood, another eminent figure in the study of liquid systems, remarked a short time after that this was an epoch-making innovation: all the cavities used before (e.g., Maxwell and Lorentz cavities, BÖTTCHER [1973], BÖTTCHER and BORDEWIJK [1978]) were just mathematical devices. Onsager was well aware of it. In its quoted papers he paid attention to the shape of the cavity, to its dependence on the thermally induced volume changes, to the problems related to the possible occurrence of hydrogen bonds. These are problems amply treated in the more recent versions of continuum models.

The second concept introduced by Onsager is that of the reaction field. He spoke of the reaction field because its solute model was a dipole, now we are speaking more generally of a reaction potential. Here again he paid attention to the physical content of this concept: he analyzed some aspects of the nonideal behavior the homogeneous continuum dielectric may have in real systems, like the phenomena of nonlinearity that can be phenomenologically related to dielectric saturation and to electrostriction effects.

Onsager also introduced a third concept, that of the cavity field **G**, occurring when the liquid sample is subjected to an external electric field **E**. **G** is related to **E** and to the geometrical factors defining size and shape of the cavity. The field **R** is the father of the electrostatic solute–solvent interaction potentials we are using in QM continuum models, **G** plays an important role in the very recent extensions of the continuum model to compute molecular properties. The development of these basic ideas will be discussed in the following pages. Here we remark that the original Onsager model continues to be amply employed, especially to get a rationale of experimentally observed trends in chemical properties. Some modifications of this models introduced new concepts. We quote its application to the description of solvent shifts in electronic spectra.

An important development was the translation of this model in a QM language. In the models called Onsager–SCRF (or simply SCRF, were the acronym stays for self-consistent reaction field) the solute is described at the QM level, put in a spherical cavity and subjected to the action of a **R** field having as its origin the dipolar contribution of the charge distribution of the solute. This model is quite simple to implement and to use when a code for the calculation of QM molecular wavefunctions is available. For this reason the first implementation of the Onsager–SCRF model (TAPIA and GOSCINSKI [1975]) was done forty years after the original Onsager paper. Actually few years before (1973) a proposal was made by Rivail's group in Nancy to introduce in the model other terms of the multipole expansion of the solute charge distribution. The proposal was expressed in a better way and documented by results in RIVAIL and RINALDI [1976]. This innovation eliminates at a good extent a defect of the too simple description of the electrostatic terms given in the original version: the use of the dipole only may produce serious deformations in the description of the solvent effects for molecules with a complex shape. A typical example is that of solutes having two or more identical polar groups, which are spatially arranged in the molecule to give a net value of the total dipole equal to zero: in this case the Onsager–SCRF model gives zero solvent effects. The original SCRF model has been continuously refined: we quote here the extension to ellipsoidal cavities and to cavities with a molecular shape. SCRF models are easy to implement and to use when the cavity has a constant curvature (sphere, ellipsoid). At present, they may also be used for more realistic cavities with a shape modeled on that of the molecule, but the elaboration of the model and in particular its use is a bit more delicate. The multipole expansion still presents some limits for complex molecules, that could be partially eliminated by using segmental local expansions, with expansion centers placed at opportune sites of the molecule. There exists programs able to do it, but they are rarely employed. The SCRF method has been further generalized to high-level QM approaches (like MCSCF and Coupled-Cluster) by MIKKELSEN, CESAR, ÅGREN and JENSEN [1995] who implemented the spherical version of the SCRF model in the Dalton quantum mechanical code (HELGAKER ET AL. [2001]).

An alternative approach to SCRF-like continuum models was proposed by our group. The first paper is of MIERTUŠ, SCROCCO and TOMASI [1981]. In this approach the multipole expansion used in all the previous models was replaced by another way of solving the electrostatic problem posed by the model. The solution of the Laplace and Poisson equations requested by the model was expressed in term of an apparent charge distribution, defined by applying theorems of classical electrostatics, and spread on the cavity surface. The solvent reaction potential was expressed in terms of this apparent charge properly discretized into point charges and introduced in the Hamiltonian of the solute as a solute–solvent interaction potential. In such a way the multipole expansions of both the molecular and the reaction potentials, rather cumbersome in the case of cavities of molecular shape, was avoided. The method was presented within a QM formalism for computing molecular wavefunctions of ab initio type (the previously quoted SCRF methods were at that time all at the semi-empirical level, probably because it was not easy to compute high multipole integrals with ab initio codes). The procedure we devised was the application to a QM problem of the boundary element method (BEM). We realized it later; actually at that time the name was not yet diffused in the literature. Afterwards it was called Polarizable Continuum Model (PCM) and it continues to keep this name even if many important improvements have been introduced in the years with respect to the original version (CAMMI and TOMASI [1995b], CANCÈS and MENNUCCI [1998b], CANCÈS, MENNUCCI and TOMASI [1997], MENNUCCI, CANCÈS and TOMASI [1997]).

We shall consider later more details and extensions of PCM. Other methods sharing some similarity with PCM appeared in the nineties: among them we quote the so-called COSMO model and the methods making explicit use of other mathematical techniques, as the Finite Element method (FEM) and the finite difference (FD) approach. In the last two cases the solvent response function is obtained by sampling the solvent electric potential at a relatively large number of points inside the bulk dielectric. One among these methods has gained wide popularity in a semi-classical version (the solute charge distributions is reduced to a set of point charges on the nuclei) to study large biomolecules (CHEN and HONIG [1997]). In the method known with the acronym COSMO, the model is different as the screening effects in the dielectric are replaced by the screening effects in a conductor (KLAMT and SCHÜÜRMANN [1993], TRUONG and STEFANOVICH [1995]). In other words, COSMO method is a solution of the Poisson equation designed for the case of very high ε, and it takes advantage of the analytic solution for the limit case of a conductor ($\varepsilon = \infty$), for which the boundary condition reduces to $V = 0$ on the surface. Anyway, apart from these differences in the theoretical background, COSMO can be described exactly in the same way as PCM.

The last solvation approach we are here quoting adopts a different strategy. It is an extension of a simple model Born employed many years ago (1921) to describe the solvation energy of a simple atomic ion. The approach is often called GB (generalized Born) approach (CRAMER and TRUHLAR [1991]). The charge distribution of the molecule is reduced to atomic point charges; the Born formula is generalized taking into account that at each atom a sphere is assigned and that such spheres partially overlap each other. The procedures following this approach contain a number of parameters that have to be empirically fixed. A version of this approach, with a line of versions progressively improving the method, has been accurately parametrized and it is widely used, first in semi-empirical versions, and now also with ab initio codes

(ZHU, LI, HAWKINS, CRAMER and TRUHLAR [1998]). The QM part of the procedure is necessary to update during the calculations the solute atomic charges, modified by the reaction potential. We have summarized the methods we consider more important for the applications the have had in the chemical domain, but several other exists, reference is made to a detailed review by TOMASI and PERSICO [1994] that can be partially adjourned by looking at CRAMER and TRUHLAR [1999].

In the following we shall made reference to the PCM-like methods only, because they are the methods that have had more extensions. We have had until now considered electrostatic models for molecules in the bulk of homogeneous liquids. The structure of PCM makes easy to introduce further boundary conditions similar to the cavity surface. Without important changes in the general structure, it is easy to divide the whole space into nonoverlapping portions, each having a given dielectric function and containing, or not, an electric charge distribution to be treated classically or at the desired QM level. This opens the possibility to study important aspects of many phenomena of chemical interest, as the description of molecular encounters giving origin to associates or to new compounds, to follow the decomposition of a molecule into separate portions, to describe the interactions among molecules separated by the solvent.

We have here recalled other phenomena occurring in the bulk of the solvent, but the extension to other types of liquids is also immediate. We quote as examples the description of the partition of a solute between two immiscible liquid phase (a procedure which has very old traditions in chemistry and many applications in drug design), the chemical processes occurring within a droplet dispersed in another liquid or suspended in the atmosphere, the behavior of molecules within a membrane, a vesicle or a molecular host (great attention is now paid to the technological exploitation of that change in the molecular properties induced by these restricted systems). The formal transformation of PCM conceived for bulk is easy, but to have a proper modelistic description of the related phenomena there is to consider other aspects of the basic solvation models we have not yet introduced. Another subject of interest is the behavior of liquids, and of solutes at the liquid surface. A lot of chemistry is given at the surfaces: liquid/gas, liquid/liquid. liquid/solid. It is an enormous field in which important chapters of basic science are intermingled with very important practical problems. It is not possible to give a correct impression just quoting few names of phenomena or of sub-fields, so we prefer to limit ourselves to this indication, adding the remark that the work in this field is at its very beginning, and that continuum electrostatic methods have to be strongly supplemented to give reliable results.

We may add here at the list of extensions of PCM, those regarding salt solutions and liquid crystals. Both have introduced in PCM by a renewed version, called IEF (integral equation formalism) (CANCÈS and MENNUCCI [1998b], CANCÈS, MENNUCCI and TOMASI [1997], MENNUCCI, CANCÈS and TOMASI [1997]) on which more will be said in the following. Here, we add only few remarks. Both slat solution and liquid crystals are of considerable scientific and practical importance. Biological processes solely occur in saline solutions, and the chemical physics of salt solutions has been the object of continuous studies and modelizations since almost 150 years. The liquid crystals are relatively newcomers in the realm of chemistry, but have rapidly gained popularity because their very peculiar properties, exploited in many ways: optical display, to give an example are almost all composed by liquid crystals subjected to an external electric field.

3. The energy of the system and the solvation energy

The energy of dissolution of a chemical into a solvent is one of the basic quantities in chemistry. Precise measurements are required in many fields of chemistry and chemical engineering. Actually to get accurate experimental values is a very hard task and, after more than one hundred years of efforts, intrinsic solvation energies are known for a very limited number of substances in a few solvents, and often with large error bars. According to the experimental setup solvation free energies or enthalpies are obtained. We recall that the former has a wider interest, because it governs the chemical transformations, the change of liquid phase, the electrochemical processes, and all the other processes in which chemistry is involved: enthalpies are not sufficient to such purposes. We also recall that thermodynamic quantities always require reference values. To formulate the solvation energy problem in computational methods it is convenient to make use of canonical ensembles. We shall consider now how this problem has been formulated in the several versions of the continuum PCM methods.

The PCM continuum model considered in this chapter uses ab initio QM methods to describe the solute. In this framework the reference energy (the zero value in the final outcome) is that of a system composed by the pure unperturbed liquid at the given (N, P, T) conditions (alternatively to the N, V, T conditions, according to the already expressed remarks) supplemented by the opportune number of electrons and nuclei, necessary to describe the solute, not interacting and at zero kinetic energy. We shall now consider two different ways of defining the total energy of the system and the energy difference corresponding to the solvation energy.

(A) The solvation process can be split into two parts: (1) the formation of a cavity of suitable shape and size within the homogeneous liquid, and (2) a charging process of formation the solute, starting from the electrons and nuclei, within the cavity. During this process the solute–solvent interactions play an active role contributing to give the final result in which both solvent and solute have reached an equilibrium including mutual interaction effects. The energy related to the first process must be computed apart.

In this scheme we have implicitly made use of one of the physical approaches to the statistical description of liquids that we have introduced before. This method, called Scaled Particle Theory (SPT), gives exact analytical expressions for the formation of a cavity of spherical shape in a liquid composed by spherical units (PIEROTTI [1976]). For coherence with the other terms this process is quantified in terms of the corresponding cavitation free energy, G_{cav}. In the SPT model, the solvent distribution function is modified only in the region corresponding to the cavity without disturbing the remainder, as requested by our two-step process. Actually, the procedure requires some checks, because in realistic applications neither the solute nor the solvent has a spherical shape. Checks have been done via MC and MD simulations, with satisfactory results, but others would be advisable.

The main part of process (2) is given by the ab initio PCM calculation in which the electrostatic, dispersion and repulsion interaction terms are coupled to the inner electrostatic interactions giving rise to the molecule. In doing this, a problem appears. In the various replicas of the canonical ensemble, the point where the cavity has been formed and the molecule has been built, occurs at different positions, being not fixed by the model. This fact gives rise to an additional entropic contribution, called communal entropy; it corresponds to the fact that the solute molecule is free to wander within the whole volume V of the system. Actually, it is a not complete freedom, because V is occupied at a large extent by solvent molecules. From here it arises the concept of free volume, on which fifty years have been spent in discussions. The solution we consider to be definitive, has been given by BEN-NAIM [1974], BEN-NAIM [1987] with the introduction of the concept of *liberation free energy*, the free energy spent to allow to solute to pass from a given position to others within V. No details will be given on the elaboration of this idea, which has been accepted in PCM. The free energy G given by PCM actually corresponds to a molecule with nuclei frozen at a given geometry (the Born–Oppenheimer approximation is used in the ab initio calculations). The contribution of nuclear motions are computed with statistical mechanics techniques and may be divided into vibrational and rotational contributions: G_{vib} and G_{rot}. To do it, formally one has to factorize the canonical partition function into molecular components q. This factorization is not exact; there are couplings, especially in the rotational component, a fact which prompts to improve further the model.

(B) A second possible approach is divided into three steps: (1) cavity formation, (2) formation in vacuo of the molecule with a charging process, and (3) insertion of the molecule in the cavity. With this second approach we may give an expression of the solvation energy based on the passage of the solute molecules from the gas to the liquid phase:

$$G_{\text{sol}} = \Delta G + G_{\text{cav}} + \Delta G_{\text{vib}} + \Delta G_{\text{rot}},$$

where $\Delta G = G - G^0$ is the difference between the free energies given by the two QM calculations (in the BO approximation), G^0 in the ideal gas phase and G in solution. The translational contribution to the free energy has been omitted, because of the use of the Ben-Naim treatment, based on a definition of the standard concentration in terms of the volume, and not of the pressure as it is usual for the gas phase.

The definition of the energy becomes a bit more complex when one studies (i) an association process, where two solutes A and B, merges into a unique entity AB with the consequence that six degrees of rotational and vibrational type of the two molecules become internal degrees of vibrations; (ii) in the case of liquid crystal, in which G depends on the orientation of the solute with respect to the axes of the dielectric tensor; (iii) in the case of molecules near a boundary, where G depends on the distance (and orientation) with respect to the boundary. We will not dwell on these problems and others similar.

The whole charging process done to compute G can be partitioned into three separate charging processes when the solute–solvent interaction potential is separated into three parts: $V = V_1 + V_2 + V_3$, where 1, 2, 3 stay for electrostatic, repulsion, and dispersion. The charging processes can be done separately, if one neglects couplings. The expression to use is based on the fact that $\ln Q = -kT(G + \Delta G)$, where G is the free energy of the system in absence of solute-solvent interactions and:

$$\Delta G = \sum_{m=1}^{3} \int_0^1 d\lambda_m \int dr\, \rho V_m g'(r; \lambda_m)$$

in which λ_m are three charging parameters (assimilated respectively to charges, overlaps and dynamic polarizabilities) going from 0 to 1, and $g'(r; \lambda_m)$ is the reduced solute-solvent distribution function when the interaction potential is $\lambda_m V_m$. Actually, this last condition is satisfied only for the electrostatic term, being the others computed at constant distribution function. The formula is of course approximate, but comparison with solvation energies computed with the approach B (in which couplings are at least in part considered), and the analysis of a few computer simulations performed ad hoc, indicate that in normal cases the errors are quite limited.

Dispersion and repulsion contributions

In general, the modeling of dispersion and repulsion interactions in solution is based either on a discrete molecular description of the liquid or on a continuum model.

The discrete approach is generally based on the use of pair potentials related to atoms or groups of atoms of the solvent S (here indicated with l) and the solute M (here indicated with m):

$$V_{\text{dis-rep}} = \sum_{m \in M} \sum_{l \in S} V_{ml}(r_{ml}) \quad \Longleftrightarrow \quad V_{ml}(r_{ml}) = \sum_n \frac{d_{ml}^{(n)}}{r_{ml}^n}$$

the dispersion ($n = 6, 8, 10$) and the repulsion ($n = 12$) coefficients are taken from the literature. Often an alternative exponential expression, more related to the physical interpretation of the interaction, is used for the repulsion term: $c_{ml} \exp(-\gamma_{ml} r_{ml})$. The related approximate expression for the dispersion–repulsion contribution to G is derived in terms of continuous distribution functions $\rho_{ml}(\mathbf{r}_{ml}) = N_l n_S g_{ml}(\mathbf{r}_{ml})$, where N_l is the number of groups of type l in each solvent molecule, n_S the solvent macroscopic density, and $g_{ml}(\mathbf{r}_{ml})$ a correlation function depending on the position of l with respect to m (we use here the bold character to indicate vectors). As a final result we can write:

$$G_{\text{dis-rep}} = \sum_{l \in S} \sum_{m \in M} \int \rho_{ml}(\mathbf{r}_{ml}) V_{ml}(\mathbf{r}_{ml})\, d\mathbf{r}_{ml}. \tag{3.1}$$

It should be clear that to go further other approximations are needed; in particular the integral operation can be simplified by defining for each l an appropriate portion of space in which there are no l centers, and $g_{ml}(\mathbf{r}_{ml})$ is always zero. This portion of space can be identified with the cavity C_l related to the Van der Waals spheres centered on the nuclei of the solute and enlarged to take into account the correspondent radius of the solvent. When the set of C_l cavities is known, we may replace the volume integrals in Eq. (3.1) with surface integrals over the surface Σ_l of the cavities C_l (FLORIS, TOMASI and PASCUAL-AHUIR [1991]) by introducing auxiliary vector functions defined in terms of the pair potential and the correlation functions. Simple analytic expressions of these functions can be obtained by reducing each $g_{ml}(\mathbf{r}_{ml})$ to a step function 0 inside and 1 outside C_l (this is often called the "uniform approximation", and is congruent with the properties of solutions at infinite dilution) and by performing a simple one-dimensional integration over r. It has to be noted that such methodology is completely independent on solute charge distribution; it thus does not involve any QM description of the system, but it only leads to an additional term in the total free energy.

The second possible approach we have mentioned to get $G_{\text{dis-rep}}$ is based on a continuum model; in this case the two contributions are treated separately. The starting point is the general expression derived from the intermolecular forces. In the previous discrete approach however, this equation were applied to calculated or estimated potentials V_{rep}^{AB} available from the literature. We now substitute V_{rep}^{AB} for a suitable expression taken directly from the theory considering that, as originated from the Pauli exclusion principle, the repulsion forces between two interacting molecules increase with the overlap of the two distributions and are strictly related to the density of electrons with the same spin (AMOVILLI and MENNUCCI [1997])

$$G_{\text{rep}} = \frac{1}{2}\rho_B \int g_{AB}(\mathbf{r})\, d\mathbf{r} \int \frac{d\mathbf{r}_1\, d\mathbf{r}_2}{r_{12}} P_A(\mathbf{r}_1; \mathbf{r}_2) P_B(\mathbf{r}|\mathbf{r}_2; \mathbf{r}_1).$$

Here the label A refers to the solute, B to the solvent, $\mathbf{r}$ is an appropriate set of coordinates defining the internal geometry of the complex AB, ρ_B is the number density and g_{AB} is a correlation function which is 0 inside the solute cavity ($\mathbf{r} \in C$) and 1 outside ($\mathbf{r} \notin C$). At this point, because in the continuum approach here exploited the electron density of the solvent is not given, it is useful to make the following two assumptions: (i) each valence electron pair of the solvent molecules can be localized in bond and lone pair regions and (ii) each pair, owing to the thermal motion of the solvent molecules, will have the same probability to be found at any point of the solution not occupied by the solute. The resulting solvent density $P_{\text{pair}}(\mathbf{R}|\mathbf{r}_2; \mathbf{r}_1)$ (here $\mathbf{R}$ is the coordinate of the centroid of the localized orbital containing the pair in a reference frame fixed on the solute molecule) can be represented in terms of a Gaussian representation of localized orbitals; this yields the simple expression

$$G_{\text{rep}} = \alpha \int_{\mathbf{r} \notin C} d\mathbf{r}\, P_A(\mathbf{r}), \tag{3.2}$$

where α is a suitable constant defined by some selected properties of the solvent.

We note that in Eq. (3.2) the repulsion energy is proportional to the fraction of solute electrons outside the cavity; as a consequence the inclusion of this contribution provides an automatic confinement of the electronic cloud of the solute.

The continuum approach to the dispersion term has a much longer history than for the repulsion, and different procedures have been developed.

The original formal theory is expressed in terms of quantum electrodynamics and the continuum medium characterized by its spectrum of complex dielectrics frequencies. A successive formulation derived from this theory, is based on the extension of the reaction field concept to a dipole subject to fluctuations exclusively electric in origin. In this framework, the dispersion free energy of a molecule placed in a cavity immersed in the solvent is related to the molecular polarizability and the complex dielectric constant. A couple of methods now in use (RINALDI, COSTA-CABRAL and RIVAIL [1986], AGUILAR and OLIVARES DEL VALLE [1986]), have extended this treatment, in origin limited to the dipolar approximation and a spherical cavity, to a full multipolar expansion and to any cavity shape. Actually, the two methods are quite different either in the solvation model they exploit or in the practical implementation; anyway an important aspect is common to both: the inclusion of the dispersive term in the solute effective Hamiltonian which allows a real influence of these interaction forces on the electronic structure of the solvated molecule (i.e., on the description of its wavefunction). In 1997 an alternative procedure (AMOVILLI and MENNUCCI [1997]) has been formulated starting, as for the repulsion contribution, from the theory of intermolecular forces. In this framework, the expression of the dispersion energy between two molecular systems A and B is given in terms of generalized frequency dependent polarizabilities. Following a scheme commonly exploited to derive the electrostatic contribution to the interaction energy, the molecule B is substituted by a continuum medium, the solvent, described by a surface charge density $\sigma_B[\varepsilon_B(i\omega), P_A(0K|\mathbf{r})]$ induced by the electric field of the solute A and spreading on the cavity surface. In defining this surface charge density the transition density $P_A(0K|\mathbf{r})$ for the solute A (for transition to state K) has to be computed as well as the solvent dielectric constant calculated at imaginary frequencies, $\varepsilon_B(i\omega)$ has to be known. The reduction of the general expression to an equation containing only terms related the solute ground state can be obtained through some simplifications on the form of σ_B and the nature of the excited states to be considered.

Electrostatic contributions

The electrostatic interactions between the charge distributions which compose the molecular system (point nuclei and electronic cloud in quantum chemistry, points charges and multipoles in molecular mechanics) are affected by the presence of the dielectric. Indeed, in solvation continuum models in which the molecular system is placed in a volume (called molecular cavity) surrounded by a continuum dielectric, the interaction energy between the two charge distributions ρ_1 and ρ_2 reads

$$E_s(\rho_1, \rho_2) = \int_{\mathbb{R}^3} \rho_1 V_2 = \int_{\mathbb{R}^3} \rho_2 V_1 = \int_{\mathbb{R}^3} \varepsilon \nabla V_1 \cdot \nabla V_2,$$

where the electrostatic potential V_k generated by ρ_k satisfies

$$-\operatorname{div}\big(\varepsilon(x)\nabla V_k(x)\big) = 4\pi\rho_k(x) \tag{3.3}$$

with $\varepsilon(x) = 1$ inside the cavity Ω and $\varepsilon(x) = \varepsilon$ outside (ε denotes the macroscopic dielectric constant of the solvent).

It is useful to decompose the potential V in the sum of the electrostatic potential

$$\phi := \rho \star \frac{1}{|x|}$$

generated by the charge distribution ρ in vacuo and the "reaction potential"

$$V^r := V - \phi.$$

If we indicate as $G(x, y) = \frac{1}{|x-y|}$ the Green kernel of the operator $-\frac{1}{4\pi}\Delta$, $G^s(x, y)$ the Green kernel of the operator $-\frac{1}{4\pi}\operatorname{div}(\varepsilon\nabla\cdot)$ and $G^r(x, y) := G^s(x, y) - G(x, y)$, the following relations yield

$$V(x) = \int_{\mathbb{R}^3} G^s(x, y)\rho(y)\,dy,$$

$$\phi(x) = \int_{\mathbb{R}^3} G(x, y)\rho(y)\,dy,$$

$$V^r(x) = \int_{\mathbb{R}^3} G^r(x, y)\rho(y)\,dy.$$

Following this scheme the total interaction energy $E_s(\rho_1, \rho_2)$ is usually split into two terms

$$E_s(\rho_1, \rho_2) = \mathcal{D}(\rho_1, \rho_2) + E_r(\rho_1, \rho_2),$$

where $\mathcal{D}(\rho_1, \rho_2)$ and $E_r(\rho_1, \rho_2)$, respectively, denote the interaction energy in the vacuum, and the so-called reaction-field contribution to the energy

$$\mathcal{D}(\rho_1, \rho_2) = \int_{\mathbb{R}^3}\int_{\mathbb{R}^3} \frac{\rho_1(x)\rho_2(y)}{|x-y|}\,dx\,dy,$$

$$E^r(\rho_1, \rho_2) := \int_{\mathbb{R}^3} \rho_1 V_2^r = \int_{\mathbb{R}^3} \rho_2 V_1^r = \int_{\mathbb{R}^3}\int_{\mathbb{R}^3} \rho_1(x) G^r(x, y)\,\rho_2(y)\,dx\,dy.$$

Beyond the standard continuum model

The electrostatic problem of charge distributions embedded in a cavity surrounded by a continuum dielectric can be extended from the standard homogeneous isotropic dielectrics, characterized by a constant scalar permittivity, ε to more complex systems such as homogeneous anisotropic dielectrics, characterized by a constant tensorial permittivity, ε, systems formed by two or more different dielectrics separated by well-defined boundaries, and solutions in which the dissolved electrolyte charges are free to move in the surrounding medium.

Anisotropic solvent. To extend the continuum model defined by Eq. (3.3) to anisotropic solvents in which the dielectric permittivity is represented by a tensor, we have to solve the anisotropic Poisson equation

$$-\operatorname{div}\big(\varepsilon(x)\nabla V(x)\big) = 4\pi\rho(x). \tag{3.4}$$

The dielectric permittivity $\varepsilon(x)$ is no more a scalar quantity but a 3×3 tensor of the type

$$\varepsilon(x) = \begin{cases} \mathbf{I}_3 & \text{if } x \in \Omega, \\ \varepsilon_s & \text{if } x \in \mathbb{R}^3 \setminus \overline{\Omega}. \end{cases}$$

($\mathbf{I}_3$ is here the unity 3×3 tensor.) For physical reasons the tensor ε is symmetric. The model of anisotropic continuum well represent liquid crystalline phases and crystal matrices

Ionic solutions. The ionic solutions (i.e., solutions in which electrolyte charges are dissolved) well represent biological environments and in particular physiological solutions. An ionic solution can be described through a continuum model if the Poisson equation (3.3) is replaced by the (nonlinear) Poisson–Boltzmann equation:

$$-\operatorname{div}\big(\varepsilon(x)\nabla V(x)\big) + \varepsilon(x)\kappa^2(x)kT\sinh\big(V(x)/kT\big) = 4\pi\rho(x) \tag{3.5}$$

with

$$\varepsilon(x) = \begin{cases} 1 & \text{if } x \in \Omega, \\ \varepsilon_s & \text{if } x \in \mathbb{R}^3 \setminus \overline{\Omega}, \end{cases}$$

and

$$\kappa(x) = \begin{cases} 0 & \text{if } x \in \Omega, \\ \kappa_s & \text{if } x \in \mathbb{R}^3 \setminus \overline{\Omega}. \end{cases}$$

The constant κ accounts for the ion screening: its inverse is known as the Debye length.

The hyperbolic sine function in Eq. (3.5) derives from a treatment of statistical physics in the thermodynamical equilibrium approximation, namely it represents the distribution of the mobile ions in the field of the nonlinear electrostatic potential V. If we represents such function in terms of a Taylor expansion and we take the linear term only, we obtain the so-called linearized PB equation (LPB).

$$-\operatorname{div}\big(\varepsilon(x)\nabla V(x)\big) + \varepsilon(x)\kappa^2(x)V(x) = 4\pi\rho(x).$$

The LPB equation is a result of the Debye–Hückel approximation applicable in the case of low potentials, a condition approached at low concentrations. For electrolytes of high-valence type and for electrolytes in media of low dielectric constant, the same approximation becomes not good.

4. Computation of the electrostatic contribution

Let us now focus on the calculation of the reaction field interaction energy

$$E^r(\rho, \rho') = \int_{\mathbb{R}^3} \rho' V^r$$

with $V^r = V - \phi$ and $\phi = \rho \star \frac{1}{|x|}$, V denoting the unique solution to Eq. (3.3) zeroing at infinity. The problem whose V^r is solution has the following characteristics: (i) it is posed on $\mathbb{R}^3$; (ii) it exhibits an interface; (iii) inside (resp. outside) the cavity the partial differential equation is linear and the differential operator has constant coefficients. Those three characteristics make it natural to resort to an integral method: the original three-dimensional problem (3.3) posed on an unbounded domain ($\mathbb{R}^3$) can thus be replaced by a two-dimensional problem posed on a bounded manifold (the interface $\Gamma = \partial\Omega$).

Basics on integral equations

For the reader convenience, we state here some basic results on integral equations which are used below. For more details, the reader is referred to HACKBUSH [1995], NÉDÉLEC and PLANCHARD [1973], NÉDÉLEC [1994]. Let us consider a function V satisfying

$$\begin{cases} -\Delta V = 0 & \text{in } \Omega, \\ -\Delta V = 0 & \text{in } \mathbb{R}^3 \setminus \overline{\Omega}, \\ V \to 0 & \text{at infinity,} \end{cases}$$

and whose interior (V_i, $\frac{\partial V}{\partial n}|_i$) and exterior ($V_e$, $\frac{\partial V}{\partial n}|_e$) traces on $\Gamma = \partial\Omega$ are well-defined and continuous. Denoting by

$$[V] := V_i - V_e \quad \text{and} \quad \left[\frac{\partial V}{\partial n}\right] := \left.\frac{\partial V}{\partial n}\right|_i - \left.\frac{\partial V}{\partial n}\right|_e,$$

the following *representation formulae* can be stated: the function V satisfies for any $x \notin \Gamma$,

$$V(x) = \int_\Gamma \frac{1}{4\pi|x-y|}\left[\frac{\partial V}{\partial n}\right](y)\,dy - \int_\Gamma \frac{\partial}{\partial n_y}\left(\frac{1}{4\pi|x-y|}\right)[V](y)\,dy \tag{4.1}$$

and for any $x \in \Gamma$,

$$\frac{V_i(x) + V_e(x)}{2} = \int_\Gamma \frac{1}{4\pi|x-y|}\left[\frac{\partial V}{\partial n}\right](y)\,dy - \int_\Gamma \frac{\partial}{\partial n_y}\left(\frac{1}{4\pi|x-y|}\right)[V](y)\,dy. \tag{4.2}$$

For $x \in \Gamma$, one has in addition

$$\frac{1}{2}\left(\frac{\partial V}{\partial n}\bigg|_i + \frac{\partial V}{\partial n}\bigg|_e\right)(x) = \int_\Gamma \frac{\partial}{\partial n_x}\left(\frac{1}{4\pi|x-y|}\right)\left[\frac{\partial V}{\partial n}\right](y)\,dy \tag{4.3}$$

$$- \int_\Gamma \frac{\partial^2}{\partial n_x \partial n_y}\left(\frac{1}{4\pi|x-y|}\right)[V](y)\,dy. \tag{4.4}$$

The two last relations suggest to introduce the operators S, D, D^* and N formally defined for $\sigma : \Gamma \to \mathbb{R}$ and $x \in \Gamma$ by

$$(S \cdot \sigma)(x) = \int_\Gamma \frac{1}{|x-y|}\sigma(y)\,dy, \tag{4.5}$$

$$(D \cdot \sigma)(x) = \int_\Gamma \frac{\partial}{\partial n_y}\left(\frac{1}{|x-y|}\right)\sigma(y)\,dy, \tag{4.6}$$

$$\left(D^* \cdot \sigma\right)(x) = \int_\Gamma \frac{\partial}{\partial n_x}\left(\frac{1}{|x-y|}\right)\sigma(y)\,dy, \tag{4.7}$$

$$(N \cdot \sigma)(x) = \int_\Gamma \frac{\partial^2}{\partial n_x \partial n_y}\left(\frac{1}{|x-y|}\right)\sigma(y)\,dy. \tag{4.8}$$

When the surface Γ is regular (C^1 at least), the Green kernels of the operators S, D and D^* present integrable singularities on the surface Γ; it is easy to check that they behave as $\frac{1}{|x-y|}$ when y goes to x (then $|(x-y)\cdot n_x| \sim |(x-y)\cdot n_y| \sim |x-y|^2$ when y is close to x). On the other hand, the Green kernel of the operator N is hypersingular (it behaves as $\frac{1}{|x-y|^3}$ when y is close to x) so that the notation (4.8) is only formal: even for regular σ, the integral $\int_\Gamma \frac{\partial^2}{\partial n_x \partial n_y}(\frac{1}{|x-y|})\sigma(y)\,dy$ has to be given a sense of Cauchy principal value.

The operators S, D, D^* and N satisfy the following relations (which follow from the properties of Calderon operators (HACKBUSH [1995])): (i) the operators S and N are self-adjoint on $L^2(\Gamma)$ and D^* is the adjoint of D; (ii) one has $DS = SD^*$, $DN = ND^*$, $D^2 - SN = 4\pi^2$, $D^{*2} - NS = 4\pi^2$.

Let us end up this section with the definition of the so-called single-layer and double-layer potentials.

A *single-layer potential* is a function V which can be written as

$$V(x) = \int_\Gamma \frac{\sigma(y)}{|x-y|}\,dy, \quad \forall x \in \mathbb{R}^3,$$

with $\sigma \in H^{-1/2}(\mathbb{R}^3)$. A single-layer potential is well-defined and continuous on $\mathbb{R}^3$ (in particular $[V] = 0$). Its normal derivative presents a discontinuity at the crossing of the interface Γ given by the formula

$$\left[\frac{\partial V}{\partial n}\right] = \frac{\partial V}{\partial n}\bigg|_i - \frac{\partial V}{\partial n}\bigg|_e = 4\pi\sigma.$$

The density σ is solution to the integral equation on Γ

$$S \cdot \sigma = V.$$

A *double-layer potential* is a function V which can be written as

$$V(x) = \int_\Gamma \frac{\partial}{\partial n_y}\left(\frac{1}{|x-y|}\right)\sigma(y)\,dy, \quad \forall x \in \mathbb{R}^3,$$

with $\sigma \in H^{-1/2}(\mathbb{R}^3)$. A double layer potential is continuous on $\mathbb{R}^3 \setminus \Gamma$ but presents a discontinuity at the crossing of the interface Γ given by

$$[V] = V|_i - V|_e = 4\pi\sigma.$$

On the other hand, its normal derivative is continuous at the crossing of Γ. The density σ is solution to the integral equation on Γ

$$N \cdot \sigma = -\frac{\partial V}{\partial n}.$$

Numerical aspects

Usually, integral equations arising in potential theory are numerically solved by either a *collocation* or a *Galerkin* method; in the latter case, boundary elements are most often used. Let us detail both methods on the example of a linear integral equation

$$A \cdot \sigma = g, \tag{4.9}$$

where the unknown σ belongs to $H^s(\Gamma)$, the right hand side g is in $H^{s'}(\Gamma)$, and the integral operator $A \in \mathcal{L}(H^s(\Gamma), H^{s'}(\Gamma))$ is characterized by the Green kernel $a(x, y)$:

$$(A \cdot \sigma)(x) = \int_\Gamma a(x, y)\sigma(y)\,dy, \quad \forall x \in \Gamma.$$

Let us consider a mesh $(T_i)_{1 \leqslant i \leqslant n}$ on Γ, that will be considered as drawn on the curved surface Γ; let us denote by x_i a representative point of the element T_i (e.g., its "center"). Theoretical and numerical details on the methodologies developed to determine surface meshes will be given in Section 6.

The P_0 collocation and Galerkin methods for solving Eq. (4.9) provide two approximations of σ in the space V_h of piecewise constant functions whose restriction to each element T_i is constant: in the collocation method, σ^c is the solution to

$$\int_\Gamma a(x_i, y)\sigma^c(y)\,dy = g(x_i), \quad \forall 1 \leqslant i \leqslant n;$$

whereas in the Galerkin method, σ^g satisfies

$$\forall \tau \in V_h, \quad \langle A \cdot \sigma^g, \tau \rangle_\Gamma = \langle g, \tau \rangle_\Gamma.$$

These two methods, respectively, lead to the matrix equations

$$[A]^c \cdot [\sigma]^c = [g]^c \quad \text{and} \quad [A]^g \cdot [\sigma]^g = [g]^g,$$

where

$$[A]^c_{ij} = \int_{T_j} a(x_i, y)\, dy, \quad [g]^c_i = g(x_i),$$

$$[A]^g_{ij} = \int_{T_j} \int_{T_j} a(x, y)\, dx\, dy, \quad [g]^g_i = \int_{T_i} g,$$

$[\sigma]^c_i$ and $[\sigma]^g_i$ denoting, respectively, the values of σ on T_i under the collocation and Galerkin approximations. The collocation method is more natural and easier to implement (at least at first sight); for these reasons, it is often used by Chemists; on the other hand, the Galerkin method leads to a *symmetric* linear system when the operator A is itself symmetric, which may appreciably simplify the numerical resolution of the linear system (QUARTERONI and VALLI [1997]).

The boundary element method (BEM) follows the standard finite element method (FEM). The only difference proceeds from the fact that in usual applications of the FEM, the operator is *local* (typically a Laplacian) whereas it is *nonlocal* in most applications of the BEM. Consequently, the stiffness matrix $[A]$ is generally sparse for FEM and full for BEM.

In many applications, the surface Γ is partitioned in small patches called tesserae T_i. This approximation renders easier the computation of the coefficients of the stiffness matrices

$$[S]_{ij} = \int_{T_i} \int_{T_j} \frac{1}{|x - y|}\, dx\, dy \quad \text{and} \quad [D]_{ij} = \int_{T_i} \left(\int_{T_j} \frac{\partial}{\partial n_y} \left(\frac{1}{|x - y|} \right) dy \right) dx.$$

In the Galerkin approximation, the exterior integration can be performed with an adaptive Gaussian integration method, the number of integration points depending on the distance and on the relative orientation of the elements T_i and T_j.

The case of an interior charge

Let us recall that in the standard case, the external medium is modelled by a homogeneous and isotropic dielectric whose dielectric constant ε_s equals the macroscopic permittivity of the solvent. Let us denote by ρ and ρ' two charge distributions of the generic form $\sum_{k=1}^{M} q_k \delta_{\bar{x}_k} - \rho_{\mathcal{D}}$ with $\bar{x}_k \in \Omega$ and $\rho_{\mathcal{D}} \in C^{0,1}(\overline{\Omega})$. By extension, the duality brackets for which the integrals below make sense are also denoted by the symbol $\int$.

Our goal is to compute the energy

$$E^r(\rho,\rho') = \int_{\mathbb{R}^3} \rho' V^r,$$

where the reaction potential V^r generated by ρ is uniquely defined by

$$V^r := V - \phi, \quad -\nabla\big(\varepsilon(x)\nabla V(x)\big) = 4\pi\rho(x), \quad -\Delta\phi = 4\pi\rho,$$

with $\varepsilon(x) = 1$ inside the cavity Ω and $\varepsilon(x) = \varepsilon_s$ in the external domain $\mathbb{R}^3 \setminus \overline{\Omega}$. It is easy to check that V^r is C^2 in $\overline{\Omega}$ and in $\mathbb{R}^3 \setminus \Omega$ and satisfies

$$\begin{cases} -\Delta V^r = 0 & \text{in } \Omega, \\ -\Delta V^r = 0 & \text{in } \mathbb{R}^3 \setminus \overline{\Omega}, \\ \left[V^r\right] = 0 & \text{on } \Gamma, \\ V^r \to 0 & \text{at infinity.} \end{cases}$$

The representation formulae (4.1) and (4.2) therefore enable ones to write down the reaction potential V^r as a single layer potential

$$V^r(x) = \int_\Gamma \frac{\sigma(y)}{|x-y|}\,dy, \quad \forall x \in \mathbb{R}^3,$$

with $\sigma = \frac{1}{4\pi}[\frac{\partial V^r}{\partial n}]$. In order to get the apparent surface charge distribution σ, suffices it to use the relations

$$\left.\frac{\partial V^r}{\partial n}\right|_i - \left.\frac{\partial V^r}{\partial n}\right|_e = 4\pi\sigma,$$

$$\frac{1}{2}\left[\left.\frac{\partial V^r}{\partial n}\right|_i + \left.\frac{\partial V^r}{\partial n}\right|_e\right] = D^* \cdot \sigma$$

and the jump condition on the interface Γ

$$0 = \left.\frac{\partial V^s}{\partial n}\right|_i - \varepsilon_s \left.\frac{\partial V^s}{\partial n}\right|_e \tag{4.10}$$

$$= \left.\frac{\partial V^r}{\partial n}\right|_i - \varepsilon \left.\frac{\partial V^r}{\partial n}\right|_e + (1-\varepsilon)\frac{\partial \phi}{\partial n}, \tag{4.11}$$

which lead by simple algebraic manipulations to the integral equation

$$\left(2\pi\frac{\varepsilon_s+1}{\varepsilon_s-1} - D^*\right)\cdot\sigma = \frac{\partial\phi}{\partial n}. \tag{4.12}$$

It can be shown that a solution σ of Eq. (4.12) exists and it is unique.

This technique for calculating the reaction potential, based on integral equations, is referred in chemistry by the acronym ASC (apparent surface charge). It is by far the most often used method. In particular, the expression (4.12) is the one used in the original version of the PCM method introduced in the previous sections. The energy $E^r(\rho,\rho')$ can then be obtained in the following way

$$\begin{aligned} E^r(\rho,\rho') &= \int_{\mathbb{R}^3} \rho' V^r = \int_{\mathbb{R}^3} \rho(x)\left(\int_\Gamma \frac{\sigma(y)}{|x-y|}\,dy\right)dx \\ &= \int_\Gamma \sigma(y)\left(\int_{\mathbb{R}^3} \frac{\rho'(x)}{|x-y|}\,dx\right)dy = \int_\Gamma \sigma\phi' \end{aligned}$$

with $\phi' = \rho' \star \frac{1}{|x|}$.

This method presents a difficulty as the electronic charge distribution is not entirely supported in Ω (the electrons are delocalized in the whole space $\mathbb{R}^3$). Rigorously speaking, it is not possible to used the technique detailed above to compute this term. However, as the part of the electronic cloud which lays outside the cavity is generally small, an accurate approximation of this interaction term can be obtained by a slightly more sophisticated integral method.

The escaped charge problem

In the derivation of Eq. (4.12) we have assumed that ρ is supported inside the cavity, in this case the interaction energy $E^r(\rho,\rho')$ equals the exact energy; otherwise the two quantities differ. Besides, if ρ (or ρ') is not supported inside the cavity, the symmetry property $E^r(\rho,\rho') = E^r(\rho',\rho)$ is broken: this is an important problems of ASC continuum models when coupled to quantum mechanical calculations. In QM in fact, although most of the electronic density ρ_{el} lays inside the cavity, electronic tails always spread outside the cavity: this is the so-called *escaped charge*. When this happens, integral equations methods like PCM can only provide an approximation of the reaction-field energy; an exact computation requires a 3D calculation. Whereas the escaped charge is usually small (a fraction of atomic units), it can nevertheless affect the evaluation of energies and, even more significantly, of properties of the solute. For this reason, it is important to take into account the escaped charge in quantum chemistry calculations.

As reported in the previous sections, the PCM method has been reformulated in a new version known as Integral Equation Formalism (IEF) (CANCÈS and MENNUCCI [1998b], CANCÈS, MENNUCCI and TOMASI [1997], MENNUCCI, CANCÈS and TOMASI [1997]). IEF method can be seen as an improvement of the standard PCM method; in particular it appears to be more efficient in the way the escaped charge is accounted for. The IEF energy reads

$$E_r^{\text{IEF}}(\rho,\rho') = \int_\Gamma \sigma^{\text{IEF}}\phi', \tag{4.13}$$

where σ^{IEF} is the solution to the integral equation

$$\left(2\pi\frac{\varepsilon+1}{\varepsilon-1}-D\right)S\sigma^{\mathrm{IEF}}=-(2\pi-D)\phi.$$

Contrary to the PCM energy, the IEF energy E_r^{IEF} is actually symmetric in the two arguments ρ and ρ' since, denoting by $\langle\cdot,\cdot\rangle_\Gamma$ the scalar product on Γ

$$\int_\Gamma \sigma^{\mathrm{IEF}}\phi'=\langle L\phi,\phi'\rangle_\Gamma=\langle\phi,L\phi'\rangle_\Gamma=\int_\Gamma \sigma'^{\mathrm{IEF}}\phi$$

the self-adjointness of the linear operator

$$L=\left[\left(2\pi\frac{\varepsilon+1}{\varepsilon-1}-D\right)S\right]^{-1}(-2\pi+D)$$

being a consequence of the commutation property $DS=SD^*$.

Anisotropic dielectrics and ionic solutions

In Section 3 we have presented the extension of the standard electrostatic problem of a charge distribution immersed in a continuum dielectric to more complex systems such as anisotropic dielectrics and ionic solutions; here we present the solutions of the corresponding problems. In these cases the Green kernels to be defined are, respectively:

$$G_e(x,y)=\begin{cases}\left(\sqrt{\det\varepsilon}\right)^{-1}\left[\left(\varepsilon^{-1}(x-y)\right)\cdot(x-y)\right]^{-1/2} & \text{(anisotropic)}\\ \exp\left(-\kappa|x-y|\right)\left(\varepsilon|x-y|\right)^{-1} & \text{(ionic solution)}\end{cases}\quad \text{if } x\notin C,$$

where C indicates the cavity.

In a parallel way as that we have used to define the operators (4.5)–(4.8) we can define two other operators, S_e and D_e, by replacing G with the corresponding function G_e defined in the outer space. In this case the derivative operator means $\partial_y G_e(x,y)=(\varepsilon\cdot\nabla_y G_e(x,y))\cdot n(y)$, where the dielectric matrix reduces to the scalar dielectric constant for ionic solutions.

For brevity's sake, here we do not report the formal derivation that, exploiting specific characteristics of these operators, leads to the definition of the surface charge σ, but we only say the latter is the unique solution to the equation:

$$A\cdot\sigma=-g \tag{4.14}$$

where:

$$A=(2\pi-D_e)S_i+S_e\left(2\pi+D_i^*\right), \tag{4.15}$$

$$g=(2\pi-D_e)\phi+S_e\frac{\partial\phi}{\partial n} \tag{4.16}$$

being I the unit operator.

5. Geometry optimization

As shown in the previous section the electrostatic interactions between the charge distributions which compose the molecular system are affected by the presence of the dielectric. Indeed, in solvation continuum models, the interaction energy between two charge distributions ρ_1 and ρ_2 can be redefined as the interaction energy we obtain in vacuo for a new system composed by ρ_1, ρ_2, and the further apparent charge ρ_1^a whose density is the surface density σ we have defined in the previous section; namely we have:

$$E\big(\rho_1(\lambda), \rho_2(\lambda)\big) = \mathcal{D}(\rho_1, \rho_2) + \mathcal{D}\big(\rho_1^a, \rho_2\big), \tag{5.1}$$

where

$$\mathcal{D}(\rho, \rho') = \iint_{\mathbb{R}^3\times\mathbb{R}^3} \left[\frac{\rho(x)\rho'(y)}{|x-y|}\right] dx\, dy$$

denotes the interaction energy between any charge distributions ρ and ρ' in vacuo. To preserve all generality, we have considered that the charge distributions ρ_1, ρ_2, depend on n real parameters $(\lambda_1, \dots, \lambda_n)$ and that the cavity also depends on the λ_i. We note in a geometry optimization algorithm the set of parameters λ_i represent the nuclear coordinates of the atoms which constitute the molecular system.

Passing now to the partial derivatives $\partial E/\partial\lambda_i$, in standard solvation methods the usual way to proceed is the following. It consists in differentiating the interaction energy expression (5.1), which leads to

$$\frac{\partial}{\partial\lambda_j}\big(E\big(\rho_1(\lambda), \rho_2(\lambda)\big)\big) = \frac{\partial}{\partial\lambda_j}\mathcal{D}(\rho_1, \rho_2) + \mathcal{D}\left(\rho_1^a, \frac{\partial\rho_2}{\partial\lambda_j}\right) + \mathcal{D}\left(\frac{\partial\rho_1^a}{\partial\lambda_j}, \rho_2\right)$$

and in computing the derivative $\partial\rho_1^a(\lambda)/\partial\lambda_j$ (the only term for which problems may arise) by differentiating the basic integral equation (4.12) after approximation by the boundary element method quoted above. Recently we have suggested an alternative way to proceed (CANCÈS and MENNUCCI [1998a], CANCÈS, MENNUCCI and TOMASI [1998]), which in particular avoids computing any of the derivatives $\partial\rho_k^a(\lambda)/\partial\lambda_j$. Globally speaking, the new approach consists in deriving first the electrostatic equation so as to write $\partial V_k/\partial\lambda$ as a solution of the differentiated equation, and next inserting it in the derivative of E.

In this scheme, we obtain the following derivative formula:

$$\begin{aligned}\frac{\partial E}{\partial\lambda_j}(\lambda) &= \left\langle\frac{\partial\rho_1}{\partial\lambda_j}(\lambda), V_2(\lambda)\right\rangle + \left\langle\frac{\partial\rho_2}{\partial\lambda_j}(\lambda), V_1(\lambda)\right\rangle \\ &\quad + \frac{1}{4\pi}\int_{\Gamma(\lambda)} \tau(\lambda)\big(U^j_{\Sigma(\lambda)}\cdot n_{\Sigma(\lambda)}\big)\end{aligned} \tag{5.2}$$

with

$$\tau(\lambda) = \frac{16\pi^2\varepsilon}{\varepsilon - 1}\sigma_1(\lambda)\sigma_2(\lambda) + (\varepsilon - 1)\big(\nabla V_1(\lambda)\big)_{\parallel}\big(\nabla V_2(\lambda)\big)_{\parallel}. \tag{5.3}$$

In Eq. (5.3) $(\nabla V_k(\lambda))_{\parallel}$ is the projection of $\nabla V_k(\lambda)$ on Γ, whereas in Eq. (5.2) we denote with U_Γ^j the derivative of the interface Γ when λ varies. This quantity assumes a very simple form for standard cavities given as union of spheres, each of them centered on a solute nucleus.

Each of the three terms on the r.h.s. of (5.2) has a clear meaning: the first two are due to the variations of the charges ρ_1 and ρ_2, respectively, while the third one comes from the deformation of the cavity. From a numerical point of view, the second term of (5.3) is not easy to deal with. Fortunately, its contribution appears to be small in practical cases and can therefore be neglected. This behavior can be formally proved only for smooth cavities and for dielectrics with high permittivities, but numerical tests performed so far have shown that this is still true for all practical cases encountered in chemistry, even for low dielectric permittivity ($\varepsilon \simeq 2$). In view of these arguments, Eq. (5.2) can be always reduced to the simplified formula

$$\frac{\partial E}{\partial \lambda_j} \simeq \left\langle \frac{\partial \rho_1}{\partial \lambda_j}, V_2 \right\rangle + \left\langle \frac{\partial \rho_2}{\partial \lambda_j}, V_1 \right\rangle + \frac{4\pi\varepsilon}{\varepsilon - 1}\int_\Gamma \sigma_1\sigma_2\big(U_\Gamma^j \cdot n_\Sigma\big).$$

6. Molecular surface meshing

In the solvation continuum methods presented in this chapter, the solute is modeled by a domain Ω, called *molecular cavity* or *molecular volume,* in the tridimensional space $\mathbb{R}^3$. Its boundary $\Gamma = \partial\Omega$ is referred to as a *molecular surface*. As shown before, physical problems can be formulated in terms of partial differential equations (PDE) and solved using for instance the finite element method (FEM) or the boundary element method (BEM). These methods are based on a spatial discretization, or *mesh*, of the surface Γ (and also the volume Ω in some cases). It is thus necessary to construct a partition of the molecular surface Γ in geometrically simple elements $\{T_i\}$ such as triangles or quadrilaterals, sometimes called herein *tesserae.*

The generation of such a mesh raises several issues that are considered in this section. First of all, a geometric definition of a molecular surface must be given precisely, in a way which makes sense in computational chemistry. This is detailed in the first subsection which shows that, in fact, different kinds of surfaces can be defined for a given molecule: VWS, SAS and SES. Then, a boundary representation of this molecular surface must be obtained, so that a mesh generator can use this model as a geometric support (see the second subsection for VWS and SAS, and the third subsection for SES). Finally, a surface mesh must be generated to simulate the chemical phenomena numerically. In such a context of numerical simulation, it is now clearly established that mesh quality has a strong influence on solution accuracy, as also convergence and speed of the computing scheme. Consequently, it is essential that the mesh elements are as regular as possible (i.e., almost equilateral), while conforming to a user-specifiable size map and closely approximating the surface (see the fourth subsection). To show some applications of the methods presented, several examples of molecular surface meshes are provided in the fifth subsection, and a brief conclusion is given at the end.

Geometric definitions of molecular surfaces: VWS, SAS, SES

To model a molecule, the basic idea is to assimilate each constituting atom to a ball B_i whose size is determined by its Van der Waals radius. Let $\mathcal{B} = \bigcup B_i$ be the union of all these possibly overlapping balls. Then, it is possible to define different kinds of surfaces, the most commonly used being the Van der Waals surface (VWS), the solvent-accessible surface (SAS) and the solvent-excluded surface (SES), as explained in a review by CONNOLLY [1996].

The VWS is simply the boundary of the union $\mathcal{B}$ (see Fig. 6.1).

The SAS, as introduced by LEE and RICHARDS [1971], involves a sphere of radius r_p, called the *probe*, which represents a single solvent molecule, for instance a water molecule. When the probe sphere rolls on the VWS, the locus of its center defines the SAS. In fact, this definition amounts to the previous one, by increasing the radius of each atom by the constant value r_p (see Fig. 6.2).

The SES is made up of two parts, the "contact surface" and the "reentrant surface". Fig. 6.3 shows a bidimensional diagram where the probe (red circle) rolls on a simplified 4-atom molecule, tracing the contact surface (green arcs) and the reentrant surface (cyan arcs). In three dimensions, the contact surface is the part of the VWS that can be touched by the probe sphere, following an idea which was introduced by RICHMOND and RICHARDS [1978]. The reentrant surface, as introduced by RICHARDS [1977], is obtained when the probe sphere is in contact with two atoms or more. As shown on Fig. 6.4, the SES consists of spherical patches lying either on the atoms (red) or on the probe (green), and toroidal patches defined when the probe rolls over a pair of atoms (cyan). As we will see later, the SES is G^1 continuous (i.e., not folded) in most cases, depending on the probe size, and this is an important property for many

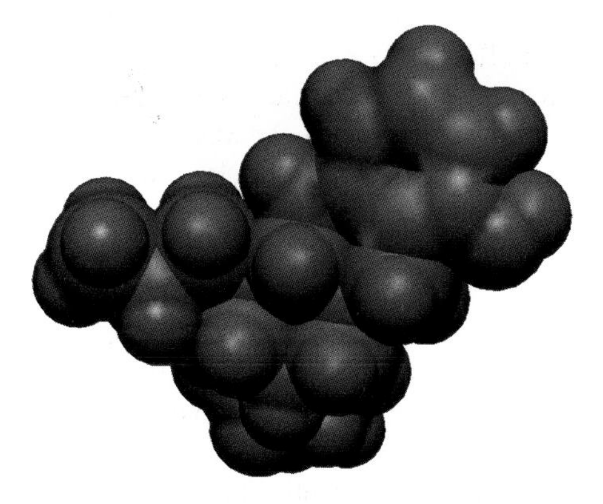

FIG. 6.1. Van der Waals surface (VWS).

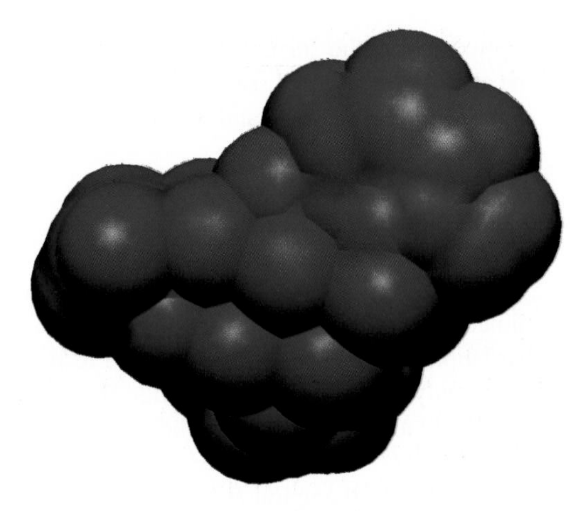

FIG. 6.2. Solvent-accessible surface (SAS).

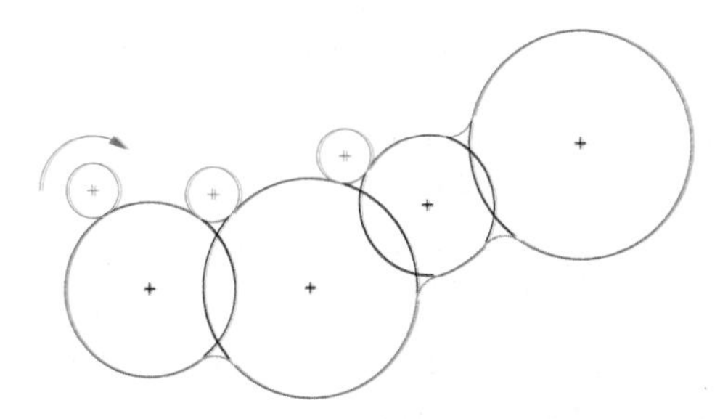

FIG. 6.3. Bidimensional diagram showing the SES of a simplified 4-atom molecule.

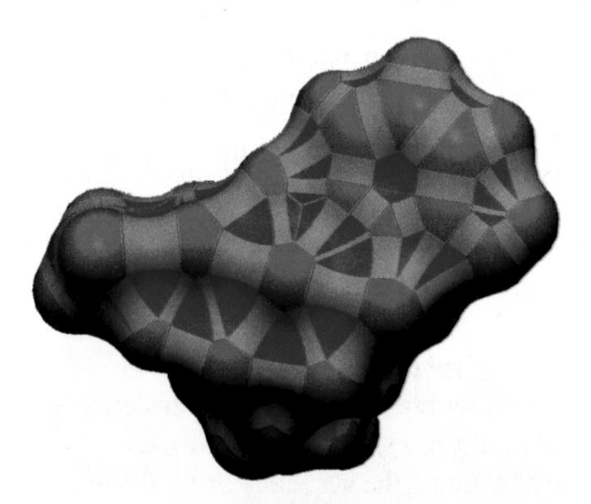

FIG. 6.4. Solvent-excluded surface (SES).

computation models. In the literature, the SES is also called "smooth molecular surface" or "Connolly surface".

At present, a molecular surface being given, it is necessary to obtain its boundary representation (B-rep), i.e., a set of patches and the topological relations between them.

The next subsections deal firstly with VWS and SAS (containing spherical patches only), and secondly with SES. (These descriptions, as well as mesh generation methods, can also be found in LAUG and BOROUCHAKI [2001].)

Boundary representation of VWS and SAS

In the case of *Van der Waals* or *solvent-accessible* surfaces (VWS or SAS), the problem can be mathematically stated as follows:

Given a set of arbitrary spheres $\{S_i\}$ *(several spheres may intersect), where each sphere* S_i *is defined by its center* (x_i, y_i, z_i) *and its radius* r_i*, determine the envelope* E *of the union of these spheres.* By definition, the envelope E is the topological boundary of the union of balls $\mathcal{B} = \bigcup B_i$, where each ball B_i is bounded by the sphere S_i.

To allow for meshing, the envelope E will be determined in the form of a composite parametric surface, i.e., the union of several patches in the 3D space. Two different patches may have a common boundary curve, called an *interface curve*, or may be totally disjoint. Each patch is the image of a planar domain, called a *parametric space*, which is defined from its boundary curves. A single 3D interface curve may be the image of several planar boundary curves.

To give a parametric representation of the envelope E, *first* its interface curves can be obtained by using the following algorithm:

(1) Compute the intersections of spheres $\{S_i\}$ two by two, giving a set of circles $\{C_j\}$.
(2) Compute the intersections of circles $\{C_j\}$ two by two, giving a set of points $\{P_k\}$.
(3) The different points $\{P_k\}$ on a circle C_j define a partition of the circle into a set of arcs. Extract the set of arcs that are outside the spheres, giving the desired interface curves.

Second, mapping functions $\{\sigma_i\}$ on parametric spaces $\{\Omega_i\}$, whose images form the envelope E, are defined. An efficient solution is to use the inverse function of a projection sometimes used in cartography.

Third, the projections of the interface curves of E give the boundaries of the parametric domains $\{\Omega_i\}$ in a 2D space. Here, the curve discretization plays an important role for the validity and shape quality of the planar mesh.

All the above steps are detailed in the following paragraphs.

Computing the intersections of spheres

We want to determine the circle representing the intersection of two spheres (S_1, S_2) with centers (C_1, C_2) and radii (r_1, r_2), if this circle exists.

Let us consider a plane containing the vertical axis $\overrightarrow{C_1 Z}$ and passing through point C_2. In this plane, it is easy to find points A and B at the intersection of two circles with centers (C_1, C_2) and radii (r_1, r_2). Actually, if $AB < \varepsilon$, where ε is a given small value, it is generally preferable to ignore the intersection because of the floating point errors and to avoid a mesh which would be locally too fine.

The intersection of spheres S_1 and S_2 in $\mathbb{R}^3$ is a circle with diameter $[AB]$. The circle lies in a plane defined by $ax + by + cz + d = 0$, where $\overrightarrow{C_1 C_2} = (a, b, c)$ is a normal vector. The molecule surface does not contain the part of sphere S_1 (resp. S_2) satisfying the inequality $ax + by + cz + d > 0$ (resp. < 0), and this property will be used later (paragraph entitled "determining the external arcs").

Computing the intersections of circles

Let us consider two circles drawn on the same sphere with center O and radius r. Each circle lies in a plane whose normal vector is known. Let $\vec{v_1}$ and $\vec{v_2}$ be these normal vectors. Let us determine the intersection points of these two circles, if they exist.

The idea is to consider a plane whose normal vector is perpendicular to both $\vec{v_1}$ and $\vec{v_2}$. The projection of each circle on this plane reduces to a straight segment. It is then easy to compute their intersection. In practice, if the two circles are nearly intersecting, or if the intersection points are very close, only one intersection point can be retained, to avoid again a locally too fine mesh.

While computing the intersections of the circles, the same points should not be computed several times. Actually, if three spheres S_1, S_2 and S_3 intersect each other, their intersections circles are $C_1 = S_1 \cap S_2$, $C_2 = S_2 \cap S_3$ and $C_3 = S_3 \cap S_1$. These three circles define only two intersections points, for $C_1 \cap C_2 = C_2 \cap C_3 = C_3 \cap C_1 = S_1 \cap S_2 \cap S_3$ (by associativity).

Now, let us consider a given circle C_j and the set of intersection points $\{P_k\}$ belonging to it. Some close points can be eliminated directly because of the above remarks, but others may remain. For instance, Fig. 6.5 represents the envelope of four slightly shifted spheres with radius 1 and coplanar centers $(0, 0)$, $(1, 0)$, $(1, 1)$ and $(0, 0.999)$, showing (left) many triangles, generally small and distorted, at the center of the mesh. To avoid this problem, it is sufficient to sort the points on a given circle with respect to their angles. If the difference between two consecutive angles is less than a certain threshold (for instance, $0.3°$), the corresponding points are merged provided they remain close to their defining circles (see Fig. 6.5, right).

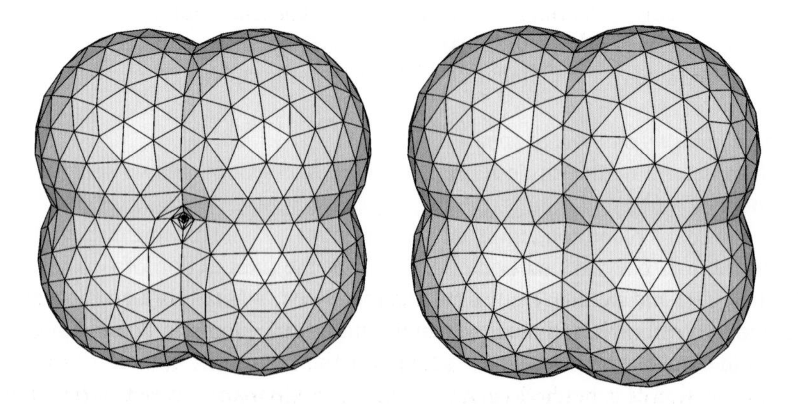

FIG. 6.5. Mesh without (*left*) and with (*right*) merging the closest points.

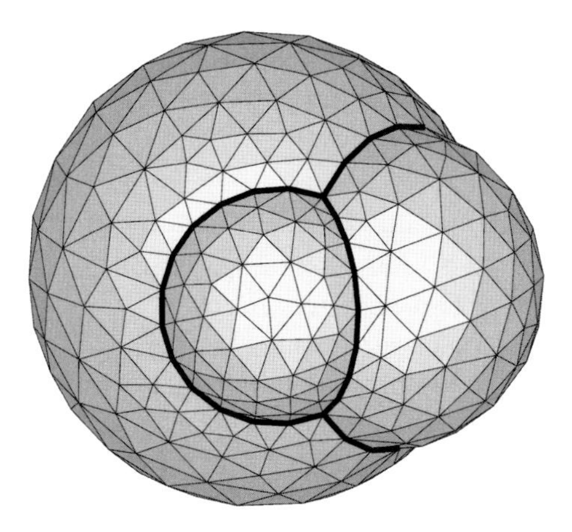

FIG. 6.6. External arcs at the intersection of three spheres.

Determining the external arcs

Let us consider again a circle C_j and the set of intersection points $\{P_k\}$ belonging to it. This set defines a partition of circle C_j into several arcs. The set of the arcs that are not inside a sphere form the interface of all the spherical surfaces (see Fig. 6.6). To determine if an arc is inside or outside, the equations of the planes containing the circles are used (see above).

More precisely, let us denote by $\{A_l\}$ the set of all the arcs found. To each arc A_l is associated a flag $A_l.ext$ with value T (true) if the arc is external and F (false) otherwise. To this end, the following algorithm written in pseudo-code can be used:

```
For each arc A_l
  A_l.ext = T
End
For each sphere S_i
  For each circle C_j lying on the sphere S_i
    (a, b, c, d) = coefficients of the plane containing C_j
    For each arc A_l
      If A_l is a part of C_j then cycle
      (x, y, z) = middle point of arc A_l
      If ax + by + cz + d > 0 then A_l.ext = F
    End
  End
End
```

Parameterizing a spherical surface

Each spherical surface must now be parameterized, which consists conversely in defining a projection from the surface to the parametric domain. We will obtain here rational polynomials, as for instance in BAJAJ, LEE, MERKERT and PASCUCCI [1997].

One of the simplest method derives from the *orthogonal projection* (see Fig. 6.7, point P_0). If $P = (x, y, z)$ is a point on a sphere with center $C = (0, 0, R)$ and radius R,

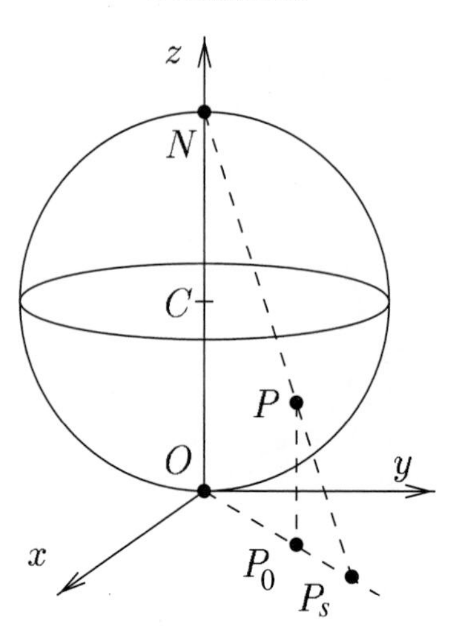

FIG. 6.7. Diagram of an orthogonal (P_0) and stereographic (P_s) projection.

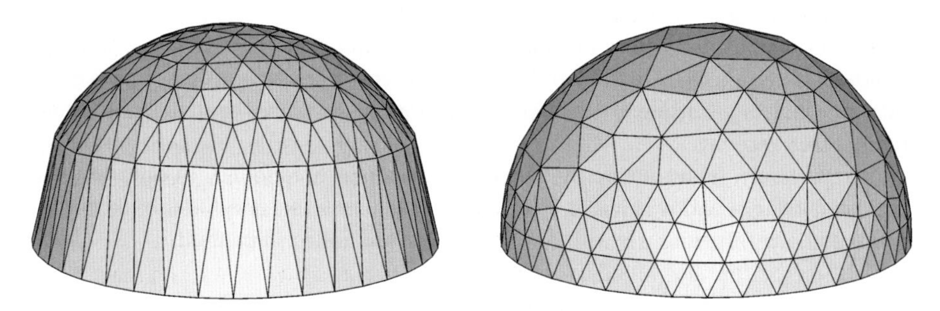

FIG. 6.8. *Left*: inverse orthogonal projection of a uniform mesh. *Right*: inverse stereographic projection of the same mesh.

its orthogonal projection is point $P_0 = (u, v)$ with $u = x$ and $v = y$. Conversely, a hemisphere can be parameterized by:

$$\sigma_0(u, v) = \begin{bmatrix} u \\ v \\ \sqrt{R^2 - u^2 - v^2} \end{bmatrix}.$$

The main drawback of this parameterization is that it can be highly unstable near the equator, for the partial derivatives become infinite. As an illustration, Fig. 6.8 (left) shows the image by the mapping function σ_0 of a uniform mesh on a plane disk Ω.

To avoid degenerate derivatives, it is more convenient to consider the *stereographic projection* (see Fig. 6.7, point P_s). If $N = (0, 0, 2R)$ is the sphere's "North Pole", the stereographic projection of point P is defined as the intersection P_s of the straight line

(NP) with the plane $z = 0$. We now have $P_s = (u, v)$ with $u = \frac{2Rx}{2R-z}$ and $v = \frac{2Ry}{2R-z}$. Conversely, the sphere without point N can be parameterized by:

$$\sigma_s(u, v) = \frac{2R}{u^2 + v^2 + 4R^2} \begin{bmatrix} 2Ru \\ 2Rv \\ u^2 + v^2 \end{bmatrix}.$$

This parameterization is stable near the equator, continuously differentiable, and preserves angles (but not distances). As a consequence, if the triangulation represents the geometry of the sphere accurately, any equilateral triangle in the parametric domain Ω is almost equilateral on the sphere, giving directly a tridimensional surface mesh with a good shape quality (see Fig. 6.8 (right)).

Since it is impossible to parameterize a whole sphere with only one domain, it can be divided into a pair of hemispheres when this case occurs. However, as far as molecular surfaces are concerned, only some parts of a sphere generally remain, hence only one domain is necessary (by choosing the sphere's "North Pole" outside these parts).

Defining and discretizing the bidimensional boundaries

We have already shown that the interface curves in 3D are circle arcs. Each arc is discretized conforming to a pre-specified size map. Finally, using a stereographic projection, a discretization of the bidimensional boundaries is obtained. However, this discretization may lead to an invalid definition of the bidimensional parametric domain, or give bad quality elements. It is thus necessary to check the crossing edges, the close edges and the adjacent edge lengths of the domain boundaries, as explained below.

(a) *Checking crossing edges.* Consider the example of Fig. 6.9. The geometric definition of the boundary is represented by dashed curves, and the initial discretization of this support by its vertices (small circles) and its edges (thin solid lines). Some edges are intersecting each other, meaning that the discretized boundary of the domain is not well defined. To rectify this, the longest crossing edges can be recursively subdivided (here, one larger circle and two thick solid lines). In fact, we present just below an algorithm which is more strict and produces a better mesh, at the cost of more calculations.

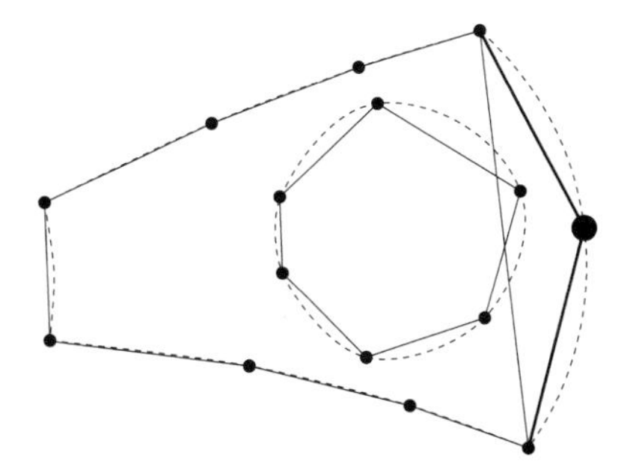

FIG. 6.9. Crossing edges.

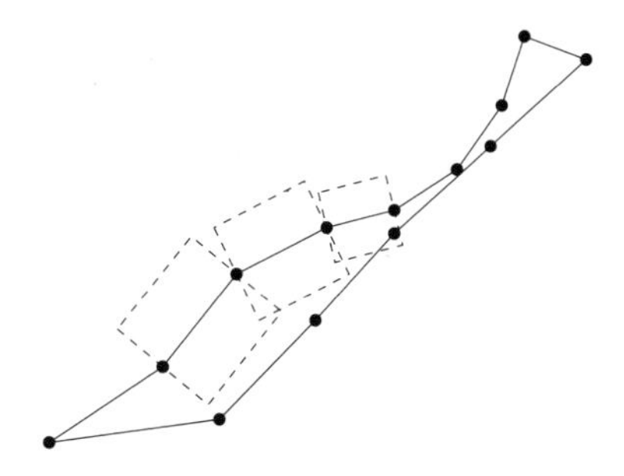

FIG. 6.10. Close edges.

FIG. 6.11. Initial mesh (left), after checking close edges (middle) and after checking edge lengths (right), thus improving the shape quality of triangles.

(b) *Checking close edges.* Fig. 6.10 shows a bidimensional boundary whose edges are very close. In this case, the domain mesh, which must comply with the boundary discretization, would contain very flat triangles. To avoid this, the idea is to associate to each edge a rectangle (dashed lines on the figure). Its length is equal to the edge length, and its width is proportional to it. If the rectangle intersects other edges, then the corresponding edge is subdivided. This process is repeated until there is no intersection. In the case where two edges make a sharp angle, this algorithm must be adapted by defining a rectangle with a smaller width.

(c) *Checking adjacent edge lengths.* Once the boundaries have been discretized (initially or after the above correction), the lengths of two adjacent edges may be very different, leading to a bad mesh quality. For instance, Fig. 6.11 presents a partial mesh in its initial state (left) and after the previous correction (middle). Then, an iterative algorithm which limits the lengths ratio of adjacent edges has been used, and the final mesh can be seen on the same figure (right).

Boundary representation of SES

So far, we have considered *Van der Waals* or *solvent-accessible* surfaces (VWS or SAS), which are made up of several spherical patches. If we now focus on *solvent-excluded* surfaces (SES), also called Connolly surfaces (cf. first subsection), different kinds of patches may be encountered:

- When the *probe sphere* (PS) is in contact with only one atom, it defines a spherical surface, which is in fact a part of the classical VWS.
- When it rolls while touching simultaneously two atoms, the part facing the molecule traces a toroidal patch. In the usual case, this patch is bounded by four arcs. However, if the two atoms become more distant compared to the probe diameter, we obtain a pair of patches bounded by three arcs each. These two 3-sided patches share a curve along which the surface is not G^1 continuous.
- When it touches three atoms at the same time, it cannot roll anymore and defines a spherical surface which is now reentrant.

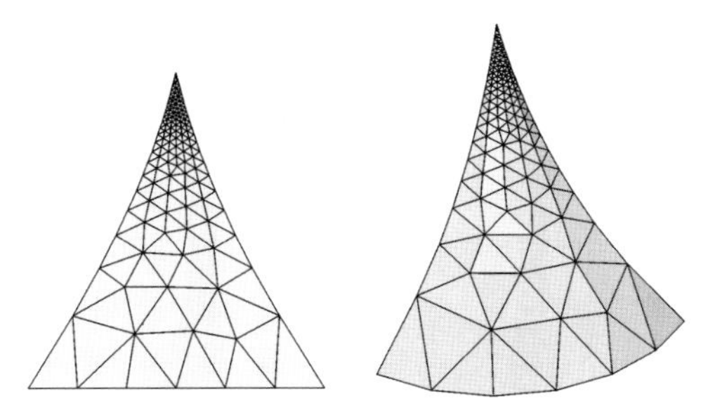

FIG. 6.12. Toroidal patch bounded by three arcs (planar and surface meshes).

Several algorithms to compute analytical models of the SES are referenced in an article by SANNER, OLSON and SPEHNER [1996], who developed an efficient program called MSMS which determines the constituting patches (spherical contact, spherical reentrant, toroidal-3 or toroidal-4). This program can either give a Connolly surface mesh suitable for visualization, or provide an intermediate description of the patches in terms of centers, radii, bounding arcs, etc. To generate a mesh from such a boundary representation, a parameterization of the obtained patches is necessary.

To parameterize a *spherical* patch (contact or reentrant), a stereographic projection can be used as before. To parameterize a *toroidal* patch, a basic idea is to consider an arc on a rotating plane. The center of this arc is defined by a radius r_1 and an angle $\varphi \in [\varphi_1, \varphi_2]$. For a given φ, a point on the arc is defined by a radius r_2 and an angle $\theta \in [\theta_1, \theta_2]$. Then, if r_1 and r_2 are known, any point on the torus is defined by the two angles φ and θ. However, in general, the parametric domain cannot be defined as the rectangle $[\varphi_1, \varphi_2] \times [\theta_1, \theta_2]$. A problem occurs when $r_1 \leqslant r_2$, i.e., when the arc intersects the axis. In this case, a toroidal patch is bounded by only three arcs (see Fig. 6.12). Then, instead of considering the two angles φ and θ, it is preferable to use curvilinear abscissae. This also gives better results when $r_1 \simeq r_2$ (but $r_1 > r_2$), for the shapes of the 2D and 3D domains become similar and the 2D triangles are not too distorted (see Fig. 6.13).

Surface meshing

Having a boundary representation and a parameterization of a molecular surface, a surface mesh must now be generated. As explained before, the size and shape of the mesh elements are a crucial point in a computational scheme. The problem of meshing, in particular surface meshing, for numerical simulation purposes is a broad and interdisciplinary field, and it is beyond the scope of this Handbook to describe the various existing algorithms suitable to produce such meshes (a comprehensive survey about mesh generation techniques for surfaces and volumes can be found in FREY and GEORGE [2000]). In short, there are two approaches to meshing parametric surfaces: direct and indirect. In the *direct* approach, the mesh is generated over the surface directly in $\mathbb{R}^3$. Among the direct approaches we can cite the octree-based method, the advancing-front-based method and the paving-based method. A direct

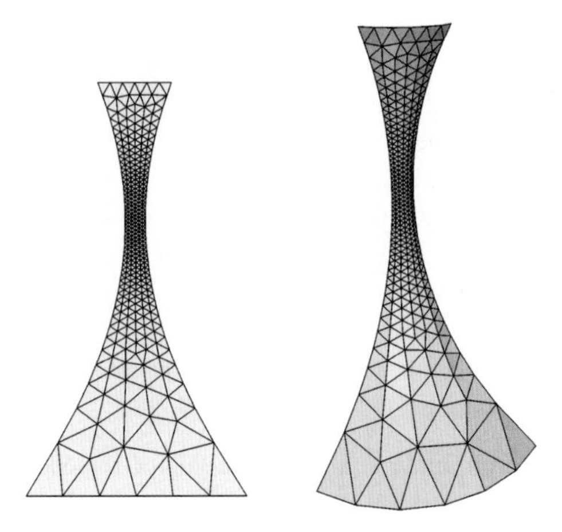

FIG. 6.13. Thin toroidal patch bounded by four arcs (planar and surface meshes).

approach, conceived mainly for molecular surface visualization, is described in AKKIRAJU and EDELSBRUNNER [1996]. The *indirect* approach consists of meshing the parametric domain and mapping the resulting mesh onto the surface. It is conceptually straightforward, as a two-dimensional mesh is generated in the parametric domain, and thus it is expected to be faster than the direct approach. However, one problem with this method is the generation of a mesh which conforms to the metric of the surface. It is briefly presented below (for more details, see BOROUCHAKI, LAUG and GEORGE [2000]).

Let Σ be such a surface parameterized by:

$$\sigma : \Omega \to \Sigma, \quad (u, v) \mapsto \sigma(u, v) = (x, y, z),$$

where Ω denotes the parametric domain. First, from the size specifications, a Riemannian metric $\mathcal{M}_3 = \frac{1}{h^2}\mathcal{I}_3$ (h being the specified size function and $\mathcal{I}_3$ the identity matrix) is defined so that the desired mesh has unit length edges with respect to the related Riemannian space (such meshes being referred to as "unit" meshes). Then, based on the intrinsic properties of the surface, namely the first fundamental form:

$$\mathcal{M}_\sigma = \begin{pmatrix} {}^t\sigma'_u\sigma'_u & {}^t\sigma'_u\sigma'_v \\ {}^t\sigma'_v\sigma'_u & {}^t\sigma'_v\sigma'_v \end{pmatrix},$$

the Riemannian structure $\mathcal{M}_3$ is induced into the parametric space as follows:

$$\widetilde{\mathcal{M}}_2 = \frac{1}{h^2}\mathcal{M}_\sigma.$$

The initial size specification is isotropic while the induced metric in parametric space is in general anisotropic, due to the variation of the tangent plane along the surface. Finally, a unit mesh is generated completely inside the parametric space such that it conforms to the induced metric $\mathcal{M}_2$. This mesh is constructed using a combined advancing-front – Delaunay approach applied within a Riemannian context: the field points are defined after an advancing-front method and are connected using a generalized Delaunay type method.

One can control explicitly the accuracy of a generated element with respect to the geometry of the surface if careful attention is paid. Indeed, a mesh of a parametric patch whose element vertices belong to the surface is "geometrically" suitable if the two following properties hold:

- each mesh element is close to the surface, and
- each mesh element is close to the tangent planes related to its vertices.

A mesh satisfying these properties is called a *geometric mesh*. The first property allows us to bound the gap between the elements and the surface. This gap measures the greatest distance between an element and the surface. The second property ensures that the surface is locally of order G^1 in terms of continuity. To obtain this, the angular gap between the element and the tangent plane at its vertices must be bounded. A sufficient condition is that the element size is locally proportional to the minimal radius of curvature. Here, we deal with two kinds of surface, namely spheres and tori. For a sphere, the minimal radius of curvature is simply its radius. In the case of a torus, this radius can be easily computed from the radii of the two main defining circles.

Note that if a given size map is specified, the two above properties can be locally violated. In fact, it is more useful to find a compromise between the geometric approximation of the surface and the size map conformity.

To give an idea of possible shapes of parametric domains, Fig. 6.14 shows two examples where several topological sides, trimming curves and multiple loops can be noticed. The first one (left) has one external and one internal boundary, each boundary being made up of several arcs. The second one (right) is constituted by three different sub-domains.

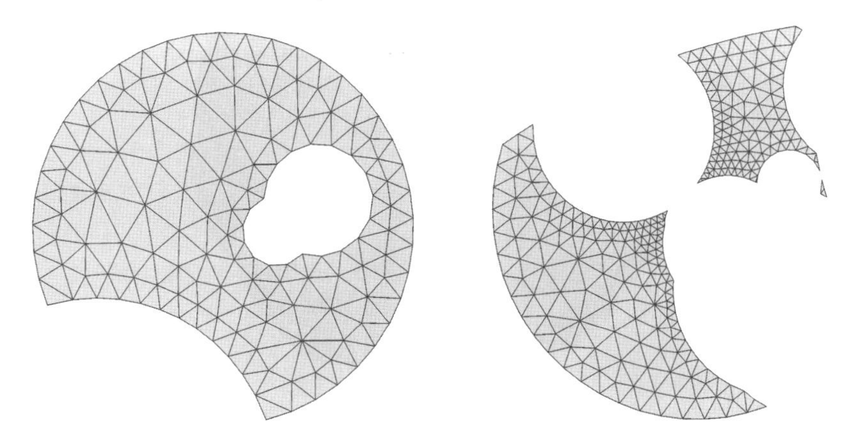

FIG. 6.14. Two examples of parametric domain meshes.

Application examples

Several sample meshes are presented in this subsection to illustrate the above methods. They all have been generated by a software package called BLMOL (LAUG and BOROUCHAKI [2002]), except in the second example (ice SES). Each molecule is represented by a list of atoms, and each atom is described by its chemical symbol, its center coordinates and its radius length. Such descriptions can be obtained for instance from the MathMol library (NEW YORK UNIVERSITY [2002]) in a PDB (Protein Data

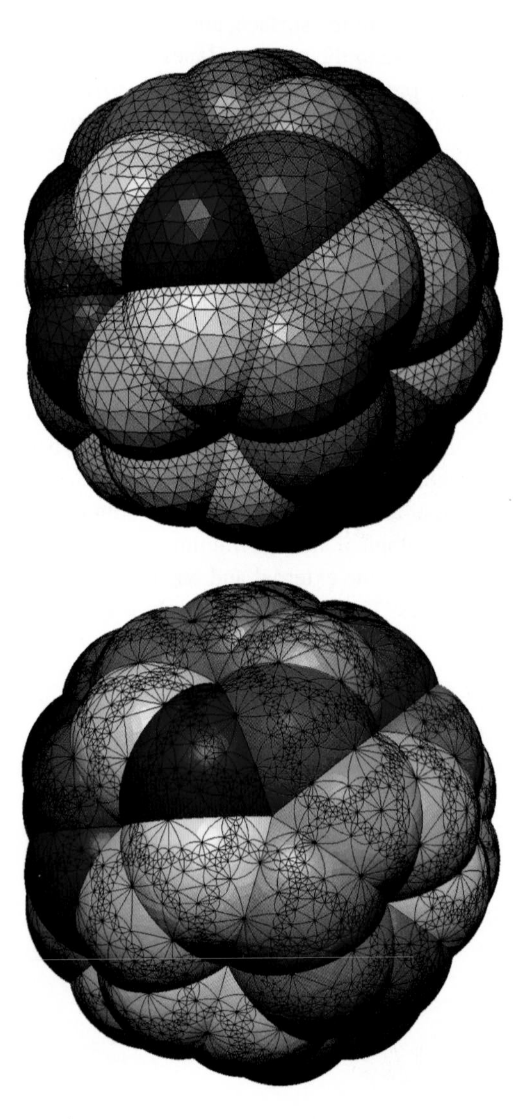

FIG. 6.15. Carbon 60. *Top*: P^1 geometric mesh. *Bottom*: P^2 mesh (curved triangles) with a given analytical field.

Bank) file. The following examples involve molecular surfaces of fullerene (C_{60}), ice, pentane, DNA and echistatin.

C_{60}. The C_{60} molecule can be modeled by 60 identical spheres whose centers are at the vertices of pentagons and hexagons. Fig. 6.15 (top) shows the default mesh generated by BLMOL. It meets the most common requirements for computational chemistry. It is a geometric mesh with a tolerance angle of 3° for the curves and 9° for the surfaces, using a gradation parameter of 1.5. The total CPU time is 5 s on a HP

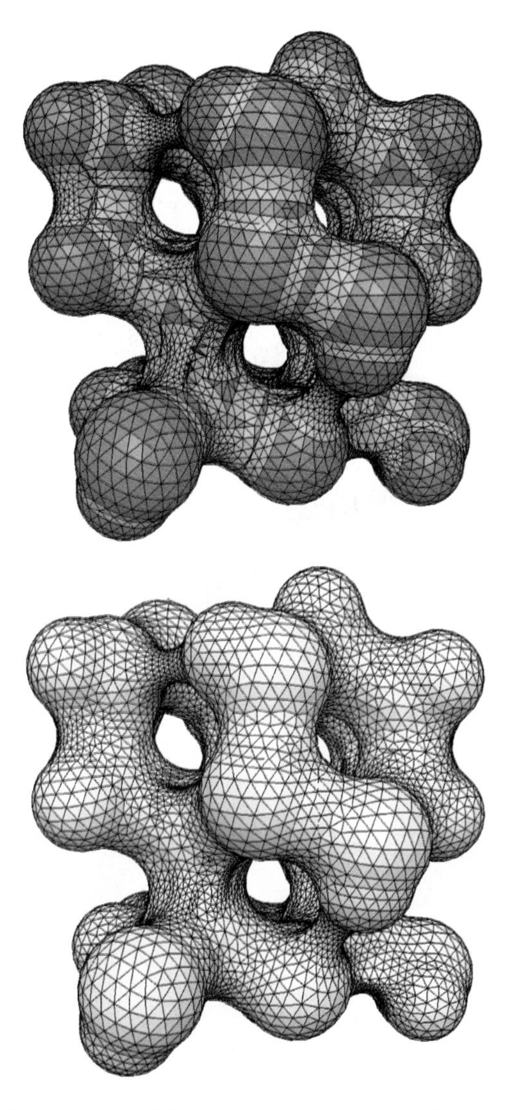

FIG. 6.16. Solvent-excluded surface (SES) of a network of hydrogen-bonded waters. *Top*: patch-dependent mesh. *Bottom*: simplified mesh.

9000/785 400 MHz for a mesh without gradation, and 12 s for the mesh shown (three iterations are necessary when a gradation is required). The number of vertices is 7,863 and the number of triangles is 15,718. Fig. 6.15 (bottom) shows a mesh which conforms to a given analytical field. Here, curved quadratic triangles (P^2) are used instead of the usual ones (P^1), thus decreasing the number of elements for a given accuracy. In this example, the mesh contains 40,776 nodes and 20,386 six-node triangles.

Ice. Fig. 6.16 shows two meshes of a solvent-excluded surface (SES) of a network of hydrogen bonded water molecules. Initially, its boundary representation has been generated by the MSMS program (SANNER, OLSON and SPEHNER [1996]), as explained in the subsection entitled "Boundary representation of SES" (a very small probe has been used to have holes in the network). Then, the BLSURF software package (LAUG and BOROUCHAKI [1999]) was used to create surface meshes. The first mesh (on the bottom) has been constructed while preserving the contours of each patch. It contains 17,446 vertices and 35,020 triangles. However, within the context of numerical simulation, it is more convenient to enhance the quality of such a mesh by merging the extremities of small edges and moving points. In the example shown on the top of the

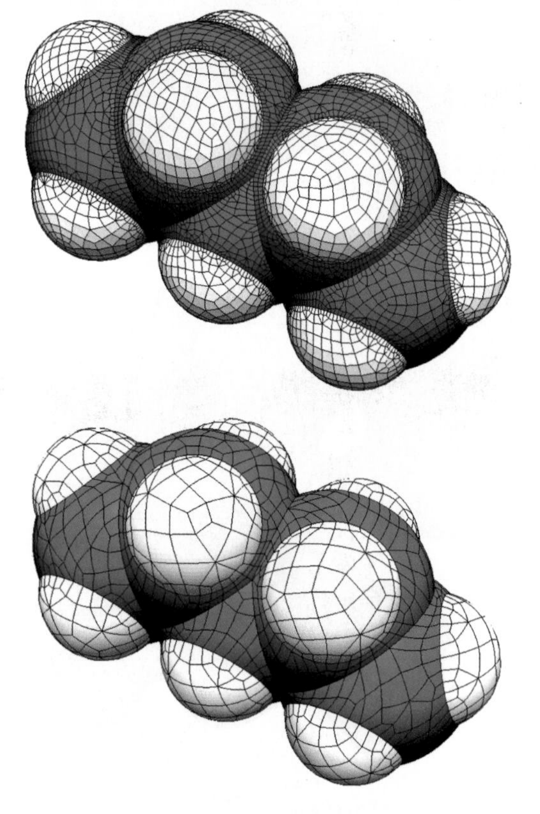

FIG. 6.17. Quadrilateral meshes of pentane. *Top*: linear (Q^1). *Bottom*: quadratic (Q^2).

FIG. 6.18. Surface mesh of DNA.

figure, the counts reduce to 16,365 vertices and 32,858 triangles with a better shape quality.

n-Pentane. The n-pentane molecule (C_5H_{12}) is made up of 17 atoms. The two meshes shown on Fig. 6.17 have quadrilateral elements that are, respectively, linear (Q^1) and quadratic (Q^2). They have been obtained by pairing adjacent triangles. Let us remark that this process have been applied to an "optimal" triangular mesh, which cannot in any case generate an optimal quadrilateral mesh. The Q^1 mesh is a geometric mesh with a tolerance angle of 3° for curves and 5° for surfaces, and a gradation parameter of 1.2. It contains 8,520 vertices and 8,518 quadrilaterals. For the Q^2 mesh, we have 10° for curves and 20° for surfaces without gradation, producing a mesh with 5,330 nodes and 1,776 eight-node quadrilaterals.

FIG. 6.19. Sphere of water containing an echistatin molecule.

DNA. Fig. 6.18 shows a mesh of a fragment of DNA strand, whose structure is a double helix. It has 637 atoms, 154,220 vertices and 309,004 triangles.

Echistatin. Finally, Fig. 6.19 shows a mesh of a sphere of water containing an echistatin molecule with 12,596 atoms, 1,329,865 vertices and 2,673,604 triangles.

Conclusion

To conclude briefly this section, some methods for generating a mesh on any molecular surface (Van der Waals, solvent-accessible or solvent-excluded surface) have been presented and examples have been shown. In computational chemistry, like in any field of computing science, mesh quality is an important issue to obtain accurate results. The choice of elements may also be taken into consideration (P^1 or P^2 triangles, Q^1 or Q^2 quadrilaterals). At present, though molecules with several thousand atoms can be addressed, a remaining challenge is to find out and implement even more efficient methods for very large molecules and dynamic problems.

CHAPTER III

Computer Simulations

Computer simulations are considered a valuable alternative to experiments to get information on the liquid state. The literature on this subject is large, with a sizeable number of detailed monographs (HANSEN and MCDONALD [1986], GRAY and GUBBINS [1984], ALLEN and TILDESLEY [1987], FRENKEL and SMIT [1987]). Here, we shall be concise, limiting our attention to the two basic approaches, Monte Carlo (MC) and molecular dynamics (MD), and giving for both a very schematic outline.

7. Monte Carlo (MC) methods

MC technically is a procedure to compute an integral with a random sampling of integrand values. It may so be used to compute the averages over the ensemble distributions necessary to get the properties in the Gibbs approach. Actually MC methods are generally limited to properties without an explicit dependence on time.

The basic formalism is simple. The value of a property X depending on the spatial $\mathbf{r}^N$ and angular ω^N set of coordinates, averaged on time, is given in the Gibbs approach by

$$\langle X\rangle = \int P\big(\mathbf{r}^N\omega^N\big)X\big(\mathbf{r}^N\omega^N\big)\,d\mathbf{r}^N\,d\omega^N = \frac{\int X(\mathbf{r}^N\omega^N)\exp[-V/kT]}{Q_N}, \tag{7.1}$$

where V indicate the potential energy of the N interacting particles. We have here used the canonical formulation but the method can be applied to other ensembles as well. In the MC approach the integrals appearing in the numerator as well as in the denominator of Eq. (7.1) are replaced by summations. Each term of the sum refers to a specific choice of the values of the coordinates, which we now collectively indicate with $\mathbf{R}$. Expression (7.1) may be so approximated by

$$\langle X\rangle = \sum_k X(\mathbf{R}_k)P_k = \frac{\sum_k X(\mathbf{R}_k)\exp[-V(\mathbf{R}_k)/kT]}{\sum_k \exp[-V(\mathbf{R}_k)/kT]}. \tag{7.2}$$

To evaluate Eq. (7.2) with the MC procedure one has to:

(1) Specify the initial coordinates of atoms $\mathbf{R}_0^N$.
(2) Generate new coordinates $\mathbf{R}_a^N$ by changing the initial coordinates at random.
(3) Compute the transition probability $W(0,a)$.

(4) Generate a uniform random number t in the range [0,1].
(5) If $W(0,a) < t$, then take the old coordinates as the new coordinates and go to step 2.
(6) Otherwise accept the new coordinates and go to step 2.

The most popular realization of the Monte Carlo method (though not the only one) for molecular systems is the Metropolis method (METROPOLIS, A.W. ROSENBLUTH, M.N. ROSENBLUTH, A.N. TELLER and E. TELLER [1953]) which consists in: (i) specifying the initial atom coordinates (e.g., from molecular mechanics geometry optimization), (ii) selecting some atom i randomly and move it by random displacement: ΔR_i; (iii) calculating the change of potential energy ΔV corresponding to this displacement. If $\Delta V < 0$, the new coordinates are accepted and the cycle is repeated. Otherwise, if $\Delta V \geqslant 0$, a random number γ in the range [0, 1] is selected and, if $e^{-\Delta V/kT} < \gamma$, the new coordinates are accepted, while if $e^{-\Delta V/kT} \geqslant \gamma$ the original coordinates are kept unchanged. In both cases the process goes back to point (iii) and iterated.

Note that the iterations are independent of one another (i.e., the system does not contain any "memory"). The possibility that the system might revert to its previous state is as probable as choosing any other state. This condition has to be satisfied if the system is to behave as a Markov process, for which the methods of calculating ensemble (i.e., statistical) averages are known. Another important condition to be satisfied is the continuity of the potential energy function in order for the system to be ergodic. In this context, ergodicity means that any state of the system can be reached from any other state.

There are several problems related to the use of this procedure. The formal elements of the Markov theory are not accompanied by stringent rules on the representativity of the chain in spanning the whole ensemble. If the random selection of new configurations is not well calibrated, the sampling will be only performed in a limited region of the conformational space, giving so a value of the property with a small statistical error, but with a large systematic error.

The move generally regards a single coordinate of a single molecule, under the form

$$q_{im}(R_{i+1}) = q_{im}(R_i) + \lambda \Delta q_m,$$

where q_m is one of the six translational and rotational coordinates of the molecule i, Δq_m is the maximal change permitted for that coordinate and λ is a random number between 0 and 1. Experience has learned what are acceptable Δq_m values for small molecules, but for large molecules as well as for cases in which many molecules exhibit positive correlation in their position (e.g., strong hydrogen networks, polymers) this definition of the moves surely is inadequate. There are now "smarter" definitions of the moves. To alleviate problems of inefficient sampling the length of the Markov chain has been greatly increased in the years: standard calculations include now several millions of accepted moves, after some extra million moves of "equilibration" used to find an initial R_i conformation not too severely biased.

The ensemble on which the simulation is performed cannot be too large for computational reasons. Generally it is defined in terms of a square box of appropriate dimensions containing a given number N of molecules. To reduce border effects the box is surrounded by other similar boxes all containing the N molecules in the same R_i conformation (translational symmetry). So when a move brings a molecule out of the box a similar molecule on the opposite side replaces it. The number N must be as large as possible, but there are limitations due to the finite computer power (the maximum for N is now a few thousands). This fact introduces new problems, especially for ionic solutions in which long range unscreened Coulomb interactions are active. The collection of boxes introduces in these cases a spurious periodic pattern of charges that severely modifies the ensemble. In any case, simulations on solutions are more sensitive to errors than simulations on pure liquids, because the low mole ratio does not permit to average local properties on many molecules.

Actually to perform calculations on a given property the general formula (7.2) is rewritten in a computationally more suitable form. For example the energy V is immediately reduced to a weighted average of the energies of the various conformations, the heat capacity at constant volume C_V to an average of the fluctuations of V around its mean value $\langle V \rangle$, that can be recovered during a single simulation run (even if with a larger error than for $\langle V \rangle$). Things are not so simple for the free energy A for which there is the need of averaging over the $\exp[-V(R)/kT]$ function. This means that to have a meaningful result for $\langle A \rangle$ one has to sample regions in the conformational space high in energy, generally less represented in the MC chain.

Generally, MC simulations work well for the canonical ensemble; one could pass from the canonical to another ensemble giving a simpler expression for the desired property. The performances of MC simulations are however worse in passing from the canonical one to other ensembles, in particular the use of the grand canonical ensemble (that it would be the best for free energy) is quite difficult.

We have dwelt upon problems and limits in the use of MC simulations, and have now to stress the great merits of this method. It is quite flexible, open to the researcher's ingenuity to derive information not limited to the thermodynamic functions: a complete listing of the successful applications of MC to condensed systems (of every type) would be very instructive and stimulating. Unfortunately we cannot review here these performances and we limit ourselves to remark a point. There are no formal limits in the use of potential functions to describe the collective or local behavior of molecules in the condensed systems. The limitations we have signaled in the use of physical methods (which are progressively reduced at the cost of great efforts) do not exist here. The only real limitation is due to the computational cost, a limitation more and more reduced by the impressive increase of computer performances. There are for example MC methods which compute the interaction potentials on the spot at the QM level, as well as MC strategies to follow the evolution of a chemical reaction along its reaction path. The limitations due to the cost are still present, however: a large part of recent studies continues to use empirical two-body potentials (now more and more expressed at the level of nonrigid molecules and with the introduction of contributions describing modifications to the potential due to nearby molecules); the use of explicit three body potentials still is a rarity.

As a result of a stochastic simulation, the large number of configurations (geometries) are accumulated and the potential energy function is calculated for each of them. These data are used to calculate thermodynamic properties of the system. The Monte Carlo method is accepted more by physicists than by chemists, probably because MC is not a deterministic method and does not offer time evolution of the system in a form suitable for viewing. It does not mean however, that for deriving the thermodynamic properties of systems molecular dynamics is better. In fact many chemical problems in statistical mechanics are approached more efficiently with MC, and some (e.g., simulations of polymers chains on a lattice) can only be done efficiently with MC. Also, for Markov chains, there are efficient methods for deriving time related quantities such as relaxation times. Currently, the stronghold of MC in chemistry is in the area of simulations of liquids and solvation processes.

8. Molecular Dynamics (MD)

The deterministic approach, called Molecular Dynamics (MD), actually simulates the time evolution of the molecular system and provides the actual trajectory of the system. The information generated from simulation methods can in principle be used to fully characterize the thermodynamic state of the system. In practice, the simulations are interrupted long before there is enough information to derive absolute values of thermodynamic functions, however the differences between thermodynamic functions corresponding to different states of the system are usually computed quite reliably. In MD, the evolution of the molecular system is studied as a series of snapshots taken at close time intervals (usually of the order of femtoseconds). For large molecular systems the computational complexity is enormous and supercomputers or special attached processors have to be used to perform simulations spanning long enough periods of time to be meaningful. Typical simulations of small proteins including surrounding solvent cover the range of tens to hundreds of picoseconds, i.e., they incorporate many thousands of elementary time steps.

Based on the potential energy function V, we can find components F_i of the force acting on an atom as:

$$F_i - = -\partial V / \partial x_i.$$

This force results in an acceleration according to Newton's equation of motion. By knowing acceleration, we can calculate the velocity of an atom in the next time step. From atom positions, velocities, and accelerations at any moment in time, we can calculate atom positions and velocities at the next time step. Integrating these infinitesimal steps yields the trajectory of the system for any desired time range. There are efficient methods for integrating these elementary steps with Verlet and leapfrog algorithms being the most commonly used.

To start the MD simulation we need an initial set of atom positions (i.e., geometry) and atom velocities. In practice, the acceptable starting state of the system is achieved by "equilibration" and "heating" runs prior to the "production" run. The initial positions of atoms are most often accepted from the prior geometry optimization with molecular mechanics. Formally, such positions correspond to the absolute zero temperature. The velocities are assigned randomly to each atom from the Maxwell distribution for some low temperature (say 20 K). The random assignment does not allocate correct velocities and the system is not at thermodynamic equilibrium. To approach the equilibrium the "equilibration" run is performed and the total kinetic energy (or temperature) of the system is monitored until it is constant. The velocities are then rescaled to correspond to some higher temperature, i.e., the heating is performed. Then the next equilibration run follows. The absolute temperature, T, and atom velocities are related through the mean kinetic energy of the system:

$$T = \frac{2}{3Nk} \sum_{i=3}^{N} \frac{m_i |v_i|^2}{2},$$

where N denotes the number of atoms in the system, m_i represents the mass of the ith atom, and k is the Boltzmann constant. By multiplying all velocities by $\sqrt{T_{\text{desired}}/T_{\text{current}}}$ we can effectively "heat" the system. Heating can also be realized by immersing the system in a "heat bath" which stochastically (i.e., randomly) accelerates the atoms of the molecular system. These cycles are repeated until the desired temperature is achieved and at this point a "production" run can start. In the actual software, the "heating" and "equilibration" stages can be introduced in a more efficient way by assigning velocities in such a way that "hot spots" (i.e., spots in which the neighboring atoms are assigned high velocities) are avoided.

Molecular dynamics for larger molecules or systems in which solvent molecules are explicitly taken into account, is a computationally intensive task even for the most powerful supercomputers, and approximations are frequently made. The most popular is the SHAKE method (RYCKAERT, CICCOTTI and BERENDSEN [1977]) which in effect freezes vibrations along covalent bonds. This method is also applied sometimes to valence angles. The major advantage of this method is not the removal of a number of degrees of freedom (i.e., independent variables) from the system, but the elimination of high frequency vibrations corresponding to "hard" bond stretching interactions. In simulations of biological molecules, these modes are usually of least interest, but their extraction allows us to increase the size of the time step, and in effect achieve a longer time range for simulations. Another approximation is the united-atom approach where hydrogen atoms which do not participate in hydrogen bonding are lumped into a heavier atom to form a pseudo-atom of larger size and mass (e.g., a CH group).

Even supercomputers have their limitations and there is always some practical limit on the size (i.e., number of atoms) of the simulated system. For situations involving solvent, the small volume of the box in which the macromolecule and solvent are contained introduces undesirable boundary effects. In fact, the results may depend sometimes more on the size and shape of the box than on the molecules involved. To circumvent this limited box size difficulty, periodic boundary conditions are used. In this approach, the original box containing a solute and solvent molecules is surrounded with identical images of itself, i.e., the positions and velocities of corresponding particles in all of the boxes are identical. The common approach is to use a cubic or rectangular parallelepiped box, but other shapes are also possible (e.g., truncated octahedron). By using this approach, we obtain what is in effect an infinite sized system. The particle (usually a solvent molecule) which escapes the box on the right side, enters it on the left side, due to periodicity. Since MD simulations are usually performed as an *NVE* (microcanonical) ensemble (i.e., at constant number of particles, constant volume, and constant total energy) or an *NVT* (canonical) ensemble, the volume of the boxes does not change during simulation, and the constancy in the number of particles is enforced by the periodicity of the lattice, e.g., a particle leaving the box on left side, enters it on the right side.

There are also techniques for performing simulations in a *NPT* (isothermal–isobaric), and *NPH* (isobaric–isoenthalpic) ensembles, where the pressure constancy during simulation is achieved by squeezing or expanding box sizes. The constant temperature is usually maintained by "coupling the system to a heat bath", i.e., by adding dissipative forces (usually Langevin, friction type forces) to the atoms of the system which as a consequence affects their velocities. However, each approximation has its price. In the case of periodic boundary conditions we are actually simulating a crystal comprised of boxes with ideally correlated atom movements. Longer simulations will be contaminated with these artificially correlated motions. The maximum length for the simulation, before artifacts start to show up, can be estimated by considering the speed of sound in water (15 Å/ps at normal conditions). This means that for a cubic cell with a side of 60 Å, simulations longer than 4 ps will incorporate artifacts due to the presence of images.

In the other popular approach, stochastic boundary conditions allow us to reduce the size of the system by partitioning the system into essentially two portions: a reaction zone and a reservoir region. The reaction zone is that portion of the system which we want to study and the reservoir region contains the portion which is inert and uninteresting. For example, in the case of an enzyme, the reaction zone should include the proximity of the active center, i.e., portions of protein, substrate, and molecules of solvent adjacent to the active center. The reservoir region is excluded from molecular dynamics calculations and is replaced by random forces whose mean corresponds to the temperature and pressure in the system. The reaction zone is then subdivided into a reaction region and a buffer region. The stochastic forces are only applied to atoms of the buffer region, in other words, the buffer region acts as a "heat buffer". There are several other approximations whose description can be found in the molecular dynamics monographs quoted here. The straightforward result of a molecular dynamics simulation, a movie showing changing atom positions as a function of time, contains a wealth of information in itself. Viewing it may shed light on the molecular mechanisms behind the biological function of the system under study.

Molecular dynamics, contrary to energy minimization in molecular mechanics, is able to climb over small energy barriers and can drive the system towards a deeper energy minimum. The use of molecular dynamics allows probing of the potential energy surface for deeper minima in the vicinity of the starting geometry. It is exploited in a *simulated annealing method* (KIRKPATRICK, GELATT JR. and VECCHI [1983]), where the molecular system is first heated to an artificially high temperature and then snapshots of the trajectory at high temperature are taken as a starting state for cooling runs. The cooled states obtained from hot frames correspond frequently to much deeper potential energy minima than the original structure taken for dynamics. It is particularly suitable for imposing geometrical constraints on the system. Such constraints may be available from experimental results. Molecular dynamics will try to satisfy these artificially imposed constraints by jumping over shallow potential energy minima – behavior which is not possible in molecular mechanics.

CHAPTER IV

Hybrid Methods

9. Quantum mechanics/molecular mechanics (QM/MM)

Each computational method has strengths and weaknesses. Molecular mechanics (MM) can model very large compounds quickly. Quantum mechanics (QM) is able to compute many properties and model chemical reactions. It is possible to combine these two methods into one calculation, which models a very large compound using molecular mechanics and one crucial portion of the molecule with quantum mechanics (QM/MM). This is designed to give results that have very good speed where only one region needs to be modeled quantum mechanically. This can also be used to model a molecule surrounded by solvent molecules. The literature on QM/MM methods is vast and rapidly updating; here we only quote a general review: GAO [1995].

The basic idea to describe molecules in solution is to build an hybrid QM/MM potential and to introduce the solvent field into quantum mechanical calculations of the solute molecule on the fly during a computer simulation such that the molecular wavefunction of the solute will be polarized by the dynamic change of the surrounding solvent molecules. The method was first described by WARSHEL and LEVITT [1976] and a detailed prescription for MO calculations was presented by FIELD, BASH and KARPLUS [1990]. The idea was to divide a condensed phase system into a QM region and a MM region, plus an appropriate boundary treatment to mimic the bulk effects. Consequently the effective Hamiltonian of the system is written as follows:

$$H_{\text{eff}} = H^0_{\text{QM}} + H_{\text{MM}} + H^{\text{elec}}_{\text{QM/MM}} + H^{\text{VdW}}_{\text{QM/MM}} \tag{9.1}$$

where H^0_{QM} describes the completely quantum mechanical part of the system, H_{MM} the purely molecular mechanics and the last two terms the QM-MM electrostatic ($H^{\text{elec}}_{\text{QM/MM}}$) and van der Waals $(H)^{\text{VdW}}_{\text{QM/MM}}$ interactions. The van der Waals term $H^{\text{VdW}}_{\text{QM/MM}}$ is to ensure that the QM and MM systems will not get too close because of a lack of electronic structural description of the solvent MM system. It turns out that the most significant part is to account for the short range electron repulsions; however, for convenience as well as for inclusion of some dispersion interactions, a Lennard-Jones term is typically used for H^{VdW}.

$$H^{\text{VdW}}_{\text{QM/MM}} = \sum_{s=1}^{S}\sum_{m=1}^{M} 4\varepsilon_{sm}\left[\left(\frac{\sigma_{sm}}{R_{sm}}\right)^{12} - \left(\frac{\sigma_{sm}}{R_{sm}}\right)^{6}\right],$$

where S and M are the number of solvent (MM) and solute (QM) atoms, and σ and ε are empirical parameters. It should be emphasized that these parameters must be carefully examined for a given QM (model, basis set, and level of theory) and MM (force fields) combination. However, on the other hand, it gives us the opportunity to optimize the performance of the hybrid QM/MM potential.

In the hybrid QM/MM scheme, solute-solvent (or QM/MM) interactions are given by

$$E_{\text{QM/MM}} = \langle \Psi | H_{\text{QM/MM}} | \Psi \rangle + E^{\text{VdW}}_{\text{QM/MM}},$$

where Ψ is the wavefunction of the solute molecule in solution, which minimizes the energy of the Hamiltonian of Eq. (9.1).

Commonly it is implicitly assumed that only point charges $\{q_s\}$ in the MM region contribute to the modified Hamiltonian in the hybrid QM/MM method through the electrostatic $H^{\text{elec}}_{\text{QM/MM}}$ term, namely:

$$H^{\text{elec}}_{\text{QM/MM}} = -\sum_{s=1}^{S}\sum_{i=1}^{2N} \frac{q_s}{r_{is}} + \sum_{s=1}^{S}\sum_{m=1}^{M} \frac{q_s Z_m}{R_{ms}},$$

where i indicate electrons, r and R are the distances of the QM electrons and nuclei from the solvent sites, respectively. This allows the wavefunction and charge distribution of the solute molecule to be polarized. However, the solvent charges are kept fixed, without responding to the solute charge reorganization as well as the solvent dipole reorientations. To improve this description and to allow both the solute and solvent polarizations on the same footing. The MM solvent molecules are represented by a set of Van der Waals parameters, a set of charges, and a set of atomic polarizabilities. Therefore, an additional term is added to the system effective Hamiltonian:

$$H_{\text{eff}} = H^0_{\text{QM}} + H_{\text{MM}} + H^{\text{elec}}_{\text{QM/MM}} + H^{\text{VdW}}_{\text{QM/MM}} + H^{\text{pol}}_{\text{QM/MM}},$$

where the last term represents the interaction between the QM region and the induced dipoles of the solvent MM region. It is given below.

$$H^{\text{pol}}_{\text{QM/MM}} = \sum_{i=1}^{N}\sum_{s=1}^{S} \frac{\mathbf{R}_{is} \cdot \mu_s}{R^3_{is}} + \sum_{a=1}^{M}\sum_{s=1}^{S} \frac{Z_a \mathbf{R}_{as} \cdot \mu_s}{R^3_{as}},$$

where μ_s are MM induced dipole vectors depending on the atomic polarizabilities. The solution of the corresponding Schrödinger equations and the convergence of MM induced dipoles are coupled, and must be solved iteratively.

If the QM and MM regions are separate molecules, having nonbonded interactions only, this scheme is sufficient. On the contrary, if the two regions are parts of the same molecule, it is necessary to describe the bond connecting the two sections. In most cases, this is done using the bonding terms in the MM method being used. This is usually done by keeping every bond, angle or torsion term that incorporates one atom from the QM region.

It is sometimes desirable to include the effect of the rest of the system, outside of the QM and MM regions. One way to do this is using periodic boundary conditions, as is done in liquid state simulations. Some researchers have defined a potential, which is intended to reproduce the effect of the bulk solvent. This solvent potential may be defined just for this type of calculation, or it may be a continuum solvation model. To represent a solid continuum, a set of point charges, called a Madelung potential, is often used.

10. Layered methods: ONIOM

An alternative formulation of QM/MM is the *energy subtraction method*. In this method, calculations are done on various regions of the molecule with various levels of theory. Then the energies are added and subtracted, to give suitable corrections. This results in computing an energy for the correct number of atoms and bonds, analogous to an isodesmic strategy. Three such methods have been proposed by FROESE and MOROKUMA [1998]. The integrated MO + MM (IMOMM) method combines an orbital based technique with an MM technique. The integrated MO + MO method (IMOMO) integrates two different orbital based techniques. The "our own n-layered integrated MO and MM" method (ONIOM) allows for three or more different techniques to be used in successive layers. The acronym ONIOM is often used to refer to all three of these methods since it is a generalization of the technique. This technique can be used to model a complete system as a small model system and the complete system. The complete (or *Real*) system would be computed using only the lower level of theory. The *model* system would be computed with both levels of theory. The energy for the complete system, combining both levels of theory, would then be

$$E = E_{\text{real}}^{\text{low}} + E_{\text{model}}^{\text{high}} - E_{\text{model}}^{\text{low}}.$$

Likewise a three layer system, could be broken into small, medium, and large regions, to be computed with low, medium and high levels of theory (L,M,H, respectively). The energy expression would then be

$$E = E_{\text{small}}^{H} + E_{\text{medium}}^{M} - E_{\text{small}}^{M} + E_{\text{large}}^{L} - E_{\text{medium}}^{L}.$$

This method has the advantage of not requiring a parameterized expression to describe the interaction of various regions. Any systematic errors in the way that the lower levels of theory describe the inner regions will be canceled out.

The definition of the model system is rather straightforward if there is no covalent bond between the layers, and is then identical to the high level layer. When covalent bond exists between the layers, the dangling bonds are saturated with link atoms, which is in fact the method of choice in many QM/MM schemes. One chooses link atoms that best mimic the substituents, and usually hydrogen atoms yield good results when carbon–carbon bonds are broken. The atoms that exist both in the model system and in the real system will have the same coordinates both systems. C_{model}–H_{link} bonds are assigned the same angular and dihedral values as the C_{model}–C_{real} bonds in the real system, with the bond lengths adjusted by scaling the C–C bond length in such way that a reasonable C_{model}–C_{real} distance yields a reasonable C_{model}–H_{link} distance. Because the geometry of the model system is a function of the geometry of the real system, the number of degrees of freedom remains 3N–6 (or 3N–5), which ensures that any method for the investigation of potential energy surfaces available for conventional methods, can be used for ONIOM as well. The derivatives of the ONIOM energy with respect to the geometrical parameters can be obtained in a similar fashion as the energy.

In addition to the energy and its geometrical derivatives, other properties are available in the ONIOM framework as well. For example, the integrated density is expressed as

$$\rho = \rho_{real}^{low} + \rho_{model}^{high} - \rho_{model}^{low}.$$

Properties related to the density, such as potentials or electric field gradients, can be expressed in the same way. Higher order or mixed properties are available as well, but the Jacobian must be employed when derivatives of the nuclear coordinates are involved.

The ONIOM method has been combines with the continuum PCM-IEF model (VREVEN, MENNUCCI, DA SILVA, MOROKUMA and TOMASI [2001]) we have described in Section 4 of Chapter II. Four versions of the method have been developed. These schemes differ mainly with respect to the level of coupling between the solute charge distribution and the continuum dielectric, which has important consequences for the computational efficiency. Any property that can be calculated by both ONIOM and PCM-IEF can also be calculated by the ONIOM-PCM method.

11. The Effective Fragment Potential (EFP) method

The effective fragment potential (EFP) method (DAY, JENSEN, GORDON, WEBB, STEVENS, KRAUSS, GARMER, BASCH and COHEN [1996]) is another QM/MM method. The EFP method has successfully been applied to the study of aqueous solvation effects, by using EFPs to represent solvent molecules while the solute molecules are treated with Hartree–Fock theory.

The basic idea behind the effective fragment potential (EFP) method is to replace the chemically inert part of a system by EFPs, while performing a regular ab initio calculation on the chemically active part. Here “inert” means that no covalent bond breaking process occurs. This “spectator region” consists of one or more “fragments”, which interact with the ab initio “active region” through nonbonded interactions, and so these EFP interactions affect the ab initio wavefunction. A simple example of an active region might be a solute molecule, with a surrounding spectator region of solvent molecules represented by fragments. Each discrete solvent molecule is represented by a single fragment potential, in marked contrast to continuum models for solvation.

The nonbonded interactions currently implemented are:

- Coulomb interaction: The charge distribution of the fragments is represented by an arbitrary number of charges, dipoles, quadrupoles, and octupoles, which interact with the ab initio Hamiltonian as well as with multipoles on other fragments.

It is possible to input a screening term that accounts for the charge penetration. Typically the multipole expansion points are located on atomic nuclei and at bond midpoints.

- Dipole polarizability: An arbitrary number of dipole polarizability tensors can be used to calculate the induced dipole on a fragment due to the electric field of the ab initio system as well as all the other fragments. These induced dipoles interact with the ab initio system as well as the other EFPs, in turn changing their electric fields. All induced dipoles are therefore iterated to self-consistency. Typically the polarizability tensors are located at the centroid of charge of each localized orbital of a fragment.
- Repulsive potential: Two different forms for the repulsive potentials are used: one for ab initio-EFP repulsion and one for EFP-EFP repulsion. The form of the potentials is empirical, and consists of distributed Gaussian or exponential functions, respectively. The primary contribution to the repulsion is the quantum mechanical exchange repulsion, but the fitting technique used to develop this term also includes the effects of charge transfer. Typically these fitted potentials are located on atomic nuclei within the fragment.

The EFP method for treating discrete solvent effects begins with the ab initio Hamiltonian of the "solute", which may include a small number of solvent molecules. The remaining solvent molecules are then treated by adding their effect on the system as one-electron terms in the ab initio Hamiltonian:

$$H = H_{\mathrm{AR}} + V,$$

where H is the Hamiltonian for the entire system, H_{AR} is the ab initio Hamiltonian of the "solute", or active region, and V represents the one-electron terms that describe the potential due to the fragment molecules. This potential includes ab initio–fragment, ab initio(nuclei)–fragment, and fragment–fragment interactions, each including the three terms mentioned above (except for the ab initio(nuclei)–fragment interaction; there are no exchange repulsion/charge transfer terms there).

Recently the EFP method has been combined with the PCM-IEF continuum model to give a new discrete/continuum solvation model (BANDYOPADHYAY, GORDON, MENNUCCI and TOMASI [2002]). The first applications have shown that EFP/PCM model gives results that are in close agreement with the much more expensive full ab initio/PCM-IEF calculation.

12. Car–Parrinello ab initio molecular dynamics: AIMD

The term ab initio molecular dynamics is used to refer to a class of methods for studying the dynamical motion of atoms, where computational work is spent in solving, as exactly as it is required, the entire quantum mechanical electronic structure problem. When the electronic wavefunctions are reliably known, it will be possible to derive the forces on the atomic nuclei using the Hellmann–Feynman theorem. The forces may then be used to move the atoms, as in standard molecular dynamics.

The most widely used theory for studying the quantum mechanical electronic structure problem of solids and large molecular systems is the density-functional theory of Hohenberg and Kohn in the local-density approximation (KOHN and VASHISHTA [1993]) (LDA). The self-consistent Schrödinger equation (or more precisely, the Kohn–Sham equations) for single-electron states is solved for the solid-state or molecular system, usually in a finite basis-set of analytical functions. The electronic ground state and its total energy is thus obtained. One widely used basis set is "plane waves", or simply the Fourier components of the numerical wavefunction with a kinetic energy less than some cutoff value. Such basis sets can only be used reliably for atomic potentials whose bound states are not too localized, and hence plane waves are almost always used in conjunction with pseudo-potentials (BACHELET, HAMANN and SCHLÜTER [1982]) that effectively represent the atomic cores as relatively smooth static effective potentials in which the valence electrons are treated.

CAR and PARRINELLO [1985] developed a method which is based upon the LDA, and uses pseudopotentials and plane wave basis sets, but they added the concept of updating iteratively the electronic wavefunctions simultaneously with the motion of atomic nuclei (electron and nucleus dynamics are coupled). This is implemented in a standard molecular dynamics paradigm, associating dynamical degrees of freedom with each electronic Fourier component (with a small but finite mass). The efficiency of this iteration scheme has allowed not only for the mentioned pseudopotential based molecular dynamics studies, but also for static calculations for far larger systems than had previously been accessible. Part of this improvement is due to the fact that some terms of the Kohn–Sham Hamiltonian can be efficiently represented in real-space, other terms in Fourier space, and that Fast Fourier Transforms (FFT) can be used to quickly transform from one representation to the other. Since the original paper by Car and Parrinello, a number of modifications (TUCKERMAN, UNGAR, VON ROSENVINGE and KLEIN [1996], PAYNE, TETER, ALLAN, ARIAS and JOANNOPOULOS [1992]) have been presented that improve significantly the efficiency of the iterative solution of the Kohn–Sham equations. The modifications include the introduction of the conjugate gradients method and a direct minimization of the total energy.

Until recently first principles electronic calculations where based on techniques which required the computationally expensive matrix diagonalization methods. Car and Parrinello formulated a new more efficient method which can be expressed in the language of molecular dynamics. The essential step was to treat the expansion coefficients of the wavefunction as dynamical variables. In a conventional molecular dynamics simulation a Lagrangian can be written in terms of the dynamical variables which normally are atomic positions $\{\mathbf{R}_I\}$ and the unit cell dimensions $\{\mathbf{B}\}$. The Car–Parrinello Lagrangian can similarly be written, but also includes a term for the electronic wavefunction. Ignoring any constraints for the moment, it is

$$\mathcal{L}' = \sum_i \big(\mu\langle\dot{\varphi}_i|\dot{\varphi}_i\rangle - E\big[\{\varphi_i\}, \{\mathbf{R}_I\}, \{\mathbf{B}\}\big]\big),$$

where μ is a fictitious mass which is associated to the expansion coefficients of the Kohn–Sham electronic wavefunctions. E is the Kohn–Sham energy functional. This is analogous to the usual form of the Lagrangian where the kinetic energy term is replaced with the fictitious dynamics of the wavefunctions and the Kohn–Sham energy functional replaces the potential energy. The Kohn–Sham electronic orbitals are subject to the orthonormal constraints

$$\int \varphi_i^*(\mathbf{r})\varphi_j(\mathbf{r})\,d\mathbf{r} = \delta_{ij}.$$

These constraints can be simply incorporated into the Car–Parrinello Lagrangian as follows:

$$\mathcal{L} = \mathcal{L}' + \sum_{i,j} \Delta_{ij}\left(\int \varphi_i^*(\mathbf{r})\varphi_j(\mathbf{r})\,d\mathbf{r} - \delta_{ij}\right),$$

where $\Delta_{i,j}$ are the Lagrange multipliers ensuring that the wavefunctions remain orthonormal. In terms of molecular dynamics, these can be thought of as additional forces on the wavefunctions which maintain orthonormality throughout the calculation. From $\mathcal{L}$, it follows that the Lagrange equations of motion

$$\frac{d}{dt}\left(\frac{\partial \mathcal{L}}{\partial \dot{\varphi}_i^*}\right) = \frac{\partial \mathcal{L}}{\partial \varphi_i^*}$$

give

$$\mu\ddot{\varphi}_i = -H\varphi_i + \sum_{i,j} \Delta_{ij}\varphi_j,$$

where H is the Kohn–Sham Hamiltonian and the force $-H\varphi_i$ is the gradient of the Kohn–Sham energy functional at the point in Hilbert space that corresponds to the wavefunction. This equation when coupled to the more standard equation of motion for the classical nuclear degrees of freedom, $\mathbb{R}_I$, fully describes the dynamics of the system.

CHAPTER V

Computing Properties

The formulation of theoretical models to describe liquids has its main application in the study of phenomena and properties involving molecular systems in solution. In this chapter we shall present results on this kind of application in two separate sections: one on approaches treating the liquid as a whole, and the other on models focussing on a subsystem of the liquid, the solute.

13. Liquids

In this section we will present some examples of how to determine properties of pure liquids by mean of the previously shown MC and MD computational procedures. Obviously the following discussion cannot be considered as a complete review of all the literature in the field, but it is rather intended to give the readers a few suggestions on how these methodologies can be used to obtain data about the most usual properties of pure liquids. From what we said in the previous section, it should be clear to the reader that continuum models cannot properly be used in order to determine properties of liquids: they are in fact concerned in the treatment of a solute in a solvent, and so they are not of direct use for the study of the solvent properties.

Calculation of equilibrium properties

Statistical mechanics concepts can supply many information from molecular dynamics (or MC) runs, since they furnish prescriptions for how to calculate the macroscopic properties of the system (e.g., thermodynamic functions like free energy and entropy changes, heats of reaction, rate constant, etc.) from statistical averages of elementary steps on molecular trajectory (or of configurations). They also allow us to uncover a correlation between the motions of groups or fragments of the system as well as provide various distribution functions. Essentially, all expressions derived below apply to both MD and MC, with the only difference being that the "time step" in MD should be changed to a "new configuration" for MC. For the moment let us choose MD for illustration, since it is more often applied to chemical problems.

Three quantities form a foundation for these calculations: the partition function Z, the Boltzmann probability functional P, and the ensemble average, $\langle B \rangle$, of a quantity B:

$$Z = \iint e^{-H(\mathbf{p},\mathbf{q})/kT}\, d\mathbf{q}\, d\mathbf{p}, \tag{13.1}$$

$$P(\mathbf{p},\mathbf{q}) = \frac{1}{Z} e^{-H(\mathbf{p},\mathbf{q})/kT},$$

$$\langle B \rangle = \int B(\mathbf{p},\mathbf{q}) P(\mathbf{p},\mathbf{q})\, d\mathbf{q}\, d\mathbf{p},$$

where **p** and **q** are all generalized momenta and coordinates of the system (e.g., angular and linear velocities and internal coordinates); $H = K + V$ is the Hamiltonian for the system, k denotes the Boltzmann constant, and T is the temperature. The integration extends over the whole range accessible to momenta and coordinates.

Free energy. For the isochoric-isothermic system of constant *NTV*, the Helmholtz free energy, $A = U - TS = G - PV$ (U, G, S and T are internal energy, Gibbs free energy, entropy and absolute temperature, respectively) for the system of unit volume is given as:

$$A = -kT \ln Z.$$

The above equation simplifies significantly when Cartesian coordinates and Cartesian momenta are used, since the kinetic energy factor of the total energy represented by the Hamiltonian in the expression for Z reduces to a constant depending upon the size, temperature and volume of the system:

$$A = -kT \ln \underbrace{\int e^{-V(\mathbf{X}^N)/kT}\, d\mathbf{X}^N}_{Z^c} + \mathrm{const}(N, V, T),$$

where the potential energy function $V(\mathbf{X}^N)$ depends only on Cartesian coordinates of atoms and the integral Z^c is called a classical configurational partition function or a configurational integral. Hence, the free energy A can in principle be calculated from the integral of the potential energy function V. In practice, however, calculation of the absolute value of the free energy is not possible, except for very simple molecules or systems in which the motion of atoms is very restricted (e.g., crystals). Note that Cartesian coordinates span the range $(-\infty, +\infty)$ and for systems which are diffusional in nature (i.e., solutions, complexes, etc.) atoms can essentially occupy any position in space and contributions from the scattered locations to the value of the partition function are not negligible.

The situation is more optimistic with differences in the free energy between two closely related states (STRAATSMA [1996]), since it can be assumed that contributions from points far apart in the configurational space will be similar, and hence will cancel each other in this case. The free energy difference between two states 0 and 1 is:

$$\Delta A_{0\to 1} = A_1 - A_0 = -kT \ln \frac{Z_1^c}{Z_0^c}, \tag{13.2}$$

$$\frac{Z_1^c}{Z_0^c} = \langle e^{-\Delta V_{01}(\mathbf{X}^N)} \rangle,$$

i.e., the ensemble average of the exponential function involving the difference between potential energy functions for state 0 and 1 calculated at coordinates obtained for the state 0. As was mentioned earlier, the difference between states 0 and 1 should be very small, i.e., a perturbation. For this reason the method is called a *free energy perturbation* (ZWANZIG [1954]). If the difference between states 0 and 1 is larger, the process has to be divided into smaller steps. In practice, the change from state 0 to 1 is represented as a function of a suitably chosen coupling parameter λ such that for state 0 $\lambda = 0$, and for state 1 $\lambda = 1$. This parameter is incorporated into the potential energy function to allow smooth transition between the two states.

Now, the path $\lambda = 0 \to \lambda = 1$ can be easily split into any number of desired substates spaced close enough to satisfy the small perturbation requirement. Assuming that the process was divided into n equal subprocesses we have: $\lambda_i = i\Delta\lambda$, where $\Delta\lambda = \lambda/n$. However, in actual calculations, the division is usually not uniform to account for the fact that the most rapid changes in free energy often occur at the beginning and at the end of the path. The simulation is run for each value of λ providing free energy changes between substates i and $i+1$, where the coordinates calculated at the substate λ_i are used to calculate ΔV between states λ_i and λ_{i+1}. Then the total change of free energy is calculated as a sum of partial contributions. The most important conclusion from the above is that the conversion between two states does not have to be conducted along a physically meaningful path. Obviously, atoms cannot be converted stepwise to one another, since they are made of elementary particles. However, basic thermodynamics, namely, the Hess's law of constant heat summation, ensures that the difference between free energy depends only upon initial and final states. Any continuous path which starts at state 0 and ends at state 1, even if it is a completely fictitious one, can be used to calculate this value.

There are other methods than perturbation of free energy to calculate the difference in free energies between states. The most accepted one is the *thermodynamic integration method* (MITCHELL and MCCAMMON [1991]) where the derivative of free energy A versus coupling parameter λ is calculated. It can be proven that the derivative of the free energy versus λ is equal to the ensemble average of the derivative of potential energy versus λ, and hence:

$$\Delta A_{0\to 1} = A_1(\lambda = 1) - A(\lambda = 0) = \int_0^1 \left\langle \frac{\partial V(\mathbf{X}^N)}{\partial \lambda} \right\rangle d\lambda.$$

In most cases, the calculations are run in both directions, to provide the value of hysteresis, i.e., the precision of the integration process. Large differences between $\Delta A_{0\to 1}$ and $\Delta A_{1\to 0}$ indicate that either the simulation was not run long enough for each of the substates, or that the number of substates is too small. While checking the hysteresis represents a proof of the precision of the calculations, it is not a proof of the accuracy of the results. Errors in estimating the free energy changes may result from an inadequate energy function and in most cases from inadequate probing of the energy surface of the system. Molecular dynamics, similar to molecular mechanics, can also suffer from the "local minimum syndrome". If atom positions corresponding to the minimum of energy in state 0 are substantially different from those for state 1, the practical limitations of the length of the computer simulation may not allow the system to explore the configuration space corresponding to the true minimum for state 1. For this reason the set of independent simulations, each starting at different atom positions should be performed to assess the adequacy in sampling of the configuration space for both states.

The methods above allow one to calculate in principle the free energies of solvation, binding, etc. In the case of free energies of solvation, for example, one solvent molecule is converted to a solute molecule in a stepwise fashion, by creating and annihilating atoms or changing their type which in effect is creating the solute molecule. Binding energies could in principle be calculated in a similar way by "creating" the ligand molecule in the enzyme cavity. In practice, these calculations are very difficult, since they involve a massive displacement of solvent, i.e., the initial and final state differ dramatically. However, in drug design for example, in most cases one is interested not in the actual free energies of solvation or binding, but in differences between these parameters for different molecules, i.e., in values of $\Delta\Delta A$ (or $\Delta\Delta G$ if an isothermic-isobaric system is being considered). The quantitative method of finding these parameters is often called a *thermodynamic cycle* approach and is becoming a routine procedure in finding the differences in free energy of binding for two different ligands or the influence of mutation in the macromolecular receptor or enzyme on binding.

Besides free energies other thermodynamic quantities can be computed applying statistical concepts to computer simulations. Here we present just two of them, namely pressure and distribution functions. Here, for simplicity's sake we shift to Monte Carlo notation.

Distribution functions. The evaluation of the pair correlation function $g(\mathbf{r})$ (either center-of-mass or site–site) is a direct application of the formula:

$$\rho g(\mathbf{r}) = \frac{1}{N}\left\langle \sum_{i \neq j} \delta\big(\mathbf{r} - (\mathbf{r}_i - \mathbf{r}_j)\big)\right\rangle.$$

For an isotropic system such as a liquid or gas, for which there is no preferred direction in space, only the magnitude r ($|\mathbf{r}| = r$) is of relevance, and thus, the function $g(r)$ depends only on the distance r between two particles and not on the relative orientation.

It is easy to see that the quantity $\rho g(\mathbf{r})\, d^3\mathbf{r}$ is proportional to the mean number of molecules which are at a distance $\mathbf{r}$ from one supposed to be in the center of the coordinate system (averaged over all the molecules in the sample) or, stated differently, the number of the pairs whose distance $\mathbf{r}$ is divided by the number of molecules. In the large $\mathbf{r}$ limit the position of the pairs are uncorrelated, so this number becomes $\rho\, d^3 r$ and we have the limit

$$\lim_{\tau \to \infty} g(\mathbf{r}) = 1$$

on the other hand, since the particles cannot occupy the same position, we have also the limit

$$\lim_{\tau \to 0} g(\mathbf{r}) = 0.$$

In general $g(\mathbf{r})$ has a large peak at short distances, and an oscillating behavior around the limiting value after the peak.

To compute $g(\mathbf{r})$ an histogram is made with slots of length dr and the number of pairs whose distance is in the range $(r; r + dr)$ is collected in analyzing the data. Since $\rho g(r)\, d^3r$ is the mean number $\mathcal{N}(r)$ of pairs whose relative distance is r, we have the equality

$$\rho g(\mathbf{r}) 4\pi r^2 \, d\mathbf{r} = \frac{1}{M} \mathcal{N}(r)$$

after M configuration have been used; from which it is easily obtained that

$$g(\mathbf{r}) = \frac{1}{\rho 4\pi r^2 \, d\mathbf{r}} \mathcal{N}(r).$$

It is interesting to notice that, according to the definition, the pair (i, j) and (j, i) (with the obvious condition $i \neq j$) must both be counted. Usually the average is performed only on the pairs such that $i > j$, and every contribution counted twice.

The function $g(\mathbf{r})$, that can be measured in neutron and X-ray difraction experiments, is important for many reasons. It tells us about the structure of complex, isotropic systems and it determines the thermodynamic quantities at the level of the pair potential approximation, such as, for example the pressure

$$P = \frac{NkT}{V} - \frac{N^2}{6V^2} \int r \frac{du}{dr} g(r) \, dr,$$

where $u(r)$ is the pair potential and V the volume.

Calculation of dynamical quantities

Molecular Dynamics simulations give access to the time evolution of the system. The time correlation function formalism provides a systematic way to study this evolution. For a MD-simulation of an equilibrium system, the time correlation function C is defined:

$$C_{AB}(\tau) = \langle A(t)B(\tau) \rangle = \lim_{T \to \infty} \frac{1}{T - \tau} \int_0^{T-\tau} A(t)B(t + \tau) \, dt,$$

where A and B are dynamical variables. In the computational practice, C can be obtained as a straightforward application of the discrete time version of such definition,

$$C_{AB}(\tau) = \frac{1}{T - \tau} \sum_{t=0}^{T-\tau} A_i(t) B_i(t + \tau).$$

One has to take care of the fact that, since we have only a finite number of time-steps saved for the analysis, the correlation function at longer times have less statistics than the correlation at small times, and so the total number of steps must be large enough.

If $A = B$ the term autocorrelation function is used. An autocorrelation function commonly calculated in simulations of atomic liquids is the velocity autocorrelation function. It is of the form $\langle \mathbf{v}_i(t) \cdot \mathbf{v}_i(0) \rangle$. If all the atoms are of the same type, then the result should be independent of i, so we may as well calculate this for each of the atoms and average.

$$\frac{1}{N}\sum_i^N \langle \mathbf{v}_i(t) \cdot \mathbf{v}_i(0) \rangle.$$

Typical behavior of a velocity autocorrelation function

- The value at $t = 0$ is $\langle \mathbf{v}^2 \rangle = 3kT/m$.
- It goes to zero for large times.
- The decay might be monotonic, or go negative and then to zero, or oscillate slightly.

The reason for interest in the velocity autocorrelation function is that it is related to the self-diffusion coefficient

$$D = \frac{1}{3}\int_0^\infty dt \langle \mathbf{v}_i(t) \cdot \mathbf{v}_i(0) \rangle$$

and it is the simplest and easiest transport coefficient to calculate. Other things, like the shear viscosity coefficient, that are more commonly measured experimentally, are much harder to calculate from simulations. Thus, if any transport or time fluctuation properties are calculated, this is the one that is chosen. To calculate the self-diffusion coefficient using this formula, it is necessary to calculate the velocity autocorrelation function for all positive times out to the time where the function is essentially zero.

There is a somewhat easier way of calculating the self-diffusion coefficient in simulations, using the mean squared displacement (MSD). The mean squared displacement is defined as

$$\langle |\mathbf{r}_i(t) - \mathbf{r}_i(0)|^2 \rangle.$$

Since all particles have the same statistical properties in a one component system, we might as well calculate

$$\frac{1}{N}\sum_i^N \langle |\mathbf{r}_i(t) - \mathbf{r}_i(0)|^2 \rangle.$$

The relationship to the self-diffusion coefficient is that the mean squared displacement is equal to $6Dt$ at long times.

The Fourier transform of velocity autocorrelation function is called the spectral density of the particle motions. By making additional assumptions it can be related to Infrared-, Raman-, and neutron inelastic scattering intensities. More in general, time autocorrelation functions are related via Fourier transformation to spectral densities or simply spectra. For example the infrared spectrum is given by

$$I(\omega) \propto \int_0^T dt \exp(i\omega\tau)\langle\mu_i(t)\cdot\mu_i(0)\rangle,$$

where the quantity in brackets is the normalized time autocorrelation function of the system dipole moment vector.

The importance of correlation functions is large. Besides being used to compute dynamical properties as shown above, they are also used to see if simulation has progressed far enough so that calculated quantities are no longer dependent on initial conditions They can be used to look at molecular order in simulation, such as dipole moment, and angular velocity (molecular orientation).

14. Molecules in solution

In this section we shall consider an important application of solvation models: the study of the interactions between molecular systems and external electromagnetic fields in the presence of a solvent. The subject is so large and the related physical phenomena so numerous – suffice it to quote the various spectroscopies – that we shall necessarily limit our exposition to some specific aspects. In particular, we shall present properties that cannot be computed by computer simulation techniques (either MC or MD) as they require accurate QM calculations. In this context, continuum solvation methods represent the most effective approach and for this reason we shall focus on these methods only; in particular, we shall describe how a specific continuum method, the PCM-IEF (see Section 4 of Chapter II), has been generalized to evaluate response properties of molecules in solution to electric and magnetic fields, or to their combination. We recall that PCM-IEF is here used as it is the method developed by our group and thus the one we know better but other continuum models can be alternatively used to get response properties; it is however worth saying that PCM-IEF represents one of the most accurate continuum model at the quantum mechanical level of calculation and certainly, the one with the largest applicability in the study of phenomena related to the interaction of solvated systems and external fields.

The generalization of continuum models to aspects of solvation going beyond the energetics has been made possible due the their specific characteristic to reduce solvent effects to a set of operators which can be cast in a physically and formally simple form. In this framework, the inclusion of the formalism of continuum models into the various approaches provided by the QM theory becomes almost straightforward. In addition, the use of an accurate representation of the solvent field through an apparent surface charge when joined to the definition of a realistic molecular cavity embedding the solute immersed in the dielectric (see previous chapters for more details), makes the continuum methods particularly suitable for this kind of studies. Finally, their recent extension to different environments, like anisotropic solvents, ionic solutions, immiscible solvents with a contact surface, etc., allows, for the first time, the analysis of important response properties also for molecular systems immersed in not standard external matrices, such as liquid crystalline phases or symmetric crystalline frames, charged solutions like those representing the natural neighborhood of proteins and other biological molecules and membranes, just to quote few examples.

Before passing to present the methodologies developed to compute electric and magnetic properties of solvated molecules, an introductory presentation of the basic aspects of the inclusion of solvation continuum models into the quantum mechanical formalism is necessary; as said we exploit here the PCM formalism as representative example of QM solvation continuum models.

The generalization of continuum models to QM calculations implies to define an *Effective Hamiltonian*, i.e., an Hamiltonian to which solute–solvent interactions are added in terms of a solvent reaction potential. The basic hypothesis of this kind of approach is that one can always define a free energy functional $\mathcal{G}(\Psi)$ depending on the solute electronic wavefunction Ψ. This energy functional can be expressed in the following general form (AMOVILLI, BARONE, CAMMI, CANCÈS, COSSI, MENNUCCI, POMELLI and TOMASI [1998]):

$$\mathcal{G}(\Psi) = \langle\Psi|\widehat{H}^0|\Psi\rangle + \langle\Psi|\hat{\rho}_r|\Psi\rangle V_r^R + \frac{1}{2}\langle\Psi|\hat{\rho}_r|\Psi\rangle \mathbf{V}_{rr'}^R\langle\Psi|\hat{\rho}_{r'}|\Psi\rangle. \tag{14.1}$$

In Eq. (14.1) the Born–Oppenheimer (BO) approximation is employed. This means a standard partition of the Hamiltonian into an electronic and a nuclear part, as well as the factorization of the wavefunction into an electronic and a nuclear component. In this approximation Eq. (14.1) refers to the electronic wavefunction with the electronic Hamiltonian dependent on the coordinates of the electrons, and, parametrically, on the coordinates of the nuclei.

The detailed description of the various terms comparing in the equation above will be given in the following, here it suffices to say that $\widehat{H}^0$ is the Hamiltonian describing the isolated molecule, $\hat{\rho}_r$ represents the operator of the solute electronic charge density, V_r^R is the *solvent permanent potential,* and $\mathbf{V}_{rr'}^R$ describes the *response function of the reaction potential* associated with the solvent. Here an extension of the Einstein convention on the sum has been exploited: the space variables r and r', appearing as repeated subscripts, imply an integration in the 3-dimensional space.

By applying the variational principle on this functional we can derive the nonlinear Schrödinger equation specific for the system under scrutiny:

$$\widehat{H}_{\mathrm{eff}}|\Psi\rangle = \left[\widehat{H}^0 + \hat{\rho}_r V_r^R + \hat{\rho}_r \mathbf{V}_{rr'}^R\langle\Psi|\hat{\rho}_{r'}|\Psi\rangle\right]|\Psi\rangle = E|\Psi\rangle \tag{14.2}$$

where E is the Lagrange multiplier introduced to fulfill the normalization condition on the electronic wavefunction. Eq. (14.2) defines the specific 'Effective Hamiltonian', $\widehat{H}_{\mathrm{eff}}$, giving the name to the whole procedure.

The first solvent term V_r^R does not lead to any difficulty, neither from the theoretical point of view, nor from the practical. Many examples are known in which an external potential is introduced in the molecular calculations. On the contrary, the treatment of the reaction potential operator $\hat{\rho}_r\mathbf{V}_{rr'}^R\langle\Psi|\hat{\rho}_{r'}|\Psi\rangle$ is rather delicate, as this term induces a nonlinear character to the solute Schrödinger equation.

By imposing that first-order variation of $\mathcal{G}$ with respect to an arbitrary variation of the solute wavefunction Ψ is zero, and following the standard scheme developed for the self consistent field theory in vacuo, we finally obtain the generalized Fock operator $\widehat{F}'$ and the corresponding equation from which the final wavefunction has to be derived. By introducing the common finite-basis approximation and a closed shell system, we can eliminate the spin dependence (the occupied spin-orbitals occur in pairs) and expand the molecular orbitals (MOs) as a linear combination of atomic orbitals (LCAO). On performing the spin integration in the equations used so far, we find, for the free energy:

$$\mathcal{G} = \mathrm{tr}\,\mathbf{Ph} + \frac{1}{2}\,\mathrm{tr}\,\mathbf{PG}(\mathbf{P}) + \mathrm{tr}\,\mathbf{Ph}^R + \frac{1}{2}\,\mathrm{tr}\,\mathbf{PX}^R(\mathbf{P}) \tag{14.3}$$

and for the generalized Fock matrix:

$$\mathbf{F}' = \mathbf{h} + \mathbf{G}(\mathbf{P}) + \mathbf{h}^R + \mathbf{X}^R(\mathbf{P}), \tag{14.4}$$

where $\mathbf{P}$ is the one-electron density matrix and $\mathbf{h}$ and $\mathbf{G}(\mathbf{P})$ are the matrices used in standard calculations *in vacuo* to collect one- and two-electron integrals, respectively. In Eqs. (14.3), (14.4) the solvent contributions, previously introduced as potential functions, have been translated into a form recalling the vacuum system, just to emphasize the parallelism between the two calculations; in fact, also solvent effects can be partitioned in one- and two-electron contributions, indicated as $\mathbf{h}^R$ and $\mathbf{X}^R(\mathbf{P})$, respectively. In general, $\mathbf{h}^R$ and $\mathbf{X}^R(\mathbf{P})$ will contain different terms related to all possible interactions (dispersive, repulsive and electrostatic) between solute and solvent. In particular for the electrostatic parts we have

$$\mathbf{h}^{\mathrm{el}} = -\sum_k \mathbf{V}(s_k)\mathrm{q}^N(s_k),$$

$$\mathbf{X}^{\mathrm{el}}(\mathbf{P}) = -\sum_k \mathbf{V}(s_k)\mathrm{q}^e(s_k),$$

where q^N and q^e represent the apparent charges induced on the cavity surface by the solute nuclei and electrons, respectively, and $\mathbf{V}$ collects the AO potential integrals on the cavity; the sum runs on all the tessera forming the cavity surface (see Sections 4 and 6 of Chapter II).

In this framework the generalized Fock equation can be solved with the same iterative procedure of the problem *in vacuo*; the only difference introduced by the presence of the continuum dielectric is that, at each SCF cycle, one has to simultaneously solve the standard quantum mechanical problem and the additional problem of the evaluation of the interaction matrices. In this scheme the apparent charges are obtained through a self-consistent technique which has to be nested in that determining the solute wavefunction; as a consequence, at the convergency, solute and solvent distribution charges are mutually equilibrated.

As final note, we observe that the nonlinear equation (14.2) is a direct consequence of the variational principle applied to $\mathcal{G}$. The (free) energy functional $\mathcal{G}$ has a privileged role in the theory, as the solution of the Schrödinger equation gives a minimum of this functional even though it is not the eigenvalue of the nonlinear Hamiltonian, here indicated as E. We stress that in the habitual linear Hamiltonians these two quantities, the Hamiltonian eigenvalue and the variational functional, coincide. The difference between E and $\mathcal{G}$ has, however, a clear physical meaning; it represents the polarization work which the solute does to create the charge density inside the solvent. It is worth remarking that this interpretation is equally valid for zero-temperature models and for those in which the thermal agitation is implicitly or explicitly taken into account.

To pass from the free energy functional $\mathcal{G}$ defined above to the thermodynamical analog G, further aspects have to be considered. As said in the introductory chapter, first we have to choose the reference state (in our case it is given by noninteracting nuclei and electrons of M, supplemented by the unperturbed, i.e., unpolarized, pure liquid S), and to take into account the contributions of the interactions not included in the reaction potential.

Electric dipole polarizabilities

In this section we are concerned with the calculation of electric properties which measure the response of a charge distribution (the solute molecule) to an external electric field. If the charge distribution is mobile, then it will redistribute itself until its energy in the external field is minimized, and this is the phenomenon of polarization. The electric moments will therefore change in the external field, and we can study the change by expanding the dipole moments as a Taylor series. The expansion terms are collected by order into the so-called polarizability tensors $\gamma^{(n)}$. We shall follow the usual convention which indicates as α, β, γ the tensors corresponding to the first three $\gamma^{(n)}$ sets of coefficients, namely the polarizability, the first and the second hyperpolarizabilities.

When the external field has an oscillatory behavior, all these quantities depend on the frequency of such oscillations; for a given $\gamma^{(n)}$ we have to consider the frequencies and phases of the various components of external fields that can be combined in all possible ways to give different electric molecular response. These elements constitute the essential part of the linear and nonlinear optics, a subject for which there is a remarkable interest to know the influence of solvation effects.

Nonlinear optics (NLO) deals with the interaction of electromagnetic fields (light) with matter to generate new electromagnetic fields, altered with respect to phase, frequency, amplitude or other propagation characteristics from the incident field. A major advantage of the use of photonics instead of electronics is the possibility to increase the speed of information processes such as photonic switching and optical computing. One of the most intensively studied nonlinear optical phenomena is second harmonic generation (SHG) or frequency doubling. By this process, near infrared laser light (frequency ω) can be converted by a nonlinear optical material to blue light (2ω). The resulting wavelength is half the incident wavelength and hence it is possible to store information with a higher density. It is obvious that the required properties of the materials depend on the application that they are used for. Traditionally, the materials used to measure second order nonlinear optical behavior were inorganic crystals. Organic materials, such as organic crystals and polymers, have been shown to offer better nonlinear optical and physical properties, such as ultrafast response times, lower dielectric constants, better processability characteristics and a remarkable resistance to optical damage, when compared to the inorganic materials. The ease of modification of organic molecular structures makes it possible to synthesize tailor-made molecules and to fine-tune the properties to the desired application. In the case of second-order nonlinear optical processes, the macroscopic nonlinearity of the material (bulk susceptibility) is derived from the microscopic molecular nonlinearity and the geometrical arrangement of the NLO-chromophores. So, optimizing a material's nonlinearity begins at the molecular structural level.

The molecular response to an electric field regards its whole charge distribution, electron and nuclei. We may introduce also for molecules in solution the usual partition of the theoretical chemistry into electronic and nuclear parts, without neglecting couplings.

Electronic contribution to dipole polarizabilities. The electronic contribution can be computed using two derivative schemes involving quantum mechanical calculations of the energy or, alternatively, of the dipole moment followed by derivatives with respect to the perturbing external field, computed at zero intensity. At Hartree–Fock (HF) or Density Functional (DF) level both approaches lead to the use of the coupled HF or Kohn–Sham theory either in its time-independent (CHF or CKS) or time-dependent (TDHF or TDKS) version according to the case (MCWEENY [1992]).

The solute Hamiltonian must be now supplemented by further terms describing the interaction with the external field; in particular, the corresponding one-electron contribution to the Fock matrix $\mathbf{F}'$ (14.4) becomes

$$\mathbf{h}' = \mathbf{h} + \mathbf{h}^R + \sum_a \mathbf{m}_a E_a,$$

where $\mathbf{m}_a$ is the matrix of the ath Cartesian component of the dipole moment operator, and E_a is the corresponding component of the electric field vector.

To obtain the various electric response functions we have to determine the density matrix of the unperturbed system $\mathbf{P}^0$ and its derivatives ($\mathbf{P}^a$, $\mathbf{P}^{ab}$, etc.) with respect to the components of the electric field, corresponding to the order of the electric response functions.

To obtain time-dependent properties, we have to shift from the basic static model to an extended version in which the solute is described thorough a time-dependent Schrödinger equation. In this extended version of the model we have also to introduce the time-dependence of the solvent polarization, which is expressed in terms of a Fourier expansion and requires the whole frequency spectrum of the dielectric permittivity $\varepsilon(\omega)$ of the solvent.

Applying the Frenkel variational principle at HF or DFT level, we arrive at the following time-dependent equation (MCWEENY [1992])

$$\left(\mathbf{F}' - i\frac{\partial}{\partial t}\right)\mathbf{T} = \mathbf{T}\varepsilon, \tag{14.5}$$

where $\mathbf{T}$ is the matrix containing the expansion coefficients of the MOs on the atomic orbital basis set, and ε is the matrix collecting the Lagrangian multipliers. The solutions of he TDHF (or TDKS) equation can be obtained in a time-dependent coupled perturbation scheme. We first expand Eq. (14.5) in its Fourier components and then each component is expanded in terms of the components of the external field. The separation by orders leads to a set of coupled perturbed equations whose Fock matrices can be written in the following form:

$$\begin{aligned}
&\mathbf{F}' = \mathbf{h}' + \mathbf{G}(\mathbf{P}_0) + \mathbf{X}_0(\mathbf{P}_0),\\
&\mathbf{F}'^{a}(\omega_\sigma;\omega) = \mathbf{m}_a + \mathbf{G}\big(\mathbf{P}^{a}(\omega_\sigma;\omega)\big) + \mathbf{X}_{\omega_\sigma}\big(\mathbf{P}^{a}(\omega_\sigma;\omega)\big),\\
&\mathbf{F}'^{ab}(\omega_\sigma;\omega_1,\omega_2) = \mathbf{G}\big(\mathbf{P}^{ab}(\omega_\sigma;\omega_1,\omega_2)\big) + \mathbf{X}_{\omega_\sigma}\big(\mathbf{P}^{ab}(\omega_\sigma;\omega_1,\omega_2)\big),\\
&\mathbf{F}'^{abc}(\omega_\sigma;\omega_1,\omega_2,\omega_3) = \mathbf{G}\big(\mathbf{P}^{abc}(\omega_\sigma;\omega_1,\omega_2,\omega_3)\big) + \mathbf{X}_{\omega_\sigma}\big(\mathbf{P}^{abc}(\omega_\sigma;\omega_1,\omega_2,\omega_3)\big),
\end{aligned} \tag{14.6}$$

where ω_σ is the frequency of the resulting wave (i.e., $\omega_\sigma = \sum_k \omega_k$, with ω_k including the sign), and matrices $\mathbf{X}_{\omega_\sigma}(\mathbf{P}^{ab\cdots}(\omega_\sigma))$ represent the interactions with the solvent which has been induced by the perturbed electron density $\mathbf{P}^{ab\cdots}(\omega_\sigma)$ oscillating at frequency ω_σ (CAMMI, COSSI, MENNUCCI and TOMASI [1996]). There is a number of equations of this type, corresponding to the different combinations of frequencies, each one related to a different phenomenon of the nonlinear optics.

Once the solution of the different TDHF (or TDKS) equations has been obtained, the dynamic (hyper)polarizabilities of interest can be expressed in the following forms:

$$\begin{aligned}
&\alpha_{ab}(\omega_\sigma;\omega) = -\,\mathrm{tr}\big[\mathbf{m}_a\mathbf{P}^{b}(\omega_\sigma;\omega)\big],\\
&\beta_{abc}(\omega_\sigma;\omega_1,\omega_2) = -\,\mathrm{tr}\big[\mathbf{m}_a\mathbf{P}^{bc}(\omega_\sigma;\omega_1,\omega_2)\big],\\
&\gamma_{abcd}(\omega_\sigma;\omega_1,\omega_2,\omega_3) = -\,\mathrm{tr}\big[\mathbf{m}_a\mathbf{P}^{bcd}(\omega_\sigma;\omega_1,\omega_2,\omega_3)\big],
\end{aligned} \tag{14.7}$$

where tr indicates a trace operation on the product of the dipole integral matrix $\mathbf{m}_a$ and the derivatives of the electronic densities $\mathbf{P}^{abc}$.

Since the frequency-dependent response of the solvent is included in the kernel of the integrals collected into $\mathbf{X}_{\omega_\sigma}$, the resulting (hyper)polarizabilities will also depend on the frequency spectrum of the dielectric function $\varepsilon(\omega)$ of the solvent. When $\varepsilon(\omega)$ is described by the Debye formula (i.e., in terms of a single relaxation mode), the resulting dispersion curve depends on the frequency corresponding to the inverse of the Debye relaxation time.

We have remarked that there is a large number of coupled Fock equations to be solved in order to get electric response properties. However, as for the case of a molecule in vacuo, the computational effort can be reduced by resorting the exploiting to so called $(2n+1)$ rule that permits to get the $(2n+1)$th response using nth derivative of $\mathbf{P}$. This is a formal property of the perturbative scheme which is known since a long time and fulfilled by all good computational codes, when applied to molecules in vacuo: the same property also holds for molecules in solution.

Nuclear contribution to electric polarizabilities. The global effect of an applied external field on a molecule involves distortions both in the electronic charge distribution and in the nuclear charge distribution; the latter leads to the so-called vibrational, or nuclear, contribution to the (hyper)polarizabilities (BISHOP [1998]).

The analysis of the vibrational components reveals the presence of the distinct components, the "curvature" related to the effect of the field vibrational motion and including the zero point vibrational correction (ZPV), and the "nuclear relaxation" (nr) originates from the shift of the equilibrium geometry induced by the field.

The nuclear relaxation is the dominant contribution and can be computed in two ways: by perturbation theory or by finite field approximation. We shall limit ourselves to the perturbation theory. In this scheme the polarizabilities which can be written in terms of a sum over vibrational states k of energy-weighted transition moments, for example:

$$\alpha_{ab} = \sum_{k \neq 0} \frac{\langle 0|\mu_a|k\rangle\langle k|\mu_b|0\rangle}{\hbar\nu_k}, \tag{14.8}$$

where $\hbar\nu_k$ is the energy of the vibrational state $|k\rangle$ relative to the ground state $|0\rangle$, which is excluded from the sum.

The vibrational transition moments can be obtained through an expansion of the electronic properties (dipole and polarizability functions) in terms of the nuclear coordinate of the molecule, analogous to the expansion of the vibrational potential within the Born–Oppenheimer approximation. For a molecule in solution the potential energy is given by $\mathcal{G}(\mathbf{R})$ as function of the nuclei configuration $\mathbf{R}$. If, we separate the components according their anharmonicity, at level of double harmonicity (electric and mechanical) only linear term are considered in the expansion of the properties and only quadratic terms are considered in the expansion of the vibrational potential. The corresponding expressions of the vibrational contribution to the electric (hyper)polarizabilities become:

$$\alpha^v_{ab} = \sum_i^{3N-6} \left(\frac{\partial\mu_a}{\partial Q_i}\right)_0 \left(\frac{\partial\mu_b}{\partial Q_i}\right)_0 \Big/ (4\pi^2\nu_i^2),$$

$$\beta^v_{abc} = \sum_i^{3N-6} \left[\left(\frac{\partial\mu_c}{\partial Q_i}\right)_0 \left(\frac{\partial\alpha_{ab}}{\partial Q_i}\right)_0 + \left(\frac{\partial\mu_b}{\partial Q_i}\right)_0 \left(\frac{\partial\alpha_{ac}}{\partial Q_i}\right)_0 + \left(\frac{\partial\mu_a}{\partial Q_i}\right)_0 \left(\frac{\alpha_{bc}}{\partial Q_i}\right)_0\right] \Big/ (4\pi^2\nu_i^2),$$

where ν_i is the harmonic frequency obtained by the eigenvalues of the nuclear Hessian computed in presence of solvent effects; its elements are the second derivatives of the free energy $\mathcal{G}(\mathbf{R})$ with respect the mass weighted nuclear Cartesian components. Q_i denotes the normal mode associated to ν_i and each partial derivative is evaluated at the equilibrium geometry of the solvated system (CAMMI, MENNUCCI and TOMASI [1998b]).

Macroscopic susceptibilities: local field effects and effective polarizabilities

The phenomenological description of the polarization of a macroscopic medium subjected to an electric field and of the connected phenomenon of linear and nonlinear response is given in terms of the macroscopic susceptibilities $\chi^{(n)}$. As of particular interest, here we focus on the optical phenomena connected with linear and nonlinear response properties of a medium subjected to an electric field given by the superposition of a static and an optical component:

$$\mathbf{E} = \mathbf{E}^0 + \mathbf{E}^\omega \cos \omega t. \tag{14.9}$$

In this case the polarization density $P(t)$ of the medium can be written in terms of Fourier components (BUTCHER and COTTER [1990]):

$$P(t) = \mathbf{P}^0 + \mathbf{P}^\omega \cos(\omega t) + \mathbf{P}^{2\omega} \cos(2\omega t) + \mathbf{P}^{3\omega} \cos(3\omega t),$$

where

$$\begin{aligned} P^0 &= \chi^{(0)} + \chi^{(1)}(0;0)\cdot\mathbf{E}^0 + \chi^{(2)}(0;0,0):\mathbf{E}^0\mathbf{E}^0 + \frac{1}{2}\chi^{(2)}(0;-\omega,\omega):\mathbf{E}^\omega\mathbf{E}^\omega \\ &\quad + \chi^{(3)}(0;0,0,0)\vdots\mathbf{E}^0\mathbf{E}^0\mathbf{E}^0 + \cdots, \\ P^\omega &= \chi^{(1)}(-\omega;\omega)\cdot\mathbf{E}^\omega + 2\chi^{(2)}(-\omega;\omega,0):\mathbf{E}^\omega\mathbf{E}^0 \\ &\quad + 3\chi^{(3)}(-\omega;\omega,0,0)\vdots\mathbf{E}^\omega\mathbf{E}^0\mathbf{E}^0 + \cdots, \\ P^{2\omega} &= \frac{1}{2}\chi^{(2)}(-2\omega;\omega,\omega):\mathbf{E}^\omega\mathbf{E}^\omega + \frac{3}{2}\chi^{(3)}(-2\omega;\omega,\omega,0)\vdots\mathbf{E}^\omega\mathbf{E}^\omega\mathbf{E}^0 + \cdots. \end{aligned} \tag{14.10}$$

In Eqs. (14.10) the argument of the susceptibilities tensors $\chi^{(n)}$ describes the nature of the frequency-dependence at the various order; in all cases the frequency of the resulting wave (which is indicated as ω_σ) is stated first, then the frequency of the incident interacting wave(s) (two in a first-order process, three in the second-order analog and four in the third-order case).

The various susceptibilities may be obtained through specific experiments in linear and nonlinear optics. The first-order static susceptibility $\chi^{(1)}(0;0)$ is related to the dielectric constant at zero frequency, $\varepsilon(0)$, while $\chi^{(1)}(-\omega;\omega)$ is the linear optical susceptibility related to the refractive index n^ω at frequency ω. Passing to nonlinear effects it is worth recalling that $\chi^{(2)}(-2\omega;\omega,\omega)$ describes frequency doubling process which is usually indicated as second harmonic generation (SHG), and $\chi^{(3)}(-2\omega;\omega,\omega,0)$ describes the influence of an external field on the SHG process, measured in the third-order process of electric-field induced second harmonic generation (EFISHG).

If we consider as macroscopic sample a liquid solution of different molecular components, each at a concentration c_J, the effects of the single components are assumed to be additive so that the global measured response becomes:

$$\chi^{(n)} = \sum_J \zeta_J^{(n)} c_J,$$

where $\zeta_J^{(n)}$ are the nth-order molar polarizabilities of the constituent J. The values of the single $\zeta_J^{(n)}$ can be extracted from measurements of $\chi^{(n)}$ at different concentrations.

The electric response properties of molecular materials, as a liquid solution, on which the macroscopic electric susceptibilities $\chi^{(n)}$ depend upon are the (hyper)polarizabilities of the constituting molecules. For the gas phase the connection between the molecular hyperpolarizabilities and the corresponding susceptibilities is given in terms of a Boltzmann thermal average process regarding the permanent and the induced molecular dipole moments. To establish the same connection for a condensed phase we have also to face the so called "Local Field Problem". The problem arises because the field acting locally on a molecule is different from the applied Maxwell field $\mathbf{E}(t)$ due to the presence of all the solvent molecules around it (WORTMANN and BISHOP [1998]).

Studies on this problem date to the beginning of the past century: 'almost' empirical recipes are based on the Onsager local field correction factor for polar liquids and the Lorentz factor for nonpolar liquids. Both factors apply to cavities of regular shape, spheres or ellipsoid (BÖTTCHER [1973], BÖTTCHER and BORDEWIJK [1978]).

In the solvation PCM-IEF model a more general description of the local field effect can be obtained (CAMMI, MENNUCCI and TOMASI [1998a], CAMMI, MENNUCCI and TOMASI [2000]). In this method the effect is described by supplementing the solute Hamiltonian by a further term describing the interaction of the solute with the apparent surface charges $\mathbf{q}^{\text{ex}}$ induced by the external field $\mathbf{E}(\omega)$. The resulting one-electron contribution to the Fock matrix is then given by:

$$\begin{aligned} \mathbf{h}' = \mathbf{h} + \mathbf{h}^R &+ \sum_a \mathbf{m}_a \left[E_a^\omega (e^{i\omega t} + e^{-i\omega t}) + E_a^0 \right] \\ &+ \sum_a \widetilde{\mathbf{m}}_a^\omega E_a^\omega (e^{i\omega t} + e^{-i\omega t}) + \sum_a \widetilde{\mathbf{m}}_a^0 E_a^0, \end{aligned} \tag{14.11}$$

where the elements of $\widetilde{\mathbf{m}}_a$ are defined in terms of the additional set of apparent charges, $\mathbf{q}^{\text{ex}}$, induced on the cavity surface by the external field, namely (CAMMI, MENNUCCI and TOMASI [1998a], CAMMI, MENNUCCI and TOMASI [2000]):

$$\widetilde{\mathbf{m}}_\alpha(\omega) = \sum_k \mathbf{V}(s_k) \frac{\partial \mathbf{q}^{\text{ex}}(\omega; s_k)}{\partial E_\alpha},$$

where we have indicated the dependence of the apparent charges and of the following $\widetilde{\mathbf{m}}$ matrix on the frequency ω of the applied field; such dependence is obtained by using a frequency-dependent permittivity $\varepsilon(\omega)$ to compute the apparent charges (CAMMI, MENNUCCI and TOMASI [2000]). The two last terms in the r.h.s. of Eq. (14.11) influence the evaluation of the derivatives of the density matrices required to get the (hyper)polarizabilities (see Eqs. (14.7)); they in fact modify the first-order derivative of the Fock matrix given in (14.6) as follows:

$$\mathbf{F}'^{a}(\omega_\sigma;\omega) = \mathbf{m}_a + \widetilde{\mathbf{m}}_a^{\omega} + \mathbf{G}\big(\widetilde{\mathbf{P}}^{a}(\omega_\sigma;\omega)\big) + \mathbf{X}_{\omega_\sigma}\big(\widetilde{\mathbf{P}}^{a}(\omega_\sigma;\omega)\big). \tag{14.12}$$

The new set of density matrix derivatives, $\widetilde{\mathbf{P}}^{ab\cdots}(\omega_\sigma)$, obtained by applying the usual procedure to this new system, can be then exploited to compute the so called *effective (hyper)polarizabilities*: (see Eqs. (14.7)) which describe directly the electric response of the molecular solute to the applied Maxwell field $\mathbf{E}(\omega)$.

The macroscopic susceptibilities can now be obtained as statistical average of the corresponding effective (hyper)polarizabilities. Expressions of the molar polarizabilities for linear processes are:

$$\zeta^{(1)}(0;0) = N_A\left(\frac{\mu^*\cdot\tilde{\mu}}{3kT} + \tilde{\alpha}_{is}(0;0)\right), \tag{14.13}$$

$$\zeta^{(1)}(-\omega;\omega) = N_A\tilde{\alpha}_{is}(-\omega;\omega) \tag{14.14}$$

and as an example for higher-order processes, we quote here the EFISHG:

$$\zeta^{(3)}_{ZZZZ}(-2\omega;\omega,\omega,0) = N_{\mathrm{A}}\left(\frac{\tilde{\beta}_{\mathbf{v}}(-2\omega;\omega,\omega)\cdot\mu^*}{15kT} + \tilde{\gamma}_s(-2\omega;\omega,\omega,0)\right), \tag{14.15}$$

where N_A is the Avogadro number and k is the Boltzmann constant. In Eqs. (14.13)–(14.14) $\tilde{\alpha}_{is}$ is $1/3$ of the trace of the effective polarizability tensor while μ^* of Eq. (14.15) is the *effective dipole moment* of the solute given by the first derivative of the free energy $\mathcal{G}$ with respect to the static component $\mathbf{E}^0$ of the field (14.9); $\tilde{\beta}_v(-2\omega;\omega,\omega)$ and $\tilde{\gamma}_s(-2\omega;\omega,\omega,0)$ of Eq. (14.15) are, respectively, the "vector part" of the second order polarizability and the "scalar part" of the third order polarizability. Parallel expressions for other NLO process can be easily formulated.

As the molar polarizabilities $\zeta_J^{(n)}$ represent an easily available 'experimental' set of data, the expressions above become important for the comparison between theoretical and experimental evaluation of bulk response properties, as we shall show in the next section.

A numerical example: susceptibilities of nitroanilines

Organic molecules that exhibit large nonlinear optical properties usually consist of a frame with a delocalized π-system, end-capped with either a donor (D) or acceptor (A) substituent or both. This asymmetry results in a high degree of intramolecular charge-transfer (ICT) interaction from the donor to acceptor, which seems to be a prerequisite for a large nonlinearity. Extensively studied classes of NLO-chromophores of this type are 1,4-disubstituted benzenes from which p-nitroaniline (pNA) is a prototypical example. The nonlinear part of the induced molecular polarization is the result of the polarizability of the π-electron system. Because of the low mass of electrons compared to that of ions in inorganic crystals, organic molecules can respond to electromagnetic fields with much higher frequencies (up to 10^{14} Hz). This faster optical response is particularly interesting for applications that depend on the speed of information processing, such as optical switching. The nonlinearity can be enhanced by using stronger donor and acceptor substituents to increase the electronic asymmetry or by increasing the conjugation length between the substituents.

Here we present results of a study on molar polarizabilities for pNA in different liquid solutions. In this presentation we collect results extracted from a research we have published on Journal of Physical Chemistry A (CAMMI, MENNUCCI and TOMASI [2000]) and unpublished data. The attention will be focussed on two specific experimental processes from which data of first and third-order molar polarizabilities have been extracted: refractometric and EFISH measurements.

The results refer to HF and DFT calculations with a Dunning double-zeta valence (DZV) basis set to which $d(0.2)$ function for C, N and O and a $p(0.1)$ function on H have been added, (the numbers in parentheses are the exponents of the extra functions). For DFT calculations the hybrid functional which mixes the Lee, Yang and Parr functional for the correlation part and Becke's three-parameter functional for the exchange (B3LYP) has been used. The solvation method used is the PCM-IEF with a molecular cavity obtained in terms of interlocking spheres centered on the six carbons of the aromatic ring, and on all the nuclei of the external groups.

The geometry of the two solutes has been optimized at B3LYP/6-311G* level in the presence of the solvent. All the effective electronic properties (both static and dynamic) have been computed with the TDHF/TDKS procedures implemented in two standard computational packages Gaussian (FRISCH ET AL. [1998]) and GAMESS (SCHMIDT, BALDRIDGE, BOATZ, ELBERT, GORDON, JENSEN, KOSEKI, MATSUNAGA, NGUYEN, SU, WINDUS, DUPUIS and MONTGOMERY [1993]) and properly modified to take into account the solvent effects.

In Table 14.1 we report the refractometric molar polarizability of pNA in three different solutions: dioxane, acetone and acetonitrile. The specific expressions to be used for the property is reported in (14.14), where the exploited frequency is that corresponding to $\lambda = 589$ nm. Three sets of computed values are reported corresponding to static, dynamic and effective calculations, respectively. Here the terms static and dynamic refer to the model exploited to describe the solvent response: namely a medium described by its static dielectric constant $\varepsilon(0)$ or by its dynamical analog $\varepsilon(\omega)$. Finally, the effective values take into account both the dynamical response of the solvent and the "local field effects" described in the previous section.

The comparison with the experimental values clearly shows the increasing accuracy of the computed results going from the static, to the dynamic and finally to the effective model. It has to be noted that exploiting a static or a dynamic model makes significant differences in the two polar solvents (acetone and acetonitrile) but not in the apolar dioxane: in apolar solvents, in fact, the static and the dynamic permittivities are very similar as the dielectric response of the solvent molecules is almost exclusively

TABLE 14.1
Computed and experimental frequency dependent first order molar polarizabilities $\zeta^{(1)}(\omega)$ of pNA in different solutions. All molar polarizabilities are in SI units (10^{-16} cm^2 V^{-1} mol^{-1}). The frequency corresponds to $\lambda = 589$ m

	Static	Dynamic	Effective	exp[a]
Dioxane	13.2	13.2	15.1	$15.7 \div 0.5$
Acetone	19.3	14.5	16.4	16.26 ± 0.5
Acetonitrile	19.8	14.6	16.5	16.28 ± 0.3

[a] WOLFF and WORTMANN [1999].

described by the electronic motions (those inducing the dynamic $\varepsilon(\omega)$) while the small or null dipolar character does not significantly contribute.

The further important improvement obtained passing from dynamic to effective values indicates that the solvation model has to take into account not only the molecular aspects of the phenomenon (through an accurate description of the solute-solvent interactions) but also the macroscopic aspects involved in the experimental measurement. This is here achieved by explicitly introducing the effect of the oscillating external field on the solvent through a further operator to be considered in the TDHF/TDKS scheme (see Eq. (14.11)).

Let us pass now to the second application we have introduced at the beginning of the section, namely that of the evaluation of EFISH second order susceptibilities.

The EFISH technique (PRASAD and WILLIAMS [1991]) is one of the most used for obtaining information on the molecular hyperpolarizability, β; here, once again, we do not report any details but just some basic notes. The operating frequency is that related to the fundamental beam of 1064 nm from a Q-switched, mode-locked Nd:YAG laser. A symmetry consideration shows that the third-order nonlinearity that is measured in the EFISH experiment for a medium which is isotropic in the absence of any external electric field has only two independent components. In order to determine these tensor elements two EFISH measurements are usually performed for two polarization conditions, the electric field vector of the fundamental being parallel ($\|$) or perpendicular ($\perp$) to the external electric field $\mathbf{E}^0$. The frequency doubled photons are detected with polarization parallel to $\mathbf{E}^0$ in both cases. The exact expressions for the corresponding two molar polarizabilities can be derived from the general equation (14.15), by taking into account also the symmetry of the molecules under examination, in our case C_{2v}.

To compute the frequency-dependent first and second hyperpolarizabilities necessary to get the third-order molar polarizabilities (14.15), HF calculations have been performed: geometries and basis set, as well as all the parameters of the PCM-IEF model are the same we have used to get first order $\zeta^{(1)}(\omega)$. In this case only a set of computed values are presented: they have obtained with the full effective model.

Once again the agreement between computed third-order molar polarizabilities and experimental EFISH data are well within the experimental error. Here, however the complex nature of the final property (given by a combination of dipole, first and second hyperpolarizabilities) is difficult to analyze in terms of clear solvent effects. Actually,

TABLE 14.2
Computed (HF) effective EFISHG third-order molar polarizabilities of pNA in dioxane. Molar polarizabilities are in SI units (10^{-36} cm^4 V^{-3} mol^{-1}). The frequency corresponds to $\lambda = 1064$ nm

	calc	exp[a]
$\zeta^{(3)}(\parallel)$	110	120 ± 11
$\zeta^{(3)}(\perp)$	36	39 ± 4

[a] WOLFF and WORTMANN [1999].

the contribution given by the second hyperpolarizability $\tilde{\gamma}$ can be neglected, being at least an order of magnitude smaller that the $\tilde{\beta} \cdot \mu^*$ term (as usually observed by the experimentalists), and thus an attempt of explanation could be based on the different effects of solvation in the calculations of dipoles and higher-order properties.

Magnetic response properties

All nuclei are surrounded by electrons. When a magnetic field is applied to an atom it induces a circulation of electrons round the nucleus. This movement of electrons produces a tiny, localized magnetic field which opposes the applied field. As a result, the nucleus experiences a reduced overall field and is described as shielded. The extent of the shielding depends on the nature of the electron density in that region. Spectrometers are sensitive enough to show the amount of shielding a nucleus experiences and this can be seen on a spectrum as the chemical shift in the Nuclear-Magnetic-Resonance (NMR) spectroscopy.

If the charge distribution about a nucleus is spherically symmetric, the induced field at the nucleus $\mathbf{B}'$ opposes the applied magnetic field $\mathbf{B}^0$. For rapidly tumbling molecules in solution, $\mathbf{B}'$ is related to the external field by the nuclear shielding tensor σ:

$$\mathbf{B}' = -\sigma \cdot \mathbf{B}^0$$

which means the local field is defined by

$$\mathbf{B}_{\mathrm{loc}} = (\mathbf{1} - \sigma) \cdot \mathbf{B}^0.$$

The components of the nuclear shielding are expressed in dimensionless units (ppm) which are independent of the applied magnetic field, and are therefore molecular characteristics.

There are several sources contributing to the secondary magnetic fields. The nuclear shielding is generally partitioned into two major components: the diamagnetic and paramagnetic shieldings. Diamagnetic nuclear shielding arises from circulation of electrons in the s orbitals and filled shells surrounding the nucleus and it depends on the ground state of the molecule. The diamagnetic contribution to nuclear shielding is normally associated with increased shielding, with the local field at the nucleus antiparallel to the external magnetic field. Paramagnetic shielding, which arises from the nonspherical orbitals, is associated with the orbital angular momentum of electrons, and it is therefore dependent on excited states of the molecule. Paramagnetic contributions normally result in deshielding, with local fields at the nuclei aligned parallel with $\mathbf{B}^0$.

The nuclear shielding is very sensitive to the molecular environment, and the NMR spectra, which are usually obtained for molecules in solution, strongly depend on the solvent. In general, the main effects on NMR spectra arise from intermolecular interactions between solute and solvent molecules.

In the past semiclassical models have been proposed to describe solvent effects on nuclear shieldings; of particular importance is the well-known analysis elaborated by Buckingham (BUCKINGHAM, SCHAEFER and SCHNEIDER [1960]). In this scheme the solvent effect on the solute shielding for nucleus λ may be partitioned as follows

$$\Delta\sigma(\lambda) = \sigma_E(\lambda) + \sigma_w(\lambda) + \sigma_a(\lambda) + \sigma_b(\lambda), \tag{14.16}$$

where σ_E is the "polar effect caused by the charge distribution in the neighboring solvent molecules, thereby perturbing its electronic structure and hence the nuclear screening constants", σ_w is "due to the Van der Waals forces between the solute and the solvent", σ_a "arises from anisotropy in the molecular susceptibility of the solvent molecules", and σ_b is the "contribution proportional to the bulk magnetic susceptibility of the medium". In the context of the subject treated in this chapter, the most important component is that due to the 'polar effect' σ_E, and in fact below we shall describe how this quantity can be evaluated within PCM-IEF solvation method. Here however we would also like to note that all terms appearing in (14.16) in principle could be analytically included into solvation models; in particular for $\sigma_E(\lambda)$, and $\sigma_w(\lambda)$ continuum approaches should be sufficient while σ_a could maybe require the inclusion of discrete solvent molecules around the target solute. A more complex analysis is that required for the bulk susceptibility effect σ_b; a possible approach could be to consider this effect as that already mentioned for other molecular response properties computed in a condensed medium. The external field acting on the molecule is modified by the presence the solvent molecules with an extra local modification which can be related to the shape and the dimension of the volume occupied by the molecule.

Nuclear magnetic shielding. The nuclear magnetic shielding for molecule *in vacuo* can be described in terms of the influence on the total energy of the molecule of the nuclear magnetic moment and of the applied uniform magnetic field. Translating this analysis to molecular solutes in the presence of solvent interactions, leads to define the components of the shielding tensor σ, as the following second derivatives of the free energy functional:

$$\sigma_{ab}^X = \frac{\partial^2 \mathcal{G}}{\partial B_a \partial \mu_b^X},$$

where B_a and μ_b^X $(a, b = x, y, z)$ are the Cartesian components of the external magnetic field $\mathbf{B}$, and of the nuclear magnetic moment μ^X (X refers to a given nucleus).

It is well known that the presence of the magnetic field introduces the problem of the definition of the origin of the corresponding vector potential. However, since σ is a molecular properties, it must be invariant with respect changes of the gauge origin. To obtain this gauge invariance in the ab initio calculation, two ways can be adopted. One is to employ a sufficiently complete basis set so that the consequences of the choice of the gauge origin on th calculated value of σ are minimal. The second method is to introduce gauge factors into either the atomic orbitals of the basis set or the molecular orbitals of a coupled Hartree–Fock calculation in such a manner that the results are independent on the gauge origin even though the calculation is approximate. Inclusion of gauge factors in the atomic orbitals may be accomplished by using gauge invariant atomic orbitals (GIAO) (DITCHFIELD [1974], WOLINSKI, HINTON and PULAY [1990]):

$$\chi_\nu(B) = \chi_\nu(0)\exp\left[-\frac{i}{2c}(B \times R_\nu)\cdot r\right], \tag{14.17}$$

where R_ν is the position vector of the basis function, and $\chi_\nu(0)$ denotes the usual field-independent basis function.

The GIAO method is used in conjunction with analytical derivative theory; in this approach the magnetic field perturbation is treated in an analogous way to the perturbation produced by changes in the nuclear coordinates. In this framework, the components of the nuclear magnetic shielding tensor are obtained as:

$$\sigma_{ab} = \mathrm{tr}\left[\mathbf{P}\mathbf{h}^{B_a\mu_b^X} + \mathbf{P}^{B_a}\mathbf{h}^{\mu_b^X}\right],$$

where $\mathbf{P}^{B_a}$ is the derivative of the density matrix with respect to the magnetic field. Matrices $\mathbf{h}^{\mu_b^X}$ and $\mathbf{h}^{B_a\mu_b^X}$ contain the first derivative of the standard one-electron Hamiltonian with respect to the nuclear magnetic moment and the second derivative with respect the magnetic field and the nuclear magnetic moment, respectively. Both terms do no contain explicit solvent-induced contributions as the latter do not depend on the nuclear magnetic moment of the solute and thus the corresponding derivatives are zero.

On the contrary, explicit solvent effects act on the first derivative of the density matrix $\mathbf{P}^{B_a}$ which can be obtained as solution of the corresponding first-order HF (or KS) equation characterized by the following derivative of the Fock matrix (CAMMI, MENNUCCI and TOMASI [1998], CAMMI, MENNUCCI and TOMASI [1999]):

$$\mathbf{F}'^{B_a} = \mathbf{h}^{B_a} + \mathbf{h}_R^{B_a} + \mathbf{G}^{B_a}(\mathbf{P}) + \mathbf{X}_R^{B_a}(\mathbf{P})$$

where $\mathbf{P}$ is the unperturbed density matrix and the solvent-induced terms, $\mathbf{h}_R^{B_a} + \mathbf{X}_R^{B_a}(\mathbf{P})$, appear due to the magnetic field dependence of the atomic orbitals.

Chiroptical solvated molecules: OR and VCD

Optical activity, and the related spectroscopy (generally called polarimetry), is one of the oldest research tools that is routinely practiced by chemists. Polarimetry is in fact a sensitive, nondestructive technique for measuring the optical activity exhibited by inorganic and organic compounds. A compound is considered to be optically active if linearly polarized light is rotated when passing through it. The amount of optical rotation is determined by the molecular structure and concentration of chiral molecules in the substance.

The polarimetric method is a simple and accurate means for determination and investigation of structure in macro, semi-micro and micro analysis of expensive and nonduplicable samples. Polarimetry is employed in quality control, process control and research in the pharmaceutical, chemical, essential oil, flavor and food industries. Research applications for polarimetry are found in industry, research institutes and universities as a means of: (i) Evaluating and characterizing optically active compounds by measuring their specific rotation and comparing this value with the theoretical values found in literature; (ii) Investigating kinetic reactions by measuring optical rotation as a function of time; (iii) Monitoring changes in concentration of an optically active component in a reaction mixture, as in enzymatic cleavage; (iv) Analyzing molecular structure by plotting optical rotatory dispersion curves over a wide range of wavelengths; (v) Distinguishing between optical isomers.

Optical rotatory dispersion (optical rotation, OR, versus wavelength) and circular dichroism (CD) are the two components of the optical activity that can be used to elucidate the absolute stereochemistry of chiral molecules. Vibrational circular dichroism (VCD), and vibrational Raman optical activity (VROA) are alternative properties that are being currently used for the same scope. The main aspect to stress here is that both sets of investigative tools are strictly related to the condensed phase, as experimental measurements are generally limited to this. Despite this, the extension of solvation models to general QM methods for predicting chiro-optical properties has been very limited. Only in this last few years some efforts (many unpublished as still in their phase of testing) have been appeared; below we shall try to give a preliminary summary of these new applications, once again with almost exclusive attention to PCM–IEF solvation scheme.

OR. As shown above chiral molecules exhibit optical rotation. With very few exceptions, optical rotation measurements are carried out in the condensed phase, most often in liquid solutions. Optical rotations of solutions of chiral molecules are solvent dependent. In the case of flexible molecules, which exhibit multiple conformations in solution, solvent effects can often be attributed predominantly to changes in conformational populations with solvent. However, in the case of rigid molecules, exhibiting a single conformation, optical rotations can still exhibit substantial solvent dependence.

According to the canonical treatment of the theory of optical rotatory power, the optical rotation at a frequency ω of an isotropic dilute solution of a chiral molecule is given by

$$\phi(\omega) = \frac{4\pi N\omega^2}{c^2}\gamma_{\mathrm{LF}}(\omega)\beta(\omega),$$

where $\phi(\omega)$ is the rotation in radians/cm, c is the velocity of light and $\beta(\omega)$ is the frequency-dependent electric dipole-magnetic dipole polarizability of the chiral molecule. In the equation above, $\gamma_{\mathrm{LF}}(\omega)$ is the "local field correction factor" (i.e., the ratio of the microscopic electric field acting on the chiral molecule to the macroscopic electric field of the light wave) of Lorentz, given by $\gamma_{\mathrm{LF}}(\omega) = (n^2(\omega) + 2)/3$, where $n^2(\omega)$ is the refractive index of the solvent. A variety of ab initio methods have recently been applied to the calculation of optical rotation, but solvent effects on $\phi(\omega)$ have been generally ignored or included using the simplified Lorentz equation. None of these calculations satisfactorily take account of solvent effects. Here, we present a theory of solvent effects on optical rotations which allows to evaluate the total solvent effects on $\phi(\omega)$ by computing an "effective" electric dipole-magnetic dipole polarizability, i.e., without the need to add a scaling parameter as that represented by γ_{LF} (MENNUCCI, TOMASI, CAMMI, CHEESEMAN, FRISCH, DEVLIN, GABRIEL and STEPHENS [2002]). In this scheme, the optical rotations are calculated using ab initio Density Functional Theory (DFT) and solvent effects are incorporated using the PCM-IEF solvation model.

The ab initio theoretical treatment of the molecular optical rotation is based on the calculation of the electric dipole-magnetic dipole polarizability tensor, given by the expression (CONDON [1937]):

$$\beta_{ab} = \frac{4\pi c}{3h} \sum_{n \neq s} \frac{1}{\omega_{ns}^2 - \omega^2} \operatorname{Im}\bigl\{\langle \Psi_s | \mu_a^{\mathrm{el}} | \Psi_n \rangle \langle \Psi_n | \mu_b^{\mathrm{mag}} | \Psi_s \rangle\bigr\}, \tag{14.18}$$

where μ_a^{el} and μ_b^{mag} are, respectively, the electric and magnetic dipole operators, and Ψ_s and Ψ_n represent the ground and excited electronic states. The angular frequency ω_{ns} and ω are those related to the $s \to n$ transition and the exciting radiation, respectively. The scalar β required to get the optical rotation $\phi(\omega)$ is obtained as $1/3$ of the trace of the tensor.

An explicit evaluation of the sum over excited states in Eq. (14.18) can be avoided by exploiting the expression based on the ground state function only:

$$\beta_{ab}(\omega) = \frac{hc}{3\pi} \operatorname{Im}\left\langle \frac{\partial \Psi(\omega)}{\partial E_a} \middle| \frac{\partial \Psi(\omega)}{\partial H_b} \right\rangle,$$

where E_a and H_b are electric and magnetic field directions, respectively.

The change in the ground state wavefunction with respect to (oscillating) applied electric or magnetic field perturbations are determined from the TDHF or TDKS procedure described in the paragraph entitled "Electric dipole polarizabilities". For the magnetic perturbation atomic basis functions χ_μ which are magnetic field dependent (GIAOs) (see Eq. (14.17)) are usually exploited.

Following what reported in the paragraph entitled "Macroscopic susceptibilities: local field effects and effective polarizabilities", in the presence of an PCM-IEF continuum dielectric, the frequency-dependent perturbation to be added to the one-electron operator of the unperturbed system can be rewritten as:

$$\begin{aligned} \mathbf{h}'(\omega) &= \frac{1}{2}\mathbf{m}^{\mathrm{elec}} \cdot \mathbf{E}\bigl(e^{-\imath\omega t} + e^{+\imath\omega t}\bigr) \\ &\quad + \frac{1}{2}\sum_s \mathbf{V}(s) \frac{\partial \mathbf{q}^{\mathrm{ex}}(s)}{\partial \mathbf{E}} \mathbf{E}\bigl(e^{-\imath\omega t} + e^{+\imath\omega t}\bigr) \end{aligned} \tag{14.19}$$

where we have assumed that the oscillating field is an electric field with strength $\mathbf{E}$ and $\mathbf{m}^{\text{elec}}$ is the electric dipole integrals matrix. In Eq. (14.19) $\mathbf{V}$ is the matrix collecting the potential integrals computed on the cavity tesserae and $\mathbf{q}^{\text{ex}}$ is the apparent charge induced on the cavity by the external oscillating field $\mathbf{E}$.

Expression (14.19) when summed to the Hamiltonian modified by the solvent terms described by $\mathbf{q}^N$ and $\mathbf{q}^e$ (see Section 14) allows one to take into account the complete reaction of the solvent to the combined action of the internal (due to the solute) and the external fields.

Approximate solutions of the time-dependent Schrödinger equation associated to the resulting effective Hamiltonian can be obtained by using the same procedures formulated for isolated systems but now the Fock operator includes explicit solvent terms. The inclusion of such additional solvent terms in the coupled perturbed equations will lead to different values of the derivatives of the ground state wavefunction for an electric perturbation.

The mixed nature of the electric dipole – magnetic dipole polarizability β requires an additional coupled perturbed procedure, this time containing a magnetic perturbation. Due to the imaginary nature of such perturbation solvent induced terms do not appear in the corresponding first-order expansion term of the Fock operator as explicit terms but only through the dependence of the atomic orbital basis set on the magnetic field (CAMMI, MENNUCCI and TOMASI [1998], CAMMI, MENNUCCI and TOMASI [1999]); as a consequence also the derivatives of the wavefunction with respect to the magnetic field will be modified by the solvent (see details in the paragraphs entitled "Electric dipole polarizabilities" and "Nuclear magnetic shielding").

It is important to note that solvent contributions obtained in terms of solvent charges which are computed using the value of the dielectric constant at the frequency of the external field. In the present case such frequency is that corresponding to the sodium D line, and thus the value for $\varepsilon(\omega)$ coincides with the so-called optical ε_{opt} defined as the square of the refractive index. For polar solvents ε_{opt} is by far smaller that the static ε_0 analog and thus introducing a solvent response determined by it instead of ε_0 means to account for only a part of the response which would appear in the presence of a static field; such situation is usually defined as *nonequilibrium* solute-solvent regime while that corresponding to a full solvent response is indicated as *equilibrium* regime.

Vibrational Circular Dichroism (VCD). After recently celebrating twenty years of development since its early years of discovery, vibrational circular dichroism (VCD) has matured to a point where the phenomenon is well understood theoretically, can be measured and calculated routinely, and is being used to uncover exciting new information about the structure of optically active molecules (STEPHENS and DEVLIN [2000]). Beyond this, VCD has been shown to be a sensitive, noninvasive diagnostic probe of chiral purity or enantiomeric separation with potential use in the synthesis and manufacture of chiral drugs and pharmaceutical products. In descriptive terms, VCD is the coupling of optical activity to infrared vibrational spectroscopy. More specifically, VCD spectra are vibrational difference spectra with respect to left and right circularly polarized radiation. The essence of VCD is to combine the stereochemical sensitivity of natural optical activity with the rich structural content of vibrational spectroscopy. The result of a VCD measurement is two vibrational spectra of a sample, the VCD and its parent infrared spectrum. These can used together to deduce information about molecular structure. The principal area of application of VCD is structure elucidation of biologically significant molecules including peptides, proteins, nucleic acids, carbohydrates, natural products and pharmaceutical molecules; also, as mentioned above, it has growing potential as a chiral diagnostic probe. VCD complements the relatively slow time scale of NMR since molecular vibrations and conformational sensitivity occur in the subpicosecond time domain. VCD is also complementary to X-ray crystallography by virtue of its applicability to molecules in gas, liquid and solution phases. Here, we present a quantum-mechanical method to simulate VCD spectra of molecules in solution.

The differential response of a chiral sample to left and right circularly polarized light can be represented by the quantity $\Delta\varepsilon$, defined as:

$$\Delta\varepsilon = \varepsilon_L - \varepsilon_R,$$

where $\varepsilon_{L,R}$ are the molar absorption coefficients for left and right circularly polarized light, respectively. Ignoring for the moment solvent effects, the differential molar absorption coefficient at frequency ν, $\Delta\varepsilon(\nu)$, is:

$$\Delta\varepsilon(\nu) = 4\gamma\nu \sum_i R_i \, \mathrm{f}(\nu_i, \nu), \tag{14.20}$$

where ν_i is the frequency of the ith transition, γ a numerical coefficient, $\mathrm{f}(\nu_i, \nu)$ the normalized line-shape function and:

$$R_i = \mathrm{Im}\big[\langle 0|\mu_{\mathrm{el}}|1\rangle_i \cdot \langle 0|\mu_{\mathrm{mag}}|1\rangle_i\big] \tag{14.21}$$

the rotational strength. In Eq. (14.21) μ_{el} and μ_{mag} are the electric and magnetic dipole moment operators, respectively, and $|0\rangle$, $|1\rangle$ are vibrational states. The computation of the transition moments in Eq. (14.21) can be avoided by resorting to a coupled perturbed HF, or KS, approach in which the derivatives of the ground state wavefunction with respect to nuclear displacement and magnetic field respectively are computed, as well as the vibrational frequencies ν_i and the normal coordinates Q_i. ν_i and Q_i are obtained simultaneously by diagonalization of the mass-weighted Cartesian force field (the Hessian).

The calculation of $\Delta\varepsilon$ in the presence of a solvent medium still relies on Eqs. (14.20), (14.21), but some refinements are needed. As for electric response properties (see the paragraph entitled "Macroscopic susceptibilities: local field effects and effective polarizabilities"), as well as for infrared intensities (CAMMI, CAPPELLI, CORNI and TOMASI [2000]), the μ_{el} operator in Eq. (14.21) has to be replaced by the sum of the dipole moment μ_{el} of the molecule and the dipole moment $\tilde{\mu}_{\mathrm{el}}$ arising from the polarization induced by the molecule on the solvent (here with the symbol $\tilde{\mu}_{\mathrm{el}}$ we indicate the operator from which the previous matrix $\tilde{\mathbf{m}}$ is derived when we shift to the representation on the atomic basis set). As already observed, $\tilde{\mu}_{\mathrm{el}}$ takes into account effects due to the field generated from the solvent response to the probing field once the cavity has been created (i.e., the cavity field). In principle also μ_{mag} should be similarly reformulated. However, by assuming the response of the solvent to magnetic perturbations to be described only in terms of its magnetic permittivity (which is usually close to unity), it is reasonable to consider that the magnetic analogous of the electric "cavity field" gives minor contributions to R_i^{sol}. Thus, the final expression for R_i^{sol} to use is (CAPPELLI, CORNI, MENNUCCI, CAMMI and TOMASI [2002]):

$$R_i^{\mathrm{sol}} = \mathrm{Im}\langle 0|(\mu_{\mathrm{el}} + \tilde{\mu}_{\mathrm{el}})|1\rangle_i \cdot \langle 0|\mu_{\mathrm{mag}}|1\rangle_i$$

where the frequencies and the normal coordinates are obtained by diagonalizing the Hessian for the molecule in solution. The calculation of the modified $\mu_{\mathrm{el}} + \tilde{\mu}_{\mathrm{el}}$ term is performed by defining an additional set of external field-induced charges $\mathbf{q}^{\mathrm{ex}}$ spread on the cavity surface: details on this point can be found in the previous Section 14. The derivatives of the solvent-modified wavefunction with respect to nuclear displacement and magnetic field can be obtained with the methods presented in the previous sections. It is also possible to account for vibrational nonequilibrium solvent effects on such derivative and on the normal modes of the molecule (CAPPELLI, CORNI, CAMMI, MENNUCCI and TOMASI [2000]). We remark that nonequilibrium effects arise since the solvent cannot instantaneously equilibrate to the charge distribution of the vibrating molecule.

CHAPTER VI

Excited States

Solvation models, used to study solvated molecules in their ground state, can be extended to electronically excited states: this extension however is neither immediate nor straightforward. There are in fact many fundamental characteristics which are specific of the molecules in their excited states and which make them unique systems to be treated with completely new techniques.

This is also true for isolated systems, for which the computational tools devoted to the excited molecules are generally different from those used for ground state systems; as well-known example suffice here to quote the Hartree–Fock approach which gives very satisfactory results for properties and reactivity of molecules in their ground state but it completely fails in treating excited systems. This specificity of the electronically excited molecules acquires an even larger importance when the molecule can interact with an external medium. In this case in fact it becomes compulsory to introduce the concepts of time and time progress, concepts which can be safely neglected in treating molecules in their ground states. In these cases in fact, and also when introducing reaction processes, one can always reduce the analysis to a completely equilibrated solute-solvent system.

On the contrary, when the attention is shifted towards dynamical phenomena as those involved in electronically transitions (absorptions and/or emissions), or towards relaxation phenomena as those which describe the time evolution of the excited state, one has to introduce new models, in which solute and solvent have proper response times which have not to be coherent or at least not before very long times, of no interest in the present research. To explain better this fundamental issue, it is useful to summarize some aspects of the dynamical behavior of the medium which have been already used in the previous chapter but without a detailed description.

In general, the dynamics of solvent can be basically characterized by two main components. One, of more immediate comprehension, is represented by the internal molecular motions inside the solvent due to changes in the charge distribution, and eventually in the geometry, of the solute system. As already shown, the solute when immersed in the solvent produces an electric field inside the bulk of the medium which can modify its structure, for example inducing phenomena of alignment and/or preferential orientation of the solvent molecules around the cavity embedding the solute. These molecular motions are characterized by specific time scales of the order of the rotational and translational times proper of the condensed phases (10^{-9}–10^{-11} s). In a analogous way, we can assume that the single solvent molecules are subjected to internal geometrical variations, i.e., vibrations, due to the changes in the solute field; once again these will be described by specific well-known time scales (10^{-14}–10^{-12} s). Both the translational, the rotational and/or the vibrational motions involve nuclear displacements and therefore, in the following, they will be collectively indicated as 'nuclear motions'. The other important component of the dynamical nature of the medium, complementary to the nuclear one, is that induced by motions of the electrons inside each solvent molecule; these motions are extremely fast (of the order of 10^{-16}–10^{-15} s) and they represent the electronic polarization of the solvent.

The dynamical phenomena we have described above, which are usually neglected in standard calculations on solvated systems. Solvation continuum models have been generalized to treat phenomena involving excited states since a long time (see the large literature in TOMASI and PERSICO [1994], CRAMER and TRUHLAR [1999]).

However, these examples are generally limited to some specific aspects of the whole problem, and/or applied to classical or low-quality QM computational levels. On the contrary, the aim is to treat the phenomenon of the excited states in solution in its totality, by using computational tools of the same quality as those, by now standard, for isolated systems.

To do that it will be necessary to introduce in the formalism of continuum models the specificities proper of the excited systems. As said, the main innovation is the dynamical nature of the phenomenon and the related time-dependent response of the solvent. The brief analysis reported above on the time scales characterizing the various motions inside the medium represents the physical bases on which to build up a theoretical model which is realistic and sensitive enough to allow distinctions among the various components of the solvent response. In fact, it is easy to see that two main components we have previously described as nuclear and electronic motions, due to their different dynamic behavior, will give rise to different effects.

In the same way, it is easy to accept that the electronic motions can be considered as instantaneous and thus the part of the solvent response they originate is always equilibrated to any change, even if fast, in the charge distribution of the solute (as in the case of vertical electronic transitions, in which no nuclear relaxations have time to happen inside the molecule which remains frozen in its initial geometry also when radical changes happen in its electronic state). On the contrary, solvent nuclear motions, by far slower than the electronic ones, can be delayed with respect to fast changes, and thus they can give origin to solute–solvent systems not completely equilibrated. This condition of nonequilibrium will successively evolve towards a more stable and completely equilibrated state in a time interval which will depend on the specific system under scrutiny.

15. Continuum solvation approaches

The PCM-like continuum models, in which the solvent response is described in terms of apparent charges on the cavity surface have been generalized to treat dynamical phenomena (AGUILAR, OLIVARES DEL VALLE and TOMASI [1993]); in particular the solvent charges can be made time dependent through the time dependence of the solute wavefunction from which they are directly originated, and also due to the time dependence of the solvent permittivity (or dielectric constant). The latter, in fact, is the numerical quantification, in terms of an electric response, of the solvent polarity. It is thus immediate to extract from its value various contributions related to different motions in the solvent. In particular, experimental measurements can give information on its purely electronic component; this coincides with the so-called optical permittivity. In general, the dielectric constant presents a time dependence which is determined by the intrinsic characteristics of the solvent and thus it will be different passing from one solvent to another, and finally not representable through a general analytic function. However, by applying theoretical models based on measurements of relaxation times (as that formulated by Debye) and by knowing by experiments the behavior of the permittivity with respect to the frequency of an external applied field (the so-called Maxwell field), it is possible to derive semiempirical formula of the dielectric constant as a function of time (BÖTTCHER [1973], BÖTTCHER and BORDEWIJK [1978]). These formula, in turn, will allow to obtain induced charges which take into account, through its dependence on the time-dependent dielectric constant, the dynamic behavior of the specific solvent they represent.

It is worth noting that in the limit case of vertical electronic transitions, in which we can assume a Franck–Condon-like response of the solvent, exactly as for the solute molecule, the nuclear motions inside and among the solvent molecules will not be able to immediately follow the fast changes in the solute charge distribution, which will instantaneously shift from that characterizing the ground state to that proper of the excited state (in an absorption process), or vice versa from that of an equilibrated excited state to that of the ground state (in an emission). During, and immediately after the transition, the solvent will thus respond to the variations only through its electronic component, while the nuclear part of the response (also indicated as inertial) will remain frozen in the state immediately previous the transition.

In the PCM framework, this will lead to a partition of the apparent charges in two separate sets (AGUILAR, OLIVARES DEL VALLE and TOMASI [1993], CAMMI and TOMASI [1995a], MENNUCCI, CAMMI and TOMASI [1988]), one related to the instantaneous electronic response, σ_{fast}, and the other to the slower response connected to the nuclear motions, σ_{slow} (their sum is just the already defined total apparent charge σ). According to what said before about the origin of the apparent charges, the electronic components will depend on the instantaneous wavefunction of the solute, on the optical dielectric constant and on the complementary set of slower apparent charges. The latter, on the contrary, will still depend on the solute wavefunction of the initial state.

From a physical point of view, we can always define the equations giving the total and the electronic surface charge by considering two parallel systems of Poisson-like equations in which both the operator and the related boundary conditions are defined in terms of ε or of ε_∞, respectively. On the contrary, the orientational surface charge has to be defined as difference of the two other charges, being the orientational contribution to the solvent polarization not related to a physically stated dielectric constant.

By applying this model to the problem of a solute electronic transition from the ground state to a given excited state, we have:

$$\begin{cases} -\Delta V = 4\pi\rho_M^{\text{fin}} & \text{within the cavity,} \\ -\varepsilon_\infty \Delta V = 0 & \text{out of the cavity,} \\ V_i - V_e = 0 & \text{on the cavity surface,} \\ (\partial V/\partial n)_i - \varepsilon_\infty(\partial V/\partial n)_e = 4\pi\sigma_{\text{slow}} & \text{on the cavity surface,} \end{cases}$$

where ρ_M^{fin} is the charge distribution of the solute in its final excited state. The second jump condition on the gradient of V has here a different form with respect to that of the equilibrium situation given before, due to the presence of the constant slow charge σ_{slow}.

As shown before for ground state equilibrated solutes, the system (15) can be solved by defining an electronic surface charge σ_{fast} which can be discretized into point charges, $\mathbf{q}_{\text{fast}}$, exploiting to the usual partition of the surface cavity into tesserae. However, in the nonequilibrium case the definition of the electrostatic free energy has to be changed; in particular the contribution of the slow polarization has to be seen an external fixed field which does not follow the variationally derived rules.

As recalled before, this is just a part of the more general process of creation and evolution of the excited states in solution. However, this aspect, even if partial, represents a valid and immediate check on the quality of the adopted model; frequencies and intensity of absorption and/or emission peaks of spectra of molecular systems in liquid solutions are in fact the most common experimental data one can compare to. Data with richer information about the dynamical aspects of the process can be derived from other experimental measurements of time-dependent fluorescence spectroscopy. In particular, the time analysis of the so-called 'time-dependent Stokes-shift', i.e., the time evolution of the difference between the maxima of emission and absorption, has a large importance for the comparison with theoretical results. Such quantity, experimentally measured, can in fact give a valid test for the models used to represent the solvent relaxation after an electronic excitation inside the solute, on the one hand, and to describe the way this relaxation interacts with that proper of the excited state, on the other hand. In fact, the time evolution of this spectroscopic quantity strongly depends on the solvent.

In the "dynamic Stokes shift" experiment, a short light pulse is used to electronically excite a probe solute and thereby change its electron distribution and hence its interactions with the solvent instantaneously (as far as solvent motions are concerned). Relaxation of the solvent is monitored by measuring the frequency shift of the solute's emission spectrum, which red-shifts in a way that directly reflects the time evolution of the solvation energy. Since solvation takes only a few picoseconds in most room-temperature solvents, emission spectra must be measured on a sub-picosecond time scale in order to see solvation taking place. To do so the technique of fluorescence upconversion is used; this affords effective time resolution of 30–50 fs. The solvation times that can be observed vary over wide ranges, with characteristic solvation times as fast as 50–100 fs in water an acetonitrile, to $\geqslant 500$ ps in long-chain alcohols. This broad range of times, and the highly nonexponential nature of the solvation relaxation are captured with reasonable fidelity using dielectric continuum models of solvation (HSU, SONG and MARCUS [1997]). In particular, recent quantum mechanical studies using PCM-IEF have shown that continuum representations of the solvent can be a fruitful way to reach a deeper understanding of solvation dynamics.

References

AGUILAR, M.A. and F.J. OLIVARES DEL VALLE (1986), A computation procedure for the dispersion component of the interaction energy in continuum solute–solvent models, *Chem. Phys.* **138**, 327–335.

AGUILAR, M.A., F.J. OLIVARES DEL VALLE and J. TOMASI (1993), Nonequilibrium solvation: An ab-initio quantum-mechanical method in the continuum cavity model approximation, *J. Chem. Phys.* **98**, 7375–7384.

AKKIRAJU, N. and H. EDELSBRUNNER (1996), Triangulating the surface of a molecule, *Discrete Appl. Math.* **71**, 5–22.

ALLEN, M.P. and D.J. TILDESLEY (1987), *Computer Simulation of Liquids* (Clarendon, Oxford).

AMOVILLI, C., V. BARONE, R. CAMMI, E. CANCÈS, M. COSSI, B. MENNUCCI, C.S. POMELLI and J. TOMASI (1998), Recent advances in the description of solvent effects with the polarizable continuum model, *Adv. Quant. Chem.* **32**, 227–261.

AMOVILLI, C. and B. MENNUCCI (1997), Self-consistent field calculation of Pauli repulsion and dispersion contributions to the solvation free energy in the polarizable continuum model, *J. Phys. Chem. B* **101**, 1051–1057.

ARRIGHINI, G.P. (1981), *Intermolecular Forces and their Evaluation by Perturbation Theory* (Springer, Berlin).

BACHELET, G.B., D.R. HAMANN and M. SCHLÜTER (1982), Pseudopotentials that work: from hydrogen to plutonium, *Phys. Rev. B* **26**, 4199–4206.

BAJAJ, C., H.Y. LEE, R. MERKERT and V. PASCUCCI (1997), NURBS based B-rep models for macromolecules and their properties, in: *Fourth Symposium on Solid Modeling and Applications, Atlanta.*

BALESCU, R. (1991), *Equilibrium and Non-Equilibrium Statistical Mechanics* (Krieger, Melbourne, FL).

BANDYOPADHYAY, P., M.S. GORDON, B. MENNUCCI and J. TOMASI (2002), An integrated effective fragment-polarizable continuum approach to solvation: Theory and application to glycine, *J. Chem. Phys.* **116**, 5023–5032.

BEN-NAIM, A. (1974), *Water and Aqueous Solutions* (Plenum, New York).

BEN-NAIM, A. (1987), *Solvation Thermodynamics* (Plenum, New York).

BISHOP, D.M. (1998), *Adv. Chem. Phys.* **104**, 1.

BOROUCHAKI, H., P. LAUG and P.L. GEORGE (2000), Parametric surface meshing using a combined advancing-front – generalized-Delaunay approach, *Internat. J. Numer. Methods. Engrg.* **49** (1–2), 233–259.

BÖTTCHER, C.J.F. (1973), *Theory of Electric Polarization*, Vol. I (Elsevier).

BÖTTCHER, C.J.F. and P. BORDEWIJK (1978), *Theory of Electric Polarization*, Vol. II (Elsevier).

BUCKINGHAM, A.D., T. SCHAEFER and W.G. SCHNEIDER (1960), Solvent effects in nuclear magnetic resonance spectra, *J. Chem. Phys.* **32**, 1227–1233.

BUTCHER, P.N. and D. COTTER (1990), *The Elements of Nonlinear Optics* (Cambridge Univ. Press, Cambridge).

CAMMI, R. (1998), The Hartree–Fock calculation of the magnetic properties of molecular solutes, *J. Chem. Phys.* **109**, 3185–3196.

CAMMI, R., C. CAPPELLI, S. CORNI and J. TOMASI (2000), On the calculation of infrared intensities in solution within the polarizable continuum model, *J. Phys. Chem. A* **104**, 9874–9879.

CAMMI, R., M. COSSI, B. MENNUCCI and J. TOMASI (1996), Analytical Hartree–Fock calculations of the dynamical polarizabilities α, β and γ of molecules in solution, *J. Chem. Phys.* **105**, 10556–10564.

CAMMI, R., B. MENNUCCI and J. TOMASI (1998a), On the calculation of local field factors for microscopic static hyperpolarizabilities of molecules in solution with the aid of quantum-mechanical methods, *J. Phys. Chem. A* **102**, 870–875.

CAMMI, R., B. MENNUCCI and J. TOMASI (1998b), Solvent effects on linear and nonlinear optical properties of Donor–Acceptor polyenes: investigation of electronic and vibrational components in terms of structure and charge distribution changes, *J. Amer. Chem. Soc.* **34**, 8834–8847.

CAMMI, R., B. MENNUCCI and J. TOMASI (1999), Nuclear magnetic shieldings in solution: Gauge invariant atomic orbital calculation using the polarizable continuum model, *J. Chem. Phys.* **110**, 7627–7638.

CAMMI, R., B. MENNUCCI and J. TOMASI (2000), An attempt to bridge the gap between computation and experiment for nonlinear optical properties: macroscopic susceptibilities in solution, *J. Phys. Chem. A* **104**, 4690–4698.

CAMMI, R. and J. TOMASI (1995a), Nonequilibrium solvation theory for the polarizable continuum model: A new formulation at the SCF level with application to the case of the frequency-dependent linear electric response function, *Int. J. Quant. Chem: Quant. Chem. Symp.* **29**, 465–474.

CAMMI, R. and J. TOMASI (1995b), Remarks in the use of the apparent surface charges (ASC) methods in solvation problems: iterative versus matrix-inversion procedures and the renormalization of the apparent charges, *J. Comp. Chem.* **16**, 1449–1458.

CANCÈS, E. and B. MENNUCCI (1998a), Analytical derivatives for geometry optimization in solvation continuum models, I: Theory, *J. Chem. Phys.* **109**, 249–259.

CANCÈS, E. and B. MENNUCCI (1998b), New applications of integral equation methods for solvation continuum models: ionic solutions and liquid crystals, *J. Math. Chem.* **23**, 309–326.

CANCÈS, E., B. MENNUCCI and J. TOMASI (1997), A new integral equation formalism for the polarizable continuum model: theoretical background and applications to isotropic and anisotropic dielectrics, *J. Chem. Phys.* **107**, 3032–3041.

CANCÈS, E., B. MENNUCCI and J. TOMASI (1998), Analytical derivatives for geometry optimization in solvation continuum models, II: Numerical applications, *J. Chem. Phys.* **109**, 260–266.

CAPPELLI, C., S. CORNI, R. CAMMI, B. MENNUCCI and J. TOMASI (2000), Nonequilibrium formulation of infrared frequencies and intensities in solution: Analytical evaluation within the polarizable continuum model, *J. Chem. Phys.* **113**, 11270–11279.

CAPPELLI, C., S. CORNI, B. MENNUCCI, R. CAMMI and J. TOMASI (2002), Vibrational circular dichroism within the polarizable continuum model: A theoretical evidence of conformation effects and hydrogen bonding for (s)-(-)-3-butyn-2-d in CCl_4 solution, *J. Phys. Chem. A* **106**, 12331–12339.

CAR, R. and M. PARRINELLO (1985), Unified approach for molecular dynamics and density functional theory, *Phys. Rev. Lett.* **55**, 2471–2474.

CHEN, S.W. and B. HONIG (1997), Monovalent and divalent salt effects on electrostatic free energies defined by the nonlinear Poisson–Boltzmann equation: Application to DNA reactions, *J. Phys. Chem. B* **101**, 9113–9118.

CONDON, E.U. (1937), Theories of optical rotatory power, *Rev. Mod. Phys.* **9**, 432–457.

CONNOLLY, M.L. (1996), Molecular surfaces: A review, http://www.netsci.org/Science/Compchem/feature14.html.

CRAMER, C.J. and D.G. TRUHLAR (1991), General parameterized SCF model for free energies of solvation in aqueous solution, *J. Amer. Chem. Soc.* **113**, 8305–8311; Molecular orbital theory calculations of aqueous solvation effects on chemical equilibria, *J. Amer. Chem. Soc.* **113**, 8552–8554.

CRAMER, C.J. and D.G. TRUHLAR (1999), Implicit solvation models: Equilibria, structure, spectra, and dynamics, *Chem. Rev.* **99**, 2161–2200.

DAY, P.N., J.H. JENSEN, M.S. GORDON, S.P. WEBB, W.J. STEVENS, M. KRAUSS, D. GARMER, H. BASCH and D. COHEN (1996), An effective fragment method for modeling solvent effects in quantum mechanical calculations, *J. Chem. Phys.* **105**, 1968–1986.

DITCHFIELD, R. (1974), Self-consistent perturbation theory of diamagnetism, I. A gauge-invariant LCAO method for N.M.R. chemical shifts, *Mol. Phys.* **27**, 789–792.

FIELD, M.J., P.A. BASH and M. KARPLUS (1990), A combined quantum mechanical and molecular mechanical potential for molecular dynamics simulations, *J. Comp. Chem.* **11**, 700–733.

FLORIS, F.M., J. TOMASI and J.L. PASCUAL-AHUIR (1991), Dispersion and repulsion contributions to the solvation energy: Refinements to a simple computational model in the continuum approximation, *J. Comput. Chem.* **12**, 784–791.

FRENKEL, D. and B. SMIT (1996), *Understanding Molecular Simulation* (Academic Press, San Diego).

FREY, P.J. and P.L. GEORGE (2000), *Mesh Generation – Application to Finite Elements* (Hermès Science Europe).

FRISCH, M.J. ET AL. (1998), *Gaussian 98,* Pittsburgh.

FROESE, R.D.J. and K. MOROKUMA (1998), in: P.v.R. Schleyer, N.L. Allinger, T. Clark, J. Gasteiger, P.A. Kollman, H.F. Schaefer III, P.R. Schreiner, eds., *The Encyclopedia of Computational Chemistry* (John Wiley) 1245–1300.

GAO, J. (1995), Methods and applications of combined quantum mechanical and molecular mechanical potentials, in: K.B. Lipkowitz and D.B. Boyd, eds., *Reviews in Computational Chemistry*, Vol. 7 (VCH) 119–185.

GRAY, C.G. and K.E. GUBBINS (1984), *Theory of Molecular Fluids, Vol. 1, Fundamentals* (Clarendon, Oxford).

HACKBUSCH, W. (1995), *Integral Equations – Theory and Numerical Treatment* (Birkhäuser, Basel).

HANSEN, J.-P. and I.R. MCDONALD (1986), *Theory of Simple Liquids* (Academic Press, San Diego).

HELGAKER, H. ET AL. (2001), Dalton, an ab initio electronic structure program, Release 1.2. See http://www.kjemi.uio.no/software/dalton/dalton.html.

HOBZA, P. and R. ZAHRADNIK (1988), *Intermolecular Complexes* (Elsevier).

HSU, C.-P., X. SONG and R.A. MARCUS (1997), Time-dependent Stokes shift and its calculation from solvent dielectric dispersion data, *J. Phys. Chem. B* **101**, 2546–2551.

KARELSON, M.M. and M.C. ZERNER (1992), A theoretical treatment of solvent effects on spectroscopy, *J. Phys. Chem.* **96**, 6949–6957.

KIRKPATRICK, S., C.D. GELATT JR. and M.P. VECCHI (1983), Optimization by simulated annealing, *Science* **220**, 671–680.

KLAMT, A. and G. SCHÜÜRMANN (1993), COSMO: A new approach to dielectric screening in solvents with explicit expressions for the screening energy and its gradient, *J. Chem. Soc. Perkin Trans.* **2**, 799–805.

KOHN, W. and P. VASHISHTA (1993), General density functional theory, in: Lundqvist and March, eds., *Theory of the Inhomogeneous Electron Gas* (Plenum, New York).

LAUG, P. and H. BOROUCHAKI (1999), BLSURF – Mesh generator for composite parametric surfaces – user's manual, INRIA Technical Report RT-0232.

LAUG, P. and H. BOROUCHAKI (2001), Molecular surface modeling and meshing, 10th International Meshing Roundtable, Newport Beach, California, USA, 31–41, to appear in: *Engineering with Computers*.

LAUG, P. and H. BOROUCHAKI (2002), BLMOL, molecular surface mesher, http://www-rocq.inria.fr/Patrick.Laug/blmol/index.html.

LEE, B. and F.M. RICHARDS (1971), The interpretation of protein structures: estimation of static accessibility, *J. Mol. Biol.* **55**, 379–400.

MARGENAU, H. and N.R. KESTNER (1971), *Theory of Intermolecular Forces* (Pergamon, Elmsford, NY).

MCQUARRIE, D.A. (1976), *Statistical Mechanics* (Harper and Row, New York).

MCWEENY, R. (1992), *Methods of Molecular Quantum Mechanics*, 2nd edn. (Academic Press, San Diego).

MENNUCCI, B., R. CAMMI and J. TOMASI (1998), Excited states and solvatochromic shifts within a nonequilibrium solvation approach: A new formulation of the integral equation formalism method at the self-consistent field, configuration interaction, and multiconfiguration self-consistent field level, *J. Chem. Phys.* **109**, 2798–2807.

MENNUCCI, B., E. CANCÈS and J. TOMASI (1997), Evaluation of solvent effects in isotropic and anisotropic dielectrics, and in ionic solutions with a unified integral equation method: theoretical bases, computational implementation and numerical applications, *J. Phys. Chem. B* **101**, 10506–10517.

MENNUCCI, B., J. TOMASI, R. CAMMI, J.R. CHEESEMAN, M.J. FRISCH, F.J. DEVLIN, S. GABRIEL and P.J. STEPHENS (2002), Polarizable continuum model (PCM) calculations of solvent effects on optical rotations of chiral molecules, *J. Phys. Chem. A* **106**, 6102–6113.

METROPOLIS, M., A.W. ROSENBLUTH, M.N. ROSENBLUTH, A.N. TELLER and E. TELLER (1953), Equation of state calculations by fast computing machines, *J. Chem. Phys.* **21**, 1087–1092.

MIERTUŠ, S., E. SCROCCO and J. TOMASI (1981), Electrostatic interaction of a solute with a continuum, *Chem. Phys.* **55**, 117–129.

MIKKELSEN, K.V., A. CESAR, H. ÅGREN and H.J.AA. JENSEN (1995), Multiconfigurational self-consistent reaction field theory for nonequilibrium solvation, *J. Chem. Phys.* **103**, 9010–9023, and references therein.

MITCHELL, M.J. and J.A. MCCAMMON (1991), Free energy difference calculations by thermodynamic integration: Difficulties in obtaining a precise value, *J. Comp. Chem.* **12**, 271–275.

NÉDÉLEC, J.C. (1994), New trends in the use and analysis of integral equations, in: *Proc. Sympos. Appl. Math.*, Vol. 48, 151–176.

NÉDÉLEC, J.C. and J. PLANCHARD (1973), Une méthode variationelle d'éléments finis pour la résolution d'un problème extérieur dans $\mathbb{R}^3$, *RAIRO* **7**, 105.

NEW YORK UNIVERSITY (2002), MathMol Library, http://www.nyu.edu/pages/mathmol/.

ONSAGER, L. (1936), Electric moments of molecules in liquids, *J. Amer. Chem. Soc.* **58**, 1486–1493.

PAYNE, M.C., M.P. TETER, D.C. ALLAN, T.A. ARIAS and J. JOANNOPOULOS (1992), Iterative minimization techniques for ab initio total energy calculations: Molecular dynamics and conjugate gradients, *Rev. Mod. Phys.* **64**, 1045–1097.

PIEROTTI, R.A. (1976), A scaled particle theory of aqueous and nonaqueous solutions, *Chem. Rev.* **76**, 717–726.

PRASAD, P.N. and D.J. WILLIAMS (1991), *Introduction to Nonlinear Optical Effects in Organic Molecules and Polymers* (Wiley).

QUARTERONI, A. and A. VALLI (1997), *Numerical Approximation of Partial Differential Equations*, 2nd edn. (Springer).

REICHL, L.E. (1980), *A Modern Course in Statistical Physics* (Edward).

RICHARDS, F.M. (1977), Areas, volumes, packing, and protein structure, *Ann. Rev. Biophys. Bioeng.* **6**, 151–176.

RICHMOND, T.J. and F.M. RICHARDS (1978), Packing of a-helices: Geometrical constraints and contact areas, *J. Mol. Biol.* **119**, 537–555.

RINALDI, D., B.J. COSTA-CABRAL and J.L. RIVAIL (1986), *Chem. Phys. Lett.* **125**, 495–500.

RIVAIL, J.-L. and D. RINALDI (1976), A quantum chemical approach to dielectric solvent effects in molecular liquids, *Chem. Phys.* **18**, 233–242.

RYCKAERT, J.P., G. CICCOTTI and H.J.C. BERENDSEN (1977), Numerical integration of the Cartesian equation of motion of a system with constraints: molecular dynamics of N-alkanes, *J. Comp. Phys.* **23**, 327–341.

SANNER, M.F., A.J. OLSON and J.C. SPEHNER (1996), Reduced surface: An efficient way to compute molecular surfaces, in: *Biopolymers*, Vol. 38, 305–320.

SCHMIDT, M.W., K.K. BALDRIDGE, J.A. BOATZ, S.T. ELBERT, M.S. GORDON, J.H. JENSEN, S. KOSEKI, N. MATSUNAGA, K.A. NGUYEN, S.J. SU, T.L. WINDUS, M. DUPUIS and J.A. MONTGOMERY (1993), General atomic and molecular electronic structure system, *J. Comp. Chem.* **14**, 1347–1363.

STEPHENS, P.J. and F.J. DEVLIN (2000), Determination of the structure of chiral molecules using ab initio vibrational circular dichroism spectroscopy, *Chirality* **12**, 172–179.

STRAATSMA, T.P. (1996), in: K.B. Lipkowitz and D.B. Boyd, eds., *Reviews in Computational Chemistry*, Vol. 9 (VCH) 81–127.

TAPIA, O. and O. GOSCINSKI (1975), Self-consistent reaction field theory of solvent effects, *Mol. Phys.* **29**, 1653–1661.

TOMASI, J., B. MENNUCCI and C. CAPPELLI (2001), Interaction in solvents and solutions, in: G. Wypych, ed., *Handbook of Solvents* (Chemtech) 387–472.

TOMASI, J. and M. PERSICO (1994), Molecular interactions in solution: An overview of methods based on continous distributions of the solvent, *Chem. Rev.* **94**, 2027–2094.

TRUONG, T.N. and E.V. STEFANOVICH (1995), A new method for incorporating solvent effect into the classical, ab initio molecular orbital and density functional theory frameworks for arbitrary shape solute, *Chem. Phys. Lett.* **240**, 253–260.

TUCKERMAN, M.E., P.J. UNGAR, T. VON ROSENVINGE and M.L. KLEIN (1996), Ab initio molecular dynamics simulations, *J. Phys. Chem.* **100**, 12878–12887.

VREVEN, T., B. MENNUCCI, C.O. DA SILVA, K. MOROKUMA and J. TOMASI (2001), The ONIOM–PCM method: Combining the hybrid molecular orbital method and the polarizable continuum model for solvation. Application to the geometry and properties of a merocyanine in solution, *J. Chem. Phys.* **115**, 62–72.

WARSHEL, A. and M. LEVITT (1976), Theoretical studies of enzymic reactions: Dielectric, electrostatic and steric stabilization of the carbonium ion in the reaction of lysozyme, *J. Mol. Biol.* **103**, 227–249.

WOLFF, J.J. and R. WORTMANN (1999), Organic materials for second-order non-linear optics, *Adv. Phys. Org. Chem.* **32**, 121–217.

WOLINSKI, K., J.F. HINTON and P. PULAY (1990), Efficient implementation of the gauge-independent atomic orbital method for NMR chemical shift calculations, *J. Amer. Chem. Soc.* **112**, 8251–8260.

WORTMANN, R. and D.M. BISHOP (1998), Effective polarizabilities and local field corrections for nonlinear optical experiments in condensed media, *J. Chem. Phys.* **108**, 1001–1007.

ZHU, J., LI, G.D. HAWKINS, C.J. CRAMER and G.G. TRUHLAR (1998), Density functional solvation model based on CM2 atomic charges, *J. Chem. Phys.* **109**, 9117–9133; errata (1999), **111**, 5624 and (2000), **113**, 3930.

ZWANZIG, R.W. (1954), High temperature equation of state by a perturbation method, I. Nonpolar gases, *J. Chem. Phys.* **22**, 1420–1426.

An Introduction to First-Principles Simulations of Extended Systems

Fabio Finocchi[a], Jacek Goniakowski[b], Xavier Gonze[c], Cesare Pisani[d]

[a]*Groupe de Physique des Solides, Universités Paris 6 – Paris 7, Tour 23, 2 place Jussieu, 75251 Paris cédex 05, France*

[b]*CRMC2–CNRS, Université Aix-Marseille II, Campus de Luminy, Case 913, 13288 Marseille cédex 9, France*

[c]*Laboratoire PCPM, 1, place Croix du Sud, Université Catholique de Louvain, B1348 Louvain La Neuve, Belgium*

[d]*Dipartimento di Chimica Inorganica, Fisica e Materiali, Università di Torino, Via P. Giuria 7, Torino, Italy*

E-mail addresses: finocchi@gps.jussieu.fr (F. Finocchi), jacek@crmc2.univ-mrs.fr (J. Goniakowski), gonze@pcpm.ucl.ac.be (X. Gonze), cesare.pisani@unito.it (C. Pisani)

1. Introduction

In this chapter, we focus on condensed matter systems with the following characteristics: (i) under the actual conditions, they are solid, which means that atomic diffusion is several order of magnitude slower that the typical experimental observation times; (ii) their linear dimensions are bigger than some tens of nm and show bulk like atomic densities. Glassy or fully disordered solids exhibiting aging phenomena do not generally satisfy condition (i). They are considered as special cases of solids, and seen as they were in equilibrium on the shorter time scales. Other systems at the nanometric scale, such as macromolecules, which do not satisfy conditions (ii), are shortly discussed in order to point out the specific differences with respect to extended systems, which constitutes the main object of the present chapter.

Computational Chemistry
Special Volume (C. Le Bris, Guest Editor) of
HANDBOOK OF NUMERICAL ANALYSIS, VOL. X
P.G. Ciarlet (Editor)
© 2003 Elsevier Science B.V. All rights reserved

We also consider these systems at the atomic-scale, starting from the knowledge of their electronic structure as the basis to the determination of their physical properties. For a solid-state computational physicist, this represents a radical choice. At the very heart of any atomic-level description of a solid, there is the question how to describe the atom–atom interactions. The so called "first-principles" methods (among which the Density Functional and the Hartree–Fock theories are being considered as well founded and practical approaches, though approximate in nature) take into account the quantum nature of the electron distribution to describe the bonds between the atoms. We thus exclude from this tutorial any method based on the so called "empirical inter-atomic potentials", in which a number of simplified assumption about the nature of the chemical bonds are used in conjunction with empirical data to provide an approximate description of many-atom systems. On the other hand, both Hartree–Fock and Density Functional theories constitute a rigorous way to reduce the many-body problem to one-electron like equations in a self-consistent mean field. Here we focus on the algorithms that make the computation of the ground state properties of real materials and their response to an external weak perturbation (such as the electric field, the applied pressure, etc.) feasible on the nowadays available computers. The emphasis is put on the mathematical framework and the advantages and drawbacks of the algorithms more than on the physical properties themselves. The latter ones are mentioned only to stress the importance of the numerical techniques that are suited for the numerical simulation of extended systems. Many applications to special physical cases are discussed in other contributions to this book and the relevant techniques presented in more detail. Some basic notions about the methods that are used to compute one- and two-particle electronic excitations as well as collective ones, such as the plasmon, are also given briefly.

The plan of the chapter is the following. The second section deals with the methods that are employed to model the real systems as a collection of a smaller number of particles. The emphasis is put on the use of specific boundary conditions and on the physical assumptions that are needed to reduce the number of degrees of freedom. In Section 3, we describe the basic features of the theories – especially the Hartree–Fock and the Density Functional methods – that are employed for the simulation of extended systems. Once the theoretical framework and the model system have been chosen, one can construct a finite representation of the Hilbert space in which the Schrödinger-like equations are solved. The fourth section aims to describe the different approaches that are commonly used to construct a basis set and provide the necessary material to discuss how they perform in various simulations.

2. Basic models of extended systems

The systems that are considered here are usually composed of many ($\geqslant 10^6$) atoms, and at the same time they are structured at the nanometer scale. Simulation techniques are being more and more used to provide information complementary to experiments, or even to investigate systems for which experiments are too heavy or practically impossible. Those systems are therefore neither small nor strictly periodic, and their computer simulation needs specific algorithms to make the problem numerically suitable. As we will see in the following, even defining the composition and the geometry of the system to be treated and the boundary conditions the electronic wavefunctions must satisfy is far from trivial; often, the differences between the various strategies of solution have their origin in the *models* they are based on. It is therefore the aim of this chapter to provide the reader with the basic characteristics of the models that are currently adopted.

First, we discuss the simplest model of all extended systems at the atomic scale, that is, the ideal crystal. On one hand, the translational symmetry can be exploited to reduce the number of degrees of freedom; on the other hand, many interesting properties and useful concepts that can be partially extended to nonstrictly periodic systems can be derived. Afterwards, we will briefly summarize the essential ingredients of the models that are especially devised to treat aperiodic systems.

2.1. *The perfect crystal model*

A perfect (ideal) crystal is a collection of atoms that are regularly placed on a three-dimensional lattice with long-range order. Because of the translational symmetry, the knowledge of the atomic structure within the elementary unit (usually called the primitive cell) suffices to know the whole crystal structure. More precisely, any lattice point $\mathbf{R_l}$ is given by linear combinations of the primitive lattice vectors $\mathbf{a}_1$, $\mathbf{a}_2$, $\mathbf{a}_3$:

$$\mathbf{R_l} = \sum_{i=1}^{3} l_i \mathbf{a}_i, \tag{2.1}$$

where the l_i are arbitrary integers, which are either unbound for an infinite crystal, or in the range $0 < l_1 < N_1$, $0 < l_2 < N_2$, $0 < l_3 < N_3$ for a crystal having $N = N_1 N_2 N_3$ unit cells. The atomic positions within the primitive cell $\mathbf{R_I}$ are defined by the basis vectors $\boldsymbol{\tau}_\mathbf{I}$. The choice of the $\{\mathbf{a}_i\}$ and the basis is arbitrary; however, one usually employs the *conventional* primitive lattice vectors. In this case, the unit cell is referred to as the Wigner–Seitz (WS) cell, of volume Ω_0. Any point $\mathbf{r}$ can thus be represented as a sum of a lattice vector and a vector ℓ belonging to the Wigner–Seitz cell: $\mathbf{r} = \sum_{i=1}^{3} l_i \mathbf{a}_i + \ell$. In addition to the translational symmetry, a perfect crystal is invariant under some point group operations. The set of the point group *and* the translational symmetry operations

TABLE 2.1

The types of symmetry elements that are consistent with the translational symmetry and the associated standard notation. The first three classes are elementary operations, while the roto-inversion is a combination of a rotation with the inversion, the roto-translation n_m (sometimes called a screw axis) is a combination of a rotation of $2\pi/n$ and a translation of m/n along a lattice vector. The glide symmetry is a combination of a reflection and a translation. For instance, $P_{\bar{4}2m}$ is a space group containing a roto-inversion fourth-fold axis, a twofold rotation axis and a mirror plane

Inversion	i
Mirror	m
Rotation $2\pi/n$	$n = 1, 2, 3, 4, 6$
Roto-inversion	$\bar{1}, \bar{2}, \bar{3}, \bar{4}, \bar{6}$
Roto-translation	n_m
Glide	a, b, c (axial); n (diagonal); d ("diamond-like")

is said to be the *space group of the crystal*. There is only a limited number (230) of three-dimensional space groups, which are usually classified according to the Hermann–Maughin notation (INTERNATIONAL TABLES FOR X-RAY CRYSTALLOGRAPHY).

The dual of the discrete linear space that is spanned by the lattice vectors is usually called *reciprocal space*. Given the vectors $\mathbf{b}_1$, $\mathbf{b}_2$, $\mathbf{b}_3$ such that $\mathbf{a}_i \cdot \mathbf{b}_j = 2\pi\delta_{ij}$, they form a basis for the reciprocal space that consists of all vectors:

$$\mathbf{G}_\mathbf{J} = \sum_{i=1}^{3} J_i \mathbf{b}_i, \tag{2.2}$$

where J_i is an arbitrary integer. Any wavevector $\mathbf{q}$ can thus be decomposed as $\mathbf{q} = \mathbf{G}_\mathbf{J} + \mathbf{k}$, where $\mathbf{k}$ belongs to the reciprocal of the Wigner–Seitz cell, that is, the Brillouin Zone (BZ), whose volume is equal to $\Omega_{\text{BZ}} = \frac{(2\pi)^3}{\Omega_0}$. Because of the duality between the direct and the reciprocal space, and the completeness of the space group, any operation of the later one is also a symmetry in the reciprocal space. The portion of the BZ that can give rise to the entire one by means of the space group operations is usually said the *irreducible Brillouin Zone*. A comprehensive account of the general properties and the symmetry-consistent representation of free-electron states in cubic crystals can be found in the work by BOUCKAERT, SMOLUCHOWSKI and WIGNER [1936] and in the book by BASSANI and PASTORI-PARRAVICINI [1975] for more general space groups.

2.1.1. *The Crystal Momentum representation*

Let us consider an elementary excitation in the perfect crystal, which can be represented through single-particle wavefunctions $\psi_m(\mathbf{r})$. In the framework of effective one-particle theories, the $\psi_m(\mathbf{r})$ may be the constituents of the Slater determinants (HF) or the Kohn–Sham orbitals (DFT). m labels the principal quantum number linked to the excitation. The basic assumption here is that the electron density ρ and any other physical observable O of the unperturbed system corresponding to a one-particle operator (for instance, $\rho = \langle \Psi | \sum_i \delta(\mathbf{r} - \mathbf{r}_i) | \Psi \rangle$) keep the crystal periodicity: $A(\mathbf{r}) = A(\mathbf{r} + \mathbf{L}_\mathbf{l})$ $(A = \rho, O)$. One can show (Bloch's theorem) that the discrete periodicity of the one-particle operator A implies the conservation of the the so-called crystal momentum $\mathbf{k}$ (ASHCROFT and MERMIN [1976]). $\mathbf{k}$ is therefore a good quantum number for the wavefunction corresponding to the one-particle excitation, which has the following property: $\psi_{m,\mathbf{k}}(\mathbf{r} + \mathbf{L}_\mathbf{l}) = e^{i\mathbf{k}\cdot\mathbf{L}_\mathbf{l}} \psi_{m,\mathbf{k}}(\mathbf{r})$. As a consequence, the wavefunction can be decomposed in a product of a phase factor by a function $u_{m,\mathbf{k}}$ having the crystal periodicity as

$$\psi_{m,\mathbf{k}}(\mathbf{r}) = (N\Omega_0)^{-\frac{1}{2}} e^{i\mathbf{k}\cdot\mathbf{r}} u_{m,\mathbf{k}}(\mathbf{r}) \quad \text{with } u_{m,\mathbf{k}}(\mathbf{r} + \mathbf{L}_\mathbf{l}) = u_{m,\mathbf{k}}(\mathbf{r}). \tag{2.3}$$

For a finite system, N is the number of unit cells. In order to simulate an infinite periodic system, the Born–von Karman (BvK) cyclic boundary conditions are used:

$$\psi_{m,\mathbf{k}}(\mathbf{r} + \mathbf{N}_\mathrm{i}\mathbf{a}_\mathrm{i}) = \psi_{m,\mathbf{k}}(\mathbf{r}), \tag{2.4}$$

m and $\mathbf{k}$ label the band and the wavevector, respectively. Because of the BvK conditions, the number of $\mathbf{k}$ wavevectors in the BZ is equal to $N = N_1 N_2 N_3$ and their density to $\Omega_0/(2\pi)^3$.

The generalization of the Bloch's theorem for many-electron wavefunctions has been given by RAJAGOPAL, NEEDS, JAMES, KENNY and FOULKES [1995]. They consider the many-electron Hamiltonian within BvK boundary conditions, which is invariant under a translation $\mathbf{L_l}$ of any electron. As for its one-particle analogous, the crystal momentum is a good quantum number and the many-electron wavefunction can be written as:

$$\Psi_{m,\mathbf{k}}\big(\{\mathbf{r}_i\}\big) = (N\Omega_0)^{-\frac{1}{2}} \exp\left(i\mathbf{k}\cdot\sum_{i=1}^{N}\mathbf{r}_i\right) U_{m,\mathbf{k}}\big(\{\mathbf{r}_i\}\big), \tag{2.5}$$

where $U_{m,\mathbf{k}}(\{\mathbf{r}_i\})$ is periodic for translations of any electron by a lattice vector and antisymmetric under electron exchange.

Eq. (2.3) is such that the orthonormalization condition of the $u_{m,\mathbf{k}}(\mathbf{r})$ functions writes

$$\langle u_{m,\mathbf{k}} | u_{n,\mathbf{k}}\rangle = \delta_{mn}, \tag{2.6}$$

where the scalar product of *periodic functions*, either represented in the real space or in the reciprocal space, is defined as

$$\langle f | g\rangle = \frac{1}{\Omega_0}\int_{\Omega_0} f^*(\mathbf{r}) g(\mathbf{r})\, d\mathbf{r}. \tag{2.7}$$

Eq. (2.6) must be fulfilled only between periodic functions characterized by the same wavevector $\mathbf{k}$. A generic operator O in the crystal momentum representation reads, consistently with Eq. (2.3):

$$O_{\mathbf{k},\mathbf{k}'} = e^{-i\mathbf{k}\cdot\mathbf{r}} O e^{i\mathbf{k}'\cdot\mathbf{r}'}. \tag{2.8}$$

Let us now restrict to the one-particle effective Hamiltonians H that can be formally derived from the Hartree–Fock or the Density Functional theories for the unperturbed, periodic, ground state. In such a case the Hamiltonian is periodic, and will not couple wavefunctions of different wavevectors. The wavefunctions $\psi_{m,\mathbf{k}}(\mathbf{r})$ are either the components of a Slater determinant (HF) or the ingredients to build up the electron density $n(\mathbf{r})$ (DFT). In both cases, the one-particle equation reads:

$$\big[H - \varepsilon_{m,\mathbf{k}}\big]\psi_{m,\mathbf{k}}(\mathbf{r}) = 0. \tag{2.9}$$

By using the Bloch theorem (Eq. (2.3)), we can write an equation for $u^{(0)}_{m,\mathbf{k}}$ in the form:

$$\big[H_{\mathbf{k},\mathbf{k}} - \varepsilon_{m,\mathbf{k}}\big]u_{m,\mathbf{k}}(\mathbf{r}) = 0, \tag{2.10}$$

where $H_{\mathbf{k},\mathbf{k}} = \exp(-i\mathbf{k}\cdot\mathbf{r}) H \exp(i\mathbf{k}\cdot\mathbf{r})$. The periodic solutions of Eq. (2.10) form a complete set of periodic functions. The Crystal Momentum representation is thus based on the eigenvectors of the one-particle Hamiltonian. Because of the periodic boundary conditions, the functions $\psi_{m,\mathbf{k}}$ and $\psi_{m,\mathbf{k}+\mathbf{G}}$, where $\mathbf{G}$ is a reciprocal lattice vector, span the same space. As a consequence, the knowledge of the eigenvectors of Eq. (2.9) where $\mathbf{k}$ is restricted to the Brillouin zone (BZ) is sufficient to build up the space of all solutions to Eq. (2.9).

The eigenvalue $\varepsilon_{m,\mathbf{k}}$ has some important properties (BOUCKAERT, SMOLUCHOWSKI and WIGNER [1936], BLOUNT [1962]): it can be demonstrated that it is a continuous function of the crystal momentum $\mathbf{k}$, in the limit of an infinite crystal. Such a property derives from considering the continuity of $H(\mathbf{k})$. For each m, the ensemble of the values $\varepsilon_{m,\mathbf{k}}$ constitutes an *energy band*. The subset of the crystal space group operations that leave $\mathbf{k}$ unchanged is called the *small group of* $\mathbf{k}$. Each energy band, regarded as a function of $\mathbf{k} \in$ BZ, possesses the full symmetry of the corresponding small group, and can thus be decomposed according to its irreducible representations. However, it may happen that two or more energy bands corresponding to distinct representations touch each other in some points of the Brillouin zone. In that case, one speaks of *accidental degeneracies* and the band characters in the vicinity of the crossing points may interchange abruptly. As a consequence, the treatment of degenerate bands is more complicated and needs a careful evaluation of their characters by inspecting the corresponding eigenvectors. The *dispersion* of a simple (noncrossing) band is given by the range of values it can have. Roughly speaking, the eigenvectors of very flat, almost nondispersive, bands often come from quasi-atomic eigenstates, whereas highly dispersive bands are the signature of eigenfunctions whose character sensitively changes as a function of $\mathbf{k}$, which are usually more delocalized.

In the following, we focus on nonmagnetic extended systems, for the sake of simplicity. We therefore assume implicitly that the spin-up and spin-down orbitals have the same spatial behavior and the corresponding eigenvalues are equal. While the definition and the mathematical properties of the energy bands $\varepsilon_{m,\mathbf{k}}$ are a direct consequence of the Bloch theorem, another concept, which can be generalized to nonperiodic and/or strongly correlated electron systems[1] is the *density of states* (DOS). The DOS is given by the number of states per unit volume at a given energy ε. For a crystal, it can be expressed by means of the Dirac delta function as:

$$N(\varepsilon) = \frac{2\Omega}{(2\pi)^3} \int_{\mathrm{BZ}} \sum_{m}^{(\mathrm{occ.})} \delta(\varepsilon - \varepsilon_{m,\mathbf{k}})\, d\mathbf{k}. \tag{2.11}$$

Then, we stress the physical implications of the Crystal Momentum representation for the calculation of the electronic structure. In both HF and DF theories, the electron density is built from the $\psi_{m,\mathbf{k}}$, under the conditions that the number of electrons per unit Wigner–Seitz cell is fixed. This is equivalent to sum up over the BZ the $u_{m,\mathbf{k}}$ corresponding to the occupied bands:

$$n(\mathbf{r}) = \frac{2}{(2\pi)^3} \int_{\mathrm{BZ}} \sum_{m}^{(\mathrm{occ.})} u_{m,\mathbf{k}}(\mathbf{r}) u_{m,\mathbf{k}}(\mathbf{r})\, d\mathbf{k}. \tag{2.12}$$

[1] In such a case, the notion of one-electron excitations is missed, since the collective behavior of the interacting electrons may dominate over the effective single-particle features.

Since we consider only spin-compensated systems for which $u^{\uparrow}_{m,\mathbf{k}}(\mathbf{r}) = u^{\downarrow}_{m,\mathbf{k}}(\mathbf{r})$, we simply put a spin degeneracy factor equal to 2 in the sums in Eqs. (2.12) and (2.13). The *Fermi energy* ε_F is fixed in such a way that there are exactly N electrons (θ is the Heaviside function):

$$N = \sum_m \frac{2\Omega}{(2\pi)^3} \int_{BZ} \theta(\varepsilon_F - \varepsilon_{m,\mathbf{k}})\, d\mathbf{k}. \tag{2.13}$$

Naively speaking, the Fermi energy ε_F is defined as the energy separating the highest occupied level from the lowest unoccupied one. However, at absolute zero temperature, such a definition is only correct when there is a continuum of states around ε_F, which is the case for metals. In that case, a certain number of bands are only partially filled (this is necessarily the case, for instance, of crystals with an odd number of electrons per unit cells without a net magnetization). The set of all manyfolds in $\mathbf{k}$ space that separate the filled from the empty bands is known as the *Fermi surface*. A qualitatively different case is provided by insulating crystals, in which a finite, nonzero gap separates the filled and empty bands. In such a case, ε_F can be uniquely defined as the limit of the chemical potential as the temperature goes to 0 K, and it can be demonstrated to fall in the middle of the gap.

Thanks to the Bloch theorem, the same set of $\mathbf{k}$-dependent equations (2.9) and (2.10) applies to all crystal unit cells. In principle, for an infinite crystal, they should be solved for all $\mathbf{k}$ vectors in the BZ, according to the BvK periodic boundary conditions (Eq. (2.4)). In practice – apart from the case in which the dependence on a particular wavevector $\mathbf{k}$ must be explicitly known – one has to compute *averages* of $\mathbf{k}$-dependent functions over the BZ:

$$\bar{f} = \frac{\Omega}{(2\pi)^3} \int_{BZ} d\mathbf{k}\, f(\mathbf{k}). \tag{2.14}$$

In this case, the average $\bar{f}$ is assumed to be a c-number, such as the expectation value of the kinetic energy operator in the DFT; in other cases the function f depends on the other coordinates – for instance, the electron density in the DFT results from a sum over the BZ of a nonnegative function defined in the real space (see Eq. (2.12)). For such cases, several schemes have been proposed to consider as few as possible $\mathbf{k}$ points in the BZ and thus reduce the computational cost. Among them, the special point technique (BALDERESCHI [1973], CHADI and COHEN [1973]) and the tetrahedron method are the most popular.

In the special point technique, the average of f over the BZ (Eq. (2.14)) can be approximated by a weighted sum over a finite number $N_{\mathbf{k}}$ of $\mathbf{k}$ points:

$$\bar{f} = \sum_{i=1}^{N_{\mathbf{k}}} w_{\mathbf{k}_i} f(\mathbf{k}_i) + E\big(\{\mathbf{k}_i\}\big), \tag{2.15}$$

where the $\sum_{i=1}^{N_{\mathbf{k}}} w_{\mathbf{k}_i} = 1$. For a given $N_{\mathbf{k}}$, the set of special points $\{\mathbf{k}_i\}$ is such that the error $E(\{\mathbf{k}_i\})$ is minimized. A demonstration can be found in the paper by CHADI and

COHEN [1973], which is based on the use of the lattice point group of the crystal and the symmetry of $f(\mathbf{k})$ for translations by reciprocal lattice vectors. In the same paper, a method for generating the special point sets is outlined and the determination of various special point sets explicitly carried out for cubic and hexagonal crystal lattices. A more general method for generating equispaced grids of **k** points for crystals of any symmetry, which is equivalent to that introduced by Chadi and Cohen in special cases, was then provided by MONKHORST and PACK [1976].

The special point technique is well suited for integrating smoothly varying functions in the BZ, particularly for insulators and semiconductors, where the number of occupied bands is independent of the wavevector **k**, which simplifies its practical implementation. In metals, the number of occupied bands can vary abruptly with **k**, as a consequence of band crossing, and the shape of the Fermi surface may be complicated and generally admits singular points. A practical way to deal with such singularities is to smoothen the Fermi surface by replacing the Heaviside function in Eq. (2.13) with a smeared occupation function $\tilde{s}(\varepsilon_F - \varepsilon_{m,\mathbf{k}}; \{\alpha\})$ where the set of parameters $\{\alpha\}$ defines the actual choice of smearing (FU and HO [1983], METHFESSEL and PAXTON [1989]). A simple example is provided by the Fermi–Dirac distribution, the width of which around ε_F is proportional to the temperature T. However, all physical quantities must generally be computed in the limit of no smearing (e.g., $T \to 0$ in the Fermi–Dirac distribution) in order to avoid the arbitrariness linked to the choice of the function $\tilde{s}$. An alternative solution is provided by the tetrahedron method (JEPSEN and ANDERSEN [1971]) that is basically a simplex method for interpolating $f(\mathbf{k})$. The irreducible part of the BZ is partitioned into tetrahedra, within which $f(\mathbf{k})$ is linearized in **k**. The linear approximation allows the integration (Eq. (2.14)) to be computed analytically, taking into account the complicated shape of the Fermi surface. While there are few doubts that at zero temperature the tetrahedron method is superior to the special point technique in metals, for insulators the latter one is believed to be computationally faster and more robust. Indeed, the partitioning of the BZ into tetrahedra may be cumbersome for low-symmetry structures, and many **k** dependent matrix elements must be available at the same time to perform the integration, which can be unfeasible when, as in plane-wave based methods (see Section 4.4), the calculation of a single matrix element of a one-particle operator is quite heavy. However, an improved tetrahedron method was proposed (BLÖCHL, JEPSEN and ANDERSEN [1994]), which is able to reduce the number of **k** points effectively. The BZ average in Eq. (2.14) is carried out through integration weights that are calculated by taking into account the curvature of $f(\mathbf{k})$, thus reducing the error $E(\{\mathbf{k}_i\})$ in Eq. (2.14). For both metals and insulators, the improved method implies a substantial reduction of computer time with respect to the original tetrahedron-based integration scheme.

A quite different case is represented by the computation of elementary excitations in crystals (see Section 3.11). In this case, it is preferable to sample the BZ by using **k** points that does not have any special symmetry. Indeed, the electronic bands may be flat or degenerate in highly symmetric **k** points, which would artificially reinforce the computed weight of such electronic transitions. A practical alternative in those cases is to generate a special point grid that is successively shifted from the BZ center by a linear combination of the reciprocal lattice vectors with small coefficients, of the order of few tenths.

2.1.2. *The Wannier representation*

We have just seen that in presence of a periodic one-electron effective Hamiltonian, a useful representation of the wavefunctions is provided by the Crystal Momentum representation (CMR), in which the Hilbert space is spanned the eigenstates $u_{m,\mathbf{k}}$. The CMR is the crystalline counterpart of the plane-wave representation for the free electron. Like the solution of the latter problem, $u_{m,\mathbf{k}}$ is delocalized, which follows directly from its periodicity over the entire crystal. In some physical situations, a representation of the electronic states that is based on localized wavefunctions can be better suited or even become compulsory when a nonperiodic potential, such as a homogeneous electric field, is considered (NUNES and VANDERBILT [1994]). A way to do that was pioneered by Wannier long time ago (WANNIER [1937]) and successively explored by KOHN [1973]. The basis functions in the Wannier representation are:

$$a_m(\mathbf{r}-\mathbf{R}) = \frac{\Omega}{(2\pi)^3}\int_{\mathrm{BZ}} d\mathbf{k}\,\psi_{m,\mathbf{k}}(\mathbf{r})\exp(-i\mathbf{k}\cdot\mathbf{r}). \tag{2.16}$$

The Wannier representation can be generalized to any function $f(\mathbf{k})$ defined on the Brillouin zone, simply by replacing $\psi_{m,\mathbf{k}}(\mathbf{r})$ with $f(\mathbf{k})$ in Eq. (2.16). From Eq. (2.16), it appears that Wannier functions $\{a_m\}$ are not unique, due to a phase indeterminacy in the Bloch orbitals $\{\Psi_m\}$. However, a rigorous proof of the fact that one can built Wannier functions $a_m(\mathbf{r}-\mathbf{R})$ which are maximally localized around a given crystalline site $\mathbf{R}$ can be found in the literature (BLOUNT [1962], MARZARI and VANDERBILT [1997]). Heuristically, one can see that if in Eq. (2.16) $|\mathbf{r}-\mathbf{R}|$ is bigger than few lattice vector moduli, then the exponential factor is rapidly varying and the corresponding integral would be small, provided that appropriate phase factors have been chosen for the Bloch orbitals.

In the past the Wannier representation (WR) was rarely adopted. Indeed, the practical construction of a set of well localized and orthonormal functions is difficult by using Eq. (2.16), since the $\psi_{m,\mathbf{k}}(\mathbf{r})$ are usually undetermined within a phase factor. A way to the practical construction of the Wannier functions was pioneered by KOHN [1973] for simple and composite bands of cubic crystals, who started from trial functions that are well localized and orthonormal, and refined them through a variational principle on the total band energy. In the last decade, there was a renewed interest for the Wannier representation, which may open the way to fast, order-N computational schemes – see other chapters in this Handbook and GALLI [2000] – due to the localization properties of the $a_m(\mathbf{r}-\mathbf{R})$. However, most of the methods recently adopted, instead of working from the very beginning in the WR, suggested practical recipes to find Wannier functions from the Bloch states (MARZARI and VANDERBILT [1997], SILVESTRELLI [1999]). Nevertheless, some interesting developments have been proposed (KOHN [1993], SHUKLA, DOLG, FULDE and STOLL [1998]), which the interested reader is referred to.

2.1.3. *Periodicity in one and two dimensions*

In the perfect crystal model, we assume the periodicity holds in any of the three spatial directions. However, there are cases in which the discrete translational property applies only for a subset of the three-dimensional space. An ideal linear polymer, for instance, fulfills the translational symmetry when considering the component of the atomic positions along the polymer axis (1D case). In this case, the direct space can be considered as a tensorial product of a 1D by a 2D space $\Re^{(1)} \bigotimes \Re^{(2)}$. In $\Re^{(1)}$, we can define a basis vector $a^{(1)}$, and a set of lattice points, let's say x_n, such that $x_n = na^{(1)}$. Therefore, any one-particle wavefunction $\psi(\mathbf{r})$ in 3D can be constructed as a product of two factors $\psi^{(1)}_{n,k_x}(x)$ and $\psi^{(2)}(y,z)$ that are defined in $\Re^{(1)}$ and $\Re^{(2)}$, respectively. The former one fulfills the Bloch theorem, so that a 1D vector k_x, which is the counterpart of the crystal momentum for the perfect crystal model, can be defined as a good quantum number. All general properties that have been illustrated in the case of the perfect crystal in 3D for the one-particle wavefunction are also valid. However, the Brillouin zone is one-dimensional and the Fermi surface reduces to two disconnected points.

An ideal surface of Miller indices (hkl) can be thought as a collection of atoms whose coordinate projections on the surface plane normal to the $[hkl]$ direction are regularly arranged. Two vectors, $\mathbf{a}_1^{(2)}$ and $\mathbf{a}_2^{(2)}$, form a basis set for the 2D lattice: $\mathbf{R}_{l_1,l_2} = l_1\mathbf{a}_1^{(2)} + l_2\mathbf{a}_2^{(2)}$. Differently from the 3D case, there are only five two-dimensional Bravais lattices (BECHSTEDT and ENDERLEIN [1988], DESJONQUÈRES and SPANJAARD [1996]). Also for the 2D case we can introduce, by means of the Bloch theorem in the periodic subspace $\Re^{(2)}$, a surface momentum $\mathbf{k}_\parallel$, and define energy bands corresponding to one-particle excitations in a 2D Brillouin zone, as a function of $\mathbf{k}_\parallel$. The Fermi surface is one-fold and may be topologically connected or not, depending on the actual surface band structure. A very useful and important concept is the *projected bulk band-structure*. Any three-dimensional vector in the bulk 3D BZ can be decomposed as a sum of two vectors, parallel and normal to the surface plane, respectively: $\mathbf{k} = \mathbf{k}_\parallel + \mathbf{k}_\perp$. The parallel component is equal to a vector $\mathbf{k}_\parallel$ belonging to the 2D Brillouin zone, within a 2D reciprocal lattice vector: $\mathbf{k}_\parallel = \mathbf{k}_\parallel^{(2)} + \mathbf{G}_\parallel^{(2)}$. The set of all energy bulk band values $\varepsilon_{n,\mathbf{k}_\parallel^{(2)}}$ for $\mathbf{k} \in$ 3D BZ, is the *projected bulk band structure* that provides a set of all possible values of bulk bands when the latter ones are considered in the 2D surface BZ. Nevertheless, the projected bulk band structure implies a fully 3D periodicity, which is not the case for the surface. In this case, some extra $\varepsilon_{n,\mathbf{k}_\parallel^{(2)}}$ may appear at values that are forbidden for the projected bulk band-structure. They are the one-particle energies of *surface states* that have a spatial localization in the region close to the surface. For an historical introduction and the recent perspectives to the theory of surface states, the interested reader can look at various review articles (BERTONI [1990], FORSTMANN [1993]).

2.1.4. *Long-range interactions*

In any extended system, it is important to distinguish between short- and long-range interactions, which must be treated differently from each other.[2] Two difficulties are akin to long-range potentials. The first one is purely mathematical: because of the

[2] Given an interaction potential energy $V(\mathbf{r})$, it is said to be long-range whenever $\int_\Omega d\mathbf{r}\,|V(\mathbf{r})|$, which is a rough measure of the total potential strength, diverges as the volume $\Omega \to \infty$. For the sake of simplicity, here we focus on *bare* interactions corresponding to an energy potential $V(\mathbf{r})$ in the Hamiltonian, and not on *effective* ones, which arise from the complex mutual interactions between particles and are thus renormalized. An example is provided by the bare Coulomb potential which is screened in the solid to yield the effective local electric fields.

divergence, care must be taken when integrating on infinite domains. The second one regards the system representation: any model of the real extended system must take into account the possibility of very long range interactions between atoms that are far apart.

As an example of treatment of long-range potentials, here we discuss how to compute the electrostatic energy E_{I-I} of a 3D periodic system of point charges, which can easily be recasted for computing the ion–ion electrostatic contribution E_{Ew} to the total energy of real systems. The problem was firstly treated by EWALD [1921] in the study of ionic crystals. We have to evaluate (atomic units $e^2 = m_e = \hbar = 1$ are used throughout):

$$E_{\mathrm{Ew}} = \frac{1}{2}\sum_{\mathbf{l}}{}' \left[\sum_{\kappa,\kappa'} \frac{Z_\kappa Z_{\kappa'}}{|\boldsymbol{\tau}_\kappa - \boldsymbol{\tau}_{\kappa'} + \mathbf{L}_\mathbf{l}|}\right], \tag{2.17}$$

where Z_κ and $\boldsymbol{\tau}_\kappa$ are the point charges and positions within the unit cell, respectively, and the sum on the lattice vectors $\mathbf{L}_\mathbf{l}$ excludes the term $\mathbf{L} = \mathbf{0}$ when $\kappa = \kappa'$. The sum is conditionally convergent for globally neutral systems. A general numerical method is therefore necessary to improve the convergence of the series and speed-up the calculation of the total electrostatic energy. The basic idea in the Ewald method is to consider a compensating charge distribution consisting of a superposition of localized smooth functions $\varrho_c(\mathbf{r})$ that are centered at the ionic sites $\boldsymbol{\tau}_\kappa$. The total charge distribution is thus:

$$\rho(\mathbf{r}) = \sum_{\mathbf{l}}\sum_{\kappa} Z_\kappa\big[\delta(\mathbf{r} - \boldsymbol{\tau}_\kappa) - \varrho_c(\mathbf{r} - \boldsymbol{\tau}_\kappa)\big] + \sum_{\mathbf{l}}\sum_{\kappa} Z_\kappa \varrho_c(\mathbf{r} - \boldsymbol{\tau}_\kappa). \tag{2.18}$$

The electrostatic potential corresponding to the first term on the left is globally neutral and rapidly converging in direct space. The distribution $\sum_\mathbf{l}\sum_\kappa \varrho_c(\mathbf{r} - \tau_\kappa)$ is smooth and may be expanded in reciprocal space conveniently. Usually, the compensating charge distribution is a Gaussian: $\varrho_c(\mathbf{r}) = (\alpha^2/\pi)^{3/2}\exp(-\alpha^2 r^2)$. Accordingly, the electrostatic energy reads:

$$\begin{aligned} E_{\mathrm{Ew}} = \frac{1}{2}\sum_{\kappa\kappa'} Z_\kappa Z'_\kappa \Bigg[\sum_{\mathbf{l}}{}' \frac{\operatorname{erfc}(\alpha|\boldsymbol{\tau}_\kappa - \boldsymbol{\tau}_{\kappa'} + \mathbf{L}_\mathbf{l}|)}{|\boldsymbol{\tau}_\kappa - \boldsymbol{\tau}_{\kappa'} + \mathbf{L}_\mathbf{l}|} - \frac{2\alpha}{\pi^{1/2}}\delta_{\kappa,\kappa'} \\ + \sum_{\mathbf{G}}{}' \frac{4\pi}{\Omega_0} \frac{e^{i\mathbf{G}\cdot(\boldsymbol{\tau}_\kappa - \boldsymbol{\tau}_{\kappa'})}\, e^{-G^2/4\alpha^2}}{G^2} - \frac{\pi}{\Omega_0 \alpha^2}\Bigg], \end{aligned} \tag{2.19}$$

where $G^2 = |\mathbf{G}|^2$, and $\operatorname{erfc}(x)$ is the complementary error function. The sum over the reciprocal lattice vectors excludes the term $\mathbf{G} = \mathbf{0}$, which directly derives from the condition of global neutrality. The parameter α can assume any value, and is adjusted to obtain the fastest convergence of both real and reciprocal space sums (first and second terms on the right of Eq. (2.19), respectively). The second right-hand term compensates the term $\kappa = \kappa'$, $\mathbf{L} = \mathbf{0}$ that appears in the real space sum, which is not automatically excluded from the $\mathbf{G}$-space summation, and is sometimes referred to as the *electrostatic self-energy* of the compensating charge distribution. Similarly, the last term on the right is a remnant of the regular part of the $\mathbf{G} = \mathbf{0}$ contribution to the reciprocal lattice sum.

In the derivation of Eq. (2.19), we have assumed that the lattice is periodic in the three dimensions. The treatment of 3D systems having a discrete periodicity in one or two dimensions is analogous,[3] but the resulting mathematical formulae and the skills to obtain a fastest convergence of the lattice sums are obviously different. For the case of surfaces, the interested reader is referred to the existing bibliography (PARRY [1975], KAWATA, MIKAMI and NAGASHIMA [2002]).

2.2. *Aperiodic systems*

Most of the models that have been devised to treat with extended aperiodic systems adopt an intermediate strategy between the perfect, infinite crystal and the isolated molecule, which are two well defined limits. In formulating the models and proposing the theoretical and computational solutions, the *locality* assumption is more or less tacitly assumed. The focus is on a given portion of the aperiodic system, which is believed to be significant for its chemical and physical behaviors. Such a portion (*the local zone*) is considered as virtually cut from the extended system, with suitable boundary conditions. Although very intuitive, such an assumption is essentially based on the consideration of the relevant length scales. Moreover, the task is not trivial at all, since the presence of the long-range Coulomb interaction makes impossible to isolate a finite portion of the real system without choosing appropriate boundary conditions that mimic the local electric field suitably. More formally, it is assumed that the geometry at the atomic scale and the one-electron density matrix vary essentially in the local zone, irrespectively of the remaining of the system. In general, one is interested in calculating *energy differences* between two or more configurations differing by geometry and/or electronic structure. On one hand, some of the errors due to the boundary conditions might compensate, so that the models can be considered as good approximations of the actual extended systems. On the other hand, such energy differences are usually tiny numbers when compared to the total energies of the various configurations, which needs much care in adopting suitable algorithms and in checking the adequacy of the model.

2.2.1. *The supercell approach*

Although periodic boundary conditions are derived under the assumption of a perfect translational symmetry, they are often used to simulate aperiodic systems. Point defects in solids, solid surfaces with and without adsorbed species, solid alloys, are among the many available examples in the literature. The local zone, consisting of N_a atoms, is then artificially repeated in space through Born–von Karman (BvK) boundary conditions (BC). The model system is thus formally equivalent to a perfect crystal having N_a atoms per unit cell, which is usually called the *supercell*. There are two main reasons underlying the choice of the supercell method. First, the use of BvK BCs permits a

[3] This must not be confused with the treatment of the Coulomb potential in D = 1 or D = 2.

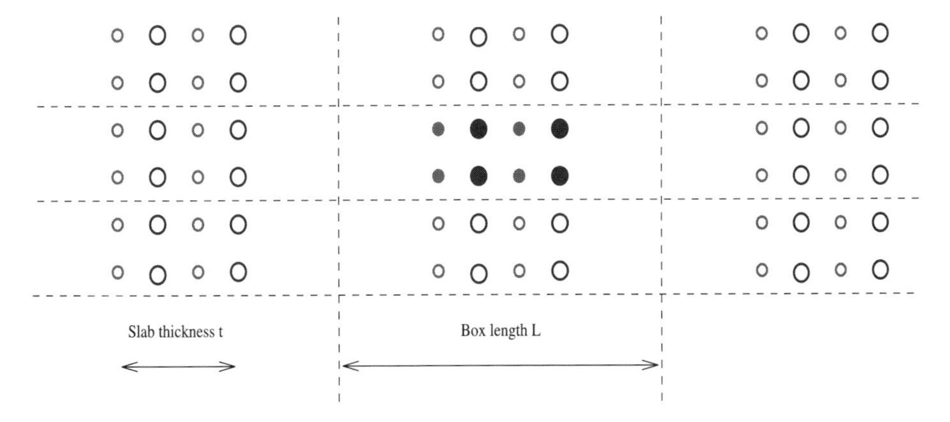

FIG. 2.1. A simple representation of a slab with periodic boundary conditions. The atoms within the unit cell are represented as full circles, while the image atoms are drawn as empty circles.

drastic reduction of the number of degrees of freedom: the knowledge of the coordinates of the atoms in one supercell are sufficient to represent the virtual periodic system through all its repeated images. Moreover, if the simulated system is bulk-like (such as a point defect, or a disordered solid) the use of BvK boundary conditions avoids to treat surface effects explicitly. Second, the mathematical framework that was developed for the perfect crystal model also applies to the virtual periodic crystal. This has important consequences on the representation of the wavefunctions that can be considered as Bloch states. Therefore, the use of periodic basis sets, such as plane waves (see Section 4.4), is made possible even for truly aperiodic systems.

As in the cluster approach that will be schematized later, the basic approximation is the choice of supercell size, which is strictly related to the number N_{a} of independent atoms. The shape and size of the supercell must be consistent with the actual characteristics of the real system (i.e., isotropy, point symmetry if any, etc.) as much as possible. Moreover, the interaction between periodic images should be monitored and evaluated. It is in principle possible to estimate finite-size effects through a fine comparison of a few calculations by adopting different cell sizes, which represents an internal check of consistency. Let us assume, for instance, a uniform variation of the supercell lattice parameters such that the supercell volume scales as $\lambda\Omega$. As $\lambda \to \infty$, the supercell model becomes exact.[4] In this limit, the Brillouin zone reduces to a single point (the Γ point $\mathbf{k} = \mathbf{0}$) and the bandwidths go to zero. Although this limit is purely hypothetical, for a finite λ a careful analysis of the band structure may give useful information on the degree of convergence of the supercell size.

The treatment of long-range forces needs a special care. A noteworthy example is the Coulomb field, whose proper treatment is a key point in any electronic structure calculation of extended systems. For instance, a localized net charge placed in an insulating system is only partially screened: this means that the resulting interaction has

[4]In absence of long-range interactions and provided that the boundary conditions are consistent with those for the real system, $\forall\lambda$.

a $1/r$ behavior at large distances. A careful evaluation of the bias introduced by a sharp cut off in direct space or by the use of BvK BCs on the computed quantities is thus a necessary previous step when setting up the model system. However, the practical treatment of the Coulomb interaction sensitively depends on the kind of periodic boundary conditions that are adopted. Many practical recipes exist, each presenting advantages and shortcomings. In the following, we specify some prototypical cases and provide the reader the useful references to look at.

Charged systems. As a prototype of nonperiodic charged extended systems the case of a charged point defect can be considered. At finite temperature, there is a small concentration of intrinsic defects in all crystalline solids. Moreover, extrinsic defects, consisting of foreign atoms (either in interstitial or in substitutional sites) are extremely frequent and technologically relevant, especially in the physics of semiconductors and insulators, and their simulation is thus a challenge. For charged defects, a compensating uniform background is often superimposed to ensure charge neutrality in the supercell (BAR-YAM and JOANNOPOULOS [1984], LESLIE and GILLAN [1985], JANSEN and SANKEY [1989]). The defect formation energy is computed via a total-energy difference of the perfect and the defective crystals. It can be shown (MARTYNA and TUCKERMAN [1999]) that energy differences between infinitely replicated periodic systems are in principle ill-defined, although the choice of a suitable reference system may remedy the situation. The treatment of the long-range Coulomb potential in reciprocal space is intimately related to the discretization of the Fourier transform into a Fourier series, which is limited to reciprocal lattice vectors. By isolating the long-range contribution, which can be treated analytically, from the short-range one, which depends on the local details of the charge distribution and can be computed numerically, some ways of dealing with the Coulomb interaction in systems with periodic boundary conditions have been devised (DELEEUW, PERRAM and SMITH [1980], MARTYNA and TUCKERMAN [1999]).

The slab model for surfaces. In this case, the surface is simulated by a film consisting of some atomic layers, plus a surrounding vacuum region, with periodic boundary conditions along the surface plane. The slab has to be thick enough to display the bulk characteristics (atomic positions, band dispersion, etc.) in the innermost layers. Despite the finite thickness, it is possible to extract convergent surface energies from this kind of calculations. The safest way, although not the most economic one as far as the computing time is concerned, is to increase the slab thickness t systematically, starting from a minimum t that allows bulk properties to be properly defined.[5] The slab total energy is then plotted as a function of the number N_l of layers, so that the average surface energy of the two slab terminations can be extrapolated at $N_l = 0$ (BOETTGER [1994], FIORENTINI and METHFESSEL [1996], BATES, KRESSE and GILLAN [1997]). In analogous way, the surface stress can be defined as the rank-2 tensor that is obtained in the limit $N_l \to 0$. A comprehensive discussion of its calculation can be found in the literature (NEEDS [1987], BOTTOMLEY and OGINO [2001]). The distinction between the two terminations, whenever they are different, as well as the evaluation of the

[5] It is noteworthy to remark that the space group of the slab is generally different from that of the infinite crystal. Much care must thus be used when considering the reference bulk state.

surface stability in equilibrium with a varying chemical environment by means of the chemical potentials is also possible (POJANI, FINOCCHI and NOGUERA [1999]). However, especially when using a plane wave basis set (see Section 4.4), periodic boundary conditions are imposed along the normal to the surface plane, which can be source of numerical errors. This is the case, for example, of slabs with a net dipole moment, which give rise to intrinsically nonperiodic electrostatic potentials. A way to overcome such difficulties still using 3D PBC has been recently discussed (BENGTSSON [1999], FINOCCHI, BOTTIN and NOGUERA [2003]).

To conclude this section, it is important to note that applying PBC to any intrinsically aperiodic system is not unobjectionable. A system that is characterized by a given charge distribution can be intrinsically different from its periodic analogous that is built by imposing a spurious periodicity, even in the limit of a very large lattice constant. The reason lies essentially in the long-range behavior of the Coulomb potential. For instance, in the Poisson equation two identical charge distributions can give rise to distinct electrostatic potentials differing by an arbitrary linear function of the spatial coordinates, which is determined by the actual boundary conditions used. The peculiarity of the Coulomb interaction shows up in many respects, including the treatment of electronic correlations: the extremely nonlocal and polarization dependency of the exchange-correlation functional in the DFT (GONZE, GHOSEZ and GODBY [1997]) is one of the numerous examples.

2.2.2. *Embedding techniques for point-defect problems*

The supercell model described in the previous section allows aperiodic problems to be formally reduced to periodic ones, while maintaining the “infinite" nature of the condensed system. The fundamental advantage which is so achieved is that one can use for the solution of the problem any one in a class of powerful computational tools that have been specifically developed for perfectly periodic solids. There are however some kinds of aperiodic problems where other solution schemes may be more convenient: metal alloys are a typical example (FAULKNER [1982]). We discuss here only the case of point defects in otherwise perfect crystals. The three- two- or one-dimensional periodicity of the host crystal (a bulk solid, a slab, a polymer) is here destroyed only by the presence of a local impurity (for instance, an interstitial atom, an adsorbed molecule at the surface of the slab, a short side chain in the polymer). We shall admit the short-range character of the effect of the impurity on the electronic structure of the host crystal (*locality assumption*); on the other hand, the latter may be changed profoundly in the local zone. We are usually interested precisely in the changes introduced by the impurity: how do atoms relax in its vicinity? what is the defect formation energy? do new electronic features appear (for instance, defect levels in a gap)? etc. The supercell model can obviously be applied to this class of problems, by designing supercells each containing a single defect, of appropriate shape and size such that the interaction between defects is negligible. However, this can require considering supercell sizes which are exceedingly large, perhaps much more than the physical range of the perturbation, especially when treating defects with a net charge or high dipole moments. In spite of all these limitations, the supercell model has achieved extraordinary success and is still one of the preferred choices in this area of studies.

Another widely adopted approach consists in mapping the problem onto one with a much smaller number of electrons, by considering a molecular cluster which includes the defect itself and a limited number of atoms of the host crystal in its vicinity. This technique has the advantage of an enormous flexibility and allows the most sophisticated tools of molecular quantum mechanics to be adopted. It is however necessary to improve on the crude basic model in order to account for the effects of the rest of the crystal that are not included directly in the quantum-mechanical simulation: they comprise finite size effects, electrostatic interactions, bond breaking, polarizability of the environment, and so on. An enormous variety of devices has been adopted for this purpose. It is outside the scopes of this chapter to illustrate these methods.

The basic idea of the "embedding techniques" to be presented here, is to recognize the quasi-crystalline nature of the problem: a perturbative approach is then adopted in the frame of Green function techniques. The solution of the comparatively simple problem of the periodic host crystal is first obtained; in a subsequent step, that solution is used so as to reduce the problem to one of the size of the "local region" within which the effects of the impurity are felt. Indeed, the term "embedding" is currently used in a wide sense, and many of the cluster techniques which introduce corrections to describe the crystalline environment are presented with this name. The advantage of the more rigorous perturbative approach over the molecular cluster model is that all effects related to the crystalline environment are taken into account accurately and in a natural way, and no artificial boundary between the defect region and its surroundings is introduced. We illustrate this approach in a very simplified way, with reference to a representative orthonormal basis set, $T \equiv \{\tau\}$. For a more adequate treatment reference is made to the relevant literature (PANTELIDES [1978], ECONOMOU [1979]).

Let us define, for any complex number z,

$$G(z) = (zI - H)^{-1} \equiv Q(z)^{-1}. \tag{2.20}$$

Here H is the Hamiltonian matrix of the system in the T representation. The *Green matrix* $G(z)$ is a function of z; it is in a one-to-one relationship to the projected-density-of-states (PDOS) matrix $\rho(\varepsilon)$, constructed with the eigenvectors $\vec{a_n}$ and eigenvalues λ_n of H (ECONOMOU [1979]):

$$\begin{cases} \rho_{\tau\tau'}(\varepsilon) \equiv \sum_n a_{n\tau} a^*_{n\tau'} \delta(\varepsilon - \lambda_n) = 1/(2\pi\imath) d/(d\varepsilon) \int_{\gamma(\varepsilon)} dz\, G_{\tau\tau'}(z), \\ G_{\tau\tau'}(z) = \int d\varepsilon\, \rho_{\tau\tau'}(\varepsilon)/(z - \varepsilon). \end{cases} \tag{2.21}$$

In the first line, $\gamma(\varepsilon)$ is a closed anti-clockwise integration path in the complex z plane which encloses all eigenvalues $\lambda_n < \varepsilon$.

Let us assume that the basis functions are localized in the sense that their spatial extension is limited; for instance, they are Gaussian type functions centered in the nuclei. Consider *any* partition of the representative space T into two subsets: $T = C \oplus D$, with $C = \{\gamma\}$, $D = \{\delta\}$. Let us call them the *cluster* and the *indented crystal* set, respectively. Generally speaking, the C set will describe the region comprising and surrounding the defect, D the rest of the infinite crystal. In a block-matrix notation, the identity $Q(z)G(z) = I$ becomes, for any z:

$$\begin{cases} Q_C G_C + Q_{CD} G_{DC} = I_C, \\ Q_{DC} G_C + Q_D G_{DC} = 0, \\ Q_C G_{CD} + Q_{CD} G_D = 0. \end{cases} \tag{2.22}$$

By solving in the second line for G_{DC} and substituting in the first one:

$$\begin{cases} G_{DC} = -(Q_D)^{-1} Q_{DC} G_C, \\ (G_C)^{-1} = Q_C - Q_{CD}(Q_D)^{-1} Q_{DC}. \end{cases} \tag{2.23}$$

Alternatively, we can obtain, from the third and first line:

$$\begin{cases} G_{DC} = G_D Q_{DC}(Q_C)^{-1}, \\ G_C = (Q_C)^{-1} - (Q_C)^{-1} Q_{CD} G_{DC}. \end{cases} \tag{2.24}$$

The use of either of the above formal results depends on the way the locality assumption is formulated. We shall distinguish two main schemes, the *standard* and the *perturbed cluster* (*PC*) ones. In both cases, reference is made to the unperturbed system (the host crystal) S^f whose solution is taken for known; the corresponding representative quantities are denoted with an f superscript (for *free*); S indicates the system *with* the defect.

2.2.3. The standard embedding approach

In the standard formulation, proposed originally by KOSTER and SLATER [1954], the following assumptions are made:

$$H_C = H_C^f + V_C; \qquad H_{CD} = H_{CD}^f; \qquad H_{DC} = H_{DC}^f; \qquad H_D = H_D^f. \tag{2.25}$$

It is here implied that the representative sets for S^f and S coincide, and that the Hamiltonian matrix is modified only in the cluster subblock CC, which can be considered to coincide with the local region. While the choice of using the same set $\{\delta\}$ of functions to describe the indented crystal with or without the defect appears sensible, the same is not true as far as the "local" $\{\gamma\}$ functions are concerned: in one case they describe only the perfect crystal atoms in the defect vicinity, and so they must coincide with the $\{\delta\}$ functions except for translations; in the other, they must describe all the new local features. This is not, however, a fundamental difficulty because it is possible to extend the results to the case where the two local subsets do not coincide. Compare now the exact relationship (line 2 of Eq. (2.23)) with the corresponding one for the perfect host system:

$$(G_C^f)^{-1} = Q_C^f - Q_{CD}^f (Q_D^f)^{-1} Q_{DC}^f. \tag{2.26}$$

If the Hamiltonian, hence the Q matrix, is perturbed only in the CC block, by subtracting the two identities we have, for any z:

$$\left(G_C^f\right)^{-1} - (G_C)^{-1} = Q_C^f - Q_C = \left(zI - H^f\right)_C - (zI - H)_C = V_C. \tag{2.27}$$

This result is formally identical to the Dyson equation (ECONOMOU [1979]). Note however that all matrices are finite since they are referred to the defect region. By substituting $G_C(z) = [(G_C^f(z))^{-1} - V_C]^{-1}$ in Eq. (2.21), the PDOS and hence all electronic properties in the local region are obtained. The procedure must be repeated to self-consistency because V_C does depend on the density matrix.

This elegant approach has had many applications in past years, both in conjunction with tight-binding schemes (PANTELIDES [1978]), and with ab initio density functional Hamiltonians (FEIBELMAN [1991]). The basic assumption (Eq. (2.25)) is usually acceptable for metallic systems where the screening of perturbations is very effective at short lengths.

2.2.4. The "perturbed-cluster" embedding technique

In the perturbed cluster approach proposed and implemented in a workable computational scheme by PISANI [1993], PISANI, DOVESI, ROETTI, CAUSÀ, ORLANDO, CASASSA and SAUNDERS [2000], the assumption is made, to be referred to as the *fundamental approximation* (f.a.), that the indented crystal block of the Green matrix is left unchanged, namely: $G_D(z) = G_D^f(z)$. By using the f.a. in the first line of Eq. (2.24) we can solve for G_{DC} in terms of Q_{DC}, Q_C and of the known quantity G_D^f; in turn, G_C can be obtained from G_{DC}, Q_{DC}, Q_C using the second line of (2.24). From a computational point of view it is convenient to work out the exact relationships (2.24) before introducing the f.a. For this purpose, it can be noticed that $(Q_C)^{-1}$ is the *cluster Green function*, which is obtained by a standard molecular program in terms of the eigenvalues $\bar{e}_j$ and eigenvectors $|\bar{v}_j\rangle$ of the cluster Hamiltonian H_C:

$$\left(Q_C(z)\right)^{-1} \equiv \overline{G}_C(z) = \sum_j \frac{|\bar{v}_j\rangle\langle\bar{v}_j|}{z - \bar{e}_j}. \tag{2.28}$$

If we adopt in the C subspace the cluster eigenvectors as a basis set ($j = \sum_\gamma \bar{v}_{\gamma j}\gamma$), this simplifies further to: $[\overline{G}_C(z)]_{jj'} = \delta_{jj'}/(z - \bar{e}_j)$. By using the mixed basis set ($\{j\} \oplus \{\delta\}$) in Eq. (2.24) and integrating over a suitable path in the complex z plane which encloses all occupied eigenvalues (PISANI [1993]), the PC equations are obtained which define the one-electron density matrix P for the S system:

$$\begin{cases} P_{\delta j} = \sum_{\delta'} M_{\delta\delta'}(\bar{e}_j) H_{\delta' j}, \\ P_{jj} = 2\theta(E_F - \bar{e}_j) - \sum_{\delta\delta'} H_{j\delta} M'_{\delta\delta'}(\bar{e}_j) H_{\delta' j}, \\ P_{j'\neq j} = \sum_\delta [H_{j'\delta} P_{\delta j} + P_{j'\delta} H_{\delta j}]/(\bar{e}_{j'} - \bar{e}_j). \end{cases} \tag{2.29}$$

A key role is played here by the *coupling matrices* $M_{DD}(e)$ with indices in the D subspace and by their energy derivatives, $M'_{DD}(e)$, whose elements are defined as follows:

$$\begin{cases} M_{\delta\delta'}(e < E_{\mathrm{F}}) = \int^{V} d\varepsilon\, \rho^{f}_{\delta\delta'}(\varepsilon)/(\varepsilon - e), \\ M_{\delta\delta'}(e > E_{\mathrm{F}}) = -\int^{O} d\varepsilon\, \rho^{f}_{\delta\delta'}(\varepsilon)/(\varepsilon - e). \end{cases} \tag{2.30}$$

The integral is extended to the virtual manifold (V) of the host crystal if e is below the Fermi energy E_{F}, to the occupied manifold (O) otherwise. The f.a. has been used only to substitute $\rho^{f}_{\delta\delta'}(\varepsilon)$ for $\rho_{\delta\delta'}(\varepsilon)$ in Eq. (2.30); otherwise, the PC equations are exact. For the PC equations to be solvable in practice, it is necessary that the nonzero part of the F_{DC} and $\rho_{\mathrm{DC}}(\varepsilon)$ matrices be finite, which means that their elements should decay rather rapidly with increasing distance between the centers of the two localized functions δ and γ. While $H_{\delta\gamma}$ normally exhibits exponential decay, the range of $\rho_{\delta\gamma}(\varepsilon)$ is much longer, especially with metallic systems (ECONOMOU [1979]), and the truncation of the infinite sums implicit in Eqs. (2.29) may become a critical issue of the method. It has to be noticed that the molecular cluster solution $\{\bar{e}_j, |\bar{v}_j\rangle\}$ is only an intermediate quantity in the computational scheme (the name of the technique, "*perturbed* cluster", describes precisely this fact); however, the analysis of the cluster eigenvalues and eigenvectors may give important information on the electronic structure in the local region.

The solution of the PC equations provides us with the complete knowledge of the one-density matrix, and with the consequent possibility to evaluate all values of one-electron observables for the ground state of the system. In principle, the method is useful also for the study of locally excited states. A review of existing applications of this technique, and an analysis of its merits and limitations has been recently provided (PISANI, DOVESI, ROETTI, CAUSÀ, ORLANDO, CASASSA and SAUNDERS [2000]).

2.2.5. *Other embedding approaches*

A common feature of the standard and the PC embedding schemes which distinguishes them from usual quantum-chemical techniques, is that they are neither variational nor charge-conserving. The Local-Space-Approach (LSA) by KIRTMAN and DE MELO [1981] is free from such problems. It is a variational HF-like approach where the energy minimum is searched as a function of arbitrary parameters that are contained in a finite matrix X. The density matrix P is made to depend on X in such a way that first, P retains its idempotent character and, second, X has the meaning of a local perturbation. These requirements give rise to a recursive set of equations which is truncated after a few iterations. It is shown that the resulting change in electron population from an initial guess (corresponding for instance to the isolated molecule plus the unperturbed semi-crystal) is zero to all orders of approximation. There seem to be serious difficulties, however, in the ab initio application of these equations to real systems.

Among embedding approaches, it is worth mentioning the Green's function matching method developed by INGLESFIELD and BENESH [1988]. Although it relies on the same approximation as the standard approach, Eq. (2.25) is not based on the Dyson equation.

Rather, it looks for a variational solution to the single particle Schrödinger equation in the C region, while imposing the continuity of the Green function and its normal derivative across the cluster boundary. This technique is particularly suited for the study of point defects at the surface of metals, where the metal is simulated by a semi-infinite helium model, and the boundary of the C region is simply a spherical surface (TRIONI, BRIVIO, CRAMPIN and INGLESFIELD [1996]).

3. The theoretical framework

3.1. Electronic and atomic degrees of freedom

Here we consider a system consisting of nuclei and electrons that interact through the Coulomb potential. By indicating the nuclear and the electronic coordinates with $\mathbf{R_I}$ and $\mathbf{r_i}$, respectively, the problem can be formulated in the framework of nonrelativistic quantum mechanics via the time-dependent Schrödinger equation (the relativistic formalism is presented and extensively discussed in another chapter of this Handbook):

$$i\hbar\frac{\partial}{\partial t}\Phi\big(\{\mathbf{r_i}\},\{\mathbf{R_I}\},t\big)=H\Phi\big(\{\mathbf{r_i}\},\{\mathbf{R_I}\},t\big). \tag{3.1}$$

In the Born–Oppenheimer approximation (BORN and HUANG [1954]), the nuclear and the electronic degrees of freedom are separated by using the following ansatz:

$$\Phi\big(\{\mathbf{r_i}\},\{\mathbf{R_I}\},t\big)=\Psi\big(\{\mathbf{r_i}\},\{\mathbf{R_I}\},t\big)\Upsilon\big(\{\mathbf{R_I}\},t\big). \tag{3.2}$$

From a physical point of view, such a separation is based on the difference between the fast electronic degrees of freedom (corresponding to a light mass m_e) and the slower nuclear ones (with masses M_I). From the mathematical point of view, the electronic wavefunction Ψ depends on the nuclear coordinates in a *parametrical way*. By considering the square modulus of Eq. (3.2), the Born–Oppenheimer approximation corresponds to factorize the probability of finding the system in a given $\{\mathbf{R_I}\}, \{\mathbf{r_i}\}$ state at time t into a probability for the ionic subsystem to be in state $\Upsilon(\{\mathbf{R_I}\},t)$ by the *conditional* probability for the electronic subsystem to be in state $\Psi(\{\mathbf{r_i}\},t)$ given $\Upsilon(\{\mathbf{R_I}\},t)$.

Within the Born–Oppenheimer approximation, one can consider the nuclei either as classical particles or as quantum objects. In the latter case, the nuclear wavefunction $\Upsilon(\{\mathbf{R_I}\},t)$ can be computed by means of various approaches (path-integrals, quantum Monte Carlo, etc.). However, in many physical situations, the De Broglie wavelengths of the nuclei is such that the nuclear wavefunctions are extremely localized and the classical treatment represents a good approximation. The nuclei are thus considered as point objects, and the set of nuclear coordinates and momenta at time t $(\{\mathbf{R_I}\},\{\mathbf{P_I}\},t)$ is sufficient to specify the state of the nuclear subsystem in the phase space.

Let us restrict to the Born–Oppenheimer approximation. A schematic account of the mathematical tools to describe the interesting physical situations for which the Born–Oppenheimer approximation is inadequate (for instance, electron–phonon coupling, level crossing (diabatic approximation), Peierls mechanism for symmetry breaking, ultra-fast dynamics with strong atom–electron coupling) may be found elsewhere. Apart from those cases, the electronic problem can be solved by considering the nuclear coordinates as parameters. The simplest case consist in finding the electronic ground state for fixed nuclei.

3.2. *The perfect crystal ground-state in Density Functional Theory*

Let $\{\xi_i\}$ the set of external parameters – such as the atomic positions, the volume, the applied fields, etc. – which defines the state of the system from the classical point of view. In this subsection, we introduce the notation for the problem of finding the electronic ground state corresponding to the $\{\xi_i\}$ that are kept fixed. Since the many-electron problem in an arbitrary external potential (which in our case is given by the superposition of Coulomb potential provided by each nucleus, and by any other potential acting on the electrons) cannot be solved analytically, suitable numerical methods have been introduced. For few-electron systems, quantum Monte Carlo is likely the most fundamental and reliable computational scheme that is nowadays able to provide a good numerical approximation to the electronic wavefunction (LESTER and HAMMOND [1990]). However, its application to systems containing many electrons is not yet feasible.[6] Less fundamental but more practical approaches to the solution of the electronic ground state for systems containing up to few thousands electrons are provided by the Hartree–Fock and the Density Functional theories. Although they have been described in the first chapter of this Handbook (CANCÉS, DEFRANCESCHI, KUTZELNIGG, LE BRIS and MADAY [2003]), here we recall the basic equations and introduce the notation that will be useful in the following.

In the DFT, the ground-state energy of the electronic system is derived from the following minimum principle (PICKETT [1989]):

$$E_{\mathrm{el}}\{\psi_\alpha\} = \sum_{\alpha}^{\mathrm{occ.}} \langle \psi_\alpha | T + v_{\mathrm{ext}} | \psi_\alpha \rangle + E_{\mathrm{Hxc}}[n], \tag{3.3}$$

where the ψ_α's are the Kohn–Sham orbitals (to be varied until the minimum is found), T is the kinetic energy operator, v_{ext} is the potential external to the electronic system that includes the one created by nuclei (or ions) and is in general completely specified in terms of the parameters ξ_i, E_{Hxc} is the Hartree and exchange-correlation energy functional of the electronic density $n(\mathbf{r})$, and the summation runs over the occupied states α. The occupied Kohn–Sham orbitals are subject to the orthonormalization constraints,

$$\int \psi_\alpha^*(\mathbf{r}) \psi_\beta(\mathbf{r})\, d\mathbf{r} = \langle \psi_\alpha | \psi_\beta \rangle = \delta_{\alpha\beta}, \tag{3.4}$$

where α and β label occupied states. This definition of scalar product for periodic functions is different from Eq. (2.6), and is valid for nonextended wavefunctions. For

[6]The reader who is interested to a up-to-date account on the Quantum Monte Carlo applications to extended systems can look to the review by Foulkes and coworkers FOULKES, MITAS, NEEDS and RAJAGOPAL [2001].

a closed-shell system (the factor 2 accounts for the spin degeneracy), the density is generated from:

$$n(\mathbf{r}) = 2\sum_{\alpha}^{\text{occ.}} \psi_\alpha^*(\mathbf{r})\psi_\alpha(\mathbf{r}). \tag{3.5}$$

The minimization of $E_{\text{el}}\{\psi\}$ under the orthonormality constraints Eq. (3.4) can be achieved using the Lagrange multiplier method. The problem turns into the minimization of

$$E_{\text{el}}^{+}\{\psi\} = \sum_{\alpha}^{\text{occ.}} \langle\psi_\alpha|T + v_{\text{ext}}|\psi_\alpha\rangle + E_{\text{Hxc}}[n] - \sum_{\alpha\beta}^{\text{occ.}} \varepsilon_{\beta\alpha}\big(\langle\psi_\alpha|\psi_\beta\rangle - \delta_{\alpha\beta}\big), \tag{3.6}$$

where $\varepsilon_{\alpha\beta}$ are the Lagrange multipliers corresponding to the set of constraints Eq. (3.4).

In the crystal momentum representation (Section 2.1.1), the ground-state wavefunctions can be obtained from the minimization of the electronic energy per unit cell; Eq. (3.3) becomes

$$E_{\text{el}}\{u\} = \frac{\Omega_0}{(2\pi)^3}\int_{\text{BZ}} \sum_{m}^{\text{occ.}} s\,\langle u_{m\mathbf{k}}|T_{\mathbf{k},\mathbf{k}} + v_{\text{ext},\mathbf{k},\mathbf{k}}|u_{m\mathbf{k}}\rangle\, d\mathbf{k} + E_{\text{Hxc}}[n]. \tag{3.7}$$

In order to keep the amount of different symbols sufficiently low, in Eq. (3.7) and in Sections 3.7, 3.8, 3.11 and 4.4 we redefine E_{el} and E_{Hxc} to be energies *per unit cell*, unlike in Eq. (3.3) where these quantities were defined *for the whole system*. We have also redefined the kinetic and potential operators, that act now on the periodic part of the Bloch functions, according to Eq. (2.8). The Euler–Lagrange equations associated with the minimization of Eq. (3.7) under the orthonormality constraint (Eq. (2.6)), followed by a unitary transformation, give

$$H_{\mathbf{k},\mathbf{k}}|u_{m,\mathbf{k}}\rangle = \varepsilon_{m,\mathbf{k}}|u_{m,\mathbf{k}}\rangle, \tag{3.8}$$

where

$$H_{\mathbf{k},\mathbf{k}} = T_{\mathbf{k},\mathbf{k}} + v_{\text{ext},\mathbf{k},\mathbf{k}} + \left.\frac{\delta E_{\text{Hxc}}}{\delta n}\right|_n. \tag{3.9}$$

3.3. The Hartree–Fock approach in solid state studies

From a formal and computational point of view, the Hartree–Fock (HF) equations for crystals are similar to the Kohn–Sham (KS) ones based on density functional theory (DFT). In both cases, we look for a set of crystalline orbitals $\{\psi_i(\mathbf{r})\}$ as eigenfunctions of a one-electron Hamiltonian, the Fock and the KS Hamiltonian, respectively. Both Hamiltonians depend on the first order density function $\gamma_1(\mathbf{r},\mathbf{r}')$ (see Chapter 1,

Section 1.3.1 for a more complete discussion), which in turn is obtained as a sum over the occupied orbitals. For a closed-shell system:

$$\gamma_1(\mathbf{r},\mathbf{r}') = 2\sum_j^{\text{occ.}} \psi_j(\mathbf{r})\psi_j^*(\mathbf{r}'). \tag{3.10}$$

Both Hamiltonians contain the kinetic and the Hartree term, and differ only in the exchange and correlation part. The Hartree–Fock Hamiltonian includes an "exact-exchange" operator $\widehat{K}$, which depends on $\gamma_1(\mathbf{r},\mathbf{r}')$ as follows:

$$\widehat{K}\phi(\mathbf{r}) = -1/2\int d\mathbf{r}'\gamma_1(\mathbf{r},\mathbf{r}')\phi(\mathbf{r}')/|\mathbf{r}-\mathbf{r}'|. \tag{3.11}$$

The Coulomb correlation of electronic motions is completely neglected. The counterpart of $\widehat{K}$ in the Kohn–Sham Hamiltonian is the exchange and correlation potential $\hat{v}_{\text{xc}}(\vec{r};[n_0]) = \frac{\delta E_{\text{xc}}}{\delta n(r)}|_{n_0}$, which depends only on the diagonal element of γ_1, the electron density $n^{(0)}(\mathbf{r}) = \gamma_1(\mathbf{r},\mathbf{r})$. This local potential accounts exactly, at least in principle, for all exchange and correlation effects. As is shown below, the nonlocal exchange term makes the HF equations somewhat more difficult to solve in crystalline applications than the KS ones, where $\hat{w}_{\text{xc}}$ is simply a multiplicative operator.

Apart from this minor complication, all tools and tricks for the solution of the crystalline problem coincide in the two approaches: **k** factorization, sampling and integration in **k**-space, Ewald treatment of long-range Coulomb interactions, techniques to improve the rate of convergence in the self-consistent field procedure, etc. In spite of that, the overwhelming majority of solid state studies has been performed, and still is, in the frame of DFT. This contrasts the almost universal acceptance of the HF approach in molecular studies. Why is it so?

The main limitation of the HF solution lies in the neglect of correlation effects. Consideration of the interaction of each electron with the average field of the others leads to an over-estimation of electron–electron repulsions: in fact, electrons move so as to keep apart from each other (otherwise stated, each electron carries along a *correlation hole*). If we are interested only in an estimate of the *dynamic* correlation energy, $E_c = E_0 - E_{\text{HF}}$, simple and powerful techniques are available, based on the precepts of DF theory (PERDEW and WANG [1992]).

It is well known that HF performs poorly for the electron gas, the prototype of simple metals: contrary to the exact result, HF predicts zero density of states at the Fermi level, and very low density in its vicinity; furthermore, electron correlation makes an essential contribution to the stability of that system at electron densities corresponding to those of the most common metals (RAIMES [1961]). In contrast to such failures, DFT performs very well in this case: as a matter of fact, the parametrization of all exchange correlation potentials in use is such as to provide the exact electron gas energy in the limit of uniform electron density. It is also shown in this same chapter that response theory, which plays a crucial role in solid state studies, is much more simply and neatly formulated in the frame of DFT than in a standard quantum mechanical approach.

These facts, and the nontrivial problem of the nonlocality of the HF exchange potential, has induced solid-state physicists to concentrate their efforts on the KS approach. As a result, very efficient and accurately parameterized computational schemes that are mainly based on the DFT have become available and spectacular successes have been achieved, which has further contributed to the prevailing opinion that a HF approach is not really needed in this field of studies. There are however some reasons for not considering this attitude as fully justified.

In recent years, the advent of powerful computers and of general purpose computational schemes (PISANI, DOVESI and ROETTI [1988], SAUNDERS, DOVESI, ROETTI, CAUSÀ, HARRISON, ORLANDO and ZICOVICH-WILSON [1998]) has made it possible to formulate a fairer judgment about the usefulness of the HF approach in solid state physics, by assessing its performance for a variety of systems: the number of applications, and the quality of the results especially in the description of ionic and covalent solids is impressive (PISANI, DOVESI, ROETTI, CAUSÀ, ORLANDO, CASASSA and SAUNDERS [2000]).

As compared to the KS approach, the unrestricted HF method has proved much more effective in the simulation of the electronic and magnetic ground state properties of transition metal ionics (oxides, sulfides, halides) (PISANI, DOVESI, ROETTI, CAUSÀ, ORLANDO, CASASSA and SAUNDERS [2000], CAUSÁ, DOVESI, ORLANDO, PISANI and SAUNDERS [1988]). In particular, for *all* investigated cases (more than 30 magnetic systems) the order between ferro- and antiferro-magnetic states is correctly reproduced. This is due to the fact that for a correct description of the super-exchange interaction between metal ions, which is responsible for magnetic order, it is essential to take into account the nonlocal nature of the exchange term.

The need for incorporating in the Hamiltonian a nonlocal exchange contribution is testified by the enormous success met in the field of molecular studies by *hybrid-exchange* recipes, such as the one proposed by BECKE [1993]: the exchange operator is there a suitably weighted mixture of HF and DFT terms. The extension of this empirical but very effective technique to periodic systems is straightforward, but requires that the evaluation of the nonlocal exchange term be feasible with the computer code adopted.

An excerpt from Fulde's book on electron correlation in solids about the value of HF computations for solids can finally be reported (FULDE [1993]):

> "Provided that the correlations are not too strong, HF calculations are a good starting point, and allow solids and molecules to be treated the same way and with the same accuracy. The development of quantum chemistry has proven that ab initio calculations based on controlled approximations capable of systematic improvement have made simpler computational schemes based on uncontrolled simplifications obsolete. Whether or not the same will eventually hold true for solid-state theory remains to be seen."

In the following, we discuss briefly the two peculiar features of the HF approach for crystals; first, the accurate evaluation of the exchange series, second, the prospective introduction of post-HF correlation corrections.

3.4. *The treatment of the exchange series*

In this section, we provide a brief discussion of the treatment of the exchange series (SAUNDERS, DOVESI, ROETTI, CAUSÀ, HARRISON, ORLANDO and ZICOVICH-WILSON [1998]); this allows us to enlighten some specificities of the exchange term in condensed systems as well as to give a feeling of the computational strategies adopted in a modern computer code for crystals.

One can express the crystalline orbitals, hence all derived quantities, as a linear combination of local functions $\chi_\nu(\mathbf{r}-\mathbf{L})$, centered in the general crystalline cell identified by the lattice vector $\mathbf{L}$. Such local functions are conveniently chosen as Gaussian type functions GTF – see Chapter 1, Section 2.2.2 of this Handbook for generalities (CANCÉS, DEFRANCESCHI, KUTZELNIGG, LE BRIS and MADAY [2003]) and BRIDDEN and JONES [2000] for their use in solid-state calculations. In particular, the one-electron density function becomes, after exploiting the translational invariance of the system:

$$\begin{aligned}\gamma_1(\mathbf{r},\mathbf{r}') &= \sum_{\lambda,\rho}\sum_{\mathbf{L}',\mathbf{M}}\left(2\sum_j^{\mathrm{occ.}} a_{j,\lambda,\mathbf{M}}a^*_{j,\rho,\mathbf{M}+\mathbf{L}'}\right)\chi_\lambda(\mathbf{r}-\mathbf{M})\chi^*_\rho(\mathbf{r}'-\mathbf{M}-\mathbf{L}')\\ &\equiv \sum_{\lambda,\rho}\sum_{\mathbf{L}'} P_{\lambda\rho}(\mathbf{L}')\sum_M \chi_\lambda(\mathbf{r}-\mathbf{M})\chi^*_\rho(\mathbf{r}'-\mathbf{M}-\mathbf{L}'),\end{aligned} \tag{3.12}$$

where $P_{\lambda\rho}(\mathbf{L}')$ is the density matrix expressed in the basis of the GTF. By substituting this expression in Eq. (3.11) we obtain, for the matrix element of the exchange operator between two GTFs:

$$\begin{aligned}K_{\mu\nu}(\mathbf{L}) &\equiv \int d\mathbf{r}\,\chi^*_\mu(\mathbf{r})\widehat{K}\chi_\nu(\mathbf{r}-\mathbf{L}) = -\frac{1}{2}\sum_{\lambda,\rho}\sum_{\mathbf{L}'} P_{\lambda\rho}(\mathbf{L}')\\ &\quad\times\sum_{\mathbf{M}}\int\int d\mathbf{r}\,d\mathbf{r}'\,\frac{\chi^*_\mu(\mathbf{r})\chi_\lambda(\mathbf{r}-\mathbf{M})\chi_\nu(\mathbf{r}'-\mathbf{L})\chi^*_\rho(\mathbf{r}'-\mathbf{M}-\mathbf{L}')}{|\mathbf{r}-\mathbf{r}'|}.\end{aligned} \tag{3.13}$$

The treatment of the exchange (as well as of the Coulomb) term, containing two infinite lattice summations requires a careful analysis and powerful and effective tools in order to ensure high numerical accuracy and computational efficiency. Here we provide an overview of an effective strategy that is based on three simple observations:

- *The overlap between two GTFs decays exponentially with distance.*
  This property can be exploited for the truncation of the $\mathbf{M}$ series in the exchange summation.
- *The electrostatic potential generated by a charge distribution at a point external to it can be approximated by a (truncated) multipolar expansion of the charge distribution.*
  The speed of convergence (and then the truncation criteria) of the expansion depends on the distance between the target point and the charge distribution, the position of the expansion centre, the size and shape of the charge distribution.

Only few of the integrals in Eq. (3.13) need be evaluated analytically; the others are grouped appropriately together and evaluated efficiently using a multipolar approximation of the two interacting distributions.

- *The density matrix decays with distance from the diagonal.*
  In a localized basis set, the elements of the density matrix of an insulator decay exponentially with the distance between the two centers, the larger the HOMO-LUMO gap, the faster the decay. The same is true for conductors, but the decay rate is much slower (PISANI, DOVESI and ROETTI [1988], CAUSÁ, DOVESI, ORLANDO, PISANI and SAUNDERS [1988]). This behavior is exploited for an efficient truncation of the exchange series: Eq. (3.13) shows that the **M** summation is limited by the exponential decay of the product $\chi_\mu^*(\mathbf{r})\chi_\lambda(\mathbf{r}-\mathbf{M})$, and for a similar reason the $|\mathbf{L}-\mathbf{M}-\mathbf{L}'|$ distance cannot be too large, otherwise the integral is negligible. These two conditions imply that the **L** and $\mathbf{L}'$ vectors cannot be very different, although their moduli could be large. However, the exponential decay of the density matrix permits us to disregard integrals involving $\mathbf{L}'$ (and, as a consequence, **L**) vectors with large moduli.

3.5. *Post-Hartree–Fock techniques for crystals*

Until very recently, the accurate treatment of electron correlation has been confined to systems where the number N of electrons is very small. The reason is that the computer requirements for classical methods scale so fast with N (of order 5 or higher) that their application to really big systems is in fact impossible, no matter how rapid the progress of computer technology is. As stated by Saebø and Pulay a decade ago, “the key to the efficient treatment of large systems is to use a localized description that allows the neglect of distant parts of the molecule” (SAEBO and PULAY [1993]). On the other hand, for two parts of the molecule to be considered as “distant” it is normally required that they are separated by a certain number of chemical bonds, or typically, that they are about 15 Bohr apart. Not very long ago, accurate calculations for molecules of this size were still prohibitively expensive. Therefore, the use of the locality assumption has become a really competitive option only recently, and a number of powerful computational tools have been implemented for the ab initio treatment of electron correlation in large molecules (SCHÜTZ, HETZER and WERNER [1999], SCHÜTZ and WERNER [2001], AYALA and SCUSERIA [1999], SAEBO and PULAY [2001]); pioneering applications to one-dimensional periodic structures have been reported latterly (SCUSERIA [2001]). The reformulation for crystalline systems of the local-correlation method as currently employed by the Stuttgart group of Theoretical Chemistry (SCHÜTZ, HETZER and WERNER [1999], SCHÜTZ and WERNER [2001]) appears particularly promising. Following Pulay’s ideas (SAEBO and PULAY [1993]), it is based on a representation of the HF wavefunction as an anti-symmetrized product of orthonormal localized orbitals, and of the virtual space in terms of atomic orbitals (AO) orthogonalized to the occupied manifold. Among its attractive features, the fact that with few variants it can be applied to all most common schemes for the treatment of correlation, that its efficiency has been tested with a variety of big molecular systems, and that asymptotically it scales linearly with molecular size. An essential step for the extension to crystalline structures of this approach is the generation of an orthonormal set of well localized, translationally equivalent functions [the so-called (*localized*) *Wannier functions*, (L)WF] which span the occupied HF manifold of the periodic system (see Section 2.1.2). Powerful and general-purpose techniques for the generation of optimally localized WFs have recently become available (MARZARI and VANDERBILT [1997], ZICOVICH-WILSON, DOVESI and SAUNDERS [2001]), and it is reasonable to expect that efficient post-HF schemes for crystals will be implemented in a near future. A particularly exciting prospect is the accurate nonparametric treatment of Van der Waals interactions in solids, which make an essential contribution to the stability and conformation of many ionic solids and of the majority of molecular crystals (GAVEZZOTTI [2002]), and seem to be out of reach of local or quasi-local DFT schemes (KRISTY and PULAY [1994]).

3.6. *First-order derivatives of the total energy: forces, stress, pressure ...*

In Sections 3.2 and 3.3, we have defined, for the perfect crystal, the equations that permit to compute the ground-state energy and wavefunctions within DFT or HF variational theories. The reference external potential v_{ext} described by the $\{\xi_i\}$ parameters, mentioned at the beginning of Section 3.2, has been considered as fixed, while the wavefunctions were determined self-consistently, so as to minimize the total energy. In the present section and the following ones, we will examine the *changes* of ground-state properties (such as the total energy and the electron density) associated with *changes* of the $\{\xi_i\}$ parameters. This set of parameters includes (among others) the atomic positions, the strength of the homogeneous electric field, or the definition of the basis vectors of the primitive cell. For the sake of conciseness, the following treatment is mainly done in the framework of DFT, although it can in principle be generalized to any variational theory, such as the Hartree–Fock one.

Suppose that the potential v_{ext} is expanded in terms of a small parameter λ (i.e., $\xi \to \xi_i + \lambda$) as follows,

$$v_{\mathrm{ext}}(\lambda) = v_{\mathrm{ext}}^{(0)} + \lambda v_{\mathrm{ext}}^{(1)} + \lambda^2 v_{\mathrm{ext}}^{(2)} + \cdots. \tag{3.14}$$

We are interested in the change of physical quantities, due to the perturbation of the external potential. We expand all of the DFT (or HF) quantities $X(\lambda)$ in the same form as $v_{\mathrm{ext}}(\lambda)$,

$$X(\lambda) = X^{(0)} + \lambda X^{(1)} + \lambda^2 X^{(2)} + \cdots, \tag{3.15}$$

where X can be E_{el}, $\psi_\alpha(\mathbf{r})$, $n(\mathbf{r})$, H, etc. The total energy E_{tot} can also be cast in terms of such an expansion. In a first approximation, E_{tot} is the sum of the electronic energy (DFT or HF) and the ion–ion (or Ewald) energy, while a more refined treatment, within the Born–Oppenheimer approximation, includes vibrational contributions. At the order zero of the expansion, all the equations developed in Section 3.2 are formally unchanged (except for a (0) superscript). For instance, the electronic energy, Eq. (3.3) writes;

$$E_{\mathrm{el}}^{(0)}\{\psi_\alpha\} = \sum_{\alpha}^{\mathrm{occ.}} \langle \psi_\alpha^{(0)} | T + v_{\mathrm{ext}}^{(0)} | \psi_\alpha^{(0)} \rangle + E_{\mathrm{Hxc}}\big[n^{(0)}\big]. \tag{3.16}$$

We now examine the first term of the total energy expansion, $E_{\text{tot}}^{(1)}$, equal to the first-order derivative of the total energy with respect to one of the $\{\xi_i\}$ parameters.

The most important of such first-order derivatives are the forces on the atoms, that the principle of virtual work identifies with minus the first derivative of the total energy with respect to an atomic displacement:

$$F_{\kappa,\alpha} = -\frac{\partial E_{\text{tot}}}{\partial \tau_{\kappa,\alpha}}. \tag{3.17}$$

This identity is valid for molecules as well as for extended systems. By contrast, the pressure, and its anisotropic generalization, the stresses, are characteristic of extended systems. The pressure is connected to the change of total energy induced by a relative change of (cell) volume, while the stresses stems from a relative change of the primitive lattice vectors (also called a strain). They can be alternatively formulated with respect to homogeneous dilatations or contractions:

$$r_{i\alpha} \to r_{i\alpha} + \Sigma_\alpha e_{\alpha\beta} r_{i\beta}. \tag{3.18}$$

Other physical properties are amenable to a formulation as first-order derivatives of the total energy: the dielectric polarization comes from the consideration of a static homogeneous electric field polarization, while the magnetization is linked to the derivative with respect to a magnetic field. However, the mathematical treatment of such perturbations for extended systems is quite complicated, as the associated change in potential breaks the periodicity of the Hamiltonian.

The computation of the first-order derivative of the total energy, neglecting vibrational contributions, include two terms, the electronic one, and the ion–ion one. The latter is computationally inexpensive. The former can be obtained quite easily, as a by-product of the calculation of the unperturbed quantities, by means of the *Hellmann–Feynman* theorem (HELLMANN [1937], FEYNMAN [1939]), although the first-order total derivative of the electronic energy, Eq. (3.6), at first sight, involves the partial derivatives with respect to both direct change of external potential and induced change of wavefunction. Indeed, Eq. (3.6) comes from a minimum principle, so that the derivative of the electronic energy with respect to the wavefunction $\frac{\partial}{\partial \psi} E_{\text{el}}^{+}$ vanishes:

$$\begin{aligned}
&\frac{d}{d\lambda} E_{\text{el}}^{+}\{\psi_\alpha(\lambda); v_{\text{ext}}(\lambda)\} \\
&\quad = \sum_{\alpha}^{\text{occ.}} \langle \psi_\alpha | \frac{\partial v_{\text{ext}}}{\partial \lambda} | \psi_\alpha \rangle + \sum_{\alpha} \int d\mathbf{r} \frac{\partial \psi_\alpha(\lambda, \mathbf{r})}{\partial \lambda} \frac{\partial E_{\text{el}}^{+}\{\psi_\alpha(\lambda, \mathbf{r}); v_{\text{ext}}(\lambda)\}}{\partial \psi_\alpha(\lambda, \mathbf{r})} \\
&\quad = \sum_{\alpha}^{\text{occ.}} \langle \psi_\alpha | \frac{\partial v_{\text{ext}}}{\partial \lambda} | \psi_\alpha \rangle.
\end{aligned} \tag{3.19}$$

In other words, $\psi^{(1)}$ is not needed in order to compute $E_{\text{tot}}^{(1)}$. This makes the computation of forces only slightly more expensive as the computation of the total energy itself, allowing for classical molecular dynamics simulations, on the basis of quantum forces.

However, the Hellman–Feynman theorem is valid if the electronic energy is completely minimized with respect to all possible variations of the wavefunction. This is not the case whenever the wavefunction is expanded in a basis set (see Section 4) and thus is represented in an incomplete Hilbert space of finite dimension. Therefore, there is an additional contribution to the force. Actually, if the basis set does *not* depends on the perturbative parameter (i.e., a plane wave basis set does not depend on the atomic coordinates), such additional contributions vanishes. In the other case (i.e., an atomic orbital basis set depends on the atomic coordinates), the additional contributions are referred to as *Pulay forces* (PULAY [1969]).

The Hellman–Feynman theorem translates into the *stress theorem* in the case of homogeneous dilatation/contraction of the space (NIELSEN and MARTIN [1985]). Its evaluation is quite easy, as in the case of forces.

3.7. *Higher derivatives of the total energy: density functional perturbation theory*

The generalization of the variational property of the energy functional of the wavefunctions, Eq. (3.6), and of the Hellman–Feynman theorem, Eq. (3.19), to higher order of perturbation reveals a particularly interesting mathematical structure (GONZE [1995]), that we will examine before introducing the physical interest of second or higher order derivatives of the total energy.

Since the zeroth- and first-order derivative of the total energy can be computed from the unperturbed wavefunction, one may wonder which derivatives of the total energy are determined by the knowledge of the nth order derivative of the wavefunctions. E_{el} satisfies a variational principle under constraints (no other condition is required). It is thus possible to derive: (1) the so-called $2n+1$ *theorem* of perturbation theory, giving the $(2n+1)$th order derivative of E_{el} from the knowledge of the derivatives of the wavefunctions up to order n and (2) a constrained variational principle for the $2n$th order derivative of E_{el} with respect to the nth order derivative of ψ_α (SINANOGLU [1961], GONZE [1995]). Explicitly, when the expansion of the wavefunction up to order n is known, the following expression gives the $(2n+1)$th order derivative of E_{el}:

$$E_{\mathrm{el}}^{(2n+1)} = \left(E_{\mathrm{el}}^{+}\left\{ \sum_{i=0}^{n} \lambda^i \psi_\alpha^{(i)} \right\} \right)^{(2n+1)}, \tag{3.20}$$

and the variational (minimum) principle for the $2n$th order derivative of E_{el} as a function of the nth order wavefunction writes

$$E_{\mathrm{el}}^{(2n)} = \min_{\psi_\alpha^{(n)}} \left(E_{\mathrm{el}}^{+}\left\{ \sum_{i=0}^{n} \lambda^i \psi_\alpha^{(i)} \right\} \right)^{(2n)}, \tag{3.21}$$

under the constraints $\sum_{i=0}^{n} \langle \psi_\alpha^{(n-i)} | \psi_\beta^{(i)} \rangle = 0$ for all occupied states α and β. This variational principle can be used in the same way as the unperturbed variational principle, namely, its solution gives the nth order wavefunction.

The explicit expressions for $E_{\text{el}}^{(2n)}$ can be worked out by introducing Eq. (3.6) into Eqs. (3.20), (3.21). We will examine more closely the case of the second-order derivative of the electronic energy: $E_{\text{el}}^{(2)}$ is the minimum of the following expression (GONZE, ALLAN and TETER [1992], SAVRASOV [1992], GONZE [1997a]):

$$\begin{aligned} E_{\text{el}}^{(2)}\{\psi^{(0)};\psi^{(1)}\} = \sum_{\alpha}^{\text{occ.}} & \big[\langle\psi_\alpha^{(1)}|H^{(0)} - \varepsilon_\alpha^{(0)}|\psi_\alpha^{(1)}\rangle \\ & + \big(\langle\psi_\alpha^{(1)}|v_{\text{ext}}^{(1)}|\psi_\alpha^{(0)}\rangle + \langle\psi_\alpha^{(0)}|v_{\text{ext}}^{(1)}|\psi_\alpha^{(1)}\rangle\big) + \langle\psi_\alpha^{(0)}|v_{\text{ext}}^{(2)}|\psi_\alpha^{(0)}\rangle\big] \\ & + \frac{1}{2}\iint \frac{\delta^2 E_{\text{Hxc}}}{\delta n(\mathbf{r})\delta n(\mathbf{r}')}\bigg|_{n^{(0)}} n^{(1)}(\mathbf{r})n^{(1)}(\mathbf{r}')\,d\mathbf{r}\,d\mathbf{r}' \\ & + \int \frac{d}{d\lambda}\frac{\delta E_{\text{Hxc}}}{\delta n(\mathbf{r})}\bigg|_{n^{(0)}} n^{(1)}(\mathbf{r})\,d\mathbf{r} + \frac{1}{2}\frac{d^2 E_{\text{Hxc}}}{d\lambda^2}\bigg|_{n^{(0)}}, \end{aligned} \tag{3.22}$$

while the first-order changes in wavefunctions $\psi_\alpha^{(1)}$ (these quantities will be referred to as the first-order wavefunctions, for brevity), are varied under the constraints (for all occupied states α and β),

$$\langle\psi_\alpha^{(0)}|\psi_\beta^{(1)}\rangle = 0, \tag{3.23}$$

and the first-order density is given by

$$n^{(1)}(\mathbf{r}) = \sum_{\alpha}^{\text{occ.}} \psi_\alpha^{*(1)}(\mathbf{r})\psi_\alpha^{(0)}(\mathbf{r}) + \psi_\alpha^{*(0)}(\mathbf{r})\psi_\alpha^{(1)}(\mathbf{r}). \tag{3.24}$$

By virtue of Eq. (3.23), the first-order wavefunctions are orthogonal to the unperturbed wavefunctions of the occupied states. Since $E_{\text{el}}^{(2)}\{\psi^{(0)};\psi^{(1)}\}$ is variational with respect to $\psi^{(1)}$, we deduce the Euler–Lagrange equations (also called self-consistent Sternheimer equations in this particular case (STERNHEIMER [1954], GONZE [1997a])),

$$P_c\big(H^{(0)} - \varepsilon_\alpha^{(0)}\big)P_c|\psi_\alpha^{(1)}\rangle = -P_c H^{(1)}|\psi_\alpha^{(0)}\rangle, \tag{3.25}$$

where P_c is the projector upon the unoccupied states (conduction bands), $H^{(0)}$, $\varepsilon_\alpha^{(0)}$, and $\psi^{(0)}$ are ground-state quantities, and the first-order Hamiltonian $H^{(1)}$ is given by

$$H^{(1)} = v_{\text{ext}}^{(1)} + v_{\text{Hxc}}^{(1)} = v_{\text{ext}}^{(1)} + \int \frac{\delta^2 E_{\text{Hxc}}}{\delta n(\mathbf{r})\delta n(\mathbf{r}')}\bigg|_{n^{(0)}} n^{(1)}(\mathbf{r}')\,d\mathbf{r}' + \frac{d}{d\lambda}\frac{\delta E_{\text{Hxc}}}{\delta n(\mathbf{r})}\bigg|_{n^{(0)}}. \tag{3.26}$$

Eq. (3.25) can be solved by different techniques. An interesting possibility is to expand the first-order wavefunction in terms of a basis set, in which case this equation reduces to an inhomogeneous systems of linear equations (GIANNOZZI, DE GIRONCOLI, PAVONE and BARONI [1991]). The self-consistency of the first-order Hamiltonian, Eq. (3.26) must also been addressed, through standard techniques. In Eq. (3.22), the contributions

$$\sum_{\alpha}^{\text{occ.}} \langle \psi_\alpha^{(0)} | v_{\text{ext}}^{(2)} | \psi_\alpha^{(0)} \rangle + \frac{1}{2} \frac{d^2 E_{\text{Hxc}}}{d\lambda^2}\bigg|_{n^{(0)}}, \tag{3.27}$$

that will be denoted by $E_{\text{nonvar}}^{(2)}$, do not depend on the first-order wavefunctions, and will not change in the course of the minimization procedure or in the self-consistent procedure. One defines also

$$v_{\text{Hxc0}}^{(1)} = \frac{d}{d\lambda} \frac{\delta E_{\text{Hxc}}}{\delta n(\mathbf{r})}\bigg|_{n^{(0)}} \tag{3.28}$$

not to be confused with $v_{\text{Hxc}}^{(1)}$, that contains one more term, see Eq. (3.26). $v_{\text{ext}}^{(1)}$ and $v_{\text{Hxc0}}^{(1)}$ do not depend on the first-order wavefunctions.

For the first-order wavefunctions that satisfy Eqs. (3.23)–(3.26), or equivalently minimize Eq. (3.22) under constraints Eq. (3.23), there are alternate formulations of $E_{\text{el}}^{(2)}$:

$$E_{\text{el}}^{(2)} = \frac{1}{2} \sum_{\alpha}^{\text{occ.}} \left(\langle \psi_\alpha^{(1)} | v_{\text{ext}}^{(1)} + v_{\text{Hxc0}}^{(1)} | \psi_\alpha^{(0)} \rangle + \langle \psi_\alpha^{(0)} | v_{\text{ext}}^{(1)} + v_{\text{Hxc0}}^{(1)} | \psi_\alpha^{(1)} \rangle \right) + E_{\text{nonvar}}^{(2)}, \tag{3.29}$$

instead of Eq. (3.22). Taking into account the time-reversal symmetry, an even simpler expressions for $E_{\text{el}}^{(2)}$ can be found:

$$E_{\text{el}}^{(2)} = \sum_{\alpha}^{\text{occ.}} \langle \psi_\alpha^{(0)} | v_{\text{ext}}^{(1)} + v_{\text{Hxc0}}^{(1)} | \psi_\alpha^{(1)} \rangle + E_{\text{nonvar}}^{(2)}, \tag{3.30}$$

or its hermitian conjugate. However, if the wavefunctions are not exactly the ones that minimize Eq. (3.22) and satisfy Eqs. (3.23)–(3.26), the error in Eqs. (3.29) and (3.30) is larger than the error in Eq. (3.22): the latter is variational, while the formers are not.

The density functional perturbation theory (DFPT) is an interesting and powerful tool. As regards the response to an atomic displacement of a molecule in its equilibrium state (that is, with respect to the state in which all forces cancels out), the second-order derivative of the total energy is equal to minus the force induced by the atomic displacement: it gives the spring constant that pushes back the atom to its equilibrium position. We examine such force constants in more details in Section 3.8. Similarly, if the perturbation is a contraction of a lattice, the second-order derivative of the total energy will be linked to the induced pressure due to the contraction, giving the elasticity (bulk) modulus of the solid. Also, the response to an applied electric field will be an induced electric polarization, with the electronic dielectric constant as proportionality coefficient.

The power of the density-functional perturbation theory is even more apparent when mixed derivatives of the total energy are considered. One can perturb a lattice by a dilatation, and examine not only the induced pressure, but also the induced forces, or the induced electric polarization. Or, one can displace an atom and examine the force induced on another atom. All these phenomena are governed by proportionality coefficients that are second-derivatives of the total energy: elastic stiffness tensor, piezoelectricity, interatomic force constants, Born effective charge, internal stresses. Before examining the formalism that allows to build such second-order derivatives of the total energy, let us also mention some physical properties linked to third-order (or higher-order) derivatives of the total energy: nonlinear dielectric susceptibility, nonlinear elasticity, Raman cross-section, Grüneisen parameters (change of force constants due to a change of lattice parameters), phonon–phonon interaction ...

We now consider two or more simultaneous Hermitian perturbations (with a set of small parameters λ_j), combined in a Taylor-like expansion of the following type (see GONZE and VIGNERON [1989] for the notation):

$$v_{\mathrm{ext}}(\boldsymbol{\lambda}) = v_{\mathrm{ext}}^{(0)} + \sum_{j_1} \lambda_{j_1} v_{\mathrm{ext}}^{j_1} + \sum_{j_1 j_2} \lambda_{j_1} \lambda_{j_2} v_{\mathrm{ext}}^{j_1 j_2} + \cdots \tag{3.31}$$

(the superscripts j_1 and j_2 are not exponents, but label the different perturbations). The mixed derivative of the energy of the electronic system is

$$E_{\mathrm{el}}^{j_1 j_2} = \frac{1}{2} \frac{\partial^2 E_{\mathrm{el}}}{\partial \lambda_{j_1} \partial \lambda_{j_2}}. \tag{3.32}$$

There are different formulas connecting the first-order responses to second-order derivatives of the energy. Some of them are *stationary* with respect to the errors made in the first-order responses. Others, inherently less accurate, have, in the case of mixed second-order derivatives of the total energy, the advantage of needing the knowledge of the first-order responses with respect to *only one* of the two perturbations. This property, in a different context, was called the "interchange theorem" (DALGARNO and STEWART [1958]). Some formulas, as well as the interchange theorem, are stationary, as a consequence of the variational principle for the total energy (PICKETT [1989]). Explicitly:

$$E_{\mathrm{el}}^{j_1 j_2} = \frac{1}{2}\big(\widetilde{E}_{\mathrm{el}}^{j_1 j_2} + \widetilde{E}_{\mathrm{el}}^{j_2 j_1}\big), \tag{3.33}$$

with

$$\begin{aligned}
&\widetilde{E}_{\mathrm{el}}^{j_1 j_2}\{\psi^{(0)}; \psi^{j_1}, \psi^{j_2}\} \\
&\quad = \sum_\alpha \Big[\big\langle \psi_\alpha^{j_1}\big| H^{(0)} - \varepsilon_\alpha^{(0)} \big| \psi_\alpha^{j_2}\big\rangle \\
&\qquad\qquad + \Big(\big\langle \psi_\alpha^{j_1}\big| v_{\mathrm{ext}}^{j_2} + v_{\mathrm{Hxc0}}^{j_2} \big| \psi_\alpha^{(0)}\big\rangle + \big\langle \psi_\alpha^{(0)}\big| v_{\mathrm{ext}}^{j_1} + v_{\mathrm{Hxc0}}^{j_1} \big| \psi_\alpha^{j_2}\big\rangle\Big) \\
&\qquad\qquad + \big\langle \psi_\alpha^{(0)}\big| v_{\mathrm{ext}}^{j_1 j_2} \big| \psi_\alpha^{(0)}\big\rangle\Big] \\
&\qquad + \frac{1}{2} \iint \frac{\delta^2 E_{\mathrm{Hxc}}}{\delta n(\mathbf{r}) \delta n(\mathbf{r}')}\bigg|_{n^{(0)}} n^{j_1}(\mathbf{r}) n^{j_2}(\mathbf{r}')\, d\mathbf{r}\, d\mathbf{r}' + \frac{1}{2} \frac{d^2 E_{\mathrm{Hxc}}}{d\lambda_{j_1} d\lambda_{j_2}}\bigg|_{n^{(0)}}.
\end{aligned} \tag{3.34}$$

It is noteworthy wondering about the numerical accuracy that one can achieve. Let us suppose that the first-order wavefunctions and densities are not exact. Eqs. (3.33) and (3.34) give an estimate of $E_{\mathrm{el}}^{j_1 j_2}$ including an error that is proportional to the *product* of errors made in the first-order quantities for the first and second perturbations. It is a *stationary* expression. If those errors are small, their product will be much smaller. However, the sign of the error is undetermined, unlike for the variational quantities (GONZE, ALLAN and TETER [1992], SAVRASOV [1992]) presented in Section 3.2. On the other hand, the following expressions do not have the stationarity property, and the numerical error is thus of the same order of those affecting the first-order wavefunctions or densities, and not their product. Nevertheless, they allows us to evaluate $E_{\mathrm{el}}^{j_1 j_2}$ from the knowledge of the derivative of wavefunctions with respect to *only one* of the perturbations:

$$
\begin{aligned}
E_{\mathrm{el}}^{j_1 j_2} &= \frac{1}{2}\sum_{\alpha}^{\mathrm{occ.}}\Big(\big\langle\psi_\alpha^{j_2}\big|v_{\mathrm{ext}}^{j_1}+v_{\mathrm{Hxc0}}^{j_1}\big|\psi_\alpha^{(0)}\big\rangle+\big\langle\psi_\alpha^{(0)}\big|v_{\mathrm{ext}}^{j_1}+v_{\mathrm{Hxc0}}^{j_1}\big|\psi_\alpha^{j_2}\big\rangle\Big)+E_{\mathrm{nonvar}}^{j_1 j_2} \\
&= \sum_{\alpha}^{\mathrm{occ.}}\big\langle\psi_\alpha^{j_2}\big|v_{\mathrm{ext}}^{j_1}+v_{\mathrm{Hxc0}}^{j_1}\big|\psi_\alpha^{(0)}\big\rangle+E_{\mathrm{nonvar}}^{j_1 j_2} \\
&= \sum_{\alpha}^{\mathrm{occ.}}\big\langle\psi_\alpha^{(0)}\big|v_{\mathrm{ext}}^{j_1}+v_{\mathrm{Hxc0}}^{j_1}\big|\psi_\alpha^{j_2}\big\rangle+E_{\mathrm{nonvar}}^{j_1 j_2},
\end{aligned}
\tag{3.35}
$$

where

$$
E_{\mathrm{nonvar}}^{j_1 j_2} = \sum_{\alpha}^{\mathrm{occ.}}\big\langle\psi_\alpha^{(0)}\big|v_{\mathrm{ext}}^{j_1 j_2}\big|\psi_\alpha^{(0)}\big\rangle+\frac{1}{2}\frac{d^2E_{\mathrm{Hxc}}}{d\lambda_{j_1}d\lambda_{j_2}}\bigg|_{n^{(0)}}.
\tag{3.36}
$$

In Eq. (3.35), the first-order wavefunctions $|\psi_\alpha^{j_1}\rangle$ are not needed, while the computation of $v_{\mathrm{ext}}^{j_1}$ and $v_{\mathrm{Hxc0}}^{j_1}$ takes little time. Similar expressions, that do not involve $|\psi_\alpha^{j_2}\rangle$ but $|\psi_\alpha^{j_1}\rangle$ are also available. The time-reversal symmetry allows to simplify further these expressions. For example,

$$
E_{\mathrm{el}}^{j_1 j_2} = \sum_{\alpha}^{\mathrm{occ.}}\big\langle\psi_\alpha^{j_2}\big|v_{\mathrm{ext}}^{j_1}+v_{\mathrm{Hxc0}}^{j_1}\big|\psi_\alpha^{(0)}\big\rangle+E_{\mathrm{nonvar}}^{j_1 j_2}.
\tag{3.37}
$$

These results (Eqs. (3.35)–(3.37)) are generalizations of the so-called "interchange theorem" (DALGARNO and STEWART [1958]).

3.8. Response calculations: wavevectors

Besides the DFPT presented in Section 3.7, the responses of crystalline solids to external perturbations, such as electric fields or atomic displacements, have been calculated within the DFT using various methods (YIN and COHEN [1980], RESTA [1985], HYBERTSEN and LOUIE [1987], ZEIN [1984], BARONI, GIANNOZZI and TESTA [1987],

GONZE, ALLAN and TETER [1992]). The simplest is a *direct approach* (YIN and COHEN [1980]) in which one freezes a finite-amplitude perturbation into the system and compare the perturbed system with the corresponding unperturbed one (e.g., the frozen-phonon technique). However, in this approach, it is impossible to handle perturbations incommensurate with the periodic lattice, or potentials linear in space (corresponding to homogeneous electric fields), while commensurate perturbations were treated through the use of supercells, sometimes with a considerable increase of computing time. Baroni, Giannozzi, and Testa (BGT) (BARONI, GIANNOZZI and TESTA [1987], GIANNOZZI, DE GIRONCOLI, PAVONE and BARONI [1991]) demonstrated the power of the *perturbative approach* (see also ZEIN [1984]), noticing its advantages to deal with perturbations that breaks the periodicity of the unperturbed lattice, like an homogeneous electric field (the corresponding potential is linear in space), or atomic displacements that differ from cell to cell. Baroni and coworkers, as well as other research groups used this formalism with plane waves and pseudopotentials, LMTO (SAVRASOV [1992], SAVRASOV [1996]) and LAPW (WANG, YU and KRAKAUER [1996]) versions of this linear-response approach have also been proposed and implemented.

We consider perturbations of the system that are incommensurate with the unperturbed periodic lattice, and characterized by a nonzero wavevector $\mathbf{q}$ (GONZE [1997a]). The *ground-state potential* operator is periodic, with:

$$v_{\text{ext}}^{(0)}(\mathbf{r}+\mathbf{R}_a) = v_{\text{ext}}^{(0)}(\mathbf{r}), \tag{3.38}$$

where $\mathbf{R}_a$ is a vector of the real space lattice, while the *perturbing potential* operator is such that

$$v_{\text{ext},\mathbf{q}}^{(1)}(\mathbf{r}+\mathbf{R}_a) = e^{i\mathbf{q}\cdot\mathbf{R}_a} v_{\text{ext},\mathbf{q}}^{(1)}(\mathbf{r}). \tag{3.39}$$

Actually, when $\mathbf{q}$ is not equal to half a vector of the reciprocal lattice, such a perturbing potential is complex, and should be always used in conjunction with its complex conjugate. However, at the level of the linear response, there is no consequence of working only with the complex $v_{\text{ext},\mathbf{q}}^{(1)}$, since the response to the sum of $v_{\text{ext},\mathbf{q}}^{(1)}$ and its hermitian conjugate is simply the sum of the response to each perturbation separately. Nevertheless we are also interested to the variational property of the second-order change in energy, for which we cannot afford a complex external potential. This difficulty is solved as follows. One considers both $v_{\text{ext},\mathbf{q}}$ and its complex conjugate – that we write $v_{\text{ext},-\mathbf{q}}$, since its wavevector is $-\mathbf{q}$) – as well as a complex expansion parameter λ, such that

$$\begin{aligned} v_{\text{ext}}(\lambda) = {} & v_{\text{ext}}^{(0)} + \left(\lambda v_{\text{ext},\mathbf{q}}^{(1)} + \lambda^* v_{\text{ext},-\mathbf{q}}^{(1)}\right) \\ & + \left(\lambda^2 v_{\text{ext},\mathbf{q},\mathbf{q}}^{(2)} + \lambda\lambda^* v_{\text{ext},\mathbf{q},-\mathbf{q}}^{(2)} + \lambda^*\lambda v_{\text{ext},-\mathbf{q},\mathbf{q}}^{(2)} + \lambda^{*2} v_{\text{ext},-\mathbf{q},-\mathbf{q}}^{(2)}\right) \\ & + \cdots. \end{aligned} \tag{3.40}$$

This definition is a generalization of Eq. (3.14). Since both $v_{\text{ext}}(\lambda)$ and $v_{\text{ext}}^{(0)}$ are real, the complex conjugates of $v_{\text{ext},\mathbf{q}}^{(1)}$ and $v_{\text{ext},\mathbf{q},\mathbf{q}}^{(2)}$ are $v_{\text{ext},-\mathbf{q}}^{(1)}$ and $v_{\text{ext},-\mathbf{q},-\mathbf{q}}^{(2)}$, respectively. One also has the freedom to impose that $v^{(2)}_{\text{ext},\mathbf{q},-\mathbf{q}}$ is real and equal to $v^{(2)}_{\text{ext},-\mathbf{q},\mathbf{q}}$. Applying a translation to the first-order wavefunctions and densities, one observes the following behaviors:

$$\psi^{(1)}_{m,\mathbf{k},\mathbf{q}}(\mathbf{r}+\mathbf{R}_a) = e^{i(\mathbf{k}+\mathbf{q})\cdot\mathbf{R}_a}\psi^{(1)}_{m,\mathbf{k},\mathbf{q}}(\mathbf{r}) \tag{3.41}$$

and

$$n^{(1)}_{\mathbf{q}}(\mathbf{r}+\mathbf{R}_a) = e^{i\mathbf{q}\cdot\mathbf{R}_a}n^{(1)}_{\mathbf{q}}(\mathbf{r}). \tag{3.42}$$

Now comes the crucial point in the treatment of perturbations characterized by a wavevector $\mathbf{q}$, like $v^{(1)}_{\text{ext},\mathbf{q}}$: It is possible to factorize the phase, in order to map the crystal subject to the incommensurate perturbation into a problem presenting the periodicity of the unperturbed crystal. For this purpose, inspired by Eqs. (3.41) and (3.42), one defines the *periodic* functions

$$u^{(1)}_{m,\mathbf{k},\mathbf{q}}(\mathbf{r}) = (N\Omega_0)^{1/2}e^{-i(\mathbf{k}+\mathbf{q})\cdot\mathbf{r}}\psi^{(1)}_{m,\mathbf{k},\mathbf{q}}(\mathbf{r}) \tag{3.43}$$

and

$$\bar{n}^{(1)}_{\mathbf{q}}(\mathbf{r}) = e^{-i\mathbf{q}\cdot\mathbf{r}}n^{(1)}_{\mathbf{q}}(\mathbf{r}). \tag{3.44}$$

Thanks to these equations, it can be shown that $E^{(2)}_{\mathbf{q},-\mathbf{q}}$ is a real quantity, variational with respect to change in the first-order wavefunctions. It only depends on *periodic* quantities. This property permits to apply the same algorithms that are employed for commensurate perturbations to the case of incommensurate perturbations.

$$\begin{aligned}
&E^{(2)}_{\text{el},-\mathbf{q},\mathbf{q}}\{u^{(0)};u^{(1)}\}\\
&\quad= \frac{\Omega_0}{(2\pi)^3}\int_{\text{BZ}}\sum_m^{\text{occ.}} s\Big(\big\langle u^{(1)}_{m\mathbf{k},\mathbf{q}}\big|H^{(0)}_{\mathbf{k}+\mathbf{q},\mathbf{k}+\mathbf{q}} - \varepsilon^{(0)}_{m\mathbf{k}}\big|u^{(1)}_{m\mathbf{k},\mathbf{q}}\big\rangle\\
&\qquad+ \big\langle u^{(1)}_{m\mathbf{k},\mathbf{q}}\big|v^{(1)}_{\text{ext},\mathbf{k}+\mathbf{q},\mathbf{k}} + v^{(1)}_{\text{Hxc0},\mathbf{k}+\mathbf{q},\mathbf{k}}\big|u^{(0)}_{m\mathbf{k}}\big\rangle\\
&\qquad+ \big\langle u^{(0)}_{m\mathbf{k}}\big|v^{(1)}_{\text{ext},\mathbf{k},\mathbf{k}+\mathbf{q}} + v^{(1)}_{\text{Hxc0},\mathbf{k}+\mathbf{q},\mathbf{k}}\big|u^{(1)}_{m\mathbf{k},\mathbf{q}}\big\rangle + \big\langle u^{(0)}_{m\mathbf{k}}\big|v^{(2)}_{\text{ext},\mathbf{k},\mathbf{k}}\big|u^{(0)}_{m\mathbf{k}}\big\rangle\Big)\,d\mathbf{k}\\
&\qquad+ \frac{1}{2}\iint_{\Omega_0}\frac{\delta^2 E_{\text{Hxc}}}{\delta n(\mathbf{r})\delta n(\mathbf{r}')}\bigg|_{n^{(0)}}\bar{n}^{(1)*}_{\mathbf{q}}(\mathbf{r})\bar{n}^{(1)}_{\mathbf{q}}(\mathbf{r}')e^{-i\mathbf{q}\cdot(\mathbf{r}-\mathbf{r}')}\,d\mathbf{r}\,d\mathbf{r}'\\
&\qquad+ \frac{1}{2}\frac{d^2 E_{\text{Hxc}}}{d\lambda d\lambda^*}\bigg|_{n^{(0)}}
\end{aligned} \tag{3.45}$$

satisfies a minimum principle with respect to variations of the first-order wavefunctions $u^{(1)}_{n,\mathbf{k},\mathbf{q}}$ under constraints

$$\big\langle u^{(0)}_{m,\mathbf{k}+\mathbf{q}}\big|u^{(1)}_{n,\mathbf{k},\mathbf{q}}\big\rangle = 0, \tag{3.46}$$

where the index n runs over the occupied states. The first-order change in density is given by:

$$\bar{n}_{\mathbf{q}}^{(1)}(\mathbf{r}) = \frac{2}{(2\pi)^3} \int_{\mathrm{BZ}} \sum_{m}^{\mathrm{occ.}} u_{m\mathbf{k}}^{(0)*}(\mathbf{r}) u_{m\mathbf{k},\mathbf{q}}^{(1)}(\mathbf{r})\, d\mathbf{k}. \tag{3.47}$$

The Euler–Lagrange equations associated with the minimization of Eq. (3.45) under constraints Eq. (3.46) are

$$P_{c,\mathbf{k}+\mathbf{q}}\big(H_{\mathbf{k}+\mathbf{q},\mathbf{k}+\mathbf{q}}^{(0)} - \varepsilon_{m,\mathbf{k}}^{(0)}\big) P_{c,\mathbf{k}+\mathbf{q}} \big|u_{m,\mathbf{k},\mathbf{q}}^{(1)}\big\rangle = -P_{c,\mathbf{k}+\mathbf{q}} H_{\mathbf{k}+\mathbf{q},\mathbf{k}}^{(1)} \big|u_{m,\mathbf{k}}^{(0)}\big\rangle \tag{3.48}$$

with

$$\begin{aligned} H_{\mathbf{k}+\mathbf{q},\mathbf{k}}^{(1)} &= v_{\mathrm{ext},\mathbf{k}+\mathbf{q},\mathbf{k}}^{(1)} + v_{\mathrm{Hxc0},\mathbf{k}+\mathbf{q},\mathbf{k}}^{(1)} \\ &\quad + \int \frac{\delta^2 E_{\mathrm{Hxc}}}{\delta n(\mathbf{r})\delta n(\mathbf{r}')}\bigg|_{n^{(0)}} \bar{n}_{\mathbf{q}}^{(1)}(\mathbf{r}') e^{-i\mathbf{q}\cdot(\mathbf{r}-\mathbf{r}')}\, d\mathbf{r}'. \end{aligned} \tag{3.49}$$

Finally, there are simpler, but nonvariational, expressions for $E_{\mathrm{el},-\mathbf{q},\mathbf{q}}^{(2)}$, in the spirit of Eq. (3.30), such as:

$$\begin{aligned} E_{\mathrm{el},-\mathbf{q},\mathbf{q}}^{(2)}\{u^{(0)}; u^{(1)}\} &= \frac{\Omega_0}{(2\pi)^3} \int_{\mathrm{BZ}} \sum_{m}^{\mathrm{occ.}} s\big(\big\langle u_{m\mathbf{k},\mathbf{q}}^{(1)}\big| v_{\mathrm{ext},\mathbf{k}+\mathbf{q},\mathbf{k}}^{(1)} + v_{\mathrm{Hxc0},\mathbf{k}+\mathbf{q},\mathbf{k}}^{(1)} \big|u_{m\mathbf{k}}^{(0)}\big\rangle \\ &\quad + \big\langle u_{m\mathbf{k}}^{(0)}\big| v_{\mathrm{ext},\mathbf{k},\mathbf{k}}^{(2)} \big|u_{m\mathbf{k}}^{(0)}\big\rangle\big)\, d\mathbf{k} + \frac{1}{2}\frac{d^2 E_{\mathrm{Hxc}}}{d\lambda d\lambda^*}\bigg|_{n^{(0)}}. \end{aligned} \tag{3.50}$$

All the quantities that appear in Eqs. (3.45)–(3.50) have the periodicity of the unperturbed lattice.

3.9. Dynamical and dielectric phenomena

We now focus on the physical significance of the response to two types of perturbations, lattice vibrations (phonons), and homogeneous, static, electric fields (GONZE [1997b]). We suppose that the theory presented in the previous two sections has been used to compute the proportionality coefficients (second-order derivatives of the total energy). In particular, as regards lattice vibrations, we consider unit displacements of atoms in the sublattice κ, along the α axis, multiplied by the infinitesimal λ (eventually, a complex quantity) and by a phase varying with the cell to which the atoms belong: the α component of their vector position is changed from $\tau_{\kappa,\alpha} + R_{a,\alpha}$ to $\tau_{\kappa,\alpha} + R_{a,\alpha} + \lambda e^{i\mathbf{q}\cdot\mathbf{R}_a}$. Atoms in the other sublattices are not displaced. We examine, briefly: (1) interatomic force constants, dynamical matrix, and phonon frequencies, linked to the second derivative of the total energy with respect to the above-mentioned atomic displacements; (2) Born effective charges, that gives the coupling between atomic displacements and an homogeneous electric field; (3) the electronic dielectric permittivity constant; (4) the long-range behavior of interatomic force constants.

The total energy of a periodic crystal with small lattice distortions from the equilibrium positions (no forces on atoms at equilibrium) can be expressed as

$$E_{\mathrm{tot}}\big(\{\Delta\boldsymbol{\tau}\}\big) = E_{\mathrm{tot}}^{(0)} + \sum_{a\kappa\alpha}\sum_{b\kappa'\beta}\frac{1}{2}\left(\frac{\partial^2 E_{\mathrm{tot}}}{\partial\tau_{\kappa\alpha}^{a}\,\partial\tau_{\kappa'\beta}^{b}}\right)\Delta\tau_{\kappa\alpha}^{a}\,\Delta\tau_{\kappa'\beta}^{b} + \cdots, \tag{3.51}$$

where $\Delta\tau_{\kappa\alpha}^{a}$ is the displacement along the α direction of the atom κ in the cell labelled a (with vector $\mathbf{R}_a$), from its equilibrium position $\boldsymbol{\tau}_\kappa$. The matrix of the interatomic force constants (IFCs) is defined as

$$C_{\kappa\alpha,\kappa'\beta}(a,b) = \left(\frac{\partial^2 E_{\mathrm{tot}}}{\partial\tau_{\kappa\alpha}^{a}\,\partial\tau_{\kappa'\beta}^{b}}\right), \tag{3.52}$$

its Fourier transform is

$$\widetilde{C}_{\kappa\alpha,\kappa'\beta}(\mathbf{q}) = \frac{1}{N}\sum_{ab} C_{\kappa\alpha,\kappa'\beta}(a,b)e^{-i\mathbf{q}\cdot(\mathbf{R}_a-\mathbf{R}_b)} = \sum_{b} C_{\kappa\alpha,\kappa'\beta}(0,b)e^{i\mathbf{q}\cdot\mathbf{R}_b}, \tag{3.53}$$

where N is the number of cells of the crystal within the Born von Karman boundary conditions (BORN and HUANG [1954]). It is connected to the dynamical matrix $\widetilde{D}_{\kappa\alpha,\kappa'\beta}(\mathbf{q})$ by

$$\widetilde{D}_{\kappa\alpha,\kappa'\beta}(\mathbf{q}) = \widetilde{C}_{\kappa\alpha,\kappa'\beta}(\mathbf{q})/(M_\kappa M_{\kappa'})^{1/2}. \tag{3.54}$$

The squares of the phonon frequencies $\omega_{m\mathbf{q}}^2$ at $\mathbf{q}$ are obtained as eigenvalues of the dynamical matrix $\widetilde{D}_{\kappa\alpha,\kappa'\beta}(\mathbf{q})$, or as solutions of the following generalized eigenvalue problem:

$$\sum_{\kappa'\beta}\widetilde{C}_{\kappa\alpha,\kappa'\beta}(\mathbf{q})U_{m\mathbf{q}}^{(\kappa'\beta)} = M_\kappa\omega_{m\mathbf{q}}^2 U_{m\mathbf{q}}^{(\kappa\alpha)}. \tag{3.55}$$

From Eqs. (3.51)–(3.53), the matrix $\widetilde{C}_{\kappa\alpha,\kappa'\beta}(\mathbf{q})$ can be linked to the second-order derivative of the total energy with respect to collective atomic displacements:

$$\widetilde{C}_{\kappa\alpha,\kappa'\beta}(\mathbf{q}) = 2E_{\mathrm{tot},-\mathbf{q},\mathbf{q}}^{\tau_{\kappa\alpha}^{*}\tau_{\kappa'\beta}}. \tag{3.56}$$

The total energy E_{tot} is made of two contributions, from the electron system and from the electrostatic interaction between the ions, which is usually computed through the Ewald method. Similarly, the $\widetilde{C}$ matrix is split in two parts:

$$\widetilde{C}_{\kappa\alpha,\kappa'\beta}(\mathbf{q}) = \widetilde{C}_{\mathrm{el},\kappa\alpha,\kappa'\beta}(\mathbf{q}) + \widetilde{C}_{\mathrm{Ew},\kappa\alpha,\kappa'\beta}(\mathbf{q}). \tag{3.57}$$

Coming now to the homogeneous (macroscopic) static electric field perturbation, one encounters two important problems. First, the potential energy of the electron, placed in such a field, is linear in space, and breaks the periodicity of the crystalline lattice:

$$v_{\mathrm{scr}}(\mathbf{r}) = \sum_{\alpha} E_{\mathrm{mac},\alpha} r_{\alpha}. \tag{3.58}$$

Secondly, the macroscopic electric field corresponds to a screened potential: the change of macroscopic electric field is the sum of an external change of field and an internal change of field, the latter being induced by the response of the electrons (the polarization of the material). In order to indicate this fact, the subscript "scr" has been used in Eq. (3.58). In the theory of classical electromagnetism (LANDAU and LIFSHITS [1960]), one writes the connection between the macroscopic displacement, electric and polarization fields as

$$D_{\mathrm{mac}}(\mathbf{r}) = E_{\mathrm{mac}}(\mathbf{r}) + 4\pi P_{\mathrm{mac}}(\mathbf{r}), \tag{3.59}$$

where $P_{\mathrm{mac}}(\mathbf{r})$ is related to the macroscopic charge density by $n_{\mathrm{mac}}(\mathbf{r}) = -\nabla \cdot P_{\mathrm{mac}}(\mathbf{r})$. It is important to emphasize that these fields are *macroscopic* fields: the microscopic fluctuations (local fields) have been averaged out in this description (LANDAU and LIFSHITS [1960]). The long-wave method is commonly used to deal with the first problem: a potential linear in space is obtained as the limit for $\mathbf{q}$ tending to $\mathbf{0}$ of

$$v(\mathbf{r}) = \lim_{\mathbf{q}\to 0} \lambda \frac{2 \sin \mathbf{q}\cdot\mathbf{r}}{|\mathbf{q}|} = \lim_{\mathbf{q}\to 0} \lambda \left(\frac{e^{i\mathbf{q}\cdot\mathbf{r}}}{i|\mathbf{q}|} - \frac{e^{-i\mathbf{q}\cdot\mathbf{r}}}{i|\mathbf{q}|} \right), \tag{3.60}$$

where $\mathbf{q}$ is in the direction of the homogeneous field.

The detailed theoretical treatment of the response to an electric field, using the long-wave method, and treating the screening adequately (in order to solve the above-mentioned second problem) is given by GONZE [1997a]. It is found that an auxiliary quantity is needed: the derivative of the ground-state wavefunctions with respect to their wavevector. Once this quantity has been obtained, the computation of the response to an homogeneous electric field per se can be performed.

In particular, the dielectric permittivity tensor, that is, the coefficient of proportionality between the macroscopic displacement field and the macroscopic electric field, in the linear regime, can be computed:

$$D_{\mathrm{mac},\alpha} = \sum_{\beta} \varepsilon_{\alpha\beta} E_{\mathrm{mac},\beta}. \tag{3.61}$$

It is obtained as

$$\varepsilon_{\alpha\beta} = \frac{\partial D_{\mathrm{mac},\alpha}}{\partial E_{\mathrm{mac},\beta}} = \delta_{\alpha\beta} + 4\pi \frac{\partial P_{\mathrm{mac},\alpha}}{\partial E_{\mathrm{mac},\beta}}. \tag{3.62}$$

For insulators, the Born effective charge tensor $Z^*_{\kappa,\beta\alpha}$ [7] is defined as the proportionality coefficient relating, at linear order, the polarization per unit cell, created along the direction β, and the displacement along the direction α of the atoms belonging to the sublattice κ, under the condition of zero electric field. The same coefficient also describes the linear relation between the force on an atom and the macroscopic electric field, because both are related to the mixed second-order derivative of the energy with respect to atomic displacements and macroscopic electric field:

$$Z^*_{\kappa,\beta\alpha} = \Omega_0 \frac{\partial P_{\mathrm{mac},\beta}}{\partial \tau_{\kappa\alpha}(\mathbf{q}=\mathbf{0})} = -\frac{\partial F_{\kappa,\alpha}}{\partial E_\beta}. \tag{3.63}$$

The Born effective charge tensor is especially important in the treatment of the dynamical matrix at short wavevector, i.e., the limit $\mathbf{q} \to \mathbf{0}$. This limit must be performed carefully. By the separate treatment of the electric field associated with collective atomic displacements in this limit, one sees that a "bare" $\mathbf{q} = \mathbf{0}$ dynamical matrix must be computed, to which a "nonanalytical" part is added, in order to reproduce correctly the $\mathbf{q} \to \mathbf{0}$ behavior along different directions. After a careful treatment, one is able to recover the important result (PICK, COHEN and MARTIN [1970], GIANNOZZI, DE GIRONCOLI, PAVONE and BARONI [1991]),

$$\widetilde{C}_{\kappa\alpha,\kappa'\beta}(\mathbf{q} \to \mathbf{0}) = \widetilde{C}_{\kappa\alpha,\kappa'\beta}(\mathbf{q}=\mathbf{0}) + \widetilde{C}^{\mathrm{NA}}_{\kappa\alpha,\kappa'\beta}(\mathbf{q} \to \mathbf{0}), \tag{3.64}$$

where the nonanalytical, direction-dependent term $\widetilde{C}^{\mathrm{NA}}_{\kappa\alpha,\kappa'\beta}(\mathbf{q} \to \mathbf{0})$ is given by

$$\widetilde{C}^{\mathrm{NA}}_{\kappa\alpha,\kappa'\beta}(\mathbf{q} \to \mathbf{0}) = \frac{4\pi}{\Omega_0} \frac{(\sum_\gamma q_\gamma Z^*_{\kappa,\gamma\alpha})(\sum_{\gamma'} q_{\gamma'} Z^*_{\kappa',\gamma'\beta})}{\sum_{\alpha\beta} q_\alpha \varepsilon^\infty_{\alpha\beta} q_\beta}. \tag{3.65}$$

Let us come back to the computation of interatomic force constants (IFCs). We have seen that the most effective way to compute the response of periodic solids to atomic displacements is to consider collective perturbations of the lattice, with a well-defined wavevector. In this case, the dynamical matrix corresponding to that wavevector can be built. If the dynamical matrices were known everywhere in the Brillouin zone, the IFCs could be built by inverting Eq. (3.53), which defines the dynamical matrix from the IFCs:

$$C_{\kappa\alpha,\kappa'\beta}(0,b) = \frac{(2\pi)^3}{\Omega_0} \int_{\mathrm{BZ}} \widetilde{C}_{\kappa\alpha,\kappa'\beta}(\mathbf{q}) e^{i\mathbf{q}\cdot\mathbf{R}_b}\, d\mathbf{q}. \tag{3.66}$$

Unfortunately, the dynamical matrices are not known everywhere in the Brillouin zone: for computational reasons they are only obtained for a small set of wavevectors. In this case, a numerical integration technique must be used to perform the integration appearing in Eq. (3.66). For that purpose, the use of a discrete Fourier transform is

[7] The star superscript of $Z^*_{\kappa,\alpha\beta}$ is not the symbol for the complex conjugation operation: $Z^*_{\kappa,\alpha\beta}$ is always a real quantity.

tempting: the dynamical matrices on a regular grid G_{lmn} of $(l \times m \times n)$ points in the Brillouin zone (MONKHORST and PACK [1976]) will generate approximate IFCs in a large box B_{lmn} made of $(l \times m \times n)$ periodic cells. Outside of this box, the IFCs are supposed to vanish:

$$C_{\kappa\alpha,\kappa'\beta}(0,b) = \begin{cases} \frac{1}{N_{\mathbf{q}}} \sum_{\mathbf{q}\in G_{lnm}} \widetilde{C}_{\kappa\alpha,\kappa'\beta}(\mathbf{q}) e^{i\mathbf{q}\cdot\mathbf{R}_b} & \text{if } \mathbf{R}_b + \boldsymbol{\tau}_\kappa - \boldsymbol{\tau}'_\kappa \in B_{lmn}, \\ 0 & \text{if } \mathbf{R}_b + \boldsymbol{\tau}_\kappa - \boldsymbol{\tau}'_\kappa \notin B_{lmn}. \end{cases} \quad (3.67)$$

The use of the *d*iscrete Fourier transform technique necessarily implies that the IFCs vanish beyond some distance. If the integrand in Eq. (3.66) were infinitely differentiable, then the IFCs would decrease exponentially fast, and this intrinsic limitation would not be a practical concern. However, for insulators with nonvanishing effective charges, Eqs. (3.64) and (3.65) shows that, close to $\mathbf{q} = \mathbf{0}$, the behavior of the dynamical matrices is strongly nonanalytical: it depends on the direction along which $\mathbf{q} = \mathbf{0}$ is attained. In the real space, it can be seen that this nonanalytical behavior corresponds to long-ranged IFCs, with an average $1/d^3$ decay (d being the distance between atoms), corresponding to dipole–dipole interactions.

Indeed, an electric dipole is often created when an atom is displaced from its original position. By definition (Eq. (3.63)), at the lowest order the dipole is equal to the atomic displacement times the Born effective charge. Even if the Born effective charge vanishes (this may be imposed by symmetry constraints, in elemental crystals), the atomic displacement will create a quadrupole or an octupole (the latter cannot be forbidden by symmetry reasons), with corresponding quadrupole–quadrupole $1/d^5$ decay, or octupole–octupole $1/d^7$ decay. However, the nonanalyticity corresponding to the dipole–dipole interaction is the strongest, and in practice, even the dipole–quadrupole interaction, with $1/d^4$ decay, is neglected. Thus, only if the Born effective charges of all atoms in a crystal vanish, can we consider that Eq. (3.67) will give adequate description of the IFCs.

For metals, the electrostatic interactions are screened for sufficiently large distances. On the other hand, Friedel oscillations, due to the abrupt change of occupation number at the Fermi level, also cause a long-ranged decay of the IFCs. In a simple isotropic model, the decay of the IFCs is given by $\cos 2k_{\mathrm{F}}d/k_{\mathrm{F}}^3 d^3$, where k_{F} is the Fermi wavevector. In more realistic situations, the decay will still be inversely proportional to the cube of the distance, but the oscillatory behavior will be more complex, and determined by the shape of the Fermi surface. In many practical applications, this long-range decay of metallic interatomic force constants in the real space, and the associated singularity in the reciprocal space, are of little importance.

For insulators with nonvanishing Born effective charges, the nonanalytical behavior of the dynamical matrices close to $\mathbf{q} = \mathbf{0}$ can nevertheless be defined once the Born effective charges and the electronic dielectric permittivity tensor are known, as shown in Eq. (3.65). This term cannot be neglected in practical applications. In a *homogeneous* material with isotropic dielectric permittivity tensor $\varepsilon\delta_{\alpha\beta}$, the dipole–dipole interaction created by the displacement of atoms with (isotropic) charges Z_κ and Z'_κ is described by the following force constants (GIANNOZZI, DE GIRONCOLI, PAVONE and BARONI

[1991]):

$$C_{\kappa\alpha,\kappa'\beta}(0,a) = \frac{Z_\kappa Z'_\kappa}{\varepsilon}\left(\frac{\delta_{\alpha\beta}}{d^3} - 3\frac{d_\alpha d_\beta}{d^5}\right), \tag{3.68}$$

where $\mathbf{d} = \mathbf{R}_a + \boldsymbol{\tau}_{\kappa'} - \boldsymbol{\tau}_\kappa$. The Fourier transform of these force constants exhibits the following nonanalytical behavior:

$$\widetilde{C}^{\mathrm{NA}}_{\kappa\alpha,\kappa'\beta}(\mathbf{q} \to \mathbf{0}) = \frac{4\pi}{\Omega_0}\frac{Z_\kappa Z'_\kappa}{\varepsilon}\frac{q_\alpha q_\beta}{q^2}. \tag{3.69}$$

Based on such theoretical considerations, the interatomic force constants of numerous crystals have been computed.

3.10. *Thermodynamic effects of lattice vibrations*

The thermodynamic functions of a solid are determined mostly by the vibrational degrees of freedom of the lattice, since in general the electronic degrees of freedom play a noticeable role only for metals at very low temperatures (LANDAU and LIFSHITS [1960]). However, the complete knowledge of the vibrational spectrum, with sufficient accuracy, is required for the calculation of these thermodynamic functions. In this section, we suppose that accurate *ab initio* interatomic force constants and phonon band structures have been computed by means of the DFPT. Here, we focus on the following thermodynamic functions: the Helmholtz free energy, the internal energy, the constant-volume specific heat and the entropy as a function of temperature. Also, the knowledge of interatomic force constants allows one to calculate the factors that describe the attenuation of X-ray diffraction intensities due to the thermal motion of the atoms. A wealth of information on these atomic temperature factors, gathered by crystallographers, is available.

The above-mentioned thermodynamic functions require summations over phonon eigenstates labelled by the phonon wavevector $\mathbf{q}$ and the phonon mode l. However, the expressions f to be evaluated at each $\mathbf{q}$ and l often depend on the latter quantities only through the frequency $\omega = \omega(\mathbf{q}, l)$. We can then turn $\sum_{\mathbf{q},l} f(\omega(\mathbf{q}, l))$ into a one-dimensional integral $3nN \int_0^{\omega_L} f(\omega)g(\omega)\,d\omega$, where n is the number of atoms per unit cell, N is the number of unit cells, ω_L is the largest phonon frequency, and $g(\omega)\,d\omega$ is defined to be the fractional number of phonon frequencies in the range between ω and $\omega + d\omega$. We normalize the phonon density of states $g(\omega)$ so that $\int_0^{\omega_L} g(\omega)\,d\omega = 1$, namely, $g(\omega) = (1/3nN)\sum_{\mathbf{q},l}\delta(\omega - \omega(\mathbf{q}, l))$. Specifically, the phonon contribution to the Helmholtz free energy ΔF, the phonon contribution to the internal energy ΔE, as well as the constant-volume specific heat C_v, and the entropy S, at temperature T have the following expressions within the harmonic approximation (MARADUDIN, MONTROLL, WEISS and IPATOVA [1971]);

$$\Delta F = 3nNk_{\mathrm{B}}T \int_0^{\omega_L} \ln\left\{2\sinh\frac{\hbar\omega}{2k_{\mathrm{B}}T}\right\} g(\omega)\,d\omega, \tag{3.70}$$

$$\Delta E = 3nN\frac{\hbar}{2}\int_0^{\omega_L} \omega \coth\left(\frac{\hbar\omega}{2k_B T}\right) g(\omega)\, d\omega, \tag{3.71}$$

$$C_v = 3nNk_B \int_0^{\omega_L} \left(\frac{\hbar\omega}{2k_B T}\right)^2 \operatorname{csch}^2\left(\frac{\hbar\omega}{2k_B T}\right) g(\omega)\, d\omega, \tag{3.72}$$

$$S = 3nNk_B \int_0^{\omega_L} \left[\frac{\hbar\omega}{2k_B T} \coth\frac{\hbar\omega}{2k_B T} - \ln\left\{2\sinh\frac{\hbar\omega}{2k_B T}\right\}\right] g(\omega)\, d\omega, \tag{3.73}$$

where k_B is the Boltzmann constant.

The diffracted intensity from a crystal depends on temperature, through the thermal vibrations of the constituent atoms. For fixed positions in the crystal, the diffracted intensity is proportional to the square of the structure factor F defined as $\sum_\kappa f_\kappa e^{2\pi i \mathbf{G}\cdot\mathbf{r}_\kappa}$, where f_κ is the scattering amplitude of the atom κ, $\mathbf{G}$ is the scattering wavevector, and $\boldsymbol{\tau}_\kappa$ is the position of the atom κ. The diffraction condition requires $\mathbf{G}$ to be a reciprocal lattice vector. At finite temperature, the atoms oscillate around their equilibrium positions and the structure factor is modified as $F_T = \sum_\kappa f_\kappa e^{-W(\kappa)} e^{2\pi i \mathbf{G}\cdot\mathbf{r}_\kappa}$ where the atomic temperature factor $e^{-W(\kappa)}$ at temperature T is defined (WILLIS and PRIOR [1975]) as

$$e^{-W(\kappa)} = \exp\left(-\frac{1}{2}\sum_{ij} B_{ij}(\kappa) G_i G_j\right) \tag{3.74}$$

with the mean-square displacement matrix $B_{ij}(\kappa)$ given by

$$B_{ij}(\kappa) = \frac{1}{N}\sum_{\mathbf{q},l} \frac{\hbar}{2\omega(\mathbf{q},l)} \coth\frac{\hbar\omega(\mathbf{q},l)}{2k_B T} e_i(\kappa|\mathbf{q},l) e_j^*(\kappa|\mathbf{q},l). \tag{3.75}$$

M_κ is the mass of the atom κ, G_i is the ith component of the scattering wavevector $\mathbf{G}$, and $e_i(\kappa|\mathbf{q},l)$ is the ith component of the displacement eigenvector associated with the atom κ and the mode l at $\mathbf{q}$ in the Cartesian coordinates, normalized such that $\sum_l M_\kappa e_i(\kappa|\mathbf{q},l) e_j^*(\kappa'|\mathbf{q},l) = \delta_{ij}\delta_{\kappa\kappa'}$. When there is only one kind of atom with sufficient local symmetry, all the $e^{-W(\kappa)}$ are identical and $|F_T|^2 = |F|^2 e^{-2W(\kappa)}$. In this case, the intensity of diffraction is reduced by a factor of $e^{-2W(\kappa)}$, which is usually called the Debye–Waller factor. For two or more kinds of atoms, there is no simple relation between $|F|^2$ and $|F_T|^2$. From Eq. (3.75), one can see that the atomic temperature factor $e^{-W(\kappa)}$ and the mean-square displacement matrix $B_{ij}(\kappa)$ cannot be calculated from $g(\omega)$ due to the explicit dependence on eigenvectors. However, it is efficient to express $B_{ij}(\kappa)$ in terms of "generalized" density of states $g_{ij}(\kappa|\omega)$ as follows,

$$B_{ij}(\kappa) = \frac{1}{M_\kappa}\int_0^{\omega_L} \frac{\hbar}{2\omega} \coth\left(\frac{\hbar\omega}{2k_B T}\right) g_{ij}(\kappa|\omega)\, d\omega, \tag{3.76}$$

where $g_{ij}(\kappa|\omega)$ is given by:

$$g_{ij}(\kappa|\omega) = \frac{1}{N}\sum_{\mathbf{q},l} M_\kappa e_i(\kappa|\mathbf{q},l)e_j^*(\kappa|\mathbf{q},l)\delta\big(\omega - \omega(\mathbf{q},l)\big). \tag{3.77}$$

The generalized density of states $g_{ij}(\kappa|\omega)$ has to be calculated only once for each atom κ and is normalized in such a way that $\int_0^{\omega_L} g_{ij}(\kappa|\omega)\,d\omega = \delta_{ij}\ \forall\kappa$.

3.11. Electronic spectra and excitation energies

Before closing this chapter that focuses on the theoretical framework for the simulation of extended systems, it is worth mentioning some peculiarities of the density-functional theory when it comes to the computation of electronic spectra, and to indicate more elaborate theories for dealing with this problem.

The DFT is, by design, a theory of the electronic ground-state. However, quite naturally in the Kohn–Sham framework, one-particle electronic eigenvalues ε_m appear (in Hartree atomic units):

$$\left[-\frac{1}{2}\Delta + v_{\text{ext}}(\mathbf{r}) + \frac{\delta E_{\text{Hxc}}}{\delta n}(\mathbf{r})\right]\psi_m(\mathbf{r}) = \varepsilon_m\psi_m(\mathbf{r}). \tag{3.78}$$

The meaning of such eigenvalues has been the subject of intense debate (see also Section 4 in Chapter 1 of this Handbook (CANCÉS, DEFRANCESCHI, KUTZELNIGG, LE BRIS and MADAY [2003])). At the experimental level, one has to differentiate between different phenomena, leading to electronic spectra and excitation energies. In one type of experiments, on extended systems one measures the energy needed to extract an electron from the energy released when one electron is captured. Typically, these corresponds to photoemission and inverse photoemission experiments. In another type of experiment such as optical measurements, the overall number of electrons in the system is preserved, although some energy is absorbed, causing excitation of an electron from an occupied state to an unoccupied state. The excited electron and the quasi-particle created by its absence (called hole) usually interact, although the electron-hole pair might dissociate.

Formally, the Kohn–Sham eigenvalues are *not linked* to any one of these experimental observables, in contrast to the total energy, or the density. Indeed, the Kohn–Sham energies have been introduced to establish a correspondance between the real system of interacting electrons and a virtual system of noninteracting ones. The total energy and the density are the same in both systems, by design. No such requirement exists for the electronic spectrum.

Yet, it has been remarked that the shape of the band structures formed from the Kohn–Sham eigenvalues are qualitatively similar to the band structures deduced from photoemission and inverse photoemission experiments. In many cases, the agreement from common approximate treatments of the DFT, using LDA or GGA functionals, is even semi-quantitative,[8] with a noticeable failure: the gap between occupied and unoccupied bands is in general much too small. Some good insulators are even described as metals in the LDA. This observation has been termed the *band-gap problem* of DFT. In finite-size systems, a practical alternative to the determination of the gap between the actual occupied and empty levels of interest is provided by the so-called Δ Self-Consistent Field (ΔSCF) (PARR and YANG [1989]): One calculates the total energy differences between two distinct states that are obtained within constrained occupation numbers, which includes the relaxation of all the Kohn–Sham orbitals when the occupation numbers are changed. In infinite systems, however, such a technique cannot be straightforwardly employed, since there is a continuum of electronic states in any allowed energy range.

A distinct treatment of the excitation spectra of electronic systems that includes the electronic relaxation is provided by the many-body perturbation theory (HEDIN and LUNDQVIST [1969]). In particular, the quasi-particles energies of the system are obtained as eigenvalues of the following equation:

$$\left[-\frac{1}{2}\Delta + v_{\text{ext}}(\mathbf{r}) + v_{\text{H}}(\mathbf{r})\right]\psi_m(\mathbf{r}) + \int \Sigma_{\text{xc}}(\mathbf{r}, \mathbf{r}'; \varepsilon_m)\psi_m(\mathbf{r}')\,d\mathbf{r}' \\ = \varepsilon_m \psi_m(\mathbf{r}). \tag{3.79}$$

The operator Σ_{xc} is called the exchange-correlation self-energy operator. It is nonlocal, energy-dependent, non-Hermitian. The eigenvalues of this equation can have an imaginary part, hence the corresponding eigenstates have a finite lifetime, which corresponds to the fact that there are internal scattering processes between the interacting electrons. In Eq. (3.78), the complex self-energy operator is replaced by a much simpler local, energy-independent, hermitian potential (the Kohn–Sham eigenvalues are real). Clearly, Σ_{xc} and V_{xc} are distinct objects. Despite the lack of theoretical justification, the formal similarity between Eqs. (3.78) and (3.79), and a connection between the exchange-correlation self-energy and the exchange-correlation potential as shown in SHAM [1985], can partially explain the good correlation that sometimes is found between the Kohn–Sham band structure and the experimentally determined spectrum in some cases. In some cases, the solution to Eq. (3.79) can be found through perturbation theory, starting from the Kohn–Sham eigenfunctions and eigenvalues. The reader who is interested in the treatment of Eq. (3.79) in both finite and infinite systems can look to recent accounts (ARYASETIAWAN and GUNNARSSON [1998], HEDIN [1999]).

Another important point to be noted is the emergence of the *Time-Dependent Density Functional Theory* (TDDFT), which is in principle devised to described globally neutral excitations (i.e., conserving the number of electrons) (GROSS, DOBSON and PETERSILKA [1996]). Despite its successes in molecules, where the experimental absorption spectra are well accounted for by the theory, only very recently the TDDFT has begun to be successfully used in crystals. This is due to the characteristics of the exchange-correlation kernel, which is frequency-dependent and long-range in nature (GONZE,

[8]This is often the case when studying the band dispersion or the relative position of electronic levels, in not too strongly correlated electron systems.

GHOSEZ and GODBY [1997]), two features that are missed by the most common local-density adiabatic approximation. A good account of the current researches about the application of the TDDFT to the computation of electronic spectra and the underlying theoretical tools has recently appeared (ONIDA, REINING and RUBIO [2002]).

4. Discretizing the Hilbert space: the use of basis sets

Section 2 focused on the choice of the particular models that are used to mimic the real materials for which the solution of the many-electron problem is searched. Section 3 described how to obtain an effective one-particle Hamiltonian to solve the many-electron problem for the ground and perturbed states. Together, those two sections allows a specific effective Hamiltonian to be built, whose solutions should be representative of those of the real materials. However, for almost all the interesting theoretical problems of materials science, analytic solutions are not available. Two main alternatives are hence possible: the first one is to build up some extremely simplified theoretical models that can be analytically treated. However, the complexity of the materials that are nowadays studied – which are often the result of a sophisticated nano-engineering – calls for systematic methods to treat the many electron problem within a large variety of the external potentials provided by the nuclei and the applied external fields. The second way is therefore to use computers to solve the established equations numerically. However, the problem must be recasted in order to do that. The main approximation, at this level, is to construct a finite representation of the Hilbert space of the solutions by means of a set of basis functions (for a comprehensive treatment, see Chapter 1, Sections 2.1 and 2.2 in this Handbook (CANCÉS, DEFRANCESCHI, KUTZELNIGG, LE BRIS and MADAY [2003])). The solutions of the one-particle Hamiltonian are thus actually built as linear combination of the basis set functions, which is generally the case, apart from some approaches that are based on finite elements. The reliability of the numerical solutions thus depends upon the ability to construct a finite basis set spanning almost entirely the Hilbert space of the exact solutions. From the computational point of view (both CPU time and memory requirements), the effectiveness of the numerical method relies on the actual size of the basis set and on the speed of the algorithms that are employed in conjunction with the use of the actual basis set. At the heart of any choice, there is necessarily a compromise between accuracy and computational fastness. The aim of this section is to provide an overview of the basis sets that are currently adopted in extended systems, illustrating the main ideas as well as the relative merits and drawbacks of each method. In particular, for most of the basis set that are discussed in the following, such as the plane waves and the LMTO methods, periodic boundary conditions are tacitly assumed, which implies that those basis set are constructed in the Crystal Momentum Representation (Section 2.1.1).

4.1. Active electrons vs. spectator electrons

In many problems in chemistry and physics, a distinction between *active* and *spectator* electronic orbitals can be made. The problem can thus be formulated in terms of the active electron wavefunctions, while the spectator electrons can be treated within suitable approximations.[9] This concept is very general and lies at the hearth of any simplified model that is used in chemistry to simulate electron systems. Among the criteria used to distinguish between active and spectator electrons, some are particularly general and relevant in most cases: (A) The spectator and active electron energy scales must differ considerably, by one order of magnitude or more; (B) They should react very unlikely to electronic perturbations; (C) Their respective density distributions are mostly localized in different regions of space.

If one is interested to the features that distinguish condensed matter systems – solids, liquids, surfaces and interfaces, large molecules and clusters – from the separated atoms, a distinction between *active* electrons and *spectator* electrons is quite natural: the outer (valence) electrons are indeed responsible for the chemical behavior, while the inner electrons can be considered, in a first approximation, as inert. Core electron orbitals lie in energy well below the valence orbitals, they are atomic-like, with a small spatial extent and large gradients. All these features imply that the core orbitals are less polarizable than the valence ones (MAHAN and SUBBASWAMY [1990]). Moreover, the chemical behavior of real materials is essentially determined by bond breaking or formation, in which the inner orbitals do not play a crucial role.

In computational chemistry, and especially when calculating the electronic structure of extended systems that consist of many atoms, it is therefore quite natural that the core electron orbitals are treated differently from those corresponding to valence electrons. Indeed, from the computational point of view, inner and valence electron states are characterized by very different length scales, which implies a careful choice of the basis set that is used to describe the Hilbert space. The natural reference for the core electrons are the atomic orbitals of the isolated atoms. They can be thus represented as a product of spherical harmonics by suitable radial functions that are centered on the nuclei. On the other hand, the valence electrons are often delocalized on many atomic sites, and may be better represented on a nearly uniform mesh in the direct space. These contrasted features are at the hearth of the choice for the Hilbert space discretization. In some methods the basis set consists of linear combination of localized functions, the tails of which can represent the valence electron wavefunctions (see Section 4.5). On the other hand, in the pseudopotential plane-wave approach, the core orbitals are frozen in their atomic state and the action of the core electrons on the valence electron wavefunctions is represented by means of suitable operators, the pseudopotentials; the valence electron orbitals are then expanded in plane waves (Section 4.4). Between those two drastic choices, some methods adopt spherical regions around the nuclei where the orbitals are expanded in spherical waves, while using plane waves or other approximated expansions to represent the extended features of the valence electron wavefunctions. These approaches will be discussed in more detail in the section devoted to the Mixed Basis Sets. Differently from the case of the pseudopotential plane-waves based approaches, the mixed and the localized basis sets are able to treat self-consistently the distribution coming from both core

[9]Here we restrict our treatment to one-electron theories such as the HF and DFT. Because of the Pauli principle, the separation of the many-body electronic wavefunction is much more delicate. A way to take into account the core-core and the core-valence correlations has been traced by SHIRLEY and MARTIN [1993].

and valence electrons self-consistently. In those cases, one speaks of *all-electron calculations*. For each kind of Hilbert space discretization, the mathematical foundations, as well as the relative merits and limits will be discussed and compared in the following.

The distinction between active and spectator electrons is also important in another respect. In self-consistent electronic structure calculations, such as in Hartree–Fock or Density Functional theories, the computational load scales as $(N_\mathrm{e})^p$ as a function of the number of active electrons N_e which are explicitly included in the self-consistent cycle or in the energy minimization. As it will be treated extensively in other contributions to this Handbook, the power p can sometimes be reduced to $p = 1$ (order-N methods). However, in most cases (and presently, in the huge majority of the calculations appearing in the literature), $p \geqslant 3$. This implies that the reduction of the number of active electrons is an effective way to lower the computational load and to make the numerical simulations of many inequivalent atom systems feasible.

Once the core-valence separation has been made, the action of the core on the valence electrons is twofold: on one side, the core orbitals contribute to the Hartree and exchange-correlation potentials felt by the valence electrons. On the other side, the valence wavefunctions must remain orthogonal to all core orbitals, thus providing a constraint which has to be fulfilled when the valence wavefunctions are allowed to vary – when minimizing the total energy, for example (see Chapter 1, Sections 2.3 and 2.4 in this Handbook (CANCÉS, DEFRANCESCHI, KUTZELNIGG, LE BRIS and MADAY [2003])). The orthogonality constraint is crucial and prevents the valence wavefunctions to collapse onto the inner orbitals during the energy minimization. At this level, three different choices may be done: in the first one, the core orbitals are explicitly and self-consistently computed (*the all-electron calculations*); in the second one, the core orbitals are taken into account as part of the total electron density, but they are frozen and not subject to the minimization of the total energy (Section 4.2). Such an option is available in most all-electron methods. The third choice is more radical: in order to save computational resources, the frozen core electrons are replaced by a suitable operator (*the pseudopotential*) which approximately reproduces their action on the valence wavefunction (Section 4.3).

4.2. The frozen-core approximation

A quite drastic choice is *the frozen-core approximation* (FCA), according to which the core orbitals are kept fixed in the calculation of the many-atom system. They can be set equal to their form in an isolated atom (a quite common choice), neutral or charged, either in its ground or excited state. Whether the FCA could be appropriate or not for describing some physical phenomena, it is a matter of (1) the intrinsic precision needed and (2) the choice of the orbitals to be considered as frozen or not. The latter issue is briefly addressed here. On the other hand, from the purely theoretical point of view, VON BARTH and GELATT [1980] discussed the validity of the FCA in the framework of the DFT. They defined an energy functional of the core and valence densities $n_c(\mathbf{r})$ and $n_v(\mathbf{r})$ separately, and derived the expression for the error on the total energy when freezing the core orbitals, which was shown to be second order in the difference between the relaxed and the frozen core densities. Core relaxation effects on the total energy and related quantities can thus be reduced, provided that enough orbitals are computed self-consistently. The answer to that question may be not trivial at all. A look at the electronic structure of the atoms in the periodic table can help us to distinguish some representative cases.

Firstly, let us consider the simple case of the F atom. Because of the energy scale (the ionization potential of the $n = 2$ shell ranges from 17.4 to 138 eV, while for the $n = 1$ shell it is comprised between 953.9 and 1103.1 eV), it is natural to consider the 1s orbitals as frozen, while the remainder 2s and 2p ones are logically treated self-consistently. Secondly, let us consider the Ti atom, which has two incomplete ($n = 3$ and $n = 4$) electronic shells. In most solid state compounds, the bonding contribution comes mainly from the 4s and 3d electrons, while the 3p orbitals remains atomic-like and are more or less perturbed, depending on the variations of the surrounding electrostatic potential. There is only a small relative gap between the fourth and the fifth ionization potentials (43.3 and 99.2 eV, respectively), and the overlap between the 3s, 3p and 3d orbitals is not negligible. Thus, most of the conditions stated above do not held for the $n = 3$ states of Ti, and consequently they should be calculated self-consistently, leaving the Ti atom with a *small core*. However, a more practical point of view might be to freeze the 3s and 3p orbitals (*large core*) and compare the computed bonding properties with those obtained within the small core approximation. Both approaches can be found in the literature – see, e.g., YAMAGUCHI and SCHAEFER III [1996]. The separation between core and valence electron might be even more difficult to disentangle when considering lanthanides or actinides. The presence of three incomplete electronic shells with localized d and f electrons, and the relevance of relativistic effects makes the task not at all obvious. In Gd, for instance, the small core choice consists in freezing the filled shells $n = 1, 2, 3$. On the other hand, one may enlarge the core by including the 4s and 4p electrons, or the 4d ones, too. Moreover, the actual spin configuration of the atomic reference system constitutes an additional degree of freedom. Two points of view can be adopted in those cases: the first and more pragmatic one consists in comparing the free and the frozen-core approaches in specific cases that are sufficiently representative of the real system, and decide which electronic states should be included in the self-consistent calculation, according to the demanded accuracy. A more rigorous approach would be to freeze only the complete electronic shells (*small core approximation*), thus fulfilling in most physical cases the requirements (A), (B) and (C) that are stated in the previous section. Such a choice appears particularly robust, which means that the small core approximation can describe different types of bonds and different chemical environments with a good accuracy. However, as it was announced in the previous section and further discussed, such a choice is numerically heavier than the large core approximation.

4.3. Pseudopotentials

4.3.1. The idea

From an historical point of view, and for understanding some concepts that will be extensively used in the following – such as *nonlocality, nonuniqueness, smoothness*, etc. – it is useful to remind the original formulation of the pseudopotential in condensed matter systems (ANTONCIK [1959], PHILLIPS and KLEINMAN [1959]). The all-electron valence orbital $|\psi_v\rangle$ is represented as a linear combination

$$|\psi_v\rangle = |\phi_v\rangle + \sum_c \alpha_{cv}|\psi_c\rangle \tag{4.1}$$

of a smooth wavefunction $|\phi_v\rangle$ and core electron orbitals $|\psi_c\rangle$ with suitable coefficients to ensure core-valence orthogonality. By using the fact that $|\psi_v\rangle$ and $|\psi_c\rangle$ are solution of the Schrödinger equation with eigenvalues ε_v and ε_c, respectively, one easily obtains the equation for $|\phi_v\rangle$:

$$\left[\widehat{H} + \sum_c (\varepsilon_v - \varepsilon_c)|\psi_c\rangle\langle\psi_c|\right]|\phi_v\rangle = \varepsilon_v|\phi_v\rangle \tag{4.2}$$

that is, the smooth valence wavefunction $|\phi_v\rangle$ is the solution of a modified Schrödinger equation with the same eigenvalue as the all-electron valence wavefunction $|\psi_v\rangle$, where $|\phi_v\rangle$ is the ground state solution of a new Hamiltonian. However, the modified Hamiltonian contains the additional projector $\hat{\wp} = \sum_c (\varepsilon_v - \varepsilon_c)|\psi_c\rangle\langle\psi_c|$ which is nonlocal. Moreover, such a term is repulsive – which can be seen by considering that the matrix element $\langle\phi_v|\hat{\wp}|\phi_v\rangle$ is positive definite – and short-range, like the core orbitals. The main advantage of such a formulation is rather clear: one can in principle replace the all-electron problem with an effective Hamiltonian acting on smooth wavefunctions representing the valence electrons. Thus, the computational load is alleviated in two respects: firstly, by reducing the number of the self-consistent wavefunctions to be determined to the valence electrons; secondly, by avoiding to represent the rapid oscillations of the all-electron valence orbitals close to the nucleus, while keeping the long tails that are mainly responsible for the formation of the chemical bonds. The actual choice of pseudopotentials that are employed to replace the action of the core electrons is therefore connected to the basis set which is used to represent the valence orbitals. This point is treated in the following.

However, one must remark other important features of such an approach: (A) $|\phi_v\rangle$ (*the pseudo-wavefunction*) is not normalized; (B) its choice is not unique, which implies that the operator $\wp$ is not unique, too. (C) $\wp = \sum_{n,l,m\in\text{core}} (\varepsilon_v - \varepsilon_{nlm})|\psi_{nlm}\rangle\langle\psi_{nlm}| = \wp_s + \wp_p + \cdots$, that is, the projector operator acts differently on the different angular momenta that are present in the core. Usually, a pseudo-Hamiltonian is constructed, analogously to Eq. (4.2), as:

$$\widehat{H}_{\text{ps}}|\phi_v\rangle = \varepsilon_v|\phi_v\rangle, \tag{4.3}$$

where some conditions are imposed on the pseudo-wavefunctions $|\phi_v\rangle$. As in the Philips–Kleinman approach, the issues (B) and (C) are shared by the pseudopotentials that are nowadays used in computational materials science. This is the reason why they cannot be treated as *true* potentials, but rather as a special class of operators, the *pseudo-potentials*.

4.3.2. Norm-conserving pseudopotentials

Historically, another approach to the construction of pseudopotentials has been devised since 1930 by Fermi, which is based on scattering theory. It consists in focusing on the matrix element of the true operator $\langle \mathbf{q}|\widehat{V}|\mathbf{q}'\rangle$ on the initial and final free-particle states, which is replaced by another operator, the pseudopotential. Its matrix elements $\langle \mathbf{q}|\widehat{V}_{ps}|\mathbf{q}'\rangle$ mimic the original ones in a as wide as possible range of energies. Such an approach has been extensively used in the seventies to reproduce the experimental electron bands and is sometimes referred to as the *empirical pseudopotential method*. The idea of reproducing the scattering properties of the original operator has been retained in the construction of the pseudopotentials that are used nowadays, through the design of optimal pseudo-wavefunctions. Let us consider the one-dimensional Schrödinger equation, that is the radial part of the effective-field equation for an isolated atom in a central potential, *for a given set of angular and azimuthal quantum numbers* l *and* m, that is, $\psi_{nlm}(\mathbf{r}) = Y_{lm}(\theta,\varphi)\chi_{nl}(r)$. Be χ_{ε_1} and χ_{ε_2} two bound solutions corresponding to eigenvalues ε_1 and ε_2, respectively. From the study of the Wronskian $W(\chi_{\varepsilon_1},\chi_{\varepsilon_2}) = \chi_{\varepsilon_1}\chi'_{\varepsilon_2} - \chi'_{\varepsilon_1}\chi_{\varepsilon_2}$, one can derive interesting mathematical properties regarding the eigenfunctions (MESSIAH [1961]). In particular, by considering the logarithmic derivative $D(x;\varepsilon) = \frac{\chi'_\varepsilon}{\chi_\varepsilon}$, one can show that the following relation between the energy derivative of $D(x;\varepsilon)$ and the norm of the corresponding eigen-solution holds:

$$\frac{\partial D(r;\varepsilon)}{\partial \varepsilon} = -\frac{1}{2\chi_\varepsilon^2(r)}\int_0^r dt\,\left|\chi_\varepsilon(t)\right|^2. \tag{4.4}$$

On the other hand, it is possible to show that the scattering by a central potential can be completely characterized in terms of the logarithmic derivatives. Therefore, if one constructs a pseudo-wavefunction whose logarithmic derivative matches the true one in wide energy and length ranges, the scattering properties of the corresponding pseudo- and true potentials on the generic l, m spherical wave would be very similar, a property which is generally referred to as *transferability*. Suppose that we replace the true wavefunction $\chi_{nl}(r)$ with a nodeless pseudo-wavefunction $\phi_l^{(\mathrm{ps})}(r)$, corresponding to two modified Schrödinger equations with the same eigenvalue, let's say ε, such that $\int_0^{r_c} dt\,|\chi_{nl}(r)|^2 = \int_0^{r_c} dt\,|\phi_l^{(\mathrm{ps})}(r)|^2$. Eq. (4.4) guarantees that, if $\chi_{nl}(r)$ and $\phi_l^{(\mathrm{ps})}(r)$ coincide $\forall r > r_c$, their logarithmic derivatives track each other at least in some energy range around ε.

Because of Eq. (4.4), the transferability has been related to the *norm conservation*, that is, given a small number η, it is possible to find a radius r_c for which the integral of the squared pseudo- and true wavefunctions differ from each other by less than η $\forall r > r_c$. The norm conservation has also a clear physical meaning: the total amount of charge inside the radius r_c is correctly given by the pseudo-wavefunction. Through the Gauss' theorem, the norm conservation ensures that the electrostatic energy associated with valence electrons is well approximated and that the long-range tail of the electrostatic potential from the nucleus plus the core electrons is correct. In the following, we give in some detail the basic principles of the construction of *norm-conserving pseudopotentials* (HAMANN, SCHLÜTER and CHIANG [1979], BACHELET, HAMANN and SCHLÜTER [1981]), which have been very widely employed in computational materials science during the last two decades. In practice, the following scheme is often used in the construction of the norm-conserving pseudopotential in the framework of the Density Functional Theory. For Hartree–Fock calculations, see MAYNAU and DAUDEY [1980] and MÜLLER, FLESCH and MEYWR [1984].

1. An all-electron calculation is carried out for a reference state of the atom. It may be done for a neutral atom in its ground state, but sometimes a ionized atom is chosen, in order to compute the all-electron orbitals $\chi_{nl}(r)$ for angular momenta l that are not populated in the ground state, up to an upper value $l_{\max}$. It is not compulsory to use the same reference atomic configuration for various l. An alternative solution, which does not need to compute the all-electron and the pseudo-wavefunctions at bound-state energies only, was proposed by HAMANN [1989]. A way to include the dependence on the atom polarization state was recently shown to be effective in treating magnetic solids (WATSON and CARTER [1998]).
2. For each l, a cutoff radius $r_c^{(l)}$ is chosen, beyond which the all-electron and the pseudo-wavefunctions are the same or exponentially converge to each other. Since the pseudo-wavefunction is the ground state solution of a pseudo-Hamiltonian, it must be nodeless. This implies that the cutoff radius $r_c^{(l)}$ must be greater than the abscissa of the last node of the all-electron wavefunction $\chi_{nl}(r)$. For $r < r_c^{(l)}$, a parametrized form is chosen for $\phi_l^{(\text{ps})}(r)$, which has to be continuous and as smooth as possible. Often, the recipe from TROULLIER and MARTINS [1991] is employed (see also Fig. 4.1).
3. $\phi_l^{(\text{ps})}(r)$ is nodeless, so that the Schrödinger equation can be inverted and the *screened* pseudopotential accordingly obtained:

$$V_l^{(\text{ps,scr})}(r) = \varepsilon_l - \frac{l(l+1)}{2r^2} + \frac{1}{2\phi_l^{(\text{ps})}(r)} \frac{d^2\phi_l^{(\text{ps})}(r)}{dr^2}. \tag{4.5}$$

4. There is a further step that is called *unscreening*. Indeed, $V_l^{(\text{ps,scr})}(r)$ in Eq. (4.5) is designed to replace the effective potential in an all-electron calculation, including both spectator and active electrons. However, in the calculations of the many-atom systems, one needs the *ionic* pseudopotential $V_l^{(\text{ps,i})}(r)$, which acts as an external potential on the valence electrons. Therefore, the ionic pseudopotential is obtained from the screened one by subtracting the contribution to the effective potential from the pseudo-wavefunctions of the valence electrons:

$$V_l^{(\text{ps,i})}(r) = V_l^{(\text{ps,scr})}\left(r; \{\chi_{\text{core}}\}, \{\phi^{(\text{ps})}\}\right) - V_l^{(\text{ps,scr})}\left(r; \{\phi^{(\text{ps})}\}\right). \tag{4.6}$$

The separation between core and pseudo-valence orbitals in the classical contribution V_H to the electron–electron interaction is trivial, since V_H fulfills the superposition principle: $V_H(\rho_c + \rho_v) = V_H(\rho_c) + V_H(\rho_c)$. The same is not true for the quantum contribution, that is the exchange-correlation term, for which the separation can be made effective only under certain approximations, which depend on the theoretical framework employed for the atomic calculation (either Hartree–Fock or DFT). The interested reader can look at the existing literature for details

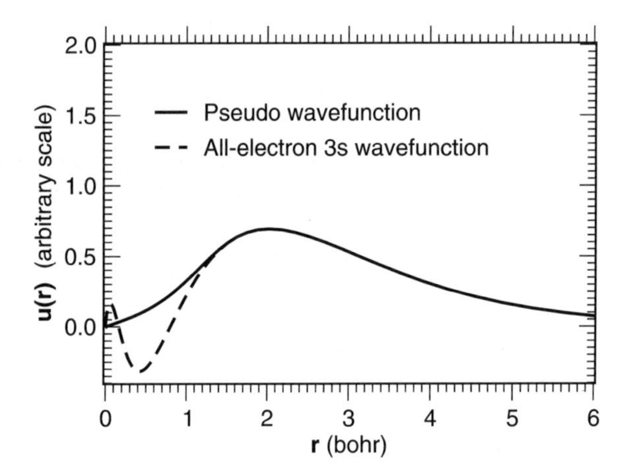

FIG. 4.1. An example of the true (all-electron) and pseudo-radial wavefunctions $u(r)$ for the 3s orbital of the Al atom. The calculations are carried out within the DFT. The pseudo-wavefunction is generated according to the scheme proposed by TROULLIER and MARTINS [1991], with a cutoff radius of about 1.5 bohr, by considering the 1s, 2s, and 2p orbitals as frozen.

(LOUIE, FROYEN and COHEN [1982], PICKETT [1989], POREZAG, PEDERSON and LIU [2000]).

5. For angular components $l > l_{\max}$, the centrifugal barrier keeps electrons out of the core region, and the pseudopotential can be taken to be the same as that corresponding to an angular momentum $l_{\text{ref}} \leqslant l_{\max}$, which is called the *reference component*.

4.3.3. Semi-local and separable pseudopotentials

All the previous considerations apply to an isolated atom, within spherical symmetry. The action of the pseudopotential on a generic wavefunction $\psi(\mathbf{r})$ is, in the real space representation:

$$|\widehat{V}^{(\text{ps})}|\psi\rangle = \int d\mathbf{r}'\, V^{(\text{ps})}(\mathbf{r}, \mathbf{r}')\psi(\mathbf{r}') \tag{4.7}$$

where

$$\begin{aligned} V^{(\text{ps})}(\mathbf{r}, \mathbf{r}') &= V^{(\text{ps})}_{l_{\text{ref}}}(r)\delta(\mathbf{r} - \mathbf{r}') \\ &\quad + \sum_{l=0}^{l_{\max}} \sum_{m=-l}^{l} \left[V^{(\text{ps})}_l(r) - V^{(\text{ps})}_{l_{\text{ref}}}(r)\right]\delta(r - r')Y_{lm}(\theta, \varphi)Y^*_{lm}(\theta', \varphi') \\ &= V^{(\text{ps})}_{l_{\text{ref}}}(r)\delta(\mathbf{r} - \mathbf{r}') + \sum_{l=0}^{l_{\max}} \sum_{m=-l}^{l} \Delta V^{(\text{ps})}_l(r)\delta(r - r')\hat{\wp}_{lm}, \end{aligned} \tag{4.8}$$

where $\hat{\wp}_{lm}$ projects out the l, m angular component from the wavefunction, while the reference component $V_{l_{\mathrm{ref}}}^{(\mathrm{ps})}(r)$ is a local operator, that is, simply multiplies the wavefunction. For $r \to \infty$, the $V_l(r)$ all behave like $-Ze^2/r$, where Z is the charge of the ionic core. The nonlocal terms $\Delta V_l^{(\mathrm{ps})}(r)$ are therefore short-range. The choice of the actual reference component may be important, as we will briefly discuss in the following. Such a pseudopotential is sometimes said to be *semi-local*, since it is local on the radial coordinates and nonlocal on the angular variables. In 1982, Kleinman and Bylander proposed the following transformation for each nonlocal component ($|\phi^{(\mathrm{ps})}\rangle$ is the radial pseudo-wavefunction) (KLEINMAN and BYLANDER [1982]):

$$\Delta \widehat{V}^{(\mathrm{ps,KB})} = \sum_{l,m} \frac{|\Delta V_l^{(\mathrm{ps})}|Y_{l,m}\phi_l^{(\mathrm{ps})}\rangle\langle Y_{l,m}\phi_l^{(\mathrm{ps})}\Delta V_l^{(\mathrm{ps})}|}{\langle Y_{l,m}\phi_l^{(\mathrm{ps})}|\Delta V_l^{(\mathrm{ps})}|Y_{l,m}\phi_l^{(\mathrm{ps})}\rangle}, \tag{4.9}$$

$\Delta \widehat{V}^{(\mathrm{ps,KB})}$ is *separable*, since it can be written as a sum of projectors:

$$\Delta \widehat{V}^{(\mathrm{ps,KB})} = \sum_{l,m} \zeta_{l,m}^*(\mathbf{r}) e_{l,m} \zeta_{l,m}(\mathbf{r}') \tag{4.10}$$

with $\zeta_{l,m}(\mathbf{r}) = |\Delta V_l^{(\mathrm{ps})}|Y_{l,m}\phi_l^{(\mathrm{ps})}\rangle$ and $e_{l,m} = [\langle Y_{l,m}\phi_l^{(\mathrm{ps})}|\Delta V_l^{(\mathrm{ps})}|Y_{l,m}\phi_l^{(\mathrm{ps})}\rangle]^{-1}$. The action of the Kleinman–Bylander pseudopotential on a generic wavefunction is, in direct space:

$$|\Delta \widehat{V}^{(\mathrm{ps,KB})}\psi\rangle = \sum_{l,m} e_{l,m}\zeta_{l,m}(\mathbf{r}) \int d^3r'\, \zeta_{l,m}^*(\mathbf{r}')\psi(\mathbf{r}'). \tag{4.11}$$

The property of separability has an important practical consequence, since in the wavevector space the matrix element between two plane waves is a simple product (see Eq. (4.25) in the following chapter). One can also easily see that the original semi-local (Eq. (4.8)) and the separable pseudopotentials (Eq. (4.10)) have the same action on the pseudo-wavefunctions $Y_{l,m}(\theta,\varphi)\phi_l^{(\mathrm{ps})}(r)$, by construction. However, the Kleinman–Bylander separable pseudopotential is not guaranteed to reproduce the semi-local pseudopotential whatever the wavefunction. Moreover, since the KB pseudopotential does not simply multiplies the radial pseudo-wavefunction, the properties of the Schrödinger equation that can be derived from the Wronskian are not longer valid. In particular, the relation between the principal quantum number and the number of nodes is lost, and spurious solutions (*ghost states*) with some nodes, but at lower energy than the pseudo-wavefunction, may happen. The mathematical conditions and the practical recipes to avoid such a problem have been firstly investigated by GONZE, STUMPF and SCHEFFLER [1991]. In practice, a careful choice of the reference angular component (see Eq. (4.8)) may avoid the appearance of ghost states.

4.3.4. *Ultra-soft pseudopotentials*

The pseudo-wavefunctions corresponding to the active electrons are usually – but not exclusively – expanded in plane waves (see next section). From the practical point of view, it is therefore important to reduce the number of plane wave coefficients in the expansion, which is directly related to the decay rate of the pseudopotential matrix elements in the wavevector space (RAPPE, RABE, KAXIRAS and JOANNOPOULOS [1990]). Pseudopotentials that need a limited amount of plane waves are usually said to be *soft*, a property which should strictly be considered in a comparative sense: while the softer of two pseudopotentials is the one corresponding to the smaller PW expansion, absolute pseudopotential softness is an ill-defined concept.

For some atoms, it is particularly difficult to improve the pseudopotential softness. This is the case of atomic 1s, 2p, 3d, 4f, etc. electrons in the periodic table, for which no effective core repulsion is present. On the other hand, a compromise between transferability and softness is often necessary: because of the norm conservation, a common way to obtain a softer pseudopotential is to increase the cutoff radius r_c, which generally worsen its transferability (FILIPPETTI, VANDERBILT, ZHONG, CAI and BACHELET [1995]). A radical alternative was traced out by VANDERBILT [1990], which is based on two main ideas. First, he let the norm conservation condition drop. Second, he introduced two or more reference energies in order to improve the pseudopotential transferability even in absence of the norm conservation condition, as also BLÖCHL [1990] independently suggested. Therefore, as a direct consequence of the lack of norm conservation, *ultra-soft* pseudopotentials can be constructed (KRESSE and HAFNER [1994]), which need a modest number of PWs for their expansion.

The reader who is interested to the practical construction of ultra-soft pseudopotentials can look at the existing bibliography (KRESSE and HAFNER [1994]). Here we would only like to stress that the ultra-soft pseudopotentials can be understood as a particular case of a dual transformation between the pseudo- and true Hilbert spaces, as pointed out recently (KRESSE and JOUBERT [1999]) and discussed in the last section of this chapter. From the practical point of view, the implementation of ultra-soft pseudopotentials in electronic structure calculations needs more care than for standard, norm-conserving ones. Indeed, the representation of the electronic density should be augmented close to the nuclei, and the mathematical expressions of a generic matrix element results more involved. However, the number of PWs needed in the expansion of the pseudo-orbitals may be reduced by a factor of two, which can be crucial for dealing with large-scale systems with transition metals and/or atoms with semi-core states treated as active electrons.

4.4. Plane waves

In the Crystal Momentum representation (see Section 2.1.1) the periodic functions corresponding to the *active* electrons can be expanded in terms of plane waves (PW) as follows,[10]

$$u_{m\mathbf{k}}^{(0)}(\mathbf{r}) = \sum_{\mathbf{G}} e^{i\mathbf{G}\cdot\mathbf{r}} u_{m\mathbf{k}}^{(0)}(\mathbf{G}), \tag{4.12}$$

[10]These notations do not distinguish among functions that are defined in real or in reciprocal space, which are Fourier transform of each other. However, no confusion should happen, since the arguments belong clearly to one space or the other.

where the sum over $\mathbf{G}$ is restricted to wavevectors such that $\frac{\hbar^2|\mathbf{k}+\mathbf{G}|^2}{2m} < E_{\text{cut}}$. E_{cut} is the maximum kinetic energy of the PW and is usually called the *cutoff energy*. The larger the cutoff energy E_{cut}, the finer the corresponding real-space grid, and the better the accuracy of the representation of the wavefunctions, the electron density and the effective potential. Therefore, the quality of the plane-wave basis set may be routinely improved by increasing the unique parameter E_{cut}. Usually, its value is chosen in such a way that the numerical values of the relevant physical quantities to be computed (such as the total energy, the atomic forces, the band structure, etc.) are little affected by raising E_{cut} further. The coefficients $u^{(0)}_{m\mathbf{k}}(\mathbf{G})$ are the Fourier transform of $u^{(0)}_{m\mathbf{k}}(\mathbf{r})$ that are defined for each vector $\mathbf{G}$ of the reciprocal lattice as:

$$u^{(0)}_{m\mathbf{k}}(\mathbf{G}) = \frac{1}{\Omega_0}\int_{\Omega_0} e^{-i\mathbf{G}\cdot\mathbf{r}} u^{(0)}_{m\mathbf{k}}(\mathbf{r})\, d\mathbf{r}. \tag{4.13}$$

The scalar products, and thus, the orthonormalization conditions can be re-expressed in terms of these Fourier coefficients:

$$\langle f|g\rangle = \frac{1}{\Omega_0}\int_{\Omega_0} f^*(\mathbf{r})g(\mathbf{r})\, d\mathbf{r} = \sum_{\mathbf{G}} f^*(\mathbf{G})g(\mathbf{G}). \tag{4.14}$$

The electron density $n^{(0)}(\mathbf{r})$ is constructed from Eqs. (2.12) and (4.12). Its Fourier transform

$$n^{(0)}(\mathbf{G}) = \frac{1}{\Omega_0}\int_{\Omega_0} n^{(0)}(\mathbf{r})\mathrm{e}^{-i\mathbf{G}\cdot\mathbf{r}}\, d\mathbf{r} \tag{4.15}$$

generally includes wavevectors such that $\frac{\hbar^2|\mathbf{G}|^2}{2m} < 4E_{\text{cut}}$, consistently with the previous equations and the cutoff energy E_{cut} for $u^{(0)}_{m\mathbf{k}}(\mathbf{G})$. The different parts of the Hamiltonian

$$H^{(0)}_{\mathbf{k},\mathbf{k}} = T^{(0)}_{\mathbf{k},\mathbf{k}} + v^{(0)}_{\text{ext},\mathbf{k},\mathbf{k}} + \left.\frac{\delta E_{\text{Hxc}}}{\delta n}\right|_{n^{(0)}} \tag{4.16}$$

are applied to wavefunctions separately. Indeed, the kinetic operator is especially simple in terms of plane wave coefficients:

$$T^{(0)}_{\mathbf{k},\mathbf{k}}(\mathbf{G},\mathbf{G}') = \frac{1}{2}(\mathbf{G}+\mathbf{k})^2\delta_{\mathbf{G}\mathbf{G}'} \tag{4.17}$$

while the local potentials are easily formulated in the real space. In applications based on plane waves, the bare nuclear potential operator is replaced by a pseudopotential, made of local and nonlocal contributions from all atoms inside each repeated cell with lattice vector $\mathbf{R}_a$:

$$v_{\text{ext}}(\mathbf{r},\mathbf{r}') = \sum_{a\kappa} v_\kappa(\mathbf{r}-\boldsymbol{\tau}_\kappa-\mathbf{R}_a, \mathbf{r}'-\boldsymbol{\tau}_\kappa-\mathbf{R}_a), \tag{4.18}$$

where $\boldsymbol{\tau}_\kappa$ is the vector position of the atoms inside the cell. The contribution of atom κ is rewritten from Eqs. (4.8) and (4.10) as:

$$v_\kappa(\mathbf{r},\mathbf{r}') = v_\kappa^{\mathrm{loc}}(\mathbf{r})\delta(\mathbf{r}-\mathbf{r}') + v_\kappa^{\mathrm{sep}}(\mathbf{r},\mathbf{r}'). \tag{4.19}$$

We consider here only nonlocal contributions of separable pseudopotentials (KLEINMAN and BYLANDER [1982], GONZE, STUMPF and SCHEFFLER [1991], PAYNE, TETER, ALLAN, ARIAS and JOANNOPOULOS [1992]),

$$v_\kappa^{\mathrm{sep}}(\mathbf{r},\mathbf{r}') = \sum_\mu e_{\mu\kappa}\zeta_{\mu\kappa}(\mathbf{r})\zeta_{\mu\kappa}^*(\mathbf{r}'), \tag{4.20}$$

where only a few separable terms, labelled by μ, are present. The functions $\zeta_{\mu\kappa}$ are short-ranged (Eq. (4.10)) and usually do not overlap between adjacent atoms (see Section 4.3.3). Because of their different mathematical expressions, the local and nonlocal parts are treated in different ways. A local potential is naturally applied on the wavefunctions in the real space, since it is a diagonal operator in that representation. A separable potential could be treated efficiently either in reciprocal space or in real space. For small systems (on the order of 10 atoms, or less), it is more efficient to apply the separable potential in the reciprocal space. The transformations of the wavefunctions between the real and reciprocal space are carried out by means of fast Fourier transforms (GALLI and PARRINELLO [1991]).

Let us first treat the local part. For every atom, it is long-ranged, with asymptotic behavior $-Z_\kappa/r$, where Z_κ is the charge of the (pseudo)ion. It is well known that, in a periodic geometry, this long-ranged part creates a divergence in the ionic potential, that must be treated together with a similar divergence in the Hartree potential (the divergences cancel each other, but give also a residue, usually incorporated in the ion–ion energy) (PICKETT [1989]). In the reciprocal space, these divergences are associated with terms at $\mathbf{G}=\mathbf{0}$, constant in real space. Thus in any case these compensating divergences are of no importance for the generation of the wavefunctions and the density, since only the mean potential is affected. Although the local potential operator as well as its derivatives are applied to the wavefunction in the real space, we will give their (simpler) expression in the reciprocal space. Their expression in the real space can be obtained by a Fourier transform, similar to Eq. (4.12). We define[11]

$$v'_{\mathrm{loc},\mathbf{k},\mathbf{k}}(\mathbf{G}) = \begin{cases} \frac{1}{\Omega_0}\sum_\kappa e^{-i\mathbf{G}\cdot\boldsymbol{\tau}_\kappa} v_\kappa^{\mathrm{loc}}(\mathbf{G}) & \text{when } \mathbf{G}\neq 0,\\ 0 & \text{when } \mathbf{G}=0. \end{cases} \tag{4.21}$$

In this expression,

$$v_\kappa^{\mathrm{loc}}(\mathbf{K}) = \int e^{-i\mathbf{K}\cdot\mathbf{r}} v_\kappa^{\mathrm{loc}}(\mathbf{r})\,d\mathbf{r}, \tag{4.22}$$

[11] The local potential $v'_{\mathrm{loc},\mathbf{k},\mathbf{k}}$ is independent of $\mathbf{k}$ and is thus simply written as v'_{loc}. Such a short notation is used for all occurrences of a $\mathbf{k}$-independent quantity. The prime symbol indicates that the $\mathbf{G}=\mathbf{0}$ term is excluded.

where the latter integral is performed throughout all the space. The limiting behavior of $v_\kappa^{\rm loc}(\mathbf{K})$ for an arbitrary wavevector $\mathbf{K}$ tending to zero diverges:

$$v_\kappa^{\rm loc}(\mathbf{K}\to 0) = -\frac{4\pi Z_\kappa}{K^2} + C_\kappa + \mathrm{O}(K^2) \tag{4.23}$$

with

$$C_\kappa = \int \left(v_\kappa^{\rm loc}(\mathbf{r}) + \frac{Z_\kappa}{r} \right) d\mathbf{r}. \tag{4.24}$$

For the separable part, one obtains:

$$v^{(0)}_{\mathrm{sep},\mathbf{k},\mathbf{k}}(\mathbf{G},\mathbf{G}') = \frac{1}{\Omega_0} \sum_{\mu\kappa} e_{\mu\kappa} \mathrm{F}_{\mu,\kappa}(\mathbf{k}) \mathrm{F}^*_{\mu,\kappa}(\mathbf{k}), \tag{4.25}$$

where

$$\mathrm{F}_{\mu,\kappa}(\mathbf{k}) = \sum_G e^{i(\mathbf{k}+\mathbf{G})\cdot\boldsymbol{\tau}_\kappa} \int d\mathbf{r}\, e^{-i(\mathbf{k}+\mathbf{G})\cdot\mathbf{r}} \zeta_{\mu\kappa}(\mathbf{r}). \tag{4.26}$$

The special form of the matrix of separable potential, Eq. (4.25), permits the fast calculation of its action onto any wavefunction. In the LDA, the exact exchange-correlation energy, which is a *f*unctional of the density, is replaced by the integral of the density $n(\mathbf{r})$ times the mean exchange-correlation energy per particle $\varepsilon_{\rm xc}(\mathbf{r})$ of the homogeneous electron gas at the point $\mathbf{r}$. However, when combined with pseudopotentials, this simple definition is to be modified, in order to take into account that only valence states are used to build the density: the contribution of the core electrons should be included, because of the nonlinear character of the exchange-correlation energy functional (LOUIE, FROYEN and COHEN [1982]). The functional then writes

$$E_{\rm xc}[n(\mathbf{r})] = \int_{\Omega_0} \big(n(\mathbf{r}) + n_c(\mathbf{r})\big) \varepsilon_{\rm xc}\big(n(\mathbf{r}) + n_c(\mathbf{r})\big)\, d\mathbf{r}, \tag{4.27}$$

where the pseudo-core density n_c is made of nonoverlapping contributions from each atom

$$n_c(\mathbf{r}) = \sum_{a\kappa} n_{c,\kappa}(\mathbf{r} - \boldsymbol{\tau}_\kappa - \mathbf{R}_a). \tag{4.28}$$

The pseudo-core density from each atom $n_{c,\kappa}$ is built at the same time as the pseudopotential (LOUIE, FROYEN and COHEN [1982]). It has spherical symmetry, and is specified by a one-dimensional radial function. The corresponding exchange-correlation potential is

$$v_{\rm xc}\big(n(\mathbf{r})\big) = \left.\frac{d((n+n_c)\varepsilon_{\rm xc}(n+n_c))}{dn}\right|_{n=n(\mathbf{r});\, n_c=n_c(\mathbf{r})}. \tag{4.29}$$

With these definitions, the electronic energy is

$$E_{\mathrm{el}}\{u^{(0)}\} = \frac{\Omega_0}{(2\pi)^3}\int_{\mathrm{BZ}}\sum_m^{\mathrm{occ.}} s\langle u_{m\mathbf{k}}^{(0)}|T_{\mathbf{k},\mathbf{k}}^{(0)} + v_{\mathrm{sep},\mathbf{k},\mathbf{k}}^{(0)}|u_{m\mathbf{k}}^{(0)}\rangle\,d\mathbf{k}$$
$$+ \int_{\Omega_0} n^{(0)}(\mathbf{r})v_{\mathrm{loc}}^{\prime(0)}(\mathbf{r})\,d\mathbf{r}$$
$$+ \int_{\Omega_0} \left(n^{(0)}(\mathbf{r}) + n_c^{(0)}(\mathbf{r})\right)\left(\varepsilon_{\mathrm{xc}}\left(n^{(0)}(\mathbf{r}) + n_c^{(0)}(\mathbf{r})\right)\right)d\mathbf{r}$$
$$+ 2\pi\Omega_0\sum_{\mathbf{G}\neq 0}\frac{|n^{(0)}(\mathbf{G})|^2}{|\mathbf{G}|^2}. \tag{4.30}$$

The Hartree energy (last term of Eq. (4.30)) can also be computed as

$$E_{\mathrm{H}} = \int_{\Omega_0} n^{(0)}(\mathbf{r})v_{\mathrm{H}}^{\prime(0)}(\mathbf{r})\,d\mathbf{r} \tag{4.31}$$

with the Hartree potential being defined as[12]

$$v_{\mathrm{H}}^{\prime(0)}(\mathbf{G}) = \begin{cases} 2\pi\Omega_0\sum_{\mathbf{G}\neq 0}\frac{n^{(0)}(\mathbf{G})}{|\mathbf{G}|^2} & \text{when } \mathbf{G}\neq 0,\\ 0 & \text{when } \mathbf{G}=0. \end{cases} \tag{4.32}$$

The Hamiltonian is given by

$$H_{\mathbf{k},\mathbf{k}}^{(0)} = T_{\mathbf{k},\mathbf{k}}^{(0)} + v_{\mathrm{sep},\mathbf{k},\mathbf{k}}^{(0)} + \left(v_{\mathrm{loc}}^{\prime(0)} + v_{\mathrm{H}}^{\prime(0)} + v_{\mathrm{xc}}^{(0)}\right). \tag{4.33}$$

The local, Hartree and exchange-correlation (XC) potentials are operators local in the real space, independent of **k** and thus possess the full lattice periodicity.

Following the Ewald summation method (see Section 2.1.4), the ion–ion contribution E_{I-I} to the unperturbed total energy per unit cell (to which the residue of the cancellation of the divergences mentioned in Eq. (4.23) must be added) is obtained as PICKETT [1989].

Such a computational scheme has several advantages: first, the plane wave basis set corresponds to a uniform real-space grid. This implies that the derivatives of the total energy with respect to the atomic displacements, i.e., the atomic forces, can be easily calculated through the Hellmann–Feynman theorem (FEYNMAN [1939]). At odds with the case of localized basis sets – often consisting of finite-range functions centered on the atomic sites – the quality and the accuracy of the description of the electron density is completely independent of the atomic positions. Therefore, when calculating the atomic forces using plane waves, one has not to evaluate Pulay forces (SAEBO and PULAY [2001]). That makes plane waves an optimal and easy-to-use basis set whenever the stable atomic configurations are not a priori known, i.e., when geometry optimization

[12]See footnote 11.

or molecular dynamics runs have to be carried out. Second, the availability of Fast Fourier Transforms (FFT) permits to perform each specific operation either in real space or in reciprocal space, according to the numerical convenience, thus avoiding any convolution product. Third, the implementation of analytical formulas in plane waves is simple and transparent, so that reading and improving plane-wave based computer codes is generally an easier task than when using a localized basis set.

On the other hand, plane waves imply some drawbacks: firstly, their number is generally very large (usually there are of the order of 10^2 or even 10^3 PWs per valence orbital). As a result, the computational effort is quite big and the memory requirements for one-hundred-atom systems may easily approach a GByte. The output files are huge and the physical quantities are hindered in such a big amount of information. In order to reduce the computational effort, PW are usually employed to describe valence electrons only, in conjunction with soft pseudopotentials. In addition, useless accuracy is employed in region of space where no electron density is present. Moreover, the use of PW generally implies PBC in the three spatial directions, which does not correspond to the physical reality in the case of isolated systems such as molecules or clusters or to semi-infinite systems such as surfaces (see Section 2.2.1).

4.5. *Localized basis set methods*

At variance with the case of plane waves, localized basis set are usually formed by a small amount of functions that are well localized in real space and usually (but not exclusively) centered on the nuclei. In order to reproduce small electron density variations, much attention is put in the choice of the type of the basis functions. Then, a number of integrals involving two, three and four basis functions, has to be computed as a function of their respective centers to calculate the total energy and the atomic forces. The related computational effort can be minimized in two respects: on one hand, by limiting the basis set size; on the other hand, by choosing suitable basis functions that make the calculation of the integrals as easy as possible. In addition to the previous issues, the basis set should be *transferable*, that is, if a given set of basis function is chosen for each chemical species, they should reproduce the chemical bonding in various compounds and possibly in aperiodic systems, such as defective crystals, surfaces, etc. From all these considerations, it is clear that the construction of a localized basis set is often an art, and that, contrarily to plane waves, its quality cannot be improved straightforwardly. Since must of the aspects linked to the practical set up of localized basis sets in atoms and molecules have already been treated in Chapter 1 of this Handbook (CANCÉS, DEFRANCESCHI, KUTZELNIGG, LE BRIS and MADAY [2003]), here we briefly remind some related issues that are peculiar for crystals.

In the Crystal Momentum representation (see Section 2.1.1), a Bloch orbital can be constructed from a linear combination of localized functions (LF) $\{\phi_{I,\xi}(\mathbf{r}-\mathbf{r_I})\}$, where $\mathbf{r_I}$ is the LFs center and ξ gathers all the quantum numbers and orbital characteristics that identify a particular function, as:

$$\psi_{m,\mathbf{k}}(\mathbf{r}) = \frac{e^{i\mathbf{k}\cdot\mathbf{r}}}{(N\Omega_0)^{\frac{1}{2}}} \sum_{I,\xi,\mathbf{L_I}} c_{m,\mathbf{k}}^{(I,\xi)} \phi_{I,\xi}(\mathbf{r}-\mathbf{r_I}-\mathbf{L_I}). \tag{4.34}$$

The minimal basis set corresponds to the exact Wannier functions (see Section 2.1.2); practically, a good choice of the LFs is a good starting point to determine the Wannier functions. In general, such LFs are not orthogonal to each other: either one works with a orthonormalized basis set that is constructed from those LFs, or solve the one-particle effective Hamiltonian in a nonorthogonal basis. In the latter case, one has to calculate the overlap matrix:

$$S_{I,\xi,J,\xi'} = \frac{1}{\Omega} \int d\mathbf{r}\, \phi^*_{I,\xi}(\mathbf{r} - \mathbf{r_I}) \phi_{J,\xi'}(\mathbf{r} - \mathbf{r_J}) \tag{4.35}$$

and express the normalization of a Bloch orbital as:

$$\int d\mathbf{r}\, \left|\psi_{m,\mathbf{k}}(\mathbf{r})\right|^2 = \sum_{I,\xi,J,\xi'} c^{*(I,\xi)}_{m,\mathbf{k}} S_{I,\xi,J,\xi'} c^{(J,\xi')}_{m,\mathbf{k}} = 1 \quad \forall m, \mathbf{k}. \tag{4.36}$$

By combining the previous equations, the electron density can be expressed as function of the LFs $\{\phi\}$ and the expansion coefficients $\{c^{(I,\xi)}_{m,\mathbf{k}}\}$. Once the LFs are set up, the latter ones are the quantities to be determined in order to compute the electron density, the total energy, etc. In the Kohn–Sham scheme (see Eq. (3.3)), this is equivalent to solve the following equation:

$$\frac{\partial}{\partial c^{*(I,\xi)}_{m,\mathbf{k}}} \left[E_{\mathrm{el}}\{c^{(J,\varsigma)}_{n,\mathbf{k}}\} - \sum_{m,\mathbf{k}} \left(\sum_{I,\xi,J,\xi'} c^{*(I,\xi)}_{m,\mathbf{k}} c^{(J,\xi')}_{m,\mathbf{k}} - 1 \right) \right] = 0 \tag{4.37}$$

which, by introducing the Hamiltonian matrix elements $\widetilde{H}_{I,\xi,J,\xi'} = \langle \phi_{I,\xi} | \widetilde{H} | \phi_{J,\xi'} \rangle$ on the LFs, results in a *generalized eigenvalue problem*:

$$\sum_{J,\xi'} \left[\widetilde{H}_{I,\xi,J,\xi'} - \varepsilon_{m,\mathbf{k}} S_{I,\xi,J,\xi'} \right] c^{(J,\xi')}_{m,\mathbf{k}} = 0. \tag{4.38}$$

From Eq. (4.38), one can see that as the overlap between two LFs approaches the unity, the determinant of the overlap matrix tends to zero and the solution of the generalized eigenvalue problem becomes unstable. Another typical drawbacks of LF basis set is the *basis set superposition errors*, which is related to the extension of the basis set when treating a problem for which the original set is markedly uncomplete. For instance, when creating vacancies in ionic solids, some electron density may be trapped at the vacancy site and not be well accounted for by the LF tails. In that case, some more LFs centered at the vacancy site can be added, but the energy differences between the two distinct configurations (i.e., the perfect solid versus the defective one) must be carefully calculated by using the most complete basis set. For more details about the use of localized basis sets in DFT, the interested reader is referred to the existing literature (SANCHEZ-PORTAL, ORDEJÓN, ARTACHO and SOLER [1997], BRIDDEN and JONES [2000]).

4.6. *Mixed basis set methods*

In the following we focus on the principal concepts common to the mixed basis set approaches, on how they appeared and became implemented in the computational schemes. For a detailed description of particular existing variants we refer the reader to the extensive bibliography.

As it has already been pointed out, for many physical applications the pseudopotential method is not entirely sufficient. On the one hand, the adequate description of certain physical problems needs the explicit treatment of core electrons. On the other hand, pseudopotentials of many elements are not weak enough to allow an efficient plane wave expansion of the pseudo-wavefunctions. Contrary to the concept of pseudo-Hilbert space used in the pseudopotential approach, the idea underlying the mixed basis set methods is to construct the basis set functions as to match the character of true crystal wavefunctions as much as possible. In order to do that, the unit cell is divided in different regions and distinct expansions of the wavefunctions are used in each of them.

4.6.1. *Construction of the basis set: Dividing up space*

The principle of the mixed basis set methods can be traced back to the original idea by SLATER [1937]. Since near the atomic nuclei the crystal potential and the crystal wavefunctions are similar to those in isolated atoms (strongly varying but nearly spherical), and in the space between the atoms both potential and wavefunctions are smooth, the space should be divided into two types of regions: atom-centered atomic spheres, and the remaining interstitial region. Different expansions should be used to represent the wavefunctions in each of these regions.

With the goal to generate basis functions resembling closely the true crystal wavefunctions, so that a small (or preferably minimal) basis set is sufficient to span the Hilbert space and to keep the related computational effort the smallest possible, a simplified form of crystal potential is used for generating the basis sets functions. Due to its resemblance to the shape of tins used for making muffins, this simplified potential is often called *muffin-tin potential*, and the related simplification, the *Muffin-Tin Approximation* (MTA).

Inside the *muffin-tin spheres* (MTs), where the crystal potential $V(\mathbf{r})$ is almost spherically symmetric, only its spherically symmetric part $V_{\mathrm{MT}}(r)$ is used for construction of partial waves $Y_L(\hat{r})\, u_l(r, \varepsilon)$ out of the numeric solutions of the radial Schrödinger equation (SE):

$$\left[-\frac{d^2}{dr^2} + \frac{l(l+1)}{r^2} + V_{\mathrm{MT}}(r) - \varepsilon\right] r u_l(r, \varepsilon) = 0. \tag{4.39}$$

In the above expression L stands for the condensed angular momentum index (l, m), and u_l are required to be well defined at $r = 0$ in order to be normalizable.

In the *interstitial region* (IR), where the potential is almost flat or slowly varying with respect to the vicinity of the nuclei, it is approximated by a constant V_{MTZ}.

The free-particle solutions of the Schrödinger equation are used for generation of the corresponding basis set expansion:

$$\left[-\nabla^2 + V_{\mathrm{MTZ}} - \varepsilon\right]\phi(\mathbf{r}, \varepsilon) = 0. \tag{4.40}$$

The basis set functions are than constructed by *augmentation* of the slowly varying *envelopes* $\phi(\mathbf{r}, \varepsilon)$ (IR) to the partial waves $Y_L(\hat{r})\ u_l(r, \varepsilon)$ (MTs). The true crystal wavefunction is represented as a linear combination of the basis set function.

In order to obtain continuous and differentiable crystal wavefunctions, the intuitive tendency is to assure these properties already at the level of the basis set functions, which can be achieved by matching smoothly the partial waves coming from the different regions at the boundary of each MT sphere. However, since in principle the basis set functions do not have to be smooth, some of the methods profit of this additional flexibility. Finally, their quality can be often improved by an increase of the basis set size.

Two important points are to be stressed here. First, the MTA is in principle used to generate the basis set functions only, whereas the true crystal wavefunction (a linear combination of basis set functions) is obtained from the full crystal potential. Second, during the iterative (self-consistent) procedure of solution of the crystal band problem, the crystal potential converges to its true, ground state form. As a consequence, the MT potential changes from one iteration to another and so do the basis set functions.

The most important difference between the various implementations of the mixed basis set methods in computational solid state physics regards the choice of the envelope functions. In this respect, practically all existing approaches can be derived from two original concepts. As suggested by its name, the Augmented Plane Wave (APW) method by Slater (SLATER [1937], SLATER [1964], LOUCKS [1967]) uses plane wave envelopes in the interstitial, whereas in the Korringa, Kohn, and Rostoker (KKR) method (KORRINGA [1947], KOHN and ROSTOKER [1954]) spherical waves are used. Historically, the KKR method was developed in the multiple-scattering formalism, whereas the Augmented Spherical Wave (ASW) method is associated with a different class of approaches (EYERT [2000]).

The character of these two types of methods is to a large extent determined by the different localization of their envelope functions: delocalized plane waves (APW) and atomic-site-centered spherical waves (KKR). Since the envelope function character determines to some extent the field of applicability of the method, and influences considerably its computational efficiency, the different localization of the envelopes has affected the historical evolution of the two kinds of approaches.

In the following we have a closer look at the practical set up of the basis set in these two cases, and point out the particularities related to each choice. For the sake of conciseness, we consider the case of a single atom per crystal unit cell, placed at the origin of the coordinate system.

4.6.1.1. Plane waves expansion. See, e.g., SINGH [1994]. Plane waves $e^{i\mathbf{Kr}}$ are the eigenfunctions of the kinetic energy operator and solutions of the SE with constant potential V_{MTZ}:

$$\mathbf{K}^2 = \varepsilon - V_{\mathrm{MTZ}}. \tag{4.41}$$

As already discussed in Section 4.4, their mathematical simplicity makes them particularly suited for representation of smooth, slowly varying periodic functions. For each vector **k** in the Brillouin zone, the crystal wavefunction $\Psi^{\mathbf{k}}(\mathbf{r})$ can be expanded in a (Fourier) series (**G** are the reciprocal lattice vectors):

$$\Psi^{\mathbf{k}}(\mathbf{r}) = \sum_{\mathbf{G}} \mathbf{c}^{\mathbf{k}}_{\mathbf{G}} \psi^{\mathbf{k}}_{\mathbf{G}}(\mathbf{r}) \tag{4.42}$$

of basis set functions $\psi^{\mathbf{k}}_{\mathbf{G}}(\mathbf{r})$ defined as plane waves in the interstitial region augmented to the partial waves inside the muffin-tin spheres:

$$\psi^{\mathbf{k}}_{\mathbf{G}}(\mathbf{r}) = \begin{cases} \sum_L a^{\mathbf{k}}_{\mathbf{G}L} u_l(r, \varepsilon) Y_L(\hat{\mathbf{r}}), & |\mathbf{r}| \leqslant R_{\mathrm{MT}}, \\ \Omega_0^{-1/2} e^{i(\mathbf{G}+\mathbf{k})\mathbf{r}}, & \mathbf{r} \in \mathrm{IR}. \end{cases} \tag{4.43}$$

Requirement of continuity of the basis set functions on the MT sphere (in the original formulation of the method the functions are continuous but not differentiable at the MTs) leads to a simple expression for the coefficients $a^{\mathbf{k}}_{\mathbf{G}L}$. Taking into account the spherical harmonic expansion of the plane waves around the origin:

$$e^{i\mathbf{G}\cdot\mathbf{r}} = 4\pi \sum_L i^l j_l(Gr) Y^{\star}_L(\widehat{\mathbf{G}}) Y_L(\hat{\mathbf{r}})$$

they can be written as:

$$a^{\mathbf{k}}_{\mathbf{G}L} = \frac{4\pi i^l}{\sqrt{\Omega_0} u_l(R_{\mathrm{MT}}, \varepsilon)} j_l(|\mathbf{k}+\mathbf{G}| R_{\mathrm{MT}}) Y^{\star}_L(\hat{\mathbf{k}}+\mathbf{G}).$$

In the above expressions, j_l stands for Bessel functions and Ω_0 is the volume of the crystal unit cell. An individual function, labelled by G and consisting of a single plane wave in the interstitial matched to radial functions in all MT spheres is the augmented plane wave, or APW. If the wavefunction of the crystal $\Psi^{\mathbf{k}}(\mathbf{r})$ is expanded in APWs the coefficients c_G become the variational parameters of the method.

Forming a complete set, plane waves reproduce correctly the behavior of Bloch states in the interstitial region. In practice, as in pseudopotential methods, the quality and the completeness of the IR part of the basis set is controlled by the energy cut-off E_{cut}, and all plane waves of kinetic energy $\hbar^2|\mathbf{G}+\mathbf{k}|^2/2m < E_{\mathrm{cut}}$ are used.

4.6.1.2. Spherical waves expansion. See, e.g., SKRIVER [1984]. Spherical waves $y_l(\kappa r)$ are the atom-centered products of spherical harmonics $Y_L(\hat{r})$ and radial solutions of the SE for a kinetic energy $\kappa^2 = \varepsilon - V_{\mathrm{MTZ}}$:

$$\left[-\frac{d^2}{dr^2} + \frac{l(l+1)}{r^2} - \kappa^2\right] r y_l(\kappa r) = 0. \tag{4.44}$$

The solutions of the Helmholtz equation (4.44) are well known as spherical Bessel j_l, Neumann n_l, and Hankel h_l functions. We bear in mind that only two of them are linearly independent since $h_l = j_l - in_l$.

Augmented to a partial wave inside the MT sphere, the so-called *muffin-tin orbital* (MTO) is conventionally defined as:

$$\psi_L(\mathbf{r}) = i^l\, Y_L(\hat{\mathbf{r}}) \begin{cases} u_l(r,\varepsilon) + \kappa \cot(n_l)\, j_l(\kappa, r), & r \leqslant \mathrm{R_{MT}}, \\ \kappa n_l(\kappa, r), & r \geqslant \mathrm{R_{MT}}. \end{cases} \tag{4.45}$$

Compared to the construction of APWs, several points are noteworthy. The Bessel function $j_l(\kappa, r)$ (*head*) is used inside the MT sphere, whereas the Neumann function $n_l(\kappa, r)$ forms the *tail* outside. The two functions are chosen to be regular at the origin (head) and at infinity (tail). If $\kappa^2 < 0$, n_l is to be substituted by Hankel function h_l, which is regular at infinity for imaginary κ. Since the conventional MTO is smooth and continuous by construction at the boundary of MT spheres, the constant $\cot(n_l)$ assures the matching of partial waves and their derivatives on the MT sphere boundary.

Basis set functions are then the Bloch functions constructed of MTOs centered in different cells of periodic crystal (**R** is the primitive translation vector):

$$\psi_L^{\mathbf{k}}(\mathbf{r}) = \sum_{\mathbf{R}} e^{i\mathbf{kR}} \psi_L(\mathbf{r} - \mathbf{R}) = \psi_L(\mathbf{r}) + \sum_{\mathbf{R} \neq \mathbf{0}} e^{i\mathbf{kR}} \psi_L(\mathbf{r} - \mathbf{R}).$$

Thanks to the lattice periodicity and to the expansion theorem $n_l(\kappa, \mathbf{r} - \mathbf{R}) = \sum_{L'} j_{l'}(\kappa, r) S_{LRL'0}(\kappa)$, that holds for any solution of Helmholtz equation (4.44): the contribution due to tails of MTOs centered at $\mathbf{R} \neq \mathbf{0}$ can be recasted in a particularly simple form:

$$\sum_{\mathbf{R} \neq \mathbf{0}} e^{i\mathbf{kR}} n_l(\kappa, \mathbf{r} - \mathbf{R}) = \sum_{L'} j_{l'}(\kappa, r) B_{LL'}^{\mathbf{k}}(\kappa),$$

where the reciprocal space *structure constants* matrix $B_{LL'}^{\mathbf{k}}(\kappa)$ is the Fourier transform of the real space structure constant matrix $S_{L0L'R}(\kappa)$.

The coefficients c_L of the linear expansion $\sum_L c_L^{\mathbf{k}} \psi_L^{\mathbf{k}}(\mathbf{r})$ are thus the variational parameters of the method. The so-called *tail cancellation condition*, according to which, inside each MT sphere local heads cancel tails coming from all other MT spheres:

$$\sum_{L'} \left[\cot(n_l) \delta_{LL'} - B_{LL'}^{\mathbf{k}}(\kappa)\right] c_L = 0$$

can be used in order to obtain the coefficients c_L.

Being localized at the atomic sites and decreasing in the interstitial regions far from their centers, the MTOs are, in principle, not an efficient expansion in the IR. In fact, they work the best when the unit cell is filled up tightly with the atomic spheres, as it is the case for face centered or body centered cubic structures. More open crystallographic structures (simple cubic, diamond), and especially surfaces, clusters, or molecules (in the supercell approach) often require use of additional basis functions centered off the actual atomic positions. This is conventionally done by the so-called *empty spheres*, which are MT spheres, placed at arbitrary positions, containing no atom, but localizing additional MTOs. In practice, the lack or an erroneous choice of empty spheres can lead to important and not always easily detectable errors. As in the case of Gaussians, moving the atoms and comparing the relative stability of different structures can be somehow tricky. However, for several physical systems (e.g., transition metals and their alloys), the MTOs provide a small but good basis set, and the resulting secular matrix is of an order of magnitude smaller compared to the APW approach. Moreover, the atomic-like character of the MTOs provides a natural basis for interpretation of the results. In the last years, the search for an optimal minimal basis set has been put in relation with the construction of the Wannier orbitals (TANK and ARCANGELI [2000]).

4.6.2. *Energy dependence of the basis set functions: Linearization*

Although we did not point it out explicitly, in both types of mixed basis set methods described above, the basis set functions are energy dependent. The two energy parameters: ε (for partial waves inside MTs), and κ^2 (or G^2) (for partial waves in the interstitial) are in principle related by Eq. (4.41) which assures that each basis set function is an exact solution of the SE for the MT potential. However, since in the case of basis set functions this requirement is of a secondary importance, the two parameters can be decoupled, giving the basis set an additional flexibility. In practice, whereas in the APW method all plane waves of kinetic energy smaller than $G^2_{\max}$ are used, in MTO methods it is more efficient to use the tails generated for one, or a few fixed values of κ chosen in the region of physical interest. Contrary to the plane wave methods, the generation of partial waves for many different κ's is not an efficient compensation of the in-completeness of the basis set related to the finite number of localization centers.

As a consequence, in both methods the solution of the band problem is defined by the value of remaining energy parameter ε and the expansion coefficients $c^{\mathbf{k}}$. Although the algebraic formulation can be different (variational principle, tail cancellation, or kink cancellation condition, etc.), the result is a set of linear, homogeneous equations $M^{\mathbf{k}}(\varepsilon)c^{\mathbf{k}}_L = 0$. In general, the secular matrix M depends on ε in a complex nonlinear way, so that the eigenvalues must be found individually by tracing the roots of its determinant. The coefficients $c^{\mathbf{k}}_L$ are than found by solving the linear equations for each ε.

Calculation of the complete band structure $\varepsilon(\mathbf{k})$ thus implies a remarkable computational effort, in contrast to methods using an energy-independent basis set, where the corresponding generalized eigenvalue problem can be solved much faster.

An approximation aiming at suppressing the energy dependence of the basis set was originally proposed by ANDERSEN [1975], who observed that in both APW and KKR methods the dependence of the valence electron wave functions on the energy is relatively weak. An approximation consisting of the two first terms in the corresponding Taylor expansion:

$$u_l(r,\varepsilon) = u_l(r,\varepsilon_l) + (\varepsilon - \varepsilon_l)\dot{u}_l(r,\varepsilon_l) + \mathrm{O}\big((\varepsilon - \varepsilon_l)^2\big) \tag{4.46}$$

gives in practice a very satisfactory result. (We bear in mind that quadratic order or error in wavefunction corresponds to fourth order in the energy.) This concept is at the origin of so-called linear (L) methods: LAPW (ANDERSEN [1975], KOELLING and ARBMAN [1975], WIMMER, KRAKAUER, WEINERT and FREEMAN [1981]) and LMTO (ANDERSEN [1975], ANDERSEN, JEPSEN and SOB [1986], SKRIVER [1984]), which in this respect can be seen as linearized descendants of APW and KKR approaches.

In practical implementations, the linearization is realized by addition of an extra partial wave $\dot{u}_l(r, \varepsilon_l)$ to the expansion inside the MT sphere:

$$Y_L(\hat{r})\left[a_L u_l(r, \varepsilon_l) + b_L \dot{u}_l(r, \varepsilon_l)\right].$$

Energies ε_l are used as additional parameters and are re-actualized during the self-consistent procedure. Their choice may seem intuitive: either from definition (Taylor expansion), or to assure the desired accuracy, one could place them around the band center of mass. Moreover, by using the information given by the appropriate moments they can be further shifted as to optimize the wavefunction or the energy. It is also possible to treat them as additional variational parameters: perform the calculations for several choices of ε_l and choose the one that gives the lowest total energy.

Beside the elimination of the energy dependence of the basis set, doubling the number of partial waves inside the sphere gives the linearized orbital an additional degree of freedom to better represent the nonspherical components of the true crystal potential, and to satisfy the boundary conditions at the MT sphere. In particular, contrary to the APS method, in most of LAPW approaches the LAPWs are made continuous and differentiable at the sphere boundary. Finally, the second term solves the asymptote problem: whereas in both types of methods, for certain energies ε_l, $u_l(R_{\mathrm{MT}}, \varepsilon_l) = 0$ and the two parts of the basis function (inside or outside the atomic spheres) become decoupled, $u_l(R_{\mathrm{MT}}, \varepsilon_l)$ and $\dot{u}_l(R_{\mathrm{MT}}, \varepsilon_l)$ rarely vanish for the same ε.

To summarize this section, the linearized methods can be seen as an extension of the LCAO approach, in which the linearized augmented partial waves are used as basis set functions. However, since the basis set evolves during the self-consistent procedure (following the changes of the MT potential), the solutions can be made exact for the MT potential, and the partial wave methods are thus by far more accurate than any fixed basis approach.

4.6.3. Core and semi-core states

As described above, the mixed basis set methods are perfectly adequate to treat self-consistently both valence and core electrons. The latter ones do not require the envelope functions, and can be satisfactorily treated within the MTA. Finally, their confinement to the MT sphere assures their orthogonality to the valence states (ANDERSEN [1975]). These elements are characteristic of the so-called all-electron methods.

However, the practical distinction between core and valence states is not always clear (see also Sections 4.1 and 4.2). Depending on the physical problem (which often determines the choice of MT spheres), some high lying core states (or *semi-core states*) may be severely influenced by the atom environment and the spherically symmetric treatment adequate for the low lying atomic states becomes insufficient. As a consequence, two (or more) partial waves with the same angular quantum number L and different principal quantum numbers n (thus, different ε_l) have to be accounted for simultaneously. Two alternative methods have been developed in order to handle the semi-core states in the band structure calculations (see, e.g., SINGH [1994]).

The *multiple panels* technique consists of choosing several nonoverlapping energy windows (panels), and treating separately states, with appropriate, independent ε_l in each of them. For a given potential and combined electronic density, this results a single diagonalization per energy window, and additional care needs to be taken to assure the orthogonality of states belonging to different panels. This technique has been implemented principally to the MTO approaches.

The *local orbital* method consists of adding an additional orbital $u_l^*(r, \varepsilon_l^*)$ to the linear combination of $u_l(r, \varepsilon_l)$ and $\dot{u}_l(r, \varepsilon_l)$ inside the sphere. The energy parameters ε_l and ε_l^* are fixed independently, and $u_l^*(r, \varepsilon_l^*)$ is handled by an appropriate choice of boundary conditions and number of nodes. Since both u_l and u_l^* enter the diagonalization, the orthogonality is assured automatically. This technique is used within the LAPW approach.

In both approaches that are described above, the orthogonality procedure may result in the so-called *ghost states*, which correspond to badly converged (core) states appearing much too high in energy. Those ghost states are not to be confused with the spurious solutions of the Kleinman–Bylander form of pseudopotentials (see Section 4.3).

4.6.4. *Atomic sphere approximation*

It is important to mention an additional approximation that leads to a particularly simple form of MTOs, and which has given origin to an entire family of mixed basis-set approaches. By setting $\kappa = 0$ in Eq. (4.45), the spherical waves behave as $r^{\pm l}$ and the corresponding MTO reads (see, e.g., SKRIVER [1984]):

$$\psi_L(\mathbf{r}) = i^l Y_L(\hat{r}) \begin{cases} u_l(r, \varepsilon) + p_l(\varepsilon)(r/R_{\mathrm{MT}})^l, & r \leqslant R_{\mathrm{MT}}, \\ (R_{\mathrm{MT}}/r)^{l+1}, & r \geqslant R_{\mathrm{MT}}, \end{cases} \tag{4.47}$$

where

$$p_l(\varepsilon) = \frac{D_l(\varepsilon) + l + 1}{D_l(\varepsilon) - l} \quad \text{and} \quad D_l(\varepsilon) = \frac{R_{\mathrm{MT}}}{u_l(R_{\mathrm{MT}}, \varepsilon)} \frac{\partial u_l(r, \varepsilon)}{\partial r}\bigg|_{r=R_{\mathrm{MT}}}.$$

This simple form of MTOs is the basic ingredient of the Atomic Sphere Approximation (ASA), in which the MT spheres overlap, and the interstitial region is effectively eliminated. This leads to factorize the matrix elements of operators into products of structure constants and radial integrals, so to improve considerably the efficiency of computer codes, and providing a convenient starting point for simplified, analytical theories of the electronic structure of solids.

The basic disadvantage of the multipole expansion appearing in the original ASA-LMTO formulation is its long range, which needs the use of Ewald technique for evaluation of lattice sums (the structure constants), and limits the applications to periodic systems. This problem was circumvented by what is called screened, or tight-binding (TB) MTOs, where an additional transformation is introduced in order to regulate the range of MTOs tails (ANDERSEN, JEPSEN and GLÖTZEL [1985], ANDERSEN, PAWLOWSKA and JEPSEN [1986]). A substantial effort has been done towards generalization and improvement of precision and flexibility of muffin-tin orbitals, resulting in what is referred to as MTOs of 2nd and 3rd generation. For further details we refer the reader to the existing reviews (see, e.g., TANK and ARCANGELI [2000]).

4.6.5. *Full-potential methods*

In the mixed basis set methods, the basis set functions are constructed from solutions of the SE obtained for the MT potential. This simplified form of potential can be further used in actual calculations of the crystal electronic structure and produce satisfactory results for many physical problems concerning principally the closed packed solids. This procedure becomes less and less reliable as the structure becomes more open, the atomic coordination numbers decrease and the symmetry lowers. In these cases, the MTA produces inaccuracies that cannot be healed simply by increasing the size of the basis set.

However, if the MTA is restricted to the construction of the basis set only, and the Hamiltonian entering the secular matrix incorporates the full, unrestricted crystal potential, the inaccuracy of the basis set functions is to a large extent alleviated by the variational principle, and the quality of solutions can be improved by increasing the basis set size. This observation is at the origin of the so-called full-potential (FP) methods, which appeared as a natural extension of simpler, MT potential approaches: FP-LAPW (WIMMER, KRAKAUER, WEINERT and FREEMAN [1981], MATTHEISS and HAMANN [1986], SOLER and WILLIAMS [1990], BLAHA, SCHWARZ, SORANTIN and TRICKEY [1990], SINGH, KRAKAUER, HAAS and LIU [1992], PETERSEN, WAGNER, HUFNAGEL, SCHEFFLER, BLAHA and SCHWARZ [2000]), and FP-LMTO (METHFESSEL, RODRIGUEZ and ANDERSEN [1989], METHFESSEL and VAN SCHILGAARDE [1993], SAVRASOV and SAVRASOV [1992], WILLIS, PRICE and COOPER [1989], BOTT, METHFESSEL, KRABS and SCHMIDT [1998]).

The additional computational difficulty of the full-potential treatment is related to the necessity of an efficient representation and handling of the nonspherical terms of the crystal potential and charge density. In practical implementations, this extension is often realized through the use of a dual basis set: in conjunction with the basis set that is employed to span the Hilbert space of wavefunctions, another one, which is independent of the first one, is introduced to expand the actual crystal potential and charge density. Since the main Coulomb contribution to the crystal potential is related to the crystal charge density (electronic plus nuclear) via the Poisson equation, in most of the calculations the charge density is expanded in a linear combination of functions that are solutions of the Poisson equation. This is relatively straightforward in the interstitial region (where the basis set functions are the eigenfunctions of kinetic energy operator), and requires an adjustment/approximation inside the MT spheres. The latter can be done, e.g., by adjusting the values of multi-poles as to reproduce those of the true electronic density (HAMANN [1979], WEINERT [1981]). Once the potential at the sphere boundary is known, a standard real space approach may be used for the integration of the Poisson equation within the MT spheres.

4.7. *Projector Augmented Wave (PAW) method*

In the preceding sections we have presented two qualitatively different concepts aiming to an efficient calculation of the band structure of periodic systems. On the one hand, the mixed basis-set methods, where an augmentation of slowly varying envelops to radial solutions of the SE inside the atomic spheres results in basis-set functions representative to true crystal wavefunctions. The price to pay for handling the true wavefunctions is a relatively heavy formalism and a computational burden. On the other hand, in the pseudopotential methods the true nuclear potential is screened by core electrons, so that the slowly varying valence pseudo-wavefunctions can be efficiently expanded in plane waves. Mathematical simplicity and computational efficiency are gained at the expense of abandoning the representation of true crystal wavefunctions in the core region.

4.7.1. *Pseudo- and true Hilbert space*

As it was pointed out by BLÖCHL [1994], if the gap between the true crystal wavefunction Ψ (which is the wanted solution of the band problem), and the well behaving pseudo-wavefunction $\widetilde{\Psi}$ (which can be computed efficiently) is bridged by a linear transformation $\Psi = T\widetilde{\Psi}$, the physical quantities (expectation values of operator $\langle A \rangle$) can be easily calculated in the pseudo-Hilbert space representation $\langle \widetilde{\Psi} | \tilde{A} | \widetilde{\Psi} \rangle$ (with $\tilde{A} = T^{\dagger} A T$), rather than directly from true wavefunctions. Similarly, the variational principle for the total energy:

$$\frac{\partial E[T|\widetilde{\Psi}\rangle]}{\partial \langle \widetilde{\Psi}|} = \varepsilon T^{\dagger} T |\widetilde{\Psi}\rangle \tag{4.48}$$

gives an equivalent of the KS equation for the pseudo-wavefunctions, and the search for the ground state can be also done in the pseudo-Hilbert space. This latter, due to the slowly varying character of the pseudo-wavefunctions can be efficiently spanned by plane waves. However, contrary to the pseudopotential method, thanks to the transformation T the pseudo-wavefunctions are directly shadowing the true crystal wavefunctions.

4.7.2. *Constructing the transformation T*

As it can be expected, the transformation T mainly regards the regions of atomic cores. In the interstitial regions, where the true crystal wavefunction varies slowly Ψ, $\widetilde{\Psi}$ can practically be built to coincide with Ψ. As a consequence, T can be seen as a sum of nonoverlapping atom-centered contributions T_R, each of them acting within the corresponding *augmentation region*, such that $T = 1 + \sum_R T_R$. By analogy with the concept of muffin-tin spheres, the transformation T can be seen as a generalization of the augmentation procedure. In the following we focus on the practical construction of the local operators T_R.

Inside each MT sphere, both the true crystal and the pseudo-wavefunctions can be represented as linear combinations of (true and pseudo-) partial waves:

$$\widetilde{\Psi} = \sum_i c_i \tilde{\psi}_i \quad \text{and} \quad \Psi = \sum_i c_i \psi_i = T\widetilde{\Psi},$$

where, since $\psi_i = T\tilde{\psi}_i$, the coefficients c_i are the same in both expansions. As a consequence:

$$\Psi = \widetilde{\Psi} + \sum_i c_i (\psi_i - \tilde{\psi}_i)$$

and the transformation T can be written as:

$$T = 1 + \sum_i (\psi_i - \tilde{\psi}_i) \langle \tilde{p}_i |.$$

In the above expression we have introduced the set of projector functions $\langle \tilde{p}_i |$, which verify the condition:

$$\widetilde{\Psi} = \sum_i c_i \tilde{\psi}_i = \sum_i \langle \tilde{p}_i | \widetilde{\Psi} \rangle \tilde{\psi}$$

and are orthogonal to the pseudo-partial waves ($\langle \tilde{p}_i | \tilde{\psi}_j \rangle = \delta_{ij}$).

Inside each augmentation region there are thus two sets of partial waves and a set of projector functions. In the following we describe briefly these three basic elements of the PAW technique.

The set of the all-electron partial waves ψ_l can be, as in the mixed basis-set methods, generated from numerical solutions of the radial SE. They can be chosen to describe the physically relevant states (valence band and semi-core states), and their number per l channel can be increased up to the desired convergence of results. Additionally, the divergent tail of the true partial wave cancels with that of the corresponding pseudo-partial wave), so that there is no constrain to the bound states only. As in linearized methods, two partial waves per valence region are often enough to span efficiently the Hilbert space. Two partial waves at two different energies, or a partial wave and its derivative, can be used.

Since the pseudo-partial waves $\tilde{\psi}_l$ are supposed to be slowly oscillating, they can be in principle generated by the techniques used in the pseudopotential approach. This procedure assures additionally the completeness of the basis set and the orthogonality to the core states. Pseudo-partial waves are then matched differentially to the all-electron partial waves at the MT sphere, so that for the same ε both functions coincide outside the augmentation region (see, e.g., KRESSE and JOUBERT [1999]).

Different practical schemes can be used to generate the set of projector functions $\tilde{p}$, dual to the set of pseudo-partial waves $\langle \tilde{p}_i | \tilde{\psi}_j \rangle = \delta_{ij}$. The Gram–Schmidt inspired scheme by BLÖCHL [1994] seems particularly well adapted for numerical implementation, assuring that the all electron and the pseudo-wave partial waves form complete sets of functions within the augmentation region.

Within the PAW formalism, the introduction of a one-center density matrix $D_{ij} = \langle\widetilde{\Psi}|\tilde{p}_i\rangle\langle\tilde{p}_j|\widetilde{\Psi}\rangle$ provides a convenient separation of the expectation value of any operator A into three terms $A = \tilde{A} + A^1 - \tilde{A}^1$:

$$
\begin{aligned}
\langle A\rangle &= \langle\widetilde{\Psi}|\tilde{A}|\widetilde{\Psi}\rangle = \langle\widetilde{\Psi}|A + \sum_{ij}|\tilde{p}_i\rangle\big(\langle\psi_i|A|\psi_j\rangle - \langle\tilde{\psi}_i|A|\tilde{\psi}_j\rangle\big)\langle\tilde{p}_j|\widetilde{\Psi}\rangle \\
&= \langle\widetilde{\Psi}|A|\widetilde{\Psi}\rangle + \sum_{ij} D_{ij}\langle\psi_i|A|\psi_j\rangle - \sum_{ij} D_{ij}\langle\tilde{\psi}_i|A|\tilde{\psi}_j\rangle.
\end{aligned}
$$

The term $\tilde{A}$ is smooth and can be evaluated on regular grid in reciprocal or in real space, whereas the two one-center contributions A^1 and $\tilde{A}^1$ are evaluated inside each augmentation region on radial grids in an angular momentum representation. This expression, with $A = |\mathbf{r}\rangle\langle\mathbf{r}|$ can be used to obtain the electronic density $n(\mathbf{r})$ necessary for evaluating the total energy functional and solving Eq. (4.48).

References

ANDERSEN, O.K. (1975), *Phys. Rev. B* **12**, 3060.

ANDERSEN, O.K., O. JEPSEN and GLÖTZEL (1985), in: F. Bassani, F. Fumi, and M.P. Tosi, eds., *Highlights of Condensed-Matter Theory* (North-Holland, New York).

ANDERSEN, O.K., O. JEPSEN and M. SOB (1986), in: M. Yussouff, ed., *Electronic Band Structure and Its Applications* (Springer-Verlag, Berlin).

ANDERSEN, O.K., Z. PAWLOWSKA and O. JEPSEN (1986), *Phys. Rev. B* **34**, 5253.

ANTONCIK, E. (1959), *J. Phys. Chem. Solids* **10**, 314.

ARYASETIAWAN, F. and O. GUNNARSSON (1998), *Rep. Prog. Phys.* **61**, 237.

ASHCROFT, N.W. and N.D. MERMIN (1976), *Solid State Physics* (Holt-Saunders, Philadelphia).

AYALA, P.Y. and G.E. SCUSERIA (1999), *J. Chem. Phys.* **110**, 3660.

BACHELET, G.B., D.R. HAMANN and M. SCHLÜTER (1982), *Phys. Rev. B* **26**, 4199.

BALDERESCHI, A. (1973), *Phys. Rev. B* **7**, 5212.

BARONI, S., P. GIANNOZZI and A. TESTA (1987), *Phys. Rev. Lett.* **58**, 1861.

BAR-YAM, Y. and J.D. JOANNOPOULOS (1984), *Phys. Rev. B* **30**, 1844.

BASSANI, F. and G. PASTORI-PARRAVICINI (1975), *Electronic States and Optical Transitions of Solids* (Pergamon Press, Oxford).

BATES, S.P., G. KRESSE and M.J. GILLAN (1997), *Surface Sci.* **385**, 386.

BECHSTEDT, F. and R. ENDERLEIN (1988), *Semiconductor Surfaces and Interfaces* (Akademie-Verlag, Berlin).

BECKE, A.D. (1993), *J. Chem. Phys.* **98**, 5648.

BENGTSSON, L. (1999), *Phys. Rev. B* **59**, 12301.

BERTONI, C.-M. (1990), in: V. Bortolani, N.H. March, and M. Tosi, eds., *Interaction of Atoms and Molecules with Solid Surfaces* (Plenum Publishing Corporation).

BLAHA, P., K. SCHWARZ, P. SORANTIN and S.B. TRICKEY (1990), *Comput. Phys. Commun.* **59**, 399.

BLÖCHL, P.E. (1990), *Phys. Rev. B* **41**, 5414.

BLÖCHL, P.E. (1994), *Phys. Rev. B* **50**, 17953.

BLÖCHL, P.E., O. JEPSEN and O.K. ANDERSEN (1994), *Phys. Rev. B* **49**, 16223.

BLOUNT, E.I. (1962), in: *Solid State Physics*, Vol. 13 (Academic Press) 305.

BOETTGER, J.C. (1994), *Phys. Rev. B* **49**, 16798.

BOTTOMLEY, D.J. and T. OGINO (2001), *Phys. Rev. B* **63**, 165412.

BOUCKAERT, L.P., R. SMOLUCHOWSKI and E. WIGNER (1936), *Phys. Rev.* **50**, 58.

BORN, M. and K. HUANG (1954), *Dynamical Theory of Crystal Lattices* (Oxford University Press, Oxford).

BOTT, E., M. METHFESSEL, W. KRABS and P.C. SCHMIDT (1998), *J. Math. Phys.* **39**, 3393.

BRIDDEN, P.R. and R. JONES (2000), *Phys. Stat. Sol. (b)* **217**, 131.

CANCÈS, E., M. DEFRANCESCHI, W. KUTZELNIGG, C. LE BRIS and Y. MADAY (2003), in: P.G. Ciarlet and C. Le Bris, eds., *Computational Chemistry*, Handbook of Numerical Analysis, Vol. X (Elsevier, Amsterdam) 3–270.

CAUSÀ, M., R. DOVESI, R. ORLANDO, C. PISANI and V.R. SAUNDERS (1988), *J. Phys. Chem.* **92**, 909.

CHADI, D.J. and M. COHEN (1973), *Phys. Rev. B* **8**, 5747.

DALGARNO, A. and A.L. STEWART (1958), *Proc. Royal Soc. London, Series A* **247**, 245.

DELEEUW, S.W., J.W. PERRAM and E.R. SMITH (1980), *Proc. Royal Soc. London, Ser. A* **373**, 27.

DESJONQUÈRES, M.-C. and D. SPANJAARD (1996), *Concepts in Surface Physics* (Springer-Verlag, Berlin).

ECONOMOU, E.N. (1979), *Green's Functions in Quantum Physics* (Springer-Verlag, Berlin).

EWALD, P.P. (1921), *Ann. Physik (Leipzig)* **64**, 253.
EYERT, V. (2000), *Int. J. Quant. Chem.* **77**, 1007.
FAULKNER, J.S. (1982), *The Modern Theory of Alloys* (Pergamon, London).
FEIBELMAN, P.J. (1991), *Phys. Rev. B* **43**, 9452.
FEYNMAN, R.P. (1939), *Phys. Rev.* **56**, 340.
FILIPPETTI, A., D. VANDERBILT, W. ZHONG, Y. CAI and G.B. BACHELET (1995), *Phys. Rev. B* **52**, 11793.
FINOCCHI, F., F. BOTTIN and C. NOGUERA (2003), in: C.R.A. Catlow and E. Kotomin, eds., *Computational Materials Science*, NATO ASI Series, in press.
FIORENTINI, V. and M. METHFESSEL (1996), *J. Phys. Condensed Matt.* **8**, 6525.
FORSTMANN, F. (1993), *Progr. Surface Sci.* **42**, 21.
FOULKES, W.M.C., L. MITAS, R.J. NEEDS and G. RAJAGOPAL (2001), *Rev. Mod. Phys.* **73**, 33.
FU, C.-L. and K.-M. HO (1983), *Phys. Rev. B* **28**, 5480.
FULDE, P. (1993), *Electron Correlations in Molecules and Solids*, Solid-State Sciences, Vol. 100 (Springer, Berlin).
GALLI, G. (2000), *Phys. Stat. Sol.* **217**, 231.
GALLI, G. and M. PARRINELLO (1991), in: M. Meyer and V. Pontikis, eds., *Computer Simulation in Materials Science*, NATO ASI Series E: Applied Sciences (Kluwer, Dordrecht) 283–304.
GAVEZZOTTI, A. (2002), *Mod. Simul. Mater. Sci. Eng.* **10** R1.
GIANNOZZI, P., S. DE GIRONCOLI, P. PAVONE and S. BARONI (1991), *Phys. Rev. B* **43**, 7231.
GONZE, X. (1995), *Phys. Rev. A* **52**, 1086; GONZE, X. (1995), *Phys. Rev. A* **52**, 1096.
GONZE, X. (1997a), *Phys. Rev. B* **55**, 10337.
GONZE, X. (1997b), *Phys. Rev. B* **55**, 10355.
GONZE, X., D.C. ALLAN and M.P. TETER (1992), *Phys. Rev. Lett.* **68**, 3603.
GONZE, X., PH. GHOSEZ and R. GODBY (1997), *Phys. Rev. B* **56**, 12811.
GONZE, X., R. STUMPF and M. SCHEFFLER (1991), *Phys. Rev. B* **44**, 8503.
GONZE, X. and J.-P. VIGNERON (1989), *Phys. Rev. B* **49**, 13120.
GROSS, E.K.U., J. DOBSON and M. PETERSILKA (1996), *Density Functional Theory* (Springer, New York).
HAMANN, D.R. (1979), *Phys. Rev. Lett.* **42**, 662.
HAMANN, D.R. (1989), *Phys. Rev. B* **40**, 2980.
HAMANN, D.R., M. SCHLÜTER and C. CHIANG (1979), *Phys. Rev. Lett.* **43**, 1494.
HEDIN, L. (1999), *J. Phys. Condensed Matt.* **11**, R489.
HEDIN, L. and S. LUNDQVIST (1969), in: H. Ehrenreich, F. Seitz, and D. Turnbull, eds., *Solid State Physics*, Vol. 23 (Academic Press, New York) 1–181.
HELLMANN, H. (1937), *Einfuhrung in die Quantumchemie* (Deuticke, Leipzig).
HYBERTSEN, M.S. and S.G. LOUIE (1987), *Phys. Rev. B* **35**, 5585.
INGLESFIELD, J.E. and G.A. BENESH (1988), *Phys. Rev. B* **37**, 6682.
INTERNATIONAL TABLES FOR X-RAY CRYSTALLOGRAPHY (1969), The Kynoch Press, Birningham, UK.
JANSEN, R.W. and O.F. SANKEY (1989), *Phys. Rev. B* **39**, 3192.
JEPSEN, O. and O.K. ANDERSEN (1971), *Solid State Comm.* **9**, 1763.
JEPSEN, O. and O.K. ANDERSEN (1984), *Phys. Rev. B* **29**, 5965.
JEPSEN, O., J. MADSEN and O.K. ANDERSEN (1982), *Phys. Rev. B* **26**, 2790.
JOHNSON, D.D. (1988), *Phys. Rev. B* **38**, 12807.
KAWATA, M., M. MIKAMI and U. NAGASHIMA (2002), *J. Chem. Phys.* **116**, 3430.
KIRTMAN, B. and C. DE MELO (1981), *J. Chem. Phys.* **75**, 4592.
KLEINMAN, L. and D.M. BYLANDER (1982), *Phys. Rev. Lett.* **48**, 1425.
KOELLING, D.D. and G.O. ARBMAN (1975), *J. Phys. F* **5**, 2041.
KOHN, W. (1973), *Phys. Rev. B* **7**, 4388.
KOHN, W. (1993), *Chem. Phys. Lett.* **208**, 167.
KOHN, W. and N. ROSTOKER (1954), *Phys. Rev.* **94**, 1111.
KORRINGA, J. (1947), *Physica* **13**, 392.
KOSTER, G.F. and J.C. SLATER (1954), *Phys. Rev.* **95**, 1167.
KRESSE, G. and J. HAFNER (1994), *J. Phys. Condensed Matt.* **6**, 8245.
KRESSE, G. and D. JOUBERT (1999), *Phys. Rev. B* **59**, 1758.

KRISTYÁN, S. and P. PULAY (1994), *Chem. Phys. Lett.* **229**, 175.

LANDAU, L.D. and E.M. LIFSHITS (1960), *Electrodynamics of Continuous Media* (Pergamon, London).

LESLIE, M. and M.J. GILLAN (1985), *J. Phys. C* **18**, 973.

LESTER, W.A. and B.L. HAMMOND (1990), *Ann. Rev. Phys. Chem.* **41**, 283.

LOUCKS, T.L. (1967), *Augmented Plane Wave Method* (Benjamin, New York).

LOUIE, S.G., S. FROYEN and M. COHEN (1982), *Phys. Rev. B* **26**, 1738.

MAHAN, G.D. and K.R. SUBBASWAMY (1990), *Local Density Theory of Polarizability* (Plenum Press, New York).

MARADUDIN, A.A., E.W. MONTROLL, G.H. WEISS and I.P. IPATOVA (1971), Theory of lattice dynamics in the harmonic approximation, in: H.E. Ehrenreich, F. Seitz and D. Turnbull, eds., *Solid State Physics*, 2nd edn. (Academic Press, New York) Chapter 4.

MARTYNA, G.J. and M.E. TUCKERMAN (1999), *J. Chem. Phys.* **110**, 2810.

MARZARI, N. and D. VANDERBILT (1997), *Phys. Rev. B* **56**, 12847.

MATTHEISS, L.F. and D.R. HAMANN (1986), *Phys. Rev. B* **33**, 823.

MAYNAU, D. and J.P. DAUDEY (1980), *Chem. Phys. Lett.* **81**, 273.

MESSIAH, A. (1961), *Quantum Mechanics* (North-Holland, Amsterdam).

METHFESSEL, M. (1988), *Phys. Rev. B* **38**, 1537.

METHFESSEL, M. and A.T. PAXTON (1989), *Phys. Rev. B* **40**, 3616

METHFESSEL, M., C.O. RODRIGUEZ and O.K. ANDERSEN (1989), *Phys. Rev. B* **40**, 2009.

METHFESSEL, M. and M. VAN SCHILFGAARDE (1993), *Phys. Rev. B* **48**, 4937.

MONKHORST, H.J. and J.D. PACK (1976), *Phys. Rev. B* **13**, 5188.

MÜLLER, W., J. FLESCH and W. MEYWR (1984), *J. Chem. Phys.* **80**, 3297.

NEEDS, R.J. (1987), *Phys. Rev. Lett.* **58**, 53.

NIELSEN, O.H. and R.M. MARTIN (1985), *Phys. Rev. B* **32**, 3780.

NOGUERA, C. (1995), *Physics and Chemistry of Oxide Surfaces* (Cambridge University Press, Cambridge).

NUNES, R.W. and D. VANDERBILT (1994), *Phys. Rev. Lett.* **73**, 712.

ONIDA, G., L. REINING and A. RUBIO (2002), *Rev. Mod. Phys.* **74**, 601.

ORDEJÓN, P., D.A. DRABOLD, R.M. MARTIN and S. ITOH (1995), *Phys. Rev. Lett.* **75**, 1324.

PANTELIDES, S.T. (1978), *Rev. Mod. Phys.* **50**, 797.

PARR, R.G. and W. YANG (1989), *Density Functional Theory of Atoms, Molecules* (Oxford Univ. Press, New York).

PARRY, D.E. (1975), *Surf. Sci.* **49**, 433.

PAYNE, M.C., M.P. TETER, D.C. ALLAN, T.A. ARIAS and J.D. JOANNOPOULOS (1992), *Rev. Mod. Phys.* **64**, 1045.

PERDEW, J.P. and Y. WANG (1992), *Phys. Rev. B* **45**, 13244.

PETERSEN, M., F. WAGNER, L. HUFNAGEL, M. SCHEFFLER, P. BLAHA and K. SCHWARZ (2000), *Comp. Phys. Comm.* **126**, 294.

PICK, R.M., M.H. COHEN and R.M. MARTIN (1970), *Phys. Rev. B* **1**, 910.

PICKETT, W.E. (1989), *Comput. Phys. Rep.* **9**, 116.

PHILLIPS, J.C. and L. KLEINMAN (1959), *Phys. Rev.* **116**, 880.

PISANI, C. (1993), *J. Molec. Catal.* **82**, 229.

PISANI, C., R. DOVESI and C. ROETTI (1988), *Hartree–Fock ab initio Treatment of Crystalline Systems*, Lecture Notes in Chem., Vol. 48 (Springer, Berlin).

PISANI, C., R. DOVESI, C. ROETTI, M. CAUSÀ, R. ORLANDO, S. CASASSA and V.R. SAUNDERS (2000), *Intern. J. Quantum Chem.* **77**, 1032.

POJANI, A., F. FINOCCHI and C. NOGUERA (1999), *Surface Sci.* **442**, 179.

POREZAG, D., M.R. PEDERSON and A.Y. LIU (2000), *Phys. Stat. Sol. (b)* **217**, 219.

PULAY, P. (1969), *Mol. Phys.* **17**, 197.

RAIMES, S. (1961), *The Wave Mechanics of Electrons in Metals* (North-Holland, Amsterdam).

RAJAGOPAL, G., R.J. NEEDS, A. JAMES, S.D. KENNY and W.M.C. FOULKES (1995), *Phys. Rev. B* **51**, 10591.

RAPPE, A.M., K.M. RABE, E. KAXIRAS and J.D. JOANNOPOULOS (1990), *Phys. Rev. B* **41**, (1990) 1227.

RESTA, R. (1985), in: P. Grosse, ed., *Festkörperprobleme*, Advances in Solid-State Physics, Vol. 25 (Vieweg, Braunschweig).

SAEBØ, S. and P. PULAY (1993), *Annu. Rev. Phys. Chem.* **44**, 213.
SAEBØ, S. and P. PULAY (2001), *J. Chem. Phys.* **115**, 3975.
SANCHEZ-PORTAL, D., P. ORDEJÓN, E. ARTACHO and J.M. SOLER (1997), *Int. J. Quantum Chem.* **65**, 453.
SAUNDERS, V.R., R. DOVESI, C. ROETTI, M. CAUSÀ, N.M. HARRISON, R. ORLANDO and C.M. ZICOVICH-WILSON (1998), *CRYSTAL98 User's Manual* (Università di Torino, Torino).
SAVRASOV, S.YU. (1992), *Phys. Rev. Lett.* **69**, 2819.
SAVRASOV, S.YU. (1996), *Phys. Rev. B* **54**, 16470.
SAVRASOV, S.Y. and D.Y. SAVRASOV (1992), *Phys. Rev. B* **46**, 12181.
SCHÜTZ, M., G. HETZER and H.-J. WERNER (1999), *J. Chem. Phys.* **111**, 5691.
SCHÜTZ, M. and H.-J. WERNER (2001), *J. Chem. Phys.* **114**, 661.
SCUSERIA, G.E. (2001), in: *Proc. 27th CHITEL*, Toulouse (France).
SHIRLEY, E.L. and R.M. MARTIN (1993), *Phys. Rev. B* **47**, 15413.
SHUKLA, A., M. DOLG, P. FULDE and H. STOLL (1998), *Phys. Rev. B* **57**, 1471.
SILVESTRELLI, P.L. (1999), *Phys. Rev. B* **59**, 9703.
SINANOGLU, O. (1961), *J. Chem. Phys.* **34**, 1237.
SINGH, D.J. (1994), *Planewaves, Pseudopotentials and the LAPW Method* (Kluwer Academic Publishers, Boston).
SINGH, D.J., H. KRAKAUER, C. HAAS and A.Y. LIU (1992), *Phys. Rev. B* **46**, 13065.
SHAM, L.J. (1985), *Phys. Rev. B* **32**, 3876.
SKRIVER, H.L. (1984), *The LMTO Method* (Springer-Verlag, Berlin).
SLATER, J.C. (1937), *Phys. Rev.* **51**, 846.
SLATER, J.C. (1964), *Advances in Quantum Chemistry* **1**, 35.
SOLER, J.M. and A.R. WILLIAMS (1990), *Phys. Rev. B* **42**, 9728.
STERNHEIMER, R.M. (1954), *Phys. Rev. B* **96**, 951.
TANK, R.W. and C. ARCANGELI (2000), *Phys. Stat. Sol. (b)* **217**, 89.
TRIONI, M.I., G.P. BRIVIO, S. CRAMPIN and J.E. INGLESFIELD (1996), *Phys. Rev. B* **53**, 8052.
TROULLIER, N. and J.L. MARTINS (1991), *Phys. Rev. B* **43**, 1993.
VANDERBILT, D. (1990), *Phys. Rev. B* **41**, 7892.
VON BARTH, U. and C.D. GELATT (1980), *Phys. Rev. B* **21**, 2222.
WANG, C.-Z., R. YU and H. KRAKAUER (1996), *Phys. Rev. B* **53**, 5430.
WANNIER, G.H. (1937), *Phys. Rev.* **52**, 191.
WATSON, S.C. and E.A. CARTER (1998), *Phys. Rev. B* **58**, R13309.
WEI, S.-H., H. KRAKAUER and M. WEINERT (1985), *Phys. Rev. B* **32**, 7792.
WEINERT, M. (1981), *J. Math. Phys.* **22**, 2433.
WILLIS, J.H., D. PRICE and B. COOPER (1989), *Phys. Rev. B* **39**, 4945; **46** (1992), 11368.
WILLIS, B.T.M. and A.W. PRYOR (1975), *Thermal Vibrations in Crystallography* (Cambridge University Press, Cambridge).
WIMMER, E., H. KRAKAUER, M. WEINERT and A.J. FREEMAN (1981), *Phys. Rev. B* **24**, 864.
YAMAGUCHI, Y. and H.F. SCHAEFER III (1996), *J. Chem. Phys.* **104**, 9841.
YIN, M.T. and M.L. COHEN (1980), *Phys. Rev. Lett.* **45**, 1004.
YU, R. and H. KRAKAUER (1994), *Phys. Rev. B* **49**, 4467.
YU, R., D.J. SINGH and H. KRAKAUER (1991), *Phys. Rev. B* **43**, 6411.
ZEIN, N.E. (1984), *Fiz. Tverd. Tela (Leningrad)* **26**, 3024 [*Sov. Phys. Solid State* **26** (1984), 1825].
ZICOVICH-WILSON, C.M., R. DOVESI and V.R. SAUNDERS (2001), *J. Chem. Phys.* **115**, 9708.

Computational Approaches of Relativistic Models in Quantum Chemistry

J.P. Desclaux[a], J. Dolbeault[b], M.J. Esteban[b], P. Indelicato[c], E. Séré[b]

[a] *15 Chemin du Billery, F-38360, Sassenage, France*

[b] *CEREMADE, Unité Mixte de Recherche du CNRS no. 7534 et Université Paris IX-Dauphine, F-75775 Paris cedex 16, France*

[c] *Laboratoire Kastler-Brossel, Unité Mixte de Recherche du CNRS no. C8552, École Normale Supérieure et Université Pierre et Marie Curie, Case 74, 4 place Jussieu, F-75252 Paris cedex 05, France*

E-mail addresses: jean-paul.desclaux@wanadoo.fr (J.P. Desclaux), dolbeaul@ceremade.dauphine.fr (J. Dolbeault), esteban@ceremade.dauphine.fr (M.J. Esteban), paul.indelicato@spectro.jussieu.fr (P. Indelicato), sere@ceremade.dauphine.fr (E. Séré)

1. Introduction

1.1. QED and relativistic models in Quantum Chemistry

It is now well known, following many experimental and theoretical results, that the use of *ab initio* relativistic calculations are mandatory if one is to obtain an accurate description of heavy atoms and ions. This is true whether one is considering highly charged ions, inner shells of neutral or quasi neutral atoms or outer shells of very heavy atoms.

Computational Chemistry
Special Volume (C. Le Bris, Guest Editor) of
HANDBOOK OF NUMERICAL ANALYSIS, VOL. X
P.G. Ciarlet (Editor)
© 2003 Elsevier Science B.V. All rights reserved

From a physics point of view, the natural formalism to treat such a system is Quantum Electrodynamics (QED), the prototype of field theories. For recent reviews of different aspects of QED in few electron ions see, e.g., EIDES, GROTCH and SHELYUTO [2001], MOHR, PLUNIEN and SOFF [1998], BEIER [2000]. Yet a direct calculation using only QED is impractical for atoms with more than one electron because of the complexity of the calculation. This is due to the slow rate of convergence of the so-called Ladder approximation $(1/Z)$, that in nonrelativistic theory amounts to a perturbation expansion using the electron–electron interaction as a perturbation. The only known method to do an accurate calculation is to attempt to treat to all orders the electron–electron interaction, and reserve QED for radiative corrections (interaction of the electron with its own radiation field, creation of virtual electron–positron pairs). The use of a naive approach however, taking a nonrelativistic Hamiltonian and replacing one-electron Schrödinger Hamiltonian by Dirac Hamiltonian fails. This approach does not take into account one of the two main features of relativity: the possibility of particle creation, and leads to severe problems as noted already in BROWN and RAVENHALL [1951] and studied in SUCHER [1980]. This theory, for example, does not preserve charge conservation in intermediate states and leads to divergence already in the second-order of perturbation expansion. The only way to derive a proper relativistic many-electron Hamiltonian is to start from QED. The Hamiltonian of a N electron system can be written formally

$$H = H_0\left[Ne^-, 0e^+\right] + H_1\left[(N+1)e^-, 1e^+\right] + H_2\left[(N+2)e^-, 2e^+\right] + \cdots. \tag{1.1.1}$$

Keeping only the first term, the so-called “no-pair” Hamiltonian reads

$$H^{\mathrm{np}} = \sum_{i=1}^{Ne^-} h_D(r_i) + \sum_{i<j} \mathcal{U}_{ij}, \tag{1.1.2}$$

where (in atomic units) $h_D(r_i) = c\boldsymbol{\alpha}\cdot\boldsymbol{p} + \beta mc^2 + V_N(\boldsymbol{r}_i)$ is a one-electron Dirac Hamiltonian in a suitable classical central potential V_N, that represents the interaction of the electron with the atomic nucleus. The speed of light is denoted by c, $\boldsymbol{\alpha}$, β are the Dirac matrices, with

$$\beta = \begin{pmatrix} \mathbb{1} & 0 \\ 0 & -\mathbb{1} \end{pmatrix}, \qquad \alpha_i = \begin{pmatrix} 0 & \sigma_i \\ \sigma_i & 0 \end{pmatrix}, \tag{1.1.3}$$

$$\sigma_1 = \begin{pmatrix} 0 & 1 \\ 1 & 0 \end{pmatrix}, \qquad \sigma_2 = \begin{pmatrix} 0 & -i \\ i & 0 \end{pmatrix}, \qquad \sigma_3 = \begin{pmatrix} 1 & 0 \\ 0 & -1 \end{pmatrix}, \tag{1.1.4}$$

$\boldsymbol{p} = -i\boldsymbol{\nabla}_i$ and

$$\mathcal{U} = \Lambda_i^+ \Lambda_j^+ V\left(|\boldsymbol{r}_i - \boldsymbol{r}_j|\right) \Lambda_i^+ \Lambda_j^+, \tag{1.1.5}$$

where Λ_i^+ is the positive spectral projection operator of a one-particle Hamiltonian similar to $h_D(r_i)$ [i.e., $\Lambda_i^+\phi = \phi$ for all eigenfunctions ϕ of $h_D(r_i)$ corresponding to positive eigenvalues]. Usually the potential used in this Hamiltonian is the direct Dirac–Fock potential (see Section 1.3). Moreover,

$$V(|\boldsymbol{r}_i - \boldsymbol{r}_j|) = \frac{1}{r_{ij}} - \frac{\boldsymbol{\alpha}_i \cdot \boldsymbol{\alpha}_j}{r_{ij}} + \left(\frac{1}{r_{ij}} - \frac{\boldsymbol{\alpha}_i \cdot \boldsymbol{\alpha}_j}{r_{ij}}\right)\left(\cos(\omega_{ij} r_{ij}/c) - 1\right)$$
$$+ c^2(\boldsymbol{\alpha}_i \cdot \boldsymbol{\nabla}_i)(\boldsymbol{\alpha}_j \cdot \boldsymbol{\nabla}_j)\frac{\cos(\omega_{ij} r_{ij}/c) - 1}{\omega_{ij}^2 r_{ij}} \tag{1.1.6}$$

is the electron–electron interaction of order 1 in $\alpha = 1/c \approx 1/137$, the fine structure constant. This expression is in Coulomb gauge, and is derived directly from QED. Here $r_{ij} = |\boldsymbol{r}_i - \boldsymbol{r}_j|$ is the inter-electronic distance, ω_{ij} is the energy of the photon exchanged between the electron i and j, which usually reduces to $\varepsilon_i - \varepsilon_j$ where the ε_i are the one-electron energies in the problem under consideration (for example, diagonal Lagrange multipliers in the case of Dirac–Fock). Note that in (1.1.6) gradient operators act only on the r_{ij} and not on the following wave functions. The presence of the ω_{ij} in this expression originates from the multi-time nature of the relativistic problem due to the finiteness of the speed of light. From this interaction, one can deduce the Breit operator, that contains retardation only to second order in $1/c$, in which the ω_{ij} can be eliminated by use of commutation relations between r and the one-particle Dirac Hamiltonian. This operator can then be readily used in the evaluation of correlation, while the higher-order in $1/c$ in the interaction (1.1.6) can only be evaluated perturbatively.

Finding bound states of (1.1.2) is difficult and requires approximations. The different methods of solution are inspired from the nonrelativistic problem. The three main categories of methods are the Relativistic Many-Body perturbation theory (RMBPT, see, e.g., LINDGREN and MORRISON [1982] for the nonrelativistic case), the Relativistic Random Phase Approximation (RRPA, see, e.g., JOHNSON and LIN [1976]), which has been heavily used for evaluation of photoionization cross-sections, and Multiconfiguration Dirac–Fock (MCDF). The RMBPT method requires the use of basis sets to sum over intermediate states. The MCDF method is a variational method.

1.2. Relativistic Many-Body perturbation theory and RRPA

In its most general version, the RMBPT method starts from a multidimensional model space and uses Rayleigh–Schrödinger perturbation theory. The concept of model space is mandatory if there are several levels of quasi-degenerate energy as in the ground state of Be-like ions ($1s^2 2s^2\ {}^1S_0$ and $1s^2 2p^2\ {}^1S_0$ are very close in energy, leading to very strong intra-shell correlation). In that case one gets would get very bad convergence of the perturbation expansion, because of the near-zero energy denominators, if building the perturbation theory on a single level.

Following LINDGREN and MORRISON [1982] we separate the Hamiltonian in a sum

$$H_T = H_0 + V_0. \tag{1.2.1}$$

We assume that we know a set of N eigenfunctions Ψ_α^0 of eigenenergies E_α^0 which are all the solutions obtained by diagonalizing H_0 on a subspace $\mathcal{P}$ (these solutions can be obtained with the Dirac–Fock method in a suitable average potential). The unperturbed Hamiltonian is then chosen as

$$H_0^N = P_0 H_0 P_0 = \sum_{\alpha=1}^{N} E_\alpha^0 |\Psi_\alpha^0\rangle\langle\Psi_\alpha^0|. \tag{1.2.2}$$

where and P_0 is the projector on $\mathcal{P}$, defined by

$$P_0 = \sum_{\alpha=1}^{N} |\Psi_\alpha^0\rangle\langle\Psi_\alpha^0|. \tag{1.2.3}$$

We define the perturbation potential by

$$V = H_T - P_0 H_0 P_0 = H_T - H_0^N. \tag{1.2.4}$$

We also define $Q_0 = 1 - P_0$ as the projector on the orthogonal space $\mathcal{Q}$. We now define the wave operator, which build the exact solution of the Hamiltonian equation (1.2.1) from the Ψ_α^0

$$\Psi_\alpha = \Omega \Psi_\alpha^0, \tag{1.2.5}$$

so that

$$H_T \Psi_\alpha = E_\alpha \Psi_\alpha, \tag{1.2.6}$$

with the property

$$P_0 \Omega P_0 = P_0. \tag{1.2.7}$$

The *exact* eigenenergies can be obtained by the application of the Model-space wave functions on the effective Hamiltonian

$$H_{\text{eff}} = P_0 H_T \Omega P_0 = P_0 H_0 P_0 + P_0 V \Omega P_0 = H_0^N + P_0 V \Omega P_0, \tag{1.2.8}$$

using Eqs. (1.2.2) and (1.2.7), $P_0^2 = P_0$ and the fact that H_0^N and P_0 commute. This operator, acting on the unperturbed wave functions give the exact eigenenergies:

$$H_{\text{eff}} \Psi_\alpha^0 = E_\alpha \Psi_\alpha^0. \tag{1.2.9}$$

The wave operator obeys the generalized Bloch equation

$$[\Omega, H_0] P_0 = V \Omega P_0 - \Omega P_0 V \Omega P_0 \tag{1.2.10}$$

using Eq. (1.2.7). This can be expanded in a series

$$\Omega = 1 + \Omega^{(1)} + \Omega^{(2)} + \cdots. \tag{1.2.11}$$

Eqs. (1.2.10) and (1.2.11) leads to the sequence of equations

$$\left[\Omega^{(1)}, H_0\right] P_0 = Q_0 V P_0, \tag{1.2.12}$$

$$\left[\Omega^{(2)}, H_0\right] P_0 = Q_0 V \Omega^{(1)} P_0 - \Omega^{(1)} P_0 V P_0. \tag{1.2.13}$$

The RRPA method is based on the solution of the Hamiltonian (1.1.2) subjected to a time-dependent perturbation (like a classical electromagnetic radiation of known frequency). This time-dependent Dirac–Fock equation is solved over a set of solutions of the unperturbed problem, leading to a set of time-dependent mixing coefficients in the usual fashion of time-dependent perturbation theory. The phases of those coefficients are approximated (leading to the name "Random Phase"), leading to differential equations very similar to the Dirac–Fock ones. This method include to all orders some classes of correlation contribution that can be easily also evaluated in the framework of RMBPT. It is mostly used for the ground state of atoms and ions to study photoionization. It is more difficult to use for excited states.

This paper is mostly devoted to the MCDF method for atoms and molecules, and to preliminary results for the linear Dirac operator.

1.3. The MCDF wave function

We first start by describing shortly the formalism used to build the Dirac–Fock solutions for a spherically-symmetric system like an isolated atom.

If we define the angular momentum operators $\boldsymbol{L} = \boldsymbol{r} \wedge \boldsymbol{p}$, $\boldsymbol{J} = \boldsymbol{L} + \frac{\sigma}{2}$, the parity Π as βP, then the total wave function is expressed in term of configuration state functions (CSF) as antisymmetric products of one-electron wave functions so that they are eigenvalues of the parity Π, the total angular momentum J and its projection M. The label ν stands for all other values (angular momentum recoupling scheme, seniority numbers, ...) necessary to define unambiguously the CSF. For a N-electron system, a CSF is thus a linear combination of Slater determinants:

$$|\nu \Pi J M\rangle = \sum_{i=1} d_i^{\nu} \begin{vmatrix} \Phi_1^{i,\nu}(r_1) & \cdots & \Phi_N^{i,\nu}(r_1) \\ \vdots & \ddots & \vdots \\ \Phi_1^{i,\nu}(r_N) & \cdots & \Phi_N^{i,\nu}(r_N) \end{vmatrix}, \tag{1.3.1}$$

all of them with the same Π and M values while the d_i's are determined by the requirement that the CSF is an eigenstate of J^2.

The total MCDF wave function is constructed as a superposition of CSFs, i.e.,

$$\Psi(\Pi J M) = \sum_{\nu=1}^{NCF} c_{\nu} |\nu \Pi J M\rangle, \tag{1.3.2}$$

where NCF is the number of configurations and the c_υ are called the configurations mixing coefficients.

The MCDF method has two variants. In one variant, one uses numerical or analytic basis sets to construct the CSF. In the other one, direct numerical solution of the MCDF equation is used. Both methods have been used in atomic and molecular physics. The numerical MCDF method is better suited for small systems, while analytic basis set techniques are better suited for cases with millions of determinants.

This chapter is organized as follows. In Section 2, different choices of basis sets for the Dirac equation are presented. In Section 3, the MCDF equations are presented, and numerical techniques adapted to the numerical MCDF method in atoms are described. In Section 4, we deal with techniques for the numerical MCDF method in molecules.

2. Linear Dirac equations

2.1. Properties of the linear Dirac operator

The unboundedness from below of the Dirac operator

$$H_0 = -i\,c\alpha \cdot \nabla + mc^2\beta \tag{2.1.1}$$

creates important difficulties when trying to find its eigenvalues. The so-called *variational collapse* is indeed related to this unboundedness property. On the other hand, finite-dimensional approximations to this problem may lead to finding *spurious solutions*: some eigenvalues of the finite-dimensional problem do not approach the eigenvalues of the Dirac operator and destroy the monotonicity of the approximated eigenvalues with respect to the basis dimension. These problems seem to be much more acute in molecular than in atomic computations, but they are already present in one-electron systems. In this section we address this difficulty for one-electron systems by describing various methods used to deal with this problem. Well-behaved approximation methods should also provide good nonrelativistic limits, that is, variational problems whose eigenvalues and eigenfunctions converge well to those of the corresponding nonrelativistic Schrödinger Hamiltonian.

A way often used to find good numerical approximations of eigenvalues of an operator A consists in projecting the eigenvalue equation

$$A\,x = \lambda\,x \tag{2.1.2}$$

over a well chosen finite-dimensional space X_N of dimension N, in order to find an approximation (λ_N, x_N) satisfying

$$A_N\,x_N = \lambda_N\,x_N, \tag{2.1.3}$$

such that (λ_N, x_N) converges to (λ, x) as $N \to \infty$. Then one looks for the eigenvalues of the $N \times N$ matrix A_N and these eigenvalues will converge either to eigenvalues of A or to points in the essential spectrum of A. As N increases, the limit set of the eigenvalues of A_N is the spectrum of A.

The difficulty with the Dirac operator is that for most physically interesting potentials V, the spectrum of $H_0 + V$ is made of its essential spectrum $(-\infty, -mc^2] \cup [mc^2, +\infty)$ and a discrete set of eigenvalues lying in the *gap* $(-mc^2, mc^2)$. Hence, the choice of the finite-dimensional space, or equivalently, of the finite basis set, is fundamental if we want to ensure that for some N large, the eigenvalues of $(H_0 + V)_N$, or at least some of them, will be approximations of the eigenvalues of $H_0 + V$ in the gap $(-mc^2, mc^2)$. The question of how to choose a good basis set has been addressed in many papers, among which DRAKE and GOLDMAN [1981], DRAKE and GOLDMAN [1988], GRANT [1989], GRANT [1982], JOHNSON, BLUNDELL and SAPIRSTEIN [1988], KUTZELNIGG [1984], LEE [2001], that we will describe with further details in Sections 2.2 and 2.3 below. In particular, Section 2.3 is devoted to the description of numerical techniques based either on discretization or on B-splines, and shows that with appropriate boundary conditions one can avoid the variational collapse.

When the operator A is bounded from below, it is often possible to characterize its spectrum by variational methods, for instance by looking for critical values of the Rayleigh quotient

$$Q(x) := \frac{(Ax, x)}{(x, x)} \tag{2.1.4}$$

over the domain of A. More concretely, when A is bounded from below, under appropriate assumptions, its ground state energy can be found by minimizing the above Rayleigh quotient. However, this cannot be done directly in the context of the Dirac operator, since it is unbounded from below (and also from above). A large number of works have been devoted to the variational resolution of this problem in view of the Dirac operator. Most of them use the approximation of an effective Hamiltonian which is bounded from below. The idea hidden behind this kind of techniques is that there is no explicit way of diagonalizing the Dirac Hamiltonian $H + V$, but this can be done at an abstract level. The diagonalized operator is then approximated via a finite expansion or an iterative procedure. These methods are therefore perturbative and contain an approximation at the operator level. They will be referred to as *perturbation theories and effective Hamiltonian methods* and will be described below (see DRAKE and GOLDMAN [1981], KUTZELNIGG [1984], GRANT [1982], JOHNSON, BLUNDELL and SAPIRSTEIN [1988] and Section 2.4 for more details).

Other variational techniques are based on a correspondence between the eigenvalues of A and those of $T(A)$, for some operator function T, like the inverse function $Tx = x^{-1}$ (see HILL and KRAUTHAUSER [1994]) or the function $Tx = x^2$ (see WALLMEIER and KUTZELNIGG [1981], BAYLISS and PEEL [1983]). Finally, some authors solve the variational problem in a subspace of the domain in which the operator is bounded from below and 'avoids' the negative continuum. Section 2.5 will be devoted to these more *direct variational approaches*, based on either linear or nonlinear constraints.

Before going into the details of the computational methods, let us start with some notations and preliminary considerations. For any ψ with values in $\mathbb{C}^4$, if we write $\psi = \binom{\varphi}{\chi}$, with φ, χ taking values in $\mathbb{C}^2$, then the eigenvalue equation

$$H\psi = (H_0 + V)\psi = \lambda\psi \tag{2.1.5}$$

is equivalent to the following system:

$$\begin{cases} R\chi = (\lambda - mc^2 - V)\varphi, \\ R\varphi = (\lambda + mc^2 - V)\chi, \end{cases} \tag{2.1.6}$$

with $R = i\,c(\vec{\sigma} \cdot \vec{\nabla}) = \sum_{j=1}^{3} i\,c\,\sigma_j \frac{\partial}{\partial x_j}$. Here σ_j, $j = 1, 2, 3$, are the Pauli matrices. As long as $\lambda + mc^2 - V \neq 0$, the system (2.1.6) can be written as

$$H^\mu\varphi := R\left(\frac{R\varphi}{g_\mu}\right) + V\varphi = \mu\varphi, \qquad \chi = \frac{R\varphi}{g_\mu}, \tag{2.1.7}$$

where $g_\mu = \mu + 2mc^2 - V$ and $\mu = \lambda - mc^2$. Note that the Hamiltonian operator H^μ is eigenvalue dependent. Reducing the 4-component spinor ψ to an equation for the 2-spinor φ is often called *partitioning*. Let us immediately notice that at least formally, the partitioned equation (2.1.7) converges to its nonrelativistic counterpart

$$-\frac{1}{2m}\Delta\varphi + V\varphi = \mu\varphi. \tag{2.1.8}$$

(see, for instance, WOOD, GRANT and WILSON [1985]). For this reason, but also because the principal part of the second order operator in (2.1.7) is semibounded for not too large potentials V, the partitioned equation has been extensively studied for finding eigenvalues of linear Dirac operators.

To end these preliminary considerations on linear Dirac equations, note that in the case of *rotationally invariant* potentials, the solutions can be put in the form

$$\psi = \frac{1}{r}\begin{pmatrix} P_\kappa(r)\chi_{\kappa m}(\theta,\varphi) \\ i\,Q_\kappa(r)\chi_{-\kappa m}(\theta,\varphi) \end{pmatrix}. \tag{2.1.9}$$

The dependence on the angular coordinates is contained in the 2-spinors $\chi_{\pm\kappa m}(\theta,\varphi)$, which are eigenfunctions of the angular momentum operators J, its third component J_z (with eigenvalues $j(j+1)$ and m, respectively) and of parity. On the other hand, the radial dependence is contained in the functions f and g which are called the upper and lower radial components of ψ.

In the ansatz defined in (2.1.9), for a given $\kappa = \pm(j + \frac{1}{2})$, with $j = \ell \mp \frac{1}{2}$, $l = 0, 1, \ldots$, the eigenvalue equation (2.1.5) is equivalent to

$$\left(H_r^\kappa + V\right)\Phi = \lambda\Phi, \tag{2.1.10}$$

with

$$H_r^\kappa = \begin{pmatrix} mc^2 & c(-\frac{d}{dr} + \kappa_j) \\ c(\frac{d}{dr} + \kappa_j) & -mc^2 \end{pmatrix}, \tag{2.1.11}$$

$\Phi = \binom{P_\kappa}{Q_\kappa}$ being a 2-vector with two scalar real components.

2.2. *Finite basis set approaches*

The choice of *finite-dimensional spaces* is essential for the discretization of the operator and the approximation of its eigenvalues. The presence of the negative continuum makes this task difficult in the case of the Dirac operator. The basic criterium to decide whether a particular space, or a generating basis set, is good, is to check that the approximated eigenvalues found are either negative and lying in the negative continuum or positive. In this case, if they are below the positive continuum, they are approximations of the discrete exact (positive) eigenvalues. Many attempts to construct finite basis sets can be found in the literature.

DRAKE and GOLDMAN [1981] introduced the so-called *Slater type orbitals* (STO):

$$\Phi(r) = r^{\gamma-1} e^{-\nu r} \sum_{i=1}^{N} r^i \left[a_i \begin{pmatrix} 1 \\ 0 \end{pmatrix} + b_i \begin{pmatrix} 0 \\ 1 \end{pmatrix} \right], \tag{2.2.1}$$

with a particular choice of γ and ν which depends on κ and V. They showed numerical evidence that such a finite basis satisfies the above properties in the case of hydrogen-like atoms. Note that the STOs exhibit the same behavior near 0 and at ∞ as the exact eigenfunctions. The properties of STO basis sets are made more explicit in GRANT [1982], where STO basis sets are replaced by orthonormal sets of *Laguerre polynomials*. The main drawback of this approach is that some of the eigenvalues of the approximated matrix are spurious roots which do not approximate any of the exact eigenvalues.

Another way to construct *basis sets* with good properties consists in imposing the so-called *kinetic balance* condition relating the *upper* and *lower components* of the functions in the basis set. See, for instance, KUTZELNIGG [1984].

Other types of basis sets proposed in the literature include those generated by *B-splines* (see JOHNSON, BLUNDELL and SAPIRSTEIN [1988]), which have very good properties since, in this approach, the matrices are very sparse: only a finite number (depending on the degree of the splines) of diagonal lines are nonzero. This kind of basis sets has been widely used in atomic and molecular computations (see Section 2.3).

The choice of a good basis set can be quite effective in some computations, but as it appears clearly in the literature that we quote, there is very often a risk of finding spurious roots or of variational collapse. In the next subsection we give some more precise examples of how to use particular basis sets in the context of Dirac operators.

2.3. Numerical basis sets

This section is devoted to the special case of basis sets whose elements are computed numerically.

Discretisation method. The Göteborg group has developed an efficient technique to obtain basis sets for the Dirac equation (SALOMONSON and ÖSTER [1989a]). The Dirac equation is discretized and solved on a grid. The atom is placed in a spherical box, large enough not to disturb the bound state wave function considered. The method provides a finite number of orbitals which is complete over the discretized space (SALOMONSON and ÖSTER [1989b]), and resemble lattice gauge field calculation (WILSON [1974]). The method enables to eliminate spurious states and preserves the Hermiticity of the discretized Hamiltonian. The appearance of spurious states in a discretized method, is traced back to the "fermion doubling", first encoutered in gauge-field lattice calculations (KOGUT [1983]). On a lattice of dimension $(D+1)$ (D spatial and one time dimensions), an equation for a massless fermion will describe not one but 2^D ones if no precaution is taken (STACEY [1982]).

As an example, let us consider a one-dimensional Dirac equation for a free fermion

$$\begin{pmatrix} mc^2 & -c\frac{d}{dx} \\ c\frac{d}{dx} & -mc^2 \end{pmatrix}\begin{pmatrix} f(x) \\ g(x) \end{pmatrix} = \left(\varepsilon + mc^2\right)\begin{pmatrix} f(x) \\ g(x) \end{pmatrix}. \tag{2.3.1}$$

The derivatives are approximated over the lattice points using

$$f_i' = \frac{f_{i+1} - f_{i-1}}{2h}, \tag{2.3.2}$$

where h is the space between adjacent lattice sites. Eliminating the large component in (2.3.1), one gets the following equation

$$-\frac{1}{2m}\left(\frac{f_{i+2} - 2f_i + f_{i-2}}{4h^2}\right) = \varepsilon\left(1 + \frac{\varepsilon}{2mc^2}\right)f_i, \tag{2.3.3}$$

in which the left-hand side is the kinetic energy operator $p_x^2/2m$ acting on f at the lattice point i. Yet this second order derivative does not connect even and odd lattice sites. The highest energy solution over the lattice is the one changing sign at each site so that

$$\cdots \approx f_{i-2} \approx -f_{i-1} \approx f_i \approx -f_{i+1} \approx \cdots. \tag{2.3.4}$$

Using the expression (2.3.3) acting on this solution gives the same results as if it had no nodes. A high-energy eigenvector thus appears as a spurious state in the low energy part of the spectrum. For low-order derivative two equivalent ways can be used (STACEY

[1982], SALOMONSON and ÖSTER [1989a]). One is to use forward derivatives for f and backward derivatives for g,

$$f_i' = \frac{f_{i+1} - f_i}{h}, \qquad g_i' = \frac{g_i - g_{i-1}}{h}. \tag{2.3.5}$$

The other consists in defining the large and small components on alternating sites on the lattice.

$$\cdots f_{i-3}\, g_{i-2}\, f_{i-1}\, g_i\, f_{i+1}\, g_{i+2} \cdots, \tag{2.3.6}$$

with h being the separation between g_{i-2} and g_i. In this case the derivative is expressed as

$$f_i' = \frac{f_{i+1} - f_i}{h}, \qquad g_j' = \frac{g_j - g_{j-1}}{h}, \tag{2.3.7}$$

with $i = 2n - 1$, $j = 2n$, $n = 1, 2, \ldots, N$. These methods reduce to the same second-order equation (STACEY [1982]).

Salomonson and Öster use a more accurate six-point formula

$$\begin{aligned} f'(x) = \frac{1}{1920h}\bigg[&-9f\Big(x - \frac{5}{2}h\Big) + 125f\Big(x - \frac{3}{2}h\Big) - 2250f\Big(x - \frac{1}{2}h\Big) \\ &+ 2250f\Big(x + \frac{1}{2}h\Big) - 125f\Big(x + \frac{3}{2}h\Big) \\ &+ 9f\Big(x + \frac{5}{2}h\Big)\bigg] + \mathcal{O}(h^6). \end{aligned} \tag{2.3.8}$$

This six-point formula combined with (2.3.6) provides a spurious-state-free solution, while using the same lattice for f and g and a forward-backward derivative scheme does not work.

In the spherical case, one needs to use a logarithmic lattice to get a good description of the wave function. The Hermiticity of the Hamiltonian must be preserved by doing the variable change

$$y(r) \to \frac{1}{\sqrt{r}}\, y(x), \quad x = \log(r). \tag{2.3.9}$$

The corresponding Dirac equation is

$$\begin{aligned} &\begin{pmatrix} V(r) & -c\big(\frac{1}{\sqrt{r}}\frac{d}{dx}\frac{1}{\sqrt{r}} - \frac{\kappa}{\sqrt{r}}\frac{1}{\sqrt{r}}\big) \\ c\big(\frac{1}{\sqrt{r}}\frac{d}{dx}\frac{1}{\sqrt{r}} + \frac{\kappa}{\sqrt{r}}\frac{1}{\sqrt{r}}\big) & V(r) - mc^2 \end{pmatrix} \begin{pmatrix} f(x) \\ g(x) \end{pmatrix} \\ &\quad = \varepsilon \begin{pmatrix} f(x) \\ g(x) \end{pmatrix}. \end{aligned} \tag{2.3.10}$$

Since the large and small component are defined on different lattices, one needs interpolation formulas to express $f(x)/\sqrt{r}$ and $g(x)/\sqrt{r}$ in the κ term.

The discretization finally provides a $2N \times 2N$ symmetric eigenvalue problem

$$\begin{pmatrix} A & {}^tD + {}^tK \\ D + K & B \end{pmatrix} \begin{pmatrix} F \\ G \end{pmatrix} = \varepsilon \begin{pmatrix} F \\ G \end{pmatrix}, \tag{2.3.11}$$

with $(F, G) = (f_1, f_3, \ldots, f_{2N-1}, g_2, g_4, \ldots, g_{2N})$. For a point nucleus, the submatrices are $A_{ii} = -Z/r_i$ and $B_{jj} = -2mc^2 - Z/r_j$, $i = 2n - 1$, $j = 2n$, $n = 1, 2, \ldots, N$. With the 6 points interpolation and derivation formulas used in SALOMONSON and ÖSTER [1989a], one obtains

$$D = \frac{c}{1920h} \begin{pmatrix} -\frac{2250}{\sqrt{r_2r_1}} & \frac{2250}{\sqrt{r_2r_3}} & -\frac{125}{\sqrt{r_2r_5}} & \frac{9}{\sqrt{r_2r_7}} & 0 & \cdots & \cdots \\ \frac{125}{\sqrt{r_4r_1}} & -\frac{2250}{\sqrt{r_4r_3}} & \frac{2250}{\sqrt{r_4r_5}} & -\frac{125}{\sqrt{r_4r_7}} & \frac{9}{\sqrt{r_4r_9}} & 0 & \cdots \\ -\frac{9}{\sqrt{r_6r_1}} & \frac{125}{\sqrt{r_6r_3}} & -\frac{2250}{\sqrt{r_6r_5}} & \frac{2250}{\sqrt{r_6r_7}} & -\frac{125}{\sqrt{r_6r_9}} & \frac{9}{\sqrt{r_6r_{11}}} & \cdots \\ 0 & -\frac{9}{\sqrt{r_8r_3}} & \frac{125}{\sqrt{r_8r_5}} & -\frac{2250}{\sqrt{r_8r_7}} & \frac{2250}{\sqrt{r_8r_9}} & -\frac{125}{\sqrt{r_8r_{11}}} & \cdots \\ \vdots & & \vdots & \vdots & \vdots & \vdots & \ddots \end{pmatrix}, \tag{2.3.12}$$

and

$$K = \frac{\kappa}{256h} \begin{pmatrix} \frac{150}{\sqrt{r_2r_1}} & \frac{150}{\sqrt{r_2r_3}} & -\frac{25}{\sqrt{r_2r_5}} & \frac{3}{\sqrt{r_2r_7}} & 0 & \cdots & \cdots \\ -\frac{25}{\sqrt{r_4r_1}} & \frac{150}{\sqrt{r_4r_3}} & \frac{150}{\sqrt{r_4r_5}} & -\frac{25}{\sqrt{r_4r_7}} & \frac{3}{\sqrt{r_4r_9}} & 0 & \cdots \\ \frac{3}{\sqrt{r_6r_1}} & -\frac{25}{\sqrt{r_6r_3}} & \frac{150}{\sqrt{r_6r_5}} & \frac{150}{\sqrt{r_6r_7}} & -\frac{25}{\sqrt{r_6r_9}} & \frac{3}{\sqrt{r_6r_{11}}} & \cdots \\ 0 & \frac{3}{\sqrt{r_8r_3}} & -\frac{25}{\sqrt{r_8r_5}} & \frac{150}{\sqrt{r_8r_7}} & \frac{150}{\sqrt{r_8r_9}} & -\frac{25}{\sqrt{r_8r_{11}}} & \cdots \\ \vdots & & \vdots & \vdots & \vdots & \vdots & \ddots \end{pmatrix}. \tag{2.3.13}$$

Eq. (2.3.11) is symmetric even though K and D are not. In the upper left corner of D, use has been made of the approximation

$$f(r) \sim r^{\gamma+1/2} + \frac{2[\gamma + \kappa - (Z\alpha)^2]}{Z\alpha^2(2\gamma + 1)} r^{\gamma+3/2} + \varepsilon \frac{[\gamma + \kappa - 2(Z\alpha)^2]}{Z(2\gamma + 1)} r^{\gamma+3/2}, \tag{2.3.14}$$

with $\gamma = \sqrt{\kappa^2 - (Z\alpha)^2}$, and the equivalent expression for g. To avoid nonlinear terms in the eigensystem (2.3.11), only the contribution independent of ε has been kept. This is a good approximation for bound states for which $\varepsilon \ll mc^2$.

Numerical basis sets based on B-splines. B-splines have been used (JOHNSON, BLUNDELL and SAPIRSTEIN [1988]) to provide numerically efficient basis sets. A knot sequence t_i is used for the radial coordinate, on which B-spline of order k provide a complete basis for piecewise polynomials of order $k-1$. This radial coordinate extends to a distance R from the origin. The solutions of the Dirac equation are expressed as linear combinations of B-splines. A *Galerkin method* is employed to obtain the solution. The Dirac equation is derived from an action principle $\delta S = 0$, with

$$\begin{aligned} S = \frac{1}{2}\int_0^R \Big\{ & cP_\kappa(r)\Big(\frac{d}{dr} - \frac{\kappa}{r}\Big)Q_\kappa(r) - cQ_\kappa(r)\Big(\frac{d}{dr} + \frac{\kappa}{r}\Big)P_\kappa(r) \\ & + V_N(r)\big[P_\kappa(r)^2 + Q_\kappa(r)^2\big] - 2mc^2 Q_\kappa(r)^2 \Big\}\, dr \\ & - \frac{1}{2}\varepsilon \int_0^R \big[P_\kappa(r)^2 + Q_\kappa(r)^2\big]\, dr \end{aligned} \tag{2.3.15}$$

using the notations of (2.1.9) (note that in this representation the gap lies between $-2mc^2$ and 0), to which suitable boundary conditions are added through

$$S' = \begin{cases} \frac{c}{4}\big[P_\kappa(R)^2 - Q_\kappa(R)^2\big] + \frac{c}{2}P_\kappa(0)^2 - \frac{c^2}{2}P_\kappa(0)Q_\kappa(0) & \text{for } \kappa < 0, \\ \frac{c}{4}\big[P_\kappa(R)^2 - Q_\kappa(R)^2\big] + c^2 P_\kappa(0)^2 - \frac{c}{2}P_\kappa(0)Q_\kappa(0) & \text{for } \kappa > 0. \end{cases} \tag{2.3.16}$$

From the point of view of the variational principle, ε is a Lagrange multiplier introduced to ensure that the solutions of the Dirac equation are normalized. The boundary constraint (2.3.16) is designed to avoid a hard boundary at the box radius R, following the idea behind the MIT bag model for quark confinement, and provides $P_\kappa(R) = Q_\kappa(R)$. Forcing $P_\kappa(R) = Q_\kappa(R) = 0$ would amount to introduce an infinite potential at the boundary and possibly leads to the Klein paradox. Other choices of boundary conditions are possible. This particular choice avoids the appearance of spurious solutions. Expanding the radial wave function as

$$P_\kappa(r) = \sum_{i=1}^{n} p_i B_{i,k}(r), \qquad Q_\kappa(r) = \sum_{i=1}^{n} q_i B_{i,k}(r), \tag{2.3.17}$$

the variational principle reduces to

$$\frac{d(S+S')}{dp_i} = 0, \qquad \frac{d(S+S')}{dq_i} = 0, \quad i = 1, 2, \ldots, n. \tag{2.3.18}$$

This leads to a $2n \times 2n$ symmetric, generalized eigenvalue equation

$$Av = \varepsilon Bv, \tag{2.3.19}$$

where $v = (p_1, p_2, \ldots, p_n, q_1, q_2, \ldots, q_n)$,

$$A = \begin{pmatrix} (V) & c\big[(D) - (\frac{\kappa}{r})\big] \\ -c\big[(D) + (\frac{\kappa}{r})\big] & (V) - 2mc^2(C) \end{pmatrix} + A' \tag{2.3.20}$$

and

$$B = \begin{pmatrix} (C) & 0 \\ 0 & (C) \end{pmatrix}. \tag{2.3.21}$$

The $2n \times 2n$ matrix A' comes from the boundary term. The $n \times n$ matrix (C) is the B-spline overlap matrix defined by

$$(C)_{ij} = \int B_{i,k}(r) B_{j,k}(r)\, dr, \tag{2.3.22}$$

(D) comes from the differential operator

$$(D)_{ij} = \int B_{i,k}(r) \frac{dB_{j,k}(r)}{dr}\, dr, \tag{2.3.23}$$

(V) is the potential term

$$(V)_{ij} = \int B_{i,k}(r) V_N(r) B_{j,k}(r)\, dr \quad \text{and} \quad \left(\frac{\kappa}{r}\right)_{ij} = \int B_{i,k}(r) \frac{\kappa}{r} B_{j,k}(r)\, dr. \tag{2.3.24}$$

Diagonalization of (2.3.20) provides $2n$ eigenvalues and eigenfunctions, n of which have energies below $-2mc^2$, a few correspond to bound states (typically 5 to 6 for $k = 7$ to 9) and the rest belongs to the positive energy continuum.

2.4. *Perturbation theory and effective Hamiltonians*

An alternative way to find the eigenvalues of the unbounded relativistic operator H consists in looking for a so-called *effective Hamiltonian* H^{eff}, which is semi-bounded, such that both Hamiltonians have *common eigenvalues* on an interval above the negative continuous spectrum. Such a Hamiltonian H^{eff} cannot usually be found in an explicit way, but can be viewed as the limit of an iterative procedure. This leads to families of Hamiltonians which approach the effective Hamiltonian and yield approximated eigenvalues for H.

One of the most popular procedure in this direction is due to FOLDY and WOUTHUYSEN [1950], whose main idea was to apply a unitary transformation Ω to $H_0 + V$ such that

$$\Omega^*(H_0 + V)\,\Omega = H^{\text{FW}} = \begin{pmatrix} H_+^{\text{FW}} & 0 \\ O & H_-^{\text{FW}} \end{pmatrix}, \tag{2.4.1}$$

so that electronic and positronic states are decoupled: electrons (resp. positrons) would be described by the eigenfunctions of H_+^{FW} (resp. H_-^{FW}). Moreover, the Hamiltonians $H_+^{\text{FW}} - mc^2$ (resp. $H_-^{\text{FW}} + mc^2$) are bounded from below (resp. above) and have correct nonrelativistic limits. Although this procedure looks very promising, the problem is that Ω is unknown in closed form, and so there is no way of diagonalizing $H_0 + V$ in an explicit way. However, approximations of Ω, and therefore of H^{FW}, can be constructed either by writing a formal series expansion for $H_{\pm}^{\mathrm{FW}}$ in the perturbation parameter c^{-2}:

$$H_{\pm}^{\mathrm{FW}} = \sum_{k=0}^{+\infty} c^{-2k} H_{2k}^{\pm}, \tag{2.4.2}$$

and cutting it at level $k \geqslant 0$, or by approaching it by an iterative procedure.

In general one identifies the effective Hamiltonian H^{eff} as a solution to a nonlinear equation $H^{\mathrm{eff}} = f(H^{\mathrm{eff}})$, which can be solved approximately in an iterative way. By instance, one can produce an equation like the above one by "eliminating" the lower component χ of the spinor as in (2.1.7), that is, by *partitioning*.

Many proposals of effective Hamiltonians for the Dirac operator can be found in the literature. Some are Hermitian, some are not, some act on 4 component spinors, others on 2-spinors. A good review about various approaches to this problem and the corresponding difficulties has been written by KUTZELNIGG [1999] (see also KUTZELNIGG [1990], RUTKOWSKI and SCHWARZ [1996], RUTKOWSKI [1999]). An important difficulty arising in this context is that most of the proposed effective Hamiltonians are quite nice when the potential V is regular, but in the case of the Coulomb potential they contain very singular terms, which are not even well defined near the nucleus. These serious singularities are avoided by a method used by CHANG, PÉLISSIER and DURAND [1986] (see also DURAND [1986], DURAND and MALRIEU [1987]), where it is proposed to use $(2mc^2 - V)^{-1}$ as an expansion parameter in the formal series defining H^{eff}, instead of c^{-2}. They obtain a 2-component Pauli-like Hamiltonian which is bounded from below, contains only well defined terms and approaches H. Similar ideas have been used by HEULLY, LINDGREN, LINDROTH, LUNDQVIST and MÅRTENSSON-PENDRILL [1986] and by VAN LENTHE, VAN LEEUWEN, BAERENDS and SNIJDERS [1994a], VAN LENTHE, VAN LEEUWEN, BAERENDS and SNIJDERS [1994b]. The latter have also made a systematic numerical analysis of this method in self-consistent calculations for the uranium atom.

2.5. *Direct variational approaches*

To begin with, let us mention two variational methods based on *nonlinear transformations of the Hamiltonian*. WALLMEIER and KUTZELNIGG [1981] look for eigenvalues of the squared Hamiltonian $(H_0 + V)^2$. The practical difficulty arises from the need to compute complicated matrix representations. HILL and KRAUTHAUSER [1994] use the Rayleigh–Ritz variational principle applied to the inverse of the Dirac Hamiltonian, $1/H$. A difficulty arises here in the computation of the matrix elements for the inverse operator. This is avoided by working in the special set of test functions defined by those which are in the image by H of a regular set of spinors. The use of these two methods can be useful in some cases, but not when the eigenvalues become close to 0.

As already noticed, the eigenvalues of the operator $H_0 + V$ are critical points of the Rayleigh quotient

$$Q_V(\psi) := \frac{((H_0 + V)\psi, \psi)}{(\psi, \psi)} \tag{2.5.1}$$

in the domain of $H_0 + V$. We are now going to describe other more sophisticated variational approaches yielding exact eigenvalues of $H_0 + V$. The particular structure of the spectrum of H_0 clearly shows that eigenvalues of $H_0 + V$ lying in the gap of the essential spectrum should be given by some kind of *min–max approach*. This had been mentioned in several papers dealing with numerical computations of Dirac eigenvalues, before it was proved in a series of papers: ESTEBAN and SÉRÉ [1997], GRIESEMER and SIEDENTOP [1999], DOLBEAULT, ESTEBAN and SÉRÉ [2000a], GRIESEMER, LEWIS and SIEDENTOP [1999], DOLBEAULT, ESTEBAN and SÉRÉ [2000b]). Basically, in all those papers, it was shown that under appropriate assumptions on the potential V, the eigenvalues are indeed characterized as a sequence of min–max values defined for Q_V on well chosen sets. A theorem in DOLBEAULT, ESTEBAN and SÉRÉ [2000b] proves that for a large class of potentials V, the ground state energy of $H_0 + V$ is given by the smallest λ in the gap $[-mc^2, mc^2]$ such that there exists φ satisfying

$$\lambda \int_{\mathbb{R}^3} |\varphi|^2 \, dx = \int_{\mathbb{R}^3} \left(\frac{|(\sigma \cdot \nabla)\varphi|^2}{1 - V + \lambda} + (1 + V)\,|\varphi|^2 \right) dx \tag{2.5.2}$$

and the corresponding eigenfunction is the spinor function

$$\psi = \begin{pmatrix} \varphi \\ -i \frac{(\sigma \cdot \nabla)\varphi}{1 - V + \lambda} \end{pmatrix}. \tag{2.5.3}$$

Note that the idea to build a semibounded energy functional had already been introduced by BAYLISS and PEEL [1983] in another context. It is closely related to previous works of DATTA and DEVIAH [1988] and TALMAN [1986], where a particular min–max procedure for the Rayleigh quotient Q_V is proposed without proof. We will not give here further details on these theoretical aspects (for tractable numerical applications, see below).

An alternative variational method has been proposed by DOLBEAULT, ESTEBAN and SÉRÉ [2000a]. It is based on rigorous results proving that for a very large class of potentials (including all those relevant in atomic models), the ground state of $H_0 + V$ can be found by a *minimization* problem posed in a class of functions defined by a *nonlinear constraint*. The main idea is to eliminate the lower component of the spinor and solve a minimization problem for the upper one. With the notations of the introduction, $\psi = \binom{\varphi}{\chi}$ is an eigenfunction of $H_0 + V$ if and only if (2.1.7) takes place. The first equation in (2.1.7) is an elliptic second order equation for the upper component φ, while the second part of (2.1.7) gives the lower component χ as a function of φ and the eigenvalue λ. The dependence of H^μ on $\lambda = \mu + mc^2$ makes this problem nonlinear, since λ is still to be found, but the difficulty of finding the unitary transformation Ω in the Foldy–Wouthuysen approach is now replaced by a much simpler problem.

We may reformulate the question as follows. Let $A(\lambda)$ be the operator defined by the quadratic form acting on 2-spinors:

$$\varphi \mapsto \int_{\mathbb{R}^3} \left(\frac{|(\sigma \cdot \nabla)\,\varphi|^2}{1 - V + \lambda} + (1 - \lambda + V)\,|\varphi|^2 \right) dx =: \big(\varphi, A(\lambda)\varphi\big) \tag{2.5.4}$$

and consider its lowest eigenvalue, $\mu_1(\lambda)$. Because of the monotonicity with respect to λ, there exists at most one λ for which $\mu_1(\lambda) = 0$. This λ is the ground state level.

An algorithm to numerically solve the above problem has been proposed in DOLBEAULT, ESTEBAN, SÉRÉ and VANBREUGEL [2000]. The idea consists in discretizing Eq. (2.5.2) in a finite-dimensional space E_n of dimension n of 2-spinor functions. The discretized version of (2.5.4) is

$$A^n(\lambda)\, x_n \cdot x_n = 0, \tag{2.5.5}$$

where $x_n \in E_n$ and $A^n(\lambda)$ is a λ-dependent $n \times n$ matrix. If E_n is generated by a basis set $\{\varphi_i, \dots \varphi_n\}$, the entries of the matrix $A^n(\lambda)$ are the numbers

$$\int_{\mathbb{R}^3} \left(\frac{\big((\sigma \cdot \nabla)\varphi_i, (\sigma \cdot \nabla)\varphi_j\big)}{1 - V + \lambda} + (1 - \lambda + V)(\varphi_i, \varphi_j) \right) dx. \tag{2.5.6}$$

The ground state energy will then be approached from above by the unique λ for which the first eigenvalue of $A^n(\lambda)$ is zero. This method has been tested on a basis of Hermite polynomials (see DOLBEAULT, ESTEBAN, SÉRÉ and VANBREUGEL [2000] for some numerical results). More efficient computations have been made recently on radially symmetric configurations with B-splines basis sets, involving very sparse matrices. Approximations from above of the excited levels can also be computed by requiring successively that the second, third, ... eigenvalues of $A^n(\lambda)$ are equal to zero.

3. The MCDF method for atoms

3.1. The Muticonfiguration Dirac–Fock (MCDF) method

The MCDF equations are obtained from (1.1.2) by a variational principle. The energy functional is written

$$E_{\text{tot}} = \frac{\langle \nu \Pi J M | H^{\text{np}} | \nu \Pi J M \rangle}{\langle \nu \Pi J M || \nu \Pi J M \rangle}. \tag{3.1.1}$$

A Hamiltonian matrix which provides the mixing coefficients by diagonalization is obtained from (3.1.1) with the help of

$$\frac{\partial}{\partial c_\nu} E_{\text{tot}} = 0, \tag{3.1.2}$$

and a set of integro-differential equations for the radial wave functions $P_\kappa(r)$ and $Q_\kappa(r)$ is obtained from the functional derivatives

$$\begin{cases} \dfrac{\delta}{\delta P_\kappa(r)} E_{\text{tot}} = 0, \\ \dfrac{\delta}{\delta Q_\kappa(r)} E_{\text{tot}} = 0. \end{cases} \tag{3.1.3}$$

One assumes the orthogonality condition (restricted Dirac–Fock)

$$\int_0^\infty \left[P_A(r)P_B(r) + Q_A(r)Q_B(r)\right] dr = \delta_{\kappa_A,\kappa_B}\delta_{n_A,n_B}, \tag{3.1.4}$$

in order to make the angular calculations possible. Eq. (3.1.3) then leads to the inhomogeneous Dirac equation for a given orbital A

$$\begin{aligned} &\begin{pmatrix} \frac{d}{dr} + \frac{\kappa_A}{r} & -\frac{2}{\alpha} + \alpha V_A(r) \\ -\alpha V_A(r) & \frac{d}{dr} - \frac{\kappa_A}{r} \end{pmatrix} \begin{pmatrix} P_A(r) \\ Q_A(r) \end{pmatrix} \\ &\quad = \alpha \sum_B \varepsilon_{A,B} \begin{pmatrix} Q_B(r) \\ -P_B(r) \end{pmatrix} + \begin{pmatrix} X_{Q_A}(r) \\ -X_{P_A}(r) \end{pmatrix}, \end{aligned} \tag{3.1.5}$$

where V_A is the sum of the nuclear potential and the direct Coulomb potential, while the exchange terms X_{P_A} and X_{Q_A} include all the two-electron interactions except for the direct Coulomb instantaneous repulsion. The constants $\varepsilon_{A,B}$ are Lagrange parameters used to enforce the orthogonality constraints of (3.1.4) and thus the summation over B runs only for orbitals with $\kappa_B = \kappa_A$. The exchange terms can be very large if the orbital A has a small effective occupation (the exchange term is a sum of exchange potentials divided by the effective occupation of the orbitals). This effective occupation is the sum

$$o_A = \sum_{i=1}^{NCF} c_\nu^2 q_\nu^{(A)}, \tag{3.1.6}$$

where $q_\nu^{(A)}$ is the number of electrons in the orbital A in the νth configuration.

The numerical MCDF methods are based on a fixed-point method, or to be precise on an iteration scheme which provides a self-consistent field (SCF) state in a way very similar to the method which is used to solve the Hartree–Fock model. Initial wave functions must be chosen, e.g., hydrogenic wave functions, wave functions in a Thomas–Fermi potential or wave functions already optimized with a smaller set of configurations. One then builds the Hamiltonian matrix (3.1.2) and obtains the mixing coefficients. Those coefficients and the initial wave functions enter the direct and exchange potential in (3.1.5), which become normal differential equations, and are solved numerically for each orbital. A new set of potential terms is then evaluated until all the wave functions are stable to a given accuracy ($\approx 10^{-2}$ in the first cycle of diagonalization to $\approx 10^{-6}$ at the last cycle, at the point where the largest variation occurs). A new Hamiltonian matrix is then built and new mixing coefficients are calculated. This process is repeated until convergence is reached. As it is a highly nonlinear process, this can be very tricky, and trial and error on the initial conditions is often required when many configuration and correlation orbitals (i.e., orbitals with very small effective occupations) are involved. All those calculations are done using direct numerical solutions of the MCDF differential equations (3.1.5), which has the advantage of providing very accurate results with relatively limited set of configurations, while MCDF methods using basis set require orders of magnitude more configurations to achieve similar accuracies.

Explicit expressions for V_A, X_{P_A} and X_{Q_A} can be found in GRANT [1987], GRANT and QUINEY [1988], DESCLAUX [1993]. All potentials can be expressed in term of the functions

$$Z^k_{i,j}(x) = \frac{1}{x^k}\int_0^x dr\, \rho_{ij}(r)\, r^k, \tag{3.1.7}$$

$$Y^k_{i,j}(x) = \frac{1}{x^k}\int_0^x dr\, \rho_{ij}(r)\, r^k + x^{k+1}\int_x^\infty dr\, \frac{\rho_{ij}(r)}{r^{k+1}}, \tag{3.1.8}$$

where $\rho_{ij}(r) = P_i(r)P_j(r) + Q_i(r)Q_j(r)$ for the Coulomb part of the interaction, to which are added terms with $\rho_{ij}(r) = P_i(r)Q_j(r)$ or $\rho_{ij}(r) = Q_i(r)P_j(r)$ when Breit retardation is included in the self-consistent field process. These potential terms can be obtained very efficiently numerically by solving a second-order differential equation (Poisson equation), as a set of two first-order differential equations, with the predictor-corrector method presented in Section 3.2.

3.2. *Numerical solution of the inhomogeneous Dirac–Fock radial equations*

In order to increase the numerical stability, the direct numerical computation of (3.1.5) is done by shooting techniques. First one chooses a change of variables to make the method more efficient because bound orbitals exhibit a rapid variation near the origin and exponential decay at large distances. One can choose either

$$t = r_0 \log(r) \quad \text{or} \quad t = r_0 \log(r) + br. \tag{3.2.1}$$

The first choice leads to a pure exponential grid, while the second leads to an exponential grid at short distances and to a linear grid at infinity, and is better suited to represent, e.g., Rydberg states. One then takes a linear grid in the new variable t, $t_n = nh$ with h ranging from 0.02 to 0.05. In order to provide the few values needed to start the numerical integration at $r = 0$, and to have accurate integrals (for evaluation of the norm for example) the wave function is represented by its series expansion at the origin, which is of the form

$$\begin{cases} P_\kappa(r) = r^\lambda(p_0 + p_1 r + \cdots), \\ Q_\kappa(r) = r^\lambda(q_0 + q_1 r + \cdots), \end{cases} \tag{3.2.2}$$

where $\lambda = \sqrt{\kappa^2 - (Z\alpha)^2}$ if $V_N(r) = -Z/r$ is a pure Coulomb potential and $\lambda = |\kappa|$ if $V_N(r)$ represents the potential of a finite charge distribution. In this case if $\kappa > 0$, $p_0 = p_2 = \cdots = 0$ and $q_1 = q_3 = \cdots = 0$, and if $\kappa < 0$, $p_1 = p_3 = \cdots = 0$, $q_0 = q_2 = \cdots = 0$.

Predictor–corrector methods. In the case of the atomic problem, the use of fancy techniques like adaptative grids is not recommended, as it is much more efficient to tabulate all wave functions over the same grid, particularly if other properties like transition probabilities are calculated as well. One then uses well proven differential equation solving techniques like predictor–corrector methods and finite difference schemes. The expansion (3.2.2) is substituted into the differential equation (3.1.5) to obtain the coefficients p_i and q_i, for $i > 0$. These coefficients are used to generate values for the wave function at the few first n points of the grid, with an arbitrary value of p_0. Then the value of the function at the next grid point is obtained using the differential equation solver. At infinity the same procedure is used. An exponential approximation of the wave function is made, and the same differential equation solver is used downward to some matching point r_m, usually chosen close to the classical turning point in the potential $V_A(r)$. In the predictor–corrector technique, an approximate value of the function at the mesh point $n+1$ is predicted from the known values at the preceding n points. This estimate is inserted in the differential equation to obtain the derivative that in turn is used to correct the first estimate, then the final value may be taken as a linear combination of the predicted and corrected values to increase the accuracy. As an example we consider the five points Adams' method that has been widely selected because of its stability properties (PRESS, FLANNERY, TEUKOLSKY and VETTERLING [1986]). The predicted, corrected and final values are given, respectively, by:

$$\begin{aligned}
p_{n+1} &= y_n + \left(1901 y'_n - 2774 y'_{n-1} + 2616 y'_{n-2} - 1274 y'_{n-3} + 251 y'_{n-4}\right)/720,\\
c_{n+1} &= y_n + \left(251 p'_{n+1} + 646 y'_n - 264 y'_{n-1} + 106 y'_{n-2} - 19 y'_{n-3}\right)/720,\\
y_{n+1} &= (475 c_{n+} + 27 p_{n+1})/502,
\end{aligned} \tag{3.2.3}$$

where p' and y' stand for the derivatives with respect to the tabulation variable. The linear combination for the final value is defined as to cancel the term of order h^6, h being the constant interval step of the mesh. In the above equations, y represents either the large or small component of the radial wave function.

Since one starts with a somewhat arbitrary energy and slope at the origin, the components of the wave function obtained by the preceding method are not continuous. A strategy must be devised to obtain the real eigenenergy and slope at the origin from the numerical solution. In the case of an homogeneous equation, one can simply make the large component continuous by multiplying the wave function by the ratio of the inward and outward values of the large component at the matching point and then change the energy until the small component is continuous, using the default in the norm. To first order the correction to the eigenvalue is

$$\delta\varepsilon = \frac{cP(r_m)[Q(r_m^-) - Q(r_m^+)]}{\int_0^\infty [P^2(u) + Q^2(u)]\,du}, \tag{3.2.4}$$

where $Q(r_m^{\pm})$ are the solutions from each side of the matching point. One then checks that the solution is the desired one by verifying that it has the right number of nodes.

In the inhomogeneous case such a strategy cannot work. In order to obtain a solution which is continuous everywhere, it is possible to proceed in the following way. One uses the well-known fact that the solution of an inhomogeneous differential equation can be written as the sum of a particular solution of the inhomogeneous equation and of the solution of the associated homogeneous equation (in the present case the equation obtained by neglecting the exchange potentials). Thus if P^o and P^i are, respectively, the outward and inward solutions for the large component, one obtains, with the same labels for the small component:

$$\begin{aligned} \left[P_{\mathrm{I}}^{\mathrm{o}} + aP_{\mathrm{H}}^{\mathrm{o}}\right]_{r=r_m^-} &= \left[P_{\mathrm{I}}^{\mathrm{i}} + bP_{\mathrm{H}}^{\mathrm{i}}\right]_{r=r_m^+}, \\ \left[Q_{\mathrm{I}}^{\mathrm{o}} + aQ_{\mathrm{H}}^{\mathrm{o}}\right]_{r=r_m^-} &= \left[Q_{\mathrm{I}}^{\mathrm{i}} + bQ_{\mathrm{H}}^{\mathrm{i}}\right]_{r=r_m^+}, \end{aligned} \tag{3.2.5}$$

where the subscripts I and H stand for the inhomogeneous and homogeneous solutions. The coefficients a and b can be obtained from the differential equation. Obviously this continuous solution will not be normalized for an arbitrary value of the diagonal parameter $\varepsilon_{A,A}$ of Eq. (3.1.5). The default in the norm is then used to modify $\varepsilon_{A,A}$ until the proper eigenvalue is found. This method is very accurate but not very efficient since it requires to solve both the inhomogeneous and the homogeneous equations to obtain a continuous solution.

Finite differences methods. As seen above, the predictor–corrector method has some disadvantages. In the nonrelativistic case the Numerov method associated with tail correction (FROESE FISCHER [1977]) provides directly a continuous approximation (the derivative remains discontinuous until the eigenvalue is found). We consider now alternative methods that easily allow to enforce the continuity of one of the two radial components. Let us define the solution at point $n+1$ as:

$$y_{n+1} = y_n + h\left(y_n' + y_{n+1}'\right) + \Delta_n, \tag{3.2.6}$$

where Δ_n is a difference correction given, in terms of central differences, by:

$$\Delta_n = \frac{-1}{12}\delta^3 y_{n+\frac{1}{2}} + \frac{1}{120}\delta^5 y_{n+\frac{1}{2}}, \tag{3.2.7}$$

with:

$$\begin{aligned} \delta^3 y_{n+\frac{1}{2}} &= y_{n+2} - 3y_{n+1} + 3y_n - y_{n-1}, \\ \delta^5 y_{n+\frac{1}{2}} &= y_{n+3} - 5y_{n+2} + 10y_{n+1} - 10y_n + 5y_{n-1} - y_{n-2}. \end{aligned} \tag{3.2.8}$$

Accurate solutions are required only when self-consistency is reached. Consequently, the difference correction Δ_n can be obtained at each iteration from the wave functions of the previous iteration as it is done for the potential terms. One can then design computationally efficient schemes (DESCLAUX, MAYERS and O'BRIEN [1971]). We define

$$
\begin{aligned}
a_n &= 1 + \frac{\kappa h}{2}\frac{r'_n}{r_n}, & u_n &= \Delta_n^P + \frac{h}{2}\left[r'_n X_n^Q + r'_{n+1} X_{n+1}^Q\right],\\
b_n &= -1 + \frac{\kappa h}{2}\frac{r'_n}{r_n}, & v_n &= \Delta_n^P + \frac{h}{2}\left[r'_n X_n^P + r'_{n+1} X_{n+1}^P\right],\\
\varphi_n &= \alpha\frac{h}{2}[\varepsilon_n - V_n]r'_n, & \theta_n &= \frac{h}{\alpha}r'_n + \theta_n,
\end{aligned}
\tag{3.2.9}
$$

where r' stands for dr/dt (to take into account the fact that the tabulation variable t is a function of r) and $X^{P(Q)} = X_{P_A(Q_A)} + \sum_{B\neq A} \varepsilon_{A,B} P_B(Q_B)$. All the functions of r are evaluated using wave functions obtained at the previous iteration. Then the system of algebraic equations:

$$
\begin{aligned}
&a_{n+1}P_{n+1} - \theta_{n+1}Q_{n+1} + b_n P_n - \theta_n Q_n = u_n,\\
&\varphi_{n+1}P_{n+1} - b_{n+1}Q_{n+1} + \varphi_n P_n - a_n Q_n = v_n,
\end{aligned}
\tag{3.2.10}
$$

determines P_{n+1} and Q_{n+1} if P_n and Q_n are known. For the outward integration, this system is solved step by step from near the origin to the matching point after getting the solution at the first point by series expansion. For the inward integration, an elimination process is used by expressing the solution in the matrix form $[M](PQ) = (uv)$ with the matrix M given by:

$$
M = \begin{bmatrix}
-a_m & \varphi_{m+1} & -b_{m+1} & & & \cdot & & & & & \\
-\theta_m & a_{m+1} & -\theta_{m+1} & & & \cdot & & & & & \\
& \varphi_{m+1} & -a_{m+1} & \varphi_{m+2} & -b_{m+2} & \cdot & & & & & \\
& b_{m+1} & -\theta_{m+1} & a_{m+2} & -\theta_{m+2} & \cdot & & & & & \\
\cdot & \cdot & \cdot & \cdot & \cdot & \cdot & \cdot & \cdot & \cdot & \cdot & \cdot \\
& & & & & \cdot & b_{N-2} & -\theta_{N-2} & a_{N-1} & \theta_{N-1} & \\
& & & & & \cdot & & & \varphi_{N-1} & -a_{N-1} & \varphi_N \\
& & & & & \cdot & & & b_{N-1} & -\theta_{N-1} & a_N
\end{bmatrix}
\tag{3.2.11}
$$

and the two column vectors (PQ) and (uv) defined as:

$$
(PQ) = \begin{bmatrix} Q_m \\ P_{m+1} \\ Q_{m+1} \\ P_{m+2} \\ \cdot \\ P_{N-1} \\ Q_{N-1} \\ P_N \end{bmatrix}, \qquad
(uv) = \begin{bmatrix} v_m - \varphi_m P_m \\ u_m - b_m P_m \\ v_{m+1} \\ u_{m+1} \\ \\ u_{N-2} \\ v_{N-1} + b_N Q_N \\ u_{N-1} + \theta_N Q_N \end{bmatrix}.
\tag{3.2.12}
$$

As displayed in Eq. (3.2.11) each row of the matrix M has at most four nonzero elements. To solve this system of equations the matrix M is decomposed into the product of two triangular matrices $M = LT$ in which L is a lower matrix with only three nonzero elements on each row and T an upper matrix with the same property. Introducing an intermediate vector (pq) it is possible to solve $L(pq) = (uv)$ for m, $m+1, \ldots, N$ and then $T(PQ) = (pq)$ for $N, N-1, \ldots, m$. The last point of tabulation N is determined by the requirement that P_N should be lower than a specified small value when assuming $Q_N = 0$. Thus the number of tabulation points of each orbital is determined automatically during the self-consistency process. This elimination process produces, as written here, a large component P that is continuous everywhere. The discontinuity of the small component at the matching point r_m can then be used to adjust the eigenvalue $\varepsilon_{A,A}$. In practice this method works very well for occupied orbitals (i.e., orbitals with effective occupations at the Dirac–Fock level $q_o^{(A)} \equiv n$, n integer larger or equal to 1). Yet it is not sufficiently accurate for correlation orbitals and leads to convergence instability. A good strategy DESCLAUX [1993] is thus to use the accurate predictor–corrector method for the outward integration and the finite differences method with the tail correction for the inward integration. However the accuracy of the inward integration is increased by computing directly the difference correction (3.2.7) from the wave function being computed rather than from the one from the previous iteration.

Diagonal Lagrange multipliers. One can use differential techniques, when the gaining of the eigenenergy ε_{AA} is difficult. Their evaluation proceeds as follows. One can obtain the first order variation of the large component P with respect to a change $\Delta\varepsilon_{AA}$ of one of the off-diagonal Lagrange multipliers by substituting the development

$$P(\varepsilon_{AA}^0 + \Delta\varepsilon_{AA}) = P(\varepsilon_{AA}^0) + \Delta\varepsilon_{AA} \left.\frac{\partial P}{\partial \varepsilon_{AA}}\right|_{\varepsilon_{AA}=\varepsilon_{AA}^0} \tag{3.2.13}$$

(and the equivalent one for the small component Q) into the differential equation (3.1.5). Defining

$$p_{AA} = \frac{\partial P}{\partial \varepsilon_{AA}}, \qquad q_{AA} = \frac{\partial Q}{\partial \varepsilon_{AA}}, \tag{3.2.14}$$

leads to the new set of differential equations

$$\begin{aligned} &\begin{pmatrix} \frac{d}{dr} + \frac{\kappa_A}{r} & -\frac{2}{\alpha} + \alpha V_A(r) \\ -\alpha V_A(r) & \frac{d}{dr} - \frac{\kappa_A}{r} \end{pmatrix} \begin{pmatrix} p_{AA}(r) \\ q_{AA}(r) \end{pmatrix} \\ &\quad = \alpha\varepsilon_{A,A} \begin{pmatrix} q_{AA}(r) \\ -p_{AA}(r) \end{pmatrix} + \alpha \begin{pmatrix} Q_B(r) \\ -P_B(r) \end{pmatrix} \end{aligned} \tag{3.2.15}$$

which is very similar to (3.1.5), with the replacement of $X_{P_A}(r)$ (resp. X_{Q_A}) by $P_B(r)$ (resp. $Q_B(r)$). This system can be solved in $p_{AA}(r)$ and $q_{AA}(r)$ by the above techniques. With this solutions $\Delta\varepsilon_{AA}$ can be calculated in first order from

$$\Delta\varepsilon_{AA} = \frac{1 - \int_0^\infty [P_A(r)P_B(r) + Q_A(r)Q_B(r)]\,dr}{2\int_0^\infty [p_{AA}(r)P_B(r) + q_{AA}(r)Q_B(r)]\,dr}. \qquad (3.2.16)$$

Note that such relations could be established to provide the change in the nondiagonal Lagrange multipliers ε_{AB} as well, if one were to solve for several orbitals of identical symmetry simultaneously.

Off-diagonal Lagrange multipliers. The self-consistent process outlined in Section 1.3 requires the evaluation of the off-diagonal Lagrange parameters to satisfy the orthonormality constraint (3.1.4). As in the nonrelativistic case, the off-diagonal Lagrange multiplier between closed[1] shells can be set to zero, which only amounts to perform a unitary transformation in the subspace of the closed shells. If the generalized occupation numbers o_A and o_B of two orbitals are different, one can use the symmetry relation

$$\varepsilon_{AB}o_A = \varepsilon_{BA}o_B \qquad (3.2.17)$$

and (3.1.5) to obtain

$$\begin{aligned} \frac{\varepsilon_{AB}(o_B - o_A)}{o_B} &= \int_0^\infty \big[V_A(r) - V_B(r)\big]\big[P_A(r)P_B(r) + Q_A(r)Q_B(r)\big]\,dr \\ &\quad - \frac{1}{\alpha}\int_0^\infty \big[X_{Q_A}(r)Q_B(r) - X_{Q_B}(r)Q_A(r) \\ &\quad + X_{P_A}(r)P_B(r) - X_{P_B}(r)P_A(r)\big]\,dr. \end{aligned} \qquad (3.2.18)$$

This equation shows that many terms will cancel out in the determination of the Lagrange multipliers (e.g., the closed shell contribution to $V_A(r)$ and $V_B(r)$) and thus provides an accurate method to calculate them provided one retains only the nonzero contributions. If $(o_B - o_A) \ll 1$, however, one must use Eqs. (3.2.15) and (3.2.16) to evaluate the Lagrange multipliers.

3.3. *Solution of the inhomogeneous Dirac–Fock equation over a basis set*

It has been found however (INDELICATO and DESCLAUX [1993], INDELICATO [1995]) that even the enhanced numerical techniques presented in Section 3.2 would not work for correlation orbitals with very small effective occupation, particularly when the contribution of the Breit interaction is used in (3.1.5). This leads to point out that in the numerical MCDF calculations, the projection operators which should be used according

[1] Closed shells are the shell filled with the maximum number of electrons as allowed by the Pauli principle, i.e., $2|\kappa|$.

to (1.1.5) are absent, as they have no explicit expression. A new method has been proposed that retains the advantages of the numerical MCDF. The idea is to expand P_A, Q_A, X_{P_A} and X_{Q_A} over a finite basis set, e.g., the one based on the B-spline calculated following the method of Section 2.3, using the full MCDF direct potential $V_A(r)$. Let us thus assume that one has a complete set of solutions $\{\phi_1^{(A)}, \ldots, \phi_{2n}^{(A)}\}$, with eigenvalues $\{\varepsilon_1^{(A)}, \ldots, \varepsilon_{2n}^{(A)}\}$ of the homogeneous equation associated to (3.1.5). One then writes

$$\begin{pmatrix} P_A(r) \\ Q_A(r) \end{pmatrix} = \sum_{i=1}^{2n} s_i^{(A)} \phi_i^{(A)}(r) \quad \text{and} \quad \begin{pmatrix} X_{P_A}(r) \\ X_{Q_A}(r) \end{pmatrix} = \sum_{i=1}^{2n} x_i^{(A)} \phi_i^{(A)}(r). \tag{3.3.1}$$

Substituting back into (3.1.5) and using the orthonormality of the basis set functions, one easily obtains

$$s_i^{(A)} = \frac{x_i^{(A)} + \sum_{B \neq A} \varepsilon_{AB}\, s_i^{(B)}}{\alpha(\varepsilon_1^{(A)} - \varepsilon_{AA})}. \tag{3.3.2}$$

The square of the norm of the solution of (3.1.5) is then easily obtained as

$$N(\varepsilon_{AA}) = \sum_{i=1}^{2n} \left(s_i^{(A)}\right)^2 = \sum_{i=1}^{2n} \left(\frac{x_i^{(A)} + \sum_{B \neq A} \varepsilon_{AB}\, s_i^{(B)}}{\alpha(\varepsilon_1^{(A)} - \varepsilon_{AA})} \right)^2. \tag{3.3.3}$$

One then can calculate the normalized solution of (3.1.5) if the off-diagonal Lagrange parameters are known, by solving $N(\varepsilon_{AA}) = 1$ for ε_{AA}. One can notice the interesting feature of (3.3.3) that the norm of the solution of the inhomogeneous equation (3.1.5) has a pole for each eigenenergy of the homogeneous equation. This method has the advantage over purely numerical techniques that by restricting the sums in (3.3.1) to positive energy eigenstates, one can explicitly implement projection operators, thus solving readily the "no-pair" Hamiltonian (1.1.2), rather than an ill-defined equation. More details on this method and on the evaluation of the off-diagonal Lagrange multipliers can be found in INDELICATO [1995].

4. Numerical relativistic methods for molecules

Most of molecular methods that include relativistic corrections are based on the expansion of the molecular orbitals in terms of basis sets (most of the time taken to be Gaussian functions). We shall not review these methods here but refer the interested reader to a book to be published soon (SCHWERDTFEGER [2002]). Let us just point out that the sometimes observed lack of convergence to upper bounds in the total energy (the so-called variational collapse) is not unambiguously related to the Dirac negative energy continuum. Indeed this attractive explanation is unfortunately unable to explain the appearance of spurious solutions. Both the existence of spurious solutions and the lack of convergence to expected levels can be traced back to originate from poor basis sets and bad finite matrix representations of the operators (in particular the kinetic energy). For an extensive discussion see DYALL, GRANT and WILSON [1984]. Numerical methods successfully used are briefly sketched in the next two paragraphs.

4.1. *Fully numerical two-dimensional method*

For diatomic molecules, the one-electron Dirac wave functions may be written as

$$\Phi = \begin{pmatrix} e^{i(m-1/2)\varphi}\,\phi_1^L(\xi,\eta) \\ e^{i(m+1/2)\varphi}\,\phi_2^L(\xi,\eta) \\ i\,e^{i(m-1/2)\varphi}\,\phi_3^S(\xi,\eta) \\ i\,e^{i(m+1/2)\varphi}\,\phi_4^S(\xi,\eta) \end{pmatrix} \tag{4.1.1}$$

where L (S) stands for the large (small) component and elliptical coordinates (ξ,η,φ) are used with:

$$\xi = (r_1 + r_2)/R, \qquad \eta = (r_1 - r_2)/R, \tag{4.1.2}$$

where r_1 and r_2 are the distances between the electron and each of the nucleus, R is the inter-nuclear distance. The third variable φ is the azimuthal angle around the axis through the nuclei.

As usual for molecular calculations, the variational collapse is avoided by defining the small component in terms of the large one (KUTZELNIGG [1984]). Starting from the Dirac equation in a local potential V one possibility is to use:

$$\phi^S = c\boldsymbol{\sigma}.\boldsymbol{p}\phi^L/\left[2c^2 + E - V\right]. \tag{4.1.3}$$

After this substitution, the large component is given as solution of a second order differential equation that can be solved using well-known relaxation methods (VARGA [1963]).

For efficiency, the distribution of integration points must be chosen as to accumulate points where the functions are rapidly varying. It was found that the transformation,

$$\mu = \operatorname{arccosh}(\xi), \qquad \nu = \operatorname{arccosh}(\eta), \tag{4.1.4}$$

which yields a quadratic distribution of points near the nuclei, is some kind of optimum to reduce the number of points needed to achieve a given accuracy. Then the derivatives of the Laplace operator are approximated by n-point finite differences. In so doing, the differential equations are replaced by a set of linear equations that can be written in a matrix form as

$$(\boldsymbol{A} - E\boldsymbol{S})X = B, \tag{4.1.5}$$

where the matrix $\boldsymbol{A}$, that represents the direct part of the Fock operator, is diagonal dominant but has nondiagonal elements arising from the discretization of the Laplace operator. Here E is the energy eigenvalue, $\boldsymbol{S}$ is the overlap diagonal matrix and B

a vector due to the exchange part of the Fock operator whose values change during iterations. Then the relaxation method can be viewed as an iterative method to find the x_i component of X such that

$$(\boldsymbol{A} - E\boldsymbol{S})x_i = b_i, \tag{4.1.6}$$

each iteration n being associated with a linear combination of the initial and final estimate of x_i at iteration $n - 1$, i.e.,

$$x_i^{\text{initial}_{n+1}} = (1 - \omega)x_i^{\text{initial}_n} + \omega x_i^{\text{final}_n}. \tag{4.1.7}$$

It was found that with overrelaxation (i.e., $\omega > 1$), the method may be slow in convergence but it is quite stable. Applications of the method outlined above may be found in SUNDHOLM, PYYKKÖ and LAAKSONEN [1987] and in references therein.

4.2. *Numerical integrations with linear combinations of atomic orbitals*

A widely used approximation in molecular calculations is to expand the molecular orbitals as a linear combination of atomic orbitals. If these atomic orbitals are chosen as the numerical solutions of some kind of Dirac–Fock atomic calculations, then small basis sets are sufficient to achieve good accuracy. The main disadvantage of this choice is that all multi-dimensional integrals have to be calculated numerically. This is compensated by two advantages: first the kinetic energy contribution can be computed by a single integral using the atomic Dirac equations (thus avoiding numerical differentiation), second, by including only positive energy atomic wave functions, no "variational collapse" will occur.

In this method, the molecular wave functions ψ are expanded in terms of symmetry molecular orbitals χ as:

$$\psi_\lambda = \sum_\nu c_\nu^\lambda \chi^\nu, \tag{4.2.1}$$

while the symmetry molecular orbitals χ are taken to be linear combinations of atomic orbitals φ:

$$\chi^\nu = \sum_i d_i^\nu \varphi^i. \tag{4.2.2}$$

The coefficients d_i^ν are given by the symmetry of the molecular orbital and are obtained from the irreducible representations of the double point groups. Computing all necessary integrals (overlaps, matrix elements of the Dirac operator, the Coulomb interaction, etc, ...) the Dirac–Fock equations are reduced to a generalized matrix eigenvalue problem that determines both the eigenvalues and the c_ν^λ coefficients of Eq. (4.2.1).

To compute the various matrix elements in the case of diatomic molecules, SEPP, KOLB, SENGLER, HARTUNG and FRICKE [1986] used Gauss–Laguerre and Gauss–Legendre integration schemes on a grid of points defined by the same variables as those of Eq. (4.1.4). Unfortunately this approach is not easy to extend beyond diatomic molecules and other methods have to be implemented. It has been shown, see for example ROSEN and ELLIS [1975], that the adaptation to molecules of the so-called Discrete Variational Method (DVM) developed for solid state calculations (ELLIS and PAINTER [1970]) may be both efficient and accurate. The DVM may be viewed as performing a multidimensional integral via a weighted sum of sampling points, i.e., to compute a matrix element $\langle f \rangle$ by:

$$\langle f \rangle = \sum_{n=1}^{N} \omega(r_i) f(r_i), \tag{4.2.3}$$

where the weight function $\omega(r_i)$ can be considered as an integration weight corresponding to a local volume per point. This function is also constrained to force the error momenta to vanish on the grid points following the work of HASELGROVE [1961]. Furthermore the set of the sampling points $[r_i]$ must be chosen to preserve the symmetries of the system under configuration (this is accomplished by taking a set of sampling points that includes all points Rr_i, R standing for operations of the symmetry group). A full description of the DVM can be found in the references given above.

References

BAYLISS, W.E. and S.J. PEEL (1983), Stable variational calculations with the Dirac Hamiltonian, *Phys. Rev. A* **28** (4), 2552–2554.

BEIER, T. (2000), The g_j factor of a bound electron and the hyperfine structure splitting in hydrogenlike ions, *Phys. Rep.* **339** (2–3), 79–213.

BROWN, G.E. and D.E. RAVENHALL (1951), On the interaction of two electrons, *Proc. R. Soc. London, Ser. A* **208** (1951), 552–559.

CHANG, CH., J.-P. PÉLISSIER and PH. DURAND (1986), Regular two-component Pauli-like effective Hamiltonians in Dirac theory, *Phys. Scripta* **34**, 394–404.

DATTA, S.N. and G. DEVIAH (1988), The minimax technique in relativistic Hartree–Fock calculations, *Pramana* **30** (5), 393–416.

DESCLAUX, J.P. (1993), A relativistic multiconfiguration Dirac–Fock package, in: E. Clementi, ed., *Methods and Techniques in Computational Chemistry: METECC-94, Vol. A: Small Systems* (STEF, Cogliari).

DESCLAUX, J.P., D.F. MAYERS and F. O'BRIEN (1971), Relativistic atomic wavefunction, *J. Phys. B: At. Mol. Opt. Phys.* **4**, 631–642.

DOLBEAULT, J., M.J. ESTEBAN and E. SÉRÉ (2000a), Variational characterization for eigenvalues of Dirac operators, *Calc. Var. and P.D.E.* **10**, 321–347.

DOLBEAULT, J., M.J. ESTEBAN and E. SÉRÉ (2000b), On the eigenvalues of operators with gaps. Application to Dirac operators, *J. Funct. Anal.* **174**, 208–226.

DOLBEAULT, J., M.J. ESTEBAN, E. SÉRÉ and M. VANBREUGEL (2000), Minimization methods for the one-particle Dirac equation, *Phys. Rev. Lett.* **85** (19), 4020–4023.

DRAKE, G.W.F. and S.P. GOLDMAN (1981), Application of discrete-basis-set methods to the Dirac equation, *Phys. Rev. A* **23**, 2093–2098.

DRAKE, G.W.F. and S.P. GOLDMAN (1988), Relativistic Sturmian and finite basis set methods in atomic physics, *Adv. Atomic Molecular Phys.* **23**, 23–29.

DURAND, PH. (1986), Transformation du Hamiltonien de Dirac en Hamiltoniens variationnels de type Pauli. Application à des atomes hydrogenoï des, *C. R. Acad. Sci. Paris, série II* **303** (2), 119–124.

DURAND, PH. and J.-P. MALRIEU (1987), Effective Hamiltonians and pseudo-potentials as tools for rigorous modelling, in: K.P. Lawley, ed., *Ab initio Methods in Quantum Chemistry I* (Wiley).

DYALL, K.G., I.P. GRANT and S. WILSON (1984), Matrix representation of operator products, *J. Phys. B: At. Mol. Phys.* **17**, 493–503.

EIDES, M.I., H. GROTCH and V.A. SHELYUTO (2001), Theory of light hydrogenlike atoms, *Phy. Rep.* **342** (2–3), 63–261.

ELLIS, D.E. and G.S. PAINTER, (1970), Discrete variational method for the energy-band problem with general crystal potentials, *Phys. Rev. B* **2** (8), 2887–2898.

ESTEBAN, M.J. and E. SÉRÉ (1997), Existence and multiplicity of solutions for linear and nonlinear Dirac problems, in: P.C. Greiner, V. Ivrii, L.A. Seco and C. Sulem, eds., *Partial Differential Equations and Their Applications*, CRM Proceedings and Lecture Notes, Vol. 12 (AMS).

FOLDY, L.L. and S.A. WOUTHUYSEN (1950), On the Dirac theory of spin-1/2 particles and its nonrelativistic limit, *Phys. Rev.* **78**, 29–36.

FROESE FISCHER, C. (1977), *The Hartree–Fock Method for Atoms* (Wiley).

GRANT, I.P. (1982), Conditions for convergence of variational solutions of Dirac's equation in a finite basis, *Phys. Rev. A* **25** (2), 1230–1232.

GRANT, I.P. (1987), Relativistic atomic structure calculations, *Meth. Comp. Chem.* **2**, 132.

GRANT, I.P. (1989), Notes on basis sets for relativistic atomic structure and QED, in: P.J. Mohr, W.R. Johnson and J. Sucher, eds., *A.I.P. Conf. Proc.*, Vol. 189, 235–253.

GRANT, I.P. and H.M. QUINEY (1988), Foundation of the relativistic theory of atomic and molecular structure, *Adv. At. Mol. Phys.* **23**, 37–86.

GRIESEMER, M., R.T. LEWIS and H. SIEDENTOP (1999), A minimax principle in spectral gaps: Dirac operators with Coulomb potentials, *Doc. Math.* **4**, 275–283 (electronic).

GRIESEMER, M. and H. SIEDENTOP (1999), A minimax principle for the eigenvalues in spectral gaps, *J. London Math. Soc. (2)* **60** (2), 490–500.

HASELGROVE, C.B. (1961), A method for numerical integration, *Math. Comput.* **15**, 323–337.

HEULLY, J.L., I. LINDGREN, E. LINDROTH, S. LUNDQVIST and A.M. MÅRTENSSON-PENDRILL (1986), Diagonalisation of the Dirac Hamiltonian as a basis for a relativistic many-body procedure, *J. Phys. B: At. Mol. Phys.* **19**, 2799–2815.

HILL, R.N. and C. KRAUTHAUSER (1994), A solution to the problem of variational collapse for the one-particle Dirac equation, *Phys. Rev. Lett.* **72** (14) (1994), 2151–2154.

INDELICATO, P. (1995), Projection operators in multiconfiguration Dirac–Fock calculations, Application to the ground state of Heliumlike ions, *Phys. Rev. A* **51** (2) (1995), 1132–1145.

INDELICATO, P. and J.P. DESCLAUX (1993), Projection operators in the multiconfiguration Dirac–Fock method, *Phys. Scr.* **T 46**, 110–114.

JOHNSON, W.R., S. BLUNDELL and J. SAPIRSTEIN (1988), Finite basis sets for Dirac equation constructed from B splines, *Phys. Rev. A* **37** (2), 307–315.

JOHNSON, W.R. and C.D. LIN, Relativistic random phase approximation applied to atoms of the He isoelectronic sequence, *Phys. Rev. A* **14** (2), 565–575.

KOGUT, J.B. (1983), Lattice gauge theory approach to quantum chromodynamics, *Rev. Mod. Phys.* **55** (3), 775–836.

KUTZELNIGG, W. (1984), Basis set expansion of the Dirac operator without variational collapse, *Int. J. Quant. Chem.* **25**, 107–129.

KUTZELNIGG, W. (1990), Perturbation theory of relativistic corrections 2. Analysis and classification of known and other possible methods, *Z. Phys. D – Atoms, Molecules and Clusters* **15**, 27–50.

KUTZELNIGG, W. (1999), Effective Hamiltonians for degenerate and quasidegenerate direct perturbation theory of relativistic effects, *J. Chem. Phys.* **110** (17), 8283–8294.

LEE, S.-H. (2001), A new basis set for the radial Dirac equation, Preprint.

VAN LEEUWEN, R., E. VAN LENTHE, E.J. BAERENDS and J.G. SNIJDERS (1994a), Exact solutions of regular approximate relativistic wave equations for hydrogen-like atoms, *J. Chem. Phys.* **101** (2), 1272–1281.

VAN LENTHE, E., R. VAN LEEUWEN, E.J. BAERENDS and J.G. SNIJDERS (1994b), Relativistic regular two-component Hamiltonians, in: R. Broek et al., eds., *New Challenges in Computational Quantum Chemistry* (Publications Dept. Chem. Phys. and Material sciences, University of Groningen).

LINDGREN, I. and J. MORRISON (1982), *Atomic Many-Body Theory* (Springer).

MOHR, P.J., G. PLUNIEN and G. SOFF (1998), QED corrections in heavy atoms, *Phys. Rep.* **293** (5&6), 227–372.

PRESS, W.H., B.P. FLANNERY, S.A. TEUKOLSKY and W.T. VETTERLING (1986), *Numerical Recipes* (Cambridge University Press).

ROSEN, A. and D.E. ELLIS (1975), Relativistic molecular calculations in the Dirac–Slater model, *J. Chem. Phys.* **62**, 3039–3049.

RUTKOWSKI, A. (1999), Iterative solution of the one-electron Dirac equation based on the Bloch equation of the 'direct perturbation theory', *Chem. Phys. Lett.* **307**, 259–264.

RUTKOWSKI, A. and W.H.E. SCHWARZ (1996), Effective Hamiltonian for near-degenerate states in direct relativistic perturbation theory. I. Formalism, *J. Chem. Phys.* **104** (21), 8546–8552.

SALOMONSON, S. and P. ÖSTER (1989a), Relativistic all-order pair functions from a discretized single-particle Dirac Hamiltonian, *Phys. Rev. A* **40** (10), 5548–5558.

SALOMONSON, S. and P. ÖSTER (1989b), Solution of the pair equation using a finite discrete spectrum, *Phys. Rev. A* **40** (10), 5559–5567.

SCHWERDTFEGER, P., ed. (2002), *Relativistic Electronic Structure Theory. Part 1: Fundamental Aspects* (Elsevier).

SEPP, W.D., D. KOLB, W. SENGLER, H. HARTUNG and B. FRICKE (1986), Relativistic Dirac–Fock–Slater program to calculate potential-energy curves for diatomic molecules, *Phys. Rev. A* **33**, 3679–3687.

STACEY, R. (1982), Eliminating lattice fermion doubling, *Phys. Rev. D* **26** (2), 468–472.

SUCHER, J. (1980), Foundation of the relativistic theory of many-electron atoms, *Phys. Rev. A* **22** (2), 348–362.

SUNDHOLM, D., P. PYYKKÖ and L. LAAKSONEN (1987), Two-dimensional fully numerical solutions of second-order Dirac equations for diatomic molecules. Part 3, *Phys. Scr.* **36**, 400–402.

TALMAN, J.D. (1986), Minimax principle for the Dirac equation, *Phys. Rev. Lett.* **57** (9), 1091–1094.

VARGA, R.S. (1963), *Matrix Iterative Analysis* (Prentice-Hall, Englewoods Cliffs, NJ).

WALLMEIER, H. and W. KUTZELNIGG (1981), Use of the squared Dirac operator in variational relativistic calculations, *Chem. Phys. Lett.* **78** (2), 341–346.

WILSON, K.G. (1974), Confinement of quarks, *Phys. Rev. D* **10** (8), 2445–2459.

WOOD, J., I.P. GRANT and S. WILSON (1985), The Dirac equation in the algebraic approximation: IV. Application of the partitioning technique, *J. Phys. B: At. Mol. Phys.* **18**, 3027–3041.

Quantum Monte Carlo Methods for the Solution of the Schrödinger Equation for Molecular Systems

Alán Aspuru-Guzik, William A. Lester, Jr.

Department of Chemistry, University of California, Berkeley, CA 94720-1460, USA

Preface

The solution of the time independent Schrödinger equation for molecular systems requires the use of modern computers, because analytic solutions are not available.

This review deals with some of the methods known under the umbrella term quantum Monte Carlo (QMC), specifically those that have been most commonly used for electronic structure. Other applications of QMC are widespread to rotational and vibrational states of molecules, such as the work of BENOIT and CLARY [2000], KWON, HUANG, PATEL, BLUME and WHALEY [2000], CLARY [2001], VIEL and WHALEY [2001], condensed matter physics (CEPERLEY and ALDER [1980], FOULKES, MITAS, NEEDS and RAJAGOPAL [2001]), and nuclear physics (WIRINGA, PIEPER, CARLSON and PANDHARIPANDE [2000], PIEPER and WIRINGA [2001]).

QMC methods have several advantages:

- Computer time scales with system size roughly as N^3, where N is the number of particles of the system. Recent developments have made possible the approach to linear scaling in certain cases.
- Computer memory requirements are small and grow modestly with system size.
- QMC computer codes are significantly smaller and more easily adapted to parallel computers than basis set molecular quantum mechanics codes.
- Basis set truncation errors are absent in the QMC formalism.

Computational Chemistry
Special Volume (C. Le Bris, Guest Editor) of
HANDBOOK OF NUMERICAL ANALYSIS, VOL. X
P.G. Ciarlet (Editor)
2003 Elsevier Science B.V.

- Monte Carlo numerical efficiency can be arbitrarily increased. QMC calculations have an accuracy dependence of $\sqrt{T}$, where T is the computer time. This enables one to choose an accuracy range and readily estimate the computer time needed for performing a calculation of an observable with an acceptable error bar.

The purpose of the present work is to present a description of the commonly used algorithms of QMC for electronic structure and to report some recent developments in the field.

The chapter is organized as follows. In Part I, we provide a short introduction to the topic, as well as enumerate some properties of wave functions that are useful for QMC applications. In Part II we describe commonly used QMC algorithms. In Part III, we briefly introduce some special topics that remain fertile research areas.

Other sources that complement and enrich the topics presented in this chapter are our previous monograph, HAMMOND, LESTER and REYNOLDS [1994] and the reviews of SCHMIDT [1986], LESTER, JR. and HAMMOND [1990], CEPERLEY and MITAS [1996], ACIOLI [1997], BRESSANINI and REYNOLDS [1998], MITAS [1998], ANDERSON [1999], LÜCHOW and ANDERSON [2000], FOULKES, MITAS, NEEDS and RAJAGOPAL [2001]. There are also chapters on QMC contained in selected computational physics texts (KOONIN and MEREDITH [1995], GOULD and TOBOCHNIK [1996], THIJSSEN [1999]). Selected applications of the method are contained in GREEFF, HAMMOND and LESTER, JR. [1996], SOKOLOVA [2000], GREEFF and LESTER, JR. [1997], FLAD, SAVIN and SCHULTHEISS [1994], GROSSMAN and MITAS [1995], FLAD and DOLG [1997], FILIPPI and UMRIGAR [1996], BARNETT, SUN and LESTER, JR. [2001], FLAD, CAFFAREL and SAVIN [1997], GROSSMAN, LESTER, JR. and LOUIE [2000], OVCHARENKO, LESTER, JR., XIAO and HAGELBERG [2001].

QMC methods that are not covered in this review are the auxiliary field QMC method (CHARUTZ and NEUHAUSER [1995], ROM, CHARUTZ and NEUHAUSER [1997], BAER, HEAD-GORDON and NEUHAUSER [1998], BAER and NEUHAUSER [2000]) and path integral methods (CEPERLEY [1995], SARSA, SCHMIDT and MAGRO [2000]).

Atomic units are used throughout, the charge of the electron e and Planck's normalized constant $\hbar$ are set to unity. In this metric system, the unit distance is the Bohr radius a_0.

Part I. Introduction

The goal of the quantum Monte Carlo (QMC) method is to solve the Schrödinger equation, which in the time independent representation is given by

$$\widehat{H}\Psi_n(\mathbf{R}) = E_n\Psi_n(\mathbf{R}). \tag{0.1}$$

Here, $\widehat{H}$ is the Hamiltonian operator of the system in state n, with wave function $\Psi_n(\mathbf{R})$ and energy E_n; $\mathbf{R}$ is a vector that denotes the $3N$ coordinates of the system of N particles (electrons and nuclei), $\mathbf{R} \equiv \{\mathbf{r}_1, \ldots, \mathbf{r}_n\}$. For molecular systems, in the absence of electric or magnetic fields, the Hamiltonian has the form $\widehat{H} \equiv \widehat{T} + \widehat{V}$, where $\widehat{T}$ is the kinetic energy operator, $\widehat{T} \equiv -\frac{1}{2}\nabla^2_{\mathbf{R}} \equiv -\frac{1}{2}\sum_i \nabla_i^2$, and $\widehat{V}$ is the potential energy operator. For atomic and molecular systems $\widehat{V}$ is the Coulomb potential between particles of charge q_i, $\widehat{V} \equiv \sum_{ij} \frac{q_{ij}}{\mathbf{r}_{ij}}$.

The first suggestion of a Monte Carlo solution of the Schrödinger equation dates back to Enrico Fermi, based on METROPOLIS and ULAM [1949]. He indicated that a solution to the stationary state equation

$$-\frac{1}{2}\nabla_{\mathbf{R}}^2\Psi(\mathbf{R}) = E\Psi(\mathbf{R}) - V(\mathbf{R})\Psi(\mathbf{R}) \tag{0.2}$$

could be obtained by introducing a wave function of the form $\Psi(\mathbf{R}, \tau) = \Psi(\mathbf{R})e^{-E\tau}$. This yields the equation

$$\frac{\partial\Psi(\mathbf{R}, \tau)}{\partial\tau} = \frac{1}{2}\nabla^2\Psi(\mathbf{R}, \tau) - V(\mathbf{R})\Psi(\mathbf{R}, \tau). \tag{0.3}$$

Taking the limit $\tau \to \infty$, Eq. (0.2) is recovered. If the second term on the right hand side of Eq. (0.3) is ignored, the equation is isomorphic with a diffusion equation, which can be simulated by a random walk (EINSTEIN [1926], COURANT, FRIEDRICHS and LEWY [1928]), where random walkers diffuse in a **R**-dimensional space. If the first term is ignored, the equation is a first-order kinetics equation with a position dependent rate constant, $V(\mathbf{R})$, which can also be interpreted as a stochastic survival probability. A numerical simulation in which random walkers diffuse through **R**-space, reproduce in regions of low potential, and die in regions of high potential leads to a stationary distribution proportional to $\Psi(\mathbf{R})$, from which expectation values can be obtained.

1. Numerical solution of the Schrödinger equation

Most efforts to solve the Schrödinger equation are wave function methods. These approaches rely exclusively on linear combinations of Slater determinants, and include configuration interaction (CI) and the multi-configuration self-consistent field (MCSCF). There are perturbation approaches including the Möller–Plesset series (MP2, MP4), and coupled cluster (CC) theory, which are presently popular computational procedures. Wave function methods suffer from scaling deficiencies. An exact calculation with a given basis set expansion requires $N!$ computer operations, where N is the number of basis functions. Competitive methods such as coupled cluster with singles and doubles, and triples perturbation treatment, CCSD(T), scale as N^7.[1]

A term that we will use later is correlation energy (CE). It is defined as the difference between the exact nonrelativistic energy, and the energy of a mean field solution of the Schrödinger equation, the Hartree–Fock method, in the limit of an infinite basis set (LÖWDIN [1959], SENATORE and MARCH [1994])

$$E_{\text{corr}} = E_{\text{exact}} - E_{\text{HF}}. \tag{1.1}$$

[1] For a more detailed analysis of the scaling of wave function based methods see, for example, HEAD-GORDON [1996], and RAGHAVACHARI and ANDERSON [1996]. For a general overview of these methods, the reader is referred to Chapter I of this book (CANCÈS, DEFRANCESCHI, KUTZELNIGG, LE BRIS and MADAY [2003]).

The CI, MCSCF, MP(N), and CC methods are all directed at generating energies that approach E_{exact}.

Other methods that have been developed include dimensional expansions (WATSON, DUNN, GERMANN, HERSCHBACH and GOODSON [1996]), and the contracted Schrödinger equation method (MAZZIOTTI [1998]). For an overview of quantum chemistry methods, see SCHLEYER, ALLINGER, CLARK, GASTEIGER, KOLLMAN, SCHAEFER III and SCHREINER [1998].

Since the pioneering work of the late forties to early sixties (METROPOLIS and ULAM [1949], DONSKER and KAC [1950], KALOS [1962]), the MC and related methods have grown in interest. QMC methods have an advantage with system size scaling, in the simplicity of algorithms and in trial wave function forms that can be used.

2. Properties of the exact wave function

The exact time independent wave function solves the eigenvalue equation (0.1). Some analytic properties of this function are very helpful in the construction of trial functions for QMC methods.

For the present discussion, we are interested in the discrete spectrum of the $\widehat{H}$ operator. In most applications the total Schrödinger equation (0.1) can be represented into an "electronic" Schrödinger equation and a "nuclear" Schrödinger equation based on the large mass difference between electrons and nuclei. This is the Born–Oppenheimer (BO) approximation. Such a representation need not be introduced in QMC but here is the practical benefit of it that the nuclei can be held fixed for electronic motion results in the simplest form of the electronic Schrödinger equation.

The wave function also must satisfy the virial, hypervirial Hellman–Feynman and generalized Hellman–Feynman theorems (HELLMAN [1937], FEYNMANN [1939], WEISSBLUTH [1978]).[2] The local energy (FROST, KELLOG and CURTIS [1960])

$$E_L(\mathbf{R}) \equiv \frac{\widehat{H}\Psi(\mathbf{R})}{\Psi(\mathbf{R})} \tag{2.1}$$

is a constant for the exact wave function.

When charged particles meet, there is a singularity in the Coulomb potential. This singularity must be compensated by a singularity in the kinetic energy, which results in a discontinuity in the first derivative, i.e., a *cusp*, in the wave function when two or more particles meet (KATO [1957], MEYERS, UMRIGAR, SETHNA and MORGAN [1991]). For one electron coalescing at a nucleus, if we focus in a one electron function or orbital $\phi(\mathbf{r}) = \chi(r)Y_l^m(\theta, \phi)$, where $\chi(r)$ is a radial function, and $Y_l^m(\theta, \phi)$ is a spherical harmonic with angular and magnetic quantum numbers l and m, the electron–nucleus cusp condition is

$$\left.\frac{1}{\eta(r)} \frac{d\eta(r)}{dr}\right|_{r=0} = -\frac{Z}{l+1}, \tag{2.2}$$

[2] The Hellman–Feynman theorem is discussed in Section 8.2.3.

where $\eta(r)$ is the radial wave function with the leading r dependence factored out, $\eta(r) \equiv \chi(r)/r^m$, and Z is the atomic number of the nucleus.

For electron–electron interactions, the cusp condition takes the form

$$\frac{1}{\eta_{ij}(r)} \frac{d\eta_{ij}(r)}{dr}\bigg|_{r_{ij}=0} = \frac{1}{2(l+1)}, \tag{2.3}$$

where $\eta_{ij}(r)$ is the r^m factored function for the electron–electron radial distribution function.

Furthermore, $\bar{\rho}(\mathbf{r})$, the spherical average of the electron density, $\rho(\mathbf{r})$,[3] must satisfy another cusp condition, namely,

$$\frac{\partial}{\partial r}\bar{\rho}(r)\bigg|_{r=0} = -2Z\bar{\rho}(r) \tag{2.4}$$

at any nucleus. Another condition on $\rho(\mathbf{r})$ is that asymptotically it decays exponentially:

$$\rho(r \to \infty) \approx e^{-2\sqrt{2I_0}r}, \tag{2.5}$$

where I_0 is the first ionization potential. This relation can be derived from consideration of a single electron at large distance. Details on these requirements can be found in MORELL, PARR and LEVY [1975] and DAVIDSON [1976].

We discuss how to impose properties of the exact wave function on QMC trial functions in Section 3.2.

2.0.1. *Approximate wave functions*

JAMES and COOLIDGE [1937] proposed three accuracy tests of a trial wave function, Ψ_T: the root mean square error in Ψ_T

$$\delta_\Psi = \left[\int (\Psi_T - \Psi_0)^2 \, d\mathbf{R}\right]^{\frac{1}{2}}, \tag{2.6}$$

the energy error

$$\delta_E = E(\Psi_T) - E_0 \tag{2.7}$$

and the root mean square local energy deviation

$$\delta_{E_L} = \left[\int \left|\left(\widehat{H} - E_0\right)\Psi\right|^2 d\mathbf{R}\right]^{\frac{1}{2}} \tag{2.8}$$

[3] If N is the number of electrons, then $\rho(r)$ is defined by

$$\rho(r) = N \int |\Psi(\mathbf{R})|^2 \, d\mathbf{R}.$$

where the local energy is defined as in Eq. (2.1).

The calculation of δ_Ψ by QMC requires sampling the exact wave function, a procedure that will be described in Section 8.1.

Several stochastic optimization schemes have been proposed for minimizing expressions (2.6)–(2.8). Most researchers have focused on (2.8), i.e., minimizing δ_{E_L}; see, for example, MCDOWELL [1981]. In Section 4 we turn to stochastic wave function optimization procedures.

Part II. Algorithms

In this part, we describe the computational procedures of QMC methods. All of these methods use MC techniques widely used in other fields, such as operations research, applied statistics, and classical statistical mechanics simulations. Techniques such as importance sampling, correlated sampling and MC optimization are similar in spirit to those described in MC treatises by BAUER [1958], HAMMERSLEY and HANDSCOMB [1964], HALTON [1970], MCDOWELL [1981], KALOS and WHITLOCK [1986], WOOD and ERPENBECK [1976], FISHMAN [1996], SOBOL [1994], MANNO [1999], DOUCET, DE FREITAS, GORDON and SMITH [2001], LIU [2001]. The reader is referred to the former for more details on the techniques described in this part.

We present the simple, yet powerful variational Monte Carlo (VMC) method, in which the Metropolis MC[4] method is used to sample a known trial function Ψ_T. We follow with the projector Monte Carlo (PMC) methods that sample the unknown ground state wave function.

3. Variational Monte Carlo

3.1. *Formalism*

Variational methods involve the calculation of the expectation value of the Hamiltonian operator using a trial wave function Ψ_T. This function is dependent on a set of parameters, Λ, that are varied to minimize the expectation value, i.e.,

$$\langle\widehat{H}\rangle = \frac{\langle\Psi_T|\widehat{H}|\Psi_T\rangle}{\langle\Psi_T|\Psi_T\rangle} \equiv E[\Lambda] \geqslant E_0. \tag{3.1}$$

The expectation value (3.1) can be sampled from a probability distribution proportional to Ψ_T^2, and evaluated from the expression

$$\frac{\int d\mathbf{R}\left[\frac{\widehat{H}\Psi_T(\mathbf{R})}{\Psi_T(\mathbf{R})}\right]\Psi_T^2(\mathbf{R})}{\int d\mathbf{R}\Psi_T^2(\mathbf{R})} \equiv \frac{\int d\mathbf{R}E_L\Psi_T^2(\mathbf{R})}{\int d\mathbf{R}\Psi_T^2(\mathbf{R})} \geqslant E_0, \tag{3.2}$$

[4] This algorithm is also known as the $M(RT)^2$, due to the full list of the authors that contributed to its development, Metropolis, Rosenbluth, Rosenbluth, Teller and Teller, see METROPOLIS, ROSENBLUTH, ROSENBLUTH, TELLER and TELLER [1953].

where E_L is the local energy defined in Section 2.0.1. The procedure involves sampling random points in **R**-space from

$$\mathcal{P}(\mathbf{R}) \equiv \frac{\Psi_T^2(\mathbf{R})}{\int d\mathbf{R}\Psi_T^2(\mathbf{R})}. \tag{3.3}$$

The advantage of using (3.3) as the probability density function is that one need not perform the averaging of the numerator and denominator of Eq. (3.2). The calculation of the ratio of two integrals with the MC method is biased by definition: the average of a quotient is not equal to the quotient of the averages, so this choice of $P(\mathbf{R})$ avoids this problem.

In general, sampling is done using the Metropolis method (METROPOLIS, ROSENBLUTH, ROSENBLUTH, TELLER and TELLER [1953]), that is well described in Chapter 3 of KALOS and WHITLOCK [1986], and briefly summarized here in Section 3.1.1.

Expectation values can be obtained using the VMC method from the following general expressions (BRESSANINI and REYNOLDS [1998]):

$$\langle\widehat{O}\rangle \equiv \frac{\int d\mathbf{R}\Psi_T(\mathbf{R})^2\widehat{O}(\mathbf{R})}{\int d\mathbf{R}\Psi_T(\mathbf{R})^2} \cong \frac{1}{N}\sum_{i=1}^{N}\widehat{O}(\mathbf{R}_i), \tag{3.4}$$

$$\langle\widehat{O}_d\rangle \equiv \frac{\int d\mathbf{R}\left[\frac{\widehat{O}_d\Psi_T(\mathbf{R})}{\Psi_T(\mathbf{R})}\right]\Psi_T(\mathbf{R})^2}{\int d\mathbf{R}\Psi_T(\mathbf{R})^2} \cong \frac{1}{N}\sum_{i=1}^{N}\frac{\widehat{O}_d\Psi_T(\mathbf{R}_i)}{\Psi_T(\mathbf{R}_i)}. \tag{3.5}$$

Eq. (3.4) is for a coordinate operator, $\widehat{O}$, and (3.5) is preferred for a differential operator, $\widehat{O}_d$.

3.1.1. The generalized Metropolis algorithm

The main idea of the Metropolis algorithm is to sample the electronic density, given hereby, $\Psi_T^2(\mathbf{R})$ using fictitious kinetics that in the limit of large simulation time yields the density at equilibrium. A coordinate move is proposed, $\mathbf{R} \to \mathbf{R}'$, which has the probability of being accepted given by

$$P(\mathbf{R} \to \mathbf{R}') = \min\left(1, \frac{T(\mathbf{R}' \to \mathbf{R})\Psi_T^2(\mathbf{R}')}{T(\mathbf{R} \to \mathbf{R}')\Psi_T^2(\mathbf{R})}\right), \tag{3.6}$$

where $T(\mathbf{R} \to \mathbf{R}')$ denotes the transition probability for a coordinate move from $\mathbf{R}$ to $\mathbf{R}'$. In the original Metropolis procedure, T was taken to be a uniform random distribution over a coordinate interval $\Delta\mathbf{R}$. Condition (3.6) is necessary to satisfy the detailed balance condition

$$T(\mathbf{R}' \to \mathbf{R})\Psi_T^2(\mathbf{R}') = T(\mathbf{R} \to \mathbf{R}')\Psi_T^2(\mathbf{R}) \tag{3.7}$$

which is necessary for $\Psi_T^2(\mathbf{R})$ to be the equilibrium distribution of the sampling process.

Several improvements to the Metropolis method have been pursued both in classical and in QMC simulations. These improvements involve new transition probability functions and other sampling procedures. See, for example, KALOS, LEVESQUE and VERLET [1974], CEPERLEY, CHESTER and KALOS [1977], RAO and BERNE [1979], PANGALI, RAO and BERNE [1979], CEPERLEY and ALDER [1980], ANDERSON [1980], BRESSANINI and REYNOLDS [1998], DEWING [2000].

A common approach for improving $T(\mathbf{R} \to \mathbf{R}')$ in VMC, is to use the quantum force,

$$\mathbf{F_q} \equiv \nabla \ln\left|\Psi_T(\mathbf{R})^2\right| \tag{3.8}$$

as a component of the transition probability. The quantum force can be incorporated by expanding $f(\mathbf{R},\tau) = |\Psi_T(\mathbf{R})^2| = e^{-\ln|\Psi_T^2(\mathbf{R})|}$ in a Taylor series in $\ln|\Psi_T^2(\mathbf{R})|$ and truncating at first order

$$T\big(\mathbf{R} \to \mathbf{R}'\big) \approx \frac{1}{N} e^{\lambda \mathbf{F_q}(\mathbf{R})\cdot(\mathbf{R}'-\mathbf{R})}, \tag{3.9}$$

where N is a normalization factor, and λ is a parameter fixed for the simulation or optimized in some fashion, for example, see STEDMAN, FOWLKES and NEKOVEE [1998]. A usual improvement is to introduce a cutoff in $\Delta\mathbf{R} = (\mathbf{R}' - \mathbf{R})$, so that if the proposed displacement is larger than a predetermined measure, the move is rejected.

A good transition probability should also contain random displacements, so that all of phase space can be sampled. The combination of the desired drift arising from the quantum force of Eq. (3.9) with a Gaussian random move, gives rise to Langevin fictitious dynamics, namely,

$$\mathbf{R}' \to \mathbf{R} + \frac{1}{2}\mathbf{F_q}(\mathbf{R}) + \mathcal{G}_{\delta\tau}, \tag{3.10}$$

where $\mathcal{G}_{\delta\tau}$ is a number sampled from a Gaussian distribution with standard deviation $\delta\tau$. The propagator or transition probability for Eq. (3.10) is

$$T_L\big(\mathbf{R} \to \mathbf{R}'\big) = \frac{1}{\sqrt{4\pi D\delta\tau}^{\,3N}} e^{-(\mathbf{R}'-\mathbf{R}-\frac{1}{2}\mathbf{F_q}(\mathbf{R})\delta\tau)^2/2\delta\tau} \tag{3.11}$$

which is a drifting Gaussian, spreading in $\delta\tau$. Using Eq. (3.10) is equivalent to finding the solution of the Fokker–Planck equation (COURANT, FRIEDRICHS and LEWY [1928])

$$\frac{\partial f(\mathbf{R},\tau)}{\partial\tau} = \frac{1}{2}\nabla\cdot(\nabla - \mathbf{F_q}) f(\mathbf{R},\tau). \tag{3.12}$$

Eq. (3.11) has proved to be a simple and effective choice for a VMC transition probability. More refined choices can be made, usually with the goal of increasing acceptance probabilities in regions of rapid change in $|\Psi_T(\mathbf{R})^2|$, such as close to nuclei. For a more detailed discussion of this formalism, the reader is directed to Chapter 2 of HAMMOND, LESTER and REYNOLDS [1994]. More elaborate transition rules can be found in UMRIGAR [1993], SUN, SOTO and LESTER, JR. [1994], STEDMAN, FOWLKES and NEKOVEE [1998], BRESSANINI and REYNOLDS [1999].

3.1.2. Statistics

Usually, VMC calculations are performed using an ensemble of $N_{\mathcal{W}}$ random walkers $\mathcal{W} \equiv \{\mathbf{R}_1, \mathbf{R}_2, \ldots, \mathbf{R_n}\}$ that are propagated following $T(\mathbf{R} \rightarrow \mathbf{R}')$ using the probability $P(\mathbf{R} \rightarrow \mathbf{R}')$ to accept or reject proposed moves for ensemble members. Statistical averaging has to take into account auto-correlation between moves that arises if the mean square displacement for the ensemble, $\Delta(\mathbf{R} \rightarrow \mathbf{R}')^2/N_{\mathcal{W}}$ is sufficiently large. In such cases, observables measured at the points $\mathbf{R}'$ will be statistically correlated with those evaluated at $\mathbf{R}$. The variance for an observable, $\widehat{O}$, measured over N_s MC steps of a random walk is

$$\sigma_{\widehat{O}} \equiv \frac{1}{N_s N_{\mathcal{W}}} \big(O_i - \langle O \rangle\big), \tag{3.13}$$

where $\langle O \rangle$ is the average of the observations, O_i, over the sample. A simple approach to remove auto-correlation between samples is to define a number of blocks, N_b, where each block is an average of N_s steps, with variance

$$\sigma_B \equiv \frac{1}{N_B N_{\mathcal{W}}} \big(O_b - \langle O \rangle\big), \tag{3.14}$$

where O_b is the average number of observations N_t in block b. If N_t is sufficiently large, σ_B is a good estimator of the variance of the observable over the random walk. The auto-correlation time is a good measure of computational efficiency, and is given by

$$T_{\mathrm{corr}} = \lim_{N_s \to \infty} N_s \left(\frac{\sigma_B^2}{\sigma_{\widehat{O}}^2} \right). \tag{3.15}$$

The efficiency of a method depends on time step (ROTHSTEIN and VBRIK [1988]). Serial correlation between sample points should vanish for an accurate estimator of the variance. For an observable $\langle O \rangle$, the serial correlation coefficient is defined as

$$\xi_k \equiv \frac{1}{(\langle O^2 \rangle - \langle O \rangle^2)(N-k)} \sum_{i=1}^{N-k} \big(O_i - \langle O \rangle\big)\big(O_{i+k} - \langle O \rangle\big), \tag{3.16}$$

where k is the number of MC steps between the points O_i and O_{i+k}. The function (3.16) decays exponentially with k. The correlation length, L, is defined as the number of steps necessary for ξ_k to decay essentially to zero. For an accurate variance estimator, blocks should be at least L steps long.

The efficiency of a simulation is inversely proportional to ξ_k. The ξ_k dependence on time step is usually strong (HAMMOND, LESTER and REYNOLDS [1994]); the larger the time step, the fewer steps/block L necessary, and the more points available for calculating the global average $\langle O\rangle$. A rule of thumb is to use an $N_t \approx 10$ times larger than the auto-correlation time to insure statistical independence of block averages, and therefore a reliable variance estimate.

The VMC method shares some of the strengths and weaknesses of traditional variational methods: the energy is an upper bound to the true ground state energy. If reasonable trial functions are used, often reliable estimates of properties can be obtained. For quantum MC applications, VMC can be used to obtain valuable results. In chemical applications, VMC is typically used to analyze and generate trial wave functions for PMC.

3.2. Trial wave functions

In contrast to wave function methods, where the wave function is constructed from linear combinations of determinants of orbitals, QMC methods can use arbitrary functional forms for the wave function subject to the requirements in Section 2. Because QMC trial wave functions are not restricted to expansions in one-electron functions (orbitals), more compact representations are routinely used. In this section, we review the forms most commonly used for QMC calculations.

Fermion wave functions must be antisymmetric with respect to the exchange of an arbitrary pair of particle coordinates. If they are constructed as the product of N functions of the coordinates, $\phi(r_1, r_2, \ldots, r_N)$, the most general wave function can be constructed enforcing explicit permutation:

$$\Psi(\mathbf{R}, \Sigma) = \frac{1}{\sqrt{(N \cdot M)!}} \sum_{n,m} (-1)^n \widehat{S}_m \widehat{P}_n \phi(r_1, \sigma_1 r_2, \sigma_2, \ldots, r_N, \sigma_N), \tag{3.17}$$

where $\widehat{P}_n$ is the nth coordinate permutation operator, $\widehat{P}_n \phi(r_1, r_2, \ldots, r_i, r_j, \ldots, r_N) = \phi(r_1, r_2, \ldots, r_j, r_i, \ldots, r_N)$, and $\widehat{S}_m \phi(\sigma_1, \sigma_2, \sigma_i, \sigma_j, \ldots, \sigma_N) = \phi(\sigma_1, \sigma_2, \sigma_j, \sigma_i, \ldots, \sigma_N)$ is the mth spin coordinate permutation operator.

If the functions ϕ_i depend only on single-particle coordinates, their antisymmetrized product can be expressed as a Slater determinant

$$D(\mathbf{R}, \Sigma) = \frac{1}{\sqrt{N!}} \det\left|\phi_1, \ldots, \phi_i(r_j, \sigma_j), \ldots, \phi_n\right|. \tag{3.18}$$

Trial wave functions constructed from orbitals scale computationally as N^3, where N is the system size, compared to $N!$ for the fully antisymmetrized form.[5] The number of evaluations can be reduced by determining which permutations contribute to a particular spin state.

[5] The evaluation of a determinant of size N requires N^2 computer operations. If the one-electron functions scale with system size as well, the scaling becomes N^3. In contrast, a fully antisymmetrized form requires the explicit evaluation of the $N!$ permutations, making the evaluation of this kind of wave functions in QMC prohibitive for systems of large N.

For QMC evaluation of properties that do not depend on spin coordinates, Σ, for a given spin state, the $M!$ configurations that arise from relabeling electrons, need not be evaluated. The reason is that the Hamiltonian of Eq. (0.1), contains no magnetic or spin operators and spin degrees of freedom remain unchanged. In this case, and for the remainder of this paper, $\sigma_\uparrow$ electrons do not permute with $\sigma_\downarrow$ electrons, so that the full Slater determinant(s) can be factored into a product of spin-up, $D^\uparrow$, and spin-down, $D^\downarrow$, determinants. The number of allowed permutations is reduced from $(N_\uparrow + N_\downarrow)!$ to $N_\uparrow!N_\downarrow!$ (CAFFAREL and CLAVERIE [1988], HUANG, SUN and LESTER, JR. [1990]).

The use of various wave function forms in QMC has been explored by ALEXANDER and COLDWELL [1997], as well as BERTINI, BRESSANINI, MELLA and MOROSI [1999]. Fully antisymmetric descriptions of the wave function are more flexible and require fewer parameters than determinants, but their evaluation is inefficient due to the $N!$ scaling.

A good compromise is to use a product wave function of a determinant or linear combination of determinants, e.g., HF, MCSCF, CASSCF, CI, multiplied by a correlation function that is symmetric with respect to particle exchange,

$$\Psi_T = \mathcal{D}\mathcal{F}. \tag{3.19}$$

Here $\mathcal{D}$ denotes the antisymmetric wave function factor and $\mathcal{F}$ is the symmetric factor. We now describe some of the forms used for $\mathcal{D}$ and then we describe forms for $\mathcal{F}$. Such products are also known as the correlated molecular orbital (CMO) wave function.

In the CMO wave functions, the antisymmetric part of the wave function is constructed as a linear combination of determinants of independent particle functions, ϕ_i (see Eq. (3.18)). The ϕ_i are usually formed as a linear combination of basis functions centered on atomic centers, $\phi_i = \sum_j c_j \chi_j$. The most commonly used basis functions in traditional *ab initio* quantum chemistry are Gaussian functions, which owe their popularity to ease of integration of molecular integrals. Gaussian basis functions take the form

$$\chi_G \equiv x^a y^b z^c e^{-\xi \mathbf{r}^2}. \tag{3.20}$$

For QMC applications, it is better to use the Slater-type basis functions

$$\chi_S \equiv x^a y^b z^c e^{-\xi \mathbf{r}}, \tag{3.21}$$

because they rigorously satisfy the electron–nuclear cusp condition (see Eq. (2.2)), and the asymptotic property of Eq. (2.5). Nevertheless, in most studies, Gaussian basis functions have been used, and corrections for enforcing the cusp conditions can be made to improve local behavior close to a nucleus. For example, in one approach (MANTEN and LÜCHOW [2001]), the region close to a nucleus is described by a Slater-type function, and a polynomial fit is used to connect the Gaussian region to the exponential. This procedure strongly reduces fluctuations of the kinetic energy of these functions, a desirable property for guided VMC and Green's function methods.

The symmetric part of the wave function is usually built as a product of terms explicitly dependent on inter-particle distance, $\mathbf{r}_{ij} = |\mathbf{r}_i - \mathbf{r}_j|$. These functions are usually constructed to reproduce the form of the wave function at electron–electron and electron–nucleus cusps. A now familiar form is that proposed by BIJL [1940], DINGLE [1949], JASTROW [1955], and known as the Jastrow *ansatz*:

$$\mathcal{F} \equiv e^{U(\mathbf{r}_{ij})} \equiv e^{\prod_{i<j} g_{ij}}, \tag{3.22}$$

where the correlation function g_{ij} is

$$g_{ij} \equiv \frac{a_{ij}\mathbf{r}_{ij}}{1 + b_{ij}\mathbf{r}_{ij}} \tag{3.23}$$

with constants specified to satisfy the cusp conditions

$$a_{ij} \equiv \begin{cases} \frac{1}{4} & \text{if } ij \text{ are like spins,} \\ \frac{1}{2} & \text{if } ij \text{ are unlike spins,} \\ -\mathcal{Z} & \text{if } ij \text{ are electron/nucleus pairs.} \end{cases} \tag{3.24}$$

Electron correlation for parallel spins is taken into account by the Slater determinants.

This simple Slater–Jastrow *ansatz* has a number of desirable properties. First, as stated above, scaling with system size for the evaluation of the trial function is N^3, where N is the number of particles in the system, the correct cusp conditions are satisfied at two-body coalescence points and the correlation function g_{ij} approaches a constant at large distances, which is the correct behavior as $\mathbf{r}_{ij} \to \infty$.

In general, the inclusion of 3-body and 4-body correlation terms has been shown to improve wave function quality. The work of HUANG, UMRIGAR and NIGHTINGALE [1997] shows that if the determinant parameters λ_D are optimized along with the correlation function parameters, λ_C, one finds that the nodal structure of the wave function does not improve noticeably in going from 3- to 4-body correlation terms, which suggests that increasing the number of determinants, N_D is more important than adding fourth- and higher-order correlation terms.

The use of Feynman–Cohen backflow correlations (FEYNMANN and COHEN [1956]), which has been suggested (SCHMIDT and MOSKOWITZ [1990]) for the inclusion of three body correlations in U, has been used in trial functions for homogeneous systems such as the electron gas (KWON, CEPERLEY and MARTIN [1993], KWON, CEPERLEY and MARTIN [1998]) and liquid helium (SCHMIDT, KALOS, LEE and CHESTER [1980], CASULLERAS and BORONAT [2000]). Feynman (FEYNMANN and COHEN [1956]) suggested replacing the orbitals by functions that include hydrodynamic *backflow* effects. His idea was based on the conservation of particle current and the variational principle. The procedure involves replacing mean field orbitals by *backflow*-corrected orbitals of the form

$$\phi_n(\mathbf{r}_i) \to \phi_n\left(\mathbf{r}_i + \sum_{j \neq i} \mathbf{r}_{ij}\nu(\mathbf{r}_{ij})\right), \tag{3.25}$$

where $\nu(\mathbf{r}_{ij})$ is the backflow function. Others (PANDHARIPANDE and ITOH [1973]) proposed that $\nu(r_{ij})$ should consist of the difference between the $l = 0$ and $l = 1$ states of an effective two-particle Schrödinger equation. Furthermore, they proposed (PANDHARIPANDE, PIEPER and WIRINGA [1986]) the inclusion of a $1/r^3$ tail, as originally suggested by Feynman and Cohen,

$$\nu(\mathbf{r}) = \lambda_\nu e^{-\left[\frac{\mathbf{r}_i - \mathbf{r}_j}{\omega_\nu}\right]^2} + \frac{\lambda_{\nu'}}{\mathbf{r}^3}, \tag{3.26}$$

where, λ_ν, $\lambda_{\nu'}$, and ω_ν are variational parameters. As recently noted by KWON, CEPERLEY and MARTIN [1993], the incorporation of the full backflow trial function into wave functions involves a power of N increase in computational expense, but yields a better DMC energy for the electron gas.[6]

Recently, one has seen the practice of taking orbitals from a mean field calculation and the inclusion of averaged backflow terms in the correlation function $\mathcal{F}$. The advantage of this approach is that orbitals are unperturbed and readily obtainable from mean field computer codes.

The correlation function form used by SCHMIDT and MOSKOWITZ [1990] is a selection of certain terms of the general form originally proposed in connection with the transcorrelated method (BOYS and HANDY [1969]):

$$\mathcal{F} = e^{\sum_{I,i<j} U_{Iij}}, \tag{3.27}$$

where

$$U_{Iij} = \sum_{k}^{N(I)} \Delta(m_{kI} n_{kI}) c_{kI} \left(g_{iI}^{m_{kI}} g_{jI}^{n_{kI}} + g_{jI}^{m_{kI}} g_{iI}^{n_{kI}}\right) g_{ij}^{o_{kI}}. \tag{3.28}$$

The sum in (3.27) goes over I nuclei, ij electron pairs, and the sum in Eq. (3.28) is over the $N(I)$ terms of the correlation function for each nucleus. The parameters m, n and o are integers. The function $\Delta(m, n)$ takes the value 1 when $m \neq n$, and $\frac{1}{2}$ otherwise. The functions g_{ij} are specified by Eq. (3.23).

This correlation function (3.27), (3.28) can be shown to have contributions to averaged backflow effects from the presence of electron–electron–nucleus correlations that correspond to values of m, n and o in Eq. (3.28) of $2, 2, 0$ and $2, 0, 2$. These contributions recover $\approx 25\%$ or more of the total correlation energy of atomic and molecular systems above that from the simple Jastrow term (SCHMIDT and MOSKOWITZ [1990]).

3.3. *The variational Monte Carlo algorithm*

The VMC algorithm is an application of the generalized Metropolis MC method. As in most applications of the method, one needs to insure that the ensemble has achieved

[6] As discussed in Section 5.6, an improved fixed-node energy is a consequence of better nodes of the trial wave function, a critically important characteristic for importance sampling functions in QMC methods.

equilibrium in the simulation sense. Equilibrium is reached when the ensemble $\mathcal{W}$ is distributed according to $\mathcal{P}(\mathbf{R})$. This is usually achieved by performing a Metropolis random walk and monitoring the trace of the observables of interest. When the trace fluctuates around a mean, it is generally safe to start averaging in order to obtain desired properties.

An implementation of the VMC algorithm follows:

(1) Equilibration stage
  (a) Generate an initial set of random walker positions, $\mathcal{W}_0$; it can be read in from a previous random walk, or generated at random.
  (b) Perform a loop over N_s steps,
    (i) For each r_i of the N_p number of particles,
      (A) Propose a move from $\Psi(\mathbf{R}) \equiv \Psi(\mathbf{r}_1, \mathbf{r}_2, \ldots, \mathbf{r}_i, \ldots, \mathbf{r}_{N_p})$ to $\Psi(\mathbf{R}') \equiv \Psi(\mathbf{r}_1, \mathbf{r}_2, \ldots, \mathbf{r}'_i, \ldots, \mathbf{r}_{N_p})$. Move from $\mathbf{r}$ to $\mathbf{r}'$ according to

$$\mathbf{r}' \leftarrow \mathbf{r} + \mathcal{G}_{\delta\tau} + \frac{1}{2}\mathbf{F_q}\delta\tau, \tag{3.29}$$

where $\mathcal{G}_{\delta\tau}$ is a Gaussian random number with standard deviation $\delta\tau$,which is a proposed step size, and $\mathbf{F_q}$ is the quantum force (see Eq. (3.9)). This is the Langevin dynamics of Eq. (3.10).

      (B) Compute the Metropolis acceptance/rejection probability

$$P(\mathbf{R} \to \mathbf{R}') = \min\left(1, \frac{T_L(\mathbf{R}' \to \mathbf{R})\Psi_T^2(\mathbf{R}')}{T_L(\mathbf{R} \to \mathbf{R}')\Psi_T^2(\mathbf{R})}\right), \tag{3.30}$$

where T_L is given by Eq. (3.11).

      (C) Compare $P(\mathbf{R} \to \mathbf{R}')$ with an uniform random number between 0 and 1, $\mathcal{U}_{[0,1]}$. If $P > \mathcal{U}_{[0,1]}$, accept the move, otherwise, reject it.
      (D) Calculate the contribution to the averages $\frac{\widehat{O}_d \Psi_T(\mathbf{R}')}{\Psi_T(\mathbf{R}')}$, and perform blocking statistics as described in Section 3.1.2.
    (ii) Continue the loop until the desired accuracy is achieved.

4. Wave function optimization

Trial wave functions $\Psi_T(\mathbf{R}, \Lambda)$ for QMC are dependent on variational parameters $\Lambda = \{\lambda_1, \ldots, \lambda_n\}$. Optimization of Λ is a key element for obtaining accurate trial functions. Importance sampling using an optimized trial function increases the efficiency of DMC simulations. There is a direct relationship between trial-function accuracy and the computer time required to calculate accurate expectation values. Some of the parameters λ_i may be fixed by imposing appropriate wave function properties, such as cusp conditions (see Section 2).

It is useful to divide Λ into groups distinguished by whether the optimization changes the nodes of the wave function. The Slater determinant parameters, $\lambda_{\mathcal{D}\uparrow\downarrow}$ and the Slater determinant weights, λ_{k_i}change wave function nodal structure (UMRIGAR, WILSON and WILKINS [1988], SCHMIDT and MOSKOWITZ [1990], SUN, BARNETT and LESTER, JR. [1992], FLAD, CAFFAREL and SAVIN [1997], BARNETT, SUN and LESTER, JR. [1997], BARNETT, SUN and LESTER, JR. [2001]). The correlation function parameters, $\lambda_{\mathcal{F}}$ do not change the nodal structure of the overall wave function, and therefore the DMC energy. For some systems, the optimization of $\lambda_{\mathcal{F}}$ is sufficient for building reliable trial functions for PMC methods, because $\mathcal{F}$ is designed in part to satisfy cusp conditions (KATO [1957], MEYERS, UMRIGAR, SETHNA and MORGAN [1991], FILIPPI and UMRIGAR [1996]).

There have been several optimization methods proposed previously. Some involve the use of analytical derivatives (HUANG, SUN and LESTER, JR. [1990], BUECKERT, ROTHSTEIN and VRBIK [1992], LÜCHOW and ANDERSON [1996], HUANG and CAO [1996], HUANG, XIE, CAO, LI, YUE and MING [1999], LIN, ZHANG and RAPPE [1976]), and others focus on the use of a fixed sample for variance minimization (CONROY [1964]), and more recently others (UMRIGAR, WILSON and WILKINS [1988], SUN, HUANG, BARNETT and LESTER, JR. [1990], NIGHTINGALE and UMRIGAR [1997]). Yet another direction is the use of histogram analysis for optimizing the energy, variance, and molecular geometry for small systems (SNAJDR, DWYER and ROTHSTEIN [1999]). In the present study, we concentrate on fixed sample optimization to eliminate stochastic uncertainty during the random walk (SUN, HUANG, BARNETT and LESTER, JR. [1990]).

Other authors optimize the trial wave function using information obtained from a DMC random walk (FILIPPI and FAHY [2000]). This approach shows promise, because usually the orbitals obtained from a mean field theory, such as HF or LDA, are frozen and used in the DMC calculation without re-optimization specifically for correlation effects within the DMC framework.

The common variance functional (VF) (UMRIGAR, WILSON and WILKINS [1988]) is given by

$$VF = \frac{\sum_{i=1}^{N} \left[\frac{\widehat{H}\Psi(\mathbf{R}_i, \Lambda)}{\Psi(\mathbf{R}_i, \Lambda)} - E_T \right]^2 w_i}{\sum_{i=1}^{N} w_i}, \tag{4.1}$$

where E_T is a trial energy, w_i is a weighting factor defined by

$$w_i(\Lambda) = \frac{\Psi_i^2(\mathbf{R}_i, \Lambda)}{\Psi_i^2(\mathbf{R}_i, \Lambda_0)}, \tag{4.2}$$

and Λ_0 is an initial set of parameters. The sum in Eq. (4.1) is over fixed sample configurations.

4.0.1. *Trial wave function quality*

The overlap of Ψ_T with the ground state wave function, $\langle \Psi_T | \Psi_0 \rangle$, by DMC methods (HORNIK, SNAJDR and ROTHSTEIN [2000]) is a very efficient way of assessing wave function quality. There is also a trend that correlates the variational energy of the wave function with associated variance in a linear relationship (KWON, CEPERLEY and

MARTIN [1993], KWON, CEPERLEY and MARTIN [1998]). This correlation is expected because both properties, δ_E, and δ_{E_L}, approach limits – E_0 and zero respectively – as wave function quality improves. Observing this correlations is a good method of validating the optimization method, as well as assessing wave function quality.

5. Projector methods

QMC methods such as DMC and GFMC are usefully called projector Monte Carlo (PMC) methods.[7] The general idea is to project out a state of the Hamiltonian by iteration of a projection operator, $\widehat{P}$. For simplicity, we assume that the desired state is the ground state, Ψ_0, but projectors can be constructed for any state,

$$\lim_{i\to\infty} \widehat{P}^i |\Psi_T\rangle \approx |\Psi_0\rangle. \tag{5.1}$$

After sufficient iterations i, the contribution of all excited states $|\Psi_i\rangle$, will be filtered out, and only the ground state is recovered.

If $|\Psi_T\rangle$ is a vector and $\widehat{P}$ is a matrix, then the procedure implied by (5.1) is the algebraic power method: If a matrix is applied iteratively to an initial arbitrary vector for a sufficient number of times, only the dominant eigenvector, $|\Psi_0\rangle$, will survive. One can see for large i,

$$\widehat{P}^i |\Psi_T\rangle = \lambda_0^i \langle\Psi_0|\Psi_T\rangle|\Psi_0\rangle + \mathcal{O}(\lambda_1^i), \tag{5.2}$$

where λ_0 is the leading eigenvalue, and λ_1 is the largest sub-leading eigenvalue.

For this approach, it is possible to obtain an estimator of the eigenvalue, as described in HAMMERSLEY and HANDSCOMB [1964], given by

$$\lambda_0 = \lim_{i\to\infty} \left(\frac{\langle\phi|\widehat{P}^{i+j}|\Psi_T\rangle}{\langle\phi|\widehat{P}^i|\Psi_T\rangle} \right)^{\frac{1}{m}}. \tag{5.3}$$

5.0.2. *Markov processes and stochastic projection*

For high-dimensional vectors, such as those encountered in molecular electronic structure, the algebraic power method described previously needs to be generalized with stochastic implementation. For this to occur, the projection operator must be symmetric, so that all eigenvalues are real. This is the case for QMC methods, because $\widehat{P}$ is a function of the Hamiltonian operator, $\widehat{H}$, which is Hermitian by construction.

A stochastic matrix is a normalized nonnegative matrix. By normalization, we mean that the stochastic matrix columns add to one, $\sum_i M_{ij} = 1$. An **R**-space representation would be a stochastic propagator $M(\mathbf{R}, \mathbf{R}')$ that satisfies the condition

$$\int M(\mathbf{R}, \mathbf{R}')\, d\mathbf{R}' = 1. \tag{5.4}$$

[7] The introductory section of this chapter follows the work of HETHERINGTON [1984], CHIN [1990], CERF and MARTIN [1995].

A Markov chain is a sequence of states obtained from subsequent transitions from state i to j with a probability related to the stochastic matrix element M_{ij}, in which the move only depends on the current state, i. For example, in $\mathbf{R}$-space, this is equivalent to the following process

$$\begin{aligned}\pi(\mathbf{R}') &= \int M(\mathbf{R}', \mathbf{R}'')\pi(\mathbf{R}'')\,d\mathbf{R}'',\\ \pi(\mathbf{R}) &= \int M(\mathbf{R}, \mathbf{R}')\pi(\mathbf{R}')\,d\mathbf{R}',\\ &\ldots\end{aligned} \tag{5.5}$$

The sequence of states $S = \{\pi(\mathbf{R}''), \pi(\mathbf{R}'), \pi(\mathbf{R}), \ldots\}$ is the Markov chain.

The propagators of QMC for electronic structure are not generally normalized, therefore they are not stochastic matrices, but we can represent them in terms of the latter by factoring,

$$\widehat{P}_{ij} = M_{ij} w_j, \tag{5.6}$$

where the weights, w_j, are defined by $w_j = \sum_i \widehat{P}_{ij}$. This definition unambiguously defines both the associated stochastic matrix M and the weight vector w.

A MC sampling scheme of $\widehat{P}_{ij}|\Psi_T\rangle$ can be generated by first performing a random walk, and keep a weight vector $W(\mathbf{R})$ for the random walkers,

$$\begin{aligned}\Psi(\mathbf{R}') &\equiv \pi(\mathbf{R}')W(\mathbf{R}') = \int P(\mathbf{R}', \mathbf{R}'')\Psi(\mathbf{R}'')\,d\mathbf{R}''\\ &= \int M(\mathbf{R}', \mathbf{R}'')B(\mathbf{R}')\Psi(\mathbf{R}'')\,d\mathbf{R}'',\\ \Psi(\mathbf{R}) &\equiv \pi(\mathbf{R})W(\mathbf{R}) = \int P(\mathbf{R}, \mathbf{R}')\Psi(\mathbf{R}')\,d\mathbf{R}'\\ &= \int M(\mathbf{R}, \mathbf{R}')B(\mathbf{R})\Psi(\mathbf{R}')\,d\mathbf{R}',\\ &\ldots\end{aligned} \tag{5.7}$$

Here, $B(\mathbf{R})$ is the function that determines the weight of the configurations at each state of the random chain. This leads to a generalized stochastic projection algorithm for unnormalized transition probabilities that forms the basis for population Monte Carlo (PopMC) algorithms, which are not only used for QMC, but they are also used for statistical information processing and robotic vision (IBA [2000]). A generalized PopMC stochastic projection algorithm, represented in $\mathbf{R}$-space follows:

(1) *Initialize*

Generate a set of $\mathbf{n}$ random walkers, located at different spatial positions, $\mathcal{W} \equiv \{\mathbf{R}_1, \mathbf{R}_2, \ldots, \mathbf{R}_\mathbf{n}\}$, where $\mathbf{R}_i$ denotes a Dirac delta function at that point in space, $\delta(\mathbf{R} - \mathbf{R}_i)$. These points are intended to sample a probability density function $\Phi(\mathbf{R})$.

(2) *Move*
   (a) Each walker j is moved independently from $\mathbf{R}$ to a new position $\mathbf{R}'$, according to the transition probability

$$T(\mathbf{R} \to \mathbf{R}') \equiv M(\mathbf{R}, \mathbf{R}'). \tag{5.8}$$

   (b) Ensure detailed balance if $T(\mathbf{R} \to \mathbf{R}') \neq T(\mathbf{R}' \to \mathbf{R})$ by using a Metropolis acceptance/rejection step as in Eq. (3.30).

(3) *Weight*
   (a) Calculate a weight vector using a weighting function $B(\mathbf{R}_i)$,

$$w_i^* = B(\mathbf{R}_i). \tag{5.9}$$

   The ideal weight function preserves normalization of $\widehat{P}(\mathbf{R}, \mathbf{R}')$ and maintains individual weights w_i close to unity.
   (b) Update the weight of the walker, multiplying the weight of the previous iteration by the weight of the new iteration,

$$w_i' = w_i^* w_i. \tag{5.10}$$

(4) *Reconfiguration*
   (a) Split walkers with large weights into multiple walkers with weights that add up to the original weight.
   (b) Remove walkers with small weight.

Step 4 is necessary to avoid statistical fluctuations in the weights. It is a form of importance sampling in the sense that makes the calculation stable over time. Some algorithms omit this step; see, for example, efforts by CAFFAREL and CLAVERIE [1988], but it has been proved that such calculations eventually diverge (ASSARAF, CAFFAREL and KHELIF [2000]). There is a slight bias associated with the introduction of step 4 together with population control methods, that will be discussed in Section 5.3. When step 4 is used, $B(\mathbf{R})$ is also referred in the literature as a *branching factor*.

It is important to recall that PopMC algorithms are not canonical Markov Chain Monte Carlo (MCMC) algorithms (BYRON and FULLER [1992], DOUCET, DE FREITAS, GORDON and SMITH [2001]), in the sense that the propagator used is not normalized, and therefore factoring the propagator into a normalized transition probability and a weighting function is required.

5.0.3. Projection operators or Green's functions

Different projection operators lead to different QMC methods. If the resolvent operator,

$$\widehat{P}(\widehat{H}) \equiv \frac{1}{1 + \delta\tau(\widehat{H} - E_R)}, \tag{5.11}$$

is used, one obtains Green's function Monte Carlo (GFMC) (KALOS [1962], CEPERLEY and KALOS [1986]). This algorithm will be described in Section 5.5.

If the imaginary time evolution operator is used, i.e.,

$$\widehat{P}(\widehat{H}) \equiv e^{-\delta\tau(\widehat{H}-E_R)}, \tag{5.12}$$

one has the DMC method (ANDERSON [1976], REYNOLDS, CEPERLEY, ALDER and LESTER, JR. [1982]), which is discussed in Section 5.1.2.

For finite $\delta\tau$, and for molecular systems, the exact projector is not known analytically. In GFMC, the resolvent of Eq. (5.11) is sampled by iteration of a simpler resolvent, whereas for DMC, the resolvent is known exactly at $\tau \to 0$, so an extrapolation to $\delta\tau \to 0$ is done.

Note that any decreasing function of $\widehat{H}$ can serve as a projector. Therefore, new QMC methods still await to be explored.

5.1. Imaginary propagator

If one transforms the time-dependent Schrödinger equation (Eq. (0.2)) to imaginary time, i.e.,

$$it \to \tau, \tag{5.13}$$

then one obtains

$$\frac{\partial}{\partial\tau}\Psi(\mathbf{R},\tau) = (\widehat{H} - E_R)\Psi(\mathbf{R},\tau). \tag{5.14}$$

Here E_R is an energy offset, called the reference energy. For real $\Psi(\mathbf{R},\tau)$, this equation has the advantage of being an equation in $\mathcal{R}^N$, whereas Eq. (0.2) has in general, complex solutions.

Eq. (5.14) can be cast into integral form,

$$\Psi(\mathbf{R},\tau+\delta\tau) = \lambda_\tau \int G(\mathbf{R},\mathbf{R}',\delta\tau)\Psi(\mathbf{R}',\tau)\,d\mathbf{R}'. \tag{5.15}$$

The Green's function, $G(\mathbf{R}',\mathbf{R},\delta\tau)$, satisfies the same boundary conditions as Eq. (5.14):

$$\frac{\partial}{\partial\tau}G(\mathbf{R},\mathbf{R}',\delta\tau) = (\widehat{H} - E_T)G(\mathbf{R},\mathbf{R}',\delta\tau) \tag{5.16}$$

with the initial conditions associated with the propagation of a Dirac delta function, namely,

$$G(\mathbf{R},\mathbf{R}',0) = \delta(\mathbf{R}-\mathbf{R}'). \tag{5.17}$$

The form of the Green's function that satisfies Eq. (5.16), subject to (5.17), is

$$G(\mathbf{R},\mathbf{R}',\delta\tau) = \langle\mathbf{R}|e^{-\tau(\widehat{H}-E_R)}|\mathbf{R}\rangle. \tag{5.18}$$

This operator can be expanded in eigenfunctions, Ψ_α, and eigenvalues E_α of the system

$$G(\mathbf{R}, \mathbf{R}', \delta\tau) = \sum_\alpha e^{-\tau(E_\alpha - E_R)} \Psi_\alpha^*(\mathbf{R}') \Psi_\alpha(\mathbf{R}). \tag{5.19}$$

For an arbitrary initial trial function, $\Psi(\mathbf{R})$, in the long term limit, $\tau \to \infty$, one has

$$\begin{aligned}\lim_{\tau\to\infty} e^{-\tau(\widehat{H} - E_T)}\Psi &= \lim_{\tau\to\infty} \int G(\mathbf{R}', \mathbf{R}, \tau)\Psi(\mathbf{R}')\, d\mathbf{R}' \\ &= \lim_{\tau\to\infty} \langle \Psi | \Psi_0 \rangle e^{-\tau(E_0 - E_R)} \phi_0,\end{aligned} \tag{5.20}$$

and only the ground state wave function Ψ_0 is obtained from any initial wave function. Therefore, the imaginary time evolution operator can be used as a projection operator as mentioned at the beginning of this chapter.

5.1.1. Diffusion Monte Carlo stochastic projection

Due to the high dimensionality of molecular systems, a MC projection procedure is used for obtaining expectation values. In this approach, the wave function is represented as an ensemble of delta functions, also known as configurations, walkers, or *psips* (psi-particles):

$$\Phi(R) \longleftrightarrow \sum_k \delta(\mathbf{R} - \mathbf{R_k}). \tag{5.21}$$

The wave function is propagated in imaginary time using the Green's function. In the continuous case, one can construct a Neumann series

$$\begin{aligned}\Psi^{(2)}(\mathbf{R}, \tau) &= \lambda_1 \int G(\mathbf{R}', \mathbf{R}, \tau_2 - \tau_1)\Psi^{(1)}(\mathbf{R}')\, d\mathbf{R}, \\ \Psi^{(3)}(\mathbf{R}, \tau) &= \lambda_2 \int G(\mathbf{R}', \mathbf{R}, \tau_3 - \tau_2)\Psi^{(2)}(\mathbf{R}')\, d\mathbf{R}, \\ &\dots\end{aligned} \tag{5.22}$$

This Neumann series is a specific case of the PopMC propagation of Section 5.0.2. The discrete Neumann series can be constructed in a similar way:

$$\Phi^{(n+1)}(\mathbf{R}, \tau + \delta\tau) \longleftrightarrow \lambda_k \sum_k G^{(n)}(\mathbf{R}, \mathbf{R}', \delta\tau). \tag{5.23}$$

Therefore, a stochastic vector of configurations $\mathcal{W} \equiv \{\mathbf{R}_1, \dots, \mathbf{R}_n\}$ is used to represent $\Psi(\mathbf{R})$ and is iterated using $G^{(n)}(\mathbf{R}, \mathbf{R}', \delta\tau)$.

5.1.2. The form of the propagator

Sampling Eq. (5.18) can not be done exactly, because the argument of the exponential is an operator composed of two terms that do not commute with each other.

In the short-time approximation (STA), the propagator $G(\mathbf{R}, \mathbf{R_k}, d\tau)$ is approximated as if the kinetic and potential energy operators commuted with each other:

$$e^{(T+V)\delta\tau} \approx e^{T\delta\tau} \cdot e^{V\delta\tau} + \mathcal{O}\big((\delta\tau)^2\big) \equiv G_{ST} \equiv G_D \cdot G_B. \tag{5.24}$$

The Green's function then becomes the product of a diffusion factor G_D and a branching factor G_B. Both propagators are known:

$$G_D = (2\pi\tau)^{-3N/2} e^{-\frac{(\mathbf{R}-\mathbf{R}')^2}{2\tau}} \tag{5.25}$$

and

$$G_B = e^{-\delta\tau(V(\mathbf{R})-2E_T)}. \tag{5.26}$$

G_D is a fundamental solution of the Fourier equation (that describes a diffusion process in wave function space) and G_B is the fundamental solution of a first-order kinetic birth–death process.

The Campbell–Baker–Hausdorff (CBH) formula,

$$e^A e^B = e^{A+B+\frac{1}{2}[A,B]+\frac{1}{12}[(A-B),[A,B]]+\cdots} \tag{5.27}$$

can help in constructing more accurate decompositions, such as an expansion with a cubic error $\mathcal{O}((\delta\tau)^3)$,

$$e^{\delta\tau(T+V)} = e^{\delta\tau(V/2)} e^{\delta\tau T} e^{\delta\tau(V/2)} + \mathcal{O}\big((\delta\tau)^3\big). \tag{5.28}$$

There are more sophisticated second order (CHIN [1990]) and fourth order (CHIN [1997], FORBERT and CHIN [2001]) expansions that reduce the error considerably and make more exact DMC algorithms at the expense of a more complex propagator.

The most common implementation using G_D as a stochastic transition probability $T(\mathbf{R} \to \mathbf{R}')$, and G_B as a weighting or branching factor, $B(\mathbf{R})$. Sampling of Eq. (5.25) can be achieved by obtaining random variates from a Gaussian distribution of standard deviation δ_τ, $\mathcal{G}_{\delta\tau}$.

5.2. Importance sampling

Direct application of the algorithm of the previous section to systems governed by the Coulomb potential leads to large population fluctuations. These arise because the potential $\widehat{V}(\mathbf{R})$ becomes unbounded and induces large fluctuations in the random walker population. A remedy, importance sampling, was first used for GFMC by KALOS [1962] and extended to the DMC method by CEPERLEY and ALDER [1980].

In importance sampling, the goal is to reduce fluctuations, by multiplying the probability distribution by a known trial function, $\Psi_T(\mathbf{R})$, that is expected to be a good approximation for the wave function of the system. Rather than $\Psi(\mathbf{R},\tau)$, one samples the product

$$f(\mathbf{R},\tau) = \Psi_T(\mathbf{R})\Psi(\mathbf{R},\tau). \tag{5.29}$$

Multiplying Eq. (5.15) by $\Psi_T(\mathbf{R})$, one obtains

$$f(\mathbf{R},\tau+d\tau) = \int K(\mathbf{R}',\mathbf{R},\delta\tau) f(\mathbf{R}',\tau)\,d\mathbf{R}', \tag{5.30}$$

where $K(\mathbf{R},\mathbf{R}',\delta\tau) \equiv e^{-\tau(\widehat{H}-E_T)}\frac{\Psi_T(\mathbf{R})}{\Psi_T(\mathbf{R}')}$. Expanding K in a Taylor series, at the $\delta\tau \to 0$ limit, one obtains the expression

$$K = Ne^{-(\mathbf{R}_2-\mathbf{R}_1+\frac{1}{2}\nabla\ln\Psi_T(\mathbf{R}_1)\delta\tau)^2/(2\delta\tau)} \times e^{-(\frac{\widehat{H}\Psi_T(\mathbf{R}_1)}{\Psi_T(\mathbf{R}_1)}-E_T)\delta\tau} \equiv K_D \times K_B. \tag{5.31}$$

Eq. (5.31) is closely associated to the product of the kernel of the Smoluchowski equation, which describes a diffusion process with drift, multiplied by a first order rate process. Here the rate process is dominated by the local energy, instead of the potential. The random walk is modified by appearance of a drift term that moves configurations to regions of high values of the wave function. This drift is the quantum force of Eq. (3.8).

The excess local energy $(E_T - E_L(\mathbf{R}))$ replaces the excess potential energy in the branching term exponent, see (5.26). The local energy has kinetic and potential energy contributions that tend to cancel each other, giving a smoother function: If $\Psi_T(\mathbf{R})$ is a reasonable function, the excess local energy will be nearly a constant. The regions where charged particles meet have to be taken care of by enforcing the cusp conditions on $\Psi_T(\mathbf{R})$ (see Section 2).

The local energy is the estimator of the energy with a lower statistical variance, so it is preferred over other possible choices for an estimator. A simple average of the local energy will yield the estimator of the energy of the quantum system,[8]

$$\begin{aligned}\langle E_L\rangle &= \int f(\mathbf{R},\tau\to\infty)E_L(\mathbf{R})\,d\mathbf{R}\Big/\int f(\mathbf{R})\,d\mathbf{R}\\ &= \int \Psi(\mathbf{R})\Psi_T(\mathbf{R})\left[\frac{\widehat{H}\Psi_T(\mathbf{R})}{\Psi_T(\mathbf{R})}\right]d\mathbf{R}\Big/\int \Psi(\mathbf{R})\Psi_T(\mathbf{R})\\ &= \int \Psi(\mathbf{R})\widehat{H}\Psi(\mathbf{R})\,d\mathbf{R}\Big/\int \Psi(\mathbf{R})\Psi_T(\mathbf{R})\,d\mathbf{R}\\ &= E_0.\end{aligned} \tag{5.32}$$

[8] For other energy estimators, refer to the discussion in CEPERLEY and KALOS [1986] and HAMMOND, LESTER and REYNOLDS [1994].

Therefore, a simple averaging of the local energy will yield the DMC energy estimator:

$$\langle E_L \rangle = \lim_{N_s \to \infty} \frac{1}{N_s} \sum_{i}^{N_w} E_L(\mathbf{R}_i). \tag{5.33}$$

Because the importance sampled propagator, $K(\mathbf{R}, \mathbf{R}', \delta\tau)$, is only exact to a certain order, for obtaining an exact estimator is necessary to extrapolate to $\delta\tau = 0$ for several values of $\langle E_L \rangle$.

Importance sampling with appropriate trial functions, such as those used for accurate VMC calculations, can increase the efficiency of the random walk by several orders of magnitude. In the limit required to obtain the exact trial function, only a single evaluation of the local energy is required to obtain the exact answer. Importance sampling has made molecular and atomic calculations feasible. Note that the quantum force present in Eq. (5.31) also moves random walkers away from the nodal regions into regions of large values of the trial wave function, reducing the number of attempted node crossings by several orders of magnitude.

5.3. *Population control*

If left uncontrolled, the population of random walkers will eventually vanish or fill all computer memory. Therefore, some form of population control is needed to stabilize the number of random walkers. Control is usually achieved by slowly changing E_T as the simulation progresses. As more walkers are produced in the procedure, one needs to lower the trial energy, E_T, or if the population starts to decrease, then one needs to raise E_T. This can be achieved by periodically changing the trial energy. One version of the adjustment is to use

$$E_T = \langle E_0 \rangle + \alpha \ln \frac{N_w^0}{N_w}, \tag{5.34}$$

where $\langle E_0 \rangle$ is the best approximation to the eigenvalue of the problem to this point, α is a parameter that should be as small as possible while still having a population control effect, N_w^0 is the number of desired random walkers, and N_w is the current number of random walkers.

This simple population control procedure has a slight bias if the population control parameter α is large, or if the population is small. The bias observed goes as $1/N_w$, and, formally a $N_w \to \infty$ extrapolation is required. The bias is absent in the limit of an infinite population.

A recently resurrected population control strategy, stochastic reconfiguration (SORELLA [1998], BUONAURA and SORELLA [1998], SORELLA and CAPRIOTTI [2000], ASSARAF, CAFFAREL and KHELIF [2000]) originally came from a the work of HETHERINGTON [1984]. In this algorithm, walkers carry a weight, but the weight is redetermined at each step to keep the population constant. The idea behind this method is to control the global weight $\overline{w}$ of the population,

$$\overline{w} = \frac{1}{N_w} \sum_{i=1}^{N_w} w_i. \tag{5.35}$$

The idea is to introduce a renormalized individual walker weight, ω_i, defined as

$$\omega_i \equiv \frac{w_i}{\overline{w}}. \tag{5.36}$$

Another stochastic reconfiguration scheme proposes setting the number of copies of walker i for the next step, proportional to the renormalized walker weight ω_i. This algorithm has shown to have less bias than the scheme of Eq. (5.34), and also has the advantage of having the same number of walkers at each step, simplifying implementations of the algorithm in parallel computers.

5.4. *Diffusion Monte Carlo algorithm*

There are several versions of the DMC algorithm. The approach presented here focuses on simplicity. For the latest developments, the reader is referred to REYNOLDS, CEPERLEY, ALDER and LESTER, JR. [1982], DEPASQUALE, ROTHSTEIN and VRBIK [1988], UMRIGAR, NIGHTINGALE and RUNGE [1993].

(1) Initialize an ensemble $\mathcal{W}$ of $N_{\mathcal{W}}$ configurations, distributed according to $P(\mathbf{R})$ for $\Psi_T(\mathbf{R})$; for example, use the random walkers obtained from a previous VMC run.
(2) For every configuration in $\mathcal{W}$:
   (a) Propose an electron move from $\Psi(\mathbf{R}) \equiv \Psi(\mathbf{r}_1, \mathbf{r}_2, \ldots, \mathbf{r}_i, \ldots, \mathbf{r}_{N_p})$ to $\Psi(\mathbf{R}') \equiv \Psi(\mathbf{r}_1, \mathbf{r}_2, \ldots, \mathbf{r}'_i, \ldots, \mathbf{r}_{N_p})$. The short-time approximation propagator, $K(\mathbf{R}, \mathbf{R}'; \delta\tau)$, has an associated stochastic move

$$\mathbf{R}' \to \mathbf{R} + \mathbf{F_q}(\mathbf{R})\delta\tau + \mathcal{G}_{\delta\tau}. \tag{5.37}$$

   (b) Enforce the fixed node constraint: if a random walker crosses a node, i.e., $\mathrm{sign}(\Psi_T(\mathbf{R})) \neq \mathrm{sign}(\Psi_T(\mathbf{R}'))$, then reject the move for the current electron and proceed to treat the next electron.
   (c) Compute the Metropolis acceptance/rejection probability

$$P(\mathbf{R} \to \mathbf{R}') = \min\left(1, \frac{K_D(\mathbf{R}, \mathbf{R}'; \delta\tau)\Psi_T^2(\mathbf{R}')}{K_D(\mathbf{R}', \mathbf{R}; \delta\tau)\Psi_T^2(\mathbf{R})}\right), \tag{5.38}$$

   where K_D is the diffusion and drift transition probability given by Eq. (5.31).
   (d) Compare $P(\mathbf{R} \to \mathbf{R}')$ with an uniform random number between 0 and 1, $\mathcal{U}_{[0,1]}$, if $P > \mathcal{U}_{[0,1]}$, accept the move, otherwise, reject it.

(3) Calculate the branching factor G_B for the current configuration

$$B(\mathbf{R}, \mathbf{R}') = e^{(E_R - \frac{1}{2}(\frac{\widehat{H}\Psi_T(\mathbf{R}')}{\Psi_T(\mathbf{R}')} + \frac{\widehat{H}\Psi_T(\mathbf{R})}{\Psi_T(\mathbf{R})}))\delta\tau}. \tag{5.39}$$

(4) Accumulate all observables, such as the energy. All contributions, O_i, are weighted by the branching factor, i.e.,

$$O_T^{(n+1)} = O_T^{(n)} + B(\mathbf{R}, \mathbf{R}') O_i(\mathbf{R}), \tag{5.40}$$

where $O_T^{(n)}$ is the cumulative sum of the observable at step n.

(5) Generate a new generation of random walkers, reproducing the existing population, creating an average $B(\mathbf{R}, \mathbf{R}')$ new walkers out of a walker at $\mathbf{R}$. The simplest procedure for achieving this goal is to generate n new copies of $\mathbf{R}$ where $n = \text{int}(B(\mathbf{R}, \mathbf{R}') + \mathcal{U}_{[0,1]})$.

(6) Perform blocking statistics (see Section 3.1.2), and apply population control (see Section 5.3)

(a) One choice is to update the reference energy E_R at the end of each accumulation block,

$$E_R \leftarrow E_R + E_R^{\omega} * E_B, \tag{5.41}$$

where E_R^{ω} is a re-weighting parameter, usually chosen to be ≈ 0.5, and E_B is the average energy for block B, $E_B = E^{\text{sum}}/N_B$.

(b) Discard a relaxation time of steps, N_{Rel}, which is of the order of a tenth of a block, because moving the reference energy induces the most bias in about one relaxation time.

(7) Continue the loop until the desired accuracy is achieved.

UMRIGAR, NIGHTINGALE and RUNGE [1993] proposed several modifications to the above algorithm to reduce time-step error. These modifications concentrate on improving the propagator in regions where the short-time approximation performs poorly; namely, near wave function nodes and Coulomb singularities. These propagator errors are expected, because the short-time approximation propagator assumes a constant potential over the move interval, which is a poor approximation in Ψ regions where the Coulomb interaction diverges.

5.5. *Green's function Monte Carlo*

The GFMC method is a QMC approach that has the advantage of having no time-step error. It has been shown to require more computer time than DMC, and therefore, has been applied less frequently than DMC to atomic and molecular systems. Good descriptions of the method can be found in KALOS [1962], CEPERLEY [1983], SKINNER, MOSKOWITZ, LEE, SCHMIDT and WHITLOCK [1985], KALOS and WHITLOCK [1986], SCHMIDT and MOSKOWITZ [1986]. The GFMC approach is a PopMC method for which the projector for obtaining the ground state Green's function is the standard resolvent for the Schrödinger equation (see Eq. (5.11)). The integral equation for this case, takes the simple form

$$\Psi^{(n+1)} = \left[\frac{E_T + E_C}{\widehat{H} + E_C}\right]\Psi^{(n)} \tag{5.42}$$

where the constant E_C is positive and fulfills the condition that $|E_C| > |E_0|$, and E_T is a trial energy. The resolvent of Eq. (5.42) is related to the DMC propagator by the one-sided Laplace transform

$$\frac{1}{\widehat{H} + E_C} = \int_0^\infty e^{-(\widehat{H}+E_C)\tau}\, d\tau. \tag{5.43}$$

This integral is evaluated by MC. After equilibration, the sampled times have a Poisson distribution with a mean of $\frac{N_s}{E_0+E_C}$ after N_s steps. E_C is a parameter that controls the average time step.

The Green's function is not known in close form, so it has to be sampled by MC. This can be done by rewriting the resolvent in the form

$$\frac{1}{\widehat{H} + E_C} = \frac{1}{\widehat{H}_U + E_C} + \frac{1}{\widehat{H}_U + E_C}\left(\widehat{H}_U - \widehat{H}\right)\frac{1}{\widehat{H} + E_C}. \tag{5.44}$$

The Hamiltonian $\widehat{H}_U$ represents a family of solvable Hamiltonians. To sample the Green's function, one samples the sum of terms on the right-hand side of Eq. (5.44). The Green's functions associated with $\widehat{H}$ and $\widehat{H}_U$ satisfy the relations

$$\left(\widehat{H} + E_C\right)G\left(\mathbf{R}, \mathbf{R}'\right) = \delta\left(\mathbf{R} - \mathbf{R}'\right), \tag{5.45}$$

$$\left(\widehat{H}_U + E_C\right)G_U\left(\mathbf{R}, \mathbf{R}'\right) = \delta\left(\mathbf{R} - \mathbf{R}'\right). \tag{5.46}$$

The most commonly used form of $\widehat{H}_U$ is

$$\widehat{H}_U = \frac{1}{2}\nabla_R^2 + U, \tag{5.47}$$

where U is a potential that is independent of $\mathbf{R}$. It is convenient to have $G_U(\mathbf{R}, \mathbf{R}')$ vanish at the domain boundary. $\widehat{H}_U$ should be a good approximation to $\widehat{H}$ in the domain to achieve good convergence.

The $\mathbf{R}$-space representation of Eq. (5.44) is

$$\begin{aligned} G\left(\mathbf{R}, \mathbf{R}'\right) &= G_U\left(\mathbf{R}, \mathbf{R}'\right) - \int_S d\mathbf{R}'' G\left(\mathbf{R}, \mathbf{R}''\right)\left[-\hat{n}\cdot\nabla G_U\left(\mathbf{R}'', \mathbf{R}\right)\right] \\ &\quad + \int_V d\mathbf{R}'' G\left(\mathbf{R}, \mathbf{R}''\right)\left[U - V\left(\mathbf{R}''\right)\right]G_U\left(\mathbf{R}'', \mathbf{R}'\right). \end{aligned} \tag{5.48}$$

5.6. *Fixed-node approximation*

We have not discussed the implications of the fermion character of $\Psi(\mathbf{R})$. It is an excited state in a manifold containing all the fermionic and bosonic states. A fermion wave function has positive and negative regions that are difficult to sample with the DMC algorithm as described in Section 5.4. Considering real wave functions, $\Psi(\mathbf{R})$ contains positive and negative regions, $\Psi^+(\mathbf{R})$, and $\Psi^-(\mathbf{R})$ that, in principle, could be represented as probabilities. The sign of the wave function could be used as an extra weight for the random walk. In practice, this is a very slowly convergent method.

Returning to the importance sampled algorithm, recall that the initial distribution, $|\Psi(\mathbf{R})|^2$, is positive. Nevertheless, the Green's function, $K(\mathbf{R}, \mathbf{R}')$, can become negative, if a random walker crosses a node of the trial wave function. Again, the sign of $K(\mathbf{R}, \mathbf{R}')$ could be used as a weight for sampling $|K(\mathbf{R}, \mathbf{R}')|$. The problem is that the statistics of this process lead to exponential growth of the variance of any observable.

The simplest approach to avoid exponential growth is to forbid moves in which the product wave function, $\Psi(\mathbf{R})\Psi_T(\mathbf{R})$, changes sign. This boundary condition on permitted moves is the defining characteristic of the fixed-node approximation (FNA). The nodes of the sampled wave function are *fixed* to be the nodes of the trial wave function. The FNA is an inherent feature of the DMC method, which is, by far, the most commonly used method for atomic and molecular MC applications (ANDERSON [1976], REYNOLDS, CEPERLEY, ALDER and LESTER, JR. [1982]).

The fixed-node energy is an upper bound to the exact energy of the system. In fact, it is the best solution for that fixed set of nodes. The DMC method has much higher accuracy than the VMC method. For atomic and molecular systems, it is common to recover 95–100% of the CE, cf. Section 1, whereas the CE recovered with the VMC approach is typically less than 80% of the total.

5.7. *Exact methods*

Probably the most important algorithmic challenge that still remains to be explored is the "node problem". Although progress has been made in systems that contain up to a dozen of electrons (ARNOW, KALOS, LEE and SCHMIDT [1982], CEPERLEY and ALDER [1984], ZHANG and KALOS [1991], ANDERSON, TRAYNOR and BOGHOSIAN [1991], LIU, ZHANG and KALOS [1994]), a stable algorithm that can sample the exact wave function without resorting to the FNA remains to be determined. In this section, we discuss a family of methods that avoid the FNA. These approaches yield exact answers, usually associated with a large increase in computational time.

The Pauli antisymmetry principle imposes a boundary condition on the wave function. It is the requirement that the exchange of like-spin electrons changes the sign of the wave function. This condition is a global condition that has to be enforced within an algorithm that only considers individual random walkers, i.e., a local algorithm. The FNA is the most commonly imposed boundary condition. It satisfies the variational principle, i.e., FN solutions approach the exact energy from above. This is an useful property, but one that does not assist the search for exact results, because there is not an easy way to parametrize the nodal surface and vary it to obtain the exact solution. We now describe methods that impose no additional boundary conditions on the wave function.

5.7.1. The release node method

The evolution operator, $e^{-\tau(\widehat{H}-E_T)}$, is symmetric and has the same form for both fermions and bosons. Straightforward application of it to an arbitrary initial wave function, $|\Psi_0\rangle$, leads to collapse to the ground state bosonic wave function, as can be seen from Eq. (5.20).

An arbitrary fermion wave function, $\Psi(\mathbf{R})$, can be separated into two functions $\Psi^+(\mathbf{R})$ and $\Psi^-(\mathbf{R})$ as follows,

$$\Psi^{\pm}(\mathbf{R},\tau) \equiv \frac{1}{2}\big[\big|\Psi(\mathbf{R},\tau)\big| \pm \Psi(\mathbf{R},\tau)\big]. \tag{5.49}$$

Note that the original trial wave function is recovered as

$$\Psi(\mathbf{R},\tau) = \Psi^+(\mathbf{R},\tau) - \Psi^+(\mathbf{R},\tau). \tag{5.50}$$

The released node (RN) algorithm involves two independent DMC calculations, using Ψ^+ and Ψ^- as the wave functions to evolve

$$\begin{aligned}\Psi(\mathbf{R},\tau) &= \int G\big(\mathbf{R},\mathbf{R}',\delta\tau\big)\Psi\big(\mathbf{R}',0\big)\,d\mathbf{R}' \\ &= \int G\big(\mathbf{R},\mathbf{R}',\tau\big)\Psi^+\big(\mathbf{R}',0\big)\,d\mathbf{R}' - \int G\big(\mathbf{R},\mathbf{R}',\tau\big)\Psi^-\big(\mathbf{R}',0\big)\,d\mathbf{R}' \\ &= \Psi^+(\mathbf{R},\tau) - \Psi^-(\mathbf{R},\tau).\end{aligned} \tag{5.51}$$

The time evolution of the system can be followed from the difference of separate simulations for $\Psi^{\pm}(\mathbf{R})$. Note that both distributions are always positive during the simulation, and that they decay to the ground state bosonic wave function. This decay is problematic because the "signal-to-noise" ratio in this method depends on the difference between these two distributions. The decay of the difference $\Psi^+(\mathbf{R},\tau) - \Psi^-(\mathbf{R},\tau)$ goes roughly as $e^{\tau(E_F-E_B)}$, where E_F is the lowest fermion state energy and E_B is the bosonic ground state energy.

For this method to be practical, one needs to start with the distribution of a good fermion trial wave function. The distribution will evolve from this starting point to the bosonic ground state at large imaginary time τ. In an intermediate "transient" regime one can collect information on the exact fermion wave function.

The energy can be estimated from the expression

$$E_{\mathrm{RN}}(\tau) = \frac{\int \Psi(\mathbf{R},\tau)\widehat{H}\Psi_T(\mathbf{R})\,d\mathbf{R}}{\int \Psi(\mathbf{R},\tau)\Psi_T\,d\mathbf{R}}$$

$$= \frac{\int \Psi^+(\mathbf{R},\tau)\widehat{H}\Psi_T(\mathbf{R})\,d\mathbf{R}}{\int[\Psi^+(\mathbf{R},\tau)-\Psi^-(\mathbf{R},\tau)]\Psi_T(\mathbf{R})\,d\mathbf{R}} - \frac{\int \Psi^-(\mathbf{R},\tau)\widehat{H}\Psi_T(\mathbf{R})\,d\mathbf{R}}{\int[\Psi^+(\mathbf{R},\tau)-\Psi^-(\mathbf{R},\tau)]\Psi_T(\mathbf{R})\,d\mathbf{R}} = E_F. \tag{5.52}$$

In the release node method (CEPERLEY and ALDER [1984]), a fixed-node distribution is propagated as usual, but now two sets of random walkers are retained: $\mathcal{W}_{\mathrm{FN}}$, the fixed node ensemble, and $\mathcal{W}_{\mathrm{RN}}$, the released node ensemble. Walkers are allowed to cross nodes, and when they do, they are transferred from $\mathcal{W}_{\mathrm{FN}}$ to $\mathcal{W}_{\mathrm{RN}}$. Also a account is made of the number of iterations that a walker has survived, $\mathcal{S}_{\mathrm{RN}} = \{s_1, \ldots, s_{N_w}\}$. This index is used to bin the walkers by age. Each time a walker crosses a node, a summation weight associated with it, $\Omega_{\mathrm{RN}} = \{\omega_1, \ldots, \omega_{N_w}\}$ changes sign. These weights determine the sign of the walker contribution to global averages.

The released node energy can be calculated using the estimator,

$$E_{\mathrm{RN}} = \frac{\sum_{i=1}^{Nw} \omega_i \frac{\Psi_T(\mathbf{R}_i)}{\Psi(\mathbf{R}_i)} E_L(\mathbf{R}_i)}{\sum_{i=1}^{Nw} \omega_i \frac{\Psi_T(\mathbf{R}_i)}{\Psi(\mathbf{R}_i)}}. \tag{5.53}$$

5.7.2. Fermion Monte Carlo

From the previous section one can infer that if a method in which the distribution does not go to the bosonic ground state, but stays in an intermediate regime, will not have the deficiency of exponential growth of "signal to noise". This leads to the fermion Monte Carlo (FMC) method. The approach by KALOS and SCHMIDT [1997], KALOS and PEDERIVA [1998], PEDERIVA and KALOS [1999], KALOS and PEDERIVA [2000a], KALOS and PEDERIVA [2000b] involves correlated random walks that achieve a constant "signal to noise".

The expectation value of Eq. (5.53) for an arbitrary distribution of signed walkers can be rewritten as

$$\langle E_{\mathrm{FMC}}\rangle = \frac{\sum_{i=1}^{Nw}\left[\frac{\widehat{H}\psi_T(\mathbf{R}_i^+)}{\Psi_G^+(R_i^+)} - \frac{\widehat{H}\psi_T(\mathbf{R}_i^-)}{\Psi_G^-(R_i^-)}\right]}{\sum_{i=1}^{Nw}\left[\frac{\psi_T(R_i^+)}{\Psi_G^+(R_i^+)}\frac{\psi_T(R_i^-)}{\Psi_G^-(R_i^-)}\right]}, \tag{5.54}$$

where $\Psi_G^{\pm}(\mathbf{R}^{\pm})$ are the guiding functions for a pair of random walkers $P_i = \{\mathbf{R}_i^+, \mathbf{R}_i^-\}$. Note that the variance of the energy estimator of Eq. (5.54) goes to infinity as the difference between the two populations goes to zero, i.e., the denominator,

$$\mathcal{D} \equiv \sum_{i=1}^{Nw}\left[\frac{\psi_T(R_i^+)}{\Psi_G^+(R_i^+)}\frac{\psi_T(R_i^-)}{\Psi_G^-(R_i^-)}\right], \tag{5.55}$$

goes to zero as the simulation approaches to the bosonic ground state. A procedure that would not change $\langle E_{\mathrm{FMC}}\rangle$ would be to cancel positive and negative random walkers whenever they meet (ARNOW, KALOS, LEE and SCHMIDT [1982]). Although random walks are guaranteed to meet in one dimension, they need not meet in several dimensions, due to the exponentially decaying walker density in **R**-space. Besides, cancellation has to be combined with other procedures to insure a stable algorithm.

Cancellation can be increased by introducing correlation between the random walkers. Recall the diffusion step in DMC, in which walkers diffuse from $\mathbf{R}$ to $\mathbf{R}'$ following G_D of Eq. (5.25). In the DMC algorithm, this is implemented stochastically by updating the coordinates of the random walkers with a random displacement taken from a Gaussian distribution with a variance of $\delta\tau$,

$$\mathbf{R}'^{+} \to \mathbf{R}^{+} + \mathcal{G}^{+}_{\delta\tau} \quad \text{and} \quad \mathbf{R}^{-} \to \mathbf{R}^{-} + \mathcal{G}^{-}_{\delta\tau}. \tag{5.56}$$

If we introduce correlation between the Gaussian vectors, $\mathcal{G}^{+}_{\delta\tau}$ and $\mathcal{G}^{-}_{\delta\tau}$, the expectation value of Eq. (5.54) is not affected, because it is linear in the density of random walkers.

An efficient cancellation scheme can be achieved if the Gaussian vectors are correlated as follows:

$$\mathcal{G}^{-}_{\delta\tau} = \mathcal{G}^{+}_{\delta\tau} U^{-} - 2\left(\mathcal{G}^{+}_{\delta\tau} \cdot \frac{(\mathbf{R}^{+} - \mathbf{R}^{-})}{|\mathbf{R}^{+} - \mathbf{R}^{-}|^{2}}\right) \cdot (\mathbf{R}^{+} - \mathbf{R}^{-}). \tag{5.57}$$

Eq. (5.57) accounts for reflection along the perpendicular bisector of the vector that connects the pair, $\mathbf{R}^{+} - \mathbf{R}^{-}$. This cancellation scheme generates a correlated random walk in one dimension along the vector $\mathbf{R}^{+} - \mathbf{R}^{-}$. This one-dimensional random walk is independent of the number of dimensions of the physical system, and therefore, overcomes the cancellation difficulties mentioned above. Walkers are guaranteed to meet under these conditions.

The modifications to the DMC algorithm mentioned to this point are necessary, but not sufficient for achieving a stable algorithm. If one were to interchange the random walker populations, $\{\mathbf{R}^{+}_{1}, \ldots, \mathbf{R}^{+}_{N_w}\} \leftrightarrow \{\mathbf{R}^{-}_{1}, \ldots, \mathbf{R}^{-}_{N_w}\}$, the fictitious dynamics would not be able to distinguish between the two populations, leading to a random walk with two degenerate ground states. Namely, a ground state in which all the positive walkers $\mathbf{R}^{+}$ are marginally on the positive region of the wave function, and vice versa, $\{\Psi^{+}(\mathbf{R}^{+}), \Psi^{-}(\mathbf{R}^{-})\}$ and $\{\Psi^{+}\{\mathbf{R}^{-}\}, \Psi^{-}(\mathbf{R}^{+})\}$. This *plus–minus* symmetry can be broken by using two distinct guiding functions. For example, the guiding function

$$\Psi^{\pm}_{G} = \sqrt{\Psi^{2}_{S}(\mathbf{R}) + c^{2}\Psi^{2}_{A}(\mathbf{R})} \pm c\Psi_{A}(\mathbf{R}), \tag{5.58}$$

where $\Psi_S(\mathbf{R})$ is a symmetric function under permutation of electron labels; $\Psi_A(\mathbf{R})$ is an antisymmetric function, and c is a small adjustable parameter. The guiding functions of Eq. (5.58) are almost equal, which provides nearly identical branching factors for the walker pair. It is positive everywhere, a requirement for the DMC algorithm, and it is symmetric under permutation of the coordinates, $\Psi^{+}_{G}(\widehat{\mathbf{P}}\mathbf{R}) = \Psi^{-}_{G}(\mathbf{R})$.

The use of different guiding functions is the last required ingredient for a stable algorithm. It breaks the *plus–minus* symmetry effectively, because the drift dynamics is different because the quantum force of Eq. (3.8) is distinct for each population.

For a complete description of the FMC algorithm, the reader is referred to KALOS and PEDERIVA [2000a].

The denominator of Eq. (5.55) is an indicator of stability of the algorithm. It is a measure of the antisymmetric component of the wave function. FMC calculations have shown stable denominators for thousands of relaxation times, indicating the stability of the fermion algorithm.

Early versions of the method (ARNOW, KALOS, LEE and SCHMIDT [1982]) do not scale well with system size, due to the use of uncorrelated cancellation schemes. Nevertheless, researchers have been applied successfully to several small molecular systems obtaining solutions to the Schrödinger equation with no systematic error (ZHANG and KALOS [1991], BHATTACHARYA and ANDERSON [1994a], BHATTACHARYA and ANDERSON [1994b], ANDERSON [2001]). This version of the FMC algorithm, with GFMC propagation and without correlated dynamics, is known as exact quantum Monte Carlo (EQMC).

5.8. Zero variance principle

An increase in computational efficiency can be achieved by improving the observables $\widehat{O}$ by renormalizing them to observables that have the same expectation value, but lower variance. Recent work by ASSARAF and CAFFAREL [1999], ASSARAF and CAFFAREL [2000] has shown that estimators for the energy and energy derivatives with respect to nuclear coordinates can be constructed.

One can propose a trial operator $\widehat{H}_V$ and auxiliary trial function Ψ_V such that the evaluation of a renormalized observable $\overline{O}$ will have a variance that is smaller than that of the original observable $\widehat{O}$, and in principle can even be suppressed.

To develop this concept, let us construct a trial operator $\widehat{H}_V$ such that,

$$\int \widehat{H}_V(\mathbf{R}, \mathbf{R}')\sqrt{\pi(\mathbf{R}')}\, d\mathbf{R}' = 0, \tag{5.59}$$

where $\pi(\mathbf{R}')$ is the MC distribution. For example, in VMC the MC distribution is the wave function squared, $\Psi_T(\mathbf{R})^2$, and in DMC it is the mixed distribution of Eq. (5.29). Next, propose a renormalized observable $\overline{O}(\mathbf{R})$ related to the observable $\widehat{O}(\mathbf{R})$ given by

$$\overline{O}(\mathbf{R}) = \widehat{O}(\mathbf{R}) + \frac{\int \widehat{H}_V(\mathbf{R}, \mathbf{R}')\psi_V(\mathbf{R}')\, d\mathbf{R}'}{\sqrt{\pi(\mathbf{R}')}}. \tag{5.60}$$

The mean of the rescaled operator is formally

$$\langle \overline{O} \rangle = \frac{\int \widehat{O}(\mathbf{R})\pi(\mathbf{R})\, d\mathbf{R} + \frac{\int\int \pi(\mathbf{R})\widehat{H}_V(\mathbf{R},\mathbf{R}')\Psi_V(\mathbf{R}')\, d\mathbf{R}\, d\mathbf{R}'}{\sqrt{\pi(\mathbf{R})}}}{\int \pi(\mathbf{R})\, d\mathbf{R}} \tag{5.61}$$

which, by property (5.59), is the same as the mean for the unnormalized operator:

$$\langle \overline{O} \rangle = \langle \widehat{O} \rangle. \tag{5.62}$$

Operator $\overline{O}$ can be used as an unbiased estimator, even though statistical errors for $\overline{O}$ and $\widehat{O}$ can be quite different. The goal of this kind of importance sampling is to reduce the fluctuations by construction of such an operator.

The implementation of the procedure requires the optimization of a set of parameters for the auxiliary trial wave function, $\Psi_V(\mathbf{R}, \Lambda_V)$, using the minimization functional

$$\int \widehat{H}_V(\mathbf{R}, \mathbf{R}')\Psi_V(\mathbf{R}, \Lambda_V)\,d\mathbf{R}' = -\left[\overline{O}(x) - \langle\overline{O}\rangle\right]\sqrt{\pi(\mathbf{R})}. \tag{5.63}$$

After the parameters Λ_V are optimized, one can run a simulation to average $\overline{O}$, instead of $\widehat{O}$. The choice of auxiliary Hamiltonian suggested by recent work of ASSARAF and CAFFAREL [1999] is

$$H_V(x, y) = -\frac{1}{2}\nabla^2_{\mathbf{R}} + \frac{1}{2\sqrt{\pi(\mathbf{R})}}\nabla^2_{\mathbf{R}}\sqrt{\pi(\mathbf{R})}. \tag{5.64}$$

Note that when Eq. (5.64) is applied to $\sqrt{\pi(\mathbf{R})}$, the $\mathbf{R}'$ integration vanishes by construction. The choice of auxiliary wave function is open, and an interesting observation is that minimization of the normalization factor of $\Psi_V(\mathbf{R})$, for any choice of auxiliary trial wave function, will reduce fluctuations in the auxiliary observable

$$\sigma(\overline{O})^2 = \sigma(\widehat{O})^2 - \frac{\left\langle \frac{\widehat{O}(\mathbf{R}) \int \widehat{H}(\mathbf{R}, \mathbf{R}')\Psi_V(\mathbf{R}')\,d\mathbf{R}'}{\sqrt{\pi(\mathbf{R})}} \right\rangle^2}{\left\langle \left[\frac{\int \widehat{H}(\mathbf{R}, \mathbf{R}')\Psi_V(\mathbf{R}')\,d\mathbf{R}'}{\sqrt{\pi(\mathbf{R})}} \right]^2 \right\rangle}, \tag{5.65}$$

because the second term on the right hand side of Eq. (5.65) always has a negative sign.

This variance reduction technique, applied to VMC and GFMC simulations, has achieved an order of magnitude reduction in computational effort (ASSARAF and CAFFAREL [1999]). It can also be used to calculate energy derivatives (ASSARAF and CAFFAREL [2000]).

Part III. Special topics

6. Fermion nodes

As discussed briefly in Section 5, the simulation of a quantum system without approximation, obtaining exact results other than the numerical integration scheme's associated error bar is still an open research topic. Several solutions have been proposed, but the challenge is to have a general method that scales favorably with system size.

For the ground state of a bosonic system, for which the wave function has the same sign everywhere, QMC provides an exact solution in a polynomial amount of computer time, i.e., is already a solved problem. The research in this field attempts to obtain an algorithm that has the same properties but that can treat wave functions that have both positive and negative regions, and therefore nodes, with the same favorable scaling.

Investigation of nodes has been pursued by CAFFAREL, CLAVERIE, MIJOULE, ANDZELM and SALAHUB [1989], CEPERLEY [1991], GLAUSER, BROWN, LESTER, JR., BRESSANINI, HAMMOND and KOSYKOWSKI [1992], CAFFAREL, KROKIDIS and MIJOULE [1992], BRESSANINI, CEPERLEY and REYNOLDS [2002] to understand the properties of the nodes of fermion wave functions.

The full nodal hyper-surfaces of a wave function, $\Psi(\mathbf{R})$, where $\mathbf{R}$ is a $3N$-dimensional vector and N is the number of fermions in the system is a $3(N-1)$-dimensional function, $\eta(\mathbf{R})$. Of that function, symmetry requirements determine a $(3N-3)$-dimensional surface, the *symmetry* sub-surface, $\sigma(\mathbf{R})$. This is unfortunate, because even though that $\sigma(\mathbf{R}) \subset \eta(\mathbf{R})$, the remainder of the nodal surface, the *peculiar* nodal surface, $\varpi\mathbf{R})$ which is a function of the specific form of the nuclear and inter-electronic potential, is difficult to be known *a priori* for an arbitrary system. Note that $\sigma(\mathbf{R}) \cup \varpi(\mathbf{R}) = \eta(\mathbf{R})$.

Understanding nodal properties is important for further development of QMC methods: these shall be exploited for bypassing the node problem.

CEPERLEY [1991] discusses general properties of wave function nodes. General properties of nodes follow.

(1) The *coincidence planes* $\pi(\mathbf{r}_i = \mathbf{r}_j)$, are located at nodes when the two electrons have the same spin, i.e., $\delta_{\sigma_{ij}} = 1$. In more than 1 dimension, $\pi(\mathbf{R})$ is a scaffolding where the complex nodal surface passes through. Note that $\pi(\mathbf{R}) \subset \sigma(\mathbf{R})$.
(2) The nodes possess all the symmetries of the ground state wave function.
(3) The nodes of the many-body wave function are distinct from orbital nodes $\phi_i(\mathbf{r})$, see Section 3.2.
(4) For degenerate wave functions, the node positions are arbitrary. For a p-fold degenerate energy level, one can pick $p-1$ points in $\mathbf{R}$ and find a linear transformation for which the transformed wave functions vanish at all but one of these points.
(5) A *nodal cell* $\Omega(\mathbf{R})$ around a point $\mathbf{R}$ is defined as the set of points that can be reached from $\mathbf{R}$ without crossing a node. For potentials of our present interest, the ground state nodal cells have the *tiling-property*: any point $\mathbf{R}'$ not on the node is related by symmetry to a point in $\Omega(\mathbf{R})$. This implies that there is only one type of nodal cell: all other cells are copies that can be accessed by relabeling the particles. This property is the generalization to fermions of the theorem that the bosonic ground state is nodewave functionless.

CEPERLEY [1991] suggests that for DMC simulations benefit from the tiling property: one only needs to sample one nodal cell, because all the other cells are copies of the first. Any trial function resulting from a strictly mean field theory will satisfy the tiling property. Such as the local density approximation (LDA) wave functions.

GLAUSER, BROWN, LESTER, JR., BRESSANINI, HAMMOND and KOSYKOWSKI [1992] showed that simple HF wave functions for the first-row atoms were shown to have four nodal regions (two nodal surfaces intersecting) instead of two. This is attributed to factorizing the wave function into two distinct Slater determinants, $D^{\uparrow}$ and $D^{\downarrow}$, each composed of two surfaces, one for the $\uparrow$ and one for the $\downarrow$ electron, as discussed in Section 3.2.

Recently, after analysis of the wave functions for He, Li and Be, it was conjectured by BRESSANINI, CEPERLEY and REYNOLDS [2002], that the wave function can be factored as follows:

$$\Psi(\mathbf{R}) = N(\mathbf{R})e^{f(\mathbf{R})}, \tag{6.1}$$

where $N(\mathbf{R})$ is antisymmetric polynomial of finite order, and $f(\mathbf{R})$ is a positive definite function. A weaker conjecture is that N may not be a polynomial, but can be closely approximated by a lower-order antisymmetric polynomial. The variables in which N should be expanded are the inter-particle coordinates.

For example, for all ^3S states of two-electron atoms, the nodal factor $N(\mathbf{R})$ in Eq. (6.1) is

$$N(\mathbf{r}_1, \mathbf{r}_2) = \mathbf{r}_1 - \mathbf{r}_2, \tag{6.2}$$

where $\mathbf{r}_1$ and $\mathbf{r}_2$ are the coordinates of the two electrons.

7. Treatment of heavy elements

So far, we have not discussed the applicability of QMC to systems with large atomic number. There is an steep computational scaling of QMC methods, with respect to the atomic number Z. The computational cost of QMC methods has been estimated to scale as $Z^{5.5-6.5}$ (HAMMOND, REYNOLDS and LESTER, JR. [1987], CEPERLEY [1986]). This has motivated the replacement of the core electrons by ECPs. With this modification, the scaling with respect to atomic number is improved to $Z^{3.4}$ (HAMMOND, REYNOLDS and LESTER, JR. [1987]). Other approaches involve the use of core-valence separation schemes (STAROVEROV, LANGFELDER, and ROTHSTEIN [1998]) and the use of model potentials (YOSHIDA and IGUCHI [1988]).

7.1. *Effective core potentials*

In the ECP method (SZASZ [1985], KRAUSS and STEVENS [1984], CHRISTIANSEN, ERMLER and PITZER [1984], BALASUBRAMANIAN and PITZER [1987], DOLG [2000]), the effect of the core electrons is simulated by an effective potential acting on the valence electrons. The effective Hamiltonian for these electrons is:

$$\mathcal{H}_{\text{val}} = \sum_i \frac{-Z_{\text{eff}}}{r_i} + \sum_{i<j} \frac{1}{r_{ij}} + \sum_i \mathcal{W}_i(\mathbf{r}), \tag{7.1}$$

where i and j designate the valence electrons, Z_{eff} is the effective nuclear charge in the absence of core electrons, and $\mathcal{W}$ is the pseudopotential operator. The latter can be written,

$$\mathcal{W}(\mathbf{r}) = \sum_{l=0}^{\infty} \mathcal{W}_l(\mathbf{r}) \sum_m |lm\rangle\langle lm|, \tag{7.2}$$

where l and m are the angular momentum and magnetic quantum numbers. The projection operator $\sum_m |lm\rangle\langle lm|$, connects the pseudopotential with the one-electron valence functions. A common approximation to this equation is to assume that the angular momentum components of the pseudopotential, $w_l(\mathbf{r})$ do not depend on l when $l > L$, the angular momentum of the core. This approximation leads to the expression

$$\mathcal{W}(\mathbf{r}) = \mathcal{W}_{L+1}(\mathbf{r}) + \sum_{l=0}^{L} \big(\mathcal{W}_l(\mathbf{r}) - \mathcal{W}_{L+1}(\mathbf{r})\big) \sum_m |lm\rangle\langle lm|. \tag{7.3}$$

The operator (7.3) can be applied to a valence orbital, i.e., pseudo-orbital, $\phi_l(\mathbf{r})$. This function is usually represented by a polynomial expansion for distances less than a cutoff radius, $r < r_c$, and by a fit to the all-electron orbital for $r > r_c$.

Rapid fluctuations in the potential terms can cause the first order approximation of Eq. (5.28) to break down, therefore, seeking a slowly varying form of ECP is relevant to QMC simulations. GREEFF and LESTER, JR. [1998] proposed the use of norm-conserving soft ECPs for QMC. Soft ECPs derive their name from the property of being finite at the nucleus, this leads to a pseudo-orbital with no singularities at the origin in the kinetic energy. The associated effective potential has no discontinuities or divergences.

7.2. Embedding methods

A commonly used approach in wave function based methods, is to use embedding schemes, in which a region of high interest of a large system is treated by an accurate procedure, and the remainder is described by a less accurate method. Recent work by FLAD, SCHAUTZ, WANG and DOLG [2002] has extended the methodology to QMC methods. In this approach, a mean field calculation, for example HF, is performed for the whole system. An electron localization procedure is performed, the orbitals to be correlated are chosen and separated from the remaining core orbitals. An effective Coulomb and exchange potential is constructed, $\widehat{V}_E$, which is added to the standard Hamiltonian of Eq. (0.1) to construct an effective Hamiltonian, $\widehat{H}_E$, that then is used in QMC calculations. Localization procedures similar to those required for ECPs are needed for representing the effect of nonlocal terms.

The effective Hamiltonian, $\widehat{H}_E$, takes the form

$$\widehat{H}_E = \widehat{H}_{\text{int}} + \widehat{V}_{\text{ext}} + \widehat{J}_{\text{ext}} + \widehat{K}_{\text{ext}} + \widehat{S}_{\text{ext}} \tag{7.4}$$

where $\widehat{H}_{\text{int}}$ is the Hamiltonian for the QMC active region, $\widehat{V}_{\text{ext}}$ is the Coulomb potential exerted by the external nuclei, $\widehat{J}_{\text{ext}}$ represents the Coulomb repulsions, The term $\widehat{K}_{\text{ext}}$ represents the exchange interactions, and $\widehat{S}_{\text{ext}}$ is a shift operator that prevents the wave function to be expanded into core orbitals, ϕ_c, by shifting their energy to infinity, and is given by

$$\widehat{S}_{\text{ext}} = \lim_{\lambda \to \infty} \lambda \sum_{\alpha}^{\text{int}} \sum_{\beta}^{\text{ext}} \big|\phi_p(\mathbf{r})\big\rangle\big\langle\phi_p(\mathbf{r})\big|\, d\mathbf{r}. \tag{7.5}$$

Here λ is an effective orbital coupling constant that is derived from considering single and double excitations into core and virtual orbitals of the system. The Coulomb term, $\widehat{J}_{\text{ext}}$, and the external Coulomb potential, $\widehat{V}_{\text{ext}}$, are local potentials, and can be evaluated within QMC without further approximation. The remaining terms require localization approximations that are discussed in detail in the original work.

8. Other properties

8.1. *Exact wave function and quantities that do not commute with the Hamiltonian*

Properties related to a trial wave function $\Psi_T(\mathbf{R})$, are readily available in VMC calculations ALEXANDER, COLDWELL, AISSING and THAKKAR [1992]. In this case, expectation values are calculated from directly from Eqs. (3.4) and (3.5). The accuracy of the results obtained with VMC depend on the quality of $\Psi_T(\mathbf{R})$.

For obtaining expectation values of operators that do not commute with the Hamiltonian in an importance sampled PMC calculation, one needs to extract the exact distribution $\Psi^2(\mathbf{R})$ from the mixed distribution $f(\mathbf{R}) = \Psi(\mathbf{R})\Psi_T(\mathbf{R})$. The expectation values for an operator $\widehat{O}$, $\langle\Psi(\mathbf{R})|\widehat{O}|\Psi_T(\mathbf{R})\rangle$ and $\langle\Psi_T(\mathbf{R})|\widehat{O}|\Psi(\mathbf{R})\rangle$, are different to each other. MC sampling requires knowledge of the exact ground state distribution: a mixed distribution does not suffice to obtain the exact answer.

If the operator $\widehat{O}$ is a multiplicative operator, then the algorithms described in this section will be pertinent. Nonmultiplicative operators, which are exemplified by forces in this chapter, are described in Section 8.2.

8.1.1. *Extrapolation method*

An approximate procedure for estimating the ground state distribution can be extrapolated from the mixed and VMC distributions. This procedure is valuable because no extra changes are needed to the canonical VMC and PMC algorithms. Being an approximate method, it can fail even in very simple cases (SARSA, SCHMIDT and MAGRO [2000]), but also has provided very accurate results in more favorable cases (LIU, KALOS and CHESTER [1974]). The mixed estimator of a coordinate operator $\widehat{O}$ is

$$\langle\widehat{O}\rangle_m = \frac{\int \Psi(\mathbf{R})\widehat{O}\Psi_T(\mathbf{R})\,d\mathbf{R}}{\int \Psi(\mathbf{R})\Psi_T(\mathbf{R})\,d\mathbf{R}}, \tag{8.1}$$

to be distinguished from the pure estimator

$$\langle\widehat{O}\rangle_p = \frac{\int \Psi(\mathbf{R})\widehat{O}\Psi(\mathbf{R})\,d\mathbf{R}}{\int \Psi(\mathbf{R})\Psi(\mathbf{R})\,d\mathbf{R}}. \tag{8.2}$$

We will label $\langle\widehat{O}\rangle_v$ the VMC estimator of Eq. (3.4). $\langle\widehat{O}\rangle_m$ can be rewritten in a Taylor series in the difference between the exact and approximate wave functions, $\delta\Psi \equiv \Psi(\mathbf{R}) - \Psi_T(\mathbf{R})$,

$$\langle\widehat{O}\rangle_m = \langle\widehat{O}\rangle_p + \int \Psi\big(\langle\widehat{O}\rangle_p - \widehat{O}(\mathbf{R})\big)\delta\Psi\,d\mathbf{R} + \mathcal{O}\big((\delta\Psi)^2\big). \tag{8.3}$$

A similar expansion can be constructed for $\langle \widehat{O} \rangle_v$,

$$\langle \widehat{O} \rangle_v = \langle \widehat{O} \rangle_p + 2 \int \Psi \left(\langle \widehat{O} \rangle_p - \widehat{O}(\mathbf{R}) \right) \delta\Psi \, d\mathbf{R} + \mathcal{O}\left((\delta\Psi)^2 \right). \tag{8.4}$$

Combining Eqs. (8.3) and (8.4), we can arrive to an expression with a second order error,

$$\langle \widehat{O} \rangle_e = 2 \langle \widehat{O} \rangle_m - \langle \widehat{O} \rangle_v = \langle \widehat{O} \rangle_p + \mathcal{O}\left((\delta\Psi)^2 \right), \tag{8.5}$$

where $\langle \widehat{O} \rangle_e$ is an extrapolation estimate readily available from VMC and PMC calculations.

8.1.2. Future walking

The future walking method can be combined with any importance sampled PMC method that leads to a mixed distribution. If one multiplies both sides of Eq. (8.1) by the ratio $\Psi(\mathbf{R})/\Psi_T(\mathbf{R})$, one can recover Eq. (8.2). The ratio is obtained from the asymptotic population of descendants of a single walker (BARNETT, REYNOLDS and LESTER, JR. [1991]).

A walker in $\mathbf{R}$-space can be represented as a sum of eigenfunctions of $\widehat{H}$:

$$\delta(\mathbf{R}' - \mathbf{R}) = \Psi(\mathbf{R}') \sum_{n=0}^{\infty} c_i(\mathbf{R}) \Psi_n(\mathbf{R}). \tag{8.6}$$

The coefficients $c_i(\mathbf{R})$ can be obtained by multiplying Eq. (8.6) by $\Psi(\mathbf{R}')/\Psi_T(\mathbf{R}')$ and integrating over $\mathbf{R}'$:

$$c_n(\mathbf{R}) = \int \delta(\mathbf{R}' - \mathbf{R}) \frac{\Psi(\mathbf{R}')}{\Psi_T(\mathbf{R}')} \, d\mathbf{R}' = \frac{\Psi(\mathbf{R})}{\Psi_T(\mathbf{R})}. \tag{8.7}$$

Clearly, we want to know the contribution to the ground state wave function, $c_0(\mathbf{R})$ of the walker at $\mathbf{R}$. If propagated for sufficiently long time, all coefficients $c_i(\mathbf{R}) \neq c_0(\mathbf{R})$ for the random walker will vanish. This can be seen from the decay in τ of Eqs. (5.19) and (5.20).

If we define $P_\infty(\mathbf{R})$ to be the asymptotic population of walkers descended from a random walker at $\mathbf{R}$, we find,

$$\begin{aligned} P_\infty(\mathbf{R}) &= \int c_0(\mathbf{R}) e^{-(E_0 - E_T)\tau} \Psi(\mathbf{R}') \Psi_T(\mathbf{R}') \, d\mathbf{R}' \\ &= \frac{\Psi(\mathbf{R})}{\Psi_T(\mathbf{R})} e^{-(E_0 - E_T)\tau} \langle \Psi(\mathbf{R}) | \Psi_T(\mathbf{R}) \rangle. \end{aligned} \tag{8.8}$$

For obtaining $P_\infty(\mathbf{R})$ in a PMC algorithm, one needs to keep a list of all the descendants of each walker $\mathbf{R}_i$ at each time step τ_j. The number of steps for which one requires to keep track of the descendants, N_d, is a critical parameter. The statistical error of the asymptotic walker population grows in the limit $N_d \to \infty$, and if only few steps are used, a bias is encountered by nonvanishing contributions from excited states $c_i(\mathbf{R}) \neq c_0(\mathbf{R})$. Efficient algorithms for keeping track of the number of descendants can be found in the literature (LIU, KALOS and CHESTER [1974], REYNOLDS, BARNETT, HAMMOND, GRIMES and LESTER, JR. [1986], EAST, ROTHSTEIN and VRBIK [1988], CAFFAREL and CLAVERIE [1988], BARNETT, REYNOLDS and LESTER, JR. [1991], HAMMOND, LESTER and REYNOLDS [1994], LANGFELDER, ROTHSTEIN and VRBIK [1997]).

The wave function overlap with the ground state can also be obtained with these methods, as shown by HORNIK, SNAJDR and ROTHSTEIN [2000]. These methods have been applied for obtaining dipole moments (SCHAUTZ and FLAD [1999]), transition dipole moments (BARNETT, REYNOLDS and LESTER, JR. [1992a]) and oscillator strengths (BARNETT, REYNOLDS and LESTER, JR. [1992b]), among other applications.

Other methods for obtaining the exact distribution that are not discussed here for reasons of space are bilinear methods (ZHANG and KALOS [1993]), and time correlation methods (CEPERLEY and BERNU [1988]).

8.2. *Force calculation*

Most QMC applications have been within the Born–Oppenheimer (BO) (BORN and OPPENHEIMER [1927]) approximation. In this approximation, the nuclear coordinates $\mathcal{R}$ are fixed at a certain position during the calculation.[9] Therefore, the wave function and energy depends parametrically on the nuclear coordinates $E(\mathcal{R})$ and $\Psi(\mathbf{R}, \mathcal{R})$. We will omit this parametric dependence for the remainder of the discussion, and simplify the symbols to E and $\Psi(\mathbf{R})$, when appropriate.

Forces are derivatives of the energy with respect to nuclear displacements:

$$F(\mathcal{R}) = \nabla_{\mathcal{R}} E(\mathcal{R}). \tag{8.9}$$

Because the stochastic nature of the algorithm, obtaining forces in QMC is a difficult task. Generally, QMC calculations at critical points, e.g., equilibrium and reaction barrier geometries have been carried out, in which the geometries are obtained with a different quantum chemical method such as density functional theory (DFT) (PARR and YANG [1989]), or wave function methods. Whereas DFT and wave function methods use the Hellman–Feynman theorem for the calculation of forces, a straightforward application of the theorem in QMC leads to estimators with very large variance.

8.2.1. *Correlated sampling*

An efficient approach for force calculation is the use of correlated sampling, which is a MC method that uses correlation between similar observations to reduce the statistical

[9]The QMC method can be used for calculations without the BO approximation, but so far the applications have been to nodeless systems, such as H_2 (TRAYNOR, ANDERSON and BOGHOSIAN [1994]).

error of the sampling. If one were to represent Eq. (8.9) in a finite difference scheme for evaluating a derivative along $\mathbf{r}_d$,

$$\frac{\partial E}{\partial \mathbf{r}_d} \approx \frac{E(\mathcal{R}+\mathbf{r}_d) - E(\mathcal{R})}{\mathbf{r}_d}, \tag{8.10}$$

then one obtains an approximate energy derivative along the $\mathbf{r}_d$. If two separate calculations are carried out, with a statistical error of the energies of σ_E, the statistical error for the difference σ_d is approximately

$$\sigma_d \approx \frac{\sigma_E}{\mathbf{r}_d}. \tag{8.11}$$

One can see that because $\mathbf{r}_d$ is a vector of a small perturbation, of ≈ 0.01 a.u., that the statistical error of the difference will be several times higher than the statistical error of the energies. If $\mathbf{r}_d$ is sufficiently small, a single random walk can be performed, while evaluating the energy at the original and perturbed geometries, $E[\Psi(\mathcal{R})]$ and $E[\Psi(\mathcal{R}+\mathbf{r}_d)]$. In this case, both the primary ($\mathcal{R}$) and secondary ($\mathcal{R}$) walks will be correlated, and therefore present lower variance than uncorrelated random walks.

If correlated sampling is used for forces in a PMC algorithm, it was recently proposed by FILIPPI and UMRIGAR [2002] to use expressions including branching factors $B(\mathbf{R})$, re-optimizing the parameters of the wave function Λ for each perturbed geometry, and perform additional coordinate transformations. Practical implementations of the correlated sampling method for derivatives are described in detail in SUN, LESTER, JR. and HAMMOND [1992] and FILIPPI and UMRIGAR [2002].

8.2.2. *Analytic derivative methods*

The calculation of analytic derivative estimators is a costly process, both for wave function based methods, and for QMC methods. Fortunately, in QMC one needs not to evaluate derivatives at each step, but rather sample points more sporadically, both for reducing computer time and serial correlation.

The local energy estimator for a DMC mixed distribution is

$$E_0 = \langle E_L \rangle = \frac{\int \Psi_0(\mathbf{R}) E_L(\mathbf{R}) \Psi_T(\mathbf{R})\, d\mathbf{R}}{\int \Psi_0(\mathbf{R}) \Psi_T(\mathbf{R})\, d\mathbf{R}}. \tag{8.12}$$

The gradient of expression (8.12) involves derivatives of the unknown exact wave function $\Psi_0(\mathbf{R})$, and the trial wave function $\Psi_T(\mathbf{R})$. Derivatives of $\Psi_0(\mathbf{R})$ have to be obtained with a method devised for sampling operators that do not commute with the Hamiltonian, which are described in Section 8.1. This can lead to an exact estimator for the derivative, but with the added computational complexity of those methods. Therefore, a simple approximation can be used, replacing the derivatives of $\Psi_0(\mathbf{R})$ with those of $\Psi_T(\mathbf{R})$ to obtain

$$\nabla_{\mathcal{R}} E_0 \approx \langle \nabla_{\mathcal{R}} E_L(\mathbf{R}) \rangle + 2\left\langle E_L \frac{\nabla_{\mathcal{R}} \Psi_T(\mathbf{R})}{\Psi_T(\mathbf{R})} \right\rangle - 2E_0 \left\langle \frac{\nabla_{\mathcal{R}} \Psi_T(\mathbf{R})}{\Psi_T(\mathbf{R})} \right\rangle. \tag{8.13}$$

The derivatives of $\Psi_T(\mathbf{R})$ are readily obtainable from the known analytic expression of $\Psi_T(\mathbf{R})$.

The expression for the exact derivative involves the cumulative weight of configuration $\mathbf{R}_i$ at time step s, $\mathbf{R}_i^s$,

$$\overline{B}_i = \prod_{s_0}^{s} B(\mathbf{R}_i^s), \tag{8.14}$$

where $s - s_0$ is the number of generations for the accumulation of the cumulative weight, and $B(\mathbf{R}_i^s)$ is the PMC branching factor of Eqs. (5.9) and (5.39). The energy expression using cumulative weights is

$$E_0 = \frac{\int \Psi_T(\mathbf{R})^2 \overline{B}(\mathbf{R}) E_L(\mathbf{R})\, d\mathbf{R}}{\int \Psi_T(\mathbf{R})^2 \overline{B}(\mathbf{R})\, d\mathbf{R}}. \tag{8.15}$$

The energy derivative of expression (8.15) leads to the following derivative expression,

$$\begin{aligned}\nabla_{\mathcal{R}} E_0 = {}& \langle \nabla_{\mathcal{R}} E_L(\mathbf{R}) \rangle + 2\left\langle E_L(\mathbf{R}) \frac{\nabla_{\mathcal{R}} \Psi_T(\mathbf{R})}{\Psi_T(\mathbf{R})} \right\rangle - 2E_0 \left\langle \frac{\nabla_{\mathcal{R}} \Psi_T(\mathbf{R})}{\Psi_T(\mathbf{R})} \right\rangle \\ & + \left\langle E_L(\mathbf{R}) \frac{\nabla_{\mathcal{R}} \overline{B}(\mathbf{R})}{\overline{B}(\mathbf{R})} \right\rangle - E_0 \left\langle \frac{\nabla_{\mathcal{R}} \overline{B}(\mathbf{R})}{\overline{B}(\mathbf{R})} \right\rangle.\end{aligned} \tag{8.16}$$

Analytic energy derivatives have been applied to H_2 (REYNOLDS, BARNETT, HAMMOND, GRIMES and LESTER, JR. [1986]), LiH and CuH (VRBIK, LAGARE and ROTHSTEIN [1990]). Higher order derivatives can be obtained as well. Details on the former can be found in VRBIK and ROTHSTEIN [1992] and BELOHOREC, ROTHSTEIN and VRBIK [1993].

8.2.3. *Hellman–Feynman derivatives and the zero variance theorem*

The Hellman–Feynman theorem states that the forces can be obtained by looking at the value of the gradient of the potential

$$\langle \nabla_{\mathcal{R}} E_0 \rangle = -\frac{\int \Psi^2(\mathbf{R}) \nabla_{\mathcal{R}} V(\mathbf{R})\, d\mathbf{R}}{\int \Psi^2(\mathbf{R})\, d\mathbf{R}}, \tag{8.17}$$

where $V(\mathbf{R})$ is the Coulomb potential for the system. A QMC estimator of the Hellman–Feynman forces, $\mathbf{F}_{\mathrm{HF}} \equiv -\nabla_{\mathcal{R}} V(\mathbf{R})$, can be constructed, but it has infinite variance. This comes from the fact that at short electron–nucleus distances, $\mathbf{r}_{iN}$, the force behaves as $\mathbf{F}_{\mathrm{HF}} \approx \frac{1}{\mathbf{r}_{iN}^2}$, therefore the variance associated with $\mathbf{F}_{\mathrm{HF}}$ depends on $\langle \mathbf{F}_{\mathrm{HF}}^2 \rangle$, which is infinite. Furthermore, the Hellman–Feynman theorem only holds for exact wave functions, and basis set errors need to be accounted for (UMRIGAR [1989]). Also, the fixed-node approximation introduces an extra requirement on the nodal surface. The former has to be independent of the position of the nuclei, or it has to be the exact one.

An elaborate discussion of this issue can be found in HUANG, NEEDS and RAJAGOPAL [2000] and SCHAUTZ and Flad [2000].

A proposed solution to the infinite variance problem is to evaluate the forces at a cutoff distance close to the origin, and then extrapolation to a cutoff distance of zero (VRBIK and ROTHSTEIN [1992]). This has the problem that the extrapolation procedure is difficult due to the increase of variance as the cutoff values decrease.

As discussed in Section 5.8, renormalized operators can be obtained, in such a way that they have the same expectation value, but lower variance. Recently ASSARAF and CAFFAREL [2000], a renormalized operator was introduced

$$\overline{\mathbf{F}}_{\mathrm{HF}} = \mathbf{F}_{\mathrm{HF}} + \left[\frac{\widehat{H}_V \Psi_V(\mathbf{R})}{\Psi_V(\mathbf{R})} - \frac{\widehat{H}_V \Psi_T(\mathbf{R})}{\Psi_T(\mathbf{R})} \right] \frac{\Psi_V(\mathbf{R})}{\Psi_T(\mathbf{R})}. \tag{8.18}$$

Here, $\Psi_V(\mathbf{R})$ are an auxiliary wave function and $\widehat{H}_V$ is an auxiliary Hamiltonian. The variance of operator (8.18) can be shown to be finite, and therefore smaller than $\mathbf{F}_{\mathrm{HF}}$. The form of $\Psi_V(\mathbf{R})$ proposed by ASSARAF and CAFFAREL [2000] is a simple form that cancels the singularities of the force in the case of a diatomic molecule. Nevertheless, general forms of the auxiliary wave function can be constructed.

8.2.4. Variational Monte Carlo dynamics

Correlated sampling can be combined with a fictitious Lagrangian technique, similar to that developed by CAR and PARRINELLO [1985] in a way first proposed by TANAKA [1994] for geometry optimization. In this approach, the expectation value of the Hamiltonian is treated as a functional of the nuclear positions and the correlation parameters:

$$\langle \widehat{H} \rangle = \frac{\langle \Psi | \widehat{H} | \Psi \rangle}{\langle \Psi | \Psi \rangle} = E\big[\{\Lambda\}, \{\mathcal{R}\}\big]. \tag{8.19}$$

With the previous functional, a fictitious Lagrangian can be constructed of the form

$$L = \sum_{\alpha} \frac{1}{2} \mu_\alpha \lambda_\alpha'^2 + \sum_I \frac{1}{2} M_I \mathcal{R}_I'^2 - E\big[\{\Lambda\}, \{\mathcal{R}\}\big], \tag{8.20}$$

where M_I are the nuclear masses and μ_a are the fictitious masses for the variational parameters, λ_α. The modified Euler–Lagrange equations can be used for generating dynamics for the sets of parameters, $\{\mathcal{R}\}$ and $\{\Lambda\}$,

$$M_I \mathcal{R}_I'' = \nabla_{R_I} E, \tag{8.21}$$

$$\mu_\alpha \lambda_\alpha'' = \frac{\partial E}{\partial \lambda_\alpha}. \tag{8.22}$$

A dissipative transformation of Eqs. (8.21) and (8.22), where the masses M_I and μ_α are replaced by damped masses $\widetilde{M_I}$ and $\widetilde{\mu_\alpha}$ can be used for geometry optimization.

A more elaborate approach that attempts to include quantum effects in the dynamics is described in TANAKA [2002].

To conclude, a method that is not described here due to reasons of space and that needs to be explored further, is the generalized re-weighting method (BORONAT and CASULLERAS [1999]).

List of symbols

We denote:

Z_j Atomic nuclear charge,

$\mathcal{R}$ Set of coordinates of clamped particles (under the Born–Oppenheimer approximation): $\mathcal{R} \equiv \{\mathcal{R}_1, \ldots, \mathcal{R}_m\}$,

r_i Electronic coordinate in a Cartesian frame $r_i = \{x_i \ldots x_d\}$, where the number of dimensions, d is 3 for most chemical applications,

$\mathbf{R}$ Set of coordinates of all n particles treated by QMC, $\mathbf{R} \equiv \{r_1, \ldots, r_n\}$,

σ_i Spin coordinate for an electron, $\sigma_\uparrow$ is spin-up and $\sigma_\downarrow$ for spin-down particles,

Σ Set of spin coordinates for particles, $\Sigma \equiv \{\sigma_1, \ldots, \sigma_N\}$,

Ψ_0 Exact ground state wave function,

Ψ_i ith exact wave function,

$\Psi_{T,i}$ Approximate trial wave function for state i,

ϕ_k Single particle molecular orbital (MO),

$D(\phi_k)$ Slater determinant of k MOs: $D = \frac{1}{\sqrt{N!}} \det|\phi_1, \ldots, \phi_N|$,

$D^\uparrow(\phi_k^\uparrow)$, $D^\downarrow(\phi_k^\downarrow)$, Spin factored Slater determinants for spin up ($\uparrow$) and spin down ($\downarrow$) electrons,

$\widehat{H}$ Hamiltonian operator: $\widehat{H} \equiv \widehat{T} + \widehat{V}$,

$\widehat{T}$ Kinetic energy operator $-\frac{1}{2}\nabla_R^2 \equiv -\frac{1}{2}\sum_i \nabla_i^2$,

$\widehat{V}$ Potential energy operator, for atomic and molecular systems: $\widehat{V} = -\sum_{ij} \frac{Z_j}{r_{ij}} + \sum_{i<j} \frac{1}{r_{ij}} + \sum_{i<j} \frac{Z_i Z_j}{r_{ij}}$,

τ Imaginary time, $\tau = it$,

$\rho(\mathbf{r})$ Electronic density,

E_L Local Energy, $\widehat{H}\Psi_T(\mathbf{R})/\Psi_T(\mathbf{R})$,

$\mathcal{U}_{[a,b]}$ Uniform random variate in the interval $[a, b]$,

$\mathcal{G}_\sigma$ Gaussian random variate of variance σ,

$\sigma_{\widehat{O}}$ Monte Carlo variance for observable $\widehat{O}$,

$\mathcal{W}$ Ensemble of random walkers, $\mathcal{W} \equiv \{\mathbf{R}_1, \mathbf{R}_2, \ldots, \mathbf{R_n}\}$,

$\mathcal{P}(\mathbf{R})$ Monte Carlo probability density function,

$P(\mathbf{R} \to \mathbf{R}')$ Monte Carlo transition probability,

$B(\mathbf{R}, \mathbf{R}')$ Branching factor for Population Monte Carlo algorithms,

$G(\mathbf{R}, \mathbf{R}'; \tau - \tau')$ Time dependent Green's function,

$G_{\mathrm{ST}}(\mathbf{R}, \mathbf{R}'; \delta\tau)$ Time dependent short time Green's function for the Schrödinger equation,

$\mathbf{F_q}$ Quantum force, $\mathbf{F_q} \equiv \nabla \ln|\Psi_T(\mathbf{R})^2|$,

$\mathcal{D}$ Denominator in fermion Monte Carlo, $\mathcal{D} \equiv \sum_{i=1}^{Nw}\left[\frac{\psi_T(R_i^+)}{\Psi_G^+(R_i^+)}\frac{\psi_T(R_i^-)}{\Psi_G^-(R_i^-)}\right]$,

M_I Ionic mass,
μ_α Fictitious parameter mass.

Acknowledgments

A.A.-G. was a holder of a Gates Millenium Fellowship during the preparation of this chapter. W.A.L. was supported by the Director, Office of Science, Office of Basic Energy Sciences, Chemical Sciences Division of the U.S. Department of Energy under Contract No. DE-AC03-76SF00098, and by the National Science Foundation through the CREST Program (HRD-9805465).

References

ACIOLI, P.H. (1997), Review of quantum Monte Carlo methods and their applications, *J. Mol. Struct. (Theochem)* **394**, 75.

ALEXANDER, S.A. and R.L. COLDWELL (1997), Atomic wave function forms, *Int. J. Quant. Chem.* **63**, 1001.

ALEXANDER, S.A., R.L. COLDWELL, G. AISSING and A.J. THAKKAR (1992), Calculation of atomic and molecular properties using variational Monte Carlo methods, *Int. J. Quant. Chem.* **26**, 213–227.

ANDERSON, J.B. (1976), Quantum chemistry by random walk, *J. Chem. Phys.* **65** (10), 4121–4127.

ANDERSON, J.B. (1980), Quantum chemistry by random walk: Higher accuracy, *J. Chem. Phys.* **73** (8), 3897–3899.

ANDERSON, J.B. (1999), Quantum Monte Carlo: Atoms, molecules, clusters, liquids and solids, in: K.B. Lipkowitz, D.B. Boyd, eds., *Reviews in Computational Chemistry*, Vol. 13 (John Wiley & Sons, New York) 133.

ANDERSON, J.B. (2001), An exact quantum Monte Carlo calculation of the helium–helium intermolecular potential. II, *J. Chem. Phys.* **115**, 4546.

ANDERSON, J.B., C.A. TRAYNOR and B.M. BOGHOSIAN (1991), Quantum chemistry by random walk: Exact treatment of many-electron systems, *J. Chem. Phys.* **95**, 7418.

ARNOW, D.M., M.H. KALOS, M.A. LEE and K.E. SCHMIDT (1982), Green's function Monte Carlo for few fermion problems, *J. Chem. Phys.* **77**, 5562.

ASSARAF, R. and M. CAFFAREL (1999), Zero-variance principle for Monte Carlo algorithms, *Phys. Rev. Lett.* **83**, 4682.

ASSARAF, R. and M. CAFFAREL (2000), Computing forces with quantum Monte Carlo, *J. Chem. Phys.* **113**, 4028.

ASSARAF, R., M. CAFFAREL and A. KHELIF (2000), Diffusion Monte Carlo methods with a fixed number of walkers, *Phys. Rev. E* **61**, 4566.

BAER, R., M. HEAD-GORDON and D. NEUHAUSER (1998), Shifted-contour auxiliary field Monte Carlo for ab initio electronic structure: Straddling the sign problem, *J. Chem. Phys.* **109**, 6219.

BAER, R. and D. NEUHAUSER (2000), Molecular electronic structure using auxiliary field Monte Carlo, plane waves and pseudopotentials, *J. Chem. Phys.* **112**, 1679.

BALASUBRAMANIAN, K. and K.S. PITZER (1987), Relativistic quantum chemistry, *Adv. Chem. Phys.* **77**, 287.

BARNETT, R.N., P.J. REYNOLDS and W.A. LESTER, JR. (1991), Monte Carlo algorithms for expectation values of coordinate operators, *J. Comp. Phys.* **96**, 258.

BARNETT, R.N., P.J. REYNOLDS and W.A. LESTER, JR. (1992a), Computation of transition dipole moments by Monte Carlo, *J. Chem. Phys.* **96**, 2141.

BARNETT, R.N., P.J. REYNOLDS and W.A. LESTER, JR. (1992b), Monte Carlo determination of the oscillator strength and excited state lifetime for the Li 2s to 2p transition, *Int. J. Quant. Chem.* **42**, 837.

BARNETT, R.N., Z. SUN and W.A. LESTER, JR. (1997), Fixed sample optimization in quantum Monte Carlo using a probability density function, *Chem. Phys. Lett.* **273**, 321.

BARNETT, R.N., Z. SUN and W.A. LESTER, JR. (2001), Improved trial functions in quantum Monte Carlo: Application to acetylene and its dissociation fragments, *J. Chem. Phys.* **114**, 2013.

BAUER, W.F. (1958), The Monte Carlo method, *J. Soc. Indust. Appl. Math.* **6** (4), 438–451. A very well described introduction to MC methods by an applied mathematician. Cites the Courant and Levy 1928 paper.

BELOHOREC, P., S.M. ROTHSTEIN and J. VRBIK (1993), Infinitesimal differential diffusion quantum Monte Carlo study of CuH spectroscopic constants, *J. Chem. Phys.* **98**, 6401.

BENOIT, D.M. and D.C. CLARY (2000), Quaternion formulation of diffusion quantum Monte Carlo for the rotation of rigid molecules in clusters, *J. Chem. Phys.* **113**, 5193.

BERTINI, L., D. BRESSANINI, M. MELLA and G. MOROSI (1999), Linear expansions of correlated functions: a variational Monte Carlo case study, *Int. J. Quant. Chem.* **74**, 23.

BHATTACHARYA, A. and J.B. ANDERSON (1994a), Exact quantum Monte Carlo calculation of the H–He interaction potential, *Phys. Rev. A* **49**, 2441.

BHATTACHARYA, A. and J.B. ANDERSON (1994b), The interaction potential of a symmetric helium trimer, *J. Chem. Phys.* **100**, 8999.

BIJL, A. (1940), *Physica* **7**, 869.

BORN, M. and J.R. OPPENHEIMER (1927), Zur Quantentheorie der Molekeln, *Ann. Phys.* **84**, 457.

BORONAT, J. and J. CASULLERAS (1999), Sampling differences in quantum Monte Carlo: A generalized reweighting method, *Comp. Phys. Comm.* **122**, 466.

BOYS, S.F. and N.C. HANDY (1969), *Proc. Roy. Soc. London Ser. A* **310**, 63.

BRESSANINI, D., D.M. CEPERLEY and P.J. REYNOLDS (2002), What do we know about wave function nodes?, in: S.M. Rothstein, W.A. Lester, Jr., S. Tanaka, eds., *Recent Advances in Quantum Monte Carlo Methods, Part II* (World Scientific, Singapore).

BRESSANINI, D. and P.J. REYNOLDS (1998), Between classical and quantum Monte Carlo methods: "Variational" QMC, *Adv. Chem. Phys.* **105**, 37.

BRESSANINI, D. and P.J. REYNOLDS (1999), Spatial-partitioning-based acceleration for variational Monte Carlo, *J. Chem. Phys.* **111**, 6180.

BUECKERT, H., S.M. ROTHSTEIN and J. VRBIK (1992), Optimization of quantum Monte Carlo wavefunctions using analytical derivatives, *Canad. J. Chem.* **70**, 366.

BUONAURA, M.C. and S. SORELLA (1998), Numerical study of the two-dimensional Heisenberg model using a Green function Monte Carlo technique with a fixed number of walkers, *Phys. Rev. B* **61**, 2559.

BYRON, F.W. and R. FULLER (1992), *Mathematics of Classical and Quantum Physics* (Dover).

CAFFAREL, M. and P. CLAVERIE (1988), Development of a pure diffusion quantum Monte Carlo method using a full generalized Feynman–Kac Formula, I. Formalism, *J. Chem. Phys.* **88**, 1088–1099.

CAFFAREL, M., P. CLAVERIE, C. MIJOULE, J. ANDZELM and D.R. SALAHUB (1989), Quantum Monte-Carlo method for some model and realistic coupled anharmonic-oscillators, *J. Chem. Phys.* **90**, 990.

CAFFAREL, M., X. KROKIDIS and C. MIJOULE (1992), On the nonconservation of the number of nodal cells of eigenfunctions, *Europhys. Lett.* **7**, 581.

CANCÈS, E., M. DEFRANCESCHI, W. KUTZELNIGG, C. LE BRIS and Y. MADAY (2003), Computational quantum chemistry: A primer, in: P.G. Ciarlet and C. Le Bris, eds., *Computational Chemistry*, Handbook of Numerical Analysis, Vol. X (Elsevier, Amsterdam) 3–270.

CAR, R. and M. PARRINELLO (1985), Unified approach for molecular dynamics and density-functional theory, *Phys. Rev. Lett.* **55**, 2471.

CASULLERAS, J. and J. BORONAT (2000), Progress in Monte Carlo calculations of Fermi systems: Normal liquid He-3, *Phys. Rev. Lett.* **84**, 3121.

CEPERLEY, D.M. (1983), The simulation of quantum systems with random walks: A new algorithm for charged systems, *J. Comp. Phys.* **51**, 404.

CEPERLEY, D.M. (1986), The statistical error of Green's function Monte Carlo, *J. Stat. Phys.* **43**, 815.

CEPERLEY, D.M. (1991), Fermion nodes, *J. Stat. Phys.* **63**, 1237–1267.

CEPERLEY, D.M. (1995), Path integral Monte Carlo in the theory of condensed matter helium, *Rev. Modern. Phys.* **67** (2), 279–356.

CEPERLEY, D.M. and B.J. ALDER (1980), Ground state of the electron gas by a stochastic method, *Phys. Rev. Lett.* **45**, 566.

CEPERLEY, D.M. and B.J. ALDER (1984), Quantum Monte Carlo for molecules: Green's function and nodal release, *J. Chem. Phys.* **81**, 5833.

CEPERLEY, D.M. and B. BERNU (1988), The calculation of excited state properties with quantum Monte Carlo, *J. Chem. Phys.* **89**, 6316.

CEPERLEY, D., G.V. CHESTER and M.H. KALOS (1977), Monte Carlo simulation of a many fermion study, *Phys. Rev. B.* **16**, 3081.

CEPERLEY, D.M. and M.H. KALOS (1986), Quantum many-body problems, in: K. Binder, ed., *Monte Carlo Methods in Statistical Physics*, 2nd edn. (Springer-Verlag, New York).
CEPERLEY, D.M. and L. MITAS (1996), Quantum Monte Carlo methods in chemistry, in: I. Prigogine, S.A. Rice, eds., *New Methods in Computational Quantum Mechanics*, Adv. Chem. Phys., Vol. XCIII (John Wiley & Sons, New York).
CERF, N.J. and O.C. MARTIN (1995), Projection Monte Carlo methods: An algorithmic analysis, *Int. J. Mod. Phys.* **6**, 693.
CHARUTZ, D.M. and D. NEUHAUSER (1995), Electronic structure via the auxiliary-field Monte-Carlo algorithm, *J. Chem. Phys.* **102**, 4495.
CHIN, S.A. (1990), Quadratic diffusion Monte Carlo algorithms for solving atomic many-body problems, *Phys. Rev. A* **42**, 6991.
CHIN, S.A. (1997), Symplectic integrators from composite operator factorizations, *Phys. Lett. A* **226**, 344.
CHRISTIANSEN, P.A., W.C. ERMLER and K.S. PITZER (1984), Relativistic effects in chemical systems, *Annu. Rev. Phys. Chem.* **35**, 357.
CLARY, D.C. (2001), Torsional diffusion Monte Carlo: A method for quantum simulations of proteins, *J. Chem. Phys.* **114**, 9725.
CONROY, H. (1964), Molecular Schrödinger equation, 2. Monte Carlo evaluation of integrals, *J. Chem. Phys.* **41**, 1331.
COURANT, R., K.O. FRIEDRICHS and H. LEWY (1928), On the partial difference equations of mathematical physics, *Math. Ann.* **100**, 32.
DAVIDSON, E.R. (1976), *Reduced Density Matrices in Quantum Chemistry* (Academic Press, New York).
DEPASQUALE, M.F., S.M. ROTHSTEIN and J. VRBIK (1988), Reliable diffusion quantum Monte Carlo, *J. Chem. Phys.* **89**, 3629.
DEWING, M. (2000), Improved efficiency with variational Monte Carlo using two level sampling, *J. Chem. Phys.* **113**, 5123.
DINGLE, R.B. (1949), *Phil. Mag.* **40**, 573.
DOLG, M. (2000), Effective core potentials, in: J. Grotendorst, ed., *Modern Methods and Algorithms of Quantum Chemistry*, Vol. 1 (John von Neumann Institute for Computing) 479–508.
DONSKER, M.D. and M. KAC (1950), A sampling method for determining the lowest eigenvalue and the principal eigenfunction of Schrödinger's equation, *J. Res. Nat. Bur. Standards* **44** (50), 551–557.
DOUCET, A., N. DE FREITAS, N. GORDON and A. SMITH, eds. (2001), *Sequential Monte Carlo Methods in Practice* (Springer–Verlag, New York).
EAST, A.L.L., S.M. ROTHSTEIN and A. VRBIK (1988), Sampling the exact electron distribution by diffusion quantum Monte Carlo, *J. Chem. Phys.* **89**, 4880.
EINSTEIN, A. (1926), *Investigations in the theory of Brownian Motion* (Metheun & Co. Ltd), English translation of Einstein's original paper.
FEYNMANN, R.P. (1939), Forces in molecules, *Phys. Rev.* **56**, 340.
FEYNMANN, R.P. and M. COHEN (1956), Energy spectrum of the excitations in liquid helium, *Phys. Rev.* **102**, 1189.
FILIPPI, C. and S. FAHY (2000), Optimal orbitals from energy fluctuations in correlated wave functions, *J. Chem. Phys.* **112**, 3532.
FILIPPI, C. and C.J. UMRIGAR (1996), Multiconfiguration wave functions for quantum Monte Carlo calculations of first-row diatomic molecules, *J. Chem. Phys.* **105**, 213.
FILIPPI, C. and C.J. UMRIGAR (2002), Interatomic forces and correlated sampling in quantum Monte Carlo, in: S.M. Rothstein, W.A. Lester Jr., S. Tanaka, eds., *Recent Advances in Quantum Monte Carlo Methods, Part II* (World Scientific, Singapore).
FISHMAN, G.S. (1996), *Monte Carlo: Concepts, Algorithms and Applications*, 1st edn. (Springer-Verlag, Berlin).
FLAD, H.-J., M. CAFFAREL and A. SAVIN (1997), Quantum Monte Carlo calculations with multi-reference trial wave functions, in: W.A. Lester, Jr., ed., *Recent Advances in Quantum Monte Carlo Methods* (World Scientific, Singapore), Chapter 5, 73–98.
FLAD, H.-J. and M. DOLG (1997), Probing the accuracy of pseudopotentials for transition metals in quantum Monte Carlo calculations, *J. Chem. Phys.* **107**, 7951–7959.

FLAD, H.-J., A. SAVIN and M. SCHULTHEISS (1994), A systematic study of fixed-node and localization error in quantum Monte Carlo calculations with pseudopotentials for group III elements, *Chem. Phys. Lett.* **222**, 274.

FLAD, H.-J., F. SCHAUTZ, Y. WANG and M. DOLG (2002), Quantum Monte Carlo study of mercury clusters, in: S.M. Rothstein, W.A. Lester, Jr., S. Tanaka, eds., *Recent Advances in Quantum Monte Carlo Methods, Part II* (World Scientific, Singapore).

FORBERT, H.A. and S.A. CHIN (2001), Fourth order diffusion Monte Carlo algorithms for solving quantum many-body problems, *Int. J. Mod. Phys. B* **15**, 1752.

FOULKES, M., L. MITAS, R. NEEDS and G. RAJAGOPAL (2001), Quantum Monte Carlo for solids, *Rev. Mod. Phys.* **73**, 33.

FROST, A.A., R.E. KELLOG and E.A. CURTIS (1960), Local energy method for electronic energy calculations, *Rev. Mod. Phys.* **32**, 313–317.

GLAUSER, W.A., W.R. BROWN, W.A. LESTER, JR., D. BRESSANINI, B.L. HAMMOND and M.L. KOSYKOWSKI (1992), Random-walk approach to mapping nodal regions of N-body wave functions: Ground-state Hartree–Fock wave functions for Li–C, *J. Chem. Phys.* **97**, 9200.

GOULD, H. and J. TOBOCHNIK (1996), *An Introduction to Computer simulation Methods: Applications to Physical Systems*, 2nd edn. (Addison–Wesley, Reading, MA).

GREEFF, C.W., B.L. HAMMOND and W.A. LESTER, JR. (1996), Electronic states of Al and Al_2 using quantum Monte Carlo with an effective core potential, *J. Chem. Phys.* **104**, 1973–1978.

GREEFF, C.W. and W.A. LESTER, JR. (1997), Quantum Monte Carlo binding energies for silicon hydrides, *J. Chem. Phys.* **106**, 6412–6417.

GREEFF, C.W. and W.A. LESTER, JR. (1998), A soft Hartree–Fock pseudopotential for carbon with application to quantum Monte Carlo, *J. Chem. Phys.* **109**, 1607–1612.

GROSSMAN, J.C., W.A. LESTER, JR. and S.G. LOUIE (2000), Quantum Monte Carlo and density functional theory characterization of 2-cyclopentenone and 3-cyclopentenone formation from O(3P) + cyclopentadiene, *J. Amer. Chem. Soc.* **122**, 703.

GROSSMAN, J.C. and L. MITAS (1995), Quantum Monte Carlo determination of electronic and structural properties of Si(n) clusters ($n < 20$), *Phys. Rev. Lett.* **74**, 1323.

HALTON, J.H. (1970), A retrospective and prospective survey of the Monte Carlo method, *SIAM Review* **12** (1), 1–53.

HAMMERSLEY, J.M. and D.C. HANDSCOMB (1964), *Monte Carlo Methods* (Methuen, London).

HAMMOND, B.L., W.A. LESTER, JR. and P.J. REYNOLDS (1994), *Monte Carlo Methods in Ab Initio Quantum Chemistry* (World Scientific, Singapore).

HAMMOND, B.L., P.J. REYNOLDS and W.A. LESTER, JR. (1987), Valence quantum Monte Carlo with ab initio effective core potentials, *J. Chem. Phys.* **87**, 1130–1136.

HEAD-GORDON, M. (1996), Quantum chemistry and molecular processes, *J. Phys. Chem.* **100**, 13213.

HELLMAN, H. (1937), *Einführung in die Quanten Theorie* (Deuticke).

HETHERINGTON, J.H. (1984), Observations on the statistical iteration of matrices, *Phys. Rev. A.* **30**, 2713.

HORNIK, M., M. SNAJDR and S.M. ROTHSTEIN (2000), Estimating the overlap of an approximate with the exact wave function by quantum Monte Carlo methods, *J. Chem. Phys.* **113**, 3496.

HUANG, C., C.J. UMRIGAR and M.P. NIGHTINGALE (1997), Accuracy of electronic wave functions in quantum Monte Carlo: The effect of high-order correlations, *J. Chem. Phys.* **107**, 3007.

HUANG, H. and Z. CAO (1996), A novel method for optimizing quantum Monte Carlo wave functions, *J. Chem. Phys.* **104**, 1.

HUANG, H., Q. XIE, Z. CAO, Z. LI, Z. YUE and L. MING (1999), A novel quantum Monte Carlo strategy: Surplus function approach, *J. Chem. Phys.* **110**, 3703.

HUANG, K.C., R.J. NEEDS and G. RAJAGOPAL (2000), Comment on "Quantum Monte Carlo study of the dipole moment of CO" [J. Chem. Phys. 110, 11700 (1999)], *J. Chem. Phys.* **112**, 4419.

HUANG, S.-Y., Z. SUN and W.A. LESTER, JR. (1990), Optimized trial functions for quantum Monte Carlo, *J. Chem. Phys.* **92**, 597.

IBA, Y. (2000), Population Monte Carlo algorithms, *Trans. Japanese Soc. Artificial Intelligence* **16** (2), 279–286.

JAMES, H.H. and A.S. COOLIDGE (1937), Criteria for goodness for approximate wave functions, *Phys. Rev.* **51**, 860–863.

JASTROW, R. (1955), Many-body problem with strong forces, *Phys. Rev.* **98**, 1479.

KALOS, M.H. (1962), Monte Carlo calculations of the ground state of three- and four-body nuclei, *Phys. Rev.* **68** (4).

KALOS, M.H., D. LEVESQUE and L. VERLET (1974), Helium at zero temperature with hard-sphere and other forces, *Phys. Rev. A* **9**, 2178.

KALOS, M.H. and F. PEDERIVA (1998), Fermion Monte Carlo, in: M.P. Nightingale, C.J. Umrigar, eds., *Quantum Monte Carlo Methods in Physics and Chemistry*, Vol. 525 (Kluwer Academic, Dordrecht) 263–286.

KALOS, M.H. and F. PEDERIVA (2000a), Exact Monte Carlo method for continuum fermion systems, *Phys. Rev. Lett.* **85**, 3547.

KALOS, M.H. and F. PEDERIVA (2000b). Fermion Monte Carlo for continuum systems, *Physica A* **279**, 236.

KALOS, M.H. and K.E. SCHMIDT (1997), Model fermion Monte Carlo with correlated pairs II, *J. Stat. Phys.* **89**, 425.

KALOS, M.H. and P.A. WHITLOCK (1986), *Monte Carlo Methods, Vol. 1: Basics* (Wiley, New York).

KATO, T. (1957), On the eigenfunctions of many-particle systems in quantum mechanics, *Comm. Pure. and Appl. Math.* **10**, 151.

KOONIN, S.E. and MEREDITH (1995), *Computational Physics, FORTRAN Version*, 3rd edn. (Addison–Wesley).

KRAUSS, M. and W.J. STEVENS (1984), Effective potentials in molecular quantum chemistry, *Annu. Rev. Phys. Chem.* **35**, 357.

KWON, Y., D.M. CEPERLEY and R.M. MARTIN (1993), Effects of three-body and backflow correlations in the two-dimensional electron gas, *Phys. Rev. B.* **48**, 12037.

KWON, Y., D.M. CEPERLEY and R.M. MARTIN (1998), Effects of backflow correlation in the three-dimensional electron gas: Quantum Monte Carlo study, *Phys. Rev. B.* **58**, 6800.

KWON, Y., P. HUANG, M.V. PATEL, D. BLUME and K.B. WHALEY (2000), Quantum solvation and molecular rotations in superfluid helium clusters, *J. Chem. Phys.* **113**, 6469.

LANGFELDER, P., S.M. ROTHSTEIN and J. VRBIK (1997), Diffusion quantum Monte Carlo calculation of non-differential properties for atomic ground states, *J. Chem. Phys.* **107**, 8525.

LESTER, W.A., JR. and B.L. HAMMOND (1990), Quantum Monte Carlo for the electronic structure of atoms and molecules, *Annu. Rev. Phys. Chem.* **41**, 283–311.

LIN, X., H. ZHANG and A.M. RAPPE (2000), Optimization of quantum Monte Carlo wave functions using analytical energy derivatives, *J. Chem. Phys.* **112**, 2650.

LIU, J.S. (2001), *Monte Carlo Strategies in Scientific Computing* (Springer–Verlag, New York).

LIU, K.S., M.H. KALOS and G.V. CHESTER (1974), Quantum hard spheres in a channel, *Phys. Rev. A* **10**, 303.

LIU, Z., S. ZHANG and M.H. KALOS (1994), Model fermion Monte Carlo method with antithetical pairs, *Phys. Rev. E* **50**, 3220.

LÖWDIN, P.-O. (1959), Correlation problem in many-electron quantum mechanics, *Adv. Chem. Phys.* **2**, 207.

LÜCHOW, A. and J.B. ANDERSON (2000), Monte Carlo methods in electronic structures for large systems, *Ann. Rev. Phys. Chem.* **51**, 501.

LÜCHOW, L. and J.B. ANDERSON (1996), First-row hydrides: Dissociation and ground state energies using quantum Monte Carlo, *J. Chem. Phys.* **105**, 7573.

MANNO, I. (1999), *Introduction to the Monte-Carlo Method* (Akademiai Kiado, Budapest).

MANTEN, S. and A. LÜCHOW (2001), On the accuracy of the fixed-node diffusion quantum Monte Carlo method, *J. Chem. Phys.* **115**, 5362.

MAZZIOTTI, D.A. (1998), Contracted Schrödinger equation: Determining quantum energies and two-particle density matrices without wavefunctions, *Phys. Rev. A* **57**, 4219.

MCDOWELL, K. (1981), Assessing the quality of a wavefunction using quantum Monte Carlo, *Int. J. Quant. Chem: Quant. Chem. Symp.* **15**, 177–181.

METROPOLIS, N., A.W. ROSENBLUTH, M.N. ROSENBLUTH, N.M. TELLER and E. TELLER (1953), Equation of state calculations by fast computing machines, *J. Chem. Phys.* **21**, 1087–1091.

METROPOLIS, N. and S. ULAM (1949), The Monte Carlo method, *J. Amer. Statist. Assoc.* **44**, 335.

MEYERS, C.R., C.J. UMRIGAR, J.P. SETHNA and J.D. MORGAN (1991), The Fock expansion, Kato's cusp conditions, and the exponential Ansatz, *Phys. Rev. A* **44**, 5537.

MITAS, L. (1998), Diffusion Monte Carlo, in: M.P. Nightingale, C.J. Umrigar, eds., *Quantum Monte Carlo Methods in Physics and Chemistry*, Vol. 525 (Kluwer Academic, Dordrecht) 247.

MORELL, M.M., R.G. PARR and M. LEVY (1975), Calculation of ionization-potentials from density matrices and natural functions, and long-range behavior of natural orbitals and electron density, *J. Chem. Phys.* **62**, 549.

NIGHTINGALE, M.P. and C.J. UMRIGAR (1997), Monte Carlo optimization of trial wave functions in quantum mechanics and statistical mechanics, in: W.A. Lester, Jr., ed., *Recent Advances in Quantum Monte Carlo Methods* (World Scientific) Chapter 12, 201–227.

OVCHARENKO, I.V., W.A. LESTER, JR., C. XIAO and F. HAGELBERG (2001), Quantum Monte Carlo characterization of small Cu-doped silicon clusters: CuSi4 and CuSi6, *J. Chem. Phys.* **114**, 9028.

PANDHARIPANDE, V.R. and N. ITOH (1973), Effective mass of 3He in liquid 4He+, *Phys. Rev. A* **8**, 2564.

PANDHARIPANDE, V.R., S.C. PIEPER and R.B. WIRINGA (1986), Variational Monte Carlo calculations of ground states of liquid 4He and 3He drops, *Phys. Rev. B* **34**, 4571.

PANGALI, C., M. RAO and B.J. BERNE (1979). On the force bias Monte Carlo simulation of water: methodology, optimization and comparison with molecular dynamics, *Mol. Phys.* **37**, 1773.

PARR, R.G. and W. YANG (1989), *Density Functional Theory of Atoms and Molecules* (Oxford Univ. Press).

PEDERIVA, F. and M.H. KALOS (1999), Fermion Monte Carlo, *Comp. Phys. Comm.* **122**, 445.

PIEPER, S.C. and R.B. WIRINGA (2001), Quantum Monte Carlo calculations of light nuclei, *Annu. Rev. Nucl. Part. Sci.* **51**, 53.

RAGHAVACHARI, K. and J.B. ANDERSON (1996), Electron correlation effects in molecules, *J. Phys. Chem.* **100**, 12960.

RAO, M. and B.J. BERNE (1979), On the force bias Monte Carlo simulation of simple liquids, *J. Chem. Phys.* **71**, 129.

REYNOLDS, P.J., R.N. BARNETT, B.L. HAMMOND, R.M. GRIMES and W.A. LESTER, JR. (1986), Quantum chemistry by quantum Monte Carlo: Beyond ground-state energy calculations, *Int. J. Quant. Chem.* **29**, 589.

REYNOLDS, P.J., D.M. CEPERLEY, B. ALDER and W.A. LESTER, JR. (1982), Fixed-node quantum Monte Carlo for molecules, *J. Chem. Phys.* **77**, 5593–5603.

ROM, N., D.M. CHARUTZ and D. NEUHAUSER (1997), Shifted-contour auxiliary-field Monte-Carlo: Circumventing the sign difficulty for electronic structure calculations, *Chem. Phys. Lett.* **270**, 382.

ROTHSTEIN, S.M. and J. VBRIK (1988), Statistical error of diffusion Monte Carlo, *J. Comp. Phys.* **74**, 127.

SARSA, A., K.E. SCHMIDT and W.R. MAGRO (2000), A path integral ground state method, *J. Chem. Phys.* **113**, 1366.

SCHAUTZ, F. and H.-J. FLAD (1999), Quantum Monte Carlo study of the dipole moment of CO, *J. Chem. Phys.* **110**, 11700.

SCHAUTZ, F. and H.-J. FLAD (2000), Response to "Comment on 'Quantum Monte Carlo study of the dipole moment of CO' " [J. Chem. Phys. 112, 4419 (2000)], *J. Chem. Phys.* **112**, 4421.

SCHLEYER, P.V.R., N.L. ALLINGER, T. CLARK, J. GASTEIGER, P.A. KOLLMAN, H.F. SCHAEFER, III and P.R. SCHREINER, eds. (1998), *The Encyclopedia of Computational Chemistry* (John Wiley & Sons, Chichester).

SCHMIDT, K., M.H. KALOS, ML. A. LEE and G.V. CHESTER (1980), Variational Monte Carlo calculations of liquid 4He with three-body correlations, *Phys. Rev. Lett.* **45**, 573.

SCHMIDT, K.E. (1986), Variational and Green's function Monte Carlo calculations of few body systems, in: *Conference on Models and Methods in Few Body Physics* (Lisbon).

SCHMIDT, K.E. and J.W. MOSKOWITZ (1986), Monte Carlo calculations of atoms and molecules, *J. Stat. Phys.* **43**, 1027.

SCHMIDT, K.E. and J.W. MOSKOWITZ (1990), Correlated Monte Carlo wave functions for the atoms He through Ne, *J. Chem. Phys.* **93** (4172).

SENATORE, G. and N.H. MARCH (1994), Recent progress in the field of electron correlation, *Rev. Mod. Phys.* **66**, 445.

SKINNER, D.W., J.W. MOSKOWITZ, M.A. LEE, K.E. SCHMIDT and P.A. WHITLOCK (1985), The solution of the Schrödinger equation in imaginary time by Green's function Monte Carlo, the rigorous sampling of the attractive Coulomb singularity, *J. Chem. Phys.* **83**, 4668.

SNAJDR, M., J.R. DWYER and S.M. ROTHSTEIN (1999), Histogram filtering: A technique to optimize wave functions for use in Monte Carlo simulations, *J. Chem. Phys.* **111**, 9971.

SOBOL, I.M. (1994), *A Primer for the Monte Carlo Method* (CRC Press).

SOKOLOVA, S.A. (2000), An ab initio study of TiC with the diffusion quantum Monte Carlo method, *Chem. Phys. Lett.* **320**, 421–424.

SORELLA, S. (1998), Green's function Monte Carlo with stochastic reconfiguration, *Phys. Rev. Lett.* **80**, 4558.

SORELLA, S. and L. CAPRIOTTI (2000), Green function Monte Carlo with stochastic reconfiguration: An effective remedy for the sign problem, *Phys. Rev. B* **61**, 2599.

STAROVEROV, V.N., P. LANGFELDER and S.M. ROTHSTEIN (1998), Monte Carlo study of core-valence separation schemes, *J. Chem. Phys.* **108**, 2873.

STEDMAN, M.L., W.M.C. FOWLKES and M. NEKOVEE (1998), An accelerated Metropolis method, *J. Chem. Phys.* **109**, 2630.

SUN, Z., R.N. BARNETT and W.A. LESTER, JR. (1992), Quantum and variational Monte Carlo interaction potentials for Li2 ($X^1\Sigma g^+$)), *Chem. Phys. Lett.* **195**, 365.

SUN, Z., S.-Y. HUANG, R.N. BARNETT and W.A. LESTER, JR. (1990), Wave function optimization with a fixed sample in quantum Monte Carlo, *J. Chem. Phys.* **93**, 5.

SUN, Z., W.A. LESTER, JR. and B.L. HAMMOND (1992), Correlated sampling of Monte Carlo derivatives with iterated-fixed sampling, *J. Chem. Phys.* **97**, 7585.

SUN, Z., M.M. SOTO and W.A. LESTER, JR. (1994), Characteristics of electron movement in variational Monte Carlo simulations, *J. Chem. Phys.* **100**, 1278.

SZASZ, L. (1985), *Pseudopotential Theory of Atoms and Molecules* (Wiley, New York).

TANAKA, S. (1994), Structural optimization in variational quantum Monte Carlo, *J. Chem. Phys.* **100**, 7416.

TANAKA, S. (2002), Ab initio approach to vibrational properties and quantum dynamics of molecules, in: S.M. Rothstein, W.A. Lester, Jr., S. Tanaka, eds., *Recent Advances in Quantum Monte Carlo Methods, Part II* (World Scientific, Singapore).

THIJSSEN, J.M. (1999), *Computational Physics* (Cambridge Univ. Press).

TRAYNOR, C.A., J.B. ANDERSON and B.M. BOGHOSIAN (1994), A quantum Monte Carlo calculation of the ground state energy of the hydrogen molecule, *J. Chem. Phys.* **94**, 3657.

UMRIGAR, C.J. (1989), Two aspects of quantum Monte Carlo: Determination of accurate wavefunctions and determination of potential energy surfaces of molecules, *Int. J. Quant. Chem.* **23**, 217.

UMRIGAR, C.J. (1993), Accelerated Metropolis method, *Phys. Rev. Lett.* **71**, 408.

UMRIGAR, C.J., M.P. NIGHTINGALE and K.J. RUNGE (1993), A diffusion Monte Carlo algorithm with very small time-step error, *J. Chem. Phys.* **99**, 2865.

UMRIGAR, C.J., K.G. WILSON and J.W. WILKINS (1988), Optimized trial wave functions for quantum Monte Carlo calculations, *Phys. Rev. Lett.* **60**, 1719–1722.

VIEL, A. and K.B. WHALEY (2001), Quantum structure and rotational dynamics of HCN in helium clusters, *J. Chem. Phys.* **115**, 10186.

VRBIK, J., D.A. LAGARE and S.M. ROTHSTEIN (1990), Infinitesimal differential diffusion quantum Monte Carlo: Diatomic molecular properties, *J. Chem. Phys.* **92**, 1221.

VRBIK, J. and S.M. ROTHSTEIN (1992), Infinitesimal differential diffusion quantum Monte Carlo study of diatomic vibrational frequencies, *J. Chem. Phys.* **96**, 2071.

WATSON, D., M. DUNN, T.C. GERMANN, D.R. HERSCHBACH and D.Z. GOODSON (1996), Dimensional expansions for atomic systems, in: C.A. Tsipis, V.S. Popov, D.R. Herschbach, J. Avery, eds., *New Methods in Quantum Theory* (Kluwer Academic, Dordrecht) 83.

WEISSBLUTH, M. (1978), *Atoms and Molecules* (Academic Press, New York) 570–572.

WIRINGA, R.B., S.C. PIEPER, J. CARLSON and V.R. PANDHARIPANDE (2000), Quantum Monte Carlo calculations of $A = 8$ nuclei, *Phys. Rev. C* **62**, 14001.

WOOD, W.W. and J.J. ERPENBECK (1976), Molecular dynamics and Monte Carlo calculations in statistical mechanics, *Ann. Rev. Phys. Chem.* **27**, 319.

YOSHIDA, T. and K. IGUCHI (1988), Quantum Monte Carlo method with the model potential, *J. Chem. Phys.* **88**, 1032.

ZHANG, S. and M.H. KALOS (1991), Exact Monte Carlo calculation for few-electron systems, *Phys. Rev. Lett.* **67**, 3074.

ZHANG, S.W. and M.H. KALOS (1993), Bilinear quantum Monte Carlo – expectation values and energy differences, *J. Stat. Phys.* **70**, 515.

Linear Scaling Methods for the Solution of Schrödinger's Equation

Stefan Goedecker

Lab. Nanostructures et Magnetisme, DRFMC/SP2M, C.E.A., 17 Avenue des Martyrs, 38054 Grenoble cedex 9, France

1. Introduction

Numerical simulations are nowadays an essential part of any investigation at the atomistic level. At this level the quantum mechanical character of the electrons that bond the atoms together is the prominent feature and truly predictive methods therefore need to be based on the many-electron equations of quantum mechanics. Realistic problems found in chemistry, physics or biology frequently involve large system sizes. As the system becomes larger the simulation time of the traditional algorithms for the solution of the quantum mechanical equations increases strongly, preventing computational methods from being applied to large systems. There has recently been a lot of progress in the development of algorithms with favorable scaling. These algorithms lead only to a moderate increase in simulation time with respect to the system size and eliminate thus the above mentioned computational bottleneck. In the best case the scaling is linear with respect to the system size and the algorithms with this property are called $O(N)$ algorithms. As a measure of the system size one typically takes the number of atoms, but one might as well take the number of electrons or the number of basis functions as all these quantities grow linearly with system size. $O(N)$ methods have already allowed the quantum mechanical simulation of systems of unprecedented size, containing many thousand atoms and are expected to have a significant impact on our understanding of large molecular systems.

Computational Chemistry
Special Volume (C. Le Bris, Guest Editor) of
HANDBOOK OF NUMERICAL ANALYSIS, VOL. X
P.G. Ciarlet (Editor)
© 2003 Elsevier Science B.V. All rights reserved

There are several review articles on $O(N)$ methods within the independent particle schemes (GOEDECKER [1999], SCUSERIA [1999], ORDEJON [1998], GALLI [1996], WU and JAYANTHI [2002]). This chapter is not intended to be yet another comprehensive review article. Instead some recent developments in the field and in particular the progress with regard to $O(N)$ methods for interacting electrons will be stressed. Nevertheless the article is self-contained.

2. Electronic structure calculations

The central equation for atomistic simulations is the many-electron Schrödinger equation. For simplicity we will in the whole chapter consider only spinless electrons. For such a system of N spinless electrons Schrödinger's equation reads:

$$\left(-\frac{1}{2}\sum_{i=1}^{N}\nabla_i^2 + \sum_{i=1}^{N}\sum_{j=i+1}^{N}\frac{1}{|\mathbf{r}_i - \mathbf{r}_j|} + \sum_{i=1}^{N} V(\mathbf{r}_i)\right)\Psi(\mathbf{r}_1, \ldots, \mathbf{r}_N)$$
$$= E\Psi(\mathbf{r}_1, \ldots, \mathbf{r}_N). \tag{2.1}$$

Atomic units were used in the above equation and will also be used throughout the text. The eigenvalue problem of Eq. (2.1) has to be solved under the constraint that the many electron wave function $\Psi(\mathbf{r}_1, \ldots, \mathbf{r}_N)$ be anti-symmetric, i.e.,

$$\Psi(\ldots, \mathbf{r}_i, \ldots, \mathbf{r}_j, \ldots) = -\Psi(\ldots, \mathbf{r}_j, \ldots, \mathbf{r}_i, \ldots).$$

The first term on the left hand side of Eq. (2.1) gives the kinetic energy, the second the electrostatic interaction energy of the electrons among themselves and the last term the interaction of the electrons with an external potential $V(\mathbf{r})$ arising for instance from the charges of the nuclei. The eigenvalue E in Eq. (2.1) is called the total energy and is the central quantity in any electronic structure calculation. Most physical observables are related to this total energy. For instance, the equilibrium geometry of an molecule can be found by varying the positions of the nuclei until the lowest total energy is found. The forces acting on the nuclei are given by the derivative of the total energy with respect to the positions of the nuclei.

Electronic structure methods (KUTZELNIGG and VON HERIGONTE [2000]) that aim at solving the full many-electron Schrödinger equation (2.1) are called wavefunction methods. With the exception of Quantum Monte Carlo methods (FOULKES, MITAS, NEEDS and RAJAGOPAL [2001], NIGHTINGALE and UMRIGAR [1999]) they are all based on an expansion of the many-electron wave function in Slater determinants formed from a set of orthonormal one-electron orbitals $\phi_i(\mathbf{r})$. Due to the mathematical properties of determinants, such an expansion leads automatically to the required antisymmetry of Ψ.

$$\Psi(\mathbf{r}_1, \ldots, \mathbf{r}_N) = \sum_I C_I \frac{1}{\sqrt{N!}} \begin{vmatrix} \phi_{i_1}(\mathbf{r}_1) & \ldots & \phi_{i_N}(\mathbf{r}_1) \\ \ldots & \ldots & \ldots \\ \phi_{i_1}(\mathbf{r}_N) & \ldots & \phi_{i_N}(\mathbf{r}_N) \end{vmatrix}. \tag{2.2}$$

A determinant is also called a configuration. The composite index $I = (i_1, \ldots, i_N)$ characterizes a configuration by specifying out of which one-electron orbitals ϕ_i it is constructed. If M orbitals ϕ_i are at our disposal we can form $M!/((M-N)!N!)$ different configurations. In the full configuration interaction (CI) method the sum in Eq. (2.2) runs over all the possible $M!/((M-N)!N!)$ terms, i.e., over the whole Hilbert space. Hence the computational effort has to scale at least as $M!/((M-N)!N!)$. Even though this scaling behavior is not strictly exponential it has exponential character in the sense that it is worse than any polynomial scaling and we will therefore stick with the usual custom of saying that the scaling of a full CI calculation is exponential with respect to the system size. The Hilbert space is invariant under unitary transformations of the orbitals ϕ_i. Let us express the original set ϕ_i in terms of a new orthonormal set $\tilde{\phi}_i$,

$$\phi_i = \sum_{j=1}^{M} U_{i,j} \tilde{\phi}_j, \tag{2.3}$$

where U is a unitary matrix. Inserting the expansion (Eq. (2.3)) into all the determinants of a wave function of the type of Eq. (2.2) and using the well-known rules for the manipulation of determinants one can the reexpress this wave function in terms of these new configurations formed by the set $\tilde{\phi}_i$ with modified coefficients $\widetilde{C}_I$.

$$\Psi(\mathbf{r}_1, \ldots, \mathbf{r}_N) = \sum_I \widetilde{C}_I \frac{1}{\sqrt{N!}} \begin{vmatrix} \tilde{\phi}_{i_1}(\mathbf{r}_1) & \ldots & \tilde{\phi}_{i_N}(\mathbf{r}_1) \\ \ldots & \ldots & \ldots \\ \tilde{\phi}_{i_1}(\mathbf{r}_N) & \ldots & \tilde{\phi}_{i_N}(\mathbf{r}_N) \end{vmatrix}. \tag{2.4}$$

The fact that different sets of orbitals can be used to obtain the same wave function will be the key to reducing the scaling of wavefunction electronic structure methods.

Because of the prohibitive exponential scaling of the traditional algorithms for the full CI method, approximate CI methods are generally used. The most popular methods at present are Coupled Cluster Methods (SZABO and OSTLUND [1982]). Conceptually truncated CI methods are however simpler and we will therefore discuss their scaling. In the CI method with single and double excitations (CISD) one includes for instance the Hartree–Fock determinant and all the "excited" determinants that one can obtain by replacing at most two occupied Hartree–Fock orbitals by virtual orbitals. This reduces the scaling from exponential to the sixth power as can be seen in the following way. Out of the N occupied Hartree–Fock orbitals we have to pick two that are to be replaced, resulting in $N(N-1)/2$ possibilities. If we have M excited orbitals at our disposal there are $M(M-1)$ ways to fill the two gaps. Hence the length of the CI expansion and therefore the dimension of the many electron Hamiltonian matrix is $(1/2)N(N-1)M(M-1)$. The number of operations in a iterative matrix diagonalization is proportional to the number of nonzero matrix elements of the CI Hamiltonian. Because of the Slater–Condon rules (MCWEENY [1989]) the Hamiltonian is a sparse matrix. These rules tell us that the matrix element coupling two configurations vanishes if the two configurations differ by more than two orbitals. If they differ exactly by two orbitals the element is given by a single electron repulsion integral, if the differ by one orbital the element is a sum of N terms and if the two

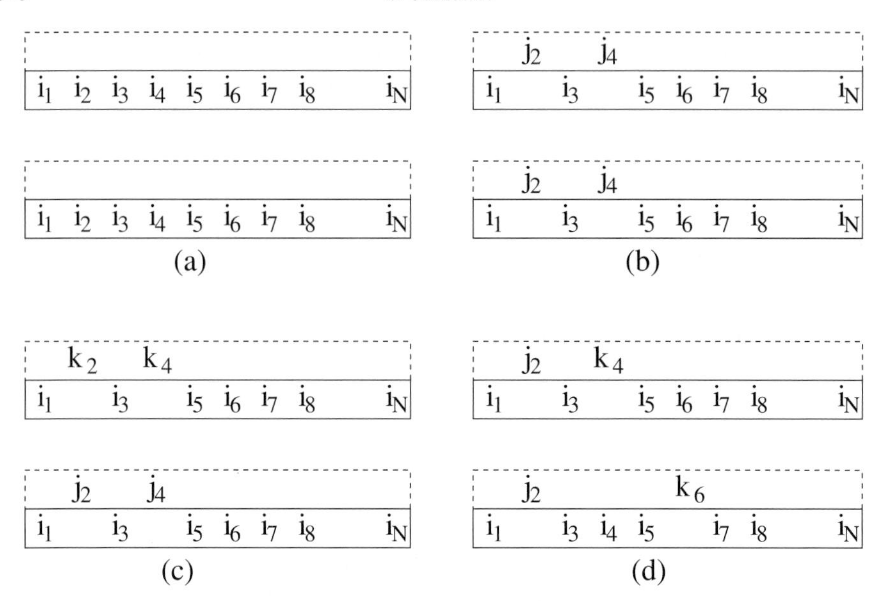

FIG. 2.1. The four different type of matrix elements in a CISD calculation; (a) matrix element of the HF configuration with itself; (b) diagonal elements of the various doubly excited configurations; (c) coupling between configurations where the same two occupied Hartree–Fock orbitals have been substituted by different excited orbitals; (d) coupling between configurations that have one excited orbital in common.

TABLE 2.1
Number of CI matrix elements, the cost to calculate one element and the total cost for the four different types of coupling shown in Fig. 2.1

	Number of elements	Cost per element	Total cost
(a)	1	$N(N-1)$	$O(N^2)$
(b)	$N(N-1)M(M-1)/2$	$N(N-1)$	$O(N^4M^2)$
(c)	$N(N-1)M(M-1)(M-2)(M-3)/4$	1	$O(N^2M^4)$
(d)	$N(N-1)(N-1)(M-1)(M-2)$	1	$O(N^3M^2)$

determinants are equal the element is a sum of $N(N-1)$ terms. Fig. 2.1 shows the four different possibilities that will result in nonzero matrix elements. Table 2.1 shows the number of floating point operations under the assumption that the electron repulsion integrals are precalculated. The dominating term is $O(N^2M^4)$, which gives the postulated sixth power scaling. This conclusions remains also valid if one examines the single excitations in the same way.

Even with these modified CI methods the scaling of the traditional algorithms is still very unfavorable and only very small systems can be treated. This is the reason why independent-electron methods, where the correlations are not taken into account explicitly but in an average way, have become highly popular for calculations of larger systems. The most prominent method in this context is the density functional method within the Kohn–Sham formulation. There the ground state is obtained by minimizing the following energy expression over all mutually orthonormal orbitals $\psi_i(\mathbf{r})$, $(\int \psi_i(\mathbf{r})\psi_j(\mathbf{r})\,d\mathbf{r} = \delta_{i,j})$:

$$E = \int \sum_{i=1}^{N} -\frac{1}{2}\psi_i(\mathbf{r})\nabla^2\psi_i(\mathbf{r}) + V_{\mathrm{H}}(\mathbf{r})\rho(\mathbf{r}) + \varepsilon_{\mathrm{xc}}\big(\rho(\mathbf{r})\big) + V(\mathbf{r})\rho(\mathbf{r})\,d\mathbf{r}, \tag{2.5}$$

where the charge density is given by

$$\rho(\mathbf{r}) = \sum_{i=1}^{N} \big|\psi_i(\mathbf{r})\big|^2. \tag{2.6}$$

The Hartree potential V_{H}

$$V_{\mathrm{H}}(\mathbf{r}) = \frac{1}{2}\int \frac{\rho(\mathbf{r})}{|\mathbf{r}-\mathbf{r}'|}\,d\mathbf{r}' \tag{2.7}$$

represents the classical electrostatic average interaction between the electrons. The so-called exchange correlation energy $\varepsilon_{\mathrm{xc}}$ is the nonclassical part of this interaction. The energy given by the minimizing orbitals is the total energy of the ground state. Since $\varepsilon_{\mathrm{xc}}$ is unknown in practice, the method is approximative. Various rather good forms of $\varepsilon_{\mathrm{xc}}$ are however available today.

There are two basic numerical approaches for obtaining the ground state within the Kohn–Sham formalism. The first is straightforward minimization of the energy expression (Eq. (2.5)) by standard minimization methods. The simplest method of this type is the steepest descent method, more sophisticated methods are the preconditioned conjugate gradient method (PAYNE, TETER, ALLAN, ARIAS and JOANNOPOULOS [1992]) or the DIIS method (HUTTER, LÜTHI and PARRINELLO [1994]). All these methods require the gradient which consists of an unconstrained part plus a correction that takes into account the orthonormality constraints. The unconstrained gradient is given by

$$\frac{\delta E}{\delta\psi_i(\mathbf{r})} = -\nabla^2\psi_k(\mathbf{r}) + 2\big(V_{\mathrm{H}}(\mathbf{r}) + \mu_{\mathrm{xc}}\big(\rho(\mathbf{r})\big) + V(\mathbf{r})\big)\psi_k(\mathbf{r}), \tag{2.8}$$

where $\mu_{\mathrm{xc}}(x) = \delta\varepsilon_{\mathrm{xc}}(x)/\delta\rho(x)$ is the exchange correlation potential. The orthonormality constraints are included by Lagrange multipliers $\lambda_{i,j} = \int \frac{\delta E}{\delta\psi_i(\mathbf{r})}\psi_j(\mathbf{r})\,d\mathbf{r}$ which leads to the effective gradient

$$\frac{\delta E}{\delta\psi_i(\mathbf{r})} - \sum_j \lambda_{i,j}\psi_j(\mathbf{r}). \tag{2.9}$$

Introducing an effective Hamiltonian operator $\mathcal{H}_{\mathrm{eff}}$ (which would be the Fock matrix in a Hartree–Fock calculation)

$$\mathcal{H}_{\mathrm{eff}} = -\frac{1}{2}\nabla^2 + \mu_{\mathrm{xc}}\big(\rho(\mathbf{r})\big) + V(\mathbf{r}), \tag{2.10}$$

the unconstrained gradient becomes

$$\frac{\delta E}{\delta \psi_i(\mathbf{r})} = 2\mathcal{H}_{\text{eff}}\psi_i(\mathbf{r})$$

and the effective gradient

$$2\mathcal{H}_{\text{eff}}\psi_i(\mathbf{r}) - 2\sum_j \lambda_{i,j}\psi_j(\mathbf{r}). \tag{2.11}$$

As in the wavefunction method case we have here again an invariance property that will turn out to be important for linear scaling. The total energy is invariant under unitary transformations of the form

$$\psi_i = \sum_j U_{i,j}\tilde{\psi}_j$$

of the N occupied minimizing orbitals. This is due to the fact that both the kinetic energy and the charge density ρ in Eq. (2.5) are invariant under this transformation. This invariance implies at the same time that the minimizing orbitals have no physical significance, in agreement with the DFT postulate that only the charge density matters.

The second numerical approach to DFT electronic structure calculations is the mixing approach. It consists of a two loop structure. In the inner loop, the total energy of Eq. (2.8) is minimized neglecting the dependence of the exchange correlation and Hartree potential on the wave functions via ρ. This means that one has to find the set of orbitals for which the gradient (Eq. (2.11)) vanishes for a fixed effective Hamiltonian $\mathcal{H}_{\text{eff}}$ constructed using a given ρ

$$H_{\text{eff}}\psi_i(\mathbf{r}) = \sum_j \lambda_{i,j}\psi_j(\mathbf{r}). \tag{2.12}$$

This equation tell us that we have to find an N-dimensional subspace that is invariant under the action of $\mathcal{H}_{\text{eff}}$, i.e., applying $\mathcal{H}_{\text{eff}}$ to one of the vectors spanning this space will give a linear combination of all the vectors spanning the space. A particular set of vectors spanning this space are the eigenvectors, in which case the matrix λ will become diagonal. The Aufbau principle (PARR and YANG [1989]) tells us that for the ground state we have to choose the space spanned by the N eigenvectors going with the N lowest eigenvalues. This inner loop can therefore be replaced by a diagonalization and this is indeed the oldest procedure for doing electronic structure calculations. However calculating the individual eigenvectors gives us more information than required. All we need is the space spanned by these lowest eigenvectors but this space can be spanned by many other sets of vectors.

The outer loop in the mixing approach is the mixing loop. The charge density obtained at the end of the inner loop from the orbitals satisfying Eq. (2.12) is different from the charge density that was used to construct $\mathcal{H}_{\text{eff}}$ when entering the inner loop.

The purpose of the outer mixing loop is now just to bring this difference between the input and output charge density to zero. Numerous methods have been developed to attain this self-consistency. The simplest method, linear mixing, constructs the new input charge density as an average of the old input and output charge density, where the weight of the output charge density is much smaller than the one of the input charge density.

Even though DFT methods are already much faster than wavefunction methods, even faster methods are required for some problems. Not surprisingly, the additional gain in speed has to be paid by a further decrease in accuracy. A method with these characteristics is the semi-empirical tight binding method. It provides us with a recipe how to construct a Hamiltonian matrix H_{TB}, given the atomic positions of a system. The dimension of this matrix is rather small, a few times the number of atoms in the system. For instance, for a silicon or carbon system H_{TB} has 4 degrees of freedom per atom corresponding to the valence s, p_x, p_y and p_z orbitals. The total energy is given by the sum of the lowest occupied eigenvalues of H_{TB} plus a classical inter-atomic interaction. The mathematical problem is thus exactly the same as the one in the inner loop of a self-consistent DFT calculation using the mixing approach. One has to find the space spanned by the lowest eigenvectors of a symmetric Hamiltonian matrix. And as in the case of self-consistent DFT calculations the problem was traditionally solved by diagonalization.

For tight binding calculations the diagonalization was typically done using a full diagonalization based on Householder transformations and the QR algorithm (PRESS, FLANNERY, TEUKOLSKY and VETTERLING [1986]). This approach gives a cubic scaling with respect to the matrix dimension. In the case of self-consistent DFT calculations with systematic basis sets such as plane waves, the matrix dimension is typically much larger than the number of required eigenvectors, since the number of basis functions is much larger than the number of electrons. Under such circumstance iterative diagonalization schemes (DAVIDSON [1975]) are more efficient. However even then some cubic terms remain. Iterative diagonalization is very similar to the direct minimization approach in DFT. Whereas in DFT one has to minimize a rather complicated functional (Eq. (2.5)), the function to be minimized in an iterative diagonalization scheme is a simple quadratic form

$$E = \sum_{i=1}^{N} \int \psi_i(\mathbf{r}) \mathcal{H} \psi_i(\mathbf{r})\, d\mathbf{r}.$$

Expanding $\psi_i(\mathbf{r})$ in terms of M orthogonal basis functions $\chi_j(\mathbf{r})$

$$\psi_i(\mathbf{r}) = \sum_j \mathbf{v}_i^j \chi_j(\mathbf{r})$$

gives

$$E = \sum_{i=1}^{N} \mathbf{v}_i^T H \mathbf{v}_i,$$

where $H_{i,j} = \int \chi_i(\mathbf{r})\mathcal{H}\chi_j(\mathbf{r})\,d\mathbf{r}$ is the Hamiltonian matrix representing the Hamiltonian operator in this basis. As in the DFT case one has an orthonormality constraint, which in its discretized version reads $\mathbf{v}_i^T\mathbf{v}_j = \delta_{i,j}$. The mere implementation of this orthogonality for N vectors of length M gives cubic scaling. Using for instance the well-known Gram–Schmidt scheme requires $4(N(N-1)/2)M$ operations. The $N(N-1)/2$ part comes from the fact that we have pairwise constraints and the factor M from the fact that in order to enforce one constraint we have to operate on vectors of length M.

3. The physical foundations of O(N) methods

Most physical quantities, such as weight, energy or heat capacity are extensive quantities, which means that they grow linearly with system size. We might therefore also hope that the computational effort will scale linearly, even though this might seem unlikely after our discussion of the traditional algorithms. The orthogonality constraint is after all related to the basic fermionic character of the electrons. Nevertheless fundamental chemistry principles strongly suggest that linear scaling is possible. To see this let us consider the two-dimensional model crystal shown in Fig. 3.1. Each atom, denoted by a small disc, forms chemical bonds with its neighbors. Let us now focus on one particular atom such as the one shown in dark black. As we have learned in an introductory chemistry course the nature and the energy of the bonds that this central atom forms with its neighbors depends on three factors. First, the number of nearest neighbors, i.e., the coordination number, second what type of atomic species the neighbor represents, i.e., whether it is for instance a carbon or an oxygen atom and, third, on the bond-length. These three properties are purely local. What is happening outside the sphere centered on the dark black atom has in this elementary picture no influence whatsoever on the bonding. A more profound analysis will show later on that

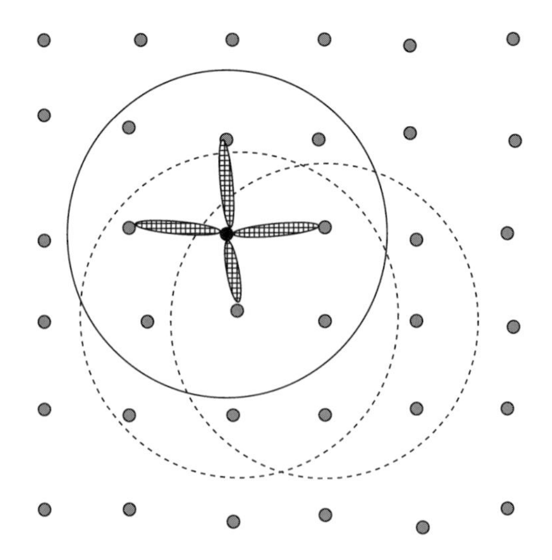

FIG. 3.1. The spherical localization regions associated with three atoms.

features outside the sphere have a certain influence on the bonds within the sphere, but this influence decays very rapidly with distance in an insulator. If one is ready to accept a small error one can therefore indeed neglect the region outside a sufficiently large sphere.

Two important conclusions can be drawn from these simple considerations. First, linear scaling of the computational effort is achievable. This comes from the fact that an electronic structure calculation in an insulator is more or less nothing else than the determination of the different bond energies. To calculate the energy of one bond, one has to take into account only a relatively small region around this bond. As the system grows, there are more bonds, but they are all embedded in spheres of the same size to do the calculation. Two more spheres are shown by dotted lines in Fig. 3.1. The total energy of the system is then obtained by simply summing all the partial bond energies. Hence the calculation scales linearly with respect to the number of bonds. The second conclusion is, that this kind of divide and conquer approach (YANG [1991]) that will result in linear scaling is only worthwhile if the system is larger than the volume of the sphere. If not, it will be more efficient to do a single electronic structure calculation, that will give the sum of all the bond energies.

Unfortunately the traditional algorithms for electronic structure calculations can not exploit this basic locality principle that was just described. In order to obtain linear scaling one has to reformulate the electronic structure problem in terms of quantities that reflect this locality and then find algorithms for calculating these new quantities. In the context of independent electron theories appropriate localized quantities are the density matrix and the Wannier functions. They are by now well characterized and will be described in the following section. In the context of wavefunction methods, the theoretical basis is much less sound. The present status will be summarized afterwards.

3.1. Locality principle for independent electrons

In this section we are interested in the decay properties of the Wannier functions and of the density matrix operator $\mathcal{F}$ for very large systems. The density matrix F is the discretized version of the operator $\mathcal{F}$

$$F_{i,j} = \int \chi_i(\mathbf{r}) \mathcal{F} \chi_j(\mathbf{r})\, d\mathbf{r}. \tag{3.1}$$

The simplest way to treat such large systems is within the context of periodic solids. There, Bloch's theorem helps us with analytical work. The general assumption is that anything that can be demonstrated for a periodic solid will also hold for a sufficiently large nonperiodic system.

The independent particle density matrix F can be defined as a matrix functional of the Hamiltonian H

$$F = f(H). \tag{3.2}$$

The function f relating the two is the Fermi distribution

$$f(x) = \frac{1}{1 + e^{(x-\mu)/(k_B T)}}, \tag{3.3}$$

where μ is the chemical potential, T the temperature and k_B Boltzmann's constant. This functional relation implies that H and F have the same eigenvectors $\mathbf{v}_i$. If

$$H\mathbf{v}_i = \varepsilon_i \mathbf{v}_i \tag{3.4}$$

then

$$F\mathbf{v}_i = n_i \mathbf{v}_i = f(\varepsilon_i)\mathbf{v}_i. \tag{3.5}$$

The eigenvalues n_i are called occupation numbers and they are obtained by evaluating the Fermi distribution at the eigenenergies ε_i of H. Since F is a hermitian matrix, it can be written down in a spectral representation,

$$F = \sum_i \mathbf{v}_i^H n_i \mathbf{v}_i, \tag{3.6}$$

where the sum runs over all the eigenvalues/vectors. The corresponding representation for the density operator is obviously

$$\mathcal{F}(\mathbf{r}, \mathbf{r}') = \sum_i \psi_i^*(\mathbf{r}) n_i \psi_i(\mathbf{r}'). \tag{3.7}$$

At zero temperature the Fermi distribution becomes a heavy-side function, being equal to one for all the states lying below the chemical potential and zero for all the states above. Consequently, in a spectral representation only the so-called occupied states below the chemical potential have to be included

$$F = \sum_{i=1}^{N} \mathbf{v}_i^H \mathbf{v}_i. \tag{3.8}$$

From the orthonormality of the eigenvectors of F it follows that

$$F^2 = F \tag{3.9}$$

which means that the zero temperature density matrix is a projection operator which projects onto the space of the occupied one-particle wave functions. As was pointed out when discussing Eq. (2.12) finding this space is what non-self-consistent electronic structure calculations at zero temperature are all about. Knowing F one has therefore the complete information about the quantum mechanical system. One can either create a set of vectors that will span the space by applying F to N linearly independent vectors or calculate the total energy directly from the density matrix in a discretized version. The latter approach requires the calculation of traces. The band-structure energy E_{BS} which would in a traditional algorithm be calculated from the eigenvalues

$$E_{\mathrm{BS}} = \sum_{i=1}^{N} \varepsilon_i$$

is for instance given by

$$E_{\mathrm{BS}} = \mathrm{Tr}[FH] \tag{3.10}$$

as can be verified immediately by calculating the trace in the eigenvector basis,

$$\mathrm{Tr}[FH] = \sum_i f(\varepsilon_i)\varepsilon_i, \tag{3.11}$$

which reduces to Eq. (3.10) at zero temperature. The other important quantity, the charge density is given by

$$\rho(\mathbf{r}) = \sum_{i,j} F_{i,j}\chi_i(\mathbf{r})\chi_j(\mathbf{r}).$$

Let us now come to the periodic case. We consider a crystal formed out of a primitive unit cell whose primitive vectors are $\mathbf{a}_1$, $\mathbf{a}_2$ and $\mathbf{a}_3$. By repeating this primitive cell n_1, n_2 and n_3 times along the three direction $\mathbf{a}_1$, $\mathbf{a}_2$ and $\mathbf{a}_3$ we form a periodic volume. It is called periodic volume because we request that all the wave functions be periodic with respect to this volume, i.e.,

$$\psi_{j,\mathbf{k}}(\mathbf{r} + ln_1\mathbf{a}_1 + mn_2\mathbf{a}_2 + nn_3\mathbf{a}_3) = \psi_{j,\mathbf{k}}(\mathbf{r})$$

for any l, m, n. Bloch's theorem (ASHCROFT and MERMIN [1976]) then tells us that the wave functions have the form

$$\psi_{j,\mathbf{k}}(\mathbf{r}) = e^{i\mathbf{kr}}u_{j,\mathbf{k}}(\mathbf{r}), \tag{3.12}$$

where $\mathbf{k}$ is the Bloch wave vector. It takes on the values

$$\mathbf{k} = \frac{m_1}{n_1}\mathbf{b}_1 + \frac{m_2}{n_2}\mathbf{b}_2 + \frac{m_3}{n_3}\mathbf{b}_3, \tag{3.13}$$

where the reciprocal lattice vectors are given by

$$\mathbf{b}_1 = 2\pi\frac{\mathbf{a}_2 \times \mathbf{a}_2}{\mathbf{a}_1 \cdot (\mathbf{a}_2 \times \mathbf{a}_2)},$$

$$\mathbf{b}_2 = 2\pi\frac{\mathbf{a}_3 \times \mathbf{a}_1}{\mathbf{a}_1 \cdot (\mathbf{a}_2 \times \mathbf{a}_3)},$$

$$\mathbf{b}_3 = 2\pi \frac{\mathbf{a}_1 \times \mathbf{a}_2}{\mathbf{a}_1 \cdot (\mathbf{a}_2 \times \mathbf{a}_3)}.$$

The function $u_{j,\mathbf{k}}(\mathbf{r})$ is periodic with respect to the primitive unit cell, i.e.,

$$u_{j,\mathbf{k}}(\mathbf{r} + l\mathbf{a}_1 + m\mathbf{a}_2 + n\mathbf{a}_3) = u_{j,\mathbf{k}}(\mathbf{r})$$

for any l, m, n and it can be determined by solving the following modified Schrödinger equation within the primitive cell

$$-\frac{1}{2}(\nabla + \mathbf{k})^2 u_{j,\mathbf{k}}(\mathbf{r}) = \varepsilon_{j,\mathbf{k}} u_{j,\mathbf{k}}(\mathbf{r}). \tag{3.14}$$

We note that the operator $(\nabla + \mathbf{k})^2$ is analytic with respect to $\mathbf{k}$. The $T = 0$ density operator for a periodic crystal is consequently given by (cf. Eqs. (3.12), (3.13), (3.7), (3.8))

$$\begin{aligned}\mathcal{F}(\mathbf{r}, \mathbf{r}') &= \sum_j \sum_{m_1=1}^{n_1} \sum_{m_2=1}^{n_2} \sum_{m_3=1}^{n_3} \psi_{j,\mathbf{k}}(\mathbf{r}) \psi^*_{j,\mathbf{k}}(\mathbf{r}') \\ &= \sum_j \sum_{m_1=1}^{n_1} \sum_{m_2=1}^{n_2} \sum_{m_3=1}^{n_3} u_{j,\mathbf{k}}(\mathbf{r}) u^*_{j,\mathbf{k}}(\mathbf{r}') e^{i\mathbf{k}(\mathbf{r}-\mathbf{r}')}.\end{aligned} \tag{3.15}$$

The Bloch vector $\mathbf{k}$ depends on m_1, m_2, m_3 through Eq. (3.13) and is contained within the Brillouin zone, which is the primitive cell of the reciprocal lattice. As the periodic volume grows towards infinity, the Bloch vectors become dense and the three spatial sums in Eq. (3.15) can be replaced by an integral over the Brillouin zone,

$$\mathcal{F}(\mathbf{r}, \mathbf{r}') = \frac{V}{(2\pi)^3} \sum_j \int d\mathbf{k}\, u_{j,\mathbf{k}}(\mathbf{r}) u^*_{j,\mathbf{k}}(\mathbf{r}') e^{i\mathbf{k}(\mathbf{r}-\mathbf{r}')}, \tag{3.16}$$

where V is the volume of the primitive cell. Since the dependence of the $\mathbf{k}$-dependent Hamiltonian on $\mathbf{k}$ in Eq. (3.14) is analytic one might hope that $u_{j,\mathbf{k}}(\mathbf{r}) u^*_{j,\mathbf{k}}(\mathbf{r}')$ is analytic as well. This is indeed the case for an isolated band j

$$\varepsilon_{j-1,\mathbf{k}} < \varepsilon_{j,\mathbf{k}} < \varepsilon_{j+1,\mathbf{k}}$$

or an isolated group of bands as was proven by DES CLOIZEAUX [1964]. Since $\mathcal{F}$ is the Fourier transform of an analytic quantity, it follows from the Paley–Wiener theorem that it decays exponentially. Since an isolated group of bands physically corresponds to an insulator we have the final result that $\mathcal{F}$ decays asymptotically exponentially, i.e., faster than any power law

$$\mathcal{F}(\mathbf{r}, \mathbf{r}') \asymp e^{-\kappa|\mathbf{r}-\mathbf{r}'|}. \tag{3.17}$$

Numerical calculations of $\mathcal{F}$ show an oscillatory behavior with an exponentially decaying amplitude. Both the wavelength of the oscillatory part and the decay length $1/\kappa$ are of the order of the inter-atomic distances. This decay is a manifestation of the locality principle in chemistry illustrated in the previous section. What is happening at a source point $\mathbf{r}$ has practically no influence at the observation point $\mathbf{r}'$ if both points are separated by a distance larger than the decay length. This statement follows from the fact that $\mathcal{F}$ completely describes our quantum mechanical system.

This strong locality principle does not hold true in a metal. For the free electron gas at $T = 0$ one can easily calculate the density operator to obtain

$$F(\mathbf{r}, \mathbf{r}') \propto \frac{\cos(k_{\mathrm{F}}|\mathbf{r} - \mathbf{r}'|)}{|\mathbf{r} - \mathbf{r}'|^2},$$

where the Fermi wave vector k_{F} is related to the electronic density (electrons/volume) by $N/V = k_{\mathrm{F}}^3/(3\pi^2)$. Being related to the discontinuity of the occupation numbers at the Fermi surface, this much slower algebraic decay is expected to hold not only for the free electron gas but also for any real metal.

Things change entirely for a metal at finite temperature. Since the Fermi surface is smoothed out one again obtains an exponential decay (GOEDECKER [1998])

$$F(\mathbf{r}, \mathbf{r}') \propto \frac{\cos(k_{\mathrm{F}}|\mathbf{r} - \mathbf{r}'|)}{|\mathbf{r} - \mathbf{r}'|^2} \exp\left(-\frac{k_{\mathrm{B}}T}{k_{\mathrm{F}}}|\mathbf{r} - \mathbf{r}'|\right). \tag{3.18}$$

Eq. (3.18) is also applicable to insulators in the case where $k_{\mathrm{B}}T$ is larger than the gap.

The second quantity that reflects the locality principle of chemistry and that can be used as the basic quantity in an independent particle electronic structure calculation are the Wannier functions $w_n(\mathbf{r} - \mathbf{R})$, where $\mathbf{R}$ points to the origin of one of the primitive cells of the periodic volume. The Wannier function are defined for an infinite periodic crystal as the Fourier transform of the Bloch functions

$$w_j(\mathbf{r} - \mathbf{R}) = \frac{V}{(2\pi)^3} \int d\mathbf{k}\, e^{-i\mathbf{k}\mathbf{R}} \psi_{j,\mathbf{k}}(\mathbf{r}). \tag{3.19}$$

Since the Fourier transformation is a unitary transformation the Wannier functions are again an orthogonal set of functions.

$$\int d\mathbf{r}\, w_j(\mathbf{r} - \mathbf{R}) w_{j'}(\mathbf{r} - \mathbf{R}') = \delta_{j,j'} \delta_{\mathbf{R},\mathbf{R}'}.$$

Wannier functions are not uniquely defined. Adding an arbitrary $\mathbf{k}$ dependent phase factor

$$\psi_{j,\mathbf{k}}(\mathbf{r}) \to \psi_{j,\mathbf{k}}(\mathbf{r}) e^{i\omega(\mathbf{k})}$$

gives another set of valid Bloch functions but changes the resulting Wannier functions. Another arbitrariness is encountered when one has nonisolated bands where on can mix the Block functions of the bands. In spite of the arbitrariness concerning the phase, NENCIU [1983] was able to demonstrate that one can find analytic and periodic Bloch functions for isolated bands. It then follows again from the Paley–Wiener theorem that the Wannier functions are exponentially localized, i.e.,

$$w_n(\mathbf{r}-\mathbf{R}) \asymp e^{-\kappa|\mathbf{r}-\mathbf{R}|}. \tag{3.20}$$

For the special cases of one-dimensional crystals and crystals with inversion symmetry the exponential localization has already been demonstrated much earlier by KOHN [1959] and DES CLOIZEAUX [1964]. HE and VANDERBILT [2001] have recently shown that the decay rate κ in Eq. (3.20) is the same as in Eq. (3.17). Despite several studies (ISMAIL-BEIGI and ARIAS [1999]) no simple rule has emerged to estimate κ for a real material. Experience tells us however that $1/\kappa$ is of the order of the inter-atomic distance.

The arbitrariness of the Wannier functions can be eliminated by determining maximally localized Wannier functions (MARZARI and VANDERBILT [1997]) which minimize the spread

$$\sum_{i=1}^{N}\left[\left\langle r^2\right\rangle_i - \langle\mathbf{r}\rangle_i^2\right].$$

This work is a generalization of the well-established Boys localization concept (BOYS [1950]) for nonperiodic systems to periodic ones. The problem to be overcome in periodic systems is that the conventional definition of the center

$$\langle\mathbf{r}\rangle_i = \int d\mathbf{r}\, \psi_i^*(\mathbf{r})\mathbf{r}\psi_i(\mathbf{r})$$

and of the spread

$$\left\langle r^2\right\rangle_i = \int d\mathbf{r}\, \psi_i^*(\mathbf{r})\mathbf{r}^2\psi_i(\mathbf{r})$$

is not valid in the periodic case because $\mathbf{r}$ and $\mathbf{r}^2$ are not bounded when arbitrary wave functions ψ_i are used. If, however, exponentially decaying Wannier functions are used the expectation values $\langle\mathbf{r}\rangle$ and $\langle\mathbf{r}^2\rangle$ are well defined. A calculation of maximally localized Wannier functions thus gives us both their centers $\mathbf{R}$, which are in general not lattice vectors, and their spread.

Numerical calculations of maximally localized Wannier functions show that they represent bonds or lone electron pairs. They are centered where one would expect bond-centers and their extension is of the order of the inter-atomic distance. Because of these properties, they have become a very valuable tool for understanding the electronic structure of various materials.

The existence of exponentially localized Wannier functions implies that within a independent particle scheme there are no nonclassical long range interactions (or more precisely only vanishingly small long range interactions) between fragments of a big system separated by a large distance. The only long range interaction is the classical electrostatic interaction. This can easily be verified by examining the total energy expressions in a Wannier basis either within Hartree–Fock or the local density DFT scheme. In particular if both fragments are neutral there is no long range interaction. The Van der Waals interactions are thus missing in independent particle schemes.

3.2. Locality principle for wavefunction methods

The usefulness of the density matrix in the independent particle context stems from the fact that it specifies the system completely and without ambiguity. The one particle reduced density matrix $D_1(\mathbf{r}_1, \mathbf{r}_1')$

$$D_1(\mathbf{r}_1, \mathbf{r}_1') = \int d\mathbf{r}_2 \cdots d\mathbf{r}_\mathbf{N}\, \Psi(\mathbf{r}_1, \mathbf{r}_2, \ldots, \mathbf{r}_\mathbf{N}) \Psi(\mathbf{r}_1', \mathbf{r}_2, \ldots, \mathbf{r}_\mathbf{N}) \tag{3.21}$$

is the generalization of the zero temperature independent particle density matrix in the many body context and it reduces to the former one in the case where the wave function consists only of a single determinant. In contrast to its independent particle counterpart, it does not entirely specify an interacting many electron system. Because of the two electron interactions, the two particle density matrix is required as well. In addition there is at present no method to calculate these two density matrices directly (COLEMAN and YUKALOV [2000]). They can only be obtained a posteriori from the many electron wave function.

The road to linear scaling is therefore based on some optimized configuration interaction expansion of the type of Eq. (2.2). Guided by their chemical intuition, which told them that correlation is a local effect, PULAY and SAEBØ [1986], SINANOGLU [1964] and many others realized a long time ago that most of the coefficients in Eq. (2.2) are vanishingly small if a set of localized one particle orbitals ϕ_i is used to construct the determinants. Their arguments can be put on a mathematically sounder basis by introducing natural Wannier functions (KOCH and GOEDECKER [2001]), which are a generalization of the ordinary Wannier functions to the many body case. This will be done in the following.

Among all the Hilbert spaces of finite dimension, the one spanned by the lowest M natural orbitals has the largest overlap with the exact many electron wave function (DAVIDSON [1976]), contained in a Hilbert space of infinite dimension. So mathematically we have here an optimality with respect to the overlap and not with respect to the physically more important energy. But calculations show that the space is also nearly optimal from an energetical point of view. The natural orbitals $\gamma_i(\mathbf{r})$ are defined as the eigenfunctions of the one particle reduced density matrix defined in Eq. (3.21),

$$\int d\mathbf{r}'\, D_1(\mathbf{r}, \mathbf{r}')\gamma_i(\mathbf{r}') = n_i \gamma_i(\mathbf{r}), \tag{3.22}$$

where n_i are the natural occupation numbers ordered in decreasing order. The lowest natural occupation numbers are nearly one and the highest tend to zero. The fact that they have to be contained in the interval [0, 1] comes from the fact that $D_1(\mathbf{r},\mathbf{r}')$ was obtained from an antisymmetric wave function (COLEMAN [1963]). Since $D_1(\mathbf{r},\mathbf{r}')$ is a hermitian matrix the natural orbitals form an orthonormal set.

The natural orbitals define only an optimal space. As was pointed out before exactly the same Hilbert space is spanned by a new set that is related to the original set by a unitary transformation. In particular one might hope that by such a transformation one might concentrate the weight of the coefficients C_I for a ground state wave function of the type of Eq. (2.2) into a few configurations which would allow to discard the majority of the other coefficients since they would be very small. The needed transformation will turn out to be the transformation to the natural Wannier functions.

As in the independent particle case we will now again consider a periodic solid. Since the one-particle density matrix has the same translational invariance properties as the independent particle Hamiltonian, the natural orbitals are Bloch functions. Consequently, the natural orbitals can be written as a product

$$\gamma_{j,\mathbf{k}}(\mathbf{r}) = e^{i\mathbf{k}\mathbf{r}} U_{j,\mathbf{k}}(\mathbf{r}), \tag{3.23}$$

where the part $U_{n,\mathbf{k}}(\mathbf{r})$ that is periodic with respect to the primitive cell is an eigenfunction of the $\mathbf{k}$ dependent one-particle density matrix $D(\mathbf{k};\mathbf{r},\mathbf{r}')$

$$\int d\mathbf{r}'\, D(\mathbf{k};\mathbf{r},\mathbf{r}') U_{j,\mathbf{k}}(\mathbf{r}') = n_j(\mathbf{k}) U_{j,\mathbf{k}}(\mathbf{r}). \tag{3.24}$$

We now postulate that $D(\mathbf{k};\mathbf{r},\mathbf{r}')$ is analytic with respect to $\mathbf{k}$ for an insulator and that it has a discontinuity for a normal metal. This postulate is physically highly plausible since many experimental characterizations of metals are based on some kind of discontinuity at the Fermi surface whereas no such discontinuity seems to exist for an insulator. The functions $U_{j,\mathbf{k}}(\mathbf{r})$ can be chosen to be analytic with respect to $\mathbf{k}$ for isolated occupation bands $n_i(\mathbf{k})$ (KOCH and GOEDECKER [2001]). It then follows again by the Paley–Wiener theorem that the natural Wannier functions defined by

$$W_j(\mathbf{r}-\mathbf{R}) = \frac{V}{(2\pi)^3} \int d\mathbf{k}\, e^{-i\mathbf{k}\mathbf{R}} \omega_{j,\mathbf{k}}(\mathbf{r}) \tag{3.25}$$

decay exponentially. We have thus found a set of orthogonal exponentially localized orbitals that span the Hilbert space associated with an arbitrary number of occupation bands of an insulator. Note the close analogy between the independent particle and the interacting case. Eq. (3.25) corresponds to Eq. (3.19), Eq. (3.24) to Eq. (3.14) and Eq. (3.23) to Eq. (3.12).

We will now proceed to show that the locality properties of an interacting system are limited by the Van der Waals interactions. Since we want to start with a reference wave function that has no long range interactions, we will take the Hartree–Fock wave function as the point of departure. The Wannier functions obtained form the

Hartree–Fock solution will be very similar to the lowest natural Wannier functions but not identical. We will now introduce a new set of exponentially localized functions $\omega_i(\mathbf{r})$ that are defined in the following way. The N lowest ω_i's are identical to the Hartree–Fock–Wannier functions. The higher ω_i's are obtained by orthogonalizing the natural Wannier functions to the N lowest ω_i's. If this orthogonalization is done by a Löwdin orthogonalization, the high ω_i's will be very similar to the high natural Wannier functions.

We will first examine the decay properties of the many body Hamiltonian in the Hilbert space spanned by the ω_i's. The matrix elements between a configuration $I = (i_1, \ldots, i_N)$ and $J = (j_1, \ldots, j_N)$ are given by the Slater–Condon rules. If the matrices differ by more than two indices in the set the matrix element is zero. For the case considered here, where the lowest N ω_i's are the Hartree–Fock orbitals, Brillouin's theorem (SZABO and OSTLUND [1982]) tell us that the matrix elements between the ground state and excited states that differ by one orbital vanish as well. So we have to consider only determinants that differ by two orbital indices. Without restriction we can assume that they differ by the first two indices, i.e., $I = (i_1, \ldots, i_N)$ and $J = (j_1, j_2, i_3, \ldots, j_N)$. Shifting one differing index to another position will only change the sign of the matrix element $\Omega_{i_1,i_2}^{j_1,j_2}$ given by

$$\Omega_{i_1,i_2}^{j_1,j_2} = \iint d\mathbf{r}\, d\mathbf{r}' \left(\frac{\omega_{i_1}(\mathbf{r})\omega_{j_1}(\mathbf{r})\omega_{i_2}(\mathbf{r}')\omega_{j_2}(\mathbf{r}')}{|\mathbf{r}-\mathbf{r}'|} - \frac{\omega_{i_1}(\mathbf{r})\omega_{j_2}(\mathbf{r})\omega_{i_2}(\mathbf{r}')\omega_{j_1}(\mathbf{r}')}{|\mathbf{r}-\mathbf{r}'|} \right). \tag{3.26}$$

The first term is a Coulombic term where a charge density that is the product of the annihilated orbital with the created orbital of the first excitation is interacting with the same charge density of the second excitation. The second term is an exchange like term where the annihilated orbital of the first excitation and the created orbital of the second excitation are forming a charge density that is interacting with the corresponding charge density formed by permuting excitation one and two. Let us denote the centers of the orbitals ω_{i_1} to ω_{j_2} by $\mathbf{R}_{i_1}, \mathbf{R}_{i_2}, \mathbf{R}_{j_1}$ and $\mathbf{R}_{j_2}$. The decay properties of the Coulomb and exchange part of $\Omega_{i_1,i_2}^{j_1,j_2}$ are summarized in Fig. 3.2.

The exponential decay of the Coulomb part with respect to the separation $|\mathbf{R}_{i_1} - \mathbf{R}_{j_1}|$ and $|\mathbf{R}_{i_2} - \mathbf{R}_{j_2}|$ comes from the exponential decay of the localized orbitals. The analogous behavior is found for the exchange part. The interaction of the two centers of the excitation $s = |\frac{1}{2}(\mathbf{R}_{i_1} + \mathbf{R}_{j_1}) - \frac{1}{2}(\mathbf{R}_{i_2} + \mathbf{R}_{j_2})|$ in the coulomb part and $s = |\frac{1}{2}(\mathbf{R}_{i_1} + \mathbf{R}_{j_2}) - \frac{1}{2}(\mathbf{R}_{i_2} + \mathbf{R}_{j_1})|$ in the exchange part is electrostatic. By virtue of the orthogonality relations none of the two interacting charges $\omega_{i_1}(\mathbf{r})\omega_{j_1}(\mathbf{r})$ and $\omega_{i_2}(\mathbf{r})\omega_{j_2}(\mathbf{r})$ can have a monopole. The longest range interaction to be found is therefore a dipole interaction that decays as $1/s^3$. This asymptotic decay will be approached whenever the overlap between the two charge distributions is small which is typically the case if the separation s is significantly larger than the inter-atomic distance.

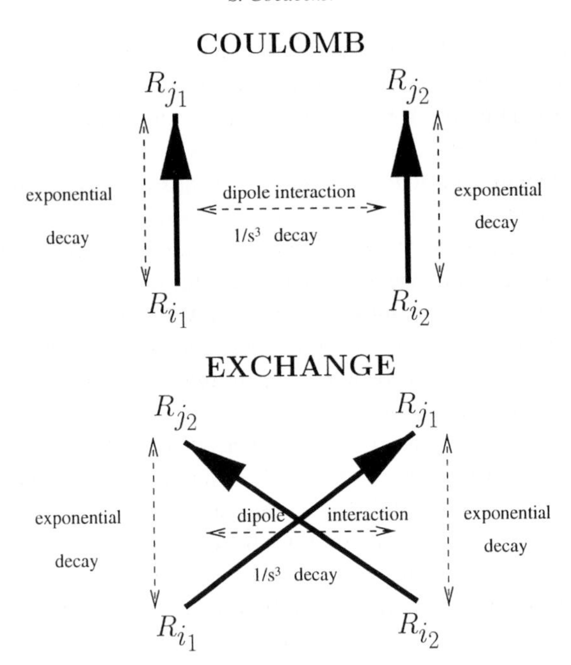

FIG. 3.2. Visualization of the decay properties of the integral $\Omega_{i_1,i_2}^{j_1,j_2}$ of Eq. (3.26).

Using perturbation theory we can now calculate the CI coefficients C_I for these double excited configurations (AYALA and SCUSERIA [1999])

$$C_{j_1,j_2,i_3,i_4,\ldots,i_N} = -\frac{\Omega_{i_1,i_2}^{j_1,j_2}}{E_0 - E_{i_1,i_2}^{j_1,j_2}}, \tag{3.27}$$

where E_0 is the Hartree–Fock ground state energy and $E_{i_1,i_2}^{j_1,j_2}$ the total energy of the excited determinant. Note that we would obtain the Möller–Plesset expression (SZABO and OSTLUND [1982]) if we had used the extended eigenfunctions of the Fock matrix instead of the localized functions ω_i obtained from the former by a unitary transformation. In this extended basis set the energy difference $E_0 - E_{i_1,i_2}^{j_1,j_2}$ is given by $\varepsilon_{i_1} - \varepsilon_{j_1} + \varepsilon_{i_2} - \varepsilon_{j_2}$. The lowest order energy contribution of the doubly excited configuration is then

$$\frac{(\Omega_{i_1,i_2}^{j_1,j_2})^2}{E_0 - E_{i_1,i_2}^{j_1,j_2}} \tag{3.28}$$

which decays as $1/s^6$ and represents a Van der Waals interaction. If we are willing to accept a result with a certain error bar, we can now discard all the negligibly small or zero contributions in the sum of Eq. (2.2). Since the decay the coefficients

$C_{j_1,j_2,i_3,i_4,\dots,i_N}$ in Eq. (3.27) is a reasonably fast with respect to the distance among the orbitals involved (Fig. 3.2), we have to keep only a small number of them and this number grows only linearly with system size. The contributions we retain arise from orbitals that are all located in a certain spatial domain and the size of this domain is independent of the number of atoms in our system if the system is sufficiently big, i.e., bigger than the domain.

Let us stress again, that the entire sequence of arguments is only valid for insulators. First, in a metal we are not granted to have exponentially localized basis functions. Second, because the basis functions may be extended, thinking in terms of separated interacting dipoles may not be justified. Finally nondegenerate perturbation theory is not applicable for a metal where the energy separation between the ground state and the excited states is vanishingly small.

In addition the arguments are only valid for insulators where second order Möller–Plesset (MP2) perturbation theory gives a qualitatively correct description of the system. The argumentation therefore would clearly break down for a strongly correlated system.

In the case of a three-dimensional solid we have to take into account that we have to sum all the Van der Waals contributions contained within a sphere with cutoff Radius r_{cut}. Hence the Van der Waals energy contribution decreases like r_{cut}^{-3}. Such a decay is not very slow. It comes to our rescue that the van der Waals energy corrections themselves are very small and that one can therefore hopefully nevertheless use a reasonably small r_{cut}. But clearly the number of doubly excited configurations within a domain is larger in a true three-dimensional solid than in a one-dimensional polymer, for example. As a consequence the prefactor of an $O(N)$ calculation for a solid will be larger than for a polymer, independently of the details of the algorithm that will be used.

The motivation for the construction of the orthogonal localized set $\omega_i(\mathbf{r})$ was the study of the localization properties in the true many body context. Unfortunately the ω_i's can not be used in an algorithm, because they are unknown at the beginning of the calculation. To construct them we need the one-particle reduced density matrix which is calculated from the solution, namely from the many determinant wave function. One can however construct other localized orbitals that will give rise to linear scaling. According to our discussion it will be necessary that the orbitals are exponentially localized and that the occupied and unoccupied spaces are separated by an energy gap. Using simply the atomic Gaussian type basis functions as orbitals would for instance not necessarily satisfy the latter condition. A localized basis set that is widely used in $O(N)$ quantum chemistry calculations (PULAY and SAEBØ [1986]) consists of orthogonal localized orbitals that were formed by unitary transformations (PIPEK and MEZEY [1989]) of the Hartree–Fock orbitals for the occupied space and of nonorthogonal Gaussian type basis functions that are projected onto the virtual Hartree–Fock space. The basis functions for the virtual space are chosen to be nonorthogonal because it turns out to be impossible to construct well localized orthogonal ones. This difficulty can be understood in the following way. In a periodic solid within the independent particle picture such as Hartree–Fock, the Hamiltonian matrix elements between Wannier functions separated by a lattice vector $\mathbf{R}$ are the Fourier components $\varepsilon_j(\mathbf{R})$ of the independent particle band structure $\varepsilon_j(\mathbf{k})$

$$\varepsilon_j(\mathbf{R}) = \frac{V}{(2\pi)^3} \int_{\mathrm{BZ}} \varepsilon_j(\mathbf{k}) e^{-i\mathbf{k}\mathbf{R}}\, d\mathbf{k}$$
$$= \frac{(2\pi)^3}{V} \int_{\mathrm{space}} W_j(\mathbf{r}') H W_j(\mathbf{r}' - \mathbf{R})\, d\mathbf{r}. \tag{3.29}$$

For high bands j, $\varepsilon_j(\mathbf{k})$ is nearly free electron like over most of the Brillouin zone, but close to the border of the Brillouin zone it has to bent over abruptly to satisfy periodicity. This abrupt bending over induces high Fourier components of large amplitude, which by Eq. (3.29) implies that the Wannier functions associated with these high bands are extended.

The difference between these kinds of localized basis sets used in practice and the $\omega_i(\mathbf{r})$ basis set is the degree of optimality. To recover a certain fraction of the correlation energy the dimension of the virtual space spanned by the ω_i's can be smaller than for any other basis set. A possible way to come closer to optimality, that has, however, not yet been explored in practice, is to construct approximative natural orbitals by using an approximate natural orbital functional (GOEDECKER and UMRIGAR [1998]) as the starting point of the wavefunction method. The natural Wannier functions constructed from the virtual natural orbitals are expected to be as well localized as the ones constructed from the occupied orbitals.

4. Obtaining sparsity from locality

Up to now we have discussed the locality properties of the density operator and the Wannier functions. The tacit assumption was that the locality of these continuous physical quantities will lead to some sparseness of their discretized counterparts, the density matrix and the vector representing a Wannier function with respect to a certain basis set. It is this sparsity that will eventually lead to a number of operations that is linear with respect to systems size. A necessary condition for obtaining sparsity from locality is evidently that the basis set be localized. In the best possible case the basis set is strictly localized which means that it is exactly zero outside a certain volume. If the basis set is in addition orthogonal the sparsity is granted as follows from the discretization prescription equation (3.1). Unfortunately, strictly localized orthogonal basis sets were deemed to be impossible until very recently when Daubechies came up with an orthogonal wavelet family with compact support (DAUBECHIES [1992]). These wavelets are not standard basis sets for electronic structure calculations. The most popular localized basis sets for electronic structure calculations are Gaussians that are not orthogonal. For nonorthogonal basis functions the discretization formula is given by

$$F_{i,j} = \int \widetilde{\chi}_i(\mathbf{r}) \mathcal{F} \widetilde{\chi}_j(\mathbf{r})\, d\mathbf{r}, \tag{4.1}$$

where $\widetilde{\chi}_j(\mathbf{r})$ is the dual of $\chi_j(\mathbf{r})$ given by

$$\widetilde{\chi}_j(\mathbf{r}) = \sum_j S^{-1}_{i,j} \chi_j(\mathbf{r})$$

and where S^{-1} is the inverse of the overlap matrix S

$$S_{i,j} = \int \chi_i(\mathbf{r}) \chi_j(\mathbf{r})\, d\mathbf{r}.$$

If S has small eigenvalues, the duals are poorly localized. As a consequence $F_{i,j}$ is not any more well localized even though $F(\mathbf{r}, \mathbf{r}')$ is.

This effect, which has up to now not been studied with mathematical rigor, is numerically illustrated in Fig. 4.1. By adjusting a simple sinusoidal potential, a one-dimensional periodic model solid was constructed with a band gap of 10 atomic units and a width of the occupied band of 1.5 a.u. These values insure that $F(\mathbf{r}, \mathbf{r}')$ decays rapidly. The occupied eigenstates of the lowest band were then calculated in a basis of equally spaced Gaussians, using 6 Gaussians per atom. The width of the Gaussians was then varied from 0.15 to 0.20 and 0.25 times the lattice constant. The width 0.20 gave an energy that was a few 10^{-4} atomic units lower than for the two neighboring values. The lowest eigenvalue of the overlap matrix for the three widths were 2×10^{-3}, 6×10^{-6} and 2×10^{-9}. Fig. 4.1 shows $F(\mathbf{r}, \mathbf{r}')$ and $F_{i,j}$ for the three cases. As is to be expected, the $F(\mathbf{r}, \mathbf{r}')$'s that were obtained from the three different basis sets are quasi identical

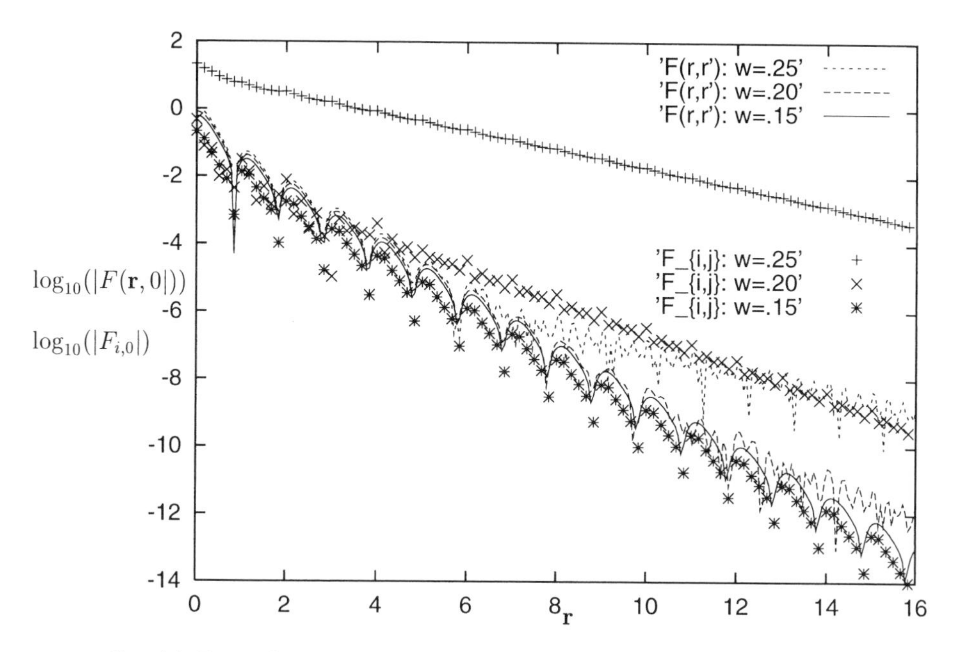

FIG. 4.1. Decay of the density operator and density matrix for nonorthogonal basis sets.

near the origin where they take on large values. However in the tail, $F(\mathbf{r}, \mathbf{r}')$ is not any more correctly described in the cases where the overlap matrix is close to being singular. For the 0.25 case the deviations start already around $r = 6$, for 0.20 only around $r = 12$. $F_{i,j}$ reflects the behavior of $F(\mathbf{r}, \mathbf{r}')$ only in the 0.15 case. For 0.20 $F_{i,j}$ tracks $F(\mathbf{r}, \mathbf{r}')$ only over a small interval and goes than over into a much slower exponential decay. For 0.25 $F_{i,j}$ is completely different form $F(\mathbf{r}, \mathbf{r}')$, being several orders of magnitude larger and decaying very slowly. Evidently, it would be very difficult to obtain some exploitable sparsity of the density matrix for the 0.25 case.

Exactly the same problems arises if one tries to represent a well localized Wannier function with a nearly overcomplete nonorthogonal basis set. The corresponding vector will not be sparse.

Numerical calculations show that an exploitable sparsity of the density matrix can be obtained with the common atom centered quantum chemistry Gaussian basis sets of moderate size in spite of the fact that they are nonorthogonal (SCUSERIA [1999]).

If one uses systematic basis sets such as finite elements or wavelets, the above described problems of nonorthogonal basis sets are not encountered either. Nevertheless the sparsity of the density matrix resulting from the physical localization properties is not sufficient for practical purposes. The storage requirement for the sparse density matrix are well above today's technical limits. As a rule of thumb, the localization region comprises some 100 atoms, which means that the density matrix will become sparse only for system sizes larger than 100 atoms. Assuming that we need 1000 basis functions per atom, the size of the density matrix would be $100\,000 \times 100\,000$, which is well beyond the memory size available today. Because of its uniqueness it would however be desirable to use the density matrix as the fundamental quantity in $O(N)$ electronic structure calculations.

This observation has prompted various research efforts to increase the sparsity of the density matrix. By representing the density matrix in a wavelet basis one can exploit not only the localization properties in real space but also in Fourier space and a significantly sparser density matrix is the result by GOEDECKER and IVANOV [1999]. Techniques related to wavelet image compression methods can also be applied to the density matrix to make it sparser (BEYLKIN, COULT and MOHLENKAMP [1999]). Another idea to obtain increased sparsity for metallic or small gap systems, not related to wavelets, is the energy renormalization group (BAER and HEAD-GORDON [1998]). It is based on a telescopic expansion of the density matrix. Let us assume we want to calculate the density matrix at some low temperature of 1/8. We can then write this density matrix in the following way

$$
\begin{aligned}
F(T = 1/8) = F(T = 1) &+ \big[F(T = 1/2) - F(T = 1)\big] \\
&+ \big[F(T = 1/4) - F(T = 1/2)\big] \\
&+ \big[F(T = 1/8) - F(T = 1/4)\big]
\end{aligned}
\tag{4.2}
$$

From Eq. (3.18) we know that $F(T = 1)$ will decay very rapidly. In a similar spirit $F(T = 1/2) - F(T = 1)$ will still decay fairly rapidly since its variation is slow as well as shown in Fig. 4.2.

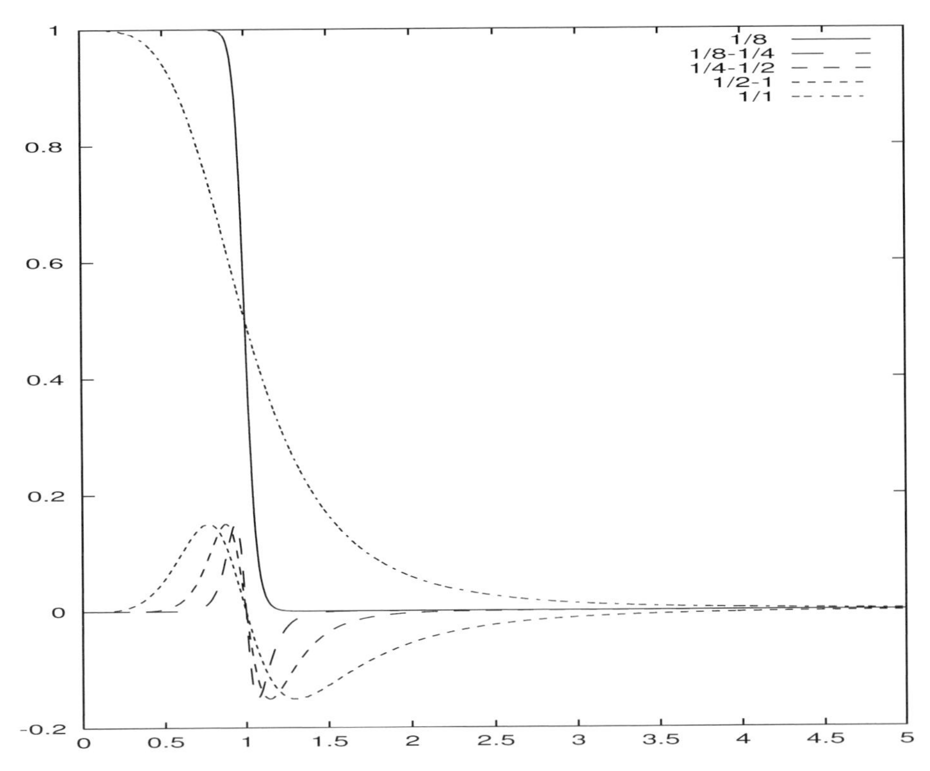

FIG. 4.2. The Fermi functions $f(T=1/8)$, $f(T=1)$, $[f(T=1/2)-f(T=1)]$, $[f(T=1/4)-f(T=1/2)]$ and $[f(T=1/8)-f(T=1/4)]$ associated with the expansion (Eq. (4.2)) of the density matrix $F(T=1/8)$.

$[F(T=1/4)-F(T=1/2)]$ decays then already much slower and $[F(T=1/8)-F(T=1/4)]$ roughly as slow as $F(T=1/8)$. So nothing seems to be gained up to now. The point however is that the effective rank of the subsequent terms of the telescopic expansion is decreasing. By effective rank we mean the number of eigenvalues that are larger in absolute value than a certain threshold. This comes form the fact that the interval where the terms in the telescopic expansion are larger than this threshold decreases (Fig. 4.2). In an appropriate basis the subsequent terms in the telescopic expansion can therefore be represented by matrices of smaller and smaller dimension. In summary, in the telescopic expansion we have first matrices of large dimension but of high sparsity, and then at the end full or nearly full matrices of small dimension.

5. Some building blocks of O(N) algorithms for the independent particle case

In this section some independent particle algorithms with attractive features are briefly introduced. A more detailed description of these algorithms, descriptions of other algorithms as well as a profound comparison of the available algorithms can be found in GOEDECKER [1999].

5.1. Purification

A purification (McWeeny [1956]) step drives an approximate density matrix closer to idempotency. After a sufficient number of steps the matrix is completely 'pure', i.e., it represents a zero temperature density matrix. The gradient G obtained from minimizing the deviation from idempotency

$$\mathrm{Tr}\big[\big(F^2 - F\big)^2\big] \tag{5.1}$$

is given by

$$G = 4F^3 - 6F^2 + 2F. \tag{5.2}$$

The simplest way to minimize the expression of Eq. (5.1) is by a steepest descent procedure, where F at the iteration $\nu + 1$ is related to F at the previous iteration ν by

$$F_{\nu+1} = F_\nu - \alpha\big(4F_\nu^3 - 6F_\nu^2 + 2F_\nu\big). \tag{5.3}$$

Even though the steepest descent procedure is usually not very efficient for higher-dimensional problems, it will turn out to be optimal in this context. For a one-dimensional problem the optimal value for α in Eq. (5.3) is the reciprocal of the curvature at the minimum. In the many-dimensional case the steepest descent method looses in general its optimality due to the presence of many different curvatures along the different principal axes. In this case, however, the curvature along all directions around the minimum is 2. Evaluating Eq. (5.1) in an eigenvector basis one obtains $\sum_j (n_j^2 - n_j)^2$. Its second derivative $\frac{\partial}{\partial n_i}$ is $2 + 12n_i(n_i - 1)$ which is 2 at any minimum where n_i is either 0 or 1. Hence our steepest descent procedure becomes

$$F_{\nu+1} = F_\nu - \frac{1}{2}\big(4F_\nu^3 - 6F_\nu^2 + 2F_\nu\big) = 2F_\nu^3 - 3F_\nu^2. \tag{5.4}$$

This formula is known as the McWeeny purification. The complete McWeeny purification algorithm for a fixed chemical potential μ with n_{it} purification steps is

$$F_0 = \big(2\max[\varepsilon_{\max} - \mu; \mu - \varepsilon_{\min}]\big)^{-1}(H - \mu I) + \frac{1}{2}I, \tag{5.5}$$

$$F_{\nu+1} = 2F_\nu^3 - 3F_\nu^2, \tag{5.6}$$

$$F \approx I - F_{n_{\mathrm{it}}}. \tag{5.7}$$

Evidently the expression of Eq. (5.1) has many minima, corresponding to projection matrices that span ground states as well as excited states with different number of electrons, i.e., with different dimensionality. Which solution will be picked in the McWeeny purification depends on the choice of F_0 in Eq. (5.5). Once we know

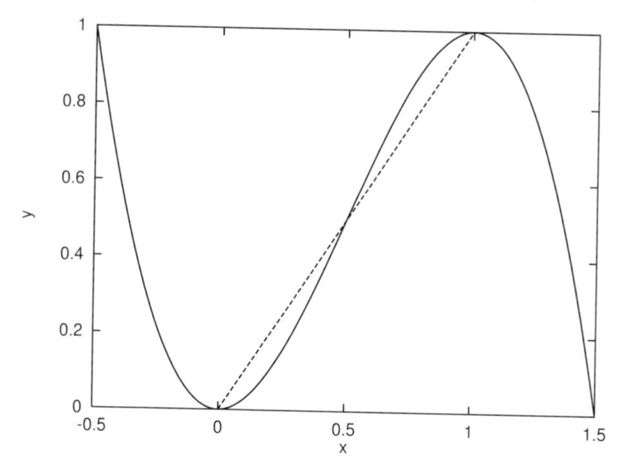

FIG. 5.1. The McWeeny (1960) purification function $3x^2 - 2x^3$.

the largest and the smallest eigenvalues $\varepsilon_{\max}$ and $\varepsilon_{\min}$ of H, we construct a scaled Hamiltonian F_0 such that the spectrum of F_0 is contained in the interval $[0, 1]$ and with a chemical potential at $\frac{1}{2}$. By looking at the plot of the McWeeny purification function (Fig. 5.1) one can convince oneself easily that the desired result is obtained. All the occupied eigenvalues of the Hamiltonian below the chemical potential are reduced in each step while the unoccupied ones tend to 1. F_ν tends to the complement of F and an approximate F is therefore given by Eq. (5.7).

The McWeeny purification iteration is equivalent to the Newton–Schultz sign matrix iteration (BEYLKIN, COULT and MOHLENKAMP [1999], NEMETH and SCUSERIA [2000])

$$F_{\nu+1} = \frac{3}{2}F_\nu - \frac{1}{2}F_\nu^3.$$

Since $\frac{3}{2}x - \frac{1}{2}x^3$ is the McWeeny function, $3x^2 - 2x^3$, shifted and stretched, such that is maps the interval $[-1, 1]$ onto $[-1, 1]$, the sign matrix iteration converges to $I - 2F$ instead of $I - F$. The F_0 of the sign matrix iteration has of course to be scaled in a slightly different way. The spectrum has to be comprised within $[-1, 1]$ with the chemical potential at 0.

Purification has an astounding rate of convergence (BEYLKIN, COULT and MOHLENKAMP [1999]). Let us assume that the distribution of the occupation numbers at a certain stage of the iteration is given by $h(x)$. The distribution in the next iteration is then given by $p_{\mathrm{MW}}(h(x))$, where $p_{\mathrm{MW}}(x) = 3x^2 - 2x^3$. The derivative of the new distribution is then $p'_{\mathrm{MW}}(h(x))h'(x)$. Evaluating this at the chemical potential $\frac{1}{2}$ we obtain $\frac{3}{2}h'(\frac{1}{2})$, which means that the slope increases by a factor of $\frac{3}{2}$ in each iteration. We have reached convergence in an insulator when the reciprocal of the slope at the Fermi level (which is something like a temperature) is less than the HOMO LUMO gap $\varepsilon_{\mathrm{gap}} = \varepsilon_{N+1} - \varepsilon_N$, i.e., if

$$\left(\frac{2}{3}\right)^{n_{\mathrm{it}}} < \frac{\varepsilon_{\mathrm{gap}}}{\varepsilon_{\max} - \varepsilon_{\min}}.$$

Hence the number of required iterations is of the order of

$$n_{\mathrm{it}} \propto \ln\left(\frac{\varepsilon_{\max} - \varepsilon_{\min}}{\varepsilon_{\mathrm{gap}}}\right) \Big/ \ln\left(\frac{3}{2}\right). \tag{5.8}$$

Due to the required matrix multiplications the purification algorithm as discussed above gives cubic scaling. In order to obtain linear scaling we have to enforce sparsity on the density matrix and to calculate only the significant matrix elements, whose number is proportional to the system size. This can be achieved by minimizing Eq. (5.1) under the constraint that $F_{k,l} = 0$ for all k, l that refer to nonsignificant matrix elements. Several criteria are imaginable to discern between significant and nonsignificant elements. The simplest criterion is to consider all matrix elements to be significant if they correspond to localized basis functions that are separated by a distance that is less than a certain cutoff distance. The sparsity constraint can be implemented using Lagrange parameters. The Lagrange parameter associated to the constraint $F_{k,l} = 0$ is given by $-G_{k,l}$ (cf. Eq. (5.2)). Hence the constrained gradient vanishes identically for all the nonsignificant matrix elements. Minimizing Eq. (5.1) under a sparsity constraint is thus very simple. All the nonsignificant matrix elements are never touched during the purification and stay therefore identically zero.

The method as described above is for a fixed chemical potential. The method has been generalized to the physically more important case of a fixed number of electrons (FALSER and MANOLOPOULOS [1998]).

5.2. *Density matrix minimization*

Minimization is one of the basic approaches in traditional electronic structure algorithms. The energy can always be written down in terms of the density matrix. For the case of a non-self-consistent problem one obtains the simple expression of Eq. (3.10). A straightforward minimization of such an energy expression with respect to the density matrix would however fail. To understand this failure let us look at the following partial minimization problem. The degrees of freedom of the density matrix can be split up in two parts, its eigenvectors and eigenvalues. We now assume that we have already found the eigenvectors and minimize only with respect to the remaining degrees of freedom, i.e., with respect to the eigenvalues n_i under the only constraint that the sum of the occupation numbers n_i give the number of electrons N. Since F and H have the same eigenvectors Eq. (3.10) becomes simply $\sum_i \varepsilon_i n_i$. This sum is not bounded from below without any further constraints. The required constraint for a fermionic system is that the occupation numbers are contained in the interval [0, 1]. This can be achieved by replacing the density matrix F in Eq. (3.10) by the purified density matrix $\widetilde{F} = 3F^2 - 2F^3$

and minimizing $E_{BS} = \text{Tr}[\widetilde{F} H]$. A second modification is still necessary. In order to get as the result a density matrix that corresponds to the occupied space below the Fermi level we have to shift the Hamiltonian to get the final Density Matrix Minimization (DMM) expression (LI, NUNES and VANDERBILT [1993])

$$\text{Tr}\big[(3F^2 - 2F^3)(H - \mu I)\big]. \tag{5.9}$$

The energy expression of Eq. (5.9) can be minimized efficiently by conjugate gradient techniques (DANIELS and SCUSERIA [1999]). The number of conjugate gradient iterations is given by GOEDECKER [1999]

$$n_{\text{it}} \propto \sqrt{\kappa} = \sqrt{\frac{\varepsilon_{\max} - \varepsilon_{\min}}{\varepsilon_{\text{gap}}}}. \tag{5.10}$$

The basic difference between purification and density matrix minimization is that in the first case the deviation from idempotency is minimized, whereas in the second case an energy is minimized. Consequently the density matrix obtained from purification is closer to idempotency whereas the DMM density matrix gives a lower energy. This is shown for a model system in Fig. 5.2.

Even though the convergence rate of the DMM method is not as fast as of the Purification method (Eq. (5.8) versus Eq. (5.10)), it has the advantage that the forces acting on the atoms can be calculated very easily due to its variational nature (GOEDECKER [1999]). By requiring that the approximate density matrix commutes

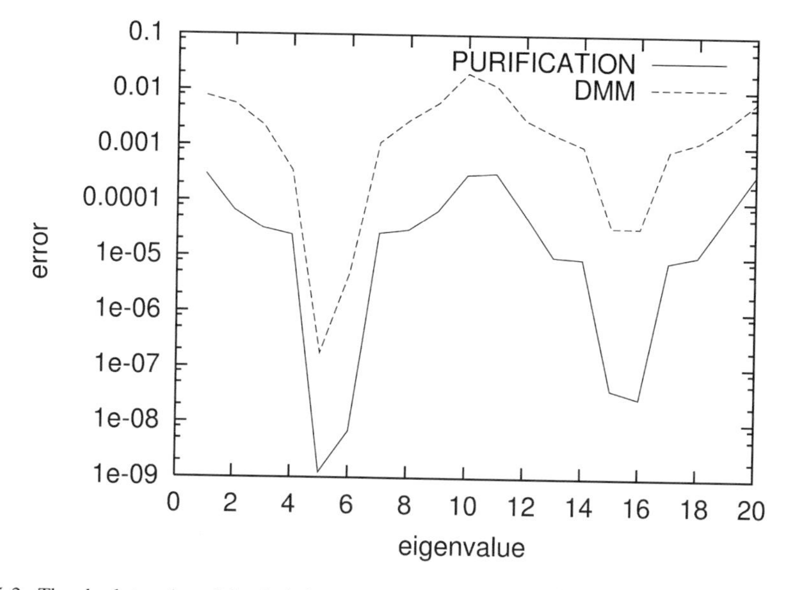

FIG. 5.2. The absolute value of the deviation of the eigenvalues of the approximate sparse density matrix from the exact values of one and zero. Shown are the results for a density matrix obtained by purification and by DMM. The latter gave a energy that was slightly lower.

with the Hamiltonian matrix during the iterations, a minimization schemes that requires a reduced number of matrix multiplies and that gives faster convergence can be obtained (CHALLACOMBE [1999]).

5.3. Projection

As was pointed out (Eq. (3.9)) the density matrix is a projection matrix that projects onto the space of the occupied independent particle orbitals. The problem in the context of O(N) algorithms is consequently to find a prescription to construct the density matrix that avoids the calculation of extended occupied orbitals. Eq. (3.2) proposes a solution to this problem, namely to calculate the density matrix directly as a matrix functional of H avoiding any reference to orbitals. In a localized basis set both H and F are sparse. The basic linear algebra operations involving H and F such as matrix times vector multiplications can therefore be done with linear scaling. Two simple computable functional representations have been proposed, a polynomial Fermi operator expansion (FOE) (GOEDECKER and COLOMBO [1994])

$$F \approx p(H) = c_0 I + c_1 H + c_2 H^2 + \cdots + c_{n_{\mathrm{pl}}} H^{n_{\mathrm{pl}}}$$

and a rational FOE (GOEDECKER [1995])

$$F \approx \frac{p(H)}{q(H)},$$

where p and q are arbitrary polynomials.

The simple-minded polynomial representation is numerically unstable. In the Fermi operator expansion this instability is avoided by formulating everything in terms of Chebyshev polynomials

$$p(H) = \frac{c_0}{2} I + \sum_{j=1}^{n_{\mathrm{pl}}} c_j T_j(H). \tag{5.11}$$

Since the Chebyshev polynomials are defined only within the interval $[-1, 1]$, we will scale H such that its spectrum falls into this interval. The Chebyshev matrix polynomials $T_j(H)$ satisfy the recursion relations

$$T_0(H) = I, \tag{5.12}$$

$$T_1(H) = H, \tag{5.13}$$

$$T_{j+1}(H) = 2H T_j(H) - T_{j-1}(H). \tag{5.14}$$

The expansion coefficients of the Chebyshev expansion can easily be determined. In an eigenfunction representation F (cf. Eq. (3.7)) is given by

$$\langle \Psi_n | F | \Psi_m \rangle = f(\varepsilon_n)\delta_{n,m}. \tag{5.15}$$

Evaluating the polynomial expansion in the same eigenfunction representation we obtain

$$\langle \Psi_n | p(H) | \Psi_m \rangle = p(\varepsilon_n)\delta_{n,m}, \tag{5.16}$$

where

$$p(\varepsilon) = \frac{c_0}{2} + \sum_{j=1}^{n_{\mathrm{pl}}} c_j T_j(\varepsilon). \tag{5.17}$$

Comparing Eqs. (5.15) and (5.16) we see that the polynomial $p(\varepsilon)$ has to approximate the Fermi distribution in the energy interval $[-1, 1]$ where the scaled and shifted Hamiltonian has its eigenvalues. In most cases one is interested in the limit of zero temperature. Nevertheless it is important to determine the Chebyshev expansion coefficients for a finite temperature to avoid Gibbs oscillations. If one is interested in the zero temperature limit it is also not necessary to use the actual Fermi distribution. Other similar functions such as a shifted and scaled error function, $f(\varepsilon) = \frac{1}{2}(1 - \mathrm{erf}(\frac{\varepsilon-\mu}{k_B T}))$ are typically better since they tend faster to zero and one away from the Fermi level for a given 'temperature'. In such a context $k_B T$ has just the significance of an energy resolution. In the case of an insulator, it has to chosen smaller than the gap, in the case of a metal smaller than the typical energy interval over which the density of states varies significantly. Since a nth order Chebyshev polynomial has n roots and thereby a resolution that is roughly proportional to $1/n$, the degree n_{pl} of the polynomial needed to represent the Fermi distribution is proportional to

$$n_{\mathrm{pl}} \propto \frac{\varepsilon_{\max} - \varepsilon_{\min}}{k_B T}, \tag{5.18}$$

where $\varepsilon_{\max}$ is the largest eigenvalue of the Hamiltonian and $\varepsilon_{\min}$ its smallest eigenvalue.

The easiest way to obtain a rational expansion is to start from Cauchy's integral formula

$$f(\varepsilon) = \frac{1}{2\pi i} \int_C \frac{dz}{\varepsilon - z}. \tag{5.19}$$

$f(\varepsilon)$ is zero if ε is outside the complex contour path C and one if it is inside. By choosing as C a circle or another path enclosing the occupied states we obtain a valid Fermi distribution for the system under consideration. Since the contour integral in Eq. (5.19) is periodic it can be approximated very efficiently by an equally spaced integration formula using some 20 integration points. The discretization has the effect of smoothing out the sharp jump of the zero temperature Fermi distribution. One has thus again some finite temperature like weight distribution. The required number of integrations points n_{pd} is given by

$$n_{\mathrm{pd}} \propto \frac{\mu - \varepsilon_{\min}}{\Delta\varepsilon}, \tag{5.20}$$

where $\Delta\varepsilon$ is again either the gap or the resolution. The implementation of the rationale Fermi Operator expansion (Rationale FOE) requires the calculation of the Greens function $(H - z_\nu)^{-1}$ at each integration point. The density matrix is then the sum of all these Green functions. Denoting $\frac{I}{H - z_\nu}$ by F_ν we have

$$(H - z_\nu)F_\nu = I, \tag{5.21}$$

$$F = \sum_\nu w_\nu F_\nu. \tag{5.22}$$

The inversion of $(H - z_\nu)$ is equivalent to the solution of M linear systems of equations, where M is the dimension of H. This can be effectuated using iterative techniques so that in the end everything can again be done by matrix times vector multiplications. The rational FOE can easily be generalized to nonorthogonal basis sets (GOEDECKER [1999]) and an efficient implementation in the context of nonorthogonal tight binding schemes has been described (BERNSTEIN [2001]).

Both the Chebyshev and the rational Fermi operator expansion can not only be used to calculate the density matrix, but also to calculate nonorthogonal Wannier functions that span the occupied one-particle space. In this case one has to solve for each integration point not M linear equations but only N. For each electron $i = 1, \ldots, N$ one has first to find an initial guess $\widetilde{V}_i$ for the nonorthogonal Wannier function based on chemical intuition, which then serves as the right hand side for the linear equation

$$(H - z_\nu)\widetilde{W}_{n,\nu} = \widetilde{V}_n. \tag{5.23}$$

The final nonorthogonal Wannier functions are then obtained by summation from the partial ones,

$$\widetilde{W}_n = \sum_\nu w_\nu \widetilde{W}_{n,\nu}. \tag{5.24}$$

Obtaining a set of orthogonal Wannier functions requires finally an orthogonalization. If this cubically scaling step should dominate in a calculation, it could either be replaced by a modified orthogonalization that retains some localization of the orbitals (and prevents therefore exact orthogonality) or by some nonorthogonal O(N) scheme (GOEDECKER [1999]).

6. Principles of O(N) methods for wavefunction methods

From the locality principles for wavefunction methods, outlined in Section 3.2 linear scaling follows straightforwardly for wavefunction methods where only doubly excited configurations are included. One simply calculates only the amplitudes C_I of the significant doubly excited configurations, whose number increases linearly with system size. A prerequisite to linear scaling is of course the use of localized orbitals with certain properties as discussed in Section 3.2.

Along these lines a linear scaling MP2 approach has been implemented (SCHÜTZ, HETZER and WERNER [1999]). Another approach to linear scaling, based on a Laplace transform of the MP2 energy expression, has also been implemented for both molecules (AYALA and SCUSERIA [1999]) and periodic solids (AYALA, KUDIN and SCUSERIA [2001]). SCHÜTZ and WERNER [2001] were able to obtain linear scaling in Coupled Cluster (SZABO and OSTLUND [1982]) calculations with single and double excitations (CCSD) as well. For best efficiency Schütz et al. have introduced a hierarchic treatment of correlation effects at different distances. Their occupied space is spanned by localized molecular orbitals (LMO's) that are obtained by applying the Pipek–Mezey localization procedure (PIPEK and MEZEY [1989]) to the Hartree–Fock molecular orbitals. The unoccupied space is spanned by projected atomic orbitals (PAOs), i.e., by the atomic orbitals whose part belonging to the occupied space was eliminated by projection. They define single and double excitation domains in the following way. A single excitation domain associated with a LMO contains all the PAO's that are centered on atoms whose Gaussian type basis functions are of considerable weight in the representation of the LMO. The exact quantification of the meaning of 'considerable' determines the actual size of the domain. A double excitation domain is simply the union of two single excitation domains. Double excitations are now treated in the following way. If the two parts of the double excitation domain are very close, the amplitudes of the doubles excitation that are possible within this domain are treated at the very accurate CCSD level. More distant pairs are treated at the MP2 level, even more distant pairs also at the MP2 level but with the involved electron repulsion integrals only evaluated approximately. Pairs far apart are finally entirely neglected. Note that the two cutoff distances that define the size and the separation of the excitation domains are directly related to the exponential and $1/s^3$ decay shown B in Fig. 3.2. CCSD contains also certain quadruple excitations, namely the disconnected quadruples. Schütz et al. have shown that these disconnected quadruple excitations have to involve also spatially close orbitals and do thus not destroy linear scaling (SCHÜTZ [2000]). It has to be stressed that linear scaling in this approach is only obtained with respect to the number of atoms and not with respect to the number of basis functions. Increasing the quality of the basis set by introducing more Gaussians per atom will lead to a very steep increase of the computational cost.

Scuseria et al. proposed another approach to linear scaling in CCSD that exploits only the localization of the atomic basis function (SCUSERIA and AYALA [1999]). By systematic thresholding the truncation error can be fully controlled in this approach.

The first step towards linear scaling has also been done within Quantum Monte Carlo (WILLIAMSON, HOOD and GROSSMAN [2001]) (QMC) methods. In QMC the wave function is written as a product of one or possibly several determinants times a Jastrow factor. The Jastrow introduces interelectronic coordinates that are necessary for a correct description of the cusp of the wave function, that occurs when two electrons are coming arbitrarily close.

$$\Psi(\mathbf{r}_1,\ldots,\mathbf{r}_N) = \begin{vmatrix} \phi_{i_1}(\mathbf{r}_1) & \ldots & \phi_{i_N}(\mathbf{r}_1) \\ \ldots & \ldots & \ldots \\ \phi_{i_1}(\mathbf{r}_N) & \ldots & \phi_{i_N}(\mathbf{r}_N) \end{vmatrix} \exp\biggl(\sum_{i,j} j\bigl(|\mathbf{r}_i - \mathbf{r}_j|\bigr)\biggr). \tag{6.1}$$

The orbitals ϕ_i can either be Hartree–Fock or density functional orbitals. QMC methods require the repeated evaluation of the wave function for arbitrary electronic positions $\mathbf{r}_1, \ldots, \mathbf{r}_N$. The fact that correlation is a local effect suggests hat the Jastrow functions j can be of finite range, which in turn means that the Jastrow part can be evaluated with linear scaling. This is actually not quite true. An analysis in NIGHTINGALE and UMRIGAR [1999] shows that the Jastrow factor decays as $1/|\mathbf{r}_i - \mathbf{r}_j|$. Nevertheless it seems to be possible to capture a significant fraction of the correlation energy with a short range Jastrow. Even if it should turn out that long range Jastrows are essential, they could be evaluated with linear or nearly linear scaling using fast summation methods. In traditional QMC calculations extended orbitals ϕ_i were used to build up the determinantal part, which means that all the N^2 entries of the determinant are nonzero (unless by chance one position coincides with a node, an event which has zero probability). Hence the scaling for the calculation of the entries of the determinant is at least quadratic. It is exactly quadratic if a localized basis set is used. Each orbital is a linear combination of all the basis functions, but if the basis functions are localized one has to sum only the contributions of the basis functions that are nonvanishing at the electronic position $\mathbf{r}_i$. It is cubic, however, if extended basis functions are used. By using localized orbitals $\phi_i(\mathbf{r})$, linear scaling is obtained since in this case the determinantal entry will be negligible unless the electronic position $\mathbf{r}_i$, is in a region close to the center of some localized orbital. We thus have to calculate only all the nonzero entries of the determinant and their number increases linearly with system size. It remains to evaluate the determinant once all its entries are known. This evaluation scales cubically, but its prefactor is much smaller than the prefactor for the calculation of the entries and therefore this part does not pose any problem in practice. In addition the cubic evaluation is only necessary at the first QMC step. In the following steps only one electron is moved. Then only one row of the determinant changes and the determinant can be updated with quadratic scaling using the Sherman–Morrison formula (CEPERLEY, CHESTER and KALOS [1977], PRESS, FLANNERY, TEUKOLSKY and VETTERLING [1986]).

7. Conclusions

Linear scaling algorithms have been found for the majority of the standard electronic structure methods both within the independent electron picture and for wavefunction methods. They all basically exploit the locality properties of weakly correlated physical and chemical systems. It remains as an unsolved problem to find linear scaling algorithms for metallic systems within density functional theory and for correlated methods with excitations of arbitrarily high order and multi-reference character. Significant improvements of the prefactors of the existing methods are certainly also possible.

Acknowledgment

Interesting discussions with P. Ayala, M. Schütz, G. Scuseria and Cyrus Umrigar are acknowledged.

References

AHLRICHS, R. and W. KUTZELNIGG (1968), *J. Chem. Phys.* **48**, 1819.
ASHCROFT, N. and N.D. MERMIN (1976), *Solid State Physics* (Saunders College, Philadelphia).
AYALA, P., K.N. KUDIN and G.E. SCUSERIA (2001), *J. Chem. Phys.* **115**, 9698.
AYALA, P. and G.E. SCUSERIA (1999), *J. Chem. Phys.* **110**, 3660.
BAER, R. and M. HEAD-GORDON (1998), *J. Phys. Chem.* **109**, 10159. BAER, R. and M. HEAD-GORDON (1998), *Phys. Rev. B* **58**, 15296.
BERNSTEIN, N. (2001), *Europhys. Lett.* **55**, 52.
BEYLKIN, G., N. COULT and M.J. MOHLENKAMP (1999), *J. Comput. Phys.* **152**, 32.
BOYS, S. (1950), *Proc. Roy Soc. London, Ser. A* **200**, 542.
CEPERLEY, D., V. CHESTER and M.H. KALOS (1977), *Phys. Rev. B* **16**, 3081.
CHALLACOMBE, M. (1999), *J. Chem. Phys.* **110**, 2332.
COLEMAN, A.J. and V.I. YUKALOV (2000), *Reduced Density Matrices: Coulson's Challenge* (Springer, New York).
COLEMAN, J. (1963), *Rev. Modern. Phys.* **35**, 668.
CSANYI, G., S. GOEDECKER and T. ARIAS (2002), *Phys. Rev. A* **65**, 032510.
DANIELS, A. and G.E. SCUSERIA (1999), *J. Chem. Phys.* **110**, 1321.
DAUBECHIES, I. (1992), *Ten Lectures on Wavelets* (SIAM, Philadelphia).
DAVIDSON, E. (1976), *Reduced Density Matrices in Quantum Chemistry* (Academic Press, New York).
DAVIDSON, E.R. (1975), *J. Comput. Phys.* **17**, 87.
DAVIDSON, E.R. (1989), *Comput. Phys. Comm.* **53**, 49.
DES CLOIZEAUX, J. (1964), *Phys. Rev.* **135**, A685. DES CLOIZEAUX, J. (1964), *Phys. Rev. A* **698**.
FOULKES, M.C., L. MITAS, R.J. NEEDS and G. RAJAGOPAL (2001), *Rev. Modern. Phys.* **73**, 33.
GALLI, G. (1996), *Current Opinion in Solid State and Materials Science* **1**, 864.
GOEDECKER, S. (1995), *J. Comput. Phys.* **118**, 261.
GOEDECKER, S. (1998), *Phys. Rev. B* **58**, 3501.
GOEDECKER, S. (1999), *Rev. Modern. Phys.* **71**, 1085–1123.
GOEDECKER, S. and L. COLOMBO (1994), *Phys. Rev. Lett.* **73**, 122.
GOEDECKER, S. and O. IVANOV (1999), *Phys. Rev. B* **59**, 7270.
GOEDECKER, S. and C. UMRIGAR (1998), *Phys. Rev. Lett.* **81**, 866.
HE, L. and D. VANDERBILT (2001), *Phys. Rev. Lett.* **86**, 5341.
HUTTER, J., H.P. LÜTHI and M. PARRINELLO (1994), *Comput. Math. Sci.* **2**, 244.
ISMAIL-BEIGI, S. and T. ARIAS (1999), *Phys. Rev. Lett.* **82**, 2127.
JENSEN, F. (1999), *Computational Chemistry* (Wiley, New York).
KOCH, E. and S. GOEDECKER (2001), *Solid. State Comm.* **119**, 105.
KOHN, W. (1959), *Phys. Rev.* **115**, 809.
KUTZELNIGG, W. and P. VON HERIGONTE (2000), *Adv. Quant. Chem.* **36**, 185.
LI, X.-P., W. NUNES and D. VANDERBILT (1993), *Phys. Rev. B* **47**, 10891.
MARZARI, N. and D. VANDERBILT (1997), *Phys. Rev. B* **56**, 12847.
MCWEENY, R. (1956), *Proc. Roy. Soc. A* **235**, 496.
MCWEENY, R. (1960), *Rev. Modern. Phys.* **32**, 335.
MCWEENY, R. (1989), *Methods of Molecular Quantum Mechanics* (Academic Press, New York).
NEMETH, K. and G. SCUSERIA (2000), *J. Chem. Phys.* **113**, 6035.
NENCIU, G. (1983), *Commun. Math. Phys.* **91**, 81.

NESBET, R.K. (1964), *Adv. Chem. Phys.* **9**, 321.
NIGHTINGALE, M.P. and C.J. UMRIGAR, eds. (1999), *Quantum Monte Carlo Methods in Physics and Chemistry*, NATO Science Series C: Mathematical and Physical Sciences, Vol. 525 (Kluwer Academic, Dordrecht).
ORDEJON, P. (1998), *Comput. Math. Sci.* **12**, 157.
PALSER, A. and D. MANOLOPOULOS (1998), *Phys. Rev. B* **58**, 12704.
PARR, R. and W. YANG (1989), *Density-Functional Theory of Atoms and Molecules* (Oxford Univ. Press, Cambridge).
PAYNE, M., M. TETER, D. ALLAN, T. ARIAS and J. JOANNOPOULOS (1992), *Rev. Modern. Phys.* **64**, 1045.
PIPEK, J. and P. MEZEY (1989), *J. Chem. Phys.* **90**, 1916.
PRESS, W., B.P. FLANNERY, S.A. TEUKOLSKY and W.T. VETTERLING (1986), *Numerical Recipes, The Art of Scientific Computing* (Cambridge Univ. Press, Cambridge).
PULAY, P. (1983), *Chem. Phys. Lett.* **100** 151.
PULAY, P. and S. SAEBØ (1986), *Theor. Chim. Acta* **69**, 357.
PURVIS, G.D. and R.J. BARTLETT (1982), *J. Chem. Phys.* **76** 1910.
SAEBØ, S. and P. PULAY (1985), *Chem. Phys. Lett.* **113**, 13.
SAEBØ, S. and P. PULAY (1987), *J. Chem. Phys.* **86**, 914.
SAEBØ, S. and P. PULAY (1988), *J. Chem. Phys.* **88**, 1884.
SCHÜTZ, M. (2000), *J. Chem. Phys.* **113**, 9986; http://www.theochem.uni-stuttgart.de/schuetz/.
SCHÜTZ, M., G. HETZER and H.-J. WERNER (1999), *J. Chem. Phys.* **111**, 5691.
SCHÜTZ, M. and H.-J. WERNER (2001), *J. Chem. Phys.* **114**, 661.
SCUSERIA, G. (1999), *J. Phys. Chem.* **103**, 4782.
SCUSERIA, G.E. and P. AYALA (1999), *J. Chem. Phys.* **111**, 8330.
SINANOGLU, O. (1964), *Adv. Chem. Phys.* **6**, 315.
STOLLHOFF, G. and P. FULDE (1980), *J. Chem. Phys.* **73**, 4548.
SZABO, A. and N. OSTLUND (1982), *Modern Quantum Chemistry* (McGraw-Hill, New York).
TARAKIN, S., D. DRABOLD and S. ELLIOT (2002), cond-mat/0110473.
WILLIAMSON, A.J., R.Q. HOOD and GROSSMAN, J.C. (2001), *Phys. Rev. Lett.* **87**, 246406.
WU, S.Y. and C.S. JAYANTHI (2002), *Phys. Rep.* **358**, 1.
YANG, W. (1991), *Phys. Rev. Lett.* **66**, 1438.
YANG, W. and T.-S. LEE (1995), *J. Chem. Phys. Rev. B* **103**, 5674.

Finite Difference Methods for Ab Initio Electronic Structure and Quantum Transport Calculations of Nanostructures

Jean-Luc Fattebert[a], Marco Buongiorno Nardelli[b]

[a]*Center for Applied Scientific Computing (CASC), Lawrence Livermore National Laboratory, P.O. Box 808, L-561, Livermore, CA 94551, USA*

[b]*Department of Physics, North Carolina State University, Raleigh, NC, and Center for Computational Sciences (CCS) and Computational Science and Mathematics Division, Oak Ridge National Laboratory, Oak Ridge, TN 37830, USA*

E-mail address: fattebert1@llnl.gov (J.-L. Fattebert)

1. Introduction

Among the numerical discretization methods used to solve the equations of density functional theory (DFT), the most widely used are linear combination of atomic orbitals (LCAO) – usually Gaussian-type orbitals (GTO) –, plane-waves (PW) and finite differences (FD). Of these three methods, FD is the most recent and less common. General fully 3D grid-based electronic structure representation using finite differences as approximate numerical schemes for partial differential operators have started being widely used in the last ten years only. However, real-space finite difference approaches have already shown to be an efficient tool in a substantial number of large scale electronic structure calculations. Among its various applications, we can cite optical properties of surfaces (SCHMIDT, BECHSTEDT and BERNHOLC [2001]), surface

Computational Chemistry
Special Volume (C. Le Bris, Guest Editor) of
HANDBOOK OF NUMERICAL ANALYSIS, VOL. X
P.G. Ciarlet (Editor)
2003 Elsevier Science B.V.

reconstruction (RAMAMOORTHY, BRIGGS and BERNHOLC [1998]), properties of GaN surfaces (BUNGARO, RAPCEWICZ and BERNHOLC [1999]), excitation energies and photoabsorption spectra of atoms and clusters (VASILIEV, OGUT and CHELIKOWSKY [1999]), diffusion of oxygen ions in SiO_2 (JIN and CHANG [2001]), first-principles molecular dynamics of carbon nanotubes (BUONGIORNO NARDELLI, YAKOBSON and BERNHOLC [1998]) and processes in solution (TAKAHASHI, HORI, HASHIMOTO and NITTA [2001], FATTEBERT and GYGI [2002]). In this chapter, we review the numerical aspects of this approach for total-energy pseudopotential calculations and show the reasons why it is becoming a method of choice. We also illustrate the method with an application to calculations of electronic structure and conductance of carbon nanotube on a metallic contact. We will limit the discussion to ab initio Density Functional Theory (DFT) models, where no parameters are fitted to experimental data.

Traditionally, chemists have mostly used LCAO methods. Atomic orbitals expressed as linear combinations of Gaussian functions provide an efficient limited basis set that allows for a good description of the electronic structure of localized finite molecules. The wide spread commercial code GAUSSIAN (Gaussian, Inc.) is based on such an approach. On the other hand, plane waves (PW) (PAYNE, TETER, ALLAN, ARIAS and JOANNOPOULOS [1992], GALLI and PASQUARELLO [1993], PARRINELLO [1997], e.g.) are very efficient to describe periodic systems. Since a plane wave represents a free electron, this approach has been very successful at describing systems with almost free electrons, like metals. By their nature, they have been mostly used and developed by solid state physicists. In this method, known as pseudospectral method in the mathematics community, the numerical basis set is completely independent of the positions of the atoms present in the simulation. It can be made as accurate as desired by systematically increasing the number of basis functions included in the basis set.

Like PW, the finite difference method is an alternative to the linear combination of atomic orbitals when highly accurate electronic wave functions are required. Both approaches try to estimate the electronic structure of a physical system without any assumption on where the atoms and electrons are located. While plane waves discretizations benefit from the long and extensive experience of many groups of solid state physicists, there are only a few groups around the world that have developed fully functional codes that allow first-principles molecular dynamics simulations on real-space grids.

In recent years, large parallel supercomputers have become an essential tool in first-principles molecular dynamics simulations. They allow not only calculations that would take months or years on single processor machines, but also calculations that would just not fit into the memory of a workstation or high-end PC. In a parallel environment, the electronic wave functions described in a PW approach can be distributed between the processors for example. Such an approach allows an efficient local application of the FFT algorithm to transform wave functions between real-space and reciprocal space (where the Laplacian is computed). However, every time a matrix element between two wave functions is required – in an orthogonalization process for example –, a huge traffic of data through the whole machine is necessary. In a real-space approach, all the expensive operations can be done locally, thanks to the real-space nature of the DFT

Hamiltonian operator and the wave functions. To compute matrix elements between wave functions, local contributions are computed on every processing element (PE) before being summed up at the end over all the PEs. This is one of the main advantages of FD over PW for nowadays simulations.

Another key element in the development of efficient grid-based FD approaches in large scale electronic structure calculations is the multigrid method (BRANDT [1977]). Indeed, real-space large-scale *ab initio* calculations involve large sparse matrices, and multigrid methods, either as solvers or as preconditioners, allow to design very efficient scalable algorithms (BRIGGS, SULLIVAN and BERNHOLC [1995], BRIGGS, SULLIVAN and BERNHOLC [1996], FATTEBERT [1996], FATTEBERT [1999], ANCILOTTO, BLANDIN and TOIGO [1999], JIN, JEONG and CHANG [1999], WANG and BECK [2000], FATTEBERT and BERNHOLC [2000], HEISKANEN, TORSTI, PUSKA and NIEMINEN [2001]).

More recently, in the context of the search for linear scaling algorithms (see GOEDECKER [2003], this Handbook, e.g.), real-space methods have appeared to be appropriate for imposing natural localization constraints on the orbitals (HERNANDEZ and GILLAN [1995], HOSHI and FUJIWARA [1997], FATTEBERT and BERNHOLC [2000]). Such an approach leads to a dramatic reduction in computer time and memory requirements for very large systems. However, since these algorithms are very recent and useful only for systems of sizes close to the limit of computing resources available today, these methods are still in development and many open questions remain.

Other advantages of FD over standard PW approaches include the possibility of introducing local mesh refinements (GYGI and GALLI [1995], MODINE, ZUMBACH and KAXIRAS [1996], FATTEBERT [1999]) and Dirichlet boundary conditions for nonperiodic systems in a very natural fashion. Some of these aspects are discussed for example in a recent review paper by BECK [2000]. If local mesh refinement is a requirement for all electrons calculations, many pseudopotential calculations can be carried out with a perfectly regular mesh. Also, local mesh refinements involve many complications such as load balancing for implementation on parallel computers, or Pulay forces. To our knowledge, no large scale first-principles molecular dynamics with local mesh refinements have been carried out so far and this aspect of FD methods will not be further discussed here.

The computational advantages of real-space approaches have also motivated some research in 3D finite elements (FE) methods for electronic structure calculations (WHITE, WILKINS and TETER [1989], MURAKAMI, SONNAD and CLEMENTI [1992], TSUCHIDA and TSUKADA [1995], KOHN, WEARE, ONG and BADEN [1997], PASK, KLEIN, FONG and STERNE [1999]). However, since electronic structure calculations in general require computation domains of very simple shapes like rectangular cells, so far FE methods have not demonstrated real advantages over FD methods for regular grids like those currently used with pseudopotentials. On the other hand, FE methods can be considerably more expensive.

Advanced numerical methods like those presented in this chapter are useful only if they allow to treat real problems in solid state physics or physical chemistry. Besides the foreseen improvement in performance for calculations of large scale systems, localized orbital adapted to their chemical environment offer the possibility of novel computational applications. In particular, the possibility of obtaining accurate localized orbital basis would be, in principles, a useful starting point for formal developments such as the semiclassical theory of electron dynamics or the theory of magnetic interactions in solids, in a natural extension of the Wannier representation of crystal wave functions (ASHCROFT and MERMIN [1976], Chapter 10, p. 187 and ff.). Among the possible applications of localized orbital methods, one of increasing technological importance is the calculation of the quantum transport properties of nanostructures.

The current limits of semiconductor electronics and the challenges for future developments involve the continuous shrinking of the physical dimensions of the devices and the attainment of higher speeds. The drive to produce smaller devices has lead the current research towards a new form of electronics in which nanoscale objects, such as clusters or molecules, replace the transistors of today's silicon technology. However, the production and integration of nanoscale individual components into easily reproducible device structures presents many challenges, both experimentally and theoretically. From the theoretical point of view, the design of such devices requires explicit modelling of quantum propagation of electrons in nanoscale systems. The quantity to be calculated is the quantum conductance, that is the measure of the ease with which electrons will transmit through a conductor, or alternatively of the resistance that electrons will encounter in their flow. As we will see in the following sections, the evaluation of conductance in nanostructures requires an electronic structure calculation of the system under consideration, the computation of its Green's function, and an accurate treatment of the coupling to and scattering at the contacts.

Recent years have witnessed a great amount of research in the field of quantum conductance in nanostructures (BEENAKKER and VAN HOUTEN [1991]). These have become the systems of choice for investigations of electrical conduction at mesoscopic scale. The improvements in nanostructured material production have stimulated developments in both experiment and theory. In particular, the formal relation between conduction and transmission, the Landauer formula (LANDAUER [1970]), has enhanced the understanding of electronic transport in extended systems and has proven to be very useful in interpreting experiments involving the conductance of nanostructures.

The problem of understanding the transport behavior of nanoscale structures cannot be effectively solved without a fully ab initio methodology. Only the latter is able to accurately describe the behavior of the electrons in the highly inhomogeneous environment of the nanoscale device, as well as account for the charge transfer and the interactions within the nano-system. Most of the existing methods to compute conductance from ab initio methods are based on the solution of the quantum scattering problem for the electronic wave functions through the conductor using a number of related techniques. Lippman–Schwinger and perturbative Green's function methods have been used to study conductance in metallic nanowires and recently in small molecular nanocontacts (LANG [1995], DI VENTRA, PANTELIDES and LANG [2000]). Conduction in nanowires, junctions and nanotube systems has been addressed using nonlocal pseudopotentials methods (CHOI and IHM [1999], CHOI, IHM, LOUIE and COHEN [2000]) and through the solution of the coupled channels equations in a scattering-theoretic approach (HIROSE and TSUKADA [1995], KOBAYASHI,

BRANDBYGE and TSUKADA [2000], LANDMAN, BARNETT, SCHERBAKOV and AVOURIS [2000]). The above methods compute ab initio transport using a plane wave representation of the electronic wave functions. This imposes severe restrictions on the size of the system because of the large number of basis functions necessary for an accurate description of the electron transmission process. Only recently real-space approaches been considered for a more efficient solution of the electronic transport problem. They are based on the use of LCAO (YOON, MAZZONI, CHOI, IHM and LOUIE [2001], TAYLOR, GUO and WANG [2000]) or Gaussian orbital bases (YALIRAKI, ROITBERG, GONZALEZ, MUJICA and RATNER [1999]). These are combined with either a scattering state solution for the transmission (YOON, MAZZONI, CHOI, IHM and LOUIE [2001]) or Green's function-based techniques (YALIRAKI, ROITBERG, GONZALEZ, MUJICA and RATNER [1999], TAYLOR, GUO and WANG [2000]).

In this chapter we do not intend to cover the wide variety of techniques that have been developed to compute quantum conductance, both from ab initio or from more phenomenological approaches. For such an exhaustive task, we refer the interested reader to the excellent monographs by DATTA [1995] and FERRY and GOODNICK [1997] and to the references at the end of this chapter. On the contrary, we have chosen to outline the main steps of the approach derived by us in the context of localized orbital methods (BUONGIORNO NARDELLI, FATTEBERT and BERNHOLC [2001]). In this chapter we will limit the discussion to the linear response regime and thus to zero bias across the conductor-lead junctions.

This chapter is organized as follows. In Section 2, we review the finite difference method in the context of the Kohn–Sham (KS) equations and pseudopotentials approach, also describing some specific features of iterative algorithms used to solve the KS equations in real-space. We also review the computation of the forces to optimize geometries and carry out first principles molecular dynamics. We finish Section 2 with some more advanced features designed to reduce the computational cost of the method in a localized orbitals representation. In Section 3, we review a numerical method to compute quantum conductance through nanostructures. This method is based on a description of the electronic structure in a basis of localized orbitals. In Section 4, we conclude this chapter by an illustration that brings together the localized grid-based orbitals and the quantum conductance algorithm to compute *ab initio* quantum conductance of a carbon nanotube on an aluminium surface.

2. Electronic structure calculation by finite differences

2.1. Kohn–Sham equations

Kohn–Sham (KS) theory (KOHN and SHAM [1965]) is a widely used model for first-principles calculations (see, e.g., CANCÈS, DEFRANCESCHI, KUTZELNIGG, LE BRIS and MADAY [2003], this volume). It states that the electronic ground state of a physical system can be described by a system of orthogonal one-particle electronic wave functions $\psi_j,\ j = 1, \dots, N$, that minimizes the KS total energy functional E_{KS}.

To simplify the discussion we neglect here the spin of the electrons by allowing double occupations of the orbitals, so that the electronic density is defined as

$$\rho_e(\vec{r}) = \sum_{i=1}^{N} f_i |\psi_i(\vec{r})|^2, \tag{2.1}$$

where $0 \leqslant f_i \leqslant 2$, $i = 1, \ldots, N$, are the occupation numbers. We also assume that the system is neutral, i.e., the total charge of the electrons neutralizes exactly the nuclei charges.

For a molecule composed of N_a atoms located at positions $\{\vec{R}_a\}_{a=1}^{N_a}$ in a computation domain Ω, the KS energy functional is given by (in atomic units)

$$\begin{aligned} E_{\mathrm{KS}}\big[\{\psi_i\}_{i=1}^{N}, \{\vec{R}_a\}_{a=1}^{N_a}\big] &= \sum_{i=1}^{N} f_i \int_\Omega \psi_i^*(\vec{r})\Big(-\frac{1}{2}\nabla^2\Big)\psi_i(\vec{r})\,d\vec{r} \\ &\quad + \frac{1}{2}\int_\Omega\int_\Omega \frac{\rho_e(\vec{r}_1)\rho_e(\vec{r}_2)}{|\vec{r}_1 - \vec{r}_2|}\,d\vec{r}_1\,d\vec{r}_2 \\ &\quad + E_{\mathrm{xc}}[\rho_e] + \int_\Omega \psi_i^*(\vec{r})(V_{\mathrm{ext}}\psi_i)(\vec{r})\,d\vec{r}. \end{aligned} \tag{2.2}$$

The first term represents the kinetic energy of the electrons. The second represents the electrostatic energy of interaction between electrons that we will note E_{es}. E_{xc} models the exchange and correlation between electrons. In this chapter, we will use the local density approximation (LDA), or the first-principles exchange-correlation functional proposed by Perdew–Burke–Ernzerhof (PBE) (PERDEW, BURKE and ERNZERHOF [1996]) which often provides results in better agreement with experiments and is appropriate for a grid-based implementation. In the last term of Eq. (2.2), the potential V_{ext} represents the total potential produced by the atomic nuclei at positions $\{\vec{R}_a\}_{a=1}^{N_a}$.

The ground state of a physical system is represented by orbitals that minimize the energy functional (2.2) under the constraints that the ψ_j are orthonormal. This minima can be found by solving the associated Euler–Lagrange equations – Kohn–Sham equations (KOHN and SHAM [1965]) –

$$H\psi_j = \left[-\frac{1}{2}\nabla^2 + v_{\mathrm{H}}(\rho_e) + \mu_{\mathrm{xc}}(\rho_e) + V_{\mathrm{ext}}\right]\psi_j = \varepsilon_j\psi_j, \tag{2.3}$$

which must be solved self-consistently for the N lowest eigenvalues ε_j, while imposing the orthonormality constraints $\langle\psi_i|\psi_j\rangle = \delta_{ij}$. We use the usual quantum mechanics notation $\langle\cdot|\cdot\rangle$ for the L^2 scalar product. The Hartree potential v_{H} represents the Coulomb potential due to the electronic charge density ρ_e, and $\mu_{\mathrm{xc}} = \delta E_{\mathrm{xc}}[\rho_e]/\delta\rho_e$ is the exchange and correlation potential.

2.2. *Finite differences approach*

In order to discretize the KS equations, we introduce a real-space rectangular grid Ω_h of mesh spacing h_x, h_y, h_z in the directions x, y, z that covers the computation domain Ω. Let M be the number of grid points. The wave functions, potentials and the electronic density are represented by their values at the grid points $\vec{r}_{ijk} = (x_i, y_i, z_k)$. Integrals over Ω are performed using the discrete summation rule

$$\int_{\Omega} u(\vec{r})\, d\vec{r} \approx h_x h_y h_z \sum_{i,j,k \in \Omega_h} u(\vec{r}_{ijk}).$$

Given the values of a function $u(\vec{r})$ on a set of nodes $\vec{r}_{i,j,k}$ the traditional finite difference approximation $w_{i,j,k}$ to the Laplacian of the function at a given node is expressed as a linear combination of values of the function at the neighboring nodes

$$w_{i,j,k} = \sum_{n=-p}^{p} c_n \big(u(x_i + nh_x, y_j, z_k) + u(x_i, y_j + nh_y, z_k) \\ + u(x_i, y_j, z_k + nh_z) \big), \tag{2.4}$$

where the coefficients $\{c_n\}$ can be computed from the Taylor expansion of u near $\vec{r}_{i,j,k}$. Such an approximation has an order of accuracy $2p$, that is for a sufficiently smooth function u, $w_{i,j,k}$ will converge at the rate $\mathrm{O}(h^{2p})$ as the mesh spacing $h \to 0$. For the second order approximation for example ($p = 1$), we have $c_0 = 2/h^2$ and $c_1 = c_{-1} = -1/h^2$. High order versions of this scheme were first used in electronic structure calculations by CHELIKOWSKY, TROUILLER and SAAD [1994].

As an alternative, one can also use a compact finite difference scheme (also called *Mehrstellenverfahren* in COLLATZ [1966]). For example, a 4th order FD scheme for the Laplacian on a cubic grid is based on the relation

$$\frac{1}{6h^2}\Bigg\{ 24u(\vec{r}_0) - 2 \sum_{\substack{\vec{r} \in \Omega_h, \\ \|\vec{r}-\vec{r}_0\| = h}} u(\vec{r}) - \sum_{\substack{\vec{r} \in \Omega_h, \\ \|\vec{r}-\vec{r}_0\| = \sqrt{2}h}} u(\vec{r}) \Bigg\} \\ = \frac{1}{72}\Bigg\{ 48(-\nabla^2 u)(\vec{r}_0) + 2 \sum_{\substack{\vec{r} \in \Omega_h, \\ \|\vec{r}-\vec{r}_0\| = h}} (-\nabla^2 u)(\vec{r}) + \sum_{\substack{\vec{r} \in \Omega_h, \\ \|\vec{r}-\vec{r}_0\| = \sqrt{2}h}} (-\nabla^2 u)(\vec{r}) \Bigg\} \\ + \mathrm{O}(h^4), \tag{2.5}$$

valid for a sufficiently differentiable function $u(\vec{r})$. For simplicity, we have assumed here that $h_x = h_y = h_z$, but this expression is easy to extend to the general case. This FD scheme requires only values at grid points within a sphere of radius $\sqrt{2}h$. Beside its good numerical properties, the compactness of this scheme reduces the amount

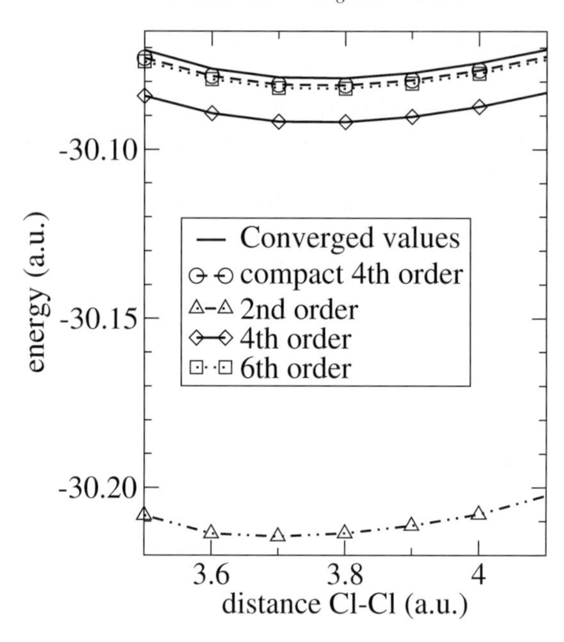

FIG. 2.1. Total energy for a Cl_2 molecule as a function of the distance Cl–Cl for several finite difference schemes, using the PBE exchange and correlation functional and pseudopotentials by HAMANN [1989]. The grid spacing used ($h = 0.34$) is sufficient to accurately compute the equilibrium bond length and binding energy of the molecule using the schemes of order 4 and 6, while it is clearly too coarse for the 2nd order scheme. The top line shows a fully converged result.

of communications in a domain-decomposition based parallel implementation. While increasing the order of the FD scheme improves the accuracy for very fine grids, it may not be the case for a given computational grid. In practice, this compact 4th order scheme consistently improves the accuracy compared to a standard 4th order scheme, as illustrated in Fig. 2.1.

REMARK 2.1. The FD method is not variational, and by refining the mesh the total energy generally increases towards convergence.

It is easy to see that a compact FD scheme like (2.5) leads to an eigenvalue problem of the form

$$(L_h + B_h V_h)\vec{\psi}_i = \varepsilon_i B_h \vec{\psi}_i, \tag{2.6}$$

where L_h represents the FD scheme on the left-hand side of Eq. (2.5) and $B_h \in \mathcal{M}_M$ is a sparse well conditioned matrix that represents the FD-like scheme on the right-hand side of Eq. (2.5). V_h represents the potential on the grid. Let $H_h = L_h + B_h V_h$. One can show that $B_h^{-1} H_h$ is symmetric (B_h and L_h commute) so that the eigenvalues ε_i are real and the eigenvectors $\vec{\psi}_i$ can be chosen orthogonals.

This type of compact FD scheme were simultaneously introduced in electronic structure calculations by BRIGGS, SULLIVAN and BERNHOLC [1995] and FATTEBERT [1996]. The simulations presented in this chapter are based on this scheme. However, to simplify the notations, we will drop the matrix B_h in the rest of this chapter. In general, from the point of view of computer time and memory requirements, FD schemes of order larger than 2 are clearly worthwhile since they allow to work with much coarser grids. This reduces the cost of operations like scalar products between trial eigenfunctions or linear combinations of trial eigenfunctions in the iterative solver (see Section 2.5).

2.3. *General formulation in nonorthogonal orbitals*

Most of the time, we are not interested in the individual eigenfunctions solutions of the KS equations, but only in the subspace spanned by these functions. It means that one can represent the subspace of the electronic orbitals by a more general basis of nonorthogonal functions, $\{\phi_1, \ldots, \phi_N\}$. We write these functions as vectors, columns of a matrix Φ,

$$\Phi = (\vec{\phi}_1, \ldots, \vec{\phi}_N).$$

An orthonormal basis of approximate eigenfunctions (Ritz functions) can be obtained by a diagonalization in this subspace of dimension N (Ritz procedure). We denote by $C \in \mathcal{M}_N$ the matrix that transforms Φ into the basis Ψ of orthonormal Ritz functions,

$$\Psi = (\vec{\psi}_1, \ldots, \vec{\psi}_N) = \Phi C. \tag{2.7}$$

The matrix C satisfies

$$CC^T = S^{-1},$$

where $S = \Phi^T \Phi$ is the overlap matrix.

In the following, for an operator A we will use the notation

$$A^{(\Phi)} = \Phi^T A \Phi.$$

We then have the relation

$$A^{(\Psi)} = C^T A^{(\Phi)} C.$$

Eq. (2.7) defines a transformation to a Ritz basis only if C is a solution of the generalized symmetric eigenvalue problem

$$H^{(\Phi)} C = SC\Lambda, \tag{2.8}$$

where $\Lambda \in \mathcal{M}_N$ is a diagonal matrix that satisfies $\Lambda = \Psi^T H \Psi$. The matrix C can actually be decomposed as a product $C = L^{-T} U$, where L is the Cholesky factorization of S,

$$S = LL^T,$$

and U is an orthogonal matrix. Knowing L, the generalized eigenvalue problem (2.8) is reduced to a standard symmetric eigenvalue problem

$$L^{-1} H^{(\Phi)} L^{-T} U = U \Lambda. \tag{2.9}$$

For a chemical potential μ, let us define $\Upsilon \in \mathcal{M}_N$ by its matrix elements

$$\Upsilon_{ij} = \delta_{ij} f\left[(\varepsilon_i - \mu)/k_{\mathrm{B}} T\right],$$

where f is a Fermi–Dirac distribution at temperature T and k_{B} is the Boltzmann constant. The density operator $\hat{\rho}$ is then defined as

$$\hat{\rho} = \Psi \Upsilon \Psi^T = \Phi C \Upsilon C^T \Phi^T.$$

For $T = 0$, $\hat{\rho}$ is a projector onto the states of eigenvalues lower than μ. The dimension of this density matrix is given by the number of degrees of freedom, i.e., the number of grid points in a grid-based approach. This number is in general so large that it is impossible to apply numerical methods that require $\rho(\vec{r}, \vec{r}')$ (matrix of size $M \times M$). However, it is useful to represent $\hat{\rho}$ in the basis Φ

$$\rho^{(\Phi)} = \Phi^T \hat{\rho} \Phi = C^{-T} \Upsilon C^{-1}.$$

Even more useful is the matrix $\bar{\rho}^{(\Phi)}$

$$\bar{\rho}^{(\Phi)} = S^{-1} \rho^{(\Phi)} S^{-1} = C \Upsilon C^T. \tag{2.10}$$

This matrix appears naturally in the expression used to compute the expectation value $\bar{A}$ of an operator A represented in the basis Φ,

$$\bar{A} = 2\,\mathrm{tr}\left(\Upsilon A^{(\Psi)}\right) = 2\,\mathrm{tr}\left(\bar{\rho}^{(\Phi)} A^{(\Phi)}\right).$$

In particular, the total number of electrons in the system is given by

$$N_e = 2\,\mathrm{tr}\left(\bar{\rho}^{(\Phi)} S\right).$$

Also, the electronic density in a nonorthogonal orbitals formulation is simply given by

$$\rho_e(\vec{r}) = 2 \sum_{j,k=1}^{N} \left(\bar{\rho}^{(\Phi)}\right)_{jk} \phi_j(\vec{r}) \phi_k(\vec{r}). \tag{2.11}$$

REMARK 2.2. If all the computed states are fully occupied, we have $\rho^{(\Phi)} = S$ and $\bar{\rho}^{(\Phi)} = S^{-1}$.

2.4. *Pseudopotentials*

On regular grids, FD methods, like plane-waves, are not very efficient to describe singular atomic potentials accurately. In particular for microcanonical molecular dynamics, it is difficult to guarantee a good conservation of the total energy of the system. These singularities can however be removed by replacing the atomic nuclei and core electrons – which can be approximated as frozen in their atomic state – by pseudopotentials. In the pseudopotential approach, the electronic structure calculation problem is reduced to the computation of a cloud of valence electrons living in a background of positive ions represented by smooth nonsingular pseudopotentials.

Accurate calculations can be performed by representing each atomic core by a nonlocal separable pseudopotential in its Kleinman–Bylander form (KLEINMAN and BYLANDER [1982])

$$V_{\text{ps}} = V_{\text{local}} + V_{\text{nl}} = v_{\text{ps}}^{\text{local}}(\vec{r}) + \sum_{\ell=0}^{\ell_{\max}} \sum_{m=-\ell}^{\ell} |v_\ell^m\rangle E_\ell^{\text{KB}} \langle v_\ell^m|, \tag{2.12}$$

where E_ℓ^{KB} are normalization coefficients. The function $v_{\text{ps}}^{\text{local}}$ contains the long range effects and is equal to $-Z/r$ outside of the core. The functions $v_\ell^m(\vec{r})$ are the product of a spherical harmonics Y_ℓ^m by a radial function $v_\ell(r)$ – centered on an atom – which vanishes beyond some critical radius. Being separable means that the matrix elements $\langle \psi_i | V_{\text{nl}} | \psi_j \rangle$ can be computed efficiently according to

$$\langle \phi_j | V_{\text{nl}} | \phi_k \rangle = \sum_{\ell=0}^{\ell_{\max}} \sum_{m=-\ell}^{\ell} E_\ell^{\text{KB}} f_{\ell m}(\phi_j) f_{\ell m}(\phi_k), \tag{2.13}$$

where the quantities $f_{\ell m}(\phi_n) = \langle \phi_n | v_\ell^m \rangle$ can be computed independently of each other. Since the functions v_ℓ^m are localized in real-space, the evaluation of all the matrix elements $\langle \psi_j | V_{\text{nl}} | \psi_k \rangle$ for a system of N atoms scales like $O(N^2)$.

In the applications presented in this chapter, we use the pseudopotentials proposed by HAMANN [1989]. An example is represented in Fig. 2.2 (Chlorine).

REMARK 2.3. All the atoms of the periodic table cannot be represented by pseudopotentials with the same degree of smoothness. To be represented accurately, atomic species like Oxygen or Nitrogen require finer discretization grids than Silicon for example. This makes the calculations of the electronic structure of a crystal of 64 Silicon atoms 5–10 times cheaper than the simulation of a cell of liquid water with 32 molecules, even if the number of valence electrons to compute is the same in both cases.

Using periodic boundary conditions, the total energy of a system is in principle invariant under spatial translations. Unlike in a PW approach, a real-space finite grid

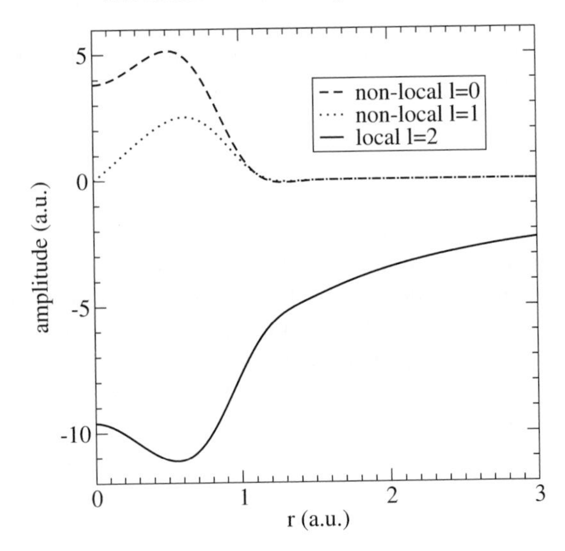

FIG. 2.2. Example of atomic pseudopotential: Chlorine for the PBE functional. The radial functions plotted are the radial components of the projector functions, v_ℓ for $\ell = 0, 1$ and the local pseudopotential $v_{\text{ps}}^{\text{local}}$ ($\ell = 2$).

representation breaks this invariance (BRIGGS, SULLIVAN and BERNHOLC [1995]). However, a discretization grid fine enough ensures that the total energy is conserved under translation within the minimal accuracy required in the calculation. Also, in order to avoid problems related to energy variations under spatial translations, the pseudopotentials can be filtered (BRIGGS, SULLIVAN and BERNHOLC [1996]). The pseudopotentials (local part and projectors) are first transformed to Fourier space by the Fourier transform

$$V_{\ell m}(\vec{k}) = \frac{1}{(2\pi)^{3/2}} \int_{\mathbb{R}^3} v_\ell^m(\vec{r}) e^{-i\vec{k}\vec{r}}\, d\vec{r} = C_{\vec{k},\ell,m} \int_0^\infty v_\ell(r) j_\ell(|\vec{k}|r) r^2\, dr. \tag{2.14}$$

The index (ℓ, m) denotes the symmetry of the functions $v_\ell^m(\vec{r}) = v_\ell(r) Y_\ell^m(\theta, \phi)$. $j_\ell(r)$ is a spherical Bessel function of order ℓ and $C_{\vec{k},\ell,m}$ is a complex factor depending on $\vec{k}, \ell, m$. For $|\vec{k}|$ larger than a cutoff k_{cut}, the coefficients $V_{\ell m}(\vec{k})$ are filtered by a Gaussian function $e^{-\beta(|\vec{k}|/k_{\text{cut}}-1)^2}$ before applying an inverse Fourier transform. One can use for example $k_{\text{cut}} = 2\pi/3h$. In practice, since the filtering depends only on $|\vec{k}|$, the coefficients $C_{\vec{k},\ell,m}$ do not need to be computed and only the 1D radial integral is required for an appropriate range of $|\vec{k}|$. The pseudopotentials lose in general their localization properties in real-space after being filtered in Fourier space and a second filtering in real-space is required to ensure that the nonlocal projectors remain confined within a limited radius. This second filtering can also be done by a Gaussian function to avoid reintroducing too many high frequency components.

Of course this filtering procedure modifies the pseudopotentials to a degree set by the grid spacing. These filtered functions will however converge towards the true pseudopotentials – together with the wave functions – as one decreases the mesh spacing. For every atomic species, one should then carefully check what grid spacing is required to ensure that the physical quantities of interest are well converged.

2.5. *Solving the eigenvalue problem*

The Kohn–Sham equations discretized by FD result in a huge 3D eigenvalue problem. Fortunately, the matrices involved are very sparse and efficient iterative methods can be used to solve this problem. In this chapter, we are going to restrict the discussion to the particularities of eigenvalue solvers for finite difference discretizations. After introducing some general features of minimization processes, we describe an appropriate multigrid preconditioner for real-space discretizations.

The multigrid full-approximation scheme (FAS) originally proposed by BRANDT [1977] is an efficient solver for nonlinear problems on a grid. In this algorithm, the entire problem has to be represented on all the grids, from the coarsest to the finest, in order to treat on equal footing all the length scales of the solution. Such an approach is not obvious to apply to the KS eigenvalue problem. Indeed, the numerous eigenfunctions of interest loose their meaning on very coarse grids, and one may be limited in the number of usable coarse grids. Only a few successful applications of the FAS algorithm for electronic structure calculations have been reported so far. They have been limited to purely academic problems (COSTINER and TAASAN [1995]) and static all electrons calculations of atoms and diatomic molecules (WANG and BECK [2000]). Recently, HEISKANEN, TORSTI, PUSKA and NIEMINEN [2001] proposed to use the Rayleigh quotient multigrid (RQMG) method (MANDEL and MCCORMICK [1989]) that avoids the coarse grid representation problem. So far it was applied only for static electronic structure calculations.

Various other methods are based on minimization schemes that make use of the steepest descent (SD) direction along which the energy functional decreases at the fastest rate. This direction is given by the gradient of the energy functional with the opposite sign. In the basis Ψ, this direction – in a space of dimension $M \times N$ – is easy to compute since it is given by the negative residual of the Kohn–Sham equations (2.3) and can be expressed as an $N \times M$ matrix

$$D^{(\Psi)} = \Psi \Lambda - H\Psi. \tag{2.15}$$

One verifies that this gradient satisfies the relation $\Psi^T D^{(\Psi)} = 0$.

For an optimum convergence rate, it is important to use the *true* steepest descent direction in algorithms expressed in nonorthogonal orbitals formulation. This direction can differ substantially from the derivative with respect to Φ if the basis Φ is highly nonorthogonal. This SD direction is easy to compute for the eigenfunctions Ψ and a simple way to obtain it in the basis Φ is to use the matrix C (from Eq. (2.7)) to derive

$$D^{(\Phi)} = D^{(\Psi)} C^{-1} = (\Psi \Lambda - H\Psi) C^{-1} = \Phi\Theta - H\Phi, \tag{2.16}$$

where $\Theta = S^{-1}H^{(\Phi)}$. In the following, we consider a SD algorithm with a linear preconditioning operator K. The basis Φ is updated according to

$$\Phi^{\text{new}} = \Phi + \eta K(\Phi\Theta - H\Phi), \tag{2.17}$$

where η is a pseudo-time step. In this algorithm all the trial wave functions are updated simultaneously. In the particular case $K = \textit{Identity}$, (2.17) is equivalent to the method proposed in GALLI and PARRINELLO [1992]. Since by definition

$$\Phi^{\text{new}} = \left(\Psi + \eta K D^{(\Psi)}\right)C^{-1} = \Psi^{\text{new}}C^{-1},$$

the same subspace is generated at each iteration, independently of the choice of the basis Φ. Note also that Eq. (2.17) does not depend on C and therefore does not require the solution of the eigenvalue problem (2.8).

REMARK 2.4. Alternatively, the above SD directions can be used in a conjugate gradient (CG) method (EDELMAN, ARIAS and SMITH [1998]). In the applications presented here, the convergence rate of the preconditioned steepest descent (PSD) algorithm with the preconditioning presented later in this Section is fast enough to make the implementation of the CG approach unnecessary.

In actual calculations, the basis functions Φ are corrected at each iteration using the PSD directions as in (2.17). A new electronic density ρ_e is then computed as well as new Hartree and exchange-correlation potentials. To avoid large oscillations of the charge distribution from step to step, these potentials can be mixed linearly with those used at the previous step. The basis Φ is refined by iterative updates until self-consistency (SC) is achieved and, at convergence accurately describe the true Kohn–Sham ground state of the system.

REMARK 2.5. For iterative algorithms based on the Ritz vectors, the transformation (2.7) is one of the most expensive part of the calculation for large scale simulations. In a nonorthogonal representation, this operation is not required anymore. The cost is however transferred to the computation of the SD directions (Eq. (2.16)).

In the particular case of a linear Hamiltonian operator and if all the orbitals are fully occupied, the Kohn–Sham functional expressed in nonorthogonal orbitals can be written as a functional without constraints

$$E_{\text{KS}} = 2\,\text{tr}\left(S^{-1}\Phi^T H\Phi\right). \tag{2.18}$$

To motivate the introduction of a preconditioner, we first estimate the expected rate of convergence of iterative algorithms based on steepest descent directions to minimize Eq. (2.18). The rate of convergence of the SD method is determined by the condition number $\chi(\mathcal{H})$ of the Hessian matrix $\mathcal{H}$ associated to the problem. To estimate $\chi(\mathcal{H})$ we compute the eigenvalues of $\mathcal{H}$. As in PFROMMER, DEMMEL and SIMON [1999], we consider electronic states $\{\phi_i\}_{i=1}^{N}$ expressed as a perturbation of the ground state eigenfunctions $\{\psi_i\}_{i=1}^{N}$

$$\phi_i = \psi_i + \sum_{l=1}^{M} c_l^{(i)} \psi_l. \tag{2.19}$$

Inserting (2.19) into (2.18), we obtain to the second order in the coefficients $c_l^{(i)}$

$$E_{\mathrm{KS}} - E_0 = 2 \sum_{i=1}^{N} \sum_{k=N+1}^{M} (\varepsilon_k - \varepsilon_i)\big(c_k^{(i)}\big)^2. \tag{2.20}$$

The coefficients $c_l^{(i)}$ for $l \leqslant N$ correspond to directions in the parameter space along which the objective function is constant. In the complementary parameter space, we obtain that the condition number of the Hessian matrix is given by

$$\chi(\mathcal{H}) = \frac{\varepsilon_M - \varepsilon_1}{\varepsilon_{N+1} - \varepsilon_N}. \tag{2.21}$$

REMARK 2.6. In general, as the number of atoms in a physical system increases, the spectrum of the Hamiltonian becomes denser, while the extreme eigenvalues (ε_M and ε_1) remain about the same.

As in PW calculations, real-space representations of the electronic wave functions require a very large number of degrees of freedom. In particular in the presence of atoms represented by very hard pseudopotentials. A very fine grid implies a quite large value for ε_M in Eq. (2.21) that negatively affects the condition number of the Hessian matrix. If Ψ is corrected at each step according to a SD algorithm without preconditioning

$$\Psi^{\mathrm{new}} = \Psi + \eta D^{(\Psi)},$$

the parameter η has to be very small – for numerical stability reasons – and the convergence can be very slow.

REMARK 2.7. Eq. (2.21) also points out at problems one can observe if the *band gap* ($\varepsilon_{N+1} - \varepsilon_N$) is very small. In practice, one can overcome this limitation by including more eigenstates than needed in the search subspace Ψ, the highest eigenstates being empty or fractionally occupied.

To introduce an appropriate preconditioner, we start by discussing the Rayleigh Quotient Iteration (RQI) method and some of its variants use in real-space electronic structure calculations. RQI is a very fast algorithm to compute one single eigenvalue of a matrix (e.g., PARLETT [1998], Chapter 4). Here we describe this method in a different form that we find more suitable for matrices of size too large to make use of direct linear solvers. Starting from an approximate eigenpair $(\varepsilon^{(k)}, \vec{\psi}^{(k)})$ of a discretized Hamiltonian matrix H, $\varepsilon^{(k)}$ given by the Rayleigh Quotient of $\vec{\psi}^{(k)}$ at step k, we look for an improved approximation $\vec{\psi}^{(k+1)}$. Specifically, we write

$$\vec{\psi}^{(k+1)} = \frac{1}{\xi}\left(\vec{\psi}^{(k)} + \delta\vec{\psi}^{(k)}\right),$$

where the correction $\delta\vec{\psi}^{(k)}$ is chosen orthogonal to $\vec{\psi}^{(k)}$. ξ is a normalization factor. Improving $\vec{\psi}^{(k)}$ by RQI requires then to find $\delta\vec{\psi}^{(k)} \perp \vec{\psi}^{(k)}$ and ξ such that

$$\left(H - \varepsilon^{(k)}\right)\frac{1}{\xi}\left(\vec{\psi}^{(k)} + \delta\vec{\psi}^{(k)}\right) = \vec{\psi}^{(k)}. \tag{2.22}$$

We eliminate ξ by projecting the whole equation onto $(\vec{\psi}^{(k)})^{\perp}$ and then rewrite it as

$$\left(I - \vec{\psi}^{(k)}(\vec{\psi}^{(k)})^T\right)\left(H - \varepsilon^{(k)}\right)\delta\vec{\psi}^{(k)} = -\left(H - \varepsilon^{(k)}\right)\vec{\psi}^{(k)}. \tag{2.23}$$

By the properties of the Rayleigh Quotient, the projector on the right hand side of the equation has been omitted.

REMARK 2.8. Eq. (2.23) is the same equation used to define the iterative corrections in the Jacobi–Davidson method (SLEIJPEN and VAN DER VORST [1996]).

This algorithm can be generalized to the simultaneous search of N eigenfunctions, with possible degeneracy of some eigenvalues (DESCLOUX, FATTEBERT and GYGI [1998], FATTEBERT [1998]). The idea is to replace the Rayleigh quotient by the Rayleigh–Ritz (RR) procedure (e.g., PARLETT [1998], Chapter 11) and to look for corrections orthogonal to the whole subspace $\Psi^{(k)}$ of trial eigenfunctions. Denoting $\Psi^{(k)} = (\vec{\psi}_1^{(k)}, \dots, \vec{\psi}_N^{(k)})$ as a matrix made of vector columns $\vec{\psi}_j^{(k)}$, we can write the following iterative algorithm:

ALGORITHM 2.1.

(1) *Let $\Psi^{(0)}$ be a trial subspace of dimension N.*
(2) *For $k = 0, 1, 2, \dots,$ do:*
   (a) *Rayleigh–Ritz for H in the subspace $\Psi^{(k)} \to \left(\varepsilon_j^{(k)}, \vec{\psi}_j^{(k)}\right)$, $j = 1, \dots, N$,*
   (b) *For $j = 1, \dots, N$, compute $\delta\vec{\psi}_j^{(k)} \perp \Psi^{(k)}$ solution of*

$$\left(I - \Psi^{(k)}\Psi^{(k)^T}\right)\left(H - \varepsilon_j^{(k)}\right)\delta\vec{\psi}_j^{(k)} = -\left(H - \varepsilon_j^{(k)}\right)\vec{\psi}_j^{(k)}, \tag{2.24}$$

   (c) *Define $\Psi^{(k+1)} = \left(\vec{\psi}_1^{(k)} + \delta\vec{\psi}_1^{(k)}, \dots, \vec{\psi}_N^{(k)} + \delta\vec{\psi}_N^{(k)}\right)$.*

An exact solution of Eq. (2.24) would lead to a locally quadratic convergence rate close to the solution of the eigenvalue problem for a non-self-consistent (i.e., *linear*) Hamiltonian as proved in FATTEBERT [1998]. Such an algorithm has actually been applied in FATTEBERT [1996], FATTEBERT [1999], where the linear systems are solved by multigrid. Slightly different versions have been proposed by JIN, JEONG and CHANG [1999] and ANCILOTTO, BLANDIN and TOIGO [1999] who omit in particular the projector in Eq. (2.24). These approaches have to deal with the difficult question of how to define meaningful potentials – and sometimes eigenfunctions – on very coarse grids. BRIGGS, SULLIVAN and BERNHOLC [1996] proposed to keep only the Laplacian in the Hamiltonian operator on the coarse grids. In this approach, the multigrid V-cycles can be seen as a preconditioner or inexact solver. Such an approach can be very efficient for self-consistent Hamiltonians, since an accurate solution for Eq. (2.24) is not always useful when the operator changes at each iteration. This point of view has been more precisely formulated in FATTEBERT and BERNHOLC [2000] where the potential operator is used only once at the beginning of the multigrid cycle to compute the residual. The main advantage of the latter approach is that the operator in Eq. (2.24) then does not depend on j and can be used in any nonorthogonal representation Φ of the trial subspace.

Let us now focus on the preconditioning approach. Looking at the correction Eq. (2.24), we note that the right hand side is the steepest descent direction for the minimization problem with orthonormality constraints associated with the KS eigenvalue problem (2.3). We note it $(-\vec{r}_j^{(k)})$, $\vec{r}_j^{(k)}$ being the residual of the eigenvalue problem. In a SD approach, $\delta\vec{\psi}_j^{(k)}$ would be given by $-\vec{r}_j^{(k)}/\varepsilon_{\max}$ where $\varepsilon_{\max}$ is the largest eigenvalue of H. However, from the point of view of the inverse iteration method, an *optimal* correction is given by

$$\delta\vec{\psi}_j^{(k)} = -\left(\left(I - \Psi^{(k)}\Psi^{(k)^T}\right)\left(H - \varepsilon_j^{(k)}\right)\big|_{\Psi^{(k)\perp}}\right)^{-1}\vec{r}_j^{(k)}. \tag{2.25}$$

Thus one can consider a preconditioner K that approximates the operator

$$\left(I - \Psi^{(k)}\Psi^{(k)^T}\right)\left(H - \varepsilon_j^{(k)}\right)\big|_{\Psi^{(k)\perp}}. \tag{2.26}$$

A close look at the Hamiltonian operator shows that for high energy states, the Laplacian is the dominant part, and the corresponding eigenfunctions are essentially similar to those of the Laplacian, i.e., plane waves perturbed by a relatively weak potential. It means that the operator $-\frac{1}{2}\nabla^2$ is a good approximation of $(H - \varepsilon_j^{(k)})$ in $\Psi^{(k)^\perp}$ for $1 \leqslant j \leqslant N$, at least close to convergence, so that one can choose

$$\begin{aligned} K &\sim \left(I - \Psi^{(k)}\Psi^{(k)^T}\right)\left(-\frac{1}{2}\nabla^2\right)\bigg|_{\Psi^{(k)\perp}} \\ &= \left(I - \Phi^{(k)}S^{-1}\Phi^{(k)^T}\right)\left(-\frac{1}{2}\nabla^2\right)\bigg|_{\Phi^{(k)\perp}}. \end{aligned} \tag{2.27}$$

In real-space, one can associate frequencies with grid resolution. Applying a single grid iterative method – like Jacobi or Gauss–Seidel – to solve a Poisson problem, one essentially obtains the high frequency components of the solution, the one that we cannot represent on a coarser grid. Using multigrid V-cycles based on such a *smoother*, we can solve the problem for components of lower frequencies by visiting coarser grids (BRANDT [1977]). Furthermore, by choosing a limited number of grids, one can *select* the components that we want to solve for. Following this heuristic argument, we define the application of the preconditioner K^{-1} to $\vec{r}_j^{(k)}$ as an iterative multigrid solver for the Poisson problem (FATTEBERT and BERNHOLC [2000])

$$-\frac{1}{2}\nabla^2\delta\vec{\psi}_j^{(k)} = -\vec{r}_j^{(k)} \tag{2.28}$$

limited to the finest grids. For practical calculations, using 2 coarse grids is often optimal. We start the process with an initial trial solution $\delta\vec{\psi}_j^{(k)} = -\alpha\vec{r}_j^{(k)}$. Its main goal is to introduce some low frequency components in the correction $\delta\vec{\psi}_j^{(k)}$. The coefficient α is defined by looking at the initial guess as the steepest descent correction one would make if the whole calculation was done on the coarse grids not visited during the V-cycles. As a smoother in the V-cycles, the Jacobi method is appropriate because of its inherent parallelism.

In Fig. 2.3, we present the convergence history of the error on the total energy for various discretization grids for an 8 atoms diamond cell self-consistent calculation. We use the PSD algorithm with the multigrid preconditioner described above, doing for each correction $\delta\vec{\psi}_j^{(k)}$ and at each SC iteration 1 V-cycle with 2 pre-smoothing and 2 post-smoothing. The grid-independence of the convergence rate is observed. All the calculations use the same coarsest grid $6 \times 6 \times 6$ for the multigrid preconditioning, and the same total number of states (16 occupied + 8 unoccupied) $N = 24$. The initial trial functions are random functions.

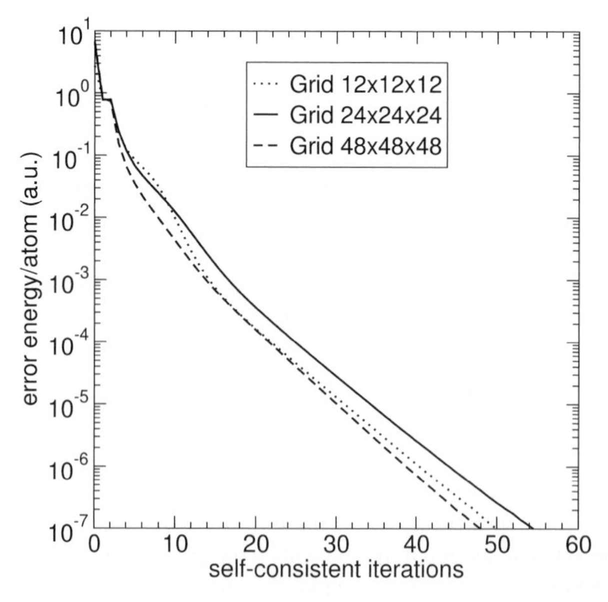

FIG. 2.3. Convergence rate for a diamond cell calculation (8 carbon atoms) for 3 different discretization grids.

REMARK 2.9. Preconditioners based on a similar idea have been developed for PW calculations (TETER, PAYNE and ALLAN [1989], FERNANDO, QUIAN, WEINERT and DAVENPORT [1989], CHETTY, WEINERT, RAHMAN and DAVENPORT [1995]). Since the numerical basis functions in PW are eigenfunctions of the Laplacian operator, efficient simple diagonal preconditioners can be designed for PW.

A different preconditioner was proposed by SAAD, STATHOPOULOS, CHELIKOWSKY, WU and OGUT [1996] in conjunction with a Lanczos algorithm. Realizing that the eigenfunctions corresponding to the lowest eigenvalues are in general smoother than the others, they proposed to apply a low frequency filter directly to the trial eigenfunctions. This is done on a single grid in real-space by an averaging of the value of a function at every grid point with the values at its neighboring points. They note however that this preconditioner is probably not always sufficient, in particular when a large number of eigenfunctions is required and the highest eigenvalues of interest correspond to eigenfunctions presenting a lower degree of smoothness.

2.6. *Energy and forces*

To optimize molecular geometries, or run molecular dynamics and measure physical quantities at finite temperature, it is important to be able to compute the forces acting on the atoms in any configuration. To derive expressions for these forces, we start by some considerations on the total energy of a physical system. In DFT, the total energy of a system can be expressed as the sum of three terms

$$E_{\mathrm{t}} = E_{\mathrm{KS}}\left[\Phi, \{\vec{R}_a\}_{a=1}^{N_a}\right] + E_{\mathrm{ions}}\left[\{\vec{R}_a\}_{a=1}^{N_a}\right] + \frac{1}{2}\sum_{a=1}^{N_a} M_a \dot{\vec{R}}_a^2, \tag{2.29}$$

where E_{ions}, the electrostatic energy between ions of charges Z_i, is given by

$$E_{\mathrm{ions}} = \frac{1}{2}\sum_{a,b=1,a\neq b}^{N_a} \frac{Z_a Z_b}{|\vec{R}_a - \vec{R}_b|}, \tag{2.30}$$

and M_a denotes the mass of the ion in $\vec{R}_a$. In the KS energy E_{KS}, the contribution due to the interaction between electrons and ions is given by the sum of two terms associated to the local and nonlocal parts of the pseudopotential

$$E_{\mathrm{ps}} = \int_\Omega v_{\mathrm{ps}}^{\mathrm{local}}(\vec{r})\rho_e(\vec{r})\,d\vec{r} + \mathrm{Tr}\left(\Phi^T V_{\mathrm{nl}}\Phi\rho^{(\Phi)}\right) = E_{\mathrm{ps}}^{\mathrm{local}} + E_{\mathrm{ps}}^{\mathrm{nl}}. \tag{2.31}$$

It is computationally more efficient to compute the electrostatic term E_{es} by solving a Poisson problem. In order to deal with a neutral charge, it is a standard procedure to add to the system smeared core charges centered at atomic sites,

$$\rho_a(\vec{r}) = -\frac{Z_a}{(\sqrt{\pi}r_c^a)^3}\exp\left(-\frac{|\vec{r} - \vec{R}_a|^2}{(r_c^a)^2}\right). \tag{2.32}$$

The sum of these charges, ρ_s, neutralizes the electronic charge by generating a total potential

$$v_s(\vec{r}) = \sum_{a=1}^{N_a} \frac{-Z_a}{|\vec{r}-\vec{R}_a|} \operatorname{erf}\left(\frac{|\vec{r}-\vec{R}_a|}{r_c^a}\right). \tag{2.33}$$

We then compute the Hartree potential v_{H} as the solution of a Poisson problem for a neutral total charge $\rho_e + \rho_s$,

$$-\nabla^2(v_{\mathrm{H}} + v_s)(\vec{r}) = 4\pi(\rho_e + \rho_s)(\vec{r}), \tag{2.34}$$

with periodic or Dirichlet boundary conditions. This problem can be efficiently solved on the discretization grid in $O(N)$ operations by the multigrid method (BRANDT [1977]).

With the introduction of smeared neutralizing core charges, one can write

$$E_{\mathrm{ions}} = \frac{1}{2}\sum_{a,b=1}^{N_a} \int_{\mathbb{R}^3} \frac{\rho_a(\vec{r}-\vec{R}_a)\rho_b(\vec{r}\,'-\vec{R}_b)}{|\vec{r}-\vec{r}\,'|}\, d\vec{r}\, d\vec{r}\,' - E_{\mathrm{self}} + E_{\mathrm{diff}}, \tag{2.35}$$

where E_{self} is the self-interaction of the core charges,

$$E_{\mathrm{self}} = \frac{1}{2}\sum_{a=1}^{N_a} \int_{\mathbb{R}^3} \frac{\rho_a(\vec{r}-\vec{R}_a)\rho_a(\vec{r}\,'-\vec{R}_a)}{|\vec{r}-\vec{r}\,'|}\, d\vec{r}\, d\vec{r}\,' = \frac{1}{\sqrt{2\pi}}\sum_{a=1}^{N_a} \frac{Z_a^2}{(r_c^a)^2} \tag{2.36}$$

and

$$\begin{aligned} E_{\mathrm{diff}} &= \frac{1}{2}\sum_{a,b=1,a\neq b}^{N_a} \left[\frac{Z_a Z_b}{|\vec{R}_a-\vec{R}_b|} - \int_{\mathbb{R}^3} \frac{\rho_a(\vec{r}-\vec{R}_a)\rho_b(\vec{r}\,'-\vec{R}_b)}{|\vec{r}-\vec{r}\,'|}\, d\vec{r}\, d\vec{r}\,'\right] \\ &= \sum_{a,b=1,a<b}^{N_a} \frac{Z_a Z_b}{|\vec{R}_a-\vec{R}_b|} \operatorname{erfc}\left(\frac{|\vec{R}_a-\vec{R}_b|}{\sqrt{(r_c^a)^2+(r_c^b)^2}}\right). \end{aligned} \tag{2.37}$$

We then have, for r_c^a sufficiently small compared to Ω,

$$\begin{aligned} &E_{\mathrm{es}} + E_{\mathrm{ps}}^{\mathrm{local}} + E_{\mathrm{ions}} \\ &= \frac{1}{2}\int_\Omega \frac{\rho_e(\vec{r})\rho_e(\vec{r}\,')}{|\vec{r}-\vec{r}\,'|}\, d\vec{r}\, d\vec{r}\,' + \int_\Omega v_{\mathrm{ps}}^{\mathrm{local}} \rho_e(\vec{r})\, d\vec{r} \\ &\quad + \frac{1}{2}\sum_{a,b=1}^{N_a} \int_{\mathbb{R}^3} \frac{\rho_a(\vec{r}-\vec{R}_a)\rho_b(\vec{r}\,'-\vec{R}_b)}{|\vec{r}-\vec{r}\,'|}\, d\vec{r}\, d\vec{r}\,' - E_{\mathrm{self}} + E_{\mathrm{diff}} \end{aligned}$$

$$
\begin{aligned}
&\approx \frac{1}{2}\int_\Omega \frac{\big(\rho_e(\vec r)+\rho_s(\vec r)\big)\big(\rho_e(\vec r\,')+\rho_s(\vec r\,')\big)}{|\vec r-\vec r\,'|}\,d\vec r\,d\vec r\,' - \int_\Omega \frac{\rho_e(\vec r)\rho_s(\vec r\,')}{|\vec r-\vec r\,'|}\,d\vec r\,d\vec r\,' \\
&\quad + \int_\Omega v_{\rm ps}^{\rm local}(\vec r)\rho_e(\vec r)\,d\vec r - E_{\rm self} + E_{\rm diff} \\
&= \frac{1}{2}\int_\Omega \big(\rho_e(\vec r)+\rho_s(\vec r)\big)(v_{\rm H}+v_s)(\vec r)\,d\vec r \\
&\quad + \int_\Omega \big(v_{\rm ps}^{\rm local}-v_s\big)(\vec r)\rho_e(\vec r)\,d\vec r - E_{\rm self} + E_{\rm diff}.
\end{aligned}
\tag{2.38}
$$

Knowing the ground state electronic structure for a given atomic configuration $\{\vec R_a\}_{a=1}^{N_a}$, one can compute the internal force acting on the ion I by deriving the total energy with respect to the atomic coordinates $\vec R_I$,

$$
\vec F_I = -\frac{d}{d\vec R_I} E_t\big(\Phi, \{\vec R_a\}_{a=1}^{N_a}\big). \tag{2.39}
$$

Using the property that Φ is the minimum of the functional E, one shows that

$$
\vec F_I = -\frac{\partial}{\partial\vec R_I} E_t\big(\Phi, \{\vec R_a\}_{a=1}^{N_a}\big) \tag{2.40}
$$

(Hellmann–Feynman forces, FEYNMAN [1939]). Since the electronic structure does not explicitly depend on the atomic positions, Eq. (2.40) means that the forces can be computed from a single ground state calculation, by deriving the atomic potentials only.

REMARK 2.10. To obtain Eq. (2.40), we also use the fact that the numerical representation of Ψ does not explicitly depend on the atomic positions since the grid is atom independent. For atom-centered orbitals moving with the atoms, this is not true anymore and additional terms (Pulay forces) have to be included.

From Eqs. (2.38) and (2.40), using Eq. (2.34), we obtain the total force acting on atom I in the form

$$
\begin{aligned}
\vec F_I &= \int_\Omega \big(v_{\rm H}(\vec r)+v_s(\vec r)\big)\frac{d}{d\vec R_I}\rho_s\,d\vec r + \int_\Omega \frac{d}{d\vec R_I}\big(v_{\rm ps}^{\rm local}-v_s\big)\rho_e(\vec r)\,d\vec r \\
&\quad + \frac{\partial}{\partial\vec R_I}{\rm Tr}\big(\Phi^T V_{\rm nl}\Phi\rho^{(\Phi)}\big) + \frac{d}{d\vec R_I}E_{\rm diff}.
\end{aligned}
\tag{2.41}
$$

Writing the forces in this form lets appear the functions $(v_{\rm ps}^{\rm local}-v_s)$ and ρ_s which are localized in real-space. This can be directly used to reduce the complexity of the computation of the forces on a grid. In principle all the derivatives with respect to $\vec R_I$ in Eq. (2.41) can be computed analytically. In practice, because of the filtering of the pseudopotentials, the derivatives have to be evaluated numerically on the filtered pseudopotentials.

REMARK 2.11. The sum of the forces over all the atoms should be zero if no external force is applied. In practice, the use of a finite grid introduces small errors (see Section 2.4) and provides an estimate of the accuracy of the forces.

2.7. *Born–Oppenheimer molecular dynamics*

To perform Born–Oppenheimer molecular dynamics simulations of quantum systems described by the KS equations, we compute the forces acting on the ions according to Eq. (2.41) and let the system evolve accordingly. The ions evolve like classical particles surrounded by quantum electrons (Born–Oppenheimer approximation). The error in the energy is second order with respect to the error in the electronic wave functions, but the error in the forces is first order. It means that one should be particularly careful in the computation of the ground state of the KS energy functional for the each atomic configuration at each iteration. It is particularly important to have accurate forces to ensure a perfect conservation of the total energy of the system in a microcanonical simulation.

As shown by JING, TROULLIER, DEAN, BINGGELI and CHELIKOWSKY [1994] and BRIGGS, SULLIVAN and BERNHOLC [1996], the computation of the forces in FD methods is accurate enough to allow for energy conserving microcanonical ab initio simulations. This is illustrated in Fig. 2.4 where we show the evolution of the energy during a molecular dynamics simulation of a Si_5 cluster. To avoid any systematic drift of the energy due to the integration scheme, the equations of motions were integrated numerically using the time reversible Verlet's second order algorithm (e.g., HEERMANN [1990], Chapter 3)

$$\vec{R}_I^{(n+1)} = 2\vec{R}_I^{(n)} - \vec{R}_I^{(n-1)} + \vec{F}_I^{(n)}(\Delta t)^2/M_I$$

A grid spacing $h = 0.56$ Bohr and a time step $\Delta t = 80$ a.u. were used for this simulation.

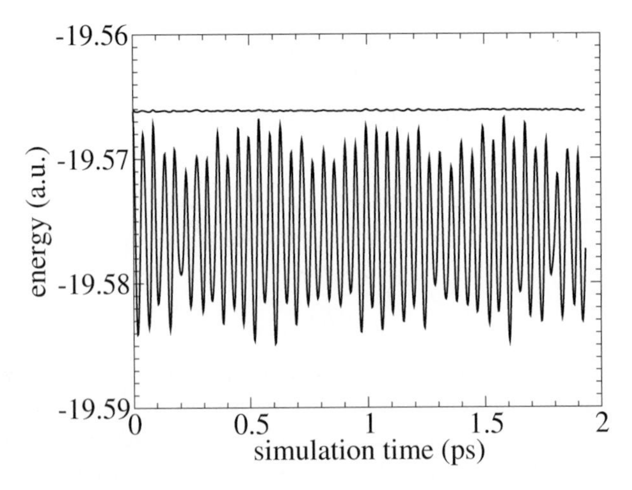

FIG. 2.4. Molecular dynamics simulation of a Si_5 cluster. The total energy (top) shows fluctuations of the order 10^{-4} a.u., but no systematic drift. The KS energy is also plotted (bottom).

2.8. *Localized orbitals*

Formulating the minimization problem in terms of nonorthogonal functions Φ instead of the Ritz functions Ψ, one can formally impose localization constraints on Φ to reduce the cost of the calculation. This is one of the most popular methods to obtain a linear scaling of the computational cost with respect to the size of the system. These so called Order N methods are discussed in another chapter (GOEDECKER [2003], this Handbook, e.g.) and we will limit here the discussion to the grid-based method exposed in this chapter.

On a real-space grid, spatial localization can be imposed by forcing each orbital to be zero outside of a spherical region centered on a particular ion (HOSHI and FUJIWARA [1997], FATTEBERT and BERNHOLC [2000]). Such a truncation will linearize the computational cost of $D^{(\Phi)}$ in Eq. (2.16) and ρ_e in Eq. (2.11), the most expensive operations in the minimization algorithm. It also reduces to $O(N)$ the storage requirements for the wave functions. Of course, this reduction in computational cost does not come for free. It introduces some approximation error that one expects to keep within a certain tolerance. It means in particular that one cannot choose the localization regions arbitrarily small.

The application of the compact FD Laplacian operator to a wave function localized in a sphere of radius R_c generates a function localized in a sphere of radius $R_c + h\sqrt{2}$, which is used as the localization radius for $H\phi_j$. This truncation suppresses some components of $H\phi_j$ that are generated by the nonlocal, short-range projectors of the pseudopotential operator. These components are lost in the correction of the wave functions since the latter should remain localized. However, they are included exactly in the computation of the matrix Θ and the total energy by writing $H^{(\Phi)}$ as the sum of two matrices

$$H^{(\Phi)} = \Phi^T (H - V_{\mathrm{nl}})\Phi + \Phi^T V_{\mathrm{nl}}\Phi,$$

which isolates the nonlocal potential V_{nl} in the second term. This second term is easily computed in $O(N)$ operations using only the nonzero terms $\langle\phi_j|v_i\rangle$. Therefore, the only approximation in this approach is the use of a localization radius to limit the spatial extent of each nonorthogonal orbital.

Since the eigenfunctions are in general not localized, the matrix C that solves Eq. (2.8) is not sparse and the computation of C requires $O(N^3)$ operations. To linearize the cost of the whole calculation, one could impose localization constraints on the density matrix, requiring $\bar{\rho}_{ij}^{(\Phi)} = 0$ if the localization regions of orbitals i and j are separated by a distance larger than a truncation radius R_ρ as in HERNANDEZ and GILLAN [1995]. This can be imposed at each step of the iterative minimization in order to achieve linear scaling. Such a truncature is justified by the exponential decay of $\rho(\vec{r}, \vec{r}')$ as $|\vec{r} - \vec{r}'| \to \infty$ in insulators or metals at a finite temperature (ISMAEL-BEIGI and ARIAS [1999]). However, for $M \gg N$, the full evaluation of S^{-1}, Θ or $\bar{\rho}^{(\Phi)}$, even with an order N^3 algorithm, constitutes a small fraction of the total calculations for a large range of system sizes. Since a good accuracy can be obtained only by keeping the number of nonzero elements in $\bar{\rho}$ much larger than in S (MILLAM and SCUSERIA [1997]), using the sparsity of $\bar{\rho}$ does not lead to much gain in this context.

REMARK 2.12. If an exact and explicit $O(N^3)$ diagonalization is performed, in Eq. (2.8), partially occupied and unoccupied orbitals can be used, which permits calculations for metallic as well as semiconducting systems. However, calculations for metals may require more localized orbitals or larger localization radii for an accurate description of their electronic structure.

For systems with $N > 1000$, solving Eq. (2.8) on a single processor becomes very expensive, if required at each self-consistent iteration. However, for fully parallel calculations, it is natural to also parallelize the $N \times N$ submatrices operations. This can be done using standard libraries, such as PBLAS (Parallel Basic Linear Algebra Subprograms) and ScaLapack (BLACKFORD, CHOI, CLEARY, D'AZEVEDO, DEMMEL, DHILLON, DONGARRA, HAMMARLING, HENRY, PETITET, STANLEY, WALKER and WHALEY [1997]).

REMARK 2.13. According to PBLAS and ScaLapack requirements, S, $H^{(\Phi)}$, Θ and $\bar{\rho}^{(\Phi)}$ are stored as full $N \times N$ matrices, distributed among the processors. Although most of the operations on these matrices can be optimized using their sparsity – except the diagonalization in Eq. (2.9) – the full storage approach is adequate for a substantial range of calculations. It is also the easiest implementation given the available standard numerical libraries for distributed memory multiprocessors computers. The solution of the eigenvalue problem (2.9) is clearly the dominant part of these $O(N^3)$ operations.

In the iterative minimization of the KS energy functional, the truncature of the orbitals modifies the correction directions in a way that can slow down the convergence process. On the other hand, the localization constraints break the invariance in the representation of the occupied subspace and may generate multiple local minima for E_{KS} (GOEDECKER [2003], this Handbook). One way to deal with these issues is to choose localization radii large enough so that one can easily end up in a minima of energy close enough to the *true* global minima – the one obtained without localization constraints. For example, FATTEBERT and BERNHOLC [2000] were able to compute accurately energy differences in a carbon nanotube using a localization radius of 8 Bohr.

Since the method described above allows to determine the eigenfunctions of the Kohn–Sham equations in a basis of localized functions – according to Eq. (2.7) –, it can be considered as a generalization of *ab initio* methods that use a linear combination of atomic orbitals (LCAO) to expand the eigenfunctions: $\psi_j = \sum_i c_i \phi_i$. The main difference is that grid-based local functions ϕ_i are defined by their values on a grid and are variationally optimized according to their environment. In particular, the functions ϕ_i have many more degrees of freedom and one can systematically increase the accuracy of the calculations by mesh refinement or expansion of the localization domain. An example of such an orbital computed for a $(5, 5)$ carbon nanotube is plotted in Fig. 2.5. The total number of basis functions in high precision calculations is thus much smaller than in LCAO approaches and minimizes the $O(N^3)$ part.

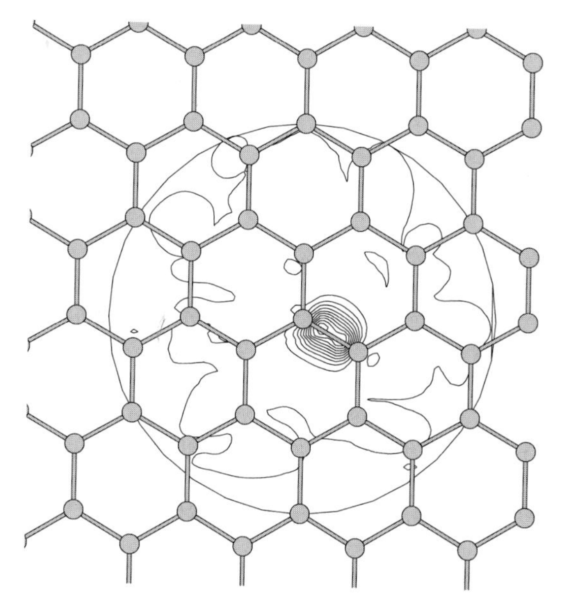

FIG. 2.5. Contour plot of the square of a typical localized orbital in the plane defined by the cylindrical surface of a (5, 5) nanotube. The external circle shows the localization region (radius 8 Bohr).

3. Quantum transport

3.1. Electron transmission and Green's functions

Let us consider a system composed of a conductor, C, connected to two semi-infinite leads, R and L, as in Fig. 3.1. A fundamental result in the theory of electronic transport is that the conductance through a region of interacting electrons (the C region in Fig. 3.1) is related to the scattering properties of the region itself via the Landauer formula (LANDAUER [1970])

$$\mathcal{C} = \frac{2e^2}{h}\mathcal{T}, \tag{3.1}$$

where $\mathcal{T}$ is the transmission function and $\mathcal{C}$ is the conductance. The former represents the probability that an electron injected at one end of the conductor will transmit to the other end. In principle, we can compute the transmission function for a coherent conductor[1] starting from the knowledge of the scattering matrix, $\mathcal{S}$. The latter is the mathematical quantity that describes the response at one lead due an excitation at another. In principle, the scattering matrix can be uniquely computed from the solution of the Schroedinger equation and would suffice to describe the transport processes we are interested in this work. However, it is a general result of conductance theory that

[1] A conductor is said to be coherent if it can be characterized by a transmission matrix that relates each of the outgoing wave amplitudes to the incoming wave amplitudes at a given energy

FIG. 3.1. A conductor described by the Hamiltonian H_C, connected to two semi-infinite leads L and R, through the coupling matrices h_{LC} and h_{CR}.

the elements of the $\mathcal{S}$-matrix can be expressed in terms of the Green's function of the conductor (DATTA [1995], FISHER and LEE [1981], MEIR and WINGREEN [1992]) which, in practice, can be sometimes simpler to compute.

Let us consider a physical system represented by an Hamiltonian H. Its Green's function for an energy E is defined by the equation

$$(E \pm i\eta - H)G(\vec{r}, \vec{r}') = \delta(\vec{r}, \vec{r}'), \tag{3.2}$$

where $i\eta > 0$ is an infinitesimal imaginary part added to the energy to incorporate the boundary conditions into the equation. The solution with $+$ sign is the retarded Green's function G^r, while the solution with $-$ sign is called advanced Green's function G^a. The transmission function can then be expressed in terms of the Green's functions of the conductors and the coupling of the conductor to the leads in a simple manner (see DATTA [1995], p. 141 and ff.)

$$\mathcal{T} = \mathrm{Tr}(\Gamma_L G_C^r \Gamma_R G_C^a), \tag{3.3}$$

where $G_C^{\{r,a\}}$ are the retarded and advanced Green's functions of the conductor, and $\Gamma_{\{L,R\}}$ are functions that describe the coupling of the conductor to the leads.

In the following we are going to restrict the discussion to discrete systems that we can describe by ordinary matrix algebra. More precisely, we are going to work with matrices representing a physical system in a basis of localized electronic orbitals centered on the atoms constituting the system. It includes in particular the tight-binding model.

For a discrete media, the Green's function is then solution of a matrix equation

$$(\varepsilon - H)G = I, \tag{3.4}$$

where $\varepsilon = E \pm i\eta$ with η arbitrarily small and I is the identity matrix. To simplify the notations, we drop the exponent $\{a, r\}$ referring to advanced and retarded functions when implicitly defined by ε. For an open system, consisting of a conductor and two semi-infinite leads (see Fig. 3.1), the above Green's function can be partitioned into sub-matrices that correspond to the individual subsystems

$$\begin{pmatrix} g_L & g_{LC} & g_{LCR} \\ g_{CL} & G_C & g_{CR} \\ g_{LRC} & g_{RC} & g_R \end{pmatrix} = \begin{pmatrix} (\varepsilon - h_L) & -h_{LC} & 0 \\ -h_{LC}^* & (\varepsilon - H_C) & -h_{CR} \\ 0 & -h_{CR}^* & (\varepsilon - h_R) \end{pmatrix}^{-1}, \tag{3.5}$$

where the matrix $(\varepsilon - H_C)$ represents the finite "isolated" conductor (with no coupling elements to the leads), $(\varepsilon - h_{\{R,L\}})$ represent the semi-infinite leads, and h_{CR} and

h_{LC} are the coupling matrices between the conductor and the leads, and h^* denotes the familiar conjugate transpose of h. As a convention, we use lower case letters for (semi-)infinite matrices and upper case for finite dimension matrices. In Eq. (3.5) we have made the assumption that there is no direct interaction between the left and right leads. From this equation it is straightforward to obtain an explicit expression for G_C

$$G_C = (\varepsilon - H_C - \Sigma_L - \Sigma_R)^{-1}, \tag{3.6}$$

where the finite-dimension matrices

$$\Sigma_L = h_{LC}^*(\varepsilon - h_L)^{-1} h_{LC}, \qquad \Sigma_R = h_{RC}(\varepsilon - h_R)^{-1} h_{RC}^* \tag{3.7}$$

are defined as the self-energy terms due to the semi-infinite leads. The self-energy terms can be viewed as effective Hamiltonians that arise from the coupling of the conductor with the leads. The coupling functions $\Gamma_{\{L,R\}}$ can then be obtained as (DATTA [1995])

$$\Gamma_{\{L,R\}} = \mathrm{i}\big[\Sigma_{\{L,R\}}^r - \Sigma_{\{L,R\}}^a\big], \tag{3.8}$$

where the advanced self-energy $\Sigma_{\{L,R\}}^a$ is the conjugate transpose of the retarded self-energy $\Sigma_{\{L,R\}}^r$. The core of the problem lies in the calculation of the self-energies of the semi-infinite leads.

It is well known that any solid (or surface) can be viewed as an infinite (semi-infinite in the case of surfaces) stack of principal layers with nearest-neighbor interactions (LEE and JOANNOPOULOS [1981a], LEE and JOANNOPOULOS [1981b]). This corresponds to transforming the original system into a linear chain of principal layers. For a lead-conductor-lead system, the conductor can be considered as one principal layer sandwiched between two semi-infinite stacks of principal layers. The next three sections are devoted to the computation of the self-energies using the principal layers approach for different geometries.

3.2. *Transmission through a bulk system*

Within the principal layer approach, the matrix elements of Eq. (3.4) between layer orbitals yield a series of matrix equations for the Green's functions

$$\begin{aligned}
(\varepsilon - H_{00})G_{00} &= I + H_{01}G_{10}, \\
(\varepsilon - H_{00})G_{10} &= H_{01}^* G_{00} + H_{01}G_{20}, \\
&\cdots \\
(\varepsilon - H_{00})G_{n0} &= H_{01}^* G_{n-1,0} + H_{01}G_{n+1,0},
\end{aligned} \tag{3.9}$$

where the finite dimension matrices H_{nm} and G_{nm} are formed by the matrix elements of the Hamiltonian and Green's function between the layer orbitals. We assume that in a bulk system $H_{00} = H_{11} = \cdots$ and $H_{01} = H_{12} = \cdots$. Following LOPEZ-SANCHO, LOPEZ-SANCHO and RUBIO [1984], LOPEZ-SANCHO, LOPEZ-SANCHO and RUBIO

[1985], this chain can be transformed in order to express the Green's function of one individual layer in terms of the Green's function of the preceding (or following) one. This is done via the introduction of the transfer matrices T and $\overline{T}$, defined such that

$$G_{10} = TG_{00}$$

and

$$G_{00} = \overline{T}G_{10}.$$

Using these definitions, we can write the bulk Green's function as (GARCIA-MOLINER and VELASCO [1992])

$$G(E) = \left(\varepsilon - H_{00} - H_{01}T - H_{01}^{*}\overline{T}\right)^{-1}. \tag{3.10}$$

The transfer matrix can be easily computed from the Hamiltonian matrix elements via an iterative procedure, as outlined in LOPEZ-SANCHO, LOPEZ-SANCHO and RUBIO [1984]. In particular T and $\overline{T}$ can be written as

$$T = t_0 + \tilde{t}_0 t_1 + \tilde{t}_0 \tilde{t}_1 t_2 + \cdots + \tilde{t}_0 \tilde{t}_1 \cdots \tilde{t}_{n-1} t_n,$$
$$\overline{T} = \tilde{t}_0 + t_0 \tilde{t}_1 + t_0 t_1 \tilde{t}_2 + \cdots + t_0 t_1 \cdots t_{n-1} \tilde{t}_n,$$

where t_i and $\tilde{t}_i$ are defined via the recursion formulas

$$t_i = \left(I - t_{i-1}\tilde{t}_{i-1} - \tilde{t}_{i-1}t_{i-1}\right)^{-1} t_{i-1}^2,$$
$$\tilde{t}_i = \left(I - t_{i-1}\tilde{t}_{i-1} - \tilde{t}_{i-1}t_{i-1}\right)^{-1} \tilde{t}_{i-1}^2$$

and

$$t_0 = (\varepsilon - H_{00})^{-1} H_{01}^{*},$$
$$\tilde{t}_0 = (\varepsilon - H_{00})^{-1} H_{01}.$$

The process is repeated until $t_n, \tilde{t}_n \leqslant \delta$ with δ arbitrarily small. Usually no more than 5 or 6 terms are required to converge the above sums.

In the hypothesis of leads and conductors being of the same material (bulk conductivity), we can identify one principal layer of the bulk system with the conductor C, so that $H_{00} \equiv H_C$. If we compare Eq. (3.10) with Eq. (3.6), we obtain the expression for the self-energies of the conductor-leads system

$$\Sigma_L = H_{01}^{*}\overline{T}, \qquad \Sigma_R = H_{01}T. \tag{3.11}$$

The coupling functions are then obtained from the sole knowledge of the transfer matrices and the coupling Hamiltonian matrix elements: $\Gamma_L = -\mathrm{Im}(H_{01}^{*}\overline{T})$ and $\Gamma_R = -\mathrm{Im}(H_{01}\overline{T})$ (BUONGIORNO NARDELLI [1999]).

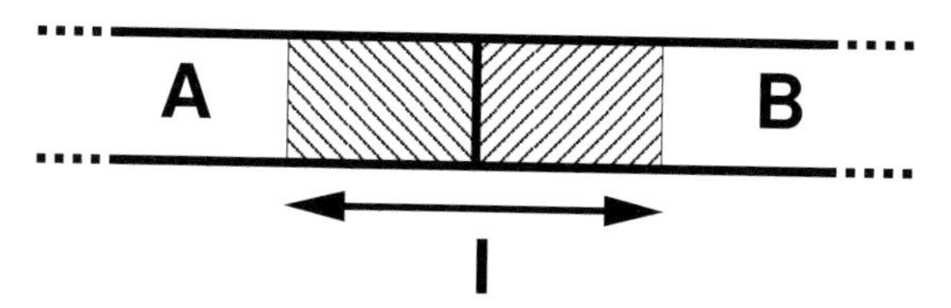

FIG. 3.2. Sketch of a system containing an interface between two media A and B. I is the interface region for which we need to compute the Green's function G_I. I is composed of two principal layers, one in each side of the interface (dashed area).

REMARK 3.1. In the application of the Landauer formula, it is customary to compute the transmission probability from one lead to the other assuming that the leads are connected to a reflectionless contact whose electron energy distribution is known (see for instance DATTA [1995], p. 59 and ff.).

3.3. *Transmission through an interface*

The procedure outlined above can also be applied in the case of electron transmission through an interface between two different media A and B. To study this case we make use of the Surface Green's Function Matching (SGFM) theory, pioneered by GARCIA-MOLINER and VELASCO [1991], GARCIA-MOLINER and VELASCO [1992].

We have to compute the Green's function G_I, where the subscript I refers to the interface region composed of two principal layers – one in each media – (see Fig. 3.2). Using the SGFM method, G_I is calculated from the bulk Green's function of the isolated systems, G_A and G_B, and the coupling between the two principal layers at the two sides of the interface, H_{AB} and H_{BA}. Via the calculation of the transmitted and reflected amplitudes of an elementary excitation that propagates from medium A to medium B, it can be shown that the interface Green's function obeys the equation (GARCIA-MOLINER and VELASCO [1992], Chapter 4)

$$G_I = \begin{pmatrix} G_{AA} & G_{AB} \\ G_{BA} & G_{BB} \end{pmatrix} = \begin{pmatrix} \varepsilon - H_{00}^A - (H_{01}^A)^* \overline{T} & -H_{AB} \\ -H_{BA} & \varepsilon - H_{00}^B - H_{01}^B T \end{pmatrix}^{-1}. \tag{3.12}$$

Once the interface Green's function is known, we can compute the transmission function in terms of block super-matrices

$$\mathcal{T}(E) = \mathrm{Tr}\big(\Gamma_A G_{AB}^r \Gamma_B G_{BA}^a\big)$$

with $\Gamma_{\{A,B\}} = \mathrm{i}\big[\Sigma_{\{A,B\}}^r - \Sigma_{\{A,B\}}^a\big]$, $\Sigma_{\{A,B\}}$ given by the analogous of Eq. (3.11) for the two semi-infinite sections, and $G_{BA}^a = (G_{AB}^r)^*$ (BUONGIORNO NARDELLI [1999]).

3.4. *Transmission through a left lead-conductor-right lead (LCR) system*

Within the SGFM framework, the approach described in the previous section can be extended to the case of multiple interfaces and superlattices (GARCIA-MOLINER and

VELASCO [1991], GARCIA-MOLINER and VELASCO [1992]) with little complication. For the calculation of conductances in realistic experimental geometry, the method can be expanded to the general configuration of a Left-lead-Conductor-Right-lead (LCR) systems – as displayed in Fig. 3.1. In the language of block matrices and principal layers, outlined in the previous sections, the LCR Green's function obeys the equation

$$
\begin{aligned}
G_{LCR} &= \begin{pmatrix} G_L & G_{LC} & G_{LR} \\ G_{CL} & G_C & G_{CR} \\ G_{RL} & G_{RC} & G_R \end{pmatrix} \\
&= \begin{pmatrix} \varepsilon - H_{00}^L - (H_{01}^L)^* \overline{T} & -H_{LC} & 0 \\ -H_{CL} & \varepsilon - H_C & -H_{CR} \\ 0 & -H_{RC} & \varepsilon - H_{00}^R - H_{01}^R T \end{pmatrix}^{-1},
\end{aligned} \tag{3.13}
$$

where $H_{nm}^{\{L,R\}}$ are the block matrices of the Hamiltonian between the layer orbitals in the left and right leads respectively, and $T_{\{L,R\}}$ and $\overline{T}_{\{L,R\}}$ are the appropriate transfer matrices. The latter are easily computed from the Hamiltonian matrix elements via the iterative procedure already described in the bulk case (Section 3.2). Correspondingly, H_{LC} and H_{CR} are the coupling matrices between the conductor and the leads principal layers in contact with the conductor.

As in Section 3.1, it is straightforward to obtain in the form of Eq. (3.6), $G_C = (\varepsilon - H_C - \Sigma_L - \Sigma_R)^{-1}$, where Σ_L and Σ_R are the self-energy terms due to the semi-infinite leads, and identify (BUONGIORNO NARDELLI and BERNHOLC [1999])

$$
\begin{aligned}
\Sigma_L &= H_{LC}^* \big(\varepsilon - H_{00}^L - (H_{01}^L)^* \overline{T}_L\big)^{-1} H_{LC}, \\
\Sigma_R &= H_{CR} \big(\varepsilon - H_{00}^R - H_{01}^R T_R\big)^{-1} H_{CR}^*.
\end{aligned} \tag{3.14}
$$

The transmission function in the LCR geometry can then be derived from Eqs. (3.3) and (3.8).

REMARK 3.2. The knowledge of the conductor's Green's function G_C gives also direct information on the electronic spectrum of the system via the spectral density of electronic states

$$N(E) = -(1/\pi)\mathrm{Im}\big[\mathrm{Tr}\big(G_C(E)\big)\big].$$

REMARK 3.3. We have assumed a truly one-dimensional chain of principal layers, which is physical only for systems like nanotubes or quantum wires that have a definite quasi-one-dimensional character. The extension to a truly three-dimensional case is straightforward using Bloch functions wave vectors $\vec{k}_\parallel$ parallel to the layers (in the directions perpendicular to the conduction). The introduction of the principal layer concept implies that along the direction of the conduction the system is described by an infinite set of wave vectors $\vec{k}_\perp$. The above procedure effectively reduces the three-dimensional system to a set of noninteracting linear-chains, one for each $\vec{k}_\parallel$ (LEE and

JOANNOPOULOS [1981a], LEE and JOANNOPOULOS [1981b]). We can then use the usual k-point summation techniques to evaluate, for instance, the quantum conductance

$$T(E) = \sum_{\vec{k}_\parallel} w_{\vec{k}_\parallel} T_{\vec{k}_\parallel}(E),$$

where $w_{\vec{k}_\parallel}$ are the relative weights of the different wave vectors $\vec{k}_\parallel$ in the irreducible wedge of the surface Brillouin zone (BALDERESCHI [1973]).

3.5. Generalization to nonorthogonal orbitals

In the previous sections we have assumed to have a Hamiltonian representation in terms of orthogonal orbitals. The expression for the Green's and transmission functions of a bulk system described by a general nonorthogonal localized-orbital Hamiltonian follows directly from the procedure outlined in Section 3.2. All the quantities can be obtained making the substitutions $(\varepsilon - H_{00}) \to (\varepsilon S_{00} - H_{00})$ and $H_{01} \to -(\varepsilon S_{01} - H_{01})$. Here, we introduce the matrices S that represent the overlap between the localized orbitals. With this recipe, the equation chain (3.9) now reads

$$\begin{aligned}
(\varepsilon S_{00} - H_{00})G_{00} &= I - (\varepsilon S_{01} - H_{01})G_{10}, \\
(\varepsilon S_{00} - H_{00})G_{10} &= -\big(\varepsilon S_{01}^* - H_{01}^*\big)G_{00} - (\varepsilon S_{01} - H_{01})G_{20}, \\
&\;\;\cdots \\
(\varepsilon S_{00} - H_{00})G_{n0} &= -\big(\varepsilon S_{01}^* - H_{01}^*\big)G_{n-1,0} - (\varepsilon S_{01} - H_{01})G_{n+1,0}.
\end{aligned}$$

From here, via the same series of algebraic manipulations as in the orthogonal case, we obtain the Green's function

$$G = \big[(\varepsilon S_{00} - H_{00}) + (\varepsilon S_{01} - H_{01})T + \big(\varepsilon S_{01}^* - H_{01}^*\big)\overline{T}\,\big]^{-1},$$

and from the latter we can identify the self-energies

$$\Sigma_L = -\big(\varepsilon S_{01}^* - H_{01}^*\big)\overline{T}, \qquad \Sigma_R = -(\varepsilon S_{01} - H_{01})T.$$

The above procedure can be extended to the case of the transmission through an interface or a LCR junction. For the latter case, we obtain

$$\begin{aligned}
\Sigma_L &= (\varepsilon S_{LC} - H_{LC})^* \\
&\quad \times \big[\varepsilon S_{00}^L - H_{00}^L + \big(\varepsilon S_{01}^L - H_{01}^L\big)^* \overline{T}_L\big]^{-1}(\varepsilon S_{LC} - H_{LC}), \\
\Sigma_R &= (\varepsilon S_{CR} - H_{CR}) \\
&\quad \times \big[\varepsilon S_{00}^R - H_{00}^R + \big(\varepsilon S_{01}^R - H_{01}^R\big)T_R\big]^{-1}(\varepsilon S_{CR} - H_{CR})^*,
\end{aligned} \tag{3.15}$$

where $H_{nm}^{\{L,R\}}$ are the matrix elements of the Hamiltonian between layer orbitals in the left and right leads, respectively. $S_{nm}^{\{L,R\}}$ are the corresponding overlap matrices and $T_{\{L,R\}}$ and $\overline{T}_{\{L,R\}}$ are the appropriate transfer matrices. The latter are easily computed from the Hamiltonian and overlap matrix elements via the usual iterative procedure (see Section 3.2). Correspondingly, H_{LC}, H_{CR}, S_{LC} and S_{CR} are the coupling and overlap matrices for the conductor-leads assembly.

4. Applications: conductivity from *ab initio* local orbital Hamiltonian

4.1. Methodology

The procedure described in Section 3 requires the knowledge of the Hamiltonian and overlap matrix elements between layer orbitals of the conductor, and the left and right leads. In *ab initio* density-functional calculations, such matrix elements can be computed using the $O(N)$-like algorithm described in Section 2.8. In this context the numerical orbitals – defined on a uniform grid in real-space – are centered on atoms and localized in spherical regions of radius R_L around the respective atoms. Since the orbitals are variationally optimized on the grid according to their environment until they accurately describe the ground state of the system, it allows us to use only a small number of orbitals per atom, much smaller than in LCAO-based calculations. The size of the matrices that enter in the quantum conductance calculation and the computational cost of the whole procedure are thus minimized. In order to ensure fast convergence and accuracy – even for metallic systems – we use both occupied and unoccupied orbitals.

The matrices that enter the electronic transport calculation of a LCR system are computed in two steps. In the first calculation, we compute the ground state of the bare leads in a supercell with periodic boundary conditions. From this calculation we extract the Hamiltonian in the basis of the localized nonorthogonal orbitals and the overlap matrices required for the computation of the self-energies. We then perform a second ground state calculation in a supercell with periodic boundary conditions containing the conductor and one principal layer of the leads. In this calculation, the orbitals in the leads are kept the same as in the bare lead calculation, in order to extract the matrix elements describing the coupling between the conductor and the leads. This procedure fully accounts for the electronic structure of the conductor and the interaction between the conductor and the leads, provided that the lead region is large enough to avoid spurious interactions between periodic images of the contacts. In order to have interactions between the nearest-neighbor principal layers only, the width of the layers has to be sufficiently large compared to the localization regions. On the other hand, the localization regions have to be large enough to ensure an accurate solution of the density-functional equations. Moreover, in the Green's function matching procedure one has to carefully align the Fermi levels of both systems in order to avoid spurious bias effects. Provided that in the conductor-lead calculation the lead region is large enough to recover bulk-like behavior far from the interfaces, we align the macroscopic average of the electrostatic potentials in the bare lead and in the conductor-lead geometry. This ensures a seamless conductor-lead geometry and prevents the spurious bias. An equivalent procedure is often used to extract band offsets in superlattice calculations

(BALDERESCHI, BARONI and RESTA [1988], BUONGIORNO NARDELLI, RAPCEWICZ and BERNHOLC [1997]).

REMARK 4.1. If a principal layer is composed of N orbitals, the calculation of the Green's function requires a matrix inversion that scales as $O(N^3)$. However, for very large systems, the localization of the orbitals allows us to divide a principal layer into thiner layers and compute the quantities of interest in largely $O(N)$ fashion (ANANTRAM and GOVINDAN [1998]).

4.2. *Example: carbon nanotube on metallic contacts*

To illustrate the above *ab initio* methodology we use the example of transport behavior of nanotube-metal contacts studied by BUONGIORNO NARDELLI, FATTEBERT and BERNHOLC [2001]. The problem of contacts in metal-carbon nanotubes assemblies is a crucial issue for technological development, and determines much of the nanoscale device characteristics. A perfect metallic nanotube behaves like a ballistic conductor: every electron injected into the nanotube at one end should come out at the other end. The basic electronic properties of metallic nanotubes imply the existence of two propagating modes for electronic transmission, independent of the diameter (BERNHOLC, BRENNER, BUONGIORNO NARDELLI, MEUNIER and ROLAND [2002]). The electronic conductance is then expected to be twice the fundamental quantum of conductance, $G_0 = 2e^2/h = 1/12.9\ (\mathrm{k}\Omega)^{-1}$. At higher energies, the electrons are able to probe different sub-bands, which gives rise to an increase in G that is proportional to the number of additional bands available for transport. Hence, G for ideal nanotubes is expected to consist of a series of "down-and-up" steps as a function of the electron energy, in which the position of the steps correlate with the band edges. An illustration of this behavior is displayed in Fig. 4.1. These considerations would suggest that nanotubes should behave as ideal device elements because of their electrical properties.

However, one of the fundamental problems that hinder a broader technological application of carbon nanotubes is the observation that most carbon nanotube devices display contact resistances of the order of $\mathrm{M}\Omega$ (TANS, DEVORET, DAI, THESS, SMALLEY, GEERLIGS and DEKKER [1997], TANS, VERSCHUEREN and DEKKER [1998], MARTEL, SCHMIDT, SHEA, HERTEL and AVOURIS [1998], BACHTOLD, HENNY, TARRIER, STRUNK, SCHONENBERGER, SALVETAT, BONARD and FORRO [1998]), rather than $\mathrm{k}\Omega$, as one would expect.

What is the physical origin behind the very high contact resistance for carbon nanotube systems? As a prototypical example, we consider the transport properties of a metallic $(5,5)$ nanotube deposited on an Al (111) surface in an idealized side-contact geometry, as shown in the inset of Fig. 4.2. In order to accurately account for the highly inhomogeneous environment of the nanowire-metal junction, and to account for the charge transfer occurring at the interface between these two dissimilar materials, it is important to use the accurate and self-consistent *ab initio* description we have previously discussed. The main characteristics of the electronic response of the system is a marked transfer of charge from the nanotube to the metal that allows the valence

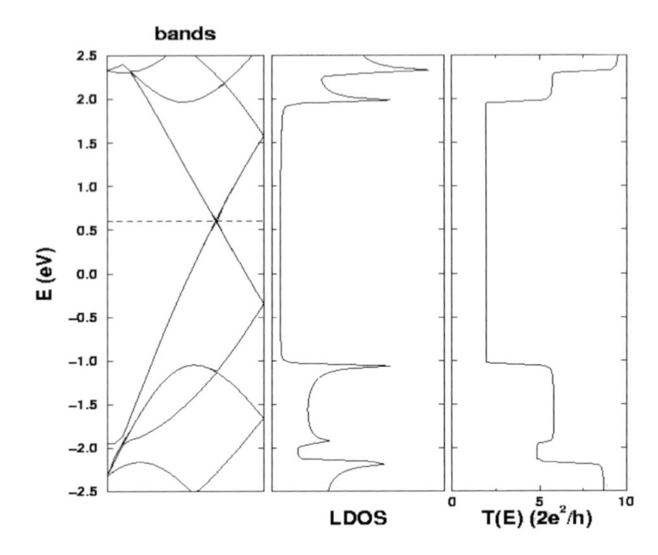

FIG. 4.1. Left: Electronic band structure of a metallic nanotube. Note the crossing of the bands at the Fermi energy. Middle: corresponding Density of States. Right: Quantum conductance spectrum. Note the metallic plateau of conductance equal to $2G_0$.

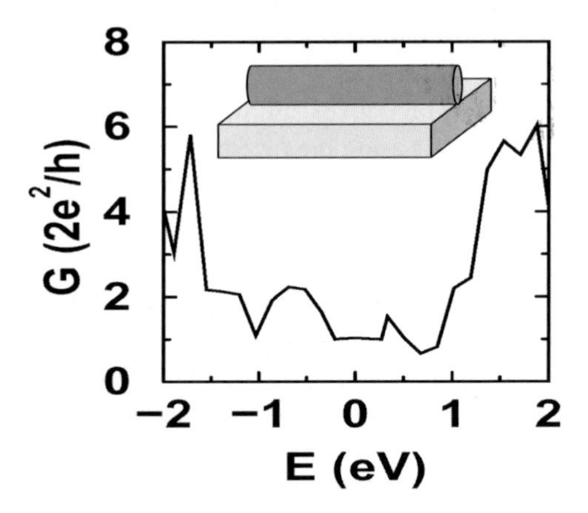

FIG. 4.2. The geometry and conductance spectrum of an infinite (5, 5) nanotube deposited on Al(111). Adapted from BUONGIORNO NARDELLI, FATTEBERT and BERNHOLC [2001].

band edge of the nanotube to align with the Fermi level of the metal electrode (XUE and DATTA [1999]).

This charge transfer, which has been already observed for other experimental systems (TANS, VERSCHUEREN and DEKKER [1998], WILDOER, VENEMA, RINZLER, SMALLEY and DEKKER [1998], MARTEL, SCHMIDT, SHEA, HERTEL and AVOURIS [1998]) and calculations (XUE and DATTA [1999], RUBIO, SANCHEZ-PORTAL, ARTACHO, ORDEJON and SOLER [1999], KONG, HAN and IHM [1999]), leads to enhanced

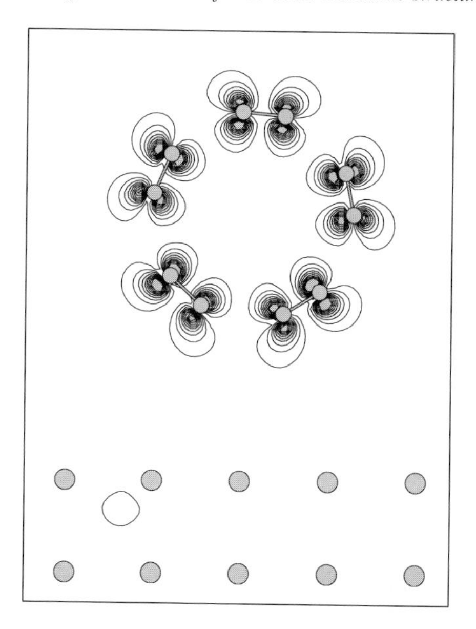

FIG. 4.3. Cross section of the square modulus of the electronic wave function corresponding to the only open eigenchannel at the Fermi level that has a sizable component on the nanotube for the system represented in Fig. 4.2. The other wave functions at the Fermi level are mostly localized on the metal. Adapted from BUONGIORNO NARDELLI, FATTEBERT and BERNHOLC [2001].

conductivity along the tube axis and gives rise to a weak ionic bonding between the tube and the metal. The conductance spectrum for the coupled nanotube is displayed in Fig. 4.2. Although the metal contact increases the resistance by a factor of two as compared to ideal isolated tubes, the transmission through the system is still substantial. To further analyze the contact resistance, we have calculated the *eigenchannels* (BRANDBYGE, SORENSEN and JACOBSEN [1997]). Among the conducting channels, i.e., those with a significant nonzero transmission coefficient, we observe a clear distinction between channels that are localized in the metal and those on the nanotube itself. This result reflects the clear separation of the individual electronic wave functions of each of the components of the system. In particular, the eigenchannel corresponding to the plateau of conductance around the Fermi energy corresponds to an individual wave function, reproduced in Fig. 4.3, almost fully localized on the nanotube. This implies that there is very little hybridization and intermixing between the nanotube and the metal in the channel responsible for conduction at the Fermi level. Thus, the conduction electron transfer between the tube and the metal in the idealized side-contact geometry considered here is very inefficient, which can explain the high contact resistance observed in nanotube-metal contacts.

This initial investigation has been extended to a geometry that more closely resembles an experimental two-terminal device, with two semi-infinite contacts connected by a nanotube bridge, 1.5 nm long. In this geometry, the system recovers the ideal conductance of an isolated tube with two conductance channels at the Fermi energy, as shown in Fig. 4.4. This behavior is induced by the alignment of the valence band edge

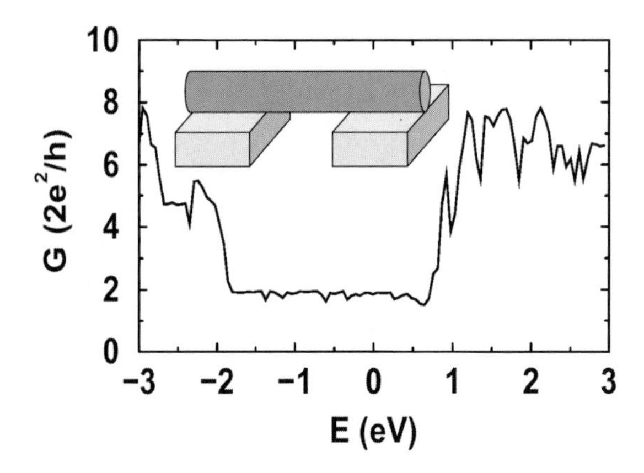

FIG. 4.4. The conductance spectrum of an ideal two-terminal device, as shown in the inset. The Fermi level is taken as a reference. From BUONGIORNO NARDELLI, FATTEBERT and BERNHOLC [2001].

of the nanotube with the Fermi energy of the metal contacts, triggered by the charge transfer in the lead regions. In this particular geometry, these conditions restore the two original eigenchannels of the nanotube and thus conserve the number of conducting channels throughout the system. It is important to note that the weak nanotube-metal interaction, responsible for the pathologically high resistance of the nanotube-metal assembly, is not strengthened.

REMARK 4.2. In this calculation, the conductor C is made of a $(5, 5)$ carbon nanotube composed of 120 atoms, while a principal layer of the leads is made of the same 120 atoms carbon nanotube and of an aluminium surface of 100 atoms. The width of a principal layer is 14.7 Å and a localization radius of 5.3 Å was used for the orbitals. The LDA exchange and correlation functional was used with the pseudopotentials by HAMANN [1989].

REMARK 4.3. The occurrence of a single channel in the first case is due to the idealized geometry of an infinite nanotube on an infinite metallic surface. The conservation of the total number of channels is ensured by the channels localized on the metal.

These examples clearly demonstrate that the weak nanotube-metal coupling is mostly responsible for the weak electron transport in the combined system, and that wave vector conservation is not a significant factor (TERSOFF [1999], DELANEY, DI VENTRA and PANTELIDES [1999]). The weak distributed coupling also explains why the measured contact resistance is inversely proportional to the contact length (TANS, DEVORET, DAI, THESS, SMALLEY, GEERLIGS and DEKKER [1997], FRANK, PONCHARAL, WANG and HEER [1998], ANANTRAM, DATTA and XUE [2000]). Although the nanotube behaves as an ideal ballistic conductor, the bonding characteristics of the nanotube-metal system prevent an efficient electron transfer mechanism from the nanotube to the Al contact. Indeed, inducing defects in the contact region, e.g., by localized electron bombardment (BACHTOLD, HENNY, TARRIER, STRUNK,

SCHONENBERGER, SALVETAT, BONARD and FORRO [1998]), drastically increases the bonding strength of the nanotube-metal assembly and greatly improve the performance of the device. Alternatively, mechanically pushing the nanotube closer to the Al surface by a small amount (≈ 1 Å, with an energy cost of ≈ 10 meV/atom) more than doubles the transmission efficiency between the metal and the nanotube. The mechanical deformation induces a small inward relaxation of the Al surface in the contact region, facilitating stronger hybridization between the nanotube and the metal contact in the conducting channels and therefore leading to a higher transmission rate.

Acknowledgments

The authors are very grateful to Prof. J. Bernholc for pointing them towards the problems of linear scaling and electronic transport, and for numerous stimulating discussions about these problems. A portion of the work was performed under the auspices of the U.S. Department of Energy by University of California Lawrence Livermore National Laboratory under contract No. W-7405-Eng-48.

References

ANANTRAM, M. and T. GOVINDAN (1998), Conductance of carbon nanotubes with disorder: a numerical study, *Phys. Rev. B* **58** (8), 4882–4887.

ANANTRAM, M.P., S. DATTA and Y.Q. XUE (2000), Coupling of carbon nanotubes to metallic contacts, *Phys. Rev. B* **61** (20), 14219–14224.

ANCILOTTO, F., P. BLANDIN and F. TOIGO (1999), Real-space full-multigrid study of the fragmentation of Li_{11}^{+} clusters, *Phys. Rev. B* **59** (12), 7868–7875.

ASHCROFT, N. and N. MERMIN (1976), *Solid state physics* (Holt–Saunders).

BACHTOLD, A., M. HENNY, C. TARRIER, C. STRUNK, C. SCHONENBERGER, J.P. SALVETAT, J.M. BONARD and L. FORRO (1998), Contacting carbon nanotubes selectively with low-ohmic contacts for four-probe electric measurements, *Appl. Phys. Lett.* **73** (2), 274–276.

BALDERESCHI, A. (1973), Mean-value point in the Brillouin zone, *Phys. Rev. B* **7**, 5212.

BALDERESCHI, A., S. BARONI and R. RESTA (1988), Band offsets in lattice-matched heterojunctions: A model and first-principles calculations for GaAs/AlAs, *Phys. Rev. Lett.* **61**, 734.

BECK, T. (2000), Real-space mesh techniques in density-functional theory, *Rev. Mod. Phys.* **72** (4), 1041–1080.

BEENAKKER, C. and H. VAN HOUTEN (1991), Quantum transport in semiconductor nanostructures, *Solid State Phys.* **44**, 1–228.

BERNHOLC, J., D. BRENNER, M. BUONGIORNO NARDELLI, V. MEUNIER and C. ROLAND (2002), Mechanical and electrical properties of nanotubes, *Annual Reviews in Materials Science*.

BLACKFORD, L.S., J. CHOI, A. CLEARY, E. D'AZEVEDO, J. DEMMEL, I. DHILLON, J. DONGARRA, S. HAMMARLING, G. HENRY, A. PETITET, K. STANLEY, D. WALKER and R.C. WHALEY (1997), *SCALAPACK User's Guide* (SIAM, Philadelphia).

BRANDBYGE, M., M. SORENSEN and K. JACOBSEN (1997), Conductance eigenchannels in nanocontacts, *Phys. Rev. B* **56**, 14956.

BRANDT, A. (1977), Multilevel adaptative solutions to boundary-value problems, *Math. Comp.* **31** (138), 333–390.

BRIGGS, E.L., D.J. SULLIVAN and J. BERNHOLC (1995), Large scale electronic structure calculations with multigrid acceleration, *Phys. Rev. B* **52** (8), R5471–5474.

BRIGGS, E.L., D.J. SULLIVAN and J. BERNHOLC (1996), Real-space multigrid-based approach to large-scale electronic structure calculations, *Phys. Rev. B* **54** (20), 14362–14375.

BUNGARO, C., K. RAPCEWICZ and J. BERNHOLC (1999), Surface sensitivity of impurity incorporation: Mg at GaN(0001) surfaces, *Phys. Rev. B* **59** (15), 9771–9774.

BUONGIORNO NARDELLI, M. (1999), Electronic transport in extended systems: Application to carbon nanotubes, *Phys. Rev. B* **60** (11), 7828–7833.

BUONGIORNO NARDELLI, M. and J. BERNHOLC (1999), Mechanical deformations and coherent transport in carbon nanotubes, *Phys. Rev. B* **60** (24), R16338–R16341.

BUONGIORNO NARDELLI, M., J.-L. FATTEBERT and J. BERNHOLC (2001), An O(N) real-space method for ab initio quantum transport calculations: application to carbon nanotube-metal contacts, *Phys. Rev. B* **64**, 245423.

BUONGIORNO NARDELLI, M., K. RAPCEWICZ and J. BERNHOLC (1997), Strain effects on the interface properties of nitride semiconductors, *Phys. Rev. B* **55**, R7323.

BUONGIORNO NARDELLI, M., B. YAKOBSON and J. BERNHOLC (1998), Mechanism of strain release in carbon nanotubes, *Phys. Rev. B* **57** (8), R4277–R4280.

CANCÈS, E., M. DEFRANCESCHI, W. KUTZELNIGG, C. LE BRIS and Y. MADAY (2003), Computational quantum chemistry: A primer, in: P.G. Ciarlet and C. Le Bris, eds., *Computational Chemistry*, Handbook of Numerical Analysis, Vol. X (Elsevier, Amsterdam) 3–270.

CHELIKOWSKY, J.R., N. TROUILLER and Y. SAAD (1994), Finite-difference-pseudopotential method: Electronic structure calculations without a basis, *Phys. Rev. Lett.* **72** (8), 1240–1243.

CHETTY, N., M. WEINERT, T.S. RAHMAN and J.W. DAVENPORT (1995), Vacancies and impurities in aluminium and magnesium, *Phys. Rev. B* **52** (9), 6313–6326.

CHOI, H. and J. IHM (1999), *Ab initio* pseudopotential method for the calculation of conductance in quantum wires, *Phys. Rev. B* **59**, 2267.

CHOI, H., J. IHM, S. LOUIE and M. COHEN (2000), Defects, quasibound states, and quantum conductance in metallic carbon nanotubes, *Phys. Rev. Lett.* **84**, 2917.

COLLATZ, L. (1966), *The Numerical Treatment of Differential Equations* (Springer, Berlin).

COSTINER, S. and S. TAASAN (1995), Simultaneous multigrid techniques for nonlinear eigenvalue problems: Solutions of the nonlinear Schrödinger–Poisson eigenvalue problem in two and three dimensions, *Phys. Rev. E* **52** (1), 1181–1192.

DATTA, S. (1995), *Electronic Transport in Mesoscopic Systems* (Cambridge University Press).

DELANEY, P., M. DI VENTRA and S.T. PANTELIDES (1999), Quantized conductance of multiwalled carbon nanotubes, *Appl. Phys. Lett.* **75** (24), 3787–3789.

DESCLOUX, J., J.-L. FATTEBERT and F. GYGI (1998), RQI (Rayleigh Quotient Iteration), an old recipe for solving modern large scale eigenvalue problems, *Comput. Phys.* **12** (1), 22–27.

DI VENTRA, M., S. PANTELIDES and N. LANG (2000), First-principles calculation of transport properties of a molecular device, *Phys. Rev. Lett.* **84**, 979.

EDELMAN, A., T. ARIAS and S. SMITH (1998), The geometry of algorithms with orthogonality constraints, *SIAM J. Matrix Anal. Appl.* **20** (2), 303–353.

FATTEBERT, J.-L. (1996), An inverse iteration method using multigrid for quantum chemistry, *BIT* **36** (3), 509–522.

FATTEBERT, J.-L. (1998), A block Rayleigh Quotient Iteration with local quadratic convergence, *ETNA* **7**, 56–74.

FATTEBERT, J.-L. (1999), Finite difference schemes and block Rayleigh Quotient Iteration for electronic structure calculations on composite grids, *J. Comput. Phys.* **149**, 75–94.

FATTEBERT, J.-L. and J. BERNHOLC (2000), Towards grid-based $O(N)$ density-functional theory methods: optimized non-orthogonal orbitals and multigrid acceleration, *Phys. Rev. B* **62** (3), 1713–1722.

FATTEBERT, J.-L. and F. GYGI (2002), Density functional theory for efficient ab initio molecular dynamics simulations in solution, *J. Comput. Chem.* **23**, 662–666.

FERNANDO, G.W., G.-X. QUIAN, M. WEINERT and J.W. DAVENPORT (1989), First-principles molecular dynamics for metals, *Phys. Rev. B* **40** (11), 7985–7988.

FERRY, F. and S. GOODNICK (1997), *Transport in Nanostructures* (University Press, Cambridge).

FEYNMAN, R. (1939), Forces in molecules, *Phys. Rev.* **56**, 340–343.

FISHER, D. and P. LEE (1981), Relation between conductivity and transmission matrix, *Phys. Rev. B* **23**, 6851.

FRANK, S., P. PONCHARAL, Z.L. WANG and W.A.D. HEER (1998), Carbon nanotube quantum resistors, *Science* **280** (5370), 1744–1746.

GALLI, G. and M. PARRINELLO (1992), Large scale electronic structure calculations, *Phys. Rev. Lett.* **69** (24), 3547–3550.

GALLI, G. and A. PASQUARELLO (1993), First-principles molecular dynamics, in: M. Allen and D. Tildesley, eds., *Computer Simulation in Chemical Physics* (Kluwer Academic Publishers) 261–313.

GARCIA-MOLINER, F. and V. VELASCO (1991), Matching methods for single and multiple interfaces: discrete and continuous media, *Phys. Rep.* **200** (3), 83–125.

GARCIA-MOLINER, F. and V. VELASCO (1992), *Theory of Single and Multiple Interfaces* (World Scientific, Singapore).

GOEDECKER, S. (2003), Linear scaling methods for the solution of Schrödinger's equation, in: P.G. Ciarlet and C. Le Bris, eds., *Computational Chemistry*, Handbook of Numerical Analysis, Vol. X (Elsevier, Amsterdam) 537–570.

GYGI, F. and G. GALLI (1995), Real-space adaptive-coordinate electronic-structure calculations, *Phys. Rev. B* **52** (4), 2229–2232.

HAMANN, D.R. (1989), Generalized norm-conserving pseudopotentials, *Phys. Rev. B* **40** (5), 2980–2987.

HEERMANN, D. (1990), *Computer Simulation Methods*, 2nd edn. (Springer-Verlag, Berlin Heidelberg).

HEISKANEN, M., T. TORSTI, M. PUSKA and R. NIEMINEN (2001), Multigrid method for electronic structure calculations, *Phys. Rev. B* **63**, 245106.

HERNANDEZ, E. and M.J. GILLAN (1995), Self-consistent first-principles technique with linear scaling, *Phys. Rev. B* **51** (15), 10157–10160.

HIROSE, K. and M. TSUKADA (1995), First-principles calculation of the electronic structure for a bielectrode junction system under strong field and current, *Phys. Rev. B* **51**, 5278.

HOSHI, T. and T. FUJIWARA (1997), Fully-selfconsistent electronic-structure calculation using nonorthogonal localized orbitals within a finite-difference real-space scheme and ultrasoft pseudopotential, *J. Phys. Soc. Jpn.* **66** (12), 3710–3713.

ISMAEL-BEIGI, S. and T.A. ARIAS (1999), Locality of the density matrix in metals, semiconductors, and insulators, *Phys. Rev. Lett.* **82** (10), 2127–2130.

JIN, Y.-G. and K. CHANG (2001), Mechanism for the enhanced diffusion of charged oxygen ions in SiO_2, *Phys. Rev. Lett.* **86** (9), 1793–1796.

JIN, Y.-G., J.-W. JEONG and K. CHANG (1999), Real-space electronic structure calculations of charged clusters and defects in semiconductors using a multigrid method, *Physica B* **273–274**, 1003–1006.

JING, X., N. TROULLIER, D. DEAN, N. BINGGELI and J. CHELIKOWSKY (1994), *Ab initio* molecular dynamics simulations of Si clusters using higher-order finite difference-pseudopotential method, *Phys. Rev. B* **50** (16), 12234–12237.

KLEINMAN, L. and D. BYLANDER (1982), Efficacious form for model pseudopotentials, *Phys. Rev. Lett.* **48** (20), 1425–1428.

KOBAYASHI, N., M. BRANDBYGE and M. TSUKADA (2000), Conduction channels at finite bias in single-atom gold contact, *Phys. Rev. B* **62**, 8430.

KOHN, S., J. WEARE, M. ONG and S. BADEN (1997), Software abstractions and computational issues in parallel structured adaptative mesh methods for electronic structure calculations, in: *Proc. of Workshop on Structured Adaptative Mesh Refinement Grid Methods* (Minneapolis).

KOHN, W. and L.J. SHAM (1965), Self-consistent equations including exchange and correlation effects, *Phys. Rev. A* **140**, 1133–1138.

KONG, K., S. HAN and J. IHM (1999), Development of an energy barrier at the metal-chain-metallic-carbon-nanotube nanocontact, *Phys. Rev. B* **60** (8), 6074–6079.

LANDAUER, R. (1970), Electrical resistance of disordered one-dimensional lattices, *Philos. Mag.* **21**, 863–867.

LANDMAN, U., R. BARNETT, A.G. SCHERBAKOV and P. AVOURIS (2000), Metal-semiconductor nanocontacts: silicon nanowires, *Phys. Rev. Lett.* **85**, 1958.

LANG, N. (1995), Resistance of atomic wires, *Phys. Rev. B* **52**, 5335.

LEE, D. and J. JOANNOPOULOS (1981a), Simple scheme for surface-band calculations. I, *Phys. Rev. B* **23** (10), 4988.

LEE, D. and J. JOANNOPOULOS (1981b), Simple scheme for surface-band calculations. II. The Green's function, *Phys. Rev. B* **23** (10), 4997.

LOPEZ-SANCHO, M., J. LOPEZ-SANCHO and J. RUBIO (1984), Quick iterative scheme for the calculation of transfer matrices: application to Mo(100), *J. Phys. F: Metal Phys.* **14**, 1205–1215.

LOPEZ-SANCHO, M., J. LOPEZ-SANCHO and J. RUBIO (1985), Highly convergent schemes for the calculation of bulk and surface Green functions, *J. Phys. F: Metal Phys.* **15**, 851–858.

MANDEL, J. and S. MCCORMICK (1989), A multilevel variational method for $Au = \lambda Bu$ on composite grids, *J. Comput. Phys.* **80** (2), 442–452.

MARTEL, R., T. SCHMIDT, H.R. SHEA, T. HERTEL and P. AVOURIS (1998), Single- and multi-wall carbon nanotube field-effect transistors, *Appl. Phys. Lett.* **73** (17), 2447–2449.

MEIR, Y. and N. WINGREEN (1992), Landauer formula for the current through an interacting electron region, *Phys. Rev. Lett.* **68** (16), 2512–2515.

MILLAM, J.M. and G.E. SCUSERIA (1997), Linear scaling conjugate gradient density matrix search as an alternative to diagonalization for first principles electronic structure calculations, *J. Chem. Phys.* **506** (13), 5569–5577.

MODINE, N.A., G. ZUMBACH and E. KAXIRAS (1996), Adaptive-coordinate real-space electronic-structure calculations on parallel computers, *Solid State Comm.* **99** (2), 57–61.

MURAKAMI, H., V. SONNAD and E. CLEMENTI (1992), A three-dimensional finite element approach towards molecular SCF computations, *Int. J. Quant. Chem.* **42**, 785–817.

PARLETT, B.N. (1998), *The Symmetric Eigenvalue Problem* (SIAM, Philadelphia).

PARRINELLO, M. (1997), From Silicon to RNA: the coming age of *ab initio* molecular dynamics, *Sol. State Comm.* **102** (2–3), 107–120.

PASK, J.E., B.M. KLEIN, C.Y. FONG and P.A. STERNE (1999), Real-space polynomial basis for solid-state electronic structure calculations: A finite-element approach, *Phys. Rev. B* **59** (19), 12352–12358.

PAYNE, M.C., M.P. TETER, D.C. ALLAN, T. ARIAS and J. JOANNOPOULOS (1992), Iterative minimization techniques for ab initio total-energy calculations: molecular dynamics and conjugate gradients, *Rev. Mod. Phys.* **64** (4), 1045–1097.

PERDEW, J., K. BURKE and M. ERNZERHOF (1996), Generalized gradient approximation made simple, *Phys. Rev. Lett.* **77** (18), 3865–3868.

PFROMMER, B., J. DEMMEL and H. SIMON (1999), Unconstraint energy functionals for electronic structure calculations, *J. Comput. Phys.* **150**, 287–298.

RAMAMOORTHY, M., E. BRIGGS and J. BERNHOLC (1998), Chemical trends in impurity incorporation into Si(100), *Phys. Rev. Lett.* **81** (8), 1642–1645.

RUBIO, A., D. SANCHEZ-PORTAL, E. ARTACHO, P. ORDEJON and J.M. SOLER (1999), Electronic states in a finite carbon nanotube: a one-dimensional quantum box, *Phys. Rev. Lett.* **82** (17), 3520–3523.

SAAD, Y., A. STATHOPOULOS, J. CHELIKOWSKY, K. WU and S. OGUT (1996), Solution of large eigenvalue problems in electronic structure calculations, *BIT* **36** (3), 563–578.

SCHMIDT, W.G., F. BECHSTEDT and J. BERNHOLC (2001), Terrace and step contributions to the optical anisotropy of Si(001) surfaces, *Phys. Rev. B* **63** (4), 045322.

SLEIJPEN, G.L.G. and H.A. VAN DER VORST (1996), A generalized Jacobi–Davidson iteration method for linear eigenvalue problem, *SIAM Matrix Anal. and Appl.* **17** (2), 401–425.

TAKAHASHI, H., T. HORI, H. HASHIMOTO and T. NITTA (2001), A hybrid QM/MM method employing real space grids for QM water in the TIP4P water solvent, *J. Comp. Chem.* **22** (12), 1252–1261.

TANS, S.J., M.H. DEVORET, H.J. DAI, A. THESS, R.E. SMALLEY, L.J. GEERLIGS and C. DEKKER (1997), Individual single-wall carbon nanotubes as quantum wires, *Nature* **386** (6624), 474–477.

TANS, S.J., A.R.M. VERSCHUEREN and C. DEKKER (1998), Room-temperature transistor based on a single carbon nanotube, *Nature* **393** (6680), 49–52.

TAYLOR, J., H. GUO and J. WANG (2000), *Ab initio* modeling of open systems: Charge transfer, electron conduction, and molecular switching of a C60 device, *Phys. Rev. B* **63**, 121104.

TERSOFF, J. (1999), Response to comment on "Contact resistance of carbon nanotubes [Appl. Phys. Lett. 75, 4028 (1999)]", *Appl. Phys. Lett.* **75** (25), 4030–4030.

TETER, M.P., M.C. PAYNE and D.C. ALLAN (1989), Solution of Schrödinger's equation for large systems, *Phys. Rev. B* **40** (18), 12255–12263.

TSUCHIDA, E. and M. TSUKADA (1995), Electronic-structure calculations based on the finite-element method, *Phys. Rev. B* **52** (8), 5573–5578.

VASILIEV, I., S. OGUT and J. CHELIKOWSKY (1999), *Ab initio* excitation spectra and collective electronic response in atoms and clusters, *Phys. Rev. Lett.* **82** (9), 1919–1922.

WANG, J. and T. BECK (2000), Efficient real-space solution of the Kohn–Sham equations with multiscale techniques, *J. Chem. Phys.* **112** (21), 9223–9228.

WHITE, S.R., J.W. WILKINS and M.P. TETER (1989), Finite-element method for electronic structure calculations, *Phys. Rev. B* **39** (9), 5819–5833.

WILDOER, J.W.G., L.C. VENEMA, A.G. RINZLER, R.E. SMALLEY and C. DEKKER (1998), Electronic structure of atomically resolved carbon nanotubes, *Nature* **391** (6662), 59–62.

XUE, Y.Q. and S. DATTA (1999), Fermi-level alignment at metal-carbon nanotube interfaces: application to scanning tunneling spectroscopy, *Phys. Rev. Lett.* **83** (23), 4844–4847.

YALIRAKI, S., A.E. ROITBERG, C. GONZALEZ, V. MUJICA and M. RATNER (1999), The injecting energy at molecule/metal interfaces: Implications for conductance of molecular junctions from an ab initio molecular description, *J. Chem. Phys.* **111**, 6997.

YOON, Y.-G., M. MAZZONI, H. CHOI, J. IHM and S. LOUIE (2001), Structural deformation and intertube conductance of crossed carbon nanotube junctions, *Phys. Rev. Lett.* **86**, 688.

Using Real Space Pseudopotentials for the Electronic Structure Problem

James R. Chelikowsky[a], Leeor Kronik[a,b], Igor Vasiliev[a,c], Manish Jain[a], Yousef Saad[d]

[a]*Department of Chemical Engineering and Materials Science, Minnesota Supercomputing Institute, University of Minnesota, Minneapolis, MN 55455, USA*
[b]*Department of Materials and Interfaces, Weizmann Institute of Science, Rehovoth 76100, Israel*
[c]*Department of Physics, New Mexico State University, Las Cruces, NM 88003, USA*
[d]*Department of Computer Science, Minnesota Supercomputing Institute, University of Minnesota, Minneapolis, MN 55455, USA*
E-mail address: jrc@msi.umn.edu, URL: jrc.cems.umn.edu (J.R. Chelikowsky)

1. Introduction

The pseudopotential model of condensed matter is one of the most promising developments within the of area computational materials science. It has led the way in providing a workable science framework for describing the properties of materials (CHELIKOWSKY and LOUIE [1996], CHELIKOWSKY and COHEN [1992]), while modern computers have provided the computational resources to implement the pseudopotential method.

The pseudopotential concept treats matter as a sea of valence electrons moving in a background of ion cores (Fig. 1.1). The cores are composed of nuclei and inert inner electrons. Within this model many of the complexities of an all-electron calculation are avoided. For example, a group IV element such as C with 6 electrons is treated in a similar fashion to Ge with 32 electrons as both elements have 4 valence electrons.

Computational Chemistry
Special Volume (C. Le Bris, Guest Editor) of
HANDBOOK OF NUMERICAL ANALYSIS, VOL. X
P.G. Ciarlet (Editor)
© 2003 Elsevier Science B.V. All rights reserved

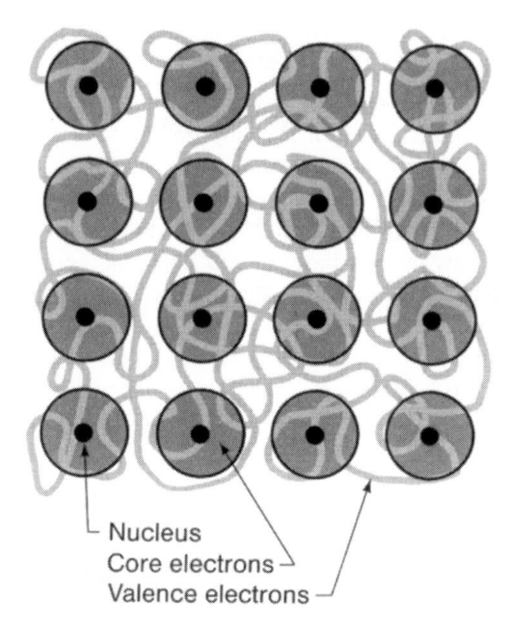

FIG. 1.1. Standard pseudopotential model of a solid. The ion cores composed of the nuclei and tightly bound core electrons are treated as chemically inert. The pseudopotential model describes only the outer, chemically active, valence electrons.

Since the pseudopotential binds only valence electron states, the resulting potential is weak and the Coulombic $1/r$ singularity at the nucleus is removed. *Without the pseudopotential approximation, real space methods would be considerably more difficult to implement, if not impossible.* Grids for the full potential must be spatially adapted to account for the rapid changes in the potential at the nuclear positions. This aspect enormously complicates the problem.

2. Constructing pseudopotentials

Contemporary pseudopotentials are predominantly based on density functional theory (DFT) (HOHENBERG and KOHN [1964], KOHN and SHAM [1965]). Within DFT, the many-body problem is mapped on to a one-electron Hamiltonian. The effects of exchange and correlation are subsumed unto a one-electron potential that depends only on the charge density. This mapping of the many body problem to a one body problem is in principle exact, but in practice approximate because the exact functional form for the one-electron exchange-correlation potential is not known. The mapping is usually accomplished via the local density approximation (LDA) (KOHN and SHAM [1965]), in which the exchange-correlation potential is assumed to be that of a homogeneous electron gas. Alternatively, the generalized gradient approximation (GGA) (PERDEW, BURKE and ERNZERHOF [1996]) incorporates some effects of electron gas inhomogeneity by considering the gradient of the charge density as well. The DFT mapping procedure, in any of its approximate forms, allows for a great simplification of the one-electron problem. Without this approach, most electronic structure methods would not be feasible for systems of more than a few dozen electrons.

The chief limitation of DFT is that it is appropriate for the ground state structure and cannot be used to describe excited states without other approximations. As such, one uses DFT to determine structural energies, compressibilities, elastic constants, vibrational modes, etc., but not band gaps (HYBERTSEN and LOUIE [1986]). However, DFT eigenvalues and wave functions can be used for computation of electronic excitations, by implementing time-dependent linear response theory on top of the standard DFT calculations.

Within the DFT approach one can write down a one-electron Hamiltonian and a corresponding one-electron Schrödinger equation, often called the *Kohn–Sham* equation (KOHN and SHAM [1965]). For an atom, the Kohn–Sham equation can be written as:

$$\left(\frac{-\hbar^2\nabla^2}{2m} - \frac{Ze^2}{r} + V_{\mathrm{H}}(\vec{r}) + V_{\mathrm{xc}}\big[\rho(\vec{r})\big]\right)\psi_n(\vec{r}) = E_n\psi_n(\vec{r}) \tag{2.1}$$

where there are Z electrons in the atom, V_{H} is the Hartree or Coulomb potential, and V_{xc} is the exchange-correlation potential. The Hartree and exchange-correlation potentials can be determined from the electronic charge density. The density is given by

$$\rho(\vec{r}) = -e \sum_{n,\,\mathrm{occup}} \left|\psi_n(\vec{r})\right|^2. \tag{2.2}$$

The summation is over all occupied states. The Hartree potential is then determined by

$$\nabla^2 V_{\mathrm{H}}(\vec{r}) = -4\pi e\rho(\vec{r}). \tag{2.3}$$

This term can be interpreted as the electrostatic interaction of an electron with the charge density of system.

The exchange-correlation potential is more problematic. The central tenant of either the LDA or the GGA is that the total exchange-correlation energy may be written as

$$E_{\mathrm{xc}}\big[\rho(\vec{r})\big] = \int \rho(\vec{r})\varepsilon_{\mathrm{xc}}\big(\rho(\vec{r})\big)\, d^3r, \tag{2.4}$$

where $\varepsilon_{\mathrm{xc}}$ is the exchange-correlation energy density (i.e., the energy per electron). $E_{\mathrm{xc}}[\rho(\vec{r})]$ and $\varepsilon_{\mathrm{xc}}(\rho(\vec{r}))$ are to be interpreted as depending on the charge density or the charge density and its gradient for LDA and GGA, respectively. The exchange-correlation potential, $V_{\mathrm{xc}}(\rho(\vec{r}))$, is then obtained as $V_{\mathrm{xc}}(\rho(\vec{r})) = \delta E_{\mathrm{xc}}[\rho(\vec{r})]/\delta\rho(\vec{r})$.

It is not difficult to solve the Kohn–Sham equation (Eq. (2.1)) for an atom. The charge density is taken to be spherically symmetric. Thus, the problem reduces to solving a one-dimensional problem. The Hartree and exchange-correlation potentials can be iterated to form a self-consistent field. Usually the process is so quick that it can be done on desktop or laptop computer in a matter of seconds. This atomic solution provides the input to construct a pseudopotential representing the effect of the core electrons and nucleus. This "ion core" pseudopotential can be transferred to other systems.

Let us consider a sodium atom for the purposes of designing an *ab initio* pseudopotential. By solving for the Na atom, we know the eigenvalue, ε_{3s}, and the corresponding wave function, $\psi_{3s}(r)$ for the valence electron. We demand several conditions for the Na pseudopotential: (1) The potential bind the valence electron, the 3s-electron for the case at hand, but not the core electrons; (2) The eigenvalue of the corresponding valence electron be identical to the full potential eigenvalue; the full potential is also called the *all-electron* potential; (3) The wave function be nodeless and *identical* to the "all-electron" wave function outside a core region.

The pseudo-wave function, $\phi_p(r)$, is chosen to identical to the all-electron wave function, $\psi_{AE}(r)$, outside the core: $\phi_p(r) = \psi_{AE}(r)$ for $r > r_c$, where r_c is the radius of a spherical core region. This attribute will guarantee that the pseudo-wave function will possess identical properties as the all-electron wave function, especially in terms of its chemical bond. For $r < r_c$, we have the opportunity to alter the all-electron wave function as we wish within certain limitations. Namely, we want the wave function in this region to be smooth and nodeless. Another very important criterion is mandated. Namely, the integral of the pseudo-charge density, i.e., square of the wave function $|\phi_p(r)|^2$, within the core should be equal to the integral of the all-electron charge density over the same spatial regime. Without this condition, the pseudo-wave function can differ by a scaling factor from the all-electron wave function, that is, $\phi_p(r) = C\psi_{AE}(r)$ for $r > r_c$ where the constant, C, may differ from unity. Since we expect the chemical bonding of an atom to be highly dependent on the tails of the valence wave functions, it is imperative that the normalized pseudo-wave function be identical to the all-electron wave functions. The criterion by which one insures $C = 1$ is called *norm conserving* (HAMANN, SCHLÜTER and CHIANG [1979]). Some of the earliest *ab initio* potentials did not incorporate this constraint (ZUNGER and COHEN [1979]). These potentials are not used for accurate computations. The chemical properties resulting from these calculations using these non-norm-conserving pseudopotentials are quite poor when compared to experiment or to the more accurate norm conserving pseudopotentials.

For our illustrative purposes, we focus on a particular construction proposed by KERKER [1980]. His method provides a straightforward procedure for constructing local density pseudopotentials that retain the *norm conserving* criterion. He suggested that the pseudo-wave function have the following form:

$$\phi_p(r) = r^l \exp\bigl(p(r)\bigr) \quad \text{for } r < r_c \tag{2.5}$$

where $p(r)$ is a simple polynomial: $p(r) = -a_0r^4 - a_1r^3 - a_2r^2 - a_3$ and

$$\phi_p(r) = \psi_{AE}(r) \quad \text{for } r > r_c. \tag{2.6}$$

This form of the pseudo-wave function for ϕ_p assures us that the function will be nodeless and have the correct behavior at large r. Kerker established criteria for fixing the parameters (a_0, a_1, a_2 and a_3). In addition to requiring that the wave function be norm conserving, the criteria include: (a) The all-electron and pseudo-wave functions have the same valence eigenvalues; (b) The pseudo-wave function be nodeless and be identical to the all-electron wave function for $r > r_c$; (c) The pseudo-wave function must be continuous as well as the first and second derivatives of the wave function at r_c. Typically, the core is taken to be less than the distance corresponding to the maximum of the valence wave function, but greater than the distance of the outermost node.

The Troullier–Martins method (TROULLIER and MARTINS [1991]), which is the one used in all our examples below, is closely related to Kerker's method, but uses a higher order polynomial $p(r)$ and imposes additional constraints, such as matching higher derivatives at the core radius.

Once the pseudo-wave function is defined, using Eqs. (2.5), (2.6) or similar well-defined "recipes", we can "invert" the Kohn–Sham equation and solve for the ion core pseudopotential, V_{ion}^{p}:

$$V_{\text{ion}}^{p}(\vec{r}) = E_n - V_{\text{H}}(\vec{r}) - V_{\text{xc}}[\rho(\vec{r})] + \frac{\hbar^2\nabla^2\phi_{p,n}}{2m\phi_{p,n}}. \tag{2.7}$$

This potential, when self-consistently screened by the pseudo-charge density, will yield an eigenvalue of E_n and a pseudo-wave function $\phi_{p,n}$. The pseudo-wave function by construction will agree with the all-electron wave function away from the core.

The pseudopotential is *state* dependent as written in Eq. (2.7), i.e., the pseudopotential is dependent on the angular momentum. It is different for an s-, p-, or d-electron. This means that the wave function needs to projected into different angular momentum quantum states, with a different pseudopotential acting on each component. The projection implies that the pseudopotential operator is nonlocal.

Nonlocality in the pseudopotential is often treated in Fourier space, but it may also be expressed in real space. The nonlocality appears in the angular dependence of the potential, but not in the radial coordinate. It is often advantageous to use a more advanced projection scheme, due to KLEINMAN and BYLANDER [1982], which is fully nonlocal. The interactions between valence electrons and pseudoionic cores may then be separated into a local potential and a non-local pseudopotential in *real space* (TROULLIER and MARTINS [1991]), which differs from zero only inside the small core region around each atom.

$$V_{\text{ion}}^{p}(\vec{r})\phi_n(\vec{r}) = \sum_a V_{\text{loc}}(|\vec{r}_a|)\phi_n(\vec{r}) + \sum_{a,n,lm} G_{n,lm}^{a}\, u_{lm}(\vec{r}_a)\Delta V_l(r_a), \tag{2.8}$$

$$K_{n,lm}^{a} = \frac{1}{\langle\Delta V_{lm}^{a}\rangle}\int u_{lm}(\vec{r}_a)\Delta V_l(r_a)\psi_n(\vec{r})\,d^3r \tag{2.9}$$

and $\langle\Delta V_{lm}^{a}\rangle$ is the normalization factor,

$$\langle\Delta V_{lm}^{a}\rangle = \int u_{lm}(\vec{r}_a)\Delta V_l(r_a)u_{lm}(\vec{r}_a)\,d^3r, \tag{2.10}$$

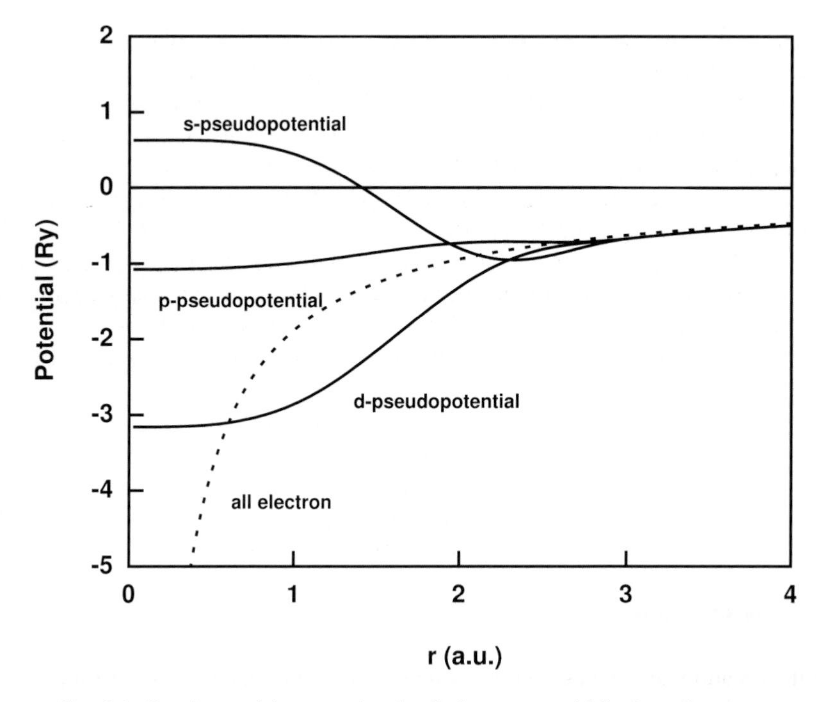

FIG. 2.1. Pseudopotential compared to the all-electron potential for the sodium atom.

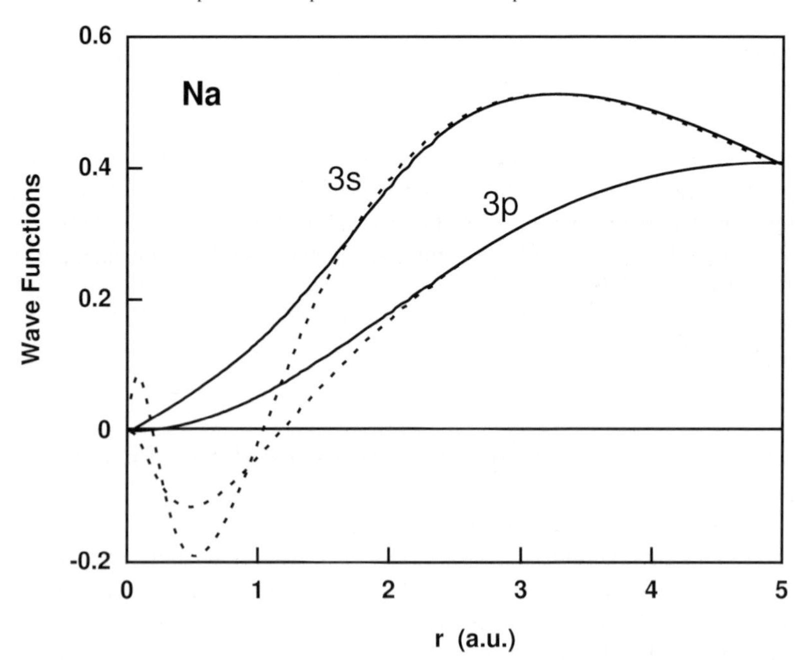

FIG. 2.2. Pseudopotential wave functions compared to all-electron wave functions for the sodium atom. The all-electron wave functions are indicated by the dashed lines.

where $\vec{r}_a = \vec{r} - \vec{R}_a$, and the u_{lm} are the atomic pseudopotential wave functions of angular momentum quantum numbers (l, m) from which the l-dependent ionic pseudopotential, $V_l(r)$, is generated. $\Delta V_l(r) = V_l(r) - V_{\text{loc}}(r)$ is the difference between the l component of the ionic pseudopotential and the local ionic potential. As a specific example, in the case of Na, we might choose the local part of the potential to replicate only the $l = 0$ component as defined by the 3s state. The nonlocal parts of the potential would then contain only the $l = 1$ and $l = 2$ components (or high l values). The choice of which angular component is chosen for the local part of the potential is somewhat arbitrary. It is often convenient to chose the local potential to correspond to the highest l-component of interest. This avoids complex projections with the high l-components. These issues can be tested by choosing different components for the local potential.

In Fig. 2.1, the ion core pseudopotential for Na is presented using the TROULLIER and MARTINS [1991] formalism for creating pseudopotentials. The nonlocality of the potential is evident by the existence of the three potentials corresponding to the s-, p- and d-states. In Fig. 2.2, we illustrate wave functions for the 3s- and 3p-states for the all-electron potential and for the Troullier–Martins pseudopotential.

3. The real space approach

Once the pseudopotential has been determined, the resulting eigenvalue problem needs to be solved for the system of interest:

$$\left(\frac{-\hbar^2\nabla^2}{2m} + V_{\text{ion}}^p(\vec{r}) + V_{\text{H}}(\vec{r}) + V_{\text{xc}}\big[\rho(\vec{r})\big]\right)\phi_n(\vec{r}) = E_n\phi_n(\vec{r}), \tag{3.1}$$

where V_{ion}^p is the ionic pseudopotential for the system. This equation is the same as Eq. (2.1), except that the true ionic potential, $-Ze^2/r$, is replaced by the pseudopotential operator corresponding to the combined effect of all ions in the system. Since the ion cores can be treated as chemically inert and highly localized, it is a simple matter to write:

$$V_{\text{ion}}^p(\vec{r}) = \sum_{\vec{R}_a} V_{\text{ion},a}^p(\vec{r} - \vec{R}_a), \tag{3.2}$$

where $V_{\text{ion},a}^p$ is the ion core pseudopotential associated with the atom, a, at a position $\vec{R}_a$.

At this point, it is worth summarizing the relevant approximations. First, the *Born–Oppenheimer approximation* is almost always invoked to separate the nuclear and electronic degrees of freedom. Second, the local density or generalized gradient approximation to DFT is used in order to map the all-electron problem into a one-electron problem. Third, the *pseudopotential* approximation is used to eliminate the core electrons and allow one to use simple bases to describe the wave functions. Of these approximations, the LDA/GGA one is probably the weakest.

Once the eigenvalue problem is solved, the total energy of the system, E_{tot}, can be evaluated from (CHELIKOWSKY and COHEN [1992])

$$E_{\text{tot}} = \sum_{\text{occup}} E_n - \frac{1}{2}\int d^3r\,\rho(\vec{r})V_{\text{H}}(\vec{r}) + \int d^3r\,\rho(\vec{r})\bigl(E_{\text{xc}}[\rho(\vec{r})] - V_{\text{xc}}[\rho(\vec{r})]\bigr) + E_{i-i}[\vec{R}_a], \qquad (3.3)$$

where E_{i-i} is the ion–ion repulsion. If one knows the behavior of the total energy as a function of atomic positions, it is possible to compute interatomic forces and perform *ab initio* molecular dynamics. It should be emphasized that LDA is a ground state theory; it is not appropriate to use this approach to describe excited state properties using the eigenvalues. It is appropriate to use this approximation for use with linear response theory to obtain excited state properties as will be outlined later.

A major difficulty in solving the eigenvalue problem in Eq. (3.1) are the length and energy scales involved. The inner (core) electrons are highly localized and tightly bound compared to the outer (valence electrons). A simple basis function approach is frequently ineffectual. For example, a plane wave basis might require 10^6 waves to represent converged wave functions for a core electron whereas only 10^2 waves are required for a valence electron. The pseudopotential overcomes this problem by removing the core states from the problem and replacing the all-electron potential by one that replicates only the chemically active, valence electron states (CHELIKOWSKY and COHEN [1992]). By construction, the pseudopotential reproduces the valence state properties such as the eigenvalue spectrum and the charge density outside the ion core.

While a plane wave basis has a number of advantages, for localized systems real space methods are clearly a superior approach. This is discussed in further detail in Section 5. The real space approach often utilizes finite difference discretization on a real space grid. A key aspect to the success of the finite difference method is the availability of *higher order finite difference expansions* for the kinetic energy operator, i.e., expansions of the Laplacian (FORNBERG and SLOAN [1994]). Higher order finite difference methods significantly improve convergence of the eigenvalue problem when compared with standard finite difference methods. If one imposes a simple, uniform grid on our system where the points are described in a finite domain by (x_i, y_j, z_k), we approximate $\frac{\partial^2\psi}{\partial x^2}$ at (x_i, y_j, z_k) by

$$\frac{\partial^2\psi}{\partial x^2} = \sum_{n=-M}^{M} C_n\psi(x_i + nh, y_j, z_k) + \mathrm{O}\bigl(h^{2M+2}\bigr), \qquad (3.4)$$

where h is the grid spacing and M is an integer. This approximation is accurate to $\mathrm{O}(h^{2M+2})$ upon the assumption that ψ can be approximated accurately by a power series in h. Algorithms are available to compute the coefficients C_n for arbitrary order in h (FORNBERG and SLOAN [1994]).

With the kinetic energy operator expanded as in Eq. (3.4), one can set up Kohn–Sham equation over a grid. One may assume a uniform grid, but this is not a necessary requirement. $\psi(x_i, y_j, z_k)$ is computed on the grid by solving the eigenvalue problem:

$$
\begin{aligned}
&-\frac{\hbar^2}{2m}\Bigg[\sum_{n_1=-M}^{M} C_{n_1}\psi_n(x_i+n_1h, y_j, z_k) + \sum_{n_2=-M}^{M} C_{n_2}\psi_n(x_i, y_j+n_2h, z_k) \\
&\qquad + \sum_{n_3=-M}^{M} C_{n_3}\psi_n(x_i, y_j, z_k+n_3h)\Bigg] \\
&\quad + \left[V_{\text{ion}}\left(x_i', y_j', z_k'\right) + V_{\text{H}}(x_i, y_j, z_k) + V_{\text{xc}}(x_i, y_j, z_k)\right]\psi_n(x_i, y_j, z_k) \\
&\quad = E_n\psi_n(x_i, y_j, z_k),
\end{aligned}
\tag{3.5}
$$

where $V_{\text{ion}}[x_i', y_j', z_k']$ is the discrete form of the pseudopotential operator. If we have L grid points, the size of the full matrix resulting from the above problem is $L \times L$.

A uniformly spaced grid in a three dimensional cube as shown in Fig. 3.1, with each grid point corresponding to a row in the matrix. However, many points in the cube are far from any atoms in the system and the wave function on these points may be replaced by zero. Special data structures may be used to discard these points and keep only those having a nonzero value for the wave function. The size of the Hamiltonian matrix is usually reduced by a factor of two to three with this strategy, which is quite important considering the large number of eigenvectors which must be saved. Further, since the Laplacian can be represented by a simple stencil, and since all local potentials sum up

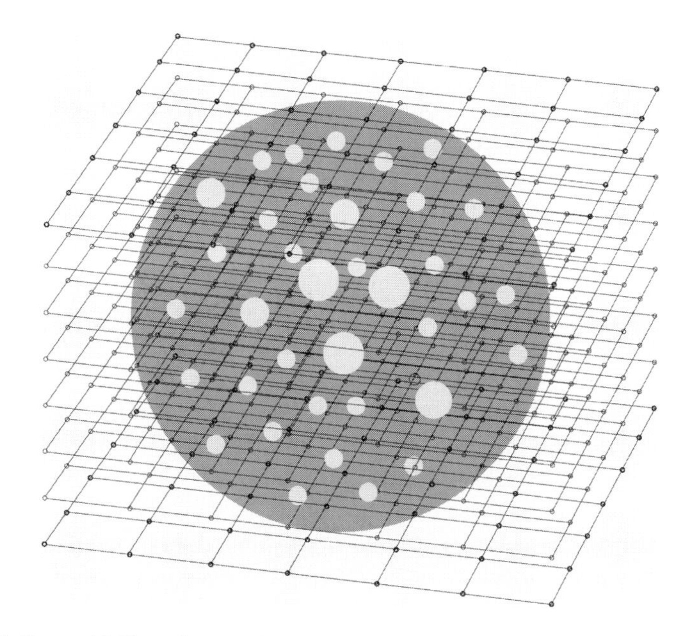

FIG. 3.1. Uniform grid illustrating a typical configuration for examining the electronic structure of a localized system. The gray sphere represents the domain where the wave functions are allowed to be nonzero. The light spheres within the domain are atoms.

to a simple diagonal matrix, the Hamiltonian need not be stored. Handling the ionic pseudopotential is complex as it consists of a local and a nonlocal term (Eqs. (2.8) and (2.9)). In the discrete form, the nonlocal term becomes a sum over all atoms, a, and quantum numbers, (l, m) of rank-one updates:

$$V_{\text{ion}} = \sum_{a} V_{\text{loc},a} + \sum_{a,l,m} c_{a,l,m} U_{a,l,m} U_{a,l,m}^{T}, \tag{3.6}$$

where $U_{a,l,m}$ are sparse vectors which are only nonzero in a localized region around each atom, $c_{a,l,m}$ are normalization coefficients.

Our discussion here is limited to fixed, uniform grids. We found the use of such grids to be highly effective and at the same time straight-forward and simple to implement. We emphasize, however, that the concepts introduced here can be used in conjunction with an adaptive grid (MODINE, ZUMBACH and KAXIRAS [1997]) or a multi-grid (BERNHOLC, BRIGGS, BUNGARO, NARDELLI, FATTEBERT, RAPCEWICZ, ROLAND, SCHMIDT and ZHAO [2000]). Various real-space approaches (including methods not based finite differences, e.g., finite-element approaches) are reviewed by BECK [2000].

4. Selected applications: Na clusters

Solving the eigenvalue problem, Eq. (3.1), provides us with complete information about the eigenvalues (energy levels) and eigenvectors (wave-functions) of the system. This information can be useful for quantitative predictions of structural, electronic, and optical properties of various materials. In this section, we demonstrate how this is achieved by computing several key properties of sodium clusters – stable and metastable isomers (a structural property), static polarizability (an electronic property), and the absorption spectrum (an optical property)

4.1. Using simulated annealing for structural properties

Perhaps the most fundamental issue in dealing with clusters is determining their structure. Before any accurate theoretical calculations can be performed for a cluster, the atomic geometry of a system must be defined. However, this can be a formidable exercise. Serious problems arise from the existence of multiple local minima in the potential-energy-surface of these systems; many similar structures can exist with vanishingly small energy differences. A complicating issue is the transcription of interatomic forces into tractable classical force fields. This transcription is especially difficult for clusters such as those involving semiconducting species. In these clusters, strong many-body forces can exist that preclude the use of pairwise forces.

A convenient method to determine the structure of small or moderate sized clusters is *simulated annealing* (BINGGELI and CHELIKOWSKY [1994]). Within this technique, atoms are randomly placed within a large cell and allowed to interact at a high (usually fictive) temperature. The atoms will sample a large number of configurations. As the system is cooled, the number of high energy configurations sampled is restricted. If the anneal is done slowly enough, the procedure should quench out structural candidates for the ground state structures.

Langevin molecular dynamics is well suited for simulated annealing methods. In Langevin dynamics, the ionic positions, $\mathbf{R}_j$, evolve according to

$$M_j \ddot{\mathbf{R}}_j = \mathbf{F}(\{\mathbf{R}_j\}) - \gamma M_j \dot{\mathbf{R}}_j + \mathbf{G}_j, \tag{4.1}$$

where $\mathbf{F}(\{\mathbf{R}_j\})$ is the interatomic force on the jth particle, and $\{M_j\}$ are the ionic masses. The last two terms on the right hand side of Eq. (4.1) are dissipation and fluctuation forces, respectively. The dissipative forces are defined by the friction coefficient, γ. The fluctuation forces are defined by random Gaussian variables, $\{\mathbf{G}_i\}$, with a white noise spectrum:

$$\langle G_i^\alpha(t)\rangle = 0 \quad \text{and} \quad \langle G_i^\alpha(t) G_j^\alpha(t')\rangle = 2\gamma M_i k_{\mathrm{B}} T \delta_{ij} \delta(t - t'), \tag{4.2}$$

where t denotes time, the angular brackets denote ensemble or time averages, α stands for the Cartesian component, T is the temperature and k_{B} is the Boltzmann constant. The coefficient of T on the right hand side of Eq. (4.2) insures that the fluctuation-dissipation theorem is obeyed, i.e., the work done on the system is dissipated by the viscous medium (RISKEN [1984]). The interatomic forces can be obtained from the Hellmann–Feynman theorem using the pseudopotential wave functions.

Choosing an initial atomic configuration for the simulation takes some care. If the atoms are too far apart, they will exhibit Brownian motion, which is appropriate for Langevin dynamics with the interatomic forces zeroed. In this case, the atoms may not form a stable cluster as the simulation proceeds. Conversely, if the atoms are too close together, they may form a metastable cluster from which the ground state may be kinetically inaccessible even at the initial high temperature. Often the initial cluster is formed by a random placement of the atoms with a constraint that any given atom must reside within 1.05 and 1.3 times the bond length from at least one atom where the bond the length is defined by the crystalline environment. The cluster in question is placed in the spherical domain shown in Fig. 3.1. In Fig. 4.1, we illustrate the simulated anneal for the Na_5 cluster. In Fig. 4.2, we present the lowest energy structures for Na_n for $9 \leqslant n \leqslant 14$ and in Fig. 4.3 the lowest energy structures for Na_n for $15 \leqslant n \leqslant 20$. All results were obtained using GGA. Often several clusters occur within a few meV of each other, indicating weak and weakly directional nature of the Na–Na bond.

4.2. *Role of temperature in the polarizabilities of simple metals*

One of the few cluster properties that can be measured with relative ease and accuracy is polarizability, i.e., the change of dipole with applied electric field. An approach for computing the polarization, which is very convenient for handling the problem for *confined* systems, like clusters, is to solve the full problem exactly within the one-electron approximation. In this approach, the external ionic potential $V_{\mathrm{ion}}(\mathbf{r})$ experienced by the electrons is modified to have an additional term given by $-e\mathbf{E}\cdot\mathbf{r}$ where $\mathbf{E}$ is the electric field vector. The Kohn–Sham equations are solved with the

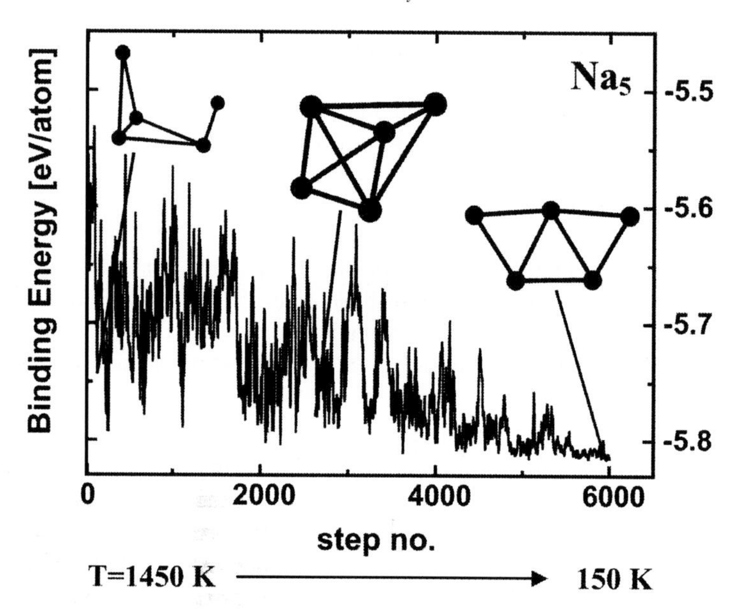

FIG. 4.1. Binding energy of Na_5 cluster during a Langevin simulation.

full external potential $V_{\text{ion}}(\mathbf{r}) - e\mathbf{E} \cdot \mathbf{r}$. For quantities like polarizability, which involve derivatives of the total energy, one can compute the energy at a few field values, and differentiate numerically. Real space methods are very suitable for confined systems, since the position operator $\mathbf{r}$ is not ill-defined, as is the case for supercell geometries in plane wave calculations (RESTA [1998]).

It is difficult to determine the polarizability of a cluster or molecule owing to the need for a complete basis in the presence of an electric field. Often polarization functions are added to complete a basis and the response of the system to the field can be sensitive to the basis required. In both real space and plane wave methods, the lack of "prejudice" with respect to the basis is a considerable asset. The real space method implemented with a uniform grid possesses a nearly "isotropic" environment with respect to the applied field. Moreover, the response can be easily checked with respect to the grid size by varying the grid spacing. Typically, the calculated electronic response of a cluster is not sensitive to the magnitude of the field over several orders of magnitude.

To illustrate this procedure, we consider a pioneering measurement of the size dependence of Na cluster polarizabilities, performed by KNIGHT, CLEMENGER, DE HEER and SAUNDERS [1985] as illustrated in Fig. 4.4. Three main features are observed in the experimental data: (a) Overall, the polarizability per atom gradually decreases from its atomic value towards its bulk value; (b) This gradual decrease is punctuated by significant "dips" at the "magic" atom numbers of 2, 8, and 18, corresponding to closed electronic shells of s-, p-, and d-like orbital character, respectively; (c) Some residual fine structure is also displayed and is usually attributed to the detailed structure of the clusters.

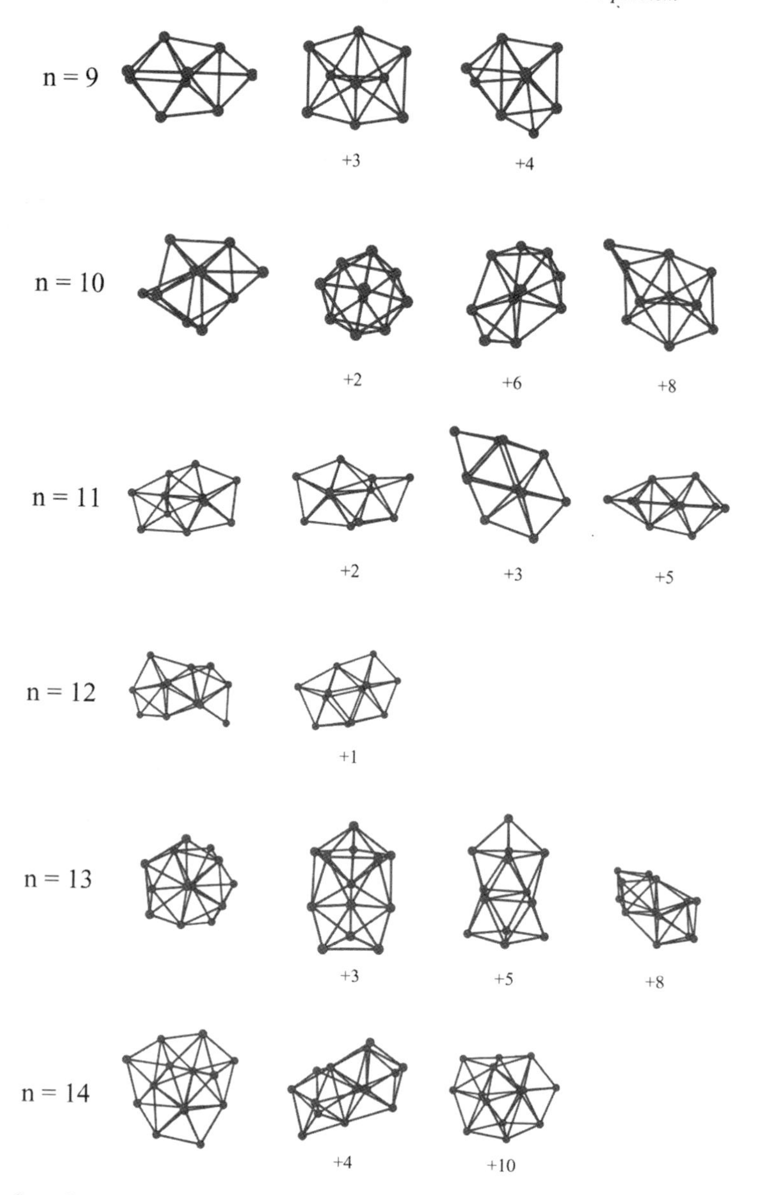

FIG. 4.2. Ground state geometries and some low-energy isomers of Na_n clusters, $n \leqslant 14$. The energy of the lower lying clusters is given in meV/atom.

Previous calculations consistently, and significantly, underestimated the experimentally observed values (RAYANE, ALLOUCHE, BENICHOU, ANTOINE, AUBERT-FRECON, DUGOURD, BROYER, RISTORI, CHANDEZON, HUBER and GUET [1999], CALAMINICI, JUG and KÖSTER [1999]). Typically the value was underestimated by $\sim$15–20%. Many workers attributed this discrepancy to a failure of DFT, which is a

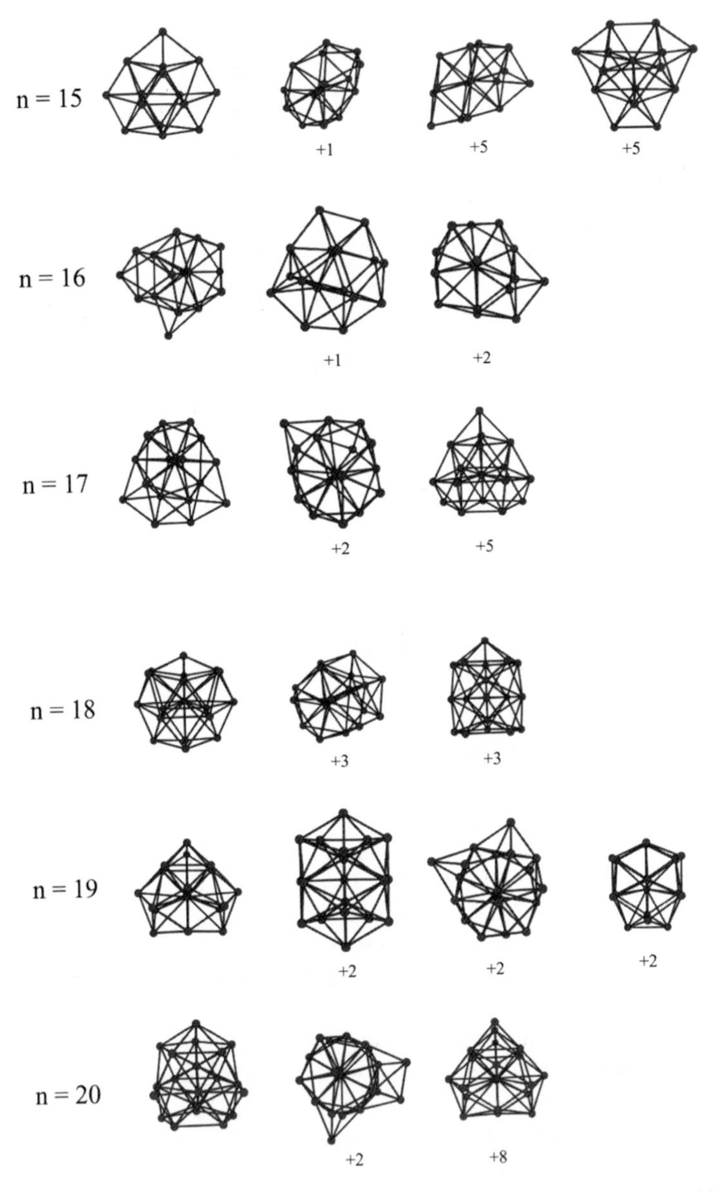

FIG. 4.3. Ground state geometries and some low-energy isomers of Na_n clusters, $15 \leqslant n \leqslant 20$. The energy of the lower lying clusters is given in meV/atom.

disconcerting conclusion given that Na is probably the element closest to replicating a homogeneous electron gas.

Workers have suggested that temperature may play an important role (KÜMMEL, AKOLA and MANNINEN [2000], KRONIK, VASILIEV and CHELIKOWSKY [2000]). One might expect that with increasing temperature, the cluster might expand in size.

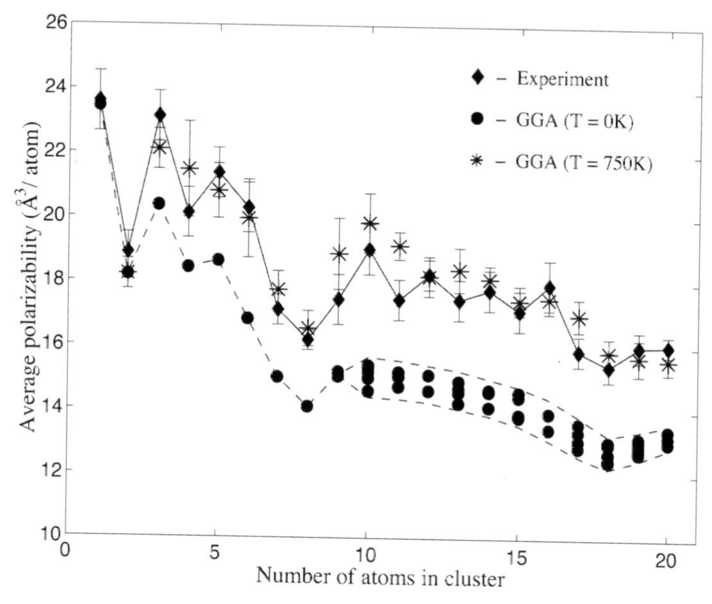

FIG. 4.4. Polarizability of Na clusters. The experiment is from KNIGHT, CLEMENGER, DE HEER and SAUNDERS [1985]. Two theoretical simulations are illustrated, both use GGA. One is at $T = 0$ K, the other is at a high temperature, $T = 750$ K as indicated.

A few percent increase in the bond length can result in a ~15–20% increase in the polarizability. This follows directly from the classical expression for the polarizability of a metallic sphere, i.e., the polarizability of such a sphere scales with its volume.

In Fig. 4.4, calculations for the polarizability of Na clusters are illustrated with and without temperature. The zero temperature results were calculated based on the low-energy structures of Figs. 4.2 and 4.3. The high-temperature results were obtained by averaging over geometries obtained by Langevin molecular dynamics with the fictive heat bath temperature fixed at 750 K (KRONIK, VASILIEV, JAIN and CHELIKOWSKY [2001]).

The zero temperature results mimic the general shape of the experimental curve, but are consistently lower than the experiment. If temperature is included, the theoretical polarizabilities are increased due to cluster expansion and distortion, and are brought into registry with experiment. This work reconciles a discrepancy that existed in the literature for a number of years.

4.3. *Optical spectra of sodium clusters*

While the theoretical background for calculating ground state properties of many-electron systems is now well established, excited state properties such as optical spectra present a challenge for DFT. Recently developed linear response theory within the time-dependent density-functional formalism provides a new tool for calculating excited states properties (CASIDA [1996]). This method, known as the time-dependent LDA

(TDLDA), allows one to compute the true excitation energies from the conventional, time independent Kohn–Sham transition energies and wave functions.

Within the TDLDA, the electronic transition energies Ω_n are obtained from the solution of the following eigenvalue problem (CASIDA [1996]):

$$\left[\omega_{ij\sigma}^2 \delta_{ik}\delta_{jl}\delta_{\sigma\tau} + 2\sqrt{f_{ij\sigma}\omega_{ij\sigma}}K_{ij\sigma,kl\tau}\sqrt{f_{kl\tau}\omega_{kl\tau}}\right]\mathbf{F}_n = \Omega_n^2\mathbf{F}_n, \tag{4.3}$$

where atomic units (a.u.) $\hbar = m = c = 1$ are used, $\omega_{ij\sigma} = \varepsilon_{j\sigma} - \varepsilon_{i\sigma}$ are the Kohn–Sham transition energies, $f_{ij\sigma} = n_{i\sigma} - n_{j\sigma}$ are the differences between the occupation numbers of the ith and jth states, the eigenvectors $\mathbf{F}_n$ are related to the transition oscillator strengths, and $K_{ij\sigma,kl\tau}$ is a coupling matrix given by:

$$K_{ij\sigma,kl\tau} = \iint \phi_{i\sigma}^*(\mathbf{r})\phi_{j\sigma}(\mathbf{r})\left(\frac{1}{|\mathbf{r}-\mathbf{r}'|} + \frac{\delta v_\sigma^{\mathrm{xc}}(\mathbf{r})}{\delta\rho_\tau(\mathbf{r}')}\right)\phi_{k\tau}(\mathbf{r}')\phi_{l\tau}^*(\mathbf{r}')\,d\mathbf{r}\,d\mathbf{r}', \tag{4.4}$$

where i, j, σ and k, l, τ are the occupied state, unoccupied state, and spin indices respectively, $\phi(\mathbf{r})$ are the Kohn–Sham wave functions, and $v^{\mathrm{xc}}(\mathbf{r})$ is the LDA exchange-correlation potential.

The TDLDA formalism is easy to implement in real space. The real-space pseudopotential code represents a natural choice for implementing TDLDA due to the real-space formulation of the general theory. With other methods, such as the plane wave approach, TDLDA calculations typically require an intermediate real-space basis. After the original plane wave calculation has been completed, all functions are transferred into that

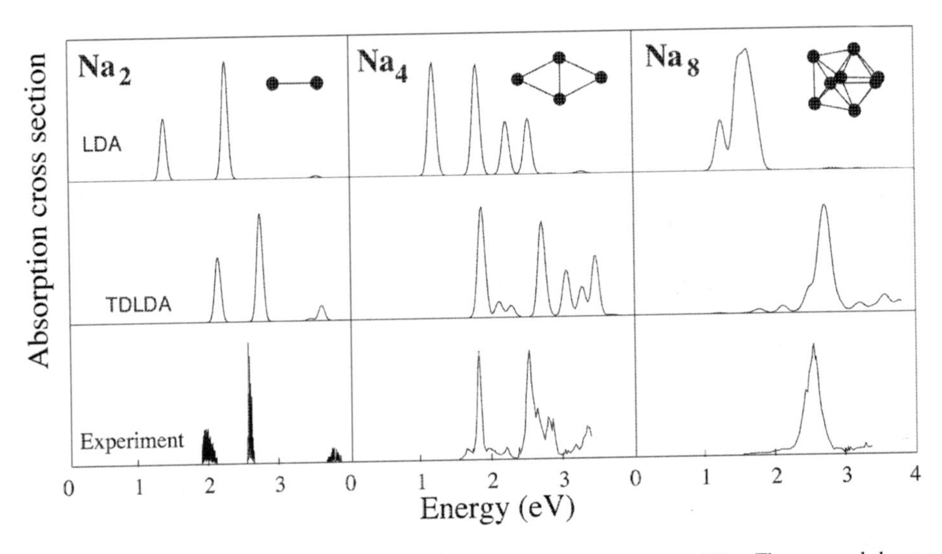

FIG. 4.5. The calculated and experimental absorption spectrum for Na_2, Na_4, and Na_8. The top panel shows a local density approximation to the spectrum using Kohn–Sham eigenvalues. The middle panel shows a TDLDA calculation. The bottom panel is experiment (WANG, POLLACK, CAMERON and KAPPES [1990], FREDRICKSON and WATSON [1927]). Also see VASILIEV, ÖĞÜT and CHELIKOWSKY [2002] and references therein.

basis, and the TDLDA response is computed in real space (BLASE, RUBIO, LOUIE and COHEN [1995]). The additional basis complicates calculations and introduces an extra error. The real-space approach simplifies implementation and allows us to perform the complete TDLDA response calculation in a single step.

We can illustrate the TDLDA technique by calculating the absorption spectra of a sodium cluster (VASILIEV, ÖĞÜT and CHELIKOWSKY [1999], VASILIEV, ÖĞÜT and CHELIKOWSKY [2002]). Accurate experimental measurements of the absorption spectra are available (WANG, POLLACK, CAMERON and KAPPES [1990], FREDRICKSON and WATSON [1927]). Since the wave functions for the unoccupied electron states are very sensitive to the boundary conditions, TDLDA calculations need to be performed within a relatively large boundary domain.

The calculated absorption spectra for Na_2 Na_4, and Na_8 are is shown in Fig. 4.5 along with experiment. In addition, we illustrate the spectrum generated by considering transitions between the LDA eigenvalues. The agreement between TDLDA and experiment is remarkable, especially when contrasted with the LDA spectrum. TDLDA correctly reproduces the experimental spectral shape, and the calculated peak positions agree with experiment within 0.1–0.2 eV. The comparison with other theoretical work demonstrates that our TDLDA absorption spectrum is as accurate as the available CI spectra (BONACIC-KOUTECKY, FANTUCCI and KOUTECKY [1990]). Furthermore, the TDLDA spectra for the Na clusters seem to be in better agreement with experiment than the calculated GW absorption spectrum (ONIDA, REINING, GODBY, DEL SOLE and ANDREONI [1995]).

5. Implementations

We begin this section by discussing the motivation for using the real-space approach instead of one based on plane wave bases. In plane wave techniques, the Laplacian term in the Hamiltonian is represented by a diagonal matrix, while the potential is represented by a dense matrix. In practice, this matrix is not computed explicitly. Instead, operations with it are performed by using fast Fourier transforms (FFT's) to move back and forth between real space and Fourier space. In contrast, matrix–vector operations in real space are trivial since the corresponding matrix is sparse. An advantage of plane wave-based methods is that they lead to natural preconditioning techniques for the eigenvalue problem, which are derived by using smaller plane wave bases, i.e., by neglecting the effect of high frequency terms. For periodic systems, these methods can be quite effective. For nonperiodic systems such as clusters, liquids or glasses, the plane wave basis must be combined with a *supercell method* (CHELIKOWSKY and COHEN [1992]), which artificially repeats the localized configuration to impose periodicity to the system. In addition to these difficulties is the issue of cost. Each of the two FFTs performed at each iteration requires on the order of $n \log n$ operations at each step versus $O(n)$ for real space methods, where n is the size of the matrix. Typically, the matrix size n is larger for real-space methods than when plane wave bases are used, but only within a constant factor. Perhaps more critical is the fact that the FFTs are not easy to implement efficiently in a parallel environment, as they tend to require an excessive amount of communication (CHELIKOWSKY, TROULLIER and SAAD [1994]). Thus, real-space finite difference methods are more general, easier to implement, and in most cases more efficient than plane wave methods.

5.1. SCF iterations

The total potential in Eq. (3.1) and the wave functions are coupled through the charge density, Eq. (2.2), the equation defining the Hartree potential, Eq. (2.3), and the equation which defines the exchange-correlation potential. Therefore, Eq. (3.1) can be viewed as a nonlinear eigenvalue problem. The solution of these problems requires the use of a Newton or an inexact Newton procedure. One procedure of this type, known as the Self Consistent Field (SCF) iteration is often used to solve the Kohn–Sham equation. The SCF iteration begins with some initial potential and solves the eigenvalue problem (3.1). The eigenvectors lead to a new potential which is 'mixed' with the one from the previous iteration and the process repeated until convergence of the potential. Here, the term 'mixing' is commonly used to describe a Broyden-type update which accelerates the convergence of the fixed point-iteration $V_{\text{tot}}^{\text{new}} = \mathbf{G}(V_{\text{tot}}^{\text{old}})$ which maps an old potential to a new one. The cost of each SCF iteration is dominated by the solution of the eigenvalue problem. Any effort in optimizing an electronic structures code should therefore focus on improving the technique for computing eigenvalues and eigenvectors.

5.2. Numerical computation of eigenvectors

From a computational point of view, the main difficulty with the eigenvalue problem Eq. (3.1) is that the number of required eigenvectors, which grows linearly with the number of atoms in the system, can be very large, reaching hundreds or even thousands. This is demanding from the point of view of computations as well as storage, since the matrices themselves will be very large, with dimensions possibly reaching hundreds of thousands to millions. In standard eigenvalue codes, a major part of the cost comes from maintaining orthogonality between computed eigenvectors. Another source of difficulty comes from the relative separation of eigenvalues which tends to deteriorate for larger matrices, leading to slower convergence. The purpose of preconditioning is to remedy this, but typical preconditioners yield limited improvements for interior eigenvalues, i.e., for eigenvalues which are located well inside the spectrum.

A popular method for computing eigenvalues and eigenvectors of large symmetric real matrices is the Davidson algorithm, which is a preconditioned variant of the Lanczos iteration, for details see, for example, PARLETT [1980], MORGAN and SCOTT [1986], DAVIDSON [1975], SAAD, STATHOPOULOS, CHELIKOWSKY, WU and ÖĞÜT [1996]. Standard preconditioning for eigenvalue problems aims as transforming the original problem into one in which the undesired eigenvectors become less dominant, leading to a fast convergence for the desired eigenvector(s). For example, instead of $H\psi = \lambda\psi$, one can solve the problem $(H - \sigma I)^{-1}\psi = \mu\psi$, whose eigenvalues $1/(\lambda - \sigma)$ are very large when $\lambda \approx \sigma$. This popular technique, known as 'shift-and-invert' (PARLETT [1980]), is not practically feasible in the present context, because of the demand to factor very large matrices for a potentially large number of shifts σ. We found that a technique based on "smoothing" or "averaging" offered a good compromise between many of the techniques we have attempted. When an approximate eigenvector is known at the points of the grid, a smoother eigenvector can be obtained by averaging the value at every point with the values of its neighboring points. For example, a low frequency filter acting on the value of the wave function ψ at the point (i, j, k) in the grid, may be described as:

$$(\psi_{i,j,k})_{\text{filtered}} = \frac{1}{12}[6\psi_{i,j,k} + \psi_{i-1,j,k} + \psi_{i,j-1,k} + \psi_{i,j,k-1} + \psi_{i+1,j,k}\psi_{i,j+1,k} + \psi_{i,j,k+1}]. \tag{5.1}$$

Filtering has the effect of reducing the highly oscillating components from the approximate eigenvector. Preconditioning 'interior' eigenvalue problems remains a difficult problem, and current solutions are not fully satisfactory.

5.3. *Parallel implementation*

In recent years, distributed memory parallel computers have emerged as the most popular high performance computing architectures in use, no doubt due to the availability of message passing software such as PVM and MPI. When implementing a real-space finite difference code on such platforms, one can think of two different ways of exploiting parallelism. The first is to take advantage of the large number of eigenpairs to be computed and to parallelize computations across these eigenvectors. The second is to rely on spatial parallelism. Our approach has adopted the latter, using a "domain decomposition" approach on the physical space.

The first task in domain decomposition techniques is to partition the physical domain in p subdomains, where p is the number of processors. It is possible to partition the domain naturally into sub-cubes, or slabs of the cube. However, zero-charge areas which must be eliminated, can be arbitrarily distributed in the domain, leading to irregular regions. These can be handled by using general purpose partitioners such as Metis (KARYPIS and KUMAR [1998]), or "manually", using a greedy approach that achieves load balancing by ordering the points and assigning the same number of points to each processor. This latter approach may lead to sub-optimal communication.

Our parallel implementation of the real-space finite difference code combines a Single Program Multiple Data (SPMD) paradigm and the master-worker paradigm. A master processor performs all the less intensive parts of the computation whose benefit from parallelization would be minimal. These include all preprocessing, computing scalar values, and the 'mixing' which obtaining the new potential after each SCF iteration. The "worker" processors compute the eigenvectors, update the charge density, and solve the Poisson equation for the Hartree potential in parallel.

Since the matrix is not actually stored, an explicit reordering can be considered so that the rows on a processor are numbered consecutively. Under this conceptually simple scheme, only a list of pointers is needed to denote where the rows of each processor start.

The nonlocal part of the matrix, which is a sum of rank-one updates, is mapped in a similar way. For each atom and for each pair of quantum numbers, a sparse vector

$u_{a,l,m}$ in (2.8) is partitioned according to the rows to which it contributes. Even though the number of nonzero elements of the u-vectors is small, their partitioning is fairly well balanced if the matrix partitioning is well balanced. With this mapping, the large storage requirements of the program are distributed.

5.4. *Parallel diagonalization*

As was already mentioned, all vectors which correspond to physical space are mapped in the same way to processors, according to the global partitioning of physical space. The Davidson algorithm will require mainly four types of operations on these vectors: (1) linear combinations of vectors; (2) matrix–vector products; (3) dot-products; and (4) preconditioning operations. Linear vector updates (saxpy operations) can be performed in parallel without communication. Dot products require a 'global reduction' of the partial inner products in each processor.

The Hamiltonian matrix in Eq. (3.1) consists of three parts. First, the main diagonal comes from the diagonal of the Laplacian and the potentials. This part is lumped with the diagonal of the Laplacian. Second, is the Laplacian term (with a modified diagonal). As in the sequential case, products with this part are performed by using the stencil information. In this operation, each processor is required to communicate with its nearest neighbors since some of the neighboring points of the local subdomain are located in the adjacent processors. Therefore, each processor must maintain a data structure which maps the local grid points to the local rows, and appends the needed interface points from other processors at the end of the local row list. The workers will extract this information and save it during the setup phase. In the second step of the matrix–vector multiplication, this interface information is exchanged among nearest neighbors and the stencil multiplication can proceed in parallel. The third part of the matrix arises from the nonlocal part of the potential. In the product with this part, each of the rank-one updates of the nonlocal components is computed as a sparse, distributed dot product. All local dot products are first computed before a global sum of their values takes place.

The preconditioning operation discussed earlier is essentially an averaging operation which uses the same stencil as that of the Laplacian. It is performed in exactly the same manner as the second step of the matrix–vector operation discussed above and therefore uses the same stencil operation. Therefore, the data structures set-up for the matrix–vector products are also used here. They are also used when solving the Poisson equation (Eq. (2.3)) for the Hartree potential with the Conjugate Gradient method and the preconditioning operation also requires the stencil and therefore they have the same communication pattern as the second step of the matrix–vector multiplication.

5.5. *Parallel TDLDA*

Our recent work has focussed on the implementation of TDLDA (VASILIEV, ÖĞÜT and CHELIKOWSKY [1999], BURDICK, SAAD, KRONIK, JAIN, VASILIEV and CHELIKOWSKY [2002]). The goal of the method is to solve the eigenvalue problem (4.3) and this leads to two distinct computational challenges. The first is the diagonalization of the matrix in Eq. (4.3) which is a large dense matrix. The second challenge, which is even bigger, lies in the computation of the coupling matrix (4.4) itself. Finally, note also that a prerequisite to the calculation of the K-matrix is the calculation of a set of eigenfunctions of Eq. (3.1), which is typically a much larger set than that of the ground states alone.

The expression (4.4) shows that there are two parts in the K-matrix. The second part, which is

$$K^{\mathrm{II}}_{ij\sigma,kl\tau} = \iint \phi_{i\sigma}(\mathbf{r})\phi_{j\sigma}(\mathbf{r}) \frac{\delta^2 E_{\mathrm{xc}}[\rho]}{\delta\rho_\sigma(\mathbf{r})\delta\rho_\tau(\mathbf{r}')} \phi_{k\tau}(\mathbf{r}')\phi_{l\tau}(\mathbf{r}')\, d\mathbf{r}\, d\mathbf{r}', \tag{5.2}$$

can be computed from a single integral thanks to the local density approximation, which inserts a delta function $\delta(\vec{r} - \vec{r}')$ in Eq. (5.2) to yield:

$$K^{\mathrm{II}}_{ij\sigma,kl\tau} = \int \phi_{i\sigma}(\mathbf{r})\phi_{j\sigma}(\mathbf{r}) \frac{\delta^2 E_{\mathrm{xc}}[\rho]}{\delta\rho_\sigma(\mathbf{r})\delta\rho_\tau(\mathbf{r})} \phi_{k\tau}(\mathbf{r})\phi_{l\tau}(\mathbf{r})\, d\mathbf{r}. \tag{5.3}$$

Computing the first term,

$$K^{\mathrm{I}}_{ij\sigma,kl\tau} = \iint \phi_{i\sigma}(\mathbf{r})\phi_{j\sigma}(\mathbf{r}) \frac{1}{|\mathbf{r} - \mathbf{r}'|} \phi_{k\tau}(\mathbf{r}')\phi_{l\tau}(\mathbf{r}')\, d\mathbf{r}\, d\mathbf{r}', \tag{5.4}$$

is more onerous. Our implementation relies on solving the Poisson's equation: $\nabla^2 \Phi_{ij\sigma}(\mathbf{r}) = -4\pi\phi_{i\sigma}(\mathbf{r})\phi_{j\sigma}(\mathbf{r})$ for each pair $\{\phi_{i\sigma}(\mathbf{r}), \phi_{j\sigma}(\mathbf{r})\}$ and then integrating in space, for each pair $\{\phi_{k\sigma}(\mathbf{r}), \phi_{l\sigma}(\mathbf{r})\}$:

$$K^{(\mathrm{I})}_{ij\sigma,kl\tau} = \int \Phi_{ij\sigma}(\mathbf{r})\phi_{k\tau}(\mathbf{r})\phi_{l\tau}(\mathbf{r})\, d\mathbf{r}. \tag{5.5}$$

The parallel code implements a master–worker paradigm just as for the eigenvalue calculation. The master sends a pair of integers i_1, i_2 with $i_1 < i_2$, to a 'worker' processor. This processor will then complete rows i_1 to i_2 of the matrix which it returns to the master. The master saves these rows to disk. This is illustrated in Fig. 5.1.

The Poisson equation is solved by the Conjugate Gradient method as was initially done in VASILIEV, ÖĞÜT and CHELIKOWSKY [1999]. One of the difficulties we encountered is that the high order approximation used does not lend itself to an efficient CG code for a number of reasons. Preconditioning was not as cost-effective as in other similar situations due to a combination of factors. First, since the 3D model does not use a fine discretization due to size limitations gains from preconditioning are limited because the unpreconditioned CG iteration performs fairly well. Second, the reduction in the CG iteration number is almost outweighed by the high cost of the preconditioning operation, which does not take advantage of block computations and data locality as well as the other operations in CG.

In order to avoid load imbalance, the master processor must itself compute a certain number of rows dynamically. Also the number of rows must be adapted as the

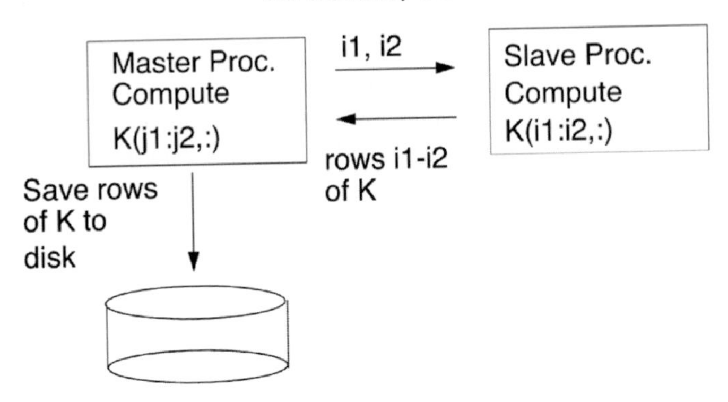

FIG. 5.1. Master–worker model for computing the coupling matrix.

computation progresses. In the early stages, larger blocks of rows must be computed by the workers and the master. Toward the end of the calculation, the size of the blocks must be reduced.

As an illustration of the matrix dimensions, for Si_{34} H_{36} the number of occupied states is $n_v = 86$ (with $E_1 \approx -1.1636$ a.u., $E_{86} \approx -0.35246$ a.u.). We have used $n_c = 154$ unoccupied states ($E_{87} \approx -0.33724$ a.u., $E_{240} \approx .202362$ a.u.). The resulting matrix is of size $N = n_v n_c = 13244$, so we need to solve a total of $N = 13244$ Poisson's equations, one for each row of the matrix.

Once the large coupling matrix is available, the eigenvalues and eigenvectors of the related matrix (4.3) must be computed. The scaLAPACK library (routine PDSYEV) can be used for this calculation, which is actually performed at a later stage. Thus, the information, i.e., the coupling matrix is saved in between the two stages. In this second stage the K-matrix is redistributed from disk to processors to prepare for a call to scalaPACK. This second stage is much less costly than the first. However, parallelism is just as important because of the memory cost. Much larger systems can be attempted if the matrix is equitably distributed among processors than otherwise.

At this point, it is useful to comment on what remains to be done and on our future plans in the area of TDLDA. Though the specific computation described above may appear to be "embarrassingly parallel", one should note that we have probably only scratched the surface on potential gains from the algorithms side. The main improvements made so far were obtained from simple techniques which exploit architectural features. In particular, an efficient use of memory allows us to deal with much bigger systems. These improvements, enabled us to perform much bigger calculations than in the original VASILIEV, ÖĞÜT and CHELIKOWSKY [2002] paper (code ≈ 3.5 times faster + much more memory-efficient). It is however still possible to reduce computational times significantly by using better algorithms. For example, we have not exploited the fact that there are several simultaneous right-hand sides in the block conjugate gradient method. Also, the conjugate gradient algorithm may be much less efficient that multigrid when solving the Poisson equation. It may even be worthwhile to resort to regular grids in order to use available Fast-Poisson Solvers, sacrificing perhaps by introducing unused points, but gaining on the side of complexity.

Acknowledgments

We would like to acknowledge support from the National Science Foundation, the United States Department of Energy and the Minnesota Supercomputing Institute.

References

BECK, T.L. (2000), Real-space mesh techniques in density-functional theory, *Rev. Mod. Phys.* **72**, 1041.

BERNHOLC, J., E.L. BRIGGS, C. BUNGARO, M.B. NARDELLI, J.L. FATTEBERT, K. RAPCEWICZ, C. ROLAND, W.G. SCHMIDT and Q. ZHAO (2000), *Phys. Stat. Solidi B* **217**, 685.

BINGGELI, N. and J.R. CHELIKOWSKY (1994), Langevin molecular dynamics with quantum forces: Application to silicon clusters, *Phys. Rev. B* **50**, 11764.

BLASE, X., A. RUBIO, S.G. LOUIE and M.L. COHEN (1995), Mixed-space formalism for the dielectric response in periodic systems, *Phys. Rev. B* **52**, R2225.

BONACIC-KOUTECKY, V., P. FANTUCCI and J. KOUTECKY (1990), Theoretical interpretation of the photoelectron detachment spectra of Na_{2-5}^- and of the absorption spectra of Na_3, Na_4 and Na_8 clusters, *J. Chem. Phys.* **93**, 3802.

An *ab initio* configuration-interaction study of the excited states of the Na_4 cluster: assignment of the absorption spectrum, *Chem. Phys. Lett.* **166**, 32.

BURDICK, W.R., Y. SAAD, L. KRONIK, M. JAIN, I. VASILIEV and J.R. CHELIKOWSKY (in preparation), Parallel implementation of time-dependent density functional theory, Technical report, Minnesota Supercomputer Institute.

CALAMINICI, P., K. JUG and KÖSTER (1999), Static polarizabilities of Na_n ($n \leqslant 9$) clusters: An all electron density functional study, *J. Chem. Phys.* **111**, 4613.

CASIDA, M.E. (1995), Time-dependent density functional response theory for molecules, in: D.P. Chong, ed., *Recent Advances in Density-Functional Methods*, Part I (World Scientific, Singapore) 155.

CASIDA, M.E. (1996), Time-dependent density functional response theory of molecular systems: theory, computational methods, and functionals, in: J.M. Seminario, ed., *Recent Developments and Applications of Modern Density Functional Theory* (Elsevier, Amsterdam) 391.

CHELIKOWSKY, J.R. and M.L. COHEN (1992), *Ab initio* pseudopotentials for semiconductors, in: P. Landsberg, ed., *Handbook on Semiconductors*, Vol. 1 (Elsevier, Amsterdam) 59.

CHELIKOWSKY, J.R. and S.G. LOUIE, eds. (1996), *Quantum Theory of Real Materials* (Kluwer Press, Boston).

CHELIKOWSKY, J.R., N. TROULLIER and Y. SAAD (1994), Finite-difference-pseudopotential method: electronic structure calculations without a basis, *Phys. Rev. Lett.* **72**, 1240–1243.

DAVIDSON, E.R. (1975), The iterative calculation of a few of the lowest eigenvalues and corresponding eigenvectors of large real-symmetric matrices, *J. Comput. Phys.* **17**, 87.

FREDRICKSON, W.R. and W.W. WATSON (1927), The sodium and potassium absorption bands, *Phys. Rev.* **30**, 429.

FORNBERG, B. and D.M. SLOAN (1994), A review of pseudospectral methods for solving partial differential equations, in: A. Iserles, ed., *Acta Numerica 94* (Cambridge University Press) 203.

HAMANN, D.R., M. SCHLÜTER and C. CHIANG (1979), Norm-conserving pseudopotentials, *Phys. Rev. Lett.* **43**, 1494.

HOHENBERG, P. and W. KOHN (1964), Inhomogeneous electron gas, *Phys. Rev.* **136**, B864.

HYBERTSEN, M. and S.G. LOUIE (1986), Electron correlation in semiconductors and insulators: Band gaps and quasiparticle energies, *Phys. Rev.* **34**, 5390.

KARYPIS, G. and V. KUMAR (1998), A fast and high-quality multi-level scheme for partitioning irregular graphs, *SIAM Journal on Scientific Computing* **20**, 359–392.

KERKER, G.P. (1980), Non-singular atomic pseudopotentials for solid state applications, *J. Phys. C* **13**, L189.

KLEINMAN, L. and D.M. BYLANDER (1982), Efficacious form for model pseudopotentials, *Phys. Rev. Lett.* **48**, 1425.

KNIGHT, W.D., K. CLEMENGER, W.A. DE HEER and W.A. SAUNDERS (1985), Polarizability of alkali clusters, *Phys. Rev. B* **31**, 2539.

KOHN, W. and L. SHAM (1965), Self-consistent equations including exchange and correlation effects, *Phys. Rev.* **140**, A1133.

KRONIK, L., I. VASILIEV and J.R. CHELIKOWSKY (2000), *Ab initio* calculations for structure and temperature effects on the polarizabilities of Na_n ($n \leqslant 20$), *Phys. Rev. B* **62**, 9992.

KRONIK, L., I. VASILIEV, M. JAIN and J.R. CHELIKOWSKY (2001), *Ab initio* structures and polarizabilities of sodium clusters, *J. Chem. Phys.* **115**, 4322.

KÜMMEL, S., J. AKOLA and M. MANNINEN (2000), Thermal expansion in small metal clusters and its impact on the electric polarizability, *Phys. Rev. Lett.* **84**, 3827.

MODINE, N.A., G. ZUMBACH and E. KAXIRAS (1997), Adaptive-coordinate real-space electronic-structure calculations for atoms, molecules, and solids, *Phys. Rev. B* **55**, 10289.

MORGAN, R.B. and D.S. SCOTT (1986), Generalizations of Davidson's method for computing eigenvalues of sparse symmetric matrices, *SIAM J. Sci. Stat. Comput.* **7**, 817.

ONIDA, G., L. REINING, R.W. GODBY, R. DEL SOLE and W. ANDREONI (1995), *Ab initio* calculations of the quasiparticle and absorption spectra of clusters: the sodium tetramer, *Phys. Rev. Lett.* **75**, 818.

PARLETT, B.N. (1980), *The Symmetric Eigenvalue Problem* (Prentice Hall, Englewood Cliffs).

PERDEW, J.P., K. BURKE and M. ERNZERHOF (1996), Generalized gradient approximation made simple, *Phys. Rev. Lett.* **77**, 3865.

RAYANE, D., A.R. ALLOUCHE, E. BENICHOU, R. ANTOINE, M. AUBERT-FRECON, PH. DUGOURD, M. BROYER, C. RISTORI, F. CHANDEZON, B.A. HUBER and C. GUET (1999), Static electric dipole polarizabilities of alkali clusters, *Eur. Phys. J. D* **9**, 243.

RESTA, R. (1998), Quantum-mechanical position operator in extended systems, *Phys. Rev. Lett.* **80**, 1800.

RISKEN, H. (1984), *The Fokker–Planck Equation* (Springer-Verlag, Berlin).

SAAD, Y. (1996), *Iterative Methods for Sparse Linear Systems* (PWS Publishing Company, Boston).

SAAD, Y., A. STATHOPOULOS, J.R. CHELIKOWSKY, K. WU and S. ÖĞÜT (1996), Solution of large eigenvalue problems in electronic structure calculations, *BIT* **36**, 563.

TONG, C.H., T.F. CHAN and C.C.J. KUO (1992), Multilevel filtering preconditioners: extensions to more general elliptic problems, *SIAM J. Sci. Stat. Comput.* **13**, 227.

TROULLIER, N. and J.L. MARTINS (1991), Efficient pseudopotentials for plane wave calculations, *Phys. Rev. B* **43**, 1993.

VASILIEV, I., S. ÖĞÜT and J.R. CHELIKOWSKY (1999), *Ab initio* excitation spectra and collective electronic response in atoms and clusters, *Phys. Rev. Lett.* **82**, 1919.

VASILIEV, I., S. ÖĞÜT and J.R. CHELIKOWSKY (2002), First-principles density-functional calculations for optical spectra of clusters and nanocrystals, *Phys. Rev. B* **65**, 15416.

WANG, C.R.C., S. POLLACK, D. CAMERON and M.M. KAPPES (1990), Molecular excited states versus collective electronic oscillations: optical absorption probes of Na_4 and Na_8, *Chem. Phys. Lett.* **93**, 3787.

ZUNGER, A. and M.L. COHEN (1979), First-principles non-local-pseudopotential approach in the density-functional formalism. II. Application to electronic and structural properties of solids, *Phys. Rev. B* **20**, 4082.

Scalable Multiresolution Algorithms for Classical and Quantum Molecular Dynamics Simulations of Nanosystems

Aiichiro Nakano[a], Timothy J. Campbell[b], Rajiv K. Kalia[a], Sanjay Kodiyalam[a], Shuji Ogata[c], Fuyuki Shimojo[d], Xiaotao Su[a], Priya Vashishta[a]

[a] *Concurrent Computing Laboratory for Materials Simulations, Biological Computation and Visualization Center, Department of Computer Science, Department of Physics & Astronomy, Louisiana State University, Baton Rouge, LA 70803, USA*

[b] *Naval Oceanographic Office, Stennis Space Center, MS 39529, USA*

[c] *Department of Applied Sciences, Yamaguchi University, Ube 755-8611, Japan*

[d] *Faculty of Integrated Arts and Sciences, Hiroshima University, Higashi-Hiroshima 739-0046, Japan*

E-mail adress: nakano@bit.csc.lsu.edu (A. Nakano)

1. Introduction

Petaflop (10^{15} floating-point operations per second) computers anticipated to be built in the near future (DONGARRA and WALKER [2001]) and metacomputing on a Grid of geographically distributed supercomputers (FOSTER and KESSELMAN [1999]) will offer tremendous opportunities for high-end computational research. However, these advanced architectures will have impacts only if applications are scalable up to the Petaflop level. Unfortunately, the majority of scientific applications are not scalable even at the current Teraflop (10^{12} floating-point operations per second) level.

To unleash the power of future Petaflop computers, a number of scalable algorithms are being developed at the forefront of scientific computing. For example, modern

Computational Chemistry
Special Volume (C. Le Bris, Guest Editor) of
HANDBOOK OF NUMERICAL ANALYSIS, VOL. X
P.G. Ciarlet (Editor)
© 2003 Elsevier Science B.V. All rights reserved

design of high-performance materials and devices focuses on controlling structures at diverse length scales from atomic to macroscopic (PECHENIK, KALIA and VASHISHTA [1999]), and a rich variety of atomistic simulation methods – ranging from empirical molecular-dynamics (MD) simulations (RAPAPORT [1995]) to *ab initio* quantum-mechanical (QM) calculations (PAYNE, TETER, ALLAN, ARIAS and JOANNOPOULOS [1992]) – are expected to play an important role in scaling down macroscopic engineering concepts to nanometer scales. Recent advances in linear-scaling simulation algorithms and scalable parallel computing frameworks have made it possible to carry out 10–100 million atom simulations of real materials and devices typically on 10–100 processors (ABRAHAM [1997], GERMAN and LOMDAHL [1999], VASHISHTA, KALIA and NAKANO [1999]). Recent benchmark tests (NAKANO, KALIA, VASHISHTA, CAMPBELL, OGATA, SHIMOJO and SAINI [2001]) have demonstrated 6.44 billion-atom MD simulation and 0.44 million-electron QM calculation based on the density functional theory (DFT) (HOHENBERG and KOHN [1964], KOHN and VASHISHTA [1983]) on a Teraflop architecture.

This chapter reviews our recent efforts to enable very large-scale atomistic simulations involving multibillion atoms by designing scalable and portable simulation algorithms. In the next section, we describe parallel linear-scaling algorithms for MD and QM calculations. Section 3 discusses software tools to support billion-atom simulations. Application of these algorithms to the study of nanoscale systems is described in Section 4, and Section 5 contains conclusions.

2. Parallel atomistic simulation algorithms

We have developed a suite of scalable MD and QM algorithms for materials simulations (SHIMOJO, CAMPBELL, KALIA, NAKANO, VASHISHTA, OGATA and TSURUTA [2000], NAKANO, KALIA, VASHISHTA, CAMPBELL, OGATA, SHIMOJO and SAINI [2001]). The linear-scaling algorithms encompass a wide spectrum of physical reality: (i) classical MD based on a many-body interatomic potential model; (ii) environment-dependent, variable-charge MD; and (iii) self-consistent QM calculation based on the density functional theory (DFT) (HOHENBERG and KOHN [1964], KOHN and VASHISHTA [1983]). These algorithms deal with the following three problems that are common in many scientific and engineering applications: (i) all-pairs function evaluation in the N-body problem; (ii) dense linear system of equations in the variable N-charge problem; and (iii) exhaustive combinatorial enumeration in the quantum N-body problem, respectively. This section describes general algorithmic techniques to obtain approximate solutions to these problems in $O(N)$ time (N is the problem size), including (i) clustering, (ii) hierarchical abstraction, and (iii) the analysis of asymptotic solution properties.

2.1. Multiresolution molecular dynamics algorithm

In the MD approach, a physical system is represented by a set of point atoms. Let $\mathbf{r}^N = \{\mathbf{r}_1, \mathbf{r}_2, \ldots, \mathbf{r}_N\}$ denote the positions of all N atoms in the system (RAPAPORT [1995]). The time evolution of $\mathbf{r}^N$ is governed by a set of coupled ordinary differential equations (Newton's equations of motion),

$$m_i \frac{d^2}{dt^2}\mathbf{r}_i = -\frac{\partial}{\partial \mathbf{r}_i} E_{\mathrm{MD}}(\mathbf{r}^N), \tag{2.1}$$

where m_i is the mass of the ith atom, and the interatomic potential energy, $E_{\mathrm{MD}}(\mathbf{r}^N)$, describes how atoms interact with each other. The MD approach obtains the phase-space trajectories of the system (positions, $\mathbf{r}_i$, and velocities, $d\mathbf{r}_i/dt$, of all atoms at all time) from the numerical solution of Eq. (2.1), by discretizing the time in unit of Δt,which is typically $\sim 10^{-15}$ second.

Accurate atomic force laws are essential for realistic simulation of devices. Mathematically, a force law is encoded in the interatomic potential energy, E_{MD}, in Eq. (2.1). In the past years, we have developed many-body interatomic potentials for a number of materials, including ceramics such as silica (SiO_2) (CAMPBELL, KALIA, NAKANO, SHIMOJO, TSURUTA and VASHISHTA [1999]), silicon nitride (Si_3N_4) (NAKANO, KALIA and VASHISHTA [1995]), and silicon carbide (SiC) (SHIMOJO, EBBSJÖ, KALIA, NAKANO, RINO and VASHISHTA [2000]), as well as semiconductors (KODIYALAM, KALIA, KIKUCHI, NAKANO, SHIMOJO and VASHISHTA [2001], SU, KALIA, MADHUKAR, NAKANO and VASHISHTA [2001]) such as gallium arsenide (GaAs), aluminum arsenide (AlAs), and indium arsenide (InAs). In our many-body interatomic potential scheme, the interatomic potential energy is expressed as an analytic function that depends on relative positions of atomic pairs and triples (VASHISHTA, KALIA, NAKANO, LI and EBBSJÖ [1996]),

$$E_{\mathrm{MD}}(\mathbf{r}^N) = \sum_{(i,j)} u_{ij}(|\mathbf{r}_{ij}|) + \sum_{(i,j,k)} v_{jik}(\mathbf{r}_{ij}, \mathbf{r}_{ik}), \tag{2.2}$$

where $\mathbf{r}_{ij} = \mathbf{r}_i - \mathbf{r}_j$. Physically, the two-body terms, u_{ij}, represent steric repulsion and electrostatic interaction due to charge transfer between atoms, and induced charge–dipole and dipole–dipole interactions that take into account the large electronic polarizability of negative ions. The three-body terms, v_{jik}, take into account covalent effects through bending and stretching of atomic bonds.

The most complex computation in the MD is the evaluation of the long-range electrostatic potential, $\Phi(\mathbf{r})$, at N destination points. Since each evaluation involves contributions from N source points, the computational cost of direct summation scales as $O(N^2)$. For scalable MD simulations on massively parallel computers, we have developed a multiresolution molecular dynamics (MRMD) algorithm (NAKANO, KALIA and VASHISHTA [1994], NAKANO, KALIA, VASHISHTA, CAMPBELL, OGATA, SHIMOJO and SAINI [2001]). This algorithm combines (i) the fast multipole method (FMM) by GREENGARD and ROKHLIN [1987] to reduce the computational complexity from $O(N^2)$ to $O(N)$ and (ii) the multiple time-scale (MTS) method (TUCKERMAN, YARNE, SAMUELSON, HUGHES and MARTYNA [2000]) to introduce multiple resolutions in time.

The first essential idea of the FMM is clustering, i.e., instead of computing interactions between all atomic pairs, atoms are clustered and cluster–cluster interactions are computed, see Fig. 2.1. At the source of interaction, cluster information is encapsulated

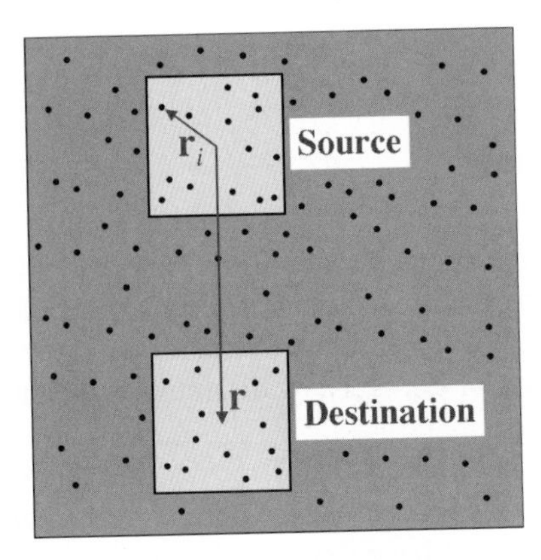

FIG. 2.1. Schematic of the clustering in the FMM (the figure shows a 2D example).

in terms of the multipoles of the charge distribution,

$$\begin{aligned}\Phi(\mathbf{r}) &= \sum_{i=1}^{N} \frac{q_i}{|\mathbf{r} - \mathbf{r}_i|} = \sum_{l=0}^{\infty} \sum_{m=-l}^{l} \left\{ \sum_{i=1}^{N} q_i r_i^l Y_l^{*m}(\theta_i, \phi_i) \right\} \frac{Y_l^m(\theta, \phi)}{r^{l+1}} \\ &= \sum_{l=0}^{\infty} \sum_{m=-l}^{l} M_l^m\left(q^N, \mathbf{r}^N\right) \frac{Y_l^m(\theta, \phi)}{r^{l+1}},\end{aligned} \tag{2.3}$$

where $\mathbf{r} = (r, \theta, \phi)$ and $\mathbf{r}_i = (r_i, \theta_i, \phi_i)$ are the spherical coordinates of the destination and source points, $Y_l^m(\theta, \phi)$ are the spherical harmonics, and the multipoles, $M_l^m(q^N, \mathbf{r}^N)$, depends on the charges, $q^N = \{q_1, \ldots, q_N\}$, and the positions, $\mathbf{r}^N$, of the source points. The multipole expansion provides a well-defined error bound. At the destination point, on the other hand, the electrostatic potential is expanded in terms of local terms, which is similar to the Taylor expansion:

$$\begin{aligned}\Phi(\mathbf{r}) &= \sum_{l=0}^{\infty} \sum_{m=-l}^{l} \left\{ \sum_{i=1}^{N} \frac{q_i Y_l^m(\theta_i, \phi_i)}{r_i^{l+1}} \right\} r^l Y_l^{*m}(\theta, \phi) \\ &= \sum_{l=0}^{\infty} \sum_{m=-l}^{l} L_l^m\left(q^N, \mathbf{r}^N\right) r^l Y_l^{*m}(\theta, \phi),\end{aligned} \tag{2.4}$$

where $L_l^m(q_i^N, \mathbf{r}_i^N)$ are the local expansion coefficients.

The second essential idea in the FMM is to use larger clusters for longer distances, in order to reduce the computational complexity as well as to keep the error constant. This is achieved by recursively subdividing the simulation box into smaller cells to form an octree data structure, see Fig. 2.2. The root of the octree is the entire simulation cell and it is divided into $2 \times 2 \times 2$ children cells of equal volume. At each level, l, of the octree, each parent cell is subdivided into 8 children cells, and this recursive subdivision is terminated at the leaf level, $l = L$, where further subdivision would make the cell size

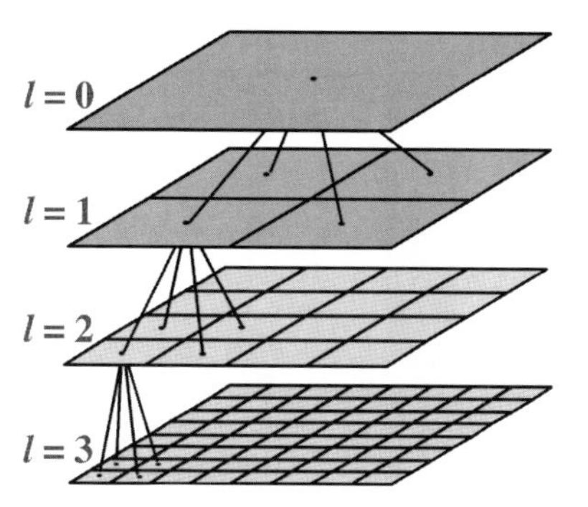

FIG. 2.2. Recursive subdivision of the simulation cell, leading to a tree structure (the figure shows a 2D example), where the lines denote parent–children relationships.

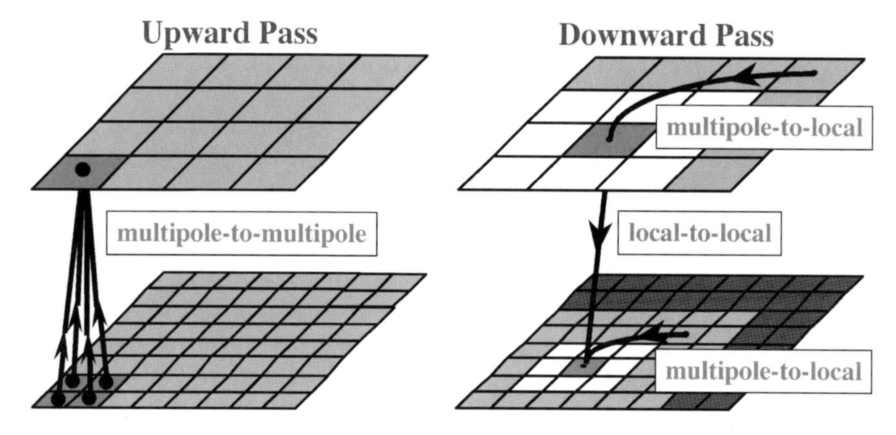

FIG. 2.3. Dual tree traversal in the FMM (the figure shows a 2D example). (Left) The upward pass uses children-to-parent shifts to compute multipoles for all cells. (Right) The downward pass translates the multipoles into local terms for all cells. For a given cell at each level, the multipoles of a constant number of interactive cells contribute to the local terms, which are added to the local terms inherited from its parent cell.

smaller than the cut-off length of the nonelectrostatic interatomic potentials (NAKANO, KALIA and VASHISHTA [1994]).

The $O(N)$ algorithm traverses this tree twice, see Fig. 2.3 (GREENGARD and ROKHLIN [1987]). In the upward pass, the multipoles are computed for all octree cells at all levels. First the multipoles of the leaf cells are computed using atomic charges and coordinates according to Eq. (2.1). The multipoles of these children cells are then combined to obtain the multipoles of the parent cells. Since the origin of the parent cell is different from that of the children cells, the multipole-to-multipole transformation formula (GREENGARD and ROKHLIN [1987]) is used to shift the origin of the multipole expansion. This procedure is repeated until the root of the octree is reached. In the downward pass, these multipoles are transformed to local expansion coefficients for all cells at all levels, starting from the root of the octree. These local expansion coefficients are computed from multipoles at different origins, using the multipole-to-local transformation formula. For a given cell at each level,

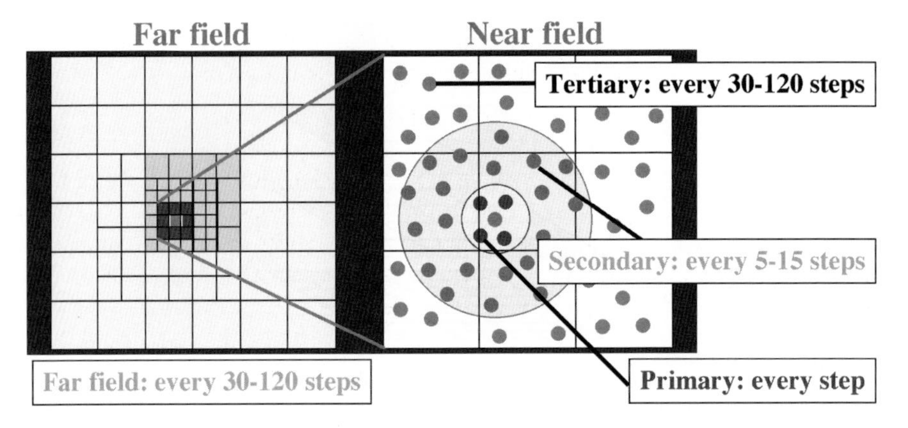

FIG. 2.4. Force update schedules used in the MTS method. The near-field interaction in the FMM is further classified into the primary, secondary and tertiary components according to the spatial distance, which are updated with progressively less frequency.

only the multipoles of a constant number of interactive cells contribute to the local expansion coefficients. Contributions from farther cells have already been computed at the previous coarse level, and they are inherited from the parent cell. (The local-to-local transformation formula is used to shift the origin of the local expansions.) On the other hand, the contributions from the nearest-neighbor cells will be computed at the next fine level. This procedure is repeated until the leaf level is reached, which concludes the computation of the far-field interaction. Finally, the nearest-neighbor-cell contributions at the leaf level (i.e., the near-field interaction) are computed by direct summation over atoms. Since constant computation is performed at each of the $O(N)$ octree nodes, the complexity of the FMM algorithm is $O(N)$.

The scalability of the parallel FMM code is further enhanced by incorporating the multiple time-scale (MTS) method (NAKANO, KALIA and VASHISHTA [1994]). The MTS method uses different force-update schedules for different force components, i.e., forces from the nearest-neighbor atoms are computed at every MD step, and forces from farther atoms are computed with less frequency, see Fig. 2.4. This not only reduces the computational cost but also enhances the data locality, and accordingly the parallel efficiency is increased.

These different force components are combined with a reversible symplectic integrator by TUCKERMAN, YARNE, SAMUELSON, HUGHES and MARTYNA [2000]. To explain their reversible reference system propagator algorithm (RESPA), let us consider a physical system in which the forces are divided into the primary and secondary components, $\mathbf{F}_i^{\mathrm{p}}$ and $\mathbf{F}_i^{\mathrm{s}}$, which vary rapidly and slowly in time, respectively. Accordingly the Liouville operator, L, of the system is decomposed into two parts, $iL = iL_{\mathrm{p}} + iL_{\mathrm{s}}$, where the primary and secondary operators are defined as

$$\begin{aligned} iL_{\mathrm{p}} &= \sum_i \dot{\mathbf{r}}_i \frac{\partial}{\partial \mathbf{r}_i} + \mathbf{F}_i^{\mathrm{p}} \frac{\partial}{\partial \mathbf{p}_i}, \\ iL_{\mathrm{s}} &= \sum_i \mathbf{F}_i^{\mathrm{s}} \frac{\partial}{\partial \mathbf{p}_i}, \end{aligned} \tag{2.5}$$

where $\mathbf{p}_N = \{\mathbf{p}_1, \ldots, \mathbf{p}_N\}$ are the momenta of the atoms. The temporal propagator for the system, $\Gamma = \{\mathbf{r}^N, \mathbf{p}^N\}$, is approximated using the Trotter formula,

$$\Gamma(t + n\Delta t) = e^{iL_s n\Delta t/2} \left(e^{iL_p \Delta t}\right)^n e^{iL_s n\Delta t/2} \Gamma(t). \tag{2.6}$$

A similar procedure is applied for multiple-part force decomposition, and the resulting algorithm consists of nested loops to use forces from different spatial regions. It has been proven that the phase-space volume occupied by atoms is a simulation-loop invariant in this algorithm (TUCKERMAN, YARNE, SAMUELSON, HUGHES and MARTYNA [2000]):

$$\frac{\partial(\mathbf{p}^N_{t+n\Delta t}, \mathbf{r}^N_{t+n\Delta t})}{\partial(\mathbf{p}^N_t, \mathbf{r}^N_t)}^T \begin{pmatrix} 0 & \mathbf{I} \\ -\mathbf{I} & 0 \end{pmatrix} \frac{\partial(\mathbf{p}^N_{t+n\Delta t}, \mathbf{r}^N_{t+n\Delta t})}{\partial(\mathbf{p}^N_t, \mathbf{r}^N_t)} = \begin{pmatrix} 0 & \mathbf{I} \\ -\mathbf{I} & 0 \end{pmatrix}, \tag{2.7}$$

where $\mathbf{I}$ is the unit matrix in the $3N$-dimensional space. This loop invariant in the RESPA has been shown to result in excellent long-time stability.

To achieve longer simulated times for stiff MD systems, we have also extended the MTS method with a hierarchical-dynamics algorithm based on fuzzy clustering (NAKANO [1997b]). This algorithm groups atoms into cohesively bound clusters, and it maps global dynamics to the rigid-body motion of these clusters (NAKANO [1999b]). A normal-mode analysis integrates the fast oscillation of each atom around its local potential minimum (ZHANG and SHLICK [1993]). The equation for the residual system is integrated using a symplectic, implicit scheme (SKEEL, ZHANG and SCHLICK [1997]). Fuzzy clustering improves the accuracy of the crisp rigid-body scheme significantly. This approach sped up a simulation of nanocluster sintering by a factor of 28 over a conventional, explicit integration scheme.

For parallelization of MD simulations, we use spatial decomposition (NAKANO, KALIA and VASHISHTA [1994]). The total volume of the system is divided into $P = P_x \times P_y \times P_z$ subsystems of equal volume, and each subsystem is assigned to a processor in an array of P processors. To calculate the nonelectrostatic as well as the near-field interaction on an atom in a subsystem, the coordinates of the atoms in the leaf cells that are at the boundaries of neighbor subsystems are "cached" from the corresponding processors, see Fig. 2.5. After updating the atomic positions due to a time-

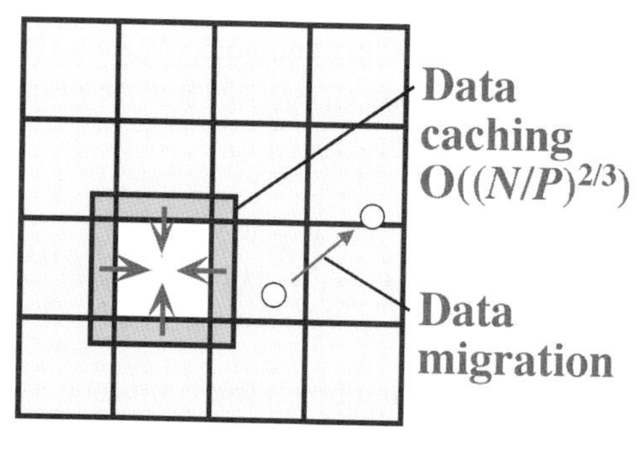

FIG. 2.5. Schematic of data caching and data migration in parallel MD algorithms. The 2-dimensional example shows an array of 4×4 subsystems.

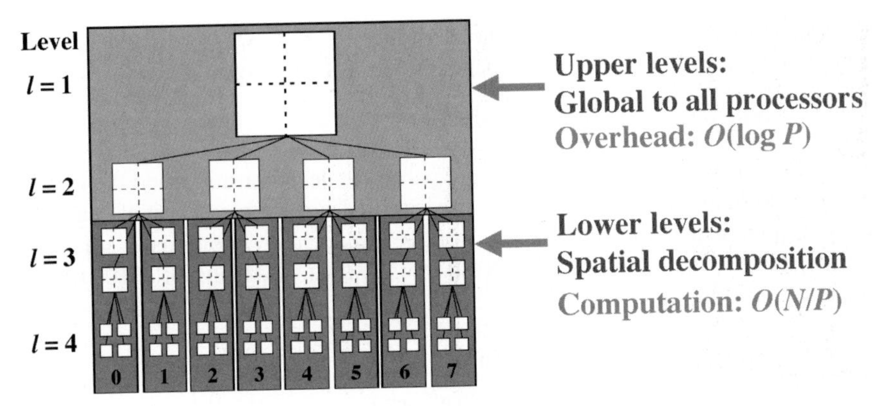

FIG. 2.6. Spatial decomposition scheme for FMM in a two-dimensional system. In the lower levels, cells are local to a node. Cell information in the upper levels is made global to all nodes.

stepping procedure, some atoms may have moved out of its subsystem. These atoms are "migrated" to the proper neighbor processors, see Fig. 2.5. To compute the far-filed interaction, upper-level octree cells are handled globally on all P processors if the number of cells is less than P, which introduces an $\mathrm{O}(\log P)$ overhead, see Fig. 2.6. At the lower octree levels, on the other hand, spatial decomposition is employed, so that the computation on each processor is $\mathrm{O}(N/P)$. The spatial decomposition also necessitates the caching of surface-cell multipoles from the nearest-neighbor processors, which involves $\mathrm{O}((N/P)^{2/3})$ communication. For typical coarse-grained applications ($N/P \sim 10^6$ and $P < 10^3$), both the global ($\log P$) and the nearest-neighbor ($(N/P)^{2/3}$) communications are negligible compared with the $\mathrm{O}(N/P)$ computation.

2.2. *Variable-charge molecular dynamics*

Physical realism of MD simulations is greatly enhanced by incorporating variable atomic charges that dynamically adapt to the local environment (CAMPBELL, KALIA, NAKANO, VASHISHTA, OGATA and RODGERS [1999], STREITZ and MINTMIRE [1994]). In this variable-charge molecular dynamics (VCMD), the interatomic potential energy of the system is expressed as a sum of nonionic, V_0, and electrostatic, V_{es}, terms: $E_{\mathrm{MD}} = V_0 + V_{\mathrm{es}}$. The electrostatic energy is a function of atomic positions, $\mathbf{r}^N$, and atomic charges, q^N:

$$V_{\mathrm{es}}\left(\mathbf{r}^N, q^N\right) = \sum_i \left(\chi_i q_i + \frac{1}{2} J_i q_i^2\right) + \sum_{i<j} \int d\mathbf{x}_1 \int d\mathbf{x}_2 \frac{\rho_i(\mathbf{x}_1; q_i)\rho_j(\mathbf{x}_2; q_j)}{|\mathbf{x}_1 - \mathbf{x}_2|}. \tag{2.8}$$

In Eq. (2.8), the first sum represents intra-atomic electrostatic energies, where χ_i and J_i denote the electronegativity and self-Coulomb repulsion of the ith atom. In the second sum in Eq. (2.8), the contributions from atomic pairs (i, j) arise from inter-atomic Coulomb interaction. Atomic charge distribution is modeled by a Slater-type orbital,

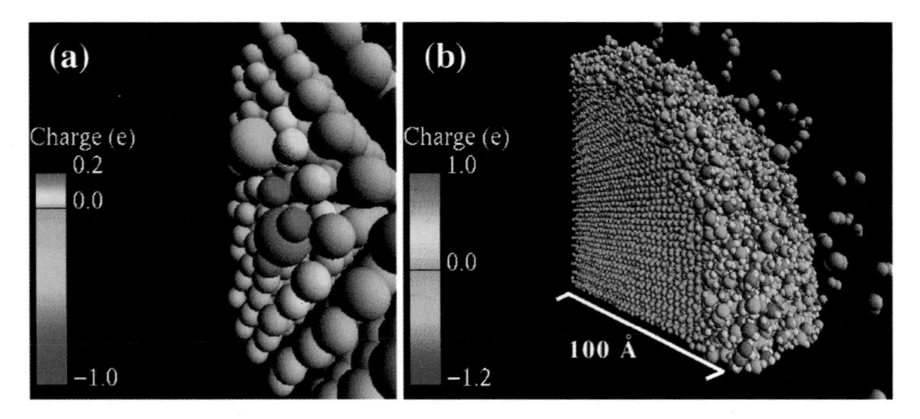

FIG. 2.7. (a) Snapshot of a variable-charge molecular-dynamics simulation of oxygen (large spheres) on an aluminum (small spheres) surface. The charge on an atom (represented by color) is allowed to vary with the environment. (b) Snapshot showing the evolution of oxidation in a wedge cut from the Al nanocluster at 30 ps of simulated time.

$\rho_i(\mathbf{x}; q_i) = (q_i\zeta_i^3/\pi)\exp(-2\zeta_i|\mathbf{x} - \mathbf{r}_i|)$, where ζ_i^{-1} is the decay length for atomic orbitals.

The increased realism of the VCMD is accompanied by increased computational complexity, $O(N^3)$, which arises as follows. In the VCMD, atomic charges, $\{q_i\}$, are determined at every MD step so as to minimize the electrostatic potential energy, V_{es}, with the charge neutrality constraint, $\Sigma_i q_i = 0$, see Fig. 2.7(a). This variable N charge problem amounts to solving a dense linear system of equations,

$$\sum_j M_{ij} q_j = \mu - \chi_i, \tag{2.9}$$

where M_{ij} denotes the Coulomb-interaction matrix, and the Lagrange's multiplier, μ, is determined to satisfy the charge neutrality constraint.

We have reduced this complexity to $O(N)$ (NAKANO [1997a], NAKANO [1997b]) by combining: (i) the fast multipole method (FMM) (GREENGARD and ROKHLIN [1987]) to perform matrix-vector multiplications with $O(N)$ operations; and (ii) an iterative minimization approach to initialize the solution using the previous MD step's charges, taking advantage of the temporal locality of the data (i.e., the atomic charges vary only slightly within one time step, Δt). This algorithm reduces the amortized computational cost averaged over simulation steps to $O(N)$.

To further accelerate the convergence of the iterative solution mentioned above, we have developed a multilevel preconditioned conjugate-gradient (MPCG) method (NAKANO [1997a], NAKANO [1997b]), by splitting the Coulomb-interaction matrix into short- and long-range components, $\mathbf{M} = \mathbf{M}_s + \mathbf{M}_l$, and using the sparse short-range matrix as a preconditioner. This decomposition of the Coulomb interaction matrix is based on the decomposition of the electrostatic energy into short- and long-range components: $V_{es} = V_s + V_l$. In the context of the FMM, they are derived from the near-field and far-field interactions, respectively. The resulting algorithm has a doubly nested loop structure. The inner loop computes a preconditioning vector $\mathbf{z}$ by solving the linear system, $\mathbf{M}_S\mathbf{z} = \mathbf{r}$, using the conjugate gradient (CG) method. Due to the short range of $\mathbf{M}_S$, this system is easier to solve than the original system. The outer loop solves the preconditioned linear system, which involves the dense matrix, $\mathbf{M}$. The extensive use of the sparse preconditioner enhances the data locality, and accordingly the parallel efficiency is increased. The computational cost of the MPCG-based VCMD code is further amortized by taking advantage of the algorithmic similarity between the MTS and the MPCG algorithms, i.e., by reusing a doubly nested loop with associated neighbor-list construction for both the MTS method for time-stepping and multilevel preconditioning for determining charges.

The MPCG algorithm has allowed us to perform the first successful MD simulation of oxidation of an Al nanoparticle (diameter 200 Å, see Fig. 2.7(b)) (CAMPBELL, KALIA, NAKANO, VASHISHTA, OGATA and RODGERS [1999]). The MD simulations are based on a reliable interaction scheme developed by STREITZ and MINTMIRE [1994], which can successfully describe a wide range of physical properties of both metallic and ceramic systems. This scheme is capable of treating bond formation and bond breakage and changes in charge transfer as the atoms move and their local environments are altered. The MD simulations provide detailed picture of the rapid evolution and culmination of the surface oxide thickness, local stresses, and atomic diffusivities. After 0.5 ns, a stable oxide scale is formed. Structural analysis reveals a 40 Å thick amorphous oxide scale on the Al nanoparticle. The thickness and structure of the oxide scale are in accordance with experimental results.

2.3. Linear-scaling quantum-mechanical calculation based on the density functional theory

Empirical interatomic potentials used in MD simulations fail to describe chemical processes. Instead, interatomic interaction in reactive regions needs to be calculated by a QM method that can describe breaking and formation of bonds (PAYNE, TETER, ALLAN, ARIAS and JOANNOPOULOS [1992]). An atom consists of a nucleus and surrounding electrons, and quantum mechanics explicitly treats the electronic degrees-of-freedom. Since each electron's wave function is a linear combination of many states, the combinatorial solution space for the many-electron problem is exponentially large. The density functional theory (DFT) avoids the exhaustive enumeration of many-electron correlations by solving M single-electron problems in a common average environment (M is the number of independent wave functions and is on the order of N). As a result, the problem is reduced to a self-consistent matrix eigenvalue problem, which can be solved with $O(M^3)$ operations (HOHENBERG and KOHN [1964], KOHN and VASHISHTA [1983]). The DFT problem can also be formulated as the minimization of the energy, $E_{\mathrm{QM}}(\mathbf{r}^N, \psi^M)$,

$$E(\mathbf{r}^N; \psi^M) = 2\sum_{n=1}^{M} \int d\mathbf{x}\, \psi_n^*(\mathbf{x}; \mathbf{r}^N)\left[-\frac{\hbar^2}{2m_e}\frac{\partial^2}{\partial \mathbf{x}^2} + V_{\mathrm{ion}}(\mathbf{x})\right]\psi_n(\mathbf{x}; \mathbf{r}^N)$$

$$+\frac{1}{2}\int d\mathbf{x}\,\rho(\mathbf{x};\mathbf{r}^N)V_{\mathrm{H}}(\mathbf{x})+E_{\mathrm{XC}}[\rho(\mathbf{x};\mathbf{r}^N)]+\sum_{i=1}^{N-1}\sum_{j=i+1}^{N}\frac{Z_iZ_je^2}{|\mathbf{r}_i-\mathbf{r}_j|}, \tag{2.10}$$

with respect to electron wave functions, $\psi^M(\mathbf{r})=\{\psi_1(\mathbf{r}),\psi_2(\mathbf{r}),\ldots,\psi_M(\mathbf{r})\}$, subject to orthonormalization constraints between the wave functions,

$$\int d\mathbf{r}\,\psi_m^*(\mathbf{r})\psi_n(\mathbf{r})=\begin{cases}1 & (m=n),\\ 0 & (m\neq n).\end{cases} \tag{2.11}$$

In Eq. (2.10), m_{e} is the mass of an electron and Z_i is the valence of the ith ion, and $V_{\mathrm{ion}}(\mathbf{x})$, $V_{\mathrm{H}}(\mathbf{x})$, and $E_{\mathrm{xc}}(\mathbf{x})$ are the valence electron–ion interaction potential, Hartree potential, and exchange-correlation energy functional, respectively. The electron number density, $\rho(\mathbf{x};\mathbf{r}^N)$, is calculated as

$$\rho(\mathbf{x};\mathbf{r}^N)=2\sum_{n=1}^{M}|\psi_{\mathrm{n}}(\mathbf{x};\mathbf{r}^N)|^2. \tag{2.12}$$

We have achieved an efficient parallel implementation of the DFT using real-space approaches based on higher-order finite differencing (CHELIKOWSKY, SAAD, ÖGÜT, VASILIEV and STATHOPOULOS [2000]) and multigrid acceleration (BECK [2000], FATTEBERT and BERNHOLC [2000]). We include electron–ion interaction, V_{ion}, using norm-conserving pseudopotentials (TROULLIER and MARTINS [1991]) and the exchange-correlation energy associated with electron–electron interaction, E_{xc}, in a generalized gradient approximation (PERDEW, BURKE and ERNZERHOF [1996]). For larger systems ($M>1{,}000$), however, the $\mathrm{O}(M^3)$ orthonormalization becomes the bottleneck.

For scalable DFT calculations, linear-scaling algorithms are essential (GOEDECKER [1999]). We have implemented (SHIMOJO, CAMPBELL, KALIA, NAKANO, VASHISHTA, OGATA and TSURUTA [2000], NAKANO, KALIA, VASHISHTA, CAMPBELL, OGATA, SHIMOJO and SAINI [2001], SHIMOJO, KALIA, NAKANO and VASHISHTA [2001]) an $\mathrm{O}(M)$ algorithm (MAURI and GALLI [1994], KIM, MAURI and GALLI [1995]) based on unconstrained minimization of a modified energy functional and a localized-basis approximation. This algorithm is based on the observation that, for most materials at most temperatures, the off-diagonal elements of the density matrix,

$$\rho(\mathbf{x},\mathbf{x}')=\sum_{n=1}^{M}\psi_n^*(\mathbf{x})\psi_n(\mathbf{x}'), \tag{2.13}$$

decays exponentially (GOEDECKER [1999]), i.e., $\rho(\mathbf{x},\mathbf{x}')\propto\exp(-C|\mathbf{x}-\mathbf{x}'|)$ for $|\mathbf{x}-\mathbf{x}'|\to\infty$ (C is a constant). Such a diagonally-dominant matrix can be represented accurately with a compact basis by maximally localizing each wave function by a unitary transformation,

$$\phi_m(\mathbf{x})=\sum_n U_{nm}\psi_n(\mathbf{x}), \tag{2.14}$$

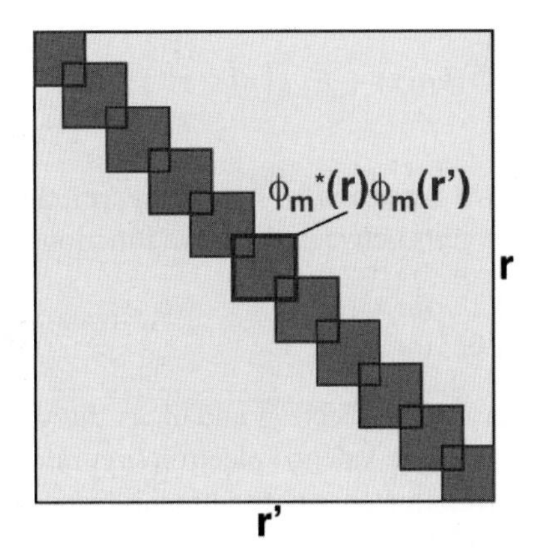

FIG. 2.8. Schematic of the localized approximation of a density matrix, $\rho(x, x')$ in one dimensional space.

and then truncating it with a finite cut-off radius, r_c, see Fig. 2.8.

Furthermore, the linear-scaling DFT algorithm uses a Lagrange-multiplier-like technique to perform unconstrained minimization of a modified energy functional to avoid the $O(M^3)$ orthonormalization procedure,

$$\widetilde{E}(\psi^M) = 2\sum_{m=1}^{M}\sum_{n=1}^{M}(2\delta_{mn} - S_{mn})(H_{nm} - \eta S_{nm}) + \eta N_{el}, \tag{2.15}$$

where

$$H_{mn} = \int d\mathbf{x}\, \psi_m^*(\mathbf{x}; \mathbf{r}^N)\widehat{H}\psi_n(\mathbf{x}; \mathbf{r}^N), \tag{2.16}$$

$$S_{mn} = \int d\mathbf{x}\, \psi_m^*(\mathbf{x}; \mathbf{r}^N)\psi_n(\mathbf{x}; \mathbf{r}^N), \tag{2.17}$$

$$\widehat{H} = -\frac{\hbar^2}{2m_e}\frac{\partial^2}{\partial \mathbf{x}^2} + V_{\text{ion}}(\mathbf{x}) + V_{\text{H}}(\mathbf{x}) + \delta E_{\text{XC}}/\delta\rho(\mathbf{x}). \tag{2.18}$$

Here, η and N_{el} are the chemical potential and the number of electrons, respectively. The modified functional, Eq. (2.15), produces the identical solution to that in constrained minimization of Eq. (2.10). It can also be proven that the minimized $\widetilde{E}$ coincides with the band structure energy. In this algorithm, each wave function is localized in a local region and therefore interacts with only those wave functions in the neighboring local regions. The energy functional is minimized iteratively using the CG method.

Spatial decomposition is used to parallelize both the real-space (RS) and the linear-scaling (LS) DFT algorithms. Grid points representing $\psi_n(\mathbf{x}; \mathbf{r}^N)$ and $\rho(\mathbf{x}; \mathbf{r}^N)$ are divided into spatial subregions, and are distributed among $P = P_x \times P_y \times P_z$ processors. Finite differencing of $\psi_n(\mathbf{x}; \mathbf{r}^N)$, as well as the preconditioning operation, necessitates interprocessor communication of the grid data because of the nonlocal operations. (We

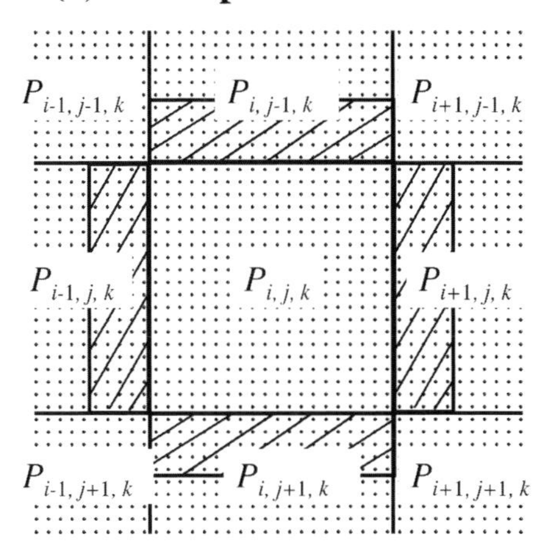

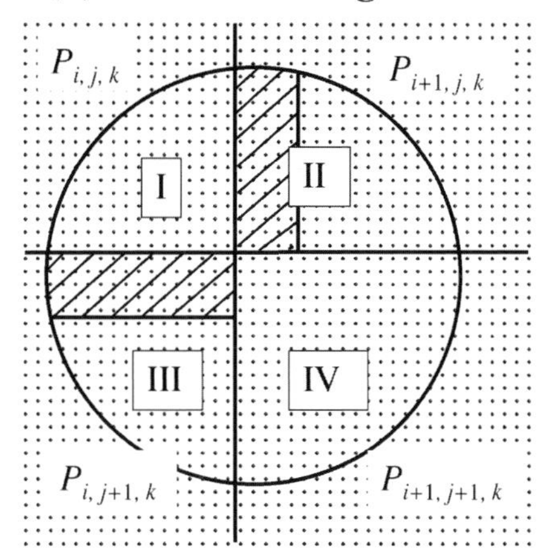

FIG. 2.9. (a) Spatial decomposition scheme for the parallel implementation of the real-space DFT algorithm. The dots show the real-space grids. The $P_{i,j,k}$ denotes the (i, j, k)th processor for $i = 1, \ldots, P_x$; $j = 1, \ldots, P_y$; $k = 1, \ldots, P_z$, with $P = P_x P_y P_z$ being the total number of processors. Each processor is assigned the corresponding sub-region. To perform the finite differencing operation in the $P_{i,j,k}$ processor, the grid data in the hatched regions have to be sent from the neighboring processors. (b) Spatial decomposition scheme for the parallel linear-scaling DFT method. The large circle shows the local region for a wave function, which is divided into four sub-regions I, II, III, and IV, and distributed to the corresponding processors. For the kinetic-energy operation in the sub-region I, the interprocessor communication of the grid data in the hatched regions is needed.

use the sixth-order expansion formula for the kinetic-energy operator.) Message passing of grid data is required only between neighboring processors, in the same way as in the classical MD algorithm.

In the spatial decomposition scheme to parallelize the RSDFT algorithm, processors are arranged as a three-dimensional array, $\{P_{i,j,k} \mid 0 \leqslant i < P_x,\ 0 \leqslant j < P_y,\ 0 \leqslant k < P_z\}$, and each processor, $P_{i,j,k}$, is assigned the corresponding subregion. To perform the finite differencing operation in the processor, $P_{i,j,k}$, the grid data in the hatched regions in Fig. 2.9(a) have to be cashed from the neighboring processors, which involves considerable communication that scales as $\mathrm{O}(N(N/P)^{2/3})$. This communication overhead can be derived as follows: The number of surface grid points that must be cached to other processors is proportional to $(N/P)^{2/3}$, and the number of communication is αM, where α is the average number of the CG iteration for each electron wave function. In addition to the surface-data caching, global communication, which scales as $\mathrm{O}(N^2 \log P)$, is necessary to evaluate the scalar products between different wave functions.

Although the basic strategy for parallelizing the LSDFT algorithm is similar to that for the RS algorithm, communication scales quite differently. Fig. 2.9(b) shows the spatial decomposition scheme for the parallel LS algorithm. The large circle denotes the local region for a wave function, which is divided into four regions I, II, III, and IV, and distributed to the corresponding processors. For the kinetic-energy operation in the region I, interprocessor communication of the grid data in the hatched regions is needed.

TABLE 2.1
Runtime (computation and communication times) requirement of the RS, and LS algorithms for an N atom problem on P processors

	Computation	Interprocessor communication	Global communication
RS	N^3/P	$N(N/P)^{2/3}$	$N^2 \log P$
LS	N/P	$(N/P)^{2/3}$	–

In the parallel LS algorithm, data structures to store discretized $\psi_n(\mathbf{x}; \mathbf{r}^N)$ are the following arrays. We use a one-dimensional array, which has sufficient size to store the total number of grid points for $\psi_n(\mathbf{x}; \mathbf{r}^N)$ assigned to each processor. Pointers are used for each wave function to specify the starting and ending array indices for grid points stored in the array. The number of surface grid points per sub-region is proportional to $(N/P)^{2/3}$ in the same way as in the RS algorithm, while the number of wave functions associated with each grid point is independent of the system size. Therefore, the communication data size between neighboring nodes scales as $O((N/P)^{2/3})$, which is much smaller than the $O(N(N/P)^{2/3})$ communication in the RS algorithm. Global communication for calculating the scalar products between different $\psi_n(\mathbf{x}; \mathbf{r}^N)$'s, which scales as $N^2 \log P$ in the RS algorithm, is unnecessary in the LS algorithm. Table 2.1 summarizes the computational and communication complexities of the RSDFT and the LSDFT algorithms.

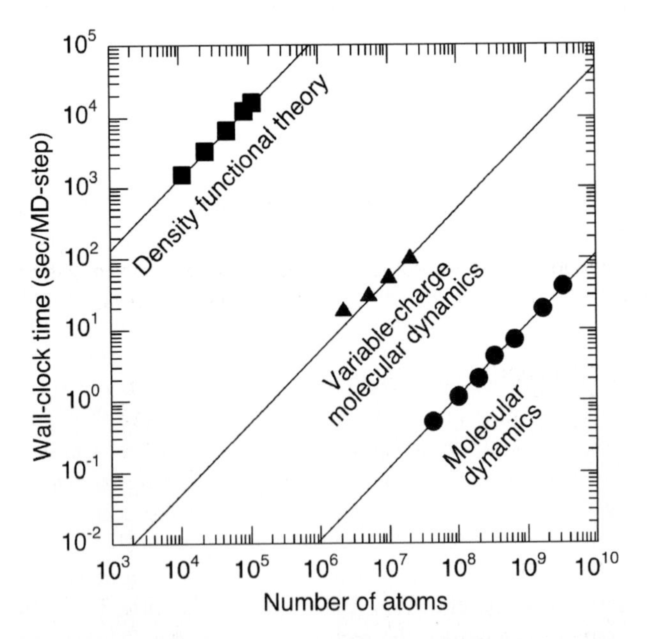

FIG. 2.10. Design-space diagram for MD and QM simulations on 1,024 IBM SP3 processors. The figure shows wall-clock time per MD step as a function of the number of atoms for three linear-scaling algorithms: Classical MD (MRMD, circles); environment-dependent variable-charge MD (VCMD, triangles); and, quantum-mechanical MD based on the DFT (LSDFT, squares). Lines show $O(N)$ scaling.

Benchmark tests of the three parallel algorithms – MRMD, VCMD, and LSDFT – have been performed on the IBM SP3 computers at the US Naval Oceanographic Office (NAVO) Major Shared Resource Center (NAKANO, KALIA, VASHISHTA, CAMPBELL, OGATA, SHIMOJO and SAINI [2001]). All the three programs are written using the Message Passing Interface (MPI) for message passing. The IBM SP3 at NAVO is configured with 375 MHz Power3 CPUs and has 334 nodes with 4 CPUs and 4 GB of memory per node.

Major design parameters for MD simulations of materials include the number of atoms in the simulated system and the methodologies to compute interatomic forces (classically in MRMD, semiempirically in VCMD, or quantum-mechanically (CAR and PARRINELLO [1985]) in LSDFT). Fig. 2.10 shows a design-space diagram for classical and quantum-mechanical MD simulations on 1,024 SP3 processors. (For LSDFT, one MD step involves 3 self-consistent DFT iterations each consisting of 20 CG steps.) The figure demonstrates linear scaling for all the three algorithms, with prefactors spanning seven-orders-of-magnitude. The largest benchmark tests in this study include 6.44-billion-atom MRMD and 111,000-atom LSDFT calculations on 1,024 SP3 processors.

3. Software tools

Practical simulations involving multibillion atoms are associated with a number of computational challenges, and this section describes software tools that address these challenges.

3.1. Wavelet-based adaptive computational-space decomposition for load balancing

Many MD simulations are characterized by irregular atomic distribution and associated load imbalance. For irregular data structures, the number of atoms assigned to each processor varies significantly, and this load imbalance degrades the parallel efficiency, see Fig. 3.1. The load-balancing problem can be stated as an optimization problem, i.e., one minimizes the load-imbalance cost as well as the size and the number of messages

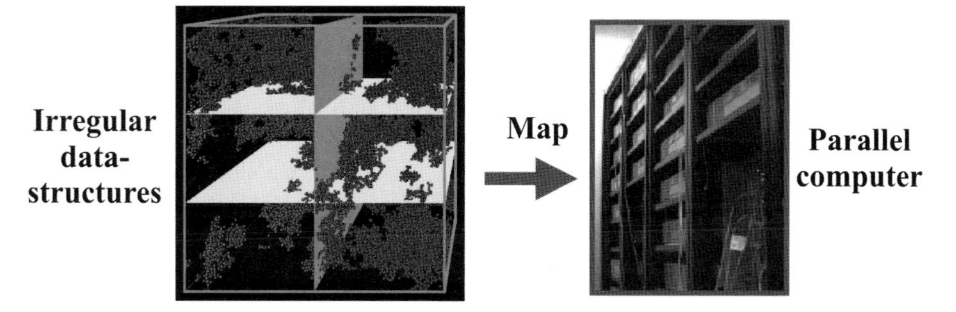

FIG. 3.1. Mapping an irregular data structure onto a parallel computer based on the regular spatial decomposition causes load imbalance and associated degradation of parallel efficiency.

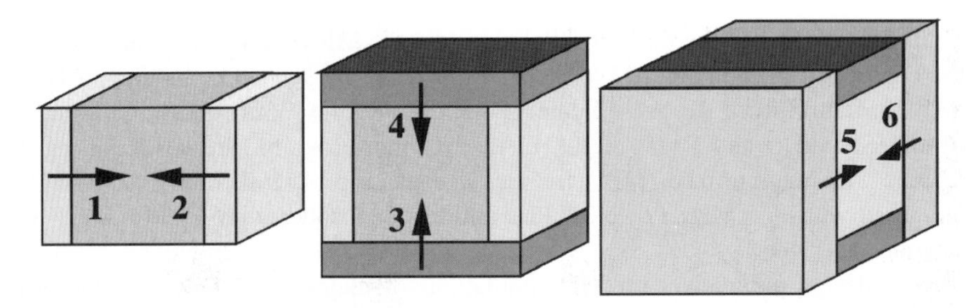

FIG. 3.2. Structured message passing in the topology-preserving computational-space decomposition.

(NAKANO and CAMPBELL [1997], NAKANO [1999a]):

$$E = t_{\text{comp}}\Big(\max_p \big|\{i\,|\,\mathbf{r}_i \in p\}\big|\Big) + t_{\text{comm}}\Big(\max_p \big|\{i\,|\,\|\mathbf{r}_i - \partial p\| < r_c\}\big|\Big) + t_{\text{latency}}\Big(\max_p \big[N_{\text{message}}(p)\big]\Big), \tag{3.1}$$

where the three terms are the load-imbalance cost, the size of messages, and the number of messages, respectively. In Eq. (3.1), ∂p and $N_{\text{message}}(p)$ denote the boundary surface of the physical volume assigned to processor p, and the number of messages per MD step for p, respectively. The expression, $\max_p f(p)$, denotes the maximum value of function $f(p)$ over all the processors, and r_c is the range of the interatomic potential. (Here, we consider only the near-field interactions, since the computation of the far-field interaction is uniform and is small in the combined FMM-MTS algorithm.) The prefactors, t_{comp}, t_{comm} and t_{latency}, are constants related to the processor speed, communication bandwidth and latency, respectively, and they are determined experimentally by test runs on the parallel computer under consideration.

To minimize the number of messages, we employ a topology-preserving spatial decomposition scheme, which uses the 3D mesh topology so that message passing is performed in a structured way in only 6 steps, see Fig. 3.2 (NAKANO, KALIA and VASHISHTA [1994]). To minimize the load imbalance cost as well as the message size, we have developed a computational-space decomposition scheme (NAKANO and CAMPBELL [1997]). The main idea of this scheme is that the computational space shrinks where the workload density is high and expands where the density is low, so that the workload is uniformly distributed in the computational space, see Fig. 3.3. To implement the curved computational space, we introduce a curvilinear coordinate transformation,

$$\xi = \mathbf{x} + \mathbf{u}(\mathbf{x}), \tag{3.2}$$

where $\mathbf{x}$ is a position in the physical Euclidean space and $\mathbf{u}(\mathbf{x})$ is a deformation field. We then use regular 3D mesh topology in the computational space, $\boldsymbol{\xi}$, to map atom i to

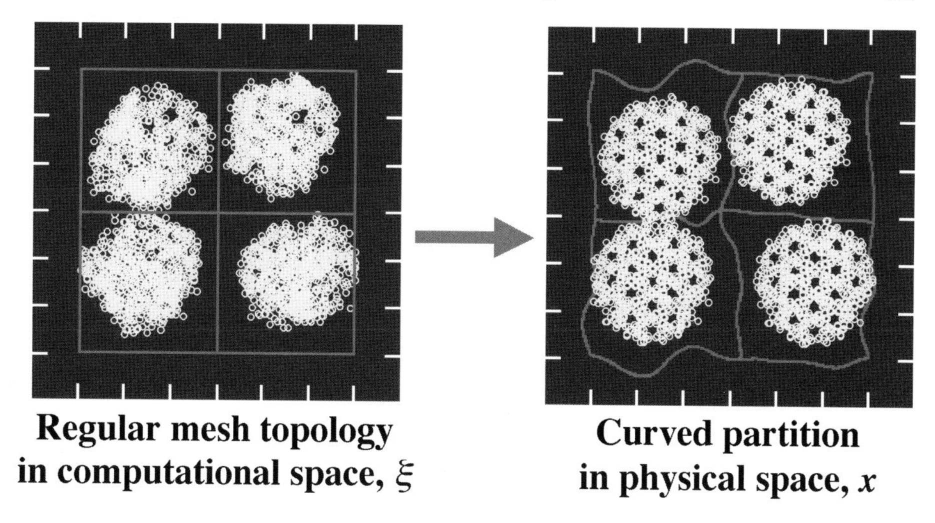

FIG. 3.3. Regular mesh decomposition in a curved computational space (left) results in curved partition boundaries in the physical Euclidean space (right).

processor p in an array of $P = P_x \times P_y \times P_z$ processors:

$$\begin{cases} p(\xi_i) = p_x(\xi_{ix}) P_y P_z + p_y(\xi_{iy}) P_z + p_z(\xi_{iz}), \\ p_\alpha(\xi_{i\alpha}) = \lfloor \xi_{i\alpha} P_\alpha / L_\alpha \rfloor \quad (\alpha = x, y, z), \end{cases} \tag{3.3}$$

where $\boldsymbol{\xi}_i = (\xi_{ix}, \xi_{iy}, \xi_{iz})$ is the coordinate of atom i and L_α is the simulation box size in the α direction in the computational space. This regular 3D mesh partition in the computational space results in curved partition boundaries in the physical space, $\mathbf{x}$. The load-imbalance and communication costs are minimized as a functional of the coordinate transformation, $\xi(\mathbf{x})$, using simulated annealing. We have found that wavelet representation leads to compact representation of curved partition boundaries, and accordingly to fast convergence of the minimization procedure (NAKANO [1999a], NAKANO [1999b]).

3.2. Spacefilling-curve-based adaptive data compression for scalable I/O

A 1.5-billion-atom MD simulation we have performed produces 150 GB of data per frame (or per minute), including atomic species, positions, velocities, and stresses. Since these data need to be transferred through network from a remote supercomputing center to a local computer for analysis and visualization, the massive data transfer has become one of the major bottlenecks of our large-scale MD simulations. For scalable input/output (I/O) and data transfer of such large datasets, we have designed a data compression algorithm (OMELTCHENKO, CAMPBELL, KALIA, LIU, NAKANO and VASHISHTA [2000]). It uses octree indexing (Fig. 3.4) and sorts atoms accordingly on the resulting spacefilling curve (Fig. 3.5). By storing differences between successive atomic coordinates, the I/O requirement for the same error tolerance level reduces from

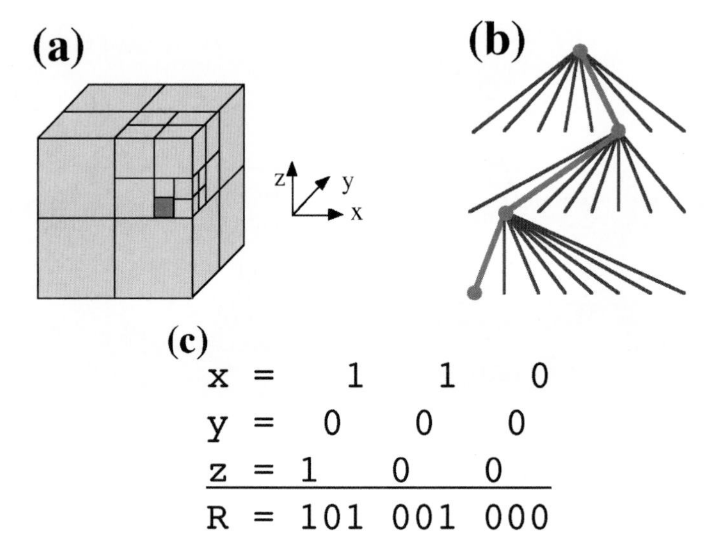

FIG. 3.4. (a) Recursively subdivision of the simulation cell into $2 \times 2 \times 2$ subsystems. (b) Each octree cell is accessed through the resulting octree data structure. (c) The octree index, R, is obtained by interleaving the binary Cartesian coordinates, x, y, and z, of the cell position.

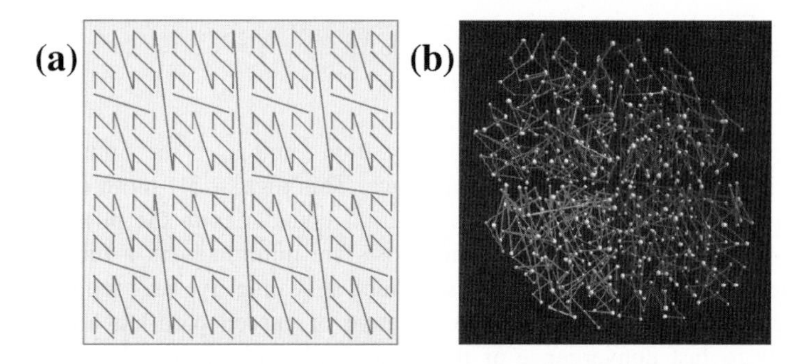

FIG. 3.5. A space-filling curve based on octree indexing maps the three-dimensional space into a sequential list, while preserving spatial proximity of consecutive list elements. (a) A two-dimensional example of the space-filling curve. (b) A three-dimensional example of atoms sorted along the space-filling curve.

$O(N \log N)$ to $O(N)$. An adaptive, variable-length encoding scheme is used to make the scheme tolerant to outliers and optimized dynamically.

Our compression algorithm for MD configurations (atomic positions, velocities, and other attributes such as atomic stress components for N atoms) consists of the following major steps:

- Quantize all double-precision data by dividing them by a user-specified error bound;
- Compute the octree index, R_i, of atomic positions for all atoms, $i = 1, \ldots, N$, by interleaving binary representations of the Cartesian coordinates, x_i, y_i, and z_i, see Fig. 3.4(c);

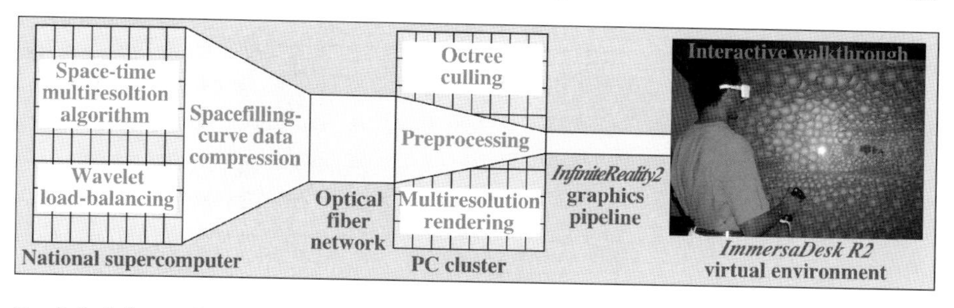

FIG. 3.6. Software pipeline for large-scale atomistic simulations. Space-time multiresolution algorithms and wavelet load balancing enable multibillion-atom simulations on supercomputers. Fractal data compression facilitates fast data transfer to LSU through OC-3 lines. Parallel visibility culling on a PC-cluster further reduces the data size to overcome the sequential bottleneck of the graphics pipeline to allow real-time walkthrough in our virtual environment.

- Sort the atoms in ascending order of R_i, see Fig. 3.5(b);
- Store differentiated $\Delta R_i = R_i - R_{i-1}$, velocities, and other data using an adaptive, variable-length encoding.

An order-of-magnitude improvement in the I/O performance was achieved for actual MD data with user-controlled error bound.

3.3. *Octree-based fast visibility culling for immersive and interactive visualization*

As discussed in the previous sections, billion-atom MD simulations are currently feasible using the space-time multiresolution algorithms and the wavelet-based load-balancing scheme at national supercomputer centers. The data are then transferred to our university using the spacefilling-curve-based data compression. The last challenge is to interactively visualize billion atoms. Interactive exploration of such large datasets is important for identifying and tracking atomic features that are responsible for macroscopic phenomena. Immersive and interactive virtual environment is an ideal platform for such explorative visualization, see Fig. 3.6.

We have developed a scalable visualization system to allow the viewer to walk through billion atoms (SHARMA, MILLER, LIU, NAKANO, KALIA, VASHISHTA, ZHAO, CAMPBELL and HAAS [2002]). To achieve interactive walkthrough speed and low access times in billion-atom systems, we employ an octree data structure and extract the regions of interest at runtime, thereby minimizing the size of the data sent to the rendering system, see Fig. 3.7. Standard visibility culling algorithms are employed to extract only the atoms that fit in the viewer's field of view. The rendering algorithm utilizes a configurable multiresolution scheme to improve the performance and visual appeal of the walkthrough experience. Each atomic entity is individually drawn at resolutions ranging from a point or spheres, each of which can be drawn using anything from 8 to 1,000 polygons. An OpenGL display list (a fast and convenient way to organize a set of OpenGL commands) for each type of atom is generated at all possible resolutions, since it is more efficient than individual function calls. At runtime, a particular resolution is called for each atom based on the distances between viewer and the object.

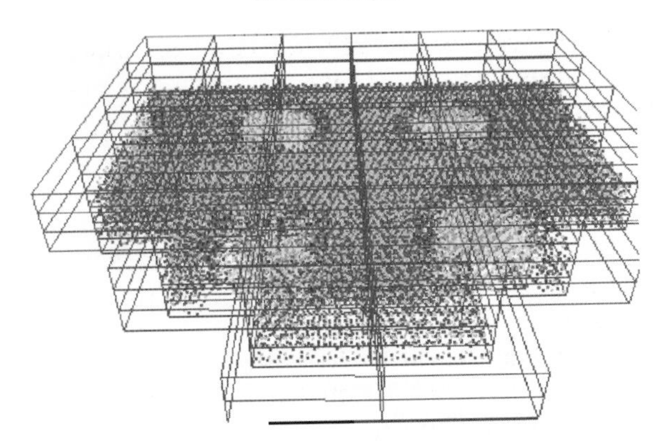

FIG. 3.7. Illustration of an octree data structure for visibility culling. Octree cells (bounded by white lines) dynamically approximate the current visible region (the position and viewing direction of the viewer is represented by the arrow). Only the visible atoms are processed for rendering, as demonstrated in the figure.

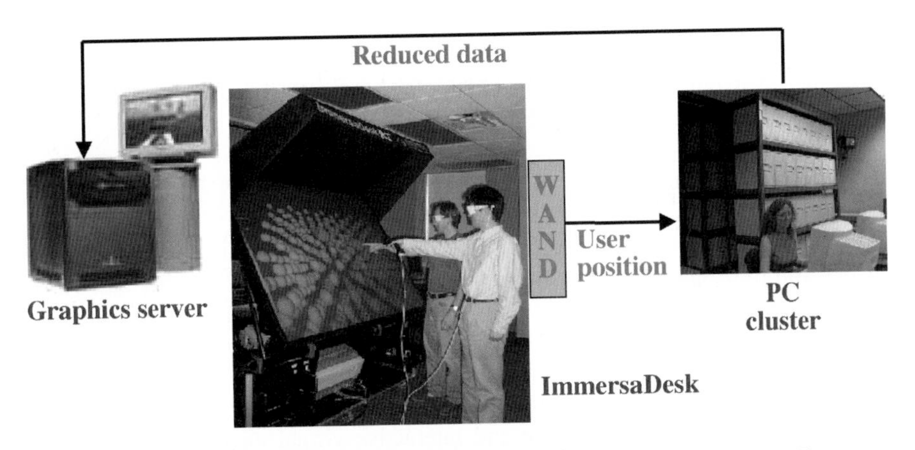

FIG. 3.8. Interactive visualization of billion atoms in an immersive visualization environment. The user uses a 3D input device (wand) to interact with the scene on the ImmersaDesk virtual environment, and the magnetic tracking system monitors the user's head movement. The user's position is sent to the PC cluster, which uses the parallel occlusion culling algorithm to select the octree cells in the user's view frustum. The reduced data are sent to the graphics server for rendering, and the resulting scene is projected on the ImmersaDesk.

To further speed up the rendering, the visibility-culling tasks described above are offloaded to a PC cluster so that the graphics server is dedicated to rendering, see Fig. 3.8. The present system is thus implemented in an *ImmersaDesk* virtual environment connected to a PC cluster, see Fig. 3.8. The *ImmersaDesk* consists of a pivotal screen, a stereoscopic projector, head-tracking system, eyewear kit, infrared emitters, wand with a tracking sensor, and wand tracking I/O subsystem. A programmable wand with three buttons and a joystick allows interactions between the viewer and simulated objects. The rendering system is an SGI Onyx with two R12000 processors, 4 GB system RAM, and an InfinityReality2 graphics pipeline. The dual-processor Onyx is connected to four PCs running Linux 6.2 each with an

800 MHz Pentium III processor and 512 MB RAM for parallel visibility culling. The resulting parallel/distributed visualization system renders a billion-atom system at nearly interactive frame rates.

4. Applications of large atomistic simulations to nanosystems

The linear-scaling simulation algorithms and the scalable parallel computing framework described in Sections 2 and 3 have been applied to the study of various properties and processes in nanoscale materials and devices. These include: Dynamic fracture in crystalline and amorphous materials (NAKANO, KALIA and VASHISHTA [1995], OMELTCHENKO, YU, KALIA and VASHISHTA [1997]); nanocomposite materials synthesized by consolidating nanometer-size ceramic particles (CHATTERJEE, KALIA, LOONG, NAKANO, OMELTCHENKO, TSURUTA, VASHISHTA, WINTERER and KLEIN [2000], KALIA, NAKANO, TSURUTA and VASHISHTA [1997], KALIA, NAKANO, OMELTCHENKO, TSURUTA and VASHISHTA [1997], CAMPBELL, KALIA, NAKANO, SHIMOJO, TSURUTA and VASHISHTA [1999]); nanoindentation to characterize the mechanical properties of surfaces (WALSH, KALIA, NAKANO, VASHISHTA and SAINI [2000]); atomic stress distribution in semiconductor/dielectric nanomesas for microelectronics (BACHLECHNER, OMELTCHENKO, NAKANO, KALIA, VASHISHTA, EBBSJÖ and MADHUKAR [2000], LIDORIKIS, BACHLECHNER, KALIA, NAKANO, VASHISHTA and VOYIADJIS [2001], OMELTCHENKO, BACHLECHNER, NAKANO, KALIA, VASHISHTA, EBBSJÖ, MADHUKAR and MESSINA [2000]); and amorphization and fracture of nanowires (LI, KALIA and VASHISHTA [1996], WALSH, KALIA, NAKANO, VASHISHTA and SAINI [2001]). This section presents some of the results from large-scale MD simulations of both epitaxical and colloidal quantum dots as building blocks for novel nanoscale devices.

4.1. *Epitaxical quantum dots*

In recent years, coherently strained 3D islands formed in semiconductor overlayers having high lattice-mismatch with underlying substrates have attracted much attention due to their importance in the study of electronic behavior in zero dimension and applications in electronic and optoelectronic devices (MADHUKAR [1996]). The role and manipulation of stress in the formation of such nanostructures have been systematically examined through a study of the growth of InAs on planar and patterned GaAs(001) substrate (these systems have a large lattice mismatch of 6.6%). On infinite planar substrates, the strain relief leads to the formation of coherent three-dimensional island structures above a critical amount, ~ 1.6 monolayers (ML), of InAs deposition. On the contrary, when InAs is deposited on $\langle 100 \rangle$ oriented GaAs square mesas of size $\leqslant 75$ nm, the island morphology is suppressed and, instead, a continuous film with flat morphology is observed. This InAs film growth is, however, self-limiting and stops at ~ 11 ML.

In order to understand the self-limiting nature of the InAs film growth, we have recently performed MD simulations of InAs/GaAs nanomesas with $\{101\}$-type sidewalls, see Fig. 4.1(a) (SU, KALIA, MADHUKAR, NAKANO and VASHISHTA

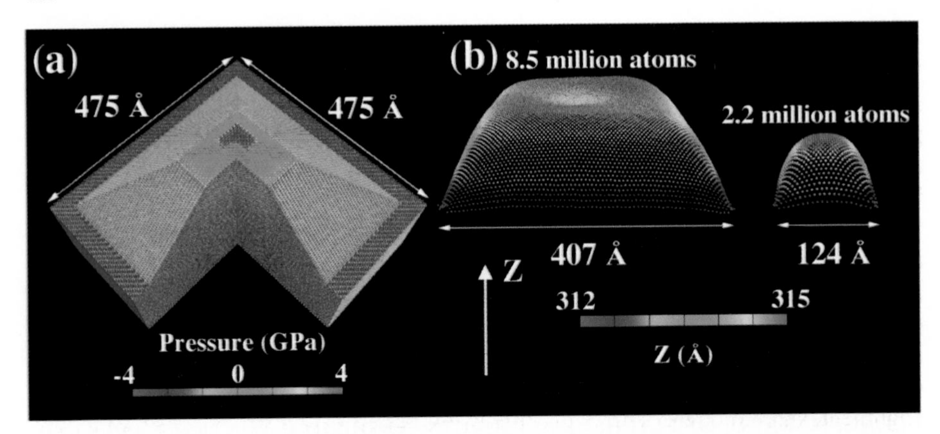

FIG. 4.1. (a) Atomic-level hydrostatic stress in an InAs/GaAs square nanomesa with a 12 ML InAs overlayer. (b) Vertical displacement of As atoms in the first As layer above the InAs/GaAs interface in the 8.5 million-atom and the 2.2 million-atom nanomesas.

[2001]). The in-plane lattice constant of InAs layers parallel to the InAs/GaAs(001) interface starts to exceed the InAs bulk value at the 12th ML and the hydrostatic stresses in InAs layers become tensile above $\sim$ 12th ML. As a result, it is not favorable to have InAs overlayers thicker than 12 ML. This may explain the experimental findings of the growth of flat InAs overlayers with self-limiting thickness of $\sim$ 11 ML on GaAs nanomesas.

Length scales are of critical significance for stress relaxation and manipulation leading to control of the island number on chosen nanoscale area arrays. For example, on stripe mesas of sub-100-nm widths on GaAs(001) substrates, deposition of InAs is shown to allow self-assembly of three, two, and single chains of InAs 3D island quantum dots selectively on the stripe mesa tops for widths decreasing from 100 nm down to 30 nm (KONKAR, MADHUKAR and CHEN [1998]).

We have recently investigated lateral size effects on the stress distribution and morphology of InAs/GaAs nanomesas using parallel MD simulations, see Fig. 4.1(b). Two mesas with the same vertical size but different lateral sizes are simulated. For the smaller mesa, a single stress domain is observed in the InAs overlayer, whereas two stress domains are found in the larger mesa (a highly compressive domain is located at the center of the InAs overlayer, whereas the peripheral region of the InAs overlayer is less compressive). This indicates the existence of a critical lateral size for domain formation in accordance with recent experimental findings. We have also studied the morphology of the InAs overlayer near the InAs/GaAs interface. For the 2.2 million-atom nanomesa, the As layer is "dome" shaped. In contrast, the As layer in the 8.5 million-atom nanomesa shows a "dimple" at the center of the mesa. This provides clear evidence that there exists a critical lateral size for such stress domain formation and the critical value is somewhere between 124 Å and 407 Å. Detailed analysis of structural correlations have revealed that the InAs overlayer in the larger mesa is laterally constrained to the GaAs bulk lattice constant but vertically relaxed to the InAs bulk lattice constant, which is consistent with the Poisson effect.

4.2. *Nanocrystalline quantum dots*

Aggregates of nanometer-size semiconductor crystals have promising applications as photovoltaics, light-emitting diodes, and single-nanocrystal, single-electron transistors (ALIVISATOS [1996]). Self-organized assembly of colloidal nanocrystals acts as an intelligent photonic-crystal material, which can be used as sensors and optical switches. Recently, self-formation of laser was demonstrated in semiconductor nanopowders due to disorder-induced photon localization mechanisms. Rod-shaped nanocrystals emit polarized light and will be useful for biological tagging applications. The most recent additions to this family of nanocrystals include tetrapods. Such systematically-controlled anisotropic shapes can be used as building blocks for self-assembly of three-dimensionally integrated nanostructures through surface-stress encoded epitaxy, self-alignment, and biological templates. Finally, these nanocrystals can be utilized as new synthetic paths to novel materials that do not exist in bulk form.

Size-dependent phase stability plays an essential role in the synthesis of nanocrystals. For example, many III–V and II–VI semiconductors transform from a four-coordinated phase to a six-coordinated phase as the pressure is increased, and the transition pressure often exhibits strong size dependence. Upon release of pressure, the metastable high-pressure phase can be kinetically trapped in nanoclusters. This is due to the large number of surface atoms such that the surface energetics essentially affects the phase stability. We may thus be able to prepare interior bonding geometries that do not occur in the known extended solid by adjusting the surface energy. In other words, it is possible to manipulate nanocrystal surfaces to trap structures that might ordinarily be unstable

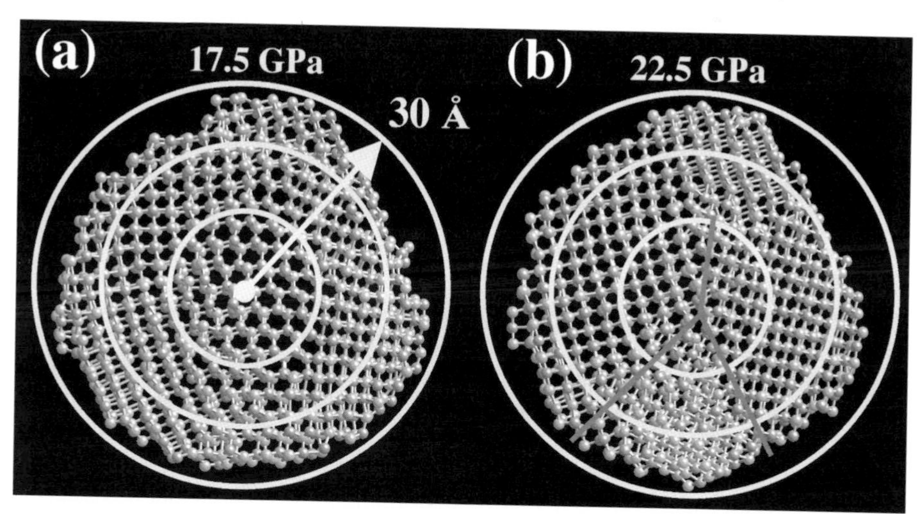

FIG. 4.2. Structural transformation in a GaAs nanocrystal from outer to inner shells. (a) An 8 Å slice of an initially spherical nanocrystal of diameter 60 Å that is partially transformed at a pressure of 17.5 GPa. Outermost shell shows the rocksalt structure (atoms making four-membered rings) while the innermost shell continues to show the zinc-blende (atoms making six-membered rings). (b) The same slice with the nanocrystal completely transformed at 22.5 GPa. The rocksalt structure can now be seen in the innermost shell. The red lines are a guide to the eye to see the differently oriented grains.

in the bulk. Nanophase engineering uses controlled pressurization and annealing to achieve new material forms, which are nonexistent in the bulk.

Molecular dynamics simulations are expected to reveal microscopic mechanisms of the nanocrystalline phase kinetics. We have performed preliminary MD simulations to investigate pressure-induced structural transformations in GaAs nanocrystals of different sizes, see Fig. 4.2 (KODIYALAM, KALIA, KIKUCHI, NAKANO, SHIMOJO and VASHISHTA [2001]). To simulate the experimental situation, the nanocrystals are immersed in a Lennard–Jones liquid so that they can be subjected to hydrostatic pressure. It is found that the transformation from four-fold (zinc-blende) to six-fold (rocksalt) coordination starts at the surfaces of nanocrystals and proceeds inwards with increasing pressure, see Fig. 4.2(a). Inequivalent nucleation of the high-pressure phase at different sites leads to an inhomogeneous deformation of the nanocrystal. For sufficiently large spherical nanocrystals, this gives rise to rocksalt structures of different orientations separated by grain boundaries, see Fig. 4.2(b).

5. Conclusions

Modern MD simulations of materials was started by RAHMAN [1964], who simulated 864 argon atoms on a CDC 3600 computer. Assuming a simple exponential growth, the number of atoms that can be simulated in classical MD simulations has doubled every 19 months to reach 6.44 billion atoms in 2000 (NAKANO, KALIA, VASHISHTA, CAMPBELL, OGATA, SHIMOJO and SAINI [2001]). Similarly, the number of atoms in DFT-based *ab initio* MD simulations (started by Roberto Car and Michelle Parrinello in 1985 for 8 Si atoms) has doubled every 12 months to 111,000 atoms in 2000 (NAKANO, KALIA, VASHISHTA, CAMPBELL, OGATA, SHIMOJO and SAINI [2001]). Petaflop computers anticipated to be built in the next ten years will maintain the growth rates in these “MD Moore’s Laws”, and we will be able to perform 10^{12}-atom classical and 10^7-atom quantum MD simulations on such computers. Multiresolution approaches described in this chapter, combined with cache-conscious techniques, will be essential to achieve scalability on petaflop architectures. Atomistic simulations have thus reached a scale such that they must be performed in a metacomputing environment of a grid of geographically-distributed multiple supercomputers. Such grid-computing efforts are also underway.

Another promising direction toward achieving scalability is the development multiscale simulations schemes that seamlessly combine QM and MD calculations with the continuum mechanics calculation based on the linear elasticity and the finite element method in a single simulation program (NAKANO, BACHLECHNER, KALIA, LIDORIKIS, VASHISHTA, VOYIADJIS, CAMPBELL, OGATA and SHIMOJO [2001], OGATA, LIDORIKIS, SHIMOJO, NAKANO, VASHISHTA and KALIA [2001]). Such a hybrid scheme enhances the scalability by performing more accurate but less scalable calculations only when and where they are needed.

Acknowledgments

This work was partially supported by AFOSR, ARL, DOE, NASA, NSF, Louisiana Board of Regents and USC–Berkeley–Princeton DURINT. Benchmark tests were performed at Department of Defense's Naval Oceanographic Office (NAVO) Major Shared Resource Center under a DoD Challenge Project. Programs have been developed using parallel computers at the Concurrent Computing Laboratory for Materials Simulations (CCLMS) at Louisiana State University. Visualization was performed using the ImmersaDesk virtual environment at the CCLMS in collaboration with Mr. Andy Haas, Dr. Xinian Liu, Mr. Paul Miller, Mr. Ashish Sharma, and Mr. Wei Zhao. Simulations of epitaxical quantum dots were performed in collaboration with Dr. Anupam Madhukar at the University of Southern California.

References

ABRAHAM, F.F. (1997), Portrait of a crack: Rapid fracture mechanics using parallel molecular dynamics, *IEEE Computational Science & Engineering* **4** (2), 66–77.

ALIVISATOS, P.A. (1996), Semiconductor clusters, nanocrystals, and quantum dots, *Science* **271**, 933–937.

BACHLECHNER, M.E., A. OMELTCHENKO, A. NAKANO, R.K. KALIA, P. VASHISHTA, I. EBBSJÖ and A. MADHUKAR (2000), Dislocation emission at silicon/silicon nitride interface – a million atom molecular dynamics simulation on parallel computers, *Phys. Rev. Lett.* **84**, 322–325.

BECK, T.L. (2000), Real-space mesh techniques in density-functional theory, *Rev. Mod. Phys.* **72**, 1041–1080.

CAMPBELL, T.J., R.K. KALIA, A. NAKANO, F. SHIMOJO, K. TSURUTA and P. VASHISHTA (1999), Structural correlations and mechanical behavior in nanophase silica glasses, *Phys. Rev. Lett.* **82**, 4018–4021.

CAMPBELL, T.J., R.K. KALIA, A. NAKANO, P. VASHISHTA, S. OGATA and S. RODGERS (1999), Dynamics of oxidation of aluminum nanoclusters using variable charge molecular-dynamics simulations on parallel computers, *Phys. Rev. Lett.* **82**, 4866–4869.

CAR, R. and M. PARRINELLO (1985), Unified approach for molecular dynamics and density functional theory, *Phys. Rev. Lett.* **55**, 2471–2474.

CHATTERJEE, A., R.K. KALIA, C.-K. LOONG, A. NAKANO, A. OMELTCHENKO, K. TSURUTA, P. VASHISHTA, M. WINTERER and S. KLEIN (2000), Sintering, structure and mechanical properties of nanophase SiC: a molecular-dynamics and neutron scattering study, *Appl. Phys. Lett.* **77**, 1132–1134.

CHELIKOWSKY, J.R., Y. SAAD, S. ÖGÜT, I. VASILIEV and A. STATHOPOULOS (2000), Electronic structure methods for predicting the properties of materials: Grids in space, *Physica Status Solidi* (*b*) **217**, 173–195.

DONGARRA J.J. and D.W. WALKER (2001), The quest for Petascale computing, *Computational Science & Engineering* **3** (3), 32–39.

FATTEBERT J.-L. and J. BERNHOLC (2000), Towards grid-based $O(N)$ density-functional theory methods: Optimized nonorthogonal orbitals and multigrid acceleration, *Phys. Rev. B* **62**, 1713–1722.

FOSTER, I. and C. KESSELMAN (1999), *The Grid: Blueprint for a New Computing Infrastructure* (Morgan Kaufmann, San Francisco).

GERMANN, T.C. and P.S. LOMDAHL (1999), Recent advances in large-scale atomistic materials simulations, *Computing in Science and Engineering* **1** (2), 10–11.

GOEDECKER, S. (1999), Linear scaling electronic structure methods, *Rev. Mod. Phys.* **71**, 1085–1123.

GREENGARD, L. and V. ROKHLIN (1987), A fast algorithm for particle simulations, *J. Comput. Phys.* **73**, 325–348.

HOHENBERG, P. and W. KOHN (1964), Inhomogeneous electron gas, *Phys. Rev.* **136**, B864–B871.

KALIA, R.K., A. NAKANO, A. OMELTCHENKO, K. TSURUTA and P. VASHISHTA (1997), Role of ultrafine microstructures in dynamic fracture in nanophase silicon nitride, *Phys. Rev. Lett.* **78**, 2144–2147.

KALIA, R.K., A. NAKANO, K. TSURUTA and P. VASHISHTA (1997), Morphology of pores and interfaces and mechanical behavior of nanocluster-assembled silicon nitride ceramic, *Phys. Rev. Lett.* **78**, 689–692.

KIM, J., F. MAURI and G. GALLI (1995), Total-energy global optimizations using nonorthogonal localized orbitals, *Phys. Rev. B* **52**, 1640–1648.

KODIYALAM, S., R.K. KALIA, H. KIKUCHI, A. NAKANO, F. SHIMOJO and P. VASHISHTA (2001), Grain boundaries in gallium arsenide nanocrystals under pressure: A parallel molecular-dynamics study, *Phys. Rev. Lett.* **86**, 55–58.

KOHN, W. and P. VASHISHTA (1983), General density functional theory, in: N.H. March and S. Lundqvist, eds., *Inhomogeneous Electron Gas* (Plenum, New York) 79–184.

KONKAR, A., A. MADHUKAR and P. CHEN (1998), Stress-engineered spatially selective self-assembly of strained InAs quantum dots on nonplanar patterned GaAs (001) substrates, *Appl. Phys. Lett.* **72**, 220–222.

LI, W., R.K. KALIA and P. VASHISHTA (1996), Amorphization and fracture in silicon diselenide nanowires: a molecular dynamics study, *Phys. Rev. Lett.* **77**, 2241–2244.

LIDORIKIS, E., M.E. BACHLECHNER, R.K. KALIA, A. NAKANO, P. VASHISHTA and G.Z. VOYIADJIS (2001), Coupling length scales for multiscale atomistic-continuum simulations: atomistically-induced stress distributions in Si/Si_3N_4 nanopixels, *Phys. Rev. Lett.* **87**, 086104:1–4.

MADHUKAR, A. (1996), A unified atomistic and kinetic framework for growth front morphology evolution and defect initiation in strained epitaxy, *J. Crystal Growth* **163**, 149–164.

MAURI, F. and G. GALLI (1994), Electronic-structure calculations and molecular-dynamics simulations with linear system-size scaling, *Phys. Rev. B* **50**, 4316–4326.

NAKANO, A. (1997a), Parallel multilevel preconditioned conjugate-gradient approach to variable-charge molecular dynamics, *Comput. Phys. Comm.* **104**, 59–69.

NAKANO, A. (1997b), Fuzzy clustering approach to hierarchical molecular dynamics simulation of multiscale materials phenomena, *Comput. Phys. Comm.* **105**, 139–150.

NAKANO, A. (1999a), Multiresolution load balancing in curved space: The wavelet representation, *Concurrency: Practice and Experience* **11**, 343–353.

NAKANO, A. (1999b), A rigid-body based multiple time-scale molecular dynamics simulation of nanophase materials, *The International Journal of High Performance Computing Applications* **13**, 154–162.

NAKANO, A., M.E. BACHLECHNER, R.K. KALIA, E. LIDORIKIS, P. VASHISHTA, G.Z. VOYIADJIS, T.J. CAMPBELL, S. OGATA and F. SHIMOJO (2001), Multiscale simulation of nanosystems, *Comput. Sci. Engrg.* **3** (4), 56–66.

NAKANO, A. and T.J. CAMPBELL (1997), An adaptive curvilinear-coordinate approach to dynamic load balancing of parallel multi-resolution molecular dynamics, *Parallel Computing* **23**, 1461–1478.

NAKANO, A., R.K. KALIA and P. VASHISHTA (1994), Multiresolution molecular dynamics algorithm for realistic materials modeling on parallel computers, *Comput. Phys. Comm.* **83**, 197–214.

NAKANO, A., R.K. KALIA and P. VASHISHTA (1995), Dynamics and morphology of brittle cracks: A molecular-dynamics study of silicon nitride, *Phys. Rev. Lett.* **75**, 3138–3141.

NAKANO, A., R.K. KALIA, P. VASHISHTA, T.J. CAMPBELL, S. OGATA, F. SHIMOJO and S. SAINI (2001), Scalable atomistic simulation algorithms for materials research, in: *Proceedings of Supercomputing 2001* (ACM, New York, NY).

OGATA, S., E. LIDORIKIS, F. SHIMOJO, A. NAKANO, P. VASHISHTA and R.K. KALIA (2001), Hybrid finite-element/molecular-dynamics/electronic-density-functional approach to materials simulations on parallel computers, *Comput. Phys. Comm.* **138**, 143–154.

OMELTCHENKO, A., M.E. BACHLECHNER, A. NAKANO, R.K. KALIA, P. VASHISHTA, I. EBBSJÖ, A. MADHUKAR and P. MESSINA (2000), Stress domains in $Si(111)/Si_3N_4(0001)$ nanopixel – 10 million-atom molecular dynamics simulations on parallel computers, *Phys. Rev. Lett.* **84**, 318–321.

OMELTCHENKO, A., T.J. CAMPBELL, R.K. KALIA, X. LIU, A. NAKANO and P. VASHISHTA (2000), Scalable I/O of large-scale molecular-dynamics simulations: A data-compression algorithm, *Comput. Phys. Comm.* **131**, 78–85.

OMELTCHENKO, A., J. YU, R.K. KALIA and P. VASHISHTA (1997), Crack front propagation and fracture in a graphite sheet, *Phys. Rev. Lett.* **78**, 2148–2151.

PAYNE, M.C., M.P. TETER, D.C. ALLAN, T.A. ARIAS and J.D. JOANNOPOULOS (1992), Iterative minimization techniques for *ab initio* total energy calculations: molecular dynamics and conjugate gradients, *Rev. Mod. Phys.* **64**, 1045–1097.

PECHENIK, A., R.K. KALIA and P. VASHISHTA (1999), *Computer-Aided Design of High-Temperature Materials* (Oxford Univ. Press, Oxford).

PERDEW, J.P., K. BURKE and M. ERNZERHOF (1996), Generalized gradient approximation made simple, *Phys. Rev. Lett.* **77**, 3865–3868.

RAHMAN, A. (1964), Correlations in the motion of atoms in liquid argon, *Phys. Rev.* **136** (2A), A405–A411.

RAPAPORT, D.C. (1995), *The Art of Molecular Dynamics Simulation* (Cambridge Univ. Press, Cambridge).

SHARMA, A., P. MILLER, X. LIU, A. NAKANO, R.K. KALIA, P. VASHISHTA, W. ZHAO, T.J. CAMPBELL and A. HAAS (2002), Immersive and interactive exploration of billion-atom systems, in: *Proceedings of IEEE Virtual Reality 2002 Conference* (IEEE Computer Society, Los Alamitos, CA) 217–223.

SHIMOJO, F., T.J. CAMPBELL, R.K. KALIA, A. NAKANO, P. VASHISHTA, S. OGATA and K. TSURUTA (2000), A scalable molecular-dynamics-algorithm suite for materials simulations: Design-space diagram on 1,024 Cray T3E processors, *Future Generation Computer Systems* **17**, 279–291.

SHIMOJO, F., I. EBBSJÖ, R.K. KALIA, A. NAKANO, J.P. RINO and P. VASHISHTA (2000), Molecular dynamics simulation of pressure induced structural transformation in silicon carbide, *Phys. Rev. Lett.* **84**, 3338–3341.

SHIMOJO, F., R.K. KALIA, A. NAKANO and P. VASHISHTA (2001), Linear-scaling density-functional-theory calculations of electronic structure based on real-space grids: design, analysis, and scalability test of parallel algorithms, *Comput. Phys. Comm.* **140**, 303–314.

SKEEL, R.D., G. ZHANG and T. SCHLICK (1997), A family of symplectic integrators, *SIAM J. Sci. Comput.* **18**, 203–222.

STREITZ, F.H. and J.W. MINTMIRE (1994), Electrostatic potentials for metal-oxide surfaces and interfaces, *Phys. Rev. B* **50**, 11996–12003.

SU, X., R.K. KALIA, A. MADHUKAR, A. NAKANO and P. VASHISHTA (2001), Million-atom molecular dynamics simulation of flat InAs overlayers with self-limiting thickness on GaAs nanomesas, *Appl. Phys. Lett.* **78**, 3717–3719.

TROULLIER, N. and J.L. MARTINS (1991), Efficient pseudopotentials for plane-wave calculations, *Phys. Rev. B* **43**, 1993–2006.

TUCKERMAN, M.E., D.A. YARNE, S.O. SAMUELSON, A.L. HUGHES and G.J. MARTYNA (2000), Exploiting multiple levels of parallelism in molecular dynamics based calculations via modern techniques and software paradigms on distributed memory computers, *Comput. Phys. Comm.* **128**, 333–376.

VASHISHTA, P., R.K. KALIA and A. NAKANO (1999), Large-scale atomistic simulation of dynamic fracture, *Comput. Sci. Engrg.* **1** (5), 56–65.

VASHISHTA, P., R.K. KALIA, A. NAKANO, W. LI and I. EBBSJÖ (1996), Molecular dynamics methods and large-scale simulations of amorphous materials, in: M.F. Thorpe and M.I. Mitkova, eds., *Amorphous Insulators and Semiconductors* (Kluwer, Dordrecht) 151–213.

WALSH, P., R.K. KALIA, A. NAKANO, P. VASHISHTA and S. SAINI (2000), Amorphization and anisotropic fracture dynamics during nanoindentation of silicon nitride – a multi-million atom molecular dynamics study, *Appl. Phys. Lett.* **77**, 4332–4334.

WALSH, P., W. LI, R.K. KALIA, A. NAKANO, P. VASHISHTA and S. SAINI (2001), Structural transformation, amorphization, and fracture in nanowires: a multi-million atom molecular dynamics study, *Appl. Phys. Lett.* **78**, 3328–3330.

ZHANG, G. and T. SCHLICK (1993), LIN: a new algorithm to simulate the dynamics of biomolecules by combining implicit-integration and normal mode techniques, *J. Comput. Chem.* **14**, 1212–1233.

Simulating Chemical Reactions in Complex Systems

M.J. Field

Laboratoire de Dynamique Moléculaire,
Institut de Biologie Structurale – Jean-Pierre Ebel,
41 rue Jules Horowitz, 38027 Grenoble cedex 1, France

1. Introduction

Chemistry is the central science, straddling as it does the physical and the biological sciences, and it has as its goal the study of the elements and the myriad compounds that they can form. Perhaps the most remarkable and the most interesting property of molecules is their ability to react or, in other words, to undergo transformations that change them into other molecular species. The evidence of chemical reaction surrounds us in our everyday life both in the inanimate – for example, in a flame – and in the animate. Reactions occur in all phases of matter, from the gas phase to the solid state, but it is in condensed phases, and particularly in solution and at surfaces, that the most diverse reactions occur.

The aim of this chapter is to give a review of modern simulation methods for the investigation of chemical reactions in the condensed phase with a bias, due to the interests of the author, towards systems of biological importance. As in many areas of molecular science, simulation approaches are proving to be essential complements to purely experimental or theoretical work but the simulation of reactions poses unique challenges because it requires a combination of a wide range of different theoretical methodologies.

The outline of this chapter is as follows. Section 2 introduces the standard classical mechanical framework for understanding reaction rate processes, Section 3 discusses

Computational Chemistry
Special Volume (C. Le Bris, Guest Editor) of
HANDBOOK OF NUMERICAL ANALYSIS, VOL. X
P.G. Ciarlet (Editor)
© 2003 Elsevier Science B.V. All rights reserved

the determination of potential energy surfaces for reacting systems and Section 4 describes how to use these surfaces for reaction studies. The chapter continues, in Section 5, with a brief outline of methods for including quantum dynamical effects in simulations of reactions and concludes, in Section 6, with some perspectives for the future.

2. Classical theories of reaction rates

This section briefly describes the theory that is necessary to calculate rate constants within the classical reactive flux formalism. It starts with a discussion of the nature of the phenomenological rate constant before going on to outline how this macroscopic quantity can be determined from microscopic theories. The arguments in this section follow closely those presented by CHANDLER [1978], CHANDLER [1987] and FIELD [1993]. Complementary reviews are given by BERNE [1985], HYNES [1985], BERNE, BORKOVEC and STRAUB [1988], ANDERSON [1995], KARPLUS [2000] and STRAUB [2001]. Due to space limitations, the discussion of this section is necessarily cursory and so readers are urged to consult these reviews for further details.

2.1. *The phenomenological rate constant*

The standard approach that chemists use when they want to predict the behavior of a set of chemical reactions is to formulate a series of phenomenological differential equations which describe how the concentrations of each reactive species change as a function of time. For simplicity, consider a unimolecular reaction that interconverts between two states, A and B, of a system:

$$\mathrm{A} \rightleftharpoons \mathrm{B}. \tag{2.1}$$

Reasonable rate equations that describe the change in the concentrations of the two states, c_{A} and c_{B} are:

$$\begin{aligned} \frac{dc_{\mathrm{A}}}{dt} &= -k_{\mathrm{A}\to\mathrm{B}}c_{\mathrm{A}}(t) + k_{\mathrm{B}\to\mathrm{A}}c_{\mathrm{B}}(t), \\ \frac{dc_{\mathrm{B}}}{dt} &= -k_{\mathrm{B}\to\mathrm{A}}c_{\mathrm{B}}(t) + k_{\mathrm{A}\to\mathrm{B}}c_{\mathrm{A}}(t), \end{aligned} \tag{2.2}$$

where t is the time and $k_{\mathrm{A}\to\mathrm{B}}$ and $k_{\mathrm{B}\to\mathrm{A}}$ are the forward and backward rate constants, respectively. There are two constraints on the concentration variables. First, the sum of the concentrations, $c_{\mathrm{A}}(t) + c_{\mathrm{B}}(t)$, must be constant at all times if the only reaction in the system is that given by Eq. (2.1) and, second, at long times the equilibrium values of the concentrations, $\langle c_{\mathrm{A}}\rangle$ and $\langle c_{\mathrm{B}}\rangle$, must obey the detailed balance condition:

$$k_{\mathrm{A}\to\mathrm{B}}\langle c_{\mathrm{A}}\rangle = k_{\mathrm{B}\to\mathrm{A}}\langle c_{\mathrm{B}}\rangle. \tag{2.3}$$

This says that the rate of transition from A to B must be equal to the rate from B to A.

The solution to these equations is:

$$\frac{c_A(t) - \langle c_A \rangle}{c_A(0) - \langle c_A \rangle} = \frac{c_B(t) - \langle c_B \rangle}{c_B(0) - \langle c_B \rangle} = \exp\left(-\frac{t}{\tau_{AB}}\right), \tag{2.4}$$

where $\tau_{AB}^{-1} = k_{A \to B} + k_{B \to A}$.

The scheme outlined above, and others like it for more complicated reactions, have been observed experimentally to work well in many cases. The problem then arises of how to relate the macroscopic parameters describing the reaction, τ_{AB}, $k_{A \to B}$ and $k_{B \to A}$, to microscopic quantities that can be calculated by simulation.

2.2. *Potential energy surfaces and molecular dynamics*

Before going on to discuss microscopic rate theories, it will be necessary to recapitulate some basic concepts concerned with how systems are described theoretically at an atomic level. As is well-known, quantum mechanics (QM) is the appropriate theory for describing molecules and one of its most fundamental equations is the time-dependent Schrödinger equation. This implies that, in principle, the behavior of any system could be determined by setting up its Schrödinger equation and then solving it. Unfortunately, this is very difficult, except in the simplest of cases, and so approximations have to be made.

The normal approach is two-fold. First of all, the time-dependent problem is transformed into a time-independent one by considering only the stationary states of the system and second, the Born–Oppenheimer approximation is invoked which has the effect of being able to treat the dynamics of the electrons and nuclei separately due to their large differences in mass. This leads to a time-independent electronic Schrödinger equation which has the following form:

$$\widehat{H}_{el}(\boldsymbol{r}, \boldsymbol{R}) \Psi_{el}(\boldsymbol{r}, \boldsymbol{R}) = E_{el}(\boldsymbol{R}) \Psi_{el}(\boldsymbol{r}, \boldsymbol{R}). \tag{2.5}$$

Here $\boldsymbol{r}$ and $\boldsymbol{R}$ are vectors of coordinates for the electrons and the nuclei, respectively, Ψ is the wavefunction that gives the distribution of electrons in the system and E_{el} is the electronic energy. $\widehat{H}_{el}$, is the nonrelativistic, electrostatic Hamiltonian for the electrons in the system, which in atomic units has the form:

$$\widehat{H}_{el} = -\frac{1}{2} \sum_i \nabla_i^2 - \sum_{i\alpha} \frac{Z_\alpha}{r_{\alpha i}} + \sum_{ij} \frac{1}{r_{ij}} + \sum_{\alpha\beta} \frac{Z_\alpha Z_\beta}{r_{\alpha\beta}}, \tag{2.6}$$

where the subscripts i and j and α and β refer to electrons and nuclei, respectively, Z is a nuclear charge and r_{xy} is the distance between particles x and y.

If the wavefunction for the system is known, the electronic energy can be written as an expectation value over the electronic Hamiltonian:

$$E_{el} = \frac{\langle \Psi | \widehat{H}_{el} | \Psi \rangle}{\langle \Psi | \Psi \rangle}. \tag{2.7}$$

In this equation the expectation value is performed with respect to the electronic variables only and not those of the nuclei because the nuclear positions are fixed and enter into Eqs. (2.5) and (2.6) parametrically. This means that the electronic energy must be solved for each distinct configuration of the nuclei of the system (defined by the vector $\boldsymbol{R}$). As this energy is a function of the nuclear coordinates, it defines a multi-dimensional surface, the potential energy surface (PES), for the system which represents the effective potential energy of interaction between the nuclei. An accurate determination of the PES of a system is a critical aspect of the study of reactions and will be discussed in Section 3.

Once the electronic problem has been solved or, in other words, once a method of obtaining the PES for the system exists, the dynamics of the nuclei can be treated. Quantum approaches do exist to do this – indeed, some will be discussed in Section 5 – but it is more usual, and certainly more straightforward, to employ classical mechanics. Probably the simplest classical mechanical algorithms start by considering the classical Hamiltonian, H, which can be written as the sum of a kinetic energy contribution, T, and a potential energy term, U. Thus:

$$H = T + U. \tag{2.8}$$

The kinetic energy term is written as:

$$T = \frac{1}{2}\boldsymbol{P}^{\mathrm{T}}\mathbf{M}^{-1}\boldsymbol{P}, \tag{2.9}$$

where $\boldsymbol{P}$ is the vector of particle momenta and $\mathbf{M}$ is the diagonal matrix of particle masses. The potential energy term is simply the electronic energy, i.e.,

$$U = E_{\mathrm{el}}. \tag{2.10}$$

The equations of motion arising from this Hamiltonian are:

$$\frac{d\boldsymbol{R}}{dt} = \mathbf{M}^{-1}\boldsymbol{P}, \tag{2.11}$$

$$\frac{d\boldsymbol{P}}{dt} = -\frac{\partial U}{\partial \boldsymbol{R}}, \tag{2.12}$$

and are equivalent to Newton's equations:

$$\boldsymbol{F} = \mathbf{M}\frac{d^2\boldsymbol{R}}{dt^2}, \tag{2.13}$$

where the forces, $\boldsymbol{F}$, are defined as:

$$\boldsymbol{F} = -\frac{\partial U}{\partial \boldsymbol{R}}. \tag{2.14}$$

The solution of these equations, or of similar ones, for the particles in the system constitutes the classical molecular dynamics method.

2.3. *Microscopic states*

The first step in describing the macroscopic process of Eqs. (2.1) and (2.2) in terms of microscopic variables is to define what the states A and B are at the atomic level. From the discussion of the preceding section, it is clear that this can be done informally by identifying which nuclear configurations or which regions of the system's PES correspond to state A and which to state B. More rigorously, the formulation is done in terms of a surface, the transition-state (TS) surface, that separates configurations of state A from those of state B.

Consider the simple case of a one-dimensional potential shown in Fig. 2.1. It is possible to choose the TS (in this case a point) anywhere along the curve that divides the regions of the surface corresponding to states A and B. For reasons that will be apparent later, though, it is customary to place the TS at the saddle point. With this definition, characteristic functions, χ_{A} and χ_{B}, may be defined for each state as:

$$\chi_{\mathrm{A}}(q) = \begin{cases} 1, & q < q^{\ddagger}, \\ 0, & q \geqslant q^{\ddagger}, \end{cases} \tag{2.15}$$

$$\chi_{\mathrm{B}}(q) = 1 - \chi_{\mathrm{A}}(q) = \begin{cases} 0, & q \leqslant q^{\ddagger}, \\ 1, & q > q^{\ddagger}, \end{cases} \tag{2.16}$$

where q is the coordinate, the reaction coordinate, whose value distinguishes the two states and $q^{\ddagger}$ is the value of the reaction coordinate at the TS.

The principles underlying the definition of the TS surface for multidimensional systems are the same as for the one-dimensional case except that the surface will have $D - 1$ dimensions if the system itself is D-dimensional. The choice of such a surface or, alternatively, of the reaction coordinate, q, that complements it, is often far from obvious, even for the simplest systems, and will be discussed in more detail in Section 4.

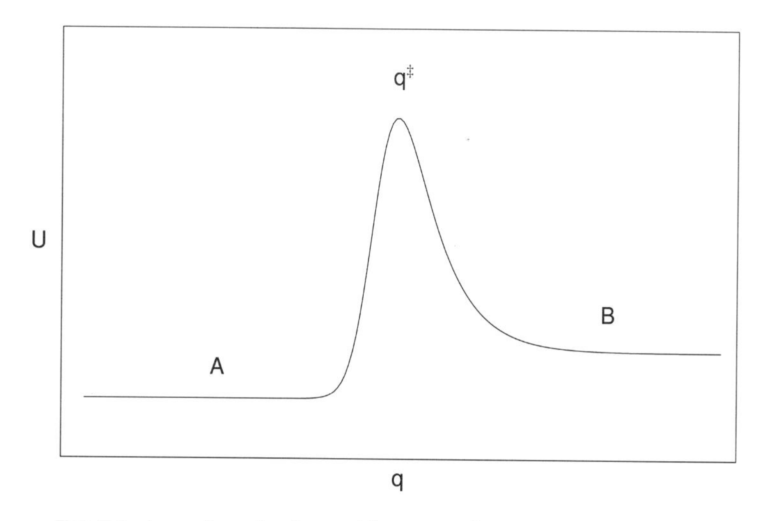

FIG. 2.1. A one-dimensional potential energy surface with two states A and B.

2.4. A rate constant expression

The connection between the macroscopic description of Section 2.1 and the microscopic states just defined is made using the equations of statistical mechanics. Suppose, for concreteness, that we consider the case of a particular thermodynamic ensemble, the canonical ensemble, in which the thermodynamic state variables are the particle number, N_{atoms}, the absolute temperature, T, and the volume, V. The ensemble average of the characteristic function for the state A is given by the following integral over the $3N_{\text{atoms}}$ momentum and $3N_{\text{atoms}}$ position variables for the system:

$$\langle \chi_{\mathrm{A}} \rangle \propto \int d\boldsymbol{P} \int d\boldsymbol{R} \exp\left[-\frac{H(\boldsymbol{P}, \boldsymbol{R})}{k_{\mathrm{B}} T} \right] \chi_{\mathrm{A}}(\boldsymbol{R}), \tag{2.17}$$

where H is the classical Hamiltonian of Eq. (2.8) and k_{B} is Boltzmann's constant. At thermodynamic equilibrium, the value of this integral is independent of time and will be proportional to the equilibrium concentration of species A, $\langle c_{\mathrm{A}} \rangle$. In fact, if A and B are the only states of the system:

$$\langle \chi_{\mathrm{A}} \rangle = \langle c_{\mathrm{A}} \rangle / \big(\langle c_{\mathrm{A}} \rangle + \langle c_{\mathrm{B}} \rangle \big). \tag{2.18}$$

Suppose that the system is perturbed a little due to a change in the Hamiltonian to, say, H', and that we want to determine how the ensemble average of the characteristic function decays back to its equilibrium value. The ensemble average of Eq. (2.17) will now depend on time so that:

$$\widetilde{\chi_{\mathrm{A}}}(t) \propto \int d\boldsymbol{P} \int d\boldsymbol{R} \exp\left[-\frac{H'}{k_{\mathrm{B}} T} \right] \chi_{\mathrm{A}}(t; \boldsymbol{R}). \tag{2.19}$$

Expanding this equation to first order in the perturbation, $H' - H$, gives the following relation:

$$\widetilde{\chi_{\mathrm{A}}}(t) - \langle \chi_{\mathrm{A}} \rangle \propto \big\langle \big(\chi_{\mathrm{A}}(t) - \langle \chi_{\mathrm{A}} \rangle \big) \big(\chi_{\mathrm{A}}(0) - \langle \chi_{\mathrm{A}} \rangle \big) \big\rangle, \tag{2.20}$$

where the right-hand side of the equation is the time autocorrelation function for the characteristic function of state A. Comparison of this equation with the solution of the macroscopic rate equations in Eq. (2.4), shows the following relation to be valid:

$$\exp\left(-\frac{t}{\tau_{\mathrm{AB}}} \right) = \frac{\langle (\chi_{\mathrm{A}}(t) - \langle \chi_{\mathrm{A}} \rangle)(\chi_{\mathrm{A}}(0) - \langle \chi_{\mathrm{A}} \rangle) \rangle}{\langle (\chi_{\mathrm{A}}(0) - \langle \chi_{\mathrm{A}} \rangle)^2 \rangle}. \tag{2.21}$$

This, greatly simplified, derivation is a consequence of the fluctuation-dissipation theorem of the linear response theory of statistical thermodynamics and allows the connection that we are seeking to be made between the macroscopic rate constant and the microscopic characteristic functions.

The expression of Eq. (2.21) is not especially convenient and it is normal to manipulate it further, the details of which will not be given here. This results in a quantity called the reactive flux correlation function, $k(t)$, whose form, for the single dimensional case of Fig. 2.1, is:

$$k(t) = \langle \dot{q}(0)\delta(q(0) - q^{\ddagger})\chi_B(q(t))\rangle \tag{2.22}$$

and in which $\dot{q}(0)$ denotes the reaction coordinate velocity at $t = 0$. The first two terms in the correlation function, $\dot{q}(0)\delta(q(0) - q^{\ddagger})$, give the flux across the transition state surface at $t = 0$ whereas the characteristic function $\chi_B(t)$ is zero unless the system is in state B at time, t.

In terms of the correlation function, $k(t)$, the rate constant for the conversion of state A to state B is:

$$k_{A\to B} \exp\left(-\frac{t}{\tau_{AB}}\right) = \frac{k(t)}{\langle \chi_A \rangle}. \tag{2.23}$$

Is this equation correct or, in other words, does $k(t)$ exhibit a simple exponential dependence? In many instances the correlation function has the form shown schematically in Fig. 2.2. At short times there is a rapid decrease in the value of the function because it takes the system a certain time, $t \sim \tau_{mol}$, to commit itself to one of the regions of the PES corresponding to the states A or B. At times shorter than τ_{mol}, the system can 'recross' the TS surface one or more times before the motions of the other degrees of freedom in the system can remove energy from the reactive motion and trap it in one of the states. At times longer than τ_{mol}, but much shorter than τ_{AB}, for which $\exp(-t/\tau_{AB}) \sim 1$, $k(t)$ appears to reach a plateau, the value of which is proportional to the rate constant $k_{A\to B}$.

The fact that the correlation function, $k(t)$, should have a plateau value for $\tau_{mol} \leqslant t \ll \tau_{AB}$ provides a criterion for the existence of a rate constant of the form given by the phenomenological model. It shows that there must be a separation of timescales

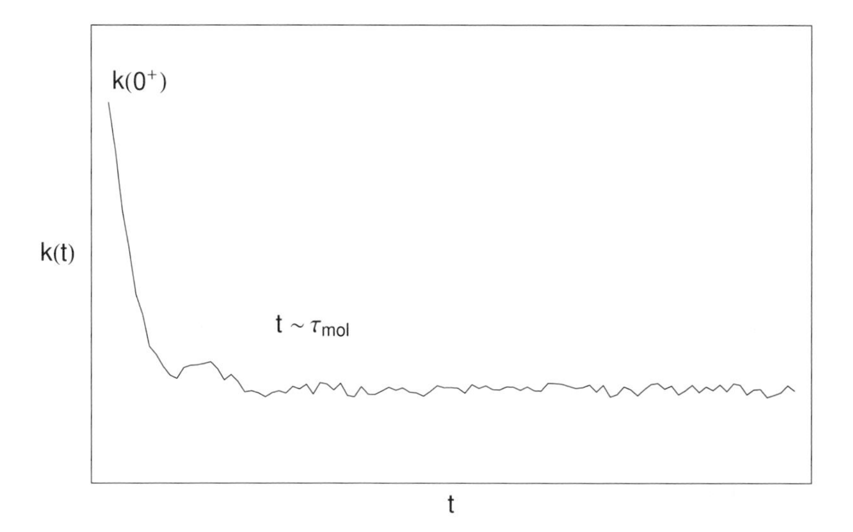

FIG. 2.2. The reactive flux correlation function.

between the 'normal', fast dynamics of the system (characterized by the time τ_{mol}) and the slower reactive dynamics (characterized by τ_{AB}). This implies that the reactive events should be rare or that the energy barrier separating the states A and B be high ($\gg k_{\text{B}}T$) so that the probability of the system existing at the transition state be small. If this is not true in a particular case it means that a phenomenological model for the reaction process of the type described in Section 2.1 is inappropriate.

2.5. *Transition state theory*

Consider the equation for the correlation function of Eq. (2.22) and suppose that the time-dependence is removed by replacing the term $\chi_{\text{B}}(q(t))$ by a term whose value is 1 if the initial velocity $\dot{q}(0)$ takes the system to state B and 0 otherwise. Denoting this modified term by $\chi_{\text{B}}(q(0^+))$ where $q(0^+)$ is the configuration of the system at an infinitesimal time after $t = 0$, gives the transition state theory (TST) rate constant, $k_{\text{A}\to\text{B}}^{\text{TST}}$:

$$k_{\text{A}\to\text{B}}^{\text{TST}} = \frac{k(0^+)}{\langle \chi_{\text{A}} \rangle}. \tag{2.24}$$

Physically what has been done by removing the time-dependence is to forbid recrossing of the TS due to the dynamics of the remaining degrees of freedom in the system. It also means that the TST rate constant will always be an upper bound to the true rate constant (see Fig. 2.2). The ratio of the true and the TST rate constants is called the transmission coefficient, κ, whence:

$$k_{\text{A}\to\text{B}} = \kappa k_{\text{A}\to\text{B}}^{\text{TST}}. \tag{2.25}$$

TST is useful because it provides a reasonably practicable way of calculating the rate constant for a rare event in a complex system. It does this by dividing the true rate constant into a statistical part, the TST rate constant, which depends uniquely upon the equilibrium properties of the system, and a dynamical part, the transmission coefficient, which is essentially equivalent to the proportion of trajectories, starting off at the transition state, that stay in the state to which they were initially directed. Each of these two terms can be evaluated independently as we shall see in Section 4.

3. Calculating condensed-phase potential energy surfaces

The evaluation of the potential energy of a reacting system as a function of its nuclear geometry is a crucial part of any simulation study. Without an accurate or, at the very least, a qualitatively-correct potential energy surface, the results of any calculation will, in most cases, be meaningless.

There are several different ways to calculate PESs. The most fundamental are the *ab initio* quantum chemical methods which try to determine the electronic energy for a system at a given nuclear configuration by solving the Schrödinger equation. At the other extreme are empirical methods that use experimental, and sometimes theoretical, data to fit a function that approximates the PES for a system in a certain region of interest. And, in between, there are methods which blend elements of both the quantum chemical and empirical approaches.

Quantum chemical methods of some type are essential for the investigation of the energetics of reacting systems because it is normally difficult to obtain sufficient empirical information to give an adequate description of a PES in the region of a reaction. It is for this reason that quantum chemical methods will be given prominence in what follows below although empirical methods, when employed in conjunction with quantum chemical potentials, are also useful tools for studying reacting systems.

3.1. Ab initio quantum chemical methods

Ab initio quantum chemical methods aim to solve the time-independent Schrödinger equation (Eq. (2.5)) directly and precisely by making as few approximations as possible. The principal methods in use are based upon density functional theory (DFT) and molecular orbital (MO) theory although valence bond (VB) and quantum Monte Carlo (QMC) approaches also exist. For many applications, it will be possible to choose a technique that will give results of the desired accuracy but for others – particularly those that involve complicated electronic structures or changes in electronic structure, such as radicals, transition metal compounds and many chemical reactions – obtaining precise results can be a challenging task.

DFT and MO-based techniques are the most common contemporary *ab initio* methods in quantum chemistry. They have been covered extensively in other chapters in this volume (see, for example, CANCÈS, DEFRANCESCHI, KUTZELNIGG, LE BRIS and MADAY [2003]), and in other references (KOCH and HOLTHAUSEN [2000], PARR and YANG [1989], KOHN [1999], POPLE [1999], SZABO and OSTLUND [1989]), and so only a few salient points concerning them will be made here.

DFT methods have revolutionized quantum chemistry over the past ten years or so and currently provide the most cost-effective means of determining the energetics and structures of reacting groups in large molecular systems. They can be implemented with efficiencies that are similar to those of the fastest MO methods based upon Hartree–Fock (HF) theory but they provide energetic and structural results for ground and excited state systems that are often as precise as much more expensive MO-based methods. They do have some limitations, notably for applications to hydrogen-bonded and weakly-bound systems and for the determination of the barrier heights for certain reactions. These problems are likely to become less marked though as better exchange-correlation functionals are developed (KOCH and HOLTHAUSEN [2000]). Another important development that should be mentioned, and which will increase the usefulness of DFT calculations, is the introduction of linear-scaling or $O(n)$ algorithms (GOEDECKER [1999]). Calculations with most existing DFT implementations have costs that scale somewhere between $O(n^2)$ and $O(n^4)$ where n is a measure of the size of the system (such as the number of electrons).

Although calculations relying upon DFT have come to dominate quantum chemistry, MO-based methods still have a role to play as they are, in principle, capable of providing results of arbitrary accuracy by systematically enhancing the quality of the basis set and including more Slater determinants in the wavefunction expansion. The simplest MO HF methods are not competitive with DFT techniques in terms of accuracy and, in any case, are often not appropriate for describing the electronic structure of a system in which bonds are being broken and formed. In contrast, the more accurate MO methods have costs that scale as $O(n^5)$ or higher which means that they are limited to systems of small size ($N_{\text{atoms}} \sim 10$).

Two other approaches deserve comment in this brief survey. The first are methods based upon VB theory in which the wavefunction for the system is constructed as a linear combination of 'resonant' structures that represent particular, idealized electronic configurations of the system. The wavefunctions of the resonant states are normally constructed from nonorthogonal spin-orbitals. *Ab initio* VB methods give results that are of similar quality to MO calculations but they are not often employed because of the extra complexity arising from the use of nonorthogonal orbitals (GALLUP, VANCE, COLLINS and NORBECK [1982], COOPER, GERRATT and RAIMONDI [1987]). The second type of approach comprise the QMC methods which make use of the similarity between the diffusion equation and the Schrödinger equation in imaginary time to statistically estimate the electronic energy of the ground state of a system (CEPERLEY and MITAS [1996]). QMC methods have the ability to produce results of very high accuracy at a much lower cost than the corresponding MO methods of equivalent precision (GROSSMAN, LESTER and LOUIE [2000]) but their use has been hampered by the absence of any widely-available general-purpose QMC programs.

To terminate this section, it should be pointed out that calculations with the quantum chemical methods of the type described here can be applied in two distinct ways. Either the calculations can be done 'on-the-fly', which means that the quantum chemical method is used directly during a simulation to obtain the electronic energy for each new configuration of a system, or the quantum chemical calculations can be done beforehand and the results employed to parametrize an analytical function that reproduces the reactive part of the PES. The latter, indirect, approach has the advantage that simulations with the parametrized surface are very fast but the disadvantage that parametrization of a surface for a reaction becomes progressively more difficult as the number of configurational degrees of freedom increases. It is for this reason that the former, direct, method is nowadays generally preferred for studies of reactions in complex systems.

3.2. Semi-empirical quantum chemical methods

Ab initio quantum chemical methods are preferable for the investigation of chemical reactions and other aspects of the electronic structure of molecules but, due to their computational expense, faster, albeit less precise, quantum chemical methods have been developed. In general, these semi-empirical methods have similar formalisms to their *ab initio* counterparts but differ in that simplifications are made that greatly speed up the more time-consuming parts of the calculation. These simplifications normally entail the replacement of analytically-determinable quantities, such as certain Hamiltonian matrix elements, by empirical functions that must be parametrized against experimental or *ab initio* data.

The great majority of semi-empirical methods in chemistry are based upon MO theory. The Hückel method was an early method that is still widely used for qualitative calculations (ALBRIGHT, BURDETT and WHANGBO [1985]). Equally successful have been the semi-empirical MO approaches developed by Dewar and co-workers. These include the AM1 and MNDO methods of Dewar himself (DEWAR and THIEL [1977], DEWAR, ZOEBISCH, HEALY and STEWART [1985]) and the PM3 parametrization of the AM1 Hamiltonian of Stewart (STEWART [1989a], STEWART [1989b]). The AM1 family of techniques are now dated and insufficiently accurate for certain problems, including many chemical reactions (THOMAS, JOURAND, BRET, AMARA and FIELD [1999]). Even so, in favorable cases, they have a precision comparable to or better than *ab initio* DFT or HF calculations performed with small basis sets (up to double-ζ) despite being substantially less expensive (DEWAR and O'CONNOR [1987]).

Given the recent success of *ab initio* DFT methods in quantum chemistry, it is perhaps not surprising that semi-empirical versions of the theory have been introduced. One promising avenue concerns the application of tight-binding methods to chemistry. These methods, commonly-used in solid-state physics, can be shown to arise from DFT by expanding the Kohn–Sham equations in terms of fluctuations in the atomic density (FOULKES and HAYDOCK [1989]). In addition to providing a firm theoretical foundation, such a link gives expressions for many of the parameters appearing in tight-binding models in terms of quantities that can be determined from *ab initio* DFT calculations. This, in principle, greatly simplifies the parametrization procedure.

An example of this type of model is the self-consistent charge tight-binding (SCCTB) method developed by Elstner, Porezag, Frauenheim and co-workers and which has been claimed to give improvements over the Dewar MO methods (ELSTNER, POREZAG, JUNGNICKEL, ELSNER, HAUGK, FRAUENHEIM, SUHAI and SEIFERT [1998], FRAUENHEIM, SEIFERT, ELSTNER, HAJNAL, JUNGNICKEL, POREZAG, SUHAI and SCHOLZ [2000], POREZAG, FRAUENHEIM, KÖHLER, SEIFERT and KASCHNER [1995]). In this method, the energy of the system, E, can be written as the sum of two terms. The first, E_0, is the typical tight-binding energy expression:

$$E_0 = \sum_i n_i \langle \psi_i | \widehat{H} | \psi_i \rangle + E_{\mathrm{rep}}, \tag{3.1}$$

where $\widehat{H}$ is the tight-binding Hamiltonian for the system, ψ_i is the ith orbital with occupation number n_i and E_{rep} is the repulsion energy between atomic cores. The second term, E_1, is:

$$E_1 = \frac{1}{2} \sum_{\alpha\beta} \gamma_{\alpha\beta} q_\alpha q_\beta, \tag{3.2}$$

where q_α is the partial charge on atom α and $\gamma_{\alpha\beta}$ is a matrix element for the interaction between two charges. The orbitals for the system are expanded in terms of atom-centered basis functions and the optimum orbitals are obtained by minimizing the total energy expression with respect to the orbital expansion coefficients. As the atomic charges are derived from a population analysis of the system's wavefunction, they depend upon the orbitals and so solution of the equations must be performed in a self-consistent fashion.

3.3. *Empirical potentials*

With quantum chemical methods, there are two possible strategies for the use of data resulting from the solution of the Schrödinger equation at a particular nuclear configuration. Either it can be used directly in a simulation or it can be employed to fit an intermediate function of arbitrary form that represents the PES of the system in the regions of interest. Once fitting is complete, this function will be used as the potential energy function in subsequent simulation. The rationale underlying empirical potentials (or, as they are also called, force fields or molecular mechanical (MM) potentials) is different. Instead, an empirical form for the energy function is chosen, each term of which encapsulates a particular aspect of the bonding properties or the interatomic interactions in the system. Once, the form of the potential has been decided, it must be parametrized so that it reproduces experimentally-observable quantities.

There are many different types of empirical force field and they are often tailored for simulations of different classes of molecule. One of the more common types and the type that is most frequently employed for simulations of biomolecular systems, such as proteins, writes the total potential energy of the system, E, as the sum of two energies, one for the covalent or bonding interactions, E_{bonding}, and one for the noncovalent or nonbonding interactions, $E_{\text{nonbonding}}$ (FIELD [1999]):

$$E = E_{\text{bonding}} + E_{\text{nonbonding}}. \tag{3.3}$$

It is normal to subdivide the bonding and nonbonding energies further. Thus, the bonding energy often consists of a sum of bond, angle, dihedral and out-of-plane terms:

$$E_{\text{bonding}} = E_{\text{bond}} + E_{\text{angle}} + E_{\text{dihedral}} + E_{\text{out-of-plane}}, \tag{3.4}$$

whereas the nonbonding energies are the sum of a electrostatic term and a Lennard-Jones term:

$$E_{\text{nonbonding}} = E_{\text{elect}} + E_{\text{LJ}}. \tag{3.5}$$

A common form for the bond energy is a sum of harmonic terms, one for each bond in the system:

$$E_{\text{bond}} = \sum_{\text{bonds}} \frac{1}{2} k_b (b - b_0)^2. \tag{3.6}$$

In this equation, b is the actual distance between the two atoms involved in the bond, b_0 is an equilibrium distance characteristic of the bond and k_b is the force constant for the bond which determines the steepness of the potential well and, hence, the bond's frequency of oscillation.

The angle energy is usually similar except that energy terms are functions of the angle, θ, subtended by three atoms:

$$E_{\text{angle}} = \sum_{\text{angles}} \frac{1}{2} k_\theta (\theta - \theta_0)^2. \tag{3.7}$$

For the dihedral or torsional energy, a harmonic form is not appropriate as the energy must be a periodic function of the torsion angle about the bond. A suitable form is an expansion in terms of trigonometric functions, i.e.,

$$E_{\text{dihedral}} = \sum_{\text{dihedrals}} \frac{1}{2} V_n \big(1 + \cos(n\phi - \delta)\big). \tag{3.8}$$

Here, V_n is the height of the torsional barrier, n is the periodicity of the term and δ is its phase.

The out-of-plane or improper torsional energy, $E_{\text{out-of-plane}}$, is used to keep atoms, such as those which are sp^2 hybridized, planar. In some force fields, a harmonic form for this energy term is employed whereas, in others, terms reminiscent of the dihedral energy are preferred.

In many force fields, the electrostatic energy, E_{elect}, is given by a simple Coulomb-type expression because the charge distribution of a molecule is represented by fixed partial charges centered on the nuclei of the atoms. The energy then has the form:

$$E_{\text{elect}} = \sum_{\alpha\beta\ \text{pairs}} \frac{q_\alpha q_\beta}{\varepsilon r_{\alpha\beta}}, \tag{3.9}$$

where q_α and q_β are the partial charges on atoms α and β and $r_{\alpha\beta}$ is the distance between them. ε is the dielectric constant for the interaction which will be unity for two atoms in vacuum.

The Lennard-Jones energy mimics the quantum mechanical exchange-repulsion interaction arising when two charge clouds overlap at short-range and the attractive dispersive inverse sixth power interaction at longer range. It has the form:

$$E_{\text{LJ}} = \sum_{\alpha\beta\ \text{pairs}} \frac{A_{\alpha\beta}}{r_{\alpha\beta}^{12}} - \frac{B_{\alpha\beta}}{r_{\alpha\beta}^{6}}, \tag{3.10}$$

where $A_{\alpha\beta}$ and $B_{\alpha\beta}$ are constants whose values depend upon the nature of the atoms α and β. The Lennard-Jones energy between two atoms will be nonzero even when they have no net charge.

The nonbonding interaction energies of Eqs. (3.9) and (3.10) are normally calculated for all pairs of atoms in the system but it is usual to exclude pairs of atoms from the sum which are separated by only one or two covalent bonds so as to avoid the overcounting that would result if both bonding and nonbonding terms were calculated for these atoms.

The calculation of the nonbonding interactions, and particularly the longer range electrostatics terms, is invariably the most expensive part of an energy calculation. This is because the number of interactions scales as the square of the number of particles whereas the number of bonding terms scales roughly linearly with the size of the system. There are a number of ways in which the cost of the nonbonding energy calculation can be reduced. One of these is the approximate truncation technique in which interactions are either neglected or tapered to zero beyond a certain cutoff distance. Other methods, which are to be preferred, attempt to calculate the full nonbonding energy of a system to a certain precision but with a cost that scales linearly with the size of the system.

The terms listed for the 'typical' force field above are by no means the only ones in use. Thus, for example, in more complicated functions there will be bonding cross-terms which couple various internal coordinate deformations as well as nonbonding polarization terms which describe the interactions due to the changes in the charge distribution of a molecule in different environments.

An advantage of empirical force fields of the type discussed here is that they are computationally efficient and can be used for the simulation of systems comprising many thousands of atoms. In addition, their analytic form is such that it is straightforward to calculate the derivatives of the energy with respect to various atomic quantities. In practical applications, the most important of these is the first derivative of the potential energy with respect to the atomic coordinates which is proportional to the force (see Eq. (2.14)). A disadvantage of force fields is that they must be parametrized to obtain values for the many parameters (b_0, k_θ, V_n, q_i, etc.) that they contain. Some applications do not require a very precise parametrization but, in most cases, it will be necessary to parametrize the force field against large amounts of data from experiments or from high quality QM calculations. Such parametrizations are laborious and can demand great effort if force fields of reasonable precision are to be obtained. Another disadvantage is that force fields are of limited flexibility as they are conceived for the simulation of particular systems in particular circumstances and will be unsuitable for studying processes outside their range of applicability. Thus, for example, it will not be possible to study reactions in which bonds are broken and formed with the force field described above because the harmonic bond term of Eq. (3.6) does not allow dissociation.

3.4. Hybrid potentials

As we have seen, the investigation of reactions in condensed-phase systems with purely quantum chemical methods is currently impractical if the system is of any reasonable size and the use of empirical force fields for studying reactions is problematical because special functions would have to be devised and parametrized for each reaction under study. To circumvent these problems, hybrid potentials have been developed in which potentials of differing accuracy are used to treat different regions of the system. For example, a chemical reaction could be studied by treating the reacting atoms and those immediately surrounding them with a QM potential and using a simpler method for the atoms of the remainder of the system. When partitioning the system in this way, the assumption is made that the reactive process is localized in the QM region – this will be

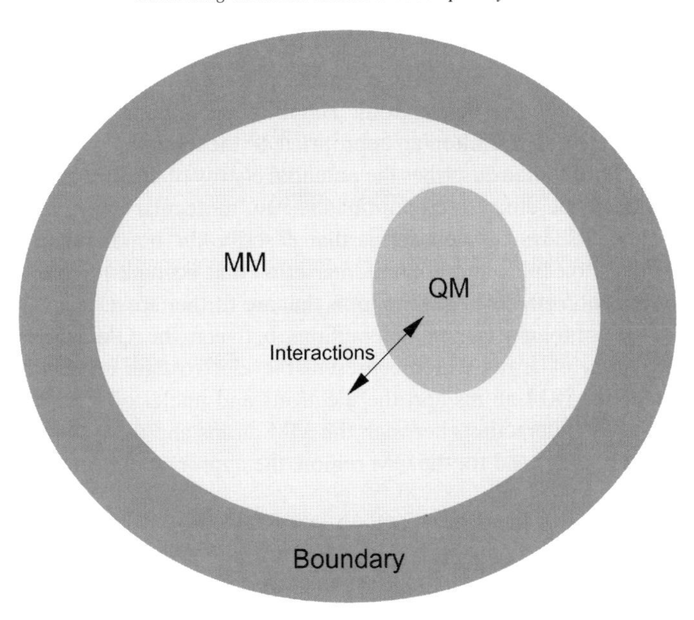

FIG. 3.1. Partitioning a system into a QM and a MM region for use with a hybrid potential.

reasonable for many reactions but it will not be valid in some instances, such as when there is a long-range electron-transfer event.

Hybrid or QM/MM potentials were first introduced for the treatment of reactions in enzymes by WARSHEL and LEVITT [1976]. Since then, many types of hybrid potential have been implemented (for reviews, see GAO [1995], AMARA and FIELD [1999], MONARD and MERZ [1999]). They differ in the number of regions into which the system is divided, the types of potential used to treat the different regions and the ways in which the interfaces between the potentials are handled. Although diverse in nature, the basic principles underlying most hybrid potentials are similar and are best illustrated by a simple example.

Consider a system divided into two regions (see Fig. 3.1), one of which is treated by a MO QM method and the other with a MM potential. The first steps in the formulation of the hybrid potential in this case are to invoke the Born–Oppenheimer approximation and to construct an effective electronic Hamiltonian, $\widehat{H}_{\mathrm{Eff}}$, which will be the sum of three terms:

$$\widehat{H}_{\mathrm{Eff}} = \widehat{H}_{\mathrm{QM}} + \widehat{H}_{\mathrm{MM}} + \widehat{H}_{\mathrm{QM/MM}}, \tag{3.11}$$

where $\widehat{H}_{\mathrm{QM}}$ is the Hamiltonian for the particles (electrons and nuclei) in the QM region of the system, $\widehat{H}_{\mathrm{MM}}$ is the Hamiltonian for the atoms in the MM region and $\widehat{H}_{\mathrm{QM/MM}}$ is the Hamiltonian for the interaction between the two regions (FIELD, BASH and KARPLUS [1990]).

The exact form for the terms in the Hamiltonian will depend upon the QM and MM potentials of the individual regions but common examples are:

$\widehat{H}_{\mathrm{QM}}$ The form for the QM Hamiltonian depends upon the approximation chosen to solve the time-independent Schrödinger equation. For an *ab initio* method this would be the standard nonrelativistic electronic Hamiltonian whereas for a semi-empirical method it would be the Hamiltonian appropriate to the method.

$\widehat{H}_{\mathrm{MM}}$ The MM Hamiltonian describes the potential energy of the MM atoms, E_{MM}, and is independent of the electronic coordinates. Any molecular mechanics force field is suitable that includes covalent terms that describe the bond, angle and dihedral energies of the molecules and nonbonding terms that account for electrostatic and dispersion/repulsion effects between atoms that are further apart.

$\widehat{H}_{\mathrm{QM/MM}}$ The Hamiltonian for the interactions between the QM and MM regions consists of a sum of terms which represent the electrostatic interactions between the charges of the MM atoms and the electrons and nuclei of the QM region and the Lennard-Jones interactions between the MM atoms and the QM nuclei. If an *ab initio* method is being used for the QM region, the appropriate form, in atomic units, would be:

$$\widehat{H}_{\mathrm{QM/MM}} = -\sum_{iM} \frac{q_M}{r_{iM}} + \sum_{\alpha M} \frac{Z_\alpha q_M}{r_{\alpha M}} + \sum_{\alpha M} \left\{ \frac{C_{\alpha M}}{r_{\alpha M}^{12}} - \frac{D_{\alpha M}}{r_{\alpha M}^{6}} \right\}, \tag{3.12}$$

where the subscripts i, α and M refer to electrons, QM nuclei and MM atoms respectively, Z is the nuclear charge and C and D are parameters for the Lennard-Jones interaction.

The energy of the system, E, and the wavefunction, Ψ, for the electrons of the QM region are determined by solving the time-independent Schrödinger equation (Eq. (2.5)) with the effective Hamiltonian defined in Eq. (3.11) as the operator. As the wavefunction and the effective Hamiltonian depends parametrically upon the positions of both the QM nuclei and the MM atoms, the Schrödinger equation must be solved at each different configuration of the QM nuclei and the MM atoms.

The hybrid potential formulated above was for a MO QM method and was done in terms of an effective Hamiltonian. For a DFT QM method, it is more appropriate to formulate the potential in terms of the electron density in the QM region but as the procedure is essentially equivalent no details will be given here. Instead, readers are referred to AMARA, VOLBEDA, FONTECILLA-CAMPS and FIELD [1999].

The treatment of the junction between the different regions is the crucial aspect of the definition of a hybrid potential and remains, in many respects, an outstanding problem. The interaction Hamiltonian of Eq. (3.12) was for the straightforward case in which there are only nonbonding interactions between the atoms of the different regions. For many applications, however, it will be necessary for a single molecule to be split between different regions which will result in there being covalent bonds between QM and MM atoms. These 'dangling' bonds must be treated in some way as otherwise the presence of broken bonds and unpaired electrons at the boundary of the QM region will dramatically change the electronic structure of the QM system. The commonest and also the simplest way of tackling this problem is to use the link-atom approximation in which a single unphysical atom is introduced into the QM region for

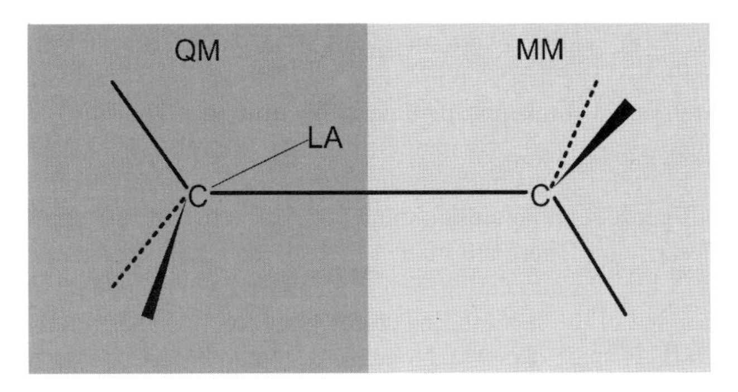

FIG. 3.2. Treating covalent bonds between atoms of the QM and MM regions of a hybrid potential using the 'link-atom' approximation.

each dangling bond (see Fig. 3.2). These atoms, which are normally hydrogens, enter into the QM calculation and serve to sate the unsatisfied valencies of the QM atoms with broken bonds. How their interactions with the MM regions are handled varies according to the implementation of the method. Apart from the link-atom method, alternatives, such as those based upon hybrid orbitals, have been proposed but all these methods are significantly more complicated to implement and, in the author's opinion, have not been shown to be consistently more accurate. In addition to the problem of covalent bonds between QM and MM atoms, the treatment of the interactions between the atoms of different regions needs improvement, especially for hybrid potentials which use *ab initio* QM methods. This is an area of on-going research.

The hybrid potentials described above have been implemented with both sophisticated *ab initio* and the simpler semi-empirical QM methods. Hybrid potentials have also been developed with even simpler QM potentials, the most notable examples being the empirical valence bond (EVB) potentials of Warshel and co-workers (WARSHEL [1991]). The EVB method is like a normal valence bond method in that the total wavefunction for the system is formed as a linear combination of wavefunctions of resonant states but differs because the matrix elements between the resonant states are parametrized using simple functional forms. Although computationally efficient and applicable to reactions, EVB potentials must be redesigned and reparametrized every time a new reaction is to be studied. Unfortunately, this requires considerable experience as there are systematic ways neither for the generation of the important quantum states that partake in the reaction nor for their parametrization. Consequently, only a relatively small number of research groups worldwide routinely apply these methods.

3.5. *Extended systems*

Experiments are usually done on systems with $O(N_{\text{Avogadro}})$ ($\sim 10^{23}$) particles which is far from the size of the system that can be studied computationally. Thus, to be feasible, a simulation of a condensed-phase system must select a certain group of atoms to treat explicitly and then approximate the effect of the remaining, neglected atoms which act as the environment to the central simulation system. There are two main ways of dealing with extended systems (apart from the easiest one of neglecting the environment entirely). The first is to use the method of periodic boundary conditions in which the entire system is taken to be a periodic array (or a 'crystal') of copies of the central simulation system (ALLEN and TILDESLEY [1987]). This method is probably the most exact as all the atoms in the system are treated explicitly. The second class of methods is to dispense with an atomic model of the environment and use some sort of approximate description. There are many of these, including the widely-used reaction field methods in which the simulation system is immersed in a medium whose dielectric constant mimics that of the surrounding atoms (see TOMASI and PERSICO [1994] and SIMONSON [2001] for nice reviews). None of the methods described in this section are specific to studies of reacting systems but no discussion of how to calculate the PES for a condensed-phase system would be complete without mentioning them.

4. Simulation methods for investigating chemical reactions

This section assumes that a method for the calculation of the PES for a reacting system is available and outlines the various simulation approaches that can be employed to obtain information about a chemical reaction. The discussion is focussed on methods for investigating reactions within the classical reactive flux framework but a more recent development is mentioned at the end.

4.1. The classical reactive flux formalism

Probably the most practical way of determining the rate constant within the reactive flux formalism is based upon Eq. (2.25) which expresses the rate constant as the product of two quantities, the TST rate constant and the transmission coefficient, which can be evaluated independently. Discussion of the transmission coefficient will be left until later, but it is convenient to manipulate the TST rate constant expression of Eq. (2.24) to put it into a more suitable form for computation. A little rearrangement allows it to be rewritten as:

$$k_{\mathrm{A}\to\mathrm{B}}^{\mathrm{TST}} = \frac{\langle \dot{q}\,\Theta(\dot{q})\,\delta(q - q^{\ddagger})\rangle}{\langle \delta(q - q^{\ddagger})\rangle} \times \frac{\rho(q^{\ddagger})}{\int_R dq\, \rho(q)}. \tag{4.1}$$

Here Θ is the Heaviside function which takes values of 1 for positive arguments and zero otherwise, $\rho(q)$ represents the probability density for the system as a function of the reaction coordinate and the integration in the denominator of the second term on the right-hand side is done over the values of the reaction coordinate that correspond to the reactant well. Not only is Eq. (4.1) easier for calculation, as we shall see below, but it also takes the same form as the well-known Arrhenius rate constant expression:

$$k_{\mathrm{A}\to\mathrm{B}}^{\mathrm{Arrhenius}} = \nu \exp\left[-\frac{\Delta G^{\ddagger}}{k_{\mathrm{B}}T}\right], \tag{4.2}$$

where ν is a frequency prefactor and $\Delta G^{\ddagger}$ is the activation free energy which is the free energy difference between the reactant and the transition state structures.

Inspection of Eq. (2.25), together with the fact that the transmission coefficient takes values between 0 and 1, indicates that the TST rate constant will always overestimate the true rate constant. As the transmission coefficient can be expensive to evaluate, especially when it takes very small values, it is most efficient to select a TS surface dividing reactants from products that minimizes the value of the TST rate constant and, hence, maximizes the value of the transmission coefficient. As Chandler has shown, this is equivalent to finding the surface with the highest free energy (CHANDLER [1978]).

Unfortunately, the selection of the TS surface of most general form and of highest free energy is a very difficult problem even for small systems. Marked simplifications arise when it is assumed that the surface dividing states A and B is a hyperplane, as in this case it is only necessary to choose the position of the hyperplane and its orientation (SCHENTER, MILLS and JÓNSSON [1994], JÓHANNESSON and JÓNSSON [2001]). Even so, calculations that make use of algorithms for finding this optimum hyperplane have been limited due to their expense. An even simpler approximation, and the one that is usually employed, is to choose a reaction coordinate that describes the reaction process between the two states, calculate the free-energy as a function of this coordinate and then take the TS as being the point of highest free-energy along the path. This approach also assumes that the dividing surface between states A and B is a hyperplane but only the location of the hyperplane is optimized – its orientation is fixed as being normal to the tangent of the reaction coordinate.

To summarize, a standard approach when using the reactive flux formalism consists of three steps. The first is the determination of a reaction coordinate that somehow represents the transition process between the reactant and product states, the second is the calculation of the TST rate constant by the estimation of the free-energy as a function of the reaction coordinate and the third is the correction of the TST rate constant by evaluating the transmission coefficient. In very general terms, each of these steps is more expensive than the previous one, so that many studies stop after only the first or second steps. This is particularly true for simulations with *ab initio* potentials as these most frequently locate a suitable TS structure for the reaction and then, at best, make a crude estimate of the TST rate constant. It should be emphasized, however, that much useful chemical information can still be gained by limited studies of this type.

4.2. Determining a reaction coordinate

The ideal reaction path is the one which minimizes the TS rate constant. Unfortunately, such paths are difficult to determine directly and so it is usual to make assumptions about what such paths look like. For the simplest reactions, it may be possible to choose a reaction path by inspection. Thus, for example, the dissociation of a diatomic molecule obviously depends upon the distance between the two atoms whereas the rotation of a methyl group or an aromatic ring within an organic molecule could be reasonably described by a torsional angle. Such choices are, however, fraught with danger as the example of the diatomic makes clear. Whereas in the gas phase the reaction coordinate is obvious, in solution it is highly probable that various solvent degrees of freedom are important for the definition of the reaction coordinate, particularly if the dissociation involves a separation of charge.

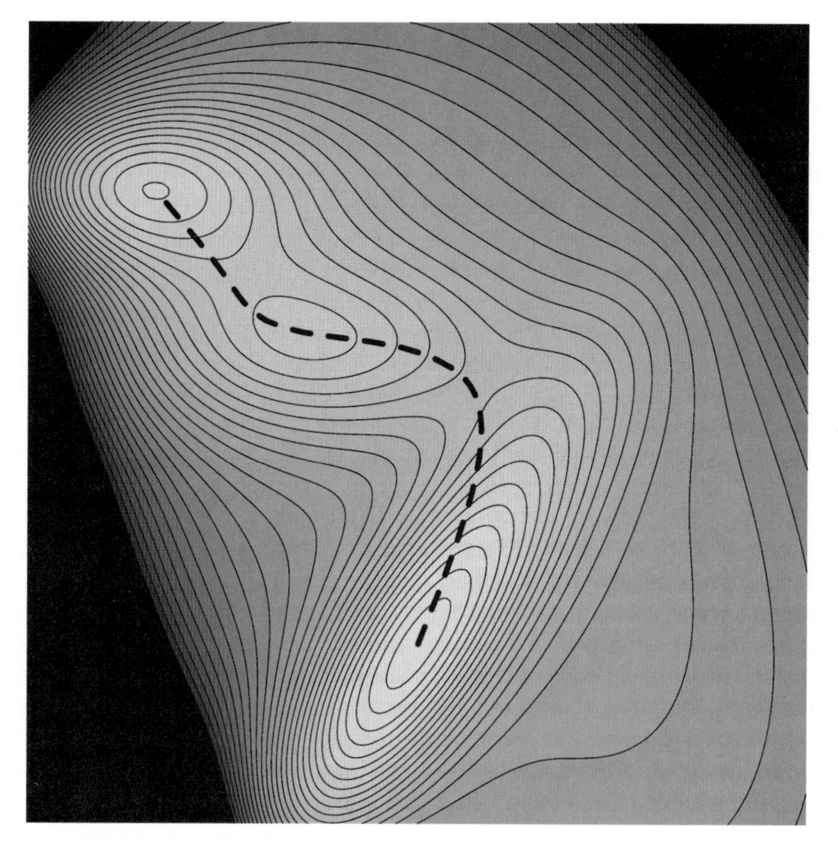

FIG. 4.1. A two-dimensional potential energy surface and a possible reaction path between two states. There is one intermediate.

A more rigorous definition of a path is one that says that the optimum path is the 'lowest-energy' path on the PES that connects in a continuous fashion the potential wells (or minima) that define the reactant and product states. This requires that the path has, as its highest-energy point, the first-order saddle point of lowest energy between the two stable states. For reactions where there are intermediate stable states between reactants and products, it is normal to define the complete path as the one made up of the individual paths between each of the intervening minima. An example is shown in Fig. 4.1.

The location of a reaction path going through saddle points is most conveniently done for small systems by first finding the saddle points and then tracing out the path downhill from the saddle point to the stable states that lie on either side. The location of saddle points is a geometry-optimization procedure with the function determining the PES as the target function and the atomic coordinates as its variables. Saddle point location is more difficult than finding minima because instead of searching for a direction which reduces the energy in every direction, it is necessary to maximize the energy in one direction and minimize in all the rest. An early, but nice introductory review of saddle-point-search algorithms has been given by BELL and CRIGHTON [1984].

Probably the most successful algorithms are the mode-walking algorithms that start off at a particular point on the surface, often a minimum, and move uphill along one of the normal modes for the system and downhill along all the rest. These algorithms were introduced by CERJAN and MILLER [1981] and have been refined by other workers, including SIMONS, JORGENSON, TAYLOR and OZMENT [1983], BAKER [1986] and WALES [1994]. The determination of the normal modes along which to walk requires the calculation or the estimation and the diagonalization of the second derivative matrix of the energy with respect to the coordinate variables. This limits the use of this algorithm to systems with a relatively small number of degrees of freedom as the calculation and storage of the second derivative matrix and its diagonalization are computationally demanding tasks.

Once the saddle point between two minima has been identified, the full path can be generated. A common definition is to follow the steepest-descent direction down from the saddle point either in Cartesian coordinates, $\boldsymbol{R}$, or in mass-weighted Cartesians, $\mathbf{M}^{-1/2}\boldsymbol{R}$ (FUKUI [1981]). A straightforward steepest-descent algorithm is often sufficient (MILLER, HANDY and ADAMS [1980], MILLER [1983], FIELD [1999]) but other algorithms have been devised if very smooth paths are desired (see, for example, DAVIS, WALES and BERRY [1990] and GONZALEZ and SCHLEGEL [1989]).

While the two-step procedure described above is suitable for finding the reaction paths for small systems, other methods, most of which do not require the use of second derivatives, have been developed for locating transition paths in systems with many degrees of freedom. A characteristic of many of these methods is that they take as input reactant and product structures, build a 'chain' of intermediate structures between them and then geometry optimize these structures until they lie on the reaction path. An early algorithm of this type was the synchronous transit algorithm of HALGREN and LIPSCOMB [1977]. A more recent series of algorithms have been developed by Elber and co-workers. These include the self-avoiding walk algorithm in which a discretized version of the average potential energy along the path, $\langle U\rangle$, is optimized (ELBER and KARPLUS [1987], CZERMINSKI and ELBER [1990]):

$$\langle U\rangle = \frac{\int_R^P dl(\boldsymbol{R})\,U(\boldsymbol{R})}{\int_R^P dl(\boldsymbol{R})}. \tag{4.3}$$

Here R and P refer to the reactant and product structures, respectively, and l is the distance along the path (which is a function of the atomic coordinates, $\boldsymbol{R}$).

The early chain-type algorithms were reasonably effective but they had some problems, including limited radii of convergence and the tendency to produce nonsmooth paths. To alleviate these problems Elber and co-workers introduced a couple of complementary algorithms to further refine the path once an initial reaction path had been determined (ULITSKY and ELBER [1990], CHOI and ELBER [1991]). A notable recent development is the nudged-elastic band algorithm that takes many elements of the Elber algorithms but considerably enhances the stability and the convergence rate of the algorithm (HENKELMAN, JÓHANNESSON and JÓNSSON [2000], HENKELMAN and JÓNSSON [2000]). The objective function of Eq. (4.3) is also not the only one that leads to reasonable reaction paths. Thus, for example, both HUO and STRAUB [1997] and ELBER and SHALLOWAY [2000] have introduced functions that allow average paths at nonzero temperature to be determined. In the work of Huo and Straub, the potential energy, U, in the integral of Eq. (4.3) is replaced by the quantity $\exp(U/k_B T)$.

4.3. Calculating the TST rate constant

As we saw in Section 4.1, the TST rate constant (Eq. (4.1)) can be expressed as the product of two factors, both of which is a configurational average. The first term is the average forward flux at the transition state surface. It can be evaluated as it appears in Eq. (4.1) by fixing the system at the transition state and then evaluating the average from a molecular dynamics simulation. Alternatively, the expression can be manipulated further by integrating over the momentum degrees of freedom to give the expression $\sqrt{k_B T/2\pi m_{\mathrm{eff}}}$ where m_{eff} is an effective or reduced mass for motion along the reaction coordinate:

$$m_{\mathrm{eff}} = \left[\frac{\left\langle \sqrt{\sum_{\alpha=1}^{N_{\mathrm{atoms}}} \frac{1}{M_\alpha} \left(\frac{\partial q}{\partial \boldsymbol{R}_\alpha} \right)^2} \delta(q - q^\ddagger) \right\rangle}{\langle \delta(q - q^\ddagger) \rangle} \right]^{-2}. \tag{4.4}$$

The effective mass can be evaluated as a configurational average too but the assumption is often made that its value is only weakly dependent on configuration and so it can be determined for a single structure that is representative of the transition state, such as a saddle point.

The second term in Eq. (4.1) is the ratio of the equilibrium populations of the system in the transition and the reactant states. It corresponds to the free-energy term of Eq. (4.2) and so allows the free-energy of activation of a reaction to be written as:

$$G^\ddagger = -k_B T \ln\left[\frac{\rho(q^\ddagger)}{\int_R dq\, \rho(q)} \right]. \tag{4.5}$$

The quantity $\mathcal{W}(q) = -k_B T \ln[\rho(q)]$ is called the potential of mean force (PMF) and gives the reversible work (or free-energy) required to displace the system along the reaction coordinate. In principle, the PMF could be calculated directly from a molecular dynamics or Monte Carlo simulation that samples the correct thermodynamic ensemble. However, in the cases that we are considering for which the energetic barrier to reaction is high, the population of the transition state relative to that of the reactant state will be extremely small. This means that very long simulations, impractically long in most cases, would be needed to adequately sample the transition state region to obtain a statistically significant estimate of the PMF.

To overcome this problem a number of techniques exist which either enhance sampling in various regions of phase space or allow the determination of the free energy of the system as a function of certain system parameters, such as the geometrical variables that define the reaction coordinate. Umbrella sampling is an example of the first type of technique and involves performing a series of simulations with constraint potentials that confine the system to different regions of configuration space. The bias introduced by the constraint potentials is then removed after the simulations when the probability density, $\rho(q)$, is constructed from the individual, biased probability densities obtained from the separate simulations. Examples of the second type of technique are the thermodynamic integration and statistical perturbation methods that rigidly constrain the system to be at a certain value of the reaction coordinate and then determine the free-energy difference as the value is changed. All these methods work well but they must be applied with care – in particular, simulations of sufficiently long duration must be performed for accurate values of the free energies to be determined. For fuller details about these techniques, readers are referred to the references by BEVERIDGE and DI CAPUA [1989], FIELD [1999], JORGENSEN [1989], KOLLMAN [1993], MARK [1998], ROUX [1995] and STRAATSMA and MCCAMMON [1992].

4.4. *Evaluating the transmission coefficient*

The transmission coefficient is the ratio of the true rate constant to the TST rate constant which, in terms of the correlation functions of Eqs. (2.22) and (2.24), can be written as:

$$\kappa \sim \frac{k(\tau_{\text{mol}})}{k(0^+)}, \tag{4.6}$$

where the plateau value of the correlation function is assumed to occur at $t \sim \tau_{\text{mol}}$.

Calculation of κ requires finding the value of the plateau for the (normalized) reactive flux correlation function displayed in Fig. 2.2 and is equivalent to determining the fraction of trajectories, originating in the reactant well, that cross the transition state and remain in the product well. There are a number of different approaches for doing this, but as they are mostly similar, only one will be outlined here. It is possible to rewrite the expression for the correlation function as:

$$k(t) \propto \big\langle \chi_{\text{B}}\big(q(t)\big)\big\rangle_{+} - \big\langle \chi_{\text{B}}\big(q(t)\big)\big\rangle_{-}. \tag{4.7}$$

The subscripts $+$ and $-$ on the angled brackets imply an averaging but using initial conditions for the position and velocity (or momentum) degrees of freedom in the system drawn from the probability distributions, P_+ and P_-, defined as follows:

$$P_{\pm} \propto \dot{q}\delta\big(q - q^{\ddagger}\big)|\dot{q}|\Theta(\pm\dot{q})\exp[-H/k_{\text{B}}T]. \tag{4.8}$$

The function P_+ forces the system to be at the transition state initially with a positive velocity that points towards the product well. The function P_- acts similarly except that the velocity points towards the reactant well.

Practically, the averaging of Eq. (4.7) is performed by starting off a large number of trajectories for the system at the transition state and with momenta that satisfy the functions of Eq. (4.8). The equations of motion for each trajectory are then integrated for a length of time that it is estimated will be sufficient for each trajectory to have committed itself to either the reactant or the product well. The average is then simply the fraction of trajectories with velocities pointing to the product well that remain in the product well minus the fraction of trajectories with velocities pointing to the reactant well that end up in the product well. This scheme works well and has been used extensively for studying complex systems (see, for example, the application to an enzyme by NERIA and KARPLUS [1997]). It can, however, be expensive to implement as many trajectories can be needed to obtain a statistically-reliable estimate of κ. In certain regimes, where there are many recrossings or where it takes a long time for the trajectories to become trapped on one side of the transition state, the method can become impractical. Alternative algorithms have been developed to cope with such circumstances but due to space limitations readers are urged to refer to STRAUB [2001] for further details.

4.5. *Transition path sampling*

The discussion of Section 2 and of the preceding part of this section has focussed on the calculation of reaction rates using the reactive flux formalism in which a reaction coordinate is defined for the reaction process, a transition state is identified at some point along the reaction coordinate and, finally, the various terms of Eqs. (4.1) and (4.6) are calculated. Such a procedure is satisfactory as long as a suitable set of variables can be found to describe the reaction coordinate and transition state. This is quite readily done, to reasonable accuracy, for simple reactions or for systems with only a few degrees of freedom by using the methods described in Section 4.2 which locate the saddle points and the associated paths that lead between reactants and products. In complex systems, however, where there can be many saddle points and paths of similar energy, these approaches are much less effective and, hence, determination of the rate constant becomes much more difficult.

To circumvent these problems, a new class of methods for investigating transitions in complex systems has recently been developed. These methods, or ones like them, are very exciting and likely to change radically the way reactions are studied in the future. The original idea was probably due to PRATT [1986] but most of the recent advances have been made by BOLHUIS, CHANDLER, DELLAGO and GEISSLER [2002]. These methods dispense with the notion of a transition state and, instead, aim to generate a representative set of transition paths that go between reactant and product configurations. Assuming that it is possible to sample a statistically significant number of trajectories and that these trajectories are consistent with a particular dynamics and thermodynamic ensemble, then dynamical quantities, such as rate constants, may be determined directly.

To give a flavor of what the methods entail, consider a transition path going between states A and B and consisting of $L + 1$ different configurations of the system. The Ith structure is specified by values of its position, $\boldsymbol{R}_I$, and its momenta, $\boldsymbol{P}_I$, and the time at which it occurs, τ_I. τ_I is equal to ΔI if the timestep between configurations is Δ. If neighboring configurations are linked by a Markovian transition probability, $p(I \rightarrow I+1)$, the probability for the whole path is given by:

$$\exp\left[-\frac{E_0}{k_B T}\right] \prod_{I=0}^{L-1} p(I \rightarrow I+1), \tag{4.9}$$

where E_0 is the total energy of the first structure. Note that it has been assumed that the structures are drawn from the canonical ensemble (hence, the Boltzmann factor) and that the transition probability conserves this distribution. To ensure that the trajectories are transition paths between the two states it is necessary to add terms corresponding to the characteristic functions for each of the states (Eqs. (2.15) and (2.16)). Thus, the probability of a path forced to start in state A is:

$$\mathcal{P}_A(\boldsymbol{R}_L) = \chi_A(\boldsymbol{R}_0) \exp\left[-\frac{E_0}{k_B T}\right] \prod_{I=0}^{L-1} p(I \rightarrow I+1), \tag{4.10}$$

whereas the probability of a path starting in state A and finishing in state B is:

$$\mathcal{P}_{AB}(\boldsymbol{R}_L) = \chi_A(\boldsymbol{R}_0) \exp\left[-\frac{E_0}{k_B T}\right] \prod_{I=0}^{L-1} p(I \rightarrow I+1) \chi_B(\boldsymbol{R}_L). \tag{4.11}$$

Once these probabilities have been defined, it is possible to generate a 'statistical mechanics' of transition paths, that is analogous in many respects to that for individual configurations in phase space, and from which expressions for time correlation functions (such as that in Eq. (2.21)) arise naturally.

Chandler and co-workers have formulated this method for different ensembles and for systems with different dynamics and they have discussed in detail how to sample effectively the transition path ensemble and calculate various dynamical quantities, such as rate constants, from it (DELLAGO, BOLHUIS, CSAJKA and CHANDLER [1998], BOLHUIS, DELLAGO and CHANDLER [1998], DELLAGO, BOLHUIS and CHANDLER [1999]). They have also applied the method to a number of different transition processes including a conformational isomerization reaction in the alanine dipeptide, which is a molecule that is often used to model the peptide bond in proteins (BOLHUIS, DELLAGO and CHANDLER [2000]), and autoionization processes in liquid water (GEISSLER, DELLAGO, CHANDLER, HUTTER and PARRINELLO [2001]). This last study was particularly interesting because it combined two of the most 'advanced' technologies available at the moment for studying reactions – molecular dynamics simulations with *ab initio* DFT QM potentials and transition path sampling. Work on similar approaches has also been done by other workers including by ZUCKERMAN and WOOLF [1999].

5. Quantum algorithms

The techniques that have been discussed so far in this chapter have assumed that the dynamics of the nuclei can be dealt with classically and QM theory has been employed solely for the calculation of the PES of the reacting system. At present, QM algorithms for the determination of rate constants in condensed-phase systems are less well developed and more complicated than their classical counterparts, but it is known that quantum dynamical effects are important and so some effort needs to be made to estimate them. This section highlights some currently active areas of research into quantum dynamical algorithms but makes no attempt at an exhaustive survey. Instead, interested readers are encouraged to consult some of the references cited below.

Perhaps the most straightforward way of including quantum effects (but not those due to spin) is with the path-integral method that can be used to calculate the equilibrium properties of a system. It is based on the observation that there is an isomorphism between a discretized Feynman path integral representation of the equilibrium QM density operator and a classical system in which the quantum particles are represented by polymers of classical particles (FEYNMAN and HIBBS [1965], CHANDLER and WOLYNES [1981]). This makes it a relatively easy method to implement in an existing classical molecular dynamics program because the simulation procedure is the same but the potential is changed as follows:

$$U_{\mathrm{PI}} = \sum_{\alpha=1}^{N_{\mathrm{atoms}}} \frac{P m_\alpha (k_{\mathrm{B}}T)^2}{2\hbar^2} \sum_{p=1}^{P} \left(\boldsymbol{R}_\alpha^{(p)} - \boldsymbol{R}_\alpha^{(p+1)}\right)^2 + \frac{1}{P} \sum_{p=1}^{P} U_{\mathrm{classical}}\left(\boldsymbol{R}_1^{(p)}, \ldots, \boldsymbol{R}_{N_{\mathrm{atoms}}}^{(p)}\right). \tag{5.1}$$

Here P is the number of beads in each polymer, m_α is the mass of atom α, and $\boldsymbol{R}_\alpha^{(p)}$ is the position of the pth bead of atom α (noting that $\boldsymbol{R}_\alpha^{(P+1)} = \boldsymbol{R}_\alpha^{(1)}$). $U_{\mathrm{classical}}$ is the normal, classical potential of the system which will be a function of the pth copy of all the quantum particles. The path-integral approach has been used successfully in many cases (see, for example, FIELD [2002] for references to applications to enzymatic reactions) but it is only valuable as a way of calculating the ensemble averages for the corresponding quantum system. Extensions of the approach to allow approximate dynamical quantities to be determined from simulations have been developed, the most notable being the path-integral centroid molecular dynamics method (see, for example, WARSHEL and CHU [1990] and VOTH [1996]).

The quantum mechanical equivalent of the classical reactive flux correlation function (Eq. (2.23)) has been known for some time. It was first derived by YAMAMOTO [1960] with later developments by a number of workers (MILLER [1974], MILLER, SCHWARTZ and TROMP [1983]). Unfortunately, applications of the formalism to real condensed-phase systems have been very limited due to its complexity although some recent work may change this (CHAKRABARTI, CARRINGTON and ROUX [1998], KOHEN and TANNOR [2000]). In the light of these problems, a very large range of approximate methodologies have been devised. These include the centroid path-integral methods mentioned above as well as methods that add quantum corrections to classical rate constants (see, for example, TRUHLAR [1998] and the references therein). Another area of active research are the surface-hopping methods that aim to study the dynamics of a system in which transitions between different electronic states can occur. Recent work in this area has been done by SHOLL and TULLY [1998], FANG and HAMMES-SCHIFFER [1999] and HACK, WENSMANN, TRUHLAR, BEN-NUN and MARTÍNEZ [2001].

6. Challenges and perspectives

The simulation of chemical reactions is a very challenging task that requires the synthesis of a number of simulation technologies from fields as diverse as molecular modeling, quantum chemistry and statistical physics. Much has been achieved and it is now possible to obtain at least qualitative insights into how many reactions occur in the condensed phase. Much, however, remains to be done. Areas of improvement are easy to identify and include (i) more accurate quantum chemical descriptions of reacting systems, whether through more precise DFT methods or faster wavefunction-based approaches, (ii) better sampling of configuration space for free-energy and transition-path-sampling simulations and (iii) effective quantum-dynamical simulation algorithms. All-in-all, though, the outlook is bright – algorithmic advances combined with increased computer power mean that precise simulation of reactions will become more straightforward. This is indeed fortunate if molecular simulation approaches are to contribute fully in the design of new materials and molecules.

Acknowledgments

The author would like to thank the Institut de Biologie Structurale – Jean-Pierre Ebel, the Commissariat à l'Energie Atomique and the Centre National de la Recherche Scientifique for financial support.

References

ALBRIGHT, T.A., J.K. BURDETT and M.-H. WHANGBO (1985), *Orbital Interactions in Chemistry* (Wiley-Interscience, New York).

ALLEN, M.P. and D.J. TILDESLEY (1987), *Computer Simulations of Liquids* (Oxford University Press, Oxford).

AMARA, P. and M.J. FIELD (1999), Hybrid potentials for large molecular systems, in: J. Leszczynski, ed., *Computational Molecular Biology* (Elsevier, Amsterdam) 1–33.

AMARA, P., A. VOLBEDA, J. FONTECILLA-CAMPS and M.J. FIELD (1999), A hybrid density functional theory/molecular mechanics study of nickel–iron hydrogenase: investigation of the active site redox states, *J. Am. Chem. Soc.* **121**, 4468–4477.

ANDERSON, J.B. (1995), Predicting rare events in molecular dynamics, *Adv. Chem. Phys.* **91**, 381–431.

BAKER, J. (1986), An algorithm for the location of transition states, *J. Comput. Chem.* **7**, 385–395.

BELL, S. and J.S. CRIGHTON (1984), Locating transition states, *J. Chem. Phys.* **80**, 2464–2475.

BERNE, B.J. (1985), Molecular dynamics and Monte Carlo simulations of rare events, in: J.U. Brackhill and B.I. Cohen, eds., *Multiple Time Scales* (Academic Press, New York) 419–436.

BERNE, B.J., M. BORKOVEC and J.E. STRAUB (1988), Classical and modern methods in reaction rate theory, *J. Phys. Chem.* **92**, 3711–3725.

BEVERIDGE, D.L. and F.M. DI CAPUA (1989), Free energy via molecular simulation: applications to biomolecular systems, *Annu. Rev. Biophys. Biophys. Chem.* **18**, 431–492.

BOLHUIS, P.G., C. DELLAGO and D. CHANDLER (1998), Sampling ensembles of deterministic transition pathways, *Faraday Discussions* **110**, 421–436.

BOLHUIS, P.G., C. DELLAGO and D. CHANDLER (2000), Reaction coordinates of biomolecular isomerization, *Proc. Natl. Acad. Sci.* **97**, 5877–5882.

BOLHUIS, P.G., D. CHANDLER, C. DELLAGO and P.L. GEISSLER (2002), Transition path sampling: throwing ropes over rough mountain passes, in the dark, *Ann. Rev. Phys. Chem.* **53**, 291–318.

CANCÈS, E., M. DEFRANCESCHI, W. KUTZELNIGG, C. LE BRIS and Y. MADAY (2003), Computational quantum chemistry: A primer, in: P.G. Ciarlet and C. Le Bris, eds., *Computational Chemistry*, Handbook of Numerical Analysis, Vol. X (Elsevier, Amsterdam) 3–270.

CEPERLEY, D.M. and L. MITAS (1996), Quantum Monte Carlo methods in chemistry, *Adv. Chem. Phys.* **93**, 1–38.

CERJAN, C.J. and W.H. MILLER (1981), On finding transition states, *J. Chem. Phys.* **75**, 2800–2806.

CHAKRABATI, N., T. CARRINGTON, JR. and B. ROUX (1998), Rate constants in quantum mechanical systems: a rigorous and practical path-integral formulation for computer simulations, *Chem. Phys. Lett.* **293**, 209–220.

CHANDLER, D. (1978), Statistical mechanics of isomerization dynamics in liquids and the transition state approximation, *J. Chem. Phys.* **68**, 2959–2970.

CHANDLER, D. (1987), *Introduction to Modern Statistical Mechanics* (Oxford University Press, Oxford).

CHANDLER, D. (1998), Barrier crossings: classical theory of rare but important events, in: B.J. Berne, G. Ciccotti and D.J. Coker, eds., *Classical and Quantum Dynamics in Condensed-Phase Simulations* (World Scientific, Singapore) 3–23.

CHANDLER, D. and P.G. WOLYNES (1981), Exploiting the isomorphism between quantum theory and classical statistical mechanics of polyatomic fluids, *J. Chem. Phys.* **74**, 4078–4095.

CHOI, C. and R. ELBER (1991), Reaction path study of helix formation in tetrapeptides: effect of side chains, *J. Chem. Phys.* **94**, 751–760.

COOPER, D.L., J. GERRATT and M. RAIMONDI (1987), Modern valence bond theory, *Adv. Chem. Phys.* **69**, 319–397.

CZERMINSKI, R. and R. ELBER (1990), Self-avoiding walk between two fixed points as a tool to calculate reaction paths in large molecular systems, *Int. J. Quant. Chem.* **24**, 167–186.

DAVIS, H.L., D.J. WALES and R.S. BERRY (1990), Exploring potential energy surfaces via transition state calculations, *J. Chem. Phys.* **92**, 4308–4319.

DELLAGO, C., P.G. BOLHUIS and D. CHANDLER (1999), On the calculation of rate constants in the transition path ensemble, *J. Chem. Phys.* **110**, 6617–6625.

DELLAGO, C., P.G. BOLHUIS, F.S. CSAJKA and D. CHANDLER (1998), Transition path sampling and the calculation of rate constants, *J. Chem. Phys.* **108**, 1964–1977.

DEWAR, M.J.S. and B.M. O'CONNOR (1987), Testing *ab initio* procedures: the 6-31G* model, *Chem. Phys. Letts.* **138**, 141–145.

DEWAR, M.J.S. and W. THIEL (1977), Ground states of molecules. 38. The MNDO method. Approximations and parameters, *J. Am. Chem. Soc.* **99**, 4899–4907.

DEWAR, M.J.S., E.G. ZOEBISCH, E.F. HEALY and J.J.P. STEWART (1985), AM1: A new general purpose quantum mechanical molecular model, *J. Am. Chem. Soc.* **107**, 3902–3909.

ELBER, R. and M. KARPLUS (1987), A method for determining reaction paths in large molecules: application to myoglobin, *Chem. Phys. Letts.* **139**, 375–380.

ELBER, R. and D. SHALLOWAY (2000), Temperature dependent reaction coordinates, *J. Chem. Phys.* **112**, 5539–5545

ELSTNER, M., D. POREZAG, G. JUNGNICKEL, J. ELSNER, M. HAUGK, TH. FRAUENHEIM, S. SUHAI and G. SEIFERT (1998), Self-consistent-charge density-functional tight-binding method for simulations of complex materials properties, *Phys. Rev. B* **58**, 7261–7268.

FANG, J.-Y. and S. HAMMES-SCHIFFER (1999), Comparison of surface hopping and mean-field approaches for model proton transfer reactions, *J. Chem. Phys.* **110**, 11166–11175.

FEYNMAN, R.P. and A.R. HIBBS (1965), *Quantum Mechanics and Path Integrals* (McGraw-Hill, New York).

FIELD, M.J. (1993), The simulation of chemical reactions, in: W.F. van Gunsteren, P.K. Weiner and A.J. Wilkinson, eds., *Computer Simulation of Biomolecular Systems*, Vol. 2 (ESCOM, Leiden) 82–123.

FIELD, M.J. (1999), *A Practical Introduction to the Simulation of Molecular Systems* (Cambridge University Press, Cambridge).

FIELD, M.J. (2002), Simulating enzyme reactions: challenges and perspectives, *J. Comput. Chem.* **23**, 48–58.

FIELD, M.J., P.A. BASH and M. KARPLUS (1990), A combined quantum mechanical and molecular mechanical potential for molecular dynamics simulations, *J. Comput. Chem.* **11**, 700–733.

FOULKES, W.M.C. and R. HAYDOCK (1989), Tight-binding models and density-functional theory, *Phys. Rev. B* **39**, 12520–12536.

FRAUENHEIM, TH., G. SEIFERT, M. ELSTNER, Z. HAJNAL, G. JUNGNICKEL, D. POREZAG, S. SUHAI and R. SCHOLZ (2000), A self-consistent charge density-functional based tight-binding method for predictive materials simulations in physics, chemistry and biology, *Phys. Stat. Sol. B* **217**, 41–62.

FUKUI, K. (1981), The path of chemical reactions – the IRC approach, *Acc. Chem. Res.* **14**, 363–368.

GALLUP, G.A., R.L. VANCE, J.R. COLLINS and J.M. NORBECK (1982), Practical valence bond calculations, *Adv. Quant. Chem.* **16**, 229–272.

GAO, J. (1995), Methods and applications of combined quantum mechanical and molecular mechanical potentials, in: K.B. Lipkowitz and D.B. Boyd, eds., *Reviews in Computational Chemistry*, Vol. 7 (VCH, New York) 119–185.

GEISSLER, P.L., C. DELLAGO, D. CHANDLER, J. HUTTER and M. PARRINELLO (2001), Autoionization in liquid water, *Science* **291**, 2121–2124.

GOEDECKER, S. (1999), Linear-scaling electronic structure methods, *Rev. Mod. Phys.* **71**, 1085–1123.

GONZALEZ, C. and H.B. SCHLEGEL (1989), An improved algorithm for reaction path following, *J. Chem. Phys.* **90**, 2154–2161.

GROSSMAN, J.C., W.A. LESTER, JR. and S.G. LOUIE (2000), Quantum Monte Carlo and density functional theory characterization of the 2-cyclopentenone and 3-cyclopentenone formation from $O(^3P)$ + cyclopentadiene, *J. Am. Chem. Soc.* **122**, 705–711.

HACK, M.D., A.M. WENSMANN, D.G. TRUHLAR, M. BEN-NUN and T.J. MARTÍNEZ (2001), Comparison of full multiple spawning, trajectory surface hopping, and converged quantum mechanics for electronically nonadiabatic dynamics, *J. Chem. Phys.* **115**, 1172–1186.

HALGREN, T.A. and W.N. LIPSCOMB (1977), The synchronous-transit method for determining reaction pathways and locating molecular transition states, *Chem. Phys. Letts.* **19**, 225–232.

HENKELMAN, G., G. JÓHANNESSON and H. JÓNSSON (2000), Methods for finding saddle points and minimum energy paths, in: S.D. Schwartz, ed., *Progress on Theoretical Chemistry and Physics* (Kluwer Academic Publishers).

HENKELMAN, G. and H. JÓNSSON (2000), Improved tangent estimate in the NEB method for finding minimum energy paths and saddle points, *J. Chem. Phys.* **113**, 9978–9985.

HUO, S. and J.E. STRAUB (1997), The MaxFlux algorithm for calculating variationally optimized reaction paths for conformational transitions in many body systems at finite temperature, *J. Chem. Phys.* **107**, 5000–5006.

HYNES, J.T. (1985), Chemical reaction dynamics in solution, *Ann. Rev. Phys. Chem.* **36**, 573–597.

JÓHANNESSON, G.H. and H. JÓNSSON (2001), Optimization of hyperplanar transition states, *J. Chem. Phys.* **115**, 9644–9656.

JORGENSEN, W.L. (1989), Free energy calculations: a breakthrough for modeling organic chemistry in solution, *Acc. Chem. Res.* **22**, 184–189.

KARPLUS, M. (2000), Aspects of protein reaction dynamics: deviations from simple behavior, *J. Phys. Chem. B* **104**, 11–27.

KOCH, W. and M.C. HOLTHAUSEN (2000), *A Chemist's Guide to Density Functional Theory* (Wiley VCH, New York).

KOHEN, D. and D.J. TANNOR (2000), Phase space approach to dissipative molecular dynamics, *Adv Chem. Phys.* **111**, 219–398.

KOHN, W. (1999), Noble lecture: electronic structure of matter – wavefunctions and density functionals, *Rev. Mod. Phys.* **71**, 1253–1266.

KOLLMAN, P.A. (1993), Free energy calculations: applications to chemical and biochemical phenomena, *Chem. Rev.* **93**, 2395–2417.

MARK, A.E. (1998), Free energy perturbation calculations, in: P. von Rague Schleyer, ed., *Encyclopedia of Computational Chemistry* (Wiley, New York) 1211–1216.

MILLER, W.H. (1974), Quantum mechanical transition state theory and a new semiclassical model for reaction rate constants, *J. Chem. Phys.* **61**, 1823–1834.

MILLER, W.H. (1983), Reaction-path dynamics for polyatomic systems, *J. Phys. Chem.* **87**, 3811–3819.

MILLER, W.H., N.C. HANDY and J.E. ADAMS (1980), Reaction path Hamiltonian for polyatomic molecules, *J. Chem Phys.* **72**, 99–112.

MILLER, W.H., S.D. SCHWARTZ and J.W. TROMP (1983), Quantum mechanical rate constants for biomolecular reactions, *J. Chem. Phys.* **79**, 4889–4898.

MONARD, G. and K.M. MERZ (1999), Combined quantum mechanical/molecular mechanical methodologies applied to biomolecular systems, *Acc. Chem. Res.* **32**, 904–911.

NERIA, E. and M. KARPLUS (1997), Molecular dynamics of an enzyme reaction: proton transfer in TIM, *Chem. Phys. Letts.* **267**, 23–30.

PARR, R.G. and W. YANG (1989), *Density-Functional Theory of Atoms and Molecules* (Oxford University Press, Oxford).

POPLE, J.A. (1999), Noble lecture: quantum chemical models, *Rev. Mod. Phys.* **71**, 1267–1274.

POREZAG, D., TH. FRAUENHEIM, TH. KÖHLER, G. SEIFERT and R. KASCHNER (1995), Construction of tight-binding-like potentials on the basis of density-functional theory: application to carbon, *Phys. Rev. B* **51**, 12947–12957.

PRATT, L.R. (1986), A statistical method for identifying transition states in high dimensional problems, *J. Chem. Phys.* **85**, 5045–5048.

ROUX, B. (1995), The calculation of the potential of mean force using computer simulations, *Comput. Phys. Commun.* **91**, 275–282.

SCHENTER, G.K., G. MILLS and H. JÓNSSON (1994), Reversible work based quantum transition state theory, *J. Chem. Phys.* **101**, 8964–8971.

SHOLL, D.S. and J.C. TULLY (1998), A generalized surface-hopping method, *J. Chem. Phys.* **109**, 7702–7710.

SIMONS, J., P. JORGENSON, H. TAYLOR and J. OZMENT (1983), Walking on potential energy surfaces, *J. Chem. Phys.* **87**, 2745–2753.

SIMONSON, T. (2001), Macromolecular electrostatics: continuum models and their growing pains, *Curr. Opin. Struct. Biol.* **11**, 243–252.

STEWART, J.J.P. (1989a), Optimization of parameters for semiempirical methods I. Method, *J. Comput. Chem.* **10**, 209–220.

STEWART, J.J.P. (1989b), Optimization of parameters for semiempirical methods I. Applications, *J. Comput. Chem.* **10**, 221–264.

STRAATSMA, T.P. and J.A. MCCAMMON (1992), Computational alchemy, *Annu. Rev. Phys. Chem.* **43**, 407–435.

STRAUB, J. (2001), Reaction rates and transition pathways, in: O.M. Becker, A.D. MacKerell, Jr., B. Roux and M. Watanabe, eds., *Computational Biochemistry and Biophysics* (Marcel Dekker, New York) 199–220.

SZABO, A. and N. OSTLUND (1989), *Modern Quantum Chemistry: Introduction to Advanced Electronic Structure Theory* (McGraw-Hill, New York).

THOMAS, A., D. JOURAND, C. BRET, P. AMARA and M.J. FIELD (1999), Is there a covalent intermediate in the viral neuraminidase reaction? A hybrid-potential free-energy study, *J. Am. Chem. Soc.* **121**, 9693–9702.

TOMASI, J. and M. PERSICO (1994), Molecular interactions in solution: an overview of methods based on continuous distributions of the solvent, *Chem. Rev.* **94**, 2027–2094.

TRUHLAR, D.G. (1998), Chemical reaction theory: summarizing remarks, *Faraday Discuss.* **110**, 521–535.

ULITSKY, A. and R. ELBER (1990), A new technique to calculate steepest descent paths in flexible polyatomic systems, *J. Chem. Phys.* **92**, 1510–1511.

VOTH, G.A. (1996), Path-integral centroid methods in quantum statistical mechanics and dynamics, *Adv. Chem. Phys.* **93**, 135–218.

WALES, D.J. (1994), Rearrangements of 55-atom Lennard-Jones and $(C_{60})_{55}$ clusters, *J. Chem. Phys.* **101**, 3750–3762.

WARSHEL, A. (1991), *Computer Modeling of Chemical Reactions in Enzymes and Solutions* (Wiley, New York).

WARSHEL, A. and Z.-T. CHU (1990), Quantum corrections for rate constants of diabatic and adiabatic reactions in solutions, *J. Chem. Phys.* **93**, 4003–4015.

WARSHEL, A. and M. LEVITT (1976), Theoretical studies of enzymic reactions: dielectric, electrostatic and steric stabilization of the carbonium ion in the reaction of lysozyme, *J. Mol. Biol.* **103**, 227–249.

YAMAMOTO, T. (1960), Quantum statistical mechanical theory of the rate of exchange chemical reactions in the gas phase, *J. Chem. Phys.* **33**, 281–289.

ZUCKERMAN, D.M. and T.B. WOOLF (1999), Dynamic reaction paths and rates through importance-sampled stochastic dynamics, *J. Chem. Phys.* **111**, 9475–9484.

Biomolecular Conformations Can Be Identified as Metastable Sets of Molecular Dynamics

Christof Schütte, Wilhelm Huisinga

Institute of Mathematics II, Department of Mathematics and Computer Science, Free University (FU) Berlin, Germany

E-mail addresses: schuette@math.fu-berlin.de (C. Schütte), huisinga@math.fu-berlin.de (W. Huisinga)

URLs: www.math.fu-berlin.de/~biocomp (C. Schütte), www.math.fu-berlin.de/~huisinga (W. Huisinga)

1. Introduction

The biochemical functions of many important biomolecules result from their *dynamical* properties, particularly from their ability to undergo so-called *conformational transitions* (cf. ZHOU, WLODEC and MCCAMMON [1998]). In a conformation, the large scale geometric structure of the molecule is understood to be conserved, whereas on smaller scales the system may well rotate, oscillate or fluctuate. Furthermore, transitions between conformations are rare events or, in other words, a typical trajectory of a molecular system stays for long periods of time within the conformation, while exits are long-term events. Hence, the term conformation includes both geometric and dynamical aspects. From the geometrical point of view, conformations are understood to represent all molecules with the same large scale geometric structure and may thus be identified with a subset of the state space. From the dynamical point of view, a conformation typically persists for long periods of time (compared to the fastest molecular motions) such that the associated subset of the state space is *metastable* and the resulting *macroscopic dynamical behavior* can be described as a flipping process between the metastable subsets.

Computational Chemistry
Special Volume (C. Le Bris, Guest Editor) of
HANDBOOK OF NUMERICAL ANALYSIS, VOL. X
P.G. Ciarlet (Editor)
© 2003 Elsevier Science B.V. All rights reserved

Understanding *conformation dynamics* – that is the statistics of the flipping process and the corresponding exit times as well as the actual transition paths between different conformations – is crucial to the understanding of biomolecular flexibility and activity. Prominent examples of conformation dynamics are the conformational changes accompanying the action of the muscle protein myosin, the light-induced conformational transition of the photo-receptor rhodopsin initializing the primary amplification cascade in vision, or the conformation conversion of human prions assumed to cause prion diseases.

The state-of-the-art biophysical explanation for the existence of conformations is as follows: The *free energy landscape* of a protein or peptide decomposes into particularly deep wells each containing huge numbers of local minima. These wells are separated by relatively large barriers – as measured on the scale of the thermal energy $k_B T$ – from each other and represent different metastable conformations. The hierarchy of barrier heights induces a hierarchy of metastable conformations (ELBER and KARPLUS [1987], FRAUENFELDER, STEINBACH and YOUNG [1989], FRAUENFELDER and MCMAHON [2000]). The corresponding hierarchy of time scales observed for conformational transitions seems to confirm the biophysical explanation for the existence of conformations (NIENHAUS, MOURANT and FRAUENFELDER [1992]). However, the entire explanation depends on the concept of the free energy landscape, whose definition is typically based on the assumption that the conformational degrees of freedom are already known in advance. In other words, the model is of minor use if conformations and conformational degrees of freedom still have to be identified by simulations.

Mathematically, the dynamical aspect of conformations is based on the concept of metastability. In this paper we will pursue two characterizations of metastability. The *exit time approach* based on *exit rates* characterizes metastability of some subset by the property that a typical trajectory will only exit the subset on macroscopically long time scales. The *ensemble dynamics approach* based on *transition probabilities* characterizes metastability of some subset in statistical terms in the following sense: the fraction of systems in an ensemble that exit from the subset during some (not necessarily long) given time span is significantly small in comparison to other subsets. It is one of the main goals of this article to discuss and compare the similarities and distinguishing aspects of these two concepts in detail (the entire Section 2 will be devoted to the conceptual differences).

We will see that the two characterizations of metastability can be formalized and studied within the unified mathematical framework of the *transfer operator approach* to metastability. This approach originated from the work of Dellnitz et al. on the identification of almost invariant sets of discrete dynamical systems with small random perturbations (DEUFLHARD, DELLNITZ, JUNGE and SCHÜTTE [1999], DELLNITZ and JUNGE [1999]) and it has been successfully applied to examine metastable behavior of deterministic Hamiltonian systems by DEUFLHARD, DELLNITZ, JUNGE and SCHÜTTE [1999]. By reformulating this idea in the context of biophysical models of molecular motion, Schütte et al. showed that biomolecular conformations can be identified via the "dominant" eigenvectors of the transfer operator associated with the dynamical model used (SCHÜTTE, FISCHER, HUISINGA and DEUFLHARD [1999], SCHÜTTE, HUISINGA and DEUFLHARD [2001], SCHÜTTE [1998], HUISINGA [2001]). It has been demonstrated that, for moderate size (bio)molecules, the eigenvectors of interest can be computed efficiently and allow to identify the desired conformations and the associated conformation dynamics in a unified setting based on simulation of the dynamical behavior of molecular systems (FISCHER, SCHÜTTE, DEUFLHARD and CORDES [2002], DEUFLHARD, HUISINGA, FISCHER and SCHÜTTE [2000], HUISINGA and SCHMIDT [2002]).

The literature on conformation dynamics is enormously rich, see, e.g., DEUFLHARD, HERMANS, LEIMKUHLER, MARK, REICH and SKEEL [1999], BERNE, CICCOTTI and COKER [1998]. However, the branch that deals with the *dynamical* aspects of conformations mainly contains approaches to the computational detection of transition paths between conformations and of the associated main transition coordinates, see, E, RAN and VANDEN-EIJNDEN [2002], FRAUENFELDER, STEINBACH and YOUNG [1989], BOLHUIS, DELLAGO, CHANDLER and GEISSLER [2001]. There are several approaches designed to bridge the time scale gap between realizably short trajectory simulations and significantly longer metastability periods of conformational substates. One example are approaches that exploit artificial accelerations of the dynamics, cf. GRUBMÜLLER [1995], HUBER, TORDA and VAN GUNSTEREN [1994]; another example is given by path integral approaches to long-term dynamics where transition paths are discretized in time using extremely large timesteps (OLENDER and ELBER [1996]).

The article will be organized as follows: First we will sharpen and conceptually complement the two characterizations of metastability (Section 2), then we will shortly summarize the different dynamical models designed to describe different aspects of the dynamical and statistical properties of molecules (Section 3). This will be followed by the presentation of the mathematical framework of the transfer operator approach to metastability (Sections 4 and 5). Within this framework we will reformulate the two concepts of metastability and justify in detail the key idea of the transfer operator approach. Moreover, it will be shown that the framework allows to incorporate almost all different dynamical models available. In Section 6 the theoretical level is left and the issues of practical realization are discussed: The concept of Galerkin discretization of transfer operators is studied which leads to the question of whether a discretization of the eigenvalue problem in huge dimensional state spaces will be possible without risking the increase of numerical effort beyond any tolerable amount. It is illustrated how this problem can be circumvented. Sections 7 and 8 conclude the article by demonstrating the application of the approach. In Section 7 the entire concept is illustrated by means of a simple but completely comprehensible test system whereas Section 8 is devoted to the application to a small oligonucleotide.

2. Conceptual preliminaries

Before we go into details about molecular dynamics, conformations, and metastability we first want to point out the fundamental principles of the approach to biomolecular conformations.

Let us assume for the remainder of this section that a mathematical model is available which, given an exact initial state, perfectly describes the true motion of the molecule under consideration in all necessary details. In general, this will be given

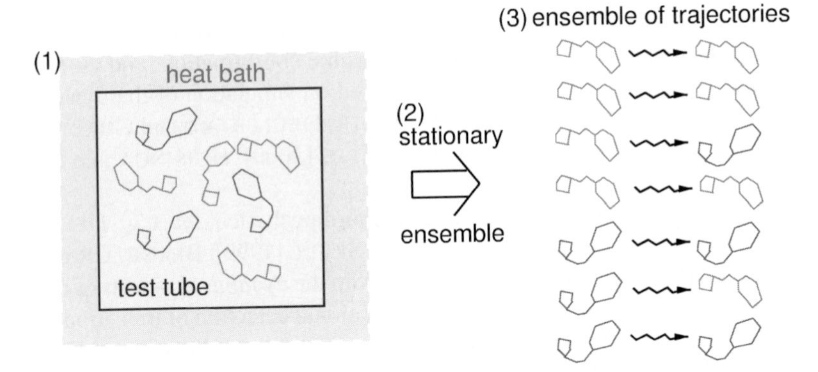

FIG. 2.1. Illustration of ensemble dynamics: (1) an ensemble of molecular systems embedded in a heat bath of constant temperature; (2) the usual assumption is that this can be modeled by some stationary probability distribution, e.g., the canonical density; (3) the dynamical behavior of each single molecule in the ensemble (here assumed to be modeled accurately by some dynamical system with flow Φ^{τ}) induces dynamical fluctuations within the ensemble without effect on the stationary distribution.

by some (discrete or continuous, deterministic or stochastic) dynamical system. In the deterministic setting, the corresponding initial value problem is thought to model the evolution of the state of a *single molecule*; its exact solution will be called "trajectory" in the following. In the stochastic setting, we use the same interpretation and wording for every single pathwise realization. Trajectories have to be distinguished from their numerical realization, which will be called "simulation" or "numerical integration".

In the context of biochemical applications one is generally not interested in single isolated molecules but in certain *molecular ensembles* that, for instance, model the collection of many identical molecules in a living cell or in a test tube with certain side conditions like, e.g., constant temperature. The molecular ensemble is represented by a statistical distribution in molecular state space. If the ensemble is assumed to be stationary, the distribution does not change in time (see Fig. 2.1). Within this setting, the *transition probability* from some sub-ensemble A to some sub-ensemble B, both specified by some subset A and B of the state space, within a pre-described time span τ is given by the fraction of systems with initial state in A at $t = 0$ and final state in B at $t = \tau$. Built upon transition probabilities we may state the

Ensemble dynamics approach: Conformations are identified as sub-ensembles/subsets, for which the fraction of systems that exit during a prescribed observation time τ is significantly smaller than for other sub-ensembles/subsets.

In order to numerically *compute* the transition probabilities from sub-ensemble A to B one has to generate (i) a sample that represents the *sub-ensemble of initial states* in A, and (ii) a sample that represents the corresponding *sub-ensemble of trajectories* starting from these initial states, as illustrated in Fig. 2.1. Within this statistical setting the ensemble dynamics approach is able to capture conformation dynamics by considering only *short-term* trajectories, since measurements on ensembles already contain information about all possible states, and short-term trajectories over time spans

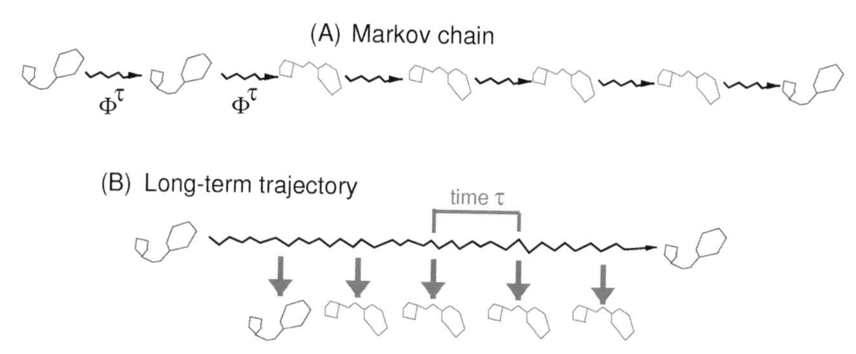

FIG. 2.2. Illustration of different algorithmic options for realizing a sample of the stationary ensemble under consideration and the induced sample the corresponding ensemble of trajectories: realization of ensemble and trajectories can be done by means of (A) specially designed Markov chains, or (B) time-τ pieces from a single ergodic long-term simulation of the corresponding dynamical system.

that are of the order of magnitude of the rapid conformational transition itself contain all transition paths from one conformation to another.

The ensemble dynamics approach has the advantage that it is based on a setting and requires information which is experimentally available: ensembles of short-term trajectories can be observed by means of femtochemistry (Nobel price 1999) (ZEWAIL [1995], ZEWAIL [1996]), a novel technique that permits us to observe the dynamical behavior of molecular systems in real-time. Boosted by the progress in laser technology, ultrashort light pulses can be generated with durations on the typical timescales of molecular vibrations, i.e., from picoseconds down to femtoseconds. The prototypical experiment follows the *pump-probe scenario*: A molecular ensemble that is initially prepared in some stationary state is excited by a first laser pulse ("*pump*") thus lifting the system into an excited state. Subsequently, a second laser pulse ("*probe*") is used to stimulate another transition (emission or absorption) that serves to generate an observable signal. By measuring the observable signal as a function of the time delay between pump and probe pulse the evolution of the system in its excited state can be monitored. Hence, pump-probe measurements allow to experimentally realize the ensemble dynamics approach.

In this chapter we will present two different algorithmic approaches to generate the data needed in steps (i) and (ii) from above. The two concepts are sketched in Fig. 2.2. On the one hand, we may use *any* method available to compute a sample that appropriately represents the stationary distribution. Given this sample one then evaluates the trajectories starting from any sample point. We will see later that the two steps may be combined into a single procedure by introducing an appropriate Markov chain. On the other hand, whenever the dynamical system under consideration is *ergodic*,[1] a single long-term trajectory represents the average behavior of molecules in the ensemble. Then, chopping the long-term trajectory into pieces of length τ will also do the job.

[1] The notion of ergodicity has several different meanings in physics and mathematics. The typical rough "definition" states that "time average equals ensemble average". We will introduce the precise meaning used in this article in Section 5.2.

This algorithmic option via long-term simulation should not be confused with another approach to conformation analysis, the

Exit time approach: Conformations are identified as subsets for which the exit time of a typical trajectory is extraordinary large in comparison to other subsets.

This approach is build upon the belief that the description of conformational transitions requires to start a long-term simulation and wait through the in general tremendously long period of time until a transition takes place. However, as we will see in the following this is not necessarily the case, and the notion of conformations in the ensemble dynamics and exit time approach in some sense turn out to be very similar.

Both approaches may exploit long-term *simulations* in the ergodic interpretation in order to generate ensembles of short-term trajectories only. Yet, the authors want to emphasize that long-term *trajectories* of the dynamical system under consideration are not necessarily needed. This is of utmost importance because the literature on predictability and sensitivity w.r.t. perturbations states that for any dynamical system and prescribed accuracy there is a certain maximal time T, up to which initial value problems are make sense. For time spans longer than T the deviation between trajectories of the dynamical system caused by perturbations may exceed the accuracy requirement. The nature of the perturbations to be considered depends on the actual application context: one may have to take into account the uncertainty of the initial value, or perturbations due to the numerical realization. The actual value of T depends on the properties of the dynamical system and on the nature of the perturbations and can be characterized by means of different estimates (e.g., by Lyapunov exponents, so-called condition numbers (DEUFLHARD and BORNEMANN [1994]), or predictability analysis (KLEEMAN [2001])). However, in the context of biomolecular dynamics all available estimates indicate that for any tolerable accuracy, the time T is many orders of magnitude smaller than the expected exit times from typical conformations. The same situation is encountered in the above mentioned pump-probe experiments, where the time spans between the pump and probe pulse are orders of magnitude smaller than typical exit times from some conformation. However, experimental observations of conformational transitions in real-time over milliseconds or even on longer scales are very limited and possible only indirectly.

Before proceeding to the algorithmic realization of the ensemble dynamics or exit time approach we may first introduce the most prominent types of dynamical systems presently discussed in the context of molecular dynamics. They can be classified in two main categories:

(MC1) Dynamical systems that are designed to model the precise motion of some molecular system, at least on short time scales.

(MC2) Dynamical systems that are designed to sample the state space of some molecular system w.r.t. some prescribed statistical distribution.

3. Description of dynamical behavior

The literature on the description of the dynamical behavior of molecular systems is extremely rich; they range from classical deterministic Hamiltonian models that try to cover the actual motion of each single molecule in the system to stochastic descriptions like Langevin dynamics or iterative schemes that only model artificial dynamics like most Markov chain Monte Carlo approaches.

3.1. Markov processes and transition functions

We now introduce the mathematical framework that subsumes both approaches, whether stochastic or deterministic.

Consider the state space $\mathbf{X} \subset \mathbf{R}^m$ for some $m \in \mathbf{N}$ equipped with the Borel σ-algebra $\mathcal{A}$ on $\mathbf{X}$.[2] The evolution of a single microscopic system is supposed to be given by a *homogeneous Markov process* $X_t = \{X_t\}_{t \in \mathbf{T}}$ in continuous or discrete time $\mathbf{T} = \mathbf{R}_0^+$ or $\mathbf{T} = \mathbf{N}$, respectively. We write $X_0 \sim \mu$, if the Markov process X_t is initially distributed according to the probability measure μ, i.e., if $\mathbf{P}[X_0 \in A] = \mu(A)$ for every $A \subset \mathbf{X}$. We use $X_0 = x$, if $X_0 \sim \delta_x$, where δ_x denotes the Dirac measure at x. The motion of X_t is given in terms of the *stochastic transition function* $p : \mathbf{T} \times \mathbf{X} \times \mathcal{B}(\mathbf{X}) \to [0, 1]$ according to

$$p(t, x, A) = \mathbf{P}[X_{t+s} \in A \mid X_s = x] \tag{3.1.1}$$

for every $t, s \in \mathbf{T}$, $x \in \mathbf{X}$ and $A \subset \mathbf{X}$. Hence, $p(t, x, A)$ describes the probability that the system moves from state x into the subset A within time t. The relation between a stochastic transition function and a homogeneous Markov process is one-to-one (MEYN and TWEEDIE [1993], Chapter 3). In the special case, where $p(t, x, A) = \delta_{\Phi(x,t)}(A)$, the Markov process is in fact a deterministic process, whose evolution is defined by the flow map $\Phi(x, t)$ in state space. Besides some more technical properties (see Appendix 9.1) the stochastic transition function fulfills the so-called Chapman–Kolmogorov equation

$$p(t + s, x, A) = \int_{\mathbf{X}} p(t, x, \mathrm{d}z) p(s, z, A), \tag{3.1.2}$$

that holds for every $t, s \in \mathbf{T}$, $x \in \mathbf{X}$ and $A \subset \mathbf{X}$ and represents the semigroup property of the Markov process. As a consequence, in the discrete time case $\mathbf{T} = \mathbf{N}$ it suffices to specify $p(x, \mathrm{d}y) = p(1, x, \mathrm{d}y)$, since the n-step transition probabilities $p^n(x, \mathrm{d}y) = p(n, x, \mathrm{d}y)$ are recursively determined by (3.1.2).

We say that the Markov process X_t admits an *invariant probability measure* μ, or μ is invariant w.r.t. X_t, if

$$\int_{\mathbf{X}} p(t, x, A) \mu(\mathrm{d}x) = \mu(A) \tag{3.1.3}$$

for every $t \in \mathbf{T}$ and $A \subset \mathbf{X}$ (MEYN and TWEEDIE [1993], Chapter 10). Note that the invariant probability measure needs not to be unique. A Markov process is called

---

[2] In the sequel every subset $C \subset \mathbf{X}$ is implicitly assumed to be measurable, i.e., we assume that additionally $C \in \mathcal{A}$ holds without further mentioning.

reversible w.r.t. an invariant probability measure μ if

$$\int_A p(t,x,B)\mu(\mathrm{d}x) = \int_B p(t,x,A)\mu(\mathrm{d}x) \tag{3.1.4}$$

for every $t \in \mathbf{T}$ and $A, B \subset \mathbf{X}$. If μ is unique, X_t is simply called reversible. For the special case of a stochastic transition function being absolutely continuous w.r.t. μ, reversibility reads $p(t,x,y) = p(t,y,x)$ for every $t \in \mathbf{T}$ and μ-a.e. $x, y \in \mathbf{X}$.

3.2. Model systems

We now turn to the most prominent examples in the context of molecular dynamics.

Let N denote the number of atoms of the system and $\Omega = \mathbf{R}^{3N}$ the position space, i.e., $q \in \Omega$ represents the vector of atomic position coordinates. Moreover, let $\xi \in \mathbf{R}^{3N}$ denote the vector of all conjugated momenta. Suppose that a differentiable potential energy function $V : \mathbf{R}^{3N} \to \mathbf{R}$ describing all interactions between the atoms is given. For each model system below we assume that the position space Ω belongs to one of the two fundamentally different cases:

Bounded systems: The potential energy function $V : \mathbf{R}^{3N} \to \mathbf{R}^{3N}$ is smooth, bounded from below, and satisfies $V \to \infty$ for $|q| \to \infty$. Such systems are called bounded, since the energy surfaces $\{(q,\xi) \in \mathbf{X}\colon\ H(q,\xi) = E\}$ are bounded subsets of Γ for every energy E.

Periodic systems: The position space Ω is some $3N$-dimensional torus and the potential energy function V is continuous on Ω and thus bounded. There is an intensive discussion concerning the question of whether V can also be assumed to be smooth as we will do herein, see SCHÜTTE [1998], Section 2, for details.

Both cases are typical for molecular dynamics applications. Periodic systems in particular include the assumption of periodic boundaries, which is by far the most popular modeling assumption for biomolecular systems.

Deterministic Hamiltonian system. The most prominent model for the dynamical behavior of molecular systems exploits classical Hamiltonian mechanics, i.e., atoms are described as mass points subject to forces that are generated by specified classical interaction potentials V. The dynamical behavior is described by some deterministic Hamiltonian system of the form

$$\dot{q} = M^{-1}\xi, \qquad \dot{\xi} = -\nabla_q V(q), \tag{3.2.1}$$

defined on the state space $\mathbf{X} = \mathbf{R}^{3N} \times \mathbf{R}^{3N}$ and M denoting the diagonal mass matrix.

Eq. (3.2.1) models an energetically closed system, whose total energy, given by the Hamiltonian

$$H(q,\xi) = \frac{1}{2}\xi^T M^{-1}\xi + V(q), \tag{3.2.2}$$

is preserved under the dynamics. For the sake of simplicity, we assume in the following that M is the identity matrix. The deterministic Hamiltonian system is typically seen as the embodiment of our class (MC1) in the context of molecular dynamics.

Let Φ^t denote the flow associated with the Hamiltonian system (3.2.1), i.e., the solution $x_t = (q_t, \xi_t)$ of (3.2.1) for the initial value $x_0 = (q_0, \xi_0)$ is given by $x_t = \Phi^t x_0$. Let $\mathbf{1}_C$ denote the characteristic function of the subset $C \subset \mathbf{X}$. Then, the stochastic transition function corresponding to (3.2.1) is given by

$$p(t, x, C) = \mathbf{1}_C(\Phi^t x) = \delta_{\Phi^t x}(C) \tag{3.2.3}$$

for every $t \in \mathbf{R}_0^+$ and $C \subset \mathbf{X}$. The Markov process $X_t = \{X_t\}_{t \in \mathbf{R}_0^+}$ induced by the stochastic transition function p coincides with the flow Φ^t; hence $X_t = \Phi^t x_0$ for the initial distribution $X_0 = x_0$.

It is well known that for every smooth function $\mathcal{F}: \mathbf{R} \to \mathbf{R}$ the probability measure $\mu(\mathrm{d}x) \propto \mathcal{F}(H)(x)\,\mathrm{d}x$ is invariant w.r.t. the Markov process X_t. The most prominent choice is the canonical density or *canonical ensemble*

$$f(x) \propto \exp(-\beta H(x))$$

for some constant $\beta > 0$ that can be interpreted as inverse temperature. The associated measure $\mu(\mathrm{d}x) \propto f(x)\,\mathrm{d}x$ is called the *canonical measure*. The canonical ensemble is often used in modeling experiments on molecular systems that are performed under the conditions of constant volume and temperature $\mathcal{T} = \frac{1}{k_B \beta}$, where k_B Boltzmann's constant. Obviously, a single solution of the Hamiltonian system (3.2.1) can never be ergodic w.r.t. the canonical measure, since it conserves the internal energy H, as defined in (3.2.2). Hence, w.r.t. the canonical measure, the deterministic Hamiltonian system is not in the class (MC2), while it might be w.r.t. to other measure such as, e.g., the microcanonical measure.

Hamiltonian system with randomized momenta. Aiming at a conformational analysis of biomolecular systems in the context of the canonical ensemble, Schütte et al. introduced a specific stochastic Hamiltonian system (SCHÜTTE, FISCHER, HUISINGA and DEUFLHARD [1999]) as a discrete time Markov chain, defined solely on the position space and derived from the deterministic Hamiltonian system by "randomizing the momenta".

For some fixed observation time span $\tau > 0$ (for comments on the choice of τ see remark below) and some inverse temperature $\beta > 0$ the stochastic transition function for the Hamiltonian system with randomized momenta is given by

$$p(q, A) = \int_{\mathbf{R}^d} \mathbf{1}_A\big(\Pi_q \Phi^\tau(q, \xi)\big)\mathcal{P}(\xi)\,\mathrm{d}\xi,$$

where $\Pi_q : (q, \xi) \mapsto q$ denotes the projection onto the position space $\Omega = \mathbf{R}^{3N}$ and $\mathcal{P}$ the canonical distribution of momenta $\mathcal{P} \propto \exp(-\beta \xi^t \xi / 2)$.

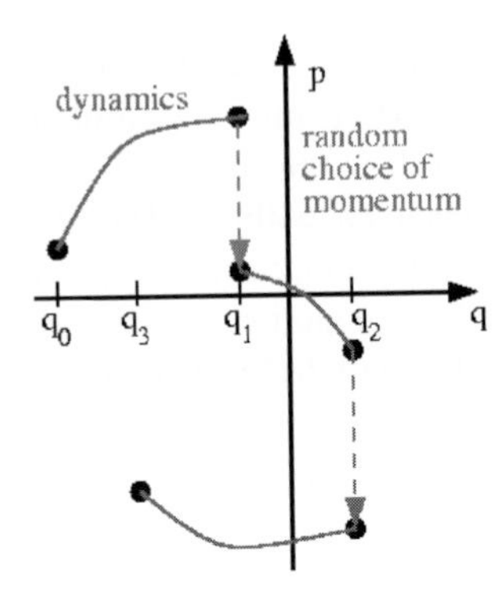

FIG. 3.1. Illustration of the Hamiltonian system with randomized momenta as defined in (3.2.4).

The associated discrete time Markov process $Q_n = \{Q_n\}_{n\in\mathbf{N}}$, defined on the state space $\mathbf{X} = \Omega$, satisfies

$$Q_{n+1} = \Pi_q \Phi^\tau(Q_n, \xi_n), \quad n \in \mathbf{N}, \tag{3.2.4}$$

where ξ_n is chosen randomly from $\mathcal{P}$ (SCHÜTTE [1998]). As it is shown in SCHÜTTE [1998] the positional canonical measure $\mu(\mathrm{d}q) \propto \exp(-\beta V(q))\,\mathrm{d}q$ is invariant w.r.t. Q_n and unique. Moreover, exploiting that Φ^τ is reversible and symplectic Q_n is shown to be reversible w.r.t. μ (SCHÜTTE [1998]).

The Hamiltonian system with randomized momenta generates an ensemble of time-τ trajectories such that each trajectory follows the deterministic Hamiltonian dynamics (3.2.1) starting at initial values distributed according to the positional canonical ensemble $f(q) \propto \exp(-\beta V(q))$ (see Fig. 2.2 for illustration). When the deterministic Hamiltonian system is believed to be contained in class (MC1), then the Hamiltonian system with randomized momenta is contained in the classes (MC1) and (MC2).

REMARK. For arbitrary, but fixed $\tau > 0$ we have defined the *one-step* transition function $p(q, D) = p^\tau(1, q, D)$. Changing the observation time to $\sigma > 0$ results in a new one-step transition function $p^\sigma(1, q, D)$. In general we will have $p^{2\tau}(1, q, D) \neq p^\tau(2, q, D)$; for an example see (SCHÜTTE [1998], Section 3.7.1).

Constant temperature molecular dynamics. One traditional aspect of molecular dynamics is the construction of (stochastic) dynamical systems that allow of sampling the canonical ensemble by means of long-term simulation. Several concepts have been discussed that all boil down to the idea to construct a Hamiltonian system in some extended state space $\widehat{\mathbf{X}}$, whose projection onto the lower dimensional state space $\mathbf{X}$ of positions and momenta allows to generate such a sampling. One of the most prominent examples is defined in terms of the Nosé Hamiltonian

$$H_{\text{Nose}}(q, \xi, s, \nu) = \underbrace{\frac{1}{2s^2}\xi^T\xi + V(q)}_{=H_s(q,\xi)} + \frac{1}{2Q}\nu^2 + \frac{1}{\beta}\log s,$$

where s is called the thermostat with conjugated momentum π and associated artificial mass Q. Let the flow of the associated Nosé Hamiltonian system be denoted Ψ^t and let Π denote the projection $(q, \xi, s, \nu) \mapsto (q, \xi)$. If Ψ^t is ergodic w.r.t. the microcanonical measure on the associated energy cell of H_{Nose}, then $\Pi\Psi^t$ is ergodic w.r.t. the canonical measure $\mu(\mathrm{d}x) \propto \exp(-\beta H_{s=1}(x))\,\mathrm{d}x$, where $x = (q, \xi)$ (BOND, BENEDICT and LEIMKUHLER [1999]). Thus, the Nosé Hamiltonian system is contained in the class (MC2) but it is at least questionable whether it is part of (MC1).

Langevin system. The most popular model for an *open system* with stochastic interaction with its environment is the so-called Langevin System (RISKEN [1996]):

$$\dot{q} = \xi, \qquad \dot{\xi} = -\nabla_q V(q) - \gamma\xi + \sigma\dot{W}_t, \tag{3.2.5}$$

defined on the state space $\mathbf{X} = \mathbf{R}^{6N}$. Here $\gamma > 0$ denotes some friction constant and $F_{\text{ext}} = \sigma\dot{W}_t$ the external forcing given by a $3N$-dimensional Brownian motion W_t. The external stochastic force is assumed to model the influence of the heat bath surrounding the molecular system. In this case, the internal energy given by the Hamiltonian H, as defined in (3.2.2), is not preserved, but the interplay between stochastic excitation and damping balances the internal energy. As a consequence, the canonical measure $\mu(\mathrm{d}x) \propto \exp(-\beta H(x))\,\mathrm{d}x$ with $x = (q, \xi)$ is invariant w.r.t. the Markov process corresponding to the Langevin system, where the noise and damping constants satisfy (RISKEN [1996]):

$$\beta = \frac{2\gamma}{\sigma^2}. \tag{3.2.6}$$

Thus, the Langevin system satisfies our expectation on (MC2) w.r.t. the canonical ensemble but simultaneously also allows to represent some essential aspects of (MC1), i.e., of the true dynamical behavior of the molecular system.

Smoluchowski system. The Smoluchowski system can be understood as an approximation to the Langevin system in the limit of high friction $\gamma \to \infty$, see HUISINGA [2001], SCHÜTTE and HUISINGA [2000] for details. While the Langevin system gives a description of molecular motion in terms of positions and momenta of all atoms in the system, the Smoluchowski system is stated in the position space only. Moreover, in contrast to the Langevin equation it defines a *reversible* Markov process that is given by the equation

$$\dot{q} = -\frac{1}{\gamma}\nabla_q V(q) + \frac{\sigma}{\gamma}\dot{W}_t. \tag{3.2.7}$$

The stochastic differential equation (3.2.7) defines a continuous time Markov process Q_t on the state space $\mathbf{X} = \Omega$ with invariant probability measure $\mu(\mathrm{d}q) \propto \exp(-\beta V(q))\,\mathrm{d}q$ (RISKEN [1996]). Thus, this dynamical model satisfies our expectation on class (MC2) but should in general not be expected to satisfy those on (MC1).

Nevertheless there is a long history of using it as a simple toolkit for investigation of dynamical behavior in complicated energy landscapes (CHANDLER [1998]). It is known that under weak conditions on the potential function V the Markov process is reversible (HUISINGA [2001]).

Markov chain Monte Carlo (MCMC). Markov chain Monte Carlo techniques are designed to sample a given probability density $f:\mathbf{R}^d \to \mathbf{R}$, particularly in highly dimensional state spaces. MCMC is an iterative realization of some specific Markov chain, whose stochastic transition function is given by

$$p(x, \mathrm{d}y) = q(x, y)\mu(\mathrm{d}y) + r(x)\delta_x(\mathrm{d}y).$$

That is, the stochastic transition function is composed of some transition kernel $q(x, y)$, which is assumed to be μ-integrable and some rejection probability

$$r(x) = 1 - \int_X q(x, y)\mu(\mathrm{d}y) \geqslant 0.$$

In almost all situations, the transition kernel q is chosen in such a way that the stochastic transition function is reversible w.r.t. μ.

In general MCMC is an artificial dynamical model that is in general understood as the embodiment of class (MC2), therefore being in general far from satisfying the properties of (MC1). However, there are MCMC methods like the popular hybrid Monte Carlo Method (HMC) that can be understood as a special realization of the Hamiltonian system with randomized momenta.

3.3. Summary

Concerning our main categories (MC1) and (MC2) the summary could be the following: constant temperature MD, MCMC, the Langevin and Smoluchowski systems clearly belong to (MC2), while the deterministic Hamiltonian system is supposed to be the incorporation of (MC1). However, this distinction is not sharp: the Langevin system is often also accepted as belonging to (MC1), while the deterministic Hamiltonian system is accepted for (MC1) only under the condition that enough details of the entire molecular system (including parts of the solute environment) are represented in atomic resolution and the interaction potential V is appropriate.

4. Metastability

Given a dynamical system, metastability of some subset of the state space is characterized by the property that the dynamical system is likely to remain within the subset for a long period of time, until it exits and hence a transition to some other region of the state space occurs. There is no unique but several definitions of metastability in literature (see, e.g., BOVIER, ECKHOFF, GAYRARD and KLEIN [2001], DAVIES [1982], SCHÜTTE, HUISINGA and DEUFLHARD [2001], SINGLETON [1984]); we will herein

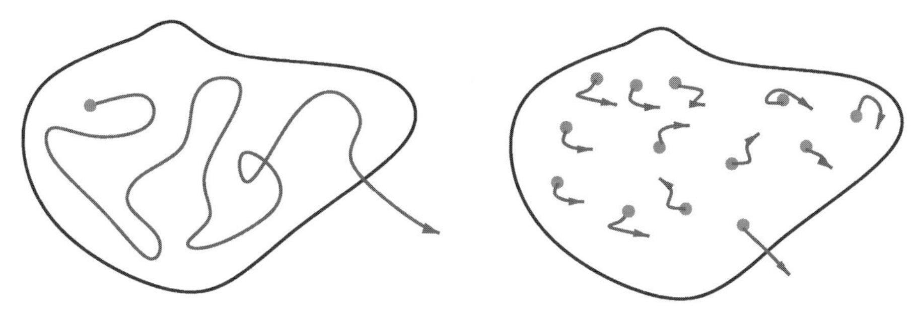

FIG. 4.1. Illustration of the different concepts of metastability: exit time approach on the left and ensemble dynamics approach on the right.

focus on two different concepts that are adapted to suit the ensemble dynamics and exit time approach, respectively, as discussed in Section 3.2.

Ensemble dynamics approach: A subset $C \subset \mathbf{X}$ of the entire state space is called metastable, if the fraction of systems in C, whose trajectory exits during some pre-described time span τ, is significantly small.

Exit time approach: A subset $C \subset \mathbf{X}$ of the entire state space is called metastable, if with high probability a typical long-term trajectory stays within C longer than some "macroscopic" time span.

Based on the discussion in section "Conceptual Preliminaries" we would expect to run into trouble when aiming at a "naive" numerical realization of the exit time approach while the ensemble dynamics approach seems to be numerically treatable for short observation time spans τ. We will come back to this question below.

In view of our biochemical application context we aim at the identification of a *decomposition of the state space into metastable subsets* and the corresponding "flipping dynamics" between these. In general, a *decomposition* $\mathcal{D} = \{D_1, \ldots, D_m\}$ of the state space $\mathbf{X}$ is a collection of subsets $D_k \subset \mathbf{X}$ with the properties:

positivity: $\mu(D_k) > 0$ for every k,
disjointness up to null sets: $\mu(D_k \cap D_l) = 0$ for $k \neq l$, and
covering property: $\bigcup_{k=1}^{m} \overline{D_k} = \mathbf{X}$.

The problem of identifying a decomposition into metastable subsets particularly poses the task of specifying the number m of subsets one is looking for. Within the transfer operator approach this is done via spectral analysis (see *key idea* on page 718).

4.1. *Ensemble dynamics approach: transition probabilities*

We aim at defining an (ensemble) transition probability from a subset B to C within some time span τ, denoted by $p(\tau, B, C)$, such that an invariant sub-ensemble C is characterized by $p(\tau, C, C) = 1$, while a metastable sub-ensemble can be characterized by $p(\tau, C, C) \approx 1$. We will see that within this approach metastability is measured w.r.t. the invariant probability measure μ of the dynamics; in biomolecular systems, the measure μ will often be defined in terms of the canonical ensemble.

We define the *transition probability* $p(t, B, C)$ from $B \subset \mathbf{X}$ to $C \subset \mathbf{X}$ within the time span t as the conditional probability

$$p(t, B, C) = \mathbf{P}_\mu[X_t \in C \mid X_0 \in B] = \frac{\mathbf{P}_\mu[X_t \in C \text{ and } X_0 \in B]}{\mathbf{P}_\mu[X_0 \in B]}, \tag{4.1.1}$$

where $\mathbf{P}_\mu$ indicates that the initially the Markov process X_t is distributed according to μ, hence $X_0 \sim \mu$. Exploiting the definition of the stochastic transition function $p(t, x, C)$ in (3.1.1) we rewrite (4.1.1) as

$$p(t, B, C) = \frac{1}{\mu(B)} \int_B p(t, x, C)\mu(\mathrm{d}x). \tag{4.1.2}$$

In other words, the transition probability quantifies the dynamical fluctuations within the stationary ensemble μ. Due to the ensemble dynamics approach to metastability we call a subset $B \subset \mathbf{X}$ *metastable* on the time scale $\tau > 0$ if

$$p(\tau, B, B^c) \approx 0, \quad \text{or equivalently,} \quad p(\tau, B, B) \approx 1,$$

where $B^c = \mathbf{X} \setminus B$ denotes the complement of B. Obviously, the approximate equalities are not sharp enough for a rigorous definition of metastability. We will come back to this problem in Section 5.2.

REMARK. It is an intrinsic property of the ensemble transition probability to depend on the observation time span τ. It is obvious from its definition that $p(\tau, C, C)$ approaches 1 for $\tau \to 0$, while it decays to $\mu(C)$ for $\tau \to \infty$, see Fig. 4.2. The most interesting

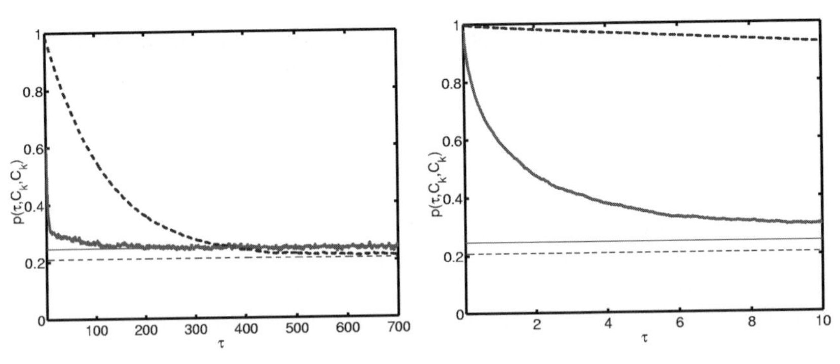

FIG. 4.2. Dependence of the transition probability $p(\tau, C, C)$ on the observation time span τ for $0 \leqslant \tau \leqslant 700$ (left) and zoomed into $0 \leqslant \tau \leqslant 10$ (right) w.r.t. two different subsets, one being metastable (dashed line) the other being much less metastable (solid line). The thick line corresponds to $p(\tau, C, C)$, while the corresponding predicted limit value $\mu(C)$ (see text) is indicated by a horizontal thin line. As can be seen from the graphics the distinction between metastable and not-metastable is clearly visible for mesoscopic observation time spans as, e.g., $1/2 \leqslant \tau \leqslant 10$. (The data are based on the Smoluchowski dynamics w.r.t. the perturbed three-well potential as illustrated in Fig. 5.1.)

phenomena occur on mesoscopic time scales τ. In the biomolecular application context, the observation time span τ will be given by the experimental setting. In the earlier mentioned pump-and-probe experiments, the values of τ range in the sub-picosecond regime.

4.2. *Exit time approach: exit rates*

The characterization of metastability within the exit time approach is related to the asymptotic decay of the distribution of exit times. Its precise formulation via exit (decay) rates requires some extended mathematical theory that hinders understanding at first reading. Therefore, we prefer to outline the fundamental idea rather than to give its mathematical justification, for which we refer to HUISINGA, MEYN and SCHÜTTE [2003].

Denote by $D \subset \mathbf{X}$ some connected open subset and consider some point $x \in D$. Then the *exit time* $\varrho_D(x)$ of the Markov process X_t from D started at $X_0 = x$ is defined as

$$\varrho_D(x) = \inf\left\{t \geqslant 0\colon \int_0^t \mathbf{1}_X(X_s \notin D)\,\mathrm{d}s > 0\right\} \tag{4.2.1}$$

and measures only exits that happen for some nonnull time interval neglecting exit events that are “singular” in time. Note that in general ϱ_D is a random variable that depends on the realization of the Markov process X_t.

The fundamental idea is a characterization of metastability in terms of the asymptotic decay of the distribution of exit times

$$F_x(s) = \mathbf{P}_x\big[\varrho_D(x) \geqslant s\big].$$

While for small values of s the function F_x may show complicated behavior, it asymptotically may decay almost exponentially, at least under certain well-established conditions. The decay rate of F_x can best be expressed by means of the conditional exit time distribution

$$F_x(s,t) = \mathbf{P}_x\big[\varrho_D(x) \geqslant s + t \mid \varrho_D(x) \geqslant t\big]$$

for $s, t \geqslant 0$ that describes the tail of the distribution, for which the exit time is larger than the so-called waiting time t. The decay rate is equal to Γ if the conditional distribution decays exponentially with rate $\Gamma > 0$, i.e.,

$$F_x(s,t) \propto \exp(-\Gamma s) \tag{4.2.2}$$

for $s \geqslant 0$ and $t \geqslant 0$. When aiming at a definition of metastability in terms of decay rates for entire *subsets*, there are two problems. Firstly, the relation in (4.2.2) will only hold for very special Markov processes (HUISINGA, MEYN and SCHÜTTE [2003]). Secondly, we would expect that the decay rate depends on the starting point, i.e., $\Gamma = \Gamma_x$.

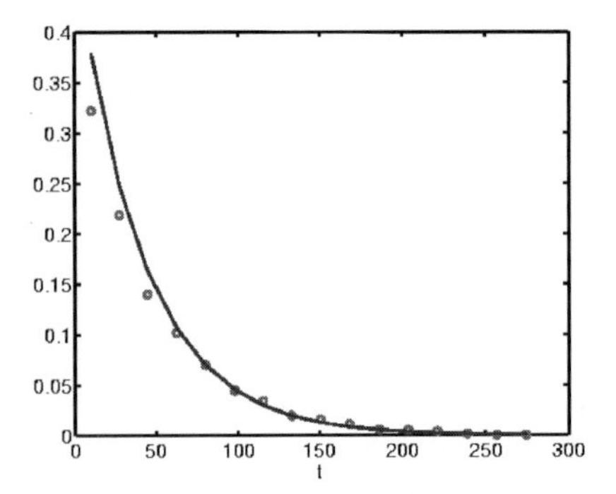

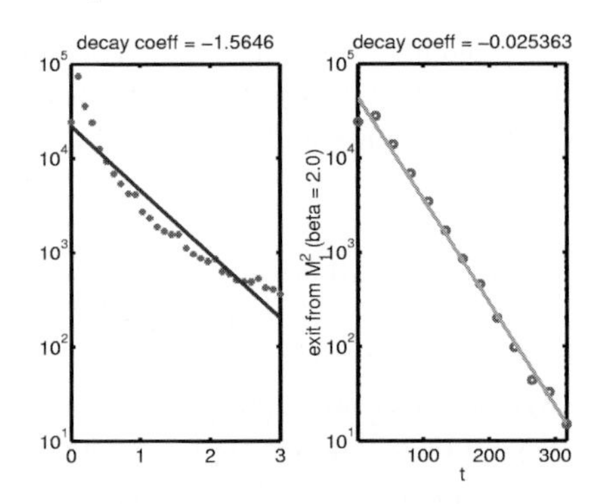

FIG. 4.3. Dependence of the exit time distribution $F_x(s)$ on the exit time s for some metastable subset D. The decay is shown for $0 \leqslant s \leqslant 300$ (left), and zoomed into $0 \leqslant s \leqslant 3$ (middle) and $10 \leqslant s \leqslant 300$ (right) on a semi-logarithmic plot. We observe regions of different decay (middle and right). Asymptotically, the decay rate of the exit time distribution is close to the predicted value of $\Gamma(D)$, i.e., $F_x(s)$ for $s > t$ decays approximately with rate $\Gamma(D)$ for $t \to \infty$. Regions of different decay rate (like, e.g., initial rapid decay followed by a much slower decay) are typically due to the fact that with high probability the process exits very rapidly, while with almost vanishing, but existing probability the process moves into some much more metastable region contained in D such that asymptotically the exit rate becomes very small.

The approach presented herein is based on the fact that there do exist subsets C, for which the decay rate is basically independent for all states $x \in C$. In a more general setting, but for a specific class of dynamical systems including, e.g., the Smoluchowski dynamics, we are able to assign a so-called *exit rate* $\Gamma = \Gamma(C)$ to an entire subset $C \subset \mathbf{X}$ rather than to single points $x \in \mathbf{X}$, thus circumventing the two above mentioned problems with (4.2.2). This exit rates may be thought of as some generalization of decay rates; see Appendix 9.2, in particular Theorem 9.2.1.

Due to the exit time approach to metastability we call a subset $B \subset \mathbf{X}$ *metastable* with exit rate $\Gamma(B)$ if

$$\Gamma(A) > \Gamma(B), \quad \text{for all open, connected sets } A \subset B,\ A \neq B. \tag{4.2.3}$$

As in the ensemble approach there will be infinitely many metastable subsets and we expect to find a hierarchy of metastability.

5. Transfer operators

We now give the mathematical foundations and the algorithmic strategies to characterize and identify a decomposition of the state space into metastable subsets. It turns out that in either of the approaches transfer operators and their generators play a crucial role.

5.1. Transfer operators and generators

Based on the assumption that the dynamical description is given by a (homogeneous) Markov process X_t we now introduce a Markov operator that allows to describe the propagation of sub-ensembles in time under the action of X_t. In so doing, we assume in the sequel that the probability measure μ is invariant w.r.t. the Markov process X_t.

The basic idea is the following: Consider all systems within the stationary ensemble, whose states are in some subset $C \subset \mathbf{X}$. This sub-ensemble is distributed according to the probability measure

$$\nu_0(A) = \frac{1}{\mu(C)} \int_A \mathbf{1}_C(x)\mu(\mathrm{d}x) = \int_A v_0(x)\mu(\mathrm{d}x)$$

corresponding to the density $v_0 = \mathbf{1}_C/\mu(C)$ w.r.t. to μ. Since every single microscopic system evolves according to the Markov process defined by its stochastic transition function p, the distribution of the sub-ensemble at time $t \in \mathbf{T}$ is given by the probability measure

$$\nu_t(A) = \int_{\mathbf{X}} v_0(x)\, p(t, x, A)\mu(\mathrm{d}x). \tag{5.1.1}$$

On the other hand, if ν_t admits the density v_t, we have

$$\nu_t(A) = \int_A v_t(x)\mu(\mathrm{d}x). \tag{5.1.2}$$

Our interest is to define a transfer operator P^t that propagates sub-ensembles in time according to

$$v_0 \mapsto v_t = P^t v_0. \tag{5.1.3}$$

This transfer operator is well-defined due to HUISINGA [2001], REVUZ [1975] and acts on the Banach spaces $L^r(\mu)$, $1 \leqslant r \leqslant \infty$, with corresponding norms $\|\cdot\|_r$, as defined in Appendix 9.1. Combining Eqs. (5.1.1) and (5.1.2), we define the *semigroup of propagators* or forward transfer operators $P^t : L^r(\mu) \to L^r(\mu)$ with $t \in \mathbf{T}$ and $1 \leqslant r < \infty$ as follows:

$$\int_A P^t v(y)\mu(\mathrm{d}y) = \int_{\mathbf{X}} v(x)p(t, x, A)\mu(\mathrm{d}x) \tag{5.1.4}$$

for $A \subset \mathbf{X}$. As a consequence of the invariance of μ, the characteristic function $\mathbf{1}_{\mathbf{X}}$ of the entire state space is an invariant density of P^t, i.e., $P^t \mathbf{1}_{\mathbf{X}} = \mathbf{1}_{\mathbf{X}}$. Furthermore, P^t is a Markov operator, i.e., P^t conserves norm: $\|P^t v\|_1 = \|v\|_1$ and positivity: $P^t v \geqslant 0$ if $v \geqslant 0$, which is a simple consequence of the definition. Due to (5.1.3), the semigroup of propagators mathematically models the physical phenomena of evolution of sub-ensembles in time.

In the theory of Markov processes another semigroup of operators is considered. We will call it the *semigroup of backward transfer operators* $T^t : L^r(\mu) \to L^r(\mu)$ with $t \in \mathbf{T}$ and $1 \leqslant r \leqslant \infty$, defined by

$$T^t u(x) = \mathsf{E}_x\big[u(X_t)\big] = \int_{\mathbf{X}} u(y)p(t, x, \mathrm{d}y). \tag{5.1.5}$$

As a consequence of property (ii) in Appendix 9.1 of the stochastic transition function, we have $T^t \mathbf{1}_\mathbf{X} = \mathbf{1}_\mathbf{X}$ for every $t \in \mathbf{T}$. Both transfer operators are closely related via the duality bracket $\langle v, u\rangle_\mu = \int_\mathbf{X} v(x)u(x)\mu(\mathrm{d}x)$ for $v \in L^1(\mu)$ and $u \in L^\infty(\mu)$, namely

$$\langle P^t v, u\rangle_\mu = \langle v, T^t u\rangle_\mu. \tag{5.1.6}$$

In discrete time $t \in \mathbf{N}$ it is convenient to use the abbreviations $P = P^1$ and $T = T^1$ corresponding to the stochastic transition function $p(x, \mathrm{d}y) = p(1, x, \mathrm{d}y)$; compare Section 3.1. Propagators associated with *reversible* Markov processes are of particular interest, since they possess additional structure on the Hilbert space $L^2(\mu)$. Such propagators will be called reversible, too.

PROPOSITION 5.1.1 (HUISINGA [2001]). *Let $P^t : L^2(\mu) \subset L^1(\mu) \to L^2(\mu)$ denote the propagator corresponding to the Markov process X_t. Then P^t is self-adjoint w.r.t. the scalar product $\langle\cdot,\cdot\rangle_\mu$ in $L^2(\mu)$, i.e.,*

$$\langle u, P^t v\rangle_\mu = \langle P^t u, v\rangle_\mu \quad t \in \mathbf{T},$$

for every $u, v \in L^2(\mu)$, if and only if the Markov process X_t is reversible.

For the semigroup of propagators $P^t : L^r(\mu) \to L^r(\mu)$ with $1 \leqslant r \leqslant \infty$ define $\mathcal{D}(\mathcal{A})$ as the set of all $v \in L^r(\mu)$ such that the strong limit

$$\mathcal{A}v = \lim_{t\to\infty} \frac{P^t v - v}{t}$$

exists. Then, the operator $\mathcal{A} : \mathcal{D}(\mathcal{A}) \to L^r(\mu)$ is called the infinitesimal *generator* corresponding to the semigroup P^t (KARATZAS and SHREVE [1991], LASOTA and MACKEY [1994]).

REMARK. Physical experiments on molecular ensembles allow to measure relative frequencies in the canonical ensemble μ. Suppose again that μ has the form

$$\mu(\mathrm{d}x) = f(x)\,\mathrm{d}x,$$

i.e., μ is absolutely continuous w.r.t. the Lebesgue measure $\mathrm{d}x$. Then physical experiments are related to densities of the form

$$v_{\text{phys}}(x) = \hat{\mathbf{1}}_C(x) f(x) \in L^1(\mathrm{d}x)$$

w.r.t. the Lebesgue measure $\mathrm{d}x$. Whenever physicists use the phrase "probability density" they refer to v_{phys} rather than to densities

$$v_{\text{math}}(x) = \hat{\mathbf{1}}_C(x) \in L^1(\mu)$$

w.r.t. probability measure μ, as we do. As it will become apparent later, it is mathematically advantageous to consider the semigroup of propagators acting on densities v_{math} rather than the semigroup of propagators acting on v_{phys}. In the former approach, we have $P^t \mathbf{1}_{\mathbf{X}} = \mathbf{1}_{\mathbf{X}}$, while in the latter this would read $P^t f = f$. However, it should be clear that results obtained in either of the two descriptions can be transformed into the other.

EXAMPLES 5.1.1. To be more specific, we now list the propagators for the different dynamical descriptions introduced in Section 3.2.

The propagator corresponding to the deterministic Hamiltonian system with flow Φ^t is known as the Frobenius–Perron operator (LASOTA and MACKEY [1994]) given by

$$P^t u(x) = u(\Phi^{-t} x).$$

For the Hamiltonian system with randomized momenta we have a kind of Frobenius–Perron operator averaged w.r.t. the momenta,

$$P^t u(q) = \int_{\mathbf{R}^d} u(\Pi_q \Phi^{-\tau}(q, \xi)) \mathcal{P}(\xi) \, \mathrm{d}\xi. \tag{5.1.7}$$

For MCMC the propagator is given by

$$Pu(y) = \int q(x, y) u(x) \, \mathrm{d}x + r(y) u(y). \tag{5.1.8}$$

For Langevin and Smoluchowski dynamics, the semigroups of propagators admit strong generators $\mathcal{A}_{\mathrm{Smo}}$ and $\mathcal{A}_{\mathrm{Lan}}$ in $L^r(\mu)$ for $1 \leqslant r < \infty$ such that the semigroups can be written as

$$P^t_{\mathrm{Smo}} = \exp(t\mathcal{A}_{\mathrm{Smo}}) \quad \text{and} \quad P^t_{\mathrm{Lan}} = \exp(t\mathcal{A}_{\mathrm{Lan}}),$$

respectively. For twice continuously differentiable $u \in L^r(\mu)$ we have the identity

$$\mathcal{A}_{\mathrm{Smo}} u = \left(\frac{\sigma^2}{2\gamma^2} \Delta_q - \frac{1}{\gamma} \nabla_q V(q) \cdot \nabla_q \right) u,$$

$$\mathcal{A}_{\mathrm{Lan}} u = \left(\frac{\sigma^2}{2} \Delta_p - p \cdot \nabla_q + \nabla_q V \cdot \nabla_p - \gamma p \cdot \nabla_p \right) u.$$

For details on $\mathcal{A}_{\mathrm{Smo}}$ and $\mathcal{A}_{\mathrm{Lan}}$ see the theory of Fokker–Planck equations and Kolmogoroff forward and backward equations (RISKEN [1996], SCHÜTTE, HUISINGA and DEUFLHARD [2001], HUISINGA [2001]).

5.2. Detecting metastability

The key idea of the transfer operator approach to metastability is to exploit the strong relation between stability properties of the Markov process and the presence of special eigenvalues in the spectrum.

Since propagators are Markov operators by definition, their spectrum is contained in the unit circle of the complex plane, i.e., the modulus of every eigenvalue is smaller or equal to 1. Suppose that some proper subset $C \subset \mathbf{X}$ is invariant under the Markov process, i.e., $p(t, x, C^c) = 0$ for all $x \in C$. Then we have:

Ensemble dynamics approach: the transition probability from C into its complement C^c is zero: $p(t, C, C^c) = 0$ for every $t \in \mathbf{T}$.
Exit rate approach: the exit rate from C is zero: $\Gamma(C) = 0$.

In literature on Markov and transfer operators, it is a well-known fact that the existence of invariant subsets has spectral consequences. Under well-established stability conditions (see (C1) and (C2) on page 719), we have:

Transfer operator: the propagator P^t exhibits an eigenvalue $\lambda_t \equiv 1$ with corresponding eigenfunction $\mathbf{1}_C$, hence $P^t \mathbf{1}_C = \mathbf{1}_C$ for every $t \in \mathbf{T}$.

Now suppose that the entire state space decomposes into exactly two invariant subsets, $X = B \cup C$. Then, the eigenvalue $\lambda = 1$ is two-fold, one corresponding to each invariant subset associated with the eigenfunctions $\mathbf{1}_B$ and $\mathbf{1}_C$. Introducing a weak coupling between the subsets B and C yields one invariant set, namely the entire state space $\mathbf{X}$, and two *weakly coupled* or *metastable* subsets, namely B and C. This has the following consequences:

Ensemble dynamics approach: the transition probability from B to C is almost zero: $p(\tau, B, C) \approx 0$ for $0 < \tau < T$ with large T. The same holds for the transition probability from C to B.
Exit rate approach: the exit rates from B and C are very small: $\Gamma(B) \approx 0$ and $\Gamma(C) \approx 0$.
Transfer operator: for $0 < \tau < T$ with large T, the propagator P^τ exhibits two dominant eigenvalues. More precisely, there exists $\eta_t \equiv 1$ corresponding to the invariant state space, and one eigenvalue $\lambda_\tau \approx 1$ corresponding to the weak coupling between the subsets B and C.

To the end, we fix some $\tau > 0$ and abbreviate $P = P^\tau$ and $p(x, C) = p(\tau, x, C)$. Hence, $(P)^n = P^{n\tau}$ corresponds to the Markov process sampled at rate τ with stochastic transition function given by $p^n(\cdot, \cdot) = p(n\tau, \cdot, \cdot)$.

The above considerations motivate the following *key idea of the transfer operator approach*:

> Metastable subsets can be detected via eigenvalues of the propagator P close to its maximal eigenvalue $\lambda = 1$; moreover they can be identified by exploiting the corresponding eigenfunctions. In doing so, the number of metastable subsets is equal to the number of eigenvalues close to 1, including $\lambda = 1$ and counting multiplicity.

The strategy mentioned above has first been proposed by DELLNITZ and JUNGE [1999] for discrete dynamical systems with weak random perturbations and has been successfully applied to molecular dynamics in different contexts (SCHÜTTE, FISCHER, HUISINGA and DEUFLHARD [1999], SCHÜTTE and HUISINGA [2000], SCHÜTTE [1998]); its justification is given in Section 5.4. The key idea requires the following two *conditions on the propagator* P (for a definition of the essential spectral radius see Appendix 9.1):

(C1) The essential spectral radius of P is less than one, i.e., $r_{\mathrm{ess}}(P) < 1$.

(C2) The eigenvalue $\lambda = 1$ of P is simple and dominant, i.e., $\eta \in \sigma(P)$ with $|\eta| = 1$ implies $\eta = 1$.

While condition (C1) allows to ensure convergence results of the numerical discretization scheme, condition (C2) excludes modeling and interpretation problems; for more details see SCHÜTTE [1998], HUISINGA [2001]. In order to proceed along the way indicated by the key idea we have to check in which situations the two conditions (C1) and (C2) may hold. In this section, we establish sufficient conditions on the Markov process that imply (C1) and (C2). Here, we mainly concentrate on reversible propagators P on $L^2(\mu)$ and refer for $L^1(\mu)$ and $L^\infty_{\mathcal{V}}$ to HUISINGA [2001] and MEYN and TWEEDIE [1993], respectively.

There are some sufficient conditions to guarantee the spectral properties of the propagator that are related to well studied stability properties of Markov processes. The $\mathcal{V}$-norm and total variation norm $\|\cdot\|_{\mathrm{TV}}$ stated in the next definition are defined in Appendix 9.1:

DEFINITION 5.2.1. Let p denote some stochastic transition function. Then

(a) p is called *geometrically ergodic* if

$$\left\| p^n(x,\cdot) - \mu \right\|_{\mathrm{TV}} \leqslant \mathcal{V}(x) q^n, \quad n \in \mathbf{N}, \tag{5.2.1}$$

for every $x \in \mathbf{X}$, some constant $q < 1$, and some integrable function $\mathcal{V}: \mathbf{X} \to \mathbf{R}$ satisfying $\mathcal{V} < \infty$ pointwise.

(b) p is called *$\mathcal{V}$-uniformly ergodic* if

$$\left\| p^n(x,\cdot) - \mu \right\|_{\mathcal{V}} \leqslant C\mathcal{V}(x) q^n, \quad n \in \mathbf{N},$$

for every $x \in \mathbf{X}$, constants $q < 1$ and $C \leqslant \infty$, and some function $\mathcal{V} \in L^1(\mu)$ satisfying $1 \leqslant \mathcal{V}$ pointwise.

The relation between geometrical and $\mathcal{V}$-uniform ergodicity is as follows: By definition, $\mathcal{V}$-uniform ergodicity implies geometric ergodicity. On the other hand, for irreducible and aperiodic stochastic transition functions geometric ergodicity implies $\mathcal{V}$-uniform ergodicity according to (ROBERTS and ROSENTHAL [1997], Proposition 2.1). Either form of ergodicity implies the properties of interest:

THEOREM 5.2.1 (HUISINGA [2001]). *Let $P: L^2(\mu) \to L^2(\mu)$ denote a reversible propagator. Then P satisfies conditions* (C1) *and* (C2) *in $L^2(\mu)$, if its stochastic transition function is geometrically or $\mathcal{V}$-uniformly ergodic.*

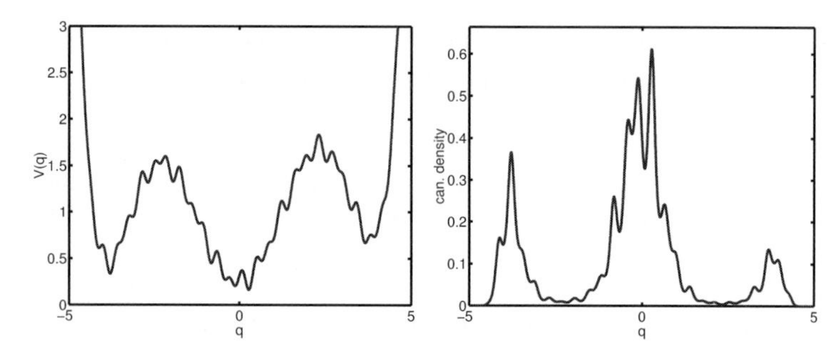

FIG. 5.1. Left: a perturbed three well potential V used for illustration. Right: canonical density corresponding to $\beta = 3.0$. Assume that the internal dynamics is given by the Smoluchowski system corresponding to $\gamma = 2.0$ and σ given by Eq. (3.2.6). The dominant spectrum of the propagator $P = P^\tau$ with $\tau = 1.0$ exhibits three eigenvalues close to 1. More precisely, we have $\lambda_1 = 1.0000$, $\lambda_2 = 0.9924$, $\lambda_3 = 0.9886$, which are separated from the remaining eigenvalues by a significant gap, since $\lambda_4 = 0.6634$. According to the key idea of the transfer operator approach we expect a decomposition of the state space into three metastable subsets, which is in agreement with our intuition for the above potential.

Conditions, under which Markov processes are geometrically or $\mathcal{V}$-uniformly ergodic are widely studied; for sufficient conditions w.r.t. the dynamical models introduced in Section 3.3, see, e.g., HUISINGA [2001], Section 6, and cited references, or MATTINGLY, STUART and HIGHAM [2003], SCHÜTTE [1998].

5.3. *Identification algorithm*

The key idea of the transfer operator approach needs to be specified regarding the actual *algorithmic identification* of metastable subsets based on the most dominant eigenvectors. The basic idea is to reduce the problem of identifying a decomposition into metastable subsets to a clustering problem, which is done by incorporating dynamical information into the process of clustering. It is therefore different from statistical clustering that is solely based on geometrical information.

Following the key idea of the transfer operator approach we are aiming at a decomposition $\mathcal{D} = \{D_1, \ldots, D_m\}$ of the state space into m metastable subsets $D_1, \ldots, D_m$ such that the number of subsets m equals the number of dominant eigenvalues. Fig. 5.2 demonstrates the mechanism of the transfer operator approach to metastability.

Based on this mechanism, virtually almost every cluster algorithm can be used to identify a decomposition of the state space into metastable subsets, *if applied to the dynamically coded sampling points*. The identification procedure introduced in DEUFLHARD, HUISINGA, FISCHER and SCHÜTTE [2000] computes this decomposition from the *sign structure* of the coded sampling points as illustrated in Fig. 5.4. Given the m dominant eigenvectors $v_1, \ldots, v_m$, we can assign to every state $x \in \mathbf{X}$ a unique *sign structure*

$$s(x) = \big(s_1(x), \ldots, s_m(x)\big) = \{+, -\}^m,$$

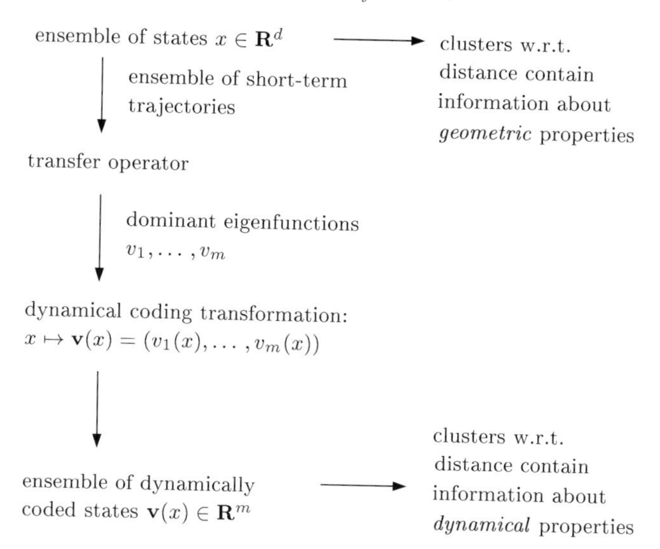

FIG. 5.2. Basic mechanism of incorporating dynamical information via dominant eigenfunctions of the transfer operator. Note that the dimension of the state space that have to be clustered reduces from d to m. In biomolecular applications, typically we have $m \ll d$.

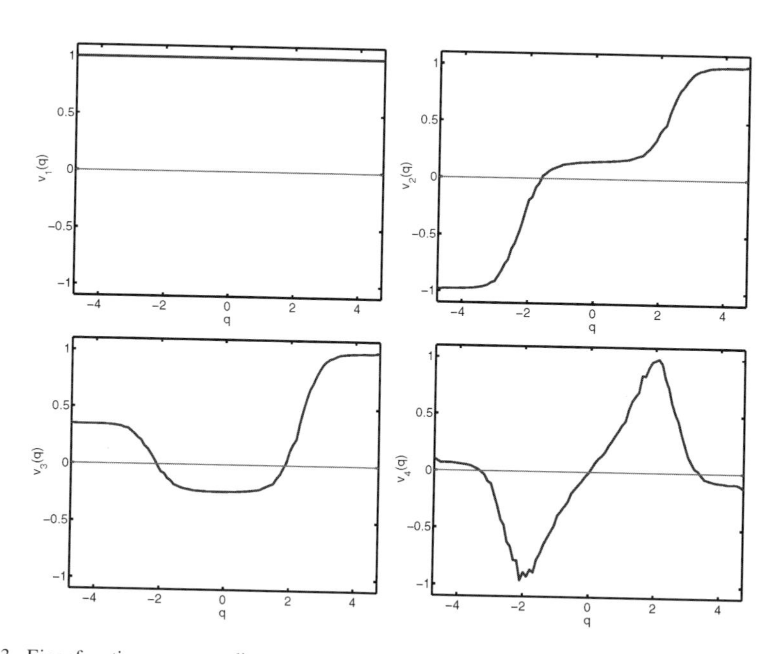

FIG. 5.3. Eigenfunction corresponding to propagator based on the Smoluchowski system in application to the perturbed three well potential V as illustrated in Fig. 5.1. Eigenfunctions corresponding to the largest eigenvalues 1.0000, 0.9924, 0.9886, 0.6634 (clockwise, starting top-left). As expected, the eigenfunction corresponding to $\lambda_1 = 1$ is constant. Note that the second and the third eigenfunction exhibit a very special structure: they are almost constant around the three wells (cf. potential in Fig. 5.1), while they show jumps near the saddle point regions.

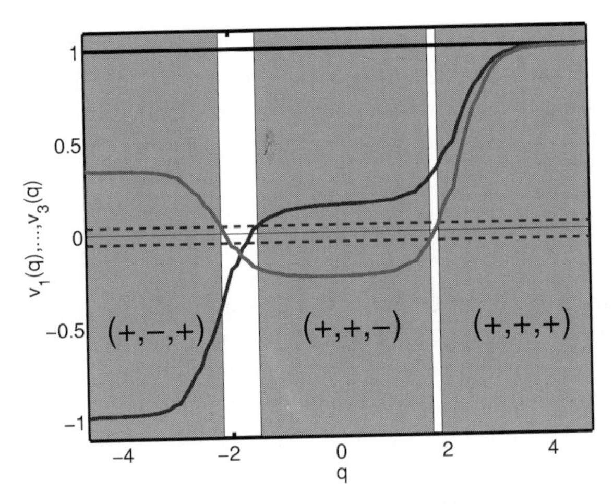

FIG. 5.4. Illustration of the identification procedure based on the perturbed three well potential presented in Fig. 5.1. The three gray shaded regions indicate the parts of the q-axis on which the three most dominant eigenvectors exhibit "unambiguous" sign structures (here $\theta = 0.05$ as indicated by dashed lines). To give an example: the gray shaded region on the very left has sign structure $(+, -, +)$ since $v_1(q), v_3(q) > \theta$ and $v_2(q) < -\theta$ for all states q on the very left.

where for some prescribed threshold value θ we define the "unambiguous" positive or negative sign $s_k(x)$ of $v_k(x)$ as

$$s_k(x) = \begin{cases} +, & v_k(x) > \theta, \\ -, & v_k(x) < \theta \end{cases}$$

as explained in Fig. 5.4. Denote by $S(\mathbf{X}) \subset \mathbf{X}$ the set of all states with at least one "unambiguous" sign s_k. If θ is large enough and the eigenfunctions are smooth, $S(\mathbf{X})$ decomposes into exactly m subsets each containing states of the same sign structure only (DEUFLHARD, HUISINGA, FISCHER and SCHÜTTE [2000]). These are the "core sets" of the metastable subsets. States with "unambiguous" sign structure are assigned to these core sets (DEUFLHARD, HUISINGA, FISCHER and SCHÜTTE [2000]) such that the resulting m metastable sets decompose the state space. This clustering algorithm has proved to be successful in many different situations; in the following subsection we give a mathematical justification of the algorithmic identification strategy from the point of view of the ensemble dynamics and the exit time approach.

5.4. *Mathematical justification*

We now give a mathematical justification of the key idea of the transfer operator approach, which illuminates the strong relation between the existence of a cluster of eigenvalues close to 1 and a possible decomposition of the state space into metastable subsets.

Justification within the ensemble dynamics approach. The investigation will be based on the following assumptions: The propagator $P : L^2(\mu) \to L^2(\mu)$ satisfies conditions

(C1), (C2) and the underlying Markov process is reversible. As a consequence the propagator P is self-adjoint due to Proposition 5.1.1.

The close relation between transition probabilities and transfer operators becomes transparent in

$$p(B, C) = \frac{\langle P\mathbf{1}_B, \mathbf{1}_C\rangle_\mu}{\langle \mathbf{1}_B, \mathbf{1}_B\rangle_\mu}. \tag{5.4.1}$$

Eq. (5.4.1) allows to give a mathematical statement relating dominant eigenvalues, the corresponding eigenfunctions and a decomposition of the state space into metastable subsets. For later reference we define the *metastability of a decomposition* $\mathcal{D}$ as the sum of the metastabilities of its subsets. The next result can be found in HUISINGA and SCHMIDT [2002]; a version for two subsets was published in HUISINGA [2001].

THEOREM 5.4.1. *Let* $P : L^2(\mu) \to L^2(\mu)$ *denote a reversible propagator satisfying* (C1) *and* (C2). *Then* P *is self-adjoint and the spectrum has the form*

$$\sigma(P) \subset [a, b] \cup \{\lambda_m\} \cup \cdots \cup \{\lambda_2\} \cup \{1\},$$

with $-1 < a \leqslant b < \lambda_m \leqslant \cdots \leqslant \lambda_1 = 1$ *and isolated, not necessarily simple eigenvalues of finite multiplicity that are counted according to multiplicity. Denote by* $v_m, \ldots, v_1$ *the corresponding eigenfunctions, normalized to* $\|v_k\|_2 = 1$. *Let* Q *be the orthogonal projection of* $L^2(\mu)$ *onto* $\mathrm{span}\{\mathbf{1}_{A_1}, \ldots, \mathbf{1}_{A_m}\}$. *The metastability of an arbitrary decomposition* $\mathcal{D} = \{A_1, \ldots, A_m\}$ *of the state space* $\mathbf{X}$ *can be bounded from above by*

$$p(A_1, A_1) + \cdots + p(A_m, A_m) \leqslant 1 + \lambda_2 + \cdots + \lambda_m,$$

while it is bounded from below according to

$$1 + \kappa_2\lambda_2 + \cdots + \kappa_m\lambda_m + c \leqslant p(A_1, A_1) + \cdots + p(A_m, A_m),$$

where $\kappa_j = \|Qv_j\|^2_{L^2(\mu)}$ *and* $c = a(1 - \kappa_2) \cdots (1 - \kappa_n)$.

Theorem 5.4.1 highlights the strong relation between a decomposition of the state space into metastable subsets and dominant eigenvalues close to 1. It states that the metastability of an arbitrary decomposition $\mathcal{D}$ cannot be larger than $1 + \lambda_2 + \cdots + \lambda_m$, while it is at least $1 + \kappa_2\lambda_2 + \cdots + \kappa_m\lambda_m + c$, which is close to the upper bound whenever the dominant eigenfunctions $v_2, \ldots, v_m$ are almost constant on the metastable subsets $A_1, \ldots, A_m$ implying $\kappa_j \approx 1$ and $c \approx 0$. The term c can be interpreted as a correction that is small, whenever $a \approx 0$ or $\kappa_j \approx 1$. It is demonstrated in HUISINGA and SCHMIDT [2002] that the lower and upper bounds are sharp and asymptotically exact.

In view of Theorem 5.4.1, it is natural to ask, whether there is an *optimal* decomposition with highest possible metastability. The answer is illustrated by Fig. 5.5: *Even if there exists an optimal decomposition, the problem of finding it might be ill-conditioned.* The graph shows the metastability of a family of decompositions. It

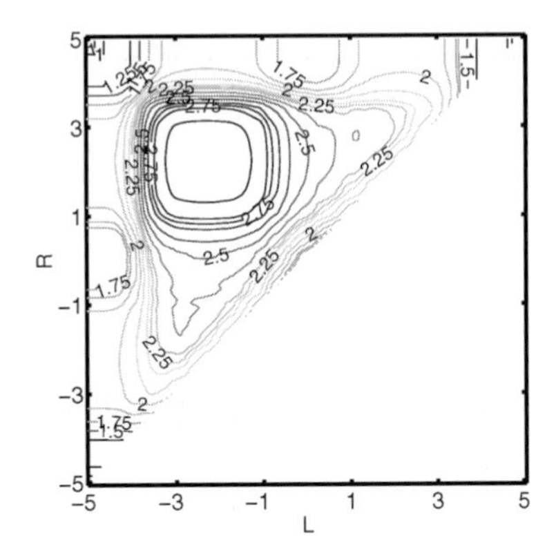

FIG. 5.5. Metastability of the decomposition of the state space $\mathbf{X} = \mathbf{R}$ into three subsets $\mathcal{D} = \{A, B, C\}$ with $A = (-\infty, L)$, $B = [L, R)$ and $C = [R, \infty)$ parameterized by $L < R \in \mathbf{R}$. The problem of finding the maximal value corresponding to the optimal decomposition is often ill-conditioned. There is a large region roughly characterized by $-3 < L < -1$ and $1 < R < 3$, in which the metastability of the corresponding decomposition is almost maximal. This region is more or less a flat plateau, for which the absolute maximum is hard to identify. The identification algorithm based on the sign structure identifies a decomposition with $L = -2.13$ and $R = 1.84$ with corresponding metastability 2.9558, which is quite close to the upper bound of 2.9916 resulting from Theorem 5.4.1. (Data based on Smoluchowski dynamics as illustrated in Fig. 5.1.)

is based on the propagator P corresponding to the Smoluchowski dynamics for the perturbed three-well potential. We identify a flat plateau of decompositions that are nearly optimal. In this case the problem of finding the maximum is ill-conditioned. We also observe that the decomposition suggested by our identification algorithm is nearly optimal. The phenomenon illustrated by Fig. 5.5 is believed to be typical in our application context, which is due to the fact that the state space admits large regions corresponding to almost vanishing statistical weight (see also HUISINGA and SCHMIDT [2002]).

Justification within the exit time approach. The justification has been worked out in HUISINGA, MEYN and SCHÜTTE [2003] based on recent literature on Markov chains. The most important restriction of this approach is the restriction to Markov processes with continuous sample paths that admit a generator.

We will herein formulate the result for general Smoluchowski systems of type (3.2.7) from Section 3.2, because this allows us to remain within the framework of self-adjoint propagators with real-valued eigenvalues and eigenfunctions. The generalization to Langevin systems seems to be possible but requires immense technical effort.

THEOREM 5.4.2. *Assume that there is some continuous Lyapunov function $\mathcal{V}$ such that the Markov process is $\mathcal{V}$-uniformly ergodic, and that there exists a twice continuously differentiable eigenfunction $v : \mathbf{X} \to \mathbf{R}$ of the Smoluchowski generator $\mathcal{A} = \mathcal{A}_{\mathrm{Smo}}$. Hence, there exists some eigenvalue $\Lambda < 0$ such that*

$$\mathcal{A}v = \Lambda v. \tag{5.4.2}$$

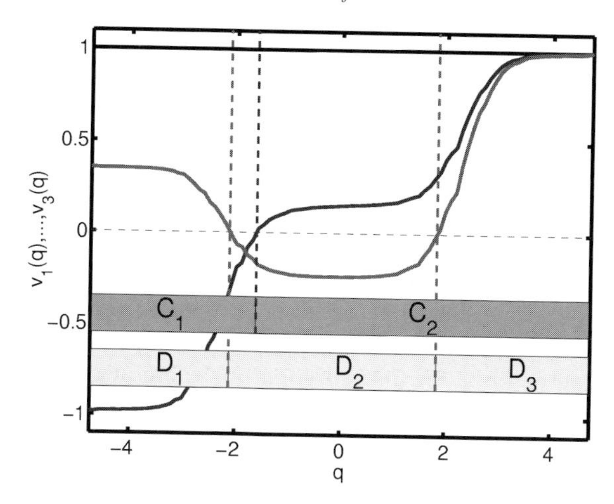

FIG. 5.6. Illustration of Theorem 5.4.2 for the perturbed three well potential illustrated in Fig. 5.1. The lowest eigenfunctions of the generator corresponding to the Smoluchowski dynamics are $\Lambda_1 = 0.0000$, $\Lambda_2 = -0.0076$, $\Lambda_3 = -0.0114$, while $\Lambda_4 = -0.4104$. The eigenfunctions v_2 and v_3 exhibit sign changes and allow to decompose the state space into subsets, on which the eigenfunction is either positive or negative. Since v_2 has exactly one zero a thereon based decomposition of the state space yields the subsets C_1 and C_2 with common exit rate $\Gamma_2 = -\Lambda_2 = 0.0076$. Analogously, v_3 defines the subsets D_1, D_2 and D_3 with common exit rate $\Gamma_3 = -\Lambda_3 = 0.0114$.

Suppose moreover that the set C under consideration and the eigenfunction v satisfy the following conditions:

(1) $v(x) > 0$ *or* $v(x) < 0$ *for all* $x \in C$,

(2) $v(x) = 0$ *and* $(\nabla v(x))^T(\nabla v(x)) > 0$ *for* $x \in \partial C$,

(3) $K_n = \{x \in \mathbf{X}\colon \mathcal{V}(x) \leqslant n v(x)\}$ *is a compact subset of* $\mathbf{X}$ *for all* $n \geqslant 1$.

Then, the set C is metastable in the sense of definition (4.2.3) *with exit rate* $\Gamma(C) = -\Lambda$, *where* Λ *is the eigenvalue associated with* v.

This theorem has the following intriguing interpretation: if there is an eigenvalue $-\Gamma_0$ of the generator close to zero, then there is an eigenvalue $\exp(-\tau\Gamma_0)$ of the propagator P^τ close to $\lambda = 1$. The corresponding eigenfunction v of the generator is also the eigenfunction of the propagator. The set of zeros of this eigenfunction decomposes the state space into *open, connected subsets* C_k, restricted to each of which the eigenfunction is either positive or negative. Now Theorem 5.4.2 states that each of this subsets C_k is metastable with the same rate $\Gamma(C_k) = \Gamma_0$ which is considerably small since Γ_0 is.

The identification algorithm is based on the *collection* of the most dominant eigenfunctions. In seemingly contrast, Theorem 5.4.2 indicates that *each* single eigenfunction induces a metastable decomposition, in particular that different eigenfunctions might induce different metastable decompositions. We illustrate this for the three-well potential in Fig. 5.6. The second and third eigenfunctions v_2 and v_3 allow application of Theorem 5.4.2 and induce two different decompositions, namely $\{C_1, C_2\}$ and $\{D_1, D_2, D_3\}$. The rate of metastability of the first is superior ($\Gamma(C_k) = 0.022$); the latter reproduces the three-well structure of the potential but its metastable subsets show less significant metastability ($\Gamma(D_k) = 0.036$). Conclusively, the result is a hierarchy of metastable decompositions with decreasingly significant metastability of the subset. However, the identification algorithm will result in a very similar hierarchy if iteratively applied to the set $\{v_1, v_2\}$ of eigenfunctions first and then to the set $\{v_1, v_2, v_3\}$.

6. Numerical realization

Identification of metastable subsets necessitates the computation of the most dominant eigenfunctions of the propagator $P = P^\tau$ for some fixed observation time span $\tau > 0$. In the following we describe the discretization procedure of the eigenvalue problem $Pv = \lambda v$. Throughout this section we assume that P satisfies the conditions (C1) and (C2) defined in Section 5.2. Part of this section follows from SCHÜTTE [1998], SCHÜTTE, FISCHER, HUISINGA and DEUFLHARD [1999].

6.1. Galerkin discretization

Let $\chi = \{\chi_1, \ldots, \chi_n\} \subset L^2(\mu)$ denote a set of *nonnegative* functions with the property to yield a partition of unity, i.e.,

$$\sum_{k=1}^{n} \chi_k = \mathbf{1}_{\mathbf{X}}.$$

The *Galerkin projection* $\Pi_n : L^2(\mu) \to \mathcal{S}_n$ onto the associated finite-dimensional ansatz space $\mathcal{S}_n = \operatorname{span}\{\chi_1, \ldots, \chi_n\}$ is defined by

$$\Pi_n v = \sum_{k=1}^{n} \frac{\langle v, \chi_k \rangle_\mu}{\langle \chi_k, \chi_k \rangle_\mu} \chi_k.$$

Application of the Galerkin projection to $Pv = \lambda v$ yields an eigenvalue problem for the discretized propagator $\Pi_n P \Pi_n$ acting on the finite-dimensional space $\mathcal{S}_n$. The matrix representation of this finite-dimensional operator is given by the $n \times n$ *transition matrix* $S = (S_{kl})$, whose entries are given by

$$S_{kl} = \frac{\langle P\chi_k, \chi_l \rangle_\mu}{\langle \chi_k, \chi_k \rangle_\mu} = \frac{\langle \chi_k, T\chi_l \rangle_\mu}{\langle \chi_k, \chi_k \rangle_\mu}, \tag{6.1.1}$$

where T denotes the adjoint transfer operator.

Properties of discretization matrix. Since P is a Markov operator and χ a partition of unity, the Galerkin discretization S is a (row) stochastic matrix, i.e., $S_{kl} \geqslant 0$ and $\sum_{l=1}^{n} S_{kl} = 1$ for $k = 1, \ldots, n$. Consequently, all its eigenvalues λ satisfy $|\lambda| \leqslant 1$. Moreover, we have the following three important properties (SCHÜTTE [1998], SCHÜTTE, FISCHER, HUISINGA and DEUFLHARD [1999]):

(1) The row vector $\pi = (\pi_1, \dots, \pi_n)$ with $\pi_k = \langle \chi_k, \chi_k \rangle_\mu = \|\chi_k\|_\mu^2$ represents the discretized invariant probability measure μ. It is a left eigenvector corresponding to the eigenvalue $\lambda = 1$, thus $\pi S = \pi$.

(2) S is *irreducible* and *aperiodic*. As a consequence, the eigenvalue $\lambda = 1$ is *simple* and *dominant*, hence $\lambda \in \sigma(S)$ implies $\lambda = 1$ or $|\lambda| < 1$. In particular, the discretized invariant density π is the *unique* invariant density of S.

(3) If P is reversible then S is *self-adjoint* w.r.t. the discrete scalar product $\langle u, v \rangle_\pi = \sum u_i \bar{v}_i \pi_i$. Equivalently, S satisfies the *detailed balance condition* $\pi_k S_{kl} = \pi_l S_{lk}$ for every $k, l \in \{1, \dots, n\}$. Hence, all eigenvalues of S are real-valued and contained in the interval $(-1, 1]$.

Partitions of unity that are defined in terms of some decomposition $\mathcal{D} = \{D_1, \dots, D_n\}$ of the state space are of particular interest. Any decomposition $\mathcal{D}$ defines a partition of unity $\chi = \{\chi_1, \dots, \chi_n\}$ with $\chi_k = \mathbf{1}_{D_k}$. Galerkin discretization based on this *box discretization* then yields a transition matrix that in addition to the above properties has the advantageous property that its entries are one-step transition probabilities from D_k to D_l

$$S_{kl} = \frac{\langle P\mathbf{1}_{D_k}, \mathbf{1}_{D_l} \rangle_\mu}{\langle \mathbf{1}_{D_k}, \mathbf{1}_{D_k} \rangle_\mu} = p(D_k, D_l).$$

Another prominent example is given by partitions of unity χ whose elements $\chi_k \in \chi$ are so-called mollifies, i.e., nonnegative C^∞-functions of compact support that approximate indicator functions. In this case, the discretization often is called *fuzzy set discretization*.

Summarizing, the discretization of the propagator can be interpreted as a *coarse graining* procedure, especially in the case of box discretization: Coarse graining the state space $\{x \in \mathbf{X}\} \to \{D_1, \dots, D_n\}$ results in a coarse graining of the propagator $P \to S$ corresponding to a coarse graining of the Markov process $p(x, D) \to p(D_k, D_l)$ with invariant measures $\mu \to \pi$. In doing so, the discretization inherits the most important properties of the propagator.

REMARK. It is important to notice that – unless in very special situations – the discretization process does not commute with the semigroup property of the transfer operator P. Hence, in general we have

$$(S^2)_{kl} \neq p^2(D_k, D_l),$$

where S^2 denotes the square of the transition matrix S obtained from discretization P, and p^2 denotes the stochastic transition function corresponding to P^2.

6.2. *The eigenvalue problem*

We restrict our considerations to the important class of *reversible* propagators $P : L^2(\mu) \to L^2(\mu)$ satisfying the conditions (C1) and (C2).

Under these assumptions, convergence results for the eigenvalues are simple consequences of, e.g., the Rayleigh–Ritz theory (see SCHÜTTE [1998] for details). Obviously we have to require that the sequence of the Galerkin ansatz spaces $\mathcal{S}_{n_1} \subset \mathcal{S}_{n_2} \subset \cdots$ is dense in $L^2(\mu)$, and the corresponding partitions of unity $\chi^{(1)}, \chi^{(2)}, \ldots$ are getting gradually finer, e.g., $\max_{\Phi \in \chi^{(k)}} \operatorname{diam}(\operatorname{supp} \Phi) \to 0$ as $n_k \to \infty$ with $k \to \infty$, where we assume that the functions in the kth partition of unity $\chi^{(k)}$ have compact support.

For the explicit numerical approximation of the dominant eigenvalues and eigenvectors different settings are available. Whenever one is interested in the dominant eigenvalues and eigenvectors of the transition matrix for a given discretization one may use iterative eigenvalue solvers, see, e.g., LEHOUCQ, SORENSEN and YANG [1998]. These allow to compute the desired information even for matrices of size $10^6 \times 10^6$, for example. Whenever one is interested in subsequent refinements of the discretization to achieve approximations with high precision, one should apply multigrid techniques with optimal efficiency even for very fine grids like those constructed in DEUFLHARD, FRIESE and SCHMIDT [1997], for example. The mentioned preconditions for efficient convergence of the techniques perfectly fit to the scenario of metastability: it is required that the dominant eigenvalues are separated from the remainder of the spectrum by a significant gap. Then the convergence rate of those methods only depends on the spectral gap and is in principle independent of the size of the stochastic transition matrix. The numerical effort is given mainly by the effort of matrix-vector multiplications.

We emphasize, however, that already for small molecules the dimension of state space X is so high as to make the transfer operator approach computationally infeasible if naively applied directly in X. This problem is sometimes called the *curse of dimensionality*. To some extend it can be ameliorated by the use of adaptive algorithms such as those in DELLNITZ and JUNGE [1998] but, for many problems in high dimensions, combination of the transfer operator approach with clustering approaches, and/or with mathematical modeling such as the exploitation of fast/slow time-scale separation, is needed to circumvent the curse of dimensionality. We will address this problem in Section 6.5.

6.3. *Evaluation of transition matrix*

We consider the evaluation of the stochastic transition matrix S obtained from discretizing $P = P^\tau$. The corresponding discrete time Markov chain is denoted by $X_n = \{X_n\}_{n \in \mathbf{N}}$. Consider two elements χ_k, χ_l of the partition of unity used for discretization. Combining $T^\tau \chi_l(x) = \mathsf{E}_x[\chi_l(X_\tau)]$ with (6.1.1) yields

$$S_{kl} = \frac{1}{\langle \chi_k, \chi_k \rangle_\mu} \int_{\mathbf{X}} \chi_k(x) \mathsf{E}_x\big[\chi_l(X_\tau)\big] \mu(\mathrm{d}x),$$

which can be approximated within two steps:

(A1) approximation of the integral

$$\int_B g(x) \mu(\mathrm{d}x) \approx \sum_{k=1}^N \alpha_k g(x_k)$$

by some deterministic or stochastic integration scheme with partition points or random variables $x_1, \dots, x_N$, respectively, and weights $\alpha_1, \dots, \alpha_N$ (DEUFLHARD and HOHMANN [1993], GILKS, RICHARDSON and SPIEGELHALTER [1997]);

(A2) approximation of the expectation value

$$\mathsf{E}_x\big[\chi_l(X_\tau)\big] \approx \frac{1}{M}\sum_{j=1}^{M} \chi_l\big(X_\tau(\omega_j, x)\big)$$

by relative frequencies, where $X_\tau(\omega_k, x)$ denotes a realization of the Markov process with initial distribution $X_0 = x$ (MEYN and TWEEDIE [1993], Chapter 17).

A combination of (A1) and (A2) with $g(x) = \chi_k(x)\mathsf{E}_x[\chi_l(X_\tau)]$ results in

$$S_{kl} \approx \frac{1}{M}\sum_{k=1}^{N}\sum_{j=1}^{M} \alpha_k \chi_k(x) \cdot \chi_l\big(X_\tau(\omega_{kj}, x_k)\big);$$

hence, for each initial point x_k, the Markov process X_τ is realized M times. The approximation quality depends on the interplay between the two approximation steps (A1) and (A2). Numerical experiments in low dimensions show that it is even possible to choose small M, if the number of partition points N is chosen in such a way that the number of points in the support of any of the χ_k is reasonably large. For high-dimensional problems, we will in general be forced to use stochastic integration schemes such as Monte Carlo methods in order to approximate the integral in (A1). For the analysis of biomolecules Monte Carlo based techniques have been applied successfully (SCHÜTTE, FISCHER, HUISINGA and DEUFLHARD [1999], HUISINGA, BEST, ROITZSCH, SCHÜTTE and CORDES [1999]).

Whenever the discrete time Markov process $X_n = \{X_n\}_{n\in\mathbf{N}}$ is ergodic w.r.t. μ, and a statistically representative realization of X_n is available, we may also combine (A1) and (A2):

EXAMPLE 6.3.1. Let $x_0, \dots, x_N$ denote a sequence of sampling points obtained from a realization of the discrete time Markov process X_n. Then

$$S_{kl} \approx S_{kl}^{(N)} = \frac{\sum_{j=1}^{N} \chi_k(x_j) \cdot \chi_l(x_{j+1})}{\sum_{j=1}^{N} \chi_k(x_j)^2}, \tag{6.3.1}$$

where convergence is guaranteed for μ-a.e. initial points x_0 by conditions (C1) and (C2) and the law of large numbers (MEYN and TWEEDIE [1993]).

6.4. *Trapping problem*

The *rate of convergence* of $S_{kl}^{(N)} \to S_{kl}$ depends on the smoothness of the partition functions χ_k as well as on the mixing properties of the Markov chain X_n (LEZAUD

[2001]). The latter property is crucial here: The convergence is geometric with a rate constant $\lambda_1 - \lambda_2 = 1 - \lambda_2$ where λ_2 denotes the second largest eigenvalue (in modulus). That is, in case of metastability and thus λ_2 being very close to $\lambda_1 = 1$, we will have dramatically slow convergence. As this is the main problem for all approaches to biomolecular dynamics and statistics, this is also a bottleneck of the transfer operator approach. An entire bunch of the literature aims at tackling this problem that is often called the *trapping problem* (BERNE and TRAUB [1997], FERGUSON, SIEPMANN and TRUHLAR [1999]). In the framework of the transfer operator approach presented herein, A. Fischer recently designed a hierarchical approach tailored to accelerate convergence called *uncoupling-coupling* (UC) (FISCHER [2000], FISCHER, SCHÜTTE, DEUFLHARD and CORDES [2002]). Its *key assumption* is the following property of all stochastic systems designed to sample the canonical ensemble: Decreasing temperature induces stronger metastabilities, while metastability vanishes for large temperatures. Thus heating can help to reduce trapping which reappears if the system is annealed to lower temperatures. The uncoupling-coupling approach has been designed to circumvent this problems by combining the idea of domain decomposition with bridge sampling techniques.

In Section 5.3 a special bridge sampling technique called *Adaptive Temperature HMC* (ATHMC) (FISCHER, CORDES and SCHÜTTE [1998]) is used. Based on the Hybrid Monte Carlo procedure ATHMC reduces trapping problems by allowing to adapt the temperature during the sampling such that it can be increased to induce exits from metastable subsets. It was demonstrated that this procedure can be described by means of a generalized ensemble such that all parts of the sample (with different temperatures) can be reweighted to the temperature of interest (FISCHER, CORDES and SCHÜTTE [1998]).

6.5. Discretization in higher dimensions

Any discretization will suffer from the *curse of dimensionality* whenever it were based on uniform partition of all of the hundreds or thousands of degrees of freedom in a typical biomolecular system. Fortunately, chemical observations reveal that – even for larger biomolecules – the curse of dimensionality can be circumvented by exploiting the hierarchical structure of the dynamical and statistical properties of biomolecular systems: Firstly, only relatively few *conformational* or *essential degrees of freedom* are needed to describe the conformational transitions (LINSSEN and BERENDSEN [1993]). Furthermore, the canonical density has a rich spatial multiscale structure induced by the rich structure of the potential energy landscape. This structure induces a hierarchical cluster structure of the sampling data that can be identified and used to define a multilevel discretization adapted to the structures of the statistical data.

These observations give rise to a collection of approaches to the construction of *structure-adapted discretizations*:

Essential degrees of freedom. In the (low dimensional) subspace of essential degrees of freedom most of the positional fluctuations occur, while in the remaining degrees of freedom the motion can be considered as "physically constrained". Based on the available sampling, we may determine essential degrees of freedom either in the position space according to LINSSEN and BERENDSEN [1993] or in the space of internal degrees of freedom, e.g., dihedral angles, by statistical analysis of circular data (HUISINGA, BEST, ROITZSCH, SCHÜTTE and CORDES [1999]). Either case is based on a principal component analysis of the sampling. As shown in HUISINGA, BEST, ROITZSCH, SCHÜTTE and CORDES [1999], this procedure may results in a enormous reduction of the number of degrees of freedom and, consequently, in a moderate number of subsets within the decomposition when discretizing the essential variables only. The principal component analysis is a linear approach to essential degrees of freedom. A characterization and identification of more general nonlinear essential degrees of freedom presently is a topic of further investigation.

Clustering algorithms. Another approach of decomposing the state space is based on clustering the sampling data by means of clustering algorithms (see, e.g., JAIN and DUBES [1988] and cited references). These methods cluster the sampling data according to structural similarity: The set of sampling points is partitioned into disjoint subsets with the property that two states belonging to the same subset are in some sense structural closer to each other than two states belonging to different subsets. A crucial question is the design of appropriate measures of structural similarity. In the biomolecular application context, these measures may either be based on the Cartesian coordinates or on the internal degrees of freedom. In contrast to the former the latter approach is invariant under rotations and translations of the entire molecule.

A novel promising approach to the above type of clustering problem uses self-organizing maps, a special kind of neural networks. Self-organizing maps allow to cluster the sampling data by assigning each sampling point to its nearest "neurons", each of them representing a subset of the decomposition. We have demonstrated its successful application to sampling data of biomolecular systems in GALLIAT, HUISINGA and DEUFLHARD [2000]. More advanced extensions, such as "box-neurons" and a hierarchical embedding, have recently been designed (GALLIAT and DEUFLHARD [2000], GALLIAT, DEUFLHARD, ROITZSCH and CORDES [2002]).

Whenever the statistical distribution allows to be clustered into a limited but significant number of clusters (e.g., a few thousand at most), these clusters can be used to define a statistics-adapted discretization as by fuzzy partitions of unity or by introducing discretization "boxes" such that each box contains a single cluster; for an application to biomolecular systems see HUISINGA, BEST, ROITZSCH, SCHÜTTE and CORDES [1999], WEBER and GALLIAT [2002].

7. Illustrative numerical experiments

We now want to illustrate the transfer operator approach to metastability in application to different dynamical descriptions that are based on the earlier introduced perturbed three-well potential (see Fig. 5.1, left). Further investigations, including dependence on parameters and discretization, can be found in HUISINGA [2001]. The theoretical justification of the transfer operator approach via conditions (C1) and (C2) for the below dynamical descriptions can be found in SCHÜTTE [1998], HUISINGA [2001].

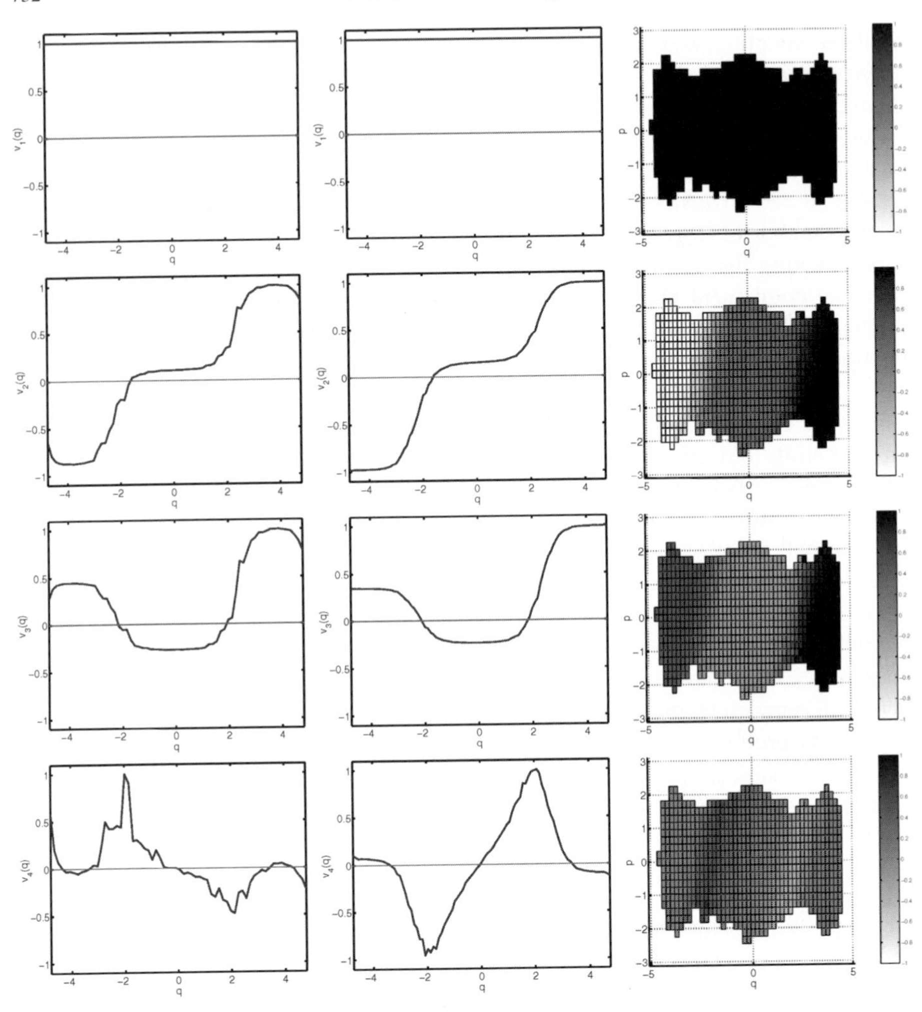

FIG. 7.1. The four dominant eigenfunctions of the propagator P_τ for different model systems. Left: Hamiltonian system with randomized momenta corresponding to the eigenvalues 1.0000, 0.9939, 0.9912, 0.6946 (from top to bottom). Middle: Smoluchowski equation for $\gamma = 2.0$ corresponding to the eigenvalues 1.0000, 0.9925, 0.9893, 0.6585. Right: Langevin equation for $\gamma = 2.0$ corresponding to the eigenvalues 1.0000, 0.9948, 0.9924, 0.6796.

Parameters. The perturbed three-well potential is defined by

$$V(q) = \frac{1}{100}\big(q^6 - 30q^4 + 234q^2 + 14q + 100 + 30\sin(17q) + 26\cos(11q)\big).$$

Below, we analyze the ensemble dynamics based on the Hamiltonian system with randomized momenta, the Langevin system and the Smoluchowski system. We choose $\beta = 2.0$ for the inverse temperature and $\gamma = 2.0$ for the friction constant. Then, σ is defined via the relation $\beta = 2\gamma/\sigma^2$ as stated in Eq. (3.2.6). For the observation time span we take $\tau = 1.0$.

Sampling. We sample the Markov processes given by the Hamiltonian system with randomized momenta with periodic boundary conditions on the positional state space $[-5, +5]$, the Langevin systems on $\mathbf{X} = \mathbf{R} \times \mathbf{R}$ and the Smoluchowski systems on $\mathbf{X} = \mathbf{R}$. The sampling length is $n = 300000$ points.

Discretization. We discretize the positional state space into $n = 70$ equally sized intervals (boxes). For the Langevin dynamics, we additionally have to discretize the space of momenta. Examining the canonical density of momenta reveals that it is very unlikely to stay outside the interval $[-3, +3]$. Hence, we discretize the state space of momenta $\mathbf{R}$ by discretizing $[-3, +3]$ into 28 equally sized intervals and adding the two infinite intervals $(-\infty, -3)$ and $(3, \infty)$. Applying the discretization methods described in Section 5.1, we end up with an $n \times n$ stochastic transition matrix S with $n = 50 \times 30$ for the Langevin dynamics and $n = 70$ otherwise. For the Hamiltonian system with randomized momenta and the Smoluchowski dynamics, the transition matrix is self-adjoint.

Hamiltonian system with randomized momenta. Solving the eigenvalue problem for S yields:

λ_1	λ_2	λ_3	λ_4	λ_5	λ_6	...
1.000	0.9939	0.9912	0.6946	0.6481	0.5952	...

Application of our identification algorithm yields a decomposition of the state space $\mathcal{D} = \{C_1, C_2, C_3\}$ with $C_1 = \{q \leqslant -2.13\}$, $C_2 = \{-2.13 < q \leqslant 1.99\}$ and $C_3 = \{1.99 < q\}$. The statistical weights $\mu(C_k)$ within the positional canonical ensemble μ and the metastabilities $p(C_k, C_k)$ are given by the following table:

metastable subset	C_1	C_2	C_3
statistical weight	0.2018	0.6979	0.1003
metastability	0.9883	0.9938	0.9800

The essential statistical behavior, i.e., the probability of transitions between the metastable subsets, is described by the coupling matrix $C = (c_{jk})_{j,k=1,2,3}$ with $c_{jk} = p(C_j, C_k)$. For our example, we obtain

$$C = \begin{pmatrix} 0.9883 & 0.0117 & 0 \\ 0.0034 & 0.9938 & 0.0029 \\ 0 & 0.0200 & 0.9800 \end{pmatrix}.$$

Langevin system. Solving the eigenvalue problem for S yields:

λ_1	λ_2	λ_3	λ_4	λ_5	...
1.0000	0.9948	0.9924	0.6796	0.5365	...

Application of our identification algorithm yields a decomposition of the state space $\mathcal{D} = \{C_1, C_2, C_3\}$. The statistical weights $\mu(C_k)$ within the canonical ensemble μ and the metastabilities $p(C_k, C_k)$ are given by the following table:

metastable subset	C_1	C_2	C_3
statistical weight	0.1918	0.7012	0.1070
metastability	0.9887	0.9946	0.9851

(7.1)

Calculating the coupling matrix yields

$$C = \begin{pmatrix} 0.9887 & 0.0113 & 0 \\ 0.0031 & 0.9946 & 0.0023 \\ 0 & 0.0149 & 0.9851 \end{pmatrix}. \tag{7.2}$$

Smoluchowski system. Solving the eigenvalue problem for S yields:

λ_1	λ_2	λ_3	λ_4	λ_5	λ_6	...
1.0000	0.9925	0.9893	0.6585	0.5496	0.4562	...

Application of our identification algorithm yields a decomposition of the state space $\mathcal{D} = \{C_1, C_2, C_3\}$ with $C_1 = \{q \leqslant -2.04\}$, $C_2 = \{-2.04 < q \leqslant 1.94\}$ and $C_3 = \{1.94 < q\}$. The statistical weights $\mu(C_k)$ within the positional canonical ensemble μ and the metastabilities $p(C_k, C_k)$ are given by the following table:

metastable subset	C_1	C_2	C_3
statistical weight	0.2083	0.6936	0.0981
metastability	0.9863	0.9926	0.9770

The coupling matrix $C = (c_{jk})_{j,k=1,2,3}$ is given by

$$C = \begin{pmatrix} 0.9863 & 0.0137 & 0 \\ 0.0041 & 0.9926 & 0.0033 \\ 0 & 0.0230 & 0.9770 \end{pmatrix}.$$

Determining a decomposition of the state space following the exit time approach and partitioning according to the zeros of the second or third eigenfunction v_2 and v_3, respectively, we obtain:

> Second eigenfunction: $\mathcal{D} = \{C_1, C_2\}$ with $C_1 = \{q \leqslant -1.32\}$ and $C_2 = \{-1.32 < q\}$ corresponding to the exit rate $\Gamma(C_i) = -\log(\lambda_2)/\tau = 0.0076$.
> Third eigenfunction: $\mathcal{D} = \{D_1, D_2, D_3\}$ with $D_1 = \{q \leqslant -2.04\}$, $D_2 = \{-2.04 < q \leqslant 1.94\}$ and $D_3 = \{1.94 < q\}$ corresponding to the exit rate $\Gamma(D_i) = -\log(\lambda_3)/\tau = 0.0107$.

Hence, in this case the decomposition induced by h_3 is identical to the decomposition obtained via the identification algorithm.

8. Application to biomolecular systems

In this section we demonstrate that the algorithmic strategy presented in Section 5.2 can be applied to identify biomolecular conformations even for large systems as, for instance, small biomolecules with hundreds of atoms. For large systems, we have to face two particular problems:

(1) How to generate a sample of the stationary distribution in a high-dimensional space?
(2) How to decompose the highly-dimensional state space in order to discretize the propagator?

We will address these problems in the following.

Analyzing a small biomolecule. This section illustrates the performance of the algorithmic approach to the tri-ribonucleotide adenylyl(3′-5′)cytidylyl(3′-5′)cytidin (r(ACC)) model system in vacuum, see Fig. 8.1. Its physical representation is based on the GROMOS96 extended atom force field (VAN GUNSTEREN, BILLETER, EISING, HÜNENBERGER, KRÜGER, MARK, SCOTT and TIRONI [1996]), resulting in $N = 70$ atoms, hence $\Omega = \mathbf{R}^{210}$ and $\Gamma = \mathbf{R}^{420}$. The internal fluctuations are modeled w.r.t. the Hamiltonian system with randomized momenta. For details see HUISINGA, BEST, ROITZSCH, SCHÜTTE and CORDES [1999].

The sampling of the canonical ensemble was generated using an adaptive temperature hybrid Monte Carlo method (FISCHER, CORDES and SCHÜTTE [1998]) at $T = 300$ K resulting in the sampling sequence $q_1, \ldots, q_{32000} \in \Omega$. The dynamical fluctuations within the canonical ensemble were approximated by integrating $M = 4$ short trajectories of length $\tau = 80$ fs starting from each sampling point. To facilitate transitions, analogous to the ATHMC sampling, the momenta were chosen according to the canonical ensemble of momenta corresponding to four different temperatures

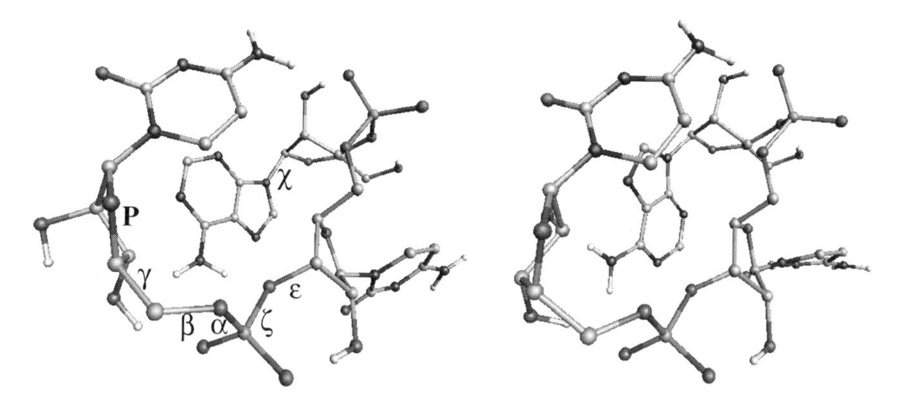

FIG. 8.1. Two representatives of different conformations of r(ACC). Left: The χ angle around the first glycosidic bond is in *anti* position (-175 degrees) and the terminal ribose pucker P is in C(3′)endo C(2′)exo conformation. Right: The χ angle is in *syn* position (19 degrees) and the terminal ribose in C(2′)endo C(3′)exo conformation. Visualization by AMIRA (KONRAD-ZUSE-ZENTRUM [2000]).

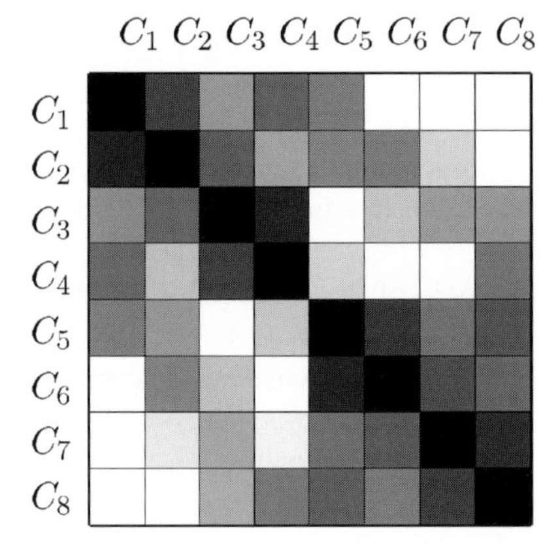

FIG. 8.2. Schematical visualization of the transition probabilities $p(\tau, C_i, C_j)$ between the conformation C_i (row) and C_j (column). The colors are chosen according to the logarithm of the corresponding entries: from $p \approx 0$ (light) to $p \approx 1$ (dark).

between 300–400 K and reweighted afterwards. This resulted in a total of $4 \times 32.000 = 128.000$ trajectories.

The configurational space was discretized using all four essential degrees of freedom, which were identified by means of a statistical analysis of the sampling data (see HUISINGA, BEST, ROITZSCH, SCHÜTTE and CORDES [1999]), resulting in $d = 36$ discretization subsets. Then the 36×36 stochastic transition matrix S was computed based on the 128.000 transitions taking the different weighting factors into account. The computation of the eigenvalues of S close to 1 yielded a cluster of eight eigenvalues with a significant gap to the remaining part of the spectrum, as shown in the following table:

k	1	2	3	4	5	6	7	8	9	...
λ_k	1.00	0.99	0.98	0.97	0.96	0.95	0.93	0.90	0.81	...

Finally, we computed conformations based on the corresponding eight eigenvectors of S via the identification algorithm presented in Section 5.3. We identified eight conformations; their statistical weights and metastabilities are shown in the following table:

conformations	C_1	C_2	C_3	C_4	C_5	C_6	C_7	C_8
statistical weight	0.11	0.01	0.12	0.03	0.32	0.04	0.29	0.10
metastability	0.99	0.94	0.96	0.89	0.99	0.95	0.98	0.96

The transition probabilities between the different conformations are visualized schematically in Fig. 8.2. The matrix allows to define a hierarchy between the clusters: on the top level, there are two clusters, one consisting of the conformations $C_1, \ldots, C_4$ and the other consisting of the conformations $C_5, \ldots, C_8$. This structure corresponds to the two 4×4 blocks on the diagonal. On the next level, each of these clusters splits up into two subclusters yielding four conformations $\{C_1, C_2\}$, $\{C_3, C_4\}$, $\{C_5, C_6\}$, $\{C_7, C_8\}$. On the bottom level, each cluster is further divided resulting in eight conformations.

9. Appendix

9.1. Some mathematical aspects of transfer operators

Consider the state space $\mathbf{X} \subset \mathbf{R}^m$ for some $m \in \mathbf{N}$ equipped with the Borel σ-algebra $\mathcal{A}$ on $\mathbf{X}$. The evolution of a single microscopic system is supposed to be given by a homogeneous Markov process $X_t = \{X_t\}_{t \in \mathbf{T}}$ in continuous or discrete time with $\mathbf{T} = \mathbf{R}_0^+$ or $\mathbf{T} = \mathbf{N}$, respectively. The motion of X_t is given in terms of a stochastic transition function p according to

$$p(t, x, A) = \mathbf{P}[X_{t+s} \in A \mid X_s = x],$$

for every $t, s \in \mathbf{T}$, $x \in \mathbf{X}$ and $A \subset \mathbf{X}$. The map $p : \mathbf{T} \times \mathbf{X} \times \mathcal{B}(\mathbf{X}) \to [0, 1]$ has the following properties

(i) $x \mapsto p(t, x, A)$ is measurable for every $t \in \mathbf{T}$ and $A \in \mathcal{B}(\mathbf{X})$;
(ii) $A \mapsto p(t, x, A)$ is a probability measure for every $t \in \mathbf{T}$ and $x \in \mathbf{X}$;
(iii) $p(0, x, \mathbf{X} \setminus \{x\}) = 0$ for every $x \in \mathbf{X}$;
(iv) the Chapman–Kolmogorov equation

$$p(t + s, x, A) = \int_{\mathbf{X}} p(t, x, \mathrm{d}z) p(s, z, A)$$

holds for every $t, s \in \mathbf{T}$, $x \in \mathbf{X}$ and $A \subset \mathbf{X}$.

The relation between Markov processes and stochastic transition functions is one-to-one, i.e., every homogeneous Markov process defines a stochastic transition function satisfying properties (i) to (iv), and vice versa (MEYN and TWEEDIE [1993], Chapter 3).

To introduce the transfer operators consider the Banach spaces of equivalence classes of measurable functions

$$L^r(\mu) = \left\{ u : \mathbf{X} \to \mathbf{C} \colon \int_{\mathbf{X}} |u(x)|^r \mu(\mathrm{d}x) < \infty \right\} \tag{9.1.1}$$

for $1 \leqslant r < \infty$ and

$$L^\infty(\mu) = \left\{ u : \mathbf{X} \to \mathbf{C} \colon \mu\text{-}\operatorname*{ess\,sup}_{x \in \mathbf{X}} |u(x)| < \infty \right\}$$

with corresponding norms $\|\cdot\|_r$ and $\|\cdot\|_\infty$, respectively. Due to Hölder's inequality we have $L^r(\mu) \subset L^s(\mu)$ for every $1 \leqslant s \leqslant r \leqslant \infty$. The propagators or forward transfer operators $P^t : L^1(\mu) \to L^1(\mu)$ are defined by

$$P^t v(y) \mu(\mathrm{d}y) = \int_{\mathbf{X}} p(t, x, \mathrm{d}y) v(x) \mu(\mathrm{d}x),$$

while the backward transfer operators $T^t : L^\infty(\mu) \to L^\infty(\mu)$ are given by

$$T^t u(x) = \mathsf{E}_x\big[u(X_t)\big] = \int_{\mathbf{X}} u(y) p(t, x, \mathrm{d}y).$$

The assumed invariance of μ w.r.t. the stochastic transition function guarantees that the forward transfer operators is well-defined (REVUZ [1975], Chapter 4) and may be consider as acting on $L^r(\mu)$ for $1 \leqslant r$. Moreover the backward transfer operator may be extended to any $L^r(\mu)$ with $1 \leqslant r$.

Due to the properties of the stochastic transition function p both definitions define semigroups of Markov operators, i.e., we especially have

$$T^t T^s = T^{t+s} \quad \text{and} \quad P^t P^s = P^{s+t},$$

and both operators, $A = T^t$ or $A = P^t$, conserve norm $\|Av\|_1 = \|A\|_1$ and positivity $Av \geqslant 0$ if $v \geqslant 0$. With the duality bracket defined in (5.1.6) the backward transfer operator is the adjoint of the propagator: $(P^t)^* = T^t$. Since $L^1(\mu)$ is a proper subset of the dual of $L^\infty(\mu)$, we have $P^t \subsetneq (T^t)^*$, hence P^t is not the adjoint of T^t. As a consequence, it is much easier to relate properties of P^t to T^t than vice versa.

Spectral properties. Consider a complex Banach space E with norm $\|\cdot\|$ and denote the spectrum[3] of a bounded linear operator $P : E \to E$ by $\sigma(P)$. For an eigenvalue $\lambda \in \sigma(P)$, the multiplicity of λ is defined as the dimension of the generalized eigenspace; see, e.g., KATO [1995], Chapter III.6. Eigenvalues of multiplicity 1 are called simple. The set of all eigenvalues $\lambda \in \sigma(P)$ that are isolated and of finite multiplicity is called the *discrete spectrum*, denoted by $\sigma_{\text{discr}}(P)$. The *essential spectral radius* $r_{\text{ess}}(P)$ of P is defined as the smallest real number, such that outside the ball of radius $r_{\text{ess}}(P)$, centered at the origin, are only discrete eigenvalues, i.e.,

$$r_{\text{ess}}(P) = \inf\big\{r \geqslant 0\colon\ \lambda \in \sigma(P) \text{ with } |\lambda| > r \text{ implies } \lambda \in \sigma_{\text{discr}}(P)\big\}.$$

This definition of $r_{\text{ess}}(P)$ is unusual in the sense that it does not involve any definition of the essential spectrum. This is owed to the surprising fact that although there are many different definitions of essential spectra, the associated spectral radii coincide (HUISINGA [2001], LEBOW and SCHECHTER [1971]) and are therefore somehow independent of the specific definition of essential spectra.

PROPOSITION 9.1.1 (HUISINGA [2001]). *Consider the* Lebesgue decomposition *of the stochastic transition function*

$$p(x, \mathrm{d}y) = p_a(x, y)\mu(\mathrm{d}y) + p_s(x, \mathrm{d}y),$$

where p_a and p_s represent the absolutely continuous and the singular part w.r.t. μ, respectively. Assume that

[3] For common functional analytical terminology see, e.g., DUNFORD and SCHWARTZ [1957], HEUSER [1986], KATO [1995], WERNER [1997].

(i) *the inequality*

$$\int_{\mathbf{X}} \int_{\mathbf{X}} p_a(x, y)^2 \mu(\mathrm{d}x)\mu(\mathrm{d}y) < \infty$$

holds, and

(ii) *there exists some* $\eta < 1$ *such that*

$$\eta = \sup p_s(x, \mathbf{X}) = 1 - \inf \int_{\mathbf{X}} p_a(x, y)\mu(\mathrm{d}y)$$

for μ*-a.e.* $x \in \mathbf{X}$.

Then, the essential spectrum is uniformly bounded away from 1, *more precisely, we have* $r_{\mathrm{ess}}(P) < \sqrt{\eta} < 1$. *In particular, condition* (C1) *is fulfilled.*

As stated in Section 5.2 the probabilistic interpretation of condition (C2) is that the Markov process admits a *unique* invariant probability measure and is *aperiodic*, hence does not show any periodic behavior. Consequently, the state space $\mathbf{X}$ can neither be decomposed into noninteracting (invariant) subsets nor into so-called cyclic subset such that the Markov process cycles with probability 1 along these cyclic subset. To interpret condition (C1), consider the Lebesgue decomposition of the stochastic transition function p. For simplicity assume that the invariant measure μ is absolutely continuous w.r.t. the Lebesgue measure as, for instance, the measure induced by the canonical ensemble. Then, as shown in Proposition 9.1.1, the essential spectral radius $r_{\mathrm{ess}}(P)$ is related to regularity conditions on the stochastic transition function. The essential spectral radius is close to one, if (a) the singular part p_s is close to one, or (b) the absolutely continuous part p_a shows a singularity like behavior, e.g., grows too fast at infinity. There is a rapidly growing literature on testable conditions, which imply that neither (a) nor (b) are valid and thus the spectral radius is strictly bounded away from 1. We only want to make the following remarks: Firstly, for the special case of the deterministic Hamiltonian system, the absolutely continuous part p_a vanishes such that $r_{\mathrm{ess}}(P) = 1$ via condition (a). Then, for the case of Langevin or Smoluchowski dynamics with smooth potentials, the singular part vanishes and the validity of condition (C1) depends on the growths of p_a at infinity only; therefore Lyapunov conditions on p_a suffice to prove (C1) in these cases. Finally, we can safely exclude problems with condition (a) whenever the transition function p allows to reach an open set with positive measure w.r.t. μ from any point $x \in \mathbf{X}$. This shows that for the Hamiltonian system with randomized momenta we only have to exclude that there is an initial position q from which τ-trajectories with arbitrary initial momentum always end up in some discrete set of positions. This typically is not the case such that condition (a) will typically be no problem for the Hamiltonian system with randomized momenta whereas it is for the deterministic Hamiltonian system, see SCHÜTTE [1998] for details.

Implications of $\mathcal{V}$-uniform ergodicity. The main body of ergodic theory and spectral theory of transfer operators is based upon another vector space setting developed in

MEYN and TWEEDIE [1993], Chapter 16. Let $\mathcal{V}:\mathbf{X}\to[1,\infty)$, finite a.e., be a given Lyapunov function, and denote by $L^\infty(\mathcal{V})$ the vector space of measurable functions $h:\mathbf{X}\to\mathbf{C}$ satisfying

$$\|h\|_{\mathcal{V}}=\sup_{x\in\mathbf{X}}\frac{|h(x)|}{\mathcal{V}(x)}<\infty.$$

Let $\|\cdot\|_{\mathcal{V}}$ also denote the operator norm induced by this $\mathcal{V}$-norm. The $\mathcal{V}$-norm on measures (as it is used in Section 5.3) is defined by

$$\|\nu\|_{\mathcal{V}}=\sup_{|v|\leqslant\mathcal{V}}\left|\int_{\mathbf{X}}v(x)\nu(\mathrm{d}x)\right|,$$

where $|v|\leqslant\mathcal{V}$ is understood to hold pointwise for every measurable function v and every $x\in\mathbf{X}$; here $\mathcal{V}$ needs to be integrable. For the special case $\mathcal{V}\equiv1$, the $\mathcal{V}$-norm coincides with the total variation norm $\|\cdot\|_{\mathrm{TV}}$.

Consider the backward transfer operator acting on the function space $L^\infty(\mathcal{V})$. We have seen in Section 5.4 that the assumption of $\mathcal{V}$-ergodicity is crucial for our approach. Then, from Theorem 5.2 of DOWN, MEYN and TWEEDIE [1995] and the results of HUISINGA, MEYN and SCHÜTTE [2003] it follows

THEOREM 9.1.1. *If the stochastic transition function p is $\mathcal{V}$-uniformly ergodic then,*

(1) *there is an invariant probability measure μ, and the semigroup T^t strongly converges in $L^\infty_{\mathcal{V}}$;*
(2) *T^t admits a spectral gap in $L^\infty(\mathcal{V})$, i.e., the set $\sigma(T^t)\cap\{z\in\mathbf{C}\colon |z-1|\leqslant\varepsilon\}$ is finite for sufficiently small $\varepsilon>0$;*
(3) *for any $B\in\mathcal{B}$ with $\mu(B)>0$, there exists $\bar{\Gamma}_B>0$ and $b<\infty$ such that*

$$\mathbf{P}_x\{\varrho_B\geqslant t\}\leqslant b\mathcal{V}(x)e^{-\bar{\Gamma}_B t},\quad x\in\mathbf{X}. \tag{9.1.2}$$

The last properties nicely illustrates that there is a deep connection between $\mathcal{V}$-ergodicity, the existence of a spectral gap, and the exponential decay of the exit time distribution. This is important for the definition of exit rates in the next subsection.

9.2. *Definition of exit rates*

Consider the Markov process X_t with transition function p. As outlined above, p defines the semigroup of transfer operators via $T^t u(x)=\mathsf{E}_x[u(X_t)]$. We now turn our attention to some open connected set B and define the *restricted process* on B induced by the process X_t via the semigroup of restricted transfer operators

$$T^t_B u(x)=\mathsf{E}_x\big[u(X_t)\mathbf{1}_X(\varrho_{B^c}\geqslant t)\big],$$

where ϱ_{B^c} denotes the exit time from B^c as introduced in (4.2.1). Thus, the new process lives on B only; considering the original process exits from B are killed. The family

$\{T_B^t\}$ is a semigroup of positive operators that in general are no longer Markov operators. We define the spectral radius for the family $\{T_B^t\}$ by

$$r(\{T_B^t\}) = \lim_{t\to\infty} (\|T_B^t\|_{\mathcal{V}})^{1/t},$$

and define the $\mathcal{V}$*-exit rate* from the set B by

$$\Gamma(B) = -\log r(\{T_B^t\}).$$

Now, one may raise the question whether the so-defined exit rate really measures the asymptotic decay of the distribution of exit times in the sense of Section 4.2. The answer is yes: If for the set B the exit time distribution decays due to (4.2.2) for some $\Gamma > 0$ that is constant in B, then $\|T^t\|_{\mathcal{V}}$ decays asymptotically as $c\exp(-\Gamma t)$ such that $\Gamma(B) = -\log r(\{T_B^t\}) = \Gamma$. The other way around and in terms of a rigorous statement for the situation described in Section 5.4, we have:

THEOREM 9.2.1. *Suppose that the conditions of Theorem* 5.4.2 *hold implying that the subset C is metastable with exit rate $\Gamma > 0$. Then there exists $\delta_0 > 0$ such that for all $s, T > 0$ the conditional distribution of exit times satisfies*

$$F_x(s,T) = e^{-\Gamma s}\left[\frac{1 + \mathrm{O}(G_x(s))}{1 + \mathrm{O}(e^{-T\delta_0}G_x(s))}\right] \quad (T \to \infty)$$

with $G_x(s) = e^{-\delta_0 s}\mathcal{V}(x)/h(x)$.

In HUISINGA, MEYN and SCHÜTTE [2003], δ_0 is some well-defined computable constant and it is shown that in typical situations we have $\delta_0 \gg \Gamma$.

References

AMADEI, A., A.B.M. LINSSEN and H.J.C. BERENDSEN (1993), Essential dynamics of proteins, *Proteins* **17**, 412–425.

BERNE, B.J., G. CICCOTTI and D.F. COKER, eds. (1998), *Classical and Quantum Dynamics in Condensed Phase Simulations* (World Scientific, Singapore).

BERNE, B.J. and J.E. STRAUB (1997), Novel methods of sampling phase space in the simulation of biological systems, *Curr. Opinion in Struct. Biol.* **7**, 181–189.

BOLHUIS, P.G., C. DELLAGO, D. CHANDLER and P. GEISSLER (2001), Transition path sampling: Throwing ropes over mountain passes, in the dark, *Ann. Rev. of Phys. Chem.*, in press.

BOND, S.D., B.B.L. BENEDICT and J. LEIMKUHLER (1999), The Nosé–Poincaré method for constant temperature molecular dynamics, *JCP* **151** (1), 114–134.

BOVIER, A., M. ECKHOFF, V. GAYRARD and M. KLEIN (2001), Metastability in stochastic dynamics of disordered mean-field models, *Probab. Theor. Rel. Fields* **119**, 99–161.

CHANDLER, D. (1998), Finding transition pathways: throwing ropes over rough mountain passes, in the dark, in: B. Berne, G. Ciccotti and D. Coker, eds., *Classical and Quantum Dynamics in Condensed Phase Simulations* (World Scientific, Singapore) 51–66.

DAVIES, E.B. (1982), Metastable states of symmetric Markov semigroups I, *Proc. London Math. Soc.* **45** (3), 133–150.

DELLNITZ, M. and O. JUNGE (1998), An adaptive subdivision technique for the approximation of attractors and invariant measures, *Comput. Visual. Sci.* **1**, 63–68.

DELLNITZ, M. and O. JUNGE (1999), On the approximation of complicated dynamical behavior, *SIAM J. Num. Anal.* **36** (2), 491–515.

DEUFLHARD, P. and F. BORNEMANN (1994), *Numerische Mathematik II* (de Gruyter).

DEUFLHARD, P., M. DELLNITZ, O. JUNGE and C. SCHÜTTE (1999), Computation of essential molecular dynamics by subdivision techniques, in: P. Deuflhard, J. Hermans, B. Leimkuhler, A.E. Mark, S. Reich and R.D. Skeel, eds., *Computational Molecular Dynamics: Challenges, Methods, Ideas*, Lecture Notes in Computational Science and Engineering, Vol. 4 (Springer) 98–115.

DEUFLHARD, P., T. FRIESE and F. SCHMIDT (1997), A nonlinear multigrid eigenproblem solver for the complex Helmholtz equation, Preprint SC-97-55, Konrad-Zuse-Zentrum, Berlin, Available via http://www.zib.de/bib/pub/pw/.

DEUFLHARD, P., J. HERMANS, B. LEIMKUHLER, A.E. MARK, S. REICH and R.D. SKEEL, eds. (1999), *Computational Molecular Dynamics: Challenges, Methods, Ideas*, Lecture Notes in Computational Science and Engineering, Vol. 4 (Springer).

DEUFLHARD, P. and A. HOHMANN (1993), *Numerische Mathematik I* (de Gruyter, Berlin).

DEUFLHARD, P., W. HUISINGA, A. FISCHER and C. SCHÜTTE (2000), Identification of almost invariant aggregates in reversible nearly uncoupled Markov chains, *Lin. Alg. Appl.* **315**, 39–59.

DOWN, D., S.P. MEYN and R.L. TWEEDIE (1995), Exponential and uniform ergodicity of Markov processes, *Ann. Prob.* **23**, 1671–1691.

DUNFORD, N. and J.T. SCHWARTZ (1957), *Linear Operators, Part I: General Theory*, Pure and Applied Mathematics, Vol. VII (Interscience, New York).

E, W., W. RAN and E. VANDEN-EIJNDEN (2002), Probing multiscale energy landscapes using the string method, *Phys. Rev. Lett.*, submitted.

ELBER, R. and M. KARPLUS (1987), Multiple conformational states of proteins: A molecular dynamics analysis of Myoglobin, *Science* **235**, 318–321.

FERGUSON, D.M., J.I. SIEPMANN and D.G. TRUHLAR, eds. (1999), *Monte Carlo Methods in Chemical Physics*, Advances in Chemical Physics, Vol. 105 (Wiley, New York).

FISCHER, A. (2000), An uncoupling–coupling technique for Markov chains Monte Carlo methods, Report 00-04, Konrad-Zuse-Zentrum, Berlin.

FISCHER, A., F. CORDES and C. SCHÜTTE (1998), Hybrid Monte Carlo with adaptive temperature in a mixed-canonical ensemble: Efficient conformational analysis of RNA, *J. Comput. Chem.* **19**, 1689–1697.

FISCHER, A., C. SCHÜTTE, P. DEUFLHARD and F. CORDES (2002), Hierarchial uncoupling–coupling of metastable conformations, in: T. Schlick and H.H. Gan, eds., *Computational Methods for Macromolecules: Challenges and Applications, Proceedings of the 3rd International Workshop on Algorithms for Macromolecular Modelling, New York, Oct. 12–14, 2000*, Lecture Notes in Computational Science and Engineering, Vol. 24 (Springer).

FRAUENFELDER, H. and B.H. MCMAHON (2000), Energy landscape and fluctuations in proteins, *Ann. Phys. (Leipzig)* **9** (9–10), 655–667.

FRAUENFELDER, H., P.J. STEINBACH and R.D. YOUNG (1989), Conformational relaxation in proteins. *Chem. Soc.* **29**A (145–150).

GALLIAT, T. and P. DEUFLHARD (2000), Adaptive hierarchical cluster analysis by self-organizing box maps, Report SC-00-13, Konrad-Zuse-Zentrum, Berlin.

GALLIAT, T., P. DEUFLHARD, R. ROITZSCH and F. CORDES (2002), Automatic identification of metastable conformations via self-organized neural networks, in: T. Schlick and H.H. Gan, eds., *Computational Methods for Macromolecules: Challenges and Applications, Proceedings of the 3rd International Workshop on Algorithms for Macromolecular Modelling, New York, Oct. 12–14, 2000*, Lecture Notes in Computational Science and Engineering, Vol. 24 (Springer).

GALLIAT, T., W. HUISINGA and P. DEUFLHARD (2000), Self-organizing maps combined with eigenmode analysis for automated cluster identification, in: H. Bothe and R. Rojas, eds., *Neural Computation* (ICSC Academic Press) 227–232.

GILKS, W., S. RICHARDSON and D. SPIEGELHALTER, eds. (1997), *Markov chain Monte-Carlo in Practice* (Chapman and Hall, London).

GRUBMÜLLER, H. (1995), Predicting slow structural transitions in macromolecular system: Conformational flooding, *Phys. Rev. E.* **52**, 2893–2906.

HEUSER, H. (1986), *Funktionalanalysis* (Teubner, Stuttgart).

HUBER, T., A.E. TORDA and W.F. VAN GUNSTEREN (1994), Local elevation: A method for improving the searching properties of molecular dynamics simulation, *JCAMD* **8**, 695–708.

HUISINGA, W. (2001), Metastability of Markovian systems: A transfer operator approach in application to molecular dynamics, PhD thesis, Free University Berlin.

HUISINGA, W., C. BEST, R. ROITZSCH, C. SCHÜTTE and F. CORDES (1999), From simulation data to conformational ensembles: Structure and dynamic based methods, *J. Comp. Chem.* **20** (16), 1760–1774.

HUISINGA, W., S. MEYN and C. SCHÜTTE (2003), Phase transitions and metastability in Markovian and molecular systems, accepted in *Ann. Appl. Probab.*

HUISINGA, W. and B. SCHMIDT (2002), Metastability and dominant eigenvalues of transfer operators, in preparation.

JAIN, A.K. and R.C. DUBES (1988), *Algorithms for Clustering Data* (Prentice Hall, New Jersey), Advanced reference series edition.

KARATZAS, I. and S.E. SHREVE (1991), *Brownian Motion and Stochastic Calculus*, Graduate Texts in Mathematics (Springer, New York).

KATO, T. (1995), *Perturbation Theory for Linear Operators* (Springer, Berlin), reprint of the 1980 edn.

KLEEMAN, R. (2001), Measuring dynamical prediction utility using relative entropy, *J. Atmos. Sci.*, accepted.

KONRAD-ZUSE-ZENTRUM (2000), AMIRA – advanced visualization, data analysis and geometry reconstruction, user's guide and reference manual (Konrad-Zuse-Zentrum für Informationstechnik Berlin (ZIB), Indeed-Visual Concepts GmbH and TGS Template Graphics Software Inc.)

LASOTA, A. and M.C. MACKEY (1994), *Chaos, Fractals and Noise*, Applied Mathematical Sciences, Vol. 97 (Springer, New York), 2nd edn.

LEBOW, A. and M. SCHECHTER (1971), Semigroups of operators and measures of noncompactness, *J. Funct. Anal.* **7**, 1–26.

LEHOUCQ, R.B., D.C. SORENSEN and C. YANG (1998), *ARPACK User's Guide: Solution of Large Eigenvalue Problems by Implicit Restartet Arnoldi Methods* (Rice University, Houston).

LEZAUD, P. (2001), Chernoff and Berry–Esséen inequalities for Markov processes, *ESIAM: P & S* **5**, 183–201.

MATTINGLY, J., A.M. STUART and D.J. HIGHAM (2003), Ergodicity for SDEs and approximations: Locally Lipschitz vector fields and degenerated noise, to appear in *Stoch. Proc. Appl.*

MEYN, S. and R. TWEEDIE (1993), *Markov Chains and Stochastic Stability* (Springer, Berlin).

NIENHAUS, G.U., J.R. MOURANT and H. FRAUENFELDER (1992), Spectroscopic evidence for conformational relaxation in Myoglobin, *PNAS* **89**, 2902–2906.

OLENDER, R. and R. ELBER (1996), Calculation of classical trajectories with a very large time step: Formalism and numerical examples, *J. Chem. Phys.* **105**, 9299–9315.

REVUZ, D. (1975), *Markov Chains* (North-Holland, Amsterdam, Oxford).

RISKEN, H. (1996), *The Fokker–Planck Equation* (Springer, New York), 2nd edn.

ROBERTS, G.O. and J.S. ROSENTHAL (1997), Geometric ergodicity and hybrid Markov chains, *Elect. Comm. in Probab.* **2**, 13–25.

SCHÜTTE, C. (1998), Conformational dynamics: modelling, theory, algorithm, and application to biomolecules, Habilitation Thesis, Fachbereich Mathematik und Informatik, Freie Universität Berlin.

SCHÜTTE, C., A. FISCHER, W. HUISINGA and P. DEUFLHARD (1999), A direct approach to conformational dynamics based on hybrid Monte Carlo, *J. Comput. Phys., Special Issue on Computational Biophysics* **151**, 146–168.

SCHÜTTE, C. and W. HUISINGA (2000), On conformational dynamics induced by Langevin processes, in: B. Fiedler, K. Gröger and J. Sprekels, eds., *EQUADIFF 99 – International Conference on Differential Equations*, Vol. 2 (World Scientific, Singapore) 1247–1262.

SCHÜTTE, C., W. HUISINGA and P. DEUFLHARD (2001), Transfer operator approach to conformational dynamics in biomolecular systems, in: B. Fiedler, ed., *Ergodic Theory, Analysis and Efficient Simulation of Dynamical Systems* (Springer) 191–223.

SINGLETON, G. (1984), Asymptotically exact estimates for metastable Markov semigroups, *Quart. J. Math. Oxford* **35** (2), 321–329.

VAN GUNSTEREN, W.F., S.R. BILLETER, A.A. EISING, P.H. HÜNENBERGER, P. KRÜGER, A.E. MARK, W.R.P. SCOTT and I.G. TIRONI (1996), *Biomolecular Simulation: The GROMOS96 Manual and User Guide* (vdf Hochschulverlag AG, ETH Zürich).

WEBER, M. and T. GALLIAT (2002), Characterization of transition states in conformational dynamics using Fuzzy sets, Technical Report 02-12, Konrad-Zuse-Zentrum (ZIB), Berlin.

WERNER, D. (1997), *Funktionalanalysis* (Springer, Berlin), 2nd edn.

ZEWAIL, A.H. (1995), Der Augenblick der Molekülbildung, *Spektrum der Wissenschaft, Digest 2: Moderne Chemie*, 18–26.

ZEWAIL, A.H. (1996), Femtochemistry: Recent progress in studies of dynamics and control of reactions and their transition state, *J. Phys. Chem.* **100** (31), 12701–12724.

ZHOU, H.X., S.T. WLODEC and J.A. MCCAMMON (1998), Conformation gating as a mechanism for enzyme specificity, *Proc. Nat. Acad. Sci. USA* **95**, 9280–9283.

Theory of Intense Laser-Induced Molecular Dissociation: From Simulation to Control

O. Atabek[a], R. Lefebvre[a,b], T.T. Nguyen-Dang[c]

[a]*Laboratoire de Photophysique Moléculaire du CNRS, Bâtiment 213, Campus d'Orsay, 91405 Orsay, France*

[b]*UFR de Physique Fondamentale et Appliquée, UPMC, 75005, Paris, France*

[c]*Département de Chimie, Université Laval, Québec, Canada*

1. Introduction

The detailed description of matter-electromagnetic field interactions within the framework of molecular alignment and fragmentation provides the key for the interpretation of photochemical reactions (BANDRAUK [1994]). Not only can structural and dynamical properties of molecules be analyzed or the nature and the role of transient species be probed, but the reaction itself can be optimally controlled by appropriately using different sources of radiation. Intense laser fields apply forces to molecules that are strong enough to produce important distortions and lead to the discovery of nonlinear, nonperturbative effects such as above-threshold dissociation (ATD) and laser-induced avoided crossings in resonant electronic excitation (HE, ATABEK and GIUSTI-SUZOR [1990], BUCKSBAUM, ZAVRIYEV, MULLER and SCHUMACHER [1990]) or pendular states in rotational excitation (FRIEDRICH and HERSCHBACH [1995], SEIDEMAN [1995]) that are discussed in this chapter. Through these effects, the laser offers the possibility to control both internal (total and partial dissociation rates as well as fragment kinetic and angular distributions) and external motions (angular positioning of the molecule with respect to the laser polarization vector). The study of laser-induced molecular

Computational Chemistry
Special Volume (C. Le Bris, Guest Editor) of
HANDBOOK OF NUMERICAL ANALYSIS, VOL. X
P.G. Ciarlet (Editor)
© 2003 Elsevier Science B.V. All rights reserved

alignment is motivated by the interpretation of the observed anisotropy in the angular distribution of fragments in photodissociation experiments (ZAVRIYEV, BUCKSBAUM, MULLER and SCHUMACHER [1990], NORMAND, LOMPRE and CORNAGGIA [1992], CHANDLER, NEYER and HECK [1998]) and also by possible laser manipulations of molecules as a goal. Besides this, alignment and orientation of molecules allow orders of magnitude enhancements in collisional cross-sections of photochemical reactivity (LOESCH and REMSCHEID [1990]) or femtochemistry experiments (ZEWAIL [1988]), to trap molecules (SEIDEMAN [1997]), and to control laser-focused molecular beams (SEIDEMAN [1997], STAPELFELDT, SAKAI, CONSTANT and CORKUM [1997]) to achieve nanoscale design (such as molecular quantum wires and dots). Highly-nonlinear field-induced barrier lowering or suppression and stabilization (trapping or quenching) mechanisms both in the Visible-Ultra-Violet (Vis-UV) and the Infra-Red (IR) frequency regimes underly chemical bond softening or hardening processes, respectively. Their interplay through the adequate adjustment of laser characteristics such as laser frequency, amplitude, phase, polarization, and temporal shape, provides control opportunities. The present chapter reviews these aspects of intense-field molecular dynamics and discusses their pertinence to laser-control theory. The chapter is organized as followed. The theory of laser–molecule interactions is presented first in Section 2, starting from a general multicharged system considered in both quantized and semiclassical field descriptions. In each case, the Hamiltonian of the laser-driven system is given in both velocity and length gauges (LOUDON [1973], REISS [1980], COHEN-TANNOUDJI, DUPONT-ROC and GRYNBERG [1987], BANDRAUK [1994]) as well as in the Bloch–Nordsieck (BLOCH and NORDSIECK [1937], PAULI and FIERZ [1938], NGUYEN-DANG and BANDRAUK [1983]) or Kramers–Henneberger representations (HENNEBERGER [1968], GAVRILA [1992]). For strong fields, the equivalence between the quantized and the semiclassical descriptions is established through the Mollow transformation (MOLLOW [1975], COHEN-TANNOUDJI, DUPONT-ROC and GRYNBERG [1987], KELLER and ATABEK [1993]). The section continues with a thorough analysis of multichannel field-dressed molecular models within the Born–Oppenheimer approximation (BORN and OPPENHEIMER [1927]) (for the molecular part) and the Floquet formalism (FLOQUET [1883], CHU [1981]) (for the radiative part). Section 3 is devoted to numerical methodologies for solving the problem either in a time independent frame, where emphasis is placed on field-induced resonances, or in a time dependent context, highlighting time-resolved features. Basically, the numerical techniques that are retained are the Fox–Goodwin propagator (FOX [1957]) using complex rotation (MOISEYEV [1998]) of the spatial coordinate (time independent approach), or the split-operator formula coupled to Fast Fourier Transform algorithms (FEIT, FLECK and STEIGER [1982]) adequately modified to take into account Volkov type solutions (KELLER [1995]) in asymptotic regions (time dependent approach). Dynamical processes with their underlying mechanisms and control strategies are presented in Section 4. The first example deals with the dissociation dynamics of the molecular ion H_2^+ to which numerous experimental and theoretical investigations have recently been devoted (GIUSTI-SUZOR, HE, ATABEK and MIES [1990], ZAVRIYEV, BUCKSBAUM, MULLER and SCHUMACHER [1990], CHANDLER, NEYER and HECK [1998]). The results of wavepacket dynamics simulation are analyzed in terms of complementary basic mechanisms, i.e., bond-softening and vibrational trapping for UV excitation or, equivalently, barrier lowering and dissociation quenching for IR excitation. This leads to the interpretation of some unexpected features in experimentally observed angularly resolved photofragment kinetic energy distributions, namely the production of less aligned fragments at the higher kinetic energies of the multiphoton ATD spectrum, when increasing the field strength. The second example, considered in Section 5 is devoted to the study of the alignment dynamics of HCN undergoing a nondissociative radiative excitation under an IR laser pulse. The results obtained from the Time Dependent Schrödinger Equation (TDSE) are interpreted through a high-frequency approximation applied to rigid-rotor pendular states, and through a discussion of adiabatic versus sudden excitation mechanisms. Orienting a molecule (i.e., with a specific direction along the laser polarization vector) is even a more challenging objective. An optimal control scheme leading to a kick mechanism for the momentum transfer to the molecule is discussed. The feasibility of orientation with experimentally reachable laser pulses is numerically illustrated using genetic algorithms.

2. General theory of laser–molecule interactions

2.1. Molecular Hamiltonian in the presence of a laser field

Many forms of the Hamiltonian operator describing a system interacting with a radiation field are currently in use (LOUDON [1973], LOUISELL [1973], COHEN-TANNOUDJI, DUPONT-ROC and GRYNBERG [1987]). These are defined within either of two descriptions: One is fully quantized and treats both the laser field and the molecular system quantum mechanically. The other is semiclassical and assumes at the outset a classical field while the charged particles are treated quantum mechanically. The present section reviews the various forms that the Hamiltonian of a laser-driven molecule may assume and serves to lay the ground work for the construction of the models encountered in the subsequent sections. Although we shall consider a system of N point-charges (electrons and nuclei) q_k of position vectors $\mathbf{r}_k$, $k = 1, \ldots, N$, the following review will be given for a single charge and a single mode. This is to avoid crowded notations and to concentrate on the essentials of the derivations. Nevertheless, for future references, the final form of the Hamiltonian for a multicharge system will be given after the single-charge version has been derived in each gauge or representation. We will start with the quantized-field picture, then give the corresponding semiclassical formulation. We then recall how the correspondence between the fully quantized formulation and the semiclassical one can be established. Field attributes that are operators in the fully quantized description will be represented by a symbol carrying a caret. This serves to distinguish these operator-valued quantities from classical functions representing the same attributes of the field in the semiclassical theory. Since no confusion is possible for the material part, which is always described quantum mechanically, no such distinction at the level of notations is necessary.

2.1.1. Fully quantal formulations

Radiation-field (RF) or velocity gauge. In the so-called Coulomb radiation-field (RF) or velocity (v) gauge, the total Hamiltonian describing the interacting single particle +

field system is:

$$H_Q^v = \frac{1}{2m}\left(\mathbf{p} - q\hat{\mathbf{A}}(\mathbf{r})\right)^2 + V(\mathbf{r}) + H_{\text{mode}}. \tag{2.1}$$

For a general multicharge system (e.g., a molecule), the Hamiltonian is:

$$H_Q^v = \sum_k \frac{1}{2m_k}\left(\mathbf{p}_k - q_k\hat{\mathbf{A}}(\mathbf{r}_k)\right)^2 + V\left(\{\mathbf{r}_k\}\right) + H_{\text{mode}}. \tag{2.2}$$

The single-mode (monochromatic), quantized, radiation field, of wavevector $\mathbf{k}$, polarization vector $\boldsymbol{\varepsilon}$, frequency $\omega = c|\mathbf{k}|$, is described by the vector potential (operator) $\hat{\mathbf{A}}(\mathbf{r})$ given by

$$\hat{\mathbf{A}}(\mathbf{r}) = \left(\frac{\hbar}{2\varepsilon_0\omega V}\right)^{1/2}\left\{\hat{a}e^{i\mathbf{k}\cdot\mathbf{r}} + \hat{a}^\dagger e^{-i\mathbf{k}\cdot\mathbf{r}}\right\}\boldsymbol{\varepsilon}. \tag{2.3}$$

The corresponding expression for the operator describing the electric field is

$$\widehat{\mathbf{E}}(\mathbf{r}) = -\frac{i}{\hbar}\left[\hbar\omega\hat{a}^\dagger\hat{a}, \hat{\mathbf{A}}(\mathbf{r})\right] = i\left(\frac{\hbar\omega}{2\varepsilon_0 V}\right)^{1/2}\left\{\hat{a}e^{i\mathbf{k}\cdot\mathbf{r}} - \hat{a}^\dagger e^{-i\mathbf{k}\cdot\mathbf{r}}\right\}\boldsymbol{\varepsilon}. \tag{2.4}$$

In Eqs. (2.3) and (2.4) V is a finite volume leading to a discretization of the laser modes. It will be shown at the end of this section that in the present context the dynamics of the molecule depends only on the photon density N/V for a laser mode involving N photons. In the above, $\hat{a}^\dagger$ and $\hat{a}$ are the creation and annihilation operators associated with the field mode. They satisfy the commutation relation:

$$\left[\hat{a}, \hat{a}^\dagger\right] = 1. \tag{2.5}$$

In terms of these operators, the single mode Hamiltonian is

$$H_{\text{mode}} = \hbar\omega\left(\hat{a}^\dagger\hat{a} + \frac{1}{2}\right) \tag{2.6}$$

and its effect on a field-eigenstate $|n\rangle$ with n photons is:

$$H_{\text{mode}}|n\rangle = \left(n + \frac{1}{2}\right)\hbar\omega|n\rangle. \tag{2.7}$$

The operators $\hat{a}$ and $\hat{a}^\dagger$ act on a photon state as:

$$\hat{a}^\dagger|n\rangle = \sqrt{n+1}|n+1\rangle, \qquad \hat{a}|n\rangle = \sqrt{n}\,|n-1\rangle. \tag{2.8}$$

The constant zero-point energy of the field $\frac{1}{2}\hbar\omega$ will be ignored, as this contributes only to the global phase factor of the dressed system's statevector and is thus dynamically irrelevant. The statevector of the system will be denoted in this representation $\Psi^v(t)\rangle$, which is expanded in terms of tensorial products involving molecular eigenfunctions times the field-states $|n\rangle$. The superscript v is a reminder that this is associated with the original velocity gauge (we will consider transformations of the Hamiltonian and the state vector to other representations). It satisfies the Schrödinger equation:

$$i\hbar\frac{\partial|\Psi^v(t)\rangle}{\partial t} = H_Q^v\big|\Psi^v(t)\big\rangle. \tag{2.9}$$

In fact, since H_Q^v is time independent, the wavefunctions solution of Eq. (2.9) can be written:

$$\big|\Psi^v(t)\big\rangle = \exp\left[-\frac{iE_{tot}t}{\hbar}\right]\big|\Phi^v\big\rangle. \tag{2.10}$$

Eq. (2.9) can be replaced by the stationary-state version:

$$H_Q^v\big|\Phi^v\big\rangle = E_{tot}\big|\Phi^v\big\rangle, \tag{2.11}$$

E_{tot} is the total energy of the particle + field (conservative) system. The wavefunctions defined by Eq. (2.10) constitute a basis to follow the motion of an arbitrary initial state, which can always be represented as a linear combination of these functions (wavepacket).

For an optical field, and in addressing its effect on internal molecular motions which are limited to molecular dimensions (as opposed to external, translational motions which can extend to large distances), it is justified to invoke the so-called long-wavelength approximation (LWA) in which one assumes that the wavelength $\lambda = 1/k$ is much larger than the extensions of the molecular motions. This implies that the wavevector of the radiation field is such that

$$\mathbf{k}\cdot\mathbf{r} \simeq 0, \tag{2.12}$$

so that the (now position-independent) vector potential and the electric-field operators reduce to

$$\hat{\mathbf{A}} = \beta\{\hat{a} + \hat{a}^\dagger\}\boldsymbol{\varepsilon}, \tag{2.13}$$

$$\widehat{\mathbf{E}} = i\beta\omega\{\hat{a} - \hat{a}^\dagger\}\boldsymbol{\varepsilon} \tag{2.14}$$

with

$$\beta = \left(\frac{\hbar}{2\varepsilon_0\omega V}\right)^{1/2}. \tag{2.15}$$

The RF gauge Hamiltonian of Eq. (2.1) becomes

$$H_Q^v = \frac{1}{2m}\left(\mathbf{p} - q\beta\{\hat{a} + \hat{a}^\dagger\}\boldsymbol{\varepsilon}\right)^2 + V(\mathbf{r}) + H_{mode}. \tag{2.16}$$

In this equation $\mathbf{r}$ and $\mathbf{p}$ act in the vectorial space of the particle, whereas $\hat{a}$ and $\hat{a}^\dagger$ operate over field-states. $V(\mathbf{r})$, which represents the potential energy (deriving from external and/or internal force fields) is a local operator, while $\mathbf{p} = -i\hbar\boldsymbol{\nabla}_\mathbf{r}$, the linear momentum, is a nonlocal operator.

Electric-field (EF) or length gauge. Within the LWA, a different form of dressed Hamiltonian is obtained through the following unitary transformation, a proper gauge transformation (COHEN-TANNOUDJI, DUPONT-ROC and GRYNBERG [1987]):

$$\widehat{\mathcal{T}}_{L\leftarrow v} = \exp\left(-\frac{i}{\hbar} q\mathbf{r}\cdot\hat{\mathbf{A}}\right) = \exp\left(-\frac{i}{\hbar}\beta q\mathbf{r}\cdot\boldsymbol{\varepsilon}\left[\hat{a} + \hat{a}^\dagger\right]\right). \tag{2.17}$$

This leaves the position vector $\mathbf{r}$ invariant and displaces the momentum by $q\hat{\mathbf{A}}$

$$\widehat{\mathcal{T}}_{L\leftarrow v}\mathbf{r}\widehat{\mathcal{T}}^\dagger_{L\leftarrow v} = \mathbf{r}, \qquad \widehat{\mathcal{T}}_{L\leftarrow v}\mathbf{p}\widehat{\mathcal{T}}^\dagger_{L\leftarrow v} = \mathbf{p} + q\hat{\mathbf{A}}, \tag{2.18}$$

while it transforms the field operators $\hat{a}, \hat{a}^\dagger$ into

$$\widehat{\mathcal{T}}_{L\leftarrow v}\hat{a}\widehat{\mathcal{T}}^\dagger_{L\leftarrow v} = \hat{a} + \frac{i}{\hbar}q\beta\mathbf{r}\cdot\boldsymbol{\varepsilon}, \qquad \widehat{\mathcal{T}}_{L\leftarrow v}\hat{a}^\dagger\widehat{\mathcal{T}}^\dagger_{L\leftarrow v} = \hat{a}^\dagger - \frac{i}{\hbar}q\beta\mathbf{r}\cdot\boldsymbol{\varepsilon}. \tag{2.19}$$

Correspondingly the statevector of the dressed system is changed into

$$\left|\Psi^L(t)\right\rangle = \widehat{\mathcal{T}}_{L\leftarrow v}\left|\Psi^v(t)\right\rangle \tag{2.20}$$

and the Schrödinger equation (2.9) is transformed into

$$i\hbar\frac{\partial|\Psi^L(t)\rangle}{\partial t} = H_\mathrm{Q}^L\left|\Psi^L(t)\right\rangle, \tag{2.21}$$

with

$$H_\mathrm{Q}^L = \widehat{\mathcal{T}}_{L\leftarrow v}H_\mathrm{Q}^v\widehat{\mathcal{T}}^\dagger_{L\leftarrow v} = \frac{1}{2m}\mathbf{p}^2 + V(\mathbf{r}) - \mathbf{d}\cdot\widehat{\mathbf{E}} + \frac{\omega}{\hbar}q^2\beta^2(\mathbf{r}\cdot\boldsymbol{\varepsilon})^2 + H_\mathrm{mode}, \tag{2.22}$$

where $\mathbf{d} = q\mathbf{r}$ is the dipole moment operator of the particle.

For a multicharge system, this expression of the transformed Hamiltonian reads

$$H_\mathrm{Q}^L = \widehat{\mathcal{T}}_{L\leftarrow v}H_\mathrm{Q}^v\widehat{\mathcal{T}}^\dagger_{L\leftarrow v} = \sum_k \frac{1}{2m_k}\mathbf{p}_k^2 + V(\{\mathbf{r}_k\}) - \mathbf{d}\cdot\widehat{\mathbf{E}} + \frac{\omega}{\hbar}\sum_k q_k^2\beta^2(\mathbf{r}_k\cdot\boldsymbol{\varepsilon})^2 + H_\mathrm{mode} \tag{2.23}$$

with $\mathbf{d} = \sum_k q_k\mathbf{r}_k$.

This new form of the dressed system's Hamiltonian contains the field-matter interaction through the scalar product of the electric field operator and the position ('length') vector $\mathbf{r}$. For this reason, the new representation is often called the 'length' gauge or 'Electric Field' (EF) gauge.

Bloch–Nordsieck (BN) representation. Another form of the dressed molecule Hamiltonian is associated with the so-called Bloch–Nordsieck (BLOCH and NORDSIECK [1937]), also called the Pauli–Fierz (PAULI and FIERZ [1938]) or space-translation representation. This is obtained from the RF or velocity gauge form within the LWA, by the following construction. Introducing a unitary operator

$$\widehat{\mathcal{T}}_{\mathrm{BN}} = \exp\left\{\frac{q\beta'}{m\hbar\omega}\mathbf{p}\cdot\boldsymbol{\varepsilon}\left[\hat{a} - \hat{a}^{\dagger}\right]\right\} \tag{2.24}$$

with

$$\beta' = \frac{\beta}{\left(1 + \frac{2q^2\beta^2}{m\hbar\omega}\right)}, \tag{2.25}$$

the $\hat{\mathbf{A}}\cdot\mathbf{p}$ coupling term originally found in H_{Q}^{v} is cancelled out exactly, yielding

$$\begin{aligned}\widehat{\mathcal{T}}_{\mathrm{BN}} H_{\mathrm{Q}}^{v} \widehat{\mathcal{T}}_{\mathrm{BN}}^{\dagger} = {} & \frac{1}{2m}\left[\left(1 - \frac{2q^2\beta\beta'}{m\hbar\omega}\right)^2 + \frac{2q^2\beta'^2}{m\hbar\omega}\right](\mathbf{p}\cdot\boldsymbol{\varepsilon})^2 + \frac{1}{2m}\mathbf{p}_{\perp}^2 \\ & + V\left(\mathbf{r} - i\frac{q\beta'}{m\omega}\boldsymbol{\varepsilon}\left[\hat{a} - \hat{a}^{\dagger}\right]\right) + \hbar\omega\hat{a}^{\dagger}\hat{a} + \frac{q^2\beta^2}{2m}\left(\hat{a}^{\dagger} + \hat{a}\right)^2.\end{aligned} \tag{2.26}$$

We can bring this result to a lot simpler (exact) form if we introduce a unitary operator of the form

$$\widehat{\mathcal{T}}_{\mathrm{sc}} = \exp\left[\sigma\left(\hat{a}^2 - \hat{a}^{\dagger 2}\right)\right] \tag{2.27}$$

with

$$\sigma = -\frac{1}{8}\ln\left(\frac{2q^2\beta^2}{m\hbar\omega} + 1\right). \tag{2.28}$$

Defining a normalized mass $m' = m/[(1 - 2q^2\beta\beta'/m\hbar\omega)^2 + \frac{2q^2\beta'^2}{m\hbar\omega}]$ and frequency $\omega' = \omega(2q^2\beta^2/m\hbar\omega + 1)^{1/2}$, the result is:

$$\begin{aligned}H_{\mathrm{Q}}^{\mathrm{BN}} = \widehat{\mathcal{T}}_{\mathrm{sc}}\widehat{\mathcal{T}}_{\mathrm{BN}} H_{\mathrm{Q}}^{v} \widehat{\mathcal{T}}_{\mathrm{BN}}^{\dagger} \widehat{T}_{\mathrm{sc}}^{\dagger} = {} & \frac{1}{2m'}(\mathbf{p}\cdot\boldsymbol{\varepsilon})^2 + \frac{1}{2m}\mathbf{p}_{\perp}^2 \\ & + V\left(\mathbf{r} - i\frac{q\beta'}{m\omega}e^{-2\sigma}\boldsymbol{\varepsilon}\left[\hat{a} - \hat{a}^{\dagger}\right]\right) + \hbar\omega'\hat{a}^{\dagger}\hat{a}.\end{aligned} \tag{2.29}$$

Due to the smallness of the dimensionless quantity $2q^2\beta^2/m\hbar\omega$ (this is typically of order 10^{-24} for an electron or nucleus in an optical field), the scaling transformation $\widehat{\mathcal{T}}_{\text{sc}}$, as well as the renormalization of the electron mass and that of the field frequency can simply be ignored. In this approximation, β' reduces to β and we can rewrite Eq. (2.29) in the final form:

$$H_{\text{Q}}^{\text{BN}} = \widehat{\mathcal{T}}_{\text{BN}} H_{\text{Q}}^{v} \widehat{\mathcal{T}}_{\text{BN}}^{\dagger} = \frac{\mathbf{p}^2}{2m} + V\left(\mathbf{r} - i\frac{q\beta}{m\omega}\boldsymbol{\varepsilon}\left[\hat{a} - \hat{a}^{\dagger}\right]\right) + \hbar\omega\hat{a}^{\dagger}\hat{a}. \tag{2.30}$$

For a multicharge system, one has:

$$H_{\text{Q}}^{\text{BN}} = \sum_k \frac{\mathbf{p}_k^2}{2m_k} + V\left(\left\{\mathbf{r}_k - i\frac{q_k\beta}{m\omega}\boldsymbol{\varepsilon}\left[\hat{a} - \hat{a}^{\dagger}\right]\right\}\right) + \hbar\omega\hat{a}^{\dagger}\hat{a}. \tag{2.31}$$

The development of this approach for a molecular system has been made by NGUYEN-DANG and BANDRAUK [1983], NGUYEN-DANG and BANDRAUK [1984] and BANDRAUK and NGUYEN-DANG [1985].

2.1.2. *Semiclassical formulations*

Each of the Hamiltonians for the dressed particle mentioned above, the one in the velocity gauge, Eq. (2.16), the Bloch–Nordsieck Hamiltonian, Eq. (2.30), or the Hamiltonian written in the length gauge, Eq. (2.22), has a counterpart in semiclassical theory where the field is considered to be an external force field described by a time dependent classical vector potential function $\mathbf{A}(\mathbf{r}, t)$, which is further considered independent of $\mathbf{r}$ in the LWA. The semiclassical formulations will be shown to be valid approximations to the quantal ones, for strong fields. They offer the advantage of simplicity in the implementation of wavepacket propagation algorithms to be used in Sections 4 and 5.

Radiation-field (RF) or velocity gauge. The Hamiltonian describing, in the RF or velocity gauge, a single-particle interacting with the field is

$$H_{\text{cl}}^{v}(t) = \frac{1}{2m}\left(\mathbf{p} - q\mathbf{A}(t)\right)^2 + V(\mathbf{r}). \tag{2.32}$$

It differs from the corresponding Hamiltonian given in Eq. (2.16) by the absence of a field Hamiltonian, as here the field is not quantized and is described by the classical function $\mathbf{A}(t)$. For a multicharge system the corresponding Hamiltonian is:

$$H_{\text{cl}}^{v}(t) = \sum_k \frac{1}{2m_k}\left(\mathbf{p}_k - q_k\mathbf{A}(t)\right)^2 + V\left(\{\mathbf{r}_k\}\right). \tag{2.33}$$

Electric-field (EF) or length gauge. The semiclassical length gauge Hamiltonian is obtained from $H_{\rm cl}^{v}(t)$ by the time dependent unitary transformation

$$\widehat{\mathcal{T}}_{L\leftarrow v}(t) = \exp\left(-\frac{i}{\hbar}q\mathbf{r}\cdot\mathbf{A}(t)\right), \tag{2.34}$$

operating on the velocity-gauge statevector $|\Psi^{v}(t)\rangle$.

The RF gauge Schrödinger equation (2.9), rewritten with the semiclassical, explicitly time dependent Hamiltonian $H_{\rm cl}^{v}$, is transformed into

$$i\hbar\frac{\partial|\Psi^{L}(t)\rangle}{\partial t} = H_{\rm cl}^{L}(t)\big|\Psi^{L}(t)\big\rangle \tag{2.35}$$

with $H_{\rm cl}^{L}(t)$ now being given by:

$$H_{\rm cl}^{L}(t) = \widehat{\mathcal{T}}_{L\leftarrow v}H_{\rm cl}^{v}(t)\widehat{\mathcal{T}}_{L\leftarrow v}^{\dagger} - i\hbar\widehat{\mathcal{T}}_{L\leftarrow v}\frac{\partial\widehat{\mathcal{T}}_{L\leftarrow v}^{\dagger}}{\partial t}. \tag{2.36}$$

As compared to Eq. (2.22), this contains an additional term which arises from the time-dependence of the unitary transformation $\widehat{\mathcal{T}}_{L\leftarrow v}(t)$. Carrying out the prescribed unitary transformation (it displaces the momentum $\mathbf{p}$ by $q\mathbf{A}(t)$) and evaluating the indicated time-derivative,

$$\widehat{\mathcal{T}}_{L\leftarrow v}\frac{\partial\widehat{\mathcal{T}}_{L\leftarrow v}^{\dagger}}{\partial t} = \frac{i}{\hbar}q\mathbf{r}\cdot\frac{\partial\mathbf{A}(t)}{\partial t} = -\frac{i}{\hbar}q\mathbf{r}\cdot\mathbf{E}(t), \tag{2.37}$$

we obtain

$$H_{\rm cl}^{L}(t) = \frac{1}{2m}\mathbf{p}^{2} + V(\mathbf{r}) - \mathbf{d}\cdot\mathbf{E}(t), \tag{2.38}$$

i.e., the same form of length-gauge Hamiltonian as in Eq. (2.22), but without the field Hamiltonian, and the $(\mathbf{r}\cdot\boldsymbol{\varepsilon})^{2}$ term, which is estimated to be of the order of $10^{-28}\hbar\omega$.[1] For a multicharge system, this reads

$$H_{\rm cl}^{L}(t) = \sum_{k}\frac{1}{2m_{k}}\mathbf{p}_{k}^{2} + V\big(\{\mathbf{r}_{k}\}\big) - \mathbf{d}\cdot\mathbf{E}(t). \tag{2.39}$$

Space-translation representation. The space-translation representation is obtained from the semiclassical velocity gauge Hamiltonian by the time dependent unitary transformation

$$\widehat{\mathcal{T}}_{\rm KH}(t) = \exp\left\{\frac{iq}{m\hbar}\mathbf{p}\cdot\int^{t}dt'\,\mathbf{A}(t')\right\}\exp\left\{-\frac{iq^{2}}{2m\hbar}\int^{t}dt'\,\mathbf{A}^{2}(t')\right\}. \tag{2.40}$$

[1] This estimate is for a quantization volume $V = 1\ \mathrm{m}^{3}$, a quantized field mode of frequency $\omega = 1000\ \mathrm{cm}^{-1}$, and for an electron $q = -e$, $m = m_{e}$ at a distance of $1\ a_{0} \simeq 0.5$ Å from an atomic nucleus.

This transformation, called the Kramers–Henneberger (KH) transformation (KRAMERS [1956], HENNEBERGER [1968]), is the semiclassical counterpart of the Bloch–Nordsieck or Pauli–Fierz transformation in the quantized field formalism. The second exponential in this operator, a pure phase factor in the LWA, serves to remove the $\mathbf{A}^2$ term, while the first exponential in this has a time derivative which exactly cancels the $\mathbf{A} \cdot \mathbf{p}$ coupling originally found in the velocity gauge Hamiltonian. In detail,

$$-i\hbar\widehat{\mathcal{T}}_{\mathrm{KH}}\frac{\partial\widehat{\mathcal{T}}^{\dagger}_{\mathrm{KH}}}{\partial t} = \frac{q}{m}\mathbf{p} \cdot \mathbf{A}(t) - \frac{q^2}{2m}\mathbf{A}^2(t). \tag{2.41}$$

As found in Eq. (2.30) for the quantized-field Bloch–Nordsieck transformation, the semiclassical unitary transformation $\widehat{\mathcal{T}}_{\mathrm{KH}}(t)$ displaces the charged particle's position vector, but now by a time dependent vector,

$$\boldsymbol{\alpha}(t) = \frac{q}{m}\int^{t} dt'\,\mathbf{A}(t'), \tag{2.42}$$

leaving the momentum $\mathbf{p}$ invariant. The new state vector evolves under the transformed Hamiltonian

$$H^{\mathrm{KH}}_{\mathrm{cl}}(t) = \widehat{\mathcal{T}}_{\mathrm{KH}}H^{v}(t)\widehat{\mathcal{T}}^{\dagger}_{\mathrm{KH}} - i\hbar\widehat{\mathcal{T}}_{\mathrm{KH}}\frac{\partial\widehat{\mathcal{T}}^{\dagger}_{\mathrm{KH}}}{\partial t} = \frac{\mathbf{p}^2}{2m} + V\big(\mathbf{r} - \boldsymbol{\alpha}(t)\big). \tag{2.43}$$

The multicharge version of this is

$$H^{\mathrm{KH}}_{\mathrm{cl}}(t) = \sum_{k}\frac{\mathbf{p}_k^2}{2m_k} + V\left(\left\{\mathbf{r}_k - \frac{q_k}{m_k}\int^{t} dt'\,\mathbf{A}(t')\right\}\right). \tag{2.44}$$

2.1.3. Mollow transformation

For monochromatic fields, it is well-known that the fully quantized field description and the semiclassical one are equivalent at high field intensities (SHIRLEY [1965]). One way to view this is to consider a high intensity classical, monochromatic field $\mathbf{E}(t)$ to correspond to a highly excited photon number state $|n\rangle$

$$H_{\mathrm{mode}}|n\rangle = n\hbar\omega|n\rangle \tag{2.45}$$

of the associated quantized field mode, with n being an extremely large number. The essence of this construct will become clear later, after the review on the Floquet theory to be found below. A more satisfactory link between the quantized formalism and the semiclassical theory is obtained by invoking the concept of coherent state and is established by a time dependent unitary transformation, henceforth called the Mollow transformation (MOLLOW [1975], COHEN-TANNOUDJI, DUPONT-ROC and GRYNBERG [1987]). Let us introduce the complex scalar quantity

$$\lambda(t) = -\frac{i}{2\beta\omega}E_0\exp\{-i\omega t\}, \tag{2.46}$$

and suppose that initially the field mode is in the coherent state

$$\left|\lambda(0)\right\rangle = \exp\left\{\lambda(0)\hat{a}^{\dagger} - \lambda^{*}(0)\hat{a}\right\}|0\rangle \tag{2.47}$$

corresponding to a high electric-field amplitude $E_0 \propto I^{1/2}$ (I is the field intensity) and to a high average photon number $\langle n\rangle_0$. In fact, we have the relations

$$\operatorname{Re}\lambda(t) = \frac{1}{2\beta}\mathbf{A}_{\mathrm{cl}}(t)\cdot\boldsymbol{\varepsilon} \tag{2.48}$$

and

$$\operatorname{Im}\lambda(t) = -\frac{1}{2\beta\omega}\mathbf{E}_{\mathrm{cl}}(t)\cdot\boldsymbol{\varepsilon}, \tag{2.49}$$

$\mathbf{A}_{\mathrm{cl}}(t)$ and $\mathbf{E}_{\mathrm{cl}}(t) = -\partial_t\mathbf{A}_{\mathrm{cl}}(t)$ being the harmonic functions describing the amplitudes of the vector potential and electric field components of the classical field, i.e.,

$$\mathbf{A}_{\mathrm{cl}}(t) = -\boldsymbol{\varepsilon}\frac{E_0}{\omega}\sin(\omega t), \qquad \mathbf{E}_{\mathrm{cl}}(t) = \boldsymbol{\varepsilon}E_0\cos(\omega t). \tag{2.50}$$

Now, the following unitary transformation

$$\widehat{\mathcal{T}}_M(t) = \exp\left\{\lambda^{*}(t)\hat{a} - \lambda(t)\hat{a}^{\dagger}\right\} \tag{2.51}$$

maps the initial coherent state of Eq. (2.47) into the vacuum state

$$\widehat{\mathcal{T}}_{\mathrm{M}}(0)\left|\lambda(0)\right\rangle = |0\rangle, \tag{2.52}$$

while it leaves the particle's dynamical variables $\mathbf{r}$, $\mathbf{p}$ invariant as the transformation is independent of these variables. In addition, it also leaves the field-mode Hamiltonian H_{mode} invariant in the sense

$$\widehat{\mathcal{T}}_{\mathrm{M}}(t)H_{\mathrm{mode}}\widehat{\mathcal{T}}^{\dagger}_{\mathrm{M}}(t) - i\hbar\widehat{\mathcal{T}}_{\mathrm{M}}(t)\frac{\partial\widehat{\mathcal{T}}^{\dagger}_{\mathrm{M}}(t)}{\partial t} = H_{\mathrm{mode}}. \tag{2.53}$$

This does not mean that the field operators are unaffected. In fact,

$$\widehat{\mathcal{T}}_{\mathrm{M}}(t)\hat{a}\widehat{\mathcal{T}}^{\dagger}_{\mathrm{M}}(t) = \hat{a} + \lambda(t), \tag{2.54}$$

$$\widehat{\mathcal{T}}_{\mathrm{M}}(t)\hat{a}^{\dagger}\widehat{\mathcal{T}}^{\dagger}_{\mathrm{M}}(t) = \hat{a}^{\dagger} + \lambda^{*}(t), \tag{2.55}$$

so that $\widehat{\mathbf{A}}$ and $\widehat{\mathbf{E}}$ are transformed according to:

$$\widehat{\mathcal{T}}_{\mathrm{M}}(t)\hat{\mathbf{A}}\widehat{\mathcal{T}}^{\dagger}_{\mathrm{M}}(t) = \mathbf{A}_{\mathrm{cl}}(t) + \hat{\mathbf{A}}, \tag{2.56}$$

$$\widehat{\mathcal{T}}_{\mathrm{M}}(t)\widehat{\mathbf{E}}\widehat{\mathcal{T}}^{\dagger}_{\mathrm{M}}(t) = \mathbf{E}_{\mathrm{cl}}(t) + \widehat{\mathbf{E}}. \tag{2.57}$$

We emphasize that the quantities $\hat{\mathbf{A}}$, $\widehat{\mathbf{E}}$ are field operators, as opposed to the classical time dependent functions $\mathbf{A}_{\text{cl}}(t)$ and $\mathbf{E}_{\text{cl}}(t)$. However, these field operators now act on a transformed field state which initially is a vacuum state, see Eq. (2.52). Thus, the charges now interact weakly with the quantized field, and their strong interaction with the intense laser is transferred to the classical fields $\mathbf{A}_{\text{cl}}(t)$ or $\mathbf{E}_{\text{cl}}(t)$. Indeed, the Mollow transformation maps the various forms of the time independent dressed molecule Hamiltonian onto (COHEN-TANNOUDJI, DUPONT-ROC and GRYNBERG [1987], NGUYEN-DANG, CHATEAUNEUF and MANOLI [1996]):

$$\widehat{\mathcal{T}}_{\text{M}}(t)H_{\text{Q}}^{v}\widehat{\mathcal{T}}_{\text{M}}^{\dagger}(t)-i\hbar\widehat{\mathcal{T}}_{\text{M}}(t)\frac{\partial\widehat{\mathcal{T}}_{\text{M}}^{\dagger}(t)}{\partial t}$$
$$=\frac{1}{2m}\left(\mathbf{p}-q\mathbf{A}_{\text{cl}}(t)-q\hat{\mathbf{A}}\right)^{2}+V(\mathbf{r})+H_{\text{mode}} \tag{2.58}$$

for the Hamiltonian in the velocity gauge, Eq. (2.16),

$$\widehat{\mathcal{T}}_{\text{M}}(t)H_{\text{Q}}^{L}\widehat{\mathcal{T}}_{\text{M}}^{\dagger}(t)-i\hbar\widehat{\mathcal{T}}_{\text{M}}(t)\frac{\partial\widehat{\mathcal{T}}_{\text{M}}^{\dagger}(t)}{\partial t}$$
$$=\frac{1}{2m}\mathbf{p}^{2}+V(\mathbf{r})-\mathbf{d}\cdot\left[\mathbf{E}_{\text{cl}}(t)+\widehat{\mathbf{E}}\right]+\frac{\omega}{\hbar}q^{2}\beta^{2}(\mathbf{r}\cdot\boldsymbol{\varepsilon})^{2}+H_{\text{mode}} \tag{2.59}$$

for the Hamiltonian in the length gauge, Eq. (2.22), and

$$\widehat{\mathcal{T}}_{\text{M}}(t)H_{\text{Q}}^{\text{BN}}\widehat{\mathcal{T}}_{\text{M}}^{\dagger}(t)-i\hbar\widehat{\mathcal{T}}_{\text{M}}(t)\frac{\partial\widehat{\mathcal{T}}_{\text{M}}^{\dagger}(t)}{\partial t}$$
$$=\frac{\mathbf{p}^{2}}{2m}+V\left(\mathbf{r}-\frac{q}{m\omega^{2}}\left[\mathbf{E}_{\text{cl}}(t)+\widehat{\mathbf{E}}\right]\right)+H_{\text{mode}}, \tag{2.60}$$

for the Bloch–Nordsieck form of Eq. (2.30). Noting that, for a harmonic field,

$$\frac{q}{m\omega^{2}}\mathbf{E}_{\text{cl}}(t)=\frac{q}{m}\int^{t}dt'\,\mathbf{A}_{\text{cl}}(t')=\boldsymbol{\alpha}(t), \tag{2.61}$$

we recognize on the right-hand side of Eq. (2.60), the semiclassical form of the Bloch–Nordsieck or space-translation Hamiltonian, if the displacement of the charge's position vector by the residual quantized field $\widehat{\mathbf{E}}$ can be neglected, so that the charged particle and the radiation field appear completely decoupled. Recall that, by construction, the Mollow transformation maps the initial coherent state of the intense field onto its vacuum state, so that in Eqs. (2.58), (2.60) the residual couplings between the quantized field and the molecule are negligible in comparison with the explicitly time dependent radiative couplings, which denote strong driving forces exerted by an external intense monochromatic field on the molecular system. In the purely semiclassical theory, presented in the previous paragraph, these residual quantized-field/molecule couplings are neglected so that the quantized field remains in the vacuum state, i.e., in the original representation, it remains coherent. The inclusion of these residual couplings are necessary to account for the feedback that the driven molecular dynamics could have on the quantized field (KELLER and ATABEK [1993]). The neglect of the residual quantized field (i.e., that of spontaneous emission), opens the way to a semiclassical representation of a pulse delivered by a noncontinuous wave laser (equivalent to a multimode quantal field), with each field-mode in a coherent state containing a large number of photons.

2.2. *Multichannel molecular models*

For a complete treatment of a laser-driven molecule, one must solve the many-body, multidimensional TDSE involving one of the Hamiltonians given above. This represents a tremendous task and direct wavepacket simulations of nuclear and electronic motions under an intense laser pulse is presently restricted to a few bodies (at most three or four) and/or to a model of low dimensionality (CHELKOWSKI, ZUO, BANDRAUK and ATABEK [1995]). For a more general treatment, an approximate separation of variables between electrons (fast subsystem) and nuclei (slow subsystem) is customarily made, yielding the Born–Oppenheimer approximation. To lay out the ideas underlying this approximation as adapted to field-driven molecular dynamics, we will consider from now on a molecule consisting of N_n nuclei (labeled $\alpha, \beta, \ldots$) and N_e electrons (labeled $i, j, \ldots$) in a classical field described within the EF or length gauge. Separation of the motion of the center of mass of the system from its internal motions is assumed to be made already. We will not be concerned with the laser-driven translational motion of the center of mass. We thus first rewrite the Hamiltonian of Eq. (2.39) in the form (we shall drop the subscript *cl* and superscript *L* which served to distinguish this semiclassical length-gauge Hamiltonian from the other forms):

$$H(t) = H_\mathrm{el}(t) + T_N - \boldsymbol{\mu}_N\left(\{\mathbf{R}_\alpha\}\right) \cdot \mathbf{E}(t) \tag{2.62}$$

with

$$H_\mathrm{el}(t) = \sum_i \frac{\mathbf{p}_i^2}{2m} + V\left(\{\mathbf{r}_i\}, \{\mathbf{R}_\alpha\}\right) + e\sum_i \mathbf{r}_i \cdot \mathbf{E}(t), \tag{2.63}$$

where T_N denotes the kinetic energy and $\boldsymbol{\mu}_N(\{\mathbf{R}_\alpha\})$ is the dipole moment operator associated with the nuclei and calculated in the center of mass coordinate system.

In the electronic Hamiltonian $H_\mathrm{el}(t)$, the potential function $V(\{\mathbf{r}_i\}, \{\mathbf{R}_\alpha\})$ comprises all Coulomb interactions between the charges, i.e., it is the sum of V_en, the attraction between electrons and nuclei, V_ee, the Coulomb repulsion between the electrons, as well as V_nn, the repulsion between the various nuclei of the molecular system. Spin dependent effects are negligible in the context of strong field interactions. The sum of $V(\{\mathbf{r}_i\}, \{\mathbf{R}_\alpha\})$ and the electronic kinetic energy defines the field-free electronic Hamiltonian, H_{el}^0:

$$H_{\mathrm{el}}^0 = \sum_i \frac{\mathbf{p}_i^2}{2m} + V\big(\{\mathbf{r}_i\}, \{\mathbf{R}_\alpha\}\big). \tag{2.64}$$

Its eigenfunctions,

$$H_{\mathrm{el}}^0 \Xi_I\big(\{\mathbf{r}_i\}; \{\mathbf{R}_\alpha\}\big) = \varepsilon_I\big(\{\mathbf{R}_\alpha\}\big)\Xi_I\big(\{\mathbf{r}_i\}; \{\mathbf{R}_\alpha\}\big), \tag{2.65}$$

which are supposedly orthonormal by construction, constitute a complete basis in terms of which the total molecular wavefunction can be expanded:

$$\Omega\big(\{\mathbf{r}_i\}, \{\mathbf{R}_\alpha\}, t\big) = \sum_I \Psi_I\big(\{\mathbf{R}_\alpha\}, t\big)\Xi_I\big(\{\mathbf{r}_i\}; \{\mathbf{R}_\alpha\}\big). \tag{2.66}$$

It is to be noted that the dependence of the electronic basis wavefunctions $\Xi_I(\{\mathbf{r}_i\}; \{\mathbf{R}_\alpha\})$ on the nuclear coordinates $\{\mathbf{R}_\alpha\}$ is a parametric one, imparted by the dependence of the electronic Hamiltonian H_{el}^0 on these geometric parameters (H_{el}^0 does not contain differential operators with respect to these). Since the field-free electronic Hamiltonian is time independent, the basis set is independent of time. Yet, the expansion coefficients $\Psi_I(\{\mathbf{R}_\alpha\}, t)$ in this basis have to obey the reduced coupled Schrödinger equations:

$$\begin{aligned} i\hbar\frac{\partial \Psi_I(\{\mathbf{R}_\alpha\}, t)}{\partial t} &= \big[T_N + \varepsilon_I\big(\{\mathbf{R}_\alpha\}\big) - \boldsymbol{\mu}\big(\{\mathbf{R}_\alpha\}\big)\cdot \mathbf{E}(t)\big]\Psi_I\big(\{\mathbf{R}_\alpha\}, t\big) \\ &\quad - \mathbf{E}(t)\cdot \sum_J \langle \Xi_I| - e\sum_i \mathbf{r}_i|\Xi_J\rangle \Psi_J\big(\{\mathbf{R}_\alpha\}, t\big) \\ &\quad + \sum_J \mathcal{C}_{I,J}(\alpha)\Psi_J\big(\{\mathbf{R}_\alpha\}, t\big), \end{aligned} \tag{2.67}$$

where $\mathcal{C}_{I,J}(\alpha)$ are nonadiabatic coupling operators, the exact form of which depends on the choice of nuclear coordinates. When the field is turned off (i.e., $\mathbf{E}(t) = 0$), neglect of the nonradiative (i.e., nonadiabatic) coupling terms on the third line of this equation yields the celebrated Born–Oppenheimer (BO) approximation (BORN and OPPENHEIMER [1927]). Since no coupling then exists any longer between various Ψ_I's, the expansion of Eq. (2.66) reduces to a single term denoting an approximate separation of nuclear and electronic variables. While the electronic part is described by Eq. (2.65), an effective Schrödinger equation describes the nuclear motions viewed as occurring on a single potential energy surface (PES), a single *electronic channel*, described by the electronic energy $\varepsilon_I(\{\mathbf{R}_\alpha\})$ which now plays the role of a potential energy function. The neglect of the nonradiative nonadiabatic couplings is usually justified, at least in part, by the disparity between the nuclear and the electronic masses. Moreover, they are dominated by the strong couplings between the electronic states or *channels* as induced by the intense laser field. Thus neglecting these nonradiative couplings (the last sum in Eq. (2.67)), we obtain

$$i\hbar\frac{\partial\Psi_I(\{\mathbf{R}_\alpha\},t)}{\partial t} = \big[T_N + \varepsilon_I(\{\mathbf{R}_\alpha\}) - \boldsymbol{\mu}(\{\mathbf{R}_\alpha\})\cdot\mathbf{E}(t)\big]\Psi_I(\{\mathbf{R}_\alpha\},t) - \mathbf{E}(t)\cdot\sum_J\langle\Xi_I| - e\sum_i \mathbf{r}_i|\Xi_J\rangle\Psi_J(\{\mathbf{R}_\alpha\},t). \tag{2.68}$$

This is the start of the construction of multichannel models of laser-driven molecules. In practice, one restricts to a finite number, N_{ch}, of electronic states, selected on the basis of physical relevance and Eq. (2.68) defines an N_{ch}-channel molecular model system. The 'electronic' Hamiltonian is, in this model, described by the operator

$$H_{\mathrm{el}}(\{\mathbf{R}_\alpha\},t) = \sum_{I,J}\big[\varepsilon_I(\{\mathbf{R}_\alpha\})\delta_{IJ} - \mathbf{E}(t)\cdot\boldsymbol{\mu}_{IJ}(\{\mathbf{R}_\alpha\})\big]|\Xi_I\rangle\langle\Xi_J|, \tag{2.69}$$

where

$$\boldsymbol{\mu}_{IJ}(\{\mathbf{R}_\alpha\}) = \langle\Xi_I| - e\sum_i \mathbf{r}_i|\Xi_J\rangle \tag{2.70}$$

is the transition dipole moment between the field-free electronic states $|\Xi_I\rangle$ and $|\Xi_J\rangle$. In matrix form, H_{el} is represented by

$$\widetilde{\mathbf{W}}(\{\mathbf{R}_\alpha\};t) = \begin{pmatrix} \varepsilon_1 & -\boldsymbol{\mu}_{12}\cdot\mathbf{E}(t) & \cdots & \cdots & \cdots \\ -\boldsymbol{\mu}_{21}\cdot\mathbf{E}(t) & \varepsilon_2 & -\boldsymbol{\mu}_{23}\cdot\mathbf{E}(t) & \cdots & \cdots \\ \cdots & -\boldsymbol{\mu}_{32}\cdot\mathbf{E}(t) & \varepsilon_3 & -\boldsymbol{\mu}_{34}\cdot\mathbf{E}(t) & \cdots \\ \cdots & \cdots & \cdots & \cdots & \cdots \end{pmatrix} \tag{2.71}$$

and, gathering the nuclear amplitudes $\Psi_I(\{\mathbf{R}_\alpha\},t)$ into a column vector $\Psi(\{\mathbf{R}_\alpha\},t)$ Eq. (2.68) reads:

$$i\hbar\frac{\partial\Psi(\{\mathbf{R}_\alpha\},t)}{\partial t} = \big[T_N\mathbf{1} + \widetilde{\mathbf{W}}(\{\mathbf{R}_\alpha\};t)\big]\Psi(\{\mathbf{R}_\alpha\},t). \tag{2.72}$$

2.3. *Floquet theory and dressed molecule picture*

The semiclassical forms of the laser-driven molecule's Hamiltonian discussed above, be they considered at an a priori, i.e., exact level, or within a multichannel model, are all time-periodic of period $T = 2\pi/\omega$, i.e., $H(t+T) = H(t)$, when the external field is a continuous wave (cw) field, corresponding to a coherent state of the quantized field mode. Owing to this time-periodicity, the Floquet theorem (FLOQUET [1883]) is applicable to the Schrödinger equation

$$i\hbar\frac{\partial|\Psi(t)\rangle}{\partial t} = H(t)|\Psi(t)\rangle. \tag{2.73}$$

It states that this equation admits solutions of the form:

$$\left|\Psi_E(t)\right\rangle = e^{-i\frac{Et}{\hbar}}\left|\Phi_E(t)\right\rangle \tag{2.74}$$

with $|\Phi_E(t+T)\rangle = |\Phi_E(t)\rangle$, which constitute a complete basis for a full description of the dynamics.[2] Introduction of the Floquet form into the wave equation produces for $|\Phi_E(t)\rangle$ the equation:

$$\left[H(t) - i\hbar\frac{\partial}{\partial t}\right]\left|\Phi_E(t)\right\rangle = E\left|\Phi_E(t)\right\rangle. \tag{2.75}$$

The operator acting on $|\Phi_E(t)\rangle$ is called the Floquet Hamiltonian,

$$H_{\mathrm{F}}(t) = H(t) - i\hbar\frac{\partial}{\partial t}, \tag{2.76}$$

and the time in this is to be treated as an additional dynamical variable so that Eq. (2.75) determines eigenvalues and eigenstates of the Floquet Hamiltonian in an extended space, the direct product of the usual molecular Hilbert space and the space of L^2 functions of $t \in [0, T]$ (see GUERIN and JAUSLIN [1997] for an application). A variant of this formulation is the (t, t') method developed by PESKIN and MOISEYEV [1993] and applied to the photodissociation of H_2^+ by MOISEYEV, CHRYSOS and LEFEBVRE [1995].

Often, as is the case for Eq. (2.33) in the velocity gauge and Eq. (2.39) in the length gauge, the Hamiltonian $H(t)$ can be split into its time independent part and a time dependent one representing the matter-field interaction:

$$H(t) = H_0 + V(t). \tag{2.77}$$

The vector $|\Phi_E(t)\rangle$ viewed as a function of time, and the time dependent interaction 'potential' $V(\mathbf{r}, \mathrm{t})$[3] can then be Fourier expanded in the complete orthonormal set of exponentials $e^{in\omega t}$,

$$V(\mathbf{r}, t) = \sum_n V_n(\mathbf{r})e^{in\omega t}, \tag{2.78}$$

and:

$$\left|\Phi_E(t)\right\rangle = \sum_n |U_n\rangle e^{in\omega t} \tag{2.79}$$

---

[2] The theorem applies to scattering and bound states with a real energy and to resonance states with a complex energy. When the energy is quantized (bound and resonance states), it is called a quasi-energy, for reasons to be developed below.

[3] To fix the idea, we shall henceforth consider the case this interaction term depends on $\mathbf{r}$, as in the length gauge Hamiltonian.

with n going from $-\infty$ to $+\infty$. This results in an elimination of the time variable through a multichannel expansion in terms of the aforementioned complete basis of the t space. Introduction of these developments into the wave equation (Eq. (2.75)) and identification of the coefficients of the exponentials indeed give a system of coupled *time independent* equations:

$$\big[H_0 + V_0(\mathbf{r})\big]|U_n\rangle + \sum_{n' \neq n} V_{n-n'}(\mathbf{r})|U_{n'}\rangle = (E - n\hbar\omega)|U_n\rangle. \tag{2.80}$$

It is worthwhile to give the explicit coupled equations for the cases where V denotes the $\mathbf{A}\cdot\mathbf{p}$ or $\boldsymbol{\mu}\cdot\mathbf{E}$ interaction term in the velocity gauge or length gauges, respectively. With

$$\mathbf{A}(t) = \mathbf{A}_0 \sin(\omega t), \tag{2.81}$$

$$\mathbf{E}(t) = -\dot{\mathbf{A}}(t) = -\omega\mathbf{A}_0 \cos(\omega t) = \mathbf{E}_0 \cos(\omega t), \quad \mathbf{E}_0 = -\omega\mathbf{A}_0, \tag{2.82}$$

the single-particle interaction

$$V^v(\mathbf{r}, t) = -\frac{q}{m}\mathbf{A}(t)\cdot\mathbf{p} = -\frac{q}{2im}\mathbf{A}_0\cdot\mathbf{p}\big\{e^{i\omega t} - e^{-i\omega t}\big\} \tag{2.83}$$

in the velocity gauge, and

$$V^L(\mathbf{r}, t) = -q\mathbf{r}\cdot\mathbf{E}(t) = -\frac{q}{2}\mathbf{r}\cdot\mathbf{E}_0\big[e^{i\omega t} + e^{-i\omega t}\big] \tag{2.84}$$

in the length gauge, has only two nonzero Fourier components:

$$V^v_{\pm 1} = \mp\frac{q}{2im}\mathbf{A}_0\cdot\mathbf{p}, \qquad V^L_{\pm 1} = -\frac{q}{2}\mathbf{r}\cdot\mathbf{E}_0 \tag{2.85}$$

so that the coupled equations (2.80) reduce to

$$(H_0 + n\hbar\omega)|U_n\rangle - \frac{iq}{2m}\mathbf{A}_0\cdot\mathbf{p}\big(|U_{n+1}\rangle - |U_{n-1}\rangle\big) = E|U_n\rangle \tag{2.86}$$

in the velocity gauge, and

$$(H_0 + n\hbar\omega)|U_n\rangle - \frac{q}{2}\mathbf{E}_0\cdot\mathbf{r}\big(|U_{n+1}\rangle + |U_{n-1}\rangle\big) = E|U_n\rangle \tag{2.87}$$

in the length gauge.

A number of basic properties of the Floquet states $|\Psi_E(t)\rangle$ (cf. Eq. (2.74)) can be derived easily. Let E_i be a quasi-energy, i.e., an eigenvalue of H_F. Consider an energy written:

$$E_{i,k} = E_i + k\hbar\omega. \tag{2.88}$$

If we multiply $\exp(-iE_{i,k}t/\hbar)$ by $e^{ik\omega t}$ we do not modify the Floquet eigenfunction, yet $E_{i,k}$ is to be interpreted as another quasi-energy, since $e^{ik\omega t}\Phi_{E_i}(t)$ is also a periodic function. Thus quasi-energies are defined modulo an integer multiple of $\hbar\omega$. The index k together with i defines a quasi-energy level in a definite Brillouin zone, which is specified by k alone. This is in complete analogy with the situation met in crystals, where the Floquet theorem also applies, but this time as a result of the spatial periodicity of the potential, and is called Bloch theorem (KITTEL [1996]). It is known that the Floquet states belonging to a single zone constitute a complete basis set (OKUNIEWICZ [1974]), in the case where the quasi-energy spectrum is discrete. Extension to cases involving continua and resonance states can be made heuristically by invoking a continuum discretization procedure.

The properties of Floquet eigenstates are such as to produce a very simple way of following the motion of a general wavepacket of the time dependent laser-driven system, at least in a *stroboscopic* way. Consider the evolution operator $\mathcal{U}(t+T,t)$ between times t and $t+T$. Starting at time t from a Floquet state

$$|\Psi_i(t)\rangle = e^{-i\frac{E_i t}{\hbar}}|\Phi_i(t)\rangle, \tag{2.89}$$

we have after time T:

$$\mathcal{U}(t+T,t)|\Psi_i(t)\rangle = e^{-i\frac{E_i(t+T)}{\hbar}}|\Phi_i(t+T)\rangle = e^{-i\frac{E_i(t+T)}{\hbar}}|\Phi_i(t)\rangle. \tag{2.90}$$

In particular if $t=0$:

$$\mathcal{U}(T,0)|\Psi_i(0)\rangle = e^{-i\frac{E_i T}{\hbar}}|\Phi_i(T)\rangle = e^{-i\frac{E_i T}{\hbar}}|\Phi_i(0)\rangle. \tag{2.91}$$

This shows that the time evolution is exactly like that of a stationary state of a time independent Hamiltonian, provided the probing is limited to T, or any multiple of T. Since $|\Psi_i(0)\rangle$ is equal to $|\Phi_i(0)\rangle$, Eq. (2.91) also shows that $\exp(-iE_iT/\hbar)$ is an eigenvalue of the evolution operator over one period of the field. Suppose now that we wish to follow the development in time of an arbitrary initial wavepacket $|\eta(0)\rangle$. We can expand it over the complete set of Floquet eigenfunctions of a given Brillouin zone at time $t=0$:

$$|\eta(0)\rangle = \sum_i |\Phi_i(0)\rangle\langle\Phi_i(0)|\eta(0)\rangle. \tag{2.92}$$

At time $t=NT$, the wavepacket has evolved into:

$$|\eta(NT)\rangle = \sum_i e^{-i\frac{E_i NT}{\hbar}}|\Phi_i(0)\rangle\langle\Phi_i(0)|\eta(0)\rangle. \tag{2.93}$$

This relation emphasizes again the similarity between the properties of Floquet eigenfunctions and the ordinary stationary eigenfunctions (see NGUYEN-DANG,

CHATEAUNEUF, ATABEK and HE [1995]) for more details). Although it has been proven here only for stroboscopic times, Eq. (2.93) may be written more generally by changing NT into an arbitrary t.

2.4. *Molecular resonance calculations in the Floquet formalism*

To illustrate the above concepts, we consider the one-electron diatomic H_2^+ molecule described in a two-channel approximation. The (time-periodic) cw field impinging on the molecule is considered to be linearly polarized in the laboratory z direction and the interatomic axis is assumed pre-aligned along this electric field. An explicit treatment of this alignment process is given later. Various representations can be defined for this system, following the formalism previously developed.

2.4.1. *Coupled equations in length gauge*

Using the notations defined in Section 2, the full Hamiltonian in the length gauge is:

$$H_{\rm cl}^{L}(t) = H_{\rm el}^{0} + T_N + eE_0\cos(\omega t)\boldsymbol{\varepsilon}\cdot\mathbf{r} \tag{2.94}$$

with $H_{\rm el}^0$ defined by

$$H_{\rm el}^{0} = \frac{\mathbf{p}^2}{2m} + V(\mathbf{r}, R). \tag{2.95}$$

T_N is the kinetic operator for the relative motion of the nuclei:

$$T_N = -\frac{\hbar^2}{2\mathcal{M}}\frac{\partial^2}{\partial R^2}. \tag{2.96}$$

$\mathcal{M}$ is the reduced mass of the nuclei $M_1M_2/(M_1+M_2)$. $V(\mathbf{r}, R)$ is the sum of Coulomb interactions among the three particles. The coupling of the nuclear motion to the external field is ignored. For the molecule H_2^+ the two relevant electronic states are the ground state of symmetry $^2\Sigma_g^+$, described by the wavefunction

$$\Xi_1(\mathbf{r}, R) = 1s\sigma_g(\mathbf{r}, R), \tag{2.97}$$

and the first excited state that is accessible via an electronic dipole allowed transition, of symmetry $^2\Sigma_u^+$, and described by

$$\Xi_2(\mathbf{r}, R) = 2p\sigma_u(\mathbf{r}, R). \tag{2.98}$$

Note that in designating the molecular orbitals representing these two electronic states as above, the $1s$ and $2p$ symbols refer to the united-atom limit. The R-dependent electronic energies of these two states are noted $\varepsilon_g(R)$ and $\varepsilon_u(R)$. The time dependent statefunction $\Omega^L(\mathbf{r}, R, t)$ of the system can then be expanded in the truncated basis of the two electronic wavefunctions, as done in Section 2 for a general N_{ch}-channel molecule,

$$\Omega^L(\mathbf{r}, R, t) = \Psi_1(R, t)\Xi_1(\mathbf{r}, R) + \Psi_2(R, t)\Xi_2(\mathbf{r}, R). \tag{2.99}$$

We now apply the Floquet ansatz, writing

$$\begin{pmatrix} \Psi_1(R, t) \\ \Psi_2(R, t) \end{pmatrix} = e^{-i\frac{Et}{\hbar}} \begin{pmatrix} \Phi_1(R, t) \\ \Phi_2(R, t) \end{pmatrix}. \tag{2.100}$$

After introduction of this form in the wave equation, we multiply on the left by either one of the two electronic functions and integrate over electronic coordinates. The "crude-diabatic" assumption $T_N \Xi_i(\mathbf{r}, R) = \Xi_i(\mathbf{r}, R) T_N$ is made. This produces the two coupled equations:

$$\begin{aligned} i\hbar\frac{\partial \Phi_1(R, t)}{\partial t} &= \big[T_N + \varepsilon_g(R) - E\big]\Phi_1(R, t) - \frac{E_0}{2}\mu_{gu}(R)\cos(\omega t)\Phi_2(R, t), \\ i\hbar\frac{\partial \Phi_2(R, t)}{\partial t} &= \big[T_N + \varepsilon_u(R) - E\big]\Phi_2(R, t) - \frac{E_0}{2}\mu_{ug}(R)\cos(\omega t)\Phi_1(R, t), \end{aligned} \tag{2.101}$$

where $\mu_{gu}(R)$ is the matrix element:

$$\mu_{gu}(R) = \big\langle \Xi_1(\mathbf{r}, R)\big| - ez\big|\Xi_2(\mathbf{r}, R)\big\rangle_{\mathbf{r}} = \mu_{ug}(R). \tag{2.102}$$

The subscript $\mathbf{r}$ is to recall that the integration implied in this matrix element is made only over the electronic coordinates. Note that the diagonal matrix elements $-ez$ vanishes for a homonuclear system such as H_2^+, owing to the inversion symmetry which prevails in each of the electronic states of the field-free molecule. Since $\Phi_k(R, t)$, $k = 1, 2$, is time-periodic, it is expressible in the form

$$\Phi_k(R, t) = \sum_{-\infty}^{+\infty} e^{in\omega t}\Psi_{k,n}(R). \tag{2.103}$$

Introduction of these developments into the above coupled equations leads to:

$$\begin{aligned} &\big[T_N + \varepsilon_g(R) + n\hbar\omega\big]\Psi_{1,n}(R) - \frac{E_0}{2}\mu_{gu}(R)\big[\Psi_{2,n-1}(R) + \Psi_{2,n+1}(R)\big] \\ &\quad = E\Psi_{1,n}(R), \\ &\big[T_N + \varepsilon_u(R) + n\hbar\omega\big]\Psi_{2,n}(R) - \frac{E_0}{2}\mu_{ug}(R)\big[\Psi_{1,n-1}(R) + \Psi_{1,n+1}(R)\big] \\ &\quad = E\Psi_{2,n}(R). \end{aligned} \tag{2.104}$$

Note that this pair of equations in fact represents an infinite set of coupled equations as n varies from $-\infty$ to $+\infty$. Also, the following parity selection rule applies: Once a parity of the Fourier indices ('numbers of photons') n associated with channel 1 is chosen, the nuclear amplitudes $\Psi_{1,n}$ supported by this channel (dressed by 'n photons') are coupled only to amplitudes $\Psi_{2,n'}$ supported by channel 2 with a number n' of opposite parity.

2.4.2. Coupled equations in velocity gauge

The Hamiltonian in velocity gauge is:

$$H_{\mathrm{cl}}^{v} = \frac{1}{2m}\left[\mathbf{p} + eA_0 \sin(\omega t)\frac{\boldsymbol{\varepsilon}\cdot\mathbf{r}}{|\mathbf{r}|}\right]^2 + T_N + V(\mathbf{r}, R). \tag{2.105}$$

Dropping the purely time dependent A^2 term, which can always be absorbed in a global phase factor of the state function,[4] we first rewrite this as:

$$H_{\mathrm{cl}}^{v} = H_{\mathrm{el}}^{0} + T_N + \frac{eA_0}{m}\sin(\omega t)\cdot p_z. \tag{2.106}$$

This differs from the one in length gauge, Eq. (2.94) only in the electron-field interaction term. We can thus transpose the above construction in the velocity gauge to the present case by replacing $E_0\cos(\omega t)$ by $A_0\sin(\omega t) = -(E_0/\omega)\sin(\omega t)$ (cf. Eq. (2.82)), and z by p_z. We note the relation:

$$p_z = \frac{im}{\hbar}\left[H_{\mathrm{el}}^{0}, z\right], \tag{2.107}$$

which implies

$$\left\langle \Xi_i(\mathbf{r}, R)\middle| p_z \middle| \Xi_j(\mathbf{r}, R)\right\rangle_{\mathbf{r}} = \frac{im}{\hbar}\big(\varepsilon_j(R) - \varepsilon_i(R)\big)\left\langle \Xi_i(\mathbf{r}, R)\middle| z \middle| \Xi_j(\mathbf{r}, R)\right\rangle_{\mathbf{r}} \tag{2.108}$$

for any pair of eigenkets $|\Xi_i(\mathbf{r}, R)\rangle$, $|\Xi_j(\mathbf{r}, R)\rangle$ of the field-free electronic Hamiltonian H_{el}^{0}, of energies $\varepsilon_i(R)$ and $\varepsilon_j(R)$. As a consequence, the diagonal matrix elements of p_z in this basis vanish identically, while

$$\left\langle \Xi_1(\mathbf{r}, R)\right| - ep_z\left| \Xi_2(\mathbf{r}, R)\right\rangle_{\mathbf{r}} = im\omega_{gu}(R)\mu_{gu}(R) \tag{2.109}$$

with $\omega_{gu}(R)$ given by

$$\omega_{gu}(R) = \frac{1}{\hbar}\left[\varepsilon_u(R) - \varepsilon_g(R)\right]. \tag{2.110}$$

[4] Suppose we have a Hamiltonian of the form $H = H_0 + f(t)$. If the solution of the associated TDSE is $\Psi(\mathbf{r}, t)$, then the function $\widetilde{\Psi}(\mathbf{r}, t)$:

$$\widetilde{\Psi}(\mathbf{r}, t) = \exp\left[+\frac{i}{\hbar}\int^t f(t')\,dt'\right]\Psi(\mathbf{r}, t)$$

satisfies the Schrödinger equation with H_0 as Hamiltonian.

The final result of the Floquet state expansion is a set of coupled equations of the same form as above. Maintaining the same notation for the R-dependent coefficients of the Fourier expansion of the two periodic nuclear functions, these equations read:

$$\begin{aligned}
&\big[T_N + \varepsilon_g(R) + n\hbar\omega\big]\Psi_{1,n}(R) \\
&\quad + \frac{E_0}{2}\frac{\omega_{gu}(R)}{\omega}\mu_{gu}(R)\big[\Psi_{2,n-1}(R) - \Psi_{2,n+1}(R)\big] = E\Psi_{1,n}(R), \\
&\big[T_N + \varepsilon_u(R) + n\hbar\omega\big]\Psi_{2,n}(R) \\
&\quad + \frac{E_0}{2}\frac{\omega_{gu}(R)}{\omega}\mu_{gu}(R)\big[\Psi_{1,n-1}(R) - \Psi_{1,n+1}(R)\big] = E\Psi_{2,n}(R).
\end{aligned} \tag{2.111}$$

These equations differ from the ones governing the components of the corresponding Floquet state in the length gauge in two respects: The appearance of the ratio $\omega_{gu}(R)/\omega$ in the coupling terms in the coupled equations (2.111) for the velocity gauge, and the phase change between the components $\Psi_{1,n-1}(R)$, $\Psi_{1,n+1}(R)$ in each of these equations. Although a unitary operator exactly links the velocity to the length gauge at the level of the primitive Hamiltonian (equivalent to using a nontruncated infinite electronic basis), as soon as we truncate the electronic basis to contain only the ground state and the first excited state, a complete equivalence of the two approaches is lost. A similarity of results is expected, however, if this restriction is a physically acceptable one. An application of the velocity gauge approach to H_2^+ has been made by CHU [1991].

2.4.3. Coupled equations in length gauge: quantized-field

In the fully quantized formalism, the dressed-molecule Hamiltonian is time independent and is, in length gauge (cf. Eq. (2.22)):

$$H = H_{\text{el}} + T_N + e\mathbf{r}\cdot\widehat{\mathbf{E}} + \hbar\omega\hat{a}^\dagger\hat{a}. \tag{2.112}$$

Corresponding to the Floquet states of the semiclassical theory, the wavefunctions of the stationary states of the dressed system can be expanded in the uncoupled basis of product of electronic Ξ_i and photon (field) states $|n\rangle$: The full wavefuction is written:

$$|\Omega\rangle = \sum_n \big[\Psi_{1,n}(R)\Xi_1(\mathbf{r}, R)|n\rangle + \Psi_{2,n}(R)\Xi_2(\mathbf{r}, R)|n\rangle\big]. \tag{2.113}$$

Introduction of this expansion into the stationary-state Schrödinger equation (Eq. (2.11)) for the dressed system followed by projection onto a definite (electronic-field) product basis state $\Xi_k(\mathbf{r}, R)|n\rangle$, $k = 1$ or 2, yields an infinite set of coupled equations:

$$\begin{aligned}
&\big[T_N + \varepsilon_g(R) + n\hbar\omega\big]\Psi_{1,n}(R) \\
&\quad + i\omega\beta\mu_{gu}(R)\big[\sqrt{n}\,\Psi_{2,n-1}(R) - \sqrt{n+1}\,\Psi_{2,n+1}(R)\big] = E\Psi_{1,n}(R), \\
&\big[T_N + \varepsilon_u(R) + n\hbar\omega\big]\Psi_{2,n}(R) \\
&\quad + i\omega\beta\mu_{gu}(R)\big[\sqrt{n}\,\Psi_{1,n-1}(R) - \sqrt{n+1}\,\Psi_{1,n+1}(R)\big] = E\Psi_{2,n}(R).
\end{aligned} \tag{2.114}$$

Often, one is interested in the dynamics of the dressed-system evolving out of an initial state with the molecule in a definite field-free channel Ξ_k, and the field in a state corresponding to a very large number of photons, say N. Generally, the numbers of photons effectively exchanged (δn, where $n = N \pm \delta n$) between the molecule and the field are very small as compared to N. For all *nuclear channels* involved in the actual exchanges of energy between the molecule and the field, we can therefore assume $\sqrt{n} \sim \sqrt{n+1} \sim \sqrt{N}$. Eqs. (2.114) for the stationary-state nuclear amplitudes $\Psi_{k,n}(R)$ ($k = 1, 2$) can be recognized to be exactly those found above in the semiclassical approach if we identify $E_0/2$ to $i\omega\beta\sqrt{N}$. Using the expression of β and taking the square moduli of these expressions we obtain

$$\varepsilon_0 \frac{E_0^2}{2} = \frac{N\hbar\omega}{V}. \tag{2.115}$$

We have here two definitions of the density of energy in the field, either classically or in a quantized approach. Note finally that in the quantized approach, the index n is positive, i.e., it is bounded below by zero, while in the Floquet approach of the semiclassical formulation, n goes to $-\infty$ to $+\infty$. This difference is of no consequence if the initial number of photons N is very large.

3. Numerical methodologies

Laser assisted and controlled photofragmentation dynamics can conceptually be viewed in two different ways. The time dependent approach offers a realistic time resolved dynamical picture of the process basically driven by an intense electromagnetic field, which in general, can only be achieved using short laser pulses. The spreading of the wavepacket or the possible presence of long-lived resonances which govern the dynamics may, however, lead to hard computational task when this approach is addressed. For long duration pulses (as compared to the time scales of the dynamics), the laser field can be considered as periodic, allowing thus the complete elimination of the time variable through the Floquet formalism. This gives rise to the time independent approach. This formalism, first analyzed in the following, not only provides a more direct and accurate way to calculate the resonances involved in the process, but also offers a useful and important interpretative tool in terms of the stationary field dressed molecular states. The time dependent approach will be developed in Section 3.2.

3.1. Dissociation rates in the time independent formalism

3.1.1. Scattering and half-collision solutions

We present the numerical methodology leading to the solution of Eqs. (2.104), resulting from the application of the Floquet formalism to the semiclassical version of the length gauge Hamiltonian of the molecule plus field system. The purpose in the present subsection is to calculate the dissociation lifetime and the population distribution among channels reflecting the number of absorbed photons and correlatively the available kinetic energy for the relative translational motion of the nuclei. Eqs. (2.104) contain nuclear channels belonging to two classes. For a given energy E, if $E - \varepsilon(R) - (N - n)\hbar\omega$, with $\varepsilon(R)$ being either $\varepsilon_g(R)$ or $\varepsilon_u(R)$, goes to a positive limit $E_n = E - (N - n)\hbar\omega$ as $R \to \infty$, the channel is open. This means that the two nuclei can separate, with a relative kinetic energy $E_n = \hbar^2 k_n^2 / 2\mathcal{M}$, k_n being the associated wave number given by:

$$k_n = \frac{1}{\hbar}\big[2\mathcal{M}\big(E - (N - n)\hbar\omega\big)\big]^{\frac{1}{2}}. \tag{3.1}$$

If the limit is negative, the outcome is not a dissociation, but a bound motion. The channel is closed. The coupled equations have solutions which depend on the imposed boundary conditions. Consider first the case where there is at least one open channel. Scattering boundary conditions consist in imposing in open channel n a combination of incoming ($\exp[-ik_n R]$) and outgoing ($\exp[+ik_n R]$) waves while all closed channel functions are constrained to vanish asymptotically. In a molecular situation, because of the strong interatomic repulsion for decreasing R, all channel functions decay to zero. Solutions then exist with a real arbitrary energy above a certain threshold. They describe, for example, a situation where the nuclei approach each other with a given number of photons in the field, and then separate with the same number of photons (elastic scattering), or a different number (inelastic scattering). The scattering situation is not the type of problem of interest in this review. When the molecule is supposed to be initially in a state belonging to a closed channel (e.g., a vibrational state of the ground electronic state), half-collision (or Siegert-type) boundary conditions are used. The experiment consists, for example, in monitoring the dissociation products when the molecule is exposed to a monochromatic field. We then impose to the functions in the open channels to be asymptotically only of the outgoing type (SIEGERT [1939]). The closed channel functions must vanish asymptotically as in the scattering case. This type of solution is only possible if the (quasi-)energy is quantized and complex, of the form $E = E_R - i\Gamma/2$. Such an energy characterizes a resonance. The closed or open character of a given channel is discussed as above, but with the real part of the energy. The total rate of dissociation (that is to say irrespective of which open channel describes the actual dissociation) is given by $\mathcal{K} = \Gamma/\hbar$. This can be checked from the behaviour of the exponential present in the Floquet wavefunction. Since a probability is calculated with the squared modulus of the wavefunction, this factor produces:

$$\left|\exp\left[-i\frac{Et}{\hbar}\right]\right|^2 = \exp\left[-\frac{\Gamma t}{\hbar}\right] = \exp[-\mathcal{K}t]. \tag{3.2}$$

At a time $t = \tau = \hbar/\Gamma$, the population of molecules is $1/e$ the initial population. τ is the lifetime. The asymptotic analysis of the wavefunction can produce, as we will see below, a more detailed description of the half-collision process in the sense that a partial rate associated with a particular open channel can be extracted. Another consequence of Siegert boundary conditions is that the wave numbers are complex, since Eq. (3.1) is now used with a complex energy. Thus k_n can be written as

$$k_n = k_{n0} - ik_{n1}, \tag{3.3}$$

with both k_{n0} and k_{n1} positive. Since the wavefunction in open channel n goes asymptotically as $\exp[ik_n R]$, we have:

$$\exp[ik_n R] = \exp[ik_{n0}R]\exp[k_{n1}R] \tag{3.4}$$

so that it diverges at infinity. The interpretation of this fact is the following: when measuring the outgoing flux at a distant position from the source, we are in fact interrogating the system at some time in the past. Since the source is decaying, the more distant we are from it, the more active was the source at the time of emission. It is possible to implement the Siegert boundary conditions in an explicit way in the determination of the multichannel wavefunction. An alternative approach is to change the reaction coordinate R into a complex coordinate $\rho\exp(i\vartheta)$ (MOISEYEV [1998]). The complex wave number k_n can also be written $\kappa\exp(-i\beta)$ with β positive. With this form for the wave number, the new channel function asymptotically goes to:

$$\begin{aligned}\exp[ik_n R] &= \exp\bigl[i\kappa\rho\exp\bigl(i(\vartheta-\beta)\bigr)\bigr]\\ &= \exp\bigl[i\kappa\rho\cos(\vartheta-\beta)\bigr]\exp\bigl[-\kappa\rho\sin(\vartheta-\beta)\bigr].\end{aligned} \tag{3.5}$$

If the condition $0 < (\vartheta - \beta) < \pi$ is fulfilled, the second exponential factor in the last form of $\exp[ik_n R]$ goes to zero. The channel function then behaves as that of a bound state. This means that all the methods available for bound state calculations are also applicable to resonance calculations. This can be the development of the channel functions in a basis of integrable functions. It is also possible to implement bound state boundary conditions in an algorithm for the direct solution of the coupled differential equations. This is the route followed here. A variant of the complex rotation method consists in changing the reaction coordinate only after some value, say R_0. The form given to the coordinate is $R_0 + (R - R_0)\exp(i\vartheta)$. This procedure is called exterior scaling (NICOLAIDES and BECK [1978], SIMON [1979]).

For the propagation of the multichannel wavefunction an efficient algorithm is the Fox–Goodwin method (FOX [1957], NORCROSS and SEATON [1973]), with a discretization of the differential operator T_N appearing in Eqs. (2.104). For two adjacent points R and $R+h$ on the grid, an inward matrix (labelled i) is defined as:

$$\mathbf{P}^i(R) = \mathbf{Q}^i(R+h)\bigl[\mathbf{Q}^i(R)\bigr]^{-1}. \tag{3.6}$$

For two adjacent points R and $R-h$ on the grid, we define the outward matrix (labelled o) as:

$$\mathbf{P}^o(R) = \mathbf{Q}^o(R-h)\bigl[\mathbf{Q}^o(R)\bigr]^{-1}. \tag{3.7}$$

$\mathbf{Q}^i\,(\mathbf{Q}^o)$ is a matrix of independent vector solutions of the coupled-channel equations satisfying the appropriate boundary conditions. If $R-h_1$, R and $R+h_2$ are three adjacent points on the grid, either all real, or all complex or mixed, we build the

Numerov matrices:

$$\boldsymbol{\alpha}(R) = h_2\left[1 + \frac{1}{12}(h_1^2 + h_1h_2 - h_2^2)(E\mathbf{1} - \boldsymbol{\varepsilon}(R))\right], \tag{3.8}$$

$$\boldsymbol{\beta}(R) = (h_1 + h_2)\left[1 - \frac{1}{12}(h_1^2 + 3h_1h_2 + h_2^2)(E\mathbf{1} - \boldsymbol{\varepsilon}(R))\right], \tag{3.9}$$

$$\boldsymbol{\gamma}(R) = h_1\left[1 + \frac{1}{12}(-h_1^2 + h_1h_2 + h_2^2)(E\mathbf{1} - \boldsymbol{\varepsilon}(R))\right]. \tag{3.10}$$

$\boldsymbol{\varepsilon}(R)$ is the potential matrix at point R. $\mathbf{1}$ is the unit matrix. The propagation is done with the two equations:

$$\mathbf{P}^i(R - h_1) = \left[\boldsymbol{\beta}(R) - \boldsymbol{\alpha}(R + h_2)\mathbf{P}^i(R)\right]^{-1}\boldsymbol{\gamma}(R - h_1), \tag{3.11}$$

$$\mathbf{P}^o(R + h_2) = \left[\boldsymbol{\beta}(R) - \boldsymbol{\alpha}(R - h_1)\mathbf{P}^0(R)\right]^{-1}\boldsymbol{\gamma}(R + h_2). \tag{3.12}$$

Let R and $R + h$ be the two points chosen for the matching of the inward and outward solutions. The matching condition has the form:

$$\det\left|\mathbf{P}^i(R) - \left(\mathbf{P}^o(R + h)\right)^{-1}\right| = 0. \tag{3.13}$$

For a single channel case described by a wavefunction $\Psi(r)$, this relation reduces to a transparent condition for the two functions produced by the inward and outward propagation:

$$\left[\frac{\Psi^i(r + h)}{\Psi^i(r)} - \left(\frac{\Psi^o(r)}{\Psi^o(r + h)}\right)^{-1}\right] = 0. \tag{3.14}$$

The fulfillment of Eq. (3.13) generates a complex energy if the inward matrix incorporates outgoing boundary conditions (with a real coordinate), or bound state boundary conditions (with a complex rotated coordinate). We describe now the advantage of the variable-step procedure in the present problem. It has been previously shown on some multichannel models (CHRYSOS and LEFEBVRE [1993]) that in an inward propagation the transition from the complex to the real part of the integration grid with the variable step technique produces an inward $\mathbf{P}^i$ matrix on the real part of the grid which is very close to that provided by an entirely real grid and explicit introduction of Siegert asymptotic conditions. Because the real part of the integration path can extend up to the asymptotic region, where stable and uncoupled components of the multichannel wavefunction are obtained, it is possible to obtain the partial rates by using traditional flux analysis, although methods exist to obtain rates in the context of the complex rotation method (MOISEYEV and PESKIN [1990]).

3.1.2. Calculation of the wavefunction. Total and partial widths

Once the complex Floquet eigenenergy has been determined from an iterative resolution of the implicit energy dependent equation (3.13), the multichannel wavefunction written as a column vector $\boldsymbol{\Psi}(R)$ can be calculated at the matching point because it satisfies the set of homogeneous linear equations:

$$\left[\mathbf{P}^i(R) - \left(\mathbf{P}^o(R+h)\right)^{-1}\right]\boldsymbol{\Psi}(R) = 0. \tag{3.15}$$

The wavefunction at the other grid points can be obtained from:

$$\begin{aligned}\boldsymbol{\Psi}(R+h) &= \left(\mathbf{P}^o(R+h)\right)^{-1}\boldsymbol{\Psi}(R),\\ \boldsymbol{\Psi}(R-h) &= \left(\mathbf{P}^i(R-h)\right)^{-1}\boldsymbol{\Psi}(R).\end{aligned} \tag{3.16}$$

For the calculation of the partial width in an open channel the Humblet–Rosenfeld formulation (HUMBLET and ROSENFELD [1961]) is used. We recall briefly here the derivation of the partial width expression obtained by these authors. If a set of channel functions $\Psi_i(R)$ obeys a set of coupled equations of the general form, as is the case in the length gauge (cf. Eqs. (2.104)):

$$\left[-\frac{\hbar^2}{2\mathcal{M}}\frac{\partial}{\partial R^2} + \varepsilon_{i,i}(R) - E\right]\Psi_i(R) + \sum_{j\neq i}\varepsilon_{i,j}(R)\Psi_j(R) = 0, \tag{3.17}$$

after multiplication of the ith equation by Ψ_i^*, addition of all equations and subtraction of the complex conjugate of the resulting expression, we obtain:

$$-\frac{\hbar^2}{2\mathcal{M}}\sum_i\left[\Psi_i^*(R)\Psi_i''(R) - \Psi_i(R)\Psi_i^{*\prime\prime}(R)\right] - (E - E^*)\sum_i\left|\Psi_i(R)\right|^2 = 0. \tag{3.18}$$

For a resonance energy written $E = E_R - i\Gamma/2$, we have $E - E^* = i\Gamma$. Integration from the origin (where all functions vanish) to some point R_0 in the asymptotic region, where the molecular potentials $\varepsilon_{i,i}(R)$ have nearly vanished, produces:

$$\frac{\hbar^2}{2\mathcal{M}}\sum_i\left[\Psi_i^*(R_0)\Psi_i'(R_0) - \Psi_i(R_0)\Psi_i^{*\prime}(R_0)\right] + i\Gamma\int_0^{R_0}\left|\sum_i\Psi_i(R)\right|^2 dR = 0. \tag{3.19}$$

Solving this equation for $\Gamma/2$ gives:

$$\frac{\Gamma}{2} = \frac{\hbar^2}{2\mathcal{M}}\frac{\sum_i \mathrm{Im}[\Psi_i^*(R_0)\Psi_i'(R_0)]}{\sum_i\int_0^{R_0}|\Psi_i(R)|^2\,dR}. \tag{3.20}$$

In the above equations, the primes and double-primes indicate first and second derivatives with respect to R. In the final equation (3.20), the total half-width is expressed as a normalized sum of fluxes. It is tempting to identify the terms corresponding to the open channels with the partial widths. However if the coupled equations are solved in the length gauge, as in most treatments of this problem, it is found that: (i) the terms in the sum giving $\Gamma/2$ do not stabilize as R_0 is moved outward; (ii) there are fluxes even in the closed channels. The reason for this is to be found in the form of the coupling function $\mu_{gu}(R)$, which for a diatomic ion such as H_2^+ diverges asymptotically as:

$$\lim_{R\to\infty} \mu_{gu}(R) = \frac{1}{2}\int d\mathbf{r}\big[1s(\mathbf{r}, R_1) + 1s(\mathbf{r}, R_2)\big](-ez)\big[1s(\mathbf{r}, R_1) - 1s(\mathbf{r}, R_2)\big]$$
$$= -\frac{e}{2}(R_1 - R_2) = \frac{eR}{2}, \tag{3.21}$$

$1s(\mathbf{r}, R_1)$ and $1s(\mathbf{r}, R_2)$ being hydrogenic $1s$ orbitals centered on either of the two nuclei. We are in the presence of so-called *persistent couplings*. One of the solutions is outlined now (CHRYSOS, ATABEK and LEFEBVRE [1993a], CHRYSOS, ATABEK and LEFEBVRE [1993b]). Let us consider the potential matrix of the coupled equations in length gauge (Eqs. (2.104)). It is a tridiagonal matrix with off diagonal elements all equal to $E_0\mu_{gu}(R)/2$ and elements on the diagonal which are alternatively $\varepsilon_g(R)$ and $\varepsilon_u(R)$ displaced by quanta of the field. This defines a diabatic potential matrix $\boldsymbol{\varepsilon}^d(R)$. It is possible to diagonalize this matrix for each value of R. This defines a set of dressed molecular *adiabatic* potentials. Let $\mathbf{C}(R)$ be the matrix diagonalizing the diabatic potential matrix according to:

$$\boldsymbol{\varepsilon}^a(R) = \mathbf{C}^\dagger(R)\boldsymbol{\varepsilon}^d(R)\mathbf{C}(R). \tag{3.22}$$

The superscripts d and a denote diabatic and adiabatic potential matrices, respectively. A solution of the coupled equations with these adiabatic potentials lead to new channel functions:

$$\boldsymbol{\Psi}^a(R) = \mathbf{C}^\dagger(R)\boldsymbol{\Psi}^d(R), \tag{3.23}$$

where $\boldsymbol{\Psi}^a(R)$ and $\boldsymbol{\Psi}^d(R)$ are the column vectors representing the solutions of the coupled equations in the two representations. This relation can be inverted to write the diabatic functions in terms of the adiabatic ones. Orthonormality of the eigenvectors of the potential matrix for each value of R provides the two relations involving different columns of the matrix $\mathbf{C}(R)$:

$$\mathbf{C}_j(R)\cdot\mathbf{C}_k(R) = \delta_{jk}, \qquad \mathbf{C}'_j(R)\cdot\mathbf{C}_k(R) = -\mathbf{C}_j(R)\cdot\mathbf{C}'_k(R). \tag{3.24}$$

Going back to Eq. (3.20), the replacement of the diabatic channel functions by the adiabatic ones produces:

$$\frac{\Gamma}{2} = \frac{\hbar^2}{2\mathcal{M}}\frac{\mathrm{Im}\big[\sum_i \Psi_i^{a*}(R_0)\Psi_i^{a'}(R_0) + 2\sum_{k>j}\sum_j \Psi_j^{a*}(R_0)\Psi_k^a(R_0)\tau_{jk}(R)\big]}{2\sum_j\int_0^{R_0}|\Psi_j^a(R')|^2\,dR'}, \tag{3.25}$$

with the definition:

$$\tau_{jk}(R) = \mathbf{C}_j(R) \cdot \mathbf{C}'_k(R). \tag{3.26}$$

We now show that the cross-terms containing the τ_{jk}'s vanish asymptotically. Because as $R \to \infty$ the molecular potentials $\varepsilon_g(R)$ and $\varepsilon_e(R)$ go to zero, the diabatic potential matrix takes a very simple structure: It is a tridiagonal matrix, with off-diagonal elements all equal to $eE_0R/4$ and adjacent diagonal elements all differing by $\hbar\omega$. The secular equations determining the components of the vectors $\mathbf{C}_j(R)$, with eigenenergy $j\hbar\omega$, are:

$$\frac{eE_0R}{4}C_{j,i-1} + (i-j)\hbar\omega C_{j,i} + \frac{eE_0R}{4}C_{j,i+1} = 0. \tag{3.27}$$

The solution is:

$$C_{j,i} = J_{j-i}\left(\frac{eE_0R}{2\hbar\omega}\right), \tag{3.28}$$

if the matrix is of infinite dimension. J_n is a Bessel function of integer order n. Truncation does not modify appreciably this result for the N_c physically accessible nuclear channels (that is to say effectively reached from a given initial channel) if the number of channels kept in the calculation exceeds sufficiently N_c. Because for large values of the argument the Bessel functions go to (Eq. (9.2.1) in ABRAMOVITZ and STEGUN [1972]):

$$J_n(x) \to \sqrt{\frac{2}{\pi x}}\cos\left(x - \frac{n\pi}{2} - \frac{1}{4}\right), \tag{3.29}$$

all Bessel functions vanish asymptotically. This applies also to the derivatives of these functions since they can be expressed as combinations of Bessel functions. Thus the τ_{jk}'s of Eq. (3.26) go to zero for large R and we are left with a single summation. There remains to show that the channels are asymptotically decoupled, since the fluxes observed in the closed channels of the length gauge have their origin in the persistent couplings of this gauge. The coupled equations of the length gauge can be written in matrix form as:

$$\left[T_N\mathbf{1} + \boldsymbol{\varepsilon}^d(R) - E\mathbf{1}\right]\boldsymbol{\Psi}^d(R) = 0, \tag{3.30}$$

where $\boldsymbol{\Psi}^d(R)$ is the vector representation of the diabatic channel functions of Eqs. (2.104), alternatively $\Psi_{1,n}(R)$, $\Psi_{2,n}(R)$. Using the matrix diagonalizing the diabatic potential matrix, we can write:

$$\mathbf{C}^\dagger(R)\left[T_N\mathbf{1} + \boldsymbol{\varepsilon}^d(R) - E\mathbf{1}\right]\mathbf{C}(R)\mathbf{C}^\dagger(R)\boldsymbol{\Psi}^d(R) = 0. \tag{3.31}$$

We recognize that the new coupled equations determine the functions $\mathbf{C}^{\dagger}(R)\boldsymbol{\Psi}^{d}(R)$ already met in Eq. (3.23). The adiabatic coupled equations are finally:

$$\left[T_N\mathbf{1}+\boldsymbol{\varepsilon}^a(R)-\frac{\hbar^2}{2\mathcal{M}}\left(\mathbf{C}^{\dagger}(R)\frac{d^2\mathbf{C}(R)}{dR^2}+2\mathbf{C}^{\dagger}(R)\frac{d\mathbf{C}(R)}{dR}\frac{d}{dR}\right)\right]\boldsymbol{\Psi}^a(R)$$
$$=E\mathbf{1}\boldsymbol{\Psi}^a(R). \tag{3.32}$$

The elements of the matrix $\mathbf{C}^{\dagger}(R)d\mathbf{C}(R)/dR$ are the $\tau_{jk}(R)$ already met in the expression of the total half-width. The matrix $\mathbf{C}^{\dagger}(R)d^2\mathbf{C}(R)/dR^2$ is made of elements also expressible in the asymptotic region in terms of Bessel functions which go to zero with R. Since $\boldsymbol{\varepsilon}^a(R)$ is by construction a diagonal matrix, we see that the adiabatic channel functions are asymptotically decoupled. This is a result associated to the particular form taken by the asymptotic diabatic potential matrix in this problem. In general a transformation to the Kramers–Henneberger frame (semiclassical approach) or to the Bloch–Nordsieck representation (quantized approach) is needed to ensure channel decoupling. It is also possible (MOISEYEV, ALON and RYABOY [1995]) to define these transformations at the level of the nuclear problem if the linear form for $\mu_{gu}(R)$ is used at all internuclear distances. The coupled equations (3.32) for the adiabatic channel functions can be the starting point for the determination of the dissociation rates (MANOLI and NGUYEN-DANG [1993]). Because of the presence of interchannel kinetic couplings, a special integration scheme has to be used (NGUYEN-DANG, DUROCHER and ATABEK [1989]).

3.1.3. Application to multiphoton absorption

We present an illustration (LEFEBVRE and ATABEK [1997]) of the success of the diabatic to adiabatic representations (practically equivalent for nuclear dynamics to a transformation from length gauge to Kramers–Henneberger frame). The molecule H_2^+ is exposed to a laser of wavelength 329.7 nm and intensity $I = 5.6 \times 10^{12}$ W/cm^2. The potential energies $\varepsilon_g(R)$ and $\varepsilon_u(R)$, as well as the electronic transition moment $\mu_{gu}(R)$ are taken from a fit to ab-initio data (BUNKIN and TUGOV [1973]). The absorption of one photon is enough to dissociate the molecule. This does not exclude the absorption of additional photons. This process is termed ATD for Above-Threshold-Dissociation and is reminiscent of the ATI or Above-Threshold-Ionization which can be induced in atoms or molecules by intense fields. The calculation is done with 10 channels, starting from channels $|g, N+2\rangle$ and $|u, N+1\rangle$ down to channels $|g, N-6\rangle$ and $|u, N-7\rangle$. Fig. 3.1 shows in the first column the functions of the most relevant channels for an interpretation of the results: $|g, N\rangle$, $|u, N-1\rangle$, $|g, N-2\rangle$ and $|u, N-3\rangle$. $|g, N\rangle$ is the reference channel, that is the channel defining the initial state, in the presence of N photons. The other channels correspond to absorption of one, two and three photons, respectively. We observe that only the third and fourth channels show oscillations which are indications of a Siegert-like behaviour, that is with outgoing complex exponentials. However there are beats showing that more than a single exponential is present in these channel functions. The right column of Fig. 3.1 shows the channel functions obtained after the transformation embodied in Eq. (3.23), but using at all internuclear distances

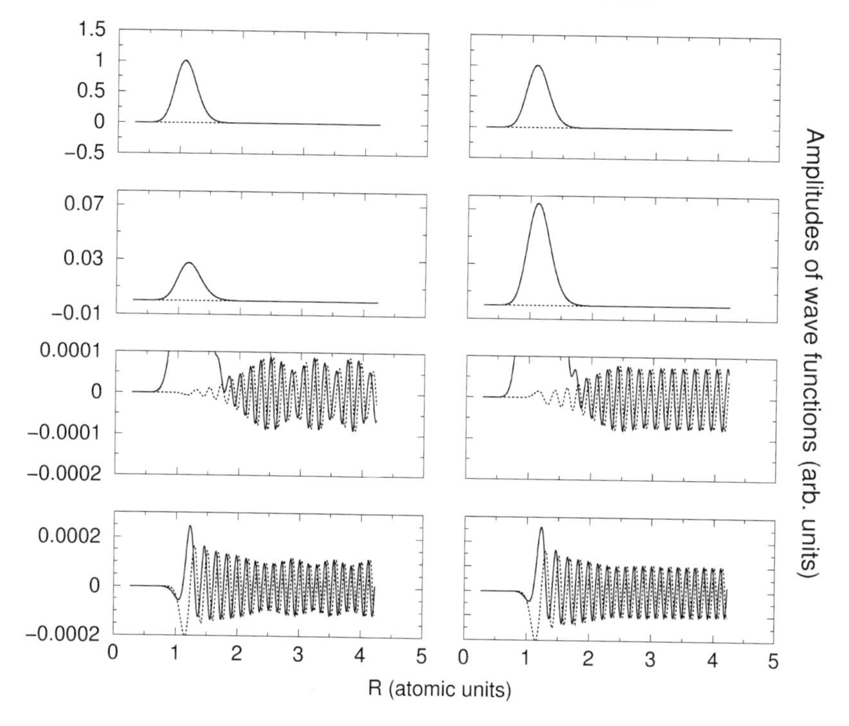

FIG. 3.1. The four channel functions giving converged rates for the dissociation of H_2^+ at an intensity $I = 5.6 \times 10^{12}$ W/cm^2 and a wavelength $\lambda = 329.7$ nm. From top to bottom the ground and the one-, two- and three-photon channels. Left column: the diabatic channel functions. Right column: the adiabatic channel functions obtained from the diabatic to adiabatic transformation valid at large interatomic distances. Solid line for the real and dotted line for the imaginary parts of the functions. Reproduced from [LEFEBVRE and ATABEK, Gauges and Fluxes in Multiphoton Absorption by H_2^+, Int. J. of Quantum Chem., **63**, 403, Copyright © (1997)], by permission of John Wiley & Sons, Inc.

the functional form taken by the matrix diagonalizing the asymptotic diabatic potential matrix. Thus only the outer parts of the functions correspond to an application of the diabatic to adiabatic transformation. However flux analysis is meaningful only at large distances. There is a striking modification of the functions, with now a single outgoing exponential in each channel. The partial width calculation leads to the conclusion that there is almost no rate for the one-photon absorption, while 27.2% of the molecules go into the two-photon channel, 70.2% into the three-photon channel and 2.3% into the four-photon channel not shown in Fig. 3.1.

3.2. *Time dependent wavepacket dynamics*

We now return to the time dependent formalism, with the purpose of solving the nuclear TDSE (cf. Eq. (2.72)), which we recall here explicitly for a diatomic multichannel molecule:

$$i\hbar \frac{\partial}{\partial t} \boldsymbol{\Psi}(R, \theta, \varphi; t) = \left[\mathbf{T}_R + \mathbf{T}_\theta + \mathbf{T}_\varphi + \widetilde{\mathbf{W}}(R, \theta; t)\right] \boldsymbol{\Psi}(R, \theta, \varphi; t), \tag{3.33}$$

where R, θ, φ are the spherical coordinates of the internuclear vector $\mathbf{R}$ in the laboratory frame. Kinetic terms are given by:

$$\mathbf{T}_R = -\frac{\hbar^2}{2\mathcal{M}}\frac{\partial^2}{\partial R^2}, \tag{3.34}$$

$$\mathbf{T}_\theta = -\frac{\hbar^2}{2\mathcal{M}}\frac{1}{R^2 \sin\theta}\frac{\partial}{\partial\theta}\left(\sin\theta\frac{\partial}{\partial\theta}\right), \tag{3.35}$$

$$\mathbf{T}_\varphi = -\frac{\hbar^2}{2\mathcal{M}}\frac{1}{R^2 \sin^2\theta}\frac{\partial^2}{\partial\varphi^2}. \tag{3.36}$$

The potential part $\widetilde{\mathbf{W}}(R, \theta; t)$ contains the field-free molecular potentials as diagonal elements and the radiative coupling in the length gauge as nondiagonal terms.

3.2.1. Propagator

Fourier-transform methodology (FEIT, FLECK and STEIGER [1982], KOSLOFF [1988]) using, for the angular variables, either spherical harmonics basis-sets expansions (KOURI and MOWREY [1987]) or grid techniques (DATEO and METIU [1991]) is one of the most popular approaches for solving Eq. (3.33). In particular, the latter authors have significantly improved the grid approach with an implementation of a unitary Cayley scheme for T_θ combined with a split operator technique using a cosine Fourier transform, resulting in a recursion formula. Apart from avoiding the numerical instabilities associated with the division by $\sin\theta$ in the kinetic energy term, the major advantage is the absence of matrix element evaluations and multiplications, the computational task being basically fast Fourier transforms. The actions of potential and kinetic operators on the wavefunction are evaluated by referring to a combination of coordinate and momentum representations, related through an exponential Fourier transform for R and a cosine transform for θ. After factorization of the azimuthal part, M_N being a good quantum number describing the invariance through rotation about the field polarization vector, the time dependent total wavepacket can be given the following compact form:

$$\boldsymbol{\Psi}_{M_N}(R, \theta, \varphi; t) = e^{iM_N\varphi}\boldsymbol{\Phi}_{M_N}(R, \theta; t), \tag{3.37}$$

where $\boldsymbol{\Phi}_{M_N}$ is to be calculated by solving the TDSE:

$$i\hbar\frac{\partial}{\partial t}\boldsymbol{\Phi}_{M_N}(R, \theta; t) = \left[\mathbf{T}_R + \mathbf{T}_\theta + \mathbf{W}_{M_N}(R, \theta; t)\right]\boldsymbol{\Phi}_{M_N}(R, \theta; t) \tag{3.38}$$

with an effective potential term

$$\mathbf{W}_{M_N}(R, \theta; t) = \widetilde{\mathbf{W}}(R, \theta; t) + \frac{\hbar^2}{2mR^2}\frac{M_N^2}{\sin^2\theta}\hat{\mathbf{1}}. \tag{3.39}$$

The essential features of the method which is based on the repeated use of the short-time propagator $U(\delta t)$:

$$\begin{aligned}\boldsymbol{\Phi}_{M_N}(R,\theta;t+\delta t) &= \mathbf{U}(\delta t)\boldsymbol{\Phi}_{M_N}(R,\theta;t)\\ &= \exp\left[-\frac{i}{\hbar}\left[\mathbf{T}_R+\mathbf{T}_\theta+\mathbf{W}_{M_N}(t)\right]\delta t\right]\boldsymbol{\Phi}_{M_N}(R,\theta;t),\end{aligned} \tag{3.40}$$

involving the value of the effective potential term at time t, are outlined in the following. Since the three operators $\mathbf{T}_R$, $\mathbf{T}_\theta$ and $\mathbf{W}_{M_N}$ do not commute, the Moyal formula is applied twice, to obtain (MOYAL [1949]):

$$\begin{aligned}\exp\left[-\frac{i}{\hbar}\left[\mathbf{T}_R+\mathbf{T}_\theta+\mathbf{W}_{M_N}(t)\right]\delta t\right] &= \exp\left[-\frac{i}{\hbar}\mathbf{W}_{M_N}\delta t/2\right]\exp\left[-\frac{i}{\hbar}\mathbf{T}_\theta\delta t/2\right]\\ &\quad\times\exp\left[-\frac{i}{\hbar}\mathbf{T}_R\delta t\right]\exp\left[-\frac{i}{\hbar}\mathbf{T}_\theta\delta t/2\right]\\ &\quad\times\exp\left[-\frac{i}{\hbar}\mathbf{W}_{M_N}\delta t/2\right]+\mathrm{O}\left(\delta t^3\right).\end{aligned} \tag{3.41}$$

All propagators appearing on the right-hand side of Eq. (3.41) are treated using the Feit–Fleck technique (FEIT and FLECK [1984]) as introduced hereafter, except the one involving $\mathbf{T}_\theta$ approximated by Cayley's formula (YOSIDA [1971]):

$$\exp\left[-\frac{i}{\hbar}\mathbf{T}_\theta\delta t/2\right]=\left[1+\left(\frac{i}{\hbar}\right)\mathbf{T}_\theta\delta t/4\right]^{-1}\left[1-\left(\frac{i}{\hbar}\right)\mathbf{T}_\theta\delta t/4\right]+\mathrm{O}\left(\delta t^3\right), \tag{3.42}$$

which maintains unitarity ($\mathbf{T}_\theta$ being self-adjoint) and is further implemented in a way that avoids matrix inversion. The action of potential and kinetic operators on the wavefunction $\boldsymbol{\Phi}_{M_N}(R,\theta;t)$ is evaluated by referring to a combination of coordinate and momentum representations, as is usual in this context. The Fourier transform theory associates with the grid points (R_p,θ_q) of the coordinate representation a wavevector grid (k_α,κ_β), and in what follows the values of the wavefunction at the grid points are labeled by the corresponding indices p,q,α and β, i.e.,

$$\boldsymbol{\Phi}_{p,q}(t)=\boldsymbol{\Phi}_{M_N}(R_p,\theta_q;t). \tag{3.43}$$

The wavefunctions $\boldsymbol{\Phi}_{p,\beta}(t)$, $\boldsymbol{\Phi}_{\alpha,q}(t)$ in a mixed momentum-coordinate representation are related to $\boldsymbol{\Phi}_{p,q}(t)$, by a cosine transform for θ:

$$\boldsymbol{\Phi}_{p,q}(t)=\sum_{\beta=0}^{N_\theta}\zeta_\beta\cos(\kappa_\beta\theta_q)\boldsymbol{\Phi}_{p,\beta}(t), \tag{3.44}$$

and a Fourier transform for R:

$$\boldsymbol{\Phi}_{p,q}(t) = \sum_{\alpha=-N_R/2}^{N_R/2-1} \exp(-ik_\alpha R_p)\boldsymbol{\Phi}_{\alpha,q}(t), \tag{3.45}$$

where N_θ and N_R are the number of grid points, $\zeta_\beta = 1/2$ for $\beta = 0$ or N_θ while $\zeta_\beta = 1$ otherwise. The application of the short-time propagator $\mathbf{U}(\delta t)$ on the wavefunction $\boldsymbol{\Phi}(t)$ can be summarized as follows:

(1) The action of the potential term which is local in the coordinate representation results into:

$$\exp\left[-\frac{i}{\hbar}\mathbf{W}_{M_N}(R_p, \theta_q; t)\delta t/2\right]\boldsymbol{\Phi}_{p,q}(t). \tag{3.46}$$

(2) The angular kinetic energy is not diagonal in the plane wave representation. After an inverse cosine transform relating $\boldsymbol{\Phi}_{p,q}$ to $\boldsymbol{\Phi}_{p,\beta}$ (inverse transform of Eq. (3.44)), a procedure based on a recurrence relation for $\boldsymbol{\Phi}_{p,\beta}(t+\delta t)$ in terms of its values $\boldsymbol{\Phi}_{p,\beta}(t)$ at an earlier time, in roughly N_θ operation, is undertaken. This recurrence relation, which was derived rigorously from Eq. (3.42), can be recast in the following compact form:

$$\begin{aligned} &\zeta_{\beta+1}a^\star_{p,\beta+1}\boldsymbol{\Phi}_{p,\beta+1}(t+\delta t) - \zeta_{\beta-1}a^\star_{p,\beta}\boldsymbol{\Phi}_{p,\beta-1}(t+\delta t) \\ &\quad - \zeta_{\beta+1}a_{p,\beta+1}\boldsymbol{\Phi}_{p,\beta+1}(t) + \zeta_{\beta-1}a_{p,\beta}\boldsymbol{\Phi}_{p,\beta-1}(t) = 0 \end{aligned} \tag{3.47}$$

(the star indicates the complex conjugate), where

$$a_{p,\beta} = \frac{2m}{\hbar}R_p^2 + i\frac{\delta t}{4}\kappa_\beta\kappa_{\beta-1} \tag{3.48}$$

for $\beta = 1, \ldots, N_\theta - 1$ and where:

$$\boldsymbol{\Phi}_{p,N_\theta}(t+\delta t) = \boldsymbol{\Phi}_{p,N_\theta-1}(t+\delta t) = 0 \tag{3.49}$$

is assumed.

(3) The action of the radial kinetic energy, diagonal in a momentum representation $\boldsymbol{\Phi}_{\alpha,\beta}$ obtained by Fourier transforming $\boldsymbol{\Phi}_{p,\beta}$ (inverse transform of Eq. (3.45)), is merely:

$$\exp\left[-\left(i\hbar k_\alpha^2/2m\right)\delta t\right]\boldsymbol{\Phi}_{\alpha,\beta}(t). \tag{3.50}$$

We are now pointing on an observation which deserves special attention. It has to be noted that in the spherical coordinate representation the polar angle θ is defined within the interval $[0, \pi]$. The even parity and 2π-periodicity of the cosine transform imply, for the angular part of the wavefunction,

$$f(\theta) = f(-\theta) = f(2\pi - \theta), \tag{3.51}$$

from which results a zero-derivative condition at both ends of the interval:

$$\left.\frac{d}{d\theta} f(\theta)\right|_{\theta=0,\pi} = 0. \tag{3.52}$$

In other words, as long as Eq. (3.52) holds, the symmetry requirements are fulfilled and the cosine Fourier representation can be worked out in an accurate way. The actual angular dependence of the wavefunction is written in terms of spherical harmonics $Y_{N,M_N}(\theta,\varphi)$ satisfying the following relation with respect to their θ-derivatives (VARSHALOVICH, MOSKALEV and KHERSONSKII [1988]):

$$\left.\frac{\partial}{\partial\theta} Y_{N,M_N}(\theta,\varphi)\right|_{\theta=\pm n\pi} = (-1)^n \left(\delta_{M_N -1} \exp^{-i\varphi} - \delta_{M_N 1} \exp^{i\varphi}\right) \times \sqrt{\frac{N(N+1)(2N+1)}{16\pi}}, \tag{3.53}$$

where $n=0$ or 1. From Eq. (3.53) it results that the θ-derivatives of the wavefunction at $\theta=0$ and π are zero except for $M_N=\pm 1$, which may be at the origin of numerical instabilities. We have to point out, however, that the cosine transform not only has been successfully tested on model problems (in DATEO and METIU [1991], for the particular case of $M_N=0$), but also served as the basic technique for the accurate calculation of angular-resolved dissociation probabilities of H_2^+, with an initial state represented by a linear combination of spherical harmonics with $M_N=0,\pm 1$ (NUMICO, KELLER and ATABEK [1995]). However, it turns out that the inaccuracies increase when the calculation of angular-resolved kinetic energy spectra in the ATD regime is attempted. The so-called contact transformation, defined by:

$$\boldsymbol{\Psi}_{M_N}(R,\theta,\varphi;t) = \frac{1}{\sqrt{\sin\theta}} \exp^{iM_N\varphi} \boldsymbol{\Phi}_{M_N}(R,\theta;t), \tag{3.54}$$

is a way out of the difficulty. Actually, the θ-derivative of $\boldsymbol{\Phi}_{M_N}$, given by

$$\frac{\partial}{\partial\theta}\boldsymbol{\Phi}_{M_N} = \left(\frac{1}{2}\frac{\cos\theta}{\sqrt{\sin\theta}}\boldsymbol{\Psi}_{M_N} + \sqrt{\sin\theta}\,\frac{d}{d\theta}\boldsymbol{\Psi}_{M_N}\right)\exp^{-iM_N\varphi}, \tag{3.55}$$

has to be evaluated at $\theta=0$ and π with an angular behavior of $\boldsymbol{\Psi}_{M_N}$ taken as a linear combination of spherical harmonics. The two limiting cases of the θ-derivatives at $\theta=0$ and π for the spherical harmonics and thus for $\boldsymbol{\Phi}_{M_N}$ lead to a cancellation at both ends of the interval $[0,\pi]$ for all positive values of M_N (except $M_N=0$). Thus numerical instabilities related with the cosine Fourier transform performed on the wavefunction $\boldsymbol{\Psi}_{M_N}$ (with $M_N=\pm 1$) can easily be removed by applying the procedure to $\boldsymbol{\Phi}_{M_N}$, through the contact transformation. It remains, however, that for $M_N=0$, where the contact transformation fails in producing zero derivatives at the ends of the θ-interval, the original procedure of DATEO and METIU [1991] provides accurate results. The conclusion that can be drawn is that the contact transformation may advantageously be combined with the cosine Fourier transform for all M_N, except $M_N = 0$, for which the cosine transform has to be used alone.

3.2.2. *Asymptotic analysis*

Until very recently the calculation of kinetic energy spectra involved the propagation of the full wavepacket on very large grids, during the total pulse duration. The basic idea leading to a restricted grid is the splitting of the wavefunction into an internal and an asymptotic part, the propagation in the latter being done analytically even during the field interaction (KELLER [1995], NUMICO, KELLER and ATABEK [1995]). A general and comprehensive overview of the method can be provided by considering two distances: $R_{\max}$, which fixes the total extension of the grid, whereas R_S delineates the boundary between the inner and the asymptotic regions. More precisely, the asymptotic region boundary R_S is defined by almost constant or negligible potential energies as compared to the kinetic energy of the fragments. The initial wavefunction $\boldsymbol{\Psi}_j(R,\theta;t=0)$ (the index M_N is dropped for simplicity and $j=1,2$ designates g and u, respectively) is sampled on the R,θ grids ($R \in [0, R_{\max}]$ and $\theta \in [0,\pi]$) and numerically propagated. This procedure is stopped at time t_i, when the probability amplitude at $R_{\max}$ reaches a given small value. Starting at that time, the asymptotic analysis proceeds by splitting wavefunctions into two parts:

$$\Psi_j(R,\theta;t_i) = \Psi_j^I(R,\theta;t_i) + \Psi_j^A(R,\theta;t_i) \quad (j=1,2) \tag{3.56}$$

(I for inner and A for asymptotic). Practically, this is done by using a smooth window function $g(R)$ of spatial extension σ, equal to 1 or 0 in the inner or asymptotic regions, respectively:

$$g(R) = \begin{cases} 1 & \text{for } R < R_S, \\ 1 - \sin^2\left(\dfrac{\pi(R-R_S)}{2\sigma}\right) & \text{for } R_S \leqslant R \leqslant R_S + \sigma, \\ 0 & \text{for } R > R_S + \sigma \end{cases} \tag{3.57}$$

and

$$\Psi_j^I(R,\theta;t_i) = g(R)\Psi_j(R,\theta;t_i), \tag{3.58}$$

$$\Psi_j^A(R,\theta;t_i) = \left[1 - g(R)\right]\Psi_j(R,\theta;t_i). \tag{3.59}$$

The further evolution is performed, on the one hand, by starting again the previous numerical procedure with the inner part of the wavepacket (3.58) as an initial wavefunction and, on the other hand, by the analysis of the asymptotic part of the wavepacket (3.59) in a way that is presented hereafter. The full time propagation consists in the iteration of this basic scheme until convergence is achieved when coherently resuming asymptotic components.

The linearity of the Schrödinger equation allows the separate propagations of Ψ_j^I and Ψ_j^A, the latter being described by:

$$i\hbar\mathbf{1}\frac{\partial}{\partial t}\begin{pmatrix}\Psi_1^A\\ \Psi_2^A\end{pmatrix}=\begin{pmatrix}T_R & -\frac{1}{2}eE(t)R\cos\theta\\ -\frac{1}{2}eE(t)R\cos\theta & T_R\end{pmatrix}\times\begin{pmatrix}\Psi_1^A\\ \Psi_2^A\end{pmatrix}, \tag{3.60}$$

where, for $R \leqslant R_S$, T_θ and W_{M_N} are neglected due to the R^{-2} factor and negligible potential terms $\varepsilon(R)$ as compared to the kinetic energy T_R, particularly important in multiphoton processes leading to high-energy fragments. The amplitude of the neglected R^{-2} terms basically depends upon the rotational excitation of the molecule, which may increase with the laser intensity (R_S is thus intensity dependent). We incidentally note that although for $\theta = 0$ or π the potential operator may diverge when using a contact transformation, it is rather its action on the accordingly decreasing wavefunction that has to be considered, the overall result being zero. The transition dipole moment between the two charge-resonant states is merely given by $\mu(R > R_S) \to \frac{1}{2}eR$. The coupled system (3.60) can be decoupled by introducing a set of wavefunctions χ, which are linear combinations of the Ψ_j^A's

$$\chi_{1(2)}(R,\theta;t)=\frac{1}{\sqrt{2}}\left[\Psi_1^A(R,\theta;t)\pm\Psi_2^A(R,\theta;t)\right] \tag{3.61}$$

and solutions of

$$i\hbar\frac{\partial}{\partial t}\chi_{1(2)}(R,\theta:t)=\left[T_R\mp\frac{1}{2}eE(t)R\cos\theta\right]\chi_{1(2)}(R,\theta;t). \tag{3.62}$$

We point out the formal analogy between these equations and the ones describing the motion of an electron in an electric field. R being formally interpreted as the electron coordinate. Volkov-type solutions (Reiss [1980]) can be obtained with the functional change

$$\chi_{1(2)}(R,\theta;t)=\exp\left[\pm\frac{i}{\hbar}R\cos\theta\,\Delta(t,-\infty)\right]\zeta_{1(2)}(R,\theta;t), \tag{3.63}$$

where

$$\Delta(\tau_2,\tau_1)=\frac{e}{2\hbar}\int_{\tau_1}^{\tau_2}E(t')\,dt'. \tag{3.64}$$

The decoupled evolution equations for $\zeta_{1(2)}$, which describe each a field-free particle, are solved in a standard way, by Fourier transforming and leads to

$$\hat{\chi}_{1(2)}(k,\theta;t)=\exp\left[-\frac{i\hbar}{2m}\int_{t_i}^{t}\left[k\mp\cos\theta\,\Delta(t,t')\right]^2dt'\right]\times\hat{\chi}_{1(2)}\left[k\mp\cos\theta\,\Delta(t,t_i),\theta;t_i\right]. \tag{3.65}$$

It has to be noted that in addition to a phase factor, $\hat{\chi}_{1(2)}(k,\theta;t)$, at time t, is obtained from $\hat{\chi}_{1(2)}(k,\theta;t_i)$ by merely shifting the momentum by the laser pulse area $\Delta(t,t_i)$ weighted by $\cos\theta$. The asymptotic wavefunctions in the momentum space result from Eq. (3.61)

$$\widehat{\Psi}_g^A(k,\theta;t) = \exp\left[-\frac{i\hbar}{2m}\int_{t_i}^{t}\left[k-\cos\theta\,\Delta(t,t')\right]^2 dt'\right]\left[\hat{\chi}_1\left[k-\cos\theta\,\Delta(t,t_i),\theta;t_i\right]\right] + \exp\left[-\frac{i\hbar}{2m}\int_{t_i}^{t}\left[k+\cos\theta\,\Delta(t,t')\right]^2 dt'\right] \times \left[\hat{\chi}_2\left[k+\cos\theta\,\Delta(t,t_i),\theta;t_i\right]\right], \tag{3.66}$$

$$\widehat{\Psi}_u^A(k,\theta;t) = \exp\left[-\frac{i\hbar}{2m}\int_{t_i}^{t}\left[k-\cos\theta\,\Delta(t,t')\right]^2 dt'\right]\left[\hat{\chi}_1\left[k-\cos\theta\,\Delta(t,t_i),\theta;t_i\right]\right] - \exp\left[-\frac{i\hbar}{2m}\int_{t_i}^{t}\left[k+\cos\theta\,\Delta(t,t')\right]^2 dt'\right] \times \left[\hat{\chi}_2\left[k+\cos\theta\,\Delta(t,t_i),\theta;t_i\right]\right]. \tag{3.67}$$

Eqs. (3.66), (3.67) are the analytical evolution of $\widehat{\Psi}_{g,u}^A(k,\theta;t)$ from t_i to t. The angular-resolved fragments probability distribution (kinetic energy spectrum) is finally given by:

$$\mathcal{P}(k,\theta)\,dk = \lim_{t\to\infty}\left[\left|\widehat{\Psi}_g^A(k,\theta;t)\right|^2 + \left|\widehat{\Psi}_u^A(k,\theta;t)\right|^2\right]dk, \tag{3.68}$$

where $t\to\infty$ means not only that the laser pulse is switched off but also that there are no outgoing continuum components of the inner-part wavepacket.

4. Dissociation dynamics of H_2^+ in intense IR and UV lasers

Todays laser technology provides a wide range of frequencies, from Infra-Red (IR), to Visible (Vis), up to Ultra-Violet (UV) domains. The intensities that are achieved (tens of TW/cm^2) are of the same order or even larger than those of Coulomb forces that tighten the molecular structure. Such strong electromagnetic fields basically induce two type of effects that monitor the dynamical behavior of the molecular system; namely, multiphoton excitations (absorptions and/or emissions) and deep modifications of the molecular internal force fields. These effects have different consequences in terms of underlying mechanisms depending on the frequency domain which is addressed.

In the *Vis-UV regime* (wavelengths within 750 nm to 40 nm), the excitation energy (a few electron-volts) is resonant with electronic transitions. Typically a single photon brings enough energy for the dissociation to occur, but due to the high intensity of the field, the molecule continues on absorbing photons even above its dissociation threshold (multiphoton ATD mechanism). The subsequent dynamics leads to fragments into different channels characterized by different kinetic energies resulting from different numbers of absorbed or emitted photons. Laser control consists in favoring some of these channels by monitoring the energy redistribution between the modes of the molecule-plus-field system. The excitation frequency (larger than 5×10^{14} Hz) is high as compared to that of molecular internal motions (vibration or rotation). As a consequence the molecule feels a time-averaged strong radiative field that deeply modifies its field-free Coulomb force field. These modifications may result into the softening of some chemical bonds (barrier suppression, BS mechanism), while others are hardened (confinement or vibrational trapping, VT mechanism). The interplay between these two complementary mechanisms opens the way to reactivity control scenarios by favoring a given reaction pathway or by producing velocity and angular selected photofragments.

In the *IR regime* (wavelengths within 750 nm to 0.1 cm), the excitation energy (less than one electron-volt) is typically resonant with nuclear vibrational motions. A single photon is not energetic enough to induce dissociation which results from a multiphoton process through the accumulation of the energy of many individual photons. In contrast with the Vis-UV regime, the IR laser frequency (10^{14} Hz) being of the same order than internal vibrational frequencies, the molecule follows the time resolved oscillations of the electromagnetic field. When properly synchronized with some of these internal nuclear motions, the laser can again be used as a tool for controlling dissociation through some IR counterparts of softening or stabilization mechanisms. Finally, it is worthwhile noting that the IR laser frequencies remain still high enough when compared to the molecular rotational motion. This gives rise to the laser control of angular distributions through time averaged pendular states which lead to alignment dynamics.

The following part of this review is devoted to small molecules, such as H_2^+ (Section 4) and HCN (Section 5), retained as illustrative examples of intense laser induced and controlled dynamical processes.

4.1. Model

The present section is concerned with the (resonant or nonresonant) electronic excitations, and concomitant large-amplitude nuclear motion, of the molecular ion H_2^+ under a UV-Vis and IR laser pulse (see for a general review GIUSTI-SUZOR, MIES, DIMAURO, CHARRON and YANG [1995]). To this end, we use the same two-channel model as introduced for this system in the previous sections. We thus solve the nuclear TDSE to generate wavepackets moving on the Born–Oppenheimer potential energy curves associated with two electronic states labeled g (ground, ${}^2\Sigma_g^+$) and u (excited, ${}^2\Sigma_u^+$),

$$i\hbar\frac{\partial}{\partial t}\begin{bmatrix}\psi_g\\ \psi_u\end{bmatrix} = \widehat{H}\begin{bmatrix}\psi_g\\ \psi_u\end{bmatrix} \tag{4.1}$$

where the Hamiltonian $\widehat{H}$ is as given in Eq. (3.33) and involves a two-by-two potential operator matrix which depend on the internuclear vector $\mathbf{R}(R,\theta,\varphi)$ and is given by

$$\widehat{W}(R,\theta,t) = \begin{pmatrix} \varepsilon_g(R) & -E_0 f(t)\mu(R)\cos\theta\cos\omega t \\ -E_0 f(t)\mu(R)\cos\theta\cos\omega t & \varepsilon_u(R) \end{pmatrix}, \tag{4.2}$$

where E_0, $f(t)$ and ω are respectively the maximum amplitude, temporal pulse shape and carrier-wave frequency of a field explicitly taken as

$$E(t) = E_0 f(t) \cos(\omega t + \delta). \tag{4.3}$$

The motion associated with the azimuthal angle φ relating **R** to the linearly polarized laser polarization vector can be separated, due to cylindrical symmetry. The remaining dynamical variables are R, the internuclear distance and θ, the polar angle between **R** and the laser polarization axis. The molecular dipole moment μ, which is the transition dipole between states g and u, does not contain any permanent part μ_0 for this homonuclear ion and is parallel to **R**. The polarizability $\boldsymbol{\alpha}$ is not explicitly introduced as, for the intensity range we are referring to, the model can be, within good approximation, strictly limited to the lowest two BO states.

In the simulations in the UV-Vis spectral range, the initial state of H_2^+ $(^2\Sigma_g^+)$ with an isotropic ensemble of para- and ortho-hydrogen in equilibrium involves two components with $J = 1$, $M_J = 0$ (statistical weight of 1/3) and $J = 1$, $M_J = \pm 1$ (statistical weight of 2/3), each of them having to be propagated separately and further combined using their relative weights. M_J is the projection of the total angular rotational momentum J on the laser polarization axis. The numerical instabilities and inaccuracies in the angular propagation associated with $M_J = \pm 1$ in the split-operator technique are removed by referring to a contact transformation (NUMICO, KELLER and ATABEK [1995]). For the simulations in the IR spectral range, the initial state involves only the single value $M_J = 0$. Once the wavepacket components $\psi_{g,u}$ are generated at a given time t, various observables can be calculated, among which the total instantaneous probabilities for the ion to remain in bound vibrational states of the ground electronic state, i.e.,

$$P_{\text{bound}}(t) = \sum_v \left| \langle v | \psi_g(t) \rangle \right|^2. \tag{4.4}$$

The dissociation probability is then simply given by

$$P_{\text{diss}}(t) = 1 - P_{\text{bound}}(t). \tag{4.5}$$

Another observable of interest is the final probability distribution for fragments specified by their momentum k and angular position θ:

$$\mathcal{P}(k, \theta) = \sum_{n=g,u} \left| \tilde{\psi}_n^A(k, \theta; t \to \infty) \right|^2, \tag{4.6}$$

where $\tilde{\psi}_{g,u}^A$ represents the Fourier transform of the asymptotic-region wavepacket components of the g and u channels, respectively.

4.2. *Multiphoton dissociation in the IR regime*

In the IR regime, the field amplitude varies typically over the same time scale as molecular vibrational motions, opening thus the possibility of a synchronization between the nuclear motion and the laser oscillations (CHATEAUNEUF, NGUYEN-DANG, OUELLET and ATABEK [1998], ABOU-RACHID, NGUYEN-DANG and ATABEK [1999]). This requires the definition of a time t_0 (or a related phase δ) when the field amplitude first reaches its maximum after the promotion of the initial wavepacket onto the excited electronic state,

$$\cos[\omega(t - t_0)] = \cos(\omega t + \delta). \tag{4.7}$$

For $t_0 = \delta = 0$, the initial wavepacket is considered to be promoted vertically and instantaneously from the ground state at maximum intensity. In contrast, if t_0 is set equal to $T/4$ ($T = 2\pi/\omega$), corresponding to $\delta = \pi/2$, the initial state preparation occurs at the start of an optical cycle, i.e., at zero field intensity. The two situations result into completely different dynamics, the former leading to dissociation quenching, while the latter is monitored by a barrier suppression mechanism. The dynamics can best be viewed as taking place on time dependent adiabatic potential surfaces $W_\pm(R, \theta, t)$ which arise from diagonalizing the potential energy operator of Eq. (4.2). The discussion is conveniently conducted by considering first the implications and results of a one-dimensional model ($\theta = 0$) representing a strictly aligned molecule (no rotational motion). Fig. 4.1 illustrates the dynamics of $W_\pm$ and of the associated nuclear probability distributions. For $\delta = 0$ the field is at its peak intensity at $t = 0$, when the initial wavepacket is prepared on the inner repulsive edge of the attractive potential W_- (close to ε_g in this region). Only

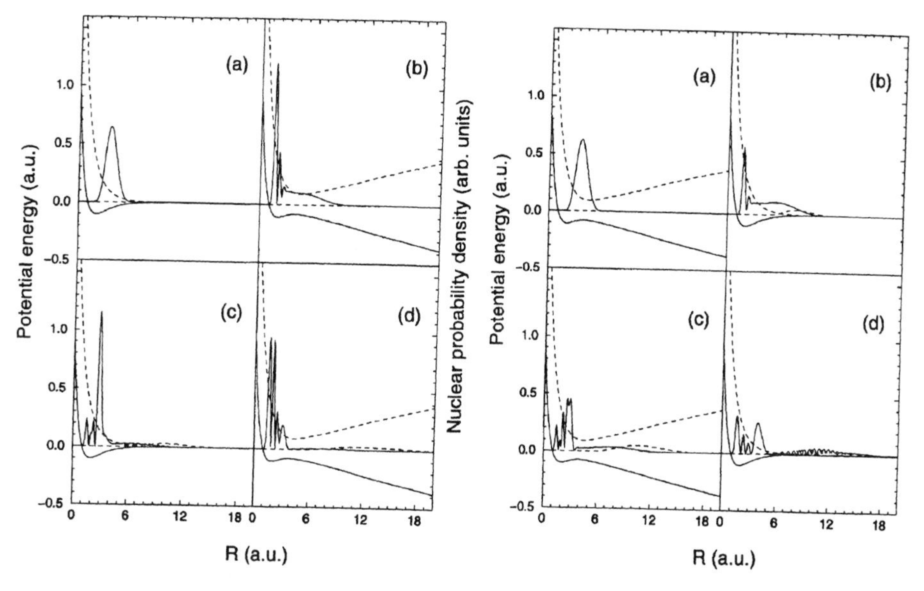

FIG. 4.1. Nuclear probability distributions associated with the wavepacket supported by the adiabatic channels $W_\pm(R, t)$ formed under a $\omega = 943.3\ \text{cm}^{-1}$, $I = 5 \times 10^{13}\ \text{W/cm}^2$ laser pulse with $\delta = 0$ on the left panel and $\delta = \pi/2$ on the right panel. The wavepackets are taken at four times within the first optical cycle: (a) $t = 1/4T$; (b) $t = 1/2T$; (c) $t = 3/4T$; (d) $t = T$. Reproduced from [CHATEAUNEUF, NGUYEN-DANG, OUELLET and ATABEK, Dynamical quenching of field-induced dissociation of H_2^+ in intense infrared lasers, J. Chem. Phys., **108**, 3974, Copyright © (1998)], by permission of the American Institute of Physics.

its tail penetrates the gap region ($R \sim 4$ a.u.) which, at that time, is widely open between W_+ and W_-. At $t = T/4$, the wavepacket components reach the gap region with a gap now closed due to the vanishing field amplitude, preventing thus any escape towards the asymptotic region. At $t = T/2$, the wavepacket components are rejected back towards the inner region and the gap is open again but without any consequence with respect to an eventual escape. During the next half cycle the wavepacket motion follows the same pattern resulting into a stabilization of the system as a consequence of a bounded vibrational motion perfectly synchronized with the opening and closing of the potential gap. This is the dynamical quenching mechanism (DDQ). A completely different wavepacket motion is associated with the case $\delta = \pi/2$. Every time the wavepacket reaches the right end of the binding potential ε_g, the gap is open permitting the escape of an important part of the wavepacket towards the asymptotic dissociative limit. This is the barrier lowering (and/or suppression) mechanism.

Fig. 4.2 shows the survival probability of H_2^+ for two laser intensities and a number of frequencies (for a phase $\delta = 0$). An optimal frequency ($\omega_{max} = 1450$ cm^{-1}) is obtained for $I = 10^{14}$ W/cm^2 leading to 70% of quenching (CHATEAUNEUF, NGUYEN-DANG, OUELLET and ATABEK [1998]). It is worthwhile noting that the field-molecule synchronization can be achieved by tuning the laser frequency and adjusting its intensity. This means that a quantitative knowledge of δ is not necessary. In particular,

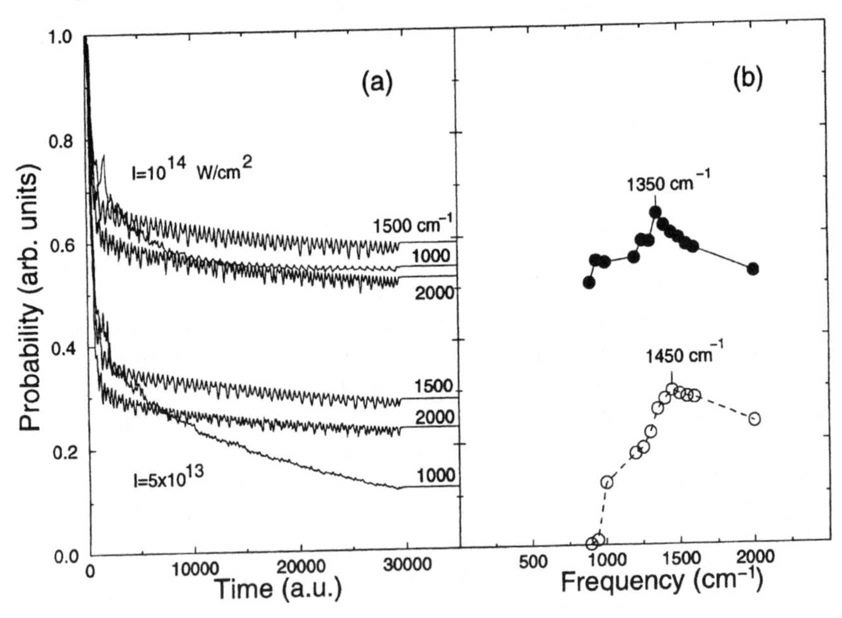

FIG. 4.2. Variations of the survival probability, P_{bound} as the field frequency varies in the range $\omega = 1000$–2000 cm^{-1}. In panel (a), P_{bound} is displayed explicitly as a function of time for three values of ω, whereas in panel (b), the survival probability evaluated at the final time, $t_f = 30000$ a.u., is shown as a function of the frequency, for $I = 5 \times 10^{13}$ W/cm^2 (full circles on solid line) and $I = 1 \times 10^{14}$ W/cm^2 (open circles on dashed line). Reproduced from [CHATEAUNEUF, NGUYEN-DANG, OUELLET and ATABEK, Dynamical quenching of field-induced dissociation of H_2^+ in intense infrared lasers, J. Chem. Phys., **108**, 3974, Copyright © (1998)], by permission of the American Institute of Physics.

a two-pulse (UV + IR) scheme can be proposed with a given time delay. An intense ultrashort UV pulse is turned on at a definite time t_0 after the onset of the long IR pulse. The UV pulse instantaneously prepares well-aligned H_2^+ ions from H_2 in its ground state. A variation of the frequency of the IR laser pulse would then allow the desired synchronization to occur (ABOU-RACHID, NGUYEN-DANG and ATABEK [2001]).

Full 2D calculations (including rotation) have recently shown that molecular alignment is not among the necessary conditions for DDQ to be sharply observed (ABOU-RACHID, NGUYEN-DANG and ATABEK [2001]). In the case where the molecule is allowed to rotate during its interaction with the field, the efficiency of DDQ is only slightly affected. A control can thus be exerted on the system by adequately combining the laser intensity (which monitors the position, the width, and the spatial spread of the gap defined as $\Delta W(R) = W_+(R) - W_-(R)$) and frequency (which monitors the gap opening and closing motions' period).

4.3. *ATD dynamics in the Vis-UV regime*

In the Vis-UV regime, the field oscillations are so fast as compared to nuclear motion that the molecular system only feels time-averaged radiatively-dressed adiabatic potentials. A transparent multiphoton interpretation to be made for the two nonlinear mechanisms affecting the chemical bond (which are the analogs of the barrier suppression and DDQ of the IR regime) is the Floquet expansion and time averaging of the molecule-plus-field Hamiltonian, valid for pulses long enough to lead to near periodic lasers. The corresponding equations (Eq. (2.104)) are written for convenience in a 1D scheme ($\theta = 0$, no rotation). The calculations and their interpretation are, of course, carried out in the full 2D space. The relatively large photon energy of a Vis-UV field not only matches the electronic transition of the molecule in a modestly elongated configuration but also approximately separates each Floquet block $[(g, n); (u, n+1)]$ from the other. Only a few channels (for a given intensity) take part for a converged calculation on the network of field-dressed diabatic potentials ($\varepsilon_l + n\hbar\omega$) leading to short distance curve crossings. This is to be contrasted to the case of IR excitation where the Floquet picture is rather inappropriate, due to single-photon crossings occurring at large distances close to the dissociative asymptotic limit, and to a multitude of crossings densely produced in the inner region of the potentials. Fig. 4.3 collects the field-dressed potential energy curves involved in the two main Floquet blocks in their diabatic and adiabatic representations. The adiabatic representation, showing avoided curve crossings, results from the diagonalization of the radiative interaction at fixed molecule-field orientations ($\theta = 0$ or π, which maximizes the couplings). Two strong field intensities are considered, namely 10^{13} W/cm^2 and 5×10^{13} W/cm^2, at a wavelength of 532 nm, which mimic the experimental situation (ZAVRIYEV, BUCKSBAUM, MULLER and SCHUMACHER [1990]). The curve crossing regions upon which the interpretation of the dynamics rests are indicated by rectangular boxes: X1 between the ground and the one-photon channels, X2 between the ground and the two-photon channels, and X3 between the two- and three-photon channels. Three initial vibrational states v are taken into account, with energies above (or below) the barriers for both intensities ($v = 5$ or $v = 2$, respectively) or in between the barriers ($v = 4$).

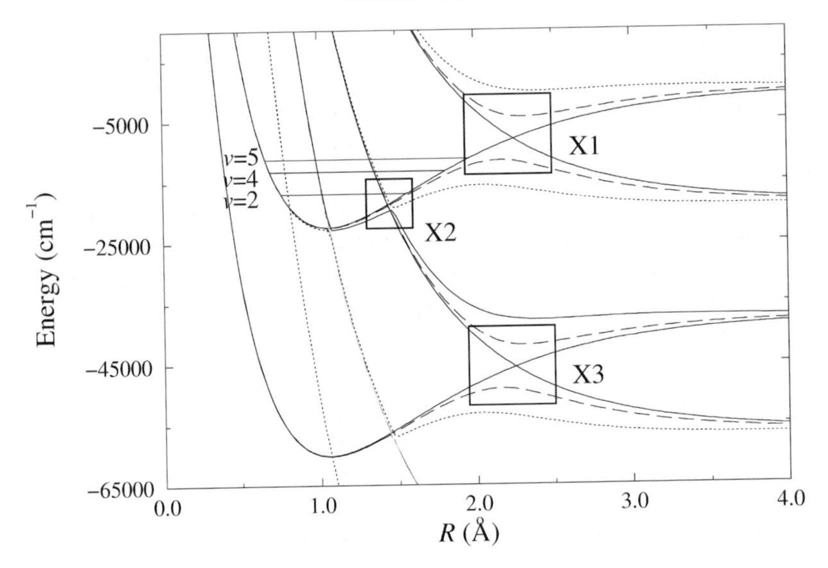

FIG. 4.3. Field-dressed potential energy curves of H_2^+ ($\lambda = 532$ nm), in the diabatic (solid lines) and adiabatic (broken lines for $I = 10^{13}$ W/cm^2 and dotted lines for $I = 5 \times 10^{13}$ W/cm^2) frames. Curve-crossing regions are outlined by rectangular boxes X1, X2, and X3. The energies of the $v = 2, 4, 5$ vibrational levels are indicated by thin horizontal lines. Reproduced from [NUMICO, KELLER and ATABEK, Intense laser-induced alignment in angularly resolved photofragment distribution of H_2^+, Phys. Rev. A **60**, 406, Copyright © (1999)], by permission of the American Physical Society.

An inspection of box X1 shows that, for this frequency regime, a strong radiative interaction may alter in a substantial way the molecular force field leading to important lowering and flattening of the stationary adiabatic potential barrier. The lowering of the adiabatic barrier for the field intensity 5×10^{13} W/cm^2 is such that H_2^+ in a vibrational state $v \geqslant 4$ dissociates almost exclusively by tunneling through the one-photon channel and leads to low energy photons. This Bond Softening (BS) mechanism has been abundantly discussed in the literature and experimentally verified (GIUSTI-SUZOR, HE, ATABEK and MIES [1990], BUCKSBAUM, ZAVRIYEV, MULLER and SCHUMACHER [1990], JOLICARD and ATABEK [1992], YAO and CHU [1993], AUBANEL, GAUTHIER and BANDRAUK [1993], AUBANEL, CONJUSTEAU and BANDRAUK [1993]). The counterpart of BS is the Vibrational Trapping (VT) mechanism, the interpretation of which is supported by the consideration of stationary field-induced upper closed adiabatic potentials that may accommodate long-lived resonances (ATABEK, CHRYSOS and LEFEBVRE [1994], NUMICO, KELLER and ATABEK [1997]). More precisely, a very narrow resonance (of zero width, in a semiclassical estimate) arises when a diabatic level is close to an adiabatic level (BANDRAUK and CHILD [1970]). Such coincidences may be obtained at will by just properly adjusting the laser wavelength (i.e., the relative energy positioning of the potentials) and intensity (i.e., the strength of the coupling) (ATABEK, CHRYSOS and LEFEBVRE [1994], ATABEK [1994]). When the molecule is properly prepared (by an adiabatic excitation, using a pulse with a sufficiently long rise time) in such a resonance state, good stabilization is expected (NUMICO, KELLER and ATABEK [1997]). An isotope separation scenario in the H_2^+/D_2^+

mixture has been devised in the literature (ATABEK, CHRYSOS and LEFEBVRE [1994]), based on the vibrational trapping mechanism. This relies on the great sensitivity of the prementioned diabatic-adiabatic level coincidences with respect to the laser and molecular characteristics. A coincidence obtained for a molecule-plus-field system does not hold when one proceeds to an isotopic substitution due to mass related energy shifts. A diabatic-adiabatic level coincidence for D_2^+ is obtained at $I = 1.1 \times 10^{13}$ W/cm^2 using a cw laser of 120 nm wavelength. For this intensity D_2^+ is efficiently trapped, while the H_2^+ resonance is affected by a large width. This clearly shows that one can proceed to a specific choice for a laser wavelength and intensity such that D_2^+ dissociation may completely be suppressed, whereas for the same field, H_2^+ is still dissociating fast. Another scenario of isotopic separation (CHARRON, GIUSTI-SUZOR and MIES [1995]), confirmed by the experiments of SHEEBY, WALKER and DIMAURO [1995], has been devised for HD^+ using the alignment dynamics favouring the production of either H^+ or D^+ in opposite directions. It has to be emphasized that the two stabilization mechanisms in consideration have completely different origins; namely, DDQ in the IR regime is based on a dynamical effect whereas VT in the Vis-UV regime is a pure stationary process. These basic mechanisms not only help in understanding and controlling fragments kinetic energy distributions but can also be referred to for the interpretation of angular distributions. Fig. 4.4 presents 3D graphs of adiabatic surfaces in the region of the X1 and X3 avoided crossings. The initial wavepacket basically prepared on the short range repulsive limit of the surface pictured in Fig. 4.4(a) is propagated towards the X1 region where the adiabatic potential barrier is lowered in an efficient way proportional to the laser–molecule coupling, i.e., to $\mu E_0 \cos\theta$. For a given field strength, this lowering is more pronounced for $\cos\theta \sim 1$ ($\theta \sim 0$ or π). For $\cos\theta \sim 0$ ($\theta \sim \pi/2$) the high potential barrier is hardly penetrable. Under the effect of the torque exerted by the laser on the molecule, the wavepacket skirts around the potential barrier and follows the minimum energy pathway of the dissociative valley (that is for directions close to $\theta = 0$ or π) resulting in aligned fragments when the BS mechanism is the leading one (NUMICO, KELLER and ATABEK [1995]). Fig. 4.4(b) represents, in the X3 region, the

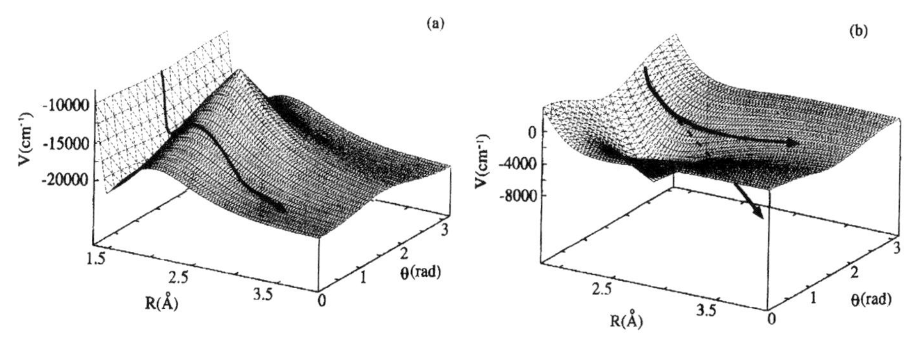

FIG. 4.4. Three-dimensional graphs of adiabatic surfaces for $I = 10^{13}$ W/cm^2 in the region of the X1 (a) or X3 (b) avoided crossings. The arrows give an illustration of the minimum energy pathway on each surface. Reproduced from [NUMICO, KELLER and ATABEK, Intense laser-induced alignment in angularly resolved photofragment distribution of H_2^+, Phys. Rev. A **60**, 406, Copyright © (1999)], by permission of the American Physical Society.

adiabatic surface leading to dissociation corresponding to the net absorption of two photons. A large portion of the initial wavepacket has to undergo a nonadiabatic transition at X2 to reach this surface. The nonadiabatic coupling responsible for the dissociation is sharply peaked around the avoided crossing position X2 with a strength much larger for $\theta \approx \pi/2$ (there is no coupling at $\theta = \pi/2$) than for $\theta = 0$ or π (NUMICO, KELLER and ATABEK [1995]). The wavepacket prepared on this surface presents an angular distribution more pronounced around $\theta \sim \pi/2$ and further dynamics on this surface also favors the $\theta \sim \pi/2$ direction as this corresponds to a potential valley. Finally, less aligned fragments are expected when nonadiabatic transitions are responsible for the dissociation of an initially trapped configuration.

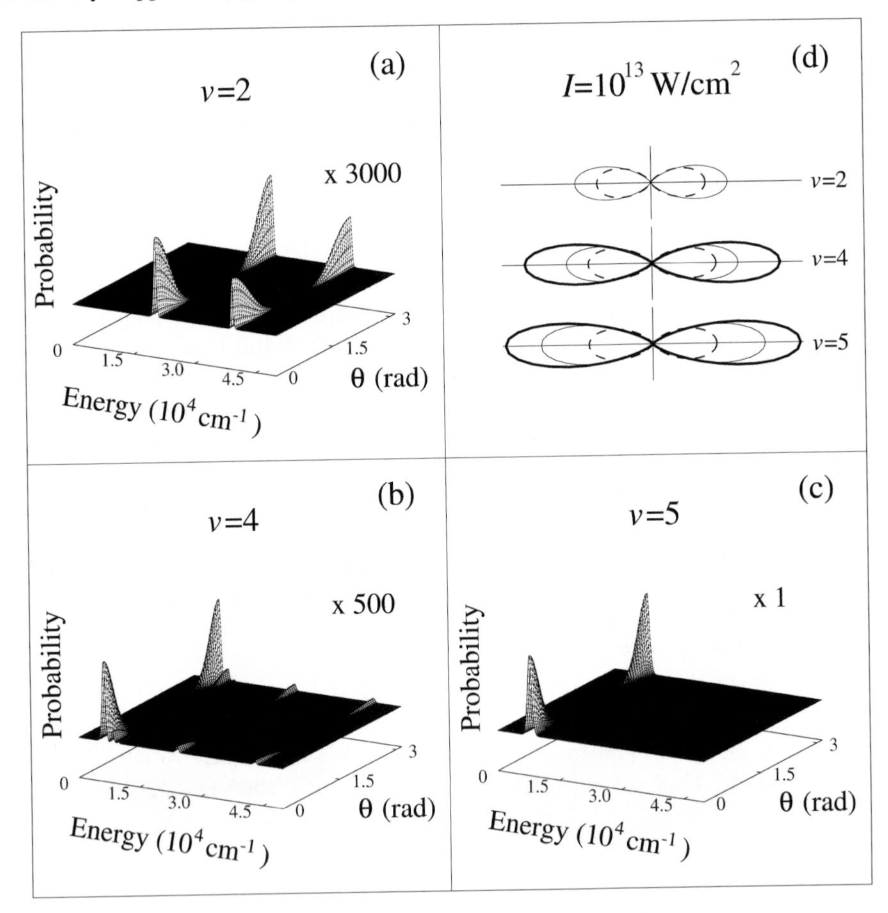

FIG. 4.5. Angular-resolved kinetic energy spectra for the initial vibrational level $v = 2$ (a), $v = 4$ (b), and $v = 5$ (c) of H_2^+ irradiated by a $\lambda = 532$ nm laser delivering an intensity of $I = 10^{13}$ W/cm^2; (d) gathers angular distributions in polar representation, for each v, of fragments contributing to the first photon peak (thick solid line), to the second photon peak (thin solid line) or the third photon peak (dashed line). Reproduced from [NUMICO, KELLER and ATABEK, Intense laser-induced alignment in angularly resolved photofragment distribution of H_2^+, Phys. Rev. A **60**, 406, Copyright © (1999)], by permission of the American Physical Society.

Angular resolved kinetic energy distribution for the three initial vibrational states of H_2^+ under consideration are given in Fig. 4.5. A multiphoton behavior, typical for ATD, is detected in terms of peaks separated by the photon energy. For $v = 2$, the potential barrier for both intensities is high enough at X1 to avoid single-photon dissociation. A third order radiative interaction at X2 results into two- or three-photon dissociation depending on the intensity: for 10^{13} W/cm^2 important nonadiabatic transitions at X3 bring into competition the two- and three-photon peaks. A completely different dissociation scenario is valid for $v = 5$; the field-induced barrier lowering at X1 favors the single photon BS process with the one-photon peak in the spectra. The case of $v = 4$ is intermediate: there is still some competition between the barrier lowering (X1) and nonadiabatic transition (X2) mechanisms with a resulting spectrum displaying three distinct peaks with a major one-photon contribution.

The alignment characteristics of fragments are given in panel (d) of Fig. 4.5 by separately displaying the contributions of each photon peaks in terms of polar plots of their normalized angular distributions $P_i(\theta)$, obtained by integrating the probability density $\mathcal{P}(k;\theta)$ over k in the peak i,

$$P_i(\theta) = \frac{\int_{k_i-\kappa_i}^{k_i+\kappa_i} dk\,\mathcal{P}(k;\theta)}{\int_0^{\pi} \sin\theta\, d\theta \int_{k_i-\kappa_i}^{k_i+\kappa_i} dk\,\mathcal{P}(k,\theta)}, \tag{4.8}$$

where k_i and $2\kappa_i$ are the maximum value and momentum range of the distribution in i. As a direct consequence of our previous analysis, the one-photon peak, resulting from the barrier lowering mechanism (X1), leads to aligned fragments, whereas the two-photon peak resulting from a single nonadiabatic transition (X2) is characterized by less aligned fragments. The three-photon peaks distributions, involving two nonadiabatic transitions (X2 and X3) are even less aligned. The intensity dependence of the alignment can also be interpreted. Stronger fields produce more efficient barrier lowering, resulting in more aligned fragments when the leading mechanism is tunneling at X1. Conversely, when the nonadiabatic transitions monitor the dynamics, distributions are less aligned at high fields (NUMICO, KELLER and ATABEK [1995]).

A more detailed analysis, in terms of the role played by the initial vibrational state v or the intensity-dependent relative contribution of the initial angular distribution (i.e., $M_J = 0$ or ± 1), is required for the interpretation of off-axis wing structures and for the discussion of some discrepancies observed in experimental findings (ZAVRIYEV, BUCKSBAUM, MULLER and SCHUMACHER [1990], CHANDLER, NEYER and HECK [1998]).

4.4. H_2 alignment in the KH frame

Another view of the alignment process makes use of the Kramers–Henneberger transformation. This change of frame leads to disappearance of the matter-field coupling, but at the expense of introducing a time dependence in the potential energy (cf. Eq. (2.44)). In the assumption of high frequency, the states of the system are supposed to be essentially governed by the cycle-averaged potential (GAVRILA [1992]). This dressing by the field

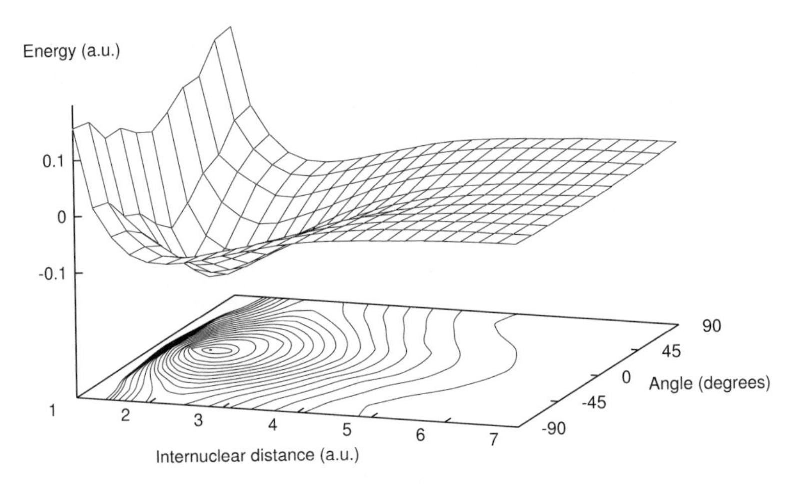

FIG. 4.6. Ground laser-dressed potential energy surface of H_2. The amplitude and the frequency of the electric field satisfy the relation $E_0/m\omega^2 = 1$ a.u.. The potential is anisotropic, with a minimum in the direction of the electric field ($\theta = 0$). Reproduced from [PEREZ DEL VALLE, LEFEBVRE and ATABEK, Dressed Potential Energy Surface of the Hydrogen Molecule in High-Frequency Floquet Theory, Int. J. of Quantum Chem., **70**, 199, Copyright © (1998)], by permission of John Wiley & Sons, Inc.

produces strong distortions of the molecular states. The method has been applied to H_2^+ (SHERTZER, CHANDLER and GAVRILA [1994]) and H_2 (PEREZ DEL VALLE, LEFEBVRE and ATABEK [1998], NGUYEN and NGUYEN-DANG [2000]). In the latter case, an interesting challenge is the introduction of electronic correlation. Standard Quantum Chemistry codes such as HONDO (DUPUIS, MARQUEZ and DAVIDSON [1995]) have been adapted (PEREZ DEL VALLE, LEFEBVRE and ATABEK [1997]) to provide the solution. Fig. 4.6 displays the field-dressed potential energy surface of H_2 ground state. There is a strong anisotropy favouring the alignment of the molecule along the electric field direction ($\theta = 0$). This treatment is a first step for the calculation of photofragmentation rates which requires to go beyond the derivation of the dressed states of Floquet high-frequency theory. This has not yet been attempted for molecules.

5. Laser-induced alignment and orientation dynamics of HCN

5.1. Model

As a second illustrative example toward a laser-induced alignment/orientation control we consider the rotational excitation of a linear HCN molecule, the intramolecular bending angle being kept fixed. The complete molecule-plus-field Hamiltonian, in BO approximation involving the ground electronic state only, is written as

$$H(\mathbf{R}, \theta, t) = \widehat{T}_{\mathbf{R}} + \widehat{T}_{\theta} + \varepsilon(\mathbf{R}) + H_{\mathrm{L}}(\mathbf{R}, \theta, t). \tag{5.1}$$

R collectively designates the lengths of the two nuclear stretching motions (R, the distance between H and the center of mass of CN, and r, the CN bond length) and θ is the polar angle of the molecular axis with respect to the laser field polarization vector. As previously, cylindrical symmetry allows for the separation of the azimuthal φ-angle motion which is not considered further. $\varepsilon(\mathbf{R})$ is the molecular ground state potential and the laser–molecule interaction H_L is given by

$$H_\mathrm{L}(\mathbf{R}, \theta, t) = -\boldsymbol{\mu}(\mathbf{R}) \cdot \mathbf{E}(t) \tag{5.2}$$

with $\mathbf{E}(t)$ the electric field vector of the linearly polarized laser and $\boldsymbol{\mu}(\mathbf{R})$ the molecular dipole moment expanded as

$$\boldsymbol{\mu}(\mathbf{R}) = \boldsymbol{\mu}_0(\mathbf{R}) + \frac{1}{2}\boldsymbol{\alpha}(\mathbf{R})\mathbf{E} \tag{5.3}$$

with $\boldsymbol{\mu}_0$ the permanent dipole moment and a laser-induced contribution involving $\boldsymbol{\alpha}$, the polarizability tensor, which, for a linear molecule, can be expressed by its components parallel, $\alpha_\parallel$, and perpendicular, $\alpha_\perp$, to the internuclear axis. The explicit expression for H_L is obtained by recasting Eq. (5.3) into Eq. (5.2),

$$H_\mathrm{L} = -\boldsymbol{\mu}_0(\mathbf{R})E(t)\cos\theta - \frac{E^2(t)}{2}\left[\alpha_\parallel(\mathbf{R})\cos^2\theta + \alpha_\perp(\mathbf{R})\sin^2\theta\right]. \tag{5.4}$$

The laser pulses used in our calculations are defined by a field envelope $f(t)$ and a field amplitude E_0 corresponding to a field intensity of 10^{12}–10^{13} W/cm^2 (below the ionization threshold).

The initial wavefunction is taken as the ground vibrational $v = 0$ and rotational $J = M_J = 0$ state. The propagated TDSE exact solution $\psi(\mathbf{R}, \theta; t)$ is obtained using the split-operator technique described in Section 2. An observation which deserves interest is that the full quantum calculation results may be interpreted in the frame of a high-frequency approximation. Actually, the laser frequencies ω, taken in the IR spectrum (i.e., $\omega = 1075$ cm^{-1} from a CO_2 laser, corresponding to a field oscillation period of 3×10^{-2} ps) are still high with respect to the molecular rotational frequencies (i.e., period of ~ 10 ps), opening the possibilities of a high-frequency Floquet approach that has been thoroughly discussed (KELLER, DION and ATABEK [2000]). It has, in particular been shown that within this approximative frame, the TDSE has a classical Lagrangian analog, given by FRIEDRICH and HERSCHBACH [1995], DION, KELLER, ATABEK and BANDRAUK [1999]:

$$\ddot{\Theta} + \Omega^2 \sin\Theta = 0 \quad (\Theta = 2\theta), \tag{5.5}$$

clearly resulting into a pendular motion of frequency

$$\Omega = \frac{E_0}{2\sqrt{I}}\left(\frac{\mu_0^2}{I\omega^2} + \alpha\right)^{1/2}, \tag{5.6}$$

I being the principal moment of inertia. It is worthwhile noting that, with the parameters of HCN,

$$\alpha \Big/ \left(\frac{\mu_0^2}{I\omega^2}\right) \approx 10^{-5}\left[\omega\left(\mathrm{cm}^{-1}\right)\right]^2, \tag{5.7}$$

i.e., the ratio of the polarizability to the permanent dipole interaction in Ω does not depend upon the field intensity and is dominated by the polarizability as soon as the frequency exceeds 315 cm^{-1}.

5.2. *Alignment*

A comprehensive summary of the results is given in Fig. 5.1 which displays the long-time behavior of the so-called half-angle $\theta_{1/2}$, retained as a quantitative measure of the alignment. $\theta_{1/2}$ corresponds to the angle within which half of the angular distribution between 0 and $\pi/2$ rad is collected (DION, KELLER, ATABEK and BANDRAUK [1999]). $\theta_{1/2} = \pi/4$ corresponds to an isotropic distribution, while smaller values describe aligned molecules. Fig. 5.1(a) corresponds to a long-duration laser pulse ($\sim$ 20 ps) which in the case of an intensity of 10^{12} $\mathrm{W/cm^2}$ corresponds to an adiabatic excitation

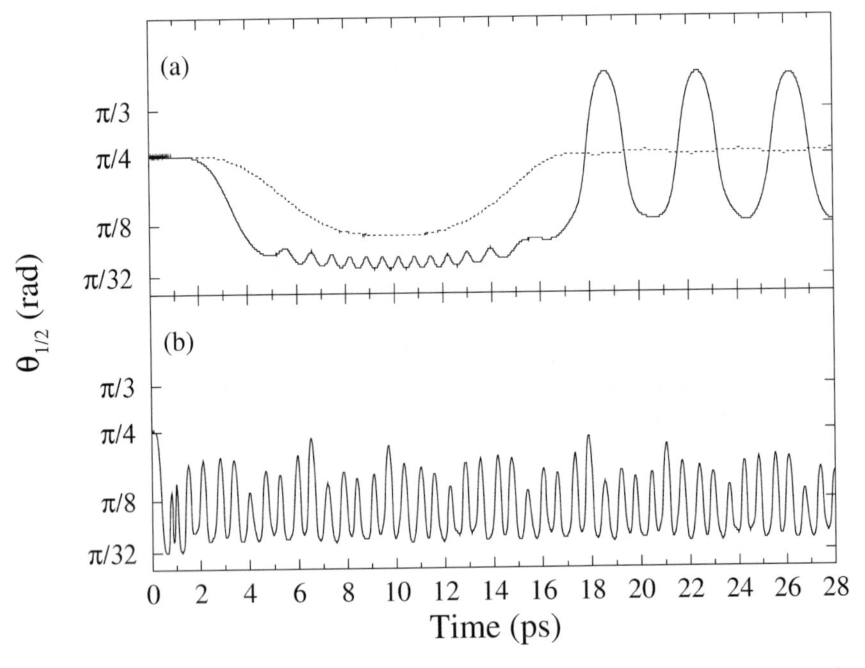

FIG. 5.1. Half angle $\theta_{1/2}$ for the HCN rigid rotor in a laser field of frequency $\hbar\omega = 1075$ cm^{-1}. (a) At intensities 10^{13} $\mathrm{W/cm^2}$ (solid line) and 10^{12} $\mathrm{W/cm^2}$ (dotted line) in a 20 ps pulse. (b) At intensity 10^{13} $\mathrm{W/cm^2}$ in a 1.7 ps pulse. Reproduced from [KELLER, DION and ATABEK, Laser-induced molecular alignment in dissociation dynamics, Phys. Rev. A **61**, 23409, Copyright © (2000)], by permission of the American Physical Society.

of the rotational motion. The initial ($J = M_J = 0$) isotropic molecular state behaves as a unique pendular state, fairly well aligned during the interaction ($\theta_{1/2} \leqslant \pi/8$) and reverting, when the laser field is turned off, to the initial isotropic distribution. A higher intensity of 10^{13} W/cm^2 induces a less adiabatic behavior: basically two pendular states are excited, as is clear from the oscillations of $\theta_{1/2}$. The alignment is better ($\theta_{1/2} \leqslant \pi/16$) due to stronger interaction. The large amplitude oscillations observed when the pulse is over correspond to the superposition of field-free $J = 0$ and $J = 2$ states on which the pendular states are expanded.

But good alignment and above all its conservation when the laser is turned off can only be obtained with sudden pulses. Fig. 5.1(b) corresponds to short-duration pulses (1.7 ps) of 10^{13} W/cm^2. Excellent alignment ($\theta_{1/2} \leqslant \pi/32$) is obtained during early dynamics which is not lost for $t > 1.7$ ps: the half-angle presents some moderate oscillations about $\pi/8$ rad.

5.3. *Orientation*

Basically, symmetry-breaking scenarios are referred to in achieving orientation: DC electric fields (LOESCH and REMSCHEID [1990], FRIEDRICH and HERSCHBACH [1991]), properly tailored microwave pulses (JUDSON, LEHMANN, RABITZ and WARREN [1990]), picosecond two-color phase-locked laser excitations (VRAKKING and STOLTE [1997]), intense linearly-polarized IR pulses combining a fundamental frequency and its second harmonic resonant with a vibrational transition (DION, BANDRAUK, ATABEK, KELLER, UMEDA and FUJIMURA [1999]) or half-cycle pulses (DION, KELLER and ATABEK [2001]), both acting on polar molecules, eventually combining permanent-dipole- and polarizability-field interactions. Among theoretical models that have so far been addressed, coherent and optimal laser control schemes have recently been proposed (JUDSON, LEHMANN, RABITZ and WARREN [1990], ORTIGOZO [1998], HOKI and FUJIMURA [2001]) as possible tools for orientation. More specifically, an optimal laser control of molecular orientation may refer to various genetic algorithms (MICHALEWICZ [1996], BEN HAJ-YEDDER, AUGER, DION, CANCÈS, KELLER, LE BRIS and ATABEK [2002], DION, BEN HAJ-YEDDER, CANCÈS, LE BRIS, KELLER and ATABEK [2002] to optimally tailor a pulse profile by properly summing up a number of individual fields characterized by their frequency, intensity, phase and temporal shape. HCN, in its ground electronic state, taken as a rigid rotor, offers an illustrative example, in a model describing both permanent dipole (μ_0) and polarizability ($\alpha_\parallel$ and $\alpha_\perp$, parallel and perpendicular components) interactions. The measure of the orientation, taken as the evaluation function for the genetic algorithm, is the expectation value of $\cos\theta$,

$$|\cos\theta\rangle(t) = \int_0^\pi \cos\theta \sin\theta \, d\theta \int_0^{2\pi} \left|\psi(\theta,\varphi;t)\right|^2 d\varphi. \tag{5.8}$$

A sum of N individual linearly-polarized pulses,

$$E(t) = \sum_n^N E_n(t) \sin(\omega_n t + \phi_n), \tag{5.9}$$

builds up the electromagnetic field through the optimization procedure. The envelope functions $E_n(t)$ are given sine-square forms, each pulse being characterized by a set of 7 adjustable parameters, namely its frequency ω_n, relative phase ϕ_n, maximum field amplitude E_{n0}, together with 4 times determining its shape (origin, rise time, plateau, and extinction time). How to tailor an electromagnetic field to reach the best possible orientation amounts to a parameter optimization problem involving $7 \times N$ variables aiming at the maximization of a target. An optimization criterion is specified as:

$$j = |\cos\theta\rangle(t_f), \tag{5.10}$$

to be minimized, namely, the best possible orientation at the final interaction time t_f. An enlightening interpretation emerges from a calculation involving only $N = 3$ pulses: it suggests a mechanism for one possible way of molecular orientation, and its simplicity leads to a discussion of experimental feasibility. Fig. 5.2 shows the orientation dynamics and the optimal laser field which monitors it. Three periods can be identified when analyzing $|\cos\theta\rangle(t)$:

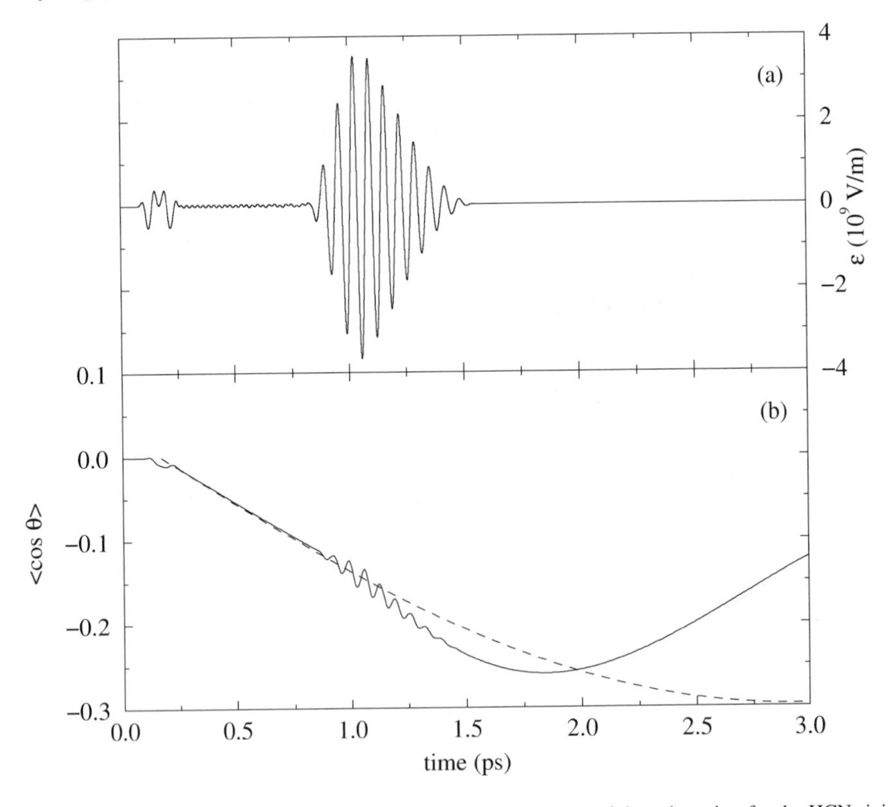

FIG. 5.2. (a) Laser field derived from the optimal control calculation of the orientation for the HCN rigid rotor. (b) Orientation expectation value with this field (solid line) and from the sudden-impact model (dashed line). Reproduced from [DION, BEN HAJ-YEDDER, CANCÈS, LE BRIS, KELLER and ATABEK, Optimal laser control of orientation: The kicked molecule, Phys. Rev. A **65**, 63408, Copyright © (2002)], by permission of the American Physical Society.

(1) An initial sudden (of approximately 0.25 ps, i.e., much shorter than the rotational period of 11 ps) and asymmetric pulse imparts a kick to the molecule that induces the dynamics of orientation in a way very similar to the one in consideration when referring to half-cycle pulses (DION, KELLER and ATABEK [2001]). The molecular response time is still not short enough for a noticeable quasi-instantaneous orientation to be observed within this period extending up to ~ 0.25 ps.

(2) The orientation continues to develop during the second period (~ 0.25 ps to ~ 1 ps) where all individual fields reach their plateau value; two of them with equal field amplitudes (within 4 digits accuracy) corresponding to rather high intensities of $\sim 3 \times 10^{12}$ W/cm^2, the third being 4 orders of magnitude smaller. But the most striking observation is that the strong field pulses present equal frequencies (within 4 digits accuracy) associated with a relative phase shift of π (within 3 digits accuracy), resulting into a quasi-absolute destructive interference that preserves the kick mechanism initiated in the first period.

(3) A dozen field oscillations within a smoother envelope before complete switch-off at $t \approx 1.5$ ps characterizes the third period. Although the molecule is solicited back and forth it continues to be oriented by the effect of the initial kick $|\langle\cos\theta(t)\rangle|$ increasing with some additional wiggles which follow the field oscillations at ~ 500 cm^{-1} frequency. The best orientation is obtained at $t = 2.2$ ps after laser switch off, and is expected to occur recurrently with the molecular rotational periodicity.

The interpretation of the kick mechanism is reinforced by the application of an impulsive "sudden-impact" model (as presented in DION, KELLER and ATABEK [2001]). This model basically reflects the fact that during the radiative interaction the molecular rotational motion can approximately be neglected. At lowest order, for a field taken from t_{ki} to t_{kf}, the fulfillment of the inequality

$$t_{kf} - t_{ki} \ll \frac{\hbar}{B\tilde{J}^2} \tag{5.11}$$

(i.e., short kick duration $t_{kf} - t_{ki}$ with respect to the rotational period) leads to the approximate solution of the TDSE:

$$\psi(\theta, \varphi; t > t_k) = \exp\left(-\frac{i}{\hbar}B\tilde{J}^2 t\right)\exp\left[i\left(\mathcal{A}\cos\theta + \mathcal{B}_\parallel \cos^2\theta + \mathcal{B}_\perp \sin^2\theta\right)\right] \times \psi(\theta, \varphi; t = t_{ki}), \tag{5.12}$$

with $t_k = (t_{kf} - t_{ki})/2$, and where the coefficients $\mathcal{A}$ and $\mathcal{B}_{\parallel,\perp}$ are obtained from the integrated electric field as

$$\mathcal{A} = \frac{\mu_0}{\hbar}\int_{t_{ki}}^{t_{kf}} E(t)\,dt, \tag{5.13}$$

$$\mathcal{B}_{\parallel,\perp} = \frac{\alpha_{\parallel,\perp}}{2\hbar}\int_{t_{ki}}^{t_{kf}} E^2(t)\,dt. \tag{5.14}$$

Freezing rotational dynamics during the kick, by use of Eq. (5.12), leads to angular distributions within fairly good accuracy, as displayed in Fig. 5.2(b). The conclusion is that the optimized field of Fig. 5.2(a) definitely offers one way to produce orientation through a kick mechanism. Although very accurately evidenced, such a mechanism is clearly not unique, other optimization criteria combining orientation efficiency and duration may presumably lead to efficient dynamics. It offers, however, apart from its physically sound and convincing interpretation, the advantage of robustness and experimental feasibility (by an appropriate duplication and recombination of a single field). Moreover, it may help as a guide to more advanced control scenarios. Namely, it suggests the use of a train of kicks acting in the same direction and progressively enhancing the orientation effect of the first, the time delay between them serving as a control parameter that basically takes into account the molecular response time to the laser excitation and the field-free evolution of the angular distribution.

6. Conclusion

Intense laser-induced molecular dissociation is definitely one of the most challenging application of Quantum Mechanics where theoretical models and their computer-based numerical simulations may help in a quantitative understanding and prediction of experimental discoveries. This has been exemplified on some simple (low-dimensional) systems, but the mechanisms that have been overviewed present a much wider application range in reactivity control scenarios. More precisely, the interplay of strong field mechanisms (barrier suppression and dynamical quenching in the IR regime or bond softening and vibrational trapping in the UV regime) are the basic tools for a detailed interpretation of experimental observations made on ATD spectra. Moreover, adequately adjusting the laser characteristics (intensity, frequency, polarization, pulse shape and phase) opens the way, by bringing into competition chemical bond softening and/or hardening mechanisms, to intense laser control scenarios of photochemical processes. Advantage can be taken of such a competition to direct the chemical reaction towards a given channel, involving the breaking of a given bond, while the others are stabilized by laser-induced trapping or quenching mechanisms. Still another possibility is to proceed to a separation in a mixture of chemical species (two isotopes, for instance) merely by dissociating some of them while stabilizing the others through the same laser excitation, the result being an enrichment of the mixture by the stabilized species. The proper modification of the chemical bond results into different photofragment distributions which may, in turn, be controlled. We have in particular shown that not only scalar observables such as total or partial dissociation rates of fragments but also vectorial observables such as fragment angular distributions can be controlled by intense lasers. Laser-induced molecular orientation and manipulations are among the future goals of such control schemes.

References

ABOU-RACHID, H., T.T. NGUYEN-DANG and O. ATABEK (1999), Dynamical quenching of laser-induced dissociations of heteronuclear diatomic molecules in intense infrared fields, *J. Chem. Phys.* **110**, 4737–4749.

ABOU-RACHID, H., T.T. NGUYEN-DANG and O. ATABEK (2001), Dynamical quenching of laser-induced dissociations of diatomic molecules in intense IR fields: Effects of molecular rotations and misalignments, *J. Chem. Phys.* **114**, 2197–2207.

ABRAMOVITZ, M. and I.A. STEGUN (1972), *Handbook of Mathematical Functions* (Dover, New York).

ATABEK, O. (1994), Isotope effects in laser-induced multiphoton molecular dynamics, *Int. J. Quantum Chem., Quantum Chemistry Symposium* **28**, 113–127.

ATABEK, O., M. CHRYSOS and R. LEFEBVRE (1994), Isotope separation using intense laser fields, *Phys. Rev. A* **49**, R8–R11.

AUBANEL, E.E., A. CONJUSTEAU and A.D. BANDRAUK (1993), Effect of rotations on stabilization in high-intensity photodissociation of H_2^+, *Phys. Rev. A* **48**, R4011–R4014.

AUBANEL, E.E., J.-M. GAUTHIER and A.D. BANDRAUK (1993), Molecular stabilization and angular distribution in photodissociation of H_2^+ in intense laser fields, *Phys. Rev. A* **48**, 2145–2152.

BANDRAUK A.D., ed. (1994), *Molecules in Laser Fields* (Dekker, New York).

BANDRAUK, A.D. and M.S. CHILD (1970), Analytic predissociation linewidths from scattering theory, *Mol. Phys.* **19**, 95–111.

BANDRAUK, A.D. and T.T. NGUYEN-DANG (1985), Molecular dynamics in intense laser fields, III. Nonadiabatic effects, *J. Chem. Phys.* **83**, 2840–2850.

BANDRAUK, A.D. and M.L. SINK (1981), Photodissociation in intense laser fields: Predissociation analogy, *J. Chem. Phys.* **74**, 1110–1117.

BEN HAJ-YEDDER, A., A. AUGER, C.M. DION, E. CANCÈS, A. KELLER, C. LE BRIS and O. ATABEK (2002), Numerical optimization of laser fields to control molecular orientation, *Phys. Rev. A* **66**, 063401-1–063401-9.

BLOCH, F. and A. NORDSIECK (1937), Note on the radiation field of the electron, *Phys. Rev.* **52**, 54–59.

BORN, M. and R. OPPENHEIMER (1927), Zur Quantentheorie der Molekeln, *Ann. Physik* **84**, 457–484.

BUCKSBAUM, P.H., A. ZAVRIYEV, H.G. MULLER and D.W. SCHUMACHER (1990), Softening of the H_2^+ bond in intense laser fields, *Phys. Rev. Lett.* **64**, 1883–1886.

BUNKIN, F.V. and I.I. TUGOV (1973), Multiphoton processes in homopolar diatomic molecules, *Phys. Rev. A* **8**, 601–612.

CHANDLER, D.W., D.W. NEYER and A.J. HECK (1998), High resolution photoelectron images and D^+ photofragment images following 532-nm photolysis of D_2, *Proc. SPIE* **3271**, 104–113.

CHARRON, E., A. GIUSTI-SUZOR and F.H. MIES (1995), Coherent control of photodissociation in intense laser fields, *J. Chem. Phys.* **103**, 7359–7373.

CHATEAUNEUF, F., T.T. NGUYEN-DANG, N. OUELLET and O. ATABEK (1998), Dynamical quenching of field-induced dissociation of H_2^+ in intense infrared lasers, *J. Chem. Phys.* **108**, 3974–3986.

CHELKOWSKI, S., T. ZUO, A.D. BANDRAUK and O. ATABEK (1995), Dissociation, ionization and Coulomb explosion in an intense laser field by numerical integration of the time-dependent Schrödinger equation, *Phys. Rev. A* **52**, 2977–2983.

CHRYSOS, M., O. ATABEK and R. LEFEBVRE (1993a), Molecular photodissociation with diverging couplings: An application to H_2^+ in intense cw laser fields, I. The single-photon problem, *Phys. Rev. A* **48**, 3845–3854.

CHRYSOS, M., O. ATABEK and R. LEFEBVRE (1993b), Molecular photodissociation with diverging couplings: An application to H_2^+ in intense cw laser fields, II. The multi-photon problem, *Phys. Rev. A* **48**, 3855–3862.

CHRYSOS, M. and R. LEFEBVRE (1993), Resonance quantization with persistent effects, *J. Phys. B: At. Mol. Opt. Phys.* **26**, 2627–2647.

CHU, S.I. (1981), Floquet theory and complex quasivibrational energy formalism for intense field molecular photodissociation, *J. Chem. Phys.* **75**, 2215–2221.

CHU, S.I. (1991), Complex quasivibrational energy formalism for intense-field multiphoton and above-threshold dissociation: Complex-scaling Fourier-grid Hamiltonian method, *J. Chem. Phys.* **94**, 7901–7909.

COHEN-TANNOUDJI, CL., J. DUPONT-ROC and G. GRYNBERG (1987), *Introduction à l'Electrodynamique Quantique*, ibid: *Processus d'interaction entre photons et atomes* (InterEditions and CNRS, Paris).

DATEO, C.E. and H. METIU (1991), Numerical solution of the time dependent Schrödinger equation in spherical coordinates by Fourier-transform methods, *J. Chem. Phys.* **95**, 7392–7400.

DION, C.M., A.D. BANDRAUK, O. ATABEK, A. KELLER, H. UMEDA and Y. FUJIMURA (1999), Two-frequency IR laser orientation of polar molecules. Numerical simulations for HCN, *Chem. Phys. Lett.* **302**, 215–223.

DION, C.M., A. BEN HAJ-YEDDER, E. CANCÈS, C. LE BRIS, A. KELLER and O. ATABEK (2002), Optimal laser control of orientation: The kicked molecule, *Phys. Rev. A* **65**, 063408-1–063408-7.

DION, C.H., A. KELLER and O. ATABEK (2001), Orienting molecules using half-cycle pulses, *Eur. Phys. J. D* **14**, 249–255.

DION, C.M., A. KELLER, O. ATABEK and A.D. BANDRAUK (1999), Laser-induced alignment dynamics of HCN: Roles of the permanent dipole moment and the polarizability, *Phys. Rev. A* **59**, 1382–1391.

DUPUIS, M., A. MARQUEZ and E.R. DAVIDSON (1995), *HONDO 95.6* (IBM Corporation, Neighborhood Road, Kingston, NY.12401).

FEIT, H.D. and J.A. FLECK Jr. (1984), Wave packet dynamics and chaos in the Hénon–Heiles system, *J. Chem. Phys.* **80**, 2578–2584.

FEIT, M.D., J.A. FLECK and A. STEIGER (1982), Solution of the Schrödinger equation by a spectral method, *J. Comput. Phys.* **47**, 412–433.

FLOQUET, G. (1883), Sur les équations différentielles linéaires à coefficients périodiques, *Ann. scient. Ecole Norm. Sup.* **12**, 47–88.

FOX, L. (1957), *The Numerical Solution of Two-Point Boundary Value Problems in Ordinary Differential Equations* (Oxford Univ. Press, London).

FRIEDRICH, B. and D. HERSCHBACH (1991), On the possibility of orienting rotationally cooled polar molecules in an electric field, *Z. f. Phys. D* **18**, 153–161.

FRIEDRICH, B. and D. HERSCHBACH (1995), Alignment and trapping of molecules in intense laser fields, *Phys. Rev. Lett.* **74**, 4623–4626.

GAVRILA, M. (1992), Atomic structure and decay in high-frequency fields, in: M. Gavrila, ed., *Atoms in Intense Laser Fields* (Academic Press, Boston) 435–508.

GIUSTI-SUZOR, A., X. HE, O. ATABEK and F.H. MIES (1990), Above-threshold dissociation of H_2^+ in intense laser fields, *Phys. Rev. Lett.* **64**, 515–518.

GIUSTI-SUZOR, A., F.H. MIES, L.F. DIMAURO, E. CHARRON and B. YANG (1995), Dynamics of H_2^+ in intense laser fields, *J. Phys. B: At. Mol. Opt. Phys.* **28**, 309–339.

GUERIN, S. and H.-R. JAUSLIN (1997), Laser-enhanced tunneling through resonant intermediate levels, *Phys. Rev. A* **55**, 1262–1275.

HE, X., O. ATABEK and A. GIUSTI-SUZOR (1990), Semidiabatic treatment of photodissociation in strong laser fields, *Phys. Rev. A* **42**, 1585–1591.

HENNEBERGER, W.C. (1968), Perturbation method for atoms in intense light beams, *Phys. Rev. Lett.* **21**, 838–841.

HOKI, K. and Y. FUJIMURA (2001), Quantum control of alignment and orientation of molecules by optimized laser pulses, *Chem. Phys.* **267**, 187–193.

HUMBLET, J. and L. ROSENFELD (1961), Theory of nuclear reactions, I. Resonant states and collision matrix, *Nuclear Phys.* **26**, 529–578.

JOLICARD, G. and O. ATABEK (1992), Above-threshold dissociation of H_2^+ with short intense laser pulses, *Phys. Rev. A* **46**, 5845–5855.

JUDSON, R.S., K.R. LEHMANN, H. RABITZ and W.S. WARREN (1990), Optimal design of external fields for controlling molecular motion: Application to rotation, *J. Mol. Struc.* **223**, 425–456.

KELLER, A. (1995), Asymptotic analysis in time-dependent calculations with divergent couplings, *Phys. Rev. A* **52**, 1450–1457.

KELLER, A. and O. ATABEK (1993), Continuum Raman scattering with short laser pulses, *Phys. Rev. A* **12**, 3741–3756.

KELLER, A., C.M. DION and O. ATABEK (2000), Laser-induced molecular alignment in dissociation dynamics, *Phys. Rev. A* **61**, 23409-1–23409-11.

KITTEL, C. (1996), *Introduction to Solid State Physics*, 7th edn. (Wiley, New York).

KOSLOFF, R. (1988), Time-dependent quantum-mechanical methods for molecular dynamics, *J. Phys. Chem.* **92**, 2087–2100.

KOURI, D.J. and R.C. MOWREY (1987), Close coupling-wave packet formalism for gas phase atom-diatom collisions, *J. Chem. Phys.* **80**, 2578–2584.

KRAMERS, H.A. (1956), *Quantum Mechanics* (North-Holland, Amsterdam).

LEFEBVRE, R. and O. ATABEK (1997), Gauges and fluxes in multiphoton absorption by H_2^+, *Int. J. of Quantum Chem.* **63**, 403–414.

LOESCH, H.J. and A. REMSCHEID (1990), Brute force in molecular reaction dynamics: A novel technique for measuring steric effects, *J. Chem. Phys.* **93**, 4779–4790.

LOUDON, R. (1973), *The Quantum Theory of Light* (Clarendon Press, Oxford).

LOUISELL, W. (1973), *Quantum Statistical Properties of Radiation* (Wiley, New York).

MANOLI, S. and T.T. NGUYEN-DANG (1993), Laser-induced-resonances calculations for the photodissociation of H_2^+ in an adiabatic electronic-field representation using the radiation-field gauge, *Phys. Rev. A* **48**, 4307–4320.

MICHALEWICZ, Z. (1996), *Genetic Algorithms + Data Structure = Evolution Programs* (Springer, Berlin).

MOISEYEV, N. (1998), Quantum theory of resonances: calculating energies, widths and cross-sections by complex scaling, *Physics Reports* **302**, 212–296.

MOISEYEV, N., O.E. ALON and V. RYABOY (1995), The (t, t') method and gauge transformations for two electronic potential surfaces: an application to the partial widths of H_2^+, *J. Phys. B: At. Mol. Opt. Phys.* **28**, 2611–2620.

MOISEYEV, N., M. CHRYSOS and R. LEFEBVRE (1995), The solution of the time dependent Schrödinger equation by the (t, t') method: application to intense field molecular photodissociation, *J. Phys. B: At. Mol. Opt. Phys.* **28**, 2599–2609.

MOISEYEV, N. and U. PESKIN (1990), Partial widths obtained by the complex resonance-scattering theory, *Phys. Rev. A* **42**, 255–260.

MOLLOW, B.R. (1975), Pure-state analysis of resonant light scattering: Radiative damping, saturation, and multiphoton effects, *Phys. Rev. A* **12**, 1919–1943.

MOYAL, J.E. (1949), Quantum mechanics as a statistical theory, *Proc. Cambridge Philos. Soc.* **45**, 99–124.

NGUYEN, N.A. and T.T. NGUYEN-DANG (2000), Molecular dichotomy within an intense high-frequency laser field, *J. Chem. Phys.* **112**, 1229–1239.

NGUYEN-DANG, T.T. and A.D. BANDRAUK (1983), Molecular dynamics in intense laser fields, I. One-dimensional systems in infrared radiation, *J. Chem. Phys.* **79**, 3256–3268.

NGUYEN-DANG, T.T. and A.D. BANDRAUK (1984), Molecular dynamics in intense laser fields, II. Adiabatic approximations and coupled equations, *J. Chem. Phys.* **80**, 4926–4939.

NGUYEN-DANG, T.T., F. CHATEAUNEUF, O. ATABEK and X. HE (1995), Time-resolved dynamics of two-channels molecular systems in cw laser fields: Wavepacket construction in the Floquet formalism, *Phys. Rev. A* **51**, 1387–1402.

NGUYEN-DANG, T.T., F. CHATEAUNEUF and S. MANOLI (1996), Molecules in high-intensity laser fields, *Can. J. Chem.* **74**, 1236–1247.

NGUYEN-DANG, T.T., S. DUROCHER and O. ATABEK (1989), Direct numerical integration of coupled equations with non-adiabatic couplings, *Chem. Phys.* **129**, 451–462.

NICOLAIDES, C.A. and D.R. BECK (1978), The variational calculation of energies and widths of resonances, *Phys. Lett. A* **65**, 11–12.

NORCROSS, D.W. and M.J. SEATON (1973), Asymptotic solutions of the coupled equations of electron-atom theory for the case of some channels closed, *J. Phys. B: At. Mol. Phys.* **6**, 604–621.

NORMAND, D., L.A. LOMPRE and G. CORNAGGIA (1992), Laser-induced molecular alignment probed by a double-pulse experiment, *J. Phys. B: At. Mol. Opt. Phys.* **25**, L497–L503.

NUMICO, R., A. KELLER and O. ATABEK (1995), Laser-induced molecular alignment in dissociation dynamics, *Phys. Rev. A* **52**, 1298–1309.

NUMICO, R., A. KELLER and O. ATABEK (1997), Nonadiabatic response to short intense laser pulses in dissociation dynamics, *Phys. Rev. A* **56**, 772–781.

NUMICO, R., A. KELLER and O. ATABEK (1999), Intense laser-induced alignment in angularly resolved photofragment distribution of H_2^+, *Phys. Rev. A* **60**, 406–413.

OKUNIEWICZ, J.M. (1974), Quasiperiodic pointwise solutions of the periodic time dependent Schrödinger equation, *J. of Math. Phys.* **15**, 1587–1595.

ORTIGOZO, J. (1998), Design of tailored microwave pulses to create rotational coherent states for an asymmetric-top molecule, *Phys. Rev. A* **57**, 4592–4599.

PAULI, W. and M. FIERZ (1938), Zur der Emission langwelliger Lichtquanten, *Nuovo Cimento* **16**, 167–188.

PEREZ DEL VALLE, C., R. LEFEBVRE and O. ATABEK (1997), Dressed states of the high-frequency Floquet theory for atoms and molecules with standard quantum chemistry programs, *J. Phys. B: At. Mol. and Opt. Phys.* **30**, 5157–5167.

PEREZ DEL VALLE, C., R. LEFEBVRE and O. ATABEK (1998), Dressed potential energy surface of the hydrogen molecule in high-frequency Floquet theory, *Int. J. of Quantum Chem.* **70**, 199–203.

PESKIN, U. and N. MOISEYEV (1993), The solution of the time-dependent Schrödinger equation by the (t, t') method: Theory, computational algorithm and applications, *J. Chem. Phys.* **99**, 4590–4596.

REISS, H.R. (1980), Effect of an intense electromagnetic field on a weakly bound system, *Phys. Rev. A* **22**, 1786–1813.

SEIDEMAN, T. (1995), Rotational excitation and molecular alignment in intense laser fields, *J. Chem. Phys.* **103**, 7887–7896.

SEIDEMAN, T. (1997), Manipulating external degrees of freedom with intense light: Laser focusing and trapping of molecules, *J. Chem. Phys.* **106**, 2881–2892; SEIDEMAN, T. (1997), Molecular optics in an intense laser field, *Phys. Rev. A* **56**, R17–R20.

SHEEBY, B., B. WALKER and L.F. DIMAURO (1995), Phase control in the two-color photodissociation of HD^+, *Phys. Rev. Lett.* **74**, 4799–4802.

SHERTZER, J., A. CHANDLER and M. GAVRILA (1994), H_2^+ in superintense laser fields: alignment and spectral restructuring, *Phys. Rev. Lett.* **73**, 2039–2042.

SHIRLEY, J.H. (1965), Solution of the Schrödinger equation with a Hamiltonian periodic in time, *Phys. Rev.* **138**, B979–B987.

SIEGERT, A.F.J. (1939), On the derivation of the dispersion formula for nuclear reactions, *Phys. Rev.* **56**, 750–752.

SIMON, B. (1979), The definition of molecular resonance curves by the method of exterior complex scaling, *Phys. Lett. A* **71**, 211–214.

STAPELFELDT, H., H. SAKAI, E. CONSTANT and P.B. CORCUM (1997), Deflection of neutral molecules using the nonresonant dipole force, *Phys. Rev. Lett.* **79**, 2787–2790.

VARSHALOVICH, A., A.N. MOSKALEV and V.K. KHERSONSKII (1988), *Quantum Theory of Angular Momentum* (World Scientific, Singapore).

VRAKKING, M.J.J. and S. STOLTE (1997), Coherent control of molecular orientation, *Chem. Phys. Lett.* **271**, 209–215.

YAO, G. and S.I. CHU (1993), Molecular-bond hardening and dynamics of molecular stabilization and trapping in intense laser pulses, *Phys. Rev. A* **48**, 485–494.

YOSIDA, K. (1971), *Functional Analysis* (Springer-Verlag, New York).

ZAVRIYEV, A., P.H. BUCKSBAUM, H.G. MULLER and D.W. SCHUMACHER (1990), Ionization and dissociation of H_2 in intense laser fields at 1.064 μn, 532 nm, and 355 nm, *Phys. Rev. A* **42**, 5500–5513.

ZEWAIL, A.H. (1988), Laser femtochemistry, *Science* **242**, 1645–1653.

Numerical Methods for Molecular Time-Dependent Schrödinger Equations – Bridging the Perturbative to Nonperturbative Regime

André D. Bandrauk, Hui-Zhong Lu

Laboratoire de Chimie Théorique, Faculté des Sciences, Université de Sherbrooke, Que. J1K 2R1, Canada

1. Introduction

Current laser technology has progressed significantly from ultrashort, i.e., duration on the attosecond (10^{-18} sec) time scale (HENTSCHEL [2001]) and intensities exceeding that of electric fields internal to atoms and molecules. Recent technological advances in ultrafast optics have permitted the generation of light wavepackets comprising only a few oscillation cycles of electric and magnetic fields (BRABEC and KRAUSZ [2000]). This revolution in laser technology is expected to also bring a revolution in experimental sciences (HAROCHE [2000]). To appreciate this new technological revolution, one has to recall the scales in atomic units (where $h/2\pi = e = m_e = 1$) the magnitude of the following important physical parameters. The period of circulation of an electron in the ground state of the hydrogen (H) atom is:

$$\tau_{\mathrm{H}}(\mathrm{a.u.}) = 24.6 \times 10^{-18}\ \mathrm{sec} = 24.6\ \mathrm{attoseconds\ (atts)}. \tag{1.1}$$

The electric field E_{H} and corresponding intensity I in that orbit are:

$$E_{\mathrm{H}}(\mathrm{a.u.}) = 5 \times 10^{9}\ \mathrm{V/cm}, \qquad I_{\mathrm{H}}(\mathrm{a.u.}) = cE^2/8\pi = 3 \times 10^{16}\ \mathrm{W/cm^2}. \tag{1.2}$$

Computational Chemistry
Special Volume (C. Le Bris, Guest Editor) of
HANDBOOK OF NUMERICAL ANALYSIS, VOL. X
P.G. Ciarlet (Editor)
© 2003 Elsevier Science B.V. All rights reserved

Due to extreme temporal and spatial confinement using current compression techniques, pulse energies in peak intensities higher than 10^{15} W/cm^2 are readily available with table top lasers (BRABEC and KRAUSZ [2000]). The resulting electric field strength approach and can exceed that of the static Coulomb field experienced by outer-shell electrons in atoms and molecules. As a consequence the laser field is strong enough to suppress the binding Coulomb potential in atoms and trigger multiphoton optical-field ionization (KELDYSH [1965], CORKUM [1993]).

Much of our understanding of intense laser field-atom interaction physics has been documented in a book where new nonperturbative phenomena such as Above Threshold Ionization, ATI (the excess absorption of laser photons after ionization of an electron), high order harmonic generation, HOHG (the emission of a large number of laser photons after recollision of an electron with its parent or neighbouring ion), have been treated theoretically and numerically from the perturbative to nonperturbative regime, as various approximations to appropriate time-dependent Schroedinger equations, TDSEs (GAVRILA [1992], EHLOTZKY [2001]). In the low frequency regime, especially with current Ti/Sap. lasers of wavelength $\lambda = 800$ nm, quasistatic tunneling models (KELDYSH [1965], CORKUM [1993], GAVRILA [1992], EHLOTZKY [2001]) have helped bridge the perturbative to nonperturbative regime in atom laser physics.

Molecules present a new challenge, due to the presence of extra degrees of freedom from nuclear motion. As an example, if one considers the frequency of vibration of the H_2 molecule, $\nu_{H_2} = 4000\ \mathrm{cm}^{-1}$ (HERTZBERG [1951]), this corresponds to a vibration period of:

$$\tau_{H_2} = (\nu_{H_2} c)^{-1} = 8.3 \times 10^{-15}\ \mathrm{sec} = 8.3\ \text{femtoseconds (fs)}, \tag{1.3}$$

where $c = 3 \times 10^{10}$ cm/s is the velocity of light. This is to be compared to the Ti/Sap. laser ($\lambda = 800$ nm) frequency of 2.7 femtoseconds, CO_2 laser ($\nu = 1000\ \mathrm{cm}^{-1}$) period of 30 femtoseconds. One finds therefore that the fastest atomic motion time, τ_p of a proton in molecules, is readily in the laser field frequency regime, i.e.,

$$3 \leqslant \tau_p \leqslant 30\ \text{femtoseconds (fs)}. \tag{1.4}$$

The proton is indeed the most elusive particle with the shortest time scale for nuclear motion in molecules, yet it is probably the most important atom in chemistry and biology (e.g., DNA code). To date no one has been able to image its motion. Recent theoretical calculations based on exact TDSEs with intense ultrashort laser pulses, $\tau_L < 10$ fs, are proposing such pulses as the ideal tool for dynamic imaging of ultrafast nuclear motion (BANDRAUK and CHELKOWSKI [2001]).

Much of theoretical work in molecules has been focused in the past fifty years on ab initio electronic structure calculations, in the Born–Oppenheimer approximation, i.e., assuming static nuclei (DEFRANCESCHI and LE BRIS [2001]). As indicated above, nuclear motion involves introducing dynamics into the structure calculations. Using Born–Oppenheimer or adiabatic molecular potentials obtained from *ab initio* calculations, considerable progress in describing nonperturbative multiphotons molecular processes has been achieved using a dressed state molecular representation. This has been summarized in various works (BANDRAUK [1994a], BANDRAUK [1994b]) and has led to the concept of laser induced avoided crossings of molecular potentials leading to laser induced molecular potentials, LIMPs, i.e., new molecular potentials created by radiative interactions, recently confirmed experimentally (WUNDERLICH, KOBLER, FIGGER and HÄNSCH [1997]) and Above Threshold Dissociation, ATD (excess laser photon absorption after photodissociation), the analogue of ATI in atoms.

Intensities approaching the atomic unit of field strength (BRABEC and KRAUSZ [2000]), leads to considerable ionization. One thus has to treat in general dissociation-ionization processes where the ionization time scale, τ_I, the nuclear motion time scale τ_p and laser time scale τ_L become comparable, i.e., one is in the few femtosecond regime where separability of electronic, nuclear and radiative processes is not possible. Thus the usual adiabatic (Born–Oppenheimer) methods of *ab initio* quantum chemistry (DEFRANCESCHI and LE BRIS [2001]) are no longer valid. Furthermore one has to deal with initial bound state (discrete spectra) to continuum state (continuum spectra) transitions. Such problems lead to the search for new numerical approaches, well beyond current quantum chemistry methods. To date only the one-electron, simplest molecule, H_2^+ has been solved exactly numerically using large grids and super-computers (CHELKOWSKI, FOISY and BANDRAUK [1998], CHELKOWSKI, CONJUSTEAU, ATABEK and BANDRAUK [1996]). Such simulations have helped discover new nonperturbative effects in molecules subjected to intense laser pulses such as Charge Resonance Enhanced Ionization, CREI (ionization induced by electron transfer) (CHELKOWSKI and BANDRAUK [1995], BANDRAUK [2000]), Laser Induced Electron Diffraction, LIED (ZUO, BANDRAUK and CORKUM [1996]) and are leading the way to suggest laser control of dissociative-ionization (ZUO and BANDRAUK [1996], CHELKOWSKI, ZAMOJSKI and BANDRAUK [2001], BANDRAUK and YU [1998], BANDRAUK, FUJIMURA and GORDON [2002]).

In summary, we would like to emphasize once again the importance of the new ultrashort laser technology to the seminal problem of laser control and manipulation of molecules. Optimization of the laser parameters, i.e., amplitude and phase, to optimize yields of interest to the chemical community has led to a new branch of photochemistry, *coherent control* of nuclear motion (see BRUMER and SHAPIRO [1994], RABITZ, TURINICI and BROWN [2003]). The present chapter focuses on a complete treatment of laser–molecule interaction, in order to arrive eventually at the laser control and manipulation of both *electronic* and *nuclear* states and their spaces, which become nonseparable, i.e., nonadiabatic processes dominate, as laser intensities are increased. Of course a complete treatment of laser–molecule interaction will require investigating numerically solutions of the full coupled TDSE and Maxwell's equations (BANDRAUK [1994b]) and finally the Time-Dependent Dirac Equation, TDDE (THALLER [1992]). TDSEs are parabolic partial differential equations, PDEs, whereas Maxwell equations and TDDEs are hyperbolic PDEs. Clearly laser–molecule interaction problems offer new numerical challenges for the current accurate description of multiphotons processes in molecules from the perturbative to nonperturbative regimes. In the present work we shall focus on methods developed by us to treat exactly one electron problems beyond the standard adiabatic Born–Oppenheimer approximation. Extension to many-electrons problems will most probably require progress in *time dependent density functional theory* methods, TDDFT (DREIZLER and GROSS [1990]), where *nonlocal* electron potentials occurring in *ab initio* quantum chemistry methods are replaced by effective local potentials. Then the one-electron numerical methods described in the present work will be easily implementable and applicable to describe completely laser–molecule interactions and the accompanying electron–nuclear dynamics for many electron systems.

2. Gauges and representations

Previous 3D calculations of intense laser–field matter interactions have relied on time-dependent Hartree–Fock, TDHF, approximations developed by KULANDER [1987] for atomic problems and applied also to H_2 (KRAUSE, SCHAFFER and KULANDER [1991]) using numerical grid methods. We have previously applied 3D Cartesian finite elements for TDHF calculations of H_2 and H_3^+ in intense laser fields (YU and BANDRAUK [1995]). Finite element bases are very flexible and are able to span bound and continuum states simultaneously. In order to go beyond TDHF method and 1D models, progress is required in exact nonperturbative numerical solution of complete TDSEs for multi-electron non-Born–Oppenheimer simulations. Furthermore molecule-radiation interactions can be represented by Hamiltonians in different gauges (GAVRILA [1992], BANDRAUK [1994b]), with each gauge having advantages and disadvantages, both numerical and physical. Thus to solve molecular TDSEs in the field of an intense laser pulse, most current numerical methods use the length gauge ($\vec{E} \cdot \vec{r}$) or the velocity gauge (Coulomb, $\vec{A} \cdot \vec{p}$) with *Euclidean* stationary coordinates systems. In the length gauge, the molecule–field interaction appears as a *potential* term whereas in the velocity gauge, this interaction appears as a *convection* term. In classical fluid mechanics, it is customary to use a *Lagrangian* (moving) coordinate system which removes the convection term so that the conservation of physical relevant observables is automatically assured (GERHART, GROSS and HOCHSTEIN [1992], HAUSBO [2000]).

As we show below, the TDSE in the Lagrangian system is a generalization for molecules of the space translation (ST) method used in intense high frequency atom-laser interaction theories. This was originally introduced by PAULI and FIERZ [1938] for vacuum corrections to atom–field interactions. To our knowledge it was first used in quantum mechanics by HUSIMI [1953] to solve the complete time-dependent harmonic oscillator in an external time-dependent force. Thus in the Lagrangian or ST system, the laser field is included in the time-dependent coordinate displacements, reducing the TDSE to a time-independent equation in some simple cases (TRUSCOTT [1993]). We show that adopting such a reference system, one obtains TDSEs with moving coordinates thus requiring adaptive grid methods to solve these equations (LU and BANDRAUK [2001]). These new equations have the advantage that for large electron–nuclear separation as occurs in dissociative ionization processes, the grid moves exactly in phase with the electron wavefunction so that the grid in asymptotic limit is an *exact adaptive* grid. We show in the present work that the Lagrangian or ST method offers many advantages over other gauges and should facilitate exact computation for nonperturbative laser–molecule interactions. Of note is that for vanishing (zero) field, the ST method contains quantum corrections to vacuum fluctuations (WELTON [1948]).

Time-dependent fields in quantum mechanics can always be transformed to suitable equivalent (isospectral) representations by appropriate unitary transformations. Less well appreciated is the fact that one can eliminate time-dependent fields by a time-dependent coordinate transformation for particular cases. We start with the usual 1D TDSE in which the potential explicitly includes a spatially uniform field that is an arbitrary function of time $f(t)$ (we use atomic units a.u.: $e = m_e = h/2\pi = 1$),

$$i\frac{\partial\psi(x,t)}{\partial t} = -\frac{1}{2}\frac{\partial^2\psi(x,t)}{\partial x^2} + \big[V(x,t) - xf(t)\big]\psi(x,t). \tag{2.1}$$

Following the procedure of HUSIMI [1953], we transform from the x, t coordinates system to a new ξ, t coordinate system where:

$$\xi = x - q(t), \tag{2.2}$$

from which it follows:

$$\frac{\partial\xi}{\partial x} = 1, \qquad \frac{\partial\xi}{\partial t} = -\dot{q}(t), \tag{2.3}$$

so that

$$\frac{\partial\psi}{\partial t} = -\dot{q}(t)\left.\frac{\partial\psi}{\partial\xi}\right|_t + \left.\frac{\partial\psi}{\partial t}\right|_\xi. \tag{2.4}$$

Introducing the transformation

$$\psi(\xi,t) = \exp(i\dot{q}\xi)\phi(\xi,t), \tag{2.5}$$

we obtain a new TDSE for $\phi(\xi,t)$:

$$\begin{aligned} i\frac{\partial\phi(\xi,t)}{\partial t} = {} & -\frac{1}{2}\frac{\partial^2\phi(\xi,t)}{\partial\xi^2} \\ & + \left[V(\xi+q,t) + (\ddot{q} - f)\xi - \left(\frac{\dot{q}^2}{2} + fq\right)\right]\phi(\xi,t). \end{aligned} \tag{2.6}$$

Setting

$$\dot{q} = \int f(s)\,ds, \qquad g(t) = \frac{\dot{q}^2}{2} + fq, \tag{2.7}$$

and performing another unitary transformation

$$\phi(\xi,t) = \exp\left[i\int g(s)\,ds\right]\Phi(\xi,t), \tag{2.8}$$

one obtains a new TDSE for $\Phi(\xi, t)$:

$$i\frac{\partial\Phi(\xi,t)}{\partial t} = -\frac{1}{2}\frac{\partial^2\Phi(\xi,t)}{\partial\xi^2} + V\big(\xi + q(t)\big)\Phi(\xi,t), \tag{2.9}$$

and the original solution of (2.1) is given by

$$\psi(x,t) = \exp\big[ix\dot{q}(t)\big]\exp\left[-\frac{i}{2}\int \dot{q}^2(t)\,dt\right]\Phi(\xi,t). \tag{2.10}$$

We note that from (2.2), ξ is a time-dependent function, i.e., it is a *moving* coordinate; however in (2.9) it is treated as a time-independent coordinate. In fact, for $V = V_0 = 0$, i.e., a constant, $\Phi(\xi,t)$ is a plane wave $e^{ik\xi} = e^{ik(x-q(t))}$ and the phase factors in (2.10) contain the ponderomotive energy $U_p = E_0^2/4\omega^2$ due to field induced oscillation of a free particle in a sinusoidal field $f(t) = E(t) = E_0\sin(\omega t)$ (GAVRILA [1992], BANDRAUK [1994b]).

In the *length* gauge, using the total 3D electronic Hamiltonian of H_2^+ for fixed nuclei at distance R in a laser field $E(t)\cos(\omega t)$ parallel to the internuclear axis, the TDSE is written (CHELKOWSKI, CONJUSTEAU, ATABEK and BANDRAUK [1996]) as:

$$\begin{aligned}
i\frac{\partial\psi(x,y,z,t)}{\partial t} &= -\frac{2m_p+1}{4m_p}\left[\frac{\partial^2}{\partial x^2} + \frac{\partial^2}{\partial y^2} + \frac{\partial^2}{\partial z^2}\right]\psi(x,y,z,t) \\
&\quad - \frac{\psi(x,y,z,t)}{\sqrt{x^2+y^2+(z\pm R/2)^2}} \\
&\quad + \frac{2m_p+2}{2m_p+1} zE(t)\cos(\omega t)\psi(x,y,z,t).
\end{aligned} \tag{2.11}$$

Here, the Hamiltonian is the exact one-electron Hamiltonian for a molecule *aligned* with the z-axis by the intense laser field with fixed protons of mass m_p (we have used a.u.: $m_e = 1$).

To propagate the TDSE in the *velocity* (Coulomb) gauge, we now use the unitary (gauge) transformation:

$$\psi \longrightarrow \psi\exp\left\{-i\frac{2m_p+1}{4m_p}\left[\int_0^t\left(\int_0^s E(\tau)\,d\tau\right)^2 ds\right] - iz\int_0^t E(s)\,ds\right\}, \tag{2.12}$$

which gives the new TDSE:

$$\begin{aligned}
i\frac{\partial\psi(x,y,z,t)}{\partial t} &= -\frac{2m_p+1}{4m_p}\left[\frac{\partial^2}{\partial x^2} + \frac{\partial^2}{\partial y^2} + \frac{\partial^2}{\partial z^2}\right]\psi(x,y,z,t) \\
&\quad - \frac{\psi(x,y,z,t)}{\sqrt{x^2+y^2+(z\pm R/2)^2}}
\end{aligned}$$

$$+ i\left[\frac{2m_p + 1}{2m_p}\int_0^t E(s)\,ds\right]\frac{\partial\psi(x,y,z,t)}{\partial z}. \tag{2.13}$$

In this equation, we see clearly the *convection* term with velocity:

$$\text{Convection velocity} = \frac{2m_p + 1}{2m_p}\int_0^t E(s)\,ds. \tag{2.14}$$

We note that Eq. (2.13) is similar to a *convection–diffusion* equation with *incompressible* convecting field but has an imaginary diffusion coefficient. Both of the two preceding formulations (2.11), (2.13) have used Eulerian (stationary) coordinates. To adopt the Lagrangian (moving) or ST coordinate system (2.2)–(2.6), we change the variable z, as suggested by Eq. (2.2),

$$z \to u = z + \frac{2m_p + 1}{2m_p}\int_0^t\int_0^s E(\tau)\,d\tau\,ds = z + \alpha(t), \tag{2.15}$$

and this transformation which is called the space translation (ST) method gives (GAVRILA [1992], BANDRAUK [1994b])

$$\begin{aligned} i\frac{\partial\psi(x,y,u,t)}{\partial t} &= -\frac{2m_p + 1}{4m_p}\left[\frac{\partial^2}{\partial x^2} + \frac{\partial^2}{\partial y^2} + \frac{\partial^2}{\partial u^2}\right]\psi(x,y,u,t) \\ &\quad - \frac{\psi(x,y,u,t)}{\sqrt{x^2 + y^2 + (u - \alpha(t) \pm R/2)^2}}. \end{aligned} \tag{2.16}$$

Comparing (2.16) with (2.11), (2.13), we note that the former is simpler: the explicit laser pulse term is removed and its effect is included in the time-dependent potential. We note that the new coordinate system is moving with respect to the fixed protons and the movement is determined by the laser strength and its phase. It is clear that in the TDSE for the dynamics of the protons, we can use the same procedure to remove the convection term with respect to the nuclei (BANDRAUK [1994b]).

Previous applications of the ST method have been done in numerical 1D simulation of H_2 (WIEDEMANN and MOSTOWSKI [1994]) and recently also in a full 3D MCSCF calculation using a Floquet expansion (NGUYEN and NGUYEN-DANG [2000]). The latter corresponds to an adiabatic-field approach. Since in general, a laser field can be characterized by the set of parameters $E = \{E_0, \omega, \phi\}$ where E_0 is the instantaneous amplitude, ω is the frequency and ϕ is the phase, then the total time derivative for any TDSE thus becomes (CHU [1989])

$$\frac{d}{dt} = \left.\frac{\partial}{\partial t}\right|_E + \dot{E}\frac{\partial}{\partial E}, \tag{2.17}$$

where

$$\dot{E}\frac{\partial}{\partial E} = \dot{E}_0\frac{\partial}{\partial E_0} + \dot{\omega}\frac{\partial}{\partial \omega} + \dot{\phi}\frac{\partial}{\partial \phi}. \tag{2.18}$$

This to be compared to Eq. (2.4) thus showing a close relation between the two approaches. All nonadiabatic field effects are contained in the time dependence of the potential via $q(t)$, Eq. (2.6). Thus accurate numerical solutions of the new TDSE, (2.6) are required to describe nonadiabatic effects expected in the short pulse limit where both amplitude E_0, frequency ω, and phase ϕ can be strongly time-dependent (BRABEC and KRAUSZ [2000]). We address next this problem by investigating various discretized propagation schemes for the TDSE of H_2^+.

3. Numerical schemes

For simplicity, we describe all numerical methods using only two space variables (y, z) in the length gauge (Eulerian coordinates). All theses schemes are easily extendable to three variables and to the Space Translation representation (Lagrangian coordinates).

A simple and efficient method is the Alternating Direction Implicit Method (ADI) (see KAWATA and KONO [1999] for a recent application of the technique). The ADI method propagates the wavefunction by the ansatz,

$$\psi(t+\delta t) = \left[1 + i\frac{\delta t}{2}A_y\right]^{-1}\left[1 + i\frac{\delta t}{2}A_z\right]^{-1} \times \left[1 - i\frac{\delta t}{2}A_z\right]\left[1 - i\frac{\delta t}{2}A_y\right]\psi(t) \tag{3.1}$$

with the notations:

$$A_y = -\frac{2m_p+1}{4m_p}\frac{\partial^2}{\partial y^2} + \frac{V_c}{2},$$

$$A_z = -\frac{2m_p+1}{4m_p}\frac{\partial^2}{\partial z^2} + \frac{V_c}{2} + \frac{2m_p+2}{2m_p+1}zE(t)\cos(\omega t). \tag{3.2}$$

Here, V_c is the Coulomb potential: In 2D, V_c is regularized to $V_c(y,z) = -1/\sqrt{0.5 + y^2 + (z \pm R/2)^2}$. In 3D, for Eulerian coordinates, the grid is chosen so that we don't need to evaluate $1/\sqrt{x^2 + y^2 + (z \pm R/2)^2}$ at its singularities but we conserve V_c in its originality. In 3D, for Lagrangian coordinates, when the laser is circularly/elliptically or linearly polarized we can adopt the same strategy as for Eulerian coordinates, but when the laser is a combination of a circular/elliptical laser and a perpendicular linear laser, we can regularize V_c by homogeneously distributing each proton's charge in a small sphere. Another way to regularize V_c in 3D is to distribute the proton's charge by a function which approximates the Dirac function and we discuss the problem of regularization next.

Regularization of Coulomb potentials to remove the singularity can be achieved by adding a positive constant to the electron–proton distance and has been used previously in 1D simulations of H_2 in the ST representation (WIEDEMANN and MOSTOWSKI [1994]). But this method has been found too inaccurate in our 3D calculation test and is abandoned. In order to conserve the 3D properties of real molecules, we investigate two regularization processes, averaging the singularity or discretizing the singularity in conformity with the grid structure. Averaging the singularity is very intuitive and is most easily achieved by distributing the charge over a small sphere of radius R_c which is generally chosen to be several times the grid cell's size and the Coulomb potential is replaced by

$$V_c(\rho, z) = \begin{cases} -\frac{3}{2R_c} + \frac{\rho^2+z^2}{2R_c^3} & \text{for } \sqrt{\rho^2+z^2} \leqslant R_c, \\ -\frac{1}{\sqrt{\rho^2+z^2}} & \text{for } \sqrt{\rho^2+z^2} \geqslant R_c. \end{cases} \tag{3.3}$$

In the discretization procedure, we replace the Coulomb potential by the atomic relation for the Laplacian $\Delta = \nabla^2$,

$$-\frac{1}{\sqrt{\rho^2+z^2}} = E_{1s} + \frac{\Delta\psi_{1s}}{\psi_{1s}}(\rho, z), \tag{3.4}$$

obtained from the Schrödinger equation of the H(1s) atomic orbital. The Laplacian is then discretized with the same finite difference schemes as in the TDSE. Alternatively, we can also use the exact relation:

$$\frac{1}{r} = \Delta\frac{r}{2}, \tag{3.5}$$

with $r = \sqrt{z^2+\rho^2}$ and discretize again the Laplacian (Δ) in conformity with the grid discretization.

We remark that the discrete propagator (3.1) is not unitary because A_y (A_z) is dependent on z (y). Furthermore, the dependence of the operator on variables of other directions make its inversion difficult to be implemented in pipelines of parallel computers. One way to remove the intercoordinate dependence in the ADI method is to use the following modified propagator involving partial exponentiation as in general split-operator (SO) methods (see BANDRAUK and SHEN [1993] for a general Fourier split-operator method with higher order precision):

$$\begin{aligned} \psi(t+\delta t) = {} & \left[1 + i\frac{\delta t}{4}S_y^1\right]^{-1} \exp\!\left(\frac{i\delta t}{4}V_c\right)\left[1 - i\frac{\delta t}{4}S_y\right] \\ & \times \left[1 + i\frac{\delta t}{4}S_z\right]^{-1} \exp\!\left(\frac{i\delta t}{4}V_c\right)\left[1 - i\frac{\delta t}{4}S_z\right] \\ & \times \left[1 + i\frac{\delta t}{4}S_z\right]^{-1} \exp\!\left(\frac{i\delta t}{4}V_c\right)\left[1 - i\frac{\delta t}{4}S_z\right] \\ & \times \left[1 + i\frac{\delta t}{4}S_y\right]^{-1} \exp\!\left(\frac{i\delta t}{4}V_c\right)\left[1 - i\frac{\delta t}{4}S_y\right]\psi(t), \end{aligned} \tag{3.6}$$

with:

$$S_y = -\frac{2m_p + 1}{4m_p}\frac{\partial^2}{\partial y^2},$$
$$S_z = -\frac{2m_p + 1}{4m_p}\frac{\partial^2}{\partial z^2} \quad \text{for (2.16)}, \tag{3.7}$$
$$S_z = -\frac{2m_p + 1}{4m_p}\frac{\partial^2}{\partial z^2} + \frac{2m_p + 2}{2m_p + 1} zE(t)\cos(\omega t) \quad \text{for (2.11)}.$$

This method (SO-1) is realizable on a pipeline of CPUs but still remains nonunitary.

To obtain a unitary propagator, one can combine the classical split operator (see BANDRAUK and SHEN [1993] for Fourier split-operator method with high precision) with the Crank–Nicholson method (HOFFMAN [1992]):

$$\begin{aligned}\psi(t+\delta t) = {} & \left[1 + i\frac{\delta t}{4}S_y\right]^{-1}\left[1 - i\frac{\delta t}{4}S_y\right]\left[1 + i\frac{\delta t}{4}S_z\right]^{-1}\left[1 - i\frac{\delta t}{4}S_z\right] \\ & \times \exp(i\delta t V_c) \\ & \times \left[1 + i\frac{\delta t}{4}S_z\right]^{-1}\left[1 - i\frac{\delta t}{4}S_z\right]\left[1 + i\frac{\delta t}{4}S_y\right]^{-1}\left[1 - i\frac{\delta t}{4}S_y\right]\psi(t).\end{aligned} \tag{3.8}$$

This third method (SO-2) is *unitary* and implementable in a series of CPUs which are serially connected.

All three preceding methods ADI (3.1), SO-1 (3.6), SO-2 (3.8) are of second order precision in time (δt^2). We will also test a method of order one in δt (the simplest propagator), SO-3,

$$\begin{aligned}\psi(t+\delta t) = {} & \left[1 + i\frac{\delta t}{2}S_y\right]^{-1}\left[1 + i\frac{\delta t}{2}S_z\right]^{-1}\exp(i\delta t V_c) \\ & \times \left[1 - i\frac{\delta t}{2}S_z\right]\left[1 - i\frac{\delta t}{2}S_y\right]\psi(t).\end{aligned} \tag{3.9}$$

Here, we have lost the unitarity of the propagator again as compared to SO-2, (3.8). For 3D calculations with the laser field parallel to the internuclear (z) axis, one replaces y by the cylindrical coordinate $\rho = \sqrt{x^2 + y^2}$ and $\frac{\partial^2}{\partial y^2}$ by $\frac{\partial^2}{\partial \rho^2} + \frac{1}{\rho}\frac{\partial}{\partial \rho}$ in (3.1)–(3.9).

For the spatial discretization, we have used a finite difference method with a *nonuniform* grid (see Fig. 3.1). Such adaptive grids are now popular in density functional, DFT, numerical calculations to treat large systems (GYGI and GALLI [1995]). The adaptive grid is obtained from a uniform grid as shown by KAWATA

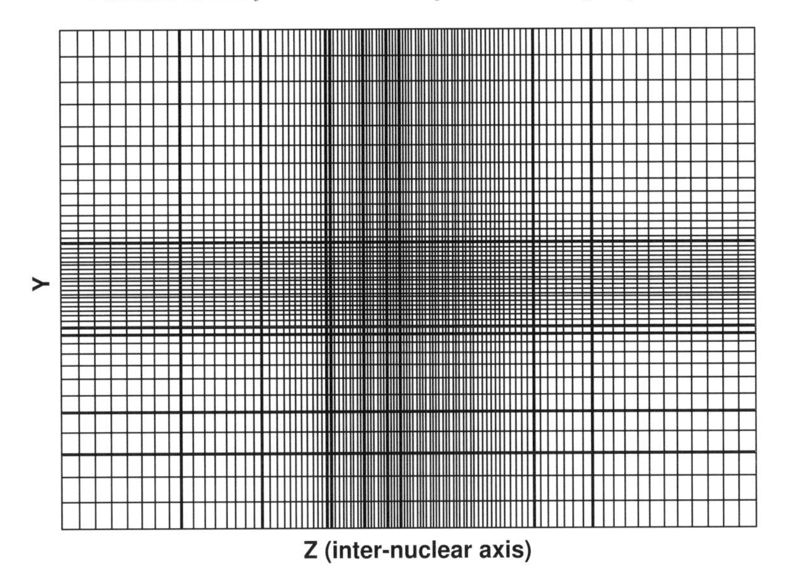

FIG. 3.1. A section of a nonuniform 3D grid.

and KONO [1999]. We use a more general transformation: $y = T(\tilde{y}), z = T(\tilde{z})$. The following is our transformation scheme:

$$T(u) = u\left(\frac{u^n + sp^n}{u^n + p^n}\right), \tag{3.10}$$

where n is an even integer, $p \geqslant 0$ representing the domain of grid refinement, and $0 \leqslant s \leqslant 1$ is set to be the minimum of $\frac{DT}{Du}$. Because $\frac{DT}{Du} = 1$ at infinity and $\frac{DT}{Du} = s$ for $T(u) = u = 0$, the generated grid is finest in the region of the singularities (neighborhood of $T(u) = 0$). In order to better conserve the wavefunction's norm, we transform it by: $\tilde{\psi} = \sqrt{T'(\tilde{y})T'(\tilde{z})}\psi$. With this spatial transformation, we use the following difference scheme to discretize the differential operator:

$$\begin{aligned}
&\frac{\partial}{\sqrt{y'}\partial\tilde{y}}\frac{\partial}{y'\partial\tilde{y}}\frac{\tilde{\psi}}{\sqrt{y'}}(\tilde{y}_i) \\
&\quad\simeq \frac{C_{i-3}+C_{i+3}}{90\delta\tilde{y}^2} - \frac{3(C_{i-2}+C_{i+2})}{20\delta\tilde{y}^2} + \frac{3(C_{i-1}+C_{i+1})}{2\delta\tilde{y}^2} \\
&\quad + \frac{1}{\delta\tilde{y}^2}\left[-\frac{D_{i+3}+D_{i-3}}{90} + \frac{3(D_{i+2}+D_{i-2})}{20} - \frac{3(D_{i+1}+D_{i-1})}{2}\right],
\end{aligned} \tag{3.11}$$

with the following notations:

$$C_j = \frac{\psi_j}{\sqrt{y_i' y_{(i+j)/2}'}\sqrt{y_j'}}, \qquad D_j = \frac{\psi_i}{y_i' y_{(i+j)/2}'}.$$

TABLE 3.1
CPU time/time-step for the four propagation methods

	ADI	SO-1	SO-2	SO-3
CPU (sec)	0.35809	0.29747	0.28205	0.17772

In the Lagrangian–ST system, z is replaced by u, Eq. (2.15). The preceding scheme is of 5th order in space, i.e., $\delta\tilde{y}^5$ and uses a stencil of 7 points. In the same way, we discretize the differential operators in other directions. The schemes for the four propagators (ADI, SO-1, SO-2, SO-3) can be easily adapted for this transformation of variable and wavefunction. We note that finite difference methods are necessary for adaptive grid methods since FFT methods as used in our previous SO methods are now difficult and inefficient (WARE [1998]).

All our 2D computations have used Cartesian coordinates with the idea of using this eventually for elliptically polarized light and the preceding discretization method without change. We have carried out a first series of 2D computations to compare different propagation methods. To show the performance of each discrete propagator, in Table 3.1 we give the CPU time of each time-step for the four methods used with a 304×688 $(y \times z)$ grid on an IBM POWER3 system (375 MHz).

Here, the CPU time for each propagator depends only on the grid point numbers, but is independent of gauge choices (Eqs. (2.11), (2.13), (2.16)). We note that the fastest method (SO-3), Eq. (3.9), needs only half of CPU time than the slowest method (ADI), Eq. (3.1), as it is of first order accuracy in δt whereas the three others, ADI, SO-1, SO-2 are of the second order accuracy. The superior performance of SO-2 (Eq. (3.8)) is to be noted due to its smaller number of exponentials and simplicity.

We next investigate numerically the above different regularization schemes (Eqs. (3.3)–(3.5)) to remove the singularities of the Coulomb potentials in a 3D simulation using cylindrical coordinates (ρ, z) for a laser field parallel to the internuclear axis. Our first calculation of ionization rates for H_2^+ in such coordinates removed the Coulomb singularity by expanding in eigenstates (Bessel functions) of the 2D cylindrical $(\rho = \sqrt{x^2 + y^2})$ Laplacian thus reducing the 3D electronic problem to 2D with large matrix for the effective Coulomb potentials (CHELKOWSKI, ZUO and BANDRAUK [1993]). The SO schemes, Eqs. (3.6)–(3.9) involve the exponentiation of the Coulomb potentials, resulting in undefined phases at the singularities. One can avoid the singularity problem partly by using nonuniform grids as in Eq. (3.10), thus enforcing that functions are zero at singularities in one direction. This results in complicated Laplacians and sometimes nonsymmetric finite difference representation (KONO [1997]). We note that in the Lagrangian–ST representation, the singularity in the Coulomb potential V_c is now moving with time due to the replacement of z by u (see Eqs. (2.15)–(2.16)). In this case, regularization is not easily implemented and we use in the exponential of the propagators the exact time dependent V_c as given in Eq. (2.16).

TABLE 3.2

Ionization rates (10^{12} s^{-1}) at $R = 4$ a.u., $I = 10^{14}$ W/cm^2, $\lambda = 1064$ nm, Z_p is defined as the position of protons on the z direction grid

(a) Original potential

Gauge, Z_p	$\delta t = 0.05$ $\delta z = \delta\rho = 0.25$	$\delta t = 0.04$ $\delta z = \delta\rho = 0.20$	$\delta t = 0.03$ $\delta z = \delta\rho = 0.15$	$\delta t = 0.02$ $\delta z = \delta\rho = 0.10$
ER, off-line	1.33423715973	1.31539308770	1.32842189536	1.34265241410
ER, on-line	0.89803248763	0.97239594933	1.07563578075	1.20825937841
ST, –	7.96097641265	4.13816007689	2.33437174921	1.53425672092

(b) Discrete potential ($\frac{1}{r} = E_{1s} - [\Delta\Psi_{1s}]/\Psi_{1s}$)

Gauge, Z_p	$\delta t = 0.05$ $\delta z = \delta\rho = 0.25$	$\delta t = 0.04$ $\delta z = \delta\rho = 0.20$	$\delta t = 0.03$ $\delta z = \delta\rho = 0.15$	$\delta t = 0.02$ $\delta z = \delta\rho = 0.10$
ER, off-line	1.30362126727	1.32835165228	1.32600533461	1.34196387133
ER, on-line	1.30125816218	1.33284154864	1.32799130791	1.33980068003
ST, –	1.63386206026	1.38552395606	1.33912138036	1.34132620691

TABLE 3.3

Ionization rates (10^{12} s^{-1}) at $R = 9.5$ a.u., $I = 10^{14}$ W/cm^2, $\lambda = 1064$ nm, Z_p is defined as the position of protons on the z direction grid

(a) Original potential

Gauge, Z_p	$\delta t = 0.05$ $\delta z = \delta\rho = 0.25$	$\delta t = 0.04$ $\delta z = \delta\rho = 0.20$	$\delta t = 0.03$ $\delta z = \delta\rho = 0.15$	$\delta t = 0.02$ $\delta z = \delta\rho = 0.10$
ER, off-line	31.4236573293	30.7844463378	29.4058863680	28.4667501641
ER, on-line	17.7299930008	19.9495445543	22.8293072187	25.2988645854
ST, –	30.1019935298	28.1778350731	27.0047899322	27.0457443269

(b) Discrete potential ($\frac{1}{r} = E_{1s} - [\Delta\Psi_{1s}]/\Psi_{1s}$)

Gauge, Z_p	$\delta t = 0.05$ $\delta z = \delta\rho = 0.25$	$\delta t = 0.04$ $\delta z = \delta\rho = 0.20$	$\delta t = 0.03$ $\delta z = \delta\rho = 0.15$	$\delta t = 0.02$ $\delta z = \delta\rho = 0.10$
ER, off-line	28.4014692441	28.4290492609	28.3075801775	28.2110591261
ER, on-line	28.4066108394	28.4327977733	28.3066810375	28.2109847369
ST, –	28.3841322200	28.3170031848	28.1807871805	28.1880178017

We have compared the ionization rates Γ (10^{12} s^{-1}) of 3D H_2^+ obtained from the time dependence of the norm of the wavefunction $N(t) = \|\psi(t)\|^2$,

$$N(t) = \left\|\psi(0)\right\|^2 e^{-\Gamma t}, \tag{3.12}$$

calculated using the SO-2 propagator (Eq. (3.8)) for the length gauge (ER) and the Lagrangian–Space Translation (ST) representation. Note each time step is unitary but absorbing boundaries reduce the norm with time. Solutions of the corresponding TDSEs, Eqs. (2.11) and (2.16), were obtained by using the: (1) original Coulomb potential ($1/r$); (2) a regularized potential averaged over a sphere; (3) a discretized

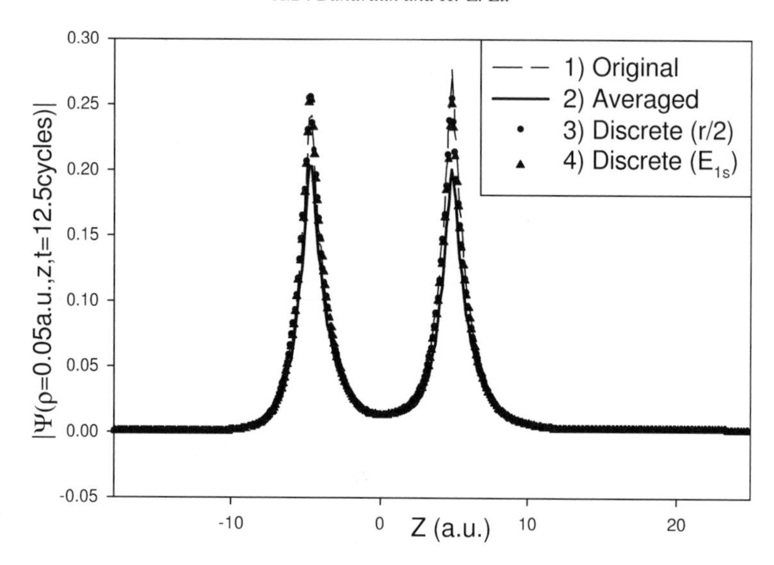

FIG. 3.2. Absolute electronic wavefunction $|\psi(\rho = 0.05\ \text{a.u.}, z, t = 12.5\ \text{cycles})|$ for H_2^+ at $I = 10^{14}$ W/cm^2, $\lambda = 1064$ nm, in the Lagrangian – Space Translation (ST) representation with (1) original V_c; (2) averaged V_c; (3) and (4) discretized V_c. Calculations were performed at $R = 9.5$ a.u. using SO-2 propagator (Eq. (3.8)) with steps: $\delta t = 0.02$, $\delta z = \delta\rho = 0.1$ a.u.

potential using Eq. (3.5) and finally (4) a discretized potential using Eq. (3.4). As illustration, we show the result for regularized potential and discretized potential using Eq. (3.4) in Tables 3.2, 3.3.

All calculations were performed for $I = 10^{14}$ W/cm^2, $\lambda = 1064$ nm ($\omega = 0.043$ a.u.), five cycle rise and then 20 cycle constant field. Table 3.2 shows results at $R = 4$ a.u., near equilibrium whereas Table 3.3 shows corresponding results at $R = 9.5$ a.u. where Charge Resonance Enhanced Ionization, CREI, has been found to occur (CHELKOWSKI and BANDRAUK [1995], BANDRAUK [2000]). All time and space discrete steps δt, δz, $\delta\rho$ are in atomic units (a.u.). The protons are situated at the middle of the ρ-grid lines in order to avoid $\rho = 0$, but are situated either on z-grid lines, i.e., with Z_p on-line, or off-line, in the latter case to avoid $z = 0$. Thus Z_p on-line results in the ER gauge are consistently poor due to the proximity of the Coulomb singularity at the proton positions Z_p. All results converge to the same rate for the smallest time and space steps, $\delta t = 0.02$, $\delta z = \delta\rho = 0.1$, thus confirming the accuracy of the propagators for very fine grids. Both discrete potential methods show remarkable stability, as seen by the smallest sensitivity of the ionization rates as a function of discretization step size. This is indicative of the importance of the same discretization of all variables in grid-methods to ensure consistency of the numerical method.

The different numerical treatments of the Coulomb singularities in the present grid method have little effect on the absolute value of the wavefunction $|\psi(z, \rho, t)|$ calculated in the Lagrangian–ST and the Eulerian-length gauge, ER methods after propagation for 12.5 cycles (Figs. 3.2, 3.4). Note that at $t = 12.5$ cycles, the field is zero and $z = u$ (Eq. (2.15)) for the Lagrangian–ST method.

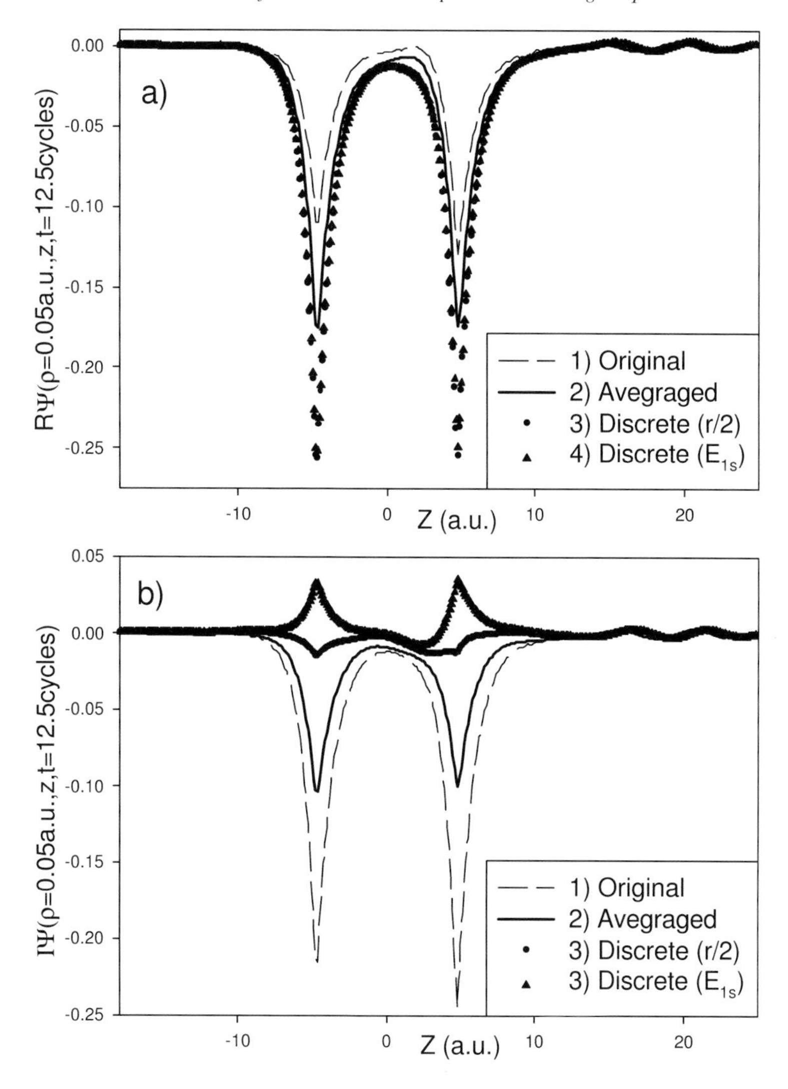

FIG. 3.3. (a) Real part – $R\psi$, (b) Imaginary part – $I\psi$ of $\psi(\rho = 0.05$ a.u., $z, t = 12.5$ cycles) for same parameters as Fig. 3.2, i.e., in ST representation.

The significant numerical differences occur in the phase of the function. In our previous careful studies of accuracies of various SO methods, we have shown that phase accuracy always lags that of amplitude accuracy by one order of magnitude in linear (BANDRAUK and SHEN [1993]) and nonlinear TDSEs (BANDRAUK [1994a]). We see this effect in Figs. 3.3 and 3.5 where using the direct ST/ER Coulomb potentials, i.e., the original potentials (1) as compared to discretized Coulomb potentials (3) and (4) in conformity with the grid properties, produces functions at the nuclei which are of

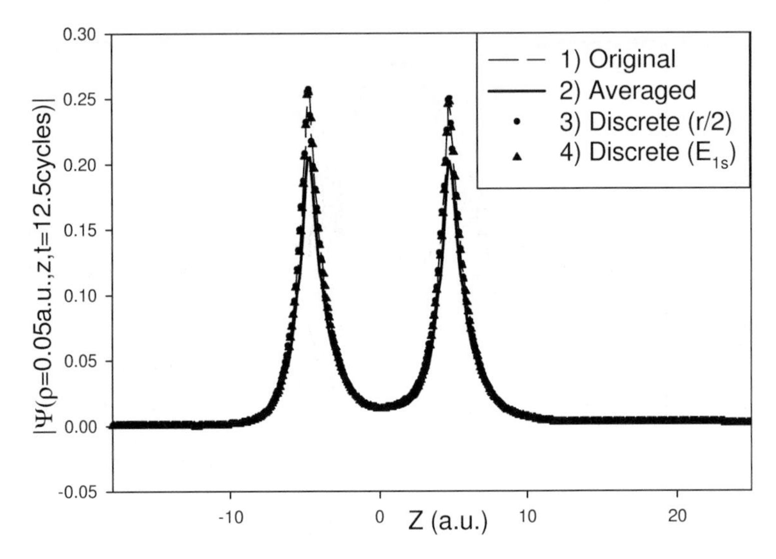

FIG. 3.4. Absolute electronic wavefunction $|\psi(\rho = 0.05\text{ a.u.}, z, t = 12.5\text{ cycles})|$ for H_2^+ at $I = 10^{14}$ W/cm^2, $\lambda = 1064$ nm, in the length gauge (ER) with (1) original V_c; (2) averaged V_c; (3) and (4) discretized V_c. Calculations were performed at $R = 9.5$ a.u. using SO-2 propagator (Eq. (3.8)) with steps: $\delta t = 0.02$, $\delta z = \delta\rho = 0.1$ a.u.

different phase for internuclear distance $R = 9.5$ a.u., i.e., especially the real part $(R\psi)$ in Fig. 3.3(a) are of opposite sign as the imaginary part $(I\psi)$ in Fig. 3.5(b).

Comparing the ST function (Figs. 3.1, 3.2) to those obtained in the ER gauge (Figs. 3.4, 3.5) using off-line proton coordinates, we remark that:

(1) In Figs. 3.5(a) and 3.5(b), the ER function obtained with the original potential has important numerical oscillation at $z = 10$ a.u. With the original potential, we have effectively approximated the wavefunction by a local polynomial which is regular in finite difference method, but using the discrete potentials the approximated function contain singular local basis. We know that the real wavefunction should be singular at the nuclei. This explains the appearance of the oscillation.

(2) Note $\psi = re^{\phi}$ the wavefunction's value at a point, the difference between two functions at a point can be written as: $\psi_1 - \psi_2 = r_2[\frac{r_1}{r_2}e^{\phi_1-\phi_2} - 1]$. Figs. 3.2 and 3.3 have shown that all the four potentials give almost the same absolute values for ST as for ER. In this case, we have

$$\left|e^{\phi_1-\phi_2} - 1\right| \simeq \sqrt{(R\psi_1 - R\psi_2)^2 + (I\psi_1 - I\psi_2)^2}. \tag{3.13}$$

According to this formula, Fig. 3.3 (ST) demonstrate an evident better convergence of wavefunction's phase than Fig. 3.5 (ER). We note furthermore that in the length gauge (ER) strong phase effects persist asymptotically and this has lead to a worse convergence of the wavefunction's phase. This is consistent with the fact that the ST wavefunction have the ponderomotive energy $U_p = E_0^2/4\omega^2$

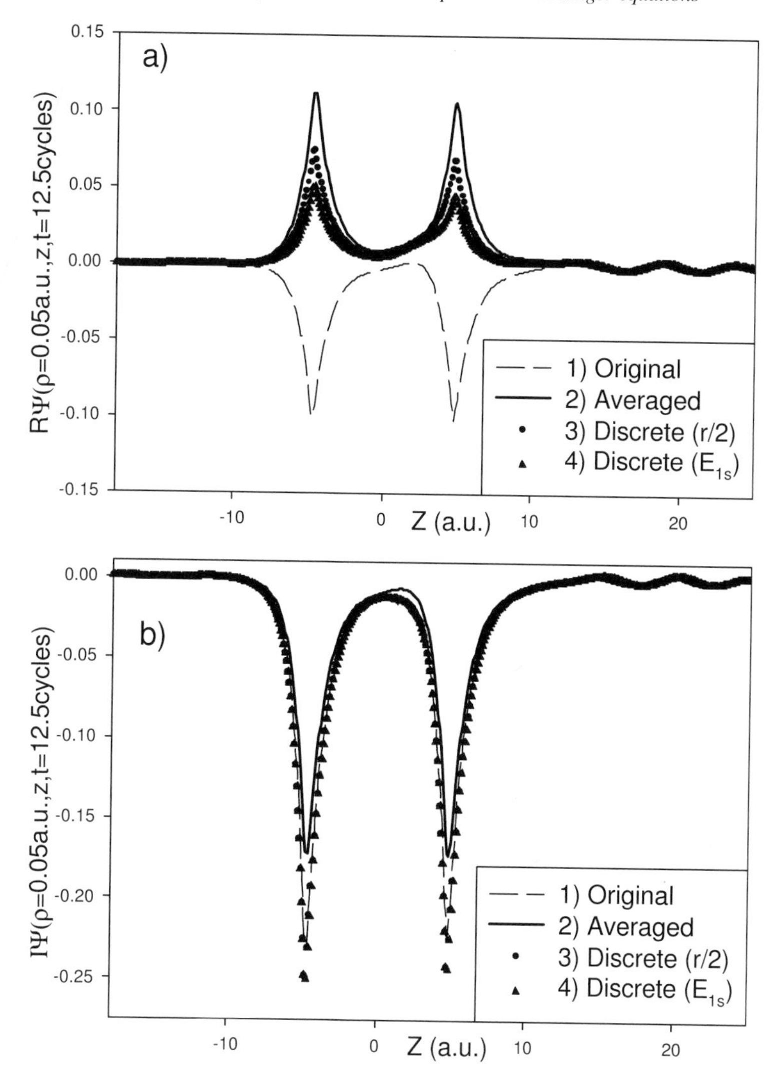

FIG. 3.5. (a) Real part – $R\psi$, (b) Imaginary part – $I\psi$ of $\psi(\rho = 0.05$ a.u., $z, t = 12.5$ cycles) for same parameters as Fig. 3.4, i.e., in ER gauge.

phase effects already removed (Eq. (2.10)). These important laser-induced phase fluctuations remain in the length gauge (ER).

4. Boundary conditions and energy spectra

As shown in the previous section, two representations, the length gauge with ER radiative coupling or the space translation (ST) method, give essentially the same accurate results for ionization rates. Nevertheless there are important differences numerically. The Ez coupling equation (2.11) diverges at large distance, especially at the grid boundaries, $|z| \to \infty$, resulting in poor convergence of the phase in that gauge. In the ST representation, the wavefunction has the ponderomotive energy removed from the phase (Eq. (2.10)). These important laser-induced phase fluctuations remain in the length (ER) gauge.

An estimate of field induced electron displacements can be obtained from the classical equation of motion of a free ionized electron driven by a laser field $E_0 \cos(\omega t)$. Such classical models reproduce remarkably well ATI (electronic) spectra (CORKUM [1993]) and HOHG spectra (BANDRAUK and SHEN [1994]). We start with the acceleration equation in the z-direction for the electron (see Fig. 3.1) and integrate twice to give the velocity $\dot{z}(t)$ and position $z(t)$, (or $\dot{q}(t)$ and $q(t)$, Eq. (2.7)):

$$\ddot{z}(t) = -E_0 \cos(\omega t + \phi), \tag{4.1}$$

$$\dot{z}(t) = -(E_0/\omega)\big[\sin(\omega t + \phi) - \sin(\phi)\big], \tag{4.2}$$

$$z(t) = -\alpha_0\big[\cos(\omega t + \phi) + \omega t \sin(\phi) - \cos(\phi)\big]. \tag{4.3}$$

ϕ is the initial laser phase at which the electron is ionized and we assume an initial velocity $\dot{z}(0) = 0$ in conformity with quasistatic tunneling models (CORKUM [1993]). α_0 is called the ponderomotive (quiver) radius and U_p is the corresponding ponderomotive energy:

$$\alpha_0(\text{a.u.}) = E_0/\omega^2, \qquad U_p = \alpha_0^2\omega^2/4 = E_0^2/4\omega^2. \tag{4.4}$$

These quantities correspond to $q(t)$ and $\dot{q}^2/2$ in the ST representation (Eq. (2.7)). These classical equations of motion can be used to predict the optimal condition for HOHG and ATI spectra. As a first hypothesis we shall look at maximum acceleration which in the classical limit should produce maximum radiation (BANDRAUK and YU [1999]). From Eq. (4.1) we obtain $\ddot{z}(\max)$ at phase $\omega t + \phi = \pi$. This yields the velocity $\dot{z} = E_0 \sin(\phi)/\omega$ with a maximum $\dot{z}(\max) = E_0/\omega$ for $\phi = \pi/2$. The resulting energy is

$$E(\max) = \dot{z}^2(\max)/2 = E_o^2/2\omega^2 = 2U_p. \tag{4.5}$$

The corresponding distance of the electron trajectory is

$$z(t) = \alpha_0(-1 + \omega t) = \alpha_0(\pi/2 - 1) = 0.57\alpha_0. \tag{4.6}$$

Thus the maximum acceleration criterion yields a maximum energy $E(\max) = 2U_p$ or HOHG order $N_m = E(\max)/\omega = 2U_p/\omega$ if an electron collides with a neighboring nucleus at distance $R = 0.57\alpha_0$. The next scenario is *recombination* of the electron with the parent ion. For maximum velocity the phase requirement is $\omega t + \phi = 3\pi/2$, thus yielding the transcendental equation $(3\pi/2 - \phi)\sin(\phi) = \cos(\phi)$ for $z(t) = 0$, i.e., recollision. The solution of this equation is $\phi = 0.08\pi$, $\dot{z}(\max) = 1.26E_0/\omega$. The resulting maximum recollision energy is

$$E_m = 3.17U_p. \tag{4.7}$$

This simple classical model explains the maximum theoretical HOHG order obtained by recollision (CORKUM [1993]), i.e.,

$$N_m = (I_p + 3.17U_p)/\omega, \tag{4.8}$$

where I_p is the ionization potential. However in molecules, as we have shown in detail (BANDRAUK and YU [1999]), electron collision can occur with neighboring ions. Then the maximum velocity condition $\omega t + \phi = 3\pi/2$ results in this velocity being from Eq. (4.2), $\dot{z} = (8E_0/\omega)(1 + \sin\phi)$ or $2E_0/\omega$ at $\phi = \pi/2$. Thus the maximum energy obtained with no constraints is

$$E_m = 2E_0^2/\omega^2 = 8U_p. \tag{4.9}$$

This is gained during a half cycle $\omega t = \pi$ and the corresponding distance travelled by the electron is

$$z(\pi/\omega) = \pi\alpha_0. \tag{4.10}$$

In summary the classical model of an ionized electron driven by a laser field predicts maximum kinetic energies $2U_p$, $3.17U_p$ and $8U_p$ at distance $z = 0.57\alpha_0$, 0 and $\pi\alpha_0$ if the initial velocity of the electron $\dot{z} = 0$. This model allows us to estimate grid size necessary to capture accurately electron dynamics upon ionization. Thus at a laser intensity $I = 3 \times 10^{14}$ W/cm^2 $= 10^{-2}$ a.u., then $E_0 = 10^{-1}$ a.u. At the wavelength 1064 nm ($\omega = 0.043$ a.u.) one obtains readily $\alpha_0 = 54$ a.u. $= 28.6 \times 10^{-8}$ cm, whereas the corresponding ponderomotive energy $U_p = 1.35$ a.u. $= 36.8$ eV. Thus electron energies up to $8U_p = 300$ eV will appear in the energy spectrum of the ionized electron and electron trajectories $\pi\alpha_0 > 160$ a.u. need to be considered. The electron radiative coupling at these distances is at least $\pi E_0\alpha_0 \simeq 12U_p \simeq 440$ eV, clearly a nonperturbative regime.

In the ST representation, one clearly needs an adaptive moving grid due to the oscillatory motion $q(t)$ of the electron with maximum displacement greater even than α_0 (e.g., Eq. (4.4)). As an example, we give in Fig. 3.1 a section of a nonuniform grid which we use regularly in our simulations. This 3D grid ($1064 \times 512 \times 512$) contains 5.4×10^8 points. Introducing up to 10 discretization points in the z-laser direction between nearest neighbor points (i.e., $\delta z = 0.1$ a.u.) and also in the x and y directions results in about $10^{12} = 1$ terabyte of memory requirements per time iteration using any one of the propagation schemes described in the previous section. This does not include yet the nuclear motion which is usually restrained to 1D, i.e., parallel to the laser field, since the strongest field–molecule interactions occur in this configuration, due to charge transfer effects between neighboring atoms (BANDRAUK [1994b]). Time iterations are performed for about 25 cycles $= 87.5$ fs (here $\tau_{\text{cycle}} = 3.5$ fs $= 142$ a.u.)

in the case of ionization, so that with time steps of $\delta t = 0.02$ a.u., one readily has $25 \times 142/0.02 \simeq 2 \times 10^5$ time iterations for calculating ionization rates.

In order to calculate ATI spectra, i.e., electron kinetic energy spectra, one must wait long after the pulse is ended in order to capture slow (low energy) electrons which are propagating to the edge of the grids. To date we have only been able to obtain 1D ATI spectra for 1D H_2^+ (CHELKOWSKI, FOISY and BANDRAUK [1998], CHELKOWSKI, ZAMOJSKI and BANDRAUK [2001]) from the exact non-Born–Oppenheimer wavefunction $\psi(z, R, t)$ where z is the electronic coordinate and R the internuclear coordinate. Simultaneous calculation of the electron (ATI) kinetic energy and proton (ATD and CE (Coulomb Explosion, the rapid explosion of completely ionized nuclei from Coulomb repulsive forces)) spectra also allows for reconstructing initial nuclear wavefunctions, both static and dynamic in the long wavelength (800 nm) (BANDRAUK and YU [1999]) and UV (30 nm) laser regime (BANDRAUK and CHELKOWSKI [2001]). This is a new application, i.e., *dynamic imaging* of nuclear motion with ultrashort intense laser pulses, which is expected to allow for time-dependent measurements of evolution of nuclear wavefunctions.

In order to obtain these kinetic energy spectra, both for electrons (ATI) and protons (ATD and CE), asymptotic projection onto exact solution of free particles are performed at large distances, z for the electron and R for the protons, i.e., at the grid edges where interparticle, Coulomb forces, are negligible. In the calculations using the length ER gauge, the exact solutions of electrons in the laser field, called Volkov states (GAVRILA [1992], BANDRAUK [1994b]) are used (CHELKOWSKI, FOISY and BANDRAUK [1998], CHELKOWSKI, ZAMOJSKI and BANDRAUK [2001]). In the case of the ST representation (see previous section) one need only project onto field-free states. Comparison of the ER gauge and ST representation ATI spectra shows identical results with the ST method which requires smaller grids, thus allowing us to finally calculate ATI spectra for the two-electron systems H_2 and H_3^+ with *moving* nuclei (KAWATA and BANDRAUK [2002]).

The asymptotic projection method used here is a special application of *domain decomposition* for numerical solution of PDEs (QUARTERONI [1991]). The large grid is subdivided near the edge into a small internal and large external box. Exact solutions from the large box are then projected onto solutions of the small box. The latter contains the Volkov (ER gauge) or free (ST representation) solution. Propagation and projection is then continued well after the pulse is over until external probabilities defined as $|\psi(x, y, z, R, T)|^2$ become constant. This method is analogous to a single domain method where the size of the grid should be the sum of both internal and external boxes. Our projection step involves simple overlap of the exact internal solution onto the asymptotic Volkov or free particle states. It should be pointed out that our overlapping domain decomposition method does not iterate on the internal interface as the well-known classical Schwarz alternating domain decomposition method. Clearly, the iteration on the interface will assure a best convergence of the numerical solutions, but it is too expensive in computing time for problem in higher dimension (> 3) aimed by our method. The numerical results of some domain decomposition methods with or without overlapping have demonstrated that the well chosen interface condition can offer very precise final results even if no iterations are made on the interface

(GUERMOND, HUBERSON and SHEN [1992], GUERMOND and LU [2000]). In our approximation, we find that the wavepackets propagate from the internal box to the external box so we have used the transfer of population from internal box to the external one to replace the interface iteration. More sophisticated methods based on Schwarz method remain to be explored in order to permit better coupling between the sub-domains without iteration on the interface.

5. Beyond the dipole approximation

We have seen in Section 2 that there are three main approaches to treating molecules – laser field interaction in the nonlinear nonperturbative regime. The standard method, which relies on principles of electrostatics, is the length ER gauge, Eq. (2.11). In this gauge radiative interactions at grid boundaries become very large thus creating numerical instabilities. Another gauge, the velocity (Coulomb) gauge, Eq. (2.13) is similar to *convection–diffusion* equations of fluid dynamics. The convection velocity is in fact called the electromagnetic potential

$$A(t) = \int_0^t E(s)\,ds, \tag{5.1}$$

and in the Coulomb gauge is *incompressible*, i.e., $\vec{\nabla} \cdot \vec{A} = 0$. Thus both $A(t)$ and $E(t)$ have no spatial dependence and this is called the *dipole approximation* which is valid whenever $r/\lambda \ll 1$, where r is the atomic-molecular dimension ($\sim$ 1–10 Å) and λ is the wavelength of the radiation field. This implies the polarization (e.g., z) is always *orthogonal* to the propagation (e.g., x, y). Any spatially independent field $E(t)$ generates automatically a spatially independent potential $A(t)$ which automatically satisfies the Coulomb gauge or incompressibility condition.

In general, A is a vector spatial – temporal function $\vec{A}(\vec{r}, t)$ so that the exact TDSE for a single particle in the velocity gauge is given by

$$i\frac{\partial\psi(\vec{r}, t)}{\partial t} = \frac{1}{2}\left[i\vec{\nabla} + \frac{1}{c}\vec{A}(\vec{r}, t)\right]^2 \psi(\vec{r}, t) + V(\vec{r})\psi(\vec{r}, t), \tag{5.2}$$

where $\vec{\nabla} = (\partial_x, \partial_y, \partial_z)$. The electromagnetic potential $\vec{A}(\vec{r}, t)$ is related to the electric and magnetic fields by $\vec{E}(\vec{r}, t) = \frac{1}{c}\frac{\partial\vec{A}(\vec{r},t)}{\partial t}$ and $\vec{B}(\vec{r}, t) = \vec{\nabla} \wedge \vec{A}$ respectively (BANDRAUK [1994b]) so that $\vec{E} \mid \vec{A}$, $\vec{B} \perp \vec{A}$. We will assume that the laser field is polarized in the z direction and propagates in the y direction:

$$\vec{A}(\vec{r}, t) = \bigl(0, 0, A_z(y, t)\bigr); \qquad \vec{E}(\vec{r}, t) = \bigl(0, 0, E_z(y, t)\bigr). \tag{5.3}$$

If one expands the potential $\vec{A}$ to first order in the space coordinate $\vec{r}(x, y, z)$ in term of the ratio $\frac{r}{\lambda}$ or $\frac{r\omega}{c}$ this results in $A_z(y, t) = A_z(t) + y\frac{\partial A_z(y,t)}{\partial y}$. This first order expansion of $\vec{A}$ corresponds to taking into account the magnetic dipole and electric quadrupole interactions (BANDRAUK [1994b]). Relativistic effects such as the mass correction and spin-orbit effects can be neglected up to $v/c \simeq 0.2$. This corresponds to an upper limit for a Ti:Sap ($\lambda = 800$ nm) laser peak intensity of $\sim 10^{17}$ W/cm^2 (WALSER, KEITEL, SCRINZI and BRABEC [2000]), since c is the vacuum light velocity = 137 a.u. (3×10^{10} cm/s) and $v \simeq E_0/\omega$, is the electron velocity in a field of peak amplitude E_0 and frequency ω (see Eq. (4.2)). Thus corrections to the dipole gauges, Section 2 must be considered whenever $\frac{r}{\lambda}$ and/or $\frac{v}{c} \simeq 1$.

For higher intensities, and lower frequencies then one must consider corrections beyond the nonrelativistic limit ($v/c \sim 1$). Thus in the latter case, one expects modification of the electron–electron Coulomb interactions (BERGOU, VARRO and FEDOROV [1981]) whereas in the former, relativistic corrections will influence interference between direct and exchange electron–electron scattering (FEDOROV and ROSHCHUPKIN [1984]).

We give next an example of such calculations, i.e., beyond the dipole approximation for intense ($I > 10^{18}$ W/cm^2), short wavelength ($\lambda = 25$ nm, $\omega = 1.8$ a.u.). Such intensities in the VUV-X-Ray regime is expected to become available soon from Free-Electron lasers. We choose the electric field amplitude polarized in the z-direction but propagating only in the y-axis, where $T_{\max} = 2n\pi/\omega$ is the number of cycles, and T is the pulse rise period,

$$\begin{aligned} E_z(y,t) &= E_0 \frac{t}{T} \sin(\omega t - \omega y/c), \quad t < T, \\ &= E_0 \sin(\omega t - \omega y/c), \quad T < t < T_{\max} - T, \\ &= E_0 \frac{T_{\max} - t}{T} \sin(\omega t - \omega y/c), \quad T_{\max} - T < t. \end{aligned} \tag{5.4}$$

Due to the electromagnetic relation (5.1), i.e., $\vec{E} = -\frac{1}{c}\frac{\partial \vec{A}}{\partial t}$, this implies the following corresponding form for the potential A,

$$\begin{aligned} &A_z(y,t) \\ &= \frac{cE_0}{\omega T}\left[t\cos\left(\omega t - \frac{\omega y}{c}\right) - \frac{1}{\omega}\sin\left(\omega t - \frac{\omega y}{c}\right) - \frac{1}{\omega}\sin\left(\frac{\omega y}{c}\right)\right], \quad t < T, \\ &= \frac{cE_0}{\omega}\cos\left(\omega t - \frac{\omega y}{c}\right), \quad T < t < T_{\max} - T, \\ &= \frac{cE_0}{\omega T}\left[(T_{\max} - t)\cos\left(\omega t - \frac{\omega y}{c}\right) + \frac{1}{\omega}\sin\left(\omega t - \frac{\omega y}{c}\right)\right. \\ &\qquad \left. + \frac{1}{\omega}\sin\left(\frac{\omega y}{c}\right)\right], \quad T_{\max} - T < t. \end{aligned} \tag{5.5}$$

The TDSE in the Coulomb (velocity) gauge is then given from Eq. (2.13),

$$i\frac{\partial \psi(x,y,z,t)}{\partial t} = \frac{2m_p + 1}{4m_p}\left[-\Delta - i\frac{2}{c}\left(\frac{2m_p + 2}{2m_p + 1}\right)A_z\frac{\partial}{\partial z}\right]\psi$$

$$+\frac{2m_p+1}{4m_p}\left(\frac{2m_p+2}{2m_p+1}\right)^2 A_z^2\psi(x,y,z,t)$$
$$+V_c\psi(x,y,z,t). \tag{5.6}$$

We now introduce a new Space Translation, in analogy with the Husimi transformation (2.2),

$$\frac{\partial\alpha}{\partial t}=\frac{1}{c}A_z(t,y=0), \tag{5.7}$$

or equivalently,

$$\begin{aligned}
\alpha(t) &= \frac{E_0}{\omega T}\left[\frac{t}{\omega}\sin(\omega t)-2\frac{1-\cos(\omega t)}{\omega^2}\right], \quad t<T,\\
&= \frac{E_0}{\omega^2}\sin(\omega t), \quad T<t<T_{\max}-T,\\
&= \frac{E_0}{\omega T}\left[\frac{T_{\max}-t}{\omega}\sin(\omega t)+2\frac{1-\cos(\omega t)}{\omega^2}\right], \quad T_{\max}-T<t.
\end{aligned} \tag{5.8}$$

Proceeding further as in the ST representation, Eq. (2.15), we introduce a coordinate displacement,

$$\begin{aligned}
z &= u+2\frac{2m_p+1}{4m_p}\frac{2m_p+2}{2m_p+1}\alpha(t),\\
\frac{\partial u}{\partial t} &= -\frac{2}{c}\frac{2m_p+1}{4m_p}\frac{2m_p+2}{2m_p+1}A_z(t,y=0).
\end{aligned} \tag{5.9}$$

In the new ST coordinate system, the TDSE now becomes:

$$\begin{aligned}
i\frac{\partial\psi(x,y,u)}{\partial t} &= \frac{2m_p+1}{4m_p}\left[-\Delta_{x,y,u}-2iB_z(t,u)\frac{\partial}{\partial u}\right]\psi(x,y,u)\\
&\quad+\frac{2m_p+1}{4m_p}\left(\frac{2m_p+2}{2m_p+1}\right)^2\frac{1}{c^2}\vec{A}^2\psi+V_c(x,y,u)\psi,
\end{aligned} \tag{5.10}$$

where

$$B_z(t,y)=\frac{1}{c}\frac{2m_p+2}{2m_p+1}\big[A_z(t,y)-A_z(t,y=0)\big]. \tag{5.11}$$

Such an equation is again an *imaginary convection–diffusion* equation in presence of a potential V_c and is a generalization of the long-wavelength, i.e., dipole approximation, ST Eq. (2.16). Thus going beyond the dipole approximation does not allow us to eliminate the convection term as in Eq. (2.16). However due to the Coulomb gauge condition $\vec{\nabla} \cdot \vec{A} = 0$, the convection coefficient $B_z(t, y)$ remains *incompressible*, i.e., $\vec{\nabla} \cdot \vec{B} = 0$.

The effect of the nondipole terms, i.e., $\frac{r}{\lambda} \sim 1$, $\frac{v}{c} \sim 1$, will be to displace electrons perpendicular to the polarization direction z, the polarization of the electric field E_z, and the propagation direction y due to a magnetic component $\vec{B}_x = (\vec{\nabla} \wedge \vec{A})_x$ along the x-axis. Nevertheless the TDSEs (5.6) and (5.10) remain symmetric with respect to x, i.e., $\psi(x) = \psi(-x)$. Due to this symmetry, one can employ a nonuniform grid Crank–Nicholson algorithm in this direction and thus reduce the number of discretization points. We use FFTs (see BANDRAUK [1994a] for recent application of FFT in Fourier split-operator method) in the polarization z-direction and the propagation y-direction in order to maintain unitarity and high accuracy of the derivatives in both these directions. The optimum splitting for Eq. (5.10) is then called FD(x) + FFT(yz) + ST(z), i.e., finite difference in x, FFTs in with Space Translation in z via the u transformation, Eq. (5.9):

$$
\begin{aligned}
\psi^{n+1} &= \exp\left\{ i\frac{\delta t}{2} \frac{2m_p+1}{4m_p} \frac{\partial^2}{\partial y^2} \right\} \\
&\quad \times \left[1 - i\frac{\delta t}{4} \frac{2m_p+1}{4m_p} \frac{\partial^2}{\partial x^2} \right]^{-1} \left[1 + i\frac{\delta t}{4} \frac{2m_p+1}{4m_p} \frac{\partial^2}{\partial x^2} \right] \\
&\quad \times \exp\left\{ -i\frac{\delta t}{2} \left[V_c + \frac{2m_p+1}{4m_p} \left(\frac{2m_p+2}{2m_p+1} \right)^2 \frac{1}{c^2} \vec{A}^2 \right] \right\} \\
&\quad \times \exp\left\{ \delta t \frac{2m_p+1}{4m_p} \left[i\frac{\partial^2}{\partial u^2} - 2B_z(t, y) \frac{\partial}{\partial u} \right] \right\} \\
&\quad \times \exp\left\{ -i\frac{\delta t}{2} \left[V_c + \frac{2m_p+1}{4m_p} \left(\frac{2m_p+2}{2m_p+1} \right)^2 \frac{1}{c^2} \vec{A}^2 \right] \right\} \\
&\quad \times \left[1 - i\frac{\delta t}{4} \frac{2m_p+1}{4m_p} \frac{\partial^2}{\partial x^2} \right]^{-1} \left[1 + i\frac{\delta t}{4} \frac{2m_p+1}{4m_p} \frac{\partial^2}{\partial x^2} \right] \\
&\quad \times \exp\left\{ i\frac{\delta t}{2} \frac{2m_p+1}{4m_p} \frac{\partial^2}{\partial y^2} \right\} \psi^n.
\end{aligned}
\tag{5.12}
$$

Replacing u by z in Eq. (5.10) would give the equivalent Coulomb (velocity) gauge propagation scheme. In other words, in Eq. (5.10) which is the *new* nondipole ST equation, we have removed the dipole field (spatially independent) $A_z(t, y = 0)$ (see Eq. (5.11)). Thus in the dipole limit, $B_z \equiv 0$, $u(t) = z$, and the TDSE (5.10) reduce to the conventional ST TDSE, Eq. (2.16).

We have calculated the influence of the nondipole terms on the electron ionization of H_2^+ using the algorithm based on Eq. (5.12) which allows to calculate the electron density $|\psi(x, y, z, t)|^2$. We illustrate in Figs. 5.1 and 5.2 the density $|\psi(0, y, z, t)|^2$ in the yz plane, i.e., in the direction of propagation (y) and polarization (z) for the

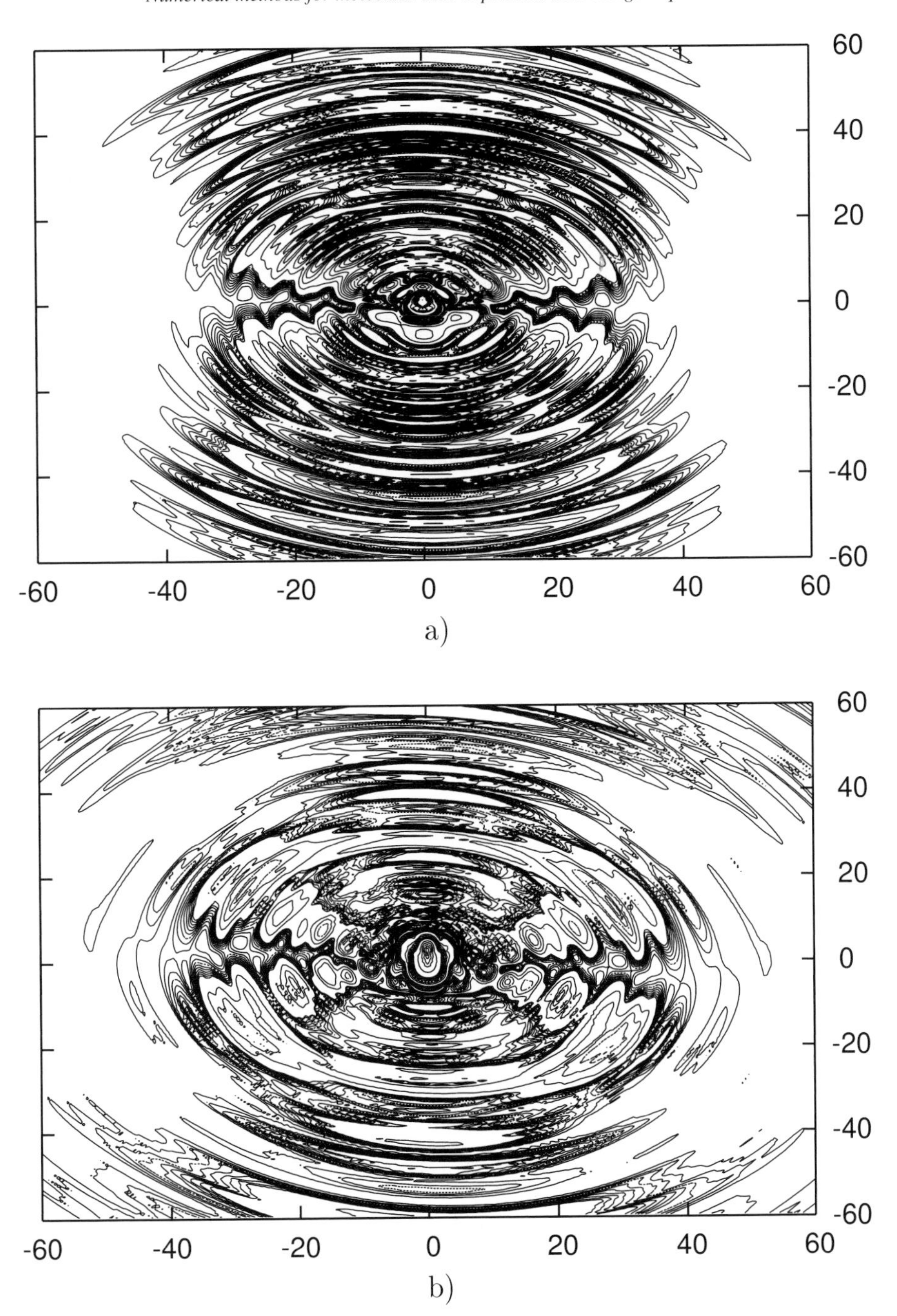

FIG. 5.1. The density $|\psi(0, y, z, t = 10 \text{ cycles})|^2$ iso-lines in the yz plane of the nondipole calculation for H_2^+ with $R = 2$ a.u. for the laser field polarized in the $z(R)$-internuclear direction and propagation in the y direction: (a) $I = 2 \times 10^{18}$ W/cm^2, $\lambda = 25$ nm; (b) $I = 10^{19}$ W/cm^2, $\lambda = 25$ nm.

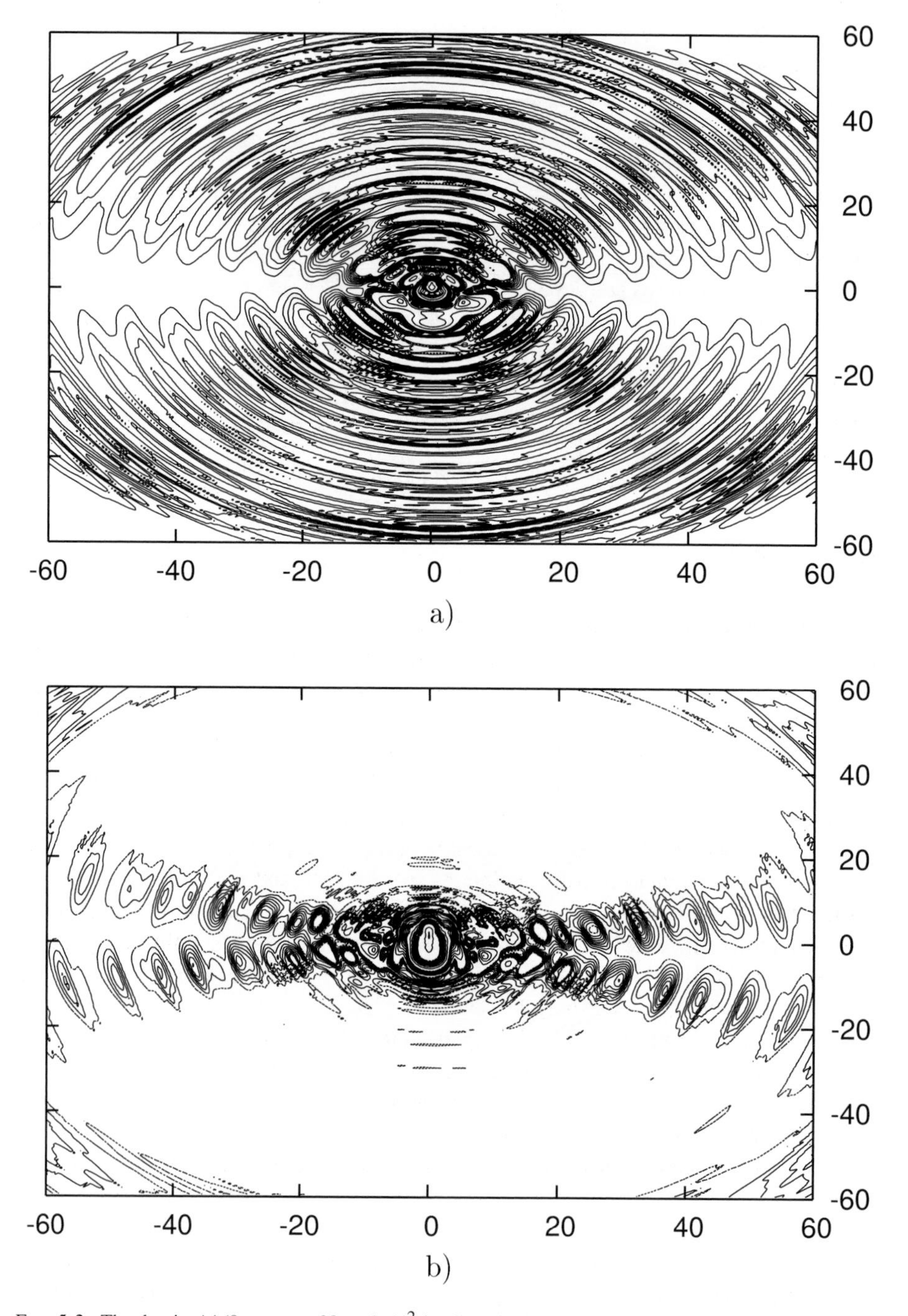

FIG. 5.2. The density $|\psi(0, y, z, t = 20 \text{ cycles})|^2$ iso-lines in the yz plane of the nondipole calculation for H_2^+ with $R = 2$ a.u. for the laser field polarized in the $z(R)$-internuclear direction and propagation in the y direction: (a) $I = 2 \times 10^{18}$ W/cm^2, $\lambda = 25$ nm; (b) $I = 10^{19}$ W/cm^2, $\lambda = 25$ nm.

intensities 2×10^{18} W/cm^2 and 10^{19} W/cm^2 at $\lambda = 25$ nm. Fig. 5.1 correspond to 10 cycles whereas Fig. 5.2 are for 20 cycles. The z polarization of the E_z field is along the internuclear axis whereas the magnetic B_x component would be along the x-axis. Thus at the lower intensity (2×10^{18} W/cm^2) (Figs. 5.1(a), 5.2(a)) one expects the electron to be emitted mainly along the z-axis as seen in both figures. The higher intensity (10^{19} W/cm^2) results, Figs. 5.1(a), 5.2(b) show a marked difference for longer times. At this intensity and wavelength the ponderomotive energy of the ionized electrons U_p (Eq. (4.4)) is 25 a.u. = 680 eV whereas at 2×10^{18} W/cm^2, $U_p = 5$ a.u. The remaining electrons (Fig. 5.2(b)) are acquiring momentum in the direction of the propagation direction, i.e., in the momentum direction of the field photons. Such effects do not appear at the lower intensity. Of further interest is that in all figures rings of electrons are created corresponding to a wavelength of about 6 a.u. $\simeq 3$ Å. As this corresponds to typical internuclear distances, one can foresee that such ionized electrons will interfere with the nuclei, leading to a new spectroscopy, LIED, Laser Induced Electron Diffraction (ZUO, BANDRAUK and CORKUM [1996]).

6. Conclusion

We have focused on methods of solving accurately TDSEs to describe laser–molecule interactions from the perturbative to nonperturbative regime taking into consideration that modern laser technology is ever evolving towards controllable, more intense and shorter laser pulses. Our emphasis has been on the simplest one-electron H_2^+ molecule with a single nuclear degree of freedom. This simple system is the *benchmark* as it allows for exact numerical treatment of competing electron–nuclear dynamics, i.e., including all non-Born–Oppenheimer corrections (see Section 4).

In practice one has the liberty of choosing various gauges or representations, which are equivalent by unitary transformations (see Section 2) and thus will yield identical results only if the algorithms are exact. We have shown in fact that the most useful representation is the Space Transformation, ST, which we have generalized beyond the dipole approximation (see Section 5) in order to take into account retardation effects and to allow one to go to short wavelengths. The advantage of the ST method is that asymptotically, at the edge of the usual large grids necessary for high intensity calculations, radiative (laser–molecule) interactions vanish so that one can project the exact numerical solutions at grid boundaries onto well known analytic free particle states, either electronic and/or nuclear.

Many-body systems unfortunately require solving multidimensional PDEs, well beyond today's super-computer technology. A promising alternative is to reduce these problems to effective one-electron systems, via new methods such as *time-dependent density functional*, TDDFT methods. The one-electron numerical methods discussed in the present work should be easily adaptable to these effective one-electron methods and these will allow for inclusion of nuclear motion, thus allowing for a complete dynamical description of molecules from the perturbative weak field limit to the nonperturbative intense field regime available with today's modern laser technology.

Acknowledgment

We wish to thank various colleagues who in the course of our work have pointed out many of the outstanding experimental and theoretical issues in laser–molecule science – P.B. Corkum (NRC, Ottawa), W. Kohn (U.C., Santa Barbara), C. Ulrich (Missouri), O. Atabek (Orsay), C. Le Bris (ENPC).

References

BANDRAUK, A.D. (1994a), *International Rev. Phys. Chem.* **13**, 123.

BANDRAUK, A.D. (1994b), *Molecules in Laser Fields* (M. Dekker Pub., New York).

BANDRAUK, A.D. (2000), in: Y. Itikawa et al., eds., *The Physics of Electronic and Atomic Collision*, AIP Conf. Proc., Vol. 500 (American Institute of Physics, New York) 102.

BANDRAUK, A.D. and S. CHELKOWSKI (2001), *Phys. Rev. Lett.* **87**, 273004.

BANDRAUK, A.D., Y. FUJIMURA and R.J. GORDON, eds. (2002), *Laser Control and Manipulation of Molecules*, ACS Symposium Series, Vol. 821 (Washington).

BANDRAUK, A.D. and H. SHEN (1993), *J. Chem. Phys.* **99**, 1185.

BANDRAUK, A.D. and H. SHEN (1994), *J. Phys. A* **27**, 7147.

BANDRAUK, A.D. and H. YU (1998), *Int. J. Mass. Spectrom.* **192**, 379.

BANDRAUK, A.D. and H. YU (1999), *Phys. Rev. A* **59**, 539.

BERGOU, J., S. VARRO and M.V. FEDOROV (1981), *J. Phys. A* **14**, 2305.

BRABEC, T. and F. KRAUSZ (2000), *Rev. Mod. Phys.* **72**, 545.

BRUMER, P. and M. SHAPIRO (1994), in: A.D. Bandrauk, ed., *Molecules in Laser Fields* (M. Dekker Pub., New York) Chapter 6.

CHELKOWSKI, S. and A.D. BANDRAUK (1995), *J. Phys. B* **28**, 1723.

CHELKOWSKI, S. and A.D. BANDRAUK (2002), *Phys. Rev. A* **65**, 0234.

CHELKOWSKI, S., A. CONJUSTEAU, O. ATABEK and A.D. BANDRAUK (1996), *Phys. Rev. A* **54**, 3235.

CHELKOWSKI, S., C. FOISY and A.D. BANDRAUK (1997), *Int. J. Quant. Chem.* **65**, 503.

CHELKOWSKI, S., C. FOISY and A.D. BANDRAUK (1998), *Phys. Rev. A* **57**, 1176.

CHELKOWSKI, S., M. ZAMOJSKI and A.D. BANDRAUK (2001), *Phys. Rev. A* **63**, 023409.

CHELKOWSKI, S., T. ZUO and A.D. BANDRAUK (1993), *Phys. Rev. A* **48**, 3837.

CHU, S.I. (1989), in: J.O. Hirschfelder, R.E. Wyatt and R.D. Coalson, eds., *Advances in Chemical Physics*, Vol. 73 (Willy Interscience, New York) Chapter 17.

CORKUM, P.B. (1993), *Phys. Rev. Lett.* **71**, 1994.

DEFRANCESCHI, M. and C. LE BRIS, eds. (2001), *Mathematical Models and Methods for Ab-Initio Quantum Chemistry*, Lecture Notes in Chemistry, Vol. 74 (Springer, New York).

DREIZLER, R.M. and E.K.U. GROSS (1990), *Density Functional Theory* (Springer, Berlin).

EHLOTZKY, F. (2001), *Phys. Repts.* **345**, 175.

FEDOROV, M.V. and S.P. ROSHCHUPKIN (1984), *J. Phys. A* **17**, 3143.

GAVRILA, M. (1992), *Atoms in Intense Laser Fields* (Academic Press, New York).

GERHART, P.M., R.J. GROSS and J.I. HOCHSTEIN (1992), *Fundamentals of Fluid Mechanics* (Addison-Wesley, New York).

GUERMOND, J.L., S. HUBERSON and W.Z. SHEN (1993), *J. Comp. Phys.* **108**, 557.

GUERMOND, J.L. and H.Z. LU (2000), *Computers and Fluids* **29**, 525.

GYGI, F. and G. GALLI (1995), *Phys. Rev. B* **52**, 2229.

HAROCHE, S. (2000), *Sciences aux Temps Ultracourts* (Académie des Sciences (Fr.), Editions TEC).

HAUSBO, P. (2000), *J. Comp. Phys.* **159**, 274.

HENTSCHEL, M. ET AL. (2001), *Nature* **414**, 509.

HERTZBERG, G. (1951), *Spectra of Diatomic Molecules* (Van Nostrand, Amsterdam).

HOFFMAN, J.D. (1992), *Numerical Methods for Engineers and Scientists* (McGraw-Hill, New York).

HUSIMI, K. (1953), *Prog. Theoret. Phys.* **9**, 381.

KAWATA, I. and A.D. BANDRAUK (2002), in preparation.

KAWATA, I. and H. KONO (1999), *J. Chem. Phys.* **111**, 9498.
KELDYSH, L.V. (1965), *Sov. Phys. JETP* **20**, 1307.
KONO, H., ET AL. (1997), *J. Comput. Phys.* **130**, 148.
KRAUSE, J.L., K.J. SCHAFFER and K.C. KULANDER (1991), *Chem. Phys. Lett.* **178**, 573.
KULANDER, K. (1987), *Phys. Rev. A* **36**, 2726.
LU, H.Z. and A.D. BANDRAUK (2001), *J. Chem. Phys.* **115**, 1670.
NGUYEN, N.A. and T.T. NGUYEN-DANG (2000), *J. Chem. Phys.* **112**, 1229.
PAULI, W. and M. FIERZ (1938), *Nuovo Cimento* **15**, 167.
QUARTERONI, A. (1991), *Surv. Math. Ind.* **1**, 75.
RABITZ, H., G. TURINICI and E. BROWN (2003), in: P.G. Ciarlet and C. Le Bris, eds., *Computational Chemistry*, Handbook of Numerical Analysis, Vol. X (Elsevier, Amsterdam), 833–887.
THALLER, B. (1992), *The Dirac Equation* (Springer-Verlag, Berlin).
TRUSCOTT, W.S. (1993), *Phys. Rev. Lett.* **70**, 1900.
WALSER, M.W., C.H. KEITEL, A. SCRINZI and T. BRABEC (2000), *Phys. Rev. Lett.* **85**, 5082.
WARE, A.F. (1998), *SIAM Rev.* **40**, 838.
WELTON, T.A. (1948), *Phys. Rev.* **74**, 1157.
WIEDEMANN, H. and T. MOSTOWSKI (1994), *Phys. Rev. A* **49**, 2719.
WUNDERLICH, C., E. KOBLER, H. FIGGER and T.W. HÄNSCH (1997), *Phys. Rev. Lett.* **78**, 2333.
YU, H. and A.D. BANDRAUK (1995), *J. Chem. Phys.* **102**, 1257.
ZUO, T. and A.D. BANDRAUK (1996), *Phys. Rev. A* **54**, 3254.
ZUO, T., A.D. BANDRAUK and P.B. CORKUM (1996), *Chem. Phys. Lett.* **259**, 313.

Control of Quantum Dynamics: Concepts, Procedures and Future Prospects

Herschel Rabitz[a], Gabriel Turinici[b], Eric Brown[c]

[a]*Department of Chemistry, Princeton University, Princeton, NJ 08544-1009, USA*
[b]*INRIA Rocquencourt, B.P. 105, 78153 Le Chesnay cedex France*
and CERMICS–ENPC, Champs sur Marne, 77455 Marne la Vallée cedex, France
[c]*Program in Applied and Computational Mathematics, 205 Fine Hall,*
Princeton University, Princeton, NJ 08544-1009, USA
E-mail addresses: hrabitz@princeton.edu (H. Rabitz),
Gabriel.Turinici@inria.fr (G. Turinici), ebrown@princeton.edu (E. Brown),
URLs: http://www.princeton.edu/~hrabitz (H. Rabitz),
http://www-rocq.inria.fr/Gabriel.Turinici (G. Turinici)

1. Introduction

1.1. Historical perspective on quantum control

Along with the development of laser technology, a rising interest has naturally appeared in connection with using lasers to influence matter at the quantum dynamical level. This new field of research, commonly designated as "quantum control", has roots that go back to the earliest days of laser development in the 1960s. What distinguishes quantum control from the traditional means of chemical manipulation is the use of delicate quantum wave interferences to alter the outcome of molecular scale dynamics phenomena. Research in quantum control accelerated in the 1980s, with the first successful results in SHI, WOODY and RABITZ [1988], TANNOR and RICE [1985], SHAPIRO

Computational Chemistry
Special Volume (C. Le Bris, Guest Editor) of
HANDBOOK OF NUMERICAL ANALYSIS, VOL. X
P.G. Ciarlet (Editor)
© 2003 Elsevier Science B.V. All rights reserved

and BRUMER [1989] driven by the recognition of the role of quantum interference and the need for rigorous design tools. The introduction of the evolutionary paradigm (see Section 4.5) in JUDSON and RABITZ [1992] and the accompanying experimental successes of ASSION, BAUMERT, BERGT, BRIXNER, KIEFER, SEYFRIED, STREHLE and GERBER [1998], LEVIS, MENKIR and RABITZ [2001], WEINACHT, AHN and BUCKSBAUM [1999], BARDEEN, YAKOVLEV, WILSON, CARPENTER, WEBER and WARREN [1997], BARDEEN, YAKOVLEV, SQUIER and WILSON [1998], HORNUNG, MEIER and MOTZKUS [2000], KUNDE, BAUMANN, ARLT, MORIER-GENOUD, SIEGNER and KELLER [2000] in the 1990s have brought an exciting new set of theoretical questions and laboratory possibilities to the field. The time scales involved lie mostly within the femtosecond (10^{-15}) – picosecond (10^{-12}) range with spatial dimensions extending from one or two atoms to large polyatomic molecules and solid state structures.

In practice the outcome of a control experiment is measured in terms of quantum observables (e.g., selective dissociation of interatomic bonds) associated with Hermitian operators acting on the system at hand. The control process consists of steering the appropriate observable or wavefunction from the initial state to a final desired state. In the laboratory this is generally done by time-varying laser fields. We refer the reader to Section 2 for a precise mathematical transcription of these concepts.

Although prospects for industrial applications motivated much of the early research on quantum control, current applications span a wide range, from high harmonic generation (BARTELS, BACKUS, ZEEK, MISOGUTI, VDOVIN, CHRISTOV, MURNANE and KAPTEYN [2000]) and fast switching in semi-conductors (KUNDE, BAUMANN, ARLT, MORIER-GENOUD, SIEGNER and KELLER [2000]) to Hamiltonian identification (see Section 5.5).

1.2. Multidisciplinarity

Although research on quantum control was initiated within the field of physical chemistry, the subject has developed to involve researchers from myriad fields, including engineering, mathematics, computer science and physics. The introduction of control tools from the engineering contexts opened the field to the influence of (applied) mathematicians; later, computer scientists also became involved through their interest in quantum computing. Besides techniques employed by control specialists working on controllability, observability and stabilization issues, the applied mathematics tools involved include the resolution of the control problem through deterministic (direct optimization or critical point equations) or stochastic (evolutionary algorithms) approaches. The variety of research paradigms contributing to quantum control has proved to be beneficial to the field and will likely be central to future advances.

1.3. Outline

This volume addresses numerical methods in various aspects of the chemical sciences and the present chapter contributes in that fashion to the subject of quantum control. However, in order to appreciate how numerical methods are relevant for quantum control it is essential to understand how they fit into the overall subject including its laboratory implementations. Thus, the chapter attempts to give a full perspective on the subject including its theoretical foundations as well as its current state in the laboratory.

The balance of this review will proceed as follows: after an introduction of the fundamental concepts in Section 2 we present in Section 3 theoretical results regarding on the controllability of bilinear quantum systems. Then, in Section 4 we present the available numerical and experimental algorithmical approaches employed to control quantum phenomena. Present topics of interest and open questions are the object of Section 5. Concluding remarks are presented in Section 6. The text draws on many works in the literature and especially a recent prospectus (BROWN and RABITZ [2002]) on the field.

2. Basic principles

2.1. Mathematical formulation of the Hamiltonian and control law

Each particular control setting requires an appropriate quantum control model. The section will be mostly devoted to the description of the bilinear dipole coupling, which is often employed in practice. However, some other control paradigms will be presented and their domain of applicability and specific capabilities discussed. Controlling a quantum system requires the introduction of external interactions. Here, we will only consider control tools that act at the atomic/molecular level, with other, more classical tools such as temperature, pressure, catalysts etc. being outside of our scope. The primary control for quantum systems discussed here will be the electric field of a laser.

Most of this paper will work under the assumption that the system to be controlled can be characterized by its state function $\psi(t)$. This is a proper representation for isolated systems starting in a pure state; the complementary case arises, for example, in collisional or condensed regimes, when a density operator $\rho(t)$ must be introduced to describe the statistical mixture of states making up the system. The density operator formulation will be discussed where relevant.

Consider a quantum system that evolves from the initial state $\psi(t=0)=\psi_0$. The objective of quantum control can generally be expressed as the desire for the expectation value

$$\langle O(t)\rangle = \langle \psi(t)|O|\psi(t)\rangle \tag{2.1.1}$$

of some predefined observable operator O to be within a prescribed target set at the final time $t=T$. More general quantum control problem formulations may deal with several observable operators $O_1,\ldots,O_k$ $(k>1)$ and additional intermediary times $0\leqslant t_j\leqslant T$ $(j=1,2,\ldots)$.

In the absence of any external control influence, evolution of the state function $\psi(t)$ under the Schrödinger equation is determined by the free Hamiltonian H_0, which by assumption does not yield dynamics producing the desired expectation values. Quantum control theory considers the addition of a laboratory accessible control law term $C(t)$ to the Hamiltonian in order to achieve these objectives, which makes $H = H_0 + C(t)$ the new Hamiltonian of the system and

$$i\hbar \frac{\partial \psi(t)}{\partial t} = \left[H_0 + C(t)\right]\psi(t) \tag{2.1.2}$$

the equation of motion. Appropriate regularity assumptions on the control law must be enforced so that the evolution of $\psi(t)$ is well defined. For example, a common control law for lasers has the form

$$C(t) = -\mu\varepsilon(t), \quad \varepsilon = (\varepsilon_i)_{i=1}^3, \quad \varepsilon_i(t) \in L^2[0, T], \quad i = 1, 2, 3,$$

where μ is the electric dipole operator, $\varepsilon(t)$ is the applied electric field, and the index i refers to spatial orientation.

REMARK 2.1. Depending on the problem, one may go beyond the first-order, bilinear term in Eq. (2.1.2) when describing the interaction between the laser and the system, cf. DION, BANDRAUK, ATABEK, KELLER, UMEDA and FUJIMURA [1999], DION, KELLER, ATABEK and BANDRAUK [1999].

Additional admissibility conditions may ensure that $\varepsilon(t)$ obeys laboratory limitations on the range of achievable laser frequencies, intensities, energy, or other criteria. In some applications, additional possibilities for $C(t)$ arise. These include

- the use of magnetic fields, in which case the control law becomes $-\mu_m B(t)$, where μ_m is the magnetic dipole operator and $B(t)$ is the magnetic field, and
- the use of materials whose design specifications themselves take the form of a control law, such as for quantum electron transport in semiconductors with variable material composition considered as the control.

Here, however, we will confine the discussion to time-dependent controls based on an external electric field $\varepsilon(t)$ coupled to the system through a dipole μ.

In some cases, it may be possible to obtain an adequate control description by replacing the Schrödinger equation with a classical representation of the system dynamics (see Section 2.6). This is especially true for interatomic phenomena, because the *de Broglie* wavelength associated with atoms is often short relative to interatomic length scales. While the relationships between classical and quantum models of molecular evolution have been extensively investigated, e.g., DAVIS and HELLER [1984], KULANDER and OREL [1981], LEFORESTIER [1986], DARDI and GRAY [1982], the implications of these relationships for control are not completely understood and will be addressed later in the context of the quantum character of the control problem.

Assuming knowledge of H_0 and a well-defined control law $C(t)$, Eq. (2.1.2) or its classical equivalent represents a complete model of the system of interest. If $C(t)$ is given a priori, the solution of Eq. (2.1.2) is a standard numerical problem in time-dependent quantum mechanics. However, the essence of the control problem is to find $C(t)$ such that the objectives in Eq. (2.1.1) are met, and since at least one of the control objectives lies in the future, this task presents some additional challenges. In particular, the Hamiltonian depends on the future state of the system through the control objectives, as can be formally represented by the expression $C(t) = C(\psi(s)\colon\ s \in [t, T])$. This noncausality introduces an entirely new set of mathematical issues which are not present in standard quantum or classical dynamics but are inherent to the theory and practice of temporal control in engineering and mathematical systems theory (cf. SONTAG [1998], BROCKETT [1970], KAILATH [1980], KHALIL [1996], NIJMEIHER and VAN DER SCHAFT [1990]). Their implications for the quantum regime are central issues. The formulation above is summarized in the following definition:

DEFINITION 2.2. The quantum control problem consists of determining a control law $C(t)$ that causes the system to optimally achieve the desired expectation values while possibly also satisfying auxiliary conditions. Quantum control theory encompasses methods of determining these control laws, their general properties, and their relationship to the underlying physical system and evolving quantum states.

2.2. *Optimal control*

Several approaches for determining control laws $C(t)$ will be discussed in this chapter. The present section introduces the concept of optimal control theory for this purpose; a more detailed algorithmic analysis will be given in Section 4. An extensive literature on optimal control theory can be found in classical engineering and mathematical systems theory (e.g., SONTAG [1998], LUENBERGER [1979], MOHLER [1983]) and increasingly in quantum mechanics (e.g., PEIRCE, DAHLEH and RABITZ [1988], ZHAO and RICE [1994], Chapter 6 of RICE and ZHAO [2000] and references therein). Considering control law design as an optimization problem is quite natural, as attaining the best possible final level of control is always the goal; further, optimization is essential when there are competing physical objectives that must simultaneously be met.

The *optimal control* approach seeks to optimize a cost functional J, which includes both terms that describe how well the objective has been met and terms that penalize undesirable effects. One simple example of a cost functional is

$$J(\varepsilon) = \langle \psi(T) | O | \psi(T) \rangle - \alpha \int_0^T \varepsilon^2(t)\, dt, \tag{2.2.1}$$

where $\alpha > 0$ is a parameter and O is the observable operator (positive semidefinite in this case) that specifies the goal. In mathematical terms, the observable O is a self-adjoint operator that acts on $\psi(T)$; in the case above, the goal is to achieve a large value $\langle \psi(T)|O|\psi(T)\rangle$. Note that in general attaining the maximum possible value of $\langle \psi(T)|O|\psi(T)\rangle$ comes at the price of a large laser fluence $\int_0^T \varepsilon^2(t)\, dt$, so that the optimum evolution will strike a balance (weighted by α) between laser fluence and operator expectation values.

2.3. *More nonlinear formulations*

The quantum control problem becomes more complex when the control law and the free Hamiltonian cannot be treated independently. An important example of this phenomenon is intense-field laser control of molecular motion, where the electric field can directly alter the dipole operator through its manipulation of the electronic degrees of freedom:

$$C(t) = -\mu\big(\varepsilon(t)\big)\varepsilon(t). \tag{2.3.1}$$

In simple cases, the relation in Eq. (2.3.1) may be expanded in terms of a low order polynomial in $\varepsilon(t)$ whose coefficients are the electric moments and polarizabilities of the system. Of special interest are situations in which the nonlinear structure may affect the controllability of the system (including the positive case in which this interaction makes a previously inaccessible target reachable).

REMARK 2.3. To date, very few mathematical studies exist to treat the situation where the control law has a nonlinear dependence on the control field; it is not clear whether successful approaches will draw upon existing methodology for proving, for example, quantum controllability (see Section 3) or upon new tools from mathematical systems theory.

The circumstances motivating Remark 2.3 can also be viewed from the larger perspective of a controllability analysis simultaneously including electronic and nuclear motion, e.g., as in CANCÈS and LE BRIS [1999]. In the latter circumstance the control field will enter the Hamiltonian linearly, but at the expense of explicitly including the electronic degrees of freedom. The same comment also applies to the performance of optimal control designs in the strong-field regime. The practical importance of investigating the latter domain has recently been demonstrated experimentally (LEVIS, MENKIR and RABITZ [2001]).

A different formulation is necessary when the back action of the quantum medium upon the propagating control field is significant. In this case the medium is called *optically dense*. This scenario has been examined experimentally for a vapor of sodium (SHEN, SHI, RABITZ, LIN, LITTMAN, HERITAGE and WEINER [1993]), and the topic is of practical importance because the controlled medium will be dense in any application directed toward collecting large amounts of product. Optically dense media can interact with the electric field to alter its phase and/or amplitude structure as it propagates. In order to model this effect, the Schrödinger equation must be coupled with Maxwell's equations, e.g., WANG and RABITZ [1996].

2.4. *Density matrix formulation*

In practical applications the medium will be at a finite temperature; in this case (and in any situation that concerns a statistical mixture of quantum states) the density operator formulation is necessary. In this formulation, the time evolution is given by the quantum Liouville equation (COHEN-TANNOUDJI, LIU and LALOË [1997]):

$$\frac{d\rho(t)}{dt} = \frac{1}{i\hbar}\big[H(t), \rho(t)\big], \tag{2.4.1}$$

and expectation values are calculated as

$$\big\langle O(t)\big\rangle = \mathrm{Tr}\big(\rho(t)O\big).$$

Upon introducing a cost functional as in Eq. (2.2.1), quantum systems described by density operators can also be treated by optimal control theory.

2.5. Special control schemes: pump-dump schemes, STIRAP

In addition to the general approach outlined below, special techniques for control law design have been developed for cases in which a priori specification of control mechanisms is possible. These methods include the weak-field regime, time-resolved "pump-dump" and simulated Raman adiabatic passage (STIRAP) schemes. Under particular quantum dynamics approximations and/or assumptions, these techniques allow for the derivation of closed form expressions for control laws that optimally accomplish certain control objectives under their specified conditions. The literature on related theoretical developments is extensive and will not be presented here (Chapters 3–5 of RICE and ZHAO [2000] and references therein, BROWN and SIBLEY [1998], SHAH and RICE [1999], MISHIMA and YAMASHITA [1999b], HOKI, OHTSUKI, KONO and FUJIMURA [1999], DE ARAUJO and WALMSLEY [1999], MISHIMA and YAMASHITA [1999a], CAO and WILSON [1997], BARDEEN, CHE, WILSON, YAKOVLEV, CONG, KOHLER, KRAUSE and MESSINA [1997], BARDEEN, CHE, WILSON, YAKOVLEV, APKARIAN, MARTENS, ZADOYAN, KOHLER and MESSINA [1997], AGARWAL [1997], GRØNAGER and HENRIKSEN [1998]). For experimental developments see Chapters 3–5 of RICE and ZHAO [2000] and references therein, BARDEEN, CHE, WILSON, YAKOVLEV, CONG, KOHLER, KRAUSE and MESSINA [1997], BARDEEN, CHE, WILSON, YAKOVLEV, APKARIAN, MARTENS, ZADOYAN, KOHLER and MESSINA [1997], MESHULACH and SILBERBERG [1998], PASTIRK, BROWN, ZHANG and DANTUS [1998].

2.6. Classical mechanics formulations

Classical modeling of quantum systems is a common and often successful technique, and it should have a level of applicability in molecular control. This section attempts to discuss its applicability, or the quantum character of molecular scale control. Nonclassical characteristics of dynamical behavior include tunneling, quantization of energy levels, and interference processes, but it is not clear which of these characterizations are most relevant to defining the quantum nature of a control problem. Understanding of this issue could be used to estimate the loss of reliability (i.e., defined upon comparison to the analogous quantum system response to the classically designed field) in resorting to the classical optimal control formulation.

Some aspects of this topic have been addressed in SCHWIETERS and RABITZ [1993], where quantum $C_q(t)$ and classical $C_c(t)$ control laws corresponding to equivalent representations of specific control problems are compared. The equations of motion analogous to Eq. (2.1.2) are:

$$\frac{dq_i^l}{dt} = \frac{\partial H}{\partial p_i^l}, \quad \frac{dp_i^l}{dt} = -\frac{\partial H}{\partial q_i^l}, \quad l = 1, \ldots, N_c, \tag{2.6.1}$$

and expectation values for the classical system are given by

$$\langle O_c \rangle = \sum_{l=1}^{N_c} \Gamma_l \bar{O}(\mathbf{q}^l, \mathbf{p}^l),$$

where H is the classical Hamiltonian of the system, i ranges over the particle coordinates, l indexes initial conditions $q^l(0)$, $p^l(0)$, and the weights Γ_l for the N_c initial conditions are chosen to mimic as best as possible the probability distribution function for the corresponding quantum system. Here, $\bar{O}$ is a classical observable corresponding to its quantum analog. It should be noted that the ordinary differential equations in Eq. (2.6.1) for some cases may be more expensive to solve than their quantum counterpart in Eq. (2.1.2). One motivation for considering classical optimal control design is for the physical insight possible with classical mechanics.

Optimal control theory has been used, e.g., in SCHWIETERS and RABITZ [1993] to separately design a control field $\varepsilon(t)$ that minimizes the difference between $\langle O \rangle$ and $\langle O_c \rangle$ and the difference between each of these expectation values and the control objectives on $[0, T]$. For the example of a Morse oscillator, it was found that an optimal control law designed in this fashion produced very close agreement between $\langle O \rangle$ and $\langle O_c \rangle$. This result suggests that in some cases classically-designed controls can also be successful as quantum controls. In related work SCHWIETERS and RABITZ [unpublished results], a method was developed for determining potentials under which evolving classical and corresponding quantum systems give similar values of classical and quantum observables; the approach met with considerable success for the control of dissociative flux and displacement. In general it is however not known for what classes of Hamiltonians and control objectives the quantum control problem can be adequately addressed using classical equations of motion. Because interference itself is a nonclassical phenomenon, this problem is related to the considerations of decoherence in Section 5.1.

3. Controllability of quantum mechanical systems

Prior to addressing the computation of the control law, it is natural to ask if the quantum control problem is well-posed such that a control law exists which will cause the objectives and auxiliary conditions to be (precisely) satisfied. Even if the answer to the latter question is negative, one may still be satisfied with achieving the control objectives (maybe only partially) through the techniques of optimal control. The fundamental importance of addressing controllability has long been recognized in engineering control applications; the broad literature on the classical aspects of the subject includes many comprehensive texts which cover linear (BROCKETT [1970], KAILATH [1980]) and nonlinear (SONTAG [1998], KHALIL [1996], NIJMEIHER and VAN DER SCHAFT [1990]) controllability. In addition, several works have considered various aspects of quantum controllability (e.g., RICE and ZHAO [2000], HUANG, TARN and CLARK [1983], JUDSON, LEHMANN, RABITZ and WARREN [1990], RAMAKRISHNA, SALAPAKA, DAHLEH, RABITZ and PIERCE [1995], TURINICI [2000c], TURINICI and RABITZ [2001a], TURINICI and RABITZ [2001b], GIRARDEAU, SCHIRMER, LEAHY and KOCH [1998], GIRARDEAU, INA, SCHIRMER and GULSRUD [1997]).

Consider a quantum system (isolated from external influences for the moment) with internal Hamiltonian H_0 that is prepared in the initial state $\psi_0(x)$ described here in the coordinate representation; its dynamics obeys the time dependent Schrödinger equation. Denoting by $\psi(x,t)$ the state at the time t one can write the evolution equations for the free system:

$$\begin{cases} i\hbar\frac{\partial}{\partial t}\psi(x,t) = H_0\psi(x,t), \\ \psi(x,t=0) = \psi_0(x), \quad \|\psi_0\|_{L^2(\mathbb{R}^\gamma)} = 1. \end{cases} \tag{3.0.1}$$

In the presence of external interactions that will be taken here as a control field amplitude $\varepsilon(t) \in \mathbb{R}$, $t \geqslant 0$, coupled to the system through a time-independent (e.g., dipole) operator μ, the (controlled) dynamical equations read:

$$\begin{cases} i\hbar\frac{\partial}{\partial t}\psi_\varepsilon(x,t) = H_0\psi_\varepsilon(x,t) - \varepsilon(t)\mu\psi_\varepsilon(x,t) = H\psi_\varepsilon(x,t), \\ \psi_\varepsilon(x,t=0) = \psi_0(x). \end{cases} \tag{3.0.2}$$

REMARK 3.1. For the sake of simplicity we have chosen in this section to treat situations with only one control law present. Extensions to many control laws representing coupling via various other operators are also available (especially for finite dimensional settings); we refer the reader to RAMAKRISHNA, SALAPAKA, DAHLEH, RABITZ and PIERCE [1995], ALBERTINI and D'ALESSANDRO [2001], TURINICI, RAMAKHRISHNA, LI and RABITZ [2002] for details.

The L^2 norm of ψ_ε is conserved throughout the evolution:

$$\left\|\psi_\varepsilon(x,t)\right\|_{L^2(\mathbb{R}^\gamma)} = \|\psi_0\|_{L^2(\mathbb{R}^\gamma)}, \quad \forall t > 0, \tag{3.0.3}$$

so the state (or wave-) function $\psi(t)$, evolves on the (complex) unit sphere $S = \{\psi \in L^2(\mathbb{R}^\gamma): \|\psi\|_{L^2(\mathbb{R}^\gamma)} = 1\}$ according to the Schrödinger equation (3.0.2) from the initial state ψ_0 to some final state $\psi(T)$.

The study of the controllability of Eq. (3.0.2) is concerned with identifying the set of final states $\psi(T)$ that can be obtained from a given initial state ψ_0 for all admissible controls. To date, very different results are available for the infinite and finite dimensional settings: while controllability is reasonably well understood for finite dimensional systems, no positive results have been obtained for infinite dimensional systems.

3.1. *Infinite dimensional controllability*

Very few results are available concerning the controllability of Eq. (3.0.2) in its infinite dimensional form. Generic *negative* results have been obtained, as is the following (Theorem 1 from TURINICI [2000c]; see also BALL, MARSDEN and SLEMROD [1982], TURINICI [2000b]):

THEOREM 3.2. *Let S be the complex unit sphere of $L^2(\mathbb{R}^\gamma)$. Let μ be a bounded operator from the Sobolev space $H_x^2(\mathbb{R}^\gamma)$ to itself and let H_0 generate a C^0 semigroup of bounded linear operators on $H_x^2(\mathbb{R}^\gamma)$. Denote by $\psi_\varepsilon(x,t)$ the solution of* (3.0.2). *Then the set of attainable states from ψ_0 defined by*

$$\mathcal{AS} = \bigcup_{T>0} \{\psi_\varepsilon(x,T);\ \varepsilon(t) \in L^2([0,T])\} \tag{3.1.1}$$

is contained in a countable union of compact subsets of $H_x^2(\mathbb{R}^\gamma)$. In particular its complement with respect to S: $\mathcal{N} = S \backslash \mathcal{AS}$ is everywhere dense on S. The same holds true for the complement with respect to $S \cap H_x^2(\mathbb{R}^\gamma)$.

The theorem implies that for any $\psi_0 \in H_x^2(\mathbb{R}^\gamma) \cap S$, within any open set around an arbitrary point $\psi \in H_x^2(\mathbb{R}^\gamma) \cap S$ there exists a state unreachable from ψ_0 with L^2 controls.

REMARK 3.3. The lack of positive controllability results to complement Theorem 3.2 should be regarded as a failure of available control theory tools to provide insight into controllability rather than as an actual restriction. It is believed that new tools and concepts will make positive results possible, especially since such results are available for the finite dimensional setting (cf. Section 3.2).

Truncating an infinite-dimensional quantum control problem to a finite-dimensional problem (i.e., so that evolution takes place in a finite dimensional vector space) changes the nature of both the control and Hamiltonian operators and the set of states available as candidate members of reachable sets. The concern is to characterize these effects by asking how a controllability result obtained in a finite-dimensional space relates to the original infinite dimensional problem from which it was derived; there are also inherently finite dimensional quantum systems (as with spins) where the latter consideration does not arise. The controllability of finite dimensional systems is the subject of the next section.

3.2. *Finite dimensional controllability*

Introduction

Let $D = \{\psi_i(x);\ i = 1, \dots, N\}$ be an orthonormal basis for a finite dimensional subspace of $L^2(\mathbb{R}^\gamma)$ of interest (for instance the vector space spanned by the first N

eigenstates of the internal Hamiltonian H_0 in Eq. (3.0.1)). Denote by M the linear space that D generates, and let A and B be the matrices of the operators $-iH_0$ and $-i\mu$ respectively, with respect to this basis. In order to exclude trivial control settings, it is supposed that $[A, B] \neq 0$ (the Lie bracket $[\cdot, \cdot]$ is defined as $[U, V] = UV - VU$). Note that since H_0 and μ are Hermitian operators, the matrices A and B are skew-Hermitian.

Let us denote by $c_\varepsilon(t) = (c_{\varepsilon i}(t))_{i=1}^N$ the coefficients of $\psi_i(x)$ in the expansion of the evolving state $\psi(t, x) = \sum_{i=1}^N c_{\varepsilon i}(t)\psi_i(x)$; then Eq. (3.0.2) reads

$$\begin{cases} \frac{d}{dt} c_\varepsilon(t) = Ac_\varepsilon(t) - \varepsilon(t)Bc_\varepsilon(t), \\ c_\varepsilon(t=0) = c_0, \end{cases} \tag{3.2.1}$$

$$c_0 = (c_{0i})_{i=1}^N, \quad c_{0i} = \langle \psi_0, \psi_i \rangle_{L^2(\mathbb{R}^\gamma)} \tag{3.2.2}$$

(where atomic units are used, i.e., we set $\hbar = 1$). Note as in Eq. (3.0.3) that the system (3.2.1) evolves on the unit sphere S_N of $L^2(\mathbb{R}^\gamma) \cap M$ which reads

$$\sum_{i=1}^N |c_{\varepsilon i}(t)|^2 = 1, \quad \forall t \geqslant 0. \tag{3.2.3}$$

Note also that the solution $c_\varepsilon(t)$ of Eq. (3.2.1) can be written

$$c_\varepsilon(t) = U_\varepsilon(t)c_0, \tag{3.2.4}$$

where the time evolution operator $U_\varepsilon(t)$ is the solution of the following

$$\begin{cases} \frac{d}{dt} U_\varepsilon(t) = AU_\varepsilon(t) - \varepsilon(t)BU_\varepsilon(t), \\ U_\varepsilon(t=0) = I_{N\times N}. \end{cases} \tag{3.2.5}$$

The matrix $U(t)$ evolves in the Lie group of unitary matrices $U(N)$, or, if both the matrices A and B have zero trace, in the Lie group of special unitary matrices $SU(N)$. Eq. (3.2.5) also prescribes the evolution of the density matrix operator: if the system starts in the mixture of states represented by the density operator ρ_0 then its evolution is given by the formula

$$\rho(t) = U(t)\rho_0 U(t)^\dagger, \tag{3.2.6}$$

where for any matrix X we denote by $X^\dagger$ its transpose-conjugate.

REMARK 3.4. Note that since $U(t)$ is unitary $\rho(t)$ has the same eigenvalues as $\rho(0)$.

We now introduce:

DEFINITION 3.5. Denote by $\mathcal{U}$ the set of all admissible control laws $\varepsilon(t)$. The system described by the state $c_\varepsilon(t)$ is called state-controllable if for any $c_i, c_f \in S_N$ there exists $0 < \tau < \infty$ and $\varepsilon \in \mathcal{U}$ such that $c_\varepsilon(t = \tau) = c_f$, where $c(t)$ satisfies (3.0.2) with $c_\varepsilon(t = 0) = c_i$.

Within the density matrix formulation the relevant definition of controllability becomes:

DEFINITION 3.6. The system described by the evolution of the density operator $\rho(t)$ satisfying Eqs. (3.2.5), (3.2.6) is called density-matrix-controllable if for any two density operators ρ_i and ρ_f compatible in the sense that there exists an unitary matrix U such that $\rho_f = U\rho_i U^\dagger$ there exists a control law $\varepsilon(t) \in \mathcal{U}$ such that given the initial condition $\rho(0) = \rho_i$ then $\rho(\tau) = \rho_f$ for some finite time τ.

REMARK 3.7. Because pure states my be represented by density matrices a system that is density-matrix-controllable is also state-controllable.

Lie group methods
Let us introduce the following

DEFINITION 3.8. A subset $\mathcal{T}$ of $U(N)$ (or $SU(N)$) is said to be *transitive* (*to act transitively*) on the sphere S_N if for any $c_i, c_f \in S_N$ there exists $g \in \mathcal{T}$ such that $c_f = gc_i$.

With this definition and considering Eq. (3.2.4) it follows that

- state-controllability of the wavefunction as in Definition 3.5 is equivalent to requiring that the set of all matrices attainable from identity $\{U_\varepsilon(t);\ 0 \leqslant t \leqslant \infty;\ \varepsilon \in \mathcal{U},\ U \text{ verify (3.2.5)}\}$ be transitive on the sphere S_N, while
- density-matrix-controllability is equivalent to requiring that the set of all matrices attainable from identity be at least $SU(N)$.

Remarkable examples of transitive subgroups of $U(N)$ are $U(N)$ itself, $SU(N)$ and, when N is even, the symplectic matrices $Sp(N/2)$. It can be proven (see ALBERTINI and D'ALESSANDRO [2001]) that, except for some special values of N and up to an isomorphism, these are the only transitive subgroups arising in quantum control.

The important result that turns this remark into a powerful tool for studying the controllability of bilinear systems is that the set of all matrices attainable from the identity via (3.2.5) is given by the connected Lie subgroup e^L of the Lie algebra L generated by A and B (when this Lie group is compact) (cf. RAMAKRISHNA, SALAPAKA, DAHLEH, RABITZ and PIERCE [1995], ALBERTINI and D'ALESSANDRO [2001]). The group e^L is called the *dynamical Lie group* of the system.

Taking $\mathcal{U}$ to be the set of all piecewise continuous functions (unconstrained in magnitude) yields the following result:

THEOREM 3.9 (RAMAKRISHNA, SALAPAKA, DAHLEH, RABITZ and PIERCE [1995]). *A sufficient condition for the density-matrix- (thus state-) controllability of the quantum system in Eq.* (3.2.1) *is that the Lie algebra L generated by A and B has dimension* N^2 *(as a vector space over the real numbers).*

Furthermore, if both A and B are traceless then a sufficient condition for the density-matrix- (thus state-) controllability of quantum system is that the Lie algebra L has dimension $N^2 - 1$.

A following result builds on Theorem 3.9 to gives the necessary and sufficient condition of state controllability:

THEOREM 3.10 (ALBERTINI and D'ALESSANDRO [2001]). *The system is state controllable if and only if the Lie algebra L generated by A and B is isomorphic (conjugate) to $sp(\frac{N}{2})$ or to $su(N)$, if the dimension N is even, or to $su(N)$, if the dimension N is odd (with or without the iI, where I is the identity matrix).*

Theorem 3.9 lends itself to algorithmic (numerical) implementation: as soon as the matrices A and B that characterize the system are given, one can compute (numerically for instance) the Lie algebra they generate and obtain its dimension. However, except for small systems, this test becomes rapidly very computationally expensive; additional results are therefore required in order to shed some light on the relationship between controllability and the structure of the A and B matrices. Two studies in this direction are available: one, from SCHIRMER, FU and SOLOMON [2001], FU, SCHIRMER and SOLOMON [2001] is presented in this section and the second, the "connectivity graph" approach, is described in the next section (cf. TURINICI [2000c], TURINICI and RABITZ [2001a], TURINICI and RABITZ [2001b], TURINICI [2000b], TURINICI [2000a]).

When the basis $D = \{\psi_i(x);\ i = 1, \ldots, N\}$ is composed of eigenstates of the internal Hamiltonian H_0 the matrix A is diagonal with purely imaginary elements $-iE_k$, where E_k are the eigenvalues of H_0, $k = 1, \ldots, N$. Let $\delta_k = E_k - E_{k+1}$ for $n = 1, \ldots, N-1$.

THEOREM 3.11 (SCHIRMER, FU and SOLOMON [2001]). *Suppose that the matrix B of the interaction operator $-i\mu$ with respect to the basis D is such that $B_{k,l} = 0$ for $|k-l| \neq 1$ and $B_{k,l} \neq 0$ for $|k-l| = 1$, $k, l = 1, \ldots, N$. Then if either*

(1) *$\delta_1 \neq 0$ and $\delta_k \neq \delta_1$ for $k = 1, \ldots, N-1$, or*

(2) *$\delta_{N-1} \neq 0$ and $\delta_{N-1} \neq \delta_k$ for $k = 1, \ldots, N-1$*

the dynamical Lie group of the system $A - \varepsilon(t)B$ with A and B as in Eq. (3.2.5) is at least $SU(N)$. If in addition $\operatorname{Tr} A \neq 0$ *then the dynamical Lie group is $U(N)$. In both cases the system is density-matrix- (thus state-) controllable.*

REMARK 3.12. Although the hypothesis of the theorem above are somewhat strong, it allows for the determination of controllability using only generic properties of the system under study, i.e., without the need to know exactly the matrices A and B. We will see later in this section (see Remark 3.17) other results concerning the controllability of the wavefunction that do not require precise evaluation of the A and B matrices.

While the results above hold in the case that the control field amplitudes are not bounded, an open question suggested in this work is the extension of the result to stricter (and more realistic) admissibility conditions: can the Lie algebraic controllability conditions (e.g., RAMAKRISHNA, SALAPAKA, DAHLEH, RABITZ and PIERCE [1995]) be extended to treat the case where both the amplitude and the frequency of the control field are bounded from above and below? This issue has practical significance as it prescribes real laboratory conditions.

REMARK 3.13. Other open problems refer to the controllability within the product state space of the coupled Schrödinger–Maxwell equations for optically dense media. While the Schrödinger–Maxwell system has a product state space representing both $\rho(t)$ (the density operator) and $\varepsilon(t)$, expectation values depend only on $\rho(t)$, which is the usual focus of controllability studies. This inspires the question: in what, if any, cases is it possible for the Schrödinger–Maxwell system to be controllable in the state space of the quantum state but possibly not controllable in that of the electric field, or vice-versa? The latter case of controlling the electric field is of importance in the allied subject of optical field propagation (e.g., XIA, MERRIAM, SHARPE, YIN and HARRIS [1997]).

REMARK 3.14. To ease the assessment of controllability using the results of this section, an automatic tool that allows computation of the dimension of the Lie algebra generated by several (skew-Hermitian) matrices is available freely on the Internet (cf. TURINICI and SCHIRMER [2001]).

Controllability analysis via the connectivity graph

This section seeks to address controllability from an analysis of the kinematic structure of the Hamiltonian. Suppose that the basis $D = \{\psi_i(x);\ i = 1, \ldots, N\}$ is composed of eigenstates of the internal Hamiltonian H_0, so that the matrix A is diagonal with purely imaginary elements $-iE_k$, where E_k are the eigenvalues of H_0, $k = 1, \ldots, N$, and that the diagonal elements of the matrix B are all zero (this is often the case in practice). We obtain the following structure:

$$A = -i\begin{pmatrix} E_1 & & & 0 \\ & E_2 & & \\ & & \ddots & \\ 0 & & & E_n \end{pmatrix}; \qquad B = -i\begin{pmatrix} 0 & b_{12} & \ldots & b_{1N} \\ b_{12}^* & 0 & b_{ij} & \vdots \\ \vdots & b_{ij}^* & \ddots & \\ b_{1N}^* & \ldots & & 0 \end{pmatrix}.$$

Assume moreover that no degenerate transitions are present, i.e.,

$$|E_i - E_j| \neq |E_k - E_l|, \quad i, j, k, l = 1, \ldots, n,\ i \neq j,\ k \neq l,\ \{i, j\} \neq \{k, l\}. \tag{3.2.7}$$

The connectivity amongst the states $\{\psi_i;\ i = 1, \ldots, n\}$ provided by the elements b_{ij}, $i, j = 1, \ldots, n$, is central to issues of controllability. The structure in B can be conveniently expressed graphically by introducing a graph $G = (V, E)$ (see CHRISTOFIDES [1975] for an introduction to graph theory): let every state be a vertex (node) of the graph G so that the set of vertices $V = \{\psi_1, \ldots, \psi_N\}$, and let there be edges between every pair of nodes ψ_i and ψ_j with $b_{ij} \neq 0$ so that the set of edges $E = \{(\psi_i, \psi_i);\ b_{ij} \neq 0\}$. Two states ψ_i and ψ_j are said to be connected by a path if there exists a connected set of edges starting in ψ_i and ending in ψ_j. The graph G is called connected if there exists a path between every pair of vertices.

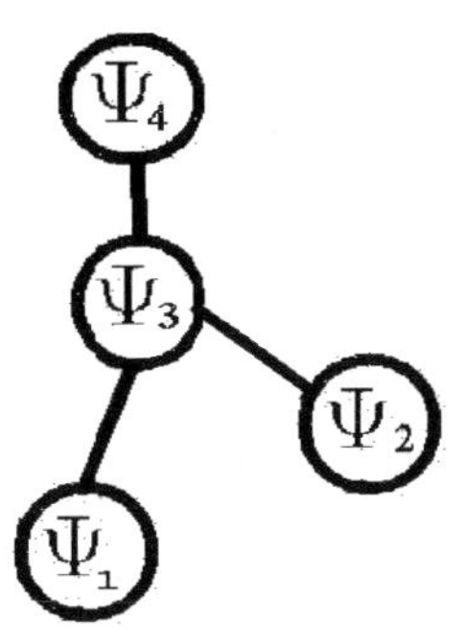

FIG. 3.1. The (connected) graph associated with the B matrix of the system in Eq. (3.2.8). Note that no direct path exits between, e.g., ψ_1 and ψ_4.

REMARK 3.15. Note that G being connected does *not* imply that any two states are necessarily *directly* connected (i.e., with a *direct* edge). One such example is the (connected) graph in Fig. 3.1 associated with the system

$$A = -i \begin{pmatrix} 1.1 & & & 0 \\ & 2.3 & & \\ & & 3.05 & \\ 0 & & & 4.6 \end{pmatrix}; \qquad B = -i \begin{pmatrix} 0 & 0 & 1 & 0 \\ 0 & 0 & 1 & 0 \\ 1 & 1 & 0 & 1 \\ 0 & 0 & 1 & 0 \end{pmatrix}. \tag{3.2.8}$$

These definitions allow formulating the following result (see also ALTAFINI [2002] for similar conclusions):

THEOREM 3.16 (TURINICI and RABITZ [2001a], TURINICI and RABITZ [2001b]). *Suppose that the graph G is connected and that the transitions of the internal Hamiltonian are nondegenerate in the sense of Eq.* (3.2.7). *Then the system in* (3.2.1) *is state-controllable.*

REMARK 3.17. The conditions of connectivity and nondegenerate transition frequencies involve only the eigenvalues of H_0 and the coefficients of B (which depend only upon which elements of B are nonzero, rather than their particular values) allowing conclusions about the controllability of the system in Eq. (3.2.1) even in the absence of quantitative information about this system.

REMARK 3.18. Controllability generally does not provide an actual control law that will achieve an objective of the form (2.1.1), but simply implies that at least one such control law exists. The proof of controllability in TURINICI [2000c] establishes an algorithm which constructively generates control laws for objective operators of the form $O_i = |\psi_i\rangle\langle\psi_i|$ in terms of sinusoidal electric fields of different fixed frequencies. Note that in this situation, the control objectives are populations of quantum states (e.g., $|\langle\psi|\psi_i\rangle|^2$).

This latter work, along with other work on special cases of constructive quantum control (e.g., HAREL and AKULIN [1999], RAMAKRISHNA [2000]), invites the question: can general constructive control solutions be developed for the quantum controllability of a broad class of objective expectation values? While the explicit construction of control solutions is generally an area of active research in control theory (NIJMEIHER and VAN DER SCHAFT [1990], SEPULCHRE, JANKOVIC and KOKOTOVIC [1998]), the case of quantum controllability of expectation values reduces this task to the specific structure of Schrödinger's equation that may be amenable to attack.

REMARK 3.19. The analysis of controllability in terms of state functions $\psi(t)$ reaches beyond what is necessary physically, as realistic objectives are expectation values of observable operators (cf. Eq. (2.1.1)). Since these quantities involve integrals of state functions, their control should generally be less demanding than that of the state itself. However, the quadratically nonlinear nature of the expectation values adds a level of additional complexity to the problem of determining controllability.

Having seen positive results for finite-dimensional wavefunction controllability we now investigate what phenomena prevent this controllability. Even if a final answer does not yet exist, available evidence suggests that conservation laws are responsible for the loss of controllability. Let us consider, as in TURINICI and RABITZ [2001a], the following simple 3-level system:

$$A = \begin{pmatrix} 1 & 0 & 0 \\ 0 & 2 & 0 \\ 0 & 0 & 3 \end{pmatrix}, \qquad B = \begin{pmatrix} 0 & 1 & 0 \\ 1 & 0 & 1 \\ 0 & 1 & 0 \end{pmatrix}, \tag{3.2.9}$$

and the corresponding evolution equations

$$i\frac{d}{dt}c_{\varepsilon 1}(t) = c_{\varepsilon 1}(t) + \varepsilon(t)c_{\varepsilon 2}(t),$$

$$i\frac{d}{dt}c_{\varepsilon 2}(t) = 2c_{\varepsilon 2}(t) + \varepsilon(t)c_{\varepsilon 1}(t) + \varepsilon(t)c_{\varepsilon 3}(t),$$

$$i\frac{d}{dt}c_{\varepsilon 3}(t) = 3c_{\varepsilon 3}(t) + \varepsilon(t)c_{\varepsilon 2}(t).$$

This system has degenerate transitions as $E_2 - E_1 = E_3 - E_2$. Upon closer examination, a "hidden symmetry" is found for this system: more precisely it is easy to prove that for any $t > 0$ and $\varepsilon(t) \in L^2([0, t])$:

$$\left| c_{\varepsilon 1}(t)c_{\varepsilon 3}(t) - \frac{c_{\varepsilon 2}(t)^2}{2} \right| = \left| c_{\varepsilon 1}(0)c_{\varepsilon 3}(0) - \frac{c_{\varepsilon 2}(0)^2}{2} \right|. \tag{3.2.10}$$

Therefore, any $\psi(t) = \sum_{i=1}^{3} c_{\varepsilon i}(t)\psi_i(x)$ that is reachable from $\psi(0)$ with $\psi(0) = \sum_{i=1}^{3} c_{\varepsilon i}(0)\psi_i(x)$ must satisfy the constraint (3.2.10). Let us consider a simple numerical example: suppose that the initial state is the ground state ψ_1 and the target is the first excited state ψ_2. We obtain for ψ_1:

$$\left| c_{\varepsilon 1}(0)c_{\varepsilon 3}(0) - \frac{c_{\varepsilon 2}(0)^2}{2} \right| = \left| 1 \cdot 0 - \frac{0^2}{2} \right| = 0$$

and for ψ_2:

$$\left| c_{\varepsilon 1}(t)c_{\varepsilon 3}(t) - \frac{c_{\varepsilon 2}(t)^2}{2} \right| = \left| 0 \cdot 0 - \frac{1^2}{2} \right| = \frac{1}{2}.$$

Since the two quantities are different, ψ_2 *is not reachable from* ψ_1 and therefore the system is not controllable, despite the fact that the connectivity assumption is satisfied.

The detailed analysis of the case $N = 3$ and a result that was communicated to us (cf. RAMAKRISHNA [2000]) suggest the conjecture that for general finite-dimensional systems, as long as no conservation laws appear (besides L^2 norm conservation) the system is controllable. This statement, if true, would have the merit of giving a result only in terms of the physical properties of the system under consideration (and independent of the mathematical transcription of the precise control situation).

REMARK 3.20. For the density matrix formulation, kinematical constraints on the controllability of systems in mixed states have been established (e.g., GIRARDEAU, SCHIRMER, LEAHY and KOCH [1998], GIRARDEAU, INA, SCHIRMER and GULSRUD [1997]), based on the eigenvalues of $\rho(0)$ and those of the objective operator.

Independent systems controllability and discrimination issues

This section introduces controllability results for independent quantum systems. Most of the text draws from TURINICI, RAMAKHRISHNA, LI and RABITZ [2002].

Successful control may be expressed as a matter of high quality discrimination, whereby the control field steers the evolving quantum system dynamics out the desired channel, while diminishing competitive flux into other undesirable channels. A potential application of quantum control techniques is to the detection of specific molecules amongst others of similar chemical/physical characteristics. We call this procedure coherent molecular discrimination. Cases of special interest for discrimination include large polyatomic molecules of similar chemical nature, whose spectra can often mask each other.

Much work needs to be done to explore and develop the concept of coherent molecular discrimination, and a basic step in this direction is to establish the criteria for independently and simultaneously controlling the dynamics of several molecular species with the same control field. Discrimination of multiple molecules is a special case of the controllability concept, where the full system consists of a set of subsystems (i.e., molecules of different type). In the simplest circumstances, the molecules may be taken as independent and noninteracting, such that the initial state $\psi(0)$ is a product $\psi(0) = \prod_{\ell=1}^{L} \psi_\ell(0)$ of states $\psi_\ell(0)$ for each of the $L \geqslant 2$ molecular species. Full controllability would correspond to the ability to simultaneously and arbitrarily steer about each of initial states $\psi_\ell(0)$ to predefined targets $\psi_\ell(T) = \psi_\ell^{\text{target}}$ under the influence of a single control laser electric field $\varepsilon(t)$, where each molecule evolves under a separate Schrödinger equation

$$i\hbar\frac{\partial}{\partial t}\psi_\ell(t) = \left[H_0^\ell - \mu^\ell \cdot \varepsilon(t)\right]\psi_\ell(t). \tag{3.2.11}$$

Here, H_0^ℓ and μ^ℓ, respectively, are the free Hamiltonian and interaction operator (dipole) of the ℓth molecule. Other milder controllability criteria might also be specified.

The purpose of this section is to state theoretical criteria for the controllability of an ensemble of L separate quantum systems in the presence of a single electric field $\varepsilon(t)$. The criteria will refer to a *finite dimensional* setting where, for each ℓ, $1 \leqslant \ell \leqslant L$, the Hamiltonian H_0^ℓ and dipole operator μ^ℓ are expressed with respect to an eigenbasis of the internal Hamiltonians, as is often the case in applications. More precisely, let $D^\ell = \{\psi_i^\ell(x);\ i = 1, \ldots, N_\ell\}$ be the set of the first N_ℓ, $N_\ell \geqslant 3$ eigenstates of the possibly infinite dimensional Hamiltonian H_0^ℓ, and let A^ℓ and B^ℓ be the matrices of the operators $-iH_0^\ell$ and $-i\mu^\ell$ respectively, with respect to this base. In order to exclude a trivial loss of controllability, it is supposed that $[A^\ell, B^\ell] \neq 0$, $\ell = 1, \ldots, L$. From the definition of the basis D^ℓ and the fact that H_0^ℓ and μ^ℓ are Hermitian operators, it follows that each A^ℓ is diagonal with purely imaginary elements and each B^ℓ is skew-Hermitian. We will suppose moreover that

For any $\ell = 1, \ldots, L$: A^ℓ has nonzero trace and B^ℓ has zero trace. (3.2.12)

With this notation, the wavefunction of the ℓth system can be written as $\psi_\ell(t) = \sum_{i=1}^{N_\ell} c_i^\ell(t)\psi_i^\ell$. The total eigenfunction $\prod_{\ell=1}^{L}\psi_\ell(t)$ will be represented as a column vector $c(t) = (c_1^1(t), \ldots, c_{N_1}^1(t), \ldots, c_1^L(t), \ldots, c_{N_L}^L(t))^T$. Denote $N = \sum_{\ell=1}^{L} N_\ell$, A be the $N \times N$ skew-Hermitian block-diagonal matrix obtained from A^ℓ, $\ell = 1, \ldots, L$, and B be the skew-Hermitian block-diagonal matrix obtained from B^ℓ, $\ell = 1, \ldots, L$:

$$A = \begin{pmatrix} A^1 & 0 & \ldots & 0 \\ 0 & A^2 & \ldots & 0 \\ \vdots & \vdots & \ddots & \vdots \\ 0 & 0 & \ldots & A^L \end{pmatrix}; \qquad B = \begin{pmatrix} B^1 & 0 & \ldots & 0 \\ 0 & B^2 & \ldots & 0 \\ \vdots & \vdots & \ddots & \vdots \\ 0 & 0 & \ldots & B^L \end{pmatrix}. \tag{3.2.13}$$

With the “atomic units” convention $\hbar = 1$, the dynamical equations read:

$$\frac{d}{dt}c(t) = Ac(t) - \varepsilon(t)Bc(t), \quad c(0) = c_0. \tag{3.2.14}$$

Recalling that each individual wavefunction $\psi_\ell(t) = \sum_{i=1}^{N_\ell} c_i^\ell(t)\psi_i^\ell$ is L^2 normalized to one, we obtain:

$$\sum_{i=1}^{N_\ell}\left|c_i^\ell(t)\right|^2 = 1, \quad \forall t \geqslant 0,\ \forall \ell = 1, \ldots, L. \tag{3.2.15}$$

Let $\mathcal{S}_{\mathbb{C}}^{k-1}$ be the complex unit sphere of $\mathbb{C}^k$. Then Eq. (3.2.15) gives:

$$c(t) \in \mathcal{S} = \prod_{\ell=1}^{L} \mathcal{S}_{\mathbb{C}}^{N_\ell - 1}, \quad \forall t \geqslant 0. \tag{3.2.16}$$

Define the admissible control set $\mathcal{U}$ as the set of all piecewise continuous functions $\varepsilon(t)$. For every $\varepsilon \in \mathcal{U}$ Eq. (3.2.14) has an (unique) solution for all $t \geqslant 0$. The system $((A^\ell, B^\ell)_{\ell=1}^{L}, \mathcal{U})$ is said to be *controllable* if for any $c_i, c_f \in \mathcal{S}$ there exists an $t_f \geqslant 0$ (possibly depending on c_i, c_f) and $\varepsilon(t) \in \mathcal{U}$ such that the solution of (3.2.14) with initial data $c(0) = c_i$ satisfy $c(t_f) = c_f$. Of course, in order for the system $((A^\ell, B^\ell)_{\ell=1}^{L}, \mathcal{U})$ to be controllable each component system $(A^\ell, B^\ell, \mathcal{U})$, $\ell = 1, \ldots, N$, taken independently has to be controllable. However, requiring that all systems be controllable *at the same time and with the same laser field* is a more demanding condition. To illustrate this statement, we will consider the simple case of two ($L = 2$) three-level systems ($N_1 = 3$, $N_2 = 3$) of TURINICI, RAMAKHRISHNA, LI and RABITZ [2002]:

$$A^1 = A^2 = -i \cdot \begin{pmatrix} 1 & 0 & 0 \\ 0 & 2 & 0 \\ 0 & 0 & 5 \end{pmatrix},$$

$$B^1 = -i \cdot \begin{pmatrix} 0 & 1 & 0 \\ 1 & 0 & 2 \\ 0 & 2 & 0 \end{pmatrix}, \qquad B^2 = -B^1. \tag{3.2.17}$$

Each system A^1, B^1 and A^2, B^2 is controllable, as can be checked by the Lie algebra criterion of RAMAKRISHNA, SALAPAKA, DAHLEH, RABITZ and PIERCE [1995] (the dimension of the Lie algebra is found to be 9).

However, denoting by (c_1^1, c_2^1, c_3^1) and by (c_1^2, c_2^2, c_3^2) the coefficients of the wavefunction of the first system and of the second system respectively, we obtain the following dynamical invariant (conservation law):

$$L(t) = \overline{c_1^1(t)} c_1^2(t) + \overline{c_3^1(t)} c_3^2(t) - \overline{c_2^1(t)} c_2^2(t) = \text{constant}, \quad \forall \varepsilon \in \mathcal{U}. \tag{3.2.18}$$

The presence of this conservation law implies that the system is not controllable. For instance, starting with both systems in the ground state

$$\left(c_1^1(0), c_2^1(0), c_3^1(0)\right) = \left(c_1^2(0), c_2^2(0), c_3^2(0)\right) = (1, 0, 0)$$

one cannot steer both to their respective first excited state

$$\left(c_1^1(T), c_2^1(T), c_3^1(T)\right) = \left(c_1^2(T), c_2^2(T), c_3^2(T)\right) = (0, 1, 0)$$

since in the ground states the dynamical invariant takes the value $L(0) = 1 + 0 - 0 = 1$ while in the first excited states the value is $L(T) = 0 + 0 - 1 = -1$.

Defining the set of attainable states

$$\mathcal{A}(c_0, T) = \{c(t);\ c(t) \text{ solution of (3.2.14)},\ t \in [0, T],\ u \in \mathcal{U}\}, \tag{3.2.19}$$

the system is controllable if (and only if) for any $c_0 \in \mathcal{S}$ the set of points attainable from c_0: $\bigcup_{t \geqslant 0} \mathcal{A}(c_0, t)$ equals $\mathcal{S}$.

We are now ready to state the following controllability result:

THEOREM 3.21 (TURINICI, RAMAKHRISHNA, LI and RABITZ [2002]). *If the dimension (computed over the scalar field $\mathbb{R}$) of the Lie algebra $\mathcal{L}(A, B)$ generated by A and B equals $1 + \sum_{\ell=1}^{L}(N_\ell^2 - 1)$ then the system $((A^\ell, B^\ell)_{\ell=1}^{L}, \mathcal{U})$ is controllable. Moreover, when the system is controllable, there exists a time $T > 0$ such that all targets can be attained before or at time T, i.e., for any $c_0 \in \mathcal{S}$, $\mathcal{A}(c_0, T) = \mathcal{S}$.*

We refer to LI, TURINICI, RAMAKHRISHNA and RABITZ [2002], TURINICI, RAMAKHRISHNA, LI and RABITZ [2002] for more general results where the hypotheses of Eq. (3.2.12) are not satisfied or when multiple $s > 1$ external fields are considered (which can be expressed by introducing multiple dipole moment operators μ_i^ℓ, $i = 1, \dots, s$).

3.3. Truncations

Little is known about the relationship between the controllability of the finite dimensional systems and that of infinite dimensional systems. We will discuss in this section some of the interesting aspects of this interplay through a list of open problems and questions.

Consider a quantum system that is controllable when its (truncated) equations of motion are expressed with respect to a particular n-dimensional basis which spans a finite dimensional space $\mathcal{H}_n$. According to Theorem 3.2, for every initial condition there must emerge a dense set of unreachable states in the limit n tends to infinity (depicted in Fig. 3.2(a)), assuming that the limiting process is well defined. In other words, for $n \to \infty$ the system may become uncontrollable in the strict sense defined above. This limit suggests the question: how does the controllability of a sequence of finite but increasingly higher dimensional quantum systems relate to the controllability of the corresponding infinite-dimensional quantum system in the limit $n \to \infty$ (if this limiting process exists)? How are the sets of unreachable states that emerge in this limit characterized?

The analysis implied in the above questions can be subtle, as evident from a simple illustration involving the emergence or disappearance of unreachable states under finite increases in the dimensionality of $\mathcal{H}_n$. For example, in hydrogenic atoms the transitions due to emission or absorption of photons must satisfy the selection rules $\Delta l = \pm 1$ and $\Delta m = \pm 1$ or 0. If the step $\mathcal{H}_n \to \mathcal{H}_{n+1}$ of the limiting process adds a basis function to which there does not exist a sequence of allowed transitions from some function ψ_1 in $\mathcal{H}_n$, the additional dimension has caused a loss of system controllability. The converse situation may also arise where the additional basis function provides a "missing pathway" between states that were mutually unreachable in $\mathcal{H}_n$: in this case, the step $\mathcal{H}_n \to \mathcal{H}_{n+1}$ might cause an uncontrollable system to become controllable. It is an open question in quantum controllability to understand how such stepwise processes may be interpreted in the infinite limit.

Now consider the related issue of controllability within a "subspace of interest" $\mathcal{H}_I$ that is contained within $\mathcal{H}_n$ (as depicted in Fig. 3.2(b)). Let $\mathcal{H}_I$ be spanned by the first I elements of the set of basis functions $\{\psi_i\colon i = 1, \dots, n\}$ spanning $\mathcal{H}_n$. Definition 3.5 may be modified to restrict analysis to the subspace of interest: controllability will be taken to mean that a system is controllable if the system can be steered between any two states ψ_1^I and ψ_2^I in $S \cap \mathcal{H}_I$. Controllability may be described as stationary within $\mathcal{H}_I$ if it remains unchanged as individual dimensions are added in any order to $\mathcal{H}_n$ until (if it exists) the limit $\lim_{n\to\infty, n\geqslant I} \mathcal{H}_n = \mathcal{H}$, is obtained. It is not known what characteristics of the Hamiltonian H_0 , the dipole or other coupling coefficients, and the spaces $\mathcal{H}_I \subset \mathcal{H}_n \subset \mathcal{H}$ are required for stationary controllability within $\mathcal{H}_I$.

The discussion above does not address the effects on the evolution of states within the truncated space $\mathcal{H}_n$ arising from states that lie outside of $\mathcal{H}_n$. This consideration also has practical consequences. For example, suppose that controllability is satisfied within $\mathcal{H}_n$ or within $\mathcal{H}_I$ for some $\mathcal{H}_n$. A realizable laboratory control might inadvertently also access states lying outside of $\mathcal{H}_n$ which might even lift the controllability in the desired subspace. Techniques from optimal control theory would be the desirable way to handle the discovery of practical fields best satisfying the assumptions under an associated controllability analysis. Following upon the latter discussion, a new class of problems is introduced if a term is added to the Schrödinger equation to represent the interaction of the remainder states that are not explicitly modeled lying in $\mathcal{H}\backslash\mathcal{H}_n$. One such term introduced in BEUMEE and RABITZ [1992], cf. also SONTAG [1998], is a n-dimensional disturbance vector w:

$$\frac{\partial \psi}{\partial t} = \left[H_0 + C(t)\right]\psi + w,$$

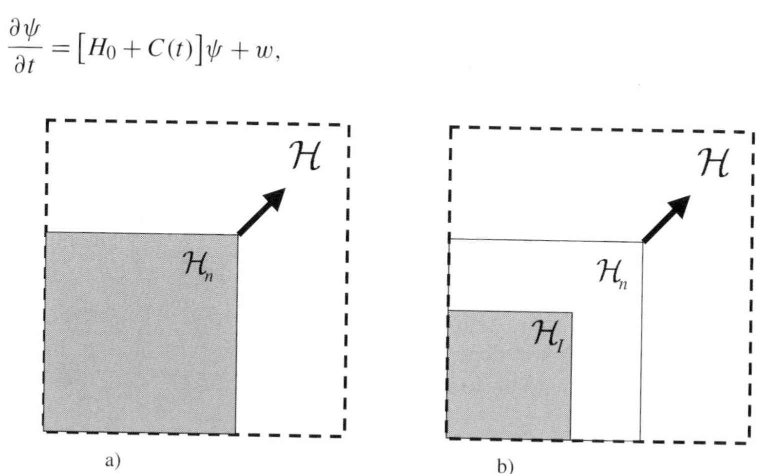

FIG. 3.2. Pictorial representation of the truncation problem. In both cases, controllability is of interest within the shaded subspace as the dimension of the truncated space H_n increases to infinity. Reprinted with permission from E. Brown and H. Rabitz, J. Math. Chem. 31, p. 23, 2002; copyright 2002, Kluwer Academic Publishers.

where $\psi \in \mathcal{H}_n$. The form and magnitude of the disturbance term w is problem-dependent and assumed to be given a priori, and generally leads to a nonunitary evolution for $\psi(t)$. In min–max optimal control theory (cf. Section 4.3), w is selected to maximize the disruptive effect of the energy-bounded disturbance (e.g., fluctuations in the laboratory environment and apparatus). In another general context, w could represent coupling to a bath external to the dynamics described by $H_0 + C(t)$. This type of coupling is important for considerations of dynamical cooling (see BARTANA and KOSLOFF [1993], BARTANA and KOSLOFF [1997], TANNOR and BARTANA [1999], TANNOR, KOSLOFF and BARTANA [1999]) in the analogous density matrix formulation. It is not yet clear what general models of dynamics exterior to $\mathcal{H}_n \subset \mathcal{H}$ can cause controllable systems to become uncontrollable and vice versa.

4. Quantum control algorithms

As analytical solutions to the quantum optimal control equations cannot generally be found, iterative numerical algorithms must be employed. The time-dependent Schrödinger equation in multiple spatial variables is computationally very expensive to solve, and, although there appear to be many opportunities to develop special control design approximations, and while some work has already been done (e.g., RICE and ZHAO [2000]), there is room for much development.

REMARK 4.1. Since the Schrödinger equation must be solved at least once (and generally many times) in most optimal control and Hamiltonian identification methods, the numerical evolution of quantum systems with many degrees of freedom must be approximated in some fashion before control design techniques can be applied. Along these lines, broad classes of quantum dynamics approximations have been developed, and in principle any of them could be applied to quantum control design. While attempting to attain designs, it is worth investigating the effects on controllability of replacing the Schrödinger equation with its various quantum dynamical approximations. Significant influence of a dynamical approximation upon a system's controllability could have serious consequences for the reliability of any resultant control designs based on the approximation.

A very important feature of quantum control is the intricate relationship between theory and the laboratory implementations, where an optimization method, usually an evolutionary (e.g., genetic) algorithm, is used to drive the experimental work which, in its turn, feeds back the optimization algorithm with necessary data. Reflecting this connection, both numerical and experimental paradigms are presented in this section, the first in Section 4.2 (extending with connected topics through Section 4.4) and the latter in Section 4.5. Common to both, the construction of the cost functionals is the object of the next section.

4.1. *Optimal control cost functional formulation*

As explained in Section 2.2, the first step in formulating the general quantum optimal control problem is to define a "cost functional" whose minimization represents the balanced achievement of control and possibly other objectives. This cost functional is given by

$$J = \sum_k J_k\bigl(C\bigl(\psi(t); t \in [0, T]\bigr)\bigr),$$

where the goal is to minimize (or maximize, as appropriately) J with respect to $C(t)$. While the specific form of the cost functional is flexible and problem-dependent, a term J_1 that addresses the achievement of the N_O optimal control objectives is always included:

$$J_1 = \sum_{j=1}^{N_O} \sum_{l=1}^{L_j} \begin{cases} \int_{\tau_j^l} W_{1,j}^l \bigl[\langle O_j(t)\rangle - \tilde{O}_j(t)\bigr]^\gamma \, dt, & \text{if } \tau_j^l \text{ is an interval,} \\ W_{1,j}^l \bigl[\langle O_j(\tau_j^l)\rangle - \tilde{O}_j(\tau_j^l)\bigr]^\gamma, & \text{if } \tau_j^l \text{ is a discrete time.} \end{cases} \tag{4.1.1}$$

Here the $W_{1,j}^l$ are positive design weights assigned to each of the objectives. The particular case $N_O = L_1 = \gamma = W_{1,1}^1 = 1$, $\tau_1^1 = T$ and $\tilde{O}_1(T) = 0$ has already been seen (cf. Eq. (2.2.1)) to give rise to a term of the form $J_1 = \langle \psi(T)|O_1|\psi(T)\rangle$.

For physically realistic control laws, the energy of the laboratory/molecular interaction must be bounded. This criteria is often included by adding the term

$$J_2 = \int_0^T W_2(t)\varepsilon(t)^2 \, dt \tag{4.1.2}$$

to the cost functional, which effectively limits the total electric field fluence. Here, $W_2(t)$ determines the time-dependent relative importance of minimizing the fluence. Note that the term J_2 does not prevent $\varepsilon(t)$ from being large in some small interval of time, although a cost on the local magnitude at any time could be introduced for this purpose.

Penalty terms may also be included, causing the minimization of the expectation values of $N_{O'}$ "undesirable" operators O_j' at the corresponding times $\tau_j'^l$:

$$J_3 = \sum_{j=1}^{N_O'} \sum_{l=1}^{L_j'} \begin{cases} \int_{\tau_j'^l} W_{3,j}^l \bigl|\langle O_j'(t)\rangle\bigr|^2 \, dt, & \text{if } \tau_j'^l \text{ is an interval,} \\ W_{3,j}^l \bigl|\langle O_j'(\tau_j'^l)\rangle\bigr|^2, & \text{if } \tau_j'^l \text{ is a discrete time.} \end{cases} \tag{4.1.3}$$

In addition to those explicitly given here, there are many other forms of J_k that could be incorporated into the cost functional. These terms could represent, for example, restrictions on the windowed Fourier transform of $\varepsilon(t)$ to a particular frequency band, minimization of sensitivity to small perturbations in the control law (as will be discussed below), or other characteristics of the desired optimal control solution.

One property of control law solutions $C(t)$ of practical import is simplicity. Several measures of simplicity could be used, such as the ability to decompose the control law into only a small number of spectral components with high accuracy (see GEREMIA, ZHU and RABITZ [2000]). However, the notion of field simplicity is best associated with the ease of stable and reliable generation in the laboratory, rather than any preconceived sense of simplicity associated with the presence of few field components. Design of simple control laws might be accomplished by introducing a term in the cost functional that favors solutions with suitable characteristics, or in a very ad hoc fashion by starting an iterative optimization algorithm with a simple control field and halting the process while some of this simplicity is still preserved but likely before complete convergence to the control objectives has been achieved. The latter suggestion follows from the observation that the final small fraction of progress toward the control objectives is often responsible for most of the complexity in the control field (e.g., see Table 6.1 of RICE and ZHAO [2000] or TERSIGNI, GASPARD and RICE [1990]). None of the above approaches has been subjected to a careful mathematical analysis, and further efforts to characterize the effects of these modifications on the optimal control process may be useful.

Thus far the terms in the cost functional have all been introduced to seek a control field that biases the objective or other goals in some specified direction. Under favorable circumstances one or more of these costs could be re-expressed as a hard demand by introducing a Lagrange multiplier. An example would be a requirement that the laser pulse energy be fixed at a specified laboratory accessible value; the reshaping would redistribute that energy as best as possible over a band of frequency components to meet the physical objective. Some absolute demands may lead to inconsistencies and resultant numerical design difficulties if the demand cannot be satisfied for some (often hidden) dynamical reason.

Once the cost functional has been defined, then the optimal control law is determined by minimizing the cost functional over the function space of admissible controls. Local or global, deterministic or stochastic optimization algorithms (e.g., gradient descent, genetic algorithms, etc.) may be used to find the minimum of the cost functional subject to satisfaction of the Schrödinger equation, possibly under suitable quantum dynamics approximations and assumptions (see KRAUSE and SCHAFER [1999], KRAUSE, REITZE, SANDERS, KUZNETSOV and STANTON [1998]). The existence of the minimum itself has been investigated in several works, including PEIRCE, DAHLEH and RABITZ [1988]; for a different formulation see CANCÈS, LE BRIS and MATHIEU [2000]. Alternatively, the Euler–Lagrange approach may be pursued, as explained below.

At the relevant minima of the cost functional, the first order variation with respect to the control law vanishes:

$$\frac{\delta J}{\delta C(t)} = 0. \tag{4.1.4}$$

Eq. (4.1.4) is subject to the dynamical constraint that $\psi(t)$ satisfies the Schrödinger equation; this may be assured through the introduction of a Lagrange multiplier function $\lambda(t)$ (cf. SONTAG [1998]). The resultant variational problem produces Euler–Lagrange equations whose solutions define the controls $C(t)$ operative at each local extrema of J. To demonstrate some of the characteristics of these equations, consider as an example a quantum optimal control problem in which there is only one objective operator O whose expectation value is to be optimized at the (single target) time T. The control law is taken to be $C(t) = -\mu\varepsilon(t)$ and the cost functional $J = J_1 + J_2$ consists of the terms given in Eqs. (4.1.1) and (4.1.2) with weights set such that J is given by Eq. (2.2.1). The extended cost functional that includes the Lagrange multiplier λ, which will be called from now on the *adjoint state*, is

$$J(\varepsilon, \psi, \lambda) = \langle \psi(T) | O | \psi(T) \rangle - \alpha \int_0^T \varepsilon^2(t)\, dt$$
$$- 2\,\mathrm{Re}\left[\int_0^T \left\langle \lambda(t) \middle| \frac{\partial}{\partial t} + i\left[H_0 - \mu\varepsilon(t)\right] \middle| \psi(t) \right\rangle dt \right]. \qquad (4.1.5)$$

With this definition we obtain first

$$\frac{\delta J(\varepsilon, \psi, \lambda)}{\delta \varepsilon} = -2\alpha\varepsilon(t) - 2\,\mathrm{Im}\{\langle \lambda(t) | \mu | \psi(t) \rangle\}. \qquad (4.1.6)$$

The derivative of J with respect to the adjoint state yields (as expected) the equation of motion for the wavefunction. In order to compute the derivative of J with respect to $\psi(t)$ one integrates by parts in the last term of $J(\varepsilon, \psi, \lambda)$ and identifies the resulting terms. The Euler–Lagrange equations then become:

$$\begin{cases} i\frac{\partial \psi(t)}{\partial t} = \left[H_0 - \mu\varepsilon(t)\right]\psi(t), \\ \psi(0) = \psi_0, \end{cases} \qquad (4.1.7)$$

$$\begin{cases} i\frac{\partial \lambda(t)}{\partial t} = \left[H_0 - \mu\varepsilon(t)\right]\lambda(t), \\ \lambda(T) = O\psi(T), \end{cases} \qquad (4.1.8)$$

$$\alpha\varepsilon(t) = -\,\mathrm{Im}\{\langle \lambda(t) | \mu | \psi(t) \rangle\}. \qquad (4.1.9)$$

If the field in Eq. (4.1.9) is substituted into Eqs. (4.1.7) and (4.1.8), the system becomes a pair of coupled nonlinear evolution equations.

REMARK 4.2. Under a mild set of assumptions the general quantum optimal control problem has been shown to possess a countable infinity of solutions, cf. DEMIRALP and RABITZ [1993]. This result has been obtained for cost functionals of the form $J = J_1 + J_2 + J_3$ having one objective operator O at a final time T and a single penalty operator O' evaluated over the entire control interval. In this work additional assumptions are that: (i) O and O' are bounded operators, (ii) O is either positive- or negative-definite, and (iii) $\mu \in (t\colon\ t \in [0, T])$ is bounded (although the proof can be extended for unbounded control terms).

4.2. *Numerical algorithms*

This section considers algorithms for determining the optimal controls based on the constrained variational problem in Eq. (4.1.4). Part of the text is drawn from the recent review ZHU and RABITZ [in press].

We will consider the quantum optimal control problem with the cost functional given in Eq. (2.2.1) and critical point Eqs. (4.1.7)–(4.1.9). The observable operator O is assumed to be positive semidefinite Hermitian. Note that Eq. (4.1.9) can also be written as

$$\begin{aligned}\varepsilon(t) &= \frac{1}{\alpha}\operatorname{Re}\big\langle\psi\big(T,\{\varepsilon\}\big)\big|O\big|\delta\psi\big(T,\{\varepsilon\}\big)/\delta\varepsilon(t)\big\rangle\\ &= -\frac{1}{\alpha}\operatorname{Im}\big\langle\psi\big(T,\{\varepsilon\}\big)\big|OU\big(T,t,\{\varepsilon\}\big)\mu\big|\psi\big(t,\{\varepsilon\}\big)\big\rangle\end{aligned}\tag{4.2.1}$$

with U being the evolution operator of the system. The presence of the argument ε on the right hand side of Eq. (4.2.1) indicates that this is actually an implicit relation for determining the control field. To solve for the control field in Eq. (4.1.9) or (4.2.1), it is evident that some type of iteration algorithm needs to be employed.

Several numerical algorithms can be used; historically, the gradient-type methods (see SHI, WOODY and RABITZ [1988], COMBARIZA, JUST, MANZ and PARAMONOV [1991]) were the first to be used. In this approach,

(1) an initial guess ε_1 is set and the iteration count k is initialized to $k = 1$;
(2) the wavefunction $\psi_k(t)$ is propagated forward in time with the field ε_k by Eq. (4.1.7);
(3) the final data for the adjoint state $\lambda_k(T)$ is derived;
(4) the wavefunction $\psi_k(t)$ and the adjoint state $\lambda_k(t)$ are propagated backward in time with the field ε_k and a new electric field $\varepsilon_{k+1} = \varepsilon_k + \gamma \cdot \frac{\delta J}{\delta \varepsilon_k}$ is computed ($\frac{\delta J}{\delta \varepsilon_k}$ is given by Eq. (4.1.6)). The constant γ is found by a linear search to optimize J in the direction of the gradient $\frac{\delta J}{\delta \varepsilon_k}$;
(5) $k \leftarrow k + 1$; step (2) is returned to and the cycle is continued until convergence.

Note that in step (4) the propagation of the wavefunction $\psi_k(t)$ is required because the storage of its trajectory (already computed in step (2)) is usually too expensive. We refer the reader to TERSIGNI, GASPARD and RICE [1990] for a conjugated gradient version of the above algorithm.

Although this algorithm proved useful in some cases, its convergence is not guaranteed (here the setting is far from the quadratic cost functional that conjugated gradient type algorithms best optimize) and it may become slow; the algorithm used by Tannor (see TANNOR, KAZAKOV and ORLOV [1992], SOMLÓI, KAZAKOV and TANNOR [1993]) based on the Krotov method (KROTOV [1973], KROTOV [1974a], KROTOV [1974b]) was designed to correct this feature. The structure of this algorithm is as follows:

(1) an initial guess ε_1 is set and the iteration count k is initialized to $k = 1$;
(2) the wavefunction $\psi_1(t)$ is propagated with the field ε_1 by Eq. (4.1.7);
(3) the final data for the adjoint state $\lambda_k(T)$ is derived;
(4) the adjoint state $\lambda_k(t)$ is propagated backward in time with the field ε_k to give $\lambda_{k0} = \lambda_k(0)$;

(5) A new field is constructed by the simultaneous resolution of the following equations:

$$\begin{cases} i\frac{\partial\psi_{k+1}(t)}{\partial t} = \left[H_0 + \frac{1}{\alpha}\mu\,\mathrm{Im}\{\langle\lambda_k(t)|\mu|\psi_{k+1}(t)\rangle\}\right]\psi_{k+1}(t), \\ \psi_{k+1}(0) = \psi_0, \end{cases} \tag{4.2.2}$$

$$\begin{cases} i\frac{\partial\lambda_{k+1}(t)}{\partial t} = \left[H_0 - \mu\varepsilon_{k+1}(t)\right]\lambda_{k+1}(t), \\ \lambda_{k+1}(T) = O\psi_{k+1}(T), \end{cases} \tag{4.2.3}$$

$$\alpha\varepsilon_{k+1}(t) = -\mathrm{Im}\{\langle\lambda_k(t)|\mu|\psi_{k+1}(t)\rangle\}; \tag{4.2.4}$$

(6) $k \leftarrow k+1$; step (3) is returned to and the cycle is continued until convergence.

As in step (4) of the gradient descent algorithm, the propagation in Eq. (4.2.3) is motivated by memory storage considerations.

In order to analyze the numerical properties of this algorithm we evaluate the difference between the value of the cost functional between two successive iterations:

$$\begin{aligned} J(\varepsilon_{k+1}) - J(\varepsilon_k) &= \langle\psi_{k+1}(T)|O|\psi_{k+1}(T)\rangle - \alpha\int_0^T \varepsilon_{k+1}(t)^2\,dt \\ &\quad - \langle\psi_k(T)|O|\psi_k(T)\rangle + \alpha\int_0^T \varepsilon_k(t)^2\,dt \\ &= \langle\psi_{k+1}(T) - \psi_k(T)|O|\psi_{k+1}(T) - \psi_k(T)\rangle \\ &\quad + 2\,\mathrm{Re}\langle\psi_{k+1}(T) - \psi_k(T)|O|\psi_k(T)\rangle \\ &\quad + \alpha\int_0^T (\varepsilon_{k+1} - \varepsilon_k)(t)^2\,dt \\ &\quad + 2\alpha\int_0^T (\varepsilon_k - \varepsilon_{k+1})(t)\varepsilon_{k+1}(t)\,dt. \end{aligned} \tag{4.2.5}$$

Since we have also:

$$\begin{aligned} &2\,\mathrm{Re}\langle\psi_{k+1}(T) - \psi_k(T)|O|\psi_k(T)\rangle = 2\,\mathrm{Re}\langle\psi_{k+1}(T) - \psi_k(T), O\psi_k(T)\rangle \\ &\quad = 2\,\mathrm{Re}\langle\psi_{k+1}(T) - \psi_k(T), \lambda_k(T)\rangle \\ &\quad = 2\,\mathrm{Re}\int_0^T \left\langle\frac{\partial(\psi_{k+1}(t) - \psi_k(t))}{\partial t}, \lambda_k(t)\right\rangle + \left\langle\psi_{k+1}(t) - \psi_k(t), \frac{\partial\lambda_k(t)}{\partial t}\right\rangle dt \\ &\quad = 2\,\mathrm{Re}\int_0^T \left\langle\frac{H_0 - \mu\varepsilon_{k+1}}{i}\psi_{k+1}(t) - \frac{H_0 - \mu\varepsilon_k}{i}\psi_k(t), \lambda_k(t)\right\rangle \\ &\qquad + \left\langle\psi_{k+1}(t) - \psi_k(t), \frac{H_0 - \mu\varepsilon_k}{i}\lambda_k(t)\right\rangle = 2\,\mathrm{Re}\int_0^T \varepsilon_{k+1}\left\langle\frac{-\mu}{i}\psi_{k+1}(t), \lambda_k(t)\right\rangle \end{aligned}$$

$$
\begin{aligned}
&- \varepsilon_k\left\langle \frac{-\mu}{i}\psi_k(t), \lambda_k(t)\right\rangle + \varepsilon_k\left\langle \psi_{k+1}(t) - \psi_k(t), \frac{-\mu}{i}\lambda_k(t)\right\rangle \\
&= 2\int_0^T \varepsilon_{k+1}\cdot\alpha\varepsilon_{k+1} - \varepsilon_k\cdot\left\langle \frac{-\mu}{i}\psi_k(t), \lambda_k(t)\right\rangle - \varepsilon_k\cdot\alpha\varepsilon_{k+1} \\
&- \varepsilon_k\cdot\left\langle \psi_k(t), \frac{-\mu}{i}\lambda_k(t)\right\rangle = 2\alpha\int_0^T \varepsilon_{k+1}(t)\cdot(\varepsilon_{k+1}-\varepsilon_k)(t)\,dt
\end{aligned}
\tag{4.2.6}
$$

we obtain thus from (4.2.5) and (4.2.6)

$$
\begin{aligned}
J(\varepsilon_{k+1}) - J(\varepsilon_k) &= \langle \psi_{k+1}(T) - \psi_k(T)|O|\psi_{k+1}(T) - \psi_k(T)\rangle \\
&\quad + \alpha\int_0^T (\varepsilon_{k+1}-\varepsilon_k)(t)^2\,dt \geqslant 0
\end{aligned}
\tag{4.2.7}
$$

because the observable O is a positive semidefinite operator. Each step of this algorithm will therefore result in an increase of the value of the cost functional; this increase is expected to be important for initial steps where the critical point equations are not fulfilled and the difference between successive fields ε_k and ε_{k+1} will be important.

This algorithm was improved with the introduction of the monotonic quadratically convergent algorithm in ZHU, BOTINA and RABITZ [1998], ZHU and RABITZ [1998] that incorporates additional feedback from the adjoint state. The following material will present some aspects of the monotonically convergent algorithms and will analyze their convergence features. In the first step, a trial field $\varepsilon^{(0)}(t)$ is used to calculate the wavefunction $\psi(t, \{\varepsilon^{(0)}\})$, then Eq. (4.2.1) can be directly applied to obtain the first iteration to the control field $\varepsilon^{(1)}(t)$ by backward propagation of the evolution operator $U(T, t, \{\varepsilon^{(1)}\})$ from T to 0. In the second step, the new control field $\varepsilon^{(2)}(t)$ also can be directly obtained by forward propagation of the wavefunction $\psi(t, \{\varepsilon^{(2)}\})$ from 0 to T. A series of control fields can be repeatedly mapped out in this way. The algorithm as an iteration sequence has the following structure:

$$
\varepsilon^{(1)}(t) = -\frac{1}{\alpha}\operatorname{Im}\langle \psi(T, \{\varepsilon^{(0)}\})|OU(T, t, \{\varepsilon^{(1)}\})\mu|\psi(t, \{\varepsilon^{(0)}\})\rangle, \tag{4.2.8}
$$

$$
\varepsilon^{(2)}(t) = -\frac{1}{\alpha}\operatorname{Im}\langle \psi(T, \{\varepsilon^{(0)}\})|OU(T, t, \{\varepsilon^{(1)}\})\mu|\psi(t, \{\varepsilon^{(2)}\})\rangle, \tag{4.2.9}
$$

$$
\vdots
$$

$$
\varepsilon^{(2i+1)}(t) = -\frac{1}{\alpha}\operatorname{Im}\langle \psi(T, \{\varepsilon^{(2i)}\})|OU(T, t, \{\varepsilon^{(2i+1)}\})\mu|\psi(t, \{\varepsilon^{(2i)}\})\rangle, \tag{4.2.10}
$$

$$
\varepsilon^{(2i+2)}(t) = -\frac{1}{\alpha}\operatorname{Im}\langle \psi(T, \{\varepsilon^{(2i)}\})|OU(T, t, \{\varepsilon^{(2i+1)}\})\mu|\psi(t, \{\varepsilon^{(2i+2)}\})\rangle. \tag{4.2.11}
$$

The introduction of the adjoint state λ allows for rewriting the above algorithm in the simplified form

(1) an initial guess ε_1 is set and the iteration count k is initialized to $k = 1$;
(2) the wavefunction $\psi_0(t)$ is propagated with the field ε_1 by Eq. (4.1.7);
(3) the final data for the adjoint state $\lambda_k(T)$ is derived;
(4) the adjoint state $\lambda_k(t)$ is propagated backward in time

$$\begin{cases} i\frac{\partial \lambda_k(t)}{\partial t} = \left[H_0 + \frac{1}{\alpha}\mu \operatorname{Im}\langle \lambda_k(t)|\mu|\psi_{k-1}(t)\rangle\right]\lambda_k(t), \\ \lambda_k(T) = O\psi_{k-1}(T). \end{cases} \tag{4.2.12}$$

Note: if the trajectory of the wavefunction $\psi_{k-1}(t)$ cannot be stored into memory, it is recomputed from the (stored) corresponding field.

(5) the wavefunction is evolved and a new field is constructed by the solution of the following equations:

$$\begin{cases} i\frac{\partial \psi_k(t)}{\partial t} = \left[H_0 + \frac{1}{\alpha}\mu \operatorname{Im}\langle \lambda_k(t)|\mu|\psi_k(t)\rangle\right]\psi_k(t), \\ \psi_k(0) = \psi_0, \end{cases} \tag{4.2.13}$$

$$\alpha\varepsilon_{k+1}(t) = -\operatorname{Im}\{\langle \lambda_k(t)|\mu|\psi_k(t)\rangle\}; \tag{4.2.14}$$

As in step (4), if needed, the trajectory of the adjoint state $\lambda_k(t)$ is recomputed.

(6) $k \leftarrow k + 1$; step (3) is returned to and the cycle is continued until convergence.

In order to analyze the convergence features of the above iteration sequence, we go back to the formulation in Eqs. (4.2.10) and (4.2.11) and consider the deviation of the objective functional between two neighboring steps (one step is defined as a pair of backward and forward propagations). Specifically, between the ith and $(i + 1)$st steps, the deviation is

$$\begin{aligned} \Delta J_{i+1,i} = {} & \left\{ \langle \psi(T, \{\varepsilon^{(2i+2)}\})|O|\psi(T, \{\varepsilon^{(2i+2)}\})\rangle - \alpha \int_0^T \left[\varepsilon^{(2i+2)}(t)\right]^2 dt \right\} \\ & - \left\{ \langle \psi(T, \{\varepsilon^{(2i)}\})|O|\psi(T, \{\varepsilon^{(2i)}\})\rangle - \alpha \int_0^T \left[\varepsilon^{(2i)}(t)\right]^2 dt \right\}. \end{aligned} \tag{4.2.15}$$

Considering the dynamical equations of the wavefunctions $\psi(t, \{\varepsilon^{(2i+2)}\})$ and $\psi(t, \{\varepsilon^{(2i)}\})$, and utilizing the field expressions in Eqs. (4.2.10) and (4.2.11), it can be derived (see ZHU and RABITZ [1998]) that:

$$\begin{aligned} \Delta J_{i+1,i} = {} & \alpha \int_0^T \left(\left[\varepsilon^{(2i+2)}(t) - \varepsilon^{(2i+1)}(t)\right]^2 + \left[\varepsilon^{(2i+1)}(t) - \varepsilon^{(2i)}(t)\right]^2 \right) dt \\ & + \langle \Delta\psi_{i+1,i}(T)|O|\Delta\psi_{i+1,i}(T)\rangle \end{aligned} \tag{4.2.16}$$

where $\Delta\psi_{i+1,i}(t) \equiv \psi(t, \{\varepsilon^{(2i+2)}\}) - \psi(t, \{\varepsilon^{(2i)}\})$. Since O is a positive semidefinite operator,

$$\langle \Delta\psi_{i+1,i}(T)|O|\Delta\psi_{i+1,i}(T)\rangle \geqslant 0, \tag{4.2.17}$$

then Eq. (4.2.16) satisfies

$$\Delta J_{i+1,i} \geqslant 0, \tag{4.2.18}$$

where the equal sign occurs only if the initial guess for the control field happens to be an exact solution of Eq. (4.2.1). Eq. (4.2.18) indicates that regardless of the choice for the initial trial field, the objective functional will monotonically converge to a local maximum of J for the iterated sequence of control fields given in Eqs. (4.2.8)–(4.2.11).

It is straightforward to show that the total gain for the objective functional after N iteration steps will be

$$\Delta J_{N,0} = \sum_{i=0}^{N-1} \langle \Delta\psi_{i+1,i}(T) | O | \Delta\psi_{i+1,i}(T) \rangle + \alpha \int_0^T \sum_{i=0}^{2N-1} \left[\Delta\varepsilon_{i+1,i}(t) \right]^2 dt, \tag{4.2.19}$$

where $\Delta\varepsilon_{i+1,i}(t) \equiv \varepsilon^{(i+1)}(t) - \varepsilon^{(i)}(t)$. Based on the above analysis, it can be concluded that a larger change of the field between neighboring steps will lead to faster convergence. As the initial guessed field usually will be far from the exact solution, the major contribution to the rapid convergence is expected to come from the first few iteration steps. An illustration of the monotonic convergence is shown in Fig. 4.1.

It is worthwhile to point out that the positive semidefiniteness of the operator O is not an intrinsic constraint on the monotonically convergent iteration algorithms: other types of monotonically convergent algorithms can lift this constraint. In the following, we will show one such procedure which is still monotonically convergent, but without the need to impose the positive semidefiniteness constraint on the operator. In the revised algorithm, the iteration sequence for the control field is slightly different from that shown in Eqs. (4.2.8)–(4.2.11). First, an initial guess for the field $\varepsilon^{(0)}(t)$ is needed to forward propagate the wavefunction $\psi(t, \{\varepsilon^{(0)}\})$. The modification of the algorithm just lies in the backward propagation. Specifically, the iteration sequence for determining the control field is (ZHU and RABITZ [1999a])

$$\varepsilon^{(1)}(t) = -\frac{1}{\alpha} \operatorname{Im}\langle \psi(t, \{\varepsilon^{(0)}\}) | U^\dagger(T, t, \{\varepsilon^{(1)}\}) O U(T, t, \{\varepsilon^{(1)}\}) \mu \times | \psi(t, \{\varepsilon^{(0)}\}) \rangle, \tag{4.2.20}$$

$$\varepsilon^{(2)}(t) = -\frac{1}{\alpha} \operatorname{Im}\langle \psi(t, \{\varepsilon^{(2)}\}) | U^\dagger(T, t, \{\varepsilon^{(1)}\}) O U(T, t, \{\varepsilon^{(1)}\}) \mu \times | \psi(t, \{\varepsilon^{(2)}\}) \rangle, \tag{4.2.21}$$

$$\vdots$$

$$\varepsilon^{(2i+1)}(t) = -\frac{1}{\alpha} \operatorname{Im}\langle \psi(t, \{\varepsilon^{(2i)}\}) | U^\dagger(T, t, \{\varepsilon^{(2i+1)}\}) O U(T, t, \{\varepsilon^{(2i+1)}\}) \mu$$

$$\times \left|\psi\left(t, \left\{\varepsilon^{(2i)}\right\}\right)\right\rangle, \tag{4.2.22}$$

$$\varepsilon^{(2i+2)}(t) = -\frac{1}{\alpha} \operatorname{Im}\left\langle\psi\left(t, \left\{\varepsilon^{(2i+2)}\right\}\right)\right|U^{\dagger}\left(T, t, \left\{\varepsilon^{(2i+1)}\right\}\right) O U\left(T, t, \left\{\varepsilon^{(2i+1)}\right\}\right)\mu \times \left|\psi\left(t, \left\{\varepsilon^{(2i+2)}\right\}\right)\right\rangle, \tag{4.2.23}$$

$$\vdots$$

REMARK 4.3. This second algorithm does not lend itself to an implementation in terms of direct and adjoint state only. The propagation of the wavefunction $\psi(t)$ and of the evolution operator $U(T, t, \{\varepsilon\})$ that is required is more costly than the propagation of $\psi(t)$ and $\lambda(t)$ of the previous monotonic convergent algorithm version because, in discrete form, it involves propagating a matrix and a vector as opposed to propagating two vectors.

To verify the monotonically convergent feature of the above algorithm, once again we evaluate the deviation of the objective functional between two neighboring steps as follows:

$$\Delta J_{i+1,i} = \left\{\left\langle\psi\left(T, \left\{\varepsilon^{(2i+2)}\right\}\right)\right|O\left|\psi\left(T, \left\{\varepsilon^{(2i+2)}\right\}\right)\right\rangle - \alpha \int_0^T \left[\varepsilon^{(2i+2)}(t)\right]^2 dt\right\} - \left\{\left\langle\psi\left(T, \left\{\varepsilon^{(2i)}\right\}\right)\right|O\left|\psi\left(T, \left\{\varepsilon^{(2i)}\right\}\right)\right\rangle - \alpha \int_0^T \left[\varepsilon^{(2i)}(t)\right]^2 dt\right\}. \tag{4.2.24}$$

Considering the dynamical equations of the wavefunction $\psi(t, \{\varepsilon^{(2i+2)}\})$ and the operator $U(T, t, \{\varepsilon^{(2i+3)}\})$, and utilizing the field expressions in Eqs. (4.2.22) and (4.2.23), it can be proven (e.g., ZHU and RABITZ [1999a]) that:

$$\Delta J_{i+1,i} = \alpha \int_0^T \left(\left[\varepsilon^{(2i+2)}(t) - \varepsilon^{(2i+1)}(t)\right]^2 + \left[\varepsilon^{(2i+1)}(t) - \varepsilon^{(2i)}(t)\right]^2\right) dt. \tag{4.2.25}$$

Comparing Eq. (4.2.25) with Eq. (4.2.16), we see that the iteration algorithm given by Eqs. (4.2.20)–(4.2.23) is still monotonically convergent for the objective functional, but without the extra constraint on the operator O. The total monotonic gain of the objective functional after N iteration steps is simply

$$\Delta J_{N,0} = \alpha \int_0^T \sum_{i=0}^{2N-1} \left[\Delta\varepsilon_{i+1,i}(t)\right]^2 dt. \tag{4.2.26}$$

We will not go further here to explore other possible monotonically convergent algorithms to iteratively solve for optimal controls. It is important to keep in mind that monotonically convergent iteration algorithms may not exist for arbitrary objective functionals. However, extended work indicates that there exist monotonically convergent iteration algorithms for most common types of the objective functionals

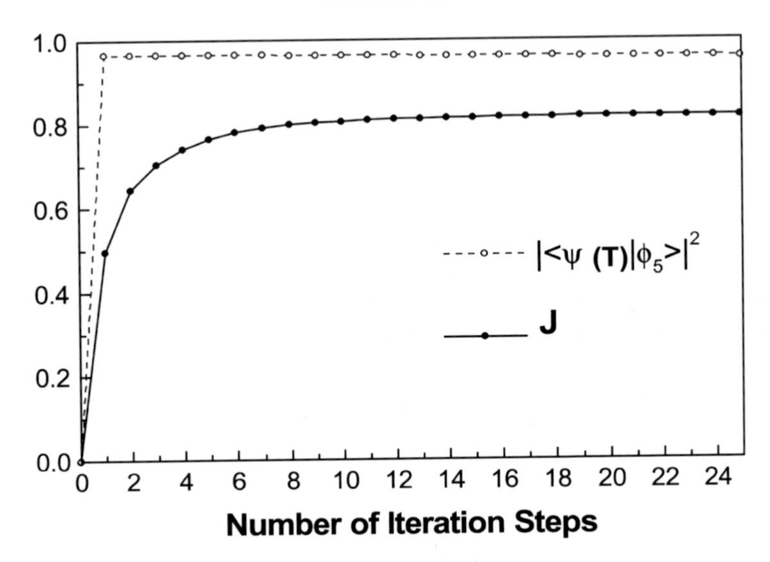

FIG. 4.1. Convergence of the cost functional to a (locally) optimal value in a numerical implementation of the quadratically convergent algorithm discussed in Section 4.2. The details of the calculation, which designs a minimum-fluence laser field to promote the $|3\rangle \to |5\rangle$ vibrational transition in an O–H bond, are given in ZHU and RABITZ [in press]. (Int. J. Quant. Chem., © copyright 2002 John Wiley & Sons Inc.)

considered in the optimal control of quantum systems (ZHU, BOTINA and RABITZ [1998], SCHIRMER, GIRARDEAU and LEAHY [2000]), and that methods similar to those presented here may be applied to the density matrix formulation (OHTSUKI, ZHU and RABITZ [1999]).

REMARK 4.4. Besides these families of algorithms that aim to solve the critical point equations directly, an alternative approach has been proposed in SHEN, DUSSAULT and BANDRAUK [1994] that uses a penalization framework. Suppose that the objective can be expressed as $\langle\psi(T)|P|\psi(T)\rangle = 1$, where P is a projection operator; a sequence of cost functionals

$$J(\varepsilon,\gamma) = \int_0^T \varepsilon^2(t)\,dt + \frac{1}{2\gamma}\big\|\langle\psi(T)\big|P\big|\psi(T)\rangle - 1\big\|^2, \tag{4.2.27}$$

(where $\psi(t)$ is the solution of Eq. (4.1.7)) is optimized with respect to ε for $\gamma \to 0$ by an inexact Newton method. The sequence of solutions of these optimization problems converges then to the solution of the initial control problem.

REMARK 4.5. The numerical resolution of the evolution equations as in Eqs. (4.1.7), (4.1.8), (4.2.2), (4.2.3), (4.2.12) or (4.2.13) requires a propagation scheme. Often used is the split-operator technique (e.g., ZHU and RABITZ [1998]) which can be written schematically: suppose that the equation to be solved is

$$i\frac{\partial}{\partial t}\chi(t) = \big(K + V(t)\big)\chi(t), \tag{4.2.28}$$

where K is the kinetic energy operator and $V(t)$ is the total potential. Then, denoting by Δt the time step, the following recurrence is used:

$$\chi(t+\Delta t) = e^{-iK\Delta t/2} e^{-iV(t)\Delta t} e^{-iK\Delta t/2} \chi(t) \tag{4.2.29}$$

which is known to be exact to second order in Δt. In order to apply the operators $e^{-iK\Delta t/2}$ and $e^{-iV(t)\Delta t}$ a dual *real space* $\leftrightarrow$ *Fourier*(*momentum*) representation is used; note that $V(t)$ is diagonal in real space while K is diagonal in momentum space; each operator is thus applied efficiently, with the transformation from one representation to the other realized by (fast) Fourier transforms.

4.3. Robust designs

Due to imperfect knowledge of system Hamiltonians and coupling operators as well as the limited precision and presence of background fluctuations inherent to any laboratory apparatus, it is impossible to perfectly reproduce either optimally designed control laws or the exact specifications under which they were designed. Hence, it is important to study the sensitivity of the control objective or cost functional to random variations or uncertainties in the operators and initial conditions describing the evolution of the system. There is extensive work on the general topic of robust optimal control in the engineering (DORATO [1987], HOSOE [1991], ACKERMAN [1985]) and quantum control literatures (DEMIRALP and RABITZ [1998], BEUMEE, SCHWIETERS and RABITZ [1990], ZHANG and RABITZ [1994]).

A general approach to assessing robustness and stability in quantum control has been considered in DEMIRALP and RABITZ [1998] based on introducing a stability operator S, the kernel of which is related to the curvature $\frac{\delta^2 J[\varepsilon]}{\delta\varepsilon(t)\delta\varepsilon(t')}$ of the cost functional with respect to the control law. Considering the curvature is necessary as the null value of the first order variation $\frac{\delta J[\varepsilon]}{\delta\varepsilon(t)} = 0$ defines the optimal solution. Conditions for robustness and optimality of the control solutions can be expressed in terms of the spectrum of S, and this analysis can also reveal qualitative relationships between the various terms in the cost functional and the robustness/optimality features of the control solutions. Work is still needed in order to find a general relationship between the dominant characteristics of a system Hamiltonian, coupling operators, and the cost functional with respect to the eigenvalues of the stability operator S.

The introduction of a penalty term of the form

$$J_3 = \int_0^T W_3 \left|\langle\psi(t)|O'|\psi(t)\rangle\right|^2 dt, \tag{4.3.1}$$

where O' is an arbitrary positive definite operator, was observed (DEMIRALP and RABITZ [1998]) to improve the robustness of optimal control solutions. The presence of J_3 can bias the system to satisfy demands tangential to the true control objectives, causing an effective "drag" along the way to the goal. Hence, the effect of J_3 may be loosely interpreted as analogous to the presence of viscous drag in stabilizing a classical mechanical system about a weakly stable point in its phase space. However, the possible stabilization mechanisms have not been carefully studied or characterized and a complete mechanism to explain how the introduction of suitable ancillary objectives may stabilize the solutions to quantum optimal control problems remains to be found.

The robustness effects of penalty operators with more specific forms than that given by Eq. (4.3.1) may be easier to intuit. For example, the term

$$J_s = \int_0^T dt \left(\frac{\delta \langle O(T) \rangle}{\delta \varepsilon(t)} \right)^2$$

(or analogous expressions with higher derivatives) may be used (KOBAYASHI [1998]) to reduce the sensitivity of the achieved control objectives at the target time T to uncertainty in control fields. Analogs of this penalty term for the sensitivity of the target objective to uncertainty in other variables were found (see BEUMEE, SCHWIETERS and RABITZ [1990]) to be capable of reducing the sensitivity to errors in force constants and other model parameters.

Design of robust quantum optimal control solutions can be achieved through the min–max procedure, which involves simultaneously maximizing the effects of an energy-bounded disturbance and minimizing the objective functional. Solutions to such min–max problems represent the best possible control under the worst possible energy-bounded disturbances. For linear dynamical systems the min–max problem becomes H_∞ control, which has an exact solution through the Ricatti equations. This procedure is well-developed in engineering control theory (see, e.g., ATHANS and FALB [1966]), and it has been applied to robust control designs for selective vibrational excitation in molecular harmonic oscillators (BEUMEE and RABITZ [1992]).

REMARK 4.6. In general, the min–max technique tends to give conservative robust solutions as it works against the worst possible bounded disturbance, and encountering this worst disturbance in practice is unlikely. This point suggests that consideration of a less extreme class of disturbances may also give useful solutions. The resulting analysis should give designs that are robust under more realistic conditions than those modeled in a worst case scenario.

The conclusions of min–max studies (ZHANG and RABITZ [1994]) reinforce the importance of this remark. For a diatomic molecule modeled as a Morse oscillator, the robustness properties of solutions to the min–max equations were compared with solutions to the standard Euler–Lagrange equations (cf. Eqs. (4.1.7)–(4.1.9)) derived without any robustness considerations. While the min–max controls performed better under the application of the worst-possible disturbance (for which they were designed), they did not necessarily outperform the standard Euler–Lagrange solutions under disturbances other than the worst case. For example, min–max control fields were demonstrated to be significantly less-robust than standard Euler–Lagrange control fields to sinusoidal perturbations with the same amplitude constraints as the worst-case disturbance. This underscores the importance of designing control laws that are robust to the particular class of disturbances most likely to occur.

Even in cases where the robustness properties of two designs are quite distinct, simulations have shown (see BEUMEE and RABITZ [1992]) that robust control designs may differ only slightly from nonrobust designs (i.e., the L^2 norm of the difference between the two control laws may be small). This similarity suggests that robustness properties in some cases may result from very subtle effects. It has also been noted (BEUMEE and RABITZ [1992]) that the relationship between the robust field and the standard design (i.e., created without robustness considerations) can take two forms: the robust field can be either a scaled, self-similar version of the standard field (which may be described as achieving robustness by "speaking louder") or can have a qualitatively different form. At present, no means exists to predict in general when either of these two cases will occur. It is suggestive that self-similar robust fields will exist for weak disturbances, but there is presently no proof of this conjecture.

4.4. Tracking

As mentioned in Remark 4.2 of Section 4, there generally exist a multiplicity of solutions to the quantum optimal control equations, suggesting that it may be possible to predefine a selected path between the initial and final conditions satisfying the control objectives. The existence of such a path exactly matching the conditions at both ends assumes that the system is controllable. The path $y(t)$ can be implicitly defined by the expectation value of a tracking operator O_{tr}:

$$y(t) = \langle \psi(t) | O_{\mathrm{tr}} | \psi(t) \rangle, \quad t \in [0, T]. \tag{4.4.1}$$

The quantum tracking control problem (e.g., GROSS, SINGH, RABITZ, HUANG and MEASE [1993], LU and RABITZ [1995], OHTSUKI, KONO and FUJIMURA [1998], ONG, HUANG, TARN and CLARK [1984]) may be viewed as a special case of optimal control theory with the target being the expectation value of O_{tr} over the entire time interval. (In some cases it may be physically attractive to only require that $\lim_{t\to T} O_{\mathrm{tr}}(t) = O$, where O is the objective operator whose expectation value is desired at T.) Given the path defined in Eq. (4.4.1), the tracking algorithm for determining the control law may be derived from the Heisenberg equation of motion

$$i\hbar \frac{d\langle \psi(t)|O_{\mathrm{tr}}|\psi(t)\rangle}{dt} = \langle \psi(t) | [H, O_{\mathrm{tr}}] | \psi(t) \rangle + \left\langle \psi(t) \left| \frac{\partial O_{\mathrm{tr}}}{\partial t} \right| \psi(t) \right\rangle. \tag{4.4.2}$$

With a control law of the form $C(t) = -\mu\varepsilon(t)$ and the assumptions that O_{tr} is independent of time along with

$$[\mu, O_{\mathrm{tr}}] \neq 0,$$

Eq. (4.4.2) can be rewritten to solve for the electric field:

$$\varepsilon(t) = \left(i\hbar \frac{dy}{dt} - \langle \psi(t) | [H_0, O_{\mathrm{tr}}] | \psi(t) \rangle \right) \Big/ \langle \psi(t) | [\mu, O_{\mathrm{tr}}] | \psi(t) \rangle. \tag{4.4.3}$$

This equation may be substituted into the Schrödinger equation, which then can be numerically solved for $\psi(t)$; substituting $\psi(t)$ back into Eq. (4.4.3) gives an explicit expression for the required control law. One important feature of this technique is that it requires only a single numerical solution of the Schrödinger equation, as opposed to the iterative methods of standard optimal control.

Given the freedom in the selection of $y(t)$, one might unknowingly choose a track that generates one or more singularities, or events at which the denominator of the control field in Eq. (4.4.3) vanishes. This type of singularity may be classified as trivial (see ZHU, SMIT and RABITZ [1999]) if it exists for all $t \in [0, T]$. Trivial singularities may be removed by formulating a tracking equation analogous to Eq. (4.4.1) for control of the kth time-derivatives of $y(t)$. A rank index may be assigned to each tracking singularity by determining the smallest order k_r for which the corresponding tracking equation has no trivial singularity; if the rank index is infinite, then the track-system pair is uncontrollable. Otherwise, any remaining (isolated) singularities may be treated as nontrivial singularities of some relative order k_{nt} (as in ZHU, SMIT and RABITZ [1999]). The magnitude of the disturbance to the trajectory resulting from a nontrivial singularity depends inversely on the magnitude of the derivatives $\frac{\partial^i y}{\partial t^i}$, $i = 1, \dots, k_{nt}$, evaluated at the singularity. This partially explains the effects of singularities on quantum tracking control and encourages the search for a noniterative algorithm to sense the occurrence of a forthcoming singularity and accordingly alter the path to avoid the momentary singularity while eventually reaching the objective.

Several extensions of exact inverse tracking which relax demands that could otherwise produce physically unreasonable fields have been developed in CHEN, GROSS, RAMAKRISHNA, RABITZ and MEASE [1995]. The first of these methods is local track generation, in which the problems associated with an a priori trajectory design are avoided by letting the track depend on the evolving quantum state: $y(t) = y(\psi(t))$. This approach is especially useful when the control objectives are not specifically defined by target operator expectation values as in Eq. (2.1.1), but rather can be expressed as the production of some qualitative change in a system. A second method is asymptotic tracking, in which the operator O_{tr} is modified to allow an asymptotic approach to possibly singular trajectories. Finally, in the competitive tracking technique a cost functional is defined whose minimization produces a solution optimally matching a number of trajectories for different tracking operators as well as minimizing the field fluence or satisfying other control objectives.

There is considerable room for further development of the tracking procedure guided by the attraction of performing only one solution of the Schrödinger equation to achieve a control design. Moreover, thus far tracking control has only been applied to the wavefunction formulation of quantum mechanics. A significant extension would be to treat mixed states in the density matrix formulation. In this context, the expectation value $\langle O_{\mathrm{tr}}(t)\rangle = \mathrm{Tr}(\rho(t) O_{\mathrm{tr}})$ would be followed and the Schrödinger equation would be replaced by Eq. (2.4.1), with the possibility of additionally including decoherence processes.

4.5. *Laboratory achievement of closed loop control*

The design of control laws poses interesting theoretical challenges, and the practical motivation for such a task is to accomplish successful control in the laboratory. This section discusses the conceptual and theoretical aspects of laboratory operations in which information about the evolving quantum systems is used to improve or define effective control laws. We will cover the technique of quantum learning control, which is increasingly proving to be the most efficient method of practically achieving many control objectives, especially in complex quantum systems; we will also discuss aspects of feedback quantum control. Learning and feedback control are closed loop experimental procedures aimed at achieving control even in the presence of Hamiltonian uncertainties and laboratory disturbances.

The computational design of a control law to meet a physical objective requires (i) explicit knowledge of the system Hamiltonian and (ii) the ability to numerically solve the quantum control equations at least once (in the case of tracking control) or many times for convergence to an optimal solution. In practice, however, these requirements can rarely be met. If the system to be controlled is sufficiently complex (e.g., a polyatomic molecule), it is likely that the Hamiltonian will be only approximately known and the corresponding quantum design equations can only be solved under serious approximations.

In light of these limitations, a completely different and practical approach to the control of quantum dynamics phenomena has been developed (see JUDSON and RABITZ [1992]). In this quantum learning control technique, the laboratory quantum system in itself serves as an analog computer to guide its own control as indicated in Fig. 4.2. This approach addresses the requirements of (i) and (ii) above: a physical quantum system can solve its Schrödinger equation in real time and with exact knowledge of its own Hamiltonian, all without computational cost to the user. Hence, the burden of knowing the Hamiltonian and solving the Schrödinger equation is shifted over to a laboratory effort with a learning algorithm guiding the control experiments. The number of physical/chemical systems treated in this way is growing rapidly, and in many cases it is easier to do the experiments than to perform the designs. However, this approach can still benefit from even approximate control designs to start the laboratory learning process, and theory also has an important role to play in introducing the proper stable and reliable algorithms to make the experiments successful.

In summarizing the methodology of quantum learning control, we will consider a simple paradigm where the state of the system is to be optimized at the final time T only. The first step is to prepare the laboratory quantum system in a convenient initial state $\psi(0) = \psi_0$, or a mixed state or a distribution of incoherent states specified by $\rho(0)$. Next, the system is allowed to evolve under its Hamiltonian and some initial trial control law C_0 applied in the laboratory. At the time T, a measurement of the corresponding control objective(s) is made. The quantum system (perturbed by this measurement) is then discarded, and the control law may be updated to C_1 based on the information gained through this measurement. The method and the frequency with which the control law is updated depends on the specific learning algorithm being used (e.g., as described below, with a genetic algorithm the control law is updated after some multiple

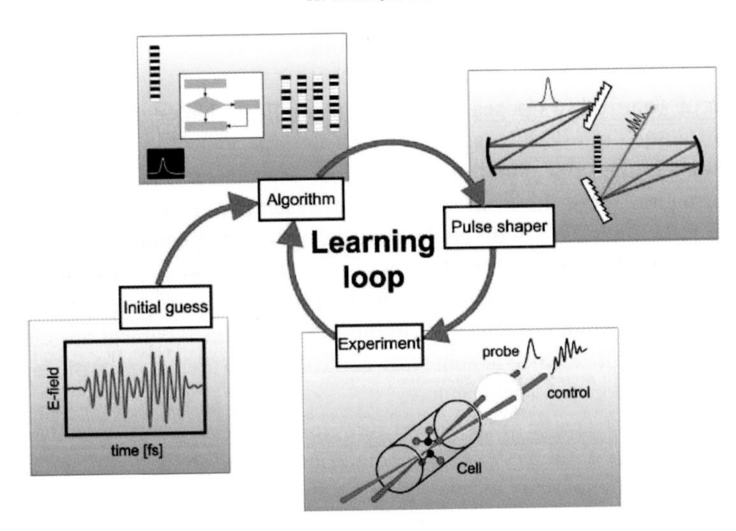

FIG. 4.2. A closed-loop process for teaching a laser to control quantum systems. The loop is entered with either an initial design estimate of even a random field in some cases. A current laser control field design is created with a pulse shaper and then applied to the sample. The action of the control is assessed, and the results are fed to a learning algorithm to suggest an improved field design for repeated excursions around the loop until the objective is satisfactory achieved. Reprinted with permission from H. Rabitz, R. de Vivie Riedle, M. Motzkus and K. Kompa, Science 288, 824 (2000). Copyright © 2000, American Association for the Advancement of Science.

of N_{pop} experiments in each time interval). This updating continues until the learning algorithm has converged to some final control law C that achieves the objectives within the convergence bounds of the learning algorithm. The methods most widely used to accomplish the updating of learning control laws are evolutionary and genetic algorithms (GAs) (e.g., GOLDBERG [1989]), although other learning algorithms could be used.

A GA involves the evolution of successive generations of control laws from their parents, in some fashion mimicking biological evolution. Each trial control field is encoded to form a gene which is part of an overall population evolved in the laboratory during the search for an optimal control. Specifically, experiments are performed in which the quantum system evolves under each of the members of a control law "population". The "fitness" of the control law is evaluated based on the degree to which the control objective(s) are achieved, and the fittest control laws are preserved and/or modified in some prescribed fashion in the next generation. This procedure is continued until the fittest members of the control law population achieve the control objectives to the required extent.

REMARK 4.7. A more general learning control setting may include a discrete (yet finite) set of objective times in $\{t_i\colon\ i = 1, \dots, n\}$ or more generally unions of discrete and continuous time intervals. When more than one target is present the problem is divided into several independent subproblems solved sequentially in time. Note that all the experiments carried out thus far (e.g., ASSION, BAUMERT, BERGT, BRIXNER, KIEFER, SEYFRIED, STREHLE and GERBER [1998], BERGT,

BRIXNER, KIEFER, STREHLE and GERBER [1999], WEINACHT, AHN and BUCKSBAUM [1999], BARDEEN, YAKOVLEV, WILSON, CARPENTER, WEBER and WARREN [1997], BARDEEN, YAKOVLEV, SQUIER and WILSON [1998], HORNUNG, MEIER and MOTZKUS [2000], LEVIS, MENKIR and RABITZ [2001]) have dealt with a single target goal.

The power of the GA and evolutionary algorithms lies in their ability to globally search the space of control laws and discover solutions that may be highly nonintuitive. This directed search takes advantage of the *ability to perform a great number of distinct control experiments in a short period of laboratory time*, and the closed loop technique has been demonstrated for a wide variety of quantum systems and control objectives. The method has also been shown to have interesting convergence properties. For example, it is observed in simulations (as in JUDSON and RABITZ [1992]) and experiments (e.g., ASSION, BAUMERT, BERGT, BRIXNER, KIEFER, SEYFRIED, STREHLE and GERBER [1998], BERGT, BRIXNER, KIEFER, STREHLE and GERBER [1999], WEINACHT, AHN and BUCKSBAUM [1999], BARDEEN, YAKOVLEV, WILSON, CARPENTER, WEBER and WARREN [1997], BARDEEN, YAKOVLEV, SQUIER and WILSON [1998], LEVIS, MENKIR and RABITZ [2001]) that the GA algorithm can converge for a set of randomly constructed initial control law populations. It is then interesting to know to what extent will the convergence properties of these algorithms be improved by incorporating trial designs into the search space.

REMARK 4.8. The choice of cost functionals (see Section 4.1) used in the experiments has the same freedom as for computational optimal control theory except that in laboratory learning control there is no direct access to the wavefunctions. At present the experiments have considered only the final target in the cost, but other criteria could be included giving rise to a possible enhancement to the procedure. Fig. 4.3 shows an application of the laboratory learning control concept.

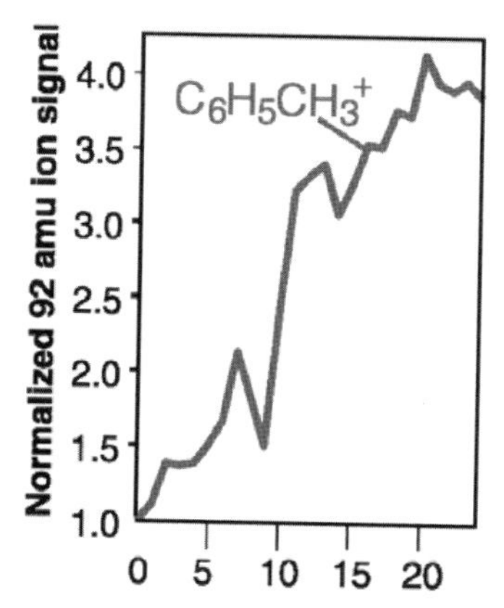

FIG. 4.3. An example the laboratory closed-loop control process. The objective, to maximize the flux of toluene (ions) from the dissociation rearrangement of acetophenone is achieved as the control law evolves over successive generations, see LEVIS, MENKIR and RABITZ [2001]. Reprinted with permission from Levis et al., Science 292, 709 (2001). Copyright © 2001, American Association for the Advancement of Science.

Simulations have considered the effects of laboratory errors (modeled as distortions, or transformations, of true input and output data) and noise upon the learning control process (GROSS, NEUHAUSER and RABITZ [1993], TÓTH, LÖRINCZ and RABITZ [1994]). In TÓTH, LÖRINCZ and RABITZ [1994], input errors were modeled by performing various functional transformations on the control laws used in simulated experiments, while output errors were represented by transforming the expectation values of the control objectives corresponding to these experiments. In general, if the input errors are systematic and the output errors are random, they may not significantly affect the ability of the learning algorithm to find an optimal solution. The fitness of the final control laws found by the GA are also demonstrated to be reasonably insensitive to noise in the control fields. These results are based on very limited studies of simple model systems, and they invite further investigation. Such an analysis could give insight into how best to operate the laboratory experiments.

In principle, any optimization algorithm can be applied to quantum learning control. For example, gradient descent and simulated annealing algorithms have been explored in simulations (GROSS, NEUHAUSER and RABITZ [1993]), but the GA outperformed them in several test cases. However, this subject has not received a thorough examination and there may exist algorithms that converge with greater efficiency or robustness than the genetic algorithm for certain classes of quantum mechanical learning control problems. In treating this topic it is important to consider the ability to perform very large numbers of quantum control experiments, which may overcome certain algorithmic shortcomings found under more common conditions. This ability is almost unprecedented in other applications of learning algorithms.

Another approach to quantum learning control is provided by the use of input $\rightarrow$ output mapping techniques as in PHAN and RABITZ [1997], PHAN and RABITZ [1999], GEREMIA, WEISS and RABITZ [2001]. These methods develop an effective map between the inputs (i.e., the parameters or features defining the control laws) and the outputs (i.e., the expectation values of objective operators). A map from the control input space $\mathcal{C}$ to the space of possible expectation values may be determined directly from the laboratory input and output data; a series of these maps may be needed to cover a sufficiently large portion of $\mathcal{C}$. The control law that optimally satisfies the objectives can be identified from these maps using a suitable learning algorithm. A central issue is establishing the efficiency of mapping techniques as compared with eliminating the maps altogether in favor of having the learning algorithm directly interfaced with the laboratory experiments. Beyond issues of efficiency, mapping techniques may offer the additional benefit of providing physical insight into control mechanisms based on the observed map structure.

5. Challenges for the future

Previous sections have presented the general framework of quantum control, the central issue of controllability, the numerical algorithms for designing control laws and the algorithms for their laboratory discovery. Along the way, several unresolved questions and topics of current interest were introduced. The purpose of the current section is to highlight some emerging areas and unresolved questions that were not a part of this overview. Most of the material is drawn from BROWN and RABITZ [2002].

5.1. Coherence and control

The robustness of coherences for controlled quantum systems in mixed states (i.e., the robustness to decay of the off-diagonal terms in the density operator (cf. Section 2.4) is a topic of special interest. The effects associated with this decay are especially important in the quantum information sciences (see CHUANG, LAFLAMME, SHOR and ZUREK [1995]), where the development of methods to curtail decoherence in information processing algorithms is an active area of research (e.g., SHOR [1995], STEANE [1998]). A relevant contribution from control might be the combination of min–max optimal control techniques with the ideas of decoherence-free subspaces (LIDAR, CHUANG and WHALEY [1998], LIDAR, BACON and WHALEY [1999], BACON, LIDAR and WHALEY [1999]), in which dynamics are invariant to interference that would otherwise cause coherences to decay (this is related to the general notion of disturbance decoupling in mathematical systems theory, see SONTAG [1998]). The result would be control solutions that maximize coherences while simultaneously minimizing an objective cost functional, which could be of significance in developing a physical understanding of the mechanisms of decoherence and its suppression. We note that other schemes are also being considered for the dynamic manipulation of decoherence and control in the presence of dissipation, such as VIOLA and LLOYD [1998], VITALI and TOMBESI [1999], CAO, MESSINA and WILSON [1997], DUAN and GUO [1999].

When the persistence of coherence is not directly important to applications, it was shown in BARDEEN, CAO, BROWN and WILSON [1999] that coherence may not be operative in the control mechanism. In an n-dimensional space $\mathcal{H}_n$ and in the presence of rapid dephasing (which implies vanishing of the off-diagonal coherence terms of the density matrix), Eq. (2.4.1) reduces to a set of rate equations for the population in the n states. A case of special interest in this regard is the control of condensed phases. Under certain conditions, BARDEEN, CAO, BROWN and WILSON [1999] demonstrates that successful controls can be designed for quantum systems whose evolution is determined by these rate equations. This raises the question of whether quantitative measures of coherence can be developed to assess its role in any quantum control problem.

The application of control methods to the cooling of quantum systems is an active area of research (BARTANA and KOSLOFF [1993], BARTANA and KOSLOFF [1997], TANNOR and BARTANA [1999], TANNOR, KOSLOFF and BARTANA [1999]) and there are several means of defining cooling on the molecular scale. One typically utilized criterion (see BARTANA and KOSLOFF [1997]) aims to minimize the von Neumann entropy $\sigma = -\sum_k p_k \log p_k$ corresponding to some observable O (such as the system Hamiltonian); here, p_k is the probability that the system is in the kth eigenstate of O. Another system cooling criterion is to increase in the Renyi entropy $\mathrm{Tr}(\rho^2)$ (as in BARTANA and KOSLOFF [1997]). With both measures, maximal cooling is achieved when all but one of the p_k are zero (i.e., achievement of a pure state). Thus, the ability to completely cool a molecular system is likely to be a challenging task in the presence of laser noise: for the purposes of molecular cooling, a laser control with noise fluctuations may be thought of as having an effective nonzero "temperature". Lower bounds related to a temperature beyond which a system cannot be cooled, if they exist, remain to be established.

5.2. *Can noise help attain control?*

In general, noise in $C(t)$ is thought of as harmful in the context of trying to achieve control objectives. However, hints from the subject of stochastic resonance (GAMMAITONI [1998]) suggest that under suitable conditions noise may possibly have beneficial effects, such as allowing the achievement of a particular level of control using smaller total field fluence than that required for the noise-free system. It remains to be demonstrated under what, if any, conditions the presence of noise can assist in the achievement of quantum control objectives, and to elucidate the possible physical mechanisms behind these effects. In addition, the process of seeking the optimal control will attempt to eliminate the deleterious influence of noise while attempting to reach the objectives.

5.3. *Qualitative behavior of optimal control solutions*

A large body of numerical studies provides examples of solutions to the quantum optimal control equations. However, none of this work has illuminated the general behavior, stability, and classes of solutions to the quantum optimal control equations (here, by classes of solutions we mean the qualitative notion of groups of solutions with particular properties, such as nondispersivity, periodicity, etc.). The possibility that unusual behavior can be expected is evident from one study which showed that the optimal control equations can be made equivalent to the standard nonlinear Schrödinger equation under suitable conditions (DEMIRALP and RABITZ [1997]) (see also in Section 5.4 the considerations on the deterministic feedback control).

5.4. *Feedback quantum control*

This section deals with the effects of real-time, laboratory measurements of the quantum system under two distinct experimental schemes, each described by a different type of feedback Schrödinger equation. The first scheme uses a sequence of repeated experiments to avoid the disturbance effects of measurement, while the second directly confronts these effects.

Deterministic feedback control

Here we consider the (deterministic) continuous-feedback Schrödinger equation

$$i\hbar\frac{\partial\psi(t)}{\partial t}=\left[H_0+C\left(\langle\psi(t)|O_C|\psi(t)\rangle\right)\right]\psi(t), \tag{5.4.1}$$

which conceptually follows from a sequence of laboratory experiments measuring some observable O_C at an increasing set of measurement times, in the limit that the intervals between these times are vanishingly small. In this approach, the probabilistic effect introduced by a measurement at each subsequent time is avoided by 'discarding' the quantum system after each measurement, using the measurement information to extend the control law over the following interval, and repeating the experiment up until the next measurement time.

Different choices of $C(\cdot)$ and O_C may result in Eq. (5.4.1) having qualitatively diverse behavior. An interesting case exists (DEMIRALP and RABITZ [1997]) under the assumptions (i) that $O_C = \delta(\mathbf{x} - \mathbf{x}')$ is the Dirac delta operator, and (ii) that the control law is $C(|\psi(\mathbf{x},t)|^2) = -\gamma|\psi(\mathbf{x},t)|^2$, where γ is a positive constant. The resulting equation

$$i\hbar\frac{\partial\psi(\mathbf{x},t)}{\partial t} = -\left[\frac{\hbar^2}{2m}\nabla^2 + \gamma\left|\psi(\mathbf{x},t)\right|^2\right]\psi(\mathbf{x},t) \tag{5.4.2}$$

admits dispersion free solutions (i.e., it preserves $|\psi(\mathbf{x} - \mathbf{v}t)|^2$) and also solitonic solutions under suitable conditions, cf. LAMB JR. [1980], DRAZIN and JOHNSON [1989]. These types of stable solutions may be significant in many applications of quantum control, including quantum information theory. Eq. (5.4.2) may also be derived from the quantum optimal control formalism under the assumptions stated above; thus, dispersion free control solutions are optimal under these same conditions. The existence of such a control law invites a search for other general classes of control Hamiltonians $H_0 + C(\langle O_C(t)\rangle)$ that exhibit nondispersive or other distinct types of qualitative behavior of practical interest.

Measurement disturbances and feedback control

This section is concerned with the effects of taking real-time measurements on a single quantum system while it is being controlled (here, "single" implies that the sequential "measure and discard" approach of the previous section is abandoned). Feedback may augment learning or optimal control methods by providing real-time information about the evolving quantum system for the stabilization of particularly sensitive objectives (e.g., locking a quantum system around an unstable point on its potential energy surface). This scenario naturally arises in the implementation of a feedback control law where measurements are taken at a discrete set of times $\{t_i\}$: the control law may be written as

$$C(t) = C\big(\langle\psi(t_i)|O_C|\psi(t_i)\rangle\big), \quad t_i \leqslant t. \tag{5.4.3}$$

There exist well-established procedures for determining feedback control laws based on measurements of evolving deterministic and stochastic classical systems in engineering control (e.g., SONTAG [1998], SOTINE and LI [1991]), and it is possible that many of these methods may be adapted to quantum mechanical control problems. Extensive consideration has been given to the effects of measurements on evolving quantum mechanical systems, including analysis in the contexts of continuous feedback and the control of quantum systems by homodyne detection (i.e., measurement of a component of the light field) cf. CARMICHAEL [1999a], CARMICHAEL [1999b], WISEMAN and MILBURN [1993b], WISEMAN and MILBURN [1993a], BRAGINSKY and KHALILI

[1992], WISEMAN [1994], WISEMAN [1993], HOFMANN, MAHLER and HESS [1998], HOFMANN, HESS and MAHLER [1998], CARMICHAEL [1996]. These works generally treat the more difficult problem of random measurement times; here, we give only an elementary discussion of ideas relevant to feedback control with measurements taken at a deterministic, discrete set of times.

A postulate of quantum mechanics states that a perfectly precise measurement of an operator O_C (with nondegenerate spectrum) on a finite-dimensional Hilbert space must both yield one of the eigenvalues λ_j of the operator and result in a disturbance such that $\psi(t)$ collapses to the associated eigenstate ψ_j. The measurement process introduces a stochastic element into the evolution of the quantum system, with the probability of collapse into ψ_j being $|\langle\psi_j|\psi(t)\rangle|^2$. Each measurement in feedback quantum control therefore involves an information trade-off: the system is perturbed away from the deterministic Schrödinger equation, but a measurement is used to update the control law. These random transitions, and the evolution they determine on intervals between the t_i via the control laws (5.4.3), determine a "stochastic quantum map" (e.g., CARMICHAEL [1999b]) between states at these times. If the measurement process does not completely determine the quantum states (or if mixed states are present for other reasons), a formulation involving a stochastic map between conditioned density operators is required, cf. CARMICHAEL [1999b].

A crucial unresolved question is what general classes of quantum problems will be assisted by incorporating feedback measurements. More specifically, it may be possible to show that taking a certain number of measurements improves control (with reasonable assumptions about the problem-dependent frequency and timing of measurements to optimize the feedback quantum control problem). A related matter is the possibility of making "weak observations" that give useful information about a quantum system while introducing only minimal perturbations. The effects of measurements on the feedback control process for systems satisfactorily described semiclassically also remain to be characterized, and this domain may be especially amenable to performing weak measurements.

Closing the loop through machine feedback

The final topic on feedback control is the possibility of closing the control loop in laboratory hardware through machine feedback. Recent work in acoustics illustrates the capability of focusing reflected waves back upon their sources cf. FINK [1999], FINK and PRADA [1996] in an iterative fashion in order to enhance the intensity in the focal volume. An analogy of this technique relevant to quantum mechanics might be "reflection" through special measurement devices that could then send modified electromagnetic waves precisely back to an emitting quantum mechanical source to better achieve the control objectives. This process may be fully quantum mechanical if carried out in a suitable optical cavity, but in general the same closed loop observation/disturbance issues raised in the previous section must be considered here. At this juncture such a machine is only a gedanken process, but its potential strongly motivates an analysis of the concept.

5.5. *Algorithms for the inversion of quantum dynamics data*

Knowledge of the potential V and the dipole μ (or other coupling coefficients) is required for control law design and is of fundamental importance to many other applications in chemistry and physics. This section concerns dynamical algorithms that invert time-dependent laboratory data to identify these operators. This problem of determining $\mu(\mathbf{x})$ or $V(\mathbf{x})$ may be related (LU and RABITZ [1995]; see also BARGHEER, DIETRICH, DONOVANG and SCHWENTNER [1999] for a different approach) to the problem of determining the control law $C(t)$ (of the form (2.3.1)) that will cause a quantum system to follow a prescribed track (see Fig. 5.1). In particular, if the expectation values $y(t) \equiv \langle O_h(t) \rangle$ of a time-independent operator O_h are established from a series of observations of an evolving quantum system, the Schrödinger equation and the Heisenberg equation of motion form the pair of coupled (forward-inverse) equations

$$i\hbar \frac{d\psi(t)}{dt} = \left[H_0 - \mu\varepsilon(t)\right]\psi(t), \tag{5.5.1}$$

$$i\hbar \frac{dy(t)}{dt} = \left\langle \psi(t) \middle| \left[H_0 - \mu\varepsilon(t), O_h\right] \middle| \psi(t) \right\rangle. \tag{5.5.2}$$

The solution of these evolution equations may in principle be attempted for any two unknowns. As knowledge of $\psi(t)$ is not available in any physical problem, the wavefunction will always be considered as one of these unknowns; the other may be chosen from either $\varepsilon(t)$, $\mu(\mathbf{x})$, or $V(\mathbf{x})$ with the complementary pair assumed as known. The first of these possibilities was treated in Section 4.4, where the fact that the expectation value on the right hand side of Eq. (5.5.2) involves only spatial integration

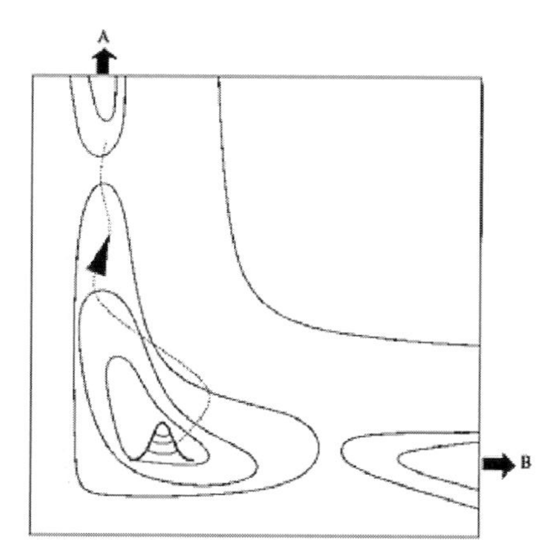

FIG. 5.1. Schematic illustration of a wavepacket track evolving on a potential surface (contours shown). For control a goal may be to find the field $\varepsilon(t)$ that will steer the track out of product channel A or B, while for inversion the track is observed in the laboratory and the goal is to determine the potential in the regions traveled by the track, cf. LU and RABITZ [1995]. Reprinted with permission from Z.M. Lu and H. Rabitz, J. Phys. Chem. 99, 13731 (1995). Copyright © 1995 American Chemical Society.

was exploited to write $\varepsilon(t)$ explicitly in the form of Eq. (4.4.3). We now turn to the solution for $\mu(\mathbf{x})$ and $V(\mathbf{x})$.

Eq. (5.5.2) may be rewritten (cf. LU and RABITZ [1995]) as a Fredholm integral equation of the second type, after regularization enforcing the physical criterion that $\mu(\mathbf{x})$ and $V(\mathbf{x})$ should decay as $x \to \infty$. The structure of this equation is

$$\int_{-\infty}^{\infty} \kappa(x, x') f(x') \, dx' + \alpha f(x) = h(x), \tag{5.5.3}$$

where $f(x)$ is the unknown (μ or V), α is the regularization parameter, and the kernel $\kappa(x, x')$ and the inhomogeneity $h(x)$ involve the data $y(t)$, the operator O_h, and the solution $\psi(t)$ to the Schrödinger equation (5.5.1). Solution of the regularized pair of Eqs. (5.5.1) and (5.5.3) may be accomplished using the tracking procedure discussed in Section 4.4, with the role of $\varepsilon(t)$ replaced by $\mu(\mathbf{x})$ or $V(\mathbf{x})$. The procedure consists of formally solving Eq. (5.5.3) for $f(x)$ and substituting the result into the Schrödinger equation (5.5.1) to solve for $\psi(t)$, which is then used to determine $f(x)$ in a final inversion step.

In the special case that the data being inverted is the probability density function $|\psi(\mathbf{x}, t)|^2$ (formally, the expectation value of the Dirac delta operators $\delta(\mathbf{x} - \mathbf{x}')$) (see KRAUSE, SCHAFER, BEN-NUN and WILSON [1997]), a direct algorithm has been developed to identify $V(\mathbf{x})$ without the expensive requirement of numerically solving the Schrödinger equation cf. ZHU and RABITZ [1999b]. The algorithm relies on Ehrenfest's relation, and appears to be difficult to generalize to other measurement operators.

In general, the goal is to solve Eqs. (5.5.1) and (5.5.2) with minimum of distortion introduced by additional criteria; in the example above, the balance between this objective and stability requirements is set by α, whose optimal value depends on the details of the particular problem and solution method. Additionally, the evolution of the quantum system which determines $\kappa(x, x')$ in (5.5.3) is in turn governed by the applied field $\varepsilon(t)$ in Eq. (5.5.1). Hence, it should be possible to determine a control law which allows inversion with maximum stability to produce optimal dynamical regularization. Note that meaningful inversion of Eqs. (5.5.1) and (5.5.2) may only be expected if the control law $\varepsilon(t)$ steers the wavefunction to be nonzero in the domain in which μ or V is to be determined; the formulation of Eqs. (5.5.1) and (5.5.2) may be extended to incorporate multiple realizations of the control law $\varepsilon_j(t)$ (see LU and RABITZ [1995]) that, taken together, may provide the desired evolution over the entire spatial domain of interest. However, for dynamical reasons the kernel $\kappa(x, x')$ may still produce a singular operator in Eq. (5.5.3) where it is significantly nonzero. The additional conditions required to resolve this problem are not immediately apparent.

A complete laboratory device may be envisioned to function as an optimal dynamics inversion machine for the efficient and automatic discovery of μ or V for diverse quantum systems (cf. RABITZ and SHI [1991], RABITZ and ZHU [2000]). This machine would operate in a closed-loop mode to take advantage of the ability to perform a very large number of high throughput control-observation experiments, and would operate through the following steps, sketched in Fig. 5.2:

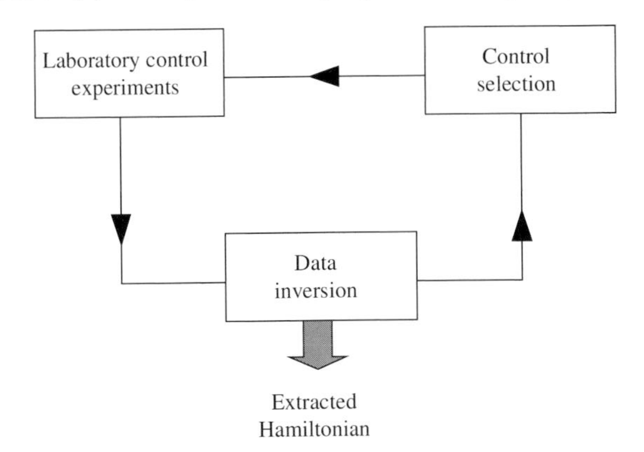

FIG. 5.2. The optimal dynamics identification machine. Reprinted with permission from E. Brown and H. Rabitz, J. Math. Chem. 31, p. 55, 2002; copyright © 2002, Kluwer Academic Publishers.

(1) Initial approximations for V and μ could be used to design an optimal control field aimed at causing the wavepacket to evolve in desired spatial areas where V and μ are being sought.
(2) Laboratory experiments using this control law would be performed to produce the data trajectory $\langle O_h(t)\rangle = y(t)$.
(3) An inversion would be performed to produce updated potential or dipole information.
(4) If the spatial domain of interest was not completely covered by the current trajectory or if the inversion quality is not adequate, the procedure would be repeated with the assistance of partial Hamiltonian information gained from step (3).

5.6. *Identification of quantum control mechanism and "rules of thumb"*

A cornerstone of chemistry is that physically similar molecules tend to exhibit similar chemical behavior. The emphasis is on "similar" and in the context of quantum control the criteria for defining similarity is not known. From the rich behavior and information content in the design, closed loop, and dynamical inversion aspects of quantum control, one can anticipate using the emerging results to provide insight or estimates for the control laws for physically related, but as yet uninvestigated, problems. The body of relationships (as just yet beginning to be observed) between quantum systems, control objectives, and control laws may be called quantum control rules of thumb. A specialized example is the explanation of the timing of the pulses used in the STIRAP control method (cf. RICE and ZHAO [2000]). However, attempts to find general rules have proved much more difficult than was at first expected. Nevertheless, the implications of these rules for both the theory of quantum control and its practical implementation are substantial: resolution of this matter may be the most important challenge ahead for the field.

A natural strategy for identifying rules of thumb might ensue from a type of quantum mechanical reverse engineering: solutions $\{C(t), \psi(t), \lambda(t), \langle O\rangle\}$ to the optimal control equations, or $C(t)$ and $\langle O\rangle$ from closed loop experiments, could provide a physical basis for understanding the mechanisms and pathways leading the quantum system from initial conditions to final control objectives. However, there exist many examples in the literature in which the structure of the final control fields and the resulting control pathways are found to be highly nonintuitive, and judging the relevance of such solutions in terms of general rules of thumb is difficult in the presence of a possibly large number of (locally) optimal solutions. Further insight into the structure of these local minima may be gained by identifying the family of locally optimal control solutions and enumerating them based on their optimality. This problem might be partially alleviated by incorporating a global search procedure in the optimization algorithm (e.g., a genetic algorithm), both for theoretical design and laboratory control. If such techniques were developed, the existence of multiple solutions could possibly be exploited as a large body of data about control behavior.

In the context of closed-loop laboratory implementation of controls, the presence of multiple solutions to the quantum optimal control problem opens up several options (GEREMIA, ZHU and RABITZ [2000]). Given that there exist many possible solutions $C(t)$ from which identification of control mechanisms could be attempted, it is important to select solutions that contain a minimum of extraneous information that detract from this task. In addition, control rules of thumb would best be developed based on solutions that are robust to realistic laboratory noise. Both the suppression of extraneous structural components in $C(t)$ and the selection of robust control fields may be accomplished through the use of appropriate cost functionals (GEREMIA, ZHU and RABITZ [2000]). This "cleanup" of control laws is likely to assist in identifying rules of thumb for the control of quantum systems.

A first step toward identifying quantum control rules of thumb involves the effective classification of similarities and differences between molecules in a context relevant to the controls directing them to certain physical objectives. This type of classification is fundamental in many fields of chemistry and physics, in which the vast numbers of molecules are categorized according to their relevant behaviors or properties. However, presently the standard measures have not been able to consistently predict the structure of control fields for particular objectives.

A three-way classification structure will be necessary, relating (i) control laws, (ii) molecular Hamiltonians and coupling terms, and (iii) control objectives. Progress toward (i) will likely involve identification of the relevant properties of control laws $C(t)$. Experience thus far suggests that the most useful features will include, but extend beyond, description of spectral components and intensities. Progress may also follow from exploring the relation between the form of the Hamiltonian and (sets of) control solutions for a fixed particular control objective, possibly through numerical optimal control calculations for a series of quantum systems whose Hamiltonians differ by small increments, but collectively cover a broad sampling of physical systems.

6. Conclusion

This article aimed to present an overview of the current state of attempts to control quantum phenomena. Special emphasis was given to the conceptual, algorithmic and numerical aspects of the subject. In envisioning further advances along these lines it is very important to explicitly consider the special capabilities of performing massive numbers of control experiments over a short laboratory time. In many respects the field of control over quantum phenomena is a field which is young with the bulk of its developments lying ahead. It is hoped that this article provides some stimulus to push the field further.

Acknowledgments

H.R. and E.B. acknowledge support from the National Science Foundation and the Department of Defense. E.B. was supported by a NSF Graduate Fellowship.

G.T. would like to thank Wusheng Zhu from the Princeton University for helpful discussions and remarks.

References

ACKERMAN, J., ed. (1985), *Uncertainty and Control, Proceedings of an International Seminar, Bonn, Germany, May 1985* (Springer).

AGARWAL, G. (1997), Nature of the quantum interference in electromagnetic-field-induced control of absorption, *Phys. Rev. A* **55**, 2467–2470.

ALBERTINI, F. and D. D'ALESSANDRO (2001), Notions of controllability for quantum mechanical systems, Preprint arXiv:quant-ph/0106128.

ALTAFINI, C. (2002), Controllability of quantum mechanical systems by root space decompositions of $su(N)$, *J. Math. Phys.* **43** (5), 2051–2062.

ASSION, A., T. BAUMERT, M. BERGT, T. BRIXNER, B. KIEFER, V. SEYFRIED, M. STREHLE and G. GERBER (1998), Control of chemical reactions by feedback-optimized phase-shaped femtosecond laser pulses, *Science* **282**, 919–922.

ATHANS, M. and P.L. FALB (1966), *Optimal Control* (McGraw-Hill).

BACON, D., D.A. LIDAR and K.B. WHALEY (1999), Robustness of decoherence-free subspaces for quantum computation, *Phys. Rev. A* **60**, 1944–1955.

BALL, J., J. MARSDEN and M. SLEMROD (1982), Controllability for distributed bilinear systems, *SIAM J. Control and Optimization* **20** (4).

BARDEEN, C., V.V. YAKOVLEV, K.R. WILSON, S.D. CARPENTER, P.M. WEBER and W.S. WARREN (1997), Feedback quantum control of molecular electronic population transfer, *Chem. Phys. Lett.* **280**, 151–158.

BARDEEN, C.J., J. CAO, F.L.H. BROWN and K.R. WILSON (1999), Using time-dependent rate equations to describe chirped pulse excitation in condensed phases, *Chem. Phys. Lett.* **302**, 405–410.

BARDEEN, C.J., J. CHE, K.R. WILSON, V.V. YAKOVLEV, V.A. APKARIAN, C.C. MARTENS, R. ZADOYAN, B. KOHLER and M. MESSINA (1997), Quantum control of I_2 in the gas phase and in condensed phase solid Kr matrix, *J. Chem. Phys.* **106**, 8486–8503.

BARDEEN, C.J., J. CHE, K.R. WILSON, V.V. YAKOVLEV, P. CONG, B. KOHLER, J.L. KRAUSE and M. MESSINA (1997), Quantum control of NaI photodissociation reaction product states by ultrafast tailored light pulses, *J. Phys. Chem. A* **101**, 3815–3822.

BARDEEN, C.J., V.V. YAKOVLEV, J.A. SQUIER and K.R. WILSON (1998), Quantum control of population transfer in green fluorescent protein by using chirped femtosecond pulses, *J. Am. Chem. Soc.* **120**, 13023–13027.

BARGHEER, M., P. DIETRICH, K. DONOVANG and N. SCHWENTNER (1999), Extraction of potentials and dynamics from condensed phase pump-probe spectra: Application to I_2 in Kr matrices, *J. Chem. Phys.* **111**, 8556–8564.

BARTANA, A. and R. KOSLOFF (1993), Laser cooling of molecular internal degrees of freedom by a series of shaped pulses, *J. Chem. Phys.* **99**, 196–210.

BARTANA, A. and R. KOSLOFF (1997), Laser cooling and internal degrees of freedom, II, *J. Chem. Phys.* **106**, 1435–1448.

BARTELS, R., S. BACKUS, E. ZEEK, L. MISOGUTI, G. VDOVIN, I.P. CHRISTOV, M.M. MURNANE and H.C. KAPTEYN (2000), Shaped-pulse optimization of coherent emission of high-harmonic soft X-rays, *Nature* **406**, 164–166.

BERGT, M., T. BRIXNER, B. KIEFER, M. STREHLE and G. GERBER (1999), Controlling the femtochemistry of $Fe(CO)_5$, *J. Phys. Chem. A* **103**, 10381–10387.

BEUMEE, J.G.B. and H. RABITZ (1992), Robust optimal control theory for selective vibrational excitation in molecules: A worst case analysis, *J. Chem. Phys.* **97**, 1353–1364.

BEUMEE, J., C. SCHWIETERS and H. RABITZ (1990), Optical control of molecular motion with robustness and application to vinylidene fluoride, *J. Opt. Soc. America B*, 1736–1747.

BRAGINSKY, V. and F. KHALILI (1992), *Quantum Measurement* (Cambridge University Press).

BROCKETT, R. (1970), *Finite Dimensional Linear Systems* (Wiley).

BROWN, E. and H. RABITZ (2002), Some mathematical and algorithmic challenges in the control of quantum dynamics phenomena, *J. Math. Chem.* **31**, 17–63.

BROWN, F. and R. SIBLEY (1998), Quantum control for arbitrary linear and quadratic potentials, *Chem. Phys. Lett.* **292**, 357–368.

CANCÈS, E. and C. LE BRIS (1999), On the time-dependent Hartree–Fock equations coupled with a classical nuclear dynamics, *Mathematical Models and Methods in Applied Sciences* **9** (7), 963–990.

CANCÈS, E., C. LE BRIS and P. MATHIEU (2000), Bilinear optimal control of a Schrödinger equation, *C. R. Acad. Sci. Paris Sér. I Math.* **7**, 567–571.

CAO, J., M. MESSINA and K. WILSON (1997), Quantum control of dissipative systems: exact solutions, *J. Chem. Phys.* **106**, 5239–5248.

CAO, J. and K. WILSON (1997), A simple physical picture for the quantum control of wave packet localization, *J. Chem. Phys.* **107**, 1441–1450.

CARMICHAEL, H. (1996), Stochastic Schrodinger equations: What they mean and what they can do, in: J. Eberly, L. Mandel and E. Wolf, eds., *Coherence and Quantum Optics VII*, 177.

CARMICHAEL, H. (1999a), *Master Equations and Fokker Plank Equations* (Springer).

CARMICHAEL, H. (1999b), *An Open Systems Approach to Quantum Optics* (Springer).

CHEN, Y., P. GROSS, V. RAMAKRISHNA, H. RABITZ and K. MEASE (1995), Competitive tracking of molecular objectives described by quantum mechanics, *J. Chem. Phys.* **102**, 8001–8010.

CHRISTOFIDES, N. (1975), *Graph Theory: An Algorithmic Approach* (Academic Press).

CHUANG, I.L., R. LAFLAMME, P.W. SHOR and W.H. ZUREK (1995), Quantum computers, factoring, and decoherence, *Science* **270**, 1633–1635.

COHEN-TANNOUDJI, C., B. LIU and F. LALOË (1997), *Quantum Mechanics* (Wiley).

COMBARIZA, J., B. JUST, J. MANZ and G. PARAMONOV (1991), Isomerization controlled by ultrashort infrared laser pulses: model simulations for the inversion of ligands (H) in the double-well potential of an organometallic compound [(C5H5)(CO)2FePH2], *J. Phys. Chem.* **95**, 10351.

DARDI, P.S. and S.K. GRAY (1982), Classical and quantum-mechanical studies of HF in an intense laser field, *J. Chem. Phys.* **77**, 1345–1353.

DAVIS, M.J. and E.J. HELLER (1984), Comparisons of classical and quantum dynamics for initially localized states, *J. Chem. Phys.* **80**, 5036–5048, this and the following three references are from the list given in note 18 of SCHWIETERS and RABITZ [1993].

DE ARAUJO, L. and I. WALMSLEY (1999), Quantum control of molecular wavepackets: an approximate analytic solution for the strong-response regime, *J. Phys. Chem. A* **103**, 10409–10416.

DEMIRALP, M. and H. RABITZ (1993), Optimally controlled quantum molecular dynamics: A perturbation formulation and the existence of multiple solutions, *Phys. Rev. A* **47**, 809–816.

DEMIRALP, M. and H. RABITZ (1997), Dispersion-free wavepackets and feedback solitonic motion in controlled quantum dynamics, *Phys. Rev. A* **55**, 673–677.

DEMIRALP, M. and H. RABITZ (1998), Assessing optimality and robustness of control over quantum dynamics, *Phys. Rev. A* **57**, 2420–2425.

DION, C., A. BANDRAUK, O. ATABEK, A. KELLER, H. UMEDA and Y. FUJIMURA (1999), Two-frequency IR laser orientation of polar molecules, numerical simulations for HCN, *Chem. Phys. Lett.* **302**, 215–223.

DION, C., A. KELLER, O. ATABEK and A. BANDRAUK (1999), Laser-induced alignment dynamics of HCN: Roles of the permanent dipole moment and the polarizability, *Phys. Rev. A* **59** (2), 1382.

DORATO, P., ed. (1987), Robust control, *IEEE Selected Reprint Series* (IEEE Press, New York).

DRAZIN, P. and R. JOHNSON (1989), *Solitons: An Introduction* (Cambridge University Press).

DUAN, L. and G. GUO (1999), Suppressing environmental noise in quantum computation through pulse control, *Phys. Lett. A* **261**, 139–144.

FINK, M. (1999), Time reversed acoustics, *Sci. Am.*, 91–97.

FINK, M. and C. PRADA (1996), Ultrasonic focusing with time-reversal mirrors, in: A. Briggs and W. Arnold, eds., *Advances in Acoustic Microscopy Series*, Vol. 2, 219–251.

FU, H., S. SCHIRMER and A. SOLOMON (2001), Complete controllability of finite-level quantum systems, *J. Phys. A* **34**, 1679–1690.

GAMMAITONI, L. (1998), Stochastic resonance, *Rev. Mod. Phys.* **70**, 223–287.

GEREMIA, J., E. WEISS and H. RABITZ (2001), Achieving the laboratory control of quantum dynamics phenomena using nonlinear functional maps, *Chem. Phys.* **267**, 209–222.

GEREMIA, J., W. ZHU and H. RABITZ (2000), Incorporating physical implementation concerns into closed loop quantum control experiments, *J. Chem. Phys.* **113**, 10841–10848.

GIRARDEAU, M.D., M. INA, S.G. SCHIRMER and T. GULSRUD (1997), Kinematical bounds on evolution and optimization of mixed quantum states, *Phys. Rev. A* **55**, R1565–R1568.

GIRARDEAU, M.D., S.G. SCHIRMER, J.V. LEAHY and R.M. KOCH (1998), Kinematical bounds on optimization of observables for quantum systems, *Phys. Rev. A* **58**, 2684–2689.

GOLDBERG, D. (1989), *Genetic Algorithms in Search, Optimization and Machine Learning* (Addison-Wesley).

GROSS, P., D. NEUHAUSER and H. RABITZ (1993), Teaching lasers to control molecules in the presence of laboratory field uncertainty and measurement imprecision, *J. Chem. Phys.* **98**, 4557–4566.

GROSS, P., J. SINGH, H. RABITZ, G. HUANG and K. MEASE (1993), Inverse quantum-mechanical control: A means for design and a test of intuition, *Phys. Rev. A* **47**, 4593–4603.

GRØNAGER, M. and N. HENRIKSEN (1998), Real-time control of electronic motion: Application to NaI, *J. Chem. Phys.* **109**, 4335–4341.

HAREL, G. and V.M. AKULIN (1999), Complete control of Hamiltonian quantum systems: Engineering of Floquet evolution, *Phys. Rev. Lett.* **82**, 1–5.

HOFMANN, H., O. HESS and G. MAHLER (1998), Quantum control by compensation of quantum fluctuations, *Optics Express* **2**, 339–346.

HOFMANN, H.F., G. MAHLER and O. HESS (1998), Quantum control of atomic systems by homodyne detection and feedback, *Phys. Rev. A* **57**, 4877–4888.

HOKI, K., Y. OHTSUKI, H. KONO and Y. FUJIMURA (1999), Quantum control of NaI predissociation in subpicosecond and several-picosecond time regimes, *J. Phys. Chem. A* **103**, 6301–6308.

HORNUNG, T., R. MEIER and M. MOTZKUS (2000), Optimal control of molecular states in a learning loop with a parameterization in frequency and time domain, *Chem. Phys. Lett.* **326**, 445–453.

HOSOE, S., ed. (1991), *Robust Control, Proceedings of a Workshop held in Tokyo, Japan, June 23–24, 1991* (Springer).

HUANG, G., T. TARN and J. CLARK (1983), On the controllability of quantum-mechanical systems, *J. Math. Phys.* **24**, 2608–2618.

JUDSON, R., K. LEHMANN, H. RABITZ and W.S. WARREN (1990), Optimal design of external fields for controlling molecular motion – application to rotation, *J. Molec. Structure* **223**, 425–456.

JUDSON, R. and H. RABITZ (1992), Teaching lasers to control molecules, *Phys. Rev. Lett.*

KAILATH, T. (1980), *Linear Systems* (Prentice Hall).

KHALIL, H.K. (1996), *Nonlinear systems* (Macmillan, New York).

KOBAYASHI, M. (1998), Mathematics makes molecules dance, *Siam News* **31** (9), 24.

KRAUSE, J.L., D.H. REITZE, G.D. SANDERS, A.V. KUZNETSOV and C.J. STANTON (1998), Quantum control in quantum wells, *Phys. Rev. B* **57**, 9024–9034.

KRAUSE, J. and K. SCHAFER (1999), Control of the emission from stark wave packets, *J. Phys. Chem. A* **103**, 10118–10125.

KRAUSE, J.L., K.J. SCHAFER, M. BEN-NUN and K.R. WILSON (1997), Creating and detecting shaped Rydberg wave packets, *Phys. Rev. Lett.* **79**, 4978–4981.

KROTOV, V. (1973), Optimization methods of control with minimax criteria, I, *Automat. Remote Control* **34**, 1863–1873; translated from *Avtomat. i Telemeh.* 1973, no. 12, 5–17.

KROTOV, V. (1974a), Optimization methods in control processes with minimax criteria, II, *Automat. Remote Control* **35**; translated from *Avtomat. i Telemeh.* 1974, no. 1, 5–15.

KROTOV, V. (1974b), Optimization methods of control with minimax criteria, III, *Automat. Remote Control* **35**, 345–353; translated from *Avtomat. i Telemeh.* 1974, no. 3, 5–14.

KULANDER, K.C. and A.E. OREL (1981), Laser-collision induced chemical-reactions – a comparison of quantum-mechanical and classical-model results, *J. Chem. Phys.* **75**, 675–680.

KUNDE, J., B. BAUMANN, S. ARLT, F. MORIER-GENOUD, U. SIEGNER and U. KELLER (2000), Adaptive feedback control of ultrafast semiconductor nonlinearities, *Appl. Phys. Lett.* **77**, 924.

LAMB JR., G. (1980), *Elements of Soliton Theory* (Wiley).

LEFORESTIER, C. (1986), Competition between dissociation and exchange processes in a collinear A + BC collision – comparison of quantum and classical results, *Chem. Phys. Lett.* **125**, 373–377.

LEVIS, R.J., G. MENKIR and H. RABITZ (2001), Selective bond dissociation and rearrangement with optimally tailored, strong-field laser pulses, *Science* **292**, 709–713.

LI, B., G. TURINICI, V. RAMAKHRISHNA and H. RABITZ (2002), Optimal dynamic discrimination of similar molecules through quantum learning control, *J. Phys. Chem. B* **106** (33), 8125.

LIDAR, D., I.L. CHUANG and K.B. WHALEY (1998), Decoherence-free subspaces for quantum computation, *Phys. Rev. Lett.* **81**, 2594–2597.

LIDAR, D.A., D. BACON and K.B. WHALEY (1999), Concatenating decoherence-free subspaces with quantum error correcting codes, *Phys. Rev. Lett.* **82**, 4556–4559.

LU, Z.M. and H. RABITZ (1995), Unified formulation for control and inversion of molecular dynamics, *J. Phys. Chem.* **99**, 13731–13735.

LUENBERGER, D. (1979), *Introduction to Dynamic Systems: Theory, Models, and Applications* (Wiley).

MESHULACH, D. and Y. SILBERBERG (1998), Coherent quantum control of two-photon transitions by a femtosecond laser pulse, *Nature* **396**, 239–241.

MISHIMA, K. and K. YAMASHITA (1999a), Quantum control of photodissociation wavepackets, *J. Molec. Structure* **461–462**, 483–491.

MISHIMA, K. and K. YAMASHITA (1999b), Theoretical study on quantum control of photodissociation and photodesorption dynamics by femtosecond chirped laser pulses, *J. Chem. Phys.* **110**, 7756–7769.

MOHLER, R. (1983), *Bilinear Control Processes* (Academic Press).

NIJMEIHER, H. and A. VAN DER SCHAFT (1990), *Nonlinear Dynamical Control Systems* (Springer).

OHTSUKI, Y., H. KONO and Y. FUJIMURA (1998), Quantum control of nuclear wave packets by locally designed optimal pulses, *J. Chem. Phys.* **109**, 9318–9331.

OHTSUKI, Y., W. ZHU and H. RABITZ (1999), Monotonically convergent algorithm for quantum optimal control with dissipation, *J. Chem. Phys.* **110**, 9825–9832.

ONG, C.K., G.M. HUANG, T.J. TARN and J.W. CLARK (1984), Invertability of quantum-mechanical control systems, *Math. Systems Theory* **17**, 335–350.

PASTIRK, I., E.J. BROWN, Q. ZHANG and M. DANTUS (1998), Quantum control of the yield of a chemical reaction, *J. Chem. Phys.* **108**, 4375–4378.

PEIRCE, A., M. DAHLEH and H. RABITZ (1988), Optimal control of quantum mechanical systems: Existence, numerical approximations, and applications, *Phys. Rev. A* **37**, 4950–4964.

PHAN, M.Q. and H. RABITZ (1997), Learning control of quantum-mechanical systems by laboratory identification of effective input-output maps, *Chem. Phys.* **217**, 389–400.

PHAN, M.Q. and H. RABITZ (1999), A self-guided algorithm for learning control of quantum-mechanical systems, *J. Chem. Phys.* **110**, 34–41.

RABITZ, H., R. DE VIVIE-RIEDLE, M. MOTZKUS and K. KOMPA (2000), Wither the future of controlling quantum phenomena?, *Science* **288**, 824–828.

RABITZ, H. and S. SHI (1991), Optimal control of molecular motion: Making molecules dance, in: J. Bowman, ed., *Advances in Molecular Vibrations and Collision Dynamics*, Vol. 1, Part A (JAI Press, Inc.) 187–214.

RABITZ, H. and W. ZHU (2000), Optimal control of molecular motion: Design, implementation, and inversion, *Acc. Chem. Res.* **33**, 572–578.

RAMAKRISHNA, V. (2000), *Private communication*.

RAMAKRISHNA, V., M. SALAPAKA, M. DAHLEH, H. RABITZ and A. PIERCE (1995), Controllability of molecular systems, *Phys. Rev. A* **51** (2), 960–966.

RAMAKRISHNA, V. ET AL. (2000), Explicit generation of unitary transformations in a single atom or molecule, *Phys. Rev. A* **61**, 032106–1–032106–6.

RICE, S. and M. ZHAO (2000), *Optical Control of Quantum Dynamics* (Wiley), many additional references to the subjects of this paper may also be found here.

SCHIRMER, S., M. GIRARDEAU and J. LEAHY (2000), Efficient algorithm for optimal control of mixed-state quantum systems, *Phys. Rev. A* **61**, 012101.

SCHIRMER, S.G., H. FU and A. SOLOMON (2001), Complete controllability of quantum systems, *Phys. Rev. A* **63**, 063410.

SCHWIETERS, C. and H. RABITZ (1993), Optimal control of classical systems with explicit quantum/classical difference reduction, *Phys. Rev. A* **48**, 2549–2457.

SCHWIETERS, C. and H. RABITZ, Designing time-independent classically equivalent potentials to reduce quantum/classical observable differences using optimal control theory, unpublished results.

SEPULCHRE, R., M. JANKOVIC and P. KOKOTOVIC (1998), *Constructive Nonlinear Control* (Springer).

SHAH, S. and S. RICE (1999), Controlling quantum wavepacket motion in reduced-dimensional spaces: reaction path analysis in optimal control of HCN isomerization, *Faraday Discuss.* **113**, 319–331.

SHAPIRO, M. and P. BRUMER (1989), Coherent chemistry: Controlling chemical reactions with lasers, *Acc. Chem. Res.* **22**, 407.

SHEN, H., J.-P. DUSSAULT and A. BANDRAUK (1994), Optimal pulse shaping for coherent control by the penalty algorithm, *Chem. Phys. Lett.* **221**, 498–506.

SHEN, L., S. SHI, H. RABITZ, C. LIN, M. LITTMAN, J.P. HERITAGE and A.M. WEINER (1993), Optimal control of the electric susceptibility of a molecular gas by designed non-resonant laser pulses of limited amplitude, *J. Chem. Phys.* **98**, 7792–7803.

SHI, S., A. WOODY and H. RABITZ (1988), Optimal control of selective vibrational excitation in harmonic linear chain molecules, *J. Chem. Phys.* **88**, 6870–6883.

SHOR, P. (1995), Scheme for reducing decoherence in quantum computer memory, *Phys. Rev. A* **52**, R2493–R2496.

SOMLÓI, J., V. KAZAKOV and D. TANNOR (1993), Controlled dissociation of I_2 via optical transitions between the X and B electronic states, *Chem. Phys.* **172**, 85–98.

SONTAG (1998), *Mathematical Control Theory* (Springer), and references within.

SOTINE, J.J. and W. LI (1991), *Applied Nonlinear Control* (Prentice Hall).

STEANE, M. (1998), Introduction to quantum error correction, *Phil. Trans. Roy. Soc. London Ser. A* **356**, 1739–1758.

TANNOR, D. and A. BARTANA (1999), On the interplay of control fields and spontaneous emission in laser cooling, *J. Chem. Phys. A* **103**, 10359–10363.

TANNOR, D., V. KAZAKOV and V. ORLOV (1992), Control of photochemical branching: Novel procedures for finding optimal pulses and global upper bounds, in: J. Broeckhove and L. Lathouwerse, eds., *Time Dependent Quantum Molecular Dynamics*, Plenum, 347–360.

TANNOR, D.J., R. KOSLOFF and A. BARTANA (1999), Laser cooling of internal degrees of freedom of molecules by dynamically trapped states, *Faraday Discuss.* **113**, 365–383.

TANNOR, D. and S. RICE (1985), Control of selectivity of chemical reaction via control of wave packet evolution, *J. Chem. Phys.* **83**, 5013–5018.

TERSIGNI, S.H., P. GASPARD and S. RICE (1990), On using shaped light-pulses to control the selectivity of product formation in a chemical-reaction – an application to a multiple level system, *J. Chem. Phys.* **93**, 1670–1680.

TÓTH, G., A. LÖRINCZ and H. RABITZ (1994), The effect of control field and measurement imprecision on laboratory feedback control of quantum systems, *J. Chem. Phys.* **101**, 3715–3722.

TURINICI, G. (2000a), Analysis of numerical methods of simulation and control in Quantum Chemistry, PhD. thesis, University of Paris VI, Paris, France.

TURINICI, G. (2000b), Controllable quantities for bilinear quantum systems, in: *Proceedings of the 39th IEEE Conference on Decision and Control, Sydney, Australia*, Vol. 2, 1364–1369.

TURINICI, G. (2000c), On the controllability of bilinear quantum systems, in: M. Defranceschi and C. Le Bris, eds., *Mathematical models and methods for ab initio Quantum Chemistry*, Lecture Notes in Chemistry, Vol. 74 (Springer) 75–92.

TURINICI, G. and H. RABITZ (2001a), Quantum wavefunction control, *Chem. Phys.* **267**, 1–9.

TURINICI, G. and H. RABITZ (2001b), Wavefunction controllability in quantum systems, *J. Phys. A*, in press.

TURINICI, G., V. RAMAKHRISHNA, B. LI and H. RABITZ (2002), Optimal discrimination of multiple quantum systems: Controllability analysis, in preparation.

TURINICI, G. and S. SCHIRMER (2001), On-line controllability calculator, www-rocq.inria.fr/Gabriel.Turinici/control/calculator.html.

VIOLA, L. and S. LLOYD (1998), Dynamical suppression of decoherence in two-state quantum systems, *Phys. Rev. A* **58**, 2733–2744.

VITALI, D. and P. TOMBESI (1999), Using parity kicks for decoherence control, *Phys. Rev. A* **59**, 4178–4185.

WANG, N. and H. RABITZ (1996), Optimal control of population transfer in an optically dense medium, *J. Chem. Phys.* 1173–1179.

WEINACHT, T., J. AHN and P. BUCKSBAUM (1999), Controlling the shape of a quantum wavefunction, *Nature* **397**, 233–235.

WISEMAN, H.M. (1993), Stochastic quantum dynamics of a continuously monitored laser, *Phys. Rev. A* **47**, 5180–5192.

WISEMAN, M. (1994), Quantum theory of continuous feedback, *Phys. Rev. A* **49**, 2133–2150.

WISEMAN, M. and G.J. MILBURN (1993a), Quantum-theory of field-quadrature measurements, *Phys. Rev. A* **47**, 643–622.

WISEMAN, M. and G.J. MILBURN (1993b), Quantum-theory of optical feedback via homodyne detection, *Phys. Rev. Lett.* **70**, 548–551.

XIA, H., A. MERRIAM, S. SHARPE, G. YIN and S.E. HARRIS (1997), Electromagnetically induced transparency in atoms with hyperfine structure, *Phys. Rev. A* **56**, R3362–R3365.

ZHANG, H. and H. RABITZ (1994), Robust optimal control of quantum molecular systems in the presence of disturbances and uncertainties, *Phys. Rev. A* **49**, 2241–2254.

ZHAO, M. and S. RICE (1994), Optimal control of product selectivity in reactions of polyatomic molecules: a reduced-space analysis, in: J. Hepburn, ed., *Laser techniques for State-selected and State-to-Chemistry II*, SPIE Vol. 2124, 246–257.

ZHU, W., J. BOTINA and H. RABITZ (1998), Rapidly convergent iteration methods for quantum optimal control of population, *J. Chem. Phys.* **108**, 1953–1963.

ZHU, W. and H. RABITZ (1998), A rapid monotonically convergent iteration algorithm for quantum optimal control over the expectation value of a positive definite operator, *J. Chem. Phys.* **109**, 385–391.

ZHU, W. and H. RABITZ (1999a), Noniterative algorithms for finding quantum optimal controls, *J. Chem. Phys.* **110**, 7142–7152.

ZHU, W. and H. RABITZ (1999b), Potential surfaces from the inversion of time dependent probability density data, *J. Chem. Phys.* **111**, 472–480.

ZHU, W. and H. RABITZ (to appear), Attaining optimal controls for manipulating quantum systems, *Int. J. Quant. Chem.*, in press.

ZHU, W., SMIT, M. and H. RABITZ (1999), Managing dynamical singular behavior in the tracking control of quantum observables, *J. Chem. Phys.* **110**, 1905–1915.

Subject Index

QA
297
H287
1990
CHEM

CHEMISTRY LIBRARY
100 Hildebrand Hall • 510-642-3753